何中虎　夏先春◎主编

CIMMYT小麦引进研究与创新利用

CIMMYT WHEAT INTRODUCTION AND UTILIZATION IN CHINA

中国农业出版社

图书在版编目（CIP）数据

CIMMYT小麦引进研究与创新利用/何中虎，夏先春主编．—北京：中国农业出版社，2016.5
ISBN 978-7-109-21538-2

Ⅰ.①C… Ⅱ.①何…②夏… Ⅲ.①小麦—作物育种—学术会议—中国—文集 Ⅳ.①S512.103-53

中国版本图书馆CIP数据核字（2016）第063104号

中国农业出版社出版
（北京市朝阳区麦子店街18号楼）
（邮政编码 100125）
责任编辑 杨天桥

中国农业出版社印刷厂印刷 新华书店北京发行所发行
2016年5月第1版 2016年5月北京第1次印刷

开本：889mm×1194mm 1/16 印张：41.25
字数：1 280千字 印数：1～800册
定价：130.00元

序

国际玉米小麦改良中心（CIMMYT）是为发展中国家服务的非赢利性国际农业研究和培训机构，专门从事玉米和小麦改良研究。早在20世纪60～70年代，在Norman Borlaug博士领导下培育的一系列矮秆、高产、抗病、广适性新小麦品种在南亚、中南美洲等国家产生了巨大影响，创立的穿梭育种方法在国际上广为应用，被誉为“绿色革命”。CIMMYT小麦育种水平一直居国际前列，并在叶锈病的成株抗性研究与利用方面形成了独特优势。因此，与CIMMYT建立长期稳定的合作关系对全面提高我国小麦育种水平具有重要战略意义。

1987年中国农业科学院与CIMMYT签署了“中国—CIMMYT小麦穿梭育种”合作协议，目的是引进并利用其优异种质，及时获得国际相关信息，同时为我国培养实用育种人才。经我推荐，何中虎有幸于1990年获得到CIMMYT做博士后的机会，在国际知名小麦育种家Sanjaya Rajaram（2014年获世界粮食奖）领导的课题组从事小麦育种研究。1993年他从CIMMYT和美国堪萨斯州立大学进修归来，“九五”期间开始主持中国农业科学院原作物育种栽培研究所（后经合并更名为作物科学研究所）小麦品质育种课题组的工作。1997年，受中国农业科学院和CIMMYT的共同委托，又增加一项任务即负责我国与CIMMYT的小麦合作育种项目。这样，他所领导的课题组工作内容包括三大方面，即小麦品质研究、中国—CIMMYT合作（以成株抗病性研究为主）和新品种培育，任务相当繁重。

经过十多年的努力，他主持完成的“中国小麦品种品质评价体系建立与分子改良技术研究”获2008年国家科技进步一等奖。在品质研究取得重要进展之后，通过与西南麦区和春麦区主要育种单位紧密合作，课题组在CIMMYT种质引进与创新利用等方面也有实质性进展。将引进筛选的有一定利用价值的1.8万份种质交国家或地方种质库保存并发放给相关育种单位利用，建立了兼抗型成株抗性育种新方法，并育成一批农艺性状优良的抗病亲本，在国内外发表论文80多篇（包括SCI论文50多篇），形成了有自己特色的优势领域，主持完成的“CIMMYT小麦引进、研究与创新利用”项目获2015年国家科技进步二等奖（注：本所地处北部冬麦区，以冬麦育种为主，CIMMYT种质的育种利用主要由合作单位完成）。

最近，课题组从上述已发表的文章中挑选48篇编辑成《CIMMYT小麦引进研究与创新利用》论文集，以便与同行进行交流，听取意见，为今后深化研究做些准备。该文集包括品种适应性分析、品质基因鉴定、成株抗性种质筛选鉴定、兼抗型成株抗性基因定位与分子检测、兼抗型成株抗性种质创新与育种应用等五个方面，其中创新性较明显的有三个方面。一是中国小麦春化基因、光周期基因鉴定及其与表型关系的研究，为了解我国小麦的冬春性和适应性提供了参考资料。二是在前一时期麦谷蛋白低分子量亚基基因特异性标记发掘的基础上，通过与CIMMYT及澳大利亚等合作，提出了鉴定25个低分子量亚基的30个国际标准品种，明确主要低分子量亚基与面筋质量和主要面食品加工品质的关系，为把低分子量亚基用于改良面筋质量提供了关键技术。三是兼抗型成株抗性鉴定、QTL定位与种质创新，共发表包括Crop Science评述性文章在内的30篇论文，占该文集论文60%以上。课题组对国内育成和引进的百农64、鲁麦21、平原50、Strampelli、Shanghai 3/Catbird等9份兼抗型成株抗性品种进行了系统的成株期抗条锈病、抗叶锈病和抗白粉病QTL定位，发现了兼抗三种病害且效应相对较大的成株抗性基因位点（QTL）5个及其相对紧密连锁的分子标记9个，在此基础上建立了分子标记与常规育种相结合的兼抗型成株抗性育种新方法，并育成农艺性状优良的兼抗型育种品系100多份（到2016年夏收进入产量比较的已超过200份），为培育兼抗型持久抗性新品种提供了遗传基础清晰的亲本、QTL、分子标记和成功范例。可以说，从育种角度来看，这一工作已基本解开持久抗性之谜（另外，国际上近几年已克隆了*Lr34*和*Lr46*两个兼抗型成株抗性基因），为初步解决锈病和白粉病抗性频繁丧失的老大难问题提供了新思路和可行方案，利用分子标记和常规技术相结合，可以做到有目的地培育持久抗性品种。这一方面的育种工作还有待继续加强，最终目标是广泛育成可大面积推广的兼抗型持久抗性品种。上述意见是否成立有待读者评说。

最后需要说明的是，课题组已将建立的品质评价体系和发掘的分子标记用于育种实践，使新品种选育逐步走出低谷（指20世纪90年代），先后育成中优9507、北京0045、中麦175和中麦895（与棉花所合作育成，已在河南、陕西推广）四个主栽品种，2016年后两个新品种的种植面积约800万亩。其中中麦175实现了高产、水肥高效、抗高温、抗寒、优良面条品质、广适性的良好结合，被选为北部冬麦区国家区试对照品种，已成为北部冬麦区第一大品种和黄淮旱肥地主栽品种，累计推广2 500万亩，预计2016年夏收面积在500万亩左右。另外，用分子标记育成的中麦1062等一批苗头品种表现突出，有望在生产中发挥作用。

总之，自1996年新课题组成立至今短短20年左右的时间，能在品质研究、兼抗型育种方法建立和新品种选育方面都有较好收获，应该说是难能可贵的。主持人何中虎胸怀全局，视野开拓，精明能干，理论联系实际，研究目标明确；又善于思考，勤奋工作，知人善任，已初步形成了一支人员和专业搭配合理、分工明确、团结协作、高效运行的研究队伍，这是研究工作能取得进展的根本原因，为此我感到由衷的高兴和鼓舞。期望继续努力，突出重点，大踏步前进，为提高我国小麦育种水平做出更大的贡献。

中国科学院院士　庄巧生

2016年3月2日

Preface

China is the largest wheat producer and consumer in the world, thus the initiation and implementation of the collaboration between International Maize and Wheat Improvement Center (CIMMYT) and China in wheat breeding is a very proud part of my career during 33 year service at CIMMMYT. The objectives of this collaboration were to introduce CIMMYT germplasm in China and then to utilize them in developing new varieties, particularly for improving disease resistance, and to train young scientists. Tremendous progress has been achieved at least in three aspects through continuous efforts of the Chinese Academy of Agricultural Science (CAAS) and other provincial academies as well as CIMMYT.

More than 18, 000 elite lines introduced from CIMMYT and other countries have been characterized and stored in the Chinese national gene bank, and over 40 leading varieties have been developed with an accumulated planting area of around 45 million ha. Genetic contribution of CIMMYT wheat to Chinese wheat acreage is 9% after 2006 as documented by Dr. Jikun Huang and his colleagues from the Chinese Academy of Sciences. Clearly, CIMMYT wheat has contributed significantly to Chinese wheat production, particularly in the provinces of Xinjiang, Gansu, Sichuan, and Yunnan. It was a great honor that three CIMMYT scientists including myself were selected to receive the prestigious Friendship Award from the State Council due to the above contributions.

Very importantly, CAAS has initiated the research and breeding program targeting adult plant resistance or slow disease based on minor genes, in collaboration with CIMMYT. They discovered five QTL conferring multi-resistance to yellow rust, leaf rust, and powdery mildew, and also found that *Yr18*/*Lr34* and *Yr29*/*Lr46* conferred resistance to powdery mildew. Furthermore, the above QTLs have been successfully used to develop elite germplasm conferring multi-disease resistance both in Beijing and Sichuan. This has provided an excellent example in utilization of minor genes based resistance for breeding program in China and other countries, as summarized and highlighted by the review article in Crop Science. I strongly encourage other breeding programs to adopt this novel approach in improving disease resistance, shifting from the traditional major gene based breeding to minor gene approach. It is a very outstanding work and has been widely recognized in the international wheat communities. Up to now, only CIMMYT and China have successful developed elite germplasm conferring multi-adult plant resistance based on minor genes approach.

Human resources development has always been a key priority for CIMMYT. This collaboration has also successfully promoted professional development of Chinese scientists through postdoctoral fellowship or the training courses. For example, Dr. Zhonghu He, has been selected as Fellow of the Crop Science Society of America and the American Society of Agronomy, the highest recognition given by the two Societies.

I am personally very satisfied with the above achievements. The Chinese team under the leadership of Dr. Zhonghu He, received the prestigious award from the Chinese State Council in 2015 for the project "CIMMYT wheat introduction, characterization, and utilization in China" . It is my great honor to write this preface for his paper collections produced from CIMMYT-China collaboration. I strongly believe that publication of these paper collections will significantly benefit wheat breeders in China as well as other countries.

The continuous supports from Professor Qiaosheng Zhuang and CAAS administration for China-CIMMYT collaboration are highly appreciated as always.

Sanjaya Rajaram

Former Wheat Program Director, CIMMYT
Former Biodiversity and Gene Management Program Director, ICARDA

前　言

经庄巧生先生推荐，我有幸于1990年获得到国际玉米小麦改良中心（CIMMYT）做博士后的机会，在国际知名小麦育种家Sanjaya Rajaram博士领导的课题组从事小麦育种研究。经过近两年学习，基本掌握了CIMMYT的育种方法与思路，其中叶锈病成株抗性（又称慢病性、部分抗性、水平抗性等）的研究与成功利用给我留下了深刻印象。在国外期间就曾设想回国后开展成株抗性遗传育种研究，1993年也曾获得过中国农业科学院院长基金的资助。1997年，受中国农业科学院和CIMMYT的共同委托，在课题组原有工作的基础上，又增加了一项任务即负责我国与CIMMYT的小麦合作育种项目。

根据国内小麦育种的需要，这一合作项目确定的主要研究任务包括：(1) CIMMYT材料引进与创新利用，明确CIMMYT品种在我国的适应性与利用途径；(2) 建立并示范成株抗性育种技术，为防止条锈病和白粉病的频繁丧失奠定基础。后来，课题组的品质分子改良技术研究取得较大进展，就用现有技术对引进的CIMMYT等品种进行了基因发掘与分子检测等工作。经过大家的共同努力，“CIMMYT小麦引进、研究与创新利用”有幸获得2015年国家科技进步二等奖。由于本所地处北部冬麦区，以冬麦育种为主，CIMMYT种质的育种利用主要由合作单位完成，在此向他们表示真诚的谢意。

为了便于向各部门同行请教和学习，全面提高我们的研究水平，同时也为了向主管部门/资助机构汇报工作进展，我们从近十年撰写的80多篇论文中精心挑选了48篇编成《CIMMYT小麦引进研究与创新利用》论文集，同时还编入了近期完成的“CIMMYT小麦引进与创新利用——进展与展望”一文，以便为读者提供一些背景资料。在这48篇论文中，有9篇是以主要合作单位河北农业大学、四川省农业科学院、新疆农业科学院等为主，与我们协作完成的。论文集包括品种适应性分析、品质基因鉴定、成株抗性种质筛选鉴定、兼抗型成株抗性基因定位与分子检测、兼抗型成株抗性种质创新与育种应用等五部分。这些论文已经在国外SCI期刊、中国农业科学、作物学报等发表。为了保证相对统一的格式和减少工作量，中文文章按《中国农业科学》的格式编排，英文文章基本保持原貌，只对文字和个别差错之处做了补充、修改。

中国与CIMMYT小麦合作育种项目一直得到中国农业科学院和作物科学研究所历任领导的大力支持，1997年之前庄巧生院士和辛志勇研究员分别担任该项目的中方负责人，为项目后期取得较快进展奠定了基础。先后获得农业部948重大国际合作项目（2003—2015）、国家自然科学基金（30471083、30060043、30821140351、31261140370、30220140636、31161140346）、国家国际科技合作专项（2011DFA31140、2012DFA32290）及国家外国专家局等的长期资助，在此表示衷心感谢。同时，向CIMMYT的主要合作者Sanjaya Rajaram、

Ravi Singh 及 Roberto Javier Peña 等表示真诚的谢意。

本论文集的编辑出版是在庄巧生院士热情鼓励下完成的，文集也是我们献给庄先生百岁华诞的一份薄礼。四川省农业科学院和新疆农业科学院等合作单位的有关专家对此项工作给予很大支持，课题组的同事和众多研究生则是具体实施者，付出了艰辛的劳动，在此一并致以谢意。感谢中国农业出版社杨天桥编审为论文集出版所做的努力。

庄巧生院士和世界粮食奖获得者、小麦育种家 Sanjaya Rajaram 博士还在百忙之中为本论文集作序，这是对我们莫大的鼓舞和鞭策，将激励我们继续努力前进。

由于时间较短，加上作者水平有限，疏漏、错误之处在所难免，敬请指教。

何中虎

2016年3月

目　录

成株抗性种质筛选鉴定

兼抗型成株抗性基因定位与分子检测

兼抗型成株抗性种质创新与育种应用

附录

CIMMYT 小麦引进研究与创新利用
——进展与展望

何中虎[1]，夏先春[1]，陈新民[1]，刘三才[1]，邹裕春[2]，
吴振录[3]，辛志勇[1]，庄巧生[1]

[1]中国农业科学院作物科学研究所/国家小麦改良中心，北京 100081；[2]四川省农业科学院作物研究所，成都 610066；[3]新疆农业科学院核技术生物技术研究所，乌鲁木齐 830000

摘要：本文系统介绍了我国在引进和利用国际玉米小麦改良中心（CIMMYT）小麦种质资源方面取得的主要进展。(1) 系统解析了主要国家品种引进后在我国主产麦区的适应性机理与利用价值，提出不同地区利用 CIMMYT 种质的具体途径和亲本组配模式，将引进筛选出的 18 165 份资源交种质库长期保存。(2) 提出鉴定 25 个低分子量亚基的 30 个国际标准品种，明确了主要亚基与面筋质量和主要食品加工品质的关系；用分子标记和常规分析对 CIMMYT 代表性种质的籽粒硬度、淀粉品质和色泽品质特性进行了评价。(3) 从引进品种及国内品种中发现了兼抗白粉、条锈和叶锈病且效应较大的成株抗性基因位点（QTL）5 个及其紧密连锁的分子标记 9 个，建立了分子标记与常规育种相结合的兼抗型成株抗性育种新方法，育成农艺性状优良的兼抗型育种品系 100 多份，为培育兼抗型持久抗性新品种提供了遗传基础清晰的亲本、基因、分子标记和成功范例。(4) 1990 年至今合作单位育成审定品种 260 个，CIMMYT 种质对提高我国小麦产量、品质和抗病性发挥了重要作用，在西北和西南地区的作用更为突出。

关键词：小麦；种质资源；适应性；成株抗性；基因定位

1 引言

20 世纪 80 年代中期，小麦育种的主要目标是提高产量和抗病性改良。条锈病是全国第一大病害，白粉病正在逐步发展成为重要病害。我国尚处在改革开放初期，对外科技交流很少，小麦育种主要存在三个问题。第一，品种抗病性频繁丧失，国内抗源又很缺乏，如何持续引进和有效利用国外抗源是育种工作面临的重要课题，培育持久抗性类型品种则是育种家的梦想，因此建立和发展稳定的国际合作平台对提高国内育种水平至关重要。第二，国内缺乏加工品质优良的品种资源，引进和评价国外优质资源十分重要。第三，由于长期封闭及语言限制，国内对国际科技动态和进展了解很少，信息滞后成为影响育种技术发展的重要因素，对科技人员进行继续培训是当务之急。

国际玉米小麦改良中心（CIMMYT）是国际农业磋商组织（CGIAR）所属、为发展中国家服务的非赢利性国际农业研究和培训机构，专门从事玉米和小麦改良研究。CIMMYT 培育的小麦品种以矮秆、高产、对光照不敏感（适应性广）和具有成株（持久）抗病性而著称，在全世界产生了巨大影响，创立的穿梭育种方法和成株抗性育种技术在国际上广为应用，被誉为绿色革命发源地。1994－2014 年间，南亚和非洲 80%品种是用 CIMMYT 种质做亲本育成的。更为重要的是，CIMMYT 与主要国家建立了开放共享的合作平台，形成了全球小麦研究网络。因此，与 CIMMYT 建立长期稳定的育种合作平台，对全面提高我国小麦育种水平具有十分重要的战略意义。

1987 年，中国农业科学院与 CIMMYT 签署了“中国－CIMMYT 小麦穿梭育种”合作协议，主要内容包括三个方面：(1) 建立中国－CIMMYT 合作育种平台，通过 CIMMYT 与其他国家建立小麦育种合作关系，系统引进国际优异资源，推动中国小麦育种研

究与世界接轨；(2) 通过双方合作育种研究，把 CIMMYT 种质的优异特性导入我国小麦，培育高产、抗病、优质、广适性新品种；(3) 资助小麦育种及相关领域科研人员到 CIMMYT 进行合作研究、参加培训和出席相关学术会议，双方交换科研信息和资料，为国内了解国际进展提供渠道，同时也向国际同行宣传我国小麦育种的相关进展。1997 年和 2007 年，中国农业科学院又与 CIMMYT 续签了该协议，合作范围和内容进一步扩大。1997 年 CIMMYT 中国办事处正式成立，在农业部 948 重大国际合作项目、国家自然科学基金会和科技部的支持下，双方的合作关系更加密切，从过去的以学习和引进为主逐步过渡到平等互利互惠的合作研究，逐步把我国小麦育种融入国际小麦研究体系。

针对我国小麦育种中存在的关键问题，结合 CIMMYT 的优势领域，我们采取国内协作与国际合作相结合、常规技术与分子标记应用相结合、品种培育与人才培养相结合的策略，组织和实施以下五个方面的工作：(1) 中国农业科学院作物科学研究所总体负责项目在国内的实施，包括制定计划、对外合作、材料引进、理论研究和人员培训等，国内主要合作单位四川省农业科学院和新疆农业科学院等分别负责当地的新品种选育和推广等。除合作单位外，还将引进材料分发给国内其他协作单位，实现材料共享，提高引进资源利用效率。同时将筛选出的可能有利用价值的优异资源交国家和地方种质库长期保存，以增加和丰富我国小麦资源的储备量。(2) 分子标记与常规技术相结合，明确引种规律和国外品种在国内的利用途径。在进行常规评价的同时，用春化 (*Vrn*)、光周期 (*Ppd*) 和株高 (*Rht*) 等基因标记鉴定引进资源，剖析影响主要国家品种在我国主产麦区适应性的机理和利用价值。通过 CIMMYT 和国内代表性试点的联合试验，明确 CIMMYT 品种在各地的利用方式和途径。(3) 充分发挥我们在品质方面的优势，利用新开发和国内外已有的基因特异性分子标记，系统研究 CIMMYT 品种的籽粒硬度、高低分子量麦谷蛋白亚基和黄色素等相关基因的分布规律，为改良我国小麦加工品质提供重要信息。(4) 重点突出兼抗型成株抗性基因发掘，为培育持久抗性品种提供新思路和新方法。在鉴定成株抗性（持久抗性）材料的基础上，针对我国的主要病害条锈病和白粉病，对国外引进和国内育成的持久抗性品种进行抗性 QTL 定位，发掘新的兼抗型成株抗性基因，为培育兼抗白粉病和条锈病的持久抗性品种提供基因、理论和方法。(5) 加强人才培养和信息交流。通过在墨西哥合作研究、参加各类培训和国际学术会议，在国内有针对性地召开研讨会、举办培训班及联合培养研究生等，为我国小麦育种持续发展提供人才储备。将 CIMMYT 相关重要文献译成中文供国内参考，将我国相关研究成果用英文在 CIMMYT 出版，扩大国际影响。下面分别对 CIMMYT 小麦品种在我国的适应性、品质性状分子鉴定、兼抗型成株抗性种质筛选与新基因发掘、新品种培育等方面的进展进行总结。

2 主要进展

2.1 小麦资源引进与品种适应性机理解析

针对高产、抗病、优质等重要目标性状，通过与 CIMMYT 合作建立的引种平台，从 CIMMYT 及主要国家系统引进各类小麦资源，包括 CIMMYT 高代稳定品系（至少进行了两年产量比较，还包括人工合成小麦及 997 份伊朗农家种）及其他 15 个国家的品种、近等基因系三套（硬度基因 *Pinb* 近等基因系、高低分子量麦谷蛋白亚基近等基因系、抗条锈基因近等基因系）和遗传研究群体（RIL 或 DH）15 个。将引进的材料分发给全国所有需求单位，根据农艺性状和抗病性进行初步筛选鉴定，利用分子标记对部分材料的高低分子量麦谷蛋白亚基、糯蛋白、兼抗基因 *Yr18* 等进行分子检测[1-5]，将筛选出的 18 165 份可能有利用价值的种质资源交国家种质库和宁夏农林科学院种质库长期保存。这些资源的系谱等信息齐全，为我国小麦育种提供了丰富的品种资源。

研究国外品种在我国的适应性，对提高引种效率、促进国外资源在国内育种中的高效利用具有重要意义。小麦的适应性主要受春化和光周期基因控制。为了明确国外主要小麦生产国品种在我国的适应性，在分析我国小麦品种春化基因与光周期基因分布的基础上[6,7]，将 20 个主要小麦生产国的部分代表性品种种植在 10 个代表性地点，通过田间评价和分子检测得出如下结论[8,9]：(1) 春化基因和光周期基因分布频率差异是导致不同国家品种适应性不同的主要原因。(2) 明确了主要国家冬麦品种的特性及其在我国的利用价值。美国品种多为冬性，光敏感型占 60%，这是晚熟的重要原因，其优良的品质特性可供国内利用；德国、英国、法国等的品种多为强冬性、光敏感型，很晚甚至不能成熟，但产量潜力高、茎秆强度好，可作为高产源利用；东欧及中东品种的冬性略弱，稍晚熟，抗条锈病和白粉病，可作为抗源利用；主要冬麦国家品种的优良高低分子量谷蛋白亚基的频率明显高于我国品种。(3) 明确了主要春性品种的特性及利用价值，意大利、印度、澳大利亚、加拿大、智利和阿根廷的品种多为春性至弱冬性，其中智利、阿根廷和加拿大多为光周期敏感型，多数品种的加工品质好，优良高低分子量谷蛋白亚基的频率高，可作为优质源利用；澳大利亚和意大利的品种抗条锈病，可作为抗源利用，澳大利亚品种的品质特性也可供国内利用。

在进行常规评价的基础上，用分子标记检测了CIMMYT小麦品种中与适应性相关基因的分布[1,2]。来自墨西哥的CIMMYT小麦为春性，光照不敏感等位基因频率高，因而适应性广；矮秆基因*Rht-B1b*频率高，同时含*Rht-B1b*和*Rht-D1b*的频率为4.6%；*Yr18*和*Yr29*等成株抗性基因频率高，表现持久抗性特点。在墨西哥4个环境和国内9个代表地点的两年联合试验表明，CIMMYT品种的穗数和穗粒数多，千粒重中等，具有广泛适应性，比我国春麦品种具有更高的产量潜力；还将国内试点与墨西哥试验环境进行聚类，为国内引种利用提供了重要信息[10,11]。CIMMYT品种引种到我国后，株高降低，抽穗期和成熟期提早；CIMMYT品种可在云南、青海和新疆直接推广种植；内蒙古、甘肃和宁夏为次适宜地区，也可以直接推广应用，但主要用作杂交亲本，单交即可取得较好效果；在黑龙江以做杂交亲本为宜，单交即可取得较好效果；在南方冬麦区主要用做亲本，最好采用三交或回交；在黄淮麦区可做亲本，必须进行三交、双交或回交才能达到预期目的[10,11]。

2.2 品质基因分子标记鉴定

通过SDS-PAGE、2-DE、MALDI-TOF-MS和PCR技术4种方法比较，提出鉴定25个低分子量亚基的30个国际标准品种，论文发表在BMC Plant Biology[12]，为国内外不同实验室的结果比较提供了可能，已在国际上广泛使用。用SDS-PAGE和分子标记等，对CIMMYT小麦品种的高低分子量亚基进行了系统鉴定，明确了优质亚基1、2*、17+18、7+8、7+9、5+10、*Glu-A3b*和*Glu-B3b*和*Glu-B3g*的分布规律[4]。在此基础上，用引进的高低分子量亚基近等基因系明确了低分子量亚基与面团流变学特性及面包、面条、馒头品质的关系，发表在BMC Plant Biology[13]。上述工作为推动低分子量亚基在育种中的应用做出了重要贡献。

将本实验室发掘的和国内外已有的基因特异性标记用于研究CIMMYT小麦的籽粒硬度、颜色和淀粉品质等基因的分布规律，为高效利用CIMMYT种质提供了重要信息。明确了CIMMYT小麦品种的*Pina*/*Pinb*基因分布，*PINA-null*占86.3%，*Pinb-D1b*占13.7%，*Pina-Dlj*/*Pinb-Dli*的籽粒硬度较*Pina-D1c*/*Pinb-Dlh*高10个单位，这是首次对CIMMYT品种籽粒硬度的研究[14]，为进一步改良磨粉品质提供了重要信息。还对制品颜色及淀粉等相关基因进行了定位和检测[3,15,16]。

2.3 兼抗型成株抗性种质筛选和新基因发掘

2.3.1 兼抗型成株抗性种质筛选

成株抗性又称慢病性、部分抗性或非小种专化抗性，具有持久抗性的特点，表现为苗期感病，成株期严重度或病害发展速率低，而非免疫或坏死反应，聚合4～5个效应相对较大的微效基因即可培育出接近免疫的成株抗性品种。CIMMYT自20世纪60年代对成株抗性的遗传和应用进行了深入系统研究，形成了一套行之有效的育种方法，育成一大批在生产上大面积长时间应用的品种。我国相关研究始于20世纪70年代，曾对条锈病和白粉病的成株抗性品种进行了鉴定，但抗性遗传研究较少，有目的地进行成株抗性育种尚未见报道。本项目针对白粉病、条锈病和叶锈病对国内品种及引进的部分品种进行了苗期分小种和成株期成株抗性鉴定，根据病害最大严重度、病程曲线面积及苗期抗性3个指标筛选出鲁麦21、平原50、Pavon76等20多份兼抗型成株抗性品种[17-20]，为抗病育种和成株抗性遗传研究提供了材料和方法。

在273份CIMMYT小麦品种中，43份材料携带*Yr18*/*Lr34*/*Pm38*成株抗性基因[2]，这些品种既抗条锈病又抗叶锈和白粉病，为兼抗性特异资源。利用该基因紧密连锁的STS标记csLV34和5个功能标记检测了536份CIMMYT小麦品种、231份国内主栽品种和422份农家种[1,21]，*Yr18*/*Lr34*/*Pm38*的频率分别为21.7%、6.1%和85.1%，但*Yr18*在一部分农家种中不表达，现已证明存在一个抑制基因[22]。由于*Yr18*/*Lr34*/*Pm38*位点在我国小麦农家种中分布广泛，推测该基因很可能起源于我国。

2.3.2 QTL定位与新基因发掘

对国内育成和引进的9份兼抗型成株抗性品种百农64、鲁麦21、平原50、Libellula、Strampelli、Fukuho-Komugi、Saar、Naxo、Shanghai3/Catbird等系统进行了抗条锈病、叶锈病和白粉病的QTL定位[23-34]，获得如下结论：(1) 证实聚合3～5个兼抗型QTL即可在病害重发区（如天水和成都）实现持久抗性；(2) 首次报道了5个效应较大的兼抗型成株抗性QTL，发现兼抗条锈和叶锈的*Yr18*/*Lr34*和*Yr29*/*Lr46*也抗白粉病，并将这两个位点对白粉病的抗性基因命名为*Pm38*和*Pm39*[24]；(3) 除加性效应外，QTL间的上位性效应对成株抗性也有重要贡献。在此基础上，将国内外已定位的条锈病和白粉病成株抗性QTL整合于同一连锁图谱，发现8个兼抗条锈病和白粉病的大基因簇，分别位于1BL、2BL、3BS、4BL、5DL、6BS和7DS，为今后鉴定和利用成株抗性基因提供了重要信息[20]。另外，在*Yr10*/6＊Avocet中克隆了1个Homeobox-like基因*TaHLRG*，成株期兼抗白粉病和条锈病，在鲁麦21/京双16、百农64/京双16和Strampelli/辉县红3个群体中可解释表型变异的9.3%～22.5%[35]。

2.3.3 育种方法建立与种质创制

针对条锈病、叶锈病和白粉病，我们进行了兼抗型成株抗性育种尝试，取得了较好效果[20,36-38]。中国农业科学院作物科学研究所在对成株抗性品种鲁麦21和百农64进行白粉病抗性QTL定位的基础上，通过单交对这两个农艺性状较好的品种进行基因聚合，育成的3个品系C181、C219和C263分别含5个成株抗性QTL，其农艺性状和抗病性均优于亲本。虽然我们以白粉病为目标，但多点鉴定证明这些材料兼抗3种病害，证实了利用连锁分子标记进行兼抗型成株抗性QTL聚合的可行性和有效性，进一步说明聚合4～5个成株抗性QTL足以在田间表达高水平的抗性[36,37]。

四川省农业科学院则以抗条锈为目标进行了类似尝试。选用CIMMYT成株抗性种质与四川小麦品种杂交，并用四川品种回交一次，目的是将成株抗性基因导入农艺性状优良、丰产性高及适应性好的小麦品种。利用成株抗性育种方法育成96份高代抗病品系，但经多环境田间鉴定及*Lr34*/*Yr18*/*Pm38*、*Lr46*/*Yr29*/*Pm39*和*Sr2*/*Yr30*分子标记检测，85份兼抗3种病害，18份含*Lr34*/*Yr18*/*Pm38*，37份含*Lr46*/*Yr29*/*Pm39*，29份含*Sr2*/*Yr30*；部分品系可能还含有未知成株抗性基因[38]。3个兼抗型成株抗性高产品系已参加了国家和四川省区试，其中川麦82兼抗条锈、叶锈、白粉病，在四川省2014年度区试中亩①产420.8kg，比对照增产13.5%，居第四组首位[38]。

在上述工作的基础上，提出了成株抗性育种的亲本选配原则和后代选择方案，即用成株抗性很高的品种与高产品种杂交并回交一次，或用两个成株抗性较高的品种杂交再用高产品种三交，适当扩大分离群体，在F_2或BC_1至F_3选择感－中感类型，F_4或F_5以后逐渐提高抗性的选择标准，至于农艺性状选择则与一般育种项目相同。高世代材料需在高度感病环境下进行鉴定，以便确认新品系的抗性[20]。通过分子标记与常规育种相结合，可有效培育兼抗型成株抗性品种，为我国小麦抗病育种提供了新思路。有关条锈病、叶锈病、白粉病成株抗性QTL整合及育种应用的评述性论文发表于*Theoretical and Applied Genetics*[39]和*Crop Science*[40]。

2.4 CIMMYT种质应用

2.4.1 技术路线

第一阶段即20世纪70年代至80年代，主要通过种子管理部门引进并大面积示范和推广CIMMYT品种，由于生育后期出现早衰等缺陷，影响了CIMMYT品种的大面积应用，为此国内育种单位开始利用我国品种改良CIMMYT小麦。在1990年前，育种单位与CIMMYT联系很少（1986年以后有所增加），国内单位间也缺乏有效组织。1973—1989年全国利用CIMMYT小麦共育成审定品92个，包括引进后直接审定品种10个，累计推广3.0亿亩，其中宁春4号、克丰3号和铁春1号等表现突出。宁春4号又名永良4号，1981年通过宁夏审定，主要特点是高产稳产、耐病性好、品质优良、适应性广，在宁夏、内蒙古、新疆和甘肃广泛种植，从1988年至今一直是春麦区第一大品种，累计推广1亿亩以上。另外，塞洛斯原名为Siete Cerros T66，与墨巴65是姊妹系，其特点是高产、矮秆、抗倒伏、中早熟、品质较好、较抗落粒，1981—1989年在新疆和云南等地累计推广约1 000万亩。

第二阶段即1990—2015年，中国农业科学院与CIMMYT建立了密切的合作关系，有组织地进行双边合作研究和CIMMYT种质利用，人员交流与合作育种同步，目标明确。CIMMYT也投入较大力量改造与利用中国小麦，1990—2005年约占其杂交组合的20%～30%。考虑到我国小麦生态类型复杂（冬性、半冬性和春性）、水浇地比例较高、后期温度高、生育期短等特点，双方合作采取三种不同的技术路线：(1) 北方冬麦区合作单位以中国农业科学院作物科学研究

① 亩为非法定计量单位。15亩=1公顷——编者注

所、河北省农林科学院、山东省农业科学院和河南省农业科学院为主，通过CIMMYT与美国冬麦区和东欧（罗马尼亚和匈亚利等）及土耳其-CIMMYT-ICARDA的冬小麦育种项目建立合作关系，主要任务是引进并利用上述地区的冬性品种，以进一步提高北方冬麦区的产量潜力、抗病性和加工品质，在黄淮麦区也可通过三交或回交改造CIMMYT的春性品种，但更重要的是通过合作研究和人员培训提高该区小麦育种的整体水平。(2) 南方冬麦区的主要合作单位包括江苏省农业科学院、湖北省农业科学院、四川省农业科学院和云南省农业科学院，由于CIMMYT小麦对赤霉病的抗性较差，后期在高温高湿环境下易早衰，采用中国亲本/CIMMYT亲本//中国亲本的三交或回交方式，把中国小麦对我国的适应性与对赤霉病的抗性和CIMMYT品种的高产、抗锈性和广泛适应性结合起来，培育适应我国生态条件的高产抗病新品种；需要说明的是，云南省的纬度和海拔与墨西哥城十分相似，因此CIMMYT小麦在云南的适应性好，能直接用于生产或用做单交亲本。(3) 春麦区的主要合作单位包括黑龙江省农业科学院、新疆维吾尔自治区农业科学院、甘肃省农业科学院、宁夏回族自治区农林科学院和内蒙古自治区农业科学院，除直接利用CIMMYT品种外，也可用其作杂交亲本。

2.4.2 新品种培育与推广应用

1973—2015年，通过引进直接审定和利用引进品种做杂交亲本，全国共育成审定品种323个（详见本书附录1），其中1990—2015年育成品种261个，包括引进后直接审定命名品种18个，用CIMMYT种质做亲本育成品种243个。由于CIMMYT小麦为春性，在西南冬麦区（云南和四川）和春麦区（新疆、甘肃、宁夏、内蒙古和黑龙江）的作用更为突出。新疆和甘肃春麦及云南品种更换主要依赖于CIMMYT种质的利用，近五年四川审定的主要品种都是CIMMYT种质的后代，用CIMMYT种质育成的邯6172等则在黄淮北片发挥了关键作用。大面积推广品种大致分为三类：(1) 抗锈高产类，四川、云南的品种多属此类，如绵农4号、川麦30、川麦42、云麦39和鄂麦18等；(2) 高产广适类，如邯6172、新春2号、新春6号等；(3) 优质高产类，如济南17、济麦19、克丰6号、川麦36和郑麦004。其中济南17、济麦19、邯6172、绵农2号、川麦42、克丰6号等分获国家科技进步二等奖，铁春1号和宁春4号分获国家科技进步三等奖。

需要说明的是，中国农业科学院地处北部冬麦区，以冬麦育种为主，对CIMMYT种质的改造利用研究较少。1990年之前曾进行春麦育种，育成的京红号品种在生产中也发挥过一定作用，筛选出的中7906后来在新疆育种中发挥了重要作用。用CIMMYT材料育成的春性优质麦中作8131-1及其冬性选系中优9507曾在生产和育种中起到一定作用。

2.4.3 CIMMYT种质对我国小麦育种的贡献

CIMMYT种质对我国小麦育种的贡献可概括为三个方面。一是提高产量潜力。选择用CIMMYT种质育成的代表性品种，连续两年的产量潜力研究表明，由于CIMMYT种质的成功应用，新疆、甘肃、云南和四川的产量潜力改良皆取得显著进展，年遗传进展分别为1.43%、0.64%、0.31%和0.73%，其中新疆高于同期黄淮麦区的遗传进展，甘肃和四川则与河南和山东遗传进展接近，云南的遗传进展相对偏低，可能与其特定的生态环境相关[41]；另外，人工合成小麦在产量改良中也发挥了重要作用[42]。二是提高抗病性。CIMMYT品种的抗条锈病性能突出，这是四川省和云南省成功应用CIMMYT种质的重要原因，近十年应用的主要抗源包括人工合成小麦、Milan和Alondra（缩写为ALD），用它们分别育成5个、7个和9个抗病品种。三是改良品质。CIMMYT种质是我国优质强筋小麦的主要亲本，如中作8131-1和临汾5064的优质源来自CIMMYT，另外宁春4号、克丰6号和郑麦004等都是知名的优质品种，其优质亲本也都来自CIMMYT。黄季焜等（2015）对CIMMYT种质在我国的利用做了系统经济分析[43]，进一步肯定了上述分析。

2.5 中国小麦品种对CIMMYT育种的贡献

为了促进双边合作，根据双方相关协议，中国农业科学院等单位先后向CIMMYT提供了500多份中国小麦品种和高代品系，CIMMYT采用CIMMYT/中国//中国和CIMMYT/中国//CIMMYT两类杂交方式，在为我国培育新种质的同时，用我国品种改良CIMMYT小麦也取得重要进展。用中国品种与CIMMYT小麦杂交育成的高代品系已向第三世界国家发放，其中GUAMVCHIL92（CATBIRD）和ARIVECHILM92（LUAN）已在墨西哥审定推广，IZGI和BAGCI 02已在土耳其审定，Norman和Iqbol在塔吉克斯坦审定。CIMMYT 20多年的育种实践表明我国小麦有两大优点：(1) 南方冬麦区的品种如川麦18、武汉1号、上海3号不仅抗赤霉病，而且对墨西哥的重要病害印度腥黑穗病及南亚等国的重要病害长蠕孢菌根腐病（*Helminthosporium* leaf blotch）、叶枯病（*Septoria tritici* 和 *Septoria nodorum*）及黑褐斑病（tan spot）也表现很好的抗性，因此在抗病育种中发挥了重要作用；(2) 北方冬麦区的冬性和半冬性品种高产、灌浆速度快、早熟，用它与 CIMMYT 小麦杂交可提高春小麦的产量潜力和耐热性。

2.6 合作平台建设与论文著作

在本项目的带动下，1997 年 CIMMYT 中国办事处在中国农业科学院正式成立，以后又相继建立了中—澳、中—美和中—日小麦联合实验室，与英国和法国等也有了实质性合作关系，为把我国融入国际小麦育种研究体系提供了平台。

以中国农业科学院为主（或参加）发表学术论文 86 篇，其中 SCI 论文 57 篇、中国农业科学和作物学报论文 25 篇（部分目录见本书附录 2），包括作物遗传育种国际核心期刊评述性论文 3 篇；出版专译著 8 部，其中《CIMMYT 的小麦育种》和《CIMMYT 麦类改良进展》较为系统地向国内介绍了其育种方法思路和具体做法，出版的 *A History of Wheat Breeding in China* 和 *Cereals in China* 则成为国际同行了解我国小麦育种的重要文献。

1990 年至 2015 年，约 250 人次到墨西哥或其他国家参加合作研究、培训班或参加学术研讨会，为我国小麦育种人才培养和学科发展起到了至关重要的作用。联合培养研究生 70 多名，在国内举办学术研讨会和培训班 20 多次，约 3000 多人次参加，为推动国内外学术交流发挥了重要作用。中国科学院农业政策研究中心于 2015 年已对人员交流与培训的效果作了全面分析（2016 年出版）。

3 展望

总之，历时 25 年的中国—CIMMYT 小麦合作育种研究实现了引进创新与自主创新的有机结合，种质引进与人才培养和信息交流的同步发展，为我国小麦育种做出了重要贡献，并在国际上产生了一定影响。随着我国实力的增强和国内科研水平的迅速提高，中国—CIMMYT 的合作形式和内容也随之发生了变化，从过去以引进为主逐步过渡到互利互惠的合作育种研究。与此同时，我们还应清醒地意识到我国的粮食安全还很严峻，由于人口不断增加和耕地持续减少，加之水和环境等问题日益突出，小麦育种面临前所未有的挑战，气候变化对农业生产的影响日益明显，继续利用国际种质、技术和人力资源仍是提高我国小麦育种整体水平的重要措施。今后 CIMMYT 仍将是我国小麦育种最重要的国际合作伙伴，预计 CIMMYT 种质在西南麦区和春麦区仍将继续发挥关键作用，在黄淮麦区也可发挥一定作用。在农业部 948 重大国际合作项目的支持下，引进和鉴定的成株抗性种质、发掘的新基因和形成的育种方法将用于主产麦区的品种选育，兼抗锈病和白粉病的持久抗性品种的培育将为防止抗性丧失提供关键技术，实现抗病育种思路的革新。在提高产量潜力、抗热和抗旱、应对气候变化、分子技术应用和人才培养等领域将与 CIMMYT 开展全面实质性合作，在现有工作基础上，在西南和西北地区建立中国—CIMMYT 合作育种基地，为我国小麦持续发展提供技术支撑，为我国培养一批国际化育种人才，为实现我国小麦育种与国际接轨提供技术支持、人才保障和合作平台。

致谢

合作单位及协作单位的许多小麦育种专家和管理人员为本项目做出了重要贡献，在此一并致谢。CIMMYT 小麦项目 4 位项目主任及小麦育种负责人 Sanjaya Rajaram 和 Ravi Singh 博士等为中国—CIMMYT 合作做出了重要贡献，先后获得我国政府友谊奖。

参考文献

[1] 梁丹，杨芳萍，何中虎，姚大年，夏先春．利用 STS 标记检测 CIMMYT 小麦品种（系）中 *Lr34*/*Yr18*、*Rht-B1b* 和 *Rht-D1b* 基因的分布．中国农业科学，2009，42：17-27.

[2] 伍玲，夏先春，朱华忠，李式昭，郑有良，何中虎．CIMMYT 273 个小麦品种抗病基因 *Lr34*/*Yr18*/*Pm38* 的分子标记检测．中国农业科学，2010，43：4553-4561.

[3] 穆培源，何中虎，徐兆华，王德森，张艳，夏先春．CIMMYT 普通小麦品系 Waxy 蛋白类型及淀粉糊化特性研究．作物学报，2006，32：1071-1075.

[4] Dan Liang，Zhonghu He，Jianwei Tang，Roberto Javier Peña，Ravi Singh，Xinyao He，Xiaoyong Shen，DanianYao，Xianchun Xia. Characterization of CIMMYT bread wheats for high-and low-molecular-weight glute-

nin subunits and other quality-related genes with SDS-PAGE, RP-HPLC and molecular markers. Euphytica, 2010, 172: 235-250.

[5] Hui Jin, Jun Yan, Roberto J. Peña, Xianchun Xia, Alexey Morgounov, Liming Han, Yong Zhang, Zhonghu He. Molecular detection of high-and low-molecular-weight glutenin subunit genes in common wheat cultivars from 20 countries using allele-specific markers. Crop & Pasture Science, 2011, 62: 746-754.

[6] X K Zhang, X C Xia, Y G Xiao, J Dubcovsky, Z H He. Allelic variation at the vernalization genes *Vrn-A1*, {*Vrn-B1*, *Vrn-D1* and *Vrn-B3* in Chinese common wheat cultivars and their association with growth habit. Crop Science, 2008, 48: 458-470.

[7] F PYang, X K Zhang, X C Xia, D A Laurie, W X Yang, Z H He. Distribution of the photoperiod insensitive *Ppd-D1a* allele in Chinese wheat cultivars. Euphytica, 2009, 165: 445-452.

[8] 杨芳萍，韩利明，阎俊，夏先春，张勇，曲延英，王忠伟，何中虎．春化和光周期基因等位变异在 23 个国家小麦品种中的分布．作物学报，2011，37：1917-1925.

[9] 杨芳萍，夏先春，张勇，张晓科，刘建军，唐建卫，杨学明，张俊儒，刘茜，李式昭，何中虎．春化、光周期和矮秆基因在不同国家小麦品种中的分布及其效应．作物学报，2012，38：1155-1166.

[10] 张勇，吴振录，张爱民，Maarten van Ginkel，何中虎．CIMMYT 小麦在中国春麦区的适应性分析．中国农业科学，2006，39：655-663.

[11] Yong Zhang, Zhonghu He, Aimin Zhang, Maarten van Ginkel, Guoyou Ye. Pattern analysis on grain yield performance of Chinese and CIMMYT spring wheat cultivars sown in China and CIMMYT. Euphytica, 2006, 147: 409-420.

[12] L Liu, T M Ikeda, G Branlard, R J Peña, W J Rogers, S E Lerner, D I María, A Kolman, X C Xia, Linhai Wang, Wujun Ma, Rudi Appels, Hisashi Yoshida, Aili Wang, Yueming Yan, Z H He. Comparison of low molecular weight glutenin subunits identified by SDS-PAGE, 2-DE, MALDI-TOF-MS and PCR in common wheat. BMC Plant Biology, 2010. 10: 124.

[13] Xiaofei Zhang, Hui Jin, Yan Zhang, Dongcheng Liu, Genying Li, Xianchun Xia, Zhonghu He, Aimin Zhang. Composition and functional analysis of low-molecular-weight glutenin alleles with Aroona near-isogenic lines of bread wheat. BMC Plant Biology, 2012, 12: 1-16.

[14] Morten Lillemo, Chen Feng, Xianchun Xia, Manilal William, Roberto J. Peña, Richard Trethowan, Zhonghu He. Puroindoline grain hardness alleles in CIMMYT bread wheat germplasm. Journal of Cereal Science, 2006, 44: 86-92.

[15] X Y He, Z H He, W Ma, R Appels, X C Xia. Allelic variants of phytoene synthase 1 (Psy1) genes in Chinese and CIMMYT wheat cultivars and development of functional markers for flour colour. Molecular Breeding, 2009, 23: 553-563.

[16] Xinyao He, Jianwu Wang, Zhonghu He, Karim Ammar, Roberto Javier Peña, Xianchun Xia. Allelic variants at the *Psy-A1* and *Psy-B1* loci in durum wheat and their associations with grain yellowness. Crop Science, 2009, 49: 2058-2064.

[17] Z L Wang, L H Li, Z H He, X Y Duan, Y L Zhou, X M Chen, M Lillemo, R P Singh, H Wang , X C Xia. Seedling and adult plant resistance to powdery mildew in Chinese bread wheat cultivars and lines. Plant Disease, 2005, 89: 457-463.

[18] Z F Li, X C Xia, X C Zhou, Y C Niu, Z H He, Y Zhang, G Q Li, A M Wan, D S Wang, X M Chen, Q L Lu, R P Singh. 2006. Seedling and slow rusting resistance to stripe rust in Chinese common wheats. Plant Disease, 90: 1302-1312.

[19] Z F Li, X C Xia, Z H He, L J Zhang, X Li, H Y Wang, Q F Meng, W X Yang, G Q Li, D Q Liu. Seedling and slow rusting resistance to leaf rust in Chinese wheat cultivars. Plant Disease, 2010, 94: 45-53.

[20] 何中虎，兰彩霞，陈新民，邹裕春，庄巧生，夏先春，小麦条锈病和白粉病成株抗性研究进展与展望．中国农业科学，2011，44：2193-2215.

[21] 杨文雄，杨芳萍，梁丹，何中虎，尚勋武，夏先春．中国小麦育成品种和农家种中慢锈基因 Lr34/Yr18 的分子检测．作物学报，2008，34：1109-1113.

[22] Ling Wu, Xianchun Xia, Garry M Rosewarne, Huazhong Zhu, Shizhao Li, Zhengyu Zhang, Zhonghu He. Stripe rust resistance gene *Yr18* and its suppressor gene in Chinese wheat landraces. Plant Breeding, 2015, 134: 634-640.

[23] S S Liang, K Suenaga, Z H He, Z L Wang, H Y Liu, D S Wang, R P Singh, P Sourdile, X C Xia. Quantitative trait loci mapping for adult-plant resistance to powdery mildew in bread wheat. Phytopathology, 2006, 96: 784-789.

[24] Morten Lillemo, Belachew Asalf, Ravi Singh, Julio Huerta-Espino, Xinmin Chen, Zhonghu He, Åsmund Bjørnstad. The adult plant rust resistance loci *Lr34*/

Yr18 and *Lr46/Yr29* are important determinants of partial resistance to powdery mildew in bread wheat line Saar. Theoretical and Applied Genetics, 2008, 116: 1155-1166.

[25] Y M Lu, C X Lan, S S Liang, X C Zhou, D Liu, Xia, Xianchun, He Zhonghu. QTL mapping for adult-plant resistance to stripe rust in Italian common wheat cultivars Libellula and Strampelli. Theoretical and Applied Genetics, 2009, 119: 1349-1359.

[26] M A Asad, B Bai, C X Lan, J Yan, X C Xia Y Zhang, Zhonghu He. Molecular mapping of quantitative trait loci for adult-plant resistance to powdery mildew in Italian wheat cultivar Libellula. Crop & Pasture Science, 2012, 63: 539-546.

[27] Asad Muhammad Azeem, Bin Bai, Caixia Lan, Jun Yan, Xianchun Xia, Yong Zhang, Zhonghu He. QTL Mapping for adult plant resistance to powdery mildew in Italian wheat cv. Strampelli. Journal of Integrate Agriculture, 2013, 12: 756-764.

[28] Yan Ren, Zhonghu He, Jia Li, Morten Lillemo, Ling Wu, Bin Bai, Qiongxian Lu, Huazhong Zhu, Gang Zhou, Jiuyuan Du, Qinglin Lu, Xianchun Xia. QTL mapping of adult-plant resistance to stripe rust in a population derived from common wheat cultivars Naxos and Shanghai 3/Catbird. Theoretical and Applied Genetics, 2012, 125: 1211-1221.

[29] Yue Zhou, Yan Ren, Morten Lillemo, Zhanjun Yao, Peipei Zhang, Xianchun Xia, Zhonghu He, Zaifeng Li, Daquan Liu. QTL mapping of adult-plant resistance to leaf rust in a RIL population derived from a cross of wheat cultivars Shanghai 3/Catbird and Naxos. Theoretical and Applied Genetics, 2014, 127: 1873-1883.

[30] Caixia Lan, Shanshan Liang, Zhulin Wang, Jun Yan, Yong Zhang, Xianchun Xia, Zhonghu He. Quantitative trait loci mapping for adult-plant resistance to powdery mildew in Chinese wheat cultivar Bainong 64. Phytopathology, 2009, 99: 1121-1126.

[31] Yan Ren, Zaifeng Li, Zhonghu He, Ling Wu, Bin Bai, Caixia Lan, Cuifen Wang, Gang Zhou, Huazhong Zhu, Xianchun Xia. QTL mapping of adult-plant resistances to stripe rust and leaf rust in Chinese wheat cultivar Bainong 64. Theoretical and Applied Genetics, 2012, 125: 1253-1262.

[32] C X Lan, X W Ni, J Yan, Y Zhang, C X Xia, X M Chen, Z H He. Quantitative trait loci mapping of adult-plant resistance to powdery mildew in Chinese wheat cultivar Lumai 21. Molecular Breeding, 2010, 25: 615-622.

[33] Yan Ren, Li Liu, Zhonghu He, Ling Wu, Bin Bai, Xianchun Xia. QTL mapping of adult-plant resistance to stripe rust in a 'Lumai 21 × Jingshuang 16' wheat population. Plant Breeding, 2015, 134: 501-507.

[34] C X Lan, S S Liang, X C Zhou, G Zhou, Q L Lu, X C Xia, Z H He. Identification of genomic regions controlling adult-plant stripe rust resistance in Chinese landrace Pingyuan 50 through bulked segregant analysis. Phytopathology, 2010, 100: 313-318.

[35] D Liu, X C Xia, Z H He, S C Xu. A novel homeobox-like gene associated with reaction to stripe rust and powdery mildew in common wheat. Phytopathology, 2008, 98: 1291-1296.

[36] B Bai, Z H He, M A Asad, C X Lan, Y Zhang, X C Xia, J Yam, X M Chen, C S Wang. Pyramiding adult plant powdery mildew resistance QTL in bread wheat. Crop & Pasture Science, 2012, 63: 606-611.

[37] 刘金栋，陈新民，何中虎，伍玲，白斌，李在峰，夏先春．小麦慢白粉病 QTL 对条锈病和叶锈病的兼抗性．作物学报，2014，40：1557-1564.

[38] 刘金栋，杨恩年，肖永贵，陈新民，伍玲，白斌，李在峰，Garry M. Rosewarne，夏先春，何中虎．兼抗型成株抗性小麦品系的培育、鉴定与分子检测．作物学报，2015，41：1472-1480.

[39] Rosewarne G M, S A Herrera-Foessel, R P Singh, J Huerta-Espino, C X Lan, Z H He. Quantitative trait loci of stripe rust resistance in wheat. Theoretical and Applied Genetics, 2013, 126: 2427-2449.

[40] Zaifeng Li, Caixia Lan, Zhonghu He, Ravi P. Singh, Garry M. Rosewarne, Xinmin Chen, Xianchun Xia. Overview and application of QTL for adult plant resistance to leaf rust and powdery mildew in wheat. Crop Science, 2014, 54: 1907-1925.

[41] 张勇，李式昭，吴振录，杨文雄，于亚雄，夏先春，何中虎．CIMMYT 种质对四川、云南、甘肃和新疆春性小麦产量遗传增益的贡献．作物学报，2011，37：1752-1762.

[42] Wuyun Yang, Dengcai Liu, Jun Li, Lianquan Zhang, Huiting Wei, Xiaorong Hu, Youliang Zheng, Zhoughu He, Yuchun Zou. Synthetic hexaploid wheat and its utilization for wheat genetic improvement in China. J. Genet. Genomics, 2009, 36: 539-546.

[43] J Huang, C Xiang, Y Wang. The impact of CIMMYT wheat germplasm on wheat productivity in China. Mexico, D. F. 2015: CGIAR Research Program on Wheat.

品种适应性分析

CIMMYT小麦在中国春麦区的适应性分析

张勇[1]，吴振录[2]，张爱民[3]，Maarten van Ginkel[4]，何中虎[1,5]

[1]中国农业科学院作物科学研究所/国家小麦改良中心/国家农作物基因资源与基因改良重大科学工程，北京 100081；[2]新疆农业科学院核技术与生物技术研究所，乌鲁木齐 830000；[3]中国科学院遗传与发育生物学研究所，北京 100101；[4]CIMMYT，Apdo. Postal 6-641，06600，Mexico，D. F.，Mexico；[5]CIMMYT中国办事处，北京 100081

摘要： 研究CIMMYT小麦在中国的适应性有助于提高春麦区的育种水平。10份CIMMYT代表性品种和15份中国春麦主栽品种于2001年和2002年种植在中国春麦区的9个试点和CIMMYT的4种不同处理环境，分析产量、产量构成因子和农艺性状的变化趋势。CIMMYT品种穗数和穗粒数多、千粒重中等，具有广泛适应性，比中国品种具有更高的产量优势；黑龙江光敏感品种植株高、抽穗和成熟晚、穗数中等、穗粒数少、千粒重和产量低；中国其他品种株高中等、抽穗和成熟早、穗数少、穗粒数中等、千粒重高、产量中等。CIMMYT品种引种到中国后，株高降低，抽穗和成熟提早，并略减产；黑龙江光敏感品种在CIMMYT种植时株高增加，抽穗和成熟推迟，千粒重降低，并显著减产；中国其他品种在CIMMYT种植时株高增加、抽穗和成熟略推迟、千粒重变化较小，并略减产。CIMMYT品种可在云南、青海和新疆直接推广种植；内蒙古、甘肃和宁夏为其次适宜地区，可以直接推广应用，但主要用作杂交亲本；在黑龙江以作杂交亲本为宜。为提高引种效率，并考虑到性状的重复力大小，在CIMMYT为中国选种时应重点选择籽粒较大的材料。为云南所选材料可略矮、适当晚熟，内蒙古、甘肃、宁夏和新疆所选材料可略高、较早熟，青海所选材料可较高、熟期相当，黑龙江应主要选择高纬度材料、植株偏高且晚熟。

关键词： 普通小麦；产量；基因型与试点互作；适应性

Adaptation of CIMMYT Wheat Germplasm in China's Spring Wheat Regions

Zhang Yong[1], Wu Zhenlu[2], Zhang Aimin[3], Maarten van Ginkel[4], He Zhonghu[1,5]

[1]*Crop Science Institute / National Wheat Improvement Center / The National Key Facilities for Crop Genetic Resources and Improvement, NFCRI, Chinese Academy of Agriculture Sciences, Beijing* 100081; [2]*Institute of Nuclear & Biological Technology, Xinjiang Academy of Agricultural Sciences, Urumqi* 830000; [3]*Institute of Genetics and Developmental Biology, Chinese Academy of Sciences, Beijing* 100101; [4]*CIMMYT, Apdo. Postal* 6-641, 06600, *Mexico, D. F., Mexico*; [5]*CIMMYT-China, C/O, Chinese Academy of Agriculture Sciences, Beijing* 100081

Abstract: Information on adaptation of CIMMYT wheat germplasm in China would greatly enhance wheat breeding efficiency in China's spring wheat regions. Twenty-five spring wheat cultivars including

原文发表在《中国农业科学》，2006，39（4）：655-663.

10 from CIMMYT and 15 from China, were sown at 9 locations in China and 4 management environments in Obregon Station at CIMMYT in 2001 and 2002 seasons, and grain yield, yield components, and other agronomic traits were investigated. CIMMYT cultivars were characterized by more spikes and grains per spike, medium thousand kernel weight, and high grain yield with broad adaptability. The photoperiod sensitive cultivars from Heilongjiang province performed tall plant height, late maturity, medium spike number, but low grains per spike, thousand-kernel weight, and grain yield. The other Chinese cultivars had medium plant height, early maturity, low spike number and medium grains per spike, but high thousand kernel weight, and medium grain yield. CIMMYT cultivars performed short plant height, earlier maturity, and a little lower grain yield when planted in China's spring wheat regions compared with that in CIMMYT. The photoperiod sensitive cultivars had taller plant height, later maturity, and much lower grain yield, while the other Chinese cultivars showed taller plant height, a little later maturity, and a little lower grain yield with almost the same thousand kernel weight when they were planted in CIMMYT. It is recommended to directly use CIMMYT wheat in production in Yunnan, Qinghai, and Xinjiang provinces. CIMMYT wheat also can be used in production in such provinces as Inner Mongolia, Gansu, and Ningxia provinces, and are more suitable as crossing parents in these areas; while only can be used as crossing parents in Heilongjiang province. In order to improve the shuttle breeding efficacy, and taking into account the genotype mean repeatability for major traits, Chinese scientists should focus on lines with large grain size when selecting materials in CIMMYT for China, choosing lines with short plant height, late maturity for Yunnan, and lines with a little tall plant height and early maturity for Inner Mongolia, Gansu, Ningxia and Xinjiang, lines with a little tall plant height and around the same maturity for Qinghai, while lines for high latitude environment with a little tall plant height and late maturity for Heilongjiang.

Key words: *T. aestivum*; Grain yield; Genotype by location interaction; Adaptation

春小麦种植面积和总产量分别约占全国小麦的14%和10%，且多分布在经济和科技相对落后的地区，生产上存在的主要问题是单产低，适应性、稳产性和品质较差。改良品种是提高春小麦竞争力的重要途径。

位于墨西哥的国际玉米小麦改良中心（CIMMYT）以春小麦育种闻名于世，育成品种丰产性好，植株较矮、株型紧凑、较抗倒伏，品质较好，抗病性强，适应性广[1]。自20世纪70年代以来，黑龙江、内蒙古、宁夏和新疆等地曾用CIMMYT小麦作杂交亲本育成了140多个品种，在生产上推广利用。我国目前春小麦品种的改良在很大程度上依赖于对CIMMYT种质的有效改造和利用[2]。

近年来，中国春小麦主要省区正与CIMMYT开展穿梭育种活动，每年有1000多份CIMMYT材料在国内种植观察，各地还常派技术人员到CIMMYT现场参与田间选择。CIMMYT用我国小麦与CIMMYT种质配制了大量组合，目的是为中国培育产量、品质和适应性更好的品种，并利用中国品种的早熟、丰产、籽粒灌浆快等特性为世界其他地区培育优良种质。但由于对CIMMYT种质引种到中国和中国品种在CIMMYT主要性状的变化规律缺乏系统研究，性状选择难以把握，从而严重制约了穿梭育种的效率。

宛秀兰根据抽穗天数和千粒重对Tanori F 71在中国春麦区的适应性进行了初步研究，认为CIMMYT小麦可在新疆和云南直接推广应用，其他春麦区可作杂交亲本[3]。姚金保和袁汉民等分别对CIMMYT小麦在中国江苏和宁夏等地区的表现进行了初步分析[4,5]。我们也对CIMMYT种质在中国春麦区的产量、品质和抗病性进行了初步报道[6]，但对CIMMYT种质和中国春麦区主栽品种在中国春麦区和CIMMYT试验站的产量、产量因子和农艺性状还未进行深入分析。为此，本文通过生育期、株高、穗数、穗粒数、千粒重和产量等性状，分析CIMMYT种质和中国春麦区主栽品种在中国春麦区和CIMMYT试验站的表现，为中国与CIMMYT穿梭育种和引种提供理论依据。

1 材料与方法

1.1 试验材料

来自CIMMYT的10份代表性品种和中国6省区的15份春麦主栽品种，于2001年和2002年分别种植在云南弥渡、黑龙江哈尔滨和克山、内蒙古呼和浩特、甘肃兰州/武威、宁夏永宁、青海西宁、新疆和静和乌鲁木齐等9个试点和CIMMYT Obregon试验站的4种不同处理环境。试验材料名称及系谱详见表1，各试点主要生态环境资料见表2。本试验选点和取材时包括云南弥渡和四川品种在内的原因是CIMMYT品种可在云南直接推广种植，而四川绵阳育成的品种在云南种植面积较大，适应性较好，是当地的主栽品种[7]。甘肃兰州和武威分别只在2001年和2002年进行了试验。Obregon试验站沙壤土，日照充足，小麦生育期降水量很少，灌溉设施优良，可模拟多种小麦生长环境，是CIMMYT小麦育种研究的主要基地，适合对产量潜力及农艺性状进行选择。

1.2 田间设计

采用Latinized alpha lattice design（拉丁方格子设计），3次重复，5行，15列。国内播期和播种量除弥渡试点（冬播150kg/hm^2）外，均为春播300kg/hm^2。小区面积4.8m^2，6行区，4m行长，0.2m行距。CIMMYT播期和播种量均为冬播130kg/hm^2。4种处理分别为：(1)灌溉、正常播期、垄作（FNB)：11月下旬播种，垄作，2垄/小区，3行/垄，3m行长，小麦生育期间5次灌溉，用水总量700mm左右，收获面积4.8m^2，模拟高产垄作最适环境[8]；（2）灌溉、正常播期、平播（FNF)：11月下旬播种，平播，8行区，3.8m行长，5次灌溉，用水总量700mm左右，收获面积3.125m^2，模拟高产平播最适环境；(3)减少灌溉、正常播期、垄作（RNB)：3次灌溉，用水总量400mm左右，在小麦抽穗后遇干旱胁迫，其余同（1)，模拟半干旱垄作环境；(4)灌溉、迟播、垄作（FLB)：1月中旬播种，垄作，2垄/小区，3行/垄，2.8m行长，5次灌溉，用水总量700mm左右，收获面积4.48m^2，在小麦生育后期遇高温胁迫，模拟热带高温环境。小区人工除草，并于抽穗至成熟期间防止鸟类危害，减少产量损失。田间管理同当地品比试验。

表1 参试品种名称、来源和系谱

Table 1 Name, origin, and pedigree of the 25 tested wheat cultivars

品 种 Cultivar	来源和主栽地区 Location of origin	系 谱 Pedigree and selection history
Seri M 82	CIMMYT	Veery, CM33027-F-15M-500Y-0M-87B-0Y-0MEX
Bacanora T 88	CIMMYT	Kauz, CM67458-4Y-1M-3Y-1M-5Y-0B-0MEX
Culiacan T 89	CIMMYT	Tui, CM74849-2M-2Y-3M-2Y-0B-46M-0Y-0MEX
Attila	CIMMYT	Nd/Vg9144//Kal/Bb/3/Yaco/4/Vee#5, CM85836-45Y-0M-0Y-4M-0Y
Turaco	CIMMYT	Cno79*2/Prl, CM90312-C-7B-2Y-3B-0X-1B-0Y
Rayon F 89	CIMMYT	Ures*2/Prl, CM90315-A-2B-2Y-1B-0Y-0MEX
Weaver	CIMMYT	Hahn*2/Prl, CM90320-A-1B-4Y-0B
Baviacora M 92	CIMMYT	Bow/Nac/Vee/3/Bjy/Coc, CM92066-J-0Y-0M-0Y-4M-0Y-0MEX
Super Seri#1	CIMMYT	Seri*4//Aga/6*Yr/3/Seri, CRG2468-I-3Y-2B-0Y
Inqalab 91	CIMMYT	Wl711/Crow, PB19545-9A-0A-0PAK
绵阳19 Mianyang 19	四川/云南 Sichuan/Yunnan	绵阳11系选 Mianyang 11 selection
绵阳20 Mianyang 20	四川/云南 Sichuan/Yunnan	绵阳11系选 Mianyang 11 selection
绵阳26 Mianyang 26	四川/云南 Sichuan/Yunnan	绵阳20/川育9号 Mianyang 20/Chuanyu 9
新克旱9 Xinkehan 9	黑龙江 Heilongjiang	克珍/克红//克69-701/3/克74F$_3$-249-3 Kezhen/Kehong//Ke69-701/3/Ke 74F$_3$-249-3

（续）

品　种 Cultivar	来源和主栽地区 Location of origin	系　谱 Pedigree and selection history
克丰 6 Kefeng 6	黑龙江 Heilongjiang	克 $85F_3$-868/克 $85F_6$-784 Ke$85F_3$-868/Ke$85F_6$-784
龙麦 19 Longmai 19	黑龙江 Heilongjiang	克 $65F_3$-196/Rulofen//克 62-348-2/Nadadores/3/龙 74-5778 Ke$65F_3$-196/Rulofen//Ke 62-348-2/Nadadores/3/Long 74-5778
龙麦 26 Longmai 26	黑龙江 Heilongjiang	沈 66-71/Tanori F71/3/松 71-175/Mexipak 66//克 74-204/4/克 88-2060-2 Shen 66-71/Tanori F71/3/Song 71-175/Mexipak 66//Ke 74-204/4/Ke88-2060-2
陇春 15 Longchun 15	甘肃 Gansu	75002/Prikumskaja 2
宁春 4 Ningchun 4	宁夏/内蒙古 Ningxia/Inner Mongolia	Sonora64/宏图 Sonora64/Hongtu
宁春 16 Ningchun 16	宁夏/内蒙古 Ningxia/Inner Mongolia	SG Ta1/宁春 4 号//宁春 4 号 SG Ta1/Ningchun 4//Ningchun 4
青春 533 Qingchun 533	青海 Qinghai	Tamworth/4/Rondine/3/幸福麦//Jubileina 2/C258/5/Alondra Tamworth/4/Rondine/3/Xingfumai//Jubileina 2/C258/5/Alondra
青春 566 Qingchun 566	青海 Qinghai	81S013/3/Abbondanza108-3-5-1-2/Yecora//75γ_2-0-6
高原 602 Plateay 602	青海 Qinghai	高原 182/3987-88 (3) Gaoyuan 182/3987-88 (3)
新春 2 Xinchun 2	新疆 Xinjiang	(Siete Cerros//新春 1 号/Orofen)^{60}Coγ (Siete Cerros//Xinchun 1/Orofen)^{60}Coγ
新春 6 Xinchun 6	新疆 Xinjiang	中 7906/新春 2 号 Zhong 7906/Xinchun 2

表 2　各试点主要生态环境资料[1)]

Table 2　Characteristics of the testing environments

试　点 Location	降水量(mm)[2)] Precipitation	最高温度(℃)[3)] Max. Temp	最低温度（℃） Min. temp	光照时数（h） Sun hour	纬度（N） Latitude	经度（E） Longitude	海拔（m） Elevation
FNB	714.8	28.8	6.9	8.8	27.3°	109.1°	39
FNF	714.8	28.8	6.9	8.8	27.3°	109.1°	39
FLB	714.8	29.3	7.5	8.9	27.3°	109.1°	39
RNB	414.8	28.4	6.6	8.6	27.3°	109.1°	39
弥渡 Midu	473.6	21.1	6.5	9.6	25.1°	100.7°	1 720
哈尔滨 Harbin	384.2	22.4	10.1	5.7	45.7°	126.7°	171
克山 Keshan	449.3	22.1	15.9	8.8	48.1°	125.5°	235
呼和浩特 Huhehaote	449.2	26	11.2	8.7	40.7°	111.7°	1 041
兰州/武威 Lanzhou/Wuwei	350.9	24.1	10.5	6	36.1°	103.9°	1 517
永宁 Yongning	389.8	30.8	7.6	8.7	38.2°	106.2°	1117
西宁 Xining	553.9	19.9	5.2	7.6	36.6°	101.8°	2 275
和静 Hejing	517.6	27.1	13.1	9.3	42.2°	86.6°	1 076
乌鲁木齐 Urumqi	692.4	25.5	8.4	9.5	43.8°	87.1°	890

1）两年资料均值；2）播种至收获期间降水量，包括灌溉用水；3）播种至收获期间平均最高和最低温度。

1）Average value of the two wheat seasons；2）Represents total precipitation and irrigation water from seeding to harvest；3）Max. temp and min. temp each represent the mean maximum and minimum temperature from seeding to harvest，respectively.

1.3 性状调查

包括抽穗期（播种至抽穗天数）、成熟期（播种至成熟天数）、株高（cm）、穗数（穗/m²）、穗粒数（20穗平均数）、千粒重（g）和产量（t/hm²）。其中穗数和穗粒数在CIMMYT的4种不同处理环境未进行调查。收获时去除两边行，测产，并折算成公顷产量。

1.4 统计分析

用合适的空间模型（spatial model）对原始试验数据进行行列效应矫正[9]，误差方差同质性检验按Bartlett的方法[10]，然后用Statistical Analysis System（SAS Institute，1997）统计分析软件进行方差及方差组分分析，计算基本统计量、性状的遗传力（重复力）和相关[11]。调用SAS PROC MIXED命令，采用品种和试点固定、年度及年度相关互作和重复随机的混合线性模型进行方差分析，并根据各效应的期望均方估计方差组分。某性状的单点单年度品种平均重复力公式为：$h_l^2=\dfrac{v_l}{v_l+\dfrac{\sigma_e^2}{n_r}}$，其中$v_l$为该性状品种方差组分，$\sigma_e^2$为误差方差组分，$n_r$为重复数；某性状的多点品种平均重复力公式为

$$h_L^2=\frac{v_L}{v_L+\frac{\sigma_{(LE)}^2}{n_E}+\frac{\sigma_{(LY)}^2}{n_Y}+\frac{\sigma_{(LEY)}^2}{n_E n_Y}+\frac{\sigma_e^2}{n_E n_Y n_r}}$$

其中，v_L为该性状品种方差组分，$\sigma_{(LE)}^2$为品种与试点互作方差组分，$\sigma_{(LY)}^2$为品种与年份互作方差组分，$\sigma_{(LEY)}^2$为品种与试点及年份互作方差组分，n_E和n_Y分别为试点和年份数，其余同单点分析。调用SAS PROC CORR命令，采用品种在多个试点性状的均值进行相关分析。由于穗数和穗粒数在CIMMYT的4种不同处理环境未加调查，故在进行与穗数和穗粒数相关的分析时，只选用了中国试点的数据。

2 结果与分析

2.1 方差分析

方差同质性检验结果表明，所有性状符合误差方差同质性（表略）。将性状方差分析结果列于表3，并将性状方差组分和重复力分析结果列于表4。

方差分析表明，除穗数的基因型与试点互作效应和千粒重的试点效应不显著外，其他性状的基因型、试点和基因型与试点互作效应分别达0.05、0.01或0.001显著水平。除穗数的年度效应达0.05显著水平外，其他性状的年度效应、基因型与年度互作效应均不显著。可以看出，千粒重大小主要受基因型影响，抽穗期、成熟期、穗数、穗粒数和产量受试点环境影响较大，株高则同时受基因型和试点的影响。

从表4可以看出，所有性状的基因型与年度互作方差组分均最小，并小于误差方差组分。株高、抽穗期、成熟期和千粒重的重复力较高，其中以株高最高（0.951），千粒重次之（0.944）；它们的基因型方差组分均大于基因型与试点互作、基因型与试点及年度互作方差组分，基因型、基因型与试点及年度互作方差组分均大于误差方差组分；其中株高和抽穗期的基因型与试点互作方差组分大于误差方差组分，成熟期和千粒重的基因型与试点互作方差组分小于误差方差组分。穗数和穗粒数的平均重复力低，分别为0.429和0.524；二者的基因型方差组分和基因型与试点互作方差组分接近，并均小于误差方差组分；基因型与试点及年度互作方差组分则均大于误差方差组分。产量的重复力中等（0.752），基因型和基因型与试点互作方差组分基本相等，且基因型与试点及年度互作方差组分大于基因型、基因型与试点互作方差组分，三者均大于误差方差组分。另据单点单年度分析表明，试点间性状重复力变异较大（资料未列出）。抽穗期和千粒重的重复力高（均为0.968），且变幅较小，分别为0.780～0.999和0.800～0.996，试点间稳定性好。其次为株高（0.966），变幅为0.541～0.999。成熟期、穗数和穗粒数的重复力较低，分别为0.889、0.841和0.876；且变幅大，分别为0.410～0.994、0.444～0.962和0.299～0.995，试点间稳定性差。产量的重复力中等（0.920），变幅中等（0.656～0.994）。

从以上分析可知，千粒重基因型方差组分是基因型与试点互作方差组分的17倍，主要受基因型影响，重复力高，试点间稳定性好。穗数和穗粒数的基因型方差组分和基因型与试点互作方差组分接近，二者的重复力低，试点间变异大。产量的基因型和基因型与试点互作方差组分基本相等，重复力中等，试点间变异较大。

表3 产量、产量因子和农艺性状方差分析

Table 3 F value of analysis of variance for grain yield, yield components and other agronomic traits

变异来源 Source of variation	自由度 DF	株高 PH	抽穗期 DH	成熟期 DM	穗数[1] SN	穗粒数 GPS	千粒重 TKW	产量 GY
基因型 Genotype (G)	24	96.66***	95.10***	28.59***	4.43***	6.37***	62.98***	16.64***
试点 Location (L)	12	6.86**	23.38***	6.58**	20.63***	4.93*	2.49	10.50***
年度 Year (Y)	1	0.39	2.30	0.59	12.78*	0.73	0.87	4.04
重复 Replication	26	4.06***	1.51*	5.08***	9.22***	5.79***	4.43***	13.36***
基因型×试点 G×L	284	2.10***	6.28***	1.61**	1.28	1.81***	1.39***	2.59***
基因型×年度 G×Y	24	1.17	1.38	0.93	1.14	1.05	1.19	1.35
试点×年度 L×Y	12	36.39***	251.34***	804.51***	3.85***	13.36***	26.92***	17.93***
基因型×试点×年度 G×L×Y	270	5.17***	6.70***	6.40***	4.26***	4.61***	9.28***	6.45***

1）穗数和穗粒数的试点、基因型与试点互作、试点与年度互作、基因型与试点及年度互作的自由度分别为8、192、8和140。PH，DH，DM，SN，GPS，TKW和GY分别表示株高，抽穗期，成熟期，穗数，穗粒数，千粒重和产量。*，**和***分别表示0.05，0.01和0.001显著水平。

1）The degree of freedom of location, genotype by location, location by year, and genotype by location by year interactions for spike number and grains per spike are 8, 192, 8, and 140, respectively. PH, DH, DM, SN, GPS, TKW, and GY each represents plant height, days to heading, days to maturity, spike number, grains per spike, thousand kernel weight, and grain yield, respectively. The same as below. *, **, and *** each indicates significant difference at 0.05, 0.01, and 0.001 probability level, respectively.

表4 产量、产量因子和农艺性状方差组分和重复力分析

Table 4 Variance components for line, line by location interaction, pooled error, and line mean repeatability across locations for grain yield, yield components, and other agronomic traits

性状 Trait	基因型 G	基因型×试点 G×L	基因型×年度 G×Y	基因型×试点×年度 G×L×Y	误差 Error	重复力 Repeatability
株高 PH	140.360	16.051	0.223	22.394	14.498	0.951
抽穗期 DH	13.664	8.389	0.172	2.875	1.554	0.906
成熟期 DM	6.073	1.200	0.000	4.165	2.034	0.865
穗数 SN	870	809	155	2 818	2 685	0.429
穗粒数 GPS	3.732	5.251	0.000	9.469	7.690	0.524
千粒重 TKW	19.750	1.165	0.126	6.519	2.196	0.944
产量 GY	0.303	0.300	0.007	0.322	0.174	0.752

2.2 CIMMYT和中国品种的产量、产量因子及其他农艺性状

将CIMMYT品种、黑龙江光敏感品种和中国其他地区来源品种在中国和CIMMYT的性状基本统计量列于表5。

从表5可以看出，CIMMYT品种、黑龙江光敏感品种和中国其他品种的多数性状在中国和CIMMYT总体表现基本一致。CIMMYT品种植株矮，穗数和穗粒数多，千粒重中等，产量高；光敏感品种植株高，抽穗和成熟晚，穗数中等，穗粒数少，千粒重和产量低；中国其他品种株高中等，抽穗和成熟早，穗数少，穗粒数中等，千粒重高，产量中等。

CIMMYT品种引种到中国后，与其在Obregon试验站相比，株高平均降低13cm，抽穗和成熟分别早6d和10d，千粒重则基本保持不变，平均每公顷减产0.46t；只有SeriM 82的产量（5.64t/hm^2）基本保持不变。与其在中国相比，光敏感品种在CIMMYT株高平均增加19cm，抽穗期和成熟期分别推迟21d和16d，千粒重显著降低（4.2g），平均每公顷减产1.57t；只有龙麦26的抽穗期和成熟期分别推迟6d和4d，千粒重基本保持不变，每公顷减产0.45t。中国其他品种在CIMMYT株高平均增加17cm，抽穗期和成熟期分别推迟6d和9d，千粒重基本保持不变，平均每公顷减产0.62t；只有绵阳26和新春6号的千粒重基本保持不变或略有增加，

每公顷分别减产 0.13t 和 0.49t。CIMMYT 品种在 CIMMYT 试点的平均产量（6.06t/hm²）比黑龙江光敏感品种（3.43t/hm²）高 76.7%，比中国其他品种（5.05t/hm²）高 20.0%。CIMMYT 品种在中国各试点的平均产量（5.60t/hm²）比黑龙江光敏感品种（5.00t/hm²）高 12.0%，与中国其他品种（5.67t/hm²）基本持平。由此可见，CIMMYT 品种在 Obregon 试验站的表现明显优于中国品种，在中国则与国内品种产量接近，能基本适应中国春麦区的环境。

表 5 CIMMYT 品种和中国品种在中国和 CIMMYT 主要性状的基本统计量

Table 5 Mean, range, and standard deviation for grain yield, yield components, and other agronomic traits of CIMMYT and Chinese wheat cultivars in China and CIMMYT

性状 Trait	试点 Location	CIMMYT (10[1]) CIMMYT wheat			光敏感品种 (4) Photoperiod sensitive wheat			中国其他品种 (11) Other Chinese wheat		
		均值 Mean	变幅 Range	标准差 SD	均值 Mean	变幅 Range	标准差 SD	均值 Mean	变幅 Range	标准差 SD
株高 (cm)	CIMMYT	90	76～100	11	122	117～124	11	101	82～123	13
PH	中国 China	77	69～87	13	103	98～106	19	84	67～96	16
抽穗期 (d)	CIMMYT	86	82～89	7	107	91～112	9	86	78～91	7
DH	中国 China	80	79～82	23	86	85～88	23	80	79～82	23
成熟期 (d)	CIMMYT	137	132～141	6	147	137～153	8	136	133～141	5
DM	中国 China	127	126～128	26	131	130～133	24	127	125～129	25
穗数 (穗/m²)	CIMMYT	—	—	—	—	—	—	—	—	—
SN	中国 China	453	410～517	143	434	378～472	137	398	377～429	122
穗粒数	CIMMYT	—	—	—	—	—	—	—	—	—
GPS	中国 China	34.0	30.3～37.5	7.0	32.0	29.7～34.8	5.4	33.1	28.6～36.3	6.9
千粒重 (g)	CIMMYT	37.8	32.8～42.8	5.2	32.7	28.1～40.8	5.5	44.2	39.2～49.5	5.8
TKW	中国 China	37.4	30.3～37.5	5.2	36.9	34.5～40.3	5.4	43.8	39.7～48.1	5.7
产量 (t/hm²)	CIMMYT	6.06	5.56～6.65	1.5	3.43	2.57～4.87	1.2	5.05	4.31～5.98	1.3
GY	中国 China	5.60	5.10～6.10	1.7	5.00	4.59～5.32	1.5	5.67	4.98～6.47	1.7

1）表示品种数；—表示数据缺失。

1）represents number of cultivars,；‘—’represents not determined.

2.3 CIMMYT 品种在中国各试点和当地主栽品种的产量比较

将 CIMMYT 品种在中国各试点和当地主栽品种的产量比较结果列于表 6。

从表 6 可以看出，CIMMYT 品种在弥渡的平均产量（4.40t/hm²）高于当地主栽品种（3.85 t/hm²），高产品种 Attila(5.04t/hm²）的产量比当地高产主栽品种(3.99t/hm²）高 26.3%。CIMMYT 品种在哈尔滨(4.98t/hm²）和克山(2.95t/hm²）的平均产量均低于当地主栽品种，高产品种 Baviacora M 92（5.62t/hm²）和 Inqalab 91（3.27t/hm²）也均低于当地所有主栽品种（5.86～6.15t/hm² 和 3.80～3.97t/hm²)。CIMMYT 品种在呼和浩特平均产量为 4.93t/hm²，高产品种 Culiacan T 89 的产量（5.73t/hm²）虽然略低于当地高产主栽品种（5.82t/hm²），但高于当地低产主栽品种（5.41t/hm²)。同样，CIMMYT 品种在兰州/武威平均产量为 5.95t/hm²，高产品种 Bacanora T 88 的产量（6.36t/hm²）接近当地主栽品种（6.44t/hm²)。CIMMYT 品种在永宁平均产量为 5.69t/hm²，高产品种 Culiacan T 89 的产量（6.43t/hm²）虽然低于当地高产主栽品种（7.17t/hm²)，但高于当地低产主栽品种（5.65t/hm²)。CIMMYT 品种在西宁平均产量为 7.40t/hm²，高产品种 Rayon F 89 的产量（8.47t/hm²）比当地高产主栽品种（8.17t/hm²）高 3.7%。CIMMYT 品种在和静和乌鲁木齐平均产量分别为 8.10t/hm² 和 5.95t/hm²，高产品种 Baviacora M 92 的产量（9.53t/hm² 和 6.8t/hm²）分别比当地高产主栽品种（9.30t/hm² 和 6.31t/hm²）高 2.5%和 7.8%。

表6 CIMMYT品种在中国各试点和当地主栽品种的产量比较

Table 6 Comparison for grain yield of CIMMYT and Chinese local cultivars in 9 Chinese locations

试点 Location	CIMMYT品种（10）CIMMYT wheat			当地主栽品种 Local cultivars		
	均值 Mean	变幅 Range	标准差 SD	均值 Mean	变幅 Range	标准差 SD
弥渡 Midu（Attila[1]）	4.40	3.56～5.04	1.0	3.85（3[2]）	3.63～3.99	1.0
哈尔滨 Harbin（Baviacora M 92）	4.98	4.39～5.62	0.8	6.01（2）	5.86～6.15	0.5
克山 Keshan（Inqalab 91）	2.95	2.32～3.27	0.5	3.89（2）	3.80～3.97	0.3
呼和浩特 Huhhot（Culiacan T 89）	4.93	3.98～5.73	1.0	5.61（2）	5.41～5.82	0.5
兰州/武威 Lanzhou/Wuwei（Bacanora T 88）	5.95	5.14～6.36	0.7	6.44（1）	6.44	0.4
永宁 Yongning（Culiacan T 89）	5.69	4.63～6.43	0.7	6.41（2）	5.65～7.17	1.2
西宁 Xining（Rayon F 89）	7.40	6.36～8.47	1.0	7.55（3）	7.11～8.17	1.6
和静 Hejing（Baviacora M 92）	8.10	7.23～9.53	0.9	9.22（2）	9.15～9.30	0.6
乌鲁木齐 Urumqi（Baviacora M 92）	5.95	5.20～6.80	0.7	6.08（2）	5.86～6.31	0.9

1）表示在该试点产量表现最高的CIMMYT品种；2）表示当地品种数。

1）represents the highest yielding CIMMYT cultivars at that location；2）represents the number of local cultivars.

从以上分析可知，CIMMYT品种在中国各试点的适应性差异较大。CIMMYT高产品种在云南、青海和新疆的产量高于当地高产主栽品种；在内蒙古、甘肃和宁夏高产品种的产量虽然低于当地高产主栽品种，但高于当地低产主栽品种；在黑龙江高产品种的产量则低于当地所有主栽品种，这主要是由于其对光照不敏感所致。

2.4 中国各地主栽品种在CIMMYT和各自来源地的性状表现

将中国各地主栽品种在CIMMYT灌溉、正常播期、平播环境（代表一般的选种环境）和各自来源地的性状表现列于表7。

表7 中国各试点主栽品种在CIMMYT和各自来源地的性状表现

Table 7 Grain yield, yield components, and other agronomic traits of Chinese cultivars in the full irrigation flat planting environment and their original locations

性状 Trait	试点 Location	云南（3[2]） Yunnan	黑龙江（4） Heilongjiang	宁夏[3]（3） Ningxia	青海（3） Qinghai	新疆（2） Xinjiang
株高（cm）	CIMMYT（98[1]）	96	126	102	123	103
PH	来源地	81	78	85	115	92
抽穗期（d）	CIMMYT（91）	94	110	86	94	87
DH	来源地	136	69	81	82	66
成熟期（d）	CIMMYT（138）	139	149	135	138	136
DM	来源地	178	107	128	150	113
千粒重（g）	CIMMYT（40.9）	49.5	35.0	48.1	42.2	49.8
TKW	来源地	43.1	38.5	42.1	39.9	50.3
产量（t/hm^2）	CIMMYT（7.74）	5.57	4.32	6.62	4.42	7.41
GY	来源地	3.85	4.79	6.06	7.55	7.65

1）表示灌溉、正常播期、平播环境中10个CIMMYT品种的平均值；2）表示当地品种数；3）包括内蒙古、甘肃和宁夏。

1）represents the mean of the 10 CIMMYT wheats in CIMMYT under full irrigation, sown normally, and planted on the flat environment；2）represents the number of local cultivars；3）includes Inner Mongolia, Gansu, and Ningxia Provinces.

从表7可以看出，与其在各自来源地相比，在CIMMYT灌溉、正常播期、平播环境，中国云南主栽品种株高平均增加15cm，比当地CIMMYT品种平均矮2cm；抽穗期和成熟期分别早42d和39d，比

CIMMYT品种分别晚3d和1d；千粒重显著增加(6.4g)，比CIMMYT品种高8.6g；平均每公顷增产1.72t，但比CIMMYT品种低2.17t。黑龙江主栽品种株高平均增加46cm，比CIMMYT品种高29cm；抽穗期和成熟期分别推迟40d和42d，比CIMMYT品种分别晚19d和11d；千粒重显著降低(3.7g)，比CIMMYT品种低5.9g；平均每公顷减产0.53t，比CIMMYT品种低3.42t。内蒙古、甘肃和宁夏主栽品种株高平均增加17cm，比CIMMYT品种高4cm；抽穗期和成熟期分别推迟5d和7d，比CIMMYT品种分别早5d和3d；千粒重显著增加(6.0g)，比CIMMYT品种高7.2g；平均每公顷增产0.56t，但比CIMMYT品种低1.12t。青海主栽品种株高平均增加8cm，比CIMMYT品种高25cm；抽穗期推迟12d，比CIMMYT品种晚3d，成熟期早12d，与CIMMYT品种相同；千粒重增加2.3g，比CIMMYT品种高1.3g；平均每公顷减产3.13t，比CIMMYT品种低3.32t。新疆主栽品种株高平均增加11cm，比CIMMYT品种高5cm；抽穗期和成熟期分别推迟21d和26d，比CIMMYT品种分别早4d和2d；千粒重基本相同，比CIMMYT品种高8.9g；平均每公顷减产0.24t，比CIMMYT品种低0.33t。

从以上分析可知，中国各地主栽品种在CIMMYT灌溉、正常播期、平播环境中性状的表现差异较大。与CIMMYT品种相比，云南品种植株略矮、抽穗和成熟略晚、千粒重高、产量低；黑龙江品种植株高、抽穗和成熟晚、千粒重和产量低；内蒙古、甘肃和宁夏品种植株略高、抽穗和成熟早、千粒重高、产量低；青海品种植株高、抽穗晚、成熟期相同、千粒重略高、产量低；新疆品种植株高、抽穗早、成熟略早、千粒重高、产量略低。

3 讨论

引种可以大大丰富中国的小麦遗传资源，拓宽遗传基础，是改良我国品种的有效途径。碧玉麦、南大2419、阿勃、阿夫、墨巴65等品种在中国的应用和推广表明，及时引进和合理利用国外种质是提高中国小麦生产和育种成效的关键[2,12]。

已有研究表明，中国春麦区的品种改良以应用CIMMYT小麦为主[2]。在国内已有研究的基础上[3-5]，本研究在Obregon试验站和中国的CIMMYT小麦适应地区同时种植CIMMYT种质和当地主栽品种，表明CIMMYT代表性品种穗数和穗粒数多，籽粒大小中等，对光周期不敏感，具有广泛适应性，比中国品种具有更高的产量潜力。另有资料表明其最新育成的品种产量潜力更高[13]，在澳大利亚也表现穗数和穗粒数较多、高产和稳产特性[14]，这充分说明了CIMMYT品种的广泛适应性。CIMMYT按降水量、温度等因素将全世界划分为12个大环境[8,15]，根据各环境的育种目标配制杂交组合，通过穿梭育种、抗病性筛选和国际多点鉴定等方法[14]，使其育成品种具有广泛适应性。因此，可以利用CIMMYT品种穗数和穗粒数多、适应性广的特性进一步改良中国春麦品种。另据研究，CIMMYT近年育成品种的蛋白质品质优于中国品种，可以利用CIMMYT品种进一步改良中国小麦的面筋强度[6]。

本试验是在宛秀兰等的研究基础上[3-5]，采用多个品种在国内多个试点和CIMMYT Obregon试验站按正规的产量比较试验进行，因此资料的代表性和可靠性高，所得结论与宛秀兰等的研究结果基本一致，对指导CIMMYT小麦在中国的应用具有一定的参考价值。CIMMYT品种在云南、青海和新疆等地可以直接推广种植，如墨巴65等品种在云南和新疆等地生产中的应用[2]，Attila已通过新疆品种审定。另有资料显示，CIMMYT品种在云南、青海和新疆等地的千粒重也较高，这可能与云南、青海等地海拔较高，日平均温度较低，生育期长有关；而新疆地区的日较差较大，日平均最高温度较高、最低温度较低，灌溉条件好。孙本普等研究表明小麦灌浆期长，日平均最高温度较高，日平均最低温度较低可使千粒重增加[16]。CIMMYT品种的光敏感性明显不如黑龙江品种，其产量和千粒重在黑龙江地区也显著低于当地品种，可见光敏感性是影响CIMMYT品种在黑龙江地区适应性的主要因素。需要说明的是，CIMMYT近年来加强了高纬度地区的育种工作，新近育成的高代品系在黑龙江的适应性明显提高。因此，黑龙江应优先利用CIMMYT高纬度环境的种质。

此外，本试验和笔者多年的观察结果表明，CIMMYT品种对叶锈病具有突出的抗性，对条锈病的抗性也优于中国大多数品种，但在白粉病的抗性方面则基本无优势。因此，中国各单位在CIMMYT选择优良种质时，在利用CIMMYT材料的品质和抗病性前提下，应参考有关性状在墨西哥和我国各地种植时的变化规律灵活掌握，以有效提高引进材料的利用效率。考虑到穗数和穗粒数重复力低，千粒重重复力高，产量重复力中等，选种时应重点选择籽粒较大的材料。云南所选材料可略矮、适当晚熟；黑龙江所选材料可较高和晚熟；内蒙古、甘肃、宁夏和新疆所选材料可略高、较早熟；青海所选材料可较高、熟期相当。

4 结论

CIMMYT品种穗数和穗粒数多，千粒重中等，具有广泛适应性；黑龙江光敏感品种植株高，抽穗和成熟晚，穗数中等，穗粒数少，千粒重和产量低；中国其他品种株高中等，抽穗和成熟早，穗数少，穗粒数中等，千粒重高，产量中等。可以利用CIMMYT品种穗数和穗粒数多、适应性广的特性进一步改良中国春麦品种。CIMMYT品种可在云南、青海和新疆直接推广种植；内蒙古、甘肃和宁夏为次适宜地区，主要用作杂交亲本；在黑龙江以作杂交亲本为宜。在CIMMYT选种时，为云南所选材料可略矮、适当晚熟，内蒙古、甘肃、宁夏和新疆所选材料可略高、较早熟，青海所选材料可较高、熟期相当，黑龙江应主要选择高纬度材料、植株偏高且晚熟。

参考文献

[1] Pingali P L (Eds). CIMMYT 1989-1999 world wheat facts and trends. Global wheat research in a changing world: challenges and achievements. Mexico, D. F.: CIMMYT. 1999.

[2] He Z H, Rajaram S. China/CIMMYT collaboration on wheat breeding and germplasm exchange: results of 10 years of shuttle breeding (1984-94). *Wheat Special Report No*. 46. Mexico, D. F.: CIMMYT. 1997.

[3] 宛秀兰．引种墨西哥小麦的初步研究—关于墨西哥小麦适应地区的分析．作物学报，1981，17（4）：249-257.

Wan X L. A preliminary study on the adaptability of Mexican wheat varieties in China. *Acta Agronomica Sinica*, 1981, 7 (4): 249-257. (in Chinese)

[4] 姚金保，周朝飞，钱存鸣，姚国才，杨学明．江苏与CIMMYT小麦穿梭育种进展与展望．麦类作物学报，1998，18（5）：14-16.

Yao J B, Tang C F, Qian C M, Yao G C, Yang X M. The progress of the shuttling breeding program between Jiangsu and CIMMYT. *Journal of Triticeae Crops*, 1998, 18 (5): 14-16. (in Chinese)

[5] 袁汉民，吴淑筠，张富国，钱晓曦．宁夏墨麦种质资源研究．宁夏农林科技，1998，(4)：8-12.

Yuan H M, Wu S J, Zhang F G, Qian XX. Studies on genetic resources of Mexican wheat in Ningxia. *Ningxia Agronomy and Forest*, 1998, (4): 8-12. (in Chinese)

[6] 吴振录，张勇，何中虎，樊哲儒，辛文利，邵立刚，李元清，杨文雄，魏亦勤，马晓刚，潘超，刘艳萍．CIMMYT小麦在我国的产量和品质表现．麦类作物学报，2004，24（3）：34-39.

Wu Z L, Zhang Y, He Z H, Fan Z R, Xin W L, Shao L G, Li Y Q, Yang W X, Wei Y Q, Ma X G, Pan C, Liu Y P. Performance on yield and quality of CIMMYT wheat in China. *Journal of Triticeae Crops*, 2004, 24 (3): 34-39. (in Chinese)

[7] 李生荣．绵阳号小麦品种改良及系谱分析//何中虎，张爱民．中国小麦育种研究进展．北京：中国科学技术出版社，2002：146-150.

Li S R. Improvement and pedigree analysis of Mianyang seris wheat. In: He Z H, Zhang A M. *Advance of Wheat Breeding in China*. Beijing: China Science and Technology Press, 2002: 146-150. (in Chinese)

[8] Rajaram S, Hettel G P (Eds.). Wheat Breeding at CIMMYT: commemorating 50 years of research in Mexico for Global wheat improvement. *Wheat Special Report . No*. 29. Mexico, D. F.: CIMMYT. 1995.

[9] Gilmour A R, Cullis B R, Verbyla A P. Accounting for natural and extraneous variation in the analysis of field experiments. *Journal of Agricultural Biological and Environmental Statistics*, 1997, 2: 269-293.

[10] Bartlett R A. Nearestneighbour models in the analysis of field experiments (with discussion). *Journal of the Royal Statistical Society*, 1978, B 40: 147-174.

[11] SAS Institute. *SAS User's Guide*: *Statistics*. SAS Institute, Cary, NC. 1997.

[12] 董玉琛，郑殿升．中国小麦遗传资源．北京：中国农业出版社，2000.

Dong Y C, Zheng D S. *Chinese Wheat Germplasm*. Beijing: China Agriculture Press, 2000. (in Chinese)

[13] Trethowan R M, van Ginkel M, Rajaram S. Progress in breeding wheat for yield and adaptation in global drought affected environments. *Crop Science*, 2002, 42: 1441-1446.

[14] Cooper M, Byth D E, Woodruff D R. An investigation of the grain yield adaptation of advanced CIMMYT wheat lines to water stress environments in

Queensland. I. Crop physiological analysis. *Australia Journal of Agricultural Research*, 1994, 45: 965-984.

[15] Dubin H J, Rajaram S. Breeding disease-resistant wheats for tropical highlands and lowlands. *Annual Review of Phytopathology*, 1996, 34: 503-526.

[16] 孙本普，王勇，李秀云，王广元，刘锋，李凤云，张金帮，王继诰，孙爱梅，王宝忠，王淑英，王峰，李萌，朱学群．气象条件对冬小麦千粒重的影响．麦类作物学报，2003，23（4）：52-56.

Sun B P, Wang Y, Li X Y, Wang G Y, Liu F, Li F Y, Zhang J B, Wang J G, Sun A M, Wang B Z, Wang S Y, Wang F, Li M, Zhu X Q. Effect of climatic elements on thousand grain weight of winter wheat. Journal of Triticeae Crops, 2003, 23 (4): 52-56. (in Chinese)

Pattern analysis on grain yield performance of Chinese and CIMMYT spring wheat cultivars sown in China and CIMMYT

Y. Zhang[1], Z. H. He[1,2], A. M. Zhang[3], M. V. Ginkel[4], and G. Y. Ye[5]

[1] *Institute of Crop Sciences/National Wheat Improvement Center*, *Chinese Academy of Agriculture Sciences* (*CAAS*), *Beijing* 100081, *China*; [2] *CIMMYT-China Office*, *c/o CAAS*, 100081 *Beijing*, *China*; [3] *Institute of Genetics and Developmental Biology*, *Chinese Academy of Sciences* (*CAS*), *Beijing* 100101, *China*; [4] *CIMMYT*, *Apdo. Postal* 6-641, 06600, *Mexico*, *D. F.*, *Mexico*; [5] *School of Land and Food*, *The University of Queensland*, *Brisbane Qld* 4072, *Australia*

Abstract: Understanding the relationships among testing environments is essential for better targeting cultivars to production environments. To identify patterns of cultivar, environment, cultivar-by-environment interactions, and opportunities for indirect selection for grain yield, a set of 25 spring wheat cultivars from China and the International Maize and Wheat Improvement Center (CIMMYT) was evaluated in nine environments in China and four management environments at CIMMYT in Cd. Obregon, Mexico, during two wheat seasons. Genetic background and original environment were the main factors influencing grain yield performance of the cultivars. Baviacora M 92, Xinchun 2 and Xinchun 6 showed relatively more stable and higher grain yields, whereas highly photoperiod sensitive cultivars Xinkehan 9, Kefeng 6 and Longmai 19 proved consistently inferior across environments, except in Harbin and Keshan, the two high latitude environments. Longmai 26, also from high latitude environments in the north-eastern Heilongjiang province, was however probably not as photoperiodicly sensitive as other cultivars from that region, and produced much higher grain yield and expressed a broader adaptation. None of the environments reported major diseases. Pattern analyses revealed that photoperiod response and planting option on beds were the two main factors underlying the observed interactions for grain yield. The production environment of planting on the flat in Mexico grouped together with Huhhot and Urumqi in both wheat seasons, indicating an indirect response to selection for grain yield in this CIMMYT managed environment could benefit the two Chinese environments. Both the environment of planting on the flat with Chinese Hejin and Yongning, and the three CIMMYT environments planting on raised beds with Chinese Yongning grouped together only in one season, showing that repeatability may not be stable in this case.

Key words: *Triticum aestivum*, bread wheat, grain yield, cultivar-by-environment interaction, pattern analysis

Introduction

Bread wheat (*Triticum aestivum* L.) is the second most important food crop following rice in China, with an annual production of more than 90 million tonnes harvested from over 23 million ha, which represents 16.5% of world wheat production. Grain yield has

Published in Euphytica, 2006, 147 (3): 409-420

long been the major objective of Chinese wheat breeding programs (He et al., 2001).

The wheat breeding program at the International Maize and Wheat Improvement Center (CIMMYT) headquartered in Mexico has a mission to develop new wheat germplasm that is widely adapted and produces high and stable yields across a broad range of environments in less developed countries including China (Rajaram, 1995). CIMMYT germplasm can be directly released for production in Xinjiang and Yunnan provinces of China, or used as crossing parents in breeding programs in such provinces as Inner Mongolia, Gansu and Ningxia (He et al., 1997). The introduced CIMMYT wheats have made significant contributions to yield improvements in China's spring-sown spring wheat regions over the past four decades (He et al., 1997). A shuttle breeding program was also established between China and Mexico to introgress the high yield potential and rust resistance of CIMMYT wheats into locally adapted Chinese wheats (He et al., 1997; Zhuang, 2003). In addition to Yunnan, which is located in the south-western autumn-sown wheat region, the targeted Chinese locations for this shuttle breeding program include five zones within the spring-sown spring wheat regions (He et al., 2001). Annually hundreds of new advanced lines from CIMMYT are received through CIMMYT's Wheat International Nurseries and/or through on-site selection by Chinese visitors and are tested in China.

It is well documented that wheat grain yield is influenced by cultivar, environment, and cultivar-by-environment interaction (Brennan & Sheppard, 1985; Cooper et al., 1994), although the relative importance of these sources differs. Knowledge on the effect of cultivar-by-environment interaction is required for the design of an efficient and economic selection strategy involving a shuttle component (Gauch & Zobel, 1997; Kang, 1998). Thus it is essential to understand the responses of CIMMYT wheats in targeted Chinese locations and the performance of Chinese wheats in various testing environments in Mexico.

Many statistical methods have been developed to analyze data from multi-environment plant variety trails to gain a better understanding of the observedcultivar-by-environment interactions and cultivar stability patterns. A worthwhile discussion on many of these methods and their relationships can be found in reviews by DeLacy et al. (1996) and Crossa et al. (2004). Pattern analysis has been applied to many multi-environment trials and shown to be very effective (Abdalla et al., 1996; Redden et al., 2000; Trethowan et al., 2001, 2002, 2003; Lillemo et al., 2004). It is based on the joint and complementary use of classification and ordination techniques (Williams, 1976). Cluster analyses summarize the complexity of the data while retaining most of the information by grouping cultivars with similar performance into relatively few environment groups that produce similar discrimination among the performance of cultivars (DeLacy et al., 1996). Relationships between cultivar performance and environment discrimination are depicted in a low dimensional representation of the original data with just a few ordination axes. Cultivars with similar performance and environments that produce similar discrimination patterns among cultivars are placed close together on the biplot (Gabriel, 1971; Kempton, 1984). Knowledge of environment classification and ordination is an important pre-requisite for effectively targeting yield trials to representative environments. This can reduce the total number of trials grown, thereby lowering trial costs while still allowing characterization of broad and/or specific adaptation patterns among cultivars.

To characterize the responses of CIMMYT wheats in targeted Chinese locations and the performances of Chinese wheats in CIMMYT's breeding environments, a multi-environment trial using major Chinese and CIMMYT spring wheat cultivars was conducted in various environments in China and Mexico in two successive wheat seasons from 2000 to 2002. The objectives of this report are: (1) to assess the interrelations between cultivars and environments; (2) to evaluate changes in grain yields of cultivars across environments; and (3) to seek opportunities for using indirect selection for grain yield in CIMMYT's management environments to identify lines relevant to China's spring wheat regions, by using the well-established pattern analysis method.

Materials and Methods

Cultivars and testing environments

Fifteen leading Chinese spring wheat cultivars including four photoperiod-sensitive lines from Heilongjiang province were selected from six provinces (Yunnan, Heilongjiang, Gansu, Ningxia, Qinghai, and Xinjiang) to cover the complete range of bread wheat types grown in China's spring wheat regions, where CIMMYT wheats can be used as cultivars directly or as crossing parents (He et al., 1997). Mianyang 19, Mianyang 20 and Mianyang 26, developed in Sichuan province, were the leading cultivars in Yunnan province and thus were also included in this trial. Ten CIMMYT cultivars were selected based on their adaptation performance across numerous international sites. Detailed cultivar information can be seen in Table 1. The nine Chinese testing environments, i. e., Midu from Yunnan, Harbin and Keshan from Heilongjiang, Huhhot from Inner Mongolia, Lanzhou/Wuwei from Gansu, Yongning from Ningxia, Xining from Qinghai, and Urumqi and Hejin from Xinjiang, ranged from 25N to 48N in latitude, and 86E to 126E in longitude, representing a wide range of environmental conditions. The four management environments at the CIMMYT experiment station in Cd. Obregon in north-western Mexico were (1) full irrigation (five irrigations), using the normal sowing date, on raised beds (FNB), (2) full irrigation, normal sowing date, on the flat (FNF), (3) reduced irrigation (three irrigations), normal sowing date, on raised beds (RNB), and (4) full irrigation, sown two months later than normal (resulting in heat stress during grain-filling), on raised beds (FLB). Detailed environmental information is shown in Table 2.

Table 1 Code, cultivar, abbreviation, origin, and mean grain yield of 25 cultivars across all 13 environments over the two seasons

Code	Cultivar	Origin	GY[a]
1	Seri M 82	CIMMYT	5.61±1.57
2	Bacanora T 88	CIMMYT	5.89±1.67
3	Culiacan T 89	CIMMYT	5.83±1.71
4	Attila	CIMMYT	5.93±1.57
5	Turaco	CIMMYT	5.31±1.53
6	Rayon F 89	CIMMYT	5.88±1.57
7	Weaver	CIMMYT	5.31±1.57
8	Baviacora M 92	CIMMYT	6.32±1.77
9	Super Seri#1	CIMMYT	5.58±1.75
10	Inqalab 91	CIMMYT	5.87±1.46
11	Mianyang 19	Sichuan/China	4.86±1.30
12	Mianyang 20	Sichuan/China	4.85±1.52
13	Mianyang 26	Sichuan/China	5.46±1.57
14	Xinkehan 9	Heilongjiang/China	4.04±1.44
15	Kefeng 6	Heilongjiang/China	4.50±1.52
16	Longmai 19	Heilongjiang/China	4.69±1.64
17	Longmai 26	Heilongjiang/China	5.24±1.47
18	Longchun 15	Gansu/China	5.21±1.32
19	Ningchun 4	Ningxia/China	5.67±1.49
20	Ningchun 16	Ningxia/China	5.91±1.75
21	Qingchun 533	Qinghai/China	5.79±1.56
22	Qingchun 566	Qinghai/China	5.18±1.49
23	Plateau 602	Qinghai/China	5.27±1.57
24	Xinchun 2	Xinjiang/China	6.09±1.87
25	Xinchun 6	Xinjiang/China	6.32±1.71

a, Mean and standard deviation for grain yield.

Table 2 Characteristics of the testing environments during the two seasons[a]

Environment	FNB	FNF	FLB	RNB	Midu[c]	Harbin	Keshan	Huhhot	Lanzhou /Wuwei	Yongning	Xining	Hejin	Urumqi
Environment code	FNB	FNF	FLB	RNB	Mid	Har	Kes	Huh	Lan /Wuw	Yon	Xin	Hej	Uru
Rainfull/irrigation[b]	4.8+700	4.8+700	4.8+700	4.8+400	173.6+300	84.2+300	149.3+100	49.2+400	350.9	39.8+400	203.9+300	17.6+600	92.4+600
	18.3+700	18.3+700	18.3+700	18.3+400	345.5+300	114.2+300	385.3	331.0+300	98.6+200	153.0+200	296.6+300	77.3+400	182.8+700
Max. temp (℃)[c]	28.6	28.6	28.8	28.3	21.9	23.3	21.7	25.7	25.1	30.6	19.8	27.9	23.2
	29	29	29.7	28.6	20.4	21.5	22.5	26.3	23.0	31.1	20.1	26.3	28.1
Min. temp	7.7	7.7	8.1	7.4	6.4	10.3	15.4	11.4	10.7	7.0	4.8	13.4	8.3
	6.1	6.1	6.8	5.9	6.7	9.9	16.3	10.9	10.2	8.1	5.6	12.7	8.5
Sun hour	8.5	8.5	8.6	8.4	9.7	6.1	8.6	8.8	6.0	9.0	7.7	8.5	9.7
	9.1	9.1	9.2	8.9	9.4	5.2	9.2	8.6	5.9	8.3	7.5	10.0	9.4
N (kg)	250	250	250	150	337	112	246	187	225	412	262	555	300
	250	250	250	150	337	75	246	225	225	450	262	600	300
P (kg)	40	40	40	40	450	150	204	285	225	187	150	240	75
	40	40	40	40	450	75	204	337	225	178	150	75	75
K (kg)	No	No	No	No	150	75	30	No	No	75	No	No	No
	No	No	No	No	150	67	30	No	75	75	No	No	No
GY (t ha^{-1})[d]	5.55±1.35	6.28±1.70	4.13±0.55	4.93±1.08	5.09±1.47	5.55±0.55	2.96±0.55	5.26±0.91	6.10±0.88	5.93±0.94	7.98±1.00	7.99±0.98	6.00±0.76
	6.08±1.62	6.70±1.60	3.93±1.19	4.37±0.73	3.82±0.80	4.61±0.79	3.56±0.35	4.22±1.08	6.02±0.94	5.27±0.85	6.12±1.23	7.82±0.98	5.28±0.96
Soil type	sandy clay	sandy clay	sandy clay	sandy clay	sandy loam	chernozem	chernozem	loam	loam	loam	sandy loam	loam	sandy loam
Latitude (N)	27.3°	27.3°	27.3°	27.3°	25.1°	45.7°	48.1°	40.7°	36.1°/ 38.1°	38.2°	36.6°	42.2°	43.8°
Longitude (E)	109.1°	109.1°	109.1°	109.1°	100.7°	126.7°	125.5°	111.7°	103.9°/ 102.4°	106.2°	101.8°	86.6°	87.1°
Elevation (m)	39	39	39	39	1720	171	235	1041	1517 / 1740	1117	2275	1076	890

a, The upper and lower line each represent the data from the 2000-2001 and 2001-2002 seasons.
b, Rainfall/irrigation represents total precipitation and irrigation water from seeding to harvest.
c, Max. temp and min. temp each represent the mean maximum and minimum temperature from seeding to harvest, respectively.
d, Mean and standard deviation for grain yield.

Experimental design and testing traits

All trials except those in Lanzhou and Wuwei were conducted during two wheat seasons from October, 2000 to August, 2002. Cultivars were sown in a latinized alpha lattice design with three replications in each environment (Barreto et al., 1997). The trials were hand-weeded, and bird scaring practices were used from anthesis to harvest to prevent grain losses. In China each plot consisted of 6 rows, 4 m long, 0.20m spacing between rows, at a sowing rate of 300kg ha^{-1} in all environments except Midu, where 150kg ha^{-1} was used. Only plants in the inner four rows over a length of 3.5m were harvested to avoid border effects. The sowing rate in the four management environments at CIMMYT was 130kg ha^{-1}. The FNB environment was sown in late November, each entry planted on two beds with three rows on top of each bed, 3m in length, followed by full furrow irrigation of about 700 mm of water in total throughout the crop cycle in the form of five separate irrigations. The harvested area was 4.8 m^2. This planting method and management type simulated bed-planting under Mega-environment 1 (ME1) (Rajaram et al., 1995). The FNF environment was also sown in late November, each entry planted on the flat (one irrigation basin per replication) as 8 single rows 3.8m in length per entry, followed by full flood irrigation of about 700mm in total. The harvested area was 3.125m^2. This planting method also simulated ME1, but on the flat as it is still practiced popularly around the world. The RNB environment was sown in late November, planted in the same fashion as FNB, but with reduced irrigation (a total of about 400mm, applied in three separate irrigations). This planting method simulated Mega-environment 4A (ME4A), in which post-flowering drought stress occurs. The FLB environment was sown in mid January to expose material to heat stress during grain-filling; each entry was planted on two beds, each containing three rows, 2.8m in length, followed by full irrigation of about 700mm in total. The harvested area was 4.48m^2. This planting method simulated Mega-environment 5 (ME5), where late heat stress occurs under conditions of full irrigation and moderate relative humidity. The harvested grain was air-dried to approximately 12% moisture before weighing, and grain yield (GY) was translated into t ha^{-1}.

Data analysis

All trials were separately analyzed by fitting an appropriate spatial model with rows and columns (Gilmour et al., 1997). The best linear unbiased predictions from the best-fit model were used as raw data for all subsequent analyses. An analysis of variance (ANOVA) was performed by treating all effects as fixed except replications, years, and interactions involving years. Pattern analyses for both cultivars and environments were performed based on standardized mean data for each environment, following Fox & Rosielle (1982). Classification of cultivars and environments was performed using an agglomerative hierarchical clustering procedure with squared Euclidean distance as the dissimilarity measure (Williams, 1976), and Ward's method, which uses incremental sums of squares as the clustering strategy (Ward, 1963). The efficacy of classification was determined by examining the partitioning of the sums of squares among sources. Cultivar classifications were truncated for the sake of brevity when 50% of the cultivar-by-environment (CE) information was retained in the reduced matrix (Byth et al., 1976). Dendrograms and cultivar group performance plots were constructed by plotting mean grain yields for cultivar groups against an environment group index based on the untransformed mean yield of the environment.

Ordination was done on the environment standardized mean yield data using a singular value decomposition algorithm with results represented by a biplot (Gabriel, 1971; Kempton, 1984). The data were modelled in two dimensions, and the efficacy of the model was determined by the proportion of the sum of squares accounted for by each vector. Within the biplot, cultivars were plotted as points and environments as direction vectors. The total variance for each environment was 1.00, and the length of the vectors indicated how well the environments were modelled in the biplot. If an environment was modelled perfectly, the length of the vector would be 1.00. Thus the angles between the vectors represented the correlation structure among the environments. The cultivar points and environment vectors allowed characterization of regions with similar cultivar performance within particular environments.

Results

Analysis ofvariance

The analysis of variance indicated that environmental main effect was the predominant source of variation, followed by cultivar-by-environment interaction (Table 3). There was a significant effect of year, and no significant cultivar-by-year interaction. The cultivar-by-environment interaction was much larger than the environment-by-year interaction and the cultivar-by-environment-by-year interaction. The proportions of the total sum of squares due to environment (E), cultivar (C), and cultivar-by-environment interaction (CE) were 68%, 10%, and 22% in the 2000-2001 season, and 60%, 15%, and 25% in the 2001-2002 season (Table 4). The ratios of the sum of squares due to cultivar-by-environment interaction and those due to cultivar main effect were 1.67 and 2.20 in the two seasons, indicating a large cultivar-by-environment interaction for grain yield.

Table 3 Sums of squares from a combined analysis of variance for grain yield of 25 cultivars grown in 13 environments across the two wheat seasons

Source	DF	SS
Y[a]	1	81.63*[b]
C	24	607.34***
E	12	2846.92***
C×E	284	825.83***
R (EY)	52	96.18***
C×Y	24	36.35
E×Y	12	241.70***
C×E×Y	272	303.37***
Error	1202	178.41

a, Y=year, C=cultivar, E=environment, C×E=cultivar-by-environment interaction, R (EY) = replication nested within environment-by-year interaction, C×Y=cultivar-by-year interaction, E×Y=environment-by-year interaction, C×L×Y=cultivar-by-environment-by-year interaction.

b, * and *** Significant at the P=0.05 and 0.001, respectively.

Table 4 Proportion of sum of squares (SS) for grain yield due to sources of cultivar (C), environment (E), and cultivar-by-environment interaction (CE); truncation level; proportion of SS retained in reduced array; proportion of CE sum of squares due to the first two vectors of singular value decomposition

Wheat season	% SS			Truncation level	% SS retained			PCA1	PCA2
	C	E	CE		C	E	CE		
2000-2001	10	68	22	7	73	47	53	44	16
2001-2002	15	60	25	7	84	55	54	49	13

Pattern analysis

Applying truncation levels where 50% or more of the sum of squares of the cultivar-by-environment interactions were retained, led to seven cultivar and environment groups in both seasons. This resulted in 85% reduction in the data array size, while retaining 73% and 84% of the sum of squares for the cultivars, and 47% and 55% of the sum of squares for the environments. The results of the classification analyses for cultivars and environments are presented in dendrograms (Figs. 1 and 2). Cultivar group performance plots are presented in Fig. 3. The ordination biplots are presented in Fig. 4, with the first two principal component vectors accounting for 60% and 62% of the total variation in the 2000-2001 and 2001-2002 seasons, respectively.

Classification

Cultivars Baviacora M 92 and Inqalab 91 were the first to separate in the dendrogram, their average yield being a high of 6.32 and 5.87t ha^{-1}, respectively (Fig. 1, Tables 1 and 5). The second group consisted of Turaco, Weaver and Super Seri # 1, with mean yields of 5.31, 5.31 and 5.58t ha^{-1}. Mianyang 19, Mianyang 20, Mianyang 26, Longmai 26 and Longchun 15 were in the next group to be separated, with mean yields ranging from 4.85 to 5.46t ha^{-1}. The fourth group consisted of Xinkehan 9, Kefeng 6 and Longmai 19, with mean yields between 4.04 and 4.69 t ha^{-1}. Ningchun 4, Qingchun 533, Qingchun 566 and Plateau 602 made up the fifth group, with mean yields varying from 5.18 to 5.79t ha^{-1}. The sixth group consisted of three cultivars, Ningchun 16, Xinchun 2 and Xinchun 6, with mean yields ranging from 5.91 to 6.32t ha^{-1}. Five cultivars, Seri M 82, Bacanora T 88, Culiacan T 89, Attila and Rayon F 89, made up the seventh and last group, with mean yields ranging from 5.61 to 5.93t ha^{-1}.

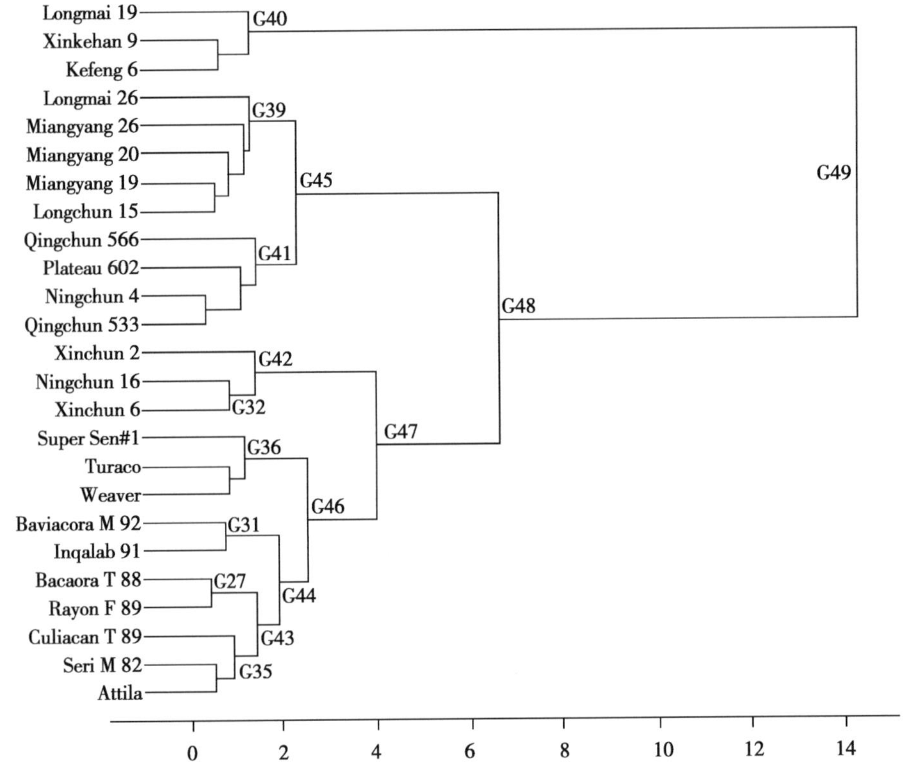

Fig. 1 Dendrogram of the classification of cultivars using Ward's method on envionnent standardised mean yield across the two seasons.

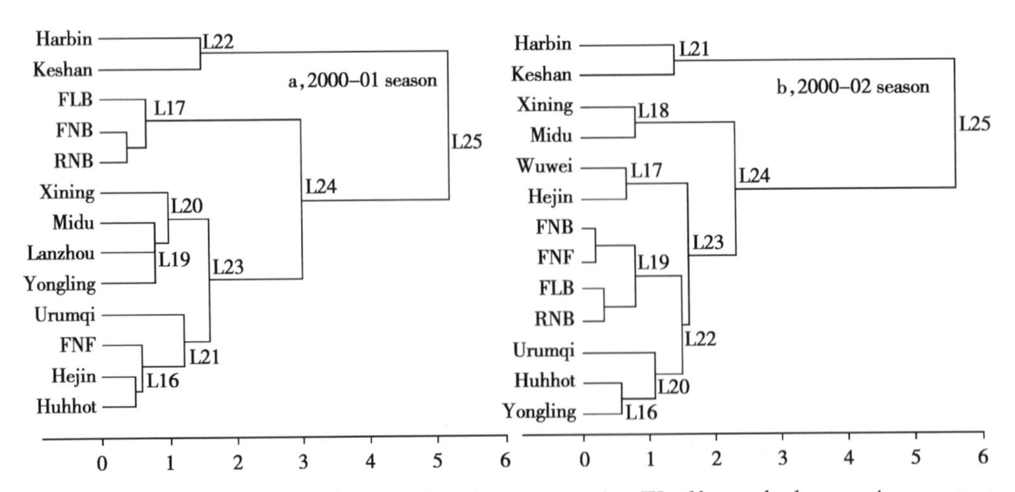

Fig. 2 Dendrograms of the classification of environments using Ward's method on environempt standardised mean yield. separately for each season (a and b).

Table 5 Cultivar group number and membership for grain yield for 25 cultivars grown in 13 environments across the two seasons

Cultivar group	Cultivar membership
G34	Baviacora M 92, Inqalab 91
G36	Turaco, Weaver, Super Seri＃1
G39	Mianyang 19, Mianyang 20, Mianyang 26, Longmai 26, Longchun 15
G40	Xinkehan 9, Kefeng 6, Longmai 19
G41	Ningchun 4, Qingchun 533, Qingchun 566, Plateau 602
G42	Ningchun 16, Xinchun 2, Xinchun 6
G43	Seri M 82, Bacanora T 88, Culiacan T 89, Attila, Rayon F 89

Environment classification indicated two major groups in the two seasons (Fig. 2). Harbin and Keshan always grouped together, with an average yield of 5.55 and 2.96t ha^{-1} in the 2000-2001 season and 4.61 and 3.56t ha^{-1} in the 2001-2002 season (Tables 2 and 6). The rest of the environments clustered in the other group. Truncating at the seven-group level, in the 2000-2001 season, Harbin, Keshan, Urumqi and Xining each separated into individual groups. CIMMYT's full irrigation (FNB) and reduced irrigation

Table 6 Environment group number and membership for grain yield for 25 cultivars grown in 13 environments during the two seasons

Wheat season	Environment group	Environment membership
2000-2001	Kesh	Keshan
	Har	Harbin
	Xin	Xining
	Uru	Urumqi
	L16	FNF, Huhhot, Hejin
	L17	FNB, FLB, RNB
	L19	Lanzhou, Yongning, Midu
2001-2002	Kesh	Keshan
	Har	Harbin
	Uru	Urumqi
	L16	Huhhot, Yongning
	L17	Wuwei, Hejin
	L18	Xining, Midu
	L19	FNB, FNF, FLB, RNB

environments sown at the normal time, planted on beds (RNB), and the full irrigation environment sown late and planted on beds (FLB) clustered together in the fifth group. The sixth group consisted of Lanzhou, Yongning, and Midu. CIMMYT's full irrigation environment, sown at the normal time and planted on the flat (FNF), Huhhot, and Hejin clustered together in the last group. In the 2001-2002 season, Harbin, Keshan and Urumqi each separated into individual groups. Huhhot and Yongning grouped together. The fifth group consisted of Xining and Midu. The four CIMMYT environments grouped together, while Wuwei and Hejin clustered together in the seventh group.

Performance plot

The response plot indicated that cultivar group 34 'Baviacora M 92 and Inqalab91' and group 39 'Mianyang 19, Mianyang 20, Mianyang 26, Longmai 26 and Longchun 15' expressed virtually no interaction with environments, and therefore can be considered to have stable yields across all environments in the 2000-2001 season (Fig. 3a, Tables 5 and 6). The first group had higher grain yields than the second group across environments. Group 36 'Turaco, Weaver and Super Seri ＃1' showed low interaction with environments and expressed low-to-medium yields across environments. This group showed the poorest yield in the high yielding environment Xining, demonstrating its lack of specific adaptation to this environment. Specific adaptation to low yielding environments Keshan and Harbin was shown by the low yielding, highly photoperiod-sensitive group 40 'Xinkehan 9, Kefeng 6 and Longmai 19', corresponding to the main split for cultivar classification (Fig. 1). This group was low yielding in almost all environments, including those with low and high mean yield, except in their region of origin in Harbin and Keshan. Group 41 'Ningchun 4, Qingchun 533, Qingchun 566 and Plateau 602' had intermediate-to-high yields in most environments and expressed low interaction with environment, and highly specific adaptation to the high yielding environment Xining. The performance plot of group 42 'Ningchun 16, Xinchun 2 and Xinchun 6', which was separated at the third split level, showed a distinct yield advantage in all high yielding environments, but poor adaptation to Harbin and Keshan. Group 43 'Seri M 82, Bacanora T 88, Culiacan T 89, Attila and Rayon F 89' also expressed poor adaptation to Keshan and Harbin, but was well adapted to the high yielding environment group 16 'FNF, Huhhot, Hejin' and environment group 17, which included all three modes of planting wheat on raised beds at CIMMYT.

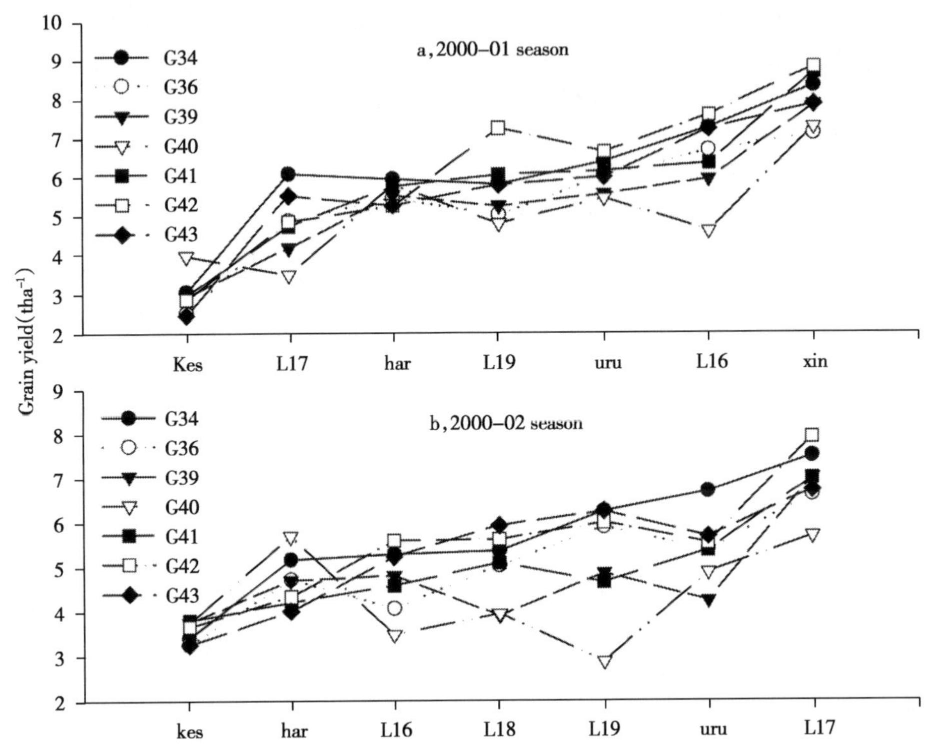

Fig. 3 The mean grain yield performance plots of cultivar groups devired from the classification using Ward's method against untransformed environment group mean yield from low to high, separately for each season (a and b).

The response plot of the seven cultivar groups across seven environment groups in the 2001-2002 season also revealed certain patterns, often similar to the previous year 2000-2001 though not always (Fig. 3b). Cultivar group 34 'Baviacora M 92 and Inqalab 91' and group 36 'Turaco, Weaver and Super Seri # 1' expressed relatively little interaction with environments, and therefore can be considered as having stable yields across all environments. The former had high yields across all environments, while the latter showed medium-to-low yield. Group 39 'Mianyang 19, Mianyang 20, Mianyang 26, Longmai 26 and Longchun 15' had medium yield in all environments except environment group 18 'Xining and Midu' and Urumqi, demonstrating its lack of specific adaptation to those environments. The performance of group 40 'Xinkehan 9, Kefeng 6 and Longmai 19' was consistently inferior across all environments except in Harbin and Keshan, to which they showed specific adaptation. Cultivar group 41 'Ningchun 4, Qingchun 533, Qingchun 566 and Plateau 602' expressed little interaction with environments, with medium-to-high yield across environments and good adaptation to Keshan. Group 42 'Ningchun 16, Xinchun 2 and Xinchun 6' produced high yield in environment group 16 'Huhhot and Yongning' and high yielding environment group 17 'Wuwei and Hejin'. This group also expressed poor yield at Harbin, as did cultivar group 43 'Seri M 82, Bacanora T 88, Culiacan T 89, Attila and Rayon F 89'. Both groups gave good yields in environment group 18 'Xining and Midu', group 19, which included all CIMMYT environments, and at Urumqi.

Ordination

The position and perpendicular projection of cultivar points relative to environment vectors could be used to determine whether a cultivar was specifically adapted to a given environment. Cultivars that were positioned further along the positive direction of a vector tended to show higher grain yield, reflecting better adaptation to that environment (Kempton, 1984). In the 2000-2001 season biplot the maximum angle among the vectors of the four CIMMYT environments was well below 90 degrees, corresponding to environments FNB and RNB (Fig. 4a), indicating they ranked cultivars similarly. Ten CIMMYT cultivars were the top yielding cultivars in these environments, with Culiacan T 89 and Rayon F 89 performing the best. Longmai 26, Ningchun 4, Ningchun 16 and Xinchun 6 were the only Chinese cultivars that performed well in these CIMMYT environments. The Chinese environments did not show the same degree of commonality as the CIMMYT types, with Huhhot and Keshan making an angle of more than 90 degrees with most other Chinese environments, indicating that they are quite different from the rest. All Chinese environments except Harbin and Keshan tended to discriminate among cultivars in a very similar fashion, with Xinchun 2 and Xinchun 6 consistently expressing high yield. Cultivars such as Mianyang 20, Xinkehan 9, Kefeng 6, Longmai 19, Qingchun 566, Plateau 602 and Xinchun 2 had high yields in Harbin and Keshan. The angle between Harbin and Keshan tended to be zero, indicative of the two locations being very similar. On average the angle between Harbin, Keshan and the other Chinese environments, the other Chinese environments and CIMMYT environments all tended towards 90 degrees. Comparing the environmental vectors for Harbin and Keshan with CIMMYT environments revealed an angle of nearly 180 degrees. Cultivar discrimination in Harbin and Keshan was thus expected to be in almost the opposite direction to that of CIMMYT environments. Thus there appeared to be three distinct groups of environments discriminating among cultivars.

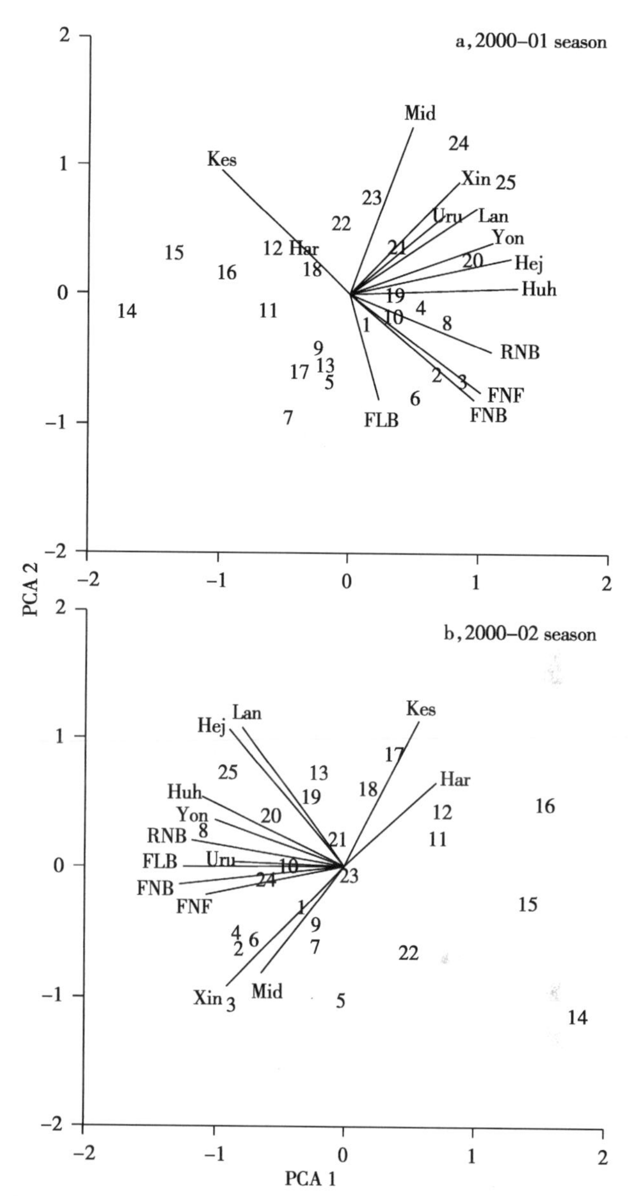

Fig. 4 Biplots using singular value decomposition of environment standardised mean yield from the ordination of environment standardised mean yield, separately for each season (a and b).

The angles among the vectors of four CIMMYT management environments were much below 90 degrees in the 2001-2002 season (Fig. 4b). Ten CIMMYT cultivars, together with Ningchun 16, Xinchun 2 and Xinchun 6 expressed top yields in these environments, with Rayon F 89 performing the best. The Chinese environments showed a large dispersion; Harbin and Keshan, in which cultivars such as Mianyang 19, Mianyang 20, Xinkehan 9, Kefeng 6, Longmai 19, Longmai 26, Longchun 15 and Ningchun 4 had high yield, tended to group together. Xining and Midu grouped together, with CIMMYT cultivars Bacanora T 88, Culiacan T 89, Attila, Turaco and Rayon F 89 performing the best. The environment vectors for Harbin and Keshan made an angle of almost 180 degrees with those of Xining and Midu. Cultivar discrimination in these environments was therefore expected to be in the opposite direction to those of Xining and Midu. All the other Chinese environments grouped together with the CIMMYT environments, and thus had similar cultivar discrimination to the CIMMYT environments. Xining and Midu were closer to CIMMYT environments than Harbin and Keshan.

Discussion and implications

In most Chinese spring wheat regions and the four CIMMYT management environments, there is a ubiquitous presence of cultivar-by-environment (CE) interaction for grain yield, which is in agreement with previous reports (Brennan & Sheppard, 1985; Cooper et al., 1994). The three principal strategies for dealing with such CE interactions are ignoring, avoiding, or exploiting them (Eisemann et al., 1990; Cooper et al., 1996). In most situations, ignoring CE interactions as a source of error or bias in assessing a cultivar (random, non-repeatable CE interactions) is not a practical strategy, since the aim is to provide farmers with adapted, but also stable, new cultivars. CE interaction can be managed by aiming for avoidance through better classification of the testing environments relative to the target population of environments (TPE). This will allow identification of sub-environments within which cultivars do not interact with environmental factors. Such cultivars can be considered as having specific adaptation. This approach of determining specifically adapted cultivars minimizes the impact of significant interactions, and allows targeting more narrowly adapted cultivars to various sub-environments (Gauch & Zobel, 1997; Kang, 1998). Exploiting the CE interactions (i. e., by selecting for broad or narrow adaptation jointly) relies on its repeatability within the TPE (DeLacy et al., 1996). The relative merit of selecting for broad and specific adaptation is dependent on the nature of the cultivar by environment interaction, particularly whether they have some degree of repeatability within the TPE (DeLacy et al., 1996).

The fact that most of thetotal variance was explained by the environment reflected a much wider range of environment main effects than cultivar main effects in these trials. Diseases were not a major factor influencing this study as none of the environments reported major diseases, or the trials were protected with fungicides. For the environments included in this study, the results suggest the existence of three environment groups for grain yield. Harbin and Keshan, both of which are high latitude environments, always clustered together at the first fusion level, expressing a high degree of association in both seasons. Specific cultivars appear to be adapted to Harbin and Keshan, suggesting that selection for specific adaptation to this region will result in faster genetic progress than selection for wide adaptation to all environments. The implication is that photoperiod sensitivity at such high latitude locations is one of the most important factors in determining adaptation. Photoperiod can be used to structure the environments into environment types, since there is a high degree of association between Harbin and Keshan. With repeatable cultivar by environment interactions across years, it is possible to structure these multiple environments. To facilitate more focused breeding for such specific environments, environments could be partitioned along mega-environmental lines. CIMMYT has taken this approach, partitioning the world into 12 mega-environments according to such factors as precipitation, temperature, soil type and photoperiod (Rajaram et al., 1995).

The three CIMMYT management environments of planting on raised beds (FNB, FLB, and RNB) clustered together and separated from the environment planting on the flat in the 2000-2001 season. This indicates that for grain yield expression planting on beds at CIMMYT results in a ranking of cultivars that does not correspond to the ranking of the same cultivars planted on the flat, highlighting significant cultivar-by-planting method interaction. This is a finding worthy of attention, since planting on the flat in basins is currently still the most common planting method in China. Collaborative work on bed-planting conducted over the past several years in China and elsewhere has indicated that at the same yield levels, inputs such as seeding rate, fertilizer, and water can be reduced by up to 10-30% when planting on beds (Wang et al., 2004; K. Sayre, personal information). Reduced diseases and lodging are also observed under bed planting conditions. The large (up to 30%) savings in irrigation water without yield penalty, and the results of this study, which indicate the existence of cultivar-by-planting method interactions, suggests that Chinese wheat breeders should start comparing planting on the flat with bed planting.

However, the four CIMMYT management environmentsall grouped together in the 2001-2002 season (Fig. 1, Table 5). This indicates that the difference between flat and bed environments is not always consistent. The only significant differences between the two seasons for the four CIMMYT environments were temperature and hours of sunshine. The hours of sunshine and mean maximum temperature were lower in the 2000-2001 season than in the 2001-2002 season, but the mean minimum temperature was higher. Therefore, these factors may need to be taken into consideration when comparisons are made between the flat and bed planting methods.

The angle between Harbin, Keshan and Midu, Xining is much lower in the 2000-2001 season biplot than that in the 2001-2002 season biplot: i. e., the angle between Harbin and Keshan, and Midu and Xining, tended towards 90 degrees in the 2000-2001 season, whereas the environment vectors for Harbin and Keshan formed an angle of almost 180 degrees with those of Xining and Midu in the 2001-2002 season. The cultivar discrimination at these environments changed between the two seasons. The only significant difference between the two seasons in these environments was in precipitation. Therefore, precipitation seems to be another factor that influences cultivar performance in these environments besides photoperiod sensitivity, since rainfall in these environments in the 2000-2001 season was much lower than in 2001-2002.

The use and study of managed environments reported in this research is an attempt to understand the effect of wheat cultivar screening regimes in CIMMYT and China on adaptation parameters. The CIMMYT planted on the flat environment grouped together with Chinese Huhhot and Urumqi in both seasons, and is much more likely to be useful for indirect selection for the two environments. Only in one of the seasons did the CIMMYT planted on the flat environment group with Chinese Yongning and Hejin, and the other three CIMMYT planting on raised beds environments group with Chinese Yongning, Therefore the level of repeatability of indirect selection of the planting on flat environment at CIMMYT with Yongning and Hejin, and of the other three CIMMYT planting on raised beds environments with Yongning, needs to be established over a number of years. The three CIMMYT environments of planting on raised beds including FLB were grouped together in both seasons, indicating that the different agronomic practices (reduced irrigation and later sowing) used to create these managed environments had only a relatively small effect on the relative cultivar performance. Previous studies indicated that it is difficult to create useful cultivar-by-environment interaction patterns by use of managed environments in a single location (Cooper et al., 1996). Managed environments can only be useful if they are created by manipulating the key biotic and/or abiotic factors underlying the GE interaction. Therefore, it may be more appropriate to first investigate the GE interaction pattern and identify the major reasons for the observed GE interaction using multi-environment trials, and to then establish a set of managed environments to measure the GE interaction.

Separation of cultivars by genetic background and site of origin is most evident in the third dendrogram split level (Fig. 1), where all CIMMYT wheat cultivars are in one group that can be further split into two groups, with Baviacora M 92 expressing the highest yield over the two seasons. Interestingly, Ningchun 16, Xinchun

2 and Xinchun 6 make up one of the groups at the third level with high yield performance, and can be further grouped together with all of the CIMMYT cultivars at the second fusion level. They were all derived from crosses of Chinese wheats with CIMMYT lines. Xinchun 2 is a parent of Xinchun 6. Xinkehan 9, Kefeng 6 and Longmai 19 all originate from Heilongjiang, are highly sensitive to photoperiod, and group together at the first fusion level. The performances of these cultivars were consistently inferior in all environments, except in Harbin and Keshan, from which they originated.

An interesting case is that of the wide adaptation of Longmai 26, from Heilongjiang province, which expresses moderately good grain yield across all environments at levels much higher than those of the other cultivars from Heilongjiang province. This may mean that it is not probably as highly photoperiodicly sensitive as the other cultivars from Heilongjiang. Seemingly the photoperiodic insensitive cultivar was selected with wider adaptation at the same time. Another cultivar with wide adaptation is Ninchun 4, which is near the center of the biplot for grain yield (Fig. 4) and hence adapted to all environments. Ninchun 4 was bred using three selection locations, i. e., Ningxia, Yunnan and Henan, during the segregating phase (Qui, personal information). In view of the above, in-depth study of the shuttle breeding methodology in China seems warranted. This approach, in which plants are selected alternatively in two environmentally contrasting sites, is the mainstay of CIMMYT's global breeding program.

Pattern analysis allowed us to summarize the results of this experiment in a sensible and useful way; it also helped to examine the natural relationships and variations in cultivar performance among different environment groups. Pattern analysis assisted in structuring environments in this study, leading to the identification of three environment groups for grain yield. Several sub-environment groups were identified within these environment groups, with photoperiod response used to structure the environments into TPEs. A number of Chinese environments have vectors located close to one another in the biplots, indicating that they discriminate among the cultivars in a similar way. Thus it may be possible to reduce the number of testing environments, thereby economizing when conducting multi-environment trials. The CIMMYT environments tend to be located close to one another and away from most Chinese locations. The dendrogram analyses indicate that one of the possible reasons may be that none of the present Chinese environments uses bed-planting, a planting method used almost exclusively at CIMMYT but now also spreading rapidly in parts of Asia, including Henan Province in China (Wang et al., 2004). Therefore an additional conclusion of this research is that bed-planting and its interactions with varietal adaptation need to be studied more.

Acknowledgments

Financial support was kindly provided by the National Natural Science Foundation (Project No. 30060043 and 39930110) and international collaborative project on wheat improvement from Ministry of Agriculture of the People's Republic of China.

References

Abdalla, O. S., J. Crossa, E. Autrique & I. H. DeLacy, 1996. Relationships among international testing sites of spring durum wheat. Crop Sci., 36: 33-40.

Barreto, H. J., G. O. Edmeades, S. C. Chapman & J. Crossa, 1997. The alpha lattice design in plant breeding and agronomy: Generation and analysis, In: G. O. Edmeades, M. Banziger, H. R. Mickelson & C. B. Pña-Valdivia (Eds.), [1997] Developing Drought- and Low N-Tolerant Maize: Proceedings of a Symposium. CIMMYT, Mexicopp. pp. 25-29. Mexico, D. F.: CIMMYT.

Brennan, P. S. & J. A. Sheppard, 1985. Retrospective assessment of environments in the determination of an objective strategy for the evaluation of the relative yield of wheat cultivars. Euphytica, 34 397-408.

Byth, D. E., R. L. Eisemann & I. H. DeLacy, 1976. Two-way pattern analysis of a large data set to evaluate genotypic adaptation. Heredity, 37: 215-230.

Cooper, M., D. E. Byth & D. R. Woodruff, 1994. A preliminary investigation of the grain yield adaptation of advanced CIMMYT wheat cultivars to water stress environments in Queensland 2. Classification analysis. Aust J Agric Res., 45: 985-1002.

Cooper, M., P. S. Brennan & J. A. Sheppard, 1996. A strategy for yield improvement of wheat which accommodates large genotype-by-environment interactions. In: M. Cooper & G. L. Hammer (Eds.) Plant Adaptation and Crop Improvement. CAB International/IRRI/ICRISAT: Wallingford, pp. 487-511.

Crossa, J. Rong-Cai Yang & P. L. Cornelius, 2004. Studying crossover genotypes x environment interaction using linear-bilinear models and mixed models. Journal of Agricultural, Biological, and Environmental Statistics, 9: 362-380.

DeLacy, I. H., K. E. Basford, M. Cooper, J. K. Bull & C. G. McLaren, 1996. Analysis of multi-environment trials: an historical perspective. In: M. Cooper &G. L. Hammer (Eds.), Plant Adaptation and Crop Improvement. CAB International/IRRI/ICRISAT: Wallingford, pp. 39-124.

Eisemann, R. L., M. Cooper & D. R. Woodruff, 1990. Beyond analytical methodology - better interpretation and exploration of genotype-by-environment interaction in breeding. In: M. S. Kang (Eds.), Genotype-by-Environment Interaction and Plant Breeding. . Louisiana State University, Baton Rouge, Louisiana, pp. 108-117.

Fox, P. N. & A. A. Rosielle, 1982. Reducing the influence of environmental main-effects on pattern analysis of plant breeding environments. Euphytica, 31: 645-656.

Gabriel, K. R., 1971. The bi-plot-graphical display of matrices with application to principal component analysis. Biometrika, 58: 453-467.

Gauch, H. G. & R. W. Zobel, 1997. Identifying mega-environments and targeting genotypes. Crop Sci., 37: 311-326.

Gilmour, A. R., B. R. Cullis & A. P. Verbyla, 1997. Accounting for natural and extraneous variation in the analysis of field experiments. J Agr Bio Env Statist, 2: 269-293.

He, Z. H. & S. Rajaram (Eds.), 1997. China/CIMMYT Collaboration on Wheat Breeding and Germplasm Exchange: Results of 10 Years of Shuttle Breeding (1984-94). Wheat Special Report No. 46. Mexico, D. F.: CIMMYT.

He, Z. H., S. Rajaram, Z. Y. Xin & G. Z. Huang (Eds.), 2001. A History of Wheat Breeding in China. Mexico, D. F.: CIMMYT. pp. 1-14.

Kang, M. S., 1998. Using genotype-by-environment interaction for crop cultivar development. Adv Agron, 62: 199-252.

Kempton, R. A., 1984. The use of bi-plots in interpreting variety by environment interactions. J Agric Sci., 103: 123-135.

Lillemo, M., M. van Ginkel, R. M. Trethowan, E. Hernandez& S. Rajaram, 2004. Associations among international CIMMYT bread wheat yield testing locations in high rainfall areas and their implications for wheat breeding. Crop Sci., 44: 1163-1169.

Rajaram, S., M. Van Ginkel & R. A. Fischer, 1995. CIMMYT's wheat breeding mega-environments (ME). In: Z. S. Li & Z. Y. Zin (Eds.), Proc 8th Int Wheat Genet Symp, China Agricultural Scientech Press, Beijing, China. pp. 1101-1106.

Redden, R. J., I. H. DeLacy, D. G. Butler & T. Usher, 2000. Analysis of line by environment interactions for yield in navy beans. 2. Pattern analysis of cultivars and environment within years. Aust J Agrci Res., 51: 607-617.

Trethowan, R. M., J. Crossa, M. van Ginkel & S. Rajaram, 2001. Relationships among bread wheat international yield testing locations in dry areas. Crop Sci., 41: 1461-1469.

Trethowan, R. M., M. van Ginkel & S. Rajaram, 2002. Progress in breeding wheat for yield and adaptation in global drought affected environments. Crop Sci., 42: 1441-1446.

Trethowan, R. M., M. van Ginkel, K. Ammar, J. Crossa, T. S. Payne, B. Cukadar, S. Rajaram & E. Hernandez, 2003. Associations among twenty years of international bread wheat yield evaluation environments. Crop Sci., 43: 1698-1711.

Wang F. H, X. Q. Wang & K. D. Sayre, 2004. Comparison of conventional, flood irrigated, flat planting with furrow irrigated, raised bed planting for winter wheat in China. Field Crops Res., 87: 35-42.

Ward, J. H., 1963. Hierarchical grouping to optimize an objective function. J Amer Statist Assoc., 58: 236-244.

Williams, W. T., 1976. Pattern Analysis in Agricultural Science. Elsevier Scientific Publishing Company, Amsterdam.

Zhuang Q. S. (Eds.), 2003. Chinese Wheat Breeding and Pedigree Analysis. China Agriculture Press, Beijing, China.

春化和光周期基因等位变异在23个国家小麦品种中的分布

杨芳萍[1,2]，韩利明[3]，阎俊[4]，夏先春[1]，张勇[1]，曲延英[3]，王忠伟[1]，何中虎[1,5]

[1]中国农业科学院作物科学研究所/国家小麦改良中心，北京 100081；[2]甘肃省农业科学院小麦研究所，兰州 730070；[3]新疆农业大学农学院，乌鲁木齐 830052；[4]中国农业科学院棉花研究所，安阳 455000；[5]CIMMYT 中国办事处，北京 100081

摘要：为促进国外资源在我国小麦育种中的有效利用，以小麦春化基因 *Vrn-A1*、*Vrn-B1*、*Vrn-D1* 和 *Vrn-B3* 及光周期位点 *Ppd-D1* 标记对23个国家的755份品种检测，同时在河南安阳秋播，观察抽穗期和成熟期。分子标记检测结果表明，*Vrn-A1*、*Vrn-B1*、*Vrn-D1* 和 *vrn-A1*+*vrn-B1*+ *vrn-D1* 的分布频率分别为13.0%、21.1%、15.6%和64.2%，显性等位变异 *Vrn-B3* 在检测材料中缺失。春化基因显性等位变异 *Vrn-A1*、*Vrn-B1* 和 *Vrn-D1* 主要分布在中国春麦区和长江中上游冬麦区、意大利、印度、日本、加拿大、墨西哥、智利、阿根廷和澳大利亚，上述地区的小麦一般为春性类型；春化位点均为隐性等位变异或 *vrn-A1*+*vrn-D1*+*Vrn-B1* 的品种主要分布在中国北方、美国中部和南部、德国、法国、挪威、乌克兰、俄罗斯、伊朗、土耳其、匈牙利、保加利亚、罗马尼亚和塞尔维亚，这些地区的小麦为冬性类型。光周期迟钝型 *Ppd-D1a* 的分布频率为55.2%。光周期敏感等位变异 *Ppd-D1b* 主要分布在纬度较高的地区，即美国各麦区以及德国、挪威、匈牙利、中国东北、加拿大、智利和阿根廷，来自其余麦区的品种均携带光周期迟钝等位变异 *Ppd-D1a*；携带 *Ppd-D1a* 的品种在河南安阳大部分能够成熟，而携带 *Ppd-D1b* 的品种在河南安阳基本不能成熟。在安阳春化显性等位变异 *Vrn-A1a* 未加速小麦抽穗，而携带 *Vrn-B1* 和 *Vrn-D1* 等位变异的部分春化需求品种能够正常抽穗，主要因安阳生长季节的温度能够满足春化需求。

关键词：小麦；春化基因；光周期基因；分子鉴定；冬春性；抽穗期

Distribution of Allelic Variation for Genes of Vernalization and Photoperiod among 755 Cultivars from 23 Wheat Countries

Yang Fangping[1,2], Han Liming[3], Yan Jun[4], Xia Xianchun[1], Zhang Yong[1], Qu Yanying[3], Wang Zhongwei[1], He Zhonghu[1,5]

[1] *Institute of Crop Sciences / National Wheat Improvement Center, Chinese Academy of Agricultural Sciences* (*CAAS*), *Beijing* 100081, *China*; [2] *Wheat Research Institute, Gansu Academy of Agricultural Sciences, Lanzhou* 730070, *China*; [3] *College of Agronomy, Xinjiang Agricultural University, Urumqi* 830052, *China*; [4] *Cotton Research Institute, CAAS, Anyang* 455000, *China*; [5] *CIMMYT China Office, Beijing* 100081, *China*

Abstract: Molecular markers for vernalization genes *Vrn-A1*, *Vrn-B1*, *Vrn-D1* and *Vrn-B3* and

原文发表在《作物学报》，2011，37（11）：1917-1925

photoperiod gene *Ppd-D1* were used to detect the presence of these genes among 755 cultivars from 23 countries. Days to heading and physiological maturity of these cultivars were also recorded in Anyang, Henan Province, China to provide information for their utilization in Chinese wheat breeding program. Frequencies of *Vrn-A1*, *Vrn-B1*, *Vrn-D1*, and *vrn-A1*+*vrn-B1*+*vrn-D1* were 13.0%, 21.1%, 15.6% and 64.2%, respectively. Dominant allele *Vrn-B3* was absent in all tested materials. Dominant vernalization alleles *Vrn-A1*, *Vrn-B1*, and *Vrn-D1* were mainly observed in Chinese spring wheat and middle and upper Yangtze Valley winter wheat regions, Italy, India, Japan, Canada, Mexico, Chile, Argentina, and Australia with spring type, while cultivars carryied all recessive alleles at the four vernalization loci. The gene recombination of *vrn-A1*, *vrn-D1*, and *Vrn-B1* was found in winter wheat regions of Northern China, middle and southern US, Germany, France, Norway, Ukraine, Russia, Turkey, Iran, Hungary, Bulgaria, Romania, and Serbia, where the wheat growth habit is winter type. The frequency of *Ppd-D1a* was 55.2%, and photoperiod sensitive allele *Ppd-D1b* was mainly observed in cultivars from higher latitude regions of US, Germany, Norway, Hungary, Northeastern China, Canada, Chile and Argentina; while photoperiod insensitive allele *Ppd-D1a* was observed in the other wheat-growing regions. Most of cultivars with photoperiod insensitive allele *Ppd-D1a* could complete physiological maturity in Anyang, whereas cultivars from Germany, Norway, Hungary, Northwestern US, Northeast China, Chile and Argentina could not mature well. In Anyang, flowering time was not speeded up by the presence of dominant vernalization allele *Vrn-A1a*, cultivars with *Vrn-B1* and *Vrn-D1* could head normally due to the completion of vernalization requirement during winter season.

Key words: Common wheat; Vernalization gene; Photoperiod gene; Molecular marker; Wheat growth habit; Heading date

小麦是世界性的重要粮食作物，广泛分布于不同国家，其适应性主要受春化、光周期和早熟性基因控制[1-3]。春化基因决定生长习性，影响开花时间，依据春化基因种类及通过春化阶段所需的温度和时间将小麦划分为冬性和春性[4,5]。小麦春化基因至少有六个，即*Vrn-A1*、*Vrn-B1*、*Vrn-D1*、*Vrn-B3*、*Vrn-D3* 和 *Vrn-D4*，分别位于 5A、5B、5D、6B、7D 和 5D 上[6-11]。*Vrn-A1*、*Vrn-B1* 和 *Vrn-D1* 统称为 *VRN1*，这三个基因对春化的影响程度不同，显性变异 *Vrn-A1* 的作用最强[12]。当 *Vrn-A1* 和 *Vrn-D1* 的任何一个为显性时，小麦的生长习性为春性；若 *VRN1* 位点三个基因全为隐性或与 *vrn-A1*、*vrn-D1* 和 *Vrn-B1* 结合时，小麦的生长习性为冬性[10]。*Vrn-A1*、*Vrn-B1*、*Vrn-D1* 和 *Vrn-B3* 已被克隆[9,13]，开发的功能标记已用于品种检测[12,14-16]，我国育成品种和地方品种中春化基因的分布状况与冬春性的一致性较高[12,14,15]。也有研究表明，春化基因 *Vrn-A1*、*Vrn-B1*、*Vrn-D1* 和 *Vrn-B3* 控制的遗传机制尚不能很好解释冬麦生长发育[10]，这可能与光周期基因的影响有关[1,2]。光周期基因包括 *Ppd-D1*（*Ppd1*）、*Ppd-B1*（*Ppd2*）和 *Ppd-A1*（*Ppd3*），分别位于染色体 2D、2B 和 2A，其中前两者的作用较强[17,18]。Tanio 等[19]利用近等基因系发现 *Ppd-B1* 和 *Ppd-D1* 均能加速短日照条件下幼穗分化，但 *Ppd-B1* 对光迟钝的作用较 *Ppd-D1* 强。Beals 等[20]根据大麦 *Ppd-H1* 基因，同源克隆到小麦 *Ppd* 基因，在 *Ppd-B1* 基因序列中没有发现与功能相关的多态性，但在 *Ppd-D1* 基因序列中发现编码序列上游存在 2 089bp 的缺失，并开发了 *Ppd-D1* 的功能标记，该标记已成功用于品种鉴定[16,21]，国内小麦材料中 *Ppd-D1a* 的分布与其对光周期的要求相吻合[21]。

为了解主要国家品种在我国主产麦区的适应性，提高国外引种的针对性，本研究采用 *VRN1* 和*VRN-B3* 位点春化基因、*PPD-D1* 位点光周期基因分子标记对来自 23 个国家的 755 份品种进行检测，并在河南安阳对其抽穗期和成熟期进行田间观察，目的是分析上述基因的分布频率，初步明确春化基因、光周期基因及其组合与抽穗期的关系。因 *Vrn-D4*、*Vrn-D3*、*Ppd-B1* 和 *Ppd-A1* 位点尚无有效的分子标记[10,11,20]，故不能对其进行基因鉴定。

1 材料和方法

1.1 供试材料

供试 755 份小麦品种（系）来自 23 个国家，其中中国品种 50 份（北部冬麦区 8 份，黄淮麦区 22 份，长江中上游冬麦区 6 份，东北春麦区 6 份，西北春麦区和新疆春麦区 8 份），引进品种包括美国 81 份（中部平原 25 份，西北部 45 份，南部 11 份），德国 75 份，法国 98 份，英国 3 份，挪威 22 份，荷兰 5 份，乌克兰 14 份，俄罗斯 23 份，伊朗 30 份，土耳其 89 份，匈牙利 10 份，保加利亚 18 份，罗马尼亚 23 份，塞尔维亚 10 份，印度 5 份，意大利 10 份，日本 10 份，加拿大 26 份，墨西哥 5 份，智利 41 份，阿根廷 33 份，澳大利亚 74 份。这 23 个国家占全球小麦总产量的 90%以上，所用材料皆为各国的主栽品种或育成的最新品系，也包括少数国际知名的历史品种，基本上反映了各国小麦生产和育种的现状。春化基因对照品种 Thatcher (*Vrn-A1a*)、中国春（*Vrn-D1*）和辽春 10 号（*Vrn-B3*）由本课题组保存。

1.2 春化和光周期基因的分子标记检测

每份材料随机取 3 粒种子，按 Lagudah 等[22]的方法提取基因组 DNA，以 3 粒种子的检测结果确认材料基因型。引物序列及其相关信息见表 1。引物由北京奥科生物技术有限公司合成。

表 1 春化基因位点 *Vrn-A1*、*Vrn-B1*、*Vrn-D1*、*Vrn-B3* 和光周期位点 *Ppd-D1* 引物序列、扩增片段及其相关信息

Table 1 Primer sequences, expected polymerase chain reaction (PCR) product sizes and related information for detected alleles at loci *Vrn-A1*, *Vrn-B1*, *Vrn-D1*, *Vrn-B3* and *Ppd-D1*

等位基因 Allele	引物名称 Primer name	引物序列 Sequence (5′-3′)	片段大小 Fragment size (bp)	退火温度 Annealing temperature (℃)	每循环延伸时间 Extending time in each cycle (s)	参考文献 Reference
Vrn-A1						
Vrn-A1a	VRN1AF	GAAAGGAAAAATTCTGCTCG	965 + 876	50	60	Yan et al. [23]
Vrn-A1b	VRN1-INT1R	GCAGGAAATCGAAATCGAAG	714			
Vrn-A1c			734			
vrn-A1			734			
Vrn-A1c	Intr1/A/F2 Intr1/A/R3	AGCCTCCACGGTTTGAAAGTAA AAGTAAGACAACACGAATGTGAGA	1 170	56	65	Fu et al. [24]
vrn-A1	Intr1/C/F Intr1/AB/R	GCACTCCTAACCCACTAACC TCATCCATCATCAAGGCAAA	1 068	58	65	Fu et al. [24]
Vrn-B1						
Vrn-B1	Intr1/B/F Intr1/B/R3	CAAGTGGAACGGTTAGGACA CTCATGCCAAAAATTGAAGATGA	709	63	43	Fu et al. [24]
vrn-B1	Intr1/B/F Intr1/B/R4	CAAGTGGAACGGTTAGGACA CAAATGAAAAGGAATGAGAGCA	1 149	58	69	Fu et al. [24]
Vrn-D1						
Vrn-D1	Intr1/D/F Intr1/D/R3	GTTGTCTGCCTCATCAAATCC GGTCACTGGTGGTCTGTGC	1 671	65	90	Fu et al. [24]
vrn-D1	Intr1/D/F Intr1/D/R4	GTTGTCTGCCTCATCAAATCC AAATGAAAAGGAACGAGAGCG	997	63	60	Fu et al. [24]
Vrn-B3						
Vrn-B3	VRN4-B-INS-F VRN4-B-INS-R	CATAATGCCAAGCCGGTGAGTAC ATGTCTGCCAATTAGCTAGC	1 200	63	70	Yan et al. [9]
vrn-B3	VRN4-B-NOINS-F VRN4-BNOINS-R	ATGCTTTCGCTTGCCATCC CTATCCCTACCGGCCATTAG	1 140	57	65	Yan et al. [9]

（续）

等位基因 Allele	引物名称 Primer name	引物序列 Sequence (5′-3′)	片段大小 Fragment size (bp)	退火温度 Annealing temperature (℃)	每循环延伸时间 Extending time in each cycle (s)	参考文献 Reference
Ppd-D1						
Ppd-D1b	Ppd-D1F Ppd-D1R1	ACGCCTCCCACTACACTG TGTTGGTTCAAACAGAGAGC	414	52	60	Beales et al. [20]
Ppd-D1a	Ppd-D1F Ppd-D1R2	ACGCCTCCCACTACACTG CACTGGTGGTAGCTGAGATT	288	52	60	Beales et al. [20]

各引物对反应体系均为 20μL，1× buffer 含 1.5mmol L^{-1} $MgCl_2$；150mmol L^{-1} dNTPs；引物浓度 VRN1AF/VRN1-INT1R 为 5pmol，其他引物对为每条引物 10pmol；Taq DNA 聚合酶 1U；模板 DNA 60～100ng。反应程序为，94℃预变性 10min，94℃变性 45s，退火温度和延伸时间见表 1，72℃延伸 1min，38 个循环。扩增产物以 2.5%琼脂糖凝胶电泳分离检测，缓冲液体系为 1×TAE 溶液，120V 电压电泳 3h，溴化乙锭染色后，用 GelDoc XR System（BIO-RAD，美国）扫描成像。

依据 Zhang 等[12]报道的方法确定不同春化基因类型，春化基因标记的扩增片段见表 1。依据 Beales 等[20]的方法确定 *Ppd-D1* 位点等位变异类型 *Ppd-D1a* 和 *Ppd-D1b*，引物和扩增片段见表 1。

1.3 生育期田间调查

除 10 份意大利和 5 份印度材料外，740 个品种（系）均于 2009 年种植在中国农业科学院作物科学研究所河南安阳试验基地，冬小麦于 10 月 8 日播种，春小麦于 10 月 22 日播种（晚播防止冻害）。为了便于观察和比较，相同来源的材料相邻种植。每份材料种 1 行，行长 2m，行距 20cm，田间管理同当地大田生产措施。分别记载各材料的抽穗期、开花期和成熟期。

1.4 冬春性和光周期敏感型的划分

依据春化基因的检测结果推测小麦的冬春性，*Vrn-A1* 和 *Vrn-D1* 的任何一个为显性时，即为春性，VRN1 位点三个基因全为隐性或 *vrn-A1*＋*vrn-D1*＋*Vrn-B1* 基因型时为冬性[10,11]。将春性品种进一步区分为春性（具有 *Vrn-A1a* 等位变异）和弱冬性（具有 *Vrn-D1* 等位变异）品种[12,14,25]，依据抽穗期的早晚，即营养生长期的长短，将冬性分为强冬性和冬性[26]，抽穗早的为冬性，抽穗晚的为强冬性。依据标记检测结果推测光周期的敏感性，携带 *Ppd-D1a* 的材料为光周期非敏感型，否则为光周期敏感型。依据抽穗期的早晚分析春化和光周期基因及其组合效应。

2 结果与分析

2.1 *Vrn-A1*、*Vrn-B1*、*Vrn-D1*、*Vrn-B3* 和 *Ppd-D1* 位点等位变异分布

在 *Vrn-A1* 位点，93 份材料扩增出与 Thatcher（*Vrn-A1a*）相同的片段（965bp 和 876bp），基因型为 *Vrn-A1a*；仅有 5 份材料扩增出 714bp 片段，基因型为 *Vrn-A1b*；657 份材料扩增出 734bp 片段，基因型可能为 *Vrn-A1c* 或 *vrn-A1*。上述 657 份材料用引物对 Intr1/C/F 和 Intr1/AB/R 扩增到 1 068bp 片段，而用 Intr1/A/F2 和 Intr1/A/R3 扩增时没有 PCR 产物出现，进一步证明它们含有 *vrn-A1* 等位变异。

在 Vrn-B1 位点，159 份材料可扩增出 706bp 片段，表明这些材料携带 *Vrn-B1* 等位变异；596 份材料用引物可扩增出 1 149bp 片段，说明携带 *vrn-B1* 等位变异。

在 *Vrn-D1* 位点，118 份材料检测到 1 671bp 片段，说明它们携带 *Vrn-D1* 显性等位变异；637 份材料检测到 994bp 片段，表明携带 *vrn-D1* 隐性等位基因。用 VrnN4-B-ISN-R/Vrn4-B-ISN-F 和 Vrn-B-NOINS-F/Vrn-B-NOINS-R 引物对扩增，所有材料中都未发现 1.2kb 带型，而全部扩增到 1.14kb 片段，说明参试材料全部含 *vrn-B3* 等位变异。

417 份材料用引物对 Ppd-D1F/Ppd-D1R2 扩增到 288bp 带，表明这些材料携带光周期非敏感基因 *Ppd-D1a*，338 份材料用引物对 Ppd-D1F/Ppd-D1R1 扩增到 414bp 片段，说明这些材料携带光周期敏感基因 *Ppd-D1b*。

2.2 春化和光周期基因在不同国家品种中的分布频率

春化基因在主要小麦生产国的品种中分布频率差异明显（表 2）。Vrn-B3 在所有检测材料中缺失；64.2%的材料在 *VRN1* 春化位点均为隐性等位变异，即为冬性类型，主要分布在中国北部冬麦区和黄淮麦区（石 4185、石新 733、衡观 33 和偃展 4110 携带 *Vrn-D1*，为弱冬性或春性）、美国、德国、法国、挪威、乌克兰、俄罗斯、土耳其、匈牙利、保加利亚、罗马尼亚和塞尔维亚；9.5%的材料携带 *vrn-A1*＋*vrn-D1*＋*Vrn-B1* 基因型，也为冬性；其余 26.3%的材料至少包含 *Vrn-A1* 或 *Vrn-D1* 显性等位变异，*Vrn-D1* 和 *Vrn-A1* 分布频率分别为 15.6%和 13.0%。显性等位变异 *Vrn-A1a* 在中国西北、东北和新疆春麦区以及加拿大、澳大利亚、阿根廷和智利的分布频率较高，在中国北方麦区、美国、德国、法国、乌克兰、俄罗斯、伊朗和土耳其品种中不存在。*Vrn-B1* 显性等位变异在印度、意大利、中国东北和新疆春麦区以及加拿大、墨西哥、智利、阿根廷和澳大利亚材料中分布频率高。*Vrn-D1* 显性等位变异在中国东北、中国南方以及日本、印度、意大利、墨西哥、智利、阿根廷和澳大利亚材料中分布频率较高。*vrn-A1*＋*Vrn-B1*＋*Vrn-D1* 基因型的分布频率为 5.4%，在 41 份材料中检出，其中伊朗 4 份，印度 3 份，意大利 3 份，墨西哥 5 份，智利 5 份，阿根廷 4 份，澳大利亚 13 份，土耳其、中国东北春麦区、中国新疆春麦区和日本各 1 份；而 *Vrn-A1*＋*Vrn-B1*＋*vrn-D1* 和 *Vrn-A1a*＋*Vrn-B1*＋*Vrn-D1* 基因型的分布频率分别为 4.8%（36 份）和 1.3%（10 份），含有 *Vrn-A1a*＋*vrn-B1*＋*Vrn-D1* 材料分布频率为 1.4%，主要分布在中国春麦区、智利、加拿大、阿根廷和澳大利亚。可见，春化基因显性等位变异主要分布于中国春麦区和南方冬麦区、加拿大、日本、意大利、印度、智利、阿根廷和澳大利亚等，说明这些地方的品种为春性；而我国北方冬麦区、欧洲各国及美国携带显性春化基因的频率极低，冬性品种占主导地位。

表 2 不同国家小麦品种的春化、光周期基因分布频率及抽穗和成熟情况

Table 2 Distribution of vernalization and photoperiod genes in cultivars from major wheat production countries and their heading and maturity situations

来源地区 Origin	品种数 No. of cultivar	基因型频率 Frequency of genotype (%)					抽穗期分布频率 Frequency of heading date (%)		能成熟材料 Maturity frequency (%)
		Vrn-A1	*Vrn-B1*	*Vrn-D1*	*vrn-A1*＋*vrn-B1*＋*vrn-D1*	*Ppd-D1a*	5 月 6 日前 Before May 6	5 月 10 日后 After May 10	
I	8	0	0	0	100.0	100.0	100.0	0	100.0
II	22	0	0	18.2	81.8	100.0	100.0	0	100.0
美国中部平原 Middle Plain US	25	0	4.0	0	96.0	40.0	36	24.0	100.0
美国西北部 North-west US	45	0	11.1	11.1	77.8	8.9	0	91.0	63.6
美国南部 Southern US	11	0	0	0	100.0	36.4	10.0	0	100.0
德国 Germany	75	0	4.0	0	96.0	1.3	0	100.0	0
法国 France	98	0	2.0	5.1	92.9	65.3	6.8	41.7	67
英国 UK	3	0	33.3	0	66.7	0	0	100.0	0
挪威 Norway	22	0	27.3	4.5	68.2	9.1	11.8	70.6	47.1
荷兰 Netherlands	5	0	0	0	100.0	0	0	100.0	0
乌克兰 Ukraine	14	0	0	7.1	85.7	64.3	7.1	35.7	71.4
俄罗斯 Russia	23	0	0	0	100.0	87.0	4.3	34.8	91.3

（续）

来源地区 Origin	品种数 No. of cultivar	基因型频率 Frequency of genotype (%)					抽穗期分布频率 Frequency of heading date (%)		能成熟材料 Maturity frequency (%)
		Vrn-A1	*Vrn-B1*	*Vrn-D1*	*vrn-A1*+*vrn-B1*+*vrn-D1*	*Ppd-D1a*	5月6日前 Before May 6	5月10日后 After May 10	
伊朗 Iran	30	0	30.0	30.0	53.3	93.3	16.6	0	100.0
土耳其 Turkey	89	0	9.0	9.0	83.1	80.9	20.2	11.2	96.6
匈牙利 Hungary	10	0	0	20.0	80.0	33.3	10.0	70.0	30.0
保加利亚 Bulgaria	18	0	33.3	0	66.7	100.0	16.7	0	100.0
罗马尼亚 Romania	23	0	0	0	100.0	82.6	0	8.7	82.6
塞尔维亚 Serbia	10	0	0	0	100.0	100.0	0	0	100.0
印度 India	5	0	100.0	60.0	0	100.0	—	—	—
意大利 Italy	10	10.0	40.0	60.0	0	100.0	—	—	—
日本 Japan	10	10.0	30.0	40.0	40.0	100.0	80.0	20.0	0
V	6	0	0	83.3	16.7	100.0	83.3	0	100.0
VI	6	50.0	83.3	50.0	0	50.0	0	100.0	16.7
VIII	4	50.0	25.0	25.0	25.0	100.0	50.0	0	100.0
X	4	50.0	75.0	25.0	0	100.0	0	0	100.0
加拿大 Canada	26	76.9	69.2	15.4	7.7	34.6	7.7	53.8	61.5
墨西哥 Mexico	5	0	100.0	100.0	0	80.0	0	0	100.0
智利 Chile	41	48.8	51.2	39.0	14.6	19.5	0	87.8	17.1
阿根廷 Argentina	33	30.3	60.6	27.3	15.2	15.2	0	84.8	24.2
澳大利亚 Australia	74	45.9	44.6	33.8	10.8	78.4	4.1	14.9	77.0

Ⅰ、Ⅱ、Ⅴ、Ⅵ、Ⅷ和Ⅸ分别代表中国北部冬麦区、黄淮冬麦区、长江中上游冬麦区、东北春麦区、西北春麦区和新疆春麦区。在*VRN-B3*位点没发现显性等位变异。—表示数据缺失（未播种）。

Ⅰ, Ⅱ, Ⅵ, Ⅶ, Ⅷ, and Ⅹ stand for northern China winter wheat, Yellow-Huaihe Rivers winter wheat, middle and upper Yangtze Valley winter wheat, Northeast China spring wheat, northwestern China spring wheat, and Xinjiang spring wheat regions, respectively. No dominant allelic variation was found at *VRN-B3* locus. "—" indicates data not available due to no sowing.

755份材料中，417份携带非敏感等位变异*Ppd-D1a*，338份含*Ppd-D1b*敏感等位变异。其中高海拔及中高纬度但对成熟期要求不严格的一年一熟地区（如美国西北部、德国、挪威、匈牙利、中国东北部、加拿大、智利和阿根廷等）*Ppd-D1b*等位变异分布频率相对高，而纬度相对较低的地区（如中国北方麦区、伊朗、土耳其、保加利亚、罗马尼亚、塞尔维亚、印度、意大利、日本、中国南方麦区和澳大利亚）*Ppd-D1a*分布频率高（表2）。法国和乌克兰品种携带Ppd-D1a的频率介于上述两类之间。据Worland等[2]报道，法国巴黎以北的地区主要种植光周期敏感材料。

2.3 不同来源品种在河南安阳的生育期表现

740份材料的抽穗时间为4月28日至5月29日，不同地区材料的成熟期分布频率存在较大差异（表2）。中国北方冬麦区、美国中部平原、日本及中国南方和西北地区的品种抽穗早（5月6日前），而美国西北部、德国、挪威、匈牙利、中国东北、加拿大、智利和阿根廷等地区的品种抽穗晚（5月10日后）；中国北方冬麦区、西北春麦区、南部冬麦区以及伊朗、土耳其、罗马尼亚、墨西哥、美国南部和美国中部平原的材料基本正常成熟，但国外大部分材料较国内品种晚熟，春性品种较冬性晚熟；来自美国西北部、德国、挪威、匈牙利、中国东北、智利和阿根廷的材料在安阳不能成熟的频率很高。

分子标记检测结果结合基于抽穗期的冬春性判定，将中国北部冬麦区全部材料（*VRN1* 位点为隐性等位变异）定为冬性；中国黄淮麦区除石 4185、石新 733、衡观 33 和偃展 4110（携带 *Vrn-D1* 显性等位变异）外，其他材料均为冬性；美国中部平原、乌克兰、俄罗斯、伊朗、土耳其、保加利亚、罗马尼亚、塞尔维亚和美国南部以冬性基因型为主，且在安阳抽穗较早（平均抽穗期为 5 月 6～10 日），将这些品种划入冬性品种；法国和美国西北部材料也以冬性基因型为主，但在安阳的抽穗期差异较大，平均抽穗期较当地生产品种晚 5～25 d，被划入冬性或强冬性类型。德国、挪威、匈牙利冬性基因型的分布频率也很高，且抽穗很晚，大部分材料在安阳不能成熟，属于强冬性品种；中国南方、意大利、墨西哥和印度材料中 *Vrn-D1* 显性等位变异分布频率高，大部分材料为弱春性；中国东北、西北和新疆地区以及加拿大、智利、阿根廷和澳大利亚的材料中，*Vrn-A1* 和 *Vrn-D1* 分布频率较高，这些品种被判定为春性或弱冬性。

2.4 小麦春化和光周期等位变异与抽穗期的关系

考察不同春化基因组合的抽穗期，发现相对于春化显性等位变异 *Vrn-A1a* 及其组合，当小麦品种的春化基因位点均为隐性（*vrn-A1*＋*vrn-B1*＋*vrn-D1*）或具有残留春化需求等位变异（*Vrn-B1* 和 *Vrn-D1*）及其组合时，其抽穗期较早（表 3），即春性强的材料在安阳秋播抽穗较晚，显性春化基因对抽穗开花没有明显的加速作用。其主要原因是安阳 2009 年 11 月至 2010 年 2 月的平均气温为 1.81～1.09℃，各月的最低气温依次为－3.17℃、－5.31℃、－7.99℃和－3.66℃，能满足冬小麦春化处理的要求，即 4～6℃保持 4～6 周[27]。另外，无论春化基因及其组合如何变化，Ppd-D1 位点的作用都很突出，光周期非敏感型 Ppd-D1a 较敏感型 Ppd-D1b 提早抽穗（表 3），说明在安阳光周期对小麦抽穗期的影响大于冬春性。

表 3 不同春化、光周期基因及其组合与抽穗期的关系（河南安阳）

Table 3 Relationship of vernalization, photoperiod genes and their combination with heading date (Anyang, Henan, China)

春化基因类型 Vernalization gene	品种数 Number of cultivar	分布频率 Frequency (%)	含 *Ppd-D1a* 品种的抽穗期 Heading date of cultivar with *Ppd-D1a*	含 *Ppd-D1b* 品种的抽穗期 Heading date of cultivar with *Ppd-D1b*
Vrn-A1	43	5.7	0506-0515	0509-0524
Vrn-B1	73	9.7	0502-0512	0507-0524
Vrn-D1	56	7.4	0428-0514	0505-0529
Vrn-A1＋*Vrn-B1*	36	4.8	0504-0513	0508-0529
Vrn-A1＋*Vrn-D1*	11	1.4	0506-0508	0509-0529
Vrn-B1＋*Vrn-D1*	41	5.4	0503-0510	0507-0522
Vrn-A1＋*Vrn-B1*＋*Vrn-D1*	10	1.3	0507-0511	0513-0522
vrn-A1＋*vrn-B1*＋*vrn-D1*	485	64.2	0430-0518	0504-0525
Total	755	100.0	0428-0518	0504-0529

在 *VRN-B3* 位点没发现显性等位变异。No dominant allelic variation was found at *VRN-B3* locus.

来自俄罗斯、保加利亚、罗马尼亚、塞尔维亚、意大利、土耳其、伊朗、印度、日本及中国北方和南方麦区的材料，其 *Vrn-A1* 位点等位变异一致，*Vrn-B1* 和 *Vrn-D1* 位点等位变异差异较大，但大部分光周期基因类型与我国黄淮麦区品种一致，以上地区的品种在安阳抽穗期差异不大，且携带非敏感型光周期基因的材料较携带敏感型等位变异的品种抽穗早。来自中高纬度的德国、荷兰、挪威、匈牙利、法国巴黎以北和乌克兰北部，以及我国东北春麦区的材料绝大部分因携带光周期敏感基因 *Ppd-D1b* 而表现抽穗晚，在安阳基本不能成熟；美国除西北部外的所有材料均携带春化位点隐性等位变异，其中西北部材料总体抽穗较晚，中部平原和南部麦区材料抽穗较早；加拿大材料抽穗和美国西北部材料相当，38.5%的材料（光周期敏感型）在安阳不能正常成熟。智利与阿根廷的材料携带春性显性等位变异（*Vrn-A1* 和 *Vrn-D1*）的频率与澳大利亚品种的分布频率差异不大，但澳大利亚的材料 78.4%为光周期迟钝型，智利和阿根廷的材料大部分为光周期敏感型，导致来自澳大利亚的大部分品种较智利和阿根廷的成熟期早。可见，以上国家的品种能否在安阳成熟与其光周期类型关系密切。

3 讨论

3.1 当地温度和纬度对小麦品种冬春性和光周期敏感程度的影响

小麦冬春性在不同国家的分布主要与冬季气温高低有关。当北半球1月平均气温介于－7℃和4℃之间时，冬季较寒冷，可满足低温春化需求，这类地区主要种植冬小麦（如德国、法国、罗马尼亚、保加利亚和我国的北方冬麦区等）；在1月气温低于－7℃的地区，如我国的黑龙江、美国的明尼苏达和加拿大等地区，因温度太低而无法种冬小麦，生产上一般种植春小麦，但美国华盛顿和俄勒冈及我国新疆等地区（1月平均温度在－10℃以下）也可种植冬麦，这主要与冬季积雪较多有关；当冬季1月平均气温高于4℃时，冬小麦难以通过春化阶段或表现晚抽穗，因而以种植秋播春小麦为主，如墨西哥、印度及中国南方冬麦区，南半球的澳大利亚、阿根廷。因此，来自欧洲国家（德国、挪威、俄罗斯、法国等）及美国的品种携带显性春化基因的频率极低，生长习性大多为强冬性或冬性，春化基因显性等位变异主要分布于中国春麦区和南方冬麦区以及加拿大、日本、意大利、印度、智利、阿根廷和澳大利亚等，这主要与当地冬季气温、纬度和降水量等密切相关[28]。本研究中一些国家春化基因显性等位变异的分布频率与 Iwaki 等[4]的报道不一致，可能与样本类型不同有关，我们以育成品种为材料，而 Iwaki 等[4]采用的是地方品种。在加拿大材料和我国西北春麦区品种中 *Vrn-A1a* 分布频率较高，分别达到 76.9%和 50.0%，这与 Iqbal 等[29]和 Zhang 等[12]的结果一致；CIMMYT 和我国长江中上游麦区品种的 *Vrn-D1* 分布频率高，这与 Van Beem 等[30]和 Zhang 等[12]的结果一致。有报道指出，日本品种携带 *Vrn-D1* 的频率较高[5,19]，但在本研究中其分布频率为 40.0%，可能与样本大小和代表性有关。

光周期敏感基因在决定小麦开花早晚和适应性方面具有重要作用，在欧洲南部光周期非敏感基因可使小麦增产 35%以上，在欧洲中部增产 15%，在欧洲西部温和地区也有增产作用，但在欧洲西部冷凉地区 *Ppd-D1a* 可导致小麦显著减产[1]。德国、挪威、匈牙利、加拿大、智利、阿根廷、美国西北部、美国中部平原、美国南部和中国东北地区品种携带光周期敏感基因 *Ppd-D1b* 的频率很高，主要与纬度高有关。来自其他国家或不同生态类型的材料携带 *Ppd-D1a* 的分布频率则高达 62.0%～100.0%，与材料原产地夏季高温干热有关，光迟钝型不仅可以早熟，避开高温影响，而且可以缩短生育期，提高复种指数，如在印度和我国长江中下游麦区及西南麦区。

3.2 春化和光周期基因与抽穗期的关系

小麦开花期早晚主要受春化、光周期和早熟基因的控制[1,2,31]。春化位点为隐性等位变异的品种需经过低温春化才能抽穗开花，若未经春化处理的品种携带 Vrn-A1a 时可提早抽穗，而携带 *Vrn-B1* 和 *Vrn-D1* 时抽穗较晚[12,16]。Eagles 等[32]研究表明，在能通过春化需求的条件下种植不同春化基因型品种，春化隐性等位变异的效应明显。在本研究中，小麦品种均秋播于安阳，其中 278 份材料的 4 个春化基因均为隐性等位变异，由于冬季气温能够满足春化需求，它们抽穗很早，而具有春化显性等位变异 *Vrn-A1a* 和 *Vrn-D1* 的大部分材料抽穗期稍晚，这也说明在安阳春化隐性等位变异作用较突出；另外，也可能与供试材料携带其他光周期和早熟性基因有关，但目前尚无法检测这些基因。本研究 755 份材料中 55.2%属于光周期非敏感型，来自同一地区的品种，光周期非敏感型的抽穗期明显偏早，这与 Iqbal 等[16]和 Eagles[32]的结果一致，也说明在安阳光周期基因是影响小麦抽穗早晚的关键因素。

Worland 等[1,2]认为，携带光周期非敏感基因的材料在长日照和短日照条件下均能提早开花，而且光周期非敏感性的效应为 *Ppd-D1*＞*Ppd-B1*，但 Tanio 等[19]认为其效应顺序为 *Ppd-D1* ×*Ppd-B1* ＞ *Ppd-B1* ＞ *Ppd-D1*。来自美国中部平原和南部地区、伊朗、日本、中国西北春麦区、新疆春麦区、南方冬麦区和北方冬麦区的绝大部分材料携带光周期非敏感基因 *Ppd-D1a*，在安阳种植抽穗期差异不大；来自德国、挪威、匈牙利和美国西北部（中高纬度）的绝大部分材料携带光周期敏感基因 *Ppd-D1b*，在安阳种植基本不能抽穗或抽穗很晚；而来自纬度相对较低的意大利、塞尔维亚、罗马利亚等国的材料引种到安阳，由于携带光周期非敏感基因 *Ppd-D1a*，能够正常成熟。这与董玉琛等[33]在河南洛阳考察欧洲小麦品种抽穗期的结果有一致性，也与我们多年在安阳的引种观察结果相符。携带 *Ppd-D1a* 的法国品种在安阳能正常抽穗，而携带 *Ppd-D1b* 的品种则抽穗偏晚7～10 d，仅有个别 *Ppd-D1b* 材料抽穗正常，这可能与光周期其他位点携带迟钝型等位变异或早熟型基因本身有关。加拿大材料大多为 *Ppd-D1b* 型，在安阳基本能成熟，也可能与 *Ppd-A1* 或 *Ppd-B1* 位点携带光周期迟钝型基因有关，由于 *Ppd-A1* 或 *Ppd-B1* 光周期位点尚无有效的分子标记[20]，目前还无法验证这一推论。低纬度墨西哥和澳大利亚材料抽穗迟也可能与早熟性基因本身有关，而智利和阿根廷材料抽穗迟主要是光周期敏感基因 *Ppd-D1b* 所致。

3.3 不同国家材料在我国的利用途径

前人对主要国家的小麦冬春性和生育期进行过一些表型观察分析，但往往没有种植在同一环境，本文则是首次从基因层面分析这些国家的小麦冬春性和光周期类型，并且在同一环境下比较其生育期，虽然表型数据仅是一年一点，但仍具有重要应用价值。根据春化基因、光周期敏感基因（*Ppd-D1b*）及抽穗期和成熟期的结果，初步提出这些材料在我国的利用途径。国外材料可为我国育种提供重要资源，以不断提高产量、改进抗病性和改良品质。以黄淮麦区为例，从英国、法国和智利引进的品种其高产性能突出，茎秆强度好，在黄淮麦区有重要利用价值，但因其携带的 *Ppd-D1b* 基因导致晚熟，可采用国内品种与引进品种杂交再回交或三交的方式，重点改造熟期，除常规选择外，还可在回交或三交一代结合分子标记辅助选择，加速育种进程。同理，引自墨西哥和澳大利亚的品种具有优良的抗性和加工品质[34,35]，但因含显性春化基因而在黄淮麦区的抗寒性较差，建议采用回交或三交方式，利用田间选择结合春化基因的相关标记选择改良品种。在黄淮北片区宜选择隐性春化基因；在黄淮南片区可保留显性春化基因 *Vrn-D1* 类型，以便利用国外资源选育高产、抗病、广适的小麦新品种。

4 结论

在755份中国和引进品种中，春化显性基因 *Vrn-B3* 缺失，显性等位变异 *Vrn-A1* 和 *Vrn-D1* 的分布频率分别为13.0%和15.6%，主要分布在春性小麦种植区，包括中国春麦区和长江中上游冬麦区、意大利、印度、日本、加拿大、墨西哥、智利、阿根廷和澳大利亚；春化位点均为隐性等位变异或 *vrn-A*＋*vrn-D1*＋*Vrn-B1* 基因型的材料主要分布在冬麦区，包括中国北方、美国中部和南部、德国、法国、挪威、乌克兰、俄罗斯、伊朗、土耳其、匈牙利、保加利亚、罗马尼亚和塞尔维亚。光周期敏感等位变异 *Ppd-D1b* 主要分布在高海拔和中高纬度地区，如美国、德国、挪威、匈牙利、中国东北、加拿大、智利和阿根廷；而光周期非敏感等位变异 *Ppd-D1a* 主要分布于纬度较低、对早熟性要求较高的地区，如中国北方冬麦区和南方麦区、伊朗、罗马尼亚、印度、意大利和澳大利亚。携带 *Ppd-D1a* 的品种在安阳大部分能够成熟，而携带 *Ppd-D1b* 的材料基本不能成熟。春化显性等位变异 *Vrn-A1a* 在安阳不能加速小麦抽穗，而携带 *Vrn-B1* 和 *Vrn-D1* 的材料因该地区温度条件满足春化要求而正常抽穗。

参考文献

[1] Worland A J. The influence of flowering time genes on environmental adaptability in European wheats. *Euphytica*, 1996, 89: 49-57.

[2] Worland A J, Borner A, Korzun V, Li W M, Petrovic S, Sayers E J. The influence of photoperiod genes on the adaptability of European winter wheats. *Euphytica*, 1998, 100: 385-394.

[3] Snape J W, Butterworth K, Whitechurch E, Worland A J. Waiting for fine times: genetics of flowering time in wheat. *Euphytica*, 2001, 119: 185-190.

[4] Iwaki K, Haruna S, Niwa T, Kato K. Adaptation and ecological differentiation in wheat with special reference to geographical variation of growth habit and *Vrn* genotype. *Plant Breeding*, 2001, 120: 107-114.

[5] Iwaki K, Nakagawa K, Kuno H, Kato K. Ecogeographical differentiation in East Asian wheat, revealed from the geographical variation of growth habit and *Vrn* genotype. *Euphytica*, 2000, 111: 137-143.

[6] Pugsley A T. A genetic analysis of the spring-winter habit of growth in wheat. *Aust J Agric Resource*. 1971, 22: 21-23.

[7] Pugsley A T. Additional genes inhibiting winter habit in wheat. *Euphytica*, 1972, 21: 547-552.

[8] McIntosh R A, Hart G E, Devos K M, Gale M D, Rogers W J. Catalogue of gene symbols for wheat, In:

Slinkard A E, ed. Proc 9th Intl Wheat Genet Symp. Vol. 5. University of Saskatchewan, Saskatoon, SK, Canada: Univ. Extension Press, 1998. pp 1-235.

[9] Yan L, Fu D, Li C, Blechl A, Tranquilli G, Bonafede M, Sanchez A, Valarik M, Yasuda S, Dubcovsky J. The wheat and barley vernalization gene *VRN3* is an orthologue of *FT*. *Proc Natl Acad Sci USA*, 2006, 103: 19581-19586.

[10] Chen Y H, Brett F, Carver, Wang S W, Cao S H, Yan L L. Genetic regulation of developmental phases in winter wheat. *Mol Breed*, 2010, 26: 573-582.

[11] Yoshida T, Nishida H, Zhu J, Nitcher R, Distelfeld A, Akashi Y, Kato K, Dubcovsky J. *Vrn-D4* is a vernalization gene located on the centromeric region of chromosome 5D in hexaploid wheat. *Theor Appl Genet*, 2010, 120: 543-552.

[12] Zhang X K, Xia X C, Xiao Y G, Zhang Y, He Z H. Allelic variation at the vernalization genes *Vrn-A1*, *Vrn-B1*, *Vrn-D1* and *Vrn-B3* in Chinese common wheat cultivars and their association with growth habit. *Crop Sci*, 2008, 48: 458-470.

[13] Yan L, Loukoianov A, Tranquilli G, Helguera M, Fahima T, Dubcovsky J. Positional cloning of the wheat vernalization gene *VRN1*. *Proc Natl Acad Sci USA*, 2003, 100: 6263-6268.

[14] Jiang Y (姜莹), Huang L-Z (黄林周), Hu Y-G (胡银岗). Distribution of vernalization genes in Chinese wheat landraces and their relationship with winter hardness. *Sci Agric Sin* (中国农业科学), 2010, 43 (13): 2619-2632. (in Chinese with English sbstract)

[15] Zhang X K (张晓科), Xia X-C (夏先春), He Z-H (何中虎), Zhou Y (周阳). Distribution of vernalization gene *Vrn-A1* in Chinese wheat cultivars detected by STS marker. *Acta Agron Sin* (作物学报), 2006, 32 (7): 1038-1043. (in Chinese with English abstract)

[16] Iqbal M, Shahzad A, Ahmed I. Allelic variation at the *Vrn-A1*, *Vrn-B1*, *Vrn-D1*, *Vrn-B3* and *Ppd-D1a* loci of Pakistani spring wheat cultivars. *Electron J Biotechnol*, DOI: 10. 2225/vol14-issue1-fulltext-6.

[17] Welsh J R, Keim D L, Pirasteh B, Richards R D. Genetic control of photoperiod response in wheat. In: Proc 4th Intl Wheat Genet Symp. University of Missouri, Columbia, 1973. pp 897-884.

[18] Law C N, Sutka J, Worland A J. A genetic study of day-length response in wheat. *Heredity*, 1978, 41: 185-191.

[19] Tanio M, Kato K. Development of near-isogenic lines for photoperiod-insensitive genes, *Ppd-B1* and *Ppd-D1*, carried by the Japanese wheat cultivars and their effect on apical development. *Breed Sci*, 2007, 57: 65-72.

[20] Beales J, Turner A, Griffiths S, Snape J W, Laurie D A. A pseudo-response regulator is misexpressed in the photoperiod insensitive *Ppd-D1a* mutant of wheat (*Triticum aestivum* L.). *Theor Appl Genet*, 2007, 115: 721-733.

[21] Yang F P, Zhang X K, Xia X C, Laurie D A, Yang W X, He Z H. Distribution of photoperiod insensitive gene *Ppd-D1a* (*Ppd1*) in Chinese common wheat. *Euphytica*, 2009, 165: 445-452.

[22] Lagudah E S, Appels R, McNeil D. The Nor-D3 locus of Triticum tauschii: Natural variation and genetic linkage to markers in chromosome 5. Genome, 1991, 34: 387-395.

[23] Yan L, Helguera M, Kato K, Fukuyama S, Sherman J, Dubcovsky J. Allelic variation at the *VRN-1* promoter region in polyploid wheat. *Theor Appl Genet*, 2004, 109: 1677-1686.

[24] Fu D, Szücs P, Yan L, Helguera M, Skinner J S, Zitzewitz J V, Hayes P M, Dubcovsky J. Large deletions within the first intron in *VRN-1* are associated with spring growth habit in barley and wheat. *Mol Genet Genomics*, 2005, 273: 54-65.

[25] Zhuang Q-S (庄巧生). Wheat Improvement and Pedigree Analysis in Chinese Wheat Cultivars (中国小麦品种改良及系谱分析). Beijing: China Agriculture Press, 2003. (in Chinese)

[26] Crofts H J. On defining a winter wheat. *Euphytica*, 1989, 44: 225-234.

[27] Porter J R, Gawith M.: Temperatures and the growth and development of wheat: a review. *Eur J Agron*, 1999, 10: 23-36.

[28] Pidwirny M, Jones S. Chapter 7. Introduction to the atmosphere: (v) Climate classification and limatic regions of the world. In: Fundamentals of Physical Geography (2nd Edn). [2010-12-03]. http://www.physicalgeography.net/fundamentals/7v.html

[29] Iqbal M, Navabi A, Salmon D F, Yang R C, Murdoch B M, Moore S S, Spaner D. Genetic analysis of flowering and maturity time in high latitude spring wheat. *Euphytica*, 2007, 154: 207-218.

[30] Van Beem J, Mohler V, Lukman R, Van Ginkel M, William M, Crossa J, Worland Anthony J. Analysis

of genetic factors influencing the developmental rate of globally important CIMMYT wheat cultivars. *Crop Sci*, 2005, 45: 2113-2119.

[31] Kato K, Yamagata H. Method for evaluation of chilling requirement and narrow-sense earliness of wheat cultivars. *Jpn J Breed*, 1988, 38: 172-186.

[32] Eagles H A, Cane K, Kuchel H, Hollamby G J, Vallance N, Eastwood R F, Gororo NN, Martin P J . Photoperiod and vernalization gene effects in southern Australian wheat. *Crop & Pasture Sci*, 2010, 61: 721-730.

[33] Dong Y-C (董玉琛), Hao C-Y (郝晨阳), Wang L-F (王兰芬), Zhang X-Y (张学勇), Gao H-T (高海涛), Zhang C-J (张灿军). Evaluation of agronomic traits of 358 wheat varieties introduced from Europe. *J Plant Genet Resour* (植物遗传资源学报), 2006, 7 (2): 129-135. (in Chinese with English abstract)

[34] Willlam H M , Singh P, Trethow R, Ginkel M, Pellegrinshi A, Huerta-Espino J, Hoisingtond D. Biotechnology applications for wheat improvement at CIMMYT. *Turk J Agric For*, 2005, 29: 113-119.

[35] Lawrence G J. The high-molecular-weight glutenin subunit composition of Australian wheat cultivars. *Aust J Agric Res*, 1986, 37: 125-133.

春化、光周期和矮秆基因在不同国家小麦品种中的分布及其效应

杨芳萍[1,2]，夏先春[1]，张勇[1]，张晓科[3]，刘建军[4]，唐建卫[5]，
杨学明[6]，张俊儒[2]，刘茜[7]，李式昭[8]，何中虎[1,9]

[1]中国农业科学院作物科学研究所/国家小麦改良中心，北京 100081；[2]甘肃省农业科学院，兰州 730070；[3]西北农林科技大学农学院，杨凌 712100；[4]山东省农业科学院，济南 250100；[5]周口市农业科学院，周口 466001；[6]江苏农业科学院，南京 210014；[7]河北省农林科学院，石家庄 050031；[8]四川农业科学院，成都 610066；[9]CIMMYT 中国办事处，北京 100081

摘要：为了促进国外种质资源在我国的有效利用，将 14 个国家的 100 份代表性小麦品种在国内的 8 个代表性地点种植，调查抽穗期、成熟期和株高，并以四个春化基因（*Vrn-A1*、*Vrn-B1*、*Vrn-D1* 和 *Vrn-B3*）、一个光周期基因（*Ppd-D1a*）及两个矮秆基因（*Rht-B1b* 和 *Rht-D1b*）的分子标记检测所有品种的基因型。春化基因 *Vrn-A1a*、*Vrn-B1*、*Vrn-D1* 和 *vrn-A1*＋*vrn-B1*＋ *vrn-D1* 的分布频率分别为 8.0%、21.0%、21.0%和 64.0%；显性等位变异 *Vrn-A1a*、*Vrn-B1* 和 *Vrn-D1* 主要存在于来自中国春麦区及意大利、印度、加拿大、墨西哥和澳大利亚的品种中，这些品种一般为春性类型；春化位点均为隐性等位变异或 *vrn-A1*＋*vrn-D1*＋*Vrn-B1* 的品种主要分布在中国冬麦区、美国冬麦区、俄罗斯冬麦区，以及英国、法国、德国、罗马尼亚、土耳其和匈牙利，这些地区的小麦均为冬性类型。秋播时，供试品种均能正常抽穗，且携带春化显性变异的材料较隐性类型抽穗早，显性等位变异表现加性效应，四个春化位点均为隐性变异的一些欧美材料因抽穗太晚在杨凌和成都不能正常成熟；而春播时，显性等位变异基因型抽穗的频率高，隐性等位变异基因型基本不能抽穗。光周期不敏感基因 *Ppd-D1a* 的分布频率为 68.0%，主要分布在中国、法国、罗马尼亚、俄罗斯、墨西哥、澳大利亚和印度，而光周期敏感等位变异 *Ppd-D1b* 主要分布在英国、德国、匈牙利和加拿大等中高纬度地区；携带 *Ppd-D1a* 的品种较 *Ppd-D1b* 品种抽穗早，大多数 *Ppd-D1a* 品种在长日照和短日照条件下均能成熟，大部分 *Ppd-D1b* 品种在短日照条件下不能成熟。*Rht-B1b* 和 *Rht-D1b* 基因的分布频率分别为 43.0%和 35.0%，其中 *Rht-B1b* 主要分布于美国、罗马尼亚、土耳其、意大利、墨西哥和澳大利亚，*Rht-D1b* 主要分布于中国、德国、英国、意大利和印度。一般来说，一个国家的品种携带 *Rht-B1b* 或 *Rht-D1b* 之一，而这两个基因在高纬度地区分布频率较低。*Rht-B1b*、*Rht-D1b* 和 *Ppd-D1a* 的降秆作用均达显著水平，*Rht-B1b* 和 *Rht-D1b* 的加性效应突出。

关键词：普通小麦；春化基因；*Ppd-D1a*；*Rht-B1b* 和 *Rht-D1b*；分子标记

原文发表在《作物学报》，2012，38（7）：1156-1166

Distribution of Allelic Variation for Vernalization, Photoperiod and Dwarfing Genes and Their Effects on Growth Period and Plant Height among Cultivars from Major Wheat Countries

Yang Fangping[1,2], Xia Xianchun[1], Zhang Yong[1], Zhang Xiaoke[3], Liu Jianjun[4], Tang Jianwei[5], Yan Xueming[6], Zhang Junru[2], Liu Qian[7], Li Shizhao[8], He Zhonghu[1,9]

[1] *Institute of Crop Sciences / National Wheat Improvement Center, Chinese Academy of Agricultural Sciences, Beijing* 100081, *China*; [2] *Gansu Academy of Agricultural Sciences, Lanzhou* 730070, *China*; [3] *College of Agronomy, Northwest A&F University, Yangling* 712100, *China*; [4] *Shandong Academy of Agricultural Sciences, Jinan* 250100, *China*; [5] *Zhoukou Academy of Agricultural Sciences, Zhoukou* 466001, *China*; [6] *Jiangsu Academy of Agricultural Sciences, Nanjing* 210014, *China*; [7] *Hebei Academy of Agriculture and Forestry Sciences, Shijiazhuang* 050031, *China*; [8] *Sichuan Academy of Agricultural Sciences, Chengdu* 610066, *China*; [9] *CIMMYT China Office, Beijing* 100081, *China*

Abstract: To efficiently use exotic resources in Chinese wheat breeding programs, the heading date, maturity, and plant height of 100 representative cultivars from 14 countries were investigated at eight locations in China, and the allelic variations of verbalization loci *VRN-1* and *VRN-B3*, photoperiod gene *Ppd-D1a*, and dwarfing genes *Rht-B1b* and *Rht-D1b* were also detected by means of molecular markers. The frequencies of verbalization loci were 8.0% for *Vrn-A1a*, 21.0% for *Vrn-B1*, 21.0% for *Vrn-D1* and 64.0% for *vrn-A1*+*vrn-B1*+ *vrn-D1*, except for the absence of dominant allele *Vrn-B3* in all tested materials. Dominant vernalization alleles *Vrn-A1a*, *Vrn-B1*, and *Vrn-D1* were mainly observed in cultivars from Chinese spring wheat region, Italy, India, Canada, Mexico, and Australia; whereas, cultivars carrying all recessive alleles at the four verbalization loci and *vrn-A1* +*vrn-D1* +*Vrn-B1* +*vrn-B3* genotype were mostly found in cultivars from Chinese winter wheat region, United States (US) winter wheat region, Russia winter wheat region, United Kingdom (UK), France, Germany, Romania, Turkey, and Hungary. All cultivars headed normally when sown in autumn. Cultivars with dominant alleles showed earlier heading date than those with recessive alleles, and genotypes with two or more dominant alleles showed additive effects. Some European and US cultivars with recessive genes at the four verbalization loci could not mature in Yangling and Chengdu. Under spring-sown condition, the cultivars with dominant verbalization alleles showed high heading frequency; in contrast, most cultivars with recessive alleles failed to head. Gene *Ppd-D1a* was distributed mainly in cultivars from China, France, Romania, Russia, Mexico, Australia, and India with the total frequency of 68%. Most cultivars with *Ppd-D1b* were from high latitude regions, such as UK, Germany, Hungary, and Canada. The *Ppd-D1a* genotypes appeared heading earlier than the *Ppd-D1b* genotypes. Daylight condition had no effect on maturity of most *Ppd-D1a* genotypes, but short daylight condition resulted in failing mature in most *Ppd-D1b* genotypes. The frequencies of dwarfing genes *Rht-B1b* and *Rht-D1b* were 43.0% and 35.0% in the cultivars tested, respectively. *Rht-B1b* was mainly observed in cultivars from US, Romania, Turkey, Italy, Mexico, and Australia, while *Rht-D1b* had high frequency in varieties from China, Germany, UK, Italy and India. Generally, cultivars from one country contain either *Rht-B1b* or *Rht-D1b*, andthe frequencies of *Rht-B1b* and *Rht-D1b* were very low in cultivars from high latitude regions. The effect of *Rht-B1b*, *Rht-D1b* and *Ppd-D1a* on reducing plant height was significant, of which *Rht-B1b* and *Rht-D1b* exhibited an additive effect.

Key words: Common wheat; Vernalization genes; *Ppd-D1a*; *Rht-B1b* and *Rht-D1b*; Molecular markers

提高产量始终是我国小麦育种的重要目标。除产量潜力本身外，产量及其稳定性还受适应性、抗病性和抗倒性等的影响。小麦的适应性主要由春化、光周期和早熟性基因决定[1-3]。国内外在春化和光周期基因的效应、定位以及分子标记开发等领域已开展了大量研究，明确了春化和光周期基因的数目及其影响小麦抽穗和成熟的分子机制[4-16]，并将相关标记应用于品种检测和辅助选择[17-20]。与抗倒性有关的矮秆基因 *Rht-B1b*、*Rht-D1b* 和 *Rht8* 广泛分布于世界各地，目前已成功克隆了 *Rht-B1b* 和 *Rht-D1b* 基因，并开发出可用于分子检测的功能标记[21-23]。

我们已利用春化、光周期和矮秆基因的功能标记分析了我国不同地区和其他国家小麦品种的生长习性及春化、光周期基因组成差异[17,20,24]，检测了不同矮秆基因在一些品种中的分布及其来源[22,23,25,26]。在上述研究中，试验材料大部分来自一个国家或麦区，尽管有些研究试验材料涉及许多国家，但试验方法仅利用分子检测或分子检测结合某一个生态区域的表型数据，缺少多点同时种植的联合观测数据，难以全面反映世界范围内小麦品种的春化、光周期和矮秆基因类型在我国的效应。鉴于此，本研究以 14 个主产国家的 100 份代表性小麦品种（系）为材料，在国内 8 个代表性育种试验点种植，观察抽穗期、成熟期和株高，并利用 *VRN-1* 和 *VRN-B3* 位点春化基因、*Ppd-D1* 位点光周期基因、*Rht-B1b* 和 *Rht-D1b* 矮秆基因的分子标记进行检测，以研究不同国家材料在我国不同地区的生育期和株高与春化、光周期和矮秆基因分布的关系，为提高引种利用效率提供理论依据。

1 材料与方法

1.1 供试材料

100 份代表性小麦品种（系）包括 10 个中国品种（7 份冬性，来自北部冬麦区和黄淮麦区；3 份春性和弱冬性，西北春麦区、西南和长江中下游冬麦区各 1 份）和 90 个引进品种。引进品种来自美国（11 份）、意大利（10 份）、英国（4 份）、法国（10 份）、德国（9 份）、罗马尼亚（9 份）、匈牙利（4 份）、俄罗斯（8 份）、土耳其（5 份）、墨西哥（5 份）、加拿大（5 份）、澳大利亚（5 份）和印度（5 份）。这 14 个国家占全球小麦总产量的 70%以上，而且所用的品种均为各国的主栽品种或最新育成品系，也包括个别国际知名的历史品种，基本上反映了各国小麦生产和育种的现状。美国和俄罗斯都是小麦生产大国，小麦品种很多，但考虑到这些国家品种生态区复杂，而我国主产区为冬性和弱冬性类型，因此只从美国和俄罗斯冬麦区选择代表性品种，其中美国品种有 6 份来自中部大平原堪萨斯州、1 份来自得克萨斯州、2 份来自南达科他州、2 份来自俄勒冈州，具有良好的冬小麦代表性。以 Thatcher（*Vrn-A1a*）、中国春（*Vrn-D1*）和辽春 10 号（*Vrn-B3*）作为春化基因的对照品种，由本课题组保存。

1.2 春化、光周期和矮秆基因的分子标记检测

每份材料随机取 3 粒种子，按 Lagudah 等[27]的方法提取单粒种子的基因组 DNA，以 3 粒种子的检测结果确认材料基因型。春化、光周期和矮秆基因的功能标记[24,25]引物由北京奥科生物技术有限公司合成。

PCR 反应体系均为 20μl，包括 1× buffer（含 1.5mmol L^{-1} $MgCl_2$）、150mmol L^{-1} dNTPs、5pmol（VRN1AF/VRN1-INT1R）或 10pmol（其他）引物、*Taq* DNA 聚合酶 1U 和模板 DNA 60～100ng。反应程序为：94℃ 预变性 10min；94℃ 变性 45s，50～65℃退火 40～60s，72℃ 延伸 40～90s，38 个循环；最后 72℃延伸 10min。扩增产物以 2.5%琼脂糖凝胶电泳分离检测，缓冲液体系为 1×TAE 溶液，120V 电压电泳 2～3h，溴化乙锭染色后，用 GelDoc XR System（BIO-RAD，美国）扫描成像。

依据 Zhang 等[17]报道的方法确定春化基因位点 *Vrn-A1*、*Vrn-B1*、*Vrn-D1* 和 *Vrn-B3* 的不同等位变异类型，依据 Beales 等[13]的方法确定 *Ppd-D1* 位点等位变异类型 *Ppd-D1a* 和 *Ppd-D1b*，按 Ellis 等[21]的方法确定 *Rht-B1* 和 *Rht-D1* 位点等位变异类型。

1.3 农艺性状调查及冬春性、光周期敏感型划分

2010—2011年度，在8个试点每份材料种1行，行长2m，行距20cm，2次重复。田间管理同当地大田生产。播种日期分别为：石家庄点，春性品种2011年3月25日，冬性品种2010年10月9日；济南点，均为2010年10月10日；河南安阳点，春性品种2011年10月28日，冬性品种2010年10月9日；河南周口点，均为2010年10月18日；南京点，均为2010年10月27日；成都点，均为2010年10月25日；陕西杨凌点，均为2010年10月23日；甘肃武威点，均为2011年3月27日。分别记载抽穗期和成熟期，并在成熟期测量株高。按我们先前报道的方法[24]记录冬春性和光周期敏感型。石家庄点仅有抽穗期数据，济南点未观测成熟期。

1.4 统计分析

用Microsoft Excel软件进行数据处理，以SAS8.0软件对矮秆基因*Rht-B1b*和*Rht-D1b*、光周期基因*Ppd-D1a*的降秆效应及不同国家品种株高差异进行显著性分析。

2 结果与分析

2.1 春化、光周期和矮秆基因等位变异类型在不同国家品种中分布

在*Vrn-A1*位点，8份材料扩增出与Thatcher相同的965bp和876bp片段，基因型为*Vrn-A1a*；92份材料扩增出734bp片段，用引物对Intr1/C/F和Intr1/AB/R能扩增到1 068bp片段，而Intr1/A/F2和Intr1/A/R3未检测到产物，说明它们含有*vrn-A1*等位变异。在*Vrn-B1*位点，21份材料扩增出709bp片段，表明这些材料携带*Vrn-B1*等位变异；78份材料用引物Intr1/B/F和Intr1/B/R4可扩增出1 149bp片段，说明携带*vrn-B1*等位变异。在*Vrn-D1*位点，21份材料检测到1 671bp片段，说明它们携带*Vrn-D1*显性等位变异；79份材料扩增到997bp片段，表明携带*vrn-D1*隐性等位基因。用VRN4-B-INS-R/VRN4-B-INS-F和VRN4-B-NOINS-F/VRN4-B-NOINS-R引物对扩增，未发现1.2kb带型，而全部扩增到1.14kb片段，因而参试材料均含*vrn-B3*等位变异。

春化基因等位变异在不同国家品种中的分布频率差异较大（表1），64%的材料在4个春化位点均为隐性等位变异，即冬性类型，主要分布在美国冬麦区、法国、德国、罗马尼亚、俄罗斯冬麦区、英国、匈牙利、土耳其及我国的北部冬麦区和黄淮冬麦区，8.0%的材料携带*vrn-A1*＋*Vrn-B1*＋*vrn-D1*基因型，也为冬性；其余28%的材料至少包含1个显性等位变异，*Vrn-A1a*和*Vrn-D1*的分布频率分别为8.0%和21.0%。显性等位变异*Vrn-A1a*在澳大利亚（80.0%）、加拿大（40.0%）和我国春性品种中（33.3%）分布频率较高，在美国冬麦区、英国、法国、德国、罗马尼亚、匈牙利、俄罗斯冬麦区、土耳其及我国北部冬麦区和黄淮麦区的品种中不存在。*Vrn-D1*显性等位变异在墨西哥（80.0%）、印度（60.0%）、意大利（60.0%）及我国春性品种中（66.7%）较高，11份材料的基因型为*vrn-A1*＋*Vrn-B1*＋*Vrn-D1*，其中，意大利3份、墨西哥5份、印度3份；加拿大品种Bluesky的基因型为*Vrn-A1*＋*Vrn-B1*＋*vrn-D1*，澳大利亚品种Amery携带*Vrn-A1*＋*Vrn-B1*＋*Vrn-D1*等位变异，*Vrn-A1*＋*vrn-B1*＋*Vrn-D1*基因型在所有材料中不存在。可见，春性显性等位变异主要分布在意大利、墨西哥、加拿大、澳大利亚、印度及我国春性麦区，这些地方的品种为春性，美国冬麦区、英国、法国、德国、罗马尼亚、俄罗斯冬麦区、土耳其、匈牙利和我国北部和黄淮麦区春性显性等位变异的分布频率极低，冬性品种占主导地位。

68份材料用引物对Ppd-D1F/Ppd-D1R2扩增到288bp片段，说明这些材料携带光周期非敏感基因*Ppd-D1a*，32份材料用引物对Ppd-D1F/Ppd-D1R1扩增到414bp片段，这些材料含光周期敏感基因*Ppd-D1b*。光周期基因等位变异在不同国家品种中的分布频率差异较大（表1）。光周期非敏感基因*Ppd-D1a*在澳大利亚、印度、墨西哥、意大利、俄罗斯（冬麦区）、罗马尼亚、法国及我国冬、春麦区等分布频率高；光周期敏感基因*Ppd-D1b*在英国、德国、匈牙利、土耳其、加拿大和美国冬麦区等国家分布频率相对较高。

利用引物对NH-BF.2与MR1对所有参试材料进行等位变异检测，42份扩增出400bp片段，说明这些材料携带*Rht-B1b*；58份材料用引物对NH-BF.2和WR1.2扩增出400bp片段，基因型为*Rht-B1a*。引物对DF和MR2在33份材料中扩增出

254bp片段，表明其基因型为 *Rht-D1b*。矮秆基因显性等位变异 *Rht-B1b* 和 *Rht-D1b* 在不同国家材料中的分布频率差异较大（表1），其中 *Rht-B1b* 在来自美国冬麦区、罗马尼亚、土耳其、匈牙利、意大利、墨西哥和澳大利亚的品种中频率较高，*Rht-D1b* 在来自英国、德国、意大利、印度和我国冬春麦区中的分布频率高，4个意大利品种中同时含 *Rht-B1b* 和 *Rht-D1b*，另外，还有28份材料不含 *Rht-B1b* 和 *Rht-D1b*，主要分布于法国（60.0%）、俄罗斯冬麦区（75.0%）、加拿大（60.0%）、德国（33.3%）、美国冬麦区（18.2%）和罗马尼亚（22.2%），可能与来源地纬度较高或为旱地有关。

表1　不同国家材料检测基因的分布频率及其冬春性

Table 1　Frequencies of detected genes and growth habit of varieties from different countries

来　源 Origin	品种数 No. of varieties	频率 Frequency (%)									冬春性 Growth habit
		Vrn-A1a	*Vrn-B1*	*Vrn-D1*	*Vrn-B1*＋*Vrn-D1*	*vrn-A1*＋*vrn-B1*＋*vrn-D1*	*Ppd-D1a*	*Rht-B1b*	*Rht-D1b*	*Rht-B1b*＋*Rht-D1b*	
中国冬麦 China winter	7	0	0	28.6	0	71.4	100	28.6	57.1	0	冬—弱冬性 W—F
美国冬麦 US winter	11	0	0	0	0	100	54.5	81.8	0	0	冬性 W
英国 UK	4	0	25.0	0	0	75.0	25.0	0	75.0	0	冬性 W
法国 France	10	0	0	0	0	100	70.0	10.0	30.0	0	冬性 W
德国 Germany	9	0	11.1	0	0	88.9	11.1	0	66.7	0	强冬性 SW
罗马尼亚 Romania	9	0	0	0	0	100	66.7	77.8	0	0	冬性 W
匈牙利 Hungary	4	0	0	25.0	0	75.0	50.0	50.0	25.0	0	冬—弱冬 W—F
俄罗斯冬麦 Russia winter	8	0	0	0	0	100	100	25.0	0	0	冬性 W
土耳其 Turkey	5	0	40.0	0	0	60.0	40.0	80.0	0	0	冬性 W
意大利 Italy	10	10.0	40.0	60.0	30.0	20.0	100	60.0	70.0	40.0	弱冬—春性 F—S
中国春性 China spring	3	33.3	0	66.7	0	0	100	33.3	66.7	0	弱冬—春性 F—S
墨西哥 Mexico	5	0	100	80.0	100	0	80.0	80.0	20.0	0	冬—春性 W—S
加拿大 Canada	5	40.0	40.0	0	0	40.0	40.0	0	40.0	0	春—冬性 S—W
澳大利亚 Australia	5	80.0	20.0	40.0	0	0	100	100	0	0	春性 S
印度 India	5	0	100	60.0	60.0	0	100	0	100	0	弱冬性 F

W：冬性；F：弱冬性；SW：强冬性；S：春性。

W：winter；F：facultative；SW：strong winter；S：spring.

2.2　不同国家品种在8个试点的抽穗期、成熟期和株高变化趋势

不同国家材料在各试点的抽穗期、成熟期及株高存在较大差异，平均抽穗期介于3月16日至6月20日之间（表2）。在武威点（春麦区），由于春播，冬性材料中除我国黄淮麦区的衡观33携带 *Vrn-D1*、美国冬麦区的 NUWEST/4/D887-74/PEW/3/LN-CR//CARSTEN/GIGANT/5/MRS/CI14482//YMH/HYS/3/RONDEZVOUS（春化位点均为隐性等位变异）和土耳其的 ARG/R16//BEZ*2/3/AGRI/KSK/5/TRK13/6/HK89（*Vrn-B1*）能抽穗外，其余均不能抽穗；而墨西哥、加拿大、澳大利亚、印度和意大利品种均为春性类型，全部可以抽穗。在石家庄点，仅英国品种 Goldentkop 不能抽穗。在杨凌、济南、安阳、周口、南京和成都点，所有参试材料均能抽穗，但抽穗期差异较大，从南向北抽穗期推迟，即在成都、南京和周口点抽穗较早（3月16日至5月9日），在杨凌、安阳、济南和石家庄点抽穗相对晚（4月22日至5月18日），这与当地春季温度变化相一致。美国冬麦区、英国、法国、德国、罗马尼亚、匈牙利和俄罗斯冬麦区的大部分品种较中国、意大利、墨西哥、加拿大、澳大利亚和印度品种抽穗晚。

表 2 不同国家材料在各试点的抽穗期和成熟期的平均值

Table 2 Means of heading date and maturity date of varieties from different countries in various locations

试点 Site	中国冬麦 China winter (n=7)	美国冬麦 US winter (n=11)	英国 UK (n=4)	法国 France (n=10)	德国 Germany (n=9)	罗马尼亚 Romania (n=9)	匈牙利 Hungary (n=4)	俄罗斯冬麦 Russia winter (n=8)	土耳其 Turkey (n=5)	意大利 Italy (n=10)	中国春麦 China spring (n=3)	墨西哥 Mexico (n=5)	加拿大 Canada (n=5)	澳大利亚 Australia (n=5)	印度 India (n=5)
抽穗期 Head date (month/day)															
武威 Wuwei	6/19 *	6/17 *	*	*	*	*	*	*	6/19 *	6/16	6/19	6/19	6/18	6/20	6/19
石家庄 Shijiazhuang	5/5	5/9	5/12	5/13	5/17	5/9	5/13	5/10	5/9	5/7	5/15	5/15	5/17	5/18	5/15
济南 Jinan	4/23	5/3	5/12	5/7	5/15	5/3	5/8	5/4	5/4	4/30	4/22	4/26	4/24	4/28	4/29
安阳 Anyang	4/28	5/4	5/14	5/10	5/16	5/6	5/8	5/6	5/5	4/30	5/2	5/5	5/6	5/5	5/4
杨凌 Yangling	4/24	4/30	5/10	5/5	5/11	5/3	5/5	5/1	5/1	4/26	4/24	4/27	4/27	4/26	4/25
周口 Zhoukou	4/19	4/26	5/4	4/30	5/8	4/29	4/29	4/27	4/27	4/20	4/16	4/18	4/20	4/19	4/16
南京 Nanjing	4/19	4/28	5/7	5/2	5/9	4/30	5/1	4/28	4/29	4/23	4/16	4/22	4/25	4/22	4/19
成都 Chengdu	4/4	4/16	4/23	4/20	4/30	4/18	4/20	4/16	4/17	3/31	3/20	3/26	3/16	3/24	3/18
成熟期 Maturity date (month/day)															
武威 Wuwei	7/27+	6/12	+	+	+	+	+	+	+	7/25	7/18	7/19	7/17	7/23	7/16
安阳 Anyang	6/5	6/9	6/12	6/11	6/17	6/11	6/13	6/9	6/11	6/6	6/7	6/9	6/8	6/12	6/9
杨凌 Yangling	6/5	6/9		6/10	+	6/12	+	6/9	6/12	6/8	6/5	6/7	6/8	6/10	6/7
周口 Zhoukou	5/28	6/2	6/9	6/5	6/10	6/4	6/6	6/2	6/5	5/30	5/29	5/30	5/30	5/31	5/29
南京 Nanjing	5/25	6/1	6/8	6/3	6/8	6/3	6/3	6/1	6/2	5/29	5/26	5/28	5/31	5/31	5/28
成都 Chengdu	5/16	5/17	5/23	5/20	+	5/18	5/19	5/18	5/18	5/16	5/16	5/18	5/17	5/17	5/16
株高 Plant height (cm)															
武威 Wuwei	50.0	—	—	—	—	—	—	—	—	55.4	64.0	70.2	97.0	68.0	86.0
济南 Jinan	76.3	85.1	86.0	85.9	76.4	78.9	78.3	80.9	83.6	60.8	66.0	72.0	92.6	70.2	70.4
安阳 Anyang	69.7	102.6	89.0	102.3	88.6	100.3	97.3	98.1	102.8	81.0	82.5	93.5	125.4	86.9	109.1
杨凌 Yangling	54.7	72.2	76.3	79.1	70.8	69.0	68.8	69.5	73.6	57.4	63.9	72.1	93.4	69.4	84.3
周口 Zhoukou	60.7	88.5	88.4	90.6	79.8	88.4	84.9	85.6	89.6	77.9	79.5	86.9	103.4	83.1	93.3
南京 Nanjing	66.9	79.5	73.6	78.9	69.3	81.4	76.0	79.4	81.6	68.7	80.3	85.6	104.7	82.5	101.6
成都 Chengdu	83.2	95.4	95.1	95.9	88.7	91.5	89.8	93.4	97.5	81.1	85.8	95.8	109.1	87.1	103.5

* 和+分别表示不能抽穗和成熟；—表示数据缺失。“ * ”and “+” indicate wheat varieties could not head and mature, respectively; “—” indicates data not available.

不同国家材料在8个试点的成熟期与抽穗期的变化趋势基本一致，来自中国、意大利、印度和墨西哥的大部分品种成熟较早，而来自英国、德国、法国和澳大利亚的大部分品种成熟较晚。在济南、周口和南京点，所有材料全部能正常成熟。在安阳点，除4份德国（Boomer、Campari、Ellivis和Maverick）和1份匈牙利（MV233-05）材料外，其余95份均能正常成熟。在杨凌点，44份材料由于抽穗太晚不能成熟，56份材料能正常成熟。在成都点，27份材料不能成熟，包括美国4份、英国3份、法国4份、德国9份、罗马尼亚3份、匈牙利2份、土耳其2份；73份材料能够正常成熟，成熟期为5月15～23日。

不同国家材料在8个试点的株高差异较大（表2），在安阳和成都点的平均株高分别为96.3cm（69.7～125.5cm）和92.9cm（83.2～109.1cm），而在杨凌和武威点的平均株高较低，分别为71.4cm（54.7～93.4cm）和72.8cm（50.0～97.0cm）。株高差异主要与试点的水分供应和光温条件等有关。加拿大、印度、法国、土耳其、罗马尼亚、美国冬麦区和俄罗斯冬麦区的品种，其植株高度显著大于我国冬麦区品种，而匈牙利、墨西哥、德国、澳大利亚和意大利的品种与我国冬、春麦区品种的株高相当（表3）。加拿大品种植株显著高于除印度以外的其他国家材料，这与加拿大属于高纬度的旱区有关，而我国冬小麦植株显著低于其他国家材料，进一步说明我国黄淮麦区水地品种在矮化育种方面取得的显著进展。不同国家材料不仅平均株高差异较大，且标准差也有较大差异，英国、法国和意大利品种的标准差较高，而美国冬麦、加拿大、德国、罗马尼亚、俄罗斯冬麦、匈牙利和土耳其品种的标准差较低（表3），说明前者品种间株高变化幅度大，后者变化幅度小。可见，即使是同一国家的品种，其株高也因生态区不同而有明显差异。

表3 不同国家材料在我国的株高比较

Table 3 Comparison of plant height of varieties from different countries

来源 Origin	品种数 No. of varieties	范围 Range (cm)	平均值±标准差 Mean ± *SD* (cm)
加拿大 Canada	5	93.0～108.7	103.7±6.6 a
印度 India	5	84.2～106.8	92.6±8.5 ab
法国 France	10	71.8～122.8	88.8±17.7 bc
土耳其 Turkey	5	75.0～93.9	88.1±7.6 bc
美国冬麦 US winter	11	78.5～97.8	87.2±6.0 bc
罗马尼亚 Romania	9	77.9～96.6	84.9±5.4 bc
英国 UK	4	65.6～125.3	84.7±27.7 bc
俄罗斯冬麦 Russia winter	8	71.7～97.5	84.5±8.1 bc
匈牙利 Hungary	4	78.6～88.0	82.5±4.0 bcd
墨西哥 Mexico	5	79.2～86.0	82.3±2.6 bcd
德国 Germany	9	70.5～91.7	78.9±6.9 bcd
澳大利亚 Australia	5	73.8～90.0	78.2±6.7 bcd
意大利 Italy	10	57.6～99.1	74.7±14.8 cd
中国春麦 China spring	3	71.4～78.6	74.6±3.7 cd
中国冬麦 China winter	7	55.5～80.3	68.2±7.2 d

具有相同字母的平均数未达到显著水平（$P<0.05$）。

Values followed by the same letter are not significantly different at 0.05 probability level.

2.3 春化、光周期等位变异与抽穗期、成熟期的关系

不同春化、光周期等位变异及其组合材料的抽穗期和成熟期在各试验点差异较大（表4）。仅携带显性春化等位变异*Vrn-A1a*的6份材料在武威均能抽穗，携带*Vrn-D1*和*Vrn-B1*等位变异的材料抽穗的频率分别为66.7%和62.5%，说明*Vrn-D1*和

*Vrn-B1*有部分春化需求，而含*vrn-A1*＋*vrn-B1*＋*vrn-D1*的材料仅4.7%能在武威抽穗；在石家庄仅1份携带*vrn-A1*＋*Vrn-B1*＋*vrn-D1*的英国材料（冬性）不能抽穗，其余各点所有试验材料全部抽穗。尽管*Vrn-A1a*对*Vrn-D1*和*Vrn-B1*具有上位作用，但在武威和石家庄没有表现出早抽穗的特点。携带*Vrn-A1*＋*Vrn-B1*＋*vrn-D1*、*vrn-A1*＋*Vrn-B1*＋*Vrn-D1*和*Vrn-A1*＋*Vrn-B1*＋*Vrn-D1*等位变异的材料在所有试点全部成熟。在秋播的济南、安阳、杨凌、周口、南京和成都试点，带多个显性春化基因的材料较带单个显性春化基因的材料早抽穗，携带春化显性等位变异的材料较4个位点均为隐性等位变异的材料（冬性）抽穗早，且显性等位变异加性效应明显。68份携带非敏感基因*Ppd-D1a*的材料在8个试验点的抽穗期和成熟期差异较大（表4），其中，在武威能抽穗的频率达到41.2%，在其余各点均能抽穗，但在杨凌和成都分别有29.4%和5.9%的材料不能成熟；32份携带*Ppd-D1b*的材料中，15.6%能在武威抽穗（大部分为春性），在其他试点均能正常抽穗，但在安阳、杨凌和成都不能成熟的频率分别达到15.6%、75.0%和71.9%。参试材料在各试点中光周期非敏感型较敏感型平均早熟1～12d，非敏感型较敏感型材料能成熟的频率明显高。无论是非敏感型或敏感型类型，在安阳、杨凌和成都不能成熟的材料基本上为欧美的隐性春化等位变异（*vrn-A1*、*vrn-B1*、*vrn-D1*和*vrn-B3*）类型，且大部分为光周期敏感型。说明春化显性等位变异可以促进早抽穗，春化隐性等位变异材料满足春化需求才能正常抽穗开花，光周期不敏感型较敏感型材料提早成熟。

表4　不同春化、光周期等位变异材料在各试点的平均抽穗期和成熟期

Table 4　Mean of heading date and maturity in tested locations for genotypes with different alleles at verbalization and photoperiod loci

试　点 Site	Only *Vrn-A1a* (*n*=6)	Only *Vrn-B1* (*n*=8)	Only *Vrn-D1* (*n*=9)	*Vrn-A1a*＋*Vrn-B1* (*n*=1)	*Vrn-B1*＋*Vrn-D1* (*n*=11)	*Vrn-A1a*＋*Vrn-B1*＋*Vrn-D1* (*n*=1)	*vrn-A1*＋*vrn-B1*＋*vrn-D1* (*n*=64)	*Ppd-D1a* (*n*=68)	*Ppd-D1b* (*n*=32)
抽穗期 Head date (month/date)									
武威 Wuwei	6/21	6/23	6/21	6/17	6/19	6/18	6/19	6/20	6/20
石家庄 Shjiazhuang	5/17	5/14	5/12	5/15	5/13	5/16	5/12	5/11	5/13
济南 Jinan	4/23	4/29	4/28	4/25	4/28	4/27	5/5	5/1	5/8
安阳 Anyang	5/4	5/8	5/4	5/3	5/3	5/5	5/7	5/4	5/10
杨凌 Yangling	4/26	5/3	4/27	4/25	4/26	4/27	5/3	4/29	5/6
周口 Zhoukou	4/19	4/25	4/20	4/17	4/18	4/17	4/30	4/23	5/1
南京 Nanjing	4/23	4/29	4/23	4/17	4/21	4/21	4/30	4/25	5/4
成都 Chengdu	3/22	4/8	4/1	3/15	3/24	3/19	4/18	4/6	4/20
成熟期 Maturity date (month/date)									
武威 Wuwei	7/23	7/19	7/17	7/17	7/21	7/18	7/18	7/21	7/18
安阳 Anyang	6/11	6/11	6/7	6/7	6/8	6/12	6/10	6/9	6/11
杨凌 Yangling	6/10	6/9	6/6	6/6	6/8	＋	6/9	6/8	6/9
周口 Zhoukou	5/31	6/3	5/29	5/29	5/30	6/1	6/4	5/31	6/6
南京 Nanjing	5/31	6/2	5/28	5/28	5/29	5/31	6/2	5/30	6/5
成都 Chengdu	5/17	5/17	5/16	5/16	5/17	5/16	5/19	5/17	5/18

＋表示不能成熟。

"＋" indicates varieties could not completely maturity.

2.4　矮秆基因与光周期*Ppd-D1*位点不同等位变异的降秆效应

矮秆基因*Rht-B1b*和*Rht-D1b*及光周期*Ppd-D1*位点等位变异对株高有较大影响（表5）。仅携带*Rht-B1b*（平均株高80.4cm）或*Rht-D1b*的材料（平均株高80.4cm）显著高于携带*Rht-B1b*和*Rht-D1b*等位变异的材料（平均65.5cm），且显著低于不含

Rht-B1b＋*Rht-D1b* 的材料（平均株高 91.2cm），说明矮秆基因 *Rht-B1b* 和 *Rht-D1b* 降秆作用显著，且两者间存在加性效应。在不含 *Rht-B1b*＋*Rht-D1b*、含 *Rht-B1b*＋*Rht-D1b* 及含其中之一的材料中，携带光周期不敏感型 *Ppd-D1a* 较敏感型 *Ppd-D1b* 材料的株高差异均达到显著水平，说明 *Ppd-D1a* 的降秆效应明显。

表 5 不同矮秆基因及其与光周期 *Ppd-D1* 位点等位变异组合的降秆效应

Table 5 Impact of dwarfing genes, *Ppd-D1* and their combination on plant height

基因型 Genotype	品种数 No. of varieties	株高 Plant height (cm)
No (*Rht-B1b*+*Rht-D1b*) +*Ppd-D1b*	9	100.1 a
No (*Rht-B1b*+*Rht-D1b*)	28	91.2 b
No (*Rht-B1b*+*Rht-D1b*) +*Ppd-D1a*	19	86.9 bc
Rht-D1b +*Ppd-D1b*	9	85.0 cd
Rht-B1b +*Ppd-D1b*	14	84.9 cd
Rht-D1b	30	80.4 ed
Rht-B1b	38	80.4 ed
Rht-D1b. +*Ppd-D1a*	21	79.6 e
Rht-B1b. + *Ppd-D1a*	24	78.2 e
Rht-B1b+*Rht-D1b*	4	65.5 f

具有相同字母的平均数未达到显著水平（$P<0.05$）。

Values followed by the same letter are not significantly different at 0.05 probability level.

不同国家品种中矮秆基因 *Rht-B1b* 或 *Rht-D1b* 的频率存在较大差异，如 9 份美国材料携带 *Rht-B1b*（81.8%），加拿大材料中 2 份携带 *Rht-D1b*（40.0%），3 份英国材料携带 *Rht-D1b*（75.0%）。我国冬小麦材料中 2 份携带 *Rht-B1b*（28.6%），4 份携带 *Rht-D1b*（57.1%）；春小麦中 2 份携带 *Rht-D1b*（66.7%）。不含 *Rht-B1b* 或 *Rht-D1b* 等位变异且株高很低的材料，如川麦 107 和 Kniish 46，可能携带其他未知矮秆基因。本课题组和 CIMMYT 的育种实践表明，我国小麦育种材料中含有其他矮秆基因。在本研究中，我们还发现一些携带 *Rht-D1b* 的材料株高均在 100cm 以上，如 GOLDENTKOP 和 Neepawa，可能与其携带的光周期敏感基因有关。

3 讨论

冬春性是决定小麦适应性的重要因素，冬小麦的 4 个位点均为隐性或仅 *Vrn-B1* 位点为显性，春小麦至少携带 1 个显性春化基因。春化显性等位变异 *Vrn-A1a*、*Vrn-B1*、*Vrn-D1* 及其组合在不同国家材料中分布频率差异较大，*Vrn-A1a* 主要分布于澳大利亚、加拿大[24,28,29]和我国西北部[17]；*Vrn-D1* 主要分布于意大利、印度[30]、墨西哥[26,31]及我国南方冬麦区和黄淮麦区等春性和弱冬性品种中[17]；冬性隐性等位变异主要分布于英国、法国、德国、罗马尼亚、土耳其、匈牙利及美国冬麦区、俄罗斯冬麦区和我国北方冬麦区品种中。这种差异与生态环境（温度和纬度）有关，是品种适应自然条件的基础[24]。春小麦携带 *Vrn-A1a* 表现早抽穗，可避免后期干热胁迫[29]；但携带 *Vrn-A1a* 的材料不适宜种植在冬季较为寒冷的区域（如我国北部冬麦区），否则冬季或早春会出现冻害[16,32]。*Vrn-B1* 或 *Vrn-D1* 单独存在时有部分春化需求，导致春播晚抽穗[15,33,34]。

小麦的适应性也受控于光周期基因，低纬度地区的品种一般为光迟钝型，而高纬度地区需要光敏感型。澳大利亚、墨西哥、意大利、罗马尼亚、法国及我国冬春麦区 *Ppd-D1a* 的分布频率均很高，而英国、德国和加拿大的品种 *Ppd-D1b* 分布频率相对较高，光周期基因的这种分布与以上国家的纬度和耕作制度要求相符合[1,24]，携带光周期非敏感和敏感基因材料在长日照条件下（如安阳、济南、周口和南京）基本能正常抽穗成熟，但在短日照条件下（成都）携带光周期敏感基因 *Ppd-D1b* 的材料抽穗期延迟或基本不能成熟[2,14]，这也是欧美多数品种在我国表现晚熟的根本原因。携带光周期不敏感基因 *Ppd-D1a* 的部分欧美冬性材料在成都不能成熟，主要是冬季温度高不能完成春化作用所致；而部分欧美冬性材料在杨凌因当年晚播（较正常年份晚播两周以上）、又逢翌年 4 月上旬和 5 月上旬连续阴雨，造成光照不足，生育后期干热风使冬性材料青干，最终不能正常成熟。需要说明的是，在本研究中美国和俄罗斯品种的抽穗期接近，但 *Ppd-D1a* 的分布频率差异很大，主要原因是现有标记只能检测一个位点，另外两个位点尚不能检测。

矮秆基因对小麦适应性也有重要作用。*Rht-B1b* 或 *Rht-D1b* 的主要作用是降低株高、提高小穗结实率和产量潜力，在气候相对温和的灌溉区如我国黄淮麦区，降低株高提高抗倒伏性是实现高产的重要途径，矮秆基因有利于增强抗倒性、提高收获指数，矮秆基因的频率很高；在后期雨水多病害重的地区，如我国长江流域，抗病是最重要的育种目标，植株过矮不仅易发病，而且产量也受影响，所以矮秆基因的频率偏低；在高纬度的干旱地区（如加拿大），对株高要求不严，少带或不含矮秆基因可提高生物学产量及抗病性，矮秆基因的分布频率较低。可见，矮秆基因的分布主要与纬度高低、生产条件（水地和旱地）及生育后期的温度有关。

Worland 等[35]研究普通小麦 2D 染色体功能时发现，与适应性有关的光周期基因、春化基因和矮秆基因之间存在互作，但春化基因不限制秋播小麦适期开花，而是限制春播小麦能否正常抽穗的重要因素[1]，本研究在不同生态区域的观测结果进一步证实这一结论。部分欧美材料在成都秋播不能成熟，与当地冬季温度高、不能完成春化有关，但光周期不敏感型较敏感型成熟频率高，说明光周期基因不同等位变异是影响其能否正常抽穗的关键因素。一般认为，春化反应影响小麦幼穗分化，随着品种冬性程度的增加，小穗数呈增加趋势，春化基因类型（本身作用大小）和数量影响小穗数量[36,37]。矮秆基因 *Rht-B1b* 和 *Rht-D1b* 对高产的贡献主要是提高小穗育性，穗粒数增加补偿了籽粒变小的不足[38]。由此看来，显性春化基因与矮秆基因对小穗数存在相反效应，即品种内显性春化基因会减弱矮秆基因增加小穗数的潜势。我国东北春麦区矮秆基因频率低，春化反应最不敏感显性基因 *Vrn-A1* 频率高，并与多个其他显性春化基因共存[24,25]，这反映了在高纬度春麦区矮秆基因与春化基因组成分布。显性春化基因和矮秆基因对小穗数的这种效应值得育种工作者关注。本研究也表明，含和不含矮秆基因的品种株高差异显著；光周期不敏感型较敏感型材料株高降低，差异也达显著水平，说明矮秆基因和光周期不敏感基因 *Ppd-D1a* 降秆作用突出。但 Sip 等[39]认为 *Ppd-D1a* 对株高没有显著效应，可能与其试验区域（中欧）的温度和降雨量有关。Addisu 等[40]认为携带 *Ppd-D1a* 的光周期不敏感型植株拔节早，导致株高降低，另外 *Ppd-D1a* 的降秆效应也可能与其紧密连锁于 *Rht8* 有关[2,41]。*Ppd-D1a* 和 *Rht8* 连锁的品种具有早熟性，可避免夏季热压获得高产，是携带 *Ppd-D1a* 和 *Rht8* 品种广泛分布于欧洲南部、东部及生态条件类似区域（如澳大利亚）的重要原因[42]。可见，在植株高度、熟性的调控上，光周期基因与矮秆基因之间可能存在一定的互作效应。

本研究采用来自 14 个国家的 100 个品种，兼有春性和冬性材料，对光周期既有敏感型，也有非敏感型，且不同国家材料矮秆基因的分布不完全一致，因而本研究结果对小麦引种和利用有重要的参考价值。由于美国和俄罗斯冬麦区分布范围广，本研究中涉及的主要基因的分布频率有一定差异，这是正常的。来自法国、美国、罗马尼亚等携带矮秆和光周期不敏感基因 *Ppd-D1a* 的冬性材料在安阳、济南、周口和南京等区域绝大部分能正常抽穗成熟，不需特别光温处理调节抽穗期，容易与当地品种进行杂交；携带 *Rht-B1b* 或 *Rht-D1b* 矮秆基因、光周期非敏感基因的春性材料（墨西哥、意大利和澳大利亚品种）可在我国高水肥春麦区或秋播春性或弱冬性麦区应用，如河西灌区、南方冬麦区及黄淮南片等；不含或少带矮秆基因且有光周期非敏感基因的半冬性材料（俄罗斯和印度品种）可应用于高海拔干旱、半干旱生态区域，如西北春麦区的定西、固原、西吉等；不含矮秆基因且光周期敏感型春性材料（加拿大品种）可在我国东北春麦区应用；其他材料的应用区域可依次类推。另外，与成都等光周期短且冬季温度又偏高类似的生态区域，需要冬性较弱或偏春性且光周期不敏感类型材料（意大利、墨西哥和澳大利亚品种）。在品种选择时，因矮秆基因 *Rht-B1b* 和 *Rht-D1b*、光周期基因 *Ppd-D1a* 对植株的降秆效应较为显著，高水肥区域应选择含以上基因的材料为主，而高海拔干旱半干旱区域则不同[43]。

品种的适宜播种期是产量潜力正常发挥的关键因素之一，春化基因不同等位变异对温度的响应时间存在较大差异，故引入我国的国外材料应依据其携带的春化基因调整播种期。秋播携带显性春化基因 *Vrn-A1a* 的国外材料时，播期可适当推迟，春播携带 *Vrn-D1* 的材料时适当提前，秋播时可适当晚播，春播携带隐性春化基因材料之前进行春化处理。欧美材料在成都不能抽穗和成熟的频率很高，故在成都等冬季温度较高的南方秋播区域，可在播种前进行适时春化处理。

4 结论

在我国 8 个育种地点，通过调查来自 14 个国家的 100 个代表性小麦品种的抽穗期、成熟期和株高，以及对春化基因位点 *VRN-1* 和 *VRN-B3*、光周期基因位点 *Ppd-D1* 和矮秆基因 *Rht-B1b* 和 *Rht-D1b* 的分子检测，发现 *Vrn-A1a*、*Vrn-B1*、*Vrn-D1* 和 *vrn-A1* + *vrn-B1* + *vrn-D1* 的分布频率分别为 8.0%、

21.0%、21.0%和 64.0%；春性品种主要分布在中国春麦区、意大利、印度、加拿大、墨西哥和澳大利亚，冬性品种主要分布在中国冬麦区、美国冬麦区、英国、法国、德国、俄罗斯冬麦区、罗马尼亚、土耳其和匈牙利。秋播时，所有供试品种均能抽穗，且携带显性春化基因的材料较隐性类型早抽穗，显性基因表现出加性效应，一些冬性欧美材料在杨凌和成都不能成熟；春播时，春性强的材料抽穗频率高，冬性强的材料基本不能抽穗。光周期非敏感材料主要来自中国、法国、罗马尼亚、俄罗斯冬麦区、墨西哥、澳大利亚和印度；光周期敏感材料主要来自英国、德国、匈牙利和加拿大等中高纬度地区。光周期不敏感材料较光周期敏感材料早抽穗，大多数光周期不敏感材料在长日照和短日照条件下均能成熟，而大部分敏感材料在短日照条件下不能成熟。矮秆基因 *Rht-B1b* 和 *Rht-D1b* 的分布频率分别为 43.0%和 35.0%，*Rht-B1b* 在美国冬麦区、罗马尼亚、土耳其、墨西哥和澳大利亚分布频率高，*Rht-D1b* 在中国、德国、英国、意大利和印度分布频率高，*Rht-B1b* 和 *Rht-D1b* 在法国、俄罗斯冬麦区和加拿大分布频率低。*Rht-B1b*、*Rht-D1b* 和 *Ppd-D1a* 的降秆作用显著，*Rht-B1b* 和 *Rht-D1b* 的加性效应明显。

参考文献

[1] Worland A J. The influence of flowering time genes on environmental adaptability in European wheats. *Euphytica*, 1996, 89: 49-57.

[2] Worland A J, Borner A, Korzun V, Li W M, Petrovic S, Sayers E J. The influence of photoperiod genes on the adaptability of European winter wheats. *Euphytica*, 1998, 100: 385-394.

[3] Snape J W, Butterworth K, Whitechurch E, Worland A J. Waiting for fine times: genetics of flowering time in wheat. *Euphytica*, 2001, 119: 185-190

[4] Pugsley A T. A genetic analysis of the spring-winter habit of growth in wheat. *Aust J Agric Res*, 1971, 22: 21-23.

[5] Pugsley A T. Additional genes inhibiting winter habit in wheat. *Euphytica*, 1972, 21: 547-552.

[6] McIntosh R A, Hart G E, Devos K M, Gale M D, Rogers W J. Catalogue of gene symbols for wheat, In: Slinkard A E, ed. In: Proc 9th Intl Wheat Genet Symp. Vol. 5. University of Saskatchewan, Saskatoon, SK, Canada: Univ. Extension Press, 1998. pp 1-235.

[7] Yan L, Fu D, Li C, Blechl A, Tranquilli G, Bonafede M, Sanchez A, Valarik M, Yasuda S, Dubcovsky J. The wheat and barley vernalization gene *VRN3* is an orthologue of *FT*. *Proc Natl Acad Sci USA*, 2006, 103: 19581-19586.

[8] Yan L, Loukoianov A, Tranquilli G, Helguera M, Fahima T, Dubcovsky J. Positional cloning of the wheat vernalization gene *VRN1*. *Proc Natl Acad Sci USA*, 2003, 100: 6263-6268.

[9] Yan L, Helguera M, Kato K, Fukuyama S, Sherman J, Dubcovsky J. Allelic variation at the *VRN-1* promoter region in polyploid wheat. *Theor Appl Genet*, 2004, 109: 1677-1686.

[10] Chen Y H, Brett F, Carver, Wang S W, Cao S H, Yan L L. Genetic regulation of developmental phases in winter wheat. *Mol Breed*, 2010, 26: 573-582.

[11] Yoshida T, Nishida H, Zhu J, Nitcher R, Distelfeld A, Akashi Y, Kato K, Dubcovsky J. *Vrn-D4* is a vernalization gene located on the centromeric region of chromosome 5D in hexaploid wheat. *Theor Appl Genet*, 2010, 120: 543-552.

[12] Welsh J R, Keim D L, Pirasteh B, Richards R D. Genetic control of photoperiod response in wheat. In: Proc 4th Intl Wheat Genet Symp. University of Missouri, Columbia, 1973. pp 897-884.

[13] Beales J, Turner A, Griffiths S, Snape J W, Laurie D A. A pseudo-response regulator is misexpressed in the photoperiod insensitive *Ppd-D1a* mutant of wheat (*Triticum aestivum* L.). *Theor Appl Genet*, 2007, 115: 721-733.

[14] Law C N, Sutka J, Worland A J. A genetic study of day-length response in wheat. *Heredity*, 1978, 41: 185-191.

[15] Eagles H A, Cane K, Kuchel H, Hollamby G J, Vallance N, Eastwood R F, Gororo N N, Martin P J. Photoperiod and vernalization gene effects in southern Australian wheat. *Crop Pasture Sci*, 2010, 61: 721-730.

[16] Tanio M, Kato K. Development of near-isogenic lines for photoperiod-insensitive genes, *Ppd-B1* and *Ppd-D1*, carried by the Japanese wheat cultivars and their effect on apical development. *Breed Sci*, 2007, 57: 65-72.

[17] Zhang X K, Xia X C, Xiao Y G, Zhang Y, He Z H. Allelic variation at the vernalization genes *Vrn-A1*,

Vrn-B1, *Vrn-D1* and *Vrn-B3* in Chinese common wheat cultivars and their association with growth habit. *Crop Sci*, 2008, 48: 458-470.

[18] Jiang Y (姜莹), Huang L-Z (黄林周), Hu Y-G (胡银岗). Distribution of vernalization genes in Chinese wheat landraces and their relationship with winter hardness. *Sci Agric Sin* (中国农业科学), 2010, 43 (13): 2619-2632 (in Chinese with English abstract).

[19] Iqbal M, Shahzad A, Ahmed I. Allelic variation at the *Vrn-A1*, *Vrn-B1*, *Vrn-D1*, *Vrn-B3* and *Ppd-D1a* loci of Pakistani spring wheat cultivars. *Electron J Biotechnol*, 2011, 14: no. 1. http: //dx. doi. org/10. 2225/vol14-issue1-fulltext-6 .

[20] Yang F P, Zhang X K, Xia X C, Laurie D A, Yang W X, He Z H. Distribution of photoperiod insensitive gene *Ppd-D1a* (*Ppd1*) in Chinese common wheat. *Euphytica*, 2009, 165: 445-452.

[21] Ellis M H, Spielmeyer W, Rebetzke G J, Richards R A. "Perfect" markers for the *Rht-B1b* and *Rht-D1b* dwarfing genes in wheat. *Theor Appl Genet*, 2002, 105: 1038-1042.

[22] Yang S-J (杨松杰), Zhang X-K (张晓科), He Z-H (何中虎), Xia X-C (夏先春), Zhou Y (周阳). Distribution of dwarfing genes *Rht-B1b* and *Rht-D1b* in Chinese bread wheats detected by STS marker. *Sci Agric Sin* (中国农业科学), 2006, 39 (8): 1680-1688 (in Chinese with English abstract).

[23] Zhang X K, Yang S J, Zhou Y, He Z H, Xia X C. Distribution of the *Rht-B1b*, *Rht-D1b* and *Rht8* reduced height genes in autumn-sown Chinese wheats detected by molecular markers. *Euphytica*, 2006, 152: 109-116.

[24] Yang F-P (杨芳萍), Han L-M (韩利明), Xia X-C (夏先春), Qu Y-Y (曲延英), Wang Z-W (王忠伟), He Z-H (何中虎). Distribution of allelic variation for genes of vernalization and photoperiod among wheat cultivars from 23 countries. *Acta Agron Sin* (作物学报), 2011, 37 (11): 1-10 (in Chinese with English abstract).

[25] Han L-M (韩利明), Yang F-P (杨芳萍), Xia X-C (夏先春), Yan J (阎俊), Zhang Y (张勇), Qu Y-Y (曲延英), Wang Z-W (王忠伟), He Z-H (何中虎). Distribution of genes related to plant height, kernel weight and disease resistance among wheat cultivars from major countries. *J Triticease Crops* (麦类作物学报), 2011, 31 (5): 1-8 (in Chinese with English abstract).

[26] Liang D (梁丹), Yang F-P (杨芳萍), He Z-H (何中虎), Yao D-N (姚大年), Xia X-C (夏先春). Characterization of *Lr34/Yr18*, *Rht-B1b*, *Rht-D1b* genes in CIMMYT wheat cultivars and advanced lines using STS markers. *Sci Agric Sin* (中国农业科学), 2009, 42 (1): 17-27 (in Chinese with English abstract).

[27] Lagudah E S, Appels R, McNeil D. The *Nor-D3* locus of *Triticum tauschii*: natural variation and genetic linkage to markers in chromosome 5. *Genome*, 1991, 34: 387-395.

[28] Iqbal M, Navabi A, Yang R C, Salmon D F, Spaner D. Molecular characterization of vernalization response genes in Canadian spring wheat. *Genome*, 2007, 50: 511-516.

[29] Iqbal M, Navabi A, Salmon D F, Yang R C, Murdoch B M, Moore S S, Spaner D. Genetic analysis of flowering and maturity time in high latitude spring wheat. *Euphytica*, 2007, 154: 207-218.

[30] Iwaki K, Haruna S, Niwa T, Kato K. Adaptation and ecological differentiation in wheat with special reference to geographical variation of growth habit and *Vrn* genotype. *Plant Breed*, 2001, 120: 107-114.

[31] Van Beem J , Mohler V, Lukman R, Van Ginkel M, William M, Crossa J, Worland Anthony J. Analysis of genetic factors influencing the developmental rate of globally important CIMMYT wheat cultivars. *Crop Sci*, 2005, 45: 2113-2119.

[32] Prasil I T, Prasilova P, Pankovak K. The relationship between vernalization requirement and frost tolerance in substitution lines of wheat. *Biol Plant*, 2005, 49: 195-200.

[33] Stelmakh A F. Genetic effects of *Vrn* genes on heading date and agronomic traits in bread wheat. *Euphytica*, 1993, 65: 53-60.

[34] Stelmakh A F. Genetic systems regulating flowering response in wheat. *Euphytica*, 1998, 100: 359-369.

[35] Worland A J, Petrovic S, Law C N. Genetic analysis of chromosome 2D of wheat: Ⅱ. The importance of the chromosome to Yugoslavian varieties. *Plant Breed*, 1988, 100: 247-259.

[36] Halse N J, Weir R N. Effects of vernalization, photoperiod, and temperature on physiological development and spikelet number of Australian wheat. *Aust J Agric Res* 1970, 21: 383-393.

[37] Flood R G, Halloran G M. The infuence of genes for vernalization response on development and growth in

wheat. *Ann Bot*, 1986, 58: 505-513.

[38] Allan R E. Agronomic comparisons between *Rht1* and *Rht2* semi-dwarf genes in winter wheat. *Crop Sci*, 1989, 29: 1103-1108.

[39] Sip V, Chrpova J, Zofajova A, Pankova K, Uzik M, Snape J W. Effects of specific *Rht* and *Ppd* alleles on agronomic traits in winter wheat cultivars grown in middle Europe. *Euphytica*, 2010, 172: 221-233 .

[40] Addisu M, Snape J W, Simmonds J R, Gooding M J. Reduced height (*Rht*) and photoperiod insensitivity (*Ppd*) allele associations with establishment and early growth of wheat in contrasting production systems. *Euphytica*, 2010, 172: 169-181.

[41] Worland A J, Korzun V, Röder M S, Ganal M W, Law C N. Genetic analysis of the dwarfing gene Rht8 in wheat: Ⅱ . The distribution and adaptive significance of allelic variants at the Rht8 locus of wheat as revealed by microsatellite screening. *Theor Appl Genet*, 1998, 96: 1110-1120.

[42] Knopf C, Becker H, Ebmeyer E, Korzun V. Occurrence of three dwarfing *Rht* genes in German winter wheat varieties. *Cereal Res Commun*, 2008, 36: 553-560.

[43] Tang N, Jiang Y, He B R, Hu Y G. The effects of dwarfing genes (*Rht-B1b*, *Rht-D1b*, and *Rht 8*) with different sensitivity to GA_3 on the coleoptile length and plant height of wheat. *Agric Sci China*, 2009, 8: 1028-1038.

Allelic variation at the vernalization genes *Vrn-A1*, *Vrn-B1*, *Vrn-D1* and *Vrn-B3* in Chinese common wheat cultivars and their association with growth habit

X. K. Zhang[1,2], Y. G. Xiao[1], Y. Zhang[1], X. C. Xia[1], J. Dubcovsky[3], and Z. H. He[1,4]

[1] *Institute of Crop Science, National Wheat Improvement Center/The National Key Facility for Crop Gene Resources and Genetic Improvement, Chinese Academy of Agricultural Sciences* (*CAAS*), 12 *Zhongguancun South Street, Beijing* 100081, *China*; [2] *College of Agronomy, Northwest Sci-Tech University of Agriculture and Forestry, Yangling, Shaanxi* 712100, *China*; [3] *Department of Plant Sciences, University of California, Davis, CA* 95615, *USA.*; [4] *CIMMYT China Office, C/O CAAS*, 12 *Zhongguancun South Street, Beijing* 100081, *China.*

Abstract: Information on the distribution of vernalization genes and their association with growth habit is crucial to understand the adaptability of wheat cultivars to different environments. In this study, 278 Chinese wheat cultivars were characterized with molecular markers for the vernalization genes *Vrn-A1*, *-B1*, *-D1*, and *-B3*. Heading time of the cultivars was evaluated in a greenhouse under long days without vernalizaton. The dominant *Vrn-D1* allele showed the highest frequency in the Chinese wheat cultivars (37.8%), followed by the dominant *Vrn-A1*, *-B1*, and *-B3* alleles. Ninety-two winter cultivars carried recessive alleles of all four vernalization loci, whereas 172 spring genotypes contained at least one dominant *Vrn* allele. All cultivars released in the North China Plain Winter Wheat Zone were winter type. Winter (53.0%), spring (36.1%) and early heading (10.9%) cultivars were grown in the Yellow and Huai River Valley Winter Zone. Most of the spring genotypes from this zone carried only the dominant *Vrn-D1* allele, which was also predominant (64.1%) in the Middle and Lower Yangtze Valley Winter Zone and Southwestern Winter Wheat Zone. In three spring-sown wheat zones, all cultivars were early heading spring types that frequently possessed the strongest dominant *Vrn-A1a* allele, and combinations with other dominant *Vrn* gene (s). The *Vrn-D1* allele is associated to the latest heading time, *Vrn-A1* the earliest and *Vrn-B1* intermediate values. The information is useful for understanding the adaptation of Chinese wheat cultivars, and also important for breeding programs in other countries with an interest in using Chinese wheats.

Introduction

Common wheat (*T. aestivum* L.) is one of the most widely cultivated food crops in the world, and is grown over a wide range of elevations, climatic conditions and soil fertility (Bushuk, 1998). The wide adaptability of wheat is largely governed by three groups of genetic factors, vernalization (*Vrn*) genes (vernalization requirement), photoperiod (*Ppd*) genes (photoperiod sensi-

Published in Crop Science, 2008, 48 (2): 458-470

tivity) and genes controlling earliness *per se* (*Eps*) (Kato and Yamagata, 1988), which act together to determine flowering time and hence the basic adaptation of a genotype for a particular environmental condition (Worland, 1996; Worland et al., 1998). Vernalization genes determine growth habits, which divide wheat into winter and spring classes. Winter cultivars are mainly adapted to areas with average January temperature between −7℃ and 4℃, whereas spring cultivars are adapted to areas with temperatures below or above this range (Iwaki et al., 2000, 2001). The different frequencies of *Vrn* alleles observed in different parts of the world suggest that these allele combinations have an adaptive value (Gotoh, 1979; Stelmakh, 1990; Iwaki et al., 2000, 2001; Goncharov, 1998; Stelmakh, 1998). For example, the dominant *Vrn-A1* allele is frequently observed in improved cultivars from Europe and Siberia, whereas higher frequencies of dominant *Vrn-D1* allele were found in commercial cultivars from countries situated nearer the equator (Stelmakh, 1990), such as in Japan (Gotoh, 1979), in the Central Asian Republic of the Former Soviet Union (Stelmakh, 1990) and in China (Iwaki et al. 2000, 2001) or in Mediterranean climates (Fu et al., 2005). Important germplasm from the International Maize and Wheat Improvement Center (CIMMYT) have also been classified for their phasic gene constitution, allowing conclusions about the frequencies of occurrence of certain gene combination (Van Beem et al., 2005). Therefore, an understanding of the vernalization genes present in wheat breeding programs is useful when developing cultivars broadly adapted to different regions.

Various studies showed that vernalization requirement is genetically controlled by at least five loci, *Vrn-A1* (formerly *Vrn1*), *Vrn-B1* (*Vrn2*), *Vrn-D1* (*Vrn3*), *Vrn4* and *Vrn-B3* (*Vrn5*) in global sets of commercial cultivars (Pugsley, 1971, 1972; McIntosh et al., 1998; Goncharov, 2003; Yan et al., 2006). The three major vernalization genes *Vrn-A1*, *Vrn-B1* and *Vrn-D1* are located on the homoeologous chromosomes 5A, 5B and 5D in common wheat, respectively (Pugsley, 1971; Law et al., 1976; Worland, 1996; Barrett et al., 2002; Yan et al., 2003), and *Vrn-B3* is located on chromosome arm 7BS (Law and Wolfe, 1966; Yan et al., 2006). The spring alleles from these genes are epistatic to the winter alleles, and therefore the winter habit is observed only when all of genes have recessive alleles (Pugsley, 1971). The *Vrn-A1a* allele is the most potent allele for spring growth habit providing complete insensitivity to vernalization, whereas *Vrn-B1*, *Vrn-D1* and *Vrn4* result in a partial elimination of the vernalization requirement (Pugsley, 1971, 1972).

Detection of vernalization genesby traditional genetic methods is time consuming. Fortunately, the recent cloning of wheat vernalization genes (Yan et al., 2003, 2006) has facilitated the development of gene specific markers or functional markers (also known as perfect or diagnostic markers). These markers provide a unique opportunity to screen large collections of wheat germplasm for allelic diversity at the *Vrn* genes. Yan et al. (2003) used diploid wheat *T. monococcum* (2n=14, A^mA^m) to clone the *Vrn-*A^m*1* gene using a positional cloning approach and to show that this gene is similar to the Arabidopsis meristem identity gene *APETALA1*. The different dominant *Vrn-A1* alleles were identified in polyploid wheats. The most abundant one in common wheat, *Vrn-A1a*, has an insertion of a foldback repetitive element and a duplicated region in the promoter (Yan et al., 2004). The less frequent *Vrn-A1b* allele shows several SNPs and deletions in the promoter region (Yan et al., 2004), whereas the rare *Vrn-A1c* allele has a large deletion in the first intron (Fu et al., 2005). The *Vrn-A1c* allele was found only in the spring hexaploid landrace IL369 from Afghanistan, but is common among tetraploid spring genotypes. The dominant *Vrn-B1* and *Vrn-D1* alleles for spring growth habit are also characterized by large deletions in the first intron of the same gene (Fu et al., 2005). A dominant allele for spring growth habit was recently identified in the cultivar Hope and was designated *Vrn-B3*, based on its orthology with the barley *Vrn-H3* gene (Yan et al., 2006). The dominant *Vrn-B3* allele has a retrotransposon insertion in the promoter region of a gene similar to Arabidopsis *FT* (Yan et al., 2006).

China is the largest wheat producer in the world. Ten major wheat agro-ecological zones have been recognized in China (Fig. 1) based on differences in wheat types, growing season, presence of major biotic and abiotic stresses, and cultivar responses to temperature and photoperiod (Zhuang, 2003). At present, autumn-sown wheats account for about 90% of production and acreage, and include zones Ⅰ (4%), Ⅱ (60%), Ⅲ (13%), Ⅳ (10%) and Ⅴ (minor area of production). Spring-sown wheats represent 7% of the wheat acreage in China and are grown in zones Ⅵ, Ⅶ and Ⅷ. Zones Ⅸ and Ⅹ cover less than 3% of the total wheat area and include both spring and fall sown wheats. Average January temperatures and wheat growing periods are greatly divergent among different Chinese wheat regions (Jin, 1986, 1997; Zhuang, 2003). The spring-sown spring wheat regions are well defined and only spring cultivars are found in these regions. However, the autumn-sown regions include both winter and spring cultivars, resulting in some confusion in cultivar classification. As an example, spring type cultivars from Zones Ⅱ and Ⅲ are sometimes planted in the northern part of Zone Ⅱ resulting in severe losses due to winter damage (Jin, 1986, 1997; Dong and Zheng, 2000; Zhuang, 2003). Therefore, information on the distribution of vernalization genes in Chinese wheats is crucially important for developing widely adapted cultivars and for providing information to extension workers and farmers.

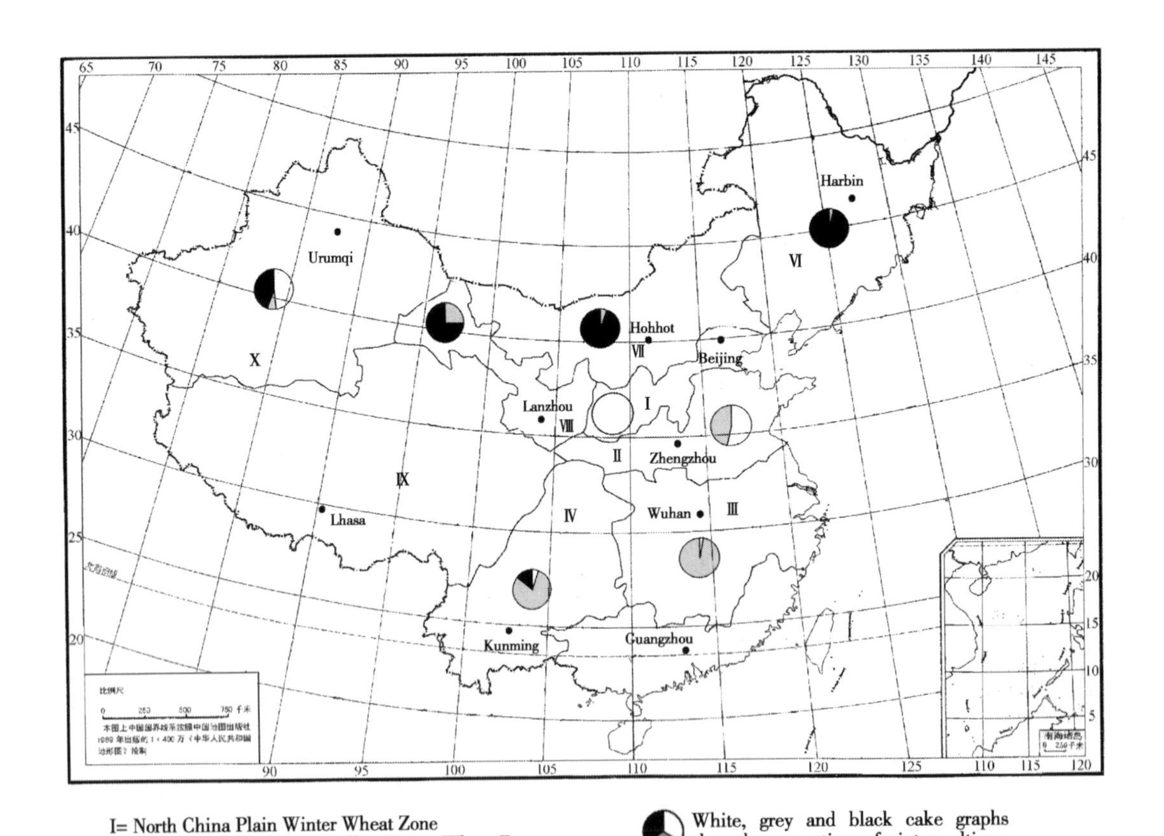

Ⅰ= North China Plain Winter Wheat Zone
Ⅱ= Yellow and Huai River Valley Winter Wheat Zone
Ⅲ = Middle and Lower Yangtze Valley Winter Wheat Zone
Ⅳ = Southwestern Winter Wheat Zone
Ⅴ = Southern Winter Wheat Zone
Ⅵ = Northeastern Spring Wheat Zone
Ⅶ = Northern Spring Wheat Zone
Ⅷ = Northwestern Spring Wheat Zone
Ⅸ = Qinghai–Tibetan Plateau Spring and Winter Wheat Zone
Ⅹ = Xinjiang Winter and Spring Wheat Zone

White, grey and black cake graphs show the proportions of winter cultivars with recessive alleles, spring cultivars without and with dominant *Vrn-A1* allele, respectively.

Fig. 1 Distribution of growth habit and vernalization allele combinations among China different wheat zones

Previously, vernalization genotypes and growth habits of 42 Chinese wheat landraces were studied by crossing them with near-isogenic lines of 'Triple Dirk' carrying different vernalization genes (Gotoh, 1979; Iwaki et al., 2000, 2001). However, the vernalization genotypes and growth habits of wheat cultivars released in China from 1960s to the present are mostly unknown. The aims of this study are (1) to characterize allelic variations at four vernalization loci *Vrn-A1*, *Vrn-B1*, *Vrn-D1* and *Vrn-B3* among the major Chinese wheat cultivars released since the 1960s using molecular markers, (2) to test heading times of these cultivars under controlled conditions, and (3) to analyze the relationships between vernalization genotypes, geographic distribution and planting times for these cultivars. This information is expected to be useful for improving the adaptation of wheat cultivars in Chinese breeding programs targeting different environments, and for foreign breeding programs interested in using Chinese wheat germplasm.

Materials and methods

Plant material

A total of 278 Chinese wheat cultivars collected from eight major zones were used to identify their vernalization genotypes using PCR methods, and the growth habits of 266 of them were assessed in the greenhouse. They include landmark landraces, leading cultivars and 12 well-known introductions, which had significant impact on Chinese wheat production and breeding after 1960. The number of entries in various zones was based on wheat acreage and number of cultivars developed by the local breeding programs (Table 1). Cultivars Thatcher (*Vrn-A1a*), Chinese Spring (*Vrn-D1*) and Hope (*Vrn-B3*) were used as controls.

Table 1 Distribution of growth habits and combination of dominant alleles at *Vrn-A1*, *Vrn-B1*, *Vrn-D1* and *Vrn-B3* loci in Chinese various wheat zones

Classification	Zone[1]								Total
	I	II	III	IV	VI	VII	VIII	X	
No. of all cultivars	32	86	29	35	24	17	27	28	278
No. of cultivars tested in the greenhouse	32	75	29	35	24	17	27	27	266
%Late heading cultivars	100	51.2	3.4	5.7	—	—	—	46.4	33.1
%Early heading cultivars	—	36.0	96.6	94.3	100	100	100	50.0	62.6
Overall % within spring cultivars									
Vrn-A1	—	—	—	15.2	95.8	94.1	74.1	85.7	44.2
Vrn-B1	—	6.7	11.1	42.4	75.0	58.8	70.4	50.0	42.4
Vrn-D1	—	93.3	96.3	78.8	50.0	29.4	14.8	28.6	61.0
Vrn-B3	—	—	—	—	8.3	—	—	—	1.2
Regional % within spring cultivars and average heading time									
Vrn-D1 alone[2] = 54d[3] (36-109d)	—	93.3	88.9	51.5	—	—	3.7	14.3	41.9
Vrn-B1 alone=47d (36-73d)	—	6.7	3.7	15.2	—	—	11.1	—	6.4
Vrn-B1 + *Vrn-D1* = 42d (36-49d)	—	—	7.4	18.1	4.2	5.9	11.1	—	7.6
Subtotal spring without *Vrn-A1*	—	100	100	84.8	4.2	5.9	25.9	14.3	55.9
Vrn-A1 alone=39d (33-46d)	—	—	—	—	12.5	23.5	26.0	35.7	11.0
Vrn-A1 + *Vrn-B1* =38d (32-49d)	—	—	—	6.1	33.2	47.1	48.1	35.7	20.9
Vrn-A1 + *Vrn-D1* =38d (36-44d)	—	—	—	6.1	12.5	17.6	—	—	4.6
Vrn-A1 + *Vrn-B1* + *Vrn-D1* =38d (33-50d)	—	—	—	3.0	29.2	5.9	—	14.3	6.4
Vrn-A1 + *Vrn-B1* + *Vrn-B3* =30d	—	—	—	—	4.2	—	—	—	0.6
Vrn-A1 +*Vrn-B1* +*Vrn-D1* +*Vrn-B3*=31d	—	—	—	—	4.2	—	—	—	0.6
Subtotal spring with *Vrn-1*	—	0	0	15.2	95.8	94.1	74.1	85.7	44.1

1, See Fig. 1.

2, Eight cultivars with the *Vrn-D1* allele alone were not tested for heading time.

3, Average heading time of tested genotypes with this genotype.

DNA extraction and molecular marker analysis

Genomic DNA was extracted from seeds following the procedure of Gale et al. (2001). Sequences of nine specific primer sets for the amplification of allelic variations at *Vrn-A1*, *Vrn-B1*, *Vrn-D1* and *Vrn-B3* loci have been published before (Yan et al., 2004, 2006; Fu et al., 2005). These primers were synthesized by Shanghai Sangon Biological Engineering Technology & Services Co. Ltd (http://www.sangon.com).

PCR was performed in an MJ Research PTC-200 thermal cycler. The PCR conditions for the primer pair VRN1AF and VRN1-INT1R were as follows: 1×PCR buffer with 1.5mM of $MgCl_2$, 150μM of each dNTPs, 2pmol of each primer, 1unit of *Taq* DNA polymerase (Tiangen Biotech Co., Ltd.) and 50-100ng of template DNA in 20μl of final volume. Thermocycing conditions were an initial denaturation at 94℃ for 10min, followed by 38 cycles of 45s at 94℃, 45s at 50℃, 1min at 72℃, and with a final extension step at 72℃ for 5min. Amplified PCR fragments were separated on a 2.5% agarose gel at 80V for 4-5h, stained with ethidium bromide, and visualized using UV light. Thatcher (*Vrn-A1a*) and Chinese Spring (*vrn-A1*) were used as controls in the test. For the other eight primer sets we used 10 pmol of each primer and the PCR conditions were similar to those for the VRN1AF/ VRN1-INT1R primer pair except for the annealing temperatures and extension times. The amplified PCR fragments were separated on a 1% agarose gel at 150V, stained with ethidium bromide, and visualized using UV light.

Greenhouse experiment

Heading times of 266 wheat cultivars were evaluated following the methods of Stelmakh (1987), Iwaki et al. (2001) and Beales et al. (2005) with minor modifications. Cultivars were grown in a greenhouse under a 16 h daylength regime, and a temperature of 18±3℃ to avoid natural vernalization. For each cultivar, five germinated seeds were sown in soil-filled containers at a space of 2.5cm (between plants in a row) and 6cm (between rows). Days to heading were recorded during six months in the greenhouse.

Results

Allelic frequencies at the *Vrn-A1*, *Vrn-B1*, *Vrn-D1* and *Vrn-B3* loci

The specific allele combinations identified in the 278 wheat cultivars are shown in the online supplemental Appendix 1. Firstly, all cultivars were tested with primers VRN1AF and VRN1-INT1R for the *Vrn-A1* promoter region. A total of 68 cultivars from Zones Ⅳ, Ⅵ, Ⅶ, Ⅷ and Ⅹ showed PCR fragments identical to those in cv. Thatcher (965bp and 876bp) indicating the presence of the dominant *Vrn-A1a* allele (Fig. 2A). Only eight cultivars (Fan 7, Chuanyu 12, and Jingmai 11 from Zone Ⅳ, Dabaipi from Zone Ⅶ, Longchun 8, Longchun 21 and Ganmai 8 from Zone Ⅷ, and Xinchun 9 from Zone Ⅹ) showed the 714-bp fragment characteristic of the *Vrn-A1b* allele (Fig. 2A). The remaining 202 cultivars exhibited the 734-bp fragment characteristic of the dominant allele *Vrn-A1c* or the recessive *vrn-A1* allele (Fig. 2A). To distinguish between these two alleles, all cultivars were tested using two primer pairs Intr1/A/F2- Intr1/A/R3 and Intr1/C/F-Intr1/AB/R for the *Vrn-A1* first intron. A 1068-bp fragment was amplified in all cultivars tested using the primer pair Intr1/C/F and Intr1/AB/R (Fig. 2C), while no PCR product was produced using primer pair Intr1/A/F2 and Intr1/A/R3 (Fig. 2B). These results indicate that the large intron 1 deletion (*Vrn-A1c* allele) was not present in the Chinese cultivars and that the 202 cultivars with the 734-bp amplification product carried the recessive *vrn-A1* allele.

A total of 73 cultivars have the dominant *Vrn-B1* allele as indicated by the amplification of a 709-bp fragment using primers Intr1/B/F and Intr1/B/R3 (Fig. 3A). The other 205 cultivars showed a 1149-bp amplification product with primers Intr1/B/F and Intr1/B/R4, which is characteristic of the recessive *vrn-B1* allele (Fig. 3B).

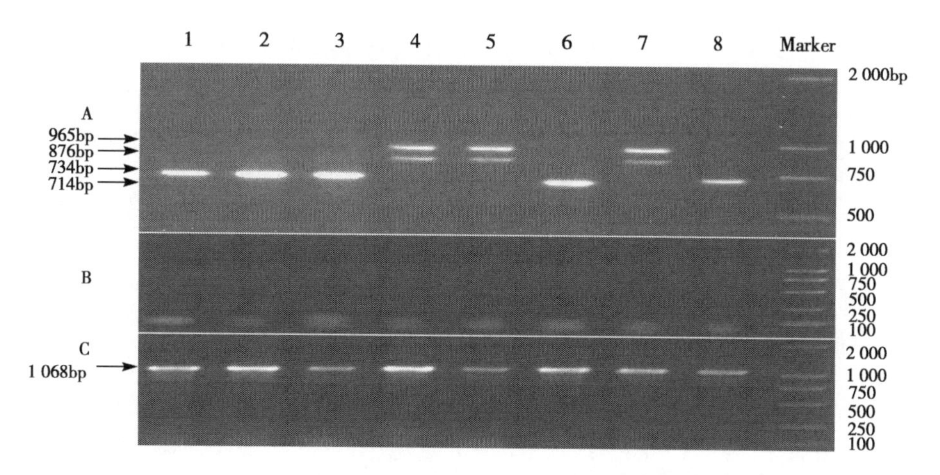

Fig. 2 PCR amplification using primer pairs VRN1AF and VRN1-INT1R (A), Intr1/A/F2 and Intr1/A/R3 (B), and Intr1/C/F and Intr1/AB/R (C) to detect alleles at the *Vrn-A1* locus.

1. Chinese Spring (*vrn-A*), 2. Jing 411 (*vrn-A*), 3. Yumai 2 (*vrn-A*), 4. Thatcher (*Vrn-A1a*), 5. Xinkehan 9 (*Vrn-A1a*), 6. Longchun 21 (*Vrn-A1b*), 7. Gan 630 (*Vrn-A1a*), 8. Shaan 229 (*vrn-A*).

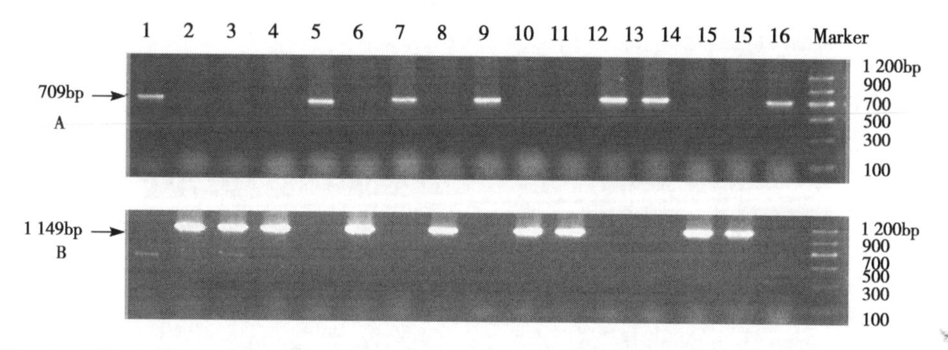

Fig. 3 PCR amplification using primer pair Intr1/B/F and Intr1/B/R3 (A), Intr1/B/F and Intr1/B/R4 (B) to detect the dominant (*Vrn-B1*) and recessive (*vrn-B1*) alleles at the *Vrn-B1* locus, respectively.

1-Abbondanza, 2-Dongfanghong 3, 3-Mentana, 4-Mazhamai, 5-Zhengmai 9023, 6-Mianyang 11, 7-Mianyang 15, 8-Miannong 4, 9-Kefeng 3, 10-Xinkehan 9, 11-Longmai 20, 12-CI12203, 13-Jinchun 14, 14-Xuzhou 25, 15-Chinese Spring, 16-Xinchun 12.

A 1671-bp fragment was generated from 105 cultivars using primer pair Intr1/D/F and Intr1/D/R3 (Fig. 4A), demonstrating that they carried the dominant *Vrn-D1* allele. Amplification of DNA from the other cultivars using primers Intr1/D/F and Intr1/D/R4 showed a 997-bp band characteristic of the recessive *vrn-D1* allele (Fig. 4B).

The dominant *Vrn-B3* allele, defined by the amplification of a 1.2-kb fragment with primers VRN4-B-INS-F and VRN4-B-INS-R, was found only in cultivars Longfumai 1 and Liaochun 10 from Zone Ⅵ (Fig. 5A). All other cultivars showed a 1.14-kb amplification fragment using primers VRN4-B-NOINS-F and VRN4-B-NOINS-R, which is characteristic of the recessive *vrn-B3* allele (Fig. 5B).

Geographic distribution of the different allele combinations

The frequencies of the different *Vrn* allele combinations were very different across different wheat agro-ecological zones (Table 1 and Fig. 1). Cultivars with recessive alleles at all the analyzed *Vrn* loci represent 38.1% of the cultivars and are mainly concentrated in Zones Ⅰ, Ⅱ and Ⅹ (Table 1). The other 61.9% includes cultivars with at least one dominant *Vrn* allele, which can be classified as spring. These cultivars are found mainly in Zones Ⅲ, Ⅳ, Ⅵ, Ⅶ, Ⅷ and Ⅹ (Table 1).

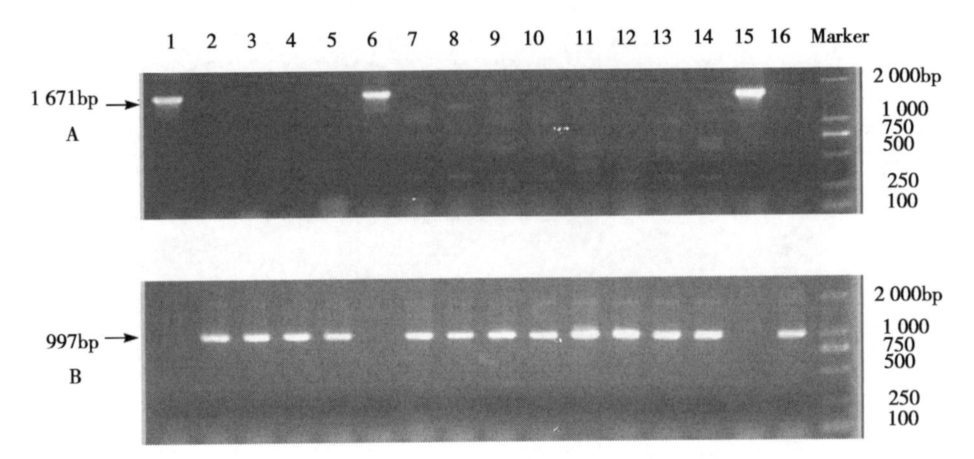

Fig. 4 PCR amplification using primer pair Intr1/D/F and Intr1/D/R3, Intr1/D/F and Intr1/D/R4 (B) detecting the dominant (*Vrn-D1*) and recessive (*vrn-D1*) alleles at the *Vrn-D1* locus, respectively.

1-Chinese Spring, 2-Beijing 10, 3-Jing 411, 4-Bima 4, 5-Abbondanza, 6-Neixiang 36, 7-Jinan 2, 8-Zhoumai 18, 9-Shijiazhuang 8, 10-Shaannong 7859, 11-Yumai 2, 12-Lumai 1, 13-Lumai 14, 14-Zhengmai 9023, 15-Mentana, 16-Emai 6.

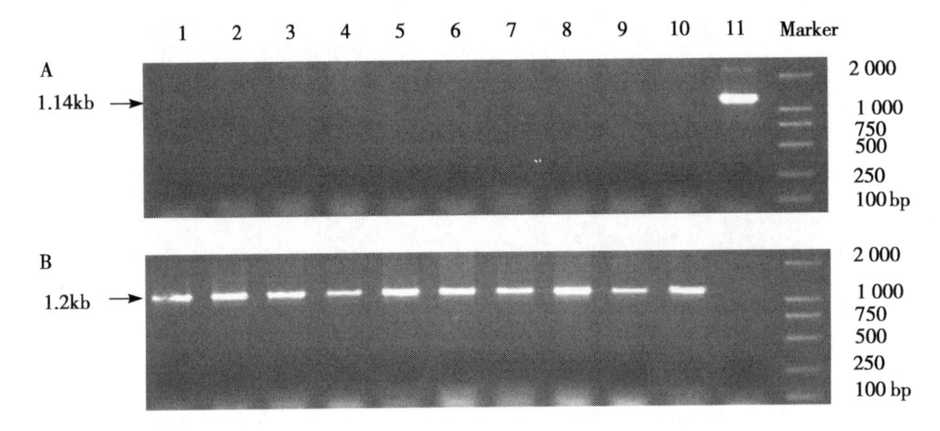

Fig. 5 PCR amplification using primer pair VRN4-B-INS-F and VRN4-B-INS-R (A), VRN4-B-NOINS-F and VRN4-B-NOINS-R (B) to detect dominant (*Vrn-B3*) and recessive (*vrn-B3*) alleles at the *Vrn-B3* locus, respectively.

1-Chinese Spring, 2-Shijiazhuang 407, 3-Lumai 21, 4-Xi'an 8, 5- Jinan 2, 6- Shaanong 7859, 7-Sumai 3, 8- Neixiang 36, 9-Neimai 19, 10-Gan 630, 11-Liaochun 10.

Among the cultivars with at least one dominant *Vrn* allele, the frequencies of the different alleles varied across regions (Table 1). The dominant *Vrn-B3* allele is present only in two cultivars from zone Ⅵ. Among the *Vrn-1* alleles *Vrn-D1* showed highest frequency, followed closely by dominant *Vrn-A1* and *Vrn-B1* alleles (Table 1). The dominant *Vrn-A1* allele is not presented in Zones Ⅰ, Ⅱ, and Ⅲ, and its frequency is low in Zone Ⅳ. However, high frequencies are observed in Zones Ⅵ, Ⅶ, Ⅷ and Ⅹ (Table 1 and Fig. 1). The dominant allele *Vrn-B1* is not presented in Zone Ⅰ, and low frequencies are observed in Zones Ⅱ and Ⅲ. However, high frequencies are observed in Zones Ⅳ, Ⅵ, Ⅶ, Ⅷ and Ⅹ. The dominant allele *Vrn-D1* is not presented in Zone Ⅰ, but is present at relatively high frequencies in Zones Ⅱ, Ⅲ, Ⅳ, Ⅵ, Ⅶ, Ⅷ and Ⅹ.

Among the four autumn-sown wheat zones (Ⅰ, Ⅱ, Ⅲ and Ⅳ) the frequency of dominant *Vrn-D1* allele is the highest, followed by *Vrn-B1* and *Vrn-A1* (*Vrn-B3* is absent) (Fig. 6). In contrast, in three spring-sown wheat zones (Ⅵ, Ⅶ and Ⅷ) the frequency of the dominant *Vrn-A1* alleleis the highest followed by *Vrn-B1* and *Vrn-D1*, respectively (Fig. 6). *Vrn-B3* frequency (2.9%) is the lowest.

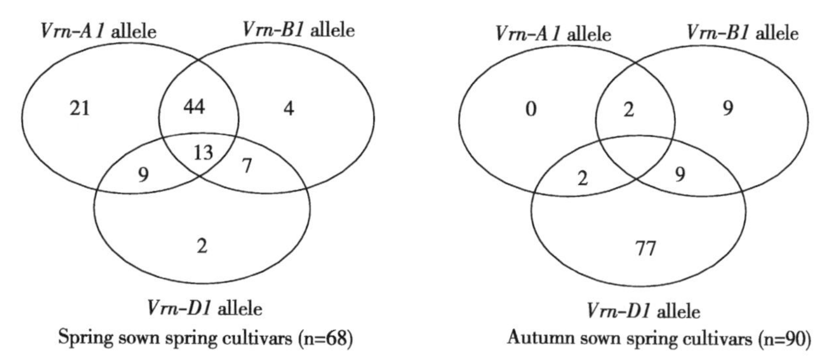

Fig. 6 Percent of different vernalization allele combinations among spring-sown spring cultivars and autumn-sown spring cultivars in China.

Frequencies of the different combinations of vernalization genes were also very different among the various wheat agro-ecological zones (Table 1). In brief, nine combinations of dominant *Vrn* alleles were identified. Among them the *Vrn-D1* allele alone was the most frequent (72 cultivars), followed by the *Vrn-A1Vrn-B1* (36 cultivars) combination. The distribution of the different allele combinations across the different zones is described in Table 1. In summary, most cultivars released in the autumn-sown wheat regions of south China (Zones Ⅲ and Ⅳ) and north China (Zone Ⅱ) possessed *Vrn-D1* as a single dominant allele. In contrast, in spring-sown wheat regions, cultivars carried the strongest dominant *Vrn-A1* alleles and the majority of them include additional dominant *Vrn* alleles at the *Vrn-B1*, *Vrn-D1* and *Vrn-B3* loci. Based on the vernalization alleles found in this study, the vernalization requirement can be ranked from strongest to weaker from Zone Ⅰ, Zone Ⅱ, Zone Ⅲ, Zone Ⅳ, with the weakest requirement in the spring-sown spring wheat regions (Zones VI, VII, and VIII).

Growth habit

Heading dates showed a continuous distribution from 30 days to in excess of six months after planting in the greenhouse. Out of 266 cultivars tested in the greenhouse, the 92 cultivars, which failed to head within 109 days all possessed recessive vernalization alleles at the four *Vrn* loci as identified by the PCR markers. Most of them were classified as winter cultivars in the literature (Jin, 1986, 1997; Zhuang, 2003). Among the 174 cultivars that headed within 109 days (early heading), 164 of them carried at least one of the tested dominant vernalization allele, and were classified as spring. The other cultivars, nine from Zones Ⅱ (Jimai 36, Taishan 1, Lumai 23, Laizhou 953, Weimai 8, Xinmai 9408, Yumai 66, Yumai 70, Xuzhou 14) and one from Zone Ⅲ (Emai 6) carried recessive alleles at the four vernalization loci. The most likely explanation for this discrepancy is the presence of an unknown allele at the four loci characterized in this study or the presence of a spring allele at the *Vrn4* locus not included in this survey because the gene is still unknown.

All 32 cultivars released in Zone Ⅰ headed after 109 days, had all three recessive *vrn-1* alleles and were classified as winter. In Zone Ⅱ, out of 75 cultivars tested in the greenhouse, 44 (58.7%) headed after 109 days. Although both winter and spring types are found in all provinces of Zone Ⅱ, the late heading cultivars were mainly cultivated in the provinces of Shandong (79.2%), Shaanxi (63.6%), and Anhui (100.0%), whereas the early heading cultivars were mostly present in the provinces of Henan, and Jiangsu. Most of the cultivars from Zones Ⅲ and Ⅳ (>94%) headed before 109 days. The frequency of early heading genotypes increased gradually from north to south in the autumn-sown regions. In Zones Ⅵ, Ⅶ and Ⅷ, all 68 cultivars tested in the greenhouse headed within 109 days. In Zone X, the frequency of early heading genotypes was 51.9%. Therefore, it was concluded that spring cultivars in China are more frequent in the high latitude regions (spring sowing) and in the low latitude area with warm winters (autumn sowing). Winter cultivars are frequently present in the middle latitude area with relatively cold winter (autumn sowing).

Relationships between *Vrn* allele combinations and growth habits

The relationships between vernalization genotypes and heading times in the absence of vernalization are shown in Table 1. The 92 late heading (winter) cultivars all carried recessive alleles at the four vernalization loci. Different combinations and proportion of vernalization alleles were found in the other 172 early heading (spring) cultivars. Single dominant alleles were observed for the *Vrn-A1* (11.0%), *Vrn-B1* (6.4%) or *Vrn-D1* (41.9%). In addition, we observed two gene combinations including *Vrn-A1* / *Vrn-B1* (20.9%), *Vrn-A1* / *Vrn-D1* (4.6%) and *Vrn-B1*/ *Vrn-D1* (7.6%); and three dominant allele combinations including *Vrn-A1Vrn-B1Vrn-D1* (6.4%), and *Vrn-A1Vrn-B1Vrn-B3* (0.6%). In addition, one very early heading cultivar (Liaochun 10) carried all four dominant alleles (*Vrn-A1Vrn-B1Vrn-D1Vrn-B3*).

Days toheading among the different combinations of dominant vernalization alleles are described in Table 1. In summary, the earliest cultivars were those carrying three to four dominant alleles including the rare *Vrn-B3* allele (average 30-31 days to heading) followed by the one, two or three gene combinations including *Vrn-A1* but not *Vrn-B3* (average 38 days to heading). Lines carrying the *Vrn-B1* /*Vrn-D1* allelle combination headed approximately 42 days after sowing, whereas those carrying only the *Vrn-B1* (average 47d) or *Vrn-D1* (average 54d) were among the latest spring cultivars. Based on these data, the strength of the dominant spring *Vrn-1* alleles can be ranked as *Vrn-A1* > *Vrn-B1* > *Vrn-D1*. *Vrn-B3* resulted in the earliest heading times in combination with other dominant *Vrn1* alleles.

Discussion

Effectivess of molecular markers for identifying vernalization alleles

Functional markers (also known as perfect markers) are derived from polymorphic sites within genes that directly affect phenotypic trait variation, and they are ideal tools for marker-assisted selection (Bagge et al., 2007). The vernalization gene markers developed by Yan et al. (2004, 2006) and Fu et al. (2005) are likely functional markers. The observed heading times in the greenhouse experiment and the growth habit determinations from the literature (Jin, 1986, 1997; Dong and Zheng, 2000; Zhuang, 2003) were consistent with the *Vrn* genotypes. However, 10 out of the 174 cultivars showed inconsistent results, with early heading in the greenhouse, but recessive alleles present at the four *Vrn* alleles characterized in this study. These exceptions are most likely due to the presence of the *Vrn4* locus which is known to be present in Chinese landraces (Iwaki et al., 2000, 2001). Unfortunately, this gene has not been cloned and no markers are currently available to screen these cultivars. An alternative possibility is the presence of new mutations at the *Vrn* loci included in this study outside of the region tested with the available primers or the presence of alleles for spring growth habit at unknown vernalization genes. Further investigation of these exceptions may provide new insights on wheat vernalization genes.

Additional exceptions were the early heading Chinese landraces Jiounong 2 and Ganmai 11 from Zone Ⅷ, which showed three amplification products with primers VRN1A and VRN1-INT1R (data not shown). This result suggests that these landraces may carry a new *Vrn-A1* allele. However, further investigation is needed to confirm this assumption. In spite of these few exceptions, the distribution of dominant *Vrn-A1*, *Vrn-B1* and *Vrn-D1* alleles based on molecular markers (Yan et al., 2004; Fu et al., 2005) among modern Chinese cultivars is similar to that reported before for a smaller set of Chinese landraces using crosses with near-isogenic lines of 'Triple Dirk' (Gotoh, 1979; Iwaki et al., 2000, 2001). These results indicate that these molecular markers can be effectively used to detect allelic variations at these four vernalization gene loci.

Reasons for different distributions of growth habit and vernalization genotypes among wheat zones of China

The average January temperatures increase from Zone I to II, from II to III, and from III to IV, whereas the length of growth periods decreases in the same direction (Dong and Zheng, 2000; Zhuang, 2003). The frequency of dominant vernalization alleles gradually increases in the same direction (Fig. 1). Only winter cultivars with recessive alleles at the four vernalization loci can survive in the cold winters of Zone I, whereas almost all cultivars released in Zones III and IV with warmer winters are spring types. Most of them carry the single dominant *Vrn-D1* allele, due to the wide utilization of some breeding parents including Mentana with *Vrn-D1* in Zones III and IV (Stelmakh, 1990; Zhuang, 2003). The *Vrn-D1* allele is the weakest of the dominant *Vrn-1* alleles and has a residual requirement for vernalization that is well suited for fall planted wheats in regions with mild winters.

Zone II is located in the middle of Zones I, III and IV (Fig. 1). The frequency of winter cultivars planted in Zone II is lower than that in Zone I, and higher than those in Zones III and IV. In general, cultivars from the northern part of Zone II have better winter hardiness or freezing resistance in comparison to cultivars from the southern part of Zone II (Zhuang, 2003). Winter wheats can not be cultivated in Zones VI, VII and VIII where the average minimum temperature in January and February is too low. In these regions spring wheats are planted in spring to avoid the colder conditions (Wilsie, 1962). The length of growth periods in spring-sown regions (VI, VII and VIII) is shorter than that in autumn-sown regions (Dong and Zheng, 2000; He et al., 2001; Zhuang, 2003). These characteristics may explain the high frequency of cultivars with the dominant *Vrn-A1* allele in these spring-sown regions, because this allele has the strongest insensitivity to vernalization conferring a very early heading time that is essential for the adaptation to short growing seasons. Heading time can be further accelerated by the presence of multiple *Vrn* alleles. Lines with multiple *Vrn* alleles are frequent in Zone VI, which has the shortest growing season in China. This region is also the only one where the very early heading *Vrn-B3* allele from Hope was found. Interestingly, Hope is not presented in the pedigrees of Liaochun 10 and Longfumai 1 (Zhuang, 2003). In Zone X, the average January temperatures are very cold, so spring cultivars are planted in spring and winter cultivars in autumn. Winter wheats can survive, largely due to snowfall in this region.

The large differences in the frequencies of dominant vernalization alleles observed across the different agro-ecological regions suggest that these distributions are largely determined by environmental factors. Particularly important are the differences in average January temperatures and the length of growth periods among different zones. Cultivars with the most suitable vernalization genes are maintained through long term natural selection and breeder selection in the wheat breeding programs. In addition to vernalization genes, other genes such as the photoperiod genes and the earliness *perse* genes play important roles in the determination of heading time in different cultivars. They are the reasons why distributions of growth habit and vernalization genotypes between Chinese and other country's wheats are different (Stelmakh, 1998; Iwaki et al., 2000, 2001; Fu et al., 2005). The effect of dominant spring *Vrn-1* alleles on heading time in this study is different with the report of Stelmakh (1993). Stelmakh (1993) used three genetic backgrouds to study genetic effects of *Vrn-1* genes on heading date in the field. It showed that the *Vrn-B1* allele was associated with the latest heading time, the *Vrn-A1* with the earliest and the *Vrn-D1* with intermediate values. The difference could be due to different genetic backgrounds of cultivars and environments. Therefore, further investigation is needed to understand the distribution of these associated genes and their interactions with the vernalization genes in the determination of the adaptability of wheat cultivars to the different agroecological regions.

Classification of autumn-sown wheat zones in China

A genetic definition of growth habit (winter vs. spring) of wheat cultivars based on vernalization genotype is superior to the customary definition based on sowing time (Crofts, 1989). Autumn-sownwheats in China are traditionally referred to as winter types (Jin, 1986, 1997; Zhuang, 2003), and therefore the classification of autumn-sown wheat zones was originally based on sowing time. Our results show that in addition to winter cultivars, the Zone II includes spring cultivars with a residual vernalization requirement (presence of *Vrn-D1*). We also showed that the majority of modern cultivars released in Zones III and IV have at least one dominant *Vrn* allele and should be classified as spring type. These results confirmed previous observations from Chinese breeders (Zhuang, 2003). We propose to re-name Zone II as the Yellow and Huai River Valley Autumn-Sown Winter and Spring Wheat Zone, and Zones III and IV as the Middle and Lower Yangtze Valley Autumn-Sown Spring Wheat Zone, and the Southwestern Autumn-Sown Spring Wheat Zone, respectively. This view is supported by He et al. (2001).

In conclusion, growth habits and distribution of dominant vernalization alleles among variouswheat zones in China were significantly different. All cultivars released in Zone I were winter types and carried recessive alleles at the four vernalization loci. In Zone II, both winter and spring cultivars were present and the later usually carried a single dominant *Vrn-D1* allele. In Zones III and IV, spring cultivars with the single dominant *Vrn-D1* allele were frequent. In spring-sown Zones VI, VII and VIII, all cultivars were spring and most of them carried the strongest dominant vernalization gene *Vrn-A1* plus other dominant gene (s). The distribution of growth habit and vernalization alleles in the different wheat zones of China were largely determined by the severity of the winter temperatures and the length of the growing season.

Acknowledgments

The authorsare grateful to Prof. Robert McIntosh from University of Sydney for kindly reviewing this manuscript. Dr Jorge Dubcovsky acknowledges support from USDA-NRI grant 2007-35301-17737. This study was supported by the National 863 Program (2006AA10Z1A7 and 2006AA100102), and the Ministry of Agriculture of China (2006-G2 and 2006BAD01A02).

References

Bagge, M., X. C. Xia, and T. Lübberstedt. 2007. Functional markers in wheat. Curr. Opin. Plant Biol., 10: 211-216.

Barrett, B., M. Bayram, and K. Kidwell. 2002. Identifying AFLP and microsatellite markers for vernalization response gene*Vrn-B1* in hexaploid wheat (*Triticum aestivum* L.) using reciprocal mapping populations. Plant Breed, 121: 400-406.

Beales, J., D. A. Laurie, and K. M. Devos. 2005. Allelic variation at the linked *AP1* and *PhyC* loci in hexaploid wheat is associated but not perfectly correlated with vernalization response. Theor. Appl. Genet., 110: 1099-1107.

Bushuk, W. 1998. Wheat breeding for end product use. Euphytica. 100: 137-145.

Crofts, H. J. 1989. On defining a winter wheat. Euphytica, 44: 225-234.

Dong, Y. S., and D. S. Zheng. 2000. Wheat Genetic Resources in China. China Agriculture Press, Beijing. (in Chinese)

Fu, D., P. Szücs, L. Yan, M. Helguera, J. S. Skinner, J. V. Zitzewitz, P. M. Hayes, and J. Dubcovsky. 2005. Large deletions within the first intron in *VRN-1* are associated with spring growth habit in barley and wheat. Mol. Gen. Genomics, 273: 54-65.

Gale, K. R., W. Ma, W. Zhang, L. Rampling, A. S. Hill, R. Appels, P. Morris, and M. Morrel. 2001. Simple high-throughput DNA markers for genotyping in wheat. p 26-31. *In*: R. Eastwood et al. (ed.) 10th Assembly Proceedings, Wheat Breeding Society of Australia Inc..

Goncharov, N. P. 1998. Genetic resources of wheat related species: the *Vrn* genes controlling growth habit (spring vs. winter). Euphytica, 100: 371-376.

Goncharov, N. P. 2003. Genetics of growth habit (spring vs. winter) in common wheat: confirmation of the existence of dominant gene *Vrn4*. Theor. Appl. Genet., 107: 768-772.

Gotoh, T. 1979. Genetic studies on growth habit of some important spring wheat cultivars in Japan, with special reference to the identification of the spring genes involved. Jap. J. Breed, 29: 133-145.

He, Z. H., S. Rajaram, Z. Y. Xin, and G. Z. Huang. 2001. A History of Wheat Breeding in China. CIMMYT, Mexico, D. F.

Iwaki, K., K. Nakagawa, H. Kuno, and K. Kato. 2000. Ecogeographical differentiation inEast Asian wheat, revealed from the geographical variation of growth habit and *Vrn* genotype. Euphytica, 111: 137-143.

Iwaki, K., S. Haruna, T. Niwa, and K. Kato. 2001. Adaptation and ecological differentiation in wheat with special reference to geographical variation of growth habit and *Vrn* genotype. Plant Breed, 120: 107-114.

Jin, S. B. 1986. Chinese Wheat Cultivars and Their Pedigrees (1962-1982). China Agriculture Press, Beijing. (in Chinese)

Jin, S. B. 1997. Chinese Wheat Cultivars and Their Pedigrees (1983-1993). China Agriculture Press, Beijing. (in Chinese)

Kato, K., and H. Yamagata. 1988. Method for evaluation of chilling requirement and narrow-sense earliness of wheat cultivars. Jap. J. Breed, 38: 172-186.

Law, C. N., A. J. Worland, and B. Giorgi. 1976. The genetic control of ear-emergence time by chromosomes 5A and 5D of wheat. Heredity, 36: 49-58.

Law, C. N., and M. S. Wolfe. 1966. Location of genetic factors for mildew resistance and ear emergence time on chromosome 7B of wheat. Can. J. Genet. Cytol., 8: 462-470.

McIntosh, R. A., G. E. Hart, K. M. Devos, M. D. Gale, and W. J. Rogers. 1998. Catalogue of gene symbols for wheat. P. 1-235. *In*: A. E. Slinkard (ed.) Proc. 9th Int. Wheat Genet. Symp., Vol. 5. University Extension Press, University of Saskatchewan, Saskatoon, Saskatchewan, Canada.

Pugsley, A. T. 1971. A genetic analysis ofthe spring-winter habit of growth in wheat. Aust. J. Agric. Res., 22: 21-23.

Pugsley, A. T. 1972. Additional genes inhibiting winter habit in wheat. Euphytica, 21: 547-552.

Stelmakh, A. F. 1987. Growth habit in common wheat (*Triticum aestivum* L. EM. Thell.). Euphytica, 36: 513-519.

Stelmakh, A. F. 1990. Geographic distribution of *Vrn* genes in landraces and improved varieties of spring bread wheat. Euphytica, 45: 113-118.

Stelmkh, A. F. 1993. Genetic effects of *Vrn* genes on heading date and agronomic traits in bread wheat. Euphytica, 65: 53-60.

Stelmakh, A. F. 1998. Genetic systems regulating flowering response in wheat. Euphytica, 100: 359-369.

Van Beem, J., V. Mohler, R. Lukman, M. van Ginkel, M. William, J Crossa, and A. J. Worland. 2005. Analysis of genetic factors influencing the developmental rate of globally important CIMMYT wheat cultivars. Crop Sci., 45: 2113-2119.

Wilsie, C. P. 1962. Crop Adaptation and Distribution. Freeman Press, San Francisco.

Worland, A. J. 1996. The influence of flowering time genes on environmental adaptability in European wheats. Euphytica, 89: 49-57.

Worland, A. J., A. Bŏrner, V. Korzun, W. M. Li, S. Petrovic, and E. J. Sayers. 1998. The influence of photoperiod genes on the adaptability of European winter wheats. Euphytica, 100: 385-394.

Yan, L., A. Loukoianov, G. Tranquilli, M. Helguera, T. Fahima, and J. Dubcovsky. 2003. Positional cloning of the wheat vernalization gene *VRN1*. Proc. Natl. Acad. Sci., USA. 100: 6263-6268.

Yan, L., M. Helguera, K. Kato, S. Fukuyama, J. Sherman, and J. Dubcovsky. 2004. Allelic variation at the *VRN-1* promoter region in polyploid wheat. Theor. Appl. Genet., 109: 1677-1686.

Yan, L., D. Fu, C. Li, A. Blechl, G. Tranquilli, M. Bonafede, A. Sanchez, M. Valarik, S. Yasuda, and J. Dubcovsky. 2006. The wheat and barley vernalization gene *VRN3* is an orthologue of *FT*. Proc. Natl. Acad. Sci., USA. 103: 19581-19586.

Zhuang, Q. S. 2003. Wheat Improvement and Pedigree Analysis in Chinese Wheat Cultivars. China Agriculture Press, Beijing. (in Chinese)

Distribution of the photoperiod insensitive *Ppd-D1a* allele in Chinese wheat cultivars

F. P. Yang[1,2], X. K. Zhang[2,3], X. C. Xia[2], D. A. Laurie[4], W. X. Yang[1], and Z. H. He[2,5]

[1] *Crop Research Institute, Gansu Academy of Agricultural Sciences, Lanzhou* 730070, *Gansu, China*; [2] *Institute of Crop Science, National Wheat Improvement Center/The National Key Facility for Crop Gene Resources and Genetic Improvement, Chinese Academy of Agricultural Sciences* (*CAAS*), 12 *Zhongguancun South Street, Beijing* 100081, *China*; [3] *College of Agronomy, Northwest Sci-Tech University of Agriculture and Forestry, Yangling, Shaanxi* 712100, *China*; [4] *Crop Genetics Department, John Innes Centre, Norwich Research Park, Colney, Norwich NR4 7UH, UK*; [5] *CIMMYT China Office, C/O CAAS, Zhongguancun South Street, Beijing* 100081, *China*

Abstract: Photoperiod response is of great importance for optimal adaptation of bread wheat cultivars to specific environments, and variation is commonly associated with allelic differences at the *Ppd-D1* locus on chromosome 2D. A total of 926 Chinese wheat landraces and improved cultivars collected from nine wheat growing zones were tested for their genotypes at the *Ppd-D1* locus using allele-specific markers. The average frequency of the photoperiod-insensitive *Ppd-D1a* allele was 66.0%, with the frequencies of 38.6% and 90.6% in landraces and improved cultivars, respectively. However, the *Ppd-D1a* allele was present in all improved cultivars released after 1970 except for spring wheats in high latitude northwestern China, and winter wheats in Gansu and Xinjiang. The presence of the *Ppd-D1a* allele in landraces and improved cultivars increased gradually from north to south, illustrating the relationship between photoperiod response and environment. *Ppd-D1a* in Chinese wheats is derived from three sources, Japanese landrace Akagomughi and Chinese landraces Mazhamai and Youzimai. The current information is important for understanding the broad adaptation of improved Chinese wheat cultivars.

Key words: Bread wheat, Photoperiod sensitive gene, *Ppd-D1*, Molecular markers

Introduction

The adaptation of common wheat (*Triticum aestivum* L.) cultivars to diverse environmental conditions is greatly influenced by flowering time (Whitechurch and Slafer, 2002), which is mainly determined by three groups of genes: vernalization response genes (*Vrn* genes), photoperiod response genes (*Ppd* genes) and developmental rate genes (earliness per se, *Eps* genes) (Snape et al., 2001). The first two groups of genes are environment-dependent, whereas the third is largely environment-independent. Photoperiod response in wheat is described as sensitive when timely flowering occurs only in long days, and insensitive when flowering occurs in either long-or short-day environments. Photoperiod response is closely associated with adaptability and grain yield in European and Canadian wheat cultivars (Martiníc, 1975; Hunt, 1979; Worland et al., 1994, 1998). In Asia, Mediterranean and North African regions, most landraces are sensitive to photoperiod, whereas all improved cultivars with high yield

Published in Euphytica, 2009, 165 (3): 445-452

potential are insensitive (Ortiz Ferrara et al., 1998). These results imply that the introduction of photoperiod insensitivity into improved wheat cultivars in these regions enhanced their adaptation to a broader range of environments than landraces with sensitivity to photoperiod.

Photoperiod response in wheat is mainly controlled by the genes*Ppd-D1* (previously designated *Ppd1*), *Ppd-B1* (*Ppd2*) and *Ppd-A1* (*Ppd3*) located on the short arms of chromosomes 2D, 2B and 2A, respectively (Welsh et al., 1973; Law et al., 1978). The dominant alleles *Ppd-D1a*, *Ppd-B1a* and *Ppd-A1a* confer photoperiod insensitivity, whereas the recessive alleles *Ppd-D1b*, *Ppd-B1b* and *Ppd-A1b* confer photoperiod sensitivity (Pugsley 1966; Dyck et al., 2004). The *Ppd-D1a* allele for photoperiod insensitivity is generally considered the most potent, followed by *Ppd-B1a* and *Ppd-A1a* (Scarth and Law, 1984). However, recent work by Tanio and Kato (2007) showed that *Ppd-B1a* could be as strong as *Ppd-D1a*.

Significant progress was recently achieved in molecular characterization of photoperiod response gene *Ppd-D1* in wheat. Beales et al. (2007) isolated homologues of barley *Ppd-H1* from a 'Chinese Spring' wheat BAC library, exploiting the collinear relationship between wheat gene *Ppd-D1* and barley gene *Ppd-H1* (Dunford et al., 2000; Turner et al., 2005). Sequence alignments from wheat cultivars with known *Ppd-D1* alleles revealed a 2, 089-bp deletion upstream of the coding region in wheat cultivars with the allele *Ppd-D1a*. Subsequently, gene-specific primer sets were developed based on the sequence deletion enabling detection of allelic variants at the locus. A 414-bp fragment was amplified for the *Ppd-D1b* allele using primer pairs Ppd-D1 _ F and Ppd-D1 _ R1, whereas a 288-bp fragment was detected for the allele *Ppd-D1a* using primer pairs Ppd-D1 _ F and Ppd-D1 _ R2 (Beales et al., 2007).

China is the largest wheat producer and consumer in the world. The wheat-growing area was divided into ten major agro-ecological zones based on differences in wheat types, growing seasons, presence of major biotic and abiotic stresses, and responses to temperature and photoperiod in different regions (Fig. 1) (He et al., 2001; Zhuang, 2003; Zhang et al., 2008). Chinese wheat is unique in several aspects. China is a secondary centre of origin for wheat, and more than 20,000 landraces are stored in the national gene bank. Cultivars are early maturing to fit the multi-cropping system, and unique products are produced and consumed. Autumn-sown winter and facultative cultivars are the most common types and are grown in the Northern China Autumn-sown Winter and Facultative Wheat Region, including the North China Plain Winter Wheat Zone (Zone Ⅰ, around 4% of the total wheat area) and the Yellow and Huai Rivers Valley Facultative Wheat Zone (Zone Ⅱ, 60%). Autumn-sown spring wheat is grown in the Southern China Autumn-sown Spring Wheat Region, comprising the Middle and Lower Yangtze Valley Autumn-Sown Spring Wheat Zone (Zone Ⅲ, 13%), the Southwestern Autumn-Sown Spring Wheat Zone (Zone Ⅳ, 10%), and Southern Autumn-Sown Spring Wheat Zone (Zone Ⅴ, < 1%). Spring-sown spring wheat (7%) is grown in Spring-sown Spring Wheat Region, including the Northeastern Spring Wheat Zone (Zone Ⅵ), the Northern Spring Wheat Zone (Zone Ⅶ), and the Northwestern Spring Wheat Zone (Zone Ⅷ). Spring-sown spring wheat and autumn-sown winter wheat regions, comprising about 6% of the area, include the Qinghai-Tibetan Plateau Spring-Winter Wheat Zone (Zone Ⅸ) and Xinjiang Winter-Spring Wheat Zone (Zone Ⅹ). There are significant differences in the length of the growth cycle, temperature and photoperiod among the different regions (Jin, 1986; 1997; Zhuang, 2003). The growth habits and genotypes at the *Vrn-A1*, *Vrn-B1*, *Vrn-D1* and *Vrn-B3* vernalization loci in Chinese wheat cultivars were documented in a previous study (Zhang et al., 2008). It was found that some cultivars from different wheat zones carried the same vernalization genotypes, but could not be introduced from one zone to another due to inappropriate maturity. For example, some spring wheat cultivars with the semi-dominant *Vrn-D1* allele planted over a large area in Zone Ⅱ were too late or too early in maturity when grown in Zone Ⅲ (Jin, 1986; 1997; Dong and Zheng, 2000; Zhuang, 2003). Therefore, characterization of photoperiod response genes in Chinese wheat cultivars is crucial to fully understanding their adaptability to regional environments, and for providing important information for wheat breeding programs elsewhere using Chinese wheat germplasm.

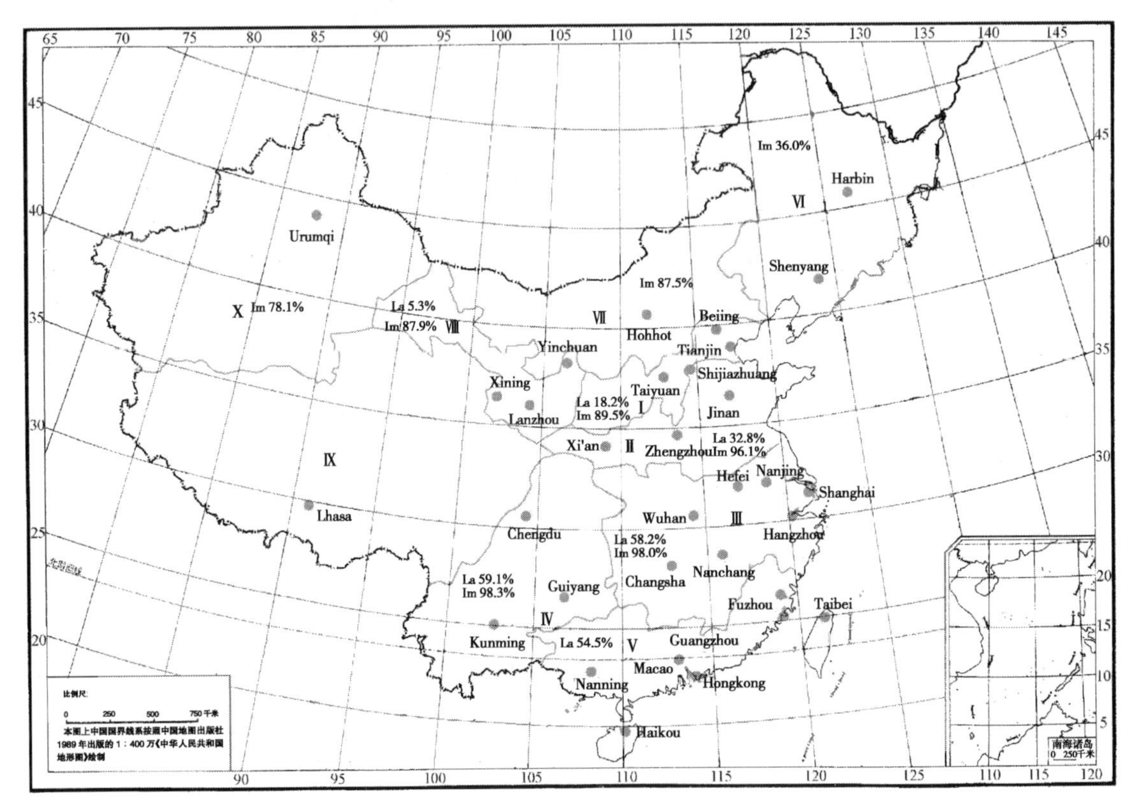

Fig. 1 Distribution of the photoperiod insensitive *Ppd-D1a* allele in landraces and improved cultivars from various wheat zones of China

Ⅰ= Northern Winter Wheat Zone, Ⅱ= Yellow and Huai Rivers Valleys Facultative Wheat Zone, Ⅲ= Middle and Lower Yangtze Valleys Autumn-Sown Spring Wheat Zone, Ⅳ= Southwestern Autumn-Sown Spring Wheat Zone, Ⅴ= Southern Autumn-Sown Spring Wheat Zone, Ⅵ= Northeastern Spring Wheat Zone, Ⅶ= Northern Spring Wheat Zone, Ⅷ= Northwestern Spring Wheat Zone, Ⅸ= Qinghai-Tibetan Plateau Spring-Winter Wheat Zone, Ⅹ= Xinjiang Winter and Spring Wheat Zone.

La and Im indicate landraces and improved cultivars, respectively.

Materials and methods

Plant materials

A total of 926 Chinese wheat cultivars collected from nine major zones were used for characterization of photoperiod genotypes at the *Ppd-D1* locus using allele-specific primers. They included 438 landraces, 475 improved cultivars and 13 introduced cultivars that had significant impacts on Chinese wheat production and breeding after the 1940s. The number of entries from each production zone was dependent on the wheat area and the number of cultivars developed by local breeding programs (Table 1, Fig. 1). All accessions are available from the National Key Facility for Crop Genetic Resources and Improvement, Institute of Crop Science, CAAS, China.

Table 1 Distribution of photoperiod insensitive allele *Ppd-D1a* in landraces and improved cultivars from various wheat zones

Zone[a]	No. of cultivars tested			No. of cultivars with *Ppd-D1a*			*Ppd-D1a* Frequency (%)		
	Landrace	Improved cultivar	Subtotal	Landrace	Improved cultivar	Subtotal	Landrace	Improved cultivar	Subtotal
Ⅰ	77	57	134	14	51	65	18.2	89.5	48.5
Ⅱ	134	206	340	44	198	242	32.8	96.1	71.2
Ⅲ	134	51	185	78	50	128	58.2	98.0	69.2
Ⅳ	22	60	82	13	59	72	59.1	98.3	87.8
Ⅴ	33	Na	33	18	na	18	54.5	na	54.5
Ⅵ	na[b]	25	25	na	9	9	na	36.0	36.0
Ⅶ	na	24	24	na	21	21	na	87.5	87.5
Ⅷ	38	33	71	2	29	31	5.3	87.9	43.7
Ⅹ	na	32	32	na	25	25	na	78.1	78.1
Total	438	488	926	169	442	611	38.6	90.6	66.0

a, Zone Ⅰ=Northern China Plain Winter Wheat Zone, Zone Ⅱ=Yellow and Huai Rivers Valleys Facultative Wheat Zone, Zone Ⅲ=Middle and Lower Yangtze Valleys Autumn-Sown Spring Wheat Zone, Zone Ⅳ=Southwestern Autumn-Sown Spring Wheat Zone, Zone Ⅴ=Southern Autumn-Sown Spring Wheat Zone, Zone Ⅵ=Northeastern Spring Wheat Zone, Zone Ⅶ=Northern Spring Wheat Zone, Zone Ⅷ=Northwestern Spring Wheat Zone, Zone Ⅹ=Xinjiang Winter-Spring Wheat Zone.

b, na=not available.

DNA extraction and molecular marker analysis

Genomic DNA was extracted from seeds following the procedure of Gale et al. (2001). PCR primers were synthesized by Beijing Augct Biological Technology Co. Ltd. (http://www.augct.com) based on the report of Beales et al. (2007).

Allelic variants at the *Ppd-D1* locus were detected using three gene-specific primers in a multiplex PCR assay, in which the primer pair Ppd-D1_F (5'-ACGCCTCCCACTACACTG-3') and Ppd-D1_R1 (5'-TGTTGGTTCAAACAGAGAGC-3') produced a 414-bp fragment in genotypes with the photoperiod sensitive allele *Ppd-D1b*, whereas primer pair Ppd-D1_F and Ppd-D1_R2 (5'-CACTGGTGGTAGCTGAGATT-3') yielded a 288-bp fragment in those with photoperiod insensitive allele *Ppd-D1a*. PCR was performed in a Peltier PTC 200 thermal cycler (Waltham, MA), with a 20-μl volume containing 1×PCR buffer with 1.5mM of $MgCl_2$, 200μM of each of dNTPs, 5pmol of the primer Ppd-D1_F and 2.5pmol each of Ppd-D1_R1 and Ppd-D1_R2, 1unit of *Taq* DNA polymerase (Takara Biotechnology Co. Ltd. Dalian, China) and 30-60ng of template DNA. Thermocycling conditions were an initial denaturation at 94℃ for 2min, followed by 40 cycles of 30s at 94℃, 30s at 52℃, 1min at 72℃, and a final extension step at 72℃ for 5min. Amplified PCR fragments were separated on a 1.2% agarose gel at 200V for 30min, stained with ethidium bromide, and visualized using UV light.

Results

Distribution of the *Ppd-D1a* allele in various wheat zones

Of 926 cultivars, 315 amplified a 414-bp PCR fragment, indicating the presence of the *Ppd-D1b* allele, and 611 genotypes gave a 288-bp fragment, indicative of the photoperiod-insensitive *Ppd-D1a* allele (Fig. 2). The overall frequency of the dominant *Ppd-D1a* allele in Chinese wheats was 66.0%, but frequencies varied across regions (Table 1, Fig. 1). For example, the highest frequency was found in the Southwestern Autumn-sown Spring Wheat Zone (Zone Ⅳ, 87.8%) and the Northern Spring Wheat Zone (Zone Ⅶ, 87.5), and the lowest was in the Northeastern Spring Wheat Zone (Zone Ⅵ, 36.0%). Among the four autumn-sown wheat zones where both landraces and improved cultivars were tested in this study, the frequency of *Ppd-D1a* in the Northern China Plain (Zone I) was much lower than that in the middle (Zone II) and southern parts (Zones III and IV).

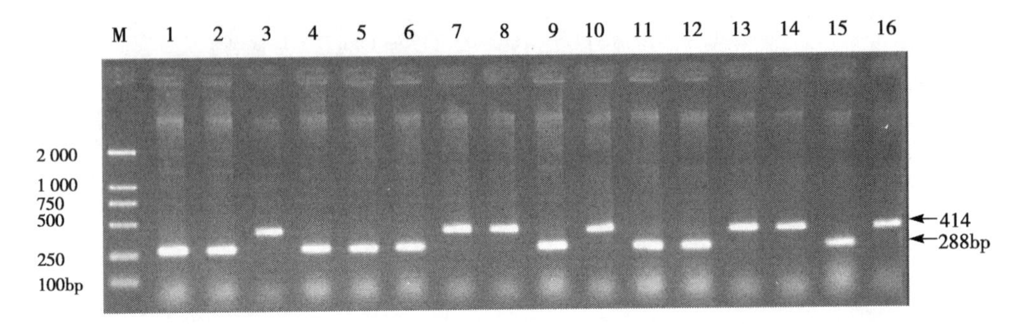

Fig. 2 Amplification products of multiplex PCR assays using primers Ppd-D1 _ F, Ppd-D1 _ R1 and Ppd-D1 _ R2 to detect recessive photoperiod sensitive (*Ppd-D1b*, 414-bp product) and dominant photoperiod insensitive (*Ppd-D1a*, 288-bp product) alleles at the *Ppd-D1* locus.

M-DNA marker 2000, 1. Opata 85, 2. Akakomugi, 3. Chinese Spring, 4. Youzimai, 5. Mazhamai, 6. Neixiang 5, 7. Biyumai (Quality), 8. Gansu 96 (CI12203), 9. Boai 7023, 10. Villa Glori, 11. Mianyang 15, 12. Fan 6, 13. Xinkehan 9, 14. Xuzhou 438, 15. Yangmai 5, 16. Neixiang 36

Comparisons of *Ppd-D1* genotypes in landraces and improved cultivars among different regions

The frequency of the *Ppd-D1a* in 263 improved cultivars released in Zones I and II (94.7%) was three fold that among 211 landraces (27.5%). The frequency of *Ppd-D1a* also increased markedly from landraces (58.3%) to improved cultivars (98.2%) in Zones III and IV. For Zones VI, VII and VIII, the frequency of *Ppd-D1a* in improved cultivars was 71.9%. Generally, the frequency of the *Ppd-D1a* in landraces (38.6%) was much lower than in improved cultivars (90.6%), a consequence of the introduction of *Ppd-D1a* to improve cultivar adaptation to various environments.

Among landraces in the autumn-sown wheat zones, the frequencies of *Ppd-D1a* in Zones I to V were 18.2%, 32.8%, 58.2%, 59.1%, and 54.5%, respectively. The frequency of *Ppd-D1a* was comparably lower (5.3%) in landraces from Zone VIII, a spring-sown spring wheat area. This indicated that the presence of *Ppd-D1a* in landraces gradually increased from north to south in the five autumn-sown wheat zones (I, II, III, IV and V).

Among improved cultivars, those in Zones II, III, and IV located in the middle and southern parts of China had high frequencies of *Ppd-D1a*, ranging from 96.1% to 98.3%, followed by Zones I (89.5%), VII (87.5%), VIII (87.9%), and X (78.1%), and the lowest frequency was observed in Zone VI (36.0%). This again showed the increasing frequency of *Ppd-D1a* in improved cultivars from north to south.

In Zone I, six improved cultivars, Dongfanghong 3, Yuandong 8585, Longdong 1, Qingxuan 1, Triumph, and Early Premium, had *Ppd-D1b*. Dongfanghong 3 released in 1968 and Yuandong 8585 developed in the 1980s were both late maturing cultivars from Beijing. Longdong 1 and Qingxuan 1 were developed and grown in the winter wheat area of Gansu province at an altitude of 1300-1600m, where one crop per year is practiced under rained conditions and late maturity is preferred for achieving high yields, thus photoperiod sensitivity is preferred. Triumph and Early Premium were introduced from Kansas State University in the 1940s and it would be expected that they carried photoperiod sensitivity alleles, especially before the green revolution. Thus, all currently improved cultivars with early maturity in Zone I carry the *Ppd-D1a* where early maturity is needed to avoid sprouting damage and to allow optimal sowing of maize after wheat.

In Zone Ⅱ, improved cultivars with the *Ppd-D1b* allele included Ji 5099 (high quality, average yield, late maturity), Shijiazhuang 407 (selected from Triumph/Yanda 1817 in 1956), Neixiang 36 (reselection of landrace Baihoumai in the 1960s), Qida 195 (reselection of a landrace), Xuzhou 438 (reselection of a landrace), Zhongliang 5, Zhongliang 11 and Quality (an Australian introduction). Zhongliang 5 and Zhongliang 11 were released in the winter area of Gansu province in the 1960s. Thus, seven of them except for Shijiazhuang 407 are expected to carry *Ppd-D1b* allele. However, further investigation is needed to understand the early maturity of Shijiazhuang 407. Therefore, all current cultivars with early maturity in this zone carry the *Ppd-D1a* allele.

In Zones Ⅲ and Ⅳ, all 109 improved cultivars with early maturity carried the *Ppd-D1a*, and only two improved cultivars (Villa Glori and Orofen) with late maturity had the *Ppd-D1b* (Jin, 1983; Zhuang, 2003). Villa Glori introduced from Italy in 1936, and Orofen introduced from Chile, were leading cultivars before the 1970s. Again, all current cultivars with early maturity carry the *Ppd-D1a* allele.

Zone Ⅵ is a high latitude environment, thus cultivars are expected to have a certain level of photoperiod sensitivity (He et al., 2001). Nine improved cultivars, Kefeng 4, Longmai 12, Longmai 19, Longmai 26, Longmai 30, Longfumai 3, Longfumai 4, Longfumai 12, and Longfumai 14, had the *Ppd-D1a*. Kefeng 4 was developed by Keshan Wheat Research Institute in Keshan of Heilongjiang province at latitude of 48°03′ in 1977. Other eight improved cultivars were developed by the Heilongjiang Academy of Agricultural Sciences in Harbin at latitude 45°41′. It is generally believed that cultivars from Harbin have less degree of photoperiod sensitivity and relative narrow adaptation in Heilongjiang in comparison with cultivars developed from Keshan (He et al., 2001). Thus, it was surprised that Kefeng 4 developed from Keshan does not have the *Ppd-D1b* allele, but it may carry other genes that affect flowering time. This needs further study.

In Zone Ⅶ, only three improved cultivars (Banong 1, Mengmai 22 and Dabaipi) had the *Ppd-D1b*. Banong 1 and Dabaipi with late maturity were released before the 1960s. Mengmai 22 with mid to late maturity was developed in 1993, but was grown only in the area close to Heilongjiang (Zone Ⅵ), and therefore was expected to carry the *Ppd-D1b*.

In Zone Ⅷ, improved cultivars, Hongnong 1, Jiunong 2, Longmai 5 and CI 12203, had the *Ppd-D1b*. Hongnong 1, Jiunong 2 and Longmai 5 were released after 1960, and further investigation is needed to understand their adaptation since all three have mid maturity. CI 12203, introduced to Gansu (Zone Ⅷ) from the USA in 1944, was named Gansu 96 in China, and was expected to carry the *Ppd-D1b* allele.

In Zone X, seven improved cultivars identified with *Ppd-D1b* included Kashize 5, Kuidong 4, Kuihua 1, Xindong 16, Xindong 17, Xindong 18 and Xindong 23. All were developed and grown in Xinjiang, and were expected to have photoperiod sensitivity.

Origin of *Ppd-D1a* in improved Chinese wheat cultivars

Four hundred and forty two improved cultivars with *Ppd-D1a* were identified in this study. Pedigree analysis showed that *Ppd-D1a* in Chinese improved cultivars was mainly derived from three primary sources: the Japanese landrace Akagomughi and Chinese landraces Mazhamai and Youzimai (Fig. 2).

Italian cultivars Mentana, Abbondanza and Funo with the *Ppd-D1a* allele were introduced into China in the 1930s and 1956, respectively. They played a crucial role in wheat production and breeding in Zones Ⅲ, Ⅳ, and Ⅷ, and 87, 110 and 98 cultivars, respectively, were developed in China using one of them as a parent (Jin, 1983; 1997). These Italian introductions were derived from crosses involving Akagomughi (Worland,

1999). Their derivatives, Boai 7023, Yangmai 1, Sumai 3, Mianyang 4, Emai 6, Hongtu, Doudu 1, Ganmai 8 and Longchun 8, became leading cultivars and important breeding parents (Zhuang , 2003).

More than 17 leading cultivars, such as Beijing 10, Shandongfu 63, Jinan 9 and Yannong 15, with *Ppd-D1a* were developed from the Henan landrace Youzimai (Zhuang, 2003). Youzimai, a leading cultivar in the 1950s, with good yield, early maturity and broad adaptation, was widely distributed in the northern part of Zone Ⅱ. More than 88 cultivars including Beijing 8, Taishan 1, Jinan 2, Bima 1 and Bima 4 were derived from the Shaanxi landrace Mazhamai, another broadly adapted high yielding genotype (Jin, 1983; 1997; Zhuang, 2003).

Discussion

Selection for earlier flowering and photoperiod insensitivity has been an important step in improving the adaptation of wheat in many environments. Wheat germplasm from the International Maize and Wheat Improvement Centre (CIMMYT) is characterized with photoperiod insensitivity and broad adaptation. It was developed through shuttle breeding between two contrasting Mexican environments, and subsequently widely used in developing countries (Trethowan et al. , 2007). A total of 257 wheat cultivars and advanced lines from CIMMYT were assayed using the same primers as in the present study, and 234 (91.1%) carried *Ppd-D1a*. The remaining 8.9%, consciously developed for high latitude environments in central Asia and northeastern China, carried *Ppd-D1b* (unpublished data). Major objectives in Chinese wheat breeding programs since 1949 have been high yield potential, broad adaptation, and early flowering and maturity to fit multi-cropping systems. Clearly the photoperiod insensitive allele *Ppd-D1a* was the target of that strategy for most areas except for spring wheat in northeastern China and winter wheat in Gansu and Xinjiang. This accounts for the significant increase in the frequency of the photoperiod insensitive *Ppd-D1a* allele.

In both the autumn-sown and spring-sown wheat regions, the frequency of *Ppd-D1a* decreased from the south where insensitivity was obligatory for early maturity to north where insensitivity is increasingly less essential due to the gradual increase of photoperiod length in the wheat growing regions (Zhuang, 2003). Compared with spring-sown regions, a higher frequency *Ppd-D1a* was found in autumn-sown regions, where the shorter photoperiod length during the wheat growing stage was present (Fig. 1). The cultivars with the strongest dominant allele removing the need for vernalization (*Vrn-A1*) combined with other dominant vernalization gene (s), in Zone Ⅵ, a spring wheat zone, are commonly considered the most sensitive to photoperiod response, and this was supported by the lowest frequency of *Ppd-D1a* among the improved cultivars released in the region (Zhuang, 2003; Zhang et al. , 2008). Conversely in the autumn-sown Zones Ⅰ, Ⅱ, Ⅲ and Ⅳ, it is important for wheat cultivars to flower and mature as early as possible in order to avoid the regular hot and desiccating summer conditions (He et al. , 2001). Combining *Ppd-D1a* with a single moderate dominant vernalization allele (*Vrn-D1*) in cultivars in Zones Ⅱ, Ⅲ and Ⅳ allows adjustment of the growth cycle to achieve higher yields, whereas *Ppd-D1a* in combination with winter genotypes in Zone Ⅰ permits avoidance of vegetative-stage frost damage as well as allowing a long grain-filling period before the onset of humid and hot summer conditions (Zhuang, 2003; Zhang et al. , 2008).

In addition to *Ppd-D1*, other genes such as *Ppd-A1*, *Ppd-B1* and earliness *per se* may also have important roles in cultivar adaptation. However, allele-specific markers are currently not available for these loci (Beales et al. , 2007). Further investigation is therefore needed to understand the distribution and importance flowering time genes in relation to the adaptation and productivity of wheat in China. For example, our previous study showed that nine early heading cultivars (Jimai 36, Taishan 1, Lumai 23, Laizhou 953, Weimai 8, Yumai 66, Yumai 70, Xuzhou 14 and Emai 6) carried recessive alleles at all four vernalization loci as determined by allele-specific assays, but all reached ear emergence in 76 to 108 days (Zhang et al., 2008). In the present study, these nine cultivars were showed to carry *Ppd-D1a* which, alone, should not cause such early flowering. These cultivars are therefore likely to carry novel alleles at one or more additional *Vrn* loci that remove the need for vernalization or additional photoperiod insensitive alleles at *Ppd-A1* and (or) *Ppd-B1* loci. The existing diagnostic assays are valuable in pin-pointing the needs of future research in understanding the genetics of adaptation.

The reduced heightgene *Rht8* located on chromosome 2D, derived from Japanese landrace Akakomugi, was introduced into Chinese wheat improved cultivars. The *Rht8* gene was found to take a dominant position with frequencies of 46.8% in 220 wheat genotypes from autumn-sown wheat regions, has less effect on reduced height than *Rht-B1b* and *Rht-D1b* (Zhang et al., 2006). It was reported that *Rht8* in close linkage with the photoperiod insensitivity gene (*Ppd-D1a*) reduced plant height by around 10% without significant negative effects on yield (Börner et al. 1993; Worland et al., 1998). The *Ppd-D1a* allele for day length insensitivity together with the less reducing height *Rht8* allele has proved highly effective in providing Chinese wheats with adaptation to specific environments, especially after the green revolution.

Acknowledgments

The authors are grateful to Prof. Robert McIntosh from the University of Sydney for kindly reviewing this manuscript. This study was supported by the National 863 Program (2006AA10Z1A7 and 2006AA100102), and International Collaboration Project from Ministry of Agriculture of China (2006-G2).

References

Beales J, Turner A, Griffiths S, Snape JW, Laurie DA (2007) A Pseudo-Response Regulator is misexpressed in the photoperiod insensitive *Ppd-D1a* mutant of wheat (*Triticum aestivum* L.). Theor Appl Genet. 115: 721-733.

Börner A, Worland AJ, Plaschke J, Schumann E, Law CN (1993) Pleiotrpic effects of genes for reduced (*Rht*) and day-length insensitivity (*Ppd*) on yield and its components for wheat grown in middle Europe. Plant Breeding. 111: 204-216.

Dong YS, Zheng DS (2000) Wheat Genetic Resources in China. China Agriculture Press, Beijing. (in Chinese)

Dunford RP, Yano M, Kurata N, Sasaki T, Huestis G, Rocheford T, Laurie DA (2000) Comparative mapping of the barley *Ppd-H1* photoperiod response gene region, which lies close to a junction between two rice linkage segments. Genetics. 161: 825-834.

Dyck JA, Matus-Cádiz MA, Hucl P, Talbert L, Hunt T, Dubuc JP, Nass H, Clayton G, Dobb J, Quick J (2004) Agronomic performance of hard red spring wheat isolines sensitive and insensitive to photoperiod. Crop Sci. 44: 1976-1981.

Gale KR, Ma W, Zhang W, Rampling L, Hill AS, Appels R, Morris P, Morrel M (2001) Simple high-throughput DNA markers for genotyping in wheat. In: Eastwood R, Hollamby G, Rathjen T, Gororo N (ed.), Proc 10th Australian Wheat Breeding Assembly, pp 26-31. Wheat Breeding Society of Australia.

He ZH, Rajaram S, Xin ZY, Huang GZ (2001) A History of Wheat Breeding in China. CIMMYT, Mexico, DF.

Hunt LA (1979) Photoperiodic responses of winter wheats from different climatic regions. Z Pflanzenzüchtung. 82: 70-80.

Jin SB (1986) Chinese Wheat Cultivars and Their Pedigrees (1962-1982). China Agriculture Press, Beijing. (in Chinese)

Jin SB (1997) Chinese Wheat Cultivars and Their Pedigrees (1983-1993). China Agriculture Press, Beijing. (in Chinese)

Law CN, Sutka J, Worland AJ (1978) A genetic study of day length response in wheat. Heredity. 41: 185-191.

Martiníc ZF (1975) Life cycle of common wheat varieties in natural environments as related to their response to shortened photoperiod. Z Pflanzenzüchtung 75: 237-251

Ortiz Ferrara G, Mosaad MG, Mahalakshmi V, Rajaram S (1998) Photoperiod and vernalisation response of Mediterranean wheats, and implications for adaptation. Euphytica. 100: 377-384.

Pugsley AT (1966) The photoperiodic sensitivity of some spring wheats with special reference to the variety Thatcher. Aust J Agric Res. 17: 591-599.

Scarth R, Law CN (1984) The control of day-length response in wheat by the group 2 chromosomes. Z Pflanzenzuchtung . 92: 140-150.

Snape JW, Butterworth K, Whitechurch E, Worland AJ (2001) Waiting for fine times: genetics of flowering time in wheat. Euphytica. 119: 185-190.

Tanio M, Kato K (2007) Development of near-isogenic lines for photoperiod-insensitive genes, *Ppd-B1* and *Ppd-D1*, carried by the Japanese wheat cultivars and their effect on apical development. Breed Sci. 57: 65-72.

Trethowan RM, Reynolds MP, Ortiz-Monasterio JI, Ortiz R (2007) The genetic basis of the green revolution in wheat production. Plant Breed Rev. 28: 39-58.

Turner A, Bearles J, Faure S, Dunford RP, Laurie DA (2005) The pseudo-response regulator *Ppd-H1* provides adaptation to photoperiod in barley. Science. 310: 1031-1034.

Welsh JR, Kein DL, Pirasteh B, Richards RD (1973) Genetic control of photoperiod response in wheat. In: Sears ER, Sears LES (ed.) Proc 4th Int Wheat Genet Symp Agric Exp Sta, pp. 879-884. University of Missouri, Columbia, USA.

Whitechurch EM, Slafer GA (2002) Contrasting *Ppd* alleles in wheat: effects on sensitivity to photoperiod in different phases. Field Crop Res. 73: 95-105.

Worland AJ (1999) The importance of Italian wheats to worldwide varietal improvement. J Genet & Breed . 53: 165-173.

Worland AJ, Appendino ML, Sayers EJ (1994) The distribution, in European winter wheats, of genes that influence ecoclimatic adaptability whist determining photoperiodic insensitivity and plant height. Euphytica. 80: 219-228.

Worland AJ, Börner A, Korzun V, Li WM, Petrovíc S, Sayers EJ (1998) The influence of photoperiod genes on the adaptability of European winter wheats. Euphytica. 100: 385-394.

Worland AJ, Korzum V, Röder MS, Ganal MW, Law CN (1998) Genetic analysis of the dwarfing gene (*Rht 8*) in wheat. Part II. The distribution and adaptive significance of allelic variants at the *Rht 8* locus of wheat as revealed by microsatellite screening. Theor Appl Genet. 96: 1110-1120.

Zhang XK, Yang SJ, Zhou Y, He ZH, Xia XC (2006) Distribution of the *Rht-B1b*, *Rht-D1b* and *Rht 8* reduced height genes in autumn-sown Chinese wheats detected by molecular markers. Euphytica. 152: 109-116.

Zhang XK, Xia XC, Xiao YG, Zhang Y, He ZH (2008) Allelic variation at the vernalization genes *Vrn-A1*, *Vrn-B1*, *Vrn-D1* and *Vrn-B3* in Chinese common wheat cultivars and their association with growth habit. Crop Sci. 48: 458-470.

Zhuang QS (2003) Wheat Improvement and Pedigree Analysis in Chinese Wheat Cultivars. China Agriculture Press, Beijing. (in Chinese)

品质基因鉴定

Puroindoline grain hardness alleles in CIMMYT bread wheat germplasm

M. Lillemo[1], F. Chen[2], X. C. Xia[2], M. William[3], R. J. Peña[3], R. Trethowan[3], and Z. H. He[2,4]

[1] *Department of Plant and Environmental Sciences, The Norwegian University of Life Sciences, P. O. Box* 5003, *N*-1432 *Ås, Norway*; [2] *Institute of Crop Sciences/National Wheat Improvement Center, Chinese Academy of Agricultural Sciences (CAAS), Zhongguancun South Street* 12, 100081, *Beijing, P.R. China*; [3] *International Maize and Wheat Improvement Center (CIMMYT), Apdo. Postal* 6-641, 06600, *México, D. F., México*; [4] *International Maize and Wheat Improvement Center (CIMMYT) China Office, c/o Chinese Academy of Agricultural Sciences, Zhongguancun South Street* 12, *Beijing* 100081, *P.R. China.*

Abstract: Grain hardness is an important quality parameter of bread wheat (*Triticum aestivum* L.) with importance for wheat classification and end use properties, and is controlled by the genes puroindoline a (*Pina*) and puroindoline b (*Pinb*). The presence of known hardness alleles were studied in a representative sample of 373 bread wheat lines from the breeding program at CIMMYT. The PINA-null mutation (*Pina-D1b*) was the most frequent hardness allele and present in 283 of the 328 lines with hard endosperm. All other hard wheats had the glycine to serine mutation in PINB (*Pinb-D1b*). A study of historically important CIMMYT bread wheat lines showed that *Pina-D1b* has been the dominating hardness allele since the inception of the wheat breeding program in Mexico. New puroindoline alleles have recently been introduced through the extensive use of synthetic hexaploid wheat, and the textural effects of various *Aegilops tauschii*-derived *Pina* and *Pinb* alleles were studied in 92 breeding lines derived from various crosses with synthetic wheat. Progeny lines with *Pina-D1j/Pinb-D1i* were on average 10 SKCS hardness units softer than those carrying the allelic combination *Pina-D1c/Pinb-D1h*. Further investigation is needed to validate the potential of such minor allelic differences for the improvement of soft wheat quality.

Key words: Bread wheat, Synthetic wheat, Grain hardness, Puroindoline

Abbreviations: ESWYT, Elite Spring Wheat Yield Trial; *Ha*, grain hardness locus; PINA, puroindoline a protein; *Pina*, puroindoline a gene; PINB, puroindoline b protein; *Pinb*, puroindoline b gene; PSI, Particle Size Index; SAWYT, Semi Arid Wheat Yield Trial; SKCS, Single Kernel Characterization System.

Introduction

Grain hardness is an important quality parameter of bread wheat (*Triticum aestivum* L.) that determines the end use properties and market classification of the grain. The main difference between soft and hard wheat lies in the binding strength between the starch granules and protein matrix in the endosperm cells (Barlow et al., 1973), which is of fundamental importance during milling and baking.

In soft wheat, the starch granules are loosely bound to the surrounding protein matrix, the grain is easy to

Published in Journal of Cereal Science, 2006, 44 (3): 86-92

mill and produces fine-textured flour with a high proportion of undamaged starch granules. In contrast, in hard wheat there is a tight linkage between the starch granules and protein matrix, more energy is needed during milling and the resulting wheat flour has a coarser texture and a higher proportion of damaged starch granules (Brennan et al., 1993). Hard wheat is preferred for yeast-leavened bread due to the higher water absorption of damaged starch, whereas soft wheat, which has low water absorption, is ideal for cookies, cakes and pastries.

Grain hardness is controlled by one major genetic factor, the hardness locus (*Ha*) on the short arm of chromosome 5D. The locus contains two closely linked genes, puroindoline a (*Pina*) and puroindoline b (*Pinb*) that confer soft endosperm when they are both in their wild-type allelic states (*Pina-D1a*/*Pinb-D1a*). Hard wheat is the result of mutations in either *Pina* or *Pinb* (reviewed in Morris, 2002). Durum wheat, which lacks both genes, is very hard. The currently known mutations in *Pina* and *Pinb* conferring hard endosperm are listed in Table 1. These mutations appear to have occurred independently from each other and some can be traced back to specific geographic areas. Whereas the *Pinb-D1b* allele prevails among the spring and winter wheats of North America, Europe, China and Australia (Cane et al., 2004; Chen et al., 2006; Lillemo and Morris, 2000; Morris et al., 2001; Xia et al., 2005), *Pinb-D1c* and *Pinb-D1d* are mostly found in western Europe (Lillemo and Morris, 2000). The three rare alleles *Pinb-D1e*, *Pinb-D1f*, *Pinb-D1g* are confined to North American cultivars (Morris et al., 2001) and *Pinb-D1p* have only been found in Chinese wheat (Chen et al., 2006; Ikeda et al., 2005; Xia et al., 2005). Likewise, the PINA null mutation (*Pina-D1b*) has been found to be the primary cause of hardness in Indian wheat (Ram et al., 2002).

Table 1 Molecular changes, phenotypes and references for the currently known *Pina* and *Pinb* hardness mutations in conventional bread wheat. A more detailed description of each allele is given in Chen et al. (2006)

Pina	*Pinb*	Molecular change	Phenotype	Reference
Pina-D1a	*Pinb-D1a*	—	Soft, wild-type	(Giroux and Morris, 1997)
Pina-D1b	*Pinb-D1a*	Unknown	Hard, PINA null	(Giroux and Morris, 1998)
Pina-D1l	*Pinb-D1a*	Base deletion in Gln-61→ frame shift.	Hard, PINA null	(Gazza et al., 2005; McIntosh et al., 2005)
Pina-D1m	*Pinb-D1a*	Pro-35 to Ser-35	Hard	(Chen et al., 2006)
Pina-D1n	*Pinb-D1a*	Trp-43 to stop codon	Hard, PINA null	(Chen et al., 2006)
Pina-D1a	*Pinb-D1b*	Gly-46 to Ser-46	Hard	(Giroux and Morris, 1997)
Pina-D1a	*Pinb-D1c*	Leu-60 to Pro-60	Hard	(Lillemo and Morris, 2000)
Pina-D1a	*Pinb-D1d*	Trp-44 to Arg-44	Hard	(Lillemo and Morris, 2000)
Pina-D1a	*Pinb-D1e*	Trp-39 to stop codon	Hard, PINB null	(Morris et al., 2001)
Pina-D1a	*Pinb-D1f*	Trp-44 to stop codon	Hard, PINB null	(Morris et al., 2001)
Pina-D1a	*Pinb-D1g*	Cys-56 to stop codon	Hard, PINB null	(Morris et al., 2001)
Pina-D1a	*Pinb-D1p*	Base deletion in Lys-42→ frame shift.	Hard, PINB null	(Ikeda et al., 2005; Xia et al., 2005)
Pina-D1a	*Pinb-D1q*	Trp-44 to Leu-44	Hard	(Chen et al., 2005)
Pina-D1a	*Pinb-D1r*	Base insertion in Glu-14→ frame shift.	Hard, PINB null	(McIntosh et al., 2005; Ram et al., 2005)
Pina-D1a	*Pinb-D1s*	Base insertion in Glu-14→ frame shift + Ala-40 to Thr-40	Hard, PINB null	(McIntosh et al., 2005; Ram et al., 2005)
Pina-D1a	*Pinb-D1t*	Gly-47 to Arg 47	Hard	(Chen et al., 2006)

The known hardness mutations all confer a big change in endosperm texture compared to soft wheat, but different alleles might confer small differences in the degree of hardness. For example, it has been shown that*Pina-D1b* gives slightly harder endosperm than *Pinb-D1b*, and is associated with lower milling yield and higher water absorption compared to *Pinb-D1b* (Cane et al., 2004; Giroux et al., 2000; Martin et al., 2001).

Additional variation on the *Ha* locus has recently been introduced into bread wheat through the artificial hybridization of durum wheat (*T. turgidum* ssp. *durum*) with different accessions of *Aegilops tauschii* to produce synthetic wheat (Mujeeb-Kazi et al., 1996). Seven additional alleles of *Pina* (*Pina-D1c*, *Pina-D1d*, *Pina-D1e*, *Pina-D1f*, *Pina-D1h*, *Pina-D1i*, *Pina-D1j*) and six different *Pinb* alleles (*Pinb-D1h*, *Pinb-D1i*, *Pinb-D1j*, *Pinb-D1m*, *Pinb-D1n*, *Pinb-D1o*) from *Ae. tauschii* were found to confer soft endosperm in synthetic wheat (Gedye et al., 2004; Massa et al., 2004).

The bread wheat breeding program of the International Maize and Wheat Improvement Center (CIMMYT) develops spring wheat lines for all major wheat growing areas in the developing world (Rajaram and van Ginkel, 2001). International nurseries with advanced breeding lines are distributed annually to collaborators in more than 60 countries, where they are tested for local adaptation and often used as parents for crossing or released directly as varieties. From its inception as a joint initiative between the Rockefeller Foundation and the Government of Mexico in 1945, and since the official establishment of CIMMYT in 1966, the breeding program has given rise to many widely adapted spring wheat cultivars. It is estimated that three quarters of the wheat growing area in developing countries is sown to cultivars that can trace their origin back to lines developed by CIMMYT (Heisey et al., 2002). The primary emphasis of CIMMYT's wheat breeding efforts is to develop cultivars with increased yield potential, wide adaptation and yield stability. However, grain processing quality has also been an important objective. As a minimum, bread wheat distributed by CIMMYT must have flat bread or chapatti quality, requiring semi-hard to hard grain and extensible dough of intermediate strength.

The purpose of the present study was to determine the distribution of puroindoline alleles in representative sets of bread wheat germplasm from CIMMYT and to study the effect on grain hardness of recently introduced puroindoline alleles from crosses with synthetic wheat.

Experimental

Germplasm, grain hardness measurements and DNA extraction procedures

Experiment 1

In all, 39 historically important varieties in the bread wheat breeding program at CIMMYT, 49 lines from the 23rd Elite Spring Wheat Yield Trial (ESWYT) and 49 lines from the 10th Semi-Arid Wheat Yield Trial (SAWYT) were planted at the experiment station of the Bayanzhuoer League of Inner Mongolia, China, in the 2003 crop season. While ESWYT represents the current elite collection of advanced breeding lines being distributed to collaborators in high-yielding areas with irrigation, SAWYT is a nursery with advanced breeding lines adapted to conditions with drought stress.

Grain hardness was measured on 300-kernel samples on Perten SKCS 4100 (Perten Instruments, Springfield, IL, USA), following the manufacturer's instructions. DNA was extracted from single kernels by the following procedure: The kernel was crushed and extracted for 30min with gentle shaking in 1ml extraction buffer (200mM Tris-HCl pH7.5; 288mM NaCl; 25mM EDTA; 0.5 % SDS). The tubes were centrifuged with 15 294g (Eppendorf Centrifuge 5417R) for 10 minutes and the supernatant transferred to a new tube and extracted in equal amounts of phenol : chloroform (1 : 1). The supernatant was transferred to a new tube, precipitated with 0.1 vol of 3M NaAc pH5.2 and 0.6vol of isopropanol, and centrifuged for 10 minutes (Eppendorf Centrifuge 5417R, 15 294g). The resulting DNA pellet was washed in 70% ethanol, air-dryed and dissolved in 50μl TE (10mM Tris pH 7.5; 1mM EDTA).

Experiment 2

In total, 236 lines from the current crossing block for the semi-arid wheat breeding program at CIMMYT were grown in the green house at El Batan in central Mexico, and DNA extracted from three weeks old leaf tissue following the procedure by Hoisington et al. (1994). Grain hardness data was obtained from the Wheat Quality lab at CIMMYT. Hardness was estimated by NIR spectroscopy (Infralyzer 400, Technicon, NY), which was calibrated following the method 39-70A of the AACC (AACC, 1995) and using the Particle Size Index method (AACC method 55-30) as reference.

Experiment 3

A total of 92 progeny lines (F_6 or higher collected from CIMMYT yield trial) were selected from 37 crosses between soft-textured synthetic or synthetic-derived bread wheats and hard-textured conventional bread wheats in the breeding program at CIMMYT. Crosses had usually been made with the primary synthetic hexaploid as the female parent and the conventional bread wheat as male parent. The F_1 had then been backcrossed to either the same or a closely related conventional bread wheat. Most crosses involving synthetic-derived bread wheats were simple crosses. The selected breeding lines were grown at the experimental station of the Bayanzhuoer League of Inner Mongolia, China, in the 2004 crop season. Grain hardness measurements and DNA extraction was the same as for experiment 1.

Detection of puroindoline alleles

Pina was amplified with the forward primer 5' CAT CTA TTC ATC TCC ACC TGC 3' and the reverse primer 5' GTG ACA GTT TAT TAG CTA GTC 3' yielding an expected PCR product of 524 bp including the coding sequence and flanking upstream and downstream sequences. *Pinb* was amplified with the forward primer 5' GAG CCT CAA CCC ATC TAT TCA TC 3' and the reverse primer 5' CAA GGG TGA TTT TAT TCA TAG 3' that amplified a PCR product of 597 bp including the coding sequence and flanking sequences at both sides. The lack of amplification by the *Pina*-specific primers from one sample and the simultaneous amplification of *Pinb* was taken as an indication of the *Pina*-null mutation (*Pina-D1b*). The *Pinb-D1b* allele was detected by the allele-specific primers developed by Giroux and Morris (1997; 1998). Detection of other *Pina* and *Pinb* alleles was done by DNA sequencing of the 524 bp and 597 bp PCR products.

Statistical analysis

Analysis of grain hardness data was carried out using SAS version 8.2 (SAS Institute, Cary, N.C., USA). PROC GLM was used for generating ANOVA tables and conducting pair-wise comparisons with the Tukey method.

Results

Frequency of puroindoline alleles in CIMMYT bread wheat

Of the 39 historically important bread wheat lines from CIMMYT, one was classified as soft, four as mixed and 34 as hard. Lerma Rojo 64, the only soft wheat sample was confirmed to have the wild-type alleles of *Pina* and *Pinb* by sequencing. Of the hard wheat samples, 30 had the PINA-null mutation (*Pina-D1b*) and four had the *Pinb-D1b* allele (Table 2). Average SKCS hardness index for the two groups of hard wheat were 88.0 and 80.3, respectively for *Pina-D1b* and *Pinb-D1b*.

Pina-D1b was by far the most common hardness allele also in the two nurseries ESWYT and SAWYT (Table 3). The only line with *Pinb-D1b* in ESWYT was the South African cultivar Kariega. The lines with this allele in SAWYT were the Syrian variety Cham 6 which is identical to Nesser, and two crosses with non-CIM-

MYT material. The *Pina-D1a/Pinb-D1i* and *Pina-D1j/Pinb-D1i* genotypes originated from crosses with synthetic wheat.

Of the 236 lines from the semi-arid wheat crossing block, 34 had soft endosperm based on data from the Quality Lab at CIMMYT and gave negative results when testing for *Pina-D1b* and *Pinb-D1b*. Seventeen of these originated from conventional bread wheat, and were genotyped as *Pina-D1a/Pinb-D1a*. The remaining lines with soft endosperm had synthetic wheat in their pedigrees, and were sequenced for proper identification of their *Pina* and *Pinb* alleles. Five haplotypes were detected, with *Pina-D1a/Pinb-D1i* and *Pina-D1c/Pinb-D1h* being the most common allelic combinations derived from synthetic wheat. Of the 202 lines with hard endosperm, 165 had the *Pina-D1b* allele and 37 lines had the *Pinb-D1b* allele (Table 3).

Table 2 Grain hardness and puroindoline alleles of historically important lines in the bread wheat breeding program at CIMMYT. Listed first are key cultivars and breeding lines developed by the breeding program in Mexico in chronological order based on the cross number, followed by cultivars based on CIMMYT germplasm that have been developed by national programs and other lines occurring frequently in the pedigrees of bread wheats from CIMMYT

Name	Selection history	Year of first release[1]	Release and impact[2]	Hardness index	Hardness class	*Pina-D1*	*Pinb-D1*
Yaqui 50	II118-2Y-1Y	1950	First stem rust resistance cultivar released in Mexico	90	H	b	a
Pitic 62	II7064-1Y-1H-1R-2M-0MEX	1962	4 countries	97	H	b	a
Penjamo T 62	II7078-1R-6M-1R-1M-0MEX	1962	4 countries	45	M	a/b[3]	a
Kalyansona	II8156-0IND	1965	4 countries, grown in several million ha	89	H	b	a
Sonora 64	II8469-2Y-6C-4C-2Y-1C-0MEX	1964	4 countries	76	H	b	a
Lerma Rojo 64	II8724	1964	Mexico, South Africa	32	S	a	a
Anza	II8739-4R-1M-1R-0USA	1973	8 countries, grown in several million ha	87	H	a	b
Sonalika	II18427-4R-1M	1967	India, grown in several million ha	69	H	a	b
Inia F 66	II19008-83M-100Y-100M-100Y-100C-0MEX	1966	Five countries	80	H	b	a
UP 301	II19008-PAU. ACC. 3132-0IND	1968	India, leading cultivar	89	H	b	a
Bluebird I5	II23584-102M-0Y-6M-0Y-1T-0T-0MEX	1970	11 countries	78	H	b	a
Yecora Rojo 76	II23584-0WM	1976	Mexico	88	H	b	a
Tanori F 71	II25717-11Y-3M-1Y-0M-0MEX	1971	Mexico	75	H	b	a
Jupateco F 73	II30842-31R-2M-2Y-0M-0MEX	1973	Mexico	88	H	b	a
Pavon F 76	CM8399-D-4M-3Y-1M-1Y-1M-0Y-0MEX	1976	13 countries, good quality	93	H	b	a
Alondra	CM11683-0DESC	1983	5 countries	87	H	b	a
Seri M 82	CM33027 - F - 15M - 500Y-0M-87B-0Y-0MEX	1982	30 countries, grown in several million ha	81	H	b	a
Veery	CM33027-F-1M-11Y-0M	—	Reselections were released in many countries	96	H	b	a

（续）

Name	Selection history	Year of first release[1]	Release and impact[2]	Hardness index	Hardness class	*Pina-D1*	*Pinb-D1*
Dashen	CM33027-F-15M-500Y-1M-0Y-0PZ-0Y-0ETH	1984	Ethiopia	—	H	b	a
Pfau	CM38212	1987	Israel, Turkey, USA	95	H	b	a
Nesser	CM40096 - 8M - 7Y - 0M-0AP-0JOR	1990	Jordan, Lebanon	88	H	a	b
Bagula	CM59123-3M-1Y-3M-2Y-1M-0Y-1M-0Y-0ECU-0Y	1988	Mexico, Pakistan, Uruguay	50	M	a/b[3]	a
Rayon F 89	CM90315-A-2B-2Y-1B-0Y-0MEX	1989	Mexico, leading cultivar for many years	103	H	b	a
Bacanora T 88	CM67458-4Y-1M-3Y-1M-5Y-0B-0MEX	1988	Mexico	92	H	b	a
San Cayetano S 97	CM67458-4Y-1M-3Y-1M-0Y-0M	1997	Mexico	96	H	b	a
PBW 343	CM85836-4Y-0M-0Y-8M-0Y-0IND	1995	3 countries, 4-5 million ha in India	82	H	b	a
Borlaug M 95	CM90320-A-1B-5Y-0B-6M-0Y-0MEX	1995	Mexico	96	H	b	a
Babax	CM92066-G-0Y-0M-0KCM-0B-0Y	1993	Mexico, Brazil, high yielding, many derivatives with top performance	79	H	b	a
Frontana	-0BRA	1950	Brazil	38	M	a/b[3]	a
Glenlea	UM714A-0CAN	1972	Canada, important parent for dough strength	79	H	b	a
WH 147	-0IND	1977	India	69	M	a/b[3]	a
UP 262	UP262-0IND	1977	India, Myanmar, Nepal, Thailand	80	H	b	a
Lok 1		1981	India, leading cultivar	77	H	a	b
Debeira	HD2172	1982	Sudan, Syria, India	89	H	b	a
Kanchan	-0BGD	1983	Bangladesh, India	88	H	b	a
HUW 234	-0IND	1984	India	93	H	b	a
HD 2329	PAU. ACC. 3079-0IND	1985	Leading cultivar inIndia	96	H	b	a
Inqalab 91	PB19545-9A-0A-0PAK	1991	Pakistan, grown in 6 million ha	90	H	b	a
UP 2338	-0IND	1993	India	87	H	b	a

1，For the lines developed in Mexico，the selection history is a unique identifier describing the selection process in each subsequent generation after the cross was made. The codes are described in Skovmand et al.（1997）.

2，This column lists countries where the particular line and other sister lines from the same cross have been released as cultivars.

3，PCR of 10 individual kernels yielded 8 *Pina-D1a* ：2 *Pina-D1b* for Penjamo T 62，2 *Pina-D1a* ：8 *Pina-D1b* for WH 147 and 9 *Pina-D1a* ：1 *Pina-D1b* for Frontana and Bagula.

Table 3 Frequency distribution of hardness alleles among the various groups of germplasm surveyed.

Hardness class	*Pina-D1*	*Pinb-D1*	Historical lines	ESWYT	SAWYT	Crossing block	Total	Frequency (%)
Mixed			4	—	—	—	4	
Soft	*a*	*a*	1	1	2	20	24	58.5
	a	*i*	—	—	2	6	8	19.5
	a	*j*	—	—	—	1	1	2.4
	c	*h*	—	—	—	5	5	12.2
	j	*i*	—	1	—	2	3	7.3
			1	2	4	34	41	100.0
Hard	*a*	*b*	4	1	3	37	45	13.7
	b	*a*	30	46	42	165	283	86.3
			34	47	45	202	328	100.0
Total			39	49	49	236	373	

Effect of puroindoline alleles introduced from synthetic wheat

Of the 92 breeding lines from crosses with synthetic wheat, 39 were soft and 53 were hard. All lines with hard endosperm had the *Pina-D1b* allele. One soft segregant from each cross was chosen for DNA sequencing to determine the corresponding alleles of *Pina* and *Pinb* inherited from the synthetic parent. Sequencing identified four different haplotypes: the wild-type *Pina-D1a*/*Pinb-D1a* and the three allelic combinations *Pina-D1a*/*Pinb-D1i*, *Pina-D1c*/*Pinb-D1h* and *Pina-D1j*/*Pinb-D1i* (Table 4). Differences in grain hardness among the different soft alleles were small, and the only significant difference in pair-wise comparisons was that between *Pina-D1c*/*Pinb-D1h* and *Pina-D1j*/*Pinb-D1i*, with the latter being about 10 SKCS hardness units softer than the former (Table 4).

Table 4 Mean SKCS hardness values of breeding lines from crosses with synthetic wheat grouped by their puroindoline alleles. Means followed by the same letter are not significantly different (α=0.05)

Genotype	Number of lines	Hardness class	Hardness Index	Standard error
Pina-D1j/*Pinb-D1i*	26	Soft	21.3a	1.8
Pina-D1a/*Pinb-D1a*	3	Soft	29.7ab	4.8
Pina-D1a/*Pinb-D1i*	3	Soft	30.5ab	4.4
Pina-D1c/*Pinb-D1h*	7	Soft	32.2b	3.2
Pina-D1b/*Pinb-D1a*	53	Hard	72.9c	1.4

Discussion

This is the first extensive survey of puroindoline alleles in CIMMYT bread wheat germplasm, and has shown that the vast majority of hard wheats in this breeding program have the PINA-null mutation (*Pina-D1b*). This is very different from the previously surveyed gene pools in North America, Europe, China and Australia where the *Pinb-D1b* allele is the primary source of grain hardness (Cane et al., 2004; Lillemo and Morris, 2000; Morris et al., 2001; Xia et al., 2005).

Data from the historical cultivars in Table 2 shows that *Pina-D1b* was already the dominating hardness allele at the beginning of the Rockefeller wheat breeding program in Mexico. The first variety to be released was Yaqui 50, which has the pedigree Newthatch/Marroqui 588, and likely inherited the *Pina-D1b* allele from Marroqui 588, which is also an important source of *Pi-*

na-D1b in North American spring wheats (Morris et al., 2001). Both ancestors of Newthatch, Hope and Thatcher, carry the *Pinb-D1b* allele (Morris et al., 2001). *Pina-D1b* was present in many of the early semi-dwarf varieties that sparked the green revolution, such as Sonora 64, Kalyansona and UP301 (sister line of Inia 66). Important exceptions are Lerma Rojo 64 with soft endosperm and Sonalika which carries the *Pinb-D1b* allele. Later, all internationally important lines such as Pavon 76, the Veery's (Seri M82, Veery, Dashen), the Kauz sister lines (Bacanora T88 and San Cayetano S97) and PBW 343 have had the *Pina-D1b* allele. Interestingly, *Pina-D1b* and *Pinb-D1b* were the only two alleles found among all the hard wheats analysed in this study.

Bread wheat lines from CIMMYT are widely used for wheat improvement around the world (Heisey et al., 2002), and CIMMYT lines have likely served as a major donor of the *Pina-D1b* allele to many other breeding programs. A survey of wheat varieties in India showed that most varieties had the *Pina-D1b* allele (Ram et al., 2002). Similarly, many important wheat varieties in Australia that carry the *Pina-D1b* allele can trace their origin back to CIMMYT germplasm (Cane et al., 2004).

Of the two puroindoline alleles found in hard wheats at CIMMYT, *Pina-D1b* tends to give harder endosperm and is associated with lower milling yield and slightly inferior bread-making quality compared to *Pinb-D1b* (Cane et al., 2004; Giroux et al., 2000; Martin et al., 2001). The biochemical basis for this difference is poorly understood, but could be due to the different membrane-affinities of PINA and PINB. Several studies have shown that the amount of PINB associated with the starch granules is strongly reduced in the absence of PINA (Capparelli et al., 2003; Corona et al., 2001; Gazza et al., 2005; Turnbull et al., 2003). Thus, while the PINA-null mutation (*Pina-D1b*) results in almost no starch-associated puroindoline, lines with the glycine to serine mutation in PINB (*Pinb-D1b*) still has a substantial, although variable amount of puroindoline associated with the starch granules (Capparelli et al., 2003). Nevertheless, increasing the frequency of more favourable hardness alleles is now an important objective in the breeding program at CIMMYT, and can be facilitated by careful choice of parents for crossing and the use of molecular markers in the selection process.

While there are many puroindoline alleles to choose from when improving hard wheat quality (Table 1), DNA sequencing has only revealed the wild type alleles *Pina-D1a* and *Pinb-D1a* in conventional bread wheat with soft endosperm (Chen et al., 2006; Lillemo and Morris, 2000; Xia et al., 2005). However, the corresponding genes in *Ae. tauschii*, the donor of the D-genome in hexaploid wheat, exhibit a high degree of sequence variability (Lillemo et al., 2002; Massa et al., 2004). The production of more than one thousand synthetic hexaploid wheats at CIMMYT by hybridizing high-yielding durum wheats (*T. turgidum* ssp. *durum*) with different accessions of *Ae. tauschii* has now made this variability easily available for wheat breeding. The new puroindoline alleles from *Ae. tauschii* all give soft endosperm, but textural differences are apparent, and can be attributed both to the durum wheat parents and the *Ae. tauschii* accessions used for making the synthetic hexaploid wheats (Gedye et al., 2004). A study of grain hardness in 75 primary synthetic wheats from CIMMYT indicated that some of the variation could be explained by allelic variation in the genes *Pina* and *Pinb* (Gedye et al., 2004).

While the previous study was done on a heterogeneous collection of primary synthetic wheats known to possess poor agronomic characters (Mujeeb-Kazi et al., 1996), the extensive use of synthetic wheats in the breeding program at CIMMYT offered the possibility of studying the effect of these new alleles in more uniform genetic backgrounds after two to three cycles of crossing with conventional bread wheat. Significant, but small differences in grain hardness were found among the *Pina* and *Pinb* alleles from synthetic wheat (Table 4). The results were contradictory to those of Gedye et al. (2004)

where *Pina-D1c* and *Pinb-D1h* had significantly softer endosperm than those with the respective wild-type alleles. The effect of the allelic combination *Pina-D1j*/*Pinb-D1i* was not tested in the primary synthetic wheats, but gave about 10 SKCS hardness units softer endosperm compared to progeny lines with *Pina-D1c*/*Pinb-D1h* in the present study of synthetic-derived breeding lines (Table 4). The tendency of *Pina-D1j*/*Pinb-D1i* to give softer endosperm than the other allelic combinations associated with soft endosperm was also observed among the synthetic derived lines in the semi-arid crossing block (data not shown). However, care should be taken when interpreting such minor differences in grain hardness since the confounding effects of other genetic factors that are known to influence grain hardness cannot be ruled out (Campbell et al., 1999; Groos et al., 2004; Lillemo and Ringlund, 2002; Perretant et al., 2000; Sourdille et al., 1996).

In conclusion, the present study has shown that the vast majority of hard wheats from CIMMYT have the *Pina-D1b* allele, which has been the dominating hardness allele since the start of the wheat breeding program in Mexico more than half a century ago. New variability at the *Ha* locus has recently been introduced through the extensive use of synthetic wheat in the breeding program, but the potential effects of the new *Ae. tauschii*-derived puroindoline alleles in conferring minor textural differences in soft wheat need further investigation.

Acknowledgements

Part of the work presented in this study was conducted during the first author's stay at CAAS, which was supported by grants from the National Basic Research Program (2002CB11300), the National Natural Science Foundation of China (30260061), and the Research Council of Norway. The technical assistance of experimental staff in Mexico and China is greatly acknowledged.

References

AACC. 1995. Official Methods of the American Association of Cereal Chemists. AACC, St. Paul, MN.

Barlow, K. K., Buttrose, M. S., Simmonds, D. H., Vesk, M.. 1973. The nature of the starch-protein interface in wheat endosperm. Cereal Chemistry, 50, 443-454.

Brennan, C. S., Sulaiman, B. D., Schofield, J. D., Vaughan, J. G.. 1993. The immunolocation of friabilin and its association with endosperm texture. Aspects of Applied Biology, 36, 69-73.

Campbell, K. G., Bergman, C. J., Gualberto, D. G., Anderson, J. A., Giroux, M. J., Hareland, G., Fulcher, R. G., Sorrells, M. E., Finney, P. L.. 1999. Quantitative trait loci associated with kernel traits in a soft x hard wheat cross. Crop Science, 39, 1184-1195.

Cane, K., Spackman, M., Eagles, H. A.. 2004. Puroindoline genes and their effects on grain quality traits in southern Australian wheat cultivars. Australian Journal of Agricultural Research, 55, 89-95.

Capparelli, R., Borriello, G., Giroux, M. J., Amoroso, M. G.. 2003. Puroindoline A-gene expression is involved in association of puroindolines to starch. Theoretical and Applied Genetics, 107, 1463-1468.

Chen, F., He, Z. H., Xia, X. C., Lillemo, M., Morris, C.. 2005. A new puroindoline b mutation present in Chinese winter wheat cultivar Jingdong 11. Journal of Cereal Science, 42, 267-269.

Chen, F., He, Z. H., Xia, X. C., Xia, L. Q., Zhang, X. Y., Lillemo, M., Morris, C.. 2006. Molecular and biochemical characterization of puroindoline a and b alleles in Chinese landraces and historical cultivars. Theoretical and Applied Genetics, 112, 400-409.

Corona, V., Gazza, L., Boggini, G., Pogna, N. E.. 2001. Variation in friabilin composition as determined by A-PAGE fractionation and PCR amplification, and its relationship to grain hardness in bread wheat. Journal of Cereal Science, 34, 243-250.

Gazza, L., Nocente, F., Ng, P. K. W., Pogna, N. E.. 2005. Genetic and biochemical analysis of common wheat cultivars lacking puroindoline a. Theoretical and Applied Genetics, 110, 470-478.

Gedye, K. R., Morris, C. F., Bettge, A. D.. 2004. Determination and evaluation of the sequence and textural effects of the puroindoline a and puroindoline b genes in a population of synthetic hexaploid wheat. Theoretical and Applied Genetics, 109, 1597-1603.

Giroux, M. J., Morris, C. F.. 1997. A glycine to serine change in puroindoline b is associated with wheat grain hardness and low levels of starch-surface friabilin. Theoretical and Applied Genetics, 95, 857-864.

Giroux, M. J., Morris, C. F.. 1998. Wheat grain hardness results from highly conserved mutations in the friabilin components puroindoline a and b. Proceedings of the National Academy of Sciences of the United States of America, 95, 6262-6266.

Giroux, M. J., Talbert, L., Habernicht, D. K., Lanning, S., Hemphill, A., Martin, J. M.. 2000. Association of puroindoline sequence type and grain hardness in hard red spring wheat. Crop Science, 40, 370-374.

Groos, C., Bervas, E., Charmet, G.. 2004. Genetic analysis of grain protein content, grain hardness and dough rheology in a hard x hard bread wheat progeny. Journal of Cereal Science, 40, 93-100.

Heisey, P. W., Lantican, M. A., Dubin, H. J.. 2002. Impacts of international wheat breeding research in developing countries, 1966-97. CIMMYT, Mexico, D. F.

Hoisington, D., Khairallah, M., Gonzalez-de-Leon, D.. 1994. Laboratory protocols: CIMMYT applied molecular genetics laboratory. 2nd ed, CIMMYT, Mexico D. F.

Ikeda, T. M., Ohnishi, N., Nagamine, T., Oda, S., Hisatomi, T., Yano, H.. 2005. Identification of new puroindoline genotypes and their relationship to flour texture among wheat cultivars. Journal of Cereal Science, 41, 1-6.

Lillemo, M., Morris, C. F.. 2000. A leucine to proline mutation in puroindoline b is frequently present in hard wheats from Northern Europe. Theoretical and Applied Genetics, 100, 1100-1107.

Lillemo, M., Ringlund, K.. 2002. Impact of puroindoline b alleles on the genetic variation for hardness in soft x hard wheat crosses. Plant Breeding, 121, 210-217.

Lillemo, M., Simeone, M. C., Morris, C. F.. 2002. Analysis of puroindoline a and b sequences from Triticum aestivum cv. 'Penawawa' and related diploid taxa. Euphytica, 126, 321-331.

Martin, J. M., Frohberg, R. C., Morris, C. F., Talbert, L. E., Giroux, M. J.. 2001. Milling and bread baking traits associated with puroindoline sequence type in hard red spring wheat. Crop Science, 41, 228-234.

Massa, A. N., Morris, C. F., Gill, B. S.. 2004. Sequence diversity of puroindoline-a, puroindoline-b, and the grain softness protein genes in Aegilops tauschii Coss. Crop Science, 44, 1808-1816.

McIntosh, R. A., Devos, K. M., Dubcovsky, J., Rogers, W. J., Morris, C. F., Appels, R., Anderson, O. D.. 2005. Catalogue of gene symbols for wheat: 2005 supplement, published online at http: //wheat. pw. usda. gov/ggpages/wgc/2005upd. html.

Morris, C. F.. 2002. Puroindolines: the molecular genetic basis of wheat grain hardness. Plant Molecular Biology, 48, 633-647.

Morris, C. F., Lillemo, M., Simeone, M. C., Giroux, M. J., Babb, S. L., Kidwell, K. K.. 2001. Prevalence of puroindoline grain hardness genotypes among historically significant North American spring and winter wheats. Crop Science, 41, 218-228.

Mujeeb-Kazi, A., Rosas, V., Roldan, S.. 1996. Conservation of the genetic variation of Triticum tauschii (Coss.) Schmalh (Aegilops squarrosa auct. non L.) in synthetic hexaploid wheats (T. turgidum L. s. lat. x T. tauschii; 2n=6x=42, AABBDD) and its potential utilization for wheat improvement. Genetic Resources and Crop Evolution, 43, 129-134.

Perretant, M. R., Cadalen, T., Charmet, G., Sourdille, P., Nicolas, P., Boeuf, C., Tixier, M. H., Branlard, G., Bernard, S., Bernard, M.. 2000. QTL analysis of bread-making quality in wheat using a doubled haploid population. Theoretical and Applied Genetics, 100, 1167-1175.

Rajaram, S., van Ginkel, M.. 2001. Mexico: 50 years of international wheat breeding, In A. P. Bonjean and W. J. Angus, (Eds.) The world wheat book. A history of wheat breeding. Lavoisier publishing, Paris. pp. 579-608.

Ram, S., Boyko, E., Giroux, M. J., Gill, B. S.. 2002. Null mutation in puroindoline a is prevalent in Indian wheats: Puroindoline genes are located in the distal part of 5DS. Journal of Plant Biochemistry and Biotechnology, 11, 79-83.

Ram, S., Jain, N., Shoran, J., Singh, R.. 2005. New frame shift mutation in puroindoline b in Indian wheat cultivars Hyb65 and NI5439. Journal of Plant Biochemistry and Biotechnology, 14, 45-48.

Skovmand, B., Villareal, R., van Ginkel, M., Rajaram, S., Ortiz-Ferrara, G.. 1997. Semidwarf bread wheats: Names, Parentages, Pedigrees, and Origins. CIMMYT, Mexico, D. F.

Sourdille, P., Perretant, M. R., Charmet, G., Leroy, P., Gautier, M. F., Joudrier, P., Nelson, J. C., Sorrells, M. E., Bernard, M.. 1996. Linkage between RFLP markers and genes affecting kernel hardness in wheat. Theoretical and Applied Genetics, 93, 580-586.

Turnbull, K. M., Marion, D., Gaborit, T., Appels, R., Rahman, S.. 2003. Early expression of grain hardness in the developing wheat endosperm. Planta, 216, 699-706.

Xia, L. Q., Chen, F., He, Z. H., Chen, X. M., Morris, C. F.. 2005. Occurrence of puroindoline alleles in Chinese winter wheats. Cereal Chemistry, 82, 38-43.

中国和 CIMMYT 小麦品种 Bx7 亚基超量表达基因（$Bx7^{OE}$）的分子检测

任妍[1,2]，梁丹[2]，张平平[2,3]，何中虎[2,4]，陈静[5]，傅体华[1]，夏先春[2]

[1]四川农业大学农学院，雅安 625014；[2]中国农业科学院作物科学研究所/国家小麦改良中心/国家农作物基因资源与基因改良重大科学工程，北京 100081；[3]江苏省农业科学院农业生物技术研究所，南京 210014；[4]CIMMYT 中国办事处，北京 100081；[5]中国科学院成都生物研究所，成都 610041

摘要：高分子量谷蛋白亚基 Bx7 的超量表达对提高小麦面筋强度有重要作用。利用反相高效液相色谱（RP-HPLC）和 STS 标记检测了 163 份中国和 CIMMYT 小麦品种（系）的高分子谷蛋白亚基 Bx7 超量表达基因（$Bx7^{OE}$）。结果表明，*TaBAC1215C06-F517/R964* 标记和 *TaBAC1215C06-F24671/R25515* 标记可分别在含有 $Bx7^{OE}$ 基因的材料中扩增出 447bp 和 844bp 的特异带，在不含 $Bx7^{OE}$ 基因的材料中无相应目标带，两个 STS 标记的检测结果完全一致。在 163 份小麦品种（系）中，11 份品种（系）含有 $Bx7^{OE}$ 基因，占总数的 6.7%。RP-HPLC 与 STS 标记检测结果一致。利用这两个 STS 标记可以方便、快速、准确地检测 $Bx7^{OE}$ 基因。

关键词：普通小麦（*Triticum aestivum* L.）；RP-HPLC；STS 标记；分子标记辅助选择；$Bx7^{OE}$

Characterization of Overexpressed Bx7 Gene ($Bx7^{OE}$) in Chinese and CIMMYT Wheats by STS Markers

Ren Yan[1,2], Lang Dan[2], Zhang Pingping[2,3], He Zhonghu[2,4], Chen Jing[5], Fu Tihua[1], Xia Xianchun[2]

[1]*College of Agronomy, Sichuan Agricultural University, Ya'an* 625014, *Sichuan*; [2]*National Wheat Improvement Center, Institute of Crop Sciences, / National Key Facility for Crop Gene Resource and Genetic Improvement, Chinese Academy of Agricultural Sciences, Beijing* 100081; [3]*Institute of Agricultural Biotechnology, Jiangsu Academy of Agricultural Sciences, Nanjing* 210014, *Jiangsu*; [4]*CIMMYT China Office, Beijing* 100081; [5]*Institute of Biology, Chinese Academy of Sciences, Chengdu* 610041, *Sichuan, China*

Abstract: Over-expression of the high molecular weight glutenin subunit (HMW-GS) Bx7 is highly associated with dough strength of wheat (*Triticum aestivum* L.) flour. A total of 163 Chinese and CIMMYT wheat cultivars and advanced lines were tested by two STS markers and RP-HPLC to understand the presence of HMW-GS gene $Bx7^{OE}$ The results indicated that the markers *TaBAC1215C06-F517/R964* and *TaBAC1215C06-F24671/R25515* could amplify a 447-bp and an 844-bp PCR fragment, respectively, in the lines with $Bx7^{OE}$, whereas no PCR products were detected in the lines without $Bx7^{OE}$. Of

原文发表在《作物学报》，2009，35（3）：403-411

the 163 cultivars and lines, specific PCR fragments were amplified in 11 genotypes by the two markers, indicating the presence of $Bx7^{OE}$ in these lines, with a frequency of 6.7%. The results obtained by RP-HPLC were consistent with those revealed by STS markers. These two STS markers could bc used to detect the presence of $Bx7^{OE}$ gene in wheat cultivars.

Key words: Common wheat (*Triticum aestivum* L.); RP-HPLC; STS marker; Marker-assisted selection; $Bx7^{OE}$

小麦谷蛋白主要由高分子量麦谷蛋白亚基（HMW-GS）和低分子量麦谷蛋白亚（LMW-GS）组成，HMW-GS与LMW-GS以二硫键结合形成聚合体，从而赋予面团独特的黏弹性，对面包烘烤品质起着重要作用[1-3]。HMW-GS分别由位于第一同源群染色体长臂的*Glu-A1*、*Glu-B1*和*Glu-D1*位点的基因（统称*Glu-1*）编码[4]。每一个*Glu-1*位点包括两个紧密连锁的等位基因，分别编码x型和y型HMW-GS[5,6]。不同HMW-GS亚基对面团特性和烘烤品质有着不同的影响。*Glu-A1*编码的1和2*亚基，*Glu-B1*编码的7+8、17+18、13+16和14+15亚基以及*Glu-D1*编码的5+10亚基均对面包加工品质有正向作用。朱金宝等[7]认为，不同亚基存在与否不能完全说明品质的变化，亚基含量和比例及其互作也对加工品质具有重要影响，即应全面考虑麦谷蛋白的质和量。最近的研究发现，*Glu-B1*位点编码的Bx7超量表达亚基与强面筋和优良烘烤品质密切相关[8-10]。发掘和利用具有Bx7超量表达亚基的基因型已成为进一步提高面筋强度的重要途径。加拿大超强筋小麦皆含有该亚基，澳大利亚等已将其列为重要的育种目标。

SDS-PAGE是分离小麦贮藏蛋白的经典方法，但对于分子量差异小的亚基很难区分或不能区分，如7*+8、7+8*、7^{OE}+8*。RP-HPLC是SDS-PAGE的重要补充，可将分子量相同或相近但疏水性不同的亚基分离，但实验成本很高，不适于大量品种鉴定。STS标记可以准确鉴定等位基因的变异，为快速、准确、廉价的检测等位基因提供了可能。

Lukow等[11]对TAA36的研究认为，Bx7亚基的超量表达可能是由于存在两个重复的基因拷贝并且同时表达造成的，而小麦品种Glenlea的Bx7亚基超量表达则依赖于基因的高效转录。D'Ovidio等[12]在对Bx7亚基超量表达品种Red River 68的研究中发现*Bx7*基因存在更强的杂交信号，支持基因复制假说。Butow等[9]发现在Bx7亚基编码区有18 bp插入，但检测结果表明18 bp的插入并不能准确说明$Bx7^{OE}$基因的存在与否。进一步研究发现，在启动子上游核基质结合区（matrix-attachment region, MAR）中43bp的插入也不能解释超量表达[13]。最近从栽培品种Glenlea中获得的一个包括*Glu-B1*位点的BAC克隆被测序，编码$Bx7^{OE}$亚基的一个10.3kb的拷贝也被确定[14]。该位点的结构分析表明，在两个重复片段中间存在一个LTR逆转座子[15]。逆转座子通过促进基因以及基因组结构改变在进化过程中发挥着重要作用，如复制或缺失[16,17]。Ragupathy等[15]根据LTR逆转座子边界与重复片段的结合区域设计了两个STS标记并检测了400多份小麦品种，发现LTR逆转座子的插入能准确判断$Bx7^{OE}$基因的存在。但这两个STS标记在国内的应用还未见报道。

本研究利用两个STS标记对163份中国和CIMMYT小麦品种（系）进行分子检测，并用RP-HPLC法对标记进行了验证，以了解$Bx7^{OE}$基因的分布状况，为我国小麦育种提供材料和分子标记辅助选择方法。

1 材料和方法

1.1 供试材料

163份供试材料（表1）中来自北部冬麦区20份，来自长江中下游冬麦区6份，来自东北春麦区10份，来自黄淮冬麦区34份，来自西南冬麦区4份，以及CIMMYT小麦品种85份、澳大利亚品种3份、加拿大品种1份。这些品种均为当地主栽品种、高代品系和常用亲本。

1.2 SDS-PAGE分析

SDS-PAGE检测参考Liu等[18]和Singh等[19]的方法。HMW-GS命名参考Payne和Lawrence[20]的方法。

表1 普通小麦品种（系）Bx7亚基的RP-HPLC定量分析与分子标记检测结果

Table 1 RP-HPLC quantification and PCR analyses of Bx7 subunit in wheat cultivars and advanced lines

Code	Cultivar	Origin	Glu-B1	Area%	M1	M2
1	ALTAR84/AE. SQUARROSA (221) //3 * BORL95/3/URES/JUN//KAUZ/4/WBLL1	CIMMYT	7+9	31.9	—	—
2	ATTILA * 2/PBW65	CIMMYT	7	31.3	—	—
3	ATTILA/3 * BCN * 2//BAV92	CIMMYT	7+9	33	—	—
4	BABAX/LR42//BABAX * 2/4/SNI/TRAP#1/3/KAUZ * 2/TRAP//KAUZ	CIMMYT	7+9	29.9	—	—
5	BABAX/LR42//BABAX * 2/4/SNI/TRAP#1/3/KAUZ * 2/TRAP//KAUZ	CIMMYT	7+9	32.5	—	—
6	BOW/NKT//CBRD/3/CBRD	CIMMYT	7+8	37.5	—	—
7	CAR//KAL/BB/3/NAC/4/VEE/PJN//2 * TUI/5/MILAN	CIMMYT	17+18	—	—	—
8	CHIBIA//PRLII/CM65531/3/SKAUZ/BAV92	CIMMYT	7+9	31.2	—	—
9	CHIBIA//PRLII/CM65531/3/SW89.5181/KAUZ	CIMMYT	7+9	36.4	—	—
10	CHUANMAI 42	CIMMYT	7+9	26.4	—	—
11	CHUANMAI 43	CIMMYT	7+9	30.9	—	—
12	CNDO/R143//ENTE/MEXI _ 2/3/AEGILOPS SQUARROSA (TAUS) /4/WEAVER/5/PASTOR	CIMMYT	7+8	33.8	—	—
13	CROC _ 1/AE. SQUARROSA (205) //FCT/3/PASTOR	CIMMYT	17+18	—	—	—
14	CROC _ 1/AE. SQUARROSA (205) //KAUZ/3/SASIA	CIMMYT	7+9	32.3	—	—
15	ELVIRA/5/CNDO/R143//ENTE/MEXI75/3/AE. SQ/4/2 * OCI	CIMMYT	17+18	—	—	—
16	FRET2 * 2/4/SNI/TRAP#1/3/KAUZ * 2/TRAP//KAUZ	CIMMYT	7+9	29.2	—	—
17	FRET2 * 2/BRAMBLING	CIMMYT	7+9	32.7	—	—
18	FRET2 * 2/KUKUNA	CIMMYT	7+9	30.2	—	—
19	FRET2/WBLL1//KAMB1	CIMMYT	17+18	—	—	—
20	HE1/3 * CNO79//2 * SERI/3/ATTILA/4/WH 542	CIMMYT	7+9	33	—	—
21	HEILO	CIMMYT	7+8	35.3	—	—
22	HEILO	CIMMYT	17+18	—	—	—
23	INIA CHURRINCHE	CIMMYT	7+8	28.4	—	—
24	INQALAB 91 * 2/TUKURU	CIMMYT	7+9	29.2	—	—
25	ITAPUA 40-OBLIGADO	CIMMYT	7+9	36.2	—	—
26	IVAN/6/SABUF/5/BCN/4/RABI//GS/CRA/3/AE. SQUARROSA (190)	CIMMYT	7+9	31	—	—
27	K6295.4A	CIMMYT	7+8	34.2	—	—
28	KANCHAN	CIMMYT	7OE +8 *	45.1	+	+
29	KAUZ//ALTAR 84/AOS/3/MILAN/KAUZ/4/HUITES	CIMMYT	17+18	—	—	—
30	KAUZ/PASTOR//PBW343	CIMMYT	7+9	35.9	—	—
31	KAUZ/PASTOR//PBW343	CIMMYT	7	34.4	—	—
32	KIRITATI//PBW65/2 * SERI.1B	CIMMYT	17+18	—	—	—
33	KIRITATI//PBW65/2 * SERI.1B	CIMMYT	17+18	—	—	—
34	KIRITATI//PBW65/2 * SERI.1B	CIMMYT	17+18	—	—	—
35	KIRITATI/3/HUW234+LR34//PRL/VEE#10	CIMMYT	7+8	38.1	—	—

（续）

Code	Cultivar	Origin	Glu-B1	Area%	M1	M2
36	KLEIN DON ENRIQUE	CIMMYT	17+18	—	—	—
37	MILAN/SHA7/3/THB/CEP7780//SHA4/LIRA/4/SHA4/CHIL	CIMMYT	7+9	33.5	—	—
38	NING MAI 9415.16//SHA4/CHIL/3/NING MAI 50	CIMMYT	7+8*	29.3	—	—
39	OR791432/VEE#3.2//MILAN	CIMMYT	17+18	—	—	—
40	PASTOR//MUNIA/ALTAR 84	CIMMYT	7+9	32.5	—	—
41	PASTOR/TEERI	CIMMYT	7+8	30.6	—	—
42	PASTOR/TEERI	CIMMYT	7+8	31	—	—
43	PASTOR/TEERI	CIMMYT	7+8	31.8	—	—
44	PBW343	CIMMYT	7	25.6	—	—
45	PBW343*2/KHVAKI	CIMMYT	7	29.3	—	—
46	PBW343*2/KUKUNA	CIMMYT	7	29.5	—	—
47	PBW343/WBLL1//PANDION	CIMMYT	17+18	—	—	—
48	PBW65/2*PASTOR	CIMMYT	7+8	31.9	—	—
49	PFAU/SERI.1B//AMAD/3/WAXWING	CIMMYT	7	29.7	—	—
50	PFAU/WEAVER*2//BRAMBLING	CIMMYT	7^{OE}+8*	42.7	+	+
51	PGO/SERI//BAV92	CIMMYT	7+9	31.6	—	—
52	PRL/2*PASTOR	CIMMYT	7+9	22.7	—	—
53	SHA5/WEAVER	CIMMYT	7+8	43.3	—	—
54	SHA8/GEN	CIMMYT	7+8	31.8	—	—
55	SOROCA	CIMMYT	7+9	32.3	—	—
56	SW03-81497	CIMMYT	7+8	33.4	—	—
57	THELIN#2//ATTILA*2/PASTOR/3/PRL/2*PASTOR	CIMMYT	7+9	32.7	—	—
58	THELIN#2/TUKURU	CIMMYT	7+9	29.5	—	—
59	THELIN//2*ATTILA*2/PASTOR	CIMMYT	17+18	—	—	—
60	THELIN//2*ATTILA*2/PASTOR	CIMMYT	17+18	—	—	—
61	THELIN/2*WBLL1	CIMMYT	7+9	30.7	—	—
62	TNMU/6/PEL74144/4/KVZ//ANE/MY64/3/PF70354/5/BR14/7/BR35	CIMMYT	7+8	33.9	—	—
63	V763.2312/V879.C8.11.11.11（36）//STAR/3/STAR	CIMMYT	7+8	32.7	—	—
64	VORB/FISCAL	CIMMYT	7	23	—	—
65	WAXWING	CIMMYT	7	24.6	—	—
66	WAXWING*2/BRAMBLING	CIMMYT	7+9	13	—	—
67	WAXWING*2/KIRITATI	CIMMYT	7	30.3	—	—
68	WAXWING*2/KIRITATI	CIMMYT	7	23	—	—
69	WAXWING*2/KIRITATI	CIMMYT	7+9	32.7	—	—
70	WAXWING*2/KIRITATI	CIMMYT	7+9	31.4	—	—
71	WAXWING*2/KUKUNA	CIMMYT	7	31.6	—	—
72	WAXWING*2/KUKUNA	CIMMYT	7	29.5	—	—
73	WAXWING*2/KUKUNA	CIMMYT	7	26.4	—	—

（续）

Code	Cultivar	Origin	Glu-B1	Area%	M1	M2
74	WAXWING * 2/KURUKU	CIMMYT	7+9	33	—	—
75	WAXWING * 2/TUKURU	CIMMYT	7	33.5	—	—
76	WAXWING * 2/TUKURU	CIMMYT	7+9	36.8	—	—
77	WAXWING * 2/VIVITSI	CIMMYT	7	34.6	—	—
78	WAXWING * 2/VIVITSI	CIMMYT	7	32.1	—	—
79	WBLL1 * 2/BRAMBLING	CIMMYT	7^{OE}+8 *	40.4	+	+
80	WBLL1 * 2/BRAMBLING	CIMMYT	7^{OE}+8 *	39.2	+	+
81	WBLL1 * 2/BRAMBLING	CIMMYT	7^{OE}+8 *	40.1	+	+
82	WBLL1 * 2/KIRITATI	CIMMYT	7+9	31.7	—	—
83	WBLL1 * 2/KIRITATI	CIMMYT	7+9	32.6	—	—
84	WBLL1 * 2/KIRITATI	CIMMYT	17+18	—	—	—
85	WBLL1/3/STAR//KAUZ/STAR/4/BAV92/RAYON	CIMMYT	7+9	31.4	—	—
86	CD87	Australia	7^{OE}+8	46.3	+	+
87	Hartog	Australia	17+18	—	—	—
88	Sunstate	Australia	17+18	—	—	—
89	Wild Cat	Canada	7^{OE}+8 *	41.1	+	+
90	CA9550	NWWR	7+8 *	20.5	—	—
91	CA9641	NWWR	7+8 *	24.2	—	—
92	CA9719	NWWR	7+8 *	21.1	—	—
93	CA9722	NWWR	7+8 *	25.4	—	—
94	京 411 Jing 411	NWWR	7+8 *	7.4	—	—
95	京冬 8 Jingdong 8	NWWR	7+9	24.5	—	—
96	农大 116 Nongda 116	NWWR	7+8	32	—	—
97	农大 152 Nongda 152	NWWR	7+8	30.9	—	—
98	农大 3197 Nongda 3197	NWWR	7+8 *	25.5	—	—
99	优选 14 Youxuan 14	NWWR	7+8	22.8	—	—
100	优选 9 Youxuan 9	NWWR	7+9	25.8	—	—
101	原冬 6 号 Yuandong 6	NWWR	7+9	21.3	—	—
102	中优 14 Zhongyou 14	NWWR	7+8	21.4	—	—
103	中优 16 Zhongyou 16	NWWR	7+8	22.8	—	—
104	中优 8 Zhongyou 8	NWWR	7+9	23.3	—	—
105	中优 9507 Zhongyou 9507	NWWR	7+9	22.2	—	—
106	中优 9701 Zhongyou 9701	NWWR	7+8	25.4	—	—
107	中优 9843 Zhongyou 9843	NWWR	7+8 *	27	—	—
108	中优 9844 Zhongyou 9844	NWWR	7+8	23.1	—	—
109	中作 8131-1 Zhongzuo 8131-1	NWWR	7+9	22.7	—	—
110	安农 91168 Annong 91168	MLYVWWR	7+8	30	—	—
111	皖麦 18 Wanmai 18	MLYVWWR	7+8	24.7	—	—
112	皖麦 19 Wanmai 19	MLYVWWR	7+9	25.1	—	—
113	皖麦 33 Wanmai 33	MLYVWWR	14+15	—	—	—

（续）

Code	Cultivar	Origin	Glu-B1	Area%	M1	M2
114	扬麦 158 Yangmai 158	MLYWWR	7+8	28	—	—
115	扬麦 5 号 Yangmai 5	MLYWWR	7+9	24.4	—	—
116	C1（克丰 6 Kefeng 6/Glenlea）	NESWR	7^{OE}+8 *	47.6	+	+
117	C5（克丰 6 Kefeng 6/Glenlea）	NESWR	7^{OE}+8 *	45.5	+	+
118	E2（龙麦 20 Longmai 20/Glenlea）	NESWR	7^{OE}+8 *	45.8	+	+
119	E3（龙麦 20 Longmai 20/Glenlea）	NESWR	17+18	—	—	—
120	E5（龙麦 20 Longmai 20/Glenlea）	NESWR	7+8	32.8	—	—
121	E6（龙麦 20 Longmai 20/Glenlea）	NESWR	7+8	31.6	—	—
122	H99（小冰麦 33 Xiaobingmai/Glenlea）	NESWR	7+8	34.2	—	—
123	克丰 3 号 Kefeng 3	NESWR	7+8	32.2	—	—
124	克丰 6 号 Kefeng 6	NESWR	7+8	31.9	—	—
125	龙麦 20 Longmai 20	NESWR	7+8	33.8	—	—
126	PH82-2-2	YHFWWR	14+15	—	—	—
127	荔垦 2 号 Liken 2	YHFWWR	7+8	20.8	—	—
128	高优 503 Gaoyou 503	YHFWWR	7+8 *	29.5	—	—
129	藁城 8901 Gaocheng 8901	YHFWWR	7+8	30.8	—	—
130	关封 2 号 Guanfeng 2	YHFWWR	7 * +8	27.6	—	—
131	淮麦 16 Huaimai 16	YHFWWR	7+9	26.5	—	—
132	济南 17 Jinan 17	YHFWWR	7+8	31.9	—	—
133	济南 19 Jinan 19	YHFWWR	7+8	31.2	—	—
134	济南 20 Jinan 20	YHFWWR	13+16	—	—	—
135	冀 5099 Ji 5099	YHFWWR	17+18	20.6	—	—
136	晋农 207 Jinong 207	YHFWWR	7+9	—	—	—
137	临汾 137 Linfen 137	YHFWWR	14+15	—	—	—
138	临旱 917 Linhan 917	YHFWWR	7+9	17.3	—	—
139	陕 160 Shaan 160	YHFWWR	7+8	21	—	—
140	陕 225 Shaan 225	YHFWWR	14+15	—	—	—
141	陕 229 Shaan 229	YHFWWR	14+15	—	—	—
142	陕 253 Shaan 253	YHFWWR	7+9	13.9	—	—
143	小堰 6 号 Xiaoyan 6	YHFWWR	14+15	—	—	—
144	小堰 54 号 Xiaoyan 54	YHFWWR	14+15	—	—	—
145	烟 239 Yan 239	YHFWWR	7+8	34.8	—	—
146	烟 2801 Yan 2801	YHFWWR	7+9	18.5	—	—
147	烟辐 188 Yanfu 188	YHFWWR	7+9	23.8	—	—
148	烟农 15 Yannong 15	YHFWWR	7+9	22.5	—	—
149	烟优 361 Yanyou 361	YHFWWR	17+18	—	—	—
150	豫麦 34 Yumai 34	YHFWWR	7+8	21.7	—	—
151	豫麦 47 Yumai 47	YHFWWR	7+8	22.1	—	—

（续）

Code	Cultivar	Origin	Glu-B1	Area%	M1	M2
152	豫麦 54 Yumai 54	YHFWWR	7+9	19.1	—	—
153	豫麦 57 Yumai 57	YHFWWR	7+9	23	—	—
154	豫麦 69 Yumai 69	YHFWWR	7+8	22.5	—	—
155	豫农 95339 Yunong 95339	YHFWWR	7+9	23.9	—	—
156	运丰早 101 Yunfengzao 101	YHFWWR	14+15	—	—	—
157	郑州 9023 Zhengzhou 9023	YHFWWR	7+8	31.6	—	—
158	郑州 992 Zhengzhou 992	YHFWWR	17+18	—	—	—
159	中育 415 Zhongyu 415	YHFWWR	14+15	—	—	—
160	川育 12 Chuanyu 12	SWWR	7+8	7.2	—	—
161	德麦 3 号 Demai 3	SWWR	7^{OE}+8 *	28	+	+
162	云麦 39 Yunmai 39	SWWR	7+9	25.9	—	—
163	中国春 Chinese Spring	SWWR	7+8	36.4	—	—

Area%：Bx7 亚基占 HMW-GS 总量的面积百分比；M1：标记 1；M2：标记 2；CIMMYT=International Maize and Wheat Improvement Center；NWWR=Northern Winter Wheat Region 北部冬麦区；MLYVWWR=Middle and Lower Yangtze Valley Winter Wheat Region 长江中下游冬麦区；NESWR=Northeastern Spring Wheat Region 东北春麦区；YHFWR=Yellow and Huai Facultative Winter Wheat Region 黄淮冬麦区；SWWR=Southwestern Winter Wheat Region 西南冬麦区.

1.3 RP-HPLC 分析

按本实验室报道的方法[21]进行反相高效液相色谱（RP-HPLC），获取 HMW-GS 各亚基峰面积相对含量。当 Bx7 亚基含量占 HMW-GS 总量的比值大于 39%时认为 Bx7 亚基超量表达，参考 Butow 等[13]的研究。

1.4 基因组 DNA 提取

采用 SDS 法[22]提取小麦基因组 DNA。每份材料分别提取 2 粒种子的 DNA，利用紫外分光光度计检测 DNA 浓度，终浓度调整至 20ng μl^{-1}。

1.5 STS 标记检测

利用 Ragupathy 等[15]开发的两个显性 STS 标记检测 *Bx7^OE^* 基因。引物由奥科生物技术有限公司合成。标记 1（重复片段与逆转座子左边结合引物）：*TaBAC1215C06-F517*：5′-ACGTGTCCAAGCTTTGGTTC-3′，*TaBAC1215C06-R964*：5′-GATTGGTGGGTGGATACAGG-3′；标记 2（重复片段与逆转座子右边结合引物）：*TaBAC1215C06-F24671*：5′-CCACTTCCAAGGTGGGACTA-3′，*TaBAC1215C06-R25515*：5′-TGCCAACACAAAAGAAGCTG-3′。

PCR 反应体系为 20μl，含 10×PCR Buffer 2μl，d NTP（A、T、C、G）各 200μmol L^{-1}，每条引物 1μl（10mmol L^{-1}），*Taq* DNA 聚合酶（TaKaRa）1U，模板 DNA 50ng。标记 1 的 PCR 反应程序为：94℃预变性 5min；94℃变性 30s，63℃退火 30s，72℃延伸 1min，35 个循环；72℃延伸 5min。标记 2 的 PCR 反应程序为：94℃预变性 5min；94℃变性 30s，58℃退火 30s，72℃延伸 45s，35 个循环；72℃延伸 5min。

PCR 扩增产物以 1.5%琼脂糖凝胶电泳分离检测，缓冲液体系为 1×TAE 溶液，180V 电压电泳 30min，溴化乙锭染色后，用 Gel Doc XR System 扫描成像并存入计算机。

2 结果与分析

2.1 RP-HPLC 分析

SDS-PAGE 分析表明，163 份材料中有 132 份含有 Bx7 亚基。利用 RP-HPLC 对 132 份材料分析，根据洗脱曲线计算 Bx7 亚基占 HMW-GS 总量的面积百分比（图 1）。在 61 份中国小麦品种中，Bx7 亚基面积百分比平均值为 26.2%，最大值是 47.6%，最小值是 7.2%，大于 39%的材料有 4 份，分别为 C1、

C5、E2和德麦3号；71份国外品种中Bx7亚基的面积百分比平均值为32.4%，最大值是46.3%，最小值是13.0%，大于39%的材料有7份，分别为CD87、Wild Cat（野猫）以及来自CIMMYT的5份材料，分别是PFAU/WEAVER * 2//BRAMBLING（1份）、KANCHAN（1份）和WBLL1 * 2/BRAMBLING（3份姐妹系）。共计11份品种（系）的面积百分比大于39%，为Bx7^{OE}类型。

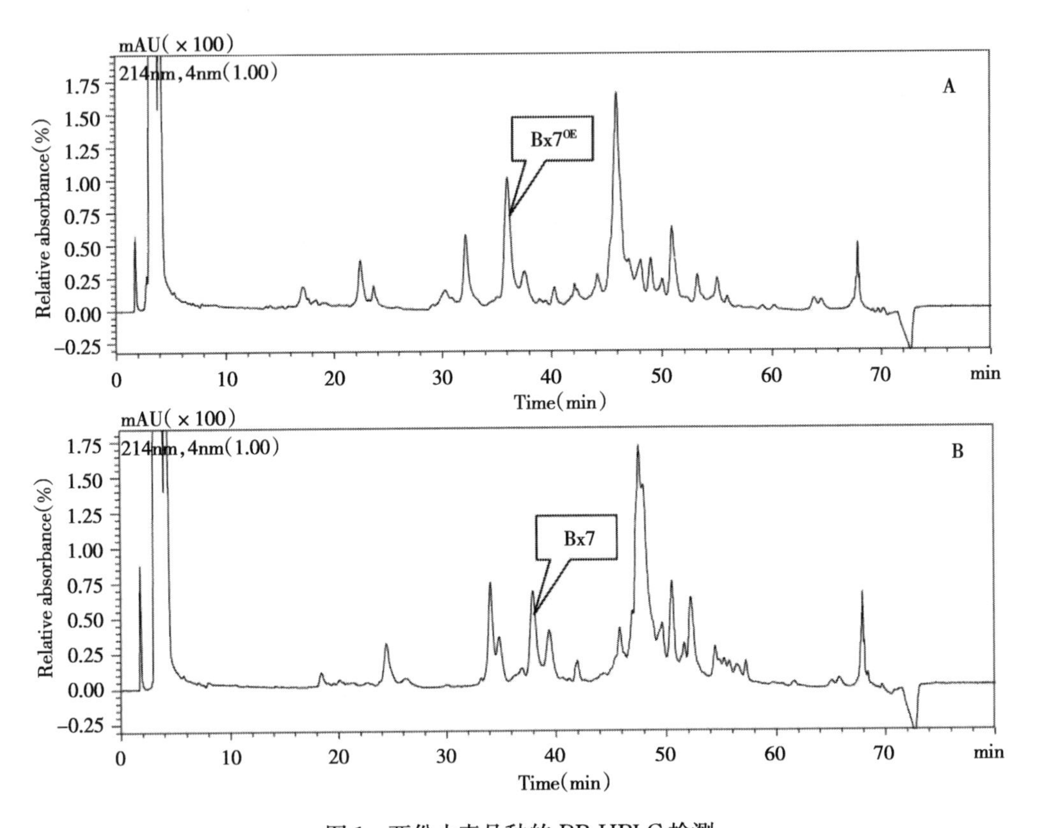

图1 两份小麦品种的RP-HPLC检测

Fig. 1 RP-HPLC profiles of two wheat accessions

A：CD87（2 *，7^{OE}+8，2+12），其Bx7亚基占全部高分子量谷蛋白亚基的46.3%；

B：H99（2 *，7+8，2+12），其Bx7亚基占全部高分子量谷蛋白亚基的34.2%.

A：CD87（2 *，7^{OE}+8，2+12），displayed HMW-GS Bx7^{OE} that represented 46.3% of its total HMW-GS composition;

B：H99（2 *，7+8，2+12），had HWM-GS Bx7 that represented 34.2% of its total HMW-GS composition.

2.2 *Bx7*^{OE}的STS标记检测

标记*TaBAC1215C06-F517/R964*在含有*Bx7*^{OE}基因的材料中可扩增出一条447bp的片段，在不含*Bx7*^{OE}基因的材料中无PCR扩增产物；标记*TaBAC1215C06-F24671/R25515*在含有*Bx7*^{OE}基因的材料中可扩增出一条844bp的片段，在不含Bx7^{OE}基因的材料中无PCR产物。两个标记的PCR产物对应出现，可以相互验证而且扩增条带清晰、重复性好（图2）。RR-HPLC检测出的面积百分比大于39%的11份品种可扩增出特异条带，因此携带Bx7^{OE}基因，其他品种则不能，说明两个STS标记的检测结果与RP-HPLC完全一致。

2.3 *Bx7*^{OE}基因的分布

159份中国和CIMMYT小麦品种（系）中仅9份材料携带*Bx7*^{OE}基因，占5.7%（表2）。4份携带*Bx7*^{OE}基因的中国材料中，3份是黑龙江农业科学院通过对加拿大品种Glenlea系选育成的材料，分别为C1、C5和E2，另一份为云南省的德麦3号。Glenlea是*Bx7*^{OE}亚基超量表达的品种，因此推测C1、C5和E2的*Bx7*^{OE}基因来自Glenlea。由于没有其亲本，德麦3号中*Bx7*^{OE}的来源尚不能确定。CIMMYT小麦品种中携带*Bx7*^{OE}基因的组合有3个，共5个品系，说明CIMMYT携带*Bx7*^{OE}基因的品种也很少。由于*Bx7*^{OE}亚基超量表达可显著提高面筋强度，因此应在今后小麦育种中充分利用 $Bx7^{OE}$ 基因。另外，澳大利亚和加拿大的材料中，携带 $Bx7^{OE}$ 基因的材料分别为 CD87 和 Wild Cat。

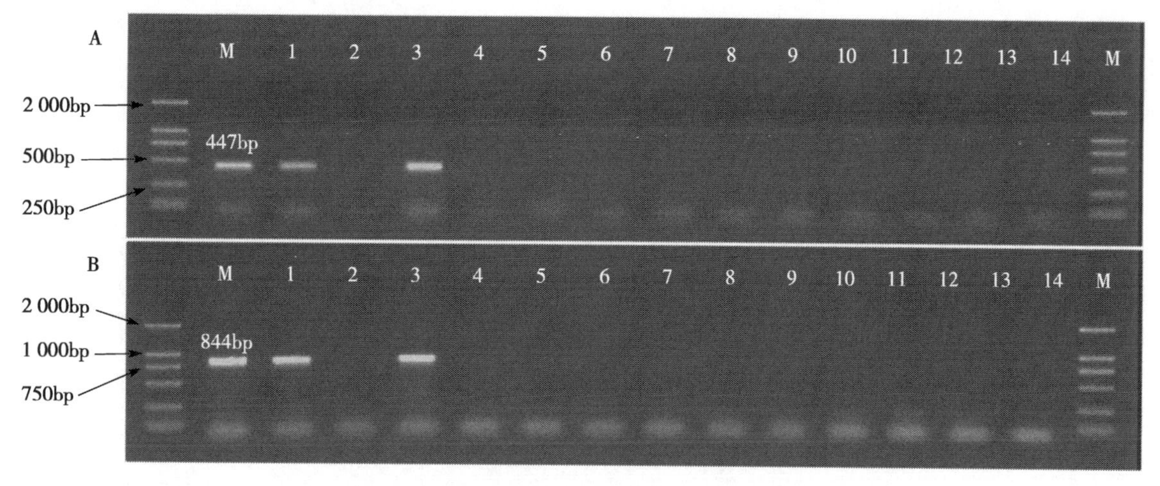

图 2 小麦品种高分子量谷蛋白亚基 $Bx7^{OE}$ 基因的分子检测

Fig. 2 Polymorphic test of PCR fragments amplified with $Bx7^{OE}$ in wheat lines

A: PCR amplified with marker *TaBAC1215C06-F517/R964*; B: PCR amplified with marker *TaBAC1215C06-F24671/R25515*. M: 2000 DNA Marker; 1: CD87 (Code: 86) 2: PFAU/WEAVER * 2//BRAMBLING (Code: 50); 3: ATTILA/3 * BCN * 2//BAV92 (Code: 3); 4: WBLL1 * 2/BRAMBLING (Code: 79); 5: BABAX/LR42//BABAX * 2/4/SNI/TRAP#1/3/KAUZ * 2/TRAP//KAUZ (Code: 4); 6: FRET2 * 2/4/SNI/TRAP#1/3/KAUZ * 2/TRAP//KAUZ (Code: 16); 7: KAUZ/PASTOR//PBW343 (Code: 30); 8: KIRITATI/3/HUW234+LR34//PRL/VEE#10 (Code: 35); 9: PBW343/WBLL1//PANDION (Code: 47); 10: PFAU/SERI. 1B//AMAD/3/WAXWING (Code: 49); 11: WAXWING * 2/TUKURU (Code: 75); 12: HEILO (Code: 21); 13: SHA8/GEN (Code: 54): 14: PASTOR/TEERI (Code: 41).

表 2 中国和 CIMMYT 小麦品种（系）$Bx7^{OE}$ 基因的分布

Table 2 Characterization of $Bx7^{OE}$ gene in Chinese and CIMMYT wheat cultivars and lines

来源 Origin	检测品种（系）数 Number of cultivars (lines) tested	含 $Bx7^{OE}$ 的品种数 Number of cultivars with $Bx7^{OE}$
中国 China	74	4
CIMMYT	85	5
澳大利亚 Australia	3	1
加拿大 Canada	1	1
合计 Total	163	11

3 讨论

分子标记辅助选择在育种中越来越受到重视，筛选并验证可靠的分子标记有助于快速准确聚合优质亚基。Ragupathy 等[15]开发了检测 $Bx7^{OE}$ 基因的两个特异性 STS 标记，但迄今为止利用分子标记检测 $Bx7^{OE}$ 基因在国内还没有报道。本试验利用这两个 STS 标记检测了 163 份小麦品种中 $Bx7^{OE}$ 基因的等位变异，结果表明该标记扩增带型清晰，稳定性好，而且两个标记可以相互对应。研究还表明，STS 标记与 RP-HPLC 的检测结果完全一致，而前者比后者的成本明显偏低。因此，$Bx7^{OE}$ 基因位点的 STS 标记是一种简单、可靠的标记，可用于小麦品种（系）$Bx7^{OE}$ 基因的检测。

对于 *Glu-A1* 位点缺失的小麦品种，将 Bx7 亚基相对于全部高分子量麦谷蛋白亚基含量面积百分比大于 39%作为携带 $Bx7^{OE}$ 基因的判断标准并不合适，原因在于 *Glu-A1* 的缺失，使 Bx7 亚基的相对含量占高分子量麦谷蛋白亚基总量的比值提高，造成偏差。例如小麦品种 SHA5/WEAVER（表 1）在 *Glu-A1* 位点缺失，虽然该品种的 Bx7 亚基相对于全部高分子麦谷蛋白亚基含量的比值为43.3%，但不携带Bx7亚基超量表达基因。

4 结论

利用两个STS标记检测了163份小麦品种（系）$Bx7^{OE}$基因的分布，并用RP-HPLC法进一步验证，两种方法检测结果完全一致。说明这两个STS标记可以快速准确地检测小麦品种是否携带$Bx7^{OE}$基因。本文报道的163份小麦品种（系）Bx7亚基超量表达基因$Bx7^{OE}$的检测结果，可以为我国小麦育种提供有用的材料以及分子标记辅助选择方法。

参考文献

[1] Payne P I, Law C N, Mudd E E. Control by homologous group 1 chromosomes of the high-molecular-weight subunits of glutenin, a major protein of wheat endosperm. *Theor Appl Genet*, 1980, 58: 113-120.

[2] Wrigley C W. Giant proteins with flour power. *Nature*, 1996, 381: 738-739.

[3] Gianibelli M C, Larroque O R, MacRitchie F. Biochemical, genetic, molecular characterization of wheat glutenin and its component subunits. *Cereal Chem*, 2001, 78: 635-646.

[4] Payne P I, Genetics of wheat storage proteins and the effect of allelic variation on bread making quality. *Ann Rev Plant Physiol*, 1987, 38: 141-153.

[5] Payne P I, Holt L M, Law C N. Structural and genetical studies on the weight subunits of wheat glutenin. *Theor Appl Genet*, 1981, 60: 229-236.

[6] Shewry P R, Halford N G, Tatham A S. High molecular weight subunits of wheat glutenin. *J Cereal Sci*, 1992, 15: 105-120.

[7] Zhu J-B (朱金宝), Liu G-T (刘广田), Zhang S-Z (张树臻). High and low molecular subunits of glutenin and their relationships with wheat quality. *Sci Agric Sin* (中国农业科学), 1996, 29 (1): 34-39. (in Chinese with English abstract)

[8] Radovanovic N, Cloutier S, Brown D, Humphreys D G, Lukow O M. Genetic variance for gluten strength contributed by high molecular weight glutenin proteins. *Cereal Chem*, 2002, 79: 843-849.

[9] Butow B J, Ma W, Gale K R, Cornish G B, Rampling L, Larroque O, Morell M K, Bekes F. Molecular discrimination of Bx7 alleles demonstrates that a highly expressed high molecular weight glutenin allele has a major impact on wheat flour dough strength. *Theor Appl Genet*, 2003, 107: 1524-1532.

[10] Vawser M J, Cornish G B. Overexpression of HMW glutenin subunitGlu-B1 7x in hexaploid wheat varieties (*Triticum aestivum*). *Aust J Agric Res*, 2004, 55: 577-588.

[11] Lukow O M, Forsyth S A, Payne P I. Over-production of HMW glutenin subunits coded on chromosome 1B in common wheat, *Triticum aestivum*. *J Genet Breed*, 1992, 46: 187-192.

[12] D' Ovidio R, Masci S, Porceddu E, Kasarda D. Duplication of the high-molecular weight glutenin subunit gene in bread wheat (*Triticum aestivum* L.) cultivar Red River 68. *Plant Breed*, 1997, 116: 525-531.

[13] Butow B J, Gale K R, Ikea J, Juhasz A, Bedo Z, Tamas L, Gianibelli M C. Dissemination of the highly expressed Bx7 glutenin subunit (*Glu-B1al* allele) in wheat as revealed by novel PCR markers and RP-HPLC. *Theor Appl Genet*, 2004, 109: 1525-1535.

[14] Cloutier S, Banks T, Nilmalgoda S, Molecular understanding of wheat evolution at the *Glu-B1* locus. In: Proceedings of the International Conference on Plant Genomics and Biotechnology: Challenges and Opportunities, Raipur, India, 2005. p 40.

[15] Ragupathy R, Naeem H A, Reimer E, Lukow O M, Sapirstein H D, Cloutier S. Evolutionary origin of the segmental duplication encompassing the wheat *Glu-B1* encoding the overexpressed Bx7 ($Bx7^{OE}$) high molecular weight glutenin subunit. *Theor Appl Genet*, 2007, 116: 283-296.

[16] Jiang N, Bao S, Zhang X, Eddy S R, Wessler S R. Pack-MULE transposable elements mediate gene evolution in plants. *Nature*, 2004, 431: 569-573.

[17] Morgante M, Brunner S, Pea G, Fengler K, Zuccolo A, Rafalski A. Gene duplication and exon shuffling by helitron-like transposon generate intraspecies diversity in Maize. *Nat Genet*, 2005, 37: 997-1002.

[18] Liu L, He Z H, Yan J, Zhang Y, Xia X C, Peña R J. Allelic variation at the *Glu-1* and *Glu-3* loci, presence of 1B. 1R translocation, and their effects on mixographic properties in Chinese bread wheats. *Euphytica*, 2005, 142: 197-204.

[19] Singh N K, Shepherd K W, Cornish G B. A simplified SDS-PAGE procedure for separating LMW subunits of glutenin. *J Cereal Sci*, 1991, 14: 203-208.

[20] Payne P I, Lawrence G J. Catalogue of alleles for the

complex loci, *Glu-A1*, *Glu-B1* and *Glu-D1*, which code for high-molecular-weight subunits of glutenin in hexaploid wheat. *Cereal Res Comm*, 1983, 11: 29-35.

[21] Zhang P-P (张平平), Zhang Y (张勇), Xia X-C (夏先春), He Z-H (何中虎). Protocol establishment of reversed-phase high-performance liquid chromatography (RP-HPLC) for analyzing wheat gluten protein. *Sci Agric Sin* (中国农业科学), 2007, 40 (期号): 1002-1009. (in Chinese with English abstract)

[22] Devos K M, Gale M D. The use of random amplified polymorphic DNA markers in wheat. *Theor Appl Genet*, 1992, 84: 567-572.

Characterization of CIMMYT bread wheats for high-and low-molecular-weight glutenin subunits and other quality-related genes with SDS-PAGE, RP-HPLC and molecular markers

D. Liang[1,2], J. W. Tang[1], R. J. Peña[3], R. P. Singh[3], X. Y. He[1], X. Y. Shen[1], D. Y. Yao[2], X. C. Xia[1], and Z. H. He[1,4]

[1] *Institute of Crop Science, National Wheat Improvement Centre/The National Key Facility for Crop Gene Resources and Genetic Improvement, Chinese Academy of Agricultural Sciences* (*CAAS*), 12 *Zhongguancun South Street, Beijing* 100081, *China*; [2] *College of Agronomy, Anhui Agricultural University, Hefei* 230036, *China*; [3] *International Maize and Wheat Improvement Centre* (*CIMMYT*), *Apdo. Postal* 6-641, 06600 *Mexico, D. F., Mexico*; [4] *International Maize and Wheat Improvement Centre* (*CIMMYT*) *China Office, c/o CAAS*, 12 *Zhongguancun South Street, Beijing* 100081, *China*

Abstract: Two hundred and seventy three CIMMYT bread wheat cultivars and advanced lines grown under irrigated conditions in Mexico during the 2005-2006 Yaqui crop cycle were characterized for quality-related genetic traits using gene-specific markers for some high-and low-molecular-weight glutenin subunit (HMW-GS and LMW-GS) genes, polyphenol oxidase (PPO), phytoene synthase (PSY), and waxy genes. Of them, 142 were analyzed for quality parameters including SDS sedimentation volume (SDS-SV), dough mixing time, and Alveograph parameters, and for HMW-GS and LMW-GS compositions using sodium-dodecyl-sulfate polyacrylamide gel electrophoresis (SDS-PAGE), and reversed-phase high-performance liquid chromatography (RP-HPLC). For the *Ppo-A1* locus tested with the marker *PPO18*, the frequencies of alleles *Ppo-A1a* and *Ppo-A1b* were 79.1% and 20.2%, respectively, and no PCR fragment was amplified in 2 lines (0.73%), whereas 227 lines (83.2%) contained the allele *Ppo-D1a* and 46 lines (16.8%) had *Ppo-D1b* detected by markers *PPO16* and *PPO29*. For the marker *YP7A*, 142 lines (52.0%) were assumed to have the allele *Psy-A1a* and 131 lines (48.0%) contained the allele *Psy-A1b*. In the case of the marker *YP7B* for the gene *Psy-B1*, the alleles *Psy-B1a* and *Psy-B1b* were detected in 155 (56.8%) and 43 (15.8%) lines, respectively, and 75 (27.4%) lines possessed the allele *Psy-B1d* detected by the marker *YP7B-3*. All 273 lines contained the alleles *Wx-A1a* and *Wx-D1a* as determined by markers *MAG264* and *MAG269*, respectively. Using the marker *Wx-B1*, 204 lines (74.7%) were presumed to have the *Wx-B1a* allele and 69 (25.3%) possessed *Wx-B1b*. The over-expressing allele of $Bx7^{OE}$ and subunit *By8* *, not clearly seen with SDS-PAGE, were detected by RP-HPLC. The numbers of lines with subunits *Ax2* *, *By8*, *By9*, *Bx17*, *Bx20*, *Dx5*, and *Glu-B3j* were 90, 16, 57, 5, 46, 118, and 33, respectively, in the 142 lines analyzed with molecular markers, and were consistent with the results obtained by SDS-PAGE, except for one line with the 1A. 1R translocation. Subunits *Ax1* and *Ax2* * at the *Glu-A1* locus showed significantly better effects on all quality parameters than subunit

Published in Euphytica, 2010, 172 (2): 235-250

Null. Subunits *5+10* gave significantly better effects for all parameters. Subunit *Glu-A3b* showed more positive effects than its alternative alleles on SDS-SV and SDS-sedimentation volume/protein content index (SPI). The allele *Glu-B3g* showed the best effect on SDS-SV and Alveograph W, whereas *Glu-B3j*, associated with the 1B. 1R translocation, exhibited a strongly negative effect on all quality parameters.

Key words: Common wheat, Quality parameters, HMW-GS, LMW-GS, SDS-PAGE, RP-HPLC, Molecular markers

Introduction

Wheat is one of the most important food crops in the world. The properties of processing quality were primarily determined by two gluten proteins in wheat flour, designated as HMW-GS and LMW-GS, which are highly associated with dough rheological properties determining bread-making quality (Branlard and Dardevet, 1985; Payne, 1987; Gupta et al., 1991, 1994; Weegels et al., 1996). Polyphenol oxidase (PPO) activity, yellow pigment (YP) content, and waxy protein have significant influence on the quality of end-use product, particularly for the Asian noodles (Yamamori et al., 1992; Zhao et al., 1998; He et al., 2004).

Bread-making quality was strongly influenced by the allelic variations of HMW-GS and LMW-GS genes (Payne, 1987; Gupta et al., 1991). The subunits *1Ax1* or *1Ax2 ** is positively associated with bread-making quality compared with null subunit at *Glu-A1* locus; the positive effect also holds true for the subunit pairs *17+18*, *7+8*, and *14+15* compared with *7+9*, and *6+8* or *7* at *Glu-B1* locus; and the *Glu-D1* group with the subunit *5+10* had more significantly positive effects on quality parameters than the *2+12* (D' Ovidio and Anderson, 1994; Gupta et al., 1991, 1994; He et al., 2004, 2005; Lawrence et al., 1988). Cornish et al. (1993) reported the gluten alleles of the *Glu-3* pattern *bbb* gave the best extensibility, particularly in combination with *Glu-1* alleles *bba*. Luo et al. (2001) considered allele *d* at *Glu-A3* locus, allele *b* at *Glu-B3* locus, and allele *b* at *Glu-D3* locus had greater quality parameters than their counterpart alleles. Recently, the subunit $Bx7^{OE}$ was found to be strongly correlated with good bread-making quality (Gianibelli et al., 2002; Butow et al., 2003).

SDS-PAGE is atraditional method for the detection of HMW-GS and LMW-GS (Payne, 1987; Vallega and Waines, 1987; Gupta et al., 1991, 1994). However, this method is time consuming, and needs experience to identify the subunits, particularly for the LMW-GS (Zhang et al., 2004). Molecular markers can also discriminate the HMW-GS and LMW-GS alleles, providing a much more convenient tool for rapid genetic analyses (Pagnotta et al., 1995). Many functional markers were developed for the glutenin loci (Ma et al., 2003, Lei et al., 2006, Butow et al., 2003, 2004, D' Ovidio and Anderson, 1994, Francis et al., 1995), and STS markers for PPO, PSY and Wax genes were also developed, and validated with Chinese and CIMMYT wheat cultivars (Nakamura et al., 2002; Liu et al., 2005; Sun et al., 2005; He et al. 2007, 2008, 2009). In addition, RP-HPLC was proved to be an effective method to discriminate subunits $Bx7^{OE}$ and $By8^{*}$ from *Bx7* and *By8*, respectively (Butow et al., 2004; Zhang et al., 2007), a distinction that was difficult to achieve by SDS-PAGE.

The 273 wheat cultivars and advanced lines employed in the present study were from the irrigated wheat breeding program, and they included leading parental lines used in the crossing program, advanced breeding lines and leading cultivars from CIMMYT and targeted national wheat breeding programs in countries such as China, India, and Pakistan. The irrigated wheat breeding program focuses on high yielding environments having irrigation and/or high rainfall, and contributing around 42 million tonnes of wheat production in those countries alone.

The present paper described molecular characterization of 273 CIMMYT wheat cultivars and advanced lines with different HMW and LMW glutenin allelic combinations, and allelic variations at the loci for PPO, PSY and waxy genes. The HMW-GS and LMW-GS of 142 CIMMYT lines were analyzed with SDS-PAGE and RP-HPLC to validate the availability of molecular markers in wheat breeding. In addition, the influences of the allelic variations at *Glu-1* and *Glu-3* loci, and the presence of the 1B. 1R translocation on quality parameters were evaluated as well. The results might provide useful information for wheat improvement at wheat breeding programs using CIMMYT wheat germplasm.

Materials and methods

Plant materials

In total, 273 cultivars and advanced lines from the CIMMYT irrigated wheat breeding program grown in Sonora, Mexico, during the 2005-2006 crop cycle, were used for the analysis of quality attributes, glutenin composition and molecular marker characterization.

Quality tests

Grain samples were milled in a Brabender Senior mill and the resulting flours analyzed for protein content by NIR (N×5.7), SDS-sedimentation volume according to Peña et al. (1990), dough mixing time in the Mixographic, and dough strength (W) and extensibility (P/L ratio) according to AACC methods 54-40A and 54-30A, respectively, (AACC, 1995).

SDS-PAGE analysis

HMW-GS and LMW-GS were separated bysodium-dodecyl-sulfate polyacrylamide gel electrophoresis (SDS-PAGE) based on the extraction method described by Singh et al. (1991), with modifications reported by Liu et al. (2005b) and He et al. (2005). The presence of the 1B. 1R translocation was determined by SDS-PAGE of alcohol-soluble and alcohol insoluble protein extracts, detecting the presence of *Sec-1* secalins in the first test and the presence of the *Glu-B3j* allele in the second.

DNA extraction and PCR reaction

Genomic DNA was extracted from seeds using a modification of the method of Sambrook et al. (1989). PCR reactions were performed in an MJ Research PTC-200 Thermal Cycler in a total volume of 20μl, including 20mM of Tris-HCl (pH 8.4), 20mM of KCl, 100mM of each of dNTPs, 1.5mM of $MgCl_2$, 5pmol of each primer, 1unit of *Taq* DNA polymerase (TIANGEN Biotech Co., Ltd., Beijing) and 50ng of template DNA. Sequences of PCR primers and fragment sizes are shown in Table 1. PCR primers were synthesized by Augct Biotechnology Company (http://www.augct.com). PCR products were stained with ethidium bromide and visualized using UV light.

Table 1 PCR primers of the molecular markers used in the study

Marker/gene	Forward and reverse primers (5′-3′)	Allele	Fragment size (bp)	Reference
Ax2 *	F: ATGACTAAGCGGTTGGTTCTT	*Glu-Ax2* *	1319	Ma et al., 2003
	R: ACCTTGCTCCCCTTGTCTTT			
ZSBy 8F5/R5	F: TTAGCGCTAAGTGCCGTCT	*Glu-By8*	527	Lei et al., 2006
	R: TTGTCCTATTTGCTGCCCTT			
ZSBy 9aF1/R3	F: TTCTCTGCATCAGTCAGGA	*Glu-By9*	707/662	Lei et al., 2006
	R: AGAGAAGCTGTGTAATGCC			
BxFp	F: CGCAACAGCCAGGACAATT	*Glu-Bx17*	675	Butow et al., 2003
	R: AGAGTTCTATCACTGCCTGGT			

(续)

Marker/gene	Forward and reverse primers (5'-3')	Allele	Fragment size (bp)	Reference
MAR	F: CCTCAGCATGCAAACATGCAGC R: CTGAAACCTTTGGCCAGTCATGTC	*Glu-Bx20*	523/560/800	Butow et al., 2004
Dx5	F: CGTCCCTATAAAAGCCTAGC R: AGTATGAAACCTGCTGCGGAC	*Glu-D1d*	450	D' Ovidio and Anderson, 1994
Glu-B3j	F: GGAGACATCATGAAACATTTG R: CTGTTGTTGGGCAGAAAG	*Glu-B3j*	1500	Francis et al., 1995
PPO16	F: TGCTGACCGACCTTGACTCC R: CTCGTCACCGTCACCCGTAT	*Ppo-D1a*	713	He et al., 2007
PPO18	F: AACTGCTGGCTCTTCTTCCCA R: AAGAAGTTGCCCATGTCCGC	*Ppo-A1*	876/685	Sun et al., 2005
PPO29	F: TGAAGCTGCCGGTCATCTAC R: AAGTTGCCCATGTCCTCGCC	*Ppo-D1b*	490	He et al., 2007
MAG264	F: CCAAAGCAAAGCAGGAAACC R: TACCTCGGAGATGACGCTGG	*Wx-A1*	336/317	Liu et al., 2005
Wx-B1	F: CTGGCCTGCTACCTCAAGAGCAACT R: CTGACGTCCATGCCGTTGACGA	*Wx-B1*	425	Nakamura et al., 2002
MAG269	F: CGAGCGGCTACTCAAGAGC R: GGCGGTCATCTGTCATTTCC	*Wx-D1*	1400/800	Liu et al., 2005
YP7A	F: GGACCTTGCTGATGACCGAG R: TGACGGTCTGAAGTGAGAATGA	*Psy-A1*	231/194	He et al., 2008
YP7B	F: GCCACAACTTGAATGTGAAAC R: ACTTCTTCCATTTGAACCCC	*Psy-B1*	151/156/Null	Unpublished
YP7B-3	F: GAGTAAGCCACCCACTGATT R: TCGCTGAGGAATGTACTGAC	*Psy-B1*	884	Unpublished

RP-HPLC analysis

The procedures for RP-HPLC analysis were according to Zhang et al. (2007). The quantity of expressed *Bx7* using the area of each subunit peak with the molecular weight value (Butow et al., 2003). Presence of subunit $Bx7^{OE}$ was assumed when the expression level of *Bx7* was higher than 39%.

Statistical analysis

Analysis of variance was conducted by PROC MIXED in the Statistical Analysis System (SAS Institute, 1997), with genotype clusters indicated by different marker alleles for one quality parameter as a categorical variable to derive the mean value and to test the significant level. Genotype cluster was treated as fixed effect, while genotypes nested in cluster as random. For the traits with only two genotype groups divided by marker alleles for a parameter, contrast command was used for the mean value comparison.

Results and discussion

Allelic variationat the *Ppo-A1*, *Ppo-D1*, *Psy-A1*, *Psy-B1* and waxy loci

Amongthe lines tested with the marker *PPO18* (Table

2), 79.1% possessed allele *Ppo-A1a* and 20.2% contained *Ppo-A1b* (Table 2), which are associated with higher and lower PPO activities, respectively (Sun et al., 2005). Two lines didn't produce any PCR amplification indicating the possibility of a third allele at this locus. At the *Ppo-D1* locus tested by markers *PPO16* and *PPO29*, 83.2% of lines contained allele *Ppo-D1a* and 16.8% had *Ppo-D1b* for lower and higher PPO activities, respectively (Table 2) (He et al., 2007). Low PPO activity is preferred for fresh Asian noodles, and therefore, selection for the alleles *Ppo-A1b* and *Ppo-D1a* is recommended in Chinese wheat breeding program.

Table 2 Allelic frequencies for polyphenol oxidase, phytoene synthase and waxy loci in 273 CIMMYT cultivars and advanced lines based on molecular markers

Allele	Marker	Fragment size (bp)	Number of accessions	Phenotype	Frequency (%)
Ppo-A1a	*PPO18*	685	216	Higher PPO content	79.1
Ppo-A1b		876	55	Lower PPO content	20.2
		—	2		0.7
Ppo-D1a	*PPO16*	713	227	Lower PPO content	83.2
Ppo-D1b	*PPO29*	490	46	Higher PPO content	16.8
Psy-A1a	*YP7A*	194	142	Higher YP content	52
Psy-A1b		231	131	Lower YP content	48
Psy-B1a	*YP7B*	151	155	Medium YP content	56.8
Psy-B1b		156	43	Lower YP content	15.8
Psy-B1d	*YP7B-3*	884	75	Not determined	27.4
Wx-A1a	*MAG264*	336	273	slightly higher amylase content	100.0
Wx-A1b		317	0	slightly lower amylase content	0.0
Wx-B1a	*Wx-B1*	425	204	Higher amylase content	74.7
Wx-B1b		—	69	Lower amylase content	25.3
Wx-D1a	*MAG269*	1400	273	slightly higher amylase content	100.0
Wx-D1b		800	0	slightly lower amylase content	0.0

"—", indicates no amplified PCR product.

At the locus*Psy-A1*, 52.0% of lines had allele *Psy-A1a* and 48.0% contained *Psy-A1b* (Table 2 and Fig. 1), and they are associated with higher and lower YP contents, respectively (He et al., 2008). Using the marker *YP7B* for the gene *Psy-B1*, 56.8% and 15.8%, respectively, of lines generated 151-bp and 156-bp fragments for alleles *Psy-B1a* and *Psy-B1b*, whereas 27.4% of lines possessed allele *Psy-B1d* detected with marker *YP7B-3* (Table 2). The alleles *Psy-B1a* and *Psy-B1b* are associated with medium and lower YP contents, respectively, in the Chinese wheat cultivars. However, no significant differences were detected for mean YP content among another set of CIMMYT wheat lines with different *Psy-B1* genotypes (He et al., 2009). High yellow pigment content is favored for durum wheat pasta, but is considered undesirable for Chinese steamed bread and white noodles (Hailu and Merker, 2007; He et al., 2004). Therefore, selection for the *Psy-A1b* and *Psy-B1d* is encouraged.

For waxy loci, all 273 lines contained the alleles *Wx-A1a* and *Wx-D1a* in the tests with the markers *MAG264* and *MAG269*, respectively. At the *Wx-B1* locus tested by marker *Wx-B1*, 74.7% of lines were predicted to have *Wx-B1a* and 25.3% to have *Wx-B1b* based on the presence and absence, respectively, of the 425-bp PCR fragment (Table 2). Yamamori and Quynh (2000) reported that the *Wx-B1b* induced lower amylase content, which was confirmed by Miura and Sugawara (1996). They also analyzed the effects of GBSS and ranked the single null genotypes as *Wx-B1b* >*Wx-D1b*>*Wx-A1b*. The results for genes *Wx-A1* and *Wx-D1* were similar to those of Yamamori et al. (1994), who found that wheat cultivars from Asia rarely contained *Wx-A1b*. They also reported that only one cultivar (Baihuomai from China) among 1, 960 tested had the allele *Wx-D1b*. Previously, 294 Chinese wheat cultivars were screened by Xu et al. (2005) for waxy protein, and the frequency of allele *Wx-B1b* was 13.2%, slightly lower than the 25.3% found in this study. Allele *Wx-B1b*, associated with a high quality of Chinese noodles, should be given more attention in wheat breeding programs.

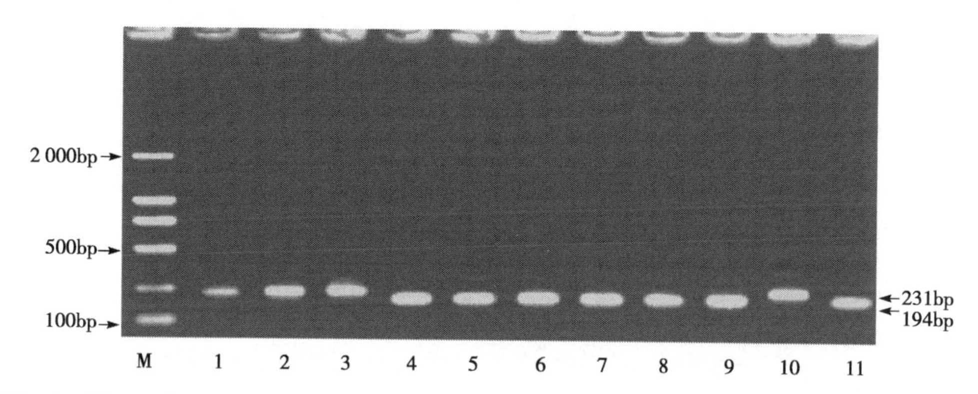

Fig. 1 Electrophoresis of PCR products amplified with *YP7A* in 11 cultivars on a 1.5 % agarose gel, yielding a 194-bp fragment for the allele *Psy-A1a* and 231-bp fragment for the allele *Psy-A1b*, respectively.

M: 2000 bp DNA ladder; 1. TOBA97/PASTOR, 2. WBLL4/KUKUNA//WBLL1, 3. ACHTAR * 3//KANZ/KS85-8-5, 4. WEEBILL1, 5. WBLL1 * 2/KIRITATI, 6. WAXWING, 7. V763.2312/V879.C8.11.11.11 (36) //STAR/3/STAR, 8. KAMB1 * 2/KIRITATI, 9. VORB/FISCAL, 10. Wanmai 33 (CK), 11. Shannong 1355 (CK).

HMW-GS and LMW-GS composition

Among the 142 cultivars and advanced lines tested by molecular markers and SDS-PAGE (Table 3), 90 (63.4%) lines gave a positive result for *Ax2* * locus and 32.4% contained subunit *Bx17* with a 675-bp fragment amplified by marker *BxFp*. The primer set *ZSBy 8F5/R5* specific to *By 8* produced a 527-bp fragment in 11.3% of lines, a 662-bp fragment indicative of subunit *By 9* was generated by marker *ZSBy 9aF1/R3* in 39.4% of lines, and 5 (3.5%) lines contained subunit *Bx20* detected by marker *MAR* with an 800-bp PCR fragment. One hundred and eighteen lines (83.1%) contained subunits *5+10* with a 450-bp PCR fragment (Fig. 2). The allele *Glu-B3j*, associated with the 1B.1R translocation, was detected in 23.3% of lines by marker *AF1/AF4*. These results were consistent with those of SDS-PAGE with one exception. Line TAM200/TUI with the 1A.1R translocation determined by SDS-PAGE was not distinguished from 1B.1R using marker *AF1/AF4*. Four lines with *By 8* * identified by RP-HPLC (Fig. 3) and molecular markers could not be distinguished from *By 8* using SDS-PAGE. Therefore, 20 lines presumed to have the subunits *By 8* or *By 8* * in SDS-PAGE (Vawser and Cornish, 2004) were separated into 16 containing the *By 8* and four with By8 * using the molecular marker *ZSBy 8F5/R5* (Lei et al., 2006). RP-HPLC showed eight lines (HEILO, K6295.4A, SHA5/WEAVER, PFAU/WEAVER * 2//BRAMBLING, WBLL1 * 2/BRAMBLING, WBLL1 * 2/BRAMBLING, NING MAI 9415.16//SHA4/CHIL/3/NING MAI 50, KANCHAN) possessing subunit $Bx7^{OE}$. Among them, HEILO, K6295.4A and SHA5/WEAVER carried $7^{OE}+8$ and the others, $7^{OE}+8^*$ (Fig. 3). The frequency of $7^{OE}+8^*$ in Argentinian and Canadian cultivars is relatively high, and is generally associated with strong gluten type (Marchylo et al., 1992; Gianibelli et al., 2002).

Table 3 Composition of HMW-GS and LMW-GS in 142 CIMMYT cultivars and advanced lines detected by SDS-PAGE, RP-HPLC and molecular markers.

No	Pedigree	Alleles detected by molecular marker (bp)				Subunits/alleles tested by SDS-PAGE and RP-HPLC				
		Glu-A1	*Glu-B1*	*Glu-D1*	1B. 1R	*Glu-A1*	*Glu-B1*	*Glu-D1*	*Glu-A3*	*Glu-B3*
1	SERI/RAYON	.	662	450	.	*1*	*7+9*	*5+10*	*c*	*b*
2	KRONSTAD F2004	1 316	662	450	.	*2 **	*7+9*	*5+10*	*c*	*h*
3	KAMBARA1	1 316	675	450	.	*2 **	*17+18*	*5+10*	*b*	*h*
4	WHEATEAR	.	675	450	.	*1*	*17+18*	*5+10*	*b*	*b*
5	WEEBILL1	1 316	662	450	.	*2 **	*7+9*	*5+10*	*e*	*h*
6	WEEBILL1	1 316	662	450	.	*2 **	*7+9*	*5+10*	*c*	*h*
7	SERI. 1B * 2/3/KAUZ * 2/BOW//KAUZ	.	675	450	.	*1*	*17+18*	*5+10*	*c*	*h*
8	ATTILA * 2/PBW65	1 316	.	450	.	*2 **	*7*	*5+10*	*c*	*h*
9	WAXWING	1 316	.	450	.	*2 **	*7*	*5+10*	*c*	*b*
10	PRL/2 * PASTOR	.	662	450	.	*1*	*7+9*	*5+10*	*d*	*g*
11	ATTILA/3 * BCN//BAV92/3/PASTOR	1 316	662	450	1 500	*2 **	*7+9*	*5+10*	*c*	*j*
12	BABAX//IRENA/KAUZ/3/HUITES	1 316	675	.	.	*2 **	*17+18*	*2+12*	*b*	*b*
13	BABAX/LR42//BABAX * 2/3/KURUKU	1 316	662	450	.	*2 **	*7+9*	*5+10*	*b*	*h*
14	BABAX/LR42//BABAX * 2/3/VIVITSI	1 316	662	450	.	*2 **	*7+9*	*5+10*	*c*	*h*
15	BABAX/LR42//BABAX * 2/3/VIVITSI	1 316	662	450	.	*2 **	*7+9*	*5+10*	*c*	*h*
16	BABAX/LR42//BABAX * 2/3/VIVITSI	1 316	662	450	.	*2 **	*7+9*	*5+10*	*c*	*h*
17	BL2064//SW89-5124 * 2/FASAN/3/TILHI	.	662	450	1 500	*0*	*7+9*	*5+10*	*c*	*j*
18	CAL/NH//H567. 71/3/SERI/4/CAL/NH//H567. 71/5/2 * KAUZ/6/PASTOR	1 316	662	.	1 500	*2 **	*7+9*	*2+12*	*c*	*j*
19	CNO79//PF70354/MUS/3/PASTOR/4/BABAX	.	662	450	.	*1*	*7+9*	*5+10*	*b*	*h*
20	FRET2 * 2/4/SNI/TRAP#1/3/KAUZ * 2/TRAP//KAUZ	1 316	662	450	.	*2 **	*7+9*	*5+10*	*c*	*h*
21	FRET2 * 2/BRAMBLING	.	662	450	.	*1*	*7+9*	*5+10*	*c*	*h*
22	FRET2 * 2/BRAMBLING	.	662	450	.	*1*	*7+9*	*5+10*	*c*	*h*
23	FRET2/TUKURU//FRET2	1 316	662	450	.	*2 **	*7+9*	*5+10*	*b*	*h*
24	FRET2/WBLL1//KAMB1	1 316	675	450	.	*2 **	*17+18*	*5+10*	*c*	*g*
25	IRENA/2 * PASTOR	.	675	450	.	*1*	*17+18*	*5+10*	*c*	*g*
26	KAMB1 * 2/BRAMBLING	1 316	675	450	.	*2 **	*17+18*	*5+10*	*b*	*h*
27	KAMB1 * 2/KIRITATI	.	675	450	.	*1*	*17+18*	*5+10*	*b*	*h*
28	KAUZ//ALTAR 84/AOS/3/MILAN/KAUZ/4/HUITES	1 316	675	450	.	*2 **	*17+18*	*5+10*	*b*	*b*
29	KAUZ//ALTAR 84/AOS/3/MILAN/KAUZ/4/HUITES	1 316	675	450	.	*2 **	*17+18*	*5+10*	*b*	*b*
30	KAUZ//ALTAR 84/AOS/3/MILAN/KAUZ/4/HUITES	1 316	675	450	.	*2 **	*17+18*	*5+10*	*b*	*b*
31	KAUZ/PASTOR//PBW343	.	.	450	1 500	*1*	*7*	*5+10*	*d*	*j*
32	KIRITATI//ATTILA * 2/PASTOR	1 316	675	450	.	*2 **	*17+18*	*5+10*	*c*	*i*
33	KIRITATI//PRL/2 * PASTOR	.	675	450	.	*1*	*17+18*	*5+10*	*e*	*i*
34	KIRITATI/WBLL1	.	675	450	.	*1*	*17+18*	*5+10*	*e*	*i*

（续）

No	Pedigree	Alleles detected by molecular marker (bp)				Subunits/alleles tested by SDS-PAGE and RP-HPLC				
		Glu-A1	*Glu-B1*	*Glu-D1*	1B. 1R	*Glu-A1*	*Glu-B1*	*Glu-D1*	*Glu-A3*	*Glu-B3*
35	MILAN/S87230//BABAX	1 316	662	450	.	2 *	7+9	5+10	*c*	*h*
36	OASIS/SKAUZ//4 * BCN * 2/3/PASTOR	.	675	450	.	1	17+18	5+10	*c*	*g*
37	OASIS/SKAUZ//4 * BCN/3/2 * PASTOR	.	675	450	.	1	17+18	5+10	*c*	*g*
38	OASIS/SKAUZ//4 * BCN/3/PASTOR/4/KAUZ * 2/YACO//KAUZ	1 316	675	450	.	2 *	17+18	5+10	*c*	*g*
39	PFAU/SERI. 1B//AMAD/3/WAXWING	1 316	.	450	.	2 *	7	5+10	*c*	*b*
40	PFAU/WEAVER * 2//BRAMBLING	.	.	450	.	1	$7^{OE}+8^{*}$	5+10	*c*	*g*
41	PFAU/WEAVER * 2//KIRITATI	.	675	450	.	1	17+18	5+10	*c*	*g*
42	PFAU/WEAVER * 2//KIRITATI	.	675	450	.	1	17+18	5+10	*e*	*i*
43	PRINIA/PASTOR	1 316	675	450	.	2 *	17+18	5+10	*c*	*g*
44	SITE/MO//PASTOR/3/TILHI	1 316	675	450	.	2 *	17+18	5+10	*c*	*g*
45	TAM200/PASTOR//TOBA97	.	662	450	1 500	1	7+9	5+10	*c*	*g*
46	THELIN/2 * WBLL1	1 316	662	450	.	2 *	7+9	5+10	*d*	*h*
47	THELIN/3/2 * BABAX/LR42//BABAX	1 316	662	450	.	2 *	7+9	5+10	*c*	*g*
48	TOBA97/PASTOR	.	675	450	.	1	17+18	5+10	*c*	*g*
49	TOBA97/PASTOR	.	675	450	.	1	17+18	5+10	*c*	*g*
50	TUKURU//BAV92/RAYON	1 316	662	450	.	2 *	7+9	5+10	*c*	*i*
51	VORB/FISCAL	1 316	.	450	.	2 *	.	5+10	*c*	*g*
52	WAXWING * 2/4/SNI/TRAP # 1/3/KAUZ * 2/TRAP//KAUZ	1 316	662	450	.	2 *	7+9	5+10	*c*	*b*
53	WAXWING * 2/KIRITATI	1 316	.	450	.	2 *	7	5+10	*c*	*b*
54	WAXWING * 2/KUKUNA	1 316	.	450	.	2 *	7	5+10	*c*	*b*
55	WAXWING * 2/KUKUNA	1 316	.	450	.	2 *	7	5+10	*c*	*b*
56	WAXWING * 2/VIVITSI	1 316	.	450	.	2 *	7	5+10	*d*	*b*
57	WAXWING * 2/VIVITSI	1 316	.	450	.	2 *	7	5+10	*c*	*b*
58	WAXWING/4/SNI/TRAP # 1/3/KAUZ * 2/TRAP//KAUZ	1 316	.	450	.	2 *	7	5+10	*b*	*b*
59	WBLL1 * 2/4/SNI/TRAP # 1/3/KAUZ * 2/TRAP//KAUZ	1 316	662	450	.	2 *	7+9	5+10	*b*	*h*
60	WBLL1 * 2/4/YACO/PBW65/3/KAUZ * 2/TRAP//KAUZ	1 316	662	450	.	2 *	7+9	5+10	*b*	*h*
61	WBLL1 * 2/BRAMBLING	1 316	662	450	.	2 *	7+9	5+10	*c*	*h*
62	WBLL1 * 2/BRAMBLING	1 316	.	450	.	2 *	$7^{OE}+8^{*}$	5+10	*c*	*h*
63	WBLL1 * 2/BRAMBLING	1 316	.	450	.	2 *	$7+8^{*}$	5+10	*c*	*h*
64	WBLL1 * 2/BRAMBLING	1 316	662	450	.	2 *	7+9	5+10	*c*	*h*
65	WBLL1 * 2/CHAPIO	1 316	662	450	.	2 *	7+9	5+10	*c*	*h*
66	WBLL1 * 2/KIRITATI	.	675	450	.	1	17+18	5+10	*c*	*h*
67	WBLL1 * 2/KUKUNA	1 316	662	450	.	2 *	7+9	5+10	*c*	*h*
68	WBLL1 * 2/KUKUNA	.	662	450	.	1	7+9	2+12	*c*	*h*
69	WBLL1 * 2/KURUKU	.	662	450	.	1	7+9	5+10	*c*	*h*

（续）

No	Pedigree	Alleles detected by molecular marker (bp)				Subunits/alleles tested by SDS-PAGE and RP-HPLC				
		Glu-A1	*Glu-B1*	*Glu-D1*	1B. 1R	*Glu-A1*	*Glu-B1*	*Glu-D1*	*Glu-A3*	*Glu-B3*
70	WBLL1 * 2/TUKURU	1 316	662	450	.	*2 **	*7+9*	*5+10*	*c*	*h*
71	WBLL1/3/STAR//KAUZ/STAR/4/BAV92/RAYON	1 316	662	450	.	*2 **	*7+9*	*5+10*	*d*	*b*
72	WBLL1/KUKUNA//KAMB1	1 316	662	450	.	*2 **	*7+9*	*5+10*	*b*	*h*
73	WBLL4/KUKUNA//WBLL1	1 316	662	450	.	*2 **	*7+9*	*5+10*	*d*	*h*
74	BOW/NKT//CBRD/3/CBRD	1 316	527	450	.	*2 **	*7+8*	*5+10*	*d*	*g*
75	CROC _ 1/AE. SQUARROSA (224) //OPATA /3/BJY/COC//PRL/BOW/4/BJY/COC//PRL/BOW	1 316	675	.	.	*2 **	*17+18*	*2+12*	*d*	*d*
76	HEILO	1 316	527	450	.	*2 **	*7^{OE}+8*	*5+10*	*d*	*g*
77	RABE/LAJ3302	1 316	675	450	.	*2 **	*17+18*	*5+10*	*c*	*b*
78	V763. 2312/V879. C8. 11. 11. 11 (36) //STAR/3/STAR	.	527	450	.	*1*	*7+8*	*5+10*	*b*	*g*
79	PAVON F 76	1 316	675	450	.	*2 **	*17+18*	*5+10*	*b*	*h*
80	JUCHI F2000	1 316	662	450	.	*2 **	*7+9*	*5+10*	*e*	*c*
81	KIRITATI	.	675	450	.	*1*	*17+18*	*5+10*	*e*	*i*
82	PFAU/WEAVER * 2//KIRITATI	.	675	450	.	*1*	*17+18*	*5+10*	*c*	*g*
83	TAM200/TUI	1 316	662	450	1 500	*2 **	*7+9*	*5+10*	*e*	*j*
84	HAR3116	.	662	450	.	*0*	*7+9*	*5+10*	*d*	*g*
85	CNDO/R143//ENTE/MEXI _ 2/3/AEGILOPS SQUARROSA (TAUS) /4/WEAVER/5/PASTOR	1 316	527	450	.	*2 **	*7+8*	*5+10*	*c*	*g*
86	K6295. 4A	.	527	.	.	*1*	*7^{OE}+8*	*2+12*	*c*	*i*
87	YANAC	1 316	800	.	.	*2 **	*Bx20*	*2+12*	*b*	*b*
88	PVN//CAR422/ANA/5/BOW/CROW//BUC/PVN/3/YR/4/TRAP#1	1 316	675	.	.	*2 **	*17+18*	*2+12*	*b*	*h*
89	BAU/TNMU	1 316	675	450	1 500	*2 **	*17+18*	*5+10*	*c*	*j*
90	CATBIRD	1 316	662	450	1 500	*2 **	*7+9*	*5+10*	*c*	*j*
91	GONDO	1 316	662	450	1 500	*2 **	*7+9*	*5+10*	*b*	*j*
92	GUAM92//PSN/BOW	.	662	450	1 500	*1*	*7+9*	*5+10*	*d*	*j*
93	IVAN/6/SABUF/5/BCN/4/RABI//GS/CRA/3/AE. SQUARROSA (190)	1 316	662	450	1 500	*2 **	*7+9*	*5+10*	*c*	*j*
94	KAUZ//TRAP#1/BOW	1 316	662	450	1 500	*2 **	*7+9*	*5+10*	*b*	*j*
95	NG8675/CBRD	.	662	450	1 500	*0*	*7+9*	*5+10*	*c*	*j*
96	NING MAI 9558	1 316	662	.	1 500	*2 **	*7+9*	*2+12*	*c*	*j*
97	SHA3/CBRD	.	800	450	.	*0*	*20*	*5+10*	*c*	*g*
98	SHA3/SERI//SHA4/LIRA	.	527	.	1 500	*0*	*7+8*	*2+12*	*e*	*j*
99	SHA5/WEAVER	.	527	450	1 500	*0*	*7^{OE}+8*	*5+10*	*e*	*j*
100	SHA8/GEN	1 316	527	450	1 500	*2 **	*7+8*	*5+10*	*e*	*j*
101	TINAMOU	1 316	675	450	1 500	*2 **	*17+18*	*5+10*	*b*	*j*
102	WUH1/VEE#5//CBRD	1 316	662	450	1 500	*2 **	*7+9*	*5+10*	*c*	*j*
103	ACHTAR * 3//KANZ/KS85-8-4	1 316	675	.	.	*2 **	*17+18*	*2+12*	*b*	*g*

（续）

No	Pedigree	Alleles detected by molecular marker (bp)				Subunits/alleles tested by SDS-PAGE and RP-HPLC				
		Glu-A1	*Glu-B1*	*Glu-D1*	1B. 1R	*Glu-A1*	*Glu-B1*	*Glu-D1*	*Glu-A3*	*Glu-B3*
104	ACHTAR * 3//KANZ/KS85-8-5	1 316	675	.	.	2 *	17+18	2+12	b	g
105	ALTAR 84/AEGILOPS SQUARROSA (TAUS) //OPATA	1 316	675	450	.	2 *	17+18	5+10	b	i
106	KANZ * 4/KS85-8-4	1 316	675	450	.	2 *	17+18	5+10	e	h
107	KANZ * 4/KS85-8-4	1 316	675	450	.	2 *	17+18	5+10	e	h
108	PRL/SARA//TSI/VEE#5	1 316	675	450	.	2 *	17+18	5+10	c	i
109	BH1146 * 3/ALD//BUC/3/DUCULA/4/DUCU-LA	1 316	662	.	.	2 *	7+9	2+12	e	h
110	PF839197/BR35//BR23/3/PASTOR	.	675	450	1 500	1	17+18	5+10	c	j
111	TNMU/6/PEL74144/4/KVZ//ANE /MY64/3/PF70354/5/BR14/7/BR35	.	527	450	.	1	7+8	5+10	b	g
112	ALTAR 84/AE. SQ//OPATA/3/2 * WH 542	1 316	662	450	1 500	2 *	7+9	5+10	d	j
113	CHUM18/BORL95//CBRD	1 316	662	450	1 500	2 *	7+9	5+10	c	j
114	CROC _ 1/AE. SQUARROSA (205) //KAUZ/ 3/SASIA	1 316	662	450	1 500	2 *	7+9	5+10	b	j
115	CROC _ 1/AE. SQUARROSA (205) //KAUZ/ 3/SASIA	1 316	662	450	1 500	2 *	7+9	5+10	b	j
116	KAUZ//ALTAR 84/AOS/3/MILAN/KAUZ	1 316	662	450	1 500	2 *	7+9	5+10	c	j
117	SW89. 3064//CMH82. 17/SERI	1 316	675	450	1 500	2 *	17+18	5+10	c	j
118	W462//VEE/KOEL/3/PEG//MRL/BUC	1 316	675	.	1 500	2 *	17+18	2+12	c	j
119	W485/HD29	1 316	527	.	.	2 *	7+8	2+12	c	g
120	CNDO/R143//ENTE/MEXI _ 2/3/AEGILOPS SQUARROSA (TAUS) /4/WEAVER/5/2 * PASTOR	.	675	450	.	1	17+18	5+10	c	g
121	PASTOR/TEERI	1 316	527	450	.	2 *	7+8	5+10	c	g
122	PASTOR/TEERI	.	527	450	.	1	7+8	5+10	c	g
123	PASTOR/TEERI	.	527	450	.	1	7+8	5+10	c	g
124	ENEIDA F94	1 316	662	450	.	2 *	7+9	5+10	b	b
125	HUW468	1 316	675	.	.	2 *	17+18	2+12	b	i
126	INQALAB 91	1 316	675	.	.	2 *	17+18	2+12	c	g
127	INQALAB 91 * 2/KUKUNA	1 316	675	.	.	2 *	17+18	2+12	c	g
128	INQALAB 91 * 2/TUKURU	.	675	.	.	1	17+18	2+12	c	g
129	KANCHAN	1 316	.	.	.	2 *	$7^{OE}+8^{*}$	2+12	d	i
130	PBW343	.	.	450	1 500	1	7	5+10	c	j
131	PBW343 * 2/KHVAKI	.	.	450	1 500	1	7	5+10	c	j
132	PBW343 * 2/KUKUNA	.	.	.	.	1	7	5+10	c	b
133	PBW343 * 2/TUKURU	.	662	450	.	1	7+9	5+10	c	i
134	58769	.	800	450	1 500	1	20	5+10	c	j
135	SW 8488 (W)	.	527	.	1 500	0	7+8	5+12	d	j
136	SW00-91382	.	662	.	1 500	0	7+9	2+12	c	j

（续）

No	Pedigree	Alleles detected by molecular marker (bp)				Subunits/alleles tested by SDS-PAGE and RP HPLC				
		Glu-A1	*Glu-B1*	*Glu-D1*	1B. 1R	*Glu-A1*	*Glu-B1*	*Glu-D1*	*Glu-A3*	*Glu-B3*
137	SW02-90137	.	527	.	.	*0*	*7+8*	*2+12*	*c*	*b*
138	SW03-81497	.	527	450	.	*1*	*7+8*	*5+10*	*c*	*b*
139	SW22725	.	662	450	.	*0*	*7+9*	*5+10*	*c*	*g*
140	YUNMAI 47	1 316	662	450	1 500	*2 **	*7+9*	*5+10*	*c*	*j*
141	80. 8	.	800	.	.	*0*	*20*	*2+12*	*c*	*i*
142	SW2148	.	800	.	1 500	*1*	*20*	*2+12*	*d*	*j*

The 1 316-bp, 527-bp, 662-bp, 675-bp, 800-bp, 450-bp, and 1 500-bp fragments detected by molecular markers indicate the presence of subunits *Ax2 **, *By8*, *By9*, *B1x17*, *Bx20*, *Dx5*, and 1B. 1R translocation, respectively. The line of number 83 was 1A. 1R translocation type.

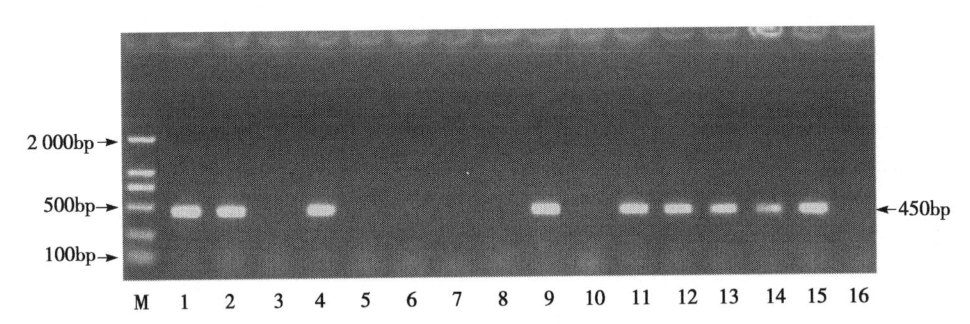

Fig. 2 Electrophoresis of PCR products amplified with the STS marker for subunit *5+10* in some CIMMYT cultivars and lines on a 1. 5% agarose gel.

M: 2000 bp DNA ladder; 1. PASTOR/TEERI, 2. ENEIDA F94, 3. HUW468, 4. PBW343, 5. INQALAB 91, 6. INQALAB 91 * 2/KUKUNA, 7. KANCHAN, 8. PBW343 * 2/KUKUNA, 9. PBW343 * 2/KHVAKI, 10. SW 8488 (W), 11. 58769, 12. PBW343 * 2/TUKURU, 13. SW03-81497, 14. SW22725, 15. CS Zhongyou 9507, 16. CS Chinese Spring.

Entries 1, 2, 4, 9, 11, 12, 13, 14 and 15 contain the subunit 5+10.

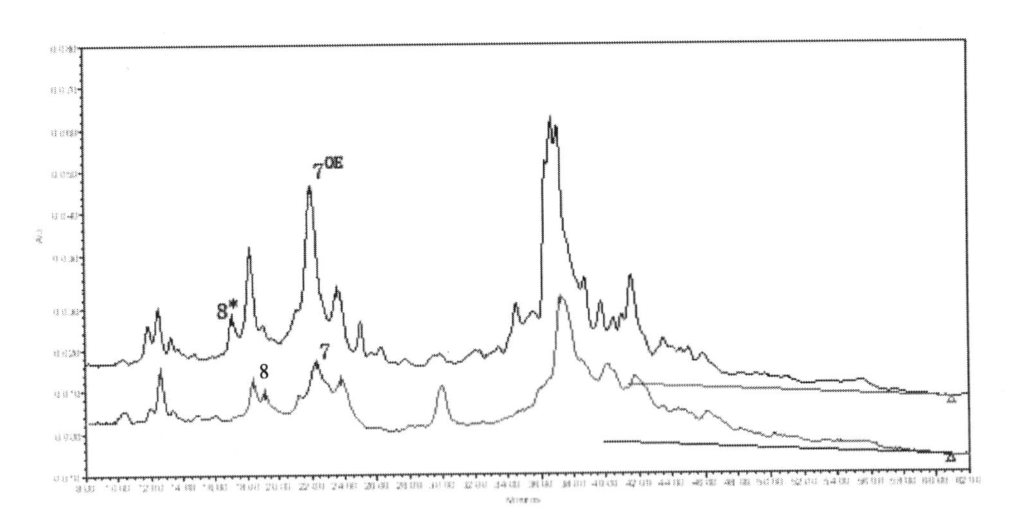

Fig. 3 Chromatograms for the identification of the subunits *Bx7*, *Bx7*^{OE}, *By8* and *By8** using RP-HPLC

In glutenin analyses using SDS-PAGE and 14-15% acrylamide running gels at CIMMYT, some genotypes showed subunit *Bx7* with a slightly slower mobility than the 'normal' one; this is called *Bx7**. RP-HPLC showed that some lines with *Bx7* or *Bx7** had the same peak resolution. Both supposed allelic variants were therefore considered to be subunit Bx*7*.

The frequencies of HMW-GS and LMW-GS among 142 lines characterized by SDS-PAGE (Table 4) indicated three allelic variants at the *Glu-A1* locus, viz., 63.4% *Ax2**, 28.9% *Ax1* and 7.7% *Null*; . nine allelic variants at *Glu-B1*, viz., *17+18*, *7+9*, *7+8*, *7*, *13+16*, *7+8**, $7^{OE}+8$, $7^{OE}+8^*$, and *20* at frequencies 32.4%, 39.4%, 9.2%, 9.2%, 0.7%, 0.7%, 2.1%, 2.1%, and 3.5%, respectively; and three subunits at *Glu-D1* viz., *5 + 10* (83.1%) (Fig. 2), *2+12* (16.3%), and *5+12* present in only one line. Four allelic variations were detected for *Glu-A3*, viz., 59.6% of lines with *Glu-A3c*, 20.6% with *Glu-A3b*, 10.6% with *Glu-A3d*, and 9.2% with *Glu-A3e*. For the *Glu-B3* locus, 7 subunits were found viz., *Glu-B3h* (26.2%), *Glu-B3g* (24.8%), *Glu-B3j* (22.7%), *Glu-B3b* (15.6%), *Glu-B3i* (9.2%), *Glu-B3c* (0.7%) and *Glu-B3d* (0.7%).

Effect of different glutenin subunits on quality parameters

The mean values of quality parameters of the lines grouped by individual glutenin subunits are shown in Table 4. At locus *Glu-A1*, the genotypic groups possessing subunits *Ax1* and *Ax2** showed a significantly higher values on SDS-SV, SPI, mixing time, and W than the group with the subunit *Null*. No significant difference was found between subunits *Ax1* and *Ax2** for the measured quality parameters. These results were similar to those of Sontag-Strohm et al. (1996) based on 95 F_6 lines from a single cross, and also those of Luo et al. (2001) who studied the quality parameter SDS-SV in five crosses among six New Zealand lines. Subunits *Ax1* and *Ax2** therefore have positive effects on the dough strength parameters and rheological properties.

Table 4 Allele frequencies and statistical analysis of the effects of HMW-GS and LMW-GS on quality parameters in the 142 lines

Locus	Subunit	Number	Frequency (%)	SDS-SV (ml)	SPI	Mixing time (min)	W (10^{-4}J)	P/L
Glu-A1	1	41	28.9	16.7a	1.6a	3.1a	360.1a	1.4b
	2*	90	63.4	16.7a	1.6a	3.0a	345.6a	1.4b
	null	11	7.7	10.8b	1.0b	2.0b	202.6b	1.6a
Glu-B1	7	13	9.2	14.2c	1.3b	2.6b	266.3b	1.2b
	7+8	20	14.0	16.7ab	1.5b	2.9b	392.3a	1.3b
	7+9	57	39.4	15.4bc	1.5b	2.9b	300.9b	1.6ab
	20	5	3.5	12.6d	1.2c	1.7c	190.8c	1.8a
	17+18	46	32.4	18.1a	1.7a	3.4a	404.3a	1.2b
Glu-D1	2+12	23	16.3	14.5b	1.3b	2.1b	260.0b	1.4a
	5+10	118	83.1	16.5a	1.6a	3.2a	354.8a	1.4a
Glu-A3	b	30	20.6	17.4a	1.6a	3.3a	383.0a	1.3a
	c	84	59.6	15.9b	1.5b	3.0a	326.8ab	1.4a
	d	15	10.6	15.9b	1.5b	2.9a	313.b	1.3a
	e	13	9.2	15.3b	1.4b	2.9a	337.6ab	1.4a
Glu-B3	b	22	15.6	16.5b	1.7a	3.0b	356.6b	1.2b
	g	35	24.8	18.6a	1.7a	3.4a	431.1a	1.2b
	h	37	26.2	17.2b	1.6b	3.2ab	334.2b	1.4b
	i	13	9.2	16.5b	1.5b	2.9b	343.1b	1.2b
	j	33	22.7	11.9c	1.1c	2.3c	234.3c	1.8a
1B.1R	non-1B.1R	109	77.3	17.4a	1.6a	3.2a	368.6a	1.3b
	1B.1R	32	22.5	11.9b	1.1b	2.6b	234.3b	1.8a

Statistical analysis in this study did not include the subunits with very small frequencies.

Different letters following the values of various quality parameters indicate significant differences at $P = 0.05$.

At *Glu-B1* locus, the genotypic group possessing subunit *17 + 18* showed superior values than those with alternative subunits on SPI and mixing time. Subunits *17+18* and *7+8* showed more significantly positive effects than subunits *7* and *20* on SDS-SV and W, and subunits *7 + 9* showed an intermediate effect on all gluten strength-related traits, whereas subunits *7+9* and *20* had negative effects on the dough extensibility value P/L. Gupta et al. (1991, 1994) found that subunit *17* had positive effects on dough strength and extensibility, whereas Shewry et al. (2003) reported that subunit *20* showed negative effects on the dough strength and rheological properties, consistent with the present results. Subunit *7+8* was also considered to have positive effects on bread-making quality (Pogna et al., 1990). Bedo et al. (1998) found that subunit *7+9* resulted in low SDS-SV, again confirmed in this study. In contrast, other reports showed that subunit *17* contributed positive effects on dough properties (Peña et al., 2005), and Branlard and Dardevet (1985) found the Zeleny sedimentation value was positively correlated with subunits *7+9*. The different results could be attributed to the different wheat germplasm used in the studies.

The *Glu-D1* group with the subunit *5 + 10* had more significantly positive effects on SDS-SV, SPI, mixing time, and W than the *2+12*, in agreement with previous studies (Payne, 1987; Luo et al., 2001), and confirming that subunits *5+10* is better for baking quality than subunits *2 + 12*. No difference in the P/L ratio was found between the two *Glu-D1* genotypic groups.

The *Glu-A3* group with *Glu-A3b* showed more positive effects on SDS-SV and SPI than the other *Glu-A3* groups, whereas no significant differences were observed regarding the effects of different allelic variants on mixing time, W and P/L. Luo et al. (2001) found that alleles *Glu-A3d* and *Glu-A3e* showed no significant differences in SDS-SV and mixing time. However, these authors found *Glu-A3d* to be significantly better than its alternative alleles in Chinese wheat cultivars, and considered *Glu-A3d* as a desirable subunit with positive effects on wheat quality (Luo et al., 2001; He et al., 2005). In the present study, 10 of 14 lines with the *Glu-A3d* carried undesirable subunits *7*, *20* and *7 + 9*, and 5 lines contained undesirable subunit *Glu-B3j*. These associations may have contributed to their weak dough strength and inferior rheological properties.

The *Glu-B3* group possessing allele *Glu-B3g* exhibited significantly positive effects on SDS-SV and W compared with the alternative alleles. The effects of *Glu-B3c* and *Glu-B3d* were not considered because of small numbers of lines. *Glu-B3* genotypic groups possessing alleles *Glu-B3b*, *Glu-B3h*, *or Glu-B3i* showed intermediate to high values for SDS-SV, SPI, mixing time and W. Previous studies showed that the allele *Glu-B3b* gave high quality values, and *Glu-B3g* was considered as a desirable allele in Chinese wheats (Luo et al., 2001). The latter is consistent with the present study indicating that *Glu-B3g* has good effects on dough strength and rheological properties. Genotypic groups *Glu-B3g*, *Glu-B3b*, *Glu-B3h*, *and Glu-B3i* showed similar low (acceptable) P/L values. *Glu-B3j*, associated with the presence of the 1B. 1R translocation, showed the most negative effect on all quality parameters evaluated in this study.

All normal 1BL. 1BS lines were significantly better than 1B. 1R translocation genotypes in SDS-SV, SPI, mixing time, and W, in agreement with the results of He et al. (2005) and Liu et al. (2005b) in studies of SDS-SV and mixing time in Chinese wheat lines. The negative effects of the 1B. 1R translocation on end-use quality can be compensated by utilizing the more desirable glutenin subunits in the same genetic backgrounds (Graybosch., 2001; Peña et al., 1990). Within the 1B. 1R translocation group there was variability in all gluten-quality related parameters (data not shown), indicating that subunits such as *7+8*, *17+18* and other *Glu-1* and *Glu-3* glutenin subunits and/or their combinations, would improve 1B. 1R translocation lines, as in entries 100, 101, 110, and 116, possessing at least intermediate gluten strength with moderate extensibility. Eight different *Glu-1*/*Glu-3* loci-combinations were analyzed in relation to mean values for the parameters SDS-SV, SPI, mixing time, W and P/L using data from the 142 lines (Table 5). Groups combining *Glu-1* subunits *1 or 2**, with *17+18* and *5+10* with any of the alleles *Glu-A3b*, *Glu-A3c*, or *Glu-A3e*, and any of the *Glu-B3* alleles *Glu-B3b*, *Glu-B3c*, *Glu-B3g*, or *Glu-B3i*, gave high mean values on SDS-SV, SPI, mixing time, and W. The groups combining *1 or 2** with *7+9*, *5+10*, either *Glu-A3b* or *Glu-A3d*, and any of *Glu-B3b*, *Glu-B3c*, *Glu-B3g*, or *Glu-B3i*, showed a intermediate values for all quality parameters, whereas the group combining of *2**, *7+9*, *2+12*, *Glu-A3c*, and *Glu-B3j*, exhibited inferior values for SDS-SV, SPI, mixing time, and W. It is important to note that the groups possessing *1* or *2**, *17+18*, *2+12*, *Glu-A3b*, and the group with *1* or *2**, *7+9*, *5+10*, with any of *Glu-B3b*, *Glu-B3c*, *Glu-B3g*, or *Glu-B3i* showed the lowest P/L ratio, meaning the greatest gluten extensibility. The overall results suggested that the combination of *1 or 2**, *17+18*, *5+10* was the best one to achieve superior overall gluten visco-elastic properties, in agreement with Lawrence et al. (1988), whereas allele *Glu-B3j* showed the largest negative effect on dough quality.

Table 5 Statistical analysis of the effects of selected genotypes quality parameters among 142 lines

Genotype					Number of accessions	Mean value				
Glu-A1	*Glu-B1*	*Glu-D1*	*Glu-A3*	*Glu-B3*		SDS-SV	SPI	mixing time	W	P/L
*1/2**	*17+18*	*5+10*	b	—	10	19.0a	1.8a	3.5ab	412.1abc	1.3bcd
*1/2**	*17+18*	*5+10*	c	—	17	18.9a	1.8a	3.7ab	438.9ab	1.3bcd
*1/2**	*17+18*	*5+10*	e	—	6	18.6ab	1.8a	3.8a	461.4a	1.3bcd
*1/2**	*17+18*	*2+12*	b	—	5	15.7c	1.5b	2.6cd	338.6bcd	1.0cd
*1/2**	*7+9*	*5+10*	b	—	6	16.9abc	1.7ab	3.6ab	348.2bcd	1.7ab
*1/2**	*7+9*	*5+10*	d	—	5	16.4bc	1.6ab	3.6ab	306.0de	1.0d
*2**	*7+9*	*5+10*	c	j	12	12.3d	1.1c	2.2d	224.8e	1.9a
*1/2**	*7+9*	*2+12*	c	h	16	17.0abc	1.7ab	3.1bc	319.6cde	1.5bc

Different letters following the values of various quality parameters indicate significant differences at $P=0.05$.
"—", indicates that any *Glu-B3* allele.

Conclusion

Wheat germplasm from CIMMYT shows genetic diversity for most of the quality-related traits that were analyzed. However, there was some predominance of specific genes or alleles, especially with respect to PPO (*Ppo-A1a*, *Ppo-D1a*) and starch-related genes *Wx-A1*, *Wx-B1*, and *Wx-D1* (*Wx-A1a*, *Wx-B1a*, and *Wx-D1a*). The frequencies of alleles *Psy-A1a* and *Psy-A1b* were 52.0% and 48.0%, respectively, and those of alleles *Psy-B1a*, *Psy-B1b* and *Psy-B1d* were 56.8%, 15.8% and 27.4%, respectively. Specific alleles at the complex *Glu-1* and *Glu-3* loci, and their combinations, contribute differently to gluten visco-elastic properties and therefore, to bread making and noodle making qualities for which different allelic effects must be considered. Although the presence of the 1B.1R translocation showed significantly negative effects on quality parameters, we confirmed that certain *Glu-1/Glu-3* allelic combinations compensate, at least in part, for the negative effects associated with the translocation.

Acknowledgments

The authors are very grateful to the critical review of this manuscript by Prof. R. A. McIntosh, Plant Breeding Institute, University of Sydney. This study was supported by the National Science Foundation of China (30671301), National Basic Research Program (2009CB118300), National 863 Program (2006AA10Z1A7 and 2006AA100102), the International Collaboration Project from the Ministry of Agriculture (2006-G2), and the earmarked fund for Modern Agro-industry Technology Research System.

References

AACC (1995) Approved Methods of the American Association of the Cereal Chemists, 9th edition. St. Paul, MN, USA.

Bedo Z, Vida G, Lang L, Karsai I, (1998) Breeding for bread-making quality using old Hungarian wheat varieties. Euphytica. 100: 179-182.

Branlard G, Dardevet M, (1985) Diversity of grain protein and bread wheat quality. II. Correlation between high-molecular-weight subunits of glutenin and flour quality characteristics. J Cereal Sci. 3: 345-354.

Butow BJ, Ma W, Gale KR, Cornish GB, Rampling L, Larroque O, Morell MK, Békés F (2003) Molecular discrimination of Bx7 alleles demonstrates that a highly expressed high-molecular-weight glutenin allele has a major impact on wheat flour dough strength. Theor Appl Genet. 107: 1524-1532.

Butow BJ, Gale KR, Ikea J, Juhász A, Bedö Z, Tamás L, Gianibelli MC (2004) Dissemination of the highly expressed *Bx7* glutenin subunit (*Glu-B1al* allele) in wheat as revealed by novel PCR markers and RP-HPLC. Theor Appl Genet. 109: 1525-1535.

Cornish GB, Burridge PM, Palmer GA, Wrigley CW (1993) Mapping the origins of some HMW and LMW glutenin subunit alleles in Australian germplasm. Proc 42nd Aust Cereal Chem Conf, Sydney, pp 255-260.

D' Ovidio R, Anderson OD (1994) PCR analysis to distinguish between alleles of a member of a multigene family correlated with wheat bread-making quality. Theor Appl Genet. 88: 759-763.

Francis HA, Leitch AR, Koebner RMD (1995) Conversion of a RAPD-generated PCR product, containing novel dispersed repetitive element, into a fast assay for the presence of rye chromatin in wheat. Theor Appl Genet. 90: 636-642.

Gianibelli MC, Echaide M, Larroque OR, Carrillo JM, Dubcovsky J (2002) Biochemical and molecular characterization of *Glu-1* loci in Argentinean wheat cultivars. Euphytica. 128: 61-73.

Graybosch RA (2001) Uneasy unions: quality effects of rye chromatin transfers to wheat. J Cereal Sci. 33: 3-16.

Gupta RB, Bekes F, Wrigley CW (1991) Prediction of physical dough properties from glutenin subunit composition in bread wheats: Correlation studies. Cereal Chem. 68: 328-333.

Gupta RB, Paul JG, Cornish GB, Palmer GA, Bekes F, Rathjen AJ (1994) Allelic variation at glutenin subunit and gliadin loci, *Glu-1*, *Glu-3* and *Gli-1*, of common wheats. 1. Its additive and interaction effects on dough properties. J Cereal Sci. 19: 9-17.

Hailu F, Merker A (2007) Variation in gluten strength and yellow pigment in Ethiopian tetraploid wheat germplasm. Genetic Resources and Crop Evolution, 10.1007/s10722-007-9233-6.

He XY, He ZH, Zhang LP, Sun DJ, Morris CF, Fuerst EP, Xia XC (2007) Allelic variation of polyphenol oxidase (PPO) genes located on chromosomes 2A and 2D and development of functional markers for the PPO genes in common wheat. Theor Appl Genet. 115: 47-58.

He XY, He ZH, Ma W, Appels R, Xia XC (2009) Allelic variants of phytoene synthase 1 (*Psy1*) genes in Chinese and CIMMYT wheat cultivars and development of functional markers for flour colour. Mol Breed. 23: 553-563.

He XY, Zhang YL, He ZH, Wu YP, Xiao YG, Ma CX, Xia XC (2008) Characterization of a phytoene synthase 1 gene (*Psy 1*) located on common wheat chromosome 7A and development of a functional marker. Theor Appl Genet. 116: 213-221.

He ZH, Yang J, Zhang Y, Quail KJ, Peña RJ (2004) Pan bread and dry white Chinese noodle quality in Chinese winter wheats. Euphytica. 139: 257-267.

He ZH, Liu L, Xia XC, Liu JJ, Peña RJ (2005) Composition of HMW and LMW glutenin subunits and their effects on dough properties, pan bread, and noodle quality of Chinese bread wheats. Cereal Chem. 82: 345-350.

Heisey PW, Lantican MA, Dubin HJ (2002) Impacts of international wheat breeding research in developing countries, 1966-97. CIMMYT, Mexico, D. F.

Lawrence GJ, MacRitchie F, Wrigley CW (1988) Dough and baking quality of wheat lines deficient in glutenin subunits controlled by the *Glu-A1*, *Glu-B1* and *Glu-D1* loci. J Cereal Sci. 7: 109-112.

Lei ZS, Gale KR, He ZH, Gianibelli C, Larroque O, Xia XC, Butow BJ, Ma W (2006) Y-type gene specific markers for enhanced discrimination of high-molecular weight glutenin alleles at the *Glu-B1* locus in hexaploid wheat. J Cereal Sci. 43: 94-101.

Liu YC, Zhu HL, Cheng SH, Ma ZQ (2005a) PCR-based molecular markers for Wx-A1 and Wx-D1 genes of wheat. J Triticeae. Crops 25: 1-5.

Liu L, He ZH, Yan J, Zhang Y, Xia XC, Peña RJ (2005b) Allelic variation at the *Glu-1* and *Glu-3* loci, presence of the 1B. 1R translocation, and their effects on mixographic properties in Chinese bread wheats. Euphytica. 142: 197-204.

Luo C, Griffin WB, Branlard G, McNeil (2001) Comparison of low- and high molecular-weight wheat glutenin alleles effects on flour quality. Theor Appl Genet. 102: 1088-1098.

Ma W, Zhang W, Gale KR (2003) Multiplex-PCR typing of high molecular weight glutenin alleles in wheat. Euphytica. 134: 51-60.

Marchylo BA, Lukow OM, Kruger JE (1992) Quantitative variation in high molecular weight glutenin subunit 7 in some Canadian wheats. J Cereal Sci. 15: 29-37.

Miura H, Sugawara A (1996) Dosage three wx genes on amylose synthesis in wheat endosperm. Theor Appl Genet. 93: 1066-1070.

Nakamura T, Yamamori M, Hirano H, Hidaka S (1993) Identification of three Wx proteins in wheat (*Triticum aestivum* L.). Biochem Genet 31: 75-78.

Nakamura T, Vrinten P, Saito M, Konda M (2002) Rapid classification of partial waxy wheats using PCR-based markers. Genome. 45: 1150-1156.

Payne PI (1987) Genetics of wheat storage proteins and the effect of allelic variation on bread-making quality. Annu. Rev. Plant. Physiol. 38: 141-153.

Pagnotta MA, NevoE, Beiles A, Porceddu E (1995) Wheat storage proteins: glutenin diversity in wild emmer, *Triticum dicoccoides*, in Israel and Turkey. 2. DNA diversity detected by PCR. Theor Appl Genet. 91: 409-414.

Peña RJ, Amaya A, Rajaram S, Mujeeb-Kazi A (1990) Variation in quality characteristics associated with some spring 1B. 1R translocation wheats. J Cereal Sci. 12: 105-112.

Peña E, Bernardo A, Soler C, Jouve N (2005) Relationship between common wheat (*Triticum aestivum* L.) gluten proteins and dough rheological properties. Euphytica. 143: 169-177.

Pogna NE, Autran JC, Mellini F, Lafiandra D, Feillet P (1990) Chromosome 1B-encoded gliadins and glutenin subunits in Durum wheat: genetics and relationship to gluten strength. J Cereal Sci. 11: 15-34.

Sambrook J, Fritsch EF, ManiatisT (1989) Molecular Cloning: a Laboratory Manual. Cold Spring Harbour Laboratory Press, New York.

Shewry PR, Gilbert SM, Savage AWJ, Tatham AS, Wan YF, Belton PS, Wellner N, D' Ovidio R, Bekes F, Halford NG (2003) Sequence and properties of HMW subunit *1Bx20* from pasta wheat (*Triticum durum*) which is associated with poor end use properties. Theor Appl Genet. 106: 744-750.

Singh NK, Shepherd KW, Cornish GB (1991) A simplified SDS-PAGE procedure for separating LMW subunits of glutenin. J Cereal Sci. 14: 203-208.

Sontag-Strohm T, Payne PI, Salovaara H (1996) Effect of allelic variation of glutenin subunits and gliadins on baking quality in the progeny on two biotypes of bread wheat cv. Ulla. J Cereal Sci. 24: 115-124.

Sun DJ, He ZH, Xia XC, Zhang LP, Morris CF, Appels R, Ma WJ, Wang H (2005) A novel STS marker for polyphenol oxidase activity in bread wheat. Molecular Breeding. 16: 209-218.

Vallega V, Waines JG (1987) High molecular weight glutenin subunit variation in *Triticum turgidum* var. *dicoccum*. Theor Appl Genet. 74: 706-710.

Vawser MJ, Cornish GB (2004) Over-expression of HMW glutenin subunit *Glu-B17x* in hexaploid wheat varieties (*Triticum aestivum*). Australian Journal of Agricultural Research. 55: 577-588.

Weegels PL, Hamer RJ, Schofield JD (1996) Functional properties of wheat glutenin. J Cereal Sci. 23: 1-18.

Xu ZH, Xia LQ, Chen XM, Xia XC, He ZH (2005) Analysis of waxy proteins in Chinese winter wheat cultivars using SDS-PAGE and molecular markers. Scientia Agricultura Sinica. 38: 1514-1521.

Yamamori M, Nakamura T, Kuroda A (1992) Variations in the content of starch-granule bound protein among several Japanese cultivars of common wheat (*Triticum aestivum* L.). Euphytica. 64: 215-219.

Yamamori M, Nakamura T, Endo TR, Nagamine T (1994) Waxy protein deficiency and chromosomal location of coding genes in common wheat. Theor Appl Genet. 89: 179-184.

Yamamori M, Quynh NT (2000) Differential effects of Wx-A1, -B1 and -D1 protein deficiencies on apparent amylose content and starch pasting properties in common wheat. Theor Appl Genet. 100: 32-38.

Zhang W, Gianibelli MC, Rampling LR, Gale KR (2004) Characterization and marker development for low molecular weight glutenin genes from *Glu-A3* alleles of bread wheat (*Triticum aestivum* L). Theor Appl Genet. 108: 1409-1419.

Zhang PP, He ZH, Chen DS, Zhang Y, Larroque OR, Xia XC (2007) Contribution of common wheat protein fractions to dough properties and quality of northen-style Chinese steamed bread. J Cereal Sci. 46: 1-10.

Zhao XC, Batey IL, Sharp PJ, CrosbieG, Barclay I, Wilson R, Morel MK, Appels R (1998) A single genetic locus associated with starch granule properties and noodle quality in wheat. J Cereal Sci. 27: 7-13.

Molecular detection of high-and low-molecular-weight glutenin subunit genes in common wheat cultivars from 20 countries using allele-specific markers

H. Jin[1], J. Yan[2], R. J. Peña[3], X. C. Xia[1], A. Morgounov[3], L. M. Han[1], Y. Zhang[1], and Z. H. He[1,3]

[1] *Institute of Crop Science, National Wheat Improvement Centre/The National Key Facility for Crop Gene Resources and Genetic Improvement, Chinese Academy of Agricultural Sciences* (*CAAS*), 12 *Zhongguancun South Street*, *Beijing* 100081, *China*; [2] *Cotton Research Institute*, *Chinese Academy of Agricultural Sciences* (*CAAS*), *Huanghedadao*, *Anyang* 455000, *Henan*, *China*; [3] *International Maize and Wheat Improvement Centre* (*CIMMYT*), *Apdo. Postal* 6-641, 06600 *Mexico*, *D. F.*, *Mexico*

Abstract: The composition and quantity of high- and low-molecular-weight glutenin subunits (HMW-GS and LMW-GS) plays an important role in determining the end-use quality of wheat products. In the present study, 718 wheat cultivars and advanced lines from 20 countries were characterized for the HMW-GS and LMW-GS with allele-specific molecular markers. For the *Glu-A1* locus, 311 cultivars (43.3%) had the allele Ax2*, which predominated in cultivars from Canada (83.3%), Romania (91.7%), Russia (72.2%) and USA (72.2%). At *Glu-B1* locus, 197 cultivars (27.4%) contained the By8 subunit and its frequency was higher in Japanese (60.0%) and Romanian (62.5%) genotypes than in those from other countries; 264 cultivars (36.8%) carried the By9 subunit, mostly existing in the cultivars from Austria (100.0%), Russia (72.2%), and Serbia (72.7%); the By16 subunit was present in 44 cultivars (6.1%), with a relatively high percentage in Chile (19.5%), whereas almost no cultivars from other countries had this subunit; the frequency of $Bx7^{OE}$ was 3.1%, and was found only in cultivars from Argentina (12.1%), Australia (4.1%), Canada (25.0%), Iran (20.0%), and Japan (30.0%). There were 446 genotypes (62.1%) with the subunit Dx5 at the *Glu-D1* locus; high frequencies of Dx5 occurred in cultivars from Hungary (90.0%), Romania (95.8%), and Ukraine (92.3%). At the *Glu-A3* locus, the frequencies of *Glu-A3a*, *b*, *c*, *d*, *e*, *f*, and *g* were 2.9, 6.8, 53.2, 12.8, 7.7, 13.8, and 2.4%, respectively. *Glu-A3a* was detected only in the cultivars from Bulgaria (13.3%), China (12.2%), Germany (2.7%), Iran (6.7%), Mexico (14.3%), Turkey (4.7%), and USA (5.1%); the high frequencies of superior alleles *Glu-A3b* and *d* were found in cultivars from Australia (39.7%) and France (24.5%); *Glu-A3c* was widely distributed in cultivars from all the countries; the high frequencies of *Glu-A3e*, *f* and *g* were detected in cultivars from Argentina (33.3%), Canada (29.2%), and Hungary (20.0%). At the *Glu-B3* locus, *Glu-B3a*, *b*, *c*, *d*, *e*, *f*, *g*, *h* and *i* were present in frequencies of 0.4, 22.3, 0.3, 2.8, 1.9, 3.9, 27.2, 18.8, and 7.1%, respectively. *Glu-B3a* was detected only in cultivars from Argentina (3.0%) and Ukrainia (15.4%) cultivars; high frequencies of *Glu-B3b* and *d* were found in the cultivars from Romania (62.5%) and Mexico (14.3%); *Glu-B3c* was detected only in Romanin (8.3%)

Published in Crop & Pasture Science, 2011, 62 (9): 746-754

genotypes; Frequencies of *e*, *f*, *h* and *i* were high in cultivars from Austria (40.0%), China (14.3%), USA (43.0%), and Argentina (33.3%); *Glu-B3g* was mostly detected in the cultivars from Germany (69.3%), Norway (77.3%), and Serbia (63.6%). The frequency of the 1B • 1R translocation was 13.4%; it occurred in cultivars from all the countries except Australia, Austria, Norway, and Serbia. The functional markers applied in this study, in agreement with the results of SDS-PAGE, were accurate and stable, and can be used effectively in wheat quality breeding.

Key words: *Triticum aestivum* L., processing quality, HMW-GS, LMW-GS, molecular markers

Introduction

The quantity and quality of glutenins are important factors in determining the dough strength and visco-elastic properties, which are responsible for bread- and noodle-making qualities of wheat (Figueroa et al., 2009; Maucher et al., 2009). Based on sodium-dodecylsulfate polyacrylamide gel electrophoresis (SDS-PAGE), gluten is classified into high-molecular-weight glutenin subunits (HMW-GS) and low-molecular-weight glutenin subunits (LMW-GS). HMW-GS are encoded by genes at three loci, designated as *Glu-A1*, *Glu-B1*, and *Glu-D1*, on the long arms of chromosomes 1A, 1B, and 1D, respectively, whereas LMW-GS are encoded by *Glu-A3*, *Glu-B3* and *Glu-D3* genes on the short arms of chromosomes of the same chromosomes (Payne et al., 1987; Singh and Shepherd 1988; D' Ovidio and Masci 2004). Previous studies indicated a close relationship between HMW-GS and bread-making quality (Payne et al., 1979, 1981; He et al., 2005; Liu et al., 2005; Meng and Cai, 2008; Pang et al., 2009). The genes at the *Glu-1* loci showed additive and epistatic effects on processing qualities of wheat-based products (Rousset et al., 1992; Luo et al., 2001; Zhang et al., 2009). Many studies indicated that the effect of the *Glu-D1* locus was the most important of the *Glu-1* loci in determining processing quality (Tabiki et al., 2006; Lei et al., 2009; Zhang et al., 2009). Of subunits at the *Glu-A1* locus, Ax1 and Ax2* showed significantly better effects on all quality parameters than the subunit Null (He et al., 2005; Liu et al., 2005; Li et al., 2010). Positive effects are also true for the subunit pairs Bx7+By8 and Bx7+By9 compared with Bx6+By8 and Bx14+By15 at the *Glu-B1* locus (Liu et al., 2005; Gobaa et al. 2008; Figueroa et al., 2009; Zhang et al., 2009). For the *Glu-D1* locus, the subunit Dx5+Dy10 had more significantly positive effects on quality parameters than the Dx2+Dy12 (Moonen et al. 1985; Payne et al., 1987; Liu et al., 2005; Tabiki et al., 2006; Gobaa et al., 2008). Recently, Bx13+By16 was shown to be positively correlated with bread-making quality (Pang and Zhang, 2008). Subunit $Bx7^{OE}$ (over-expressed Bx7 phenotype) confers improved dough strength (Butow et al., 2003; Juhász et al., 2003).

However, LMW-GS have significant effects on dough extensibility because they account for 75% of the total gluten (Brett et al., 1993; Gupta et al., 1994; He et al., 2005; Figueroa et al., 2009; Liu et al., 2009; Li et al., 2010; Tsenov et al., 2010), but their application in wheat breeding has been limited, largely due to the difficult of identifying alleles in breeding programs. Previous studies showed that *Glu-B3* played more significant roles in dough property determination than *Glu-A3* or *Glu-D3* (Liu et al., 2005; Tabiki et al., 2006; Zhang et al., 2009). Cornish et al. (1993) showed that *Glu-3* patterns with *b b b* alleles gave the best extensibility. *Glu-A3d* and *Glu-B3d* were considered better than others (He et al., 2005; Liu et al., 2005). Wiester et al., (2000) reported that the 1B • 1R translocation was superior in improving adaptation and yield in wheat, but it had significantly negative effects on the processing quality.

There are four common methods in distinguishing gluten proteins. SDS-PAGE is a traditional method, with the advantages of simple operation and low cost, but it has the disadvantages of time involved and low discrimination ability, particularly of LMW-GS (Zhang et al., 2004). RP-HPLC is an effective method with high accuracy and discrimination ability, but it is very costly and not well suited for screening large numbers of samples. Another method is capillary electrophoresis that can resolve the disadvantages of SDS-PAGE and RP-HPLC, but the preparation of samples is inconvenient and its repeatability still needs improvement. In comparison with these methods, molecular markers developed from gene sequences can effectively discriminate HMW-GS and LMW-GS alleles, providing a much more efficient tool for rapid genetic analyses (Pagnotta et al., 1995). Molecular marker assisted selection for these genes is being increasingly applied in wheat breeding.

Currently, many functional markers are available for glutenin genes. Ma et al. (2003) developed several markers for multiplex-PCR use in discriminating *Glu-1* loci. Lei et al. (2006) designed a number of markers to distinguish *Glu-B1* alleles. Two pairs of complementary primers were designed for discriminating of Bx7 and $Bx7^{OE}$ (Ragupathy et al., 2008). A set of PCR markers was developed to distinguish allelic variations at *Glu-A3*, *Glu-B3* and *Glu-D3* loci (Zhang et al., 2004; Zhao et al., 2007a, b). Recently, Wang et al. (2009, 2010) designed 10 STS markers for discriminating *Glu-B3* alleles and 7 STS markers for *Glu-A3* alleles. Sui et al. (2010) developed an allele-specific marker for *Glu-B3* alleles and established a multiplex PCR assay.

Quality improvement has become an important objective for wheat breeding in China. Recently, we introduced 669 wheat cultivars and advanced lines from 19 globally important wheat producing countries, with the objective of using them as parents for quality improvement in China. The aims of the present study were to characterize the combinations of HMW-GS and LMW-GS genes in these lines and 49 Chinese wheat cultivars using molecular markers. The results will benefit the improvement of wheat quality in China and elsewhere.

Materials and methods

Plant materials

In total, 718 wheat cultivars and advanced lines from 20 countries were used for the characterization of HMW-GS and LMW-GS. They were from Argentina (33 genotypes), Australia (73), Austria (5), Bulgaria (15), Canada (24), Chile (41), China (49), France (94), Germany (75), Hungary (10), Iran (30), Japan (10), Mexico (7), Norway (22), Romania (24), Russia (18), Serbia (11), Turkey (85), Ukraine (13), and USA (79). These cultivars and advanced lines were collected from breeders in various countries and the 17th Facultative and Winter Wheat Observing Nursery for Reduced Irrigation (FAWWON-IRR) and the 17th Facultative and Winter Wheat Observing Nursery for Semiarid Conditions (FAWWON-SA) from the Turkey-CIMMYT-ICARDA Joint International Winter Wheat Program. The number of genotypes from Mexico (CIMMYT) was smaller than other countries since the composition of HMW-GS and LMW-GS was reported previously (Liang et al., 2010). Among the tested genotypes, 512 entries were winter cultivars and advanced lines and 206 were spring genotypes, based on the presence of vernalization genes (unpublished data from our lab).

DNA extraction and PCR

Genomic DNA was extracted from seeds using the method of Sambrook et al. (1989) with minor modification. PCR were performed in an MJ Research PTC-200 Thermal Cycler in total volumes of 20μl, including 20mM of Tris - HCl (pH 8.4), 20mM of KCl, 100 mM of each of dNTPs, 1.5mM of $MgCl_2$, 5pmol of each primer, 1unit of *Taq* polymerase (TaKaRa Biotechnology Co., Ltd, Dalian, China) and 50ng of template DNA. Sequences of PCR primers and fragment sizes are shown in Table 1. PCR primers were synthesized by Augct Biotechnology Company (http://www.augct.com). PCR products were stained with ethidium bromide and visualized using UV light.

Table 1 PCR primers of molecular markers used for the detection of HMW-GS and LMW-GS

	Locus	Allele or Subunit	Sequence (5′→3′)	Fragment size (bp)	Reference
HMW-GS	*Glu-A1*	*Ax2* *	5′-ATGACTAAGCGGTTGGTTCTT-3′	1 319	Ma et al. 2003
			5′-ACCTTGCTCCCCTTGTCTTT-3′		
	Glu-B1	*By8*	5′-TTAGCGCTAAGTGCCGTCT-3′	527	Lei et al. 2006
			5′-TTGTCCTATTTGCTGCCCTT-3′		
		By9	5′-TTCTCTGCATCAGTCAGGA-3′	707/662	Lei et al. 2006
			5′-AGAGAAGCTGTGTAATGCC-3′		
		By16	5′-GCAGTACCCAGCTTCTCA A-3′	3 fragments	Lei et al. 2006
			5′-CCTTGTCTTGTTTGTTGCC-3′		
		$Bx7^{OE}$	5′-CCACTTCCAAGGTGGGACTA-3′	844	Ragupathy et al. 2008
	Glu-D1	*Dx5*	5′- CGTCCCTATAAAAGCCTAGC-3′	478	Ma et al. 2003
			5′-AGTATGAAACCTGCTGCGGAC-3′		
			5′-TGCCAACACAAAAGAAGCTG-3′		
LMW-GS	*Glu-A3*	*Glu-A3a*	5′-AAACAGAATTATTAAAGCCGG-3′	529	Wang et al. 2010
			5′-GGTTGTTGTTGTTGCAGCA-3′		
		Glu-A3ac	5′-AAACAGAATTATTAAAGCCGG-3′	573	Wang et al. 2010
			5′-GTGGCTGTTGTGAAAACGA-3′		
		Glu-A3b	5′-TTCAGATGCAGCCAAACAA-3′	894	Wang et al. 2010
			5′-GCTGTGCTTGGATGATACTCTA-3′		
		Glu-A3d	5′-TTCAGATGCAGCCAAACAA-3′	967	Wang et al. 2010
			5′-TGGGGTTGGGAGACACATA-3′		
		Glu-A3e	5′-AAACAGAATTATTAAAGCCGG-3′	158	Wang et al. 2010
			5′-GGCACAGACGAGGAAGGTT-3′		
		Glu-A3f	5′-AAACAGAATTATTAAAGCCGG-3′	552	Wang et al. 2010
			5′-GCTGCTGCTGCTGTGTAAA-3′		
		Glu-A3g	5′-AAACAGAATTATTAAAGCCGG-3′	1 345	Wang et al. 2010
			5′-AAACAACGGTGATCCAACTAA-3′		
	Glu-B3	*Glu-B3a*	5′-CACAAGCATCAAAACCAAGA-3′	1 095	Wang et al. 2009
			5′-TGGCACACTAGTGGTGGTC-3′		
		Glu-B3b	5′-ATCAGGTGTAAAAGTGATAG-3′	1 570	Wang et al. 2009
			5′-TGCTACATCGACATATCCA-3′		
		Glu-B3c	5′-CAAATGTTGCAGCAGAGA-3′	472	Wang et al. 2009
			5′-CATATCCATCGACTAAACAAA-3′		
		Glu-B3d	5′-CACCATGAAGACCTTCCTCA-3′	662	Wang et al. 2009
			5′-GTTGTTGCAGTAGAACTGGA-3′		
		Glu-B3e	5′-GACCTTCCTCATCTTCGCA-3′	669	Wang et al. 2009
			5′-GCAAGACTTTGTGGCATT-3′		
		Glu-B3fg	5′-TATAGCTAGTGCAACCTACCAT-3′	812	Wang et al. 2009
			5′-CAACTACTCTGCCACAACG-3′		
		Glu-B3g	5′-CCAAGAAATACTAGTTAACACTAGTC-3′	853	Wang et al. 2009
			5′-GTTGGGGTTGGGAAACA-3′		

（续）

Locus	Allele or Subunit	Sequence (5′→3′)	Fragment size (bp)	Reference
	Glu-B3h	5′-CCACCACAACAAACATTAA-3′	1 022	Wang et al. 2009
		5′-GTGGTGGTTCTATACAACGA-3′		
	Glu-B3i	5′-TATAGCTAGTGCAACCTACCAT-3′	621	Wang et al. 2009
		5′-TGGTTGTTGCGGTATAATTT-3′		
	Glu-B3bef	5′-GCATCAACAACAAATAGTACTAGAA-3′	750	Wang et al. 2009
		5′-GGCGGGTCACACATGACA-3′		
1B·1R	*Glu-B3j*	5′-GGAGACATCATGAAACATTTG-3′	1 500	Francis et al. 1995
		5′-CTGTTGTTGGGCAGAAAG-3′		

Results

HMW-GS and LMW-GS investigated with molecular markers

Among 718 wheat cultivars and advanced lines, 311 (43.3%) possessed the *Glu-A1* subunit Ax2*, amplifying a 1319-bp fragment with the Ax2* marker (Table 2). For *Glu-B1* locus, 197 genotypes (27.4%) with the subunit By8 amplified a 527-bp PCR fragment using ZSBy8F5/R5; 264 cultivars (36.8%) contained By9 with a 662-bp PCR fragment amplified, and *Glu-B1f* (By16) were in 44 genotypes (6.1%). Only 22 cultivars (3.1%) had Bx7^{OE}; they came from Argentina, Australia, Canada, Iran, and Japan. Four hundred and forty six genotypes (62.1%) had the subunit Dx5, amplifying a 478-bp fragment. At the *Glu-A3* locus, there were 21 cultivars (2.9%) with *Glu-A3a* which amplifies a 529-bp band using the marker *gluA3a*; 92 cultivars (12.8%) contained the allele *Glu-A3d*, generating a 967-bp PCR product with the marker *gluA3d* (Fig. 1); 49 (6.8%), 382 (53.2%), 55 (7.7%), 99 (13.8%), and 17 (2.4%) of genotypes contained *Glu-A3b*, *c*, *e*, *f* and *g*, respectively; 3cultivars (0.4%) did not produce any PCR product with the set of markers for *Glu-A3*. For *Glu-B3*, alleles *a*, *b*, *c*, *d*, *e*, *f*, *h*, and *i* were present in 3 (0.4%), 160 (22.3%), 2 (0.3%), 20 (2.8%), 14 (1.9%), 28 (3.9%), 135 (18.8%), 51 (7.1%) of genotypes using the markers gluB3a, gluB3b, gluB3c, gluB3d, gluB3e, gluB3f, gluB3h, and gluB3i, respectively; 195 cultivars (27.2%) with allele *Glu-B3g* amplified a 853-bp PCR product with the marker gluB3g (Fig. 2). The allele *Glu-B3j*, associated with 1B·1R translocation, was detected in 96 lines (13.4%) by marker AF1/AF4. Fourteen lines (1.9%) did not generate any PCR product at the *Glu-B3* locus.

Table 2 Frequencies of alleles for HMW-GS and LMW-GS in 718 cultivars and advanced lines

Locus	Alleleor Subunit	Number	Frequency (%)
Glu-A1	*Ax2**	311	43.3
	Others	407	56.7
Glu-B1	*Bx7^{OE}*	22	3.1
	By8	197	27.4
	By9	264	36.8
	By16	44	6.1
	Others	213	29.7
Glu-D1	*Dx5*	446	62.1
	Others	272	37.9
Glu-A3	*a*	21	2.9

（续）

Locus	Alleleor Subunit	Number	Frequency (%)
	b	49	6.8
	c	382	53.2
	d	92	12.8
	e	55	7.7
	f	99	13.8
	g	17	2.4
	/	3	0.4
Glu-B3	*a*	3	0.4
	b	160	22.3
	c	2	0.3
	d	20	2.8
	e	14	1.9
	f	28	3.9
	g	195	27.2
	h	135	18.8
	i	51	7.1
1B・1R	*j*	96	13.4
	/	14	1.9

"/", indicates no PCR product amplified

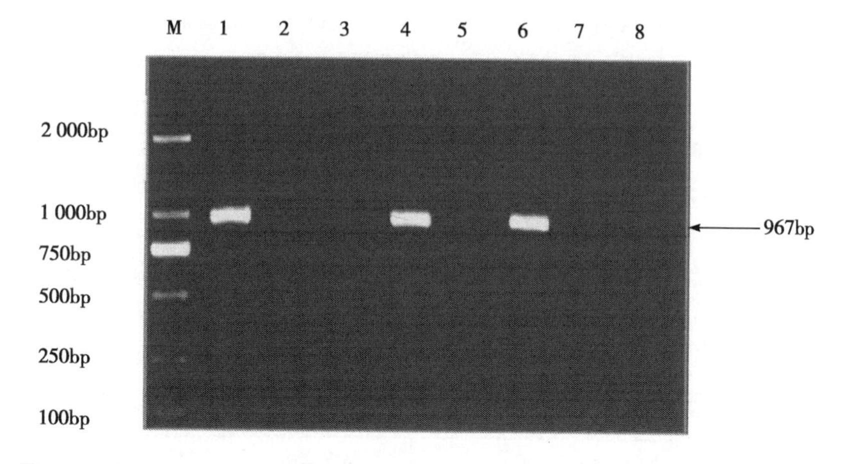

Fig. 1 Electrophoresis of PCR products amplified with the *Glu-A3d* marker on 1.5% agarose gel. M. Marker DL2000, 1: Aikang 58, 2: Adler, 3: Akratos, 4: Actros, 5: Ceganne, 6: Bruta, 7: Lasen, 8: Marke

Frequency distributions of allelic variants in cultivars from different countries

Large differences in frequency distributions for HMW-GS and LMW-GS were found in the cultivars from different countries (Tables 3 and 4). High frequencies of Ax2* were present in cultivars from Canada (83.3%), Romania (91.7%), Russia (72.2%), and USA (72.2%), whereas the corresponding frequencies were much lower in cultivars from Chile (22.0%), China (20.4%) and Germany (1.3%), and no cultivar with Ax2* was found among genotypes from Austria and Japan. The frequency of By8 was high in Japanese (60.0%) and Romanian (62.5%) materials. Frequen-

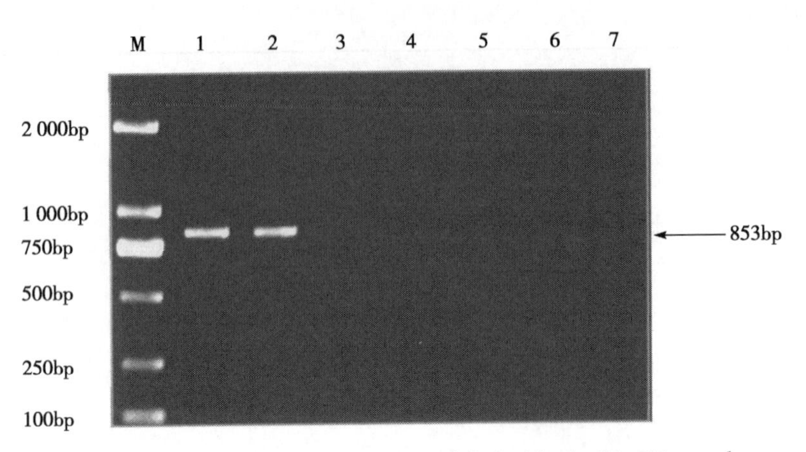

Fig. 2 Electrophoresis of PCR products amplified with the *Glu-B3g* marker on a 1.5% agarose gel. M. Marker DL2000, 1. Hengguan 33, 2. Kefeng 12, 3. Aikang 58, 4. Actros, 5. Ceganne, 6. Bainong 64, 7. Lasen

cies of By9 were high in wheats from Austria (100.0%), Russia (72.2%), and Serbia (72.7%). By16 was most frequent in cultivars from Chile (19.5%) and USA (19.0%). The $Bx7^{OE}$ allele was only present in cultivars from Argentina (12.1%), Australia (4.1%), Canada (25.0%), Iran (20.0%), and Japan (30.0%). A high frequency of Dx5 occurred in cultivars from Hungary (90.0%), Romania (95.8%), Russia (88.9%), and Ukraine (92.3%). For the *Glu-A3* locus, allele *Glu-A3a* was found only in cultivars from Bulgaria (13.3%), China (12.2%), Germany (2.7%), Iran (6.7%), Mexico (14.3%), Turkey (4.7%) and USA (5.1%); the highest frequencies of *b* and *d* were detected in the cultivars from Australia (39.7%) and France (24.5%), respectively; *c* was widely present in all materials; *e* was mostly in cultivars from Argentina (33.3%), Canada (29.2%) and Ukraine (23.1%); *f* was detected mainly in Canada (29.2%), Norway (27.3%), and Germany (26.7%) lines; *g* was only in materials from Argentina (6.1%), Canada (12.5%), Chile (4.9%), China (2.0%), France (7.4%) and Hungary (20.0%). At the *Glu-B3* locus, *a* was only in cultivars from Argentina (3.0%) and Ukraine (15.4%); the frequencies of *b*, *c*, *d* and *e* were highest in cultivars from Romania (62.5%), Romania (8.7%), Mexico (14.3%), and Austria (40.0%), respectively; the highest frequency of *f*, *h* and *i* were in cultivars from China (14.3%), USA (43.0%) and Argentina (33.3%), respectively; *g* was mainly detected in lines from Germany (69.3%), Norway (77.3%), and Serbia (63.6%), and the 1B • 1R translocation (*Glu-B3j*) was widely present in all materials except those from Australia, Austria, Norway, and Serbia.

Table 3 Allelic frequencies for HMW-GS in wheat cultivars from 20 countries

Origin	HMW-GS					
	Glu-A1 (%)	Glu-B1 (%)				Glu-D1 (%)
	Ax2*	$Bx7^{OE}$	By8	By9	By16	Dx5
Argentina	60.6	12.1	30.3	24.2	3.0	66.7
Australia	39.7	4.1	35.6	8.2	4.1	39.7
Austria	0.0	0.0	0.0	100.0	0.0	80.0
Bulgaria	53.3	0.0	33.3	53.3	6.7	80.0
Canada	83.3	25.0	29.2	50.0	0.0	87.5
Chile	22.0	0.0	24.4	14.6	19.5	68.3
China	20.4	0.0	26.5	46.9	2.0	26.5
France	31.9	0.0	25.5	35.1	2.1	72.3

（续）

Origin	HMW-GS					
	Glu-A1 (%)	Glu-B1 (%)				Glu-D1 (%)
	Ax2*	$Bx7^{OE}$	By8	By9	By16	Dx5
Germany	1.3	0.0	5.3	38.7	6.7	73.3
Hungary	60.0	0.0	40.0	60.0	0.0	90.0
Iran	46.7	20.0	46.7	33.3	6.7	26.7
Japan	0.0	30.0	60.0	30.0	0.0	10.0
Mexico	71.4	0.0	14.3	28.6	14.3	85.7
Norway	59.1	0.0	4.5	50.0	4.5	72.7
Romania	91.7	0.0	62.5	33.3	0.0	95.8
Russia	72.2	0.0	22.2	72.2	0.0	88.9
Serbia	27.3	0.0	27.3	72.7	0.0	72.7
Turkey	55.3	0.0	38.8	41.2	4.7	63.5
Ukraine	30.8	0.0	30.8	61.5	0.0	92.3
USA	72.2	0.0	16.5	38.0	19.0	51.9

Table 4 Allelic frequencies for LMW-GS in wheat cultivars from 20 countries

Origin	LMW-GS																
	Glu-A3 (%)							Glu-B3 (%)									
	a	b	c	d	e	f	g	a	b	c	d	e	f	g	h	i	j
Argentina	0.0	0.0	33.3	6.1	33.3	15.2	6.1	3.0	12.1	0.0	3.0	0.0	3.0	15.2	18.2	33.3	9.1
Australia	0.0	39.7	38.4	16.4	4.1	1.4	0.0	0.0	54.8	0.0	0.0	0.0	2.7	1.4	34.2	6.8	0.0
Austria	0.0	0.0	100.0	0.0	0.0	0.0	0.0	0.0	0.0	0.0	0.0	40.0	0.0	60.0	0.0	0.0	0.0
Bulgaria	13.3	6.7	73.3	0.0	6.7	0.0	0.0	0.0	33.3	0.0	6.7	6.7	0.0	20.0	13.3	0.0	20.0
Canada	0.0	0.0	29.2	0.0	29.2	29.2	12.5	0.0	8.3	0.0	0.0	0.0	8.3	25.0	37.5	8.3	4.2
Chile	0.0	4.9	48.8	22.0	2.4	17.1	4.9	0.0	12.2	0.0	0.0	0.0	2.4	7.3	39.0	12.2	26.8
China	12.2	2.0	59.2	10.2	10.2	4.1	2.0	0.0	2.0	0.0	12.2	0.0	14.3	26.5	16.3	0.0	28.6
France	0.0	0.0	53.2	24.5	1.1	12.8	7.4	0.0	2.1	0.0	3.2	0.0	4.3	58.5	3.2	5.3	17.0
Germany	2.7	0.0	49.3	20.0	1.3	26.7	0.0	0.0	1.3	0.0	6.7	0.0	6.7	69.3	0.0	0.0	16.0
Hungary	0.0	10.0	40.0	10.0	0.0	20.0	20.0	0.0	50.0	0.0	0.0	0.0	0.0	0.0	0.0	0.0	40.0
Iran	6.7	16.7	50.0	6.7	13.3	6.7	0.0	0.0	50.0	0.0	0.0	0.0	3.3	3.3	26.7	13.3	3.3
Japan	0.0	0.0	50.0	20.0	20.0	10.0	0.0	0.0	20.0	0.0	10.0	0.0	0.0	20.0	10.0	30.0	10.0
Mexico	14.3	28.1	57.1	0.0	0.0	0.0	0.0	0.0	14.3	0.0	14.3	0.0	0.0	14.3	38.6	0.0	28.6
Norway	0.0	0.0	63.6	9.1	0.0	27.3	0.0	0.0	9.1	0.0	0.0	0.0	0.0	77.3	0.0	13.6	0.0
Romania	0.0	0.0	62.5	0.0	20.8	16.7	0.0	0.0	62.5	8.3	0.0	0.0	0.0	4.2	12.5	0.0	12.5
Russia	0.0	11.1	77.8	0.0	0.0	11.1	0.0	0.0	50.0	0.0	0.0	0.0	0.0	0.0	16.7	5.6	22.2
Serbia	0.0	0.0	81.8	0.0	18.2	0.0	0.0	0.0	27.3	0.0	0.0	0.0	0.0	63.6	9.1	0.0	0.0
Turkey	4.7	3.5	61.2	8.2	5.9	16.5	0.0	0.0	41.2	0.0	2.4	11.8	2.4	11.8	15.3	4.7	10.6
Ukraine	0.0	0.0	53.8	15.4	23.1	7.7	0.0	15.4	30.8	0.0	0.0	0.0	0.0	15.4	7.7	7.7	7.7
USA	5.1	3.8	57.0	12.7	5.1	16.5	0.0	0.0	11.4	0.0	0.0	1.3	3.8	16.5	43.0	8.9	13.9

Differences in allelic variation between winter and spring wheats

Among the 512 winter wheat and 206 spring wheat cultivars tested, there were some differences in allelic variation in HMW-GS and LMW-GS (Table 5). In all, for HMW-GS, the difference for the frequency distribution of Ax2* and By16 were not significant. The frequencies of By9 and Dx5 were higher in winter wheat cultivars than in spring wheats. The frequencies of $Bx7^{OE}$ and By8 were higher in spring wheat cultivars than in winter wheats. For LMW-GS, frequencies of the superior alleles *Glu-A3b* and *Glu-B3b* were higher in spring wheat cultivars, whereas the frequency of the 1B·1R translocation was higher in winter wheats. The winter wheat cultivars in this study were mainly from European countries (Bulgaria, France, Germany, Hungary, Norway, Romania, Russia, Serbia, and Ukraine), Asian countries (China, Iran, Turkey) and USA. The spring wheat cultivars primarily came from the Americas (Argentina, Canada, Chile, Mexico), and Asian countries (China, Japan) and Australia. The differences in allelic variations between winter and spring wheat cultivars was attributed to the largely different sets of wheat parents used in winter and spring wheat breeding programs in different countries.

Table 5 Allelic frequencies for HMW-GS and LMW-GS in winter and spring cultivars

Locus	Allele (Subunit)	Winter wheat (%)	Spring wheat (%)
Glu-A1	*Ax2**	43.6	42.7
Glu-B1	$Bx7^{OE}$	1.2	7.8
	By8	26.0	31.1
	By9	42.4	22.8
	By16	6.1	6.3
Glu-D1	*Dx5*	65.2	54.4
Glu-A3	*a*	4.1	0.0
	b	2.9	16.5
	c	57.4	42.7
	d	12.9	12.6
	e	5.7	12.6
	f	14.8	11.2
	g	2.0	3.4
Glu-B3	*a*	0.4	0.5
	b	20.9	25.7
	c	0.4	0.0
	d	3.3	1.5
	e	2.7	0.0
	f	3.7	4.4
	g	32.6	13.6
	h	14.3	30.1
	i	4.9	12.6
1B·1R	*j*	14.6	10.2

Discussion

Effectiveness of the molecular markers

Among 718 wheat cultivars and advanced lines, 240 were tested with SDS-PAGE previously. Comparison of the results with those using allele-specific markers indicated agreements of 94.6, 93.4, 98.0, 90.0, and 91.3% for allelic variants identified for the loci *Glu-A1*, *Glu-B1*, *Glu-D1*, *Glu-A3* and *Glu-B3*, respectively. For the *Glu-A1* locus, 9 cultivars with Ax2* identified SDS-PAGE were not identified by the PCR marker. Different subunits were identified by the mobility on SDS-PAGE, but there was not always consistent agreement between molecular weight and mobility, presumably this is the main reason for the discrepancy. For the *Glu-B1* locus, the main discrepancies were in results for Bx7 and Bx7^{OE} because SDS-PAGE does not distinguish subunits Bx7 and Bx7^{OE} (Marchylo et al., 1992; Butow et al., 2004). For the *Glu-D1* locus, 3 cultivars contained Dx5 based on the PCR marker, but were not identified on SDS-PAGE, due to overestimated molecular masses and low resolution (Gao et al., 2010). For the *Glu-A3* locus, the main discrepancies were for *Glu-A3e* and *Glu-A3f* because *Glu-A3e* could not easily be distinguished from *Glu-A3f* by SDS-PAGE (Liu et al., 2010). For the *Glu-B3* locus, the main difference was between *Glu-B3b* and *Glu-B3h*, possibly due to mixed genotypes. Liu et al. (2010) concluded that PCR was the simplest, most accurate, low cost and therefore recommendable method for identification of *Glu-A3* and *Glu-B3* alleles in breeding programs after comparing SDS-PAGE, 2-DE, MALDL-TOF-MS and PCR methods. The present results are supportive of those findings.

Distributions of desirable alleles in Australian, Canadian, Chinese, and USA genotypes

Improvement of gluten quality rather than protein content is the major breeding objective in China, thus high quality wheats from Australia, Canada, USA, and CIMMYT Mexico are frequently used as crossing parents in Chinese wheat breeding programs targeting quality improvement (He et al., 2002; Liu et al., 2003). The compositions of HMW-GS and LMW-GS of CIMMYT germplasm were documented previously (Liang et al., 2010; Liu et al., 2010). Subunits Ax1 and Ax2* at the *Glu-A1* locus, Bx7+By8, Bx7+By9 at the *Glu-B1* locus, Dx5+Dy10 at the *Glu-D1*, and *Glu-A3b*, *Glu-B3b* at the *Glu-A3* and *Glu-B3* loci showed significantly positive effects on the processing quality of wheat products (He et al., 2005; Liu et al., 2005; Li et al., 2010). As shown in Table 3, the frequencies of Ax2* were 39.7, 83.3, 20.4 and 72.2% in Australian, Canadian, Chinese and USA lines, respectively. The highest frequency of By8 was in the cultivars from Australia (35.6%). For By9, Canadian cultivars had the highest frequency (50.0%). By16 is present in only 2.0% of Chinese cultivars, whereas it was found in 19.0% of cultivars from USA, 4.1% of those from Australia and not present in those from Canada. Bx7^{OE} was present only in Australian (4.1%) and Canadian (25.0%) cultivars. As shown in Table 4, the frequencies of *Glu-A3b* were 39.7, 0.0, 2.0 and 3.8 in wheats from Australia, Canada, China, and USA, respectively. Of 160 cultivars with the allele *Glu-B3b*, 40, 2, 1, 9 came from Australia, Canada, China, and USA. The 1B·1R translocation was detected in Canadian (4.2%), Chinese (28.6%) and USA (13.9%) lines, whereas 1B·1R was found among Australian cultivars. Thus, compared with the wheat cultivars from Australia, Canada, and USA, Chinese wheats have significantly lower frequencies of desirable alleles for both HMW-GS and LMW-GS, and thus have inferior processing quality, whereas the frequency of the undesirable 1B·1R translocation was much higher, consistent with our previous studies (He et al., 1992, 2005; Liu et al., 2005). Therefore it is essential to combine superior alleles such as Ax2*, Bx7+By8 or Bx13+By16, Dx5+Dy10, *Glu-A3b* and *Glu-B3b* into targeted genotypes to improve the end-use qualities of Chinese wheat cultivars. Molecular markers will be a useful tool for pyramiding those genes in wheat breeding programs aimed at quality improvement.

Development and application of functional markers for HMW-GS and LMW-GS genes

The development of functional markers (or allele-specific markers) is based on the polymorphisms of nucleotide sequences among different alleles, and the relationship between markers and phenotypes needs to be established. To date, several functional markers have been developed to identify HMW-GS Ax2* (Ma et al., 2003; Liu et al., 2008), $Bx7^{OE}$ (Ragupathy et al., 2008), Bx6+By8 (Schwarz et al., 2004), Bx17+By18 (Butow et al., 2003, 2004; Ma et al., 2003), Bx7+By8, Bx7+By8*, Bx7+By9, Bx13+By16, Bx14+By15 and Bx20 (Lei et al., 2006), Dx5+Dy10 and Dx2+Dy12 (D'Ovidio and Anderson, 1994; Smith et al., 1994; Ahmad, 2000; De Bustos et al., 2000, 2001; Radovanovic and Cloutier, 2003; Schwarz et al., 2003; Liu et al., 2008). For LMW-GS, the functional markers are available to discriminate different alleles at *Glu-A3* and *Glu-B3* loci (Zhang et al. 2004; Wang et al., 2009, 2010; Sui et al., 2010). However, no functional markers are available for some glutenin subunits, such as Bx6.8+By20, Bx7*, Bx21, Bx22, Bx23+By24, Dx1.5+Dy10.5, Dx2.2+Dy12, Dx5+Dy10.5 and *Glu-D3* alleles, due to the lack of reliable nucleotide sequences or little polymorphisms among different alleles. In addition, no information is available about the correlations between some subunits and wheat processing quality, so even though the functional marker is designed for these subunits, it has no practical use. To apply the molecular markers in wheat breeding, it is necessary to affirm the validity of markers using sufficient germplasms with corresponding glutenin subunits.

Acknowledgments

The authors are grateful to Prof. Robert McIntosh, University of Sydney, for reviewing this manuscript. This study was supported by the National Basic Research Program (2009CB118300), the National Science Foundation of China (30830072), international collaboration project from Ministry of Agriculture, and China Agriculture Research System.

References

Ahmad M. 2000. Molecular marker-assisted selection of HMW glutenin alleles related to wheat bread quality by PCR-generated DNA markers. *Theoretical and Applied Genetics*. 101, 892-896. doi: 10.1007/s001220051558

Brett GM, Mills ENC, Tatham AS, Fido RJ, Shewry PR, Morgan MRA. 1993. Immunochemical identification of LMW subunits of glutenin associated with bread-making quality of wheat flours. *Theoretical and Applied Genetics*. 86, 442-448. doi: 10.1007/BF00838559

Butow BJ, Gale KR, Ikea J, Juhász A, Bedö Z, Tamás L, Gianibelli MC. 2004. Dissemination of the highly expressed Bx7 glutenin subunit (Glu-B1al allele) in wheat as revealed by novel PCR markers and RP-HPLC. *Theoretical and Applied Genetics*. 109, 1525-1535. doi: 10.1007/s00122-004-1776-8

Butow BJ, Ma W, Gale KR, Cornish GB, Rampling L, Larroque O, Morell MK, Békés F. 2003. Molecular discrimination of Bx7 alleles demonstrates that a highly expressed high-molecular-weight glutenin allele has a major impact on wheat flour dough strength. *Theoretical and Applied Genetics*. 107, 1524-1532. doi: 10.1007/s00122-003-1396-8

Cornish GB, Burridge PM, Palmer GA, Wrigley CW. 1993. Mapping the origins of some HMW and LMW glutenin subunit alleles in Australian germplasm. Proc 42nd Aust. Cereal Chem. Conf.. 255-260, Sydney, Australia.

De Bustos A, Rubio P, Jouve N. 2000. Molecular characterization of the inactive allele of the gene *Glu-A1* and the development of a set of AS-PCR markers for HMW glutenins of wheat. *Theoretical and Applied Genetics*. 100, 1085-1094. doi: 10.1007/s001220051390

De Bustos A, Rubio P, Soler C, Garcia P, Jouve N. 2001. Marker assisted selection to improve HMW-glutenins in wheat. *Euphytica*. 119, 69-73. doi: 10.1023/A:1017534203520

D' Ovidio R, Anderson OD. 1994. PCR analysis to distinguish between alleles of a member of a multigene family correlated with wheat bread-making quality. *Theoretical and Applied Genetics*. 88, 759-763. doi: 10.1007/BF01253982

D' Ovidio R, Masci S. 2004. The low-molecular-weight glutenin subunits of wheat gluten. *Journal of Cereal Science*. 39, 321-339. doi: 10.1016/j.jcs.2003.12.002

Figueroa JDC, Maucher T, Reule W, Peña RJ . 2009. Influence of high molecular weight glutenins on viscoelastic properties of intact wheat kernel and relation to functional properties of wheat dough. *Cereal Chemistry*. 86, 139-144. doi: 10.1094/CCHEM-86-2-0139

Francis HA, Leitch AR, Koebner RMD . 1995. Conversion of a RAPD-generated PCR product, containing a novel dispersed repetitive element, into a fast and robust assay for the presence of rye chromatin in wheat. *Theoretical and Applied Genetics*. 90, 636-642. doi: 10.1007/BF00222127

Gao LY, Ma WJ, Chen J, Wang K, Li J, Wang SL, Bekes F, Appels R, Yan YM. 2010. Characterization and comparative analysis of wheat high molecular weight glutenin subunits by SDS-PAGE, RP-HPLC, HPCE, and MALDI-TOF-MS. *Journal of Agricultural and Food Chemistry*. 58, 2777-2786. doi: 10.1021/jf903363z

Gobaa S, Brabant C, Kleijer G, Stamp P . 2008. Effects of the 1BL. 1RS translocation and of the Glu-B3 variation on fifteen quality tests in a doubled haploid population of wheat (*Triticum aestivum* L.). *Journal of Cereal Science*. 48, 598-603. doi: 10.1016/j.jcs.2007.12.006

Gupta GB, MacRitchie F . 1994. Allelic variation at glutenin subunit and gliadin loci, *Glu-1*, *Glu-3* and *Gli-1* of common wheats. Ⅱ. Biochemical basis of the allelic effects on dough properties. *Journal of Cereal Science*. 19, 19-29. doi: 10.1006/jcrs.1994.1004

He ZH, Lin ZJ, Wang LJ, Xiao ZM, Wan FS, Zhuang QS. 2002. Classification on Chinese wheat regions based on quality. *Scientia Agricultura Sinica*. 35, 359-364.

He ZH, Liu L, Xia XC, Liu JJ, Peña RJ. 2005. Composition of HMW and LMW glutenin subunits and their effects on dough properties, pan bread, and noodle quality of Chinese bread wheats. *Cereal Chemistry*. 82, 345-350. doi: 10.1094/CC-82-0345

He ZH, Peña RJ, Rajaram S . 1992. High molecular weight glutenin subunit composition of Chinese bread wheats. *Euphytica*. 64, 11-20. doi: 10.1007/BF00023533

Juhász A, Larroque OR, Tamás L, Hsam SLK, Zeller FJ, Békés F, Bedő Z . 2003. Bankuti 1201-an old Hungarian wheat variety with special storage protein composition. *Theoretical and Applied Genetics*. 107, 697-704. doi: 10.1007/s00122-003-1292-2

Lei ZS, Gale KR, He ZH, Gianibelli C, Larroque O, Xia XC, Butow BJ, Ma W . 2006. Y-type gene specific markers for enhanced discrimination of high-molecular weight glutenin alleles at the Glu-B1 locus in hexaploid wheat. *Journal of Cereal Science*. 43, 94-101. doi: 10.1016/j.jcs.2005.08.003

Lei ZS, Liu L, Wang MF, Yan J, Yang P, Zhang Y, He ZH . 2009. Effect of HMW and LMW glutenin subunits on processing quality in common wheat. *Acta Agronomica Sinica*. 35, 203-210. doi: 10.3724/SP.J.1006.2009.00203

Li YL, Zhou RH, Branlard G, Jia JZ . 2010. Development of introgression lines with 18 alleles of glutenin subunits and evaluation of the effects of various alleles on quality related traits in wheat (*Triticum aestivum* L.). *Journal of Cereal Science*. 51, 127-133. doi: 10.1016/j.jcs.2009.10.008

Liang D, Tang JW, Peña RJ, Singh R, He XY, Shen XY, Yao DN, Xia XC, He ZH . 2010. Characterization of CIMMYT bread wheats for high- and low-molecular weight glutenin subunits and other quality-related genes with SDS-PAGE, RP-HPLC and molecular markers. *Euphytica*. 172, 235-250. doi: 10.1007/s10681-009-0054-x

Liu JJ, He ZH, Zhao ZD, Peña RJ, Rajaram S . 2003. Wheat quality traits and quality parameters of cooked dry white Chinese noodles. *Euphytica*. 131, 147-154. doi: 10.1023/A: 1023972032592

Liu L, He ZH, Ma WJ, Liu JJ, Xia XC, Peña RJ . 2009. Allelic variation at the Glu-D3 locus in Chinese bread wheat and effects on dough properties, pan bread and noodle qualities. *Cereal Research Communications*. 37, 57-64. doi: 10.1556/CRC.37.2009.1.7

Liu L, He ZH, Yan J, Zhang Y, Xia XC, Peña RJ . 2005. Allelic variation at the *Glu-1* and *Glu-3* loci, presence of the 1B·1R translocation, and their effects on mixographic properties in Chinese bread wheats. *Euphytica*. 142, 197-204. doi: 10.1007/s10681-005-1682-4

Liu L, Ikeda TM, Branlard G, Peña RJ, Rogers WJ , Lerner SE, Kolman MA, Xia XC, Wang LH, Ma WJ, Appels R, Yoshida H, Wang AL, Yan YM, He ZH . 2010. Comparison of low molecular weight glutenin subunits identified by SDS-PAGE, 2-DE, MALDI-TOF-MS and PCR in common wheat. *BMC Plant Biology*. 10, 124-147. doi: 10.1186/1471-2229-10-124

Liu SX, Chao SM, Anderson JA . 2008. New DNA markers for high molecular weight glutenin subunits in wheat. *Theoretical and Applied Genetics*. 118, 177-183. doi: 10.1007/s00122-008-0886-0

Luo C, Griffin WB, Branlard G, McNeil DL . 2001. Comparison of low- and high-molecular-weight wheat glutenin alleles effects on flour quality. *Theoretical and Applied*

Genetics. 102, 1088-1098. doi: 10.1007/s001220000433

Ma W, Zhang W, Gale KR . 2003. Multiplex-PCR typing of high molecular weight glutenin alleles in wheat. *Euphytica.* 134, 51-60. doi: 10.1023/A: 1026191918704

Marchylo BA, LukowOM, Kruger JE . 1992. Quantitative variation in high molecular weight glutenin subunit 7 in some Canadian wheats. *Journal of Cereal Science.* 15, 29-37. doi: 10.1016/S0733-5210 (09) 80054-4

Maucher T, Figueroa JDC, Reule W, Peña RJ . 2009. Influence of low molecular weight glutenins on viscoelastic properties of intact wheat kernels and their relation to functional properties of wheat dough. *Cereal Chemistry.* 86, 372-375. doi: 10.1094/CCHEM-86-4-0372

Meng XG, Cai SX . 2008. Association between glutenin alleles and Lanzhou alkaline stretched noodle quality of northwest China spring wheats. Ⅱ. Relationship with the variations at the Glu-1 loci. *Cereal Research Communications.* 36, 107-115. doi: 10.1556/CRC. 36. 2008. 1. 11

Moonen JHE, Zeven AC. 1985. Association between high molecular weight subunits of glutenin and bread-making quality in wheat lines derived from backcrosses between *Triticum aestivum* and *Triticum speltoides*. *Journal of Cereal Science.* 3, 97-101. doi: 10.1016/S0733-5210 (85) 80020-5

Pagnotta MA, Nevo E, Beiles A, Porceddu E . 1995. Wheat storage proteins: glutenin diversity in wild emmer, *Triticum dicoccoides*, in Israel and Turkey. 2. DNA diversity detected by PCR. *Theoretical and Applied Genetics.* 91, 409-414. doi: 10.1007/BF00222967

Pang BS, Yang YS, Wang LF, Zhang XY, Yu YJ . 2009. Complementary effect of high-molecular-weight glutenin subunits on bread-making quality in common wheat. *Acta Agronomica Sinica.* 35, 1379- 1385. doi: 10.1016/S1875-2780 (08) 60094-2

Pang BS, Zhang XY . 2008. Isolation and molecular characterization of high molecular weight glutenin subunit genes *1Bx13* and *1By16* from hexaploid wheat. J. Integ. *Plant Biology.* 50, 329-337. doi: 10.1111/j. 1744-7909. 2007. 00573. x

Payne PI, Corfield KG, Blackman, JA. 1979. Identification of a high-molecular weight subunit of glutenin whose presence correlates with breadmaking quality in wheats of related pedigree. *Theoretical and Applied Genetics.* 55, 153-159. doi: 10.1007/BF00295442

Payne PI, Corfield KG, Holt LM, Blackman JA. 1981. Correlations between the inheritance of certain high-molecular-weight subunits of glutenin and breadmaking quality in progenies of six crosses of bread wheat. *Journal of the Science of Food and Agriculture.* 32, 51-60. doi: 10.1002/jsfa. 2740320109

Payne PI, Nightingale MA, Krattiger AF, Holt LM. 1987. The relationship between HMW glutenin subunit composition and the bread-making quality of British-grown wheat varieties. *Journal of the Science of Food and Agriculture.* 40, 51-65. doi: 10.1002/jsfa. 2740400108

Radovanovic N, Cloutier S. 2003. Gene-assisted selection for high molecular weight glutenin subunits in wheat doubled haploid breeding programs. *Molecular Breeding.* 12, 51-59. doi: 10.1023/A: 1025484523771

Ragupathy R, Naeem HA, Reimer E, Lukow OM, Sapirstein HD, Cloutier S . 2008. Evolutionary origin of the segmental duplication encompassing the wheat Glu-B1 encoding the overexpressed Bx7 (Bx7OE) high molecular weight glutenin subunit. *Theoretical and Applied Genetics.* 116, 283-296. doi: 10.1007/s00122-007-0666-2

Rousset M, Carrillo JM, Qualset CO, Kasarda DD. 1992. Use of recombinant inbred lines of wheat for study of associations of high-molecular-weight glutenin subunit alleles to quantitative traits. *Theoretical and Applied Genetics.* 83, 403-412. doi: 10. 1007/BF01186074

Sambrook J, Fritsch EF, Maniatis T . 1989. Molecular cloning: a laboratory manual. 2nd ed. , Cold Spring Harbour Laboratory Press, New York.

Singh NK, Shepherd KW . 1988. Linkage mapping of genes controlling endosperm storage proteins in wheat. 1. Genes on the short arms of group 1 chromosomes. *Theoretical and Applied Genetics.* 75, 628-641. doi: 10.1007/BF00289132

Schwarz G, Sift A, Wenzel G, Mohler V . 2003. DHPLC scoring of a SNP between promoter sequences of HMW glutenin x-type alleles at the Glu-D1 locus in wheat. *Journal of Agricultural and Food Chemistry.* 51, 4263-4267. doi: 10.1021/jf0261304

Schwarz G, Felsenstein FG, Wenzel G . 2004. Development and validation of a PCR-based marker assay for negative selection of the HMW glutenin allele *Glu-B1-1d* (*Bx-6*) in wheat. *Theoretical and Applied Genetics.* 109, 1064-1069. doi: 10.1007/s00122-004-1718-5

Smith RL, Schweder ME, Barnett RD . 1994. Identification of glutenin alleles in wheat and triticale using PCR-generated DNA markers. *Crop Science.* 34, 1373-1378.

Sui XX, Wang LH, Xia XC, Wang ZL, He ZH. 2010. Development of an allele-specific marker for *Glu-B3* alleles in common wheat and establishment of a multiplex PCR assay. *Crop & Pasture Science*. 61, 978-987. doi: org/10.1071/CP10241

Tabiki T, Ikeguchi S, Ikeda TM. 2006. Effects of high-molecular-weight and low-molecular-weight glutenin subunit alleles on common wheat flour quality. *Breeding Science*. 56, 131-136. doi: 10.1270/jsbbs.56.131

Tsenov N, Atanasova D, Todorov I, Ivanova I, Stoeva I. 2010. Quality of winter common wheat advanced lines depending on allelic variation of Glu-A3. *Cereal Research Communications*. 38, 250-258. doi: 10.1556/CRC.38.2010.2.11

Wang LH, Li GY, Peña RJ, Xia XC, He ZH. 2010. Development of STS markers and establishment of multiplex PCR for Glu-A3 alleles in common wheat (*Triticum aestivum* L.). *Journal of Cereal Science*. 51, 305-312. doi: 10.1016/j.jcs.2010.01.005

Wang LH, Zhao XL, He ZH, Ma W, Appels R, Peña RJ, Xia XC. 2009. Characterization of low-molecular-weight glutenin subunit Glu-B3 genes and development of STS markers in common wheat (*Triticum aestivum* L.). *Theoretical and Applied Genetics*. 118, 525-539. doi: 10.1007/s00122-008-0918-9

Wieser H, Kieffer R, Lelley T. 2000. The influence of 1B•1R chromosome translocation on gluten protein composition and technological properties of bread wheat. *Journal of the Science of Food and Agriculture*. 80, 1640-1647. doi: 10.1002/1097-0010 (20000901)

Zhang W, Gianibelli MC, Rampling LR, Gale KR. 2004. Characterisation and marker development for low molecular weight glutenin genes from *Glu-A3* alleles of bread wheat (*Triticum aestivum* L.). *Theoretical and Applied Genetics*. 108, 1409-1419. doi: 10.1007/s00122-003-1558-8

Zhang Y, Tang JW, Yan J, Zhang YL, Zhang Y, Xia XC, He ZH. 2009. The gluten protein and interactions between components determine mixograph properties in an F6 recombinant inbred line population in bread wheat. *Journal of Cereal Science*. 50, 219-226. doi: 10.1016/j.jcs.2009.05.005

Zhao XL, Ma W, Gale KR, Lei ZS, He ZH, Sun QX, Xia XC. 2007a. Identification of SNPs and development of functional markers for LMW-GS genes at *Glu-D3* and *Glu-B3* loci in bread wheat (*Triticum aestivum* L.). *Molecular Breeding*. 20, 223-231. doi: 10.1007/s 11032-007-9085-y

Zhao XL, Xia XC, He ZH, Lei ZS, Appels R, Yang Y, Sun QX, Ma W. 2007b. Novel DNA variations to characterize low molecular weight glutenin *Glu-D3* genes and develop STS markers in common wheat. *Theoretical and Applied Genetics*. 114, 451-460. doi: 10.1007/s00122-006-0445-5

Comparison of low molecular weight glutenin subunits identified by SDS-PAGE, 2-DE, MALDI-TOF-MS and PCR in common wheat

L. Liu[1], T. M. Ikeda[2], G. Branlard[3], R. J. Peña[4], W. J. Rogers[5], S. E. Lerner[6], A. Kolman[5], X. C. Xia[1], L. H. Wang[1], W. J. Ma[7], R. Appels[7], H. Yoshida[8], Y. M. Yan[9], and Z. H. He[1,10]

[1] *Institute of Crop Science, National Wheat Improvement Center/The National Key Facility for Crop Genetic Resources and Genetic Improvement, Chinese Academy of Agricultural Sciences* (*CAAS*), 12 *Zhongguancun South Street, Beijing* 100081, *China*; [2] *National Agriculture and Food Research Organization*, 6-12-1 *Nishifukatsu, Fukuyama, Hiroshima*, 721-8514, *Japan*; [3] *INRA Station d' Amelioration des Plantes, Domaine de Crouelle*, 63039 *Clermont-Ferrand, France*; [4] *CIMMYT Mexico, Apdo, Postal*, 6-641, 06600, *Mexico, DF, Mexico*; [5] *CIISAS, CICPBA-BIOLAB AZUL, Facultad de Agronomía, Universidad Nacional del Centro de la Provincia de Buenos Aires, Av. República Italia* 780, *C.C.* 47, (7300), *Azul, Provincia de Buenos Aires, Argentina. CONICET INBA-CEBB-MdP*; [6] *CRESCAA, Facultad de Agronomía, Universidad Nacional del Centro de la Provincia de Buenos Aires, Av. República Italia* 780, *C.C.* 47, (7300), *Azul, Provincia de Buenos Aires, Argentina*; [7] *Western Australia Department of Agriculture and Food, State Agriculture Biotechnology Center, Murdoch University, Murdoch, WA* 6150, *Australia*; [8] *National Agriculture and Food Research Organization*, 3-1-1 *Kannondai, Tsukuba, Ibaraki*, 305-8517, *Japan*; [9] *Key Laboratory of Genetics and Biotechnology, College of Life Science, Capital Normal University*, 105 *Xisanhuan Beilu, Beijing* 100037, *China*; [10] *International Maize and Wheat Improvement Center* (*CIMMYT*) *China Office, c/o CAAS*, 12 *Zhongguancun South Street, Beijing* 100081, *China*

Abstract: Low-molecular-weight glutenin subunits (LMW-GS) play a crucial role in determining end-use quality of common wheat by influencing the viscoelastic properties of dough. Four different methods - sodium dodecyl sulfate polyacrylamide gel electrophoresis (SDS-PAGE), two-dimensional gel electrophoresis (2-DE, IEF × SDS-PAGE), matrix-assisted laser desorption/ionization time-of-flight mass spectrometry (MALDI-TOF-MS) and polymerase chain reaction (PCR), were used to characterize the LMW-GS composition in 103 cultivars from 12 countries. At the *Glu-A3* locus, all seven alleles could be reliably identified by 2-DE and PCR. However, the alleles *Glu-A3e* and *Glu-A3d* could not be routinely distinguished from *Glu-A3f* and *Glu-A3g*, respectively, based on SDS-PAGE, and the allele *Glu-A3a* could not be differentiated from *Glu-A3c* by MALDI-TOF-MS. At the *Glu-B3* locus, alleles *Glu-B3a*, *Glu-B3b*, *Glu-B3c*, *Glu-B3g*, *Glu-B3h* and *Glu-B3j* could be clearly identified by all four methods, whereas *Glu-B3ab*, *Glu-B3ac*, *Glu-B3ad* could only be identified by the 2-DE method. At the *Glu-D3* locus, allelic identification was problematic for the electrophoresis based methods and PCR. MALDI-TOF-MS has the potential to reliably identify the *Glu-D3* alleles. PCR is the simplest, most accurate, lowest cost, and therefore recommended method for identification of *Glu-A3* and *Glu-B3* alleles in breed-

Published in MBC Plant Biology, 2010, 10: 124

ing programs. A combination of methods was required to identify certain alleles, and would be especially useful when characterizing new alleles. A standard set of 30 cultivars for use in future studies was chosen to represent all LMW-GS allelic variants in the collection. Among them, Chinese Spring, Opata 85, Seri 82 and Pavon 76 were recommended as a core set for use in SDS-PAGE gels. *Glu-D3c* and *Glu-D3e* are the same allele. Two new alleles, namely, *Glu-D3m* in cultivar Darius, and *Glu-D3n* in Fengmai 27, were identified by 2-DE. Utilization of the suggested standard cultivar set, seed of which is available from the CIMMYT and INRA Clermont-Ferrand germplasm collections, should also promote information sharing in the identification of individual LMW-GS and thus provide useful information for quality improvement in common wheat.

Abbreviations: 2-DE, two-dimensional gel electrophoresis (IEF × SDS-PAGE); ACN, acetonitrile; BIOLAB AZUL, Laboratory of Functional Biology and Biotechnology; CAAS, Chinese Academy of Agricultural Sciences; CEBB-MdP, Biotechnology and Biodiversity Study Center, Mar del Plata, Argentina; CHAPS, 3- [(3-Cholanidopropyl) dimethylammonio] -1-propanesulfonate; CICPBA, Scientific Research Commission of the Province of Buenos Aires; CIMMYT, International Maize and Wheat Improvement Center; CIISAS, Center for Integrated Research in Sustainable Agricultural Systems, COUNTRY; CONICET, National Science and Technology Research Council, Argentina; CRESCAA, Regional Center for Systemic Study of Agro-alimentary Chains, COUNTRY; CTAB, cetyltrimethylammonium bromide; DNA, deoxyribonucleic acid; dNTPs, deoxynucleoside triphosphates; DTT, dithiothreitol; HMW-GS, high-molecular-weight glutenin subunits; HPLC, high-performance liquid chromatography; IEF, isoelectric focusing; INBA, Research Institute of Agricultural and Environmental Biosciences; INRA, National Institute for Agricultural Research; kDa, kilodalton; kVh, kilo-volt-ampere-hour; LMW-GS, low-molecular-weight glutenin subunits; MALDI-TOF-MS, matrix-assisted laser desorption/ ionization time-of-flight mass spectrometry; $MgCl_2$, magnesium chloride; NARO, National Agriculture and Food Research Organization; NIL, near-isogenic lines; PCR, polymerase chain reaction; pI, isoelectric points; SA, sinapinic acid; SDS, sodium dodecyl sulfate; SDS-PAGE, sodium dodecyl sulphate-polyacrylamide gel electrophoresis; TFA, trifluoroacetic acid; Tris-HCl, tris (hydroxymethyl) aminomethane hydrochloride.

Background

Glutenin proteins are the major factors responsible for the unique viscoelastic characteristics of wheat dough. They determine rheological properties and bread-making performance [1-3]. The polymeric glutenin proteins, with molecular weights ranging from less than 300 to more than 1, 000 kDa, are composed of two groups of subunits. These subunits include the LMW-GS, which are similar in size and structure to the γ- gliadins (30～40 kDa), and the high-molecular-weight glutenin subunits (HMW-GS) which range in molecular mass from 65 to 90 kDa[4]. The LMW-GS represent about one-third of the total seed protein and 60% of total glutenins[5], and are essential in determining dough properties, such as dough extensibility[6] and gluten strength[2]. Hence characterization of allelic variation among cultivars and investigation of their relationships with end-use quality has been a key area of research on quality improvement during the last 15 years, and is the basis for the success of using specific LMW-GS alleles in breeding programs [7-9].

The genes coding for LMW-GS are located on the short arms of homoeologous group 1 chromosomes at the *Glu-A3*, *Glu-B3* and *Glu-D3* loci, and are tightly linked to the *Gli-1* loci [10-12]. The *Glu-A3* locus on chromosome 1A encodes relatively few LMW-GS, with alleles *Glu-A3e* in hexaploid or common wheat and *Glu-*

A3h in tetraploid wheat being null alleles that do not express any *Glu-A3* product [13,14]. In contrast, there is extensive variation for LMW-GS encoded by chromosome 1B in common wheat. The *Glu-D3* locus has less variability with five alleles reported originally by Gupta and Shepherd [13], four alleles by Lerner et al. [15] and only three alleles observed by Jackson et al. [16] and Eagles et al. [17]. Nonetheless, recent studies using protein and PCR analyses have identified 11 *Glu-D3* alleles [18,19], suggesting that a reexamination should be carried out to clarify the genetic variability at this locus.

Despite the abundance of the LMW-GS, they have received much less attention than the HMW-GS, probably due to their complexity, heterogeneity and co-migration with gliadins in SDS-PAGE [19,20]. In the SDS-PAGE system, utilizing gliadins as indicators provided an indirect way to define LMW-GS alleles [16]. The 2-DE analytical process that could generate much more information than SDS-PAGE [21] was not generally recommended for use in breeding programs, due to its time-consuming procedure, high costs and skill requirements. MALDI-TOF-MS is currently the most efficient method to analyze proteins and requires only 4-5 minutes per sample. It is a high throughput technology for analyzing wheat gluten proteins [22-25], but being relatively new and expensive, few wheat breeding programs can afford to acquire such equipment. Recently, a simple, rapid and sensitive PCR approach, has proven to be a very useful tool for identifying LMW-GS composition in common wheat [19,26-28].

LMW-GS were first identified by gel filtration and starch gel electrophoresis of extracts of wheat flour[29,30]. They were classically subdivided into B, C, and D groups (no relationship to the A, B and D genomes of wheat), according to their electrophoretic mobilities in SDS-PAGE and their isoelectric points (pI) [31]. Based on the locations of cysteine residues involved in the formation of intermolecular disulfide bridges, Ikeda et al. [32] classified LMW gene sequences into six types, each containing several different groups based upon differences in their N- and C-terminal acid-amino compositions. Altogether, 12 groups were differentiated, but an additional five groups were reported by Juhász and Gianibelli [33].

The allelic nomenclature system for the LMW-GS was defined through the chromosomal location of the DNA coding regions by Gupta and Shepherd [13] and was reviewed by Jackson et al. [16]. Branlard et al. [34] proposed a schematic presentation of SDS-PAGE relative subunit mobilities to characterize the different alleles encoded at *Glu-A3*, *Glu-B3* and *Glu-D3* loci. Ikeda et al. [35] recently compared *Glu-3* allele identifications from five laboratories, confirming inconsistencies between laboratories in identifying *Glu-3* alleles due to differences between the separation and identification methods. The study also indicated new *Glu-3* alleles in a number of the cultivars analyzed.

The N-terminal sequences of LMW-GS were used to divide the protein subunits into two main groups [32,36]. The first group corresponded to typical LMW-GS, i. e., LMW-i (or i-type, first amino acid isoleucine) and LMW-m (or m-type, methionine) types, and the second group, named gliadin-like sequences [37] as these subunits have N-terminal sequences similar to α-, γ- and ω- gliadins. Most gliadins are monomeric, but some have an extra cys that allows them to be incorporated into glutenin polymers. Payne [1] termed the prominent bands observed by SDS-PAGE of reduced glutenin protein as A (HMW-GS), B (many of the LMW-GS) and C (the smaller LMW-GS). Later, other researchers also observed larger gliadin-like subunits, between the A and B bands, and they named them as D- subunits [31]. Most of the B- subunits were shown to possess i-, m- or s (serine) -type N-terminal sequences [38]. C-subunits including α-, and γ- gliadins-like subunits as well as subunits with classic LMW-GS sequences occur in large numbers, although their relative amounts are lower than those of B-subunits. Similarly, D- subunits have N-terminal sequences that correspond to ω-gliadins, another type of gliadin-like sequence [2,39,40].

The use of two distinct nomenclature systems, one based upon the relative mobilities in SDS-PAGE and the other upon N-terminal sequences, make it extremely difficult to compare work from different laboratories. The main ambiguities from these different classification systems can be summarized as follows: 1) at the *Glu-A3* locus, both *Glu-A3a* and *Glu-A3c* were reported for the same cultivar, and similarly, *Glu-A3a*, *Glu-A3b*, *Glu-A3c*, *Glu-A3d* were reported to be identical to *Glu-A3e*; 2) at the *Glu-B3* locus, results differed for *Glu-B3b* and *Glu-B3g*, and for *Glu-B3f* and *Glu-B3g* in the same cultivars; and 3) at the *Glu-D3* locus, there was ambiguity between *Glu-D3a* and *Glu-D3c*, and between *Glu-D3a* and *Glu-D3b* in the same cultivars[41]. As a consequence of these problems, reports of correlations between certain allelic forms of LMW-GS and quality parameters in common wheat have often been contradictory [7,42-45]. It is, therefore, essential to establish a simple and uniform classification through a set of standard cultivars for each LMW-GS allele.

In 2005, a cooperative program was developed among the following five laboratories to establish such a set of standard cultivars for identifying LMW-GS alleles: Chinese Academy of Agricultural Sciences (CAAS, China), International Maize and Wheat Improvement Center (CIMMYT, Mexico), National Institute for Agricultural Research (INRA, France), National Agriculture and Food Research Organization (NARO, Japan), and National University of the Center of the Province of Buenos Aires (Universidad Nacional, Argentina). A set of 103 cultivars used in various previously studies [35] in 12 countries was assembled and distributed to all laboratories, including Murdoch University as an additional laboratory, for the identification of LMW-GS alleles. Their preliminary *Glu-3* allelic assignments were summarized in a previous paper[35]. The objectives of the current paper are 1) to compare the LMW-GS compositions obtained by SDS-PAGE, 2-DE, MALDI-TOF-MS and PCR in order to clearly identify the protein compositions of cultivars in the collection; and 2) to establish a set of standard cultivars for the identification of LMW-GS alleles, enabling information regarding the effects of individual LMW-GS on gluten properties to be readily and continuously shared between laboratories and applied in breeding programs.

Materials and methods

Plant materials

One hundred and three cultivars of common wheat collected from 12 countries were used to develop a set of standard cultivars for identification of LMW-GS alleles (Table 1). They included 21 cultivars from China, 19 from Argentina, 15 from Australia, 14 from France, 10 from Japan, eight from Mexico, seven from Canada, three from the USA, two from Italy, two from the Netherlands, one from Finland and one from Germany. These cultivars were widely utilized in investigating glutenin subunit compositions and their relationships to processing quality [41].

Table 1 Compositions of LMW-GS alleles in 103 wheat cultivars identified by SDS-PAGE, 2-DE, MALDI-TOF-MS and allele-specific markers. Data from five laboratories are combined data for SDS-PAGE. — indicates data not available.

Cultivar	Origin	*Glu-A3*	*Glu-B3*	*Glu-D3*
Aca 303	Argentina	f/f/f/f*	h/h/h/h	c/c/c/-
Aca 601	Argentina	f/f/f/f	b/b/b/b	c/c/c/-
Aca 801	Argentina	c/c/a or c/c	g/ac/g/g	b/b/b/-
Buck Brasil	Argentina	f/f/f/f	g/ac/g/g	d/d/? /-
Buck Mejorpán	Argentina	f/f/f/f	b/b/b/b	c/c/c/-
Buck Pingo	Argentina	f/f/f/f	i/ad/d or i/i	c/c/c/-

（续）

Cultivar	Origin	*Glu-A3*	*Glu-B3*	*Glu-D3*
Klein Capricornio	Argentina	c/c/a or c/c	h/h/h/h	b/b/b/-
Klein Chaja	Argentina	c/c/a or c/c	h/h/h/h	b/b/b/-
Klein Flecha	Argentina	c/c/a or c/c	h/h/h/h	b/b/b/-
Klein Jabal 1	Argentina	d/d/d/g	g/g/g/g	c/c/c/-
Klein Martillo	Argentina	e/e/e/e	j/j/j/j	b/b/b/-
Klein Proteo	Argentina	g/g/e/g	g/ac/g/g	b/b/b/-
Nidera Baguette 10	Argentina	d/d/d/d	g/g/g/g	c/c/c/-
Nidera Baguette 20	Argentina	f/f/f/f	g/g/g/g	c/c/c/-
ProINTA Amanecer	Argentina	f/f/f/f	j/j/j/j	a/a/b/-
ProINTA Colibr 1	Argentina	d/d/d/d	b/b/b/b	a/a/a/-
ProINTA Isla Verde	Argentina	b/b/b/b	b/b/b/b	b/b/b/-
ProINTA Redomon	Argentina	c/c/a or c/c	h/h/h/h	c/c/c/-
Thomas Nevado	Argentina	c/c/a or c/c	j/j/j/j	b/b/b/-
Angas	Australia	c/c/a or c/c	g/g/g/g	c/c/c/-
Avocet	Australia	c/c/a or c/c	b/b/b/b	b/b/b/-
Carnamah	Australia	c/c/a or c/c	i/ad/d or i/i	c/c/c/-
Gabo	Australia	b/b/b/b	b/b/b/b	b/b/b/-
Grebe	Australia	c/c/a or c/c	j/j/j/j	b/b/b/-
Halberd	Australia	e/e/e/e	c/c/c/c	c/c/c/-
Insignia	Australia	f/f/f/f	c/c/c/c	c/c/c/-
Millewa	Australia	c/c/a or c/c	g/g/g/g	b/b/b/-
Spear	Australia	e/e/e/e	h/h/h/h	c/c/c/-
Stiletto	Australia	c/c/a or c/c	h/h/h/h	c/c/c/-
Tasman	Australia	b/b/b/b	i/ad/d or i/i	a/a/a/-
Trident	Australia	e/e/e/e	h/h/h/h	c/c/c/-
Westonia	Australia	c/c/a or c/c	h/h/h/h	c/c/c/-
Wilgoyne	Australia	d/d/d/d	h/h/h/h	b/b/b/-
AC Vista	Canada	e/e/e/e	i/ad/d or i/i	c/c/c/-
Bluesky	Canada	g/g/e/g	g/g/g/g	c/c/c/-
Glenlea	Canada	g/g/e/g	g/g/g/g	c/c/c/-
Katepwa	Canada	e/e/e/e	h/h/h/h	c/c/c/-
Marquis	Canada	e/e/e/e	b/b/b/b	a/a/a/-
Neepawa	Canada	e/e/e/e	h/h/h/h	c/c/c/-
Pioneer	Canada	e/e/e/e	i/ad/d or i/i	c/c/c/-
99G46	China	f/f/f/f	j/j/j/j	c/c/c/-
CA9641	China	d/d/d/d	h/h/h/h	c/c/c/-
CA9722	China	c/c/a or c/c	h/h/h/h	c/l/c/-
Chinese Spring	China	a/a/a or c/a	a/a/a/a	a/a/a/-

（续）

Cultivar	Origin	*Glu-A3*	*Glu-B3*	*Glu-D3*
Demai 3	China	c/c/a or c/c	i/d or i/d or i/i	b/b/b/-
Fengmai 27	China	c/c/a or c/c	f/f/f/f	a/n/a/-
Guanfeng 2	China	c/c/a or c/c	b/b/b/b	a/a/a/-
Huaimai 16	China	f/f/f/f	h/h/h/h	c/c/c/-
Jing 411	China	c/c/a or c/c	h/h/h/h	c/l/c/-
Lumai 23	China	c/c/a or c/c	d/d or i/d or i/d	c/l/c/-
Neixiang 188	China	a/a/a or c/a	j/j/j/j	a/a/a/-
Shan 229	China	c/c/a or c/c	j/j/j/j	b/b/b/-
Wanmai 33	China	d/d/d/d	g/g/g/g	a/a/b/-
Yan 239	China	c/c/a or c/c	j/j/j/j	b/b/b/-
Yangmai 158	China	c/c/a or c/c	g/g/g/g	c/c/c/-
Yumai 54	China	c/c/a or c/c	d/d or i/d or i/d	c/c/c/-
Yumai 63	China	c/c/a or c/c	d/d or i/d or i/d	c/c/c/-
Yumai 69	China	c/c/a or c/c	d/d or i/d or i/d	a/a/b/-
Zhongyou 9507	China	d/d/d/d	b/b/b/b	c/c/c/-
Zhongyou 9701	China	d/d/d/d	d/d or i/d or i/d	c/c/c/-
Zhongyu 415	China	c/c/a or c/c	d/d or i/d/d	c/c/c/-
Ruso	Finland	c/c/a or c/c	i/ad/d or i/i	a/a/a/-
Brimstone	France	c/c/a or c/c	g/g/g/g	d/d/? /-
Cappelle-Desprez	France	d/d/d/d	g/g/g/g	c/c/c/-
Chopin	France	c/c/a or c/c	h/h/h/h	c/c/c/-
Clément	France	f/f/f/f	j/j/j/j	c/c/c/-
Courtot	France	c/c/a or c/c	b/b/b/b	c/l/c/-
Darius	France	d/d/d/d	g/g/g/g	b/m/m/-
Etoilede Choisy	France	d/d/d/d	i/d or i/d or i/i	c/l/c/-
Festin	France	f/f/f/f	b/b/b/b	c/l/c/-
Magali Blondeau	France	e/e/e/e	g/g/g/f	b/b/b/-
Magdalena	France	d/d/d/d	b/b/b/b	a/a/a/-
Petrel	France	d/d/d/d	h/h/h/h	c/c/c/-
Renan	France	f/f/f/f	b/b/b/b	b/b/b/-
Soissons	France	c/c/a or c/c	b/b/b/b	c/c/c/-
Thesee	France	c/c/a or c/c	g/ac/g/g	c/l/c/-
Apollo	Germany	d/d/d/d	j/j/j/j	c/c/c/-
Manital	Italy	c/c/a or c/c	b/b/b/b	a/a/a/-
Salmone	Italy	c/c/a or c/c	c/c/c/g	c/c/c/-
Aoba-komugi	Japan	e/e/e/e	b/b/b/b	c/c/c/-
Eshimashinriki	Japan	c/c/a or c/c	d/d or i/d or i/d	a/a/a/-
Haruyutaka	Japan	c/c/a or c/c	h/h/h/h	b/b/b/-
Kanto 107	Japan	c/c/a or c/c	g/g/g/g	a/a/a/-
Kitanokaori	Japan	f/f/f/f	j/j/j/j	c/c/c/-

（续）

Cultivar	Origin	*Glu-A3*	*Glu-B3*	*Glu-D3*
Nanbu-komugi	Japan	d/d/d/d	b/ab/b/b	a/a/a/-
Norin 61	Japan	d/d/d/d	i/d or i/d or i/-	c/c/c/-
Norin 67	Japan	c/c/a or c/c	g/g/g/g	b/b/b/-
Shinchunaga	Japan	c/c/a or c/c	i/ad/d or i/-	a/a/a/-
Shirane-komugi	Japan	e/e/e/e	i/ad/d or i/i	a/a/a/-
Amadina	Mexico	e/e/e/e	j/j/j/j	c/l/c/-
Attila	Mexico	c/c/a or c/c	h/h/h/h	b/b/b/-
Heilo	Mexico	f/f/f/f	i/ad/d or i/i	c/l/c/-
Opata 85	Mexico	b/b/b/b	i/ad/d or i/i	a/a/a/-
Pastor	Mexico	c/c/a or c/c	g/g/g/g	b/b/b/-
Pavon 76	Mexico	b/b/b/b	h/h/h/h	b/b/b/-
Pitic	Mexico	c/c/a or c/c	b/b/b/b	b/b/b/-
Rebeca	Mexico	c/c/a or c/c	g/g/g/g	b/b/b/-
Seri 82	Mexico	c/c/a or c/c	j/j/j/j	b/b/b/-
Orca	Netherlands	d/d/d/d	d/d or i/d or i/d	c/c/c/-
Pepital	Netherlands	f/f/f/f	d/ d or i/d or i/d	c/l/c/-
Ernest	USA	d/d/d/d	d/d or i/d or i/d	d/? /? -
Splendor	USA	e/e/e/e	g/g/g/g	b/b/b/-
Verde	USA	f/f/f/f	h/h/h/h	c/c/c/-

*, the first, second, third and fourth symbol in each column are alleles of *Glu-3* loci identified by SDS-PAGE, 2-DE, MALDI-TOF-MS and PCR, respectively.

Protein extraction

A similar protocol was adopted for protein extraction in all five laboratories. Proteins were extracted from 100 mg whole meal according to the sequential procedure of Branlard and Bancel [53]. The samples were treated with 1.0mL of 50% propanol-1-ol (v/v) for 5min with continuous vortexing, followed by incubation (20min at 65°C), vortexing (5min), and centrifugation (5min at 10, 000g). This step was repeated three times to remove most of the gliadins. The glutenin in the pellet was reduced with 50% propanol-1-ol, 50mM Tris-HCl solution containing 1% w/v dithiothreitol (DTT), after which 1.4% v/v of 4-vinylpyridine was added, and alkylation was continued overnight at room temperature. The protein of each cultivar was extracted in three replicates.

SDS-PAGE

SDS-PAGE was performed in all five laboratories. Glutenin and gliadin protein extracts were separated using the method of Singh et al. [46] with some modifications in different laboratories to obtain the best resolution. To summarize, there were differences in three aspects. The concentrations of separation gel were 14.0% concentration (T) with 1.3% cross linker (C), 15.0% T with 1.3% C, 12.5% T with 0.97% C, 15.0% T with 1.4% C, and 13.5% T with 0.8% C in the laboratory of CAAS, CIMMYT, INRA, NARO and Universidad Nacionalm of Argentina, respectively. The pH for separation gel was pH8.8 in all laboratories except in CIMMYT with pH8.5. The currents of running gel were 16, 12.5, 30, 30 and 40mA in the laboratory of CAAS, CIMMYT, INRA, NARO and Universidad Nacionalm of Argentina laboratory, respectively. Generally, lower current results in better resolution, but we could not find the optimum conditions for maximum resolution of LMW-GS in all laboratories since each laboratory used its own optimum conditions. Details were reported by Ikeda et al. [35].

The LMW-GS compositions were identified according

to Singh et al.[46] and Jackson et al. [16] and the gliadins were used as indicators of LMW-GS based on the linkage between LMW-GS and gliadin because the gliadin composition can be screened more readily than specific LMW-GS. The nomenclature system of LMW-GS followed Gupta and Shepherd [13], Jackson et al.[16], Branlard et al. [34], Ikeda et al. [35], Appelbee et al.[19] and the catalogue of gene symbols for wheat (http://wheat.pw.usda.gov/ggpages/awn/53/Textfile/WGC.html).

2-DE procedure

The 2-DE method was only performed at CAAS and NARO. The 2-DE procedure employed to identify LMW-GS was performed with an IPGphor (GE Healthcare, Sweden) for isoelectric focusing (IEF), and an AE-6530 chamber and an AE-8450 power supply (ATTO, Japan) for SDS-PAGE. The glutenin fraction was precipitated with 80% acetone [54], and the resulting pellets containing 150μg protein were dissolved in 250μl of IEF rehydration solution [7M urea, 2M thiourea, 4% w/v CHAPS, 2% v/v IPG buffer pH 6-11 (GE Healthcare) and 20mM DTT] for very basic proteins [55]. After incubation for 30min at room temperature, samples were applied to Immobiline DryStrip pH 6-11 (13cm, GE Healthcare). The rehydration step was carried out for 12h at 20℃. IEF was performed with a step-wise protocol to 45kVh. After IEF, the strips were stored at −80℃ or prepared directly for 2-DE as follows: the gel strips were first equilibrated under gentle shaking for 15min in equilibration buffer (50mM Tris-HCl, pH 8.8, 6M urea, 30% v/v glycerol, 2% w/v SDS) with 2% w/v DTT, and then in equilibration buffer containing 1.4% v/v 4-vinylpyridine. The second dimension separations (SDS-PAGE) were carried out on 13% acrylamide constant gels and ran at 7mA/gel for 45min and then 25mA/gel for approximately 4h, until the bromophenol blue had run off the bottom of the gel [56]. After the completion of 2-DE, gels were fixed and stained with Coomassie Brilliant Blue-G250 according to Neuhoff et al.[57]. The resulting gels were scanned using an Image Scanner (GE Healthcare) and the images analyzed with ImageMaster 2D Platinum v6.0 software (GE Healthcare). At least three gel images of each sample were taken and compared. The LMW-GS compositions were identified with the distinctive spot on 2-DE gels according to Ikeda et al. [18]. The nomenclature system of LMW-GS was the same as above SDS-PAGE separation.

In some cases the 2-DE was modified where glutenin-proteins were not alkylated; 16% isopropanol was added to the IEF buffer, and IEF was performed at 18 kVh [18].

MALDI-TOF-MS protocol

MALDI-TOF-MS was performed at the State Agriculture Biotechnology Center, Murdoch University, Australia. The glutenin fraction was precipitated with 80% acetone [54], and the resulting pellets containing 100 μg protein were dissolved in 60μl acetonitrile (ACN) / H_2O (50 : 50 v/v) containing 0.05% v/v trifluoroacetic acid (TFA) for 1h at room temperature. Sample preparation was carried out according to the dried droplet method [58], using sinapinic acid (SA) as matrix. The matrix solution was prepared by dissolving SA in 50% ACN/0.05% TFA (w/w) at a concentration of 10mg/mL. A sandwich matrix/sample/matrix 1 : 1 : 1 (0.7μl) was deposited on to a 96-sample MALDI target, and dried at room temperature.

MALDI-TOF-MS was performed on a Voyager DE-PRO TOF mass spectrometer (Applied Biosystems, Foster City, CA, USA) equipped with a 337nm nitrogen laser and delayed extraction. Analyses were carried out on a positive linear ion mode at a mass range of 10,000-50,000m/z with an accelerating voltage of 25 kV and a delay time of 900ns. A low mass gate value of 10,000m/z was selected for analysis to avoid saturation of the detector. The identification of LMW-GS alleles based on MALDI-TOF-MS was established using a set of 19 near-isogenic lines (NIL) of cultivar Aroona (unpublished data, A Wang, W Ma, R Appels, Murdoch University, Australia).

DNA extraction and PCR amplification

PCR was performed onlyat CAAS. Genomic DNA was extracted from seeds using a modified CTAB procedure[59]. PCR was performed using TaKaRa (Dalian, China) *Taq* DNA polymerase (1. 0unit) in 20μl reaction volumes containing approximately 50ng of genomic DNA, 1× PCR buffer (1. 5mM $MgCl_2$), 100μM of each dNTP and 7. 5pmol of each PCR primer. Details of allele-specific markers for the discrimination of *Glu-A3* and *Glu-B3* alleles and PCR conditions were reported previously [27, 28].

Results and discussion

Analysis of LMW-GS by SDS-PAGE

The LMW-GS compositions identified in participating laboratories by SDS-PAGE were combined and listed in Table 1 (details available upon request); discrepancies among different laboratories were discussed by Ikeda et al. [35]. At the *Glu-A3* locus, alleles *Glu-A3a*, *Glu-A3b*, *Glu-A3c* and *Glu-A3f* could be readily identified using SDS-PAGE (Fig. 1). Alleles *Glu-A3d* and *Glu-A3g* could be differentiated with the aid of the gliadin SDS-PAGE gel; by the presence or absence of the *Gli-A1o* allele, which we believe is linked to *Glu-A3d*, but not to *Glu-A3g* (Fig. 2). It was difficult to distinguish *Glu-A3f* from *Glu-A3e* (null allele). In previous studies [7,46,47] both alleles tended to be detected as *Glu-A3e*.

Fig. 3 shows cultivars representing different *Glu-B3* alleles. At the *Glu-B3* locus, three alleles, *Glu-B3d*, *Glu-B3h* and *Glu-B3i*, each carried the slowest LMW-GS bands in the SDS-PAGE region B among the

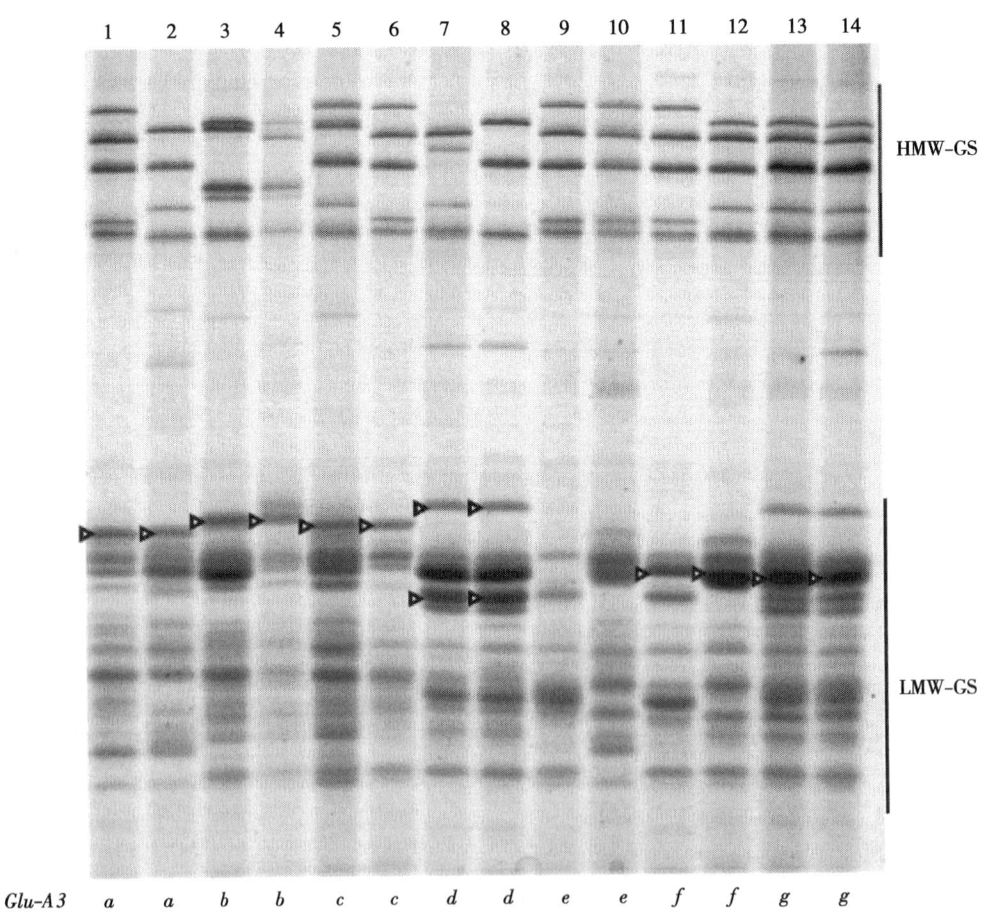

Fig. 1 SDS-PAGE of LMW-GS. The LMW-GS are propanol-insoluble fractions extracted with 50% propanol + 1% w/v DTT + 1. 4% v/v 4-vinylpyridine (The same as below). Cultivars: 1. Neixiang 188, 2. Chinese Spring, 3. Gabo, 4. Pavon 76, 5. Pitic, 6. Seri, 7. Nidera Baguette 10, 8. Cappelle-Desprez, 9. Amadina, 10. Marquis, 11. Kitanokaori, 12. Renan, 13. Bluesky, 14. Glenlea. Arrow heads indicate bands corresponding to different *Glu-A3* alleles.

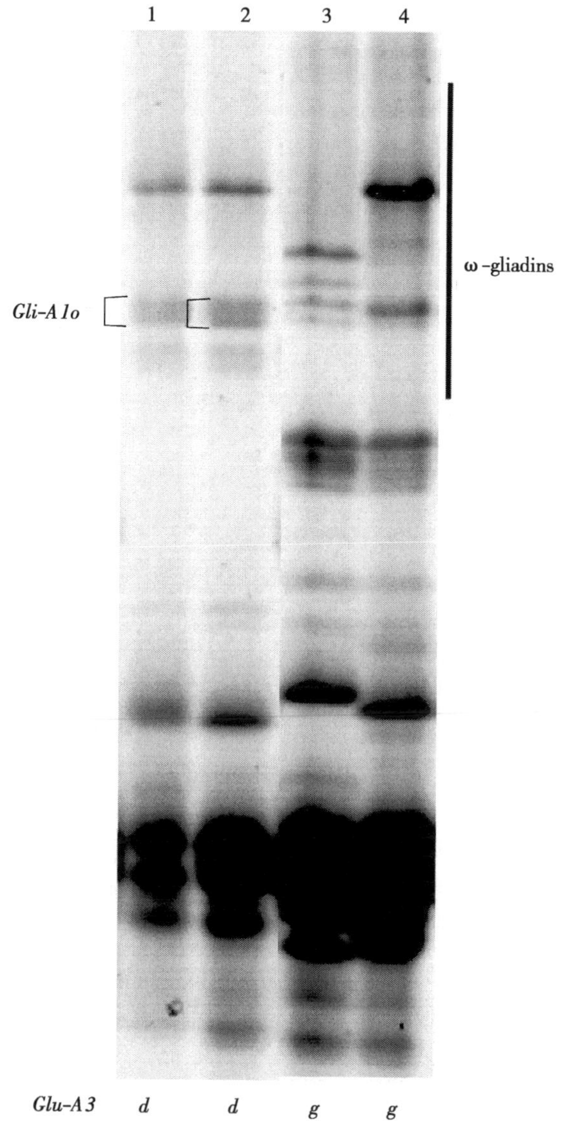

Fig. 2 SDS-PAGE of gliadins. The gliadins are 50% propanol (v/v) soluble fractions (The same as below). Cultivars in lanes 7, 8, 13, and 14, correspond to the same shown with same number in Fig. 1. The linkage between *Gli-A1o* (indicated in lanes 7 and 8 in the omega-gliadin zone) and *Glu-A3d* helps to differentiate the latter from *Glu-A3g*.

cultivars studied. The slowest *Glu-B3* band, *Glu-B3b*, almost coincided with *Glu-A3a*, but the *Glu-B3b* band was usually lighter and thinner, permitting their discrimination. Allele *Glu-B3f* could not be reliably discriminated from *Glu-B3g* since these bands had very similar mobilities, including the presence of a band in the SDS-PAGE region (Fig. 3, lanes 8-10) as previously reported [7,34,41]. However, taking advantage of the *Glu-B3/Gli-B1* linkage, one can look at the omega-gliadins region in SDS-PAGE, to identify with confidence several of the *Glu-B3* alleles (Fig. 4). Actually, differentiating between several *Glu-B3* alleles is possible only looking at both, gliadin and glutenin SDS-PAGE gels. Using this criteria, *Glu-B3* alleles in lanes 15, 16, 17, 18, and 19, seem to correspond to *Glu-B3b*, *Glu-B3g*, *Glu-B3g*, *Glu-B3i*, and *Glu-B3i*, respectively (Figs. 3 and 4), however, 2-DE analysis indicates that these genotypes correspond to new alleles provisionally designated as *Glu-B3ab*, *Glu-B3ac*, *Glu-B3ac*, *Glu-B3ad*, and *Glu-B3ad*, respectively (Fig. 3).

Fig. 5 shows cultivars representing different *Glu-D3* alleles. Although alleles *Glu-D3a*, *Glu-D3b*, *Glu-D3c* and *Glu-D3d* were frequently identified in germplasm from various origins [35], only alleles *Glu-D3a*, *Glu-D3b* and *Glu-D3d* were consistently differentiated [34]. *Glu-D3* alleles had similar mobilities to gliadins and were generally faintly stained due to the rapid diffusion of low molecular mass proteins from the gel. Thus the identification of *Glu-D3* alleles was quite difficult using only SDS-PAGE, leading to the reported discrepancies[13,19,41]. Although improvements to the SDS-PAGE protocol now allow differentiating several of the *Glu-D3* alleles with more certainty, as it is shown in Fig. 5, other methods for definitive identification of these alleles, such as 2-DE, MALDI-TOF-MS and PCR, had to be implemented to facilitate identification of *Glu-D3* alleles.

Characterization of LMW-GS by 2-DE

The identification of the LMW-GS alleles by 2-DE was consistent between the two laboratories. The discrimination between LMW-GS alleles in the collection by high resolution 2-DE is illustrated in Fig. 6-9 and the results are shown in Table 1. The *Glu-A3* alleles *Glu-A3d* [Fig. 6, (4)], *Glu-A3e* [Fig. 6, (5)], *Glu-A3f* [Fig. 6, (6)] and *Glu-A3g* [Fig. 7, (1)], were readily differentiated on the basis of protein spots with clearly different molecular masses and pI. Alleles *Glu-A3a*

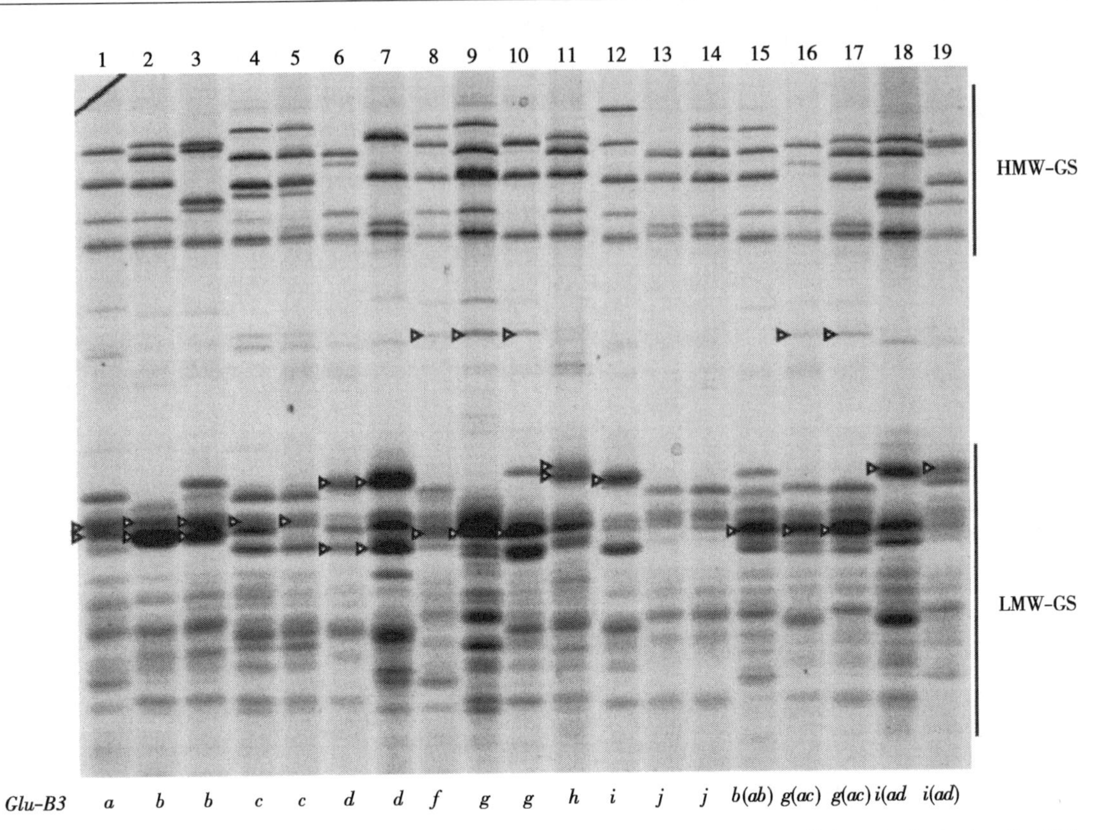

Fig. 3 SDS-PAGE of LMW-GS. Cultivars: 1. Chinese Spring, 2. Renan, 3. Gabo, 4. Insignia, 5. Halberd, 6. Pepital, 7. Ernest, 8. Fengmai 27, 9. Splendor, 10. Cappelle-Desprez, 11. Aca 303, 12. Norin 61, 13. Grebe, 14. Seri 82, 15. Nanbu-komugi, 16. Thesee, 17. Aca 801, 18. Heilo, 19. Opata. Arrow heads indicate bands corresponding to different *Glu-B3* alleles. *Glu-B3* allele designation between brackets for cultivars in lanes 15～19 correspond to provisional nomenclature as indicated by spot differences in 2-DE.

[Fig. 6, (1)], *Glu-A3b* [Fig. 6, (2)] and *Glu-A3c* [Fig. 6, (3) had identical pI but different molecular masses, making it possible to discriminate between them.

At the *Glu-B3*, the alleles *Glu-B3ab* [Fig. 7, (4)], *Glu-B3ac* [Fig. 8, (2)], *Glu-B3h* [Fig. 8, (3)], *Glu-B3ad* [Fig. 8, (4)] and *Glu-B3j* [Fig. 8, (5)] were easily differentiated by protein spots having different molecular masses and pI. Alleles *Glu-B3ab* [Fig. 7, (4)], *Glu-B3ac* [Fig. 8, (2)] and *Glu-B3ad* [Fig. 8, (4)] were each discriminated from *Glu-B3b* [Fig. 7, (3)], *Glu-B3g* [Fig. 8, (1)] and *Glu-B3i* [image not provided] by two distinct protein spots. Although the majority of the protein spots for alleles *Glu-B3b* and *Glu-B3g* had identical molecular masses and pI, they could be discriminated since allele *Glu-B3g* had one additional spot, at pH6, located between the HMW-GS and gliadins. There were no obvious differences in molecular mass or pI between alleles *Glu-B3d* [Fig. 7, (6)] and *Glu-B3i* (image not provided), or between *Glu-B3f* (image not provided) and *Glu-B3g* [Fig. 8, (1)], making differentiation by 2-DE impossible.

At the *Glu-D3*, only *Glu-D3c* [Fig. 9, (2)], *Glu-D3l* [Fig. 9, (3)] and *Glu-D3m* [Fig. 9, (4)] could be definitely identified by 2-DE. Allele *Glu-D3l* [Fig. 9, (3)] had two more distinctive spots compared to *Glu-D3c* [Fig. 9, (2)] in 2-DE separations. As expected, alleles *Glu-D3c* and *Glu-D3e* (image not provided) could not be separated by 2-DE. These alleles appeared to be the same based on SDS-PAGE and MALDI-TOF-MS in the present study as they were in a previous study[16].

2-DE did not distinguish *Glu-D3a* [Fig. 8, (6)], *Glu-*

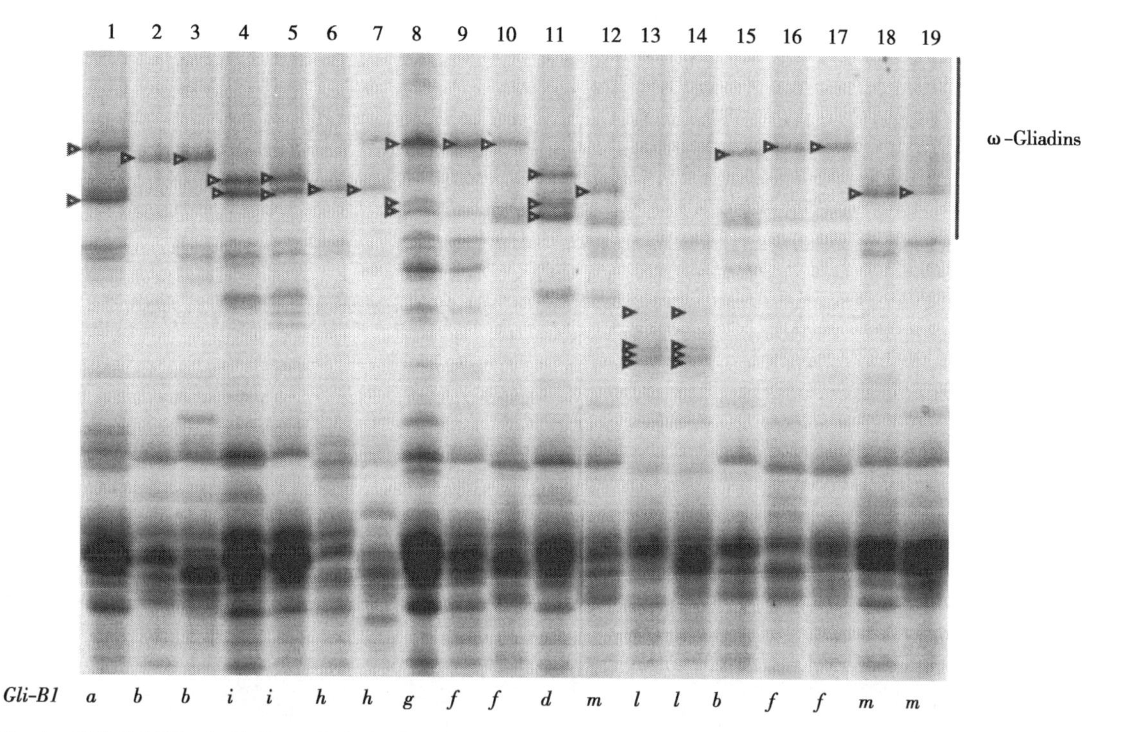

Fig. 4 SDS-PAGE of gliadins. Cultivars in lanes 1～19 are the same as in Fig. 3. Arrow heads indicate bands corresponding to different *Gli-B1* alleles. *Glu-B3* and *Gli-B1* alleles in each of the lanes 1 to 19 of Fig. 3 and 4 are tightly linked. The bands indicated with arrow heads of Fig. 4 are used as assisted bands for the identification of some *Glu-B3* alleles shown on Fig. 3 based on *Glu-B3*/*Gli-B1* linkage.

D3b [Fig. 9, (1)] and *Glu-D3d* (image not provided), hence further investigation should target discrimination of *Glu-D3* alleles by combining 2-DE with other methods such as PCR.

Identification of LMW-GS by MALDI-TOF-MS

The compositions of LMW-GS analyzed by MALDI-TOF-MS are presented in Table 1. As shown in Figures 10-13, the spectra of LMW subunits analyzed by this method consist of complex sets of peaks, consistent with the extensive diversity of the subunits. The LMW-GS exhibited molecular masses of 25-43kDa in MALDI-TOF-MS spectra, considerably lower than the corresponding molecular masses of 42-51kDa determined by SDS-PAGE and indicative of limitations of the SDS-PAGE method in determining the molecular masses of LMW glutenins [24]. Two major regions with masses from 30 to 35kDa and from 36 to 43kDa were separated in spectra of MALDI-TOF-MS (Figs. 10-13). These regions correspond to the C- LMW-GS and B- LMW-GS classified by SDS-PAGE. The region with molecular masses of 30-35kDa also corresponds in mass to the major gliadins range [1]. The results were in agreement with previous studies based on SDS-PAGE, where there were extensive overlaps between gliadins and LMW-GS with lower molecular masses[48].

MALDI-TOF-MS-based identification of LMW-GS alleles was established using a set of 19 near-isogenic lines (NIL) of cultivar Aroona (unpublished data, A Wang, W Ma, R Appels, Murdoch University, Australia). Most of the distinct peaks of the *Glu-A3* alleles exhibited higher masses in the ranges of about 41.8-42.1kDa and 43.5-43.8kDa, whereas the distinct peaks of the *Glu-D3* alleles showed lower masses of 33.2-33.7kDa. The middle masses in the ranges of about 40.1-40.2kDa and 42.8-43.3kDa corresponded to the *Glu-B3* alleles. The distributions of distinct peaks of the *Glu-3* alleles in the MALDI-TOF-MS were in agreement with their position in SDS-PAGE [34].

Compared to the other loci, *Glu-A3* was less diverse

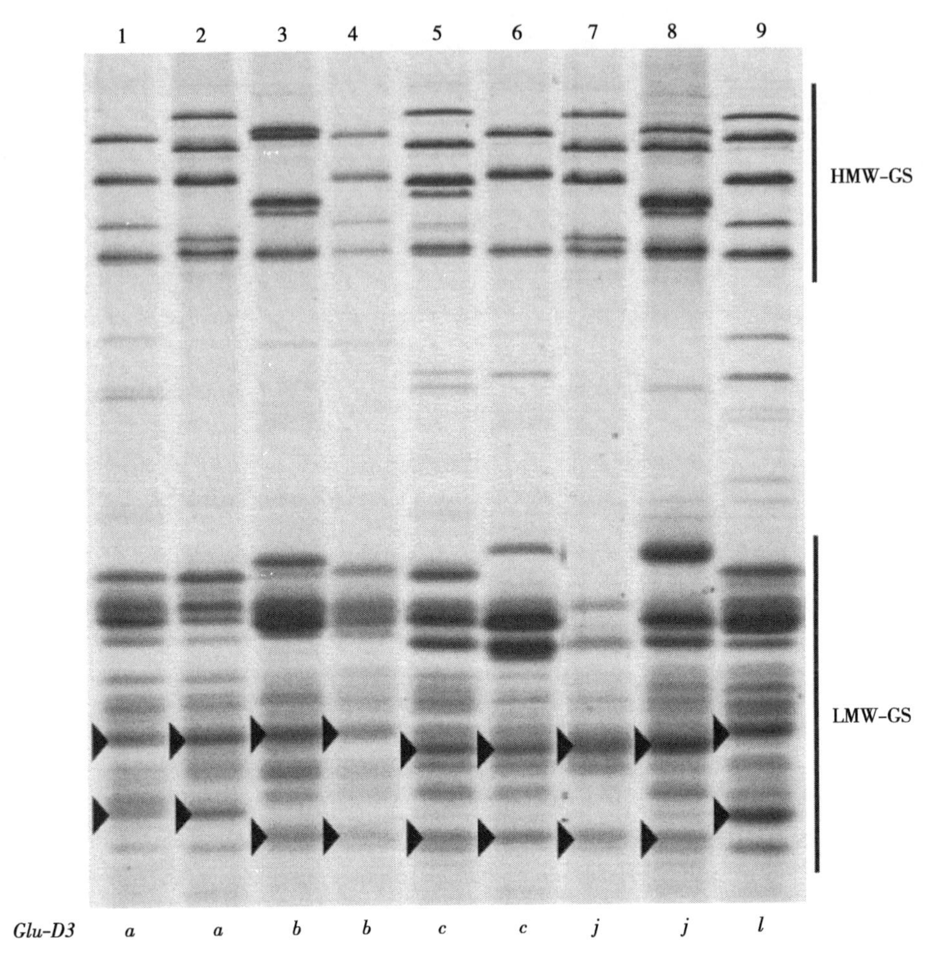

Fig. 5 SDS-PAGE of LMW-GS. Cultivars: 1. Chinese Spring, 2. Neixiang 188, 3. Gabo, 4. Avocet, 5. Insignia, 6. Cappelle- Desprez, 7. Amadina, 8. Heilo, 9. Fengmai 27. Arrow heads indicate bands corresponding to different *Glu-D3* alleles.

and most protein bands had lower mobilities, so discrimination between them using SDS-PAGE is usually feasible. Similarly, most of the distinct peaks of the *Glu-A3* alleles were well separated in MALDI-TOF-MS spectra, and alleles *Glu-A3b* [Fig. 10, (2)], *Glu-A3d* [Fig. 10, (3)], *Glu-A3e* [Fig. 10, (4)] and *Glu-A3f* [Fig. 11, (1)] were reliably discriminated.

The *Glu-B3* alleles *Glu-B3a* [Fig. 11, (2)], *Glu-B3b* [Fig. 11, (3)], *Glu-B3c* [Fig. 11, (4)], *Glu-B3h* [Fig. 12, (3)], and *Glu-B3j* [Fig. 12, (4)], as well as seven other alleles, were readily distinguished by MALDI-TOF-MS.

With regard the *Glu-D3* locus, MALDI-TOF-MS clearly differentiated the *Glu-D3a* [Fig. 13, (1)], *Glu-D3b* [Fig. 13, (2)], *Glu-D3c* [Fig. 13, (3)] and *Glu-D3m* [Fig. 13, (4)] alleles. As expected, *Glu-D3e* (image not provided) could not be discriminated from *Glu-D3c* [Fig. 13, (3)] . Improved discrimination will be achieved as calibration technology improves. In addition, it may be of value to utilize the close linkage between gliadin and LMW glutenin alleles to further improve the power of MALDI-TOF-MS in differentiating LMW glutenin alleles.

Detection of LMW-GS by allele specific PCR markers

Seven primer pairs [27], including *gluA3a*, *gluA3b*, *gluA3ac*, *gluA3d*, *gluA3e*, *gluA3f* and *gluA3g* were used to identify *Glu-A3* alleles (Table 1). The amplified fragment sizes for each marker were 529bp for *Glu-A3a*, 894bp for *Glu-A3b*, 967bp for *Glu-A3d*, 158bp for *Glu-A3e*, 552bp for *Glu-A3f*, and 1345bp

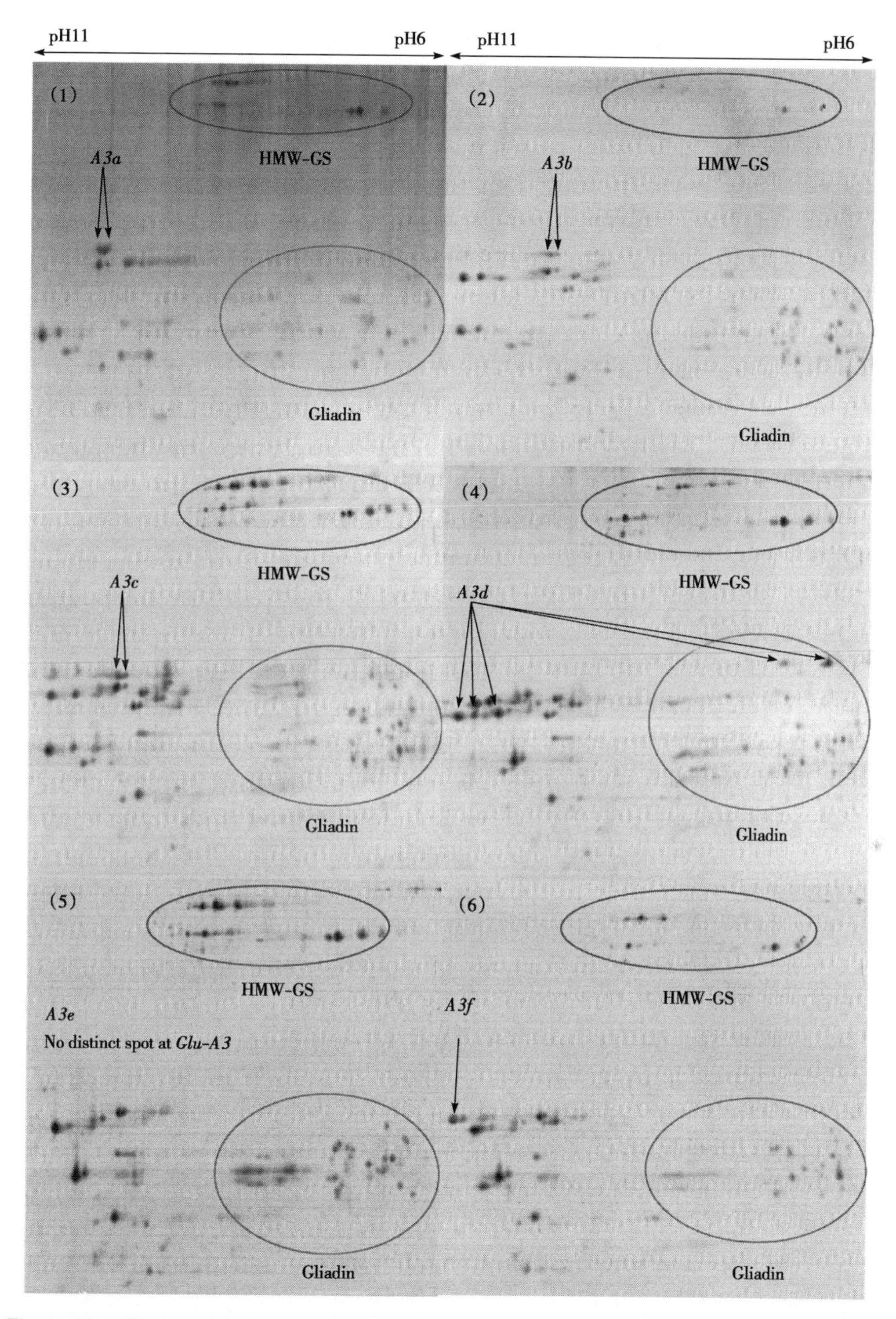

Fig. 6 Identification of LMW-GS by two-dimensional gel electrophoresis (2-DE). Discrimination of alleles *Glu-A3a*, *Glu-A3b*, *Glu-A3c*, *Glu-A3d*, *Glu-A3e* and *Glu-A3f*. Cultivars: 1. Neixiang 188, 2. Gabo, 3. Pitic, 4. Nidera Baguette 10, 5. Amadina, 6. Kitanokaori.

for *Glu-A3g*, indicating that the *Glu-A3* alleles in the collection could be readily distinguished from one another. Since no *Glu-A3c* allele-specific primer has been developed, identification of this allele required the use of the *gluA3ac* with a 573bp band in combination with the marker *gluA3a* [27].

Ten primer pairs developed by Wang et al.[28] were

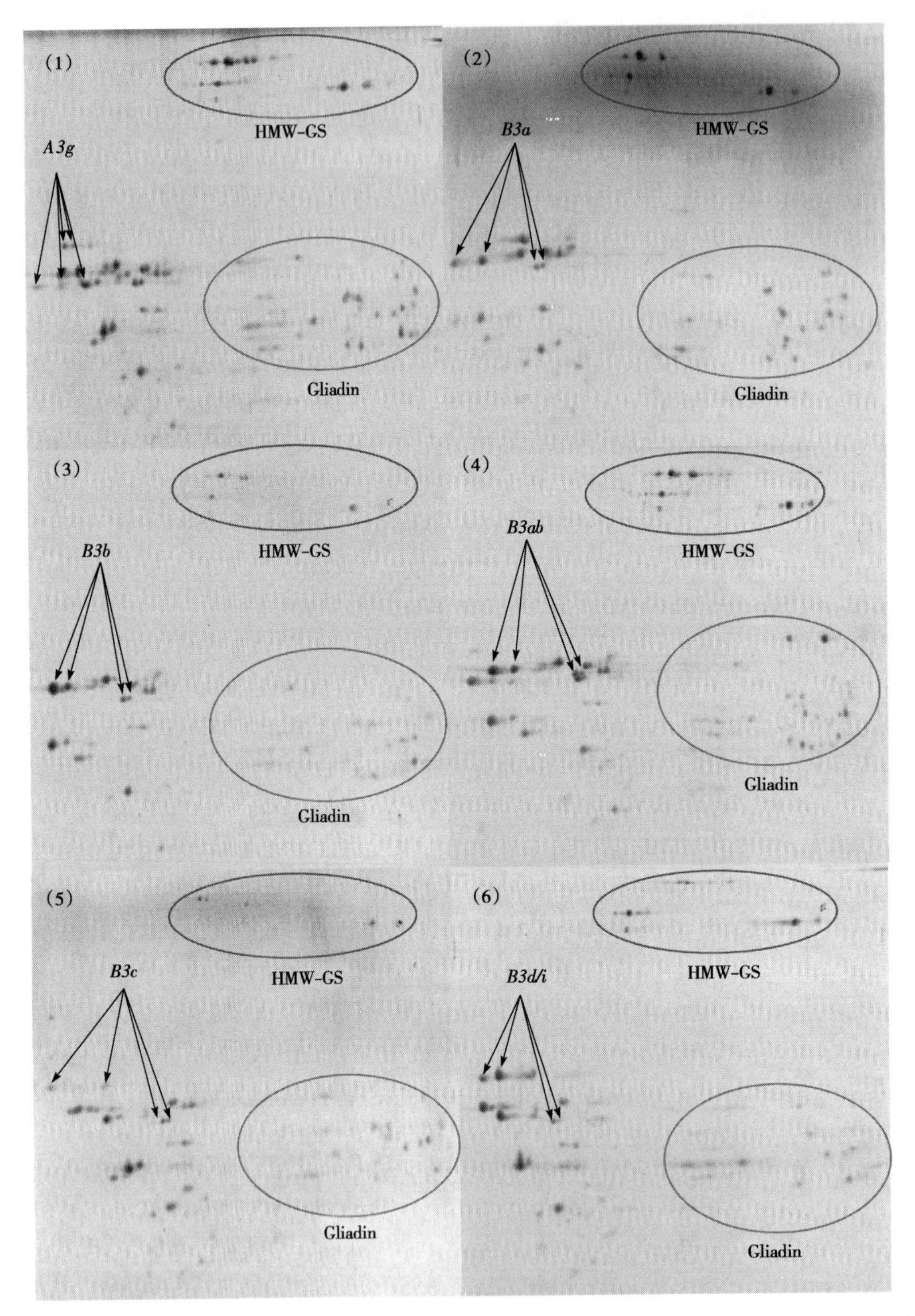

Fig. 7 Identification of LMW-GS by two-dimensional gel electrophoresis (2-DE). Discrimination of alleles *Glu-A3g*, *Glu-B3a*, *Glu-B3b*, *Glu-B3ab*, *Glu-B3c* and *Glu-B3d/i*. Cultivars: 1. Bluesky, 2. Chinese Spring, 3. Renan, 4. Nanbu-komugi, 5. Insignia, 6. Pepital. Letters preceding and following "/" indicate pairs of alleles that could not be reliably distinguished.

utilized to test for *Glu-B3* alleles and the results are summarized in Table 1. Specifically amplified fragments included 1095bp for *Glu-B3a*, 1549bp for *Glu-B3b*, 472bp for *Glu-B3c*, 662bp for *Glu-B3d*, 669bp for *Glu-B3e*, 853bp for *Glu-B3g*, 1022bp for *Glu-B3h*, and 621bp for *Glu-B3i*, indicating that the *Glu-*

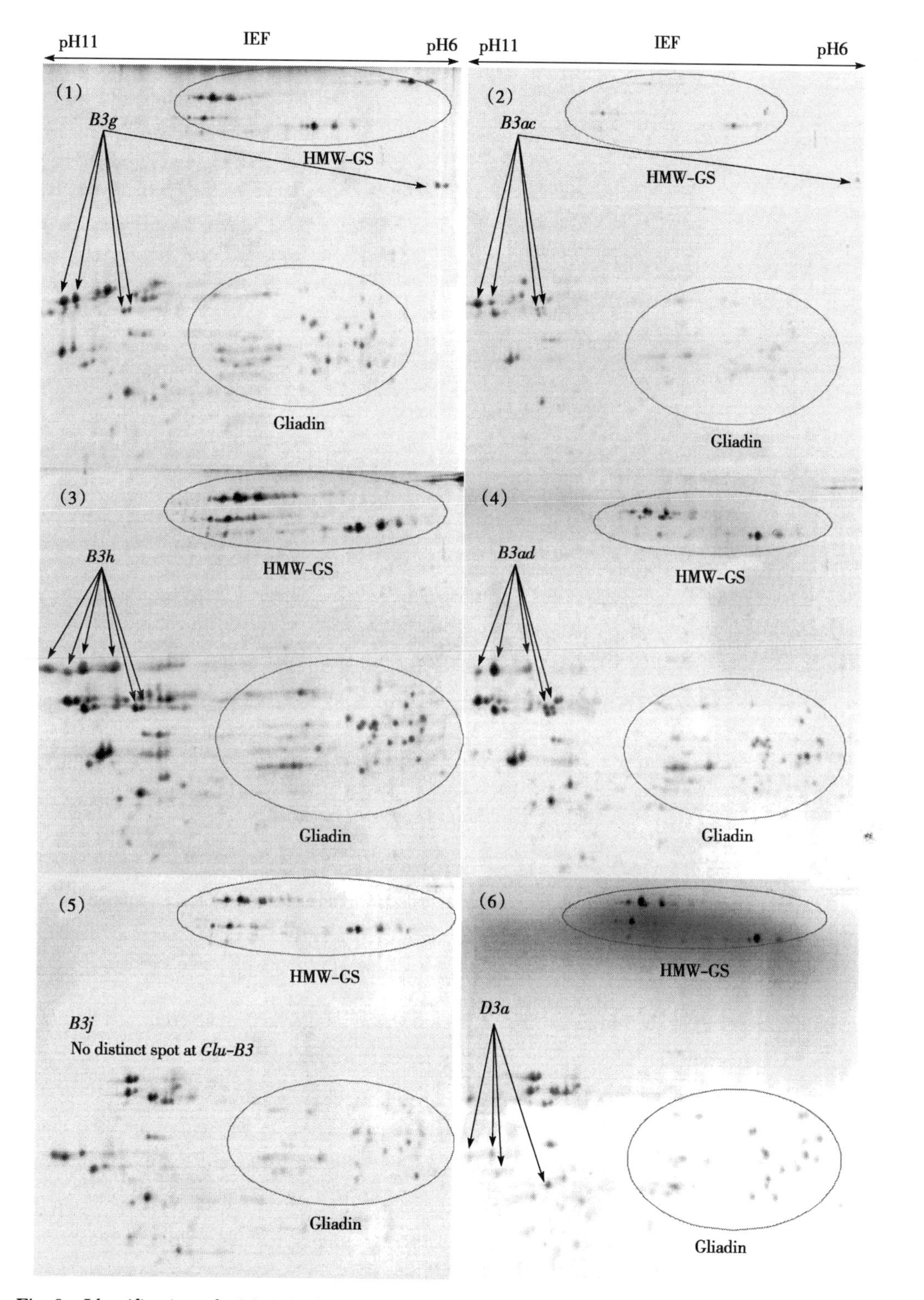

Fig. 8 Identification of LMW-GS by two-dimensional gel electrophoresis (2-DE). Discrimination of alleles *Glu-B3g*, *Glu-B3ac*, *Glu-B3h*, *Glu-B3ad*, *Glu-B3j* and *Glu-D3a*. Cultivars: 1. Splendor, 2. Thesee, 3. Aca 303, 4. Heilo, 5. Grebe, 6. Chinese Spring.

B3 alleles could be well differentiated based on corresponding markers. Detection of *Glu-B3f* required the use of the *Glu-B3fg* marker with an 812-bp marker in combination with the *Glu-B3g* marker since no *Glu-B3f* allele-specific marker has been designed. Although *Glu-B3f* could not be clearly distinguished from *Glu-B3g* by protein based methods, these alleles could be definitively differentiated by PCR. In addition,

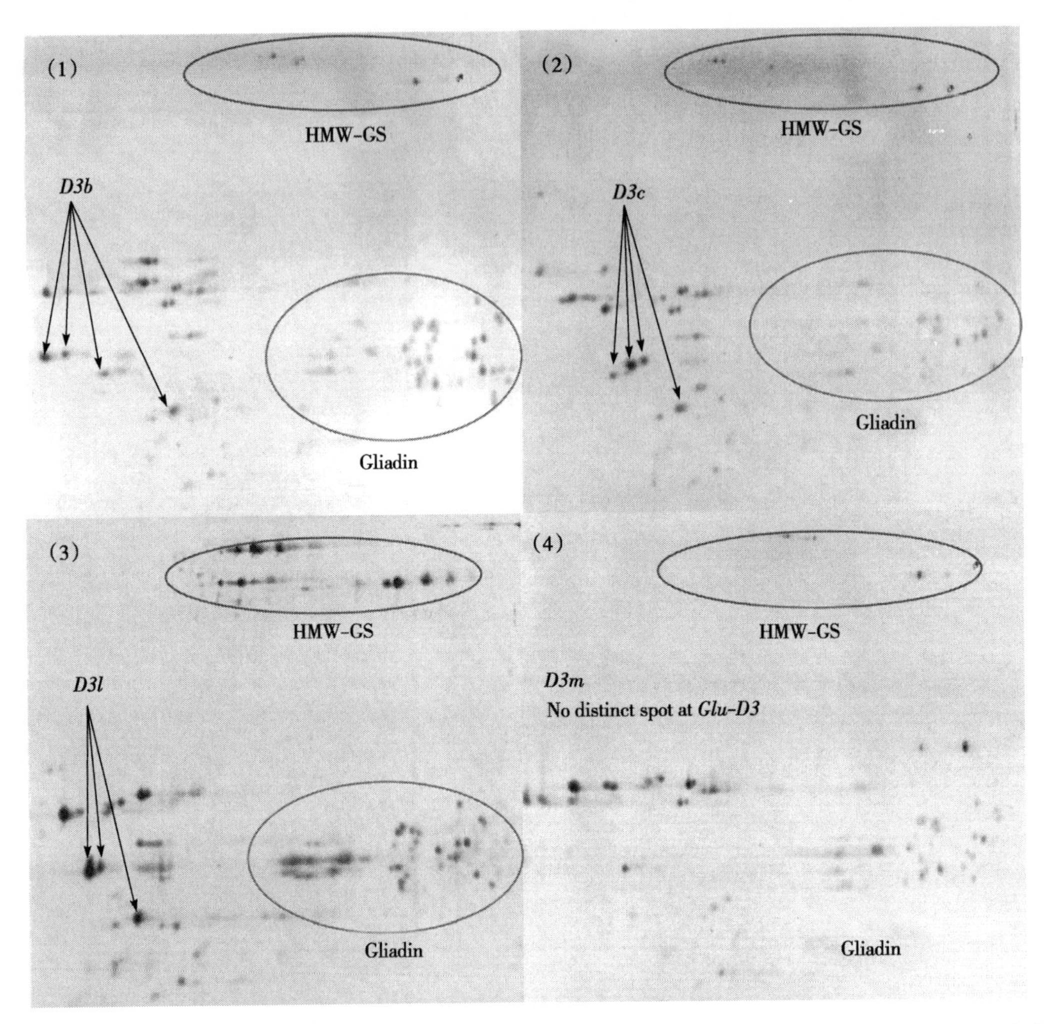

Fig. 9 Identification of LMW-GS by two-dimensional gel electrophoresis (2-DE). Discrimination of alleles *Glu-D3b*, *Glu-D3c*, *Glu-D3l* and *Glu-D3m*. Cultivars: 1. Gabo, 2. Insignia, 3. Amadina, 4. Darus.

there were obvious differences between genes *GluB3-1* and *GluB3-2* in the gene sequences of *Glu-B3f* and *Glu-B3g* [28]. The differences were firstly, the sequence length of *Glu-B3f* was 60 bp longer than that of *Glu-B3g* in the *GluB3-1* gene, and secondly, there were single base differences between *Glu-B3f* and *Glu-B3g* in both *GluB3-1* and *GluB3-2*. Therefore, alleles *Glu-B3f* and *Glu-B3g* reported in previous studies were different alleles although they could not be reliably differentiated by SDS-PAGE, 2-DE or MALDI-TOF-MS [13,34]

Glu-D3 appeared to be the most complicated locus. It contains the highest number of genes and expressed subunits compared to the other two loci, and yet most of the subunits across different alleles have similar molecular weights. Electrophoresis based methods and PCR are not efficient in differentiating *Glu-D3* alleles. The MALDI-TOF-MS based method can differentiate *Glu-D3* alleles since it is able to differentiate subtle changes in mass values. High accuracy mass calibration to remove the variations in mass measurement is the key to improve the efficiency of MALDI-TOF in differentiating these alleles.

Comparison of the four methods for identification of LMW-GS composition

The data from all five laboratories and the four methods employed showed that alleles *Glu-A3b*, *Glu-A3d* and *Glu-A3e* were consistently identified by all four meth-

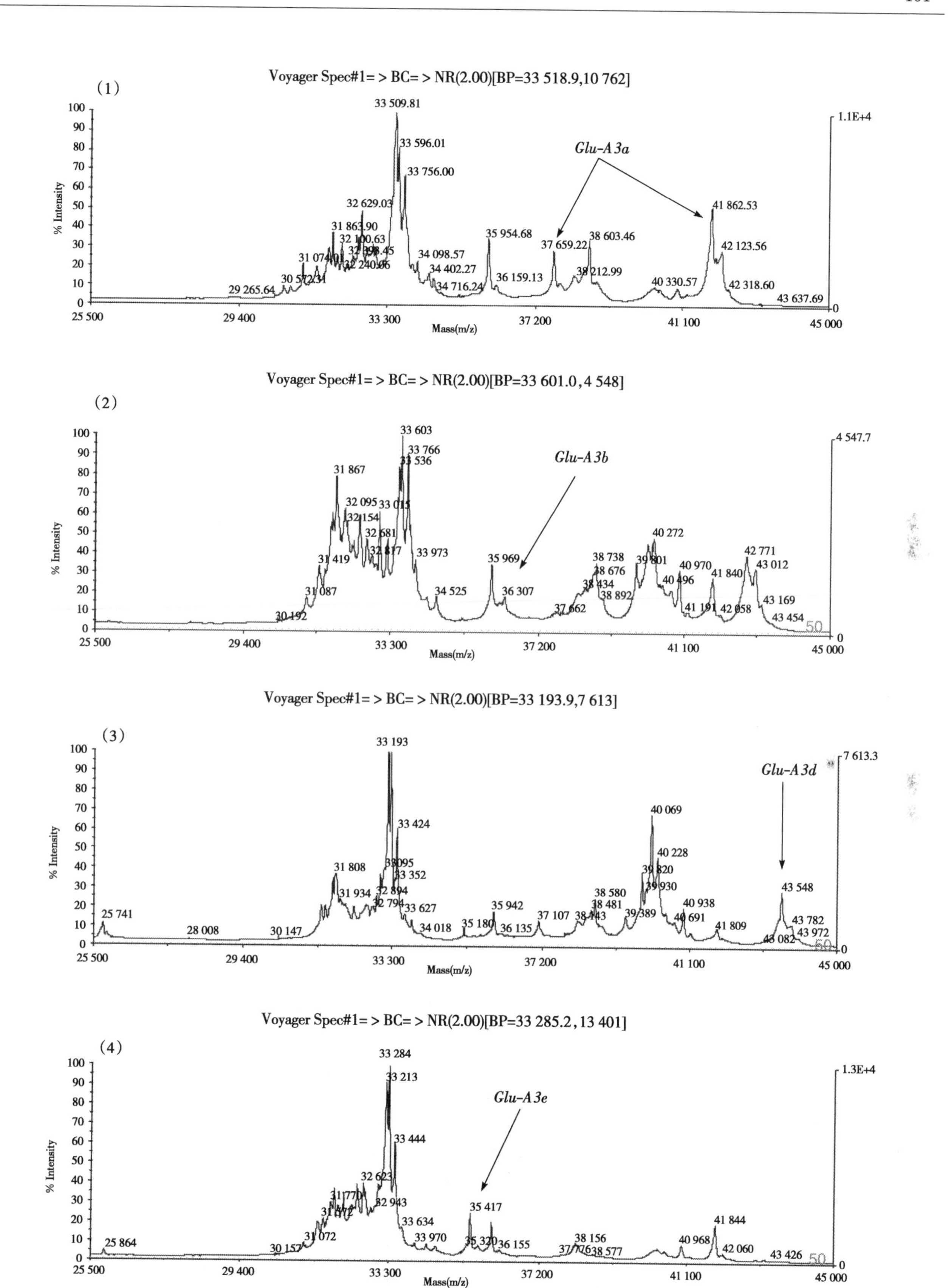

Fig. 10 Detection of LMW-GS by MALDI-TOF-MS. Identification of alleles *Glu-A3a*, *Glu-A3b*, *Glu-A3d* and *Glu-A3e*. Cultivars: 1. Neixiang 188, 2. Gabo, 3. Nidera Baguette 10, 4. Amadina.

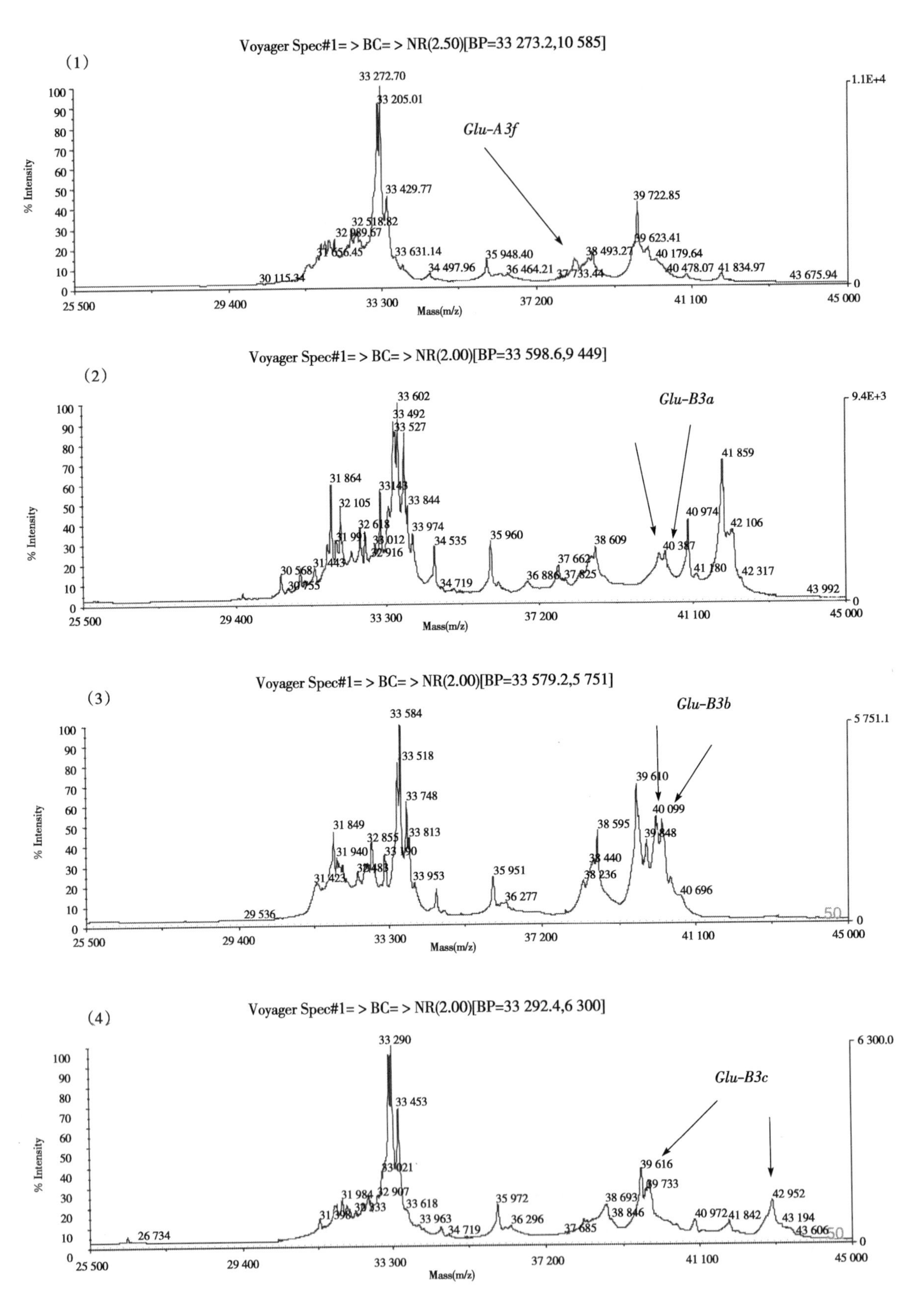

Fig. 11 Detection of LMW-GS by MALDI-TOF-MS. Identification of alleles *Glu-A3f*, *Glu-B3a*, *Glu-B3b* and *Glu-B3c*. Cultivars: 1. Kitanokaori, 2. Chinese Spring, 3. Renan, 4. Insignia.

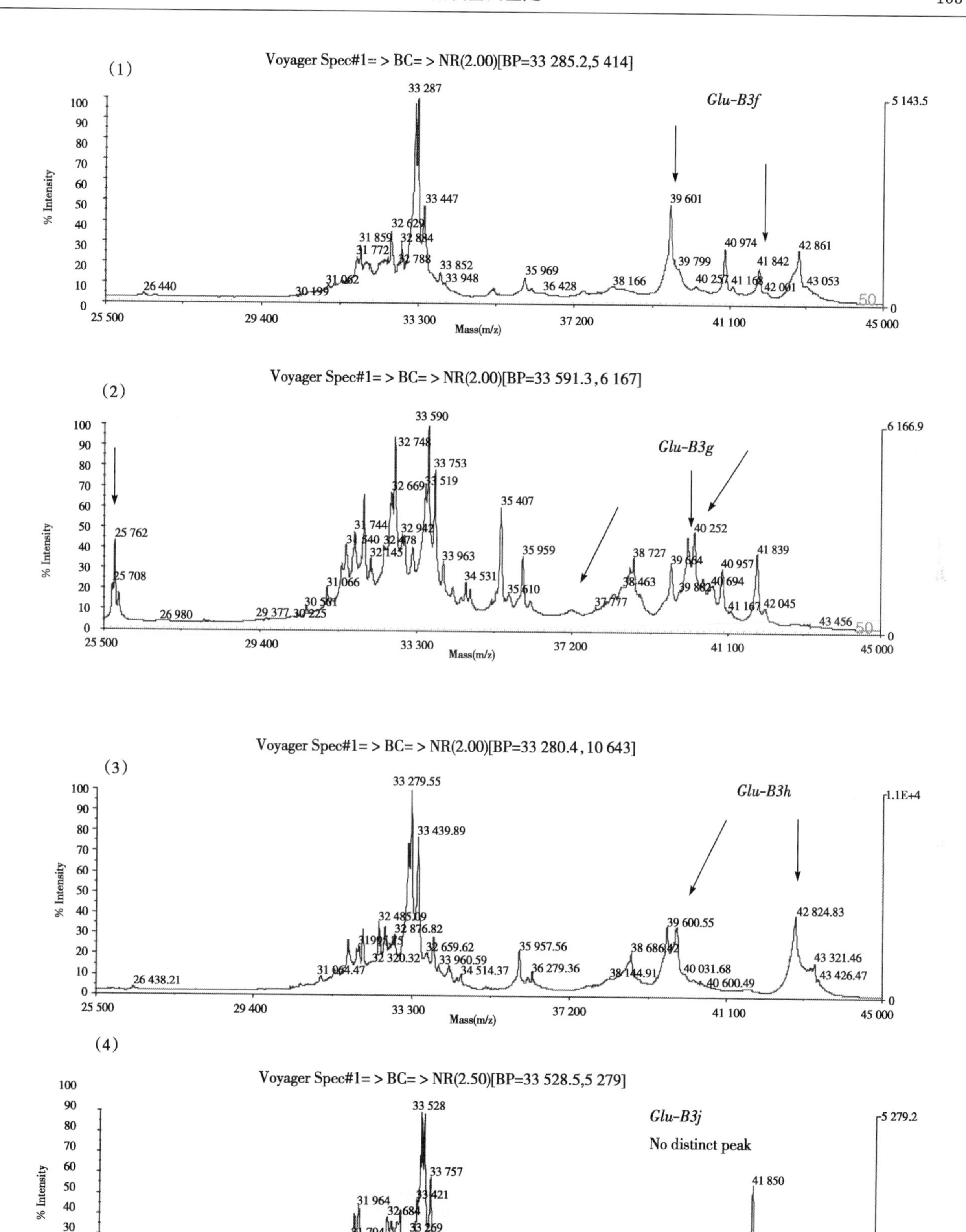

Fig. 12 Detection of LMW-GS by MALDI-TOF-MS. Identification of alleles *Glu-B3f*, *Glu-B3g*, *Glu-B3h* and *Glu-B3j*. Cultivars: 1. Pepital, 2. Splendor, 3. Aca 303, 4. Grebe.

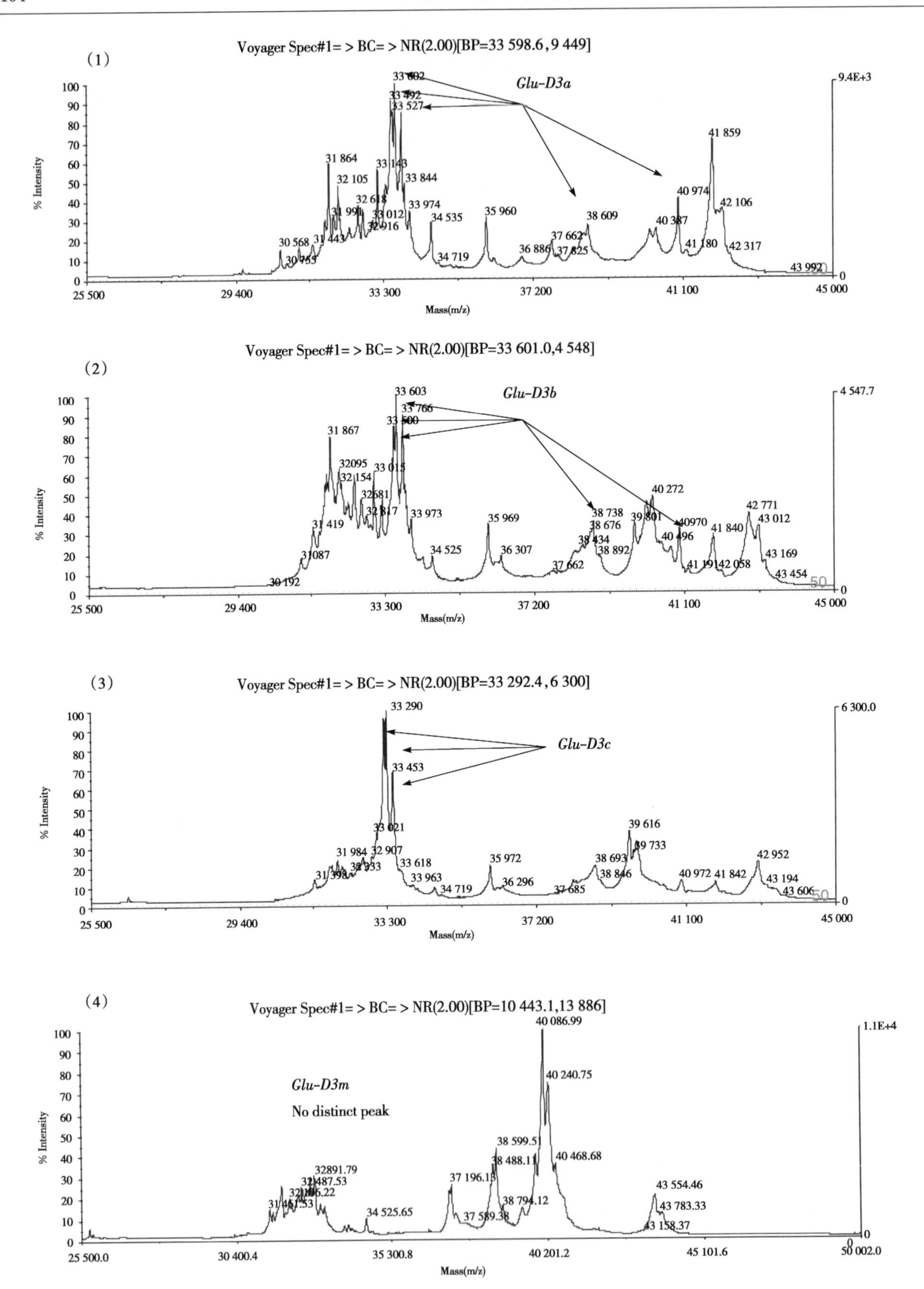

Fig. 13 Detection of LMW-GS by MALDI-TOF-MS. Identification of alleles *Glu-D3a*, *Glu-D3b*, *Glu-D3c* and *Glu-D3m*. Cultivars: 1. Chinese Spring, 2. Gabo, 3. Insignia, 4. Darus.

ods. Similarly, analyses of alleles *Glu-B3a*, *Glu-B3b*, *Glu-B3c*, *Glu-B3h* and *Glu-B3j* were in agreement for all four methods. At the *Glu-D3*, only the *Glu-D3c* allele was consistently identified by SDS-PAGE, 2-DE and MALDI-TOF-MS. The discrepancies in allelic identification using the different methods are indicated in Table 2. Alleles *Glu-A3a* and *Glu-A3c* could not be distinguished by MALDI-TOF-MS due to their nearly identical molecular masses. Similarly, these two alleles could not be reliably identified by SDS-PAGE and 2-DE due to their identical mobilities and pI. However, it was easy to differentiate them by PCR. In SDS-PAGE gels, the higher mobility patterns of alleles *Glu-B3d*, *Glu-B3h*, *Glu-B3i* overlapped with those of alleles *Glu-A3a* or *Glu-A3c*, and lower mobility patterns overlapped with those of allele *Glu-A3b*. These results were in agreement with the reports of Gupta and Shepherd [13], who concluded that ambiguous identification of subunits was possibly caused by differential staining intensity of banding patterns. The difficulty to differentiate *Glu-B3b* and *Glu-B3g* based on SDS-PAGE banding patterns arose from their similar mobilities. However, as shown in Fig. 3 and 4, several *Glu-B3* alleles could be readily discriminated using gliadins as a marker for glutenin by SDS-PAGE. These alleles had clearly different peaks or spots using MALDI-TOF-MS or 2-DE, respectively. Alleles *Glu-D3a* and *Glu-D3b* could not be reliably separated by MALDI-TOF-MS or 2-DE. It is suggested that the *Glu-D3* alleles should be differentiated by a combination of primers [49-51].

Table 2 Allelic variants of LMW-GS identified using different methods

Locus	Subunit	SDS-PAGE	2-DE	MALDI-TOF-MS	PCR
Glu-A3	*Glu-A3a*	√	√	√	√
	Glu-A3b	√	√	√	√
	Glu-A3c	√	√	√	√
	Glu-A3d	√	√	√	√
	Glu-A3e	√	√	√	√
	Glu-A3f		√	√	√
	Glu-A3g		√		√
Glu-B3	*Glu-B3a*	√	√	√	√
	Glu-B3b	√	√	√	√
	Glu-B3c	√	√	√	√
	Glu-B3d	√		√	√
	Glu-B3f			√	√
	Glu-B3g	√	√	√	√
	Glu-B3h	√	√	√	√
	Glu-B3i	√			√
	Glu-B3j	√	√	√	√
	Glu-B3ab		√		
	Glu-B3ac		√		
	Glu-B3ad		√		
Glu-D3	*Glu-D3a*	√		√	—
	Glu-D3b	√		√	—
	Glu-D3c	√	√	√	—
	Glu-D3m		√	√	—
	Glu-D3l	√	√		—
	Glu-D3n		√		

√, confirmed; —, data not available.

The 2-DE method is generally considered as the most powerful tool for identifying storage protein polymorphism of proteins in wheat[52]. However, different bands in SDS-PAGE separations were not always distinguishable in 2-DE separations. For example, alleles *Glu-B3d* and *Glu-B3i* could be identified by SDS-PAGE, but not by 2-DE. For LMW-GS identification in wheat breeding programs, PCR and/or SDS-PAGE of both gliadin and glutenin extracts should be used as the basic method, with 2-DE and MALDI-TOF-MS as complementary approaches. A combination of different methods is recommended for differentiating certain LMW-GS alleles, particularly those suspected as being novel.

Comparison of the fourmethods is presented in Table 3.

Utilization of a particular method will depend upon research objectives and the targeted materials. With appropriate classification of glutenin alleles, it is possible to improve wheat quality by selection of alleles and allelic combinations with desired quality performance. If progeny screening and cultivar development is the objective, PCR will likely be adequate for the identification of *Glu-A3* and *Glu-B3* alleles. However, if the aim is to determine the glutenin subunits of potential parents for predicting cross performance and designing crossing schemes, or to identify specific alleles such as *Glu-A3g*, *Glu-B3ab*, *Glu-B3ac*, or distinguish between the *Glu-D3* alleles, a combination of methods should be used, i. e. PCR with 2-DE or PCR with SDS-PAGE and 2-DE, in order to achieve the correct identification of LMW-GS alleles.

Table 3 Relative efficiencies of methods of gluten analysis for situations where cultivar identification is required

Subject	SDS-PAGE	2-DE	MALDI-TOF-MS	PCR
Required sample amount	40μg (Protein)	150μg (Protein)	0.04μg (Protein)	2μl (DNA)
Purity required	Low	High	High	Medium
Number of alleles	19	22	21	16
Alleles efficiently resolved	*Glu-B3b* and *Glu-B3g*, *Glu-B3d* and *Glu-B3i*,	*Glu-A3e*, *Glu-A3f*, *Glu-A3g*, *Glu-B3b*, *Glu-B3g*, *Glu-B3ab*, *Glu-B3ac*, *Glu-B3ad*, *Glu-D3l* and *Glu-D3m*	*Glu-A3e*, and *Glu-A3f*; *Glu-D3a*, *Glu-D3b*, *Glu-D3c*, and *Glu-D3m*	*Glu-A3e* and *Glu-A3f*, *Glu-B3d* and *Glu-B3i*, *Glu-B3f* and *Glu-B3g*
Mass accuracy	Inaccurate	Inaccurate	Accurate	Accurate
pI	Unknown	Known	Unknown	Unknown
Cost of equipment	≈$7, 000	≈$30, 000	≈$20, 000-400, 000	≈$5, 500
Costper sample	≈$1.0	≈$70.0	≈$0.3	≈$0.3
Number of samples analysed per day for skilled technician	30-160*	1	100	100
Automation	Not possible	Not possible	Possible	Possible
Experience required	Considerable	Considerable	Less	Less
Safety	High toxicity	High toxicity	Safe	Toxicity
False positives	No	Yes	No	Yes
Accuracy level	++	+++	++	++

* Thirty samples/day if running two gels. Up to 160 samples/day if using multi-gel (8 gels) buffer tank.

A set of standard cultivars for identification of LMW-GS

From this study of 103wheat cultivars from 12 countries we propose a set of 30 cultivars for determination of LMW-GS (Table 4) irrespective of the method to be used. Fig. 1, 3, 5 show glutenin electropherograms of 28 (missing Ernest and Darius) of the 30 genotypes presented in Table 4. They cover all LMW-GS allelic variants identified in the original set. A core set of Chinese Spring, Opata 85, Seri 82 and Pavon 76 is recommended for inclusion in all gels. Most of the common *Glu*-3 alleles are represented among this group and their distributions on gels will provide useful landmarks for comparison with other bands. In this classification, it is possible to differentiate alleles *Glu-A3g* from *Glu-A3d*, *Glu-B3ab* from *Glu-B3b*, *Glu-B3ac* from *Glu-B3g*, *Glu-B3ad* from *Glu-B3i*, and *Glu-D3l* from *Glu-D3c*. Alleles *Glu-D3e* and *Glu-D3c* are assumed to be identical. The allele in cultivar Darius, with no distinct spot in 2-DE gels, is a new allele, *Glu-D3m*. The new allele *Glu-D3n* identified in the cultivar Fengmai 27 has a distinct spot in 2-DE and different mobility in SDS-PAGE (Fig. 5). However, more work is needed to further characterize these new alleles at the *Glu-D3* locus. The other alleles were the same as those observed by Gupta and Shepherd [13].

Allele *Glu-A3g*, identified in the Canadian cultivars Bluesky and Glenlea by 2-DE in the current collection, is widely distributed in many cultivars from Canada and the U. S. A. [41]. In previous studies, allele *Glu-A3g* was frequently identified as *Glu-A3d* due to their similar SDS-PAGE patterns. The role of *Glu-A3g* in bread making quality therefore requires further study. Similarly, effects on bread making quality of alleles *Glu-B3ab*, *Glu-B3ac*, *Glu-B3ad* and *Glu-D3l*, with two additional distinct spots compared to alleles *Glu-B3b*, *Glu-B3g*, *Glu-B3i* and *Glu-D3c*, respectively, also need further investigation.

Table 4 Thirty cultivars recommended as standards for the determination of LMW-GS alleles. The core group is in bold

Locus	Allele	Standard cultivar
Glu-A3	*Glu-A3a*	Neixiang 188, **Chinese Spring**
	Glu-A3b	Gabo, **Pavon 76**
	Glu-A3c	Pitic, **Seri 82**
	Glu-A3d	Nidera Baguette 10, Cappelle-Desprez
	Glu-A3e	Amadina, Marquis
	Glu-A3f	Kitanokaori, Renan
	Glu-A3g	Bluesky, Glenlea
Glu-B3	*Glu-B3a*	**Chinese Spring**
	Glu-B3b	Renan, Gabo
	Glu-B3c	Insignia, Halberd
	Glu-B3d	Pepital, Ernest
	Glu-B3f	Fengmai 27
	Glu-B3g	Splendor, Cappelle-Desprez
	Glu-B3h	Aca303, **Pavon 76**
	Glu-B3i	Norin 61
	Glu-B3j	Grebe, **Seri 82**
	Glu-B3ab	Nanbu-komugi
	Glu-B3ac	Thesee, Aca 801
	Glu-B3ad	Heilo, **Opata 85**
Glu-D3	*Glu-D3a*	**Chinese Spring**, Neixiang 188
	Glu-D3b	Gabo, Avocet
	Glu-D3c	Insignia, Cappelle-Desprez
	Glu-D3m	Darius
	Glu-D3l	Amadina, Heilo
	Glu-D3n	Fengmai 27

Conclusions

Four methods, SDS-PAGE, 2-DE, MALDI-TOF-MS and PCR, were used for identifying the LMW-GS composition in wheat cultivars from 12 countries. All seven *Glu-A3* alleles could be identified by 2-DE and PCR, and only four and five of the seven could be differentiated by MALDI-TOF-MS and SDS-PAGE of the glutenin extract, respectively. The *Glu-B3* alleles *Glu-B3a*, *Glu-B3b*, *Glu-B3c*, *Glu-B3g*, *Glu-B3h* and *Glu-B3j* could be identified by all four methods, but alleles *Glu-B3ab*, *Glu-B3ac*, *Glu-B3ad* could only be identified by the 2-DE method. *Glu-D3* alleles were very difficult to clearly distinguish by SDS-PAGE, 2-DE and PCR. MALDI-TOF-MS was promising in reliably differentiating them. PCR is a simple, accurate, and low cost method for identifying *Glu-A3* and *Glu-B3* alleles that are currently routinely analysed by SDS-PAGE in breeding programs. However, SDS-PAGE using a multi-gel buffer chamber, and running both gliadins and glutenin extracts is also a highly reliable method. A combination of all methods will help to identify specific alleles, especially potentially new alleles.

Aset of 30 cultivars (Table 4) was recommended for identifying LMW-GS alleles. These standard cultivars cover all variants of LMW-GS in the collection investigated. Among them, Chinese Spring, Opata 85, Seri 82 and Pavon 76, are recommended as a core set to be included in each SDS-PAGE gel when identifying alleles of LMW-GS genes. The 30 cultivars have been placed in CIMMYT' s and INRA Clermont Ferrand, France germplasm banks and seed is being multiplied to make them freely available as a set upon request. Accession numbers will be assigned once the *Glu-1*/*Glu-3* allelic composition is confirmed.

Acknowledgements

The authors are grateful to Prof. R. A. McIntosh, University of Sydney, for reviewing the manuscript. The study was supported by the National Natural Science Foundation of China (30830072), National Basic Research Program (2009CB118300) and National 863 Programs (2006AA10Z1A7 and 2006AA100102).

References

[1] Payne PI: Genetics of wheat storage protein and the effect of allelic variation on pan bread quality. *Annu Rev Plant Physiol*. 1987, 38: 141-153.

[2] Gianibelli MC, Larroque OR, MacRichie F, Wrigley CW: Biochemical, genetic, and molecular characterization of wheat glutenin and its component subunits. *Cereal Chem.* 2001, 78 (6): 635-646.

[3] Shewry PR, Halford NG, Lafiandra D: Genetics of wheat gluten proteins. *Adv Genet.* 2003, 49: 111-184.

[4] Shewry PR, Tatham AS: The prolamin storage proteins of cereal seeds, structure and evolution. *Biochem J*. 1990, 267 (1): 1-12.

[5] Bietz JA, Wall JS: Isolation and characterization of gliadin-like subunits from glutelin. *Cereal Chem.* 1973, 50 (5): 537-547.

[6] Cornish GB, Békés F, Allen HM, Martin JM: Flour proteins linked to quality traits in an Australian doubled haploid wheat population. *Aust J Agric Res*. 2001, 52 (12): 1339-1348.

[7] Gupta RB, Paul JG, Cornish GB, Palmer GA, Bekes F, Rathjen AJ: Allelic variation at glutenin subunit and gliadin loci, *Glu-1*, *Glu-3* and *Gli-1*, of common wheats. I. Its additive and interaction effects on dough properties. *J Cereal Sci*. 1994, 19 (1): 9-17.

[8] He ZH, Liu L, Xia XC, Liu JJ, Peña RJ: Composition of HMW and LMW glutenin subunits and their effects on dough properties, pan bread, and noodle quality of Chinese bread wheats. *Cereal Chem.* 2005, 82 (4): 345-350.

[9] Békés F, Morell M: An integrated approach to predicting end-product quality of wheat. In: Appels R, Eastwood R, Lagudah E, Langridge P, Mackay M, McIntyre L, Sharp P (eds). Proc 11th Int Wheat Genet Symp, Sydney University Press, Sydney, Australia. 2008, O45.

[10] Singh NK, Shepherd FW: Linkage mapping of genes controlling endosperm storage proteins in wheat. 1. Genes on the short arms of group 1 chromosomes. *Theor Appl Genet.* 1988, 75 (4): 628-641.

[11] Pogna NE, Autran JC, Mellini F, Lafiandra D, Feillet

P: Chromosome 1B-encoded gliadins and glutenin subunits in durum wheat, genetics and relationship to gluten strength. *J Cereal Sci* . 1990, 11 (1): 15-34.

[12] Anderson OD, Gu YQ, Kong XY, Lazo GR, Wu JJ: The wheat ω-gliadin genes: structure and EST analysis. *Funct Integr Genomics.* 2009, 9 (3): 397-410.

[13] Gupta RB, Shepherd KW: Two-step one-dimensional SDS-PAGE analysis of LMW subunits of glutelin. 1. Variation and genetic control of the subunits in hexaploid wheats. *Theor Appl Genet.* 1990, 80 (1): 65-74.

[14] Yan Y, Hsam SL, Yu JZ, Jiang Y, Ohtsuka I, Zeller FJ: HMW and LMW glutenin alleles among putative tetraploid and hexaploid European spelt wheat (*Triticum spelta* L.) progenitors. *Theor Appl Genet.* 2003, 107 (7): 1321-1330.

[15] Lerner SE, Kolman MA, Rogers WJ: Quality and endosperm storage protein variation in Argentinean grown bread wheat. I. Allelic diversity and discrimination between cultivars. *J Cereal Sci.* 2009, 49 (3): 337-345.

[16] Jackson EA, Morel MH, Sontag-Strohm T, Branlard G, Metakovsky EV, Redaelli R: Proposal for combining the classification systems of alleles of *Gli-1* and *Glu-3* loci in bread wheat (*Triticum aestivum* L.). *J Genet Breed.* 1996, 50: 321-336.

[17] Eagles HA, Hollamby GJ, Gororo NN, Eastwood RF: Estimation and utilisation of glutenin gene effects from the analysis of unbalanced data from wheat breeding programs. *Aust J Agric Res.* 2002, 53 (4): 367-377.

[18] Ikeda TM, Araki E, Fujita Y, Yano H: Characterization of low-molecular-weight glutenin subunit genes and their protein products in common wheats. *Theor Appl Genet* . 2006, 112 (2): 327-334.

[19] Appelbee MJ, Mekuria GT, Nagasandra V, Bonneau JP, Eagles HA, Eastwood RF, Mather DE: Novel allelic variants encoded at the *Glu-D3* locus in bread wheat. *J Cereal Sci.* 2009, 49 (2): 254-261.

[20] D' Ovidio R, Masci S: The low-molecular-weight glutenin subunits of wheat gluten. *J Cereal Sci.* 2004, 39 (3): 321-339.

[21] Anderson NG, Tollaksen SL, Pascoe FH, Anderson L: Two-dimensional electrophoretic analysis of wheat seed proteins. *Crop Sci.* 1985, 25 (4): 667-674.

[22] Dworschak RG, Ens W, Standing KG, Preston KR, Marchylo BA, Nightingale MJ, Stevenson SG, Hatcher DW: Analysis of wheat gluten proteins by matrix-assisted laser desorption/ionization mass spectrometry. *J Mass Spectrom.* 1998, 33 (5): 429-435.

[23] Ghirardo A, Sørensen HA, Petersen M, Jacobsen S, Søndergaard I: Early prediction of wheat quality, analysis during grain development using mass spectrometry and multivariate data analysis. *Rapid Commun Mass Spectrom.* 2005, 19 (4): 525-532.

[24] Muccilli V, Cunsolo V, Saletti R, Foti S, Masci S, Lafiandra D: Characterization of B- and C-type low molecular weight glutenin subunits by electrospray ionization mass spectrometry and matrix-assisted laser desorption/ ionization mass spectrometry. *Proteomics.* 2005, 5 (3): 719-728.

[25] Liu L, Wang AL, Appels R, Ma JH, Xia XC, Lan P, He ZH, Bekes F, Yan YM, Ma WJ: A MALDI-TOF based analysis of high molecular weight glutenin subunits for wheat breeding. *J Cereal Sci* . 2009, 50 (2): 295-301.

[26] Zhang W, Gianibelli MC, Rampling LR, Gale KR: Characterisation and marker development for low molecular weight glutenin genes from *Glu-A3* alleles of bread wheat (*Triticum aestivum* L.). *Theor Appl Genet.* 2004, 108 (7): 1409-1419.

[27] Wang LH, Li GY, Peña RJ, Xia XC, He ZH: Development of STS markers and establishment of multiplex PCR for *Glu-A3* alleles in common wheat (*Triticum aestivum* L.). *J Cereal Sci.* 2010, 51 (3): 305-312.

[28] Wang LH, Zhao XL, He ZH, Ma W, Appels R, Peña RJ, Xia XC: Characterization of low-molecular-weight glutenin subunit *Glu-B3* genes and development of STS markers in common wheat (*Triticum aestivum* L.). *Theor Appl Genet.* 2009, 118 (3): 525-539.

[29] Beckwith AC, Nielsen HC, Wall JS, Huebner FR: Isolation and characterization of a high-molecular-weight protein from wheat gliadin. *Cereal Chem.* 1966, 43 (1): 14-28.

[30] Elton GAH, Ewart JAD: Glutenins and gliadins electrophoretic studies. *J Sci Food Agric.* 1966, 17 (1): 34-38.

[31] Jackson EA, Holt LM, Payne PI: Characterisation of high molecular weight gliadin and low-molecular-weight glutenin subunits of wheat endosperm by two-dimensional electrophoresis and the chromosomal localisation of their controlling genes. *Theor Appl Genet.* 1983, 66 (1): 29-37.

[32] Ikeda TM, Nagamine T, Fukuoka H, Yano H: Iden-

tification of new low-molecular-weight glutenin subunit genes in wheat. *Theor Appl Genet*. 2002, 104 (4): 680-687.

[33] Juhász A, Gianibelli MC: Information hidden in the low molecular weight glutenin gene sequences. In: The Gluten Proteins. Lafiandra D, Masci S, D'Ovidio R, (eds), Proc 8th Gluten Workshop, Bitervo, Italy. 2003, pp 62 - 65.

[34] Branlard G, Dardevet M, Amiour N, Igrejas G: Allelic diversity of HMW and LMW glutenin subunits and omega-gliadins in French bread wheat (*Triticum aestivum* L.). *Genet Resour Crop Evol*. 2003, 50 (7): 669-679.

[35] Ikeda TM, Branlard G, Peña RJ, Takata K, Liu L, He ZH, Lerner SE, Kolman MA, Yoshida H, Rogers WJ: International collaboration for unifying *Glu-3* nomenclature systems in common wheat. In: Appels R, Eastwood R, Lagudah E, Langridge P, Mackay M, McIntyre L, Sharp P (eds). Proc 11th Int Wheat Genet Symp, Sydney University Press, Sydney, Australia. 2008, O42.

[36] Lew EJL, Kuzmicky DD, Kasarda DD: Characterization of low molecular weight glutenin subunits by reversed-phase high performance liquid chromatography, sodium dodecyl sulphate-polyacrylamide gel electrophoresis, and N-terminal amino acid sequencing. *Cereal Chem*. 1992, 69 (5): 508-515.

[37] Tao HP, Kasarda DD: Two-dimensional gel mapping and N-terminal sequencing of LMW-glutenin subunits. *J Exp Bot*. 1989, 40 (9): 1015-1020.

[38] Masci S, Rovelli L, Kasarda DD, Vensel WH, Lafiandra D: Characterisation and chromosomal localization of C-type low-molecular-weight glutenin subunits in the bread wheat cultivar Chinese Spring. *Theor Appl Genet*. 2002, 104 (2): 422-428.

[39] Masci S, Lafiandra D, Porceddu E, Lew EJL, Tao HP, Kasarda DD: D-glutenin subunits, N-terminal sequences and evidence for the presence of cysteine. *Cereal Chem*. 1993, 70 (5): 581-585.

[40] Nieto-Taladriz MT, Rodriguez-Quijano M, Carrillo JM: Biochemical and genetic characterization of a D glutenin subunit encoded at the *Glu-B3* locus. *Genome*. 1998, 41 (2): 215-220.

[41] Wrigley CW, Bekes F, Bushuk W: The gluten composition of wheat varieties and genotypes. AACC International, ISBN 1-891127 - 51-9. 2006.

[42] Cornish GB, Burridge PM, Palmer GA, Wrigley CW: Mapping the origins of some HMW and LMW glutenin subunit alleles in Australian germplasm. In Wrigley CW (ed). Proc 43rd Aust Cereal Chem Conf, Royal Aust Chem Inst, Melbourne. 1993, pp255 - 260.

[43] Branlard G, Dardevet R, Saccomano F, Lagoutte F, Gourdon J: Genetic diversity of wheat storage proteins and bread wheat quality. *Euphytica*. 2001, 119 (1): 59-67.

[44] Luo C, Griffin WB, Branlard G, McNeil DL: Comparison of low and high molecular weight wheat glutenin allele effects on flour quality. *Theor Appl Genet*. 2001, 102 (6): 1088-1098.

[45] Vawser MJ, Cornish GB, Shepherd KW: Rheological dough properties of Aroona isolines differing in glutenin subunit composition. In Black CK, Panozzo JF, Wrigley CW, Batey IL, Larsen N (eds). Proc 52nd Aust Cereal Chem Conf, Royal Aust Chem Inst, Christchurch, New Zealand. 2002, pp53 - 58.

[46] Singh NK, Shepherd KW, Cornish GB: A simplified SDS-PAGE procedure for separating LMW subunits of glutenin. *J Cereal Sci*. 1991, 14 (3): 203-208.

[47] Liu L, He ZH, Yan J, Zhang Y, Xia XC, Peña RJ: Allelic variation at the *Glu-1* and *Glu-3* loci, presence of the 1B/1R translocation, and their effects on mixographic properties in Chinese bread wheats. *Euphytica*. 2005, 142 (3): 197-204.

[48] Marchylo BA, Handel KA, Mellish VJ: Fast horizontal sodium dodecyl sulfate gradient polyacrylamide gel electrophoresis for rapid wheat cultivar identification and analysis of high molecular weight glutenin subunits. *Cereal Chem*. 1989, 66 (3): 186-192.

[49] Zhao XL, Ma W, Gale KR, Lei ZS, He ZH, Sun QX, Xia XC: Identification of SNPs and development functional markers for LMW-GS genes at *Glu-D3* and *Glu-B3* loci in bread wheat (*Triticum aestivum* L.). *Mol Breed*. 2007, 20 (3): 223-231.

[50] Zhao XL, Xia XC, He ZH, Gale KR, Lei ZS, Appels R, Ma WJ: Characterization of three low-molecular-weight *Glu-D3* subunit genes in common wheat. *Theor Appl Genet*. 2006, 113 (7): 1247-1259.

[51] Zhao XL, Xia XC, He ZH, Lei ZS, Appels R, Yang Y, Sun QX, Ma W: Novel DNA variations to characterize low molecular weight glutenin *Glu-D3* genes and develop STS markers in common wheat. *Theor Appl Genet*. 2007, 114 (3): 451-460.

[52] Yahata E, Maruyama-Funatsuki W, Nishio Z, Tabiki T, Takata K, Yamamoto Y, Tanida M, Saruyama H:

Wheat cultivar-specific proteins in grain revealed by 2-DE and their application to cultivar identification of flour. *Proteomics*. 2005，5 (15)：3942-3953.

[53] Branlard G，Bancel E：Grain protein extraction. Plant proteomics，methods and protocols. *Methods Mol Biol*. 2006，355：15-25.

[54] Melas V，Morel MH，Autran JC，Feillet P：Simple and rapid method for purifying low molecular weight subunits of glutenin from wheat. *Cereal Chem* 1994，71 (3)：234-237.

[55] Dumur J，Jahier J，Bancel E，Laurière M，Bernard M，Branlard G：Proteomic analysis of aneuploid lines in the homeologous group 1 of the hexaploid wheat cultivar Courtot. *Proteomics*. 2004，4 (9)：2685-2695.

[56] Görg A，Obermaier C，Boguth G，Harder A，Scheibe B，Wildgruber R，Weiss W：The current state of two-dimensional electrophoresis with immobilised pH gradients. *Electrophoresis*. 2000，21 (6)：1037-1053.

[57] Neuhoff V，Arold N，Taube D，Ehrhardt W：Improved staining of proteins in polyacrylamide gels including isoelectric focusing gels with clear background at nanogram sensitivity using Coomassie Brilliant Blue G-250 and R-250. *Electrophoresis*. 1988，9 (6)：255-262.

[58] Kussmann M，Nordhoff E，Rahbek-Nielsen H，Haebel S，Rossel-Larsen M，Jakobsen L，Gobom J，Mirgorodskaya E，Kroll-Kristensen A，Palm L，Roepstorff P：MALDI-MS sample preparation techniques designed for various peptide and protein analytes. *J Mass Spectrom*. 1997，32 (6)：593-601.

[59] Gale KR，Ma W，Zhang W，Rampling L，Hill AS，Appels R，Morris P，Morell M：Simple high-throughput DNA markers for genotyping in wheat. In，Eastwood R，et al (eds). Proc 10th Australian Wheat Breeding Assembly，Wheat Breeding Soc of Australia. 2001，pp 26-31.

Composition and functional analysis of low-molecular-weight glutenin alleles with Aroona near-isogenic lines of bread wheat

X. F. Zhang[1,2], H. Jin[1], Y. Zhang[1], D. C. Liu[2], G. Y. Li[3], X. C. Xia[1], Z. H. He[1,4], and A. M. Zhang[2]

[1] *Institute of Crop Science, National Wheat Improvement Center, Chinese Academy of Agricultural Sciences* (*CAAS*), 12 *Zhongguancun South Street, Beijing* 100081, *China*; [2] *State Key Laboratory of Plant Cell and Chromosome Engineering, National Center for Plant Gene Research, Institute of Genetics and Developmental Biology, Chinese Academy of Sciences*, 1 *West Beichen Road, Beijing* 100101, *China*; [3] *Crop Research Institute, Shandong Academy of Agricultural Sciences, Jinan* 250100, *Shandong, China*; [4] *International Maize and Wheat Improvement Center* (*CIMMYT*) *China Office, c/o CAAS*, 12 *Zhongguancun South Street, Beijing* 100081, *China.*

Abstract: Low-molecular-weight glutenin subunits (LMW-GS) strongly influence the bread-making quality of bread wheat. These proteins are encoded by a multi-gene family located at the *Glu-A3*, *Glu-B3* and *Glu-D3* loci on the short arms of homoeologous group 1 chromosomes, and show high allelic variation. To characterize the genetic and protein compositions of LMW-GS alleles, we investigated 16 Aroona near-isogenic lines (NILs) using SDS-PAGE, 2D-PAGE and the LMW-GS gene marker system. Moreover, the composition of glutenin macro-polymers, dough properties and pan bread quality parameters were determined for functional analysis of LMW-GS alleles in the NILs. Using the LMW-GS gene marker system, 14-20 LMW-GS genes were identified in individual NILs. At the *Glu-A3* locus, two m-type and 2-4 i-type genes were identified and their allelic variants showed high polymorphisms in length and nucleotide sequences. The *Glu-A3d* allele possessed three active genes, the highest number among *Glu-A3* alleles. At the *Glu-B3* locus, 2-3 m-type and 1-3 s-type genes were identified from individual NILs. Based on the different compositions of s-type genes, *Glu-B3* alleles were divided into two groups, one containing *Glu-B3a*, *B3b*, *B3f* and *B3g*, and the other comprising *Glu-B3c*, *B3d*, *B3h* and *B3i*. Eight conserved genes were identified among *Glu-D3* alleles, except for *Glu-D3f*. The protein products of the unique active genes in each NIL were detected using protein electrophoresis. Among *Glu-3* alleles, the *Glu-A3e* genotype without i-type LMW-GS performed worst in almost all quality properties. *Glu-B3b*, *B3g* and *B3i* showed better quality parameters than the other *Glu-B3* alleles, whereas the *Glu-B3c* allele containing s-type genes with low expression levels had an inferior effect on bread-making quality. Due to the conserved genes at *Glu-D3* locus, *Glu-D3* alleles showed no significant differences in effects on all quality parameters. This work provided new insights into the composition and function of 18 LMW-GS alleles in bread wheat. The variation of i-type genes mainly contributed to the high diversity of *Glu-A3* alleles, and the differences among *Glu-B3* alleles were mainly derived from the high polymorphism of s-type genes. Among LMW-GS alleles, *Glu-A3e* and *Glu-B3c* represented inferior alleles for bread-making quality, whereas *Glu-A3d*, *Glu-B3b*, *Glu-B3g* and *Glu-B3i* were correlated with superior bread-making quality.

Published in MBC Plant Biology, 2012, 12 (2): 1-16

Glu-D3 alleles played minor roles in determining quality variation in bread wheat. Thus, LMW-GS alleles not only affect dough extensibility but greatly contribute to the dough resistance, glutenin macro-polymers and bread quality.

Key words: *Triticum aestivum*, *Glu-3* alleles, bread-making quality

Abbreviations: 2D-PAGE, Two-dimensional gel electrophoresis (IEF × SDS-PAGE); %UPP= UPP/ (UPP+EPP) × 100; ANOVA, Analysis of variance; CIMMYT, International Maize and Wheat Improvement Center; cMs, centimorgans; DT, Farinograph development time (min); DTT, Dithiothreitol; EA, Energy area (cm^2); EC, External color; EPP, Extractable glutenin polymeric protein; Ext, Extensograph extensibility (mm); Gli/Glu, Ratio of gliadin to glutenin; GMP, glutenin macro-polymers; HMW-GS, high-molecular-weight glutenin subunits; HPLC, High-performance liquid chromatography; IC, Inner color; IEF, Isoelectric focusing; kDa, kilodalton; KP: Kernel protein (%, 14% m. b.); LMW-GS, low-molecular-weight glutenin subunits; LV, Loaf volume (cm^3); LVS, Loaf volume score; NIL, Near-isogenic lines; ORFs, Open reading frame; PBTS, Pan bread total score; PCR, Polymerase chain reaction; pI: isoelectric points; RILs, Recombinant inbred lines; Rmax, Extensograph maximum resistance (B. U.); SDS, Sodium dodecyl sulfate; SDS-PAGE, Sodium dodecyl sulphate-polyacrylamide gel electrophoresis; Sha, Shape; Smo: Smoothness; Spr, Springiness; ST, Farinograph stability time (min); Str, Structure; TCA, trichloroacetic acid; TF, Taste flavor; Tris-HCl, Tris (hydroxymethyl) aminomethane hydrochloride; UPP, Unextractable glutenin polymeric protein; Wab, Farinograph water absorption (%); ZSV, Zeleny sedimentation value (ml).

Background

The unique viscoelastic properties conferred by gluten proteins in bread wheat are the basis of the flexible processing qualities in producing a wide range of food products for a large proportion of the world population. Gluten proteins, also named prolamins, are classically divided into gliadins and glutenins, based on different solubilities in an alcohol/water mixture[1]. The gliadins are generally monomeric proteins, divided into three groups, α/β-, γ- and ω-gliadins, based on their electrophoretic mobilities at low pH[2]. Glutenins form polymeric proteins stabilized by interchain disulfide bonds. Based on different molecular weights, glutenins can be classified into two groups, high-molecular-weight glutenin subunits (HMW-GS) and low-molecular-weight glutenin subunits (LMW-GS)[2,3]. LMW-GS are further divided into B-, C- and D-group subunits according to their mobilities in sodium dodecyl sulphate polyacrylamide-gel electrophoresis (SDS-PAGE)[4].

In bread wheat, HMW-GS are encoded by genes at the orthologous *Glu-1* loci on the long arms of chromosomes *1A*, *1B* and *1D* (*Glu-A1*, *Glu-B1* and *Glu-D1*). Each locus possesses two paralogous genes encoding one x- and one y-type subunit[5]. LMW-GS genes are located at the *Glu-A3*, *Glu-B3* and *Glu-D3* loci on the short arms of group 1 chromosomes. The LMW-GS genes at the *Glu-3* loci and the gliadin genes at the *Gli-1* loci are tightly linked and form gene clusters covering several centimorgans (cMs)[6-8]. Moreover, unlike the simple composition of HMW-GS, LMW-GS are encoded by a complex multigene family without the information of the exact number of ge nes[9, 10]. A large number of genes and abundant allelic variations at *Glu-3* loci and their tight linkage with gliadin genes make it difficult to elucidate the composition and function of LMW-GS genes in bread wheat[4].

SDS-PAGE is widely used to investigate the abundant seed storage proteins in bread wheat. Based on the mobility of proteins in SDS-PAGE gels, whole seed proteins are divided into four groups, HMW-GS, D-group, B-group, and C-group. D-group proteins are mainly composed of ω-gliadin proteins, whereas B-group mostly consists of LMW-GS proteins, and C-group comprise α, β and γ-type gliadins and several LMW-GS proteins[4]. Based on the different electrophoretic patterns, LMW-GS protein alleles encoded by *Glu-3* loci are designated alphabetically (e. g., *Glu-A3a*)[11]. However, identification of the LMW-GS composition in breeding programs remains a significant challenge because determination of LMW-GS alleles with SDS-PAGE needs much experience. This is why the functions of LMW-GS alleles are not well characterized. Gene-specific markers for *Glu-A3* and *Glu-B3* alleles were developed to identify different LMW-GS alleles. However, molecular markers for *Glu-D3* alleles are still not available due to the slight differences among alleles[12-17]. Using BAC library screening and proteomics methods, LMW-GS genes in Norin 61 (*Glu-A3d*, *Glu-B3i* and *Glu-D3c*), Glenlea (*Glu-A3g*, *Glu-B3g* and *Glu-D3c*) and Xiaoyan 54 (*Glu-A3d*, *Glu-B3d* and *Glu-D3c*) were identified and characterized[10, 18-20]. These studies greatly improved our understanding of the unique genes encoding different LMW-GS alleles in bread wheat. Recently, based on the conserved and polymorphic structure of LMW-GS genes, we developed a LMW-GS gene marker system and a full-length gene cloning method[21, 22]. They were successfully used to identify and characterize more than 16 LMW-GS genes in individual wheat varieties[21]. Both methods are helpful in elucidating the composition of *Glu-3* alleles in LMW-GS genes of bread wheat.

The effects of glutenin alleles on dough properties and processing qualities were mostly studied in two types of populations: structured populations [e. g., recombinant inbred lines (RILs) and doubled haploid lines] derived from biparental crosses, and non-structured populations, general collections of varieties and breeding lines. Due to the simple composition and easy identification of allelic variants of HMW-GS, the contributions of HMW-GS to dough properties and end-use quality were well investigated and widely used in breeding programs[23]. However, HMW-GS alone could not explain the variation in quality among wheat varieties, as LMW-GS also contributed to dough properties[24-31]. For example, LMW-GS alleles made a slightly larger contribution than HMW-GS to dough extensibility[32, 33]. Compared with HMW-GS, LMW-GS formed highly polymorphic protein complex and contain abundant allelic variation. Using SDS-PAGE and allele-specific primers, LMW-GS alleles were identified in wheat collections or structured populations, and their effects on processing quality were analyzed and discussed. However, controversies were common in regard to different kinds of populations or collections. For example, Cane et al.[34] and Eagles et al.[33] reported that *Glu-A3e* was correlated with inferior dough resistance and extensibility, whereas Zheng et al.[35] showed that *Glu-A3e* was a favorable allele for dough-mixing properties. Due to the complex composition of LMW-GS alleles and difficulties in distinguishing LMW-GS from gliadins in SDS-PAGE gels, the molecular genetic mechanisms behind the functional differences of LMW-GS alleles are not well investigated.

In the present study, a set of near-isogenic lines (NILs) containing five *Glu-A3* alleles, eight *Glu-B3* alleles and five *Glu-D3* alleles, was used to study the effects of LMW-GS on the composition of glutenin macro-polymers (GMP), dough properties, and pan bread making quality. These NILs were investigated using SDS-PAGE, the LMW-GS gene marker system and two-dimensional gel electrophoresis (2D-PAGE) for identifying the composition of LMW-GS genes and proteins in each LMW-GS allele, and analyzing their association with dough properties and bread-making quality.

Methods

Plant materials

The wheat variety Aroona and 15 near isogenic lines (NILs) were kindly provided by Dr. Marie Appelbee and Prof. Ken Shepherd, SARDI Grain Quality Research Laboratory, South Australia. Each NIL contains a unique LMW-GS allele from a donor variety added to Aroona (Supplementary Table 5). They were planted at the Xinjiang Academy of Agri-Reclamation Sciences, Shihezi, and Xinjiang Academy of Agricultural Sciences, Urumqi, Xinjiang province, in randomized complete blocks with two replications during the 2010 cropping season.

Analysis of LMW-GS genes

Genomic DNA of 16 Aroona NILs was extracted from young leaves of seedlings following Saghai-Maroof et al. [51]. The LMW-GS genes were separated by the LMW-GS gene molecular marker system[22]. With the help of the LMW-GS genes available from the Aroona NILs and some other wheat varieties[10, 13, 14, 17, 18], LMW-GS genes were characterized using Lasergene software (DNAStar; http://www.dnastar.com/), ClustalW2 (http://www.ebi.ac.uk/Tools/msa/clustalw2/), and MEGA 5 software[52].

Isolation and separation of LMW-GS proteins

Glutenin extraction was performed according to the method described by Singh et al. [11]. These proteins were separated by SDS-PAGE using the method described by Sunbrook and Russell[53]. Whole seed proteins were isolated from wheat flour based on the SDS/Phenol method[54] with some modifications. Briefly, proteins in 0.12g flour were precipitated with 10% TCA/acetone at −20℃ overnight. After centrifuging at 20, 000g at 4℃ for 15min, the pellet was washed three times with 80% acetone, then dried at 50℃. Whole proteins were extracted with SDS/Phenol buffer (50% Tris-phenol pH8.0, 30% sucrose, 2% SDS, 0.1M Tris-HCl pH8.0 and 2% DTT). The upper phenol phase was transferred into a new 2ml tube. A fivefold volume of methanol containing 0.1M ammonium acetate was added to the tube. The proteins were deposited at −20℃ for 10min or overnight. After centrifuging at 20, 000g for 5min at 4℃, the pellet was washed once with 100% methanol and twice with 80% acetone before briefly drying in air. The proteins were dissolved in isoelectric focusing (IEF) sample extraction solution and used in 2D-PAGE analysis according to Dong et al. [18]. The images of SDS-PAGE were analyzed using NIH ImageJ software program (http://rsb.info.nih.gov/ij/).

Quality testing and evaluation of pan bread

Measures of grain hardness, protein content, Zeleny sedimentation values, Farinograph and Extensograph parameters, and pan bread qualities were performed by methods reported in He et al. [55]. The glutenin macropolymer compositions were measured following Zhang et al. [56].

Statistical analysis

The SAS statistical package (SAS Institute, Cary, NC) was used for data analysis. All statistical analyses were based on averaged data from two locations.

Results

1. Separation of LMW-GS proteins in Aroona NILs using SDS-PAGE

The glutenin alleles in the flour of 16 Aroona NILs were separated by SDS-PAGE (Fig. 1a). Aroona possessed five HMW-GS proteins, viz., 1, 7+9 and 2+12, encoded by genes at *Glu-A1*, *Glu-B1* and *Glu-D1* loci, respectively. All NILs had the same HMW-GS as Aroona, and their unique LMW-GS and gliadin bands were labeled in Fig. 1a. Four *Glu-A3* NILs possessed unique LMW-GS bands at the B-group region from Aroona (*Glu-A3c*). Aroona-*Glu-A3d* and Aroona-*Glu-A3e* also contained specific gliadin bands (Fig. 1a). Among eight *Glu-B3* NILs, Aroona (*Glu-B3b*), Aroona-*Glu-B3a*, *B3f* and *B3g* shared similar B-group proteins, whereas Aroona-*Glu-B3c*, *B3d*, *B3h* and *B3i* showed another group of electrophoretic patterns (Fig. 1a). Among the latter four NILs, Aroona-*Glu-B3c* possessed the lowest quantity of B-group LMW-GS proteins, especially the protein with the largest molecular weight. Four *Glu-D3* NILs and Aroona shared the same B-group LMW-GS proteins (Fig. 1a). Aroona-*Glu-D3b* and *D3d* had the same protein bands and contained one unique LMW-GS from Aroona. Except for one gliadin protein, the electrophoretic pattern of Aroona-*Glu-D3a* was the same as those of Aroona-

Glu-D3b and *D3d*. Aroona-*Glu-D3f* produced a quite different protein pattern in the C-group region compared with other *Glu-D3* NILs. Generally, the *Glu-3* loci were tightly linked with *Gli-1* loci[6-8], and it was difficult to break this linkage through genetic recombination in conventional crosses. Thus, each *Glu-3* NIL not only possessed unique LMW-GS but also contained 1 or 2 specific gliadins.

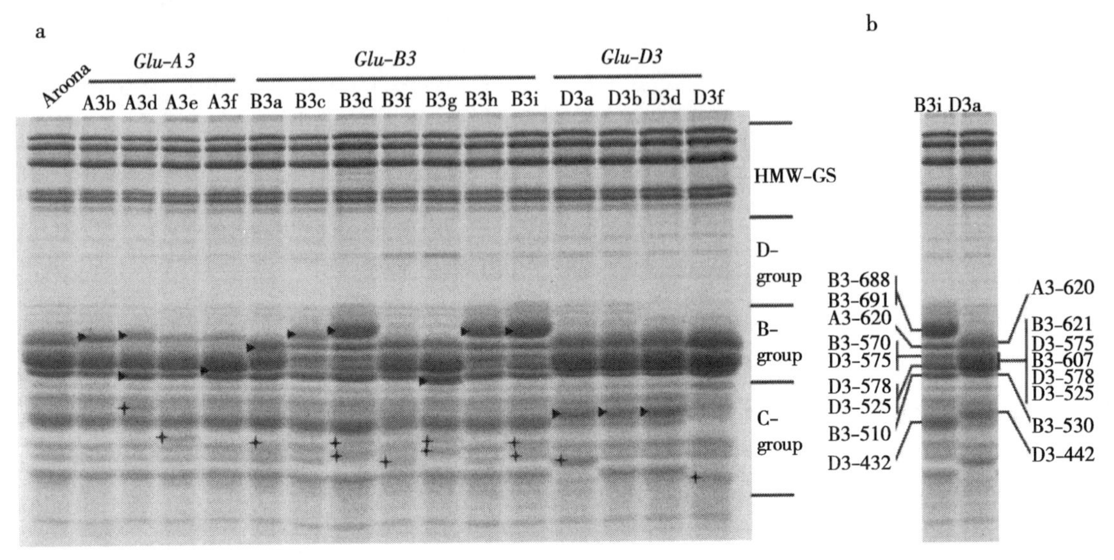

Fig. 1 a) Separation of glutenin proteins from flour of Aroona NILs. Arrowheads mark unique LMW-GS protein bands in individual NILs, and asterisks label specific gliadin bands. b) Identification of LMW-GS bands in Aroona-*Glu-A3d*, Aroona-*Glu-B3i* and Aroona-*Glu-D3a*.

2. Dissecting genes encoding LMW-GS alleles in the Aroona NILs

To analyze the genes encoding LMW-GS alleles in Aroona NILs, we investigated the 16 Aroona NILs using a previously developed LMW-GS gene marker system, that was efficient in separating LMW-GS genes in bread wheat (Fig. 2; Table 1)[21, 22]. Eighteen LMW-GS genes were identified in Aroona, including four *Glu-A3* genes (*A3-391*, *A3-400*, *A3-502-2* and *A3-620*), four *Glu-B3* genes (*B3-530-2*, *B3-578*, *B3-607* and *B3-621-1*), and eight *Glu-D3* genes (*D3-385*, *D3-393*, *D3-394*, *D3-432*, *D3-528*, *D3-575*, *D3-578-1* and *D3-591*) (Table 1). The other two novel genes corresponding to DNA fragments 388 and 410 present in all 16 lines (Table 1) were pseudogenes with premature termination codons in the CDS regions. Among 16 Aroona NILs, Aroona-*Glu-D3f* possessed 14 unique genes, with the least number of LMW-GS genes, whereas Aroona-*Glu-A3d* had the largest number of LMW-GS genes (20 genes). Comparison of the LMW-GS genes among NILs indicated that each NIL differed from Aroona at only one *Glu-3* locus. With the help of the complete gene sequences available in Aroona NILs and some other wheat varieties[10, 13, 14, 17-19, 21], the composition of LMW-GS genes in each Aroona NIL was well characterized.

For *Glu-A3* alleles, 4-6 genes were identified in Aroona (*Glu-A3c*) and four *Glu-A3* NILs (Table 1; Fig. 2). Except for Aroona-*Glu-A3d*, these NILs possessed LMW-GS genes *A3-391* and *A3-400* (Table 1; Fig. 2), which were m-type pseudogenes with premature termination codons. The other genes were i-type genes. The *A3-502* gene contained four allelic variants with unique SNPs and InDels, i. e., *A3-502-1*, *A3-502-2*, *A3-502-3* and *A3-502-4*. All *A3-502* allelic variants and the *A3-640* gene were pseudogenes, containing premature termination codons in the coding sequences.

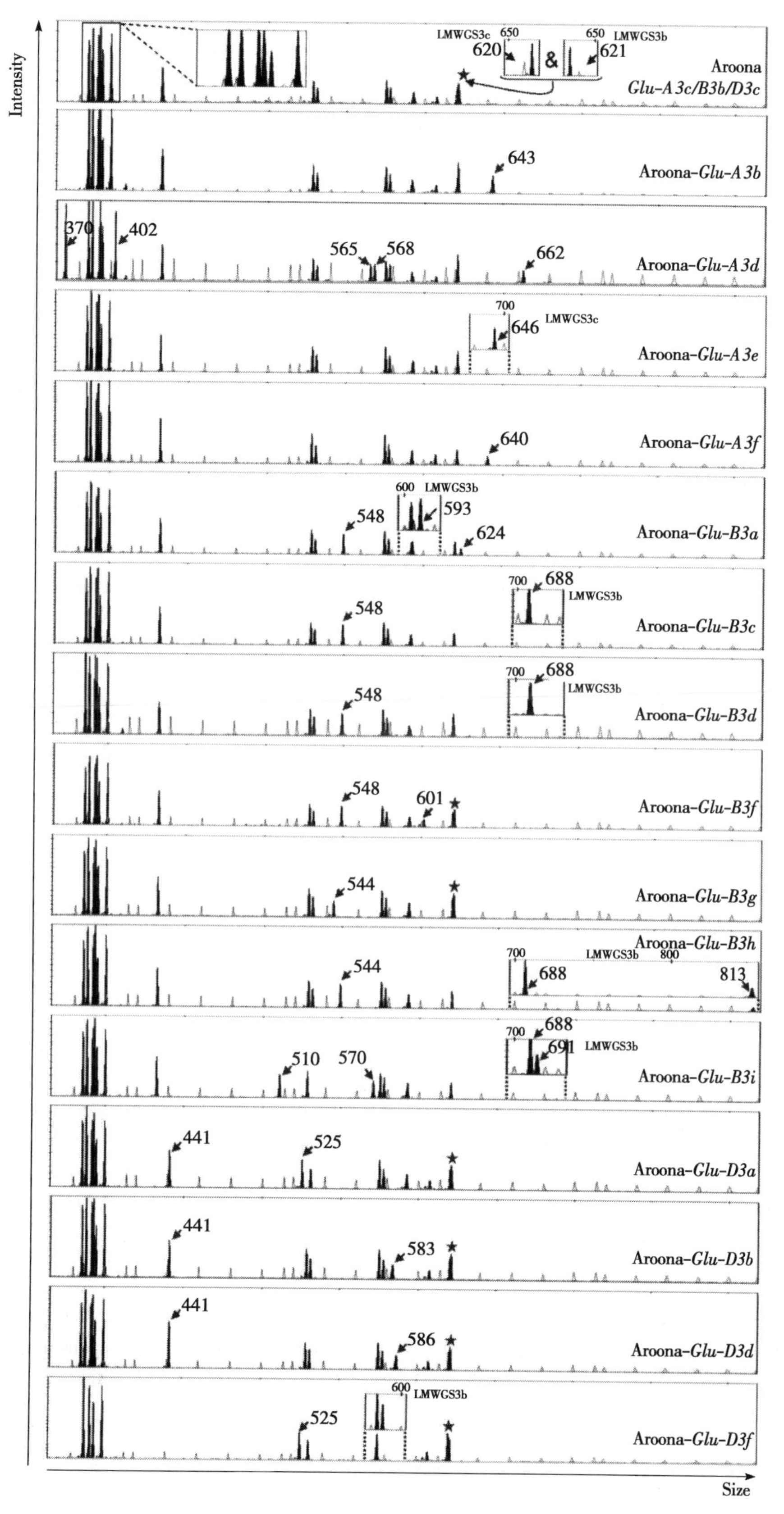

Fig. 2 Separation of LMW-GS genes in Aroona NILs using the LMW-GS gene molecular marker system. The marker system contained three sets of primers, LMWGS1, LMW-GS2 and LMWGS3. The main data were derived from primer set LMWGS1, and the added figures were from the primer LMWGS3b or LMWGS3c. The peaks marked with the asterisk comprise corresponding two genes, *A3-620* and *B3-621*. The arrows indicate unique genes in each Aroona NIL.

Table 1 LMW-GS genes identified in 16 Aroona NILs using the LMW-GS gene molecular marker system

Line	Genes at *Glu-A3* locus	Genes at *Glu-B3* locus	Genes at *Glu-D3* locus	New genes
Aroona (*Glu-A3c*, *B3b*, *D3c*)	391 400 502-2 **620**[a]	**530-2** 578 **607** **621-1**	**385** 393 **394** **432** **528** **575** **578-1** 591	388 410
Aroona-*Glu-A3b*	391 400 502-1 **643**	**530-2** 578 **607** **621-1**	**385** 393 **394** **432** **528** **575** **578-1** 591	388 410
Aroona-*Glu-A3d*	370 **402** 484 565 **568** **662**	**530-2** 578 **607** **621-1**	**385** 393 **394** **432** **528** **575** **578-1** 591	388 410
Aroona-*Glu-A3e*	391 400 502-3 **646**	**530-2** 578 **607** **621-1**	**385** 393 **394** **432** **528** **575** **578-1** 591	388 410
Aroona-*Glu-A3f*	391 400 502-4 **573** **640**	**530-2** 578 **607** **621-1**	**385** 393 **394** **432** **528** **575** **578-1** 591	388 410
Aroona-*Glu-B3a*	391 400 502-2 **620**	**530-1** 548 578 **593** **624**	**385** 393 **394** **432** **528** **575** **578-1** **591**	388 410
Aroona-*Glu-B3c*	391 400 502-2 **620**	**530-1** 548 **688-1**	**385** 393 **394** **432** **528** **575** **578-1** **591**	388 410
Aroona-*Glu-B3d*	391 400 502-2 **620**	**530-1** 548 **688-2**	**385** 393 **394** **432** **528** **575** **578-1** **591**	388 410
Aroona-*Glu-B3f*	391 400 502-2 **620**	**530-3** 548 578 **601** **621-1**	**385** 393 **394** **432** **528** **575** **578-1** **591**	388 410
Aroona-*Glu-B3g*	391 400 502-2 **620**	**530-3** 578 **544** **621-2**	**385** 393 **394** **432** **528** **575** **578-1** **591**	388 410
Aroona-*Glu-B3h*	391 400 502-2 **620**	**530-2** 548 **688-3** 813	**385** 393 394 **432** **528** **575** **578-1** **591**	388 410
Aroona-*Glu-B3i*	391 400 502-2 **620**	**510** **570** **688-4** **691**	**385** 393 **394** **432** **528** **575** **578-1** **591**	388 410
Aroona-*Glu-D3a*	391 400 502-2 **620**	**530-2** 578 **607** **621-1**	**385** 393 **394** **441** **525** **575** **578-2** **591**	388 410
Aroona-*Glu-D3b*	391 400 502-2 **620**	**530-2** 578 **607** **621-1**	**385** 393 **394** **441** **528** **575** **578-2** **583**	388 410
Aroona-*Glu-D3d*	391 400 502-2 **620**	**530-2** 578 **607** **621-1**	**385** 393 **394** **441** **528** **575** **578-2** **586**	388 410
Aroona-*Glu-D3f*	391 400 502-2 **620**	**530-2** 578 **607** **621-1**	— — **394** — **525** **575** **578** —	388 410

a. Bold numbers indicate active LMW-GS genes.

The i-type genes, *A3-643*, *A3-620*, *A3-646* and *A3-573*, were the only active genes in Aroona-*Glu-A3b*, *A3c*, *A3e* and *A3f*, respectively. Sequence alignments showed that *A3-620* and *A3-643* shared high identity (>99%) differing only in one InDel and one SNP, whereas *A3-573* and *A3-646* were greatly different from each other and the other i-type genes. A3-573 was much shorter than the other i-type proteins due to deletion of three repetitive units and several glutamine residues in repetitive regions, and lacking GTFLQPH in the C-terminal domain. Aroona-*Glu-A3d* contained six unique LMW-GS genes (Table 1; Fig. 2). For two m-type genes, *A3-370* was a pseudogene, whereas *A3-402* contained an intact coding sequence that could be expressed in the developing grain[18]. The other four genes, *A3-484*, *A3-565*, *A3-568* and *A3-662* were i-type genes, and among them *A3-568* and *A3-662* were active. Both genes contained unique SNPs in the C-terminal domain and different lengths of repetitive region compared to other i-type genes. Thus, compared to only one LMW-GS gene expressed in the other *Glu-A3* alleles, *Glu-A3d* contained three active genes, one m-type (*A3-402*) and two i-type (*A3-568* and *A3-662*) LMW-GS genes. Except for *A3-402*, the active genes at the *Glu-A3* locus were i-type genes. These i-type genes encoded B-group LMW-GS, producing different electrophoretic patterns in the B-group region among *Glu-A3* NILs (Fig. 1).

For *Glu-B3* alleles, 3-5 LMW-GS genes were identified in Aroona (*Glu-B3b*) and seven *Glu-B3* NILs. Gene *B3-530* was present in all Aroona NILs and four allelic variants (*B3-530-1*, *-2*, *-3* and *B3-510*) were identified (Table 1). Sequence alignment showed that the deduced protein sequences of the three *B3-530* allelic variants differed in only four amino acids, whereas B3-510 contained a deletion in the repetitive domain and differed by seven amino acids from the B3-530 proteins. *B3-544*, *B3-593*, *B3-601* and *B3-607* were allelic variants, sharing over 99% of identity and their deduced protein sequences differed only in the repetitive domain with the deletion of glutamine residues or one repetitive unit. *B3-621-1*, *B3-621-2* and *B3-624* genes were highly conserved and their proteins differed only in one InDel of a glutamine residue in the repetitive domain and one amino acid at the C terminal region. The *B3-688* gene was quite distinct, differing from *B3-621* and *B3-624* in several SNPs and InDels. Moreover, *B3-688* varied among Aroona-*Glu-B3c*, *B3d*, *B3h* and *B3i*, and the protein products differed in seven amino acids and two InDels. Among *Glu-B3* genes, *B3-548*, *B3-578* and *B3-813* were pseudogenes with premature termination codons, whereas the other genes contained intact ORFs (Open Reading Frame). Sequence analysis showed that *B3-530*, *B3-510*, *B3-548* and *B3-570* were m-type genes and the others were s-type genes. Based on the differences in s-type genes, *Glu-B3* NILs were classified into two groups. Aroona-*Glu-B3a*, *B3b*, *B3f* and *B3g* contained *B3-578*, *B3-544/593/601/607*, and *B3-621/624*, and the others had longer sequences of genes, *B3-688* and *B3-691/813/Null* (Table 1). This classification was consistent with the different electrophoretic patterns contrasting the two groups (Fig. 1a).

Aroona-*Glu-B3a*, *B3b*, *B3f* and *B3g* had similar LMW-GS compositions, but each of them possessed unique allelic variants (Table 1; Fig. 2). Both Aroona-*Glu-B3f* and *B3g* contained B3-530-3, but they differed in *B3-544*, *B3-601*, *B3-621-1* and *B3-621-2* genes (Table 1). Aroona-*Glu-B3f* and Aroona (*Glu-B3b*) shared the same *B3-621-1*, but possessed different genes *B3-601*/*B3-530-3* and *B3-607*/*B3-530-2*, respectively (Table 1). On the other hand, Aroona-*Glu-B3c*, *B3d*, *B3h*, and *B3i* also differed in active genes and their allelic variants. Aroona-*Glu-B3c* and *B3d* contained B3-530-1, but their B3-688 proteins differed in five amino acids (Table 1). Aroona-*Glu-B3i* was unique, containing four active LMW-GS genes *B3-510*, *B3-570*, *B3-688-4* and *B3-691* (Table 1; Fig. 2). *B3-570* was only present in *Glu-B3i*, encoding an m-type LMW-GS with the N-terminal domain METSQIPGLEKPS. Although *B3-688-4* and *B3-691* were different genes, they shared about 99% identity and *B3-691* is formally reported for the first time.

At the *Glu-D3* locus, eight genes were identified from individual NILs except Aroona-*Glu-D3f* (Table 1). *D3-393* and *D3-583/586/591* had premature termination codons. All the other genes were widely reported and were expressed in developing grains[10, 17, 18, 20]. *D3-385*, *D3-394* and *D3-575* were highly conversed among the Aroona NILs. *D3-432* and *D3-441* were allelic variants with over 99% identity, and differed in only two SNPs and two InDels. Allelic variants *D3-525* and *D3-528* were identical, except for one InDel. Two allelic variants, *D3-578-1* and *D3-578-2* differed by eight SNPs (Table 1). Compared with the active *Glu-D3* genes in Aroona (*Glu-D3c*), *D3-441* and *D3-578-2* were present in Aroona-*Glu-D3b and D3d*, which possessed the same active LMW-GS genes (Table 1; Fig. 2). Besides *D3-441* and *D3-578-2*, Aroona-*Glu-D3a* contained unique gene *D3-525* in contrast to Aroona. In Aroona-*Glu-D3f*, only four LMW-GS genes were identified; the other four genes might be absent due to deletion, substitution or other mutations (Table 1; Fig. 2).

Each Aroona NIL possessed unique LMW-GS alleles. The gene composition of these LMW-GS alleles were dissected, all the LMW-GS gene sequences were characterized, and the proteins sequences were deduced (Table 1; Fig. 2). Based on the size of the deduced LMW-GS and the LMW-GS 2D-PAGE spots identified from Xiaoyan 54, Jing 411 and Norin 61[18, 19], LMW-GS genes were assigned to protein bands in SDS-PAGE gels (Fig. 1b). The active i-type genes at the *Glu-A3* locus, all the active genes at *Glu-B3* and the genes *D3-525/528*, *D3-575* and *D3-578* at *Glu-D3* encoded B-group LMW-GS, whereas the protein products of the other active genes belonged to the C-group. Moreover, *B3-621/624*, *B3-593/601/607*, *D3-575* and *D3-578* encoded proteins with similar molecular weights (38-40 kDa), and ran synchronously on SDS-PAGE, forming the thickest protein bands in Aroona-*Glu-B3a*, *B3b* and *B3f*. *B3-688* genes encoding the longest protein conferred the unique protein pattern on Aroona-*Glu-*

B3d, *B3h* and *B3i*. (Fig. 1).

3. Comparison of whole proteins among LMW-GS NILs using 2D-PAGE

The data from SDS-PAGE and the LMW-GS marker system showed that each *Glu-3* allele contained several unique LMW-GS and gliadin genes. To analyze the composition of LMW-GS and gliadin proteins in individual NILs, we performed 2D-PAGE analysis of the whole flour proteins (Figs. 3, 4 and 5). In the previous studies, using 2D-PAGE coupled with MS or N-terminal sequencing technology, LMW-GS proteins in Xiaoyan 54, Jing 411, Norin 61 and Butte 86 were successfully identified[18, 19, 36]. These data greatly contributed to the identification of flour proteins in the Aroona NILs.

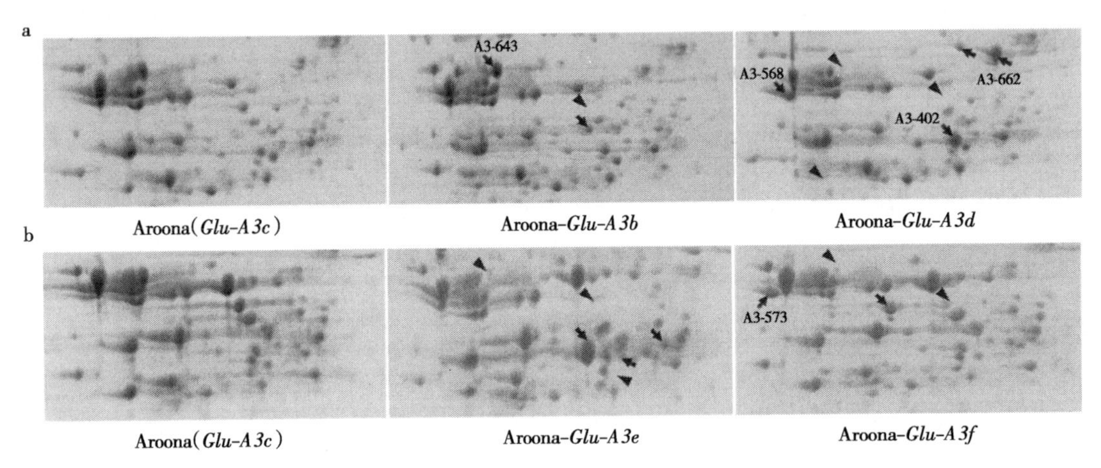

Fig. 3 Separation of flour proteins in *Glu-A3* NILs using two-dimensional gel electrophoresis (2D-PAGE). a) Comparison of storage proteins from Aroona, Aroona-*Glu-A3b* and Aroona-*Glu-A3d*. b) Comparison of storage proteins from Aroona, Aroona-*Glu-A3e* and Aroona-*Glu-A3f*. The arrows indicate the unique protein spots in each NIL, and the arrowheads show the protein spots present in Aroona but absent in other NILs. The high molecular weight glutenin subunit protein spots are the same and not shown here due to limited space.

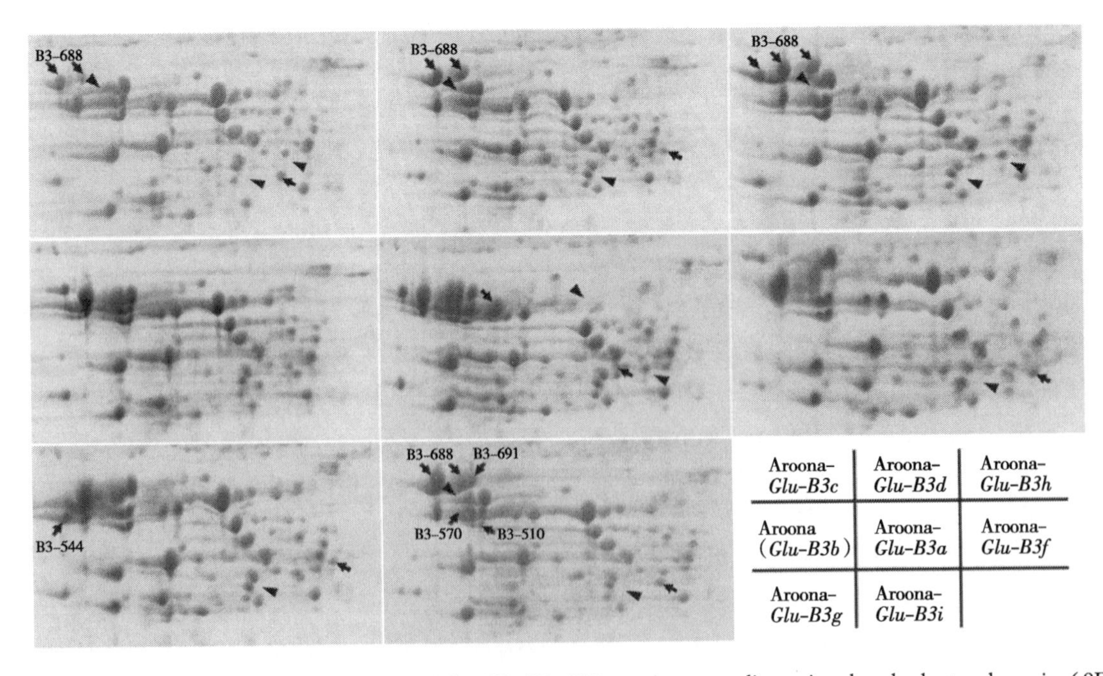

Fig. 4 Separation of flour proteins in eight *Glu-B3* NILs using two-dimensional gel electrophoresis (2D-PAGE). Arrows indicate the unique protein spots in each NIL, and the arrowheads show protein spots present in Aroona but absent in other NILs.

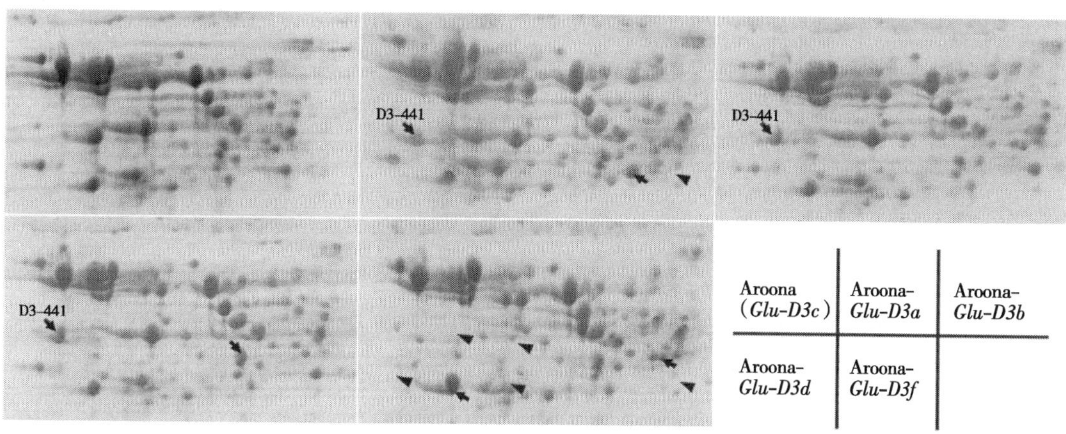

Fig. 5 Separation of flour proteins in five *Glu-D3* NILs using two-dimensional gel electrophoresis (2D-PAGE). The arrows indicate the unique protein spots in each NIL, and the arrowheads show the protein spots present in Aroona but absent in other NILs.

Among five *Glu-A3* NILs, Aroona-*Glu-A3b* and Aroona (*Glu-A3c*) shared the same protein patterns, except for one gliadin spot, which was in agreement with the similar composition of LMW-GS genes and the high identity between *A3-620* and *A3-643* (Fig. 3a). Compared with Aroona, Aroona-*Glu-A3d* contained three unique LMW-GS, A3-402, A3-568 and A3-662, and two gliadin spots were absent (Fig. 3a). For Aroona-*Glu-A3e*, the protein product of the unique i-type gene *A3-646* was too little to be detected, but three unique gliadin spots were present in 2D-PAGE (Fig. 3b). Aroona-*Glu-A3f* also contained one unique i-type protein (A3-573) and one gliadin spot different from Aroona (Fig. 3b). Thus, accompanying the unique LMW-GSs in individual *Glu-A3* NILs, generally 1-2 gliadin spots were different among them. The only exception was Aroona-*Glu-A3e* which possessed the highest number of gliadins and the lowest quantity (none) of i-type LMW-GS among the *Glu-A3* NILs (Fig. 3b).

Among eight *Glu-B3* NILs, Aroona-*Glu-B3c*, *B3d*, *B3h* and *B3i* shared similar protein spot patterns on 2D-PAGE, consistent with their similar electrophoretic patterns on SDS-PAGE (Fig. 1a and 4). Compared with the protein spot pattern of Aroona (*Glu-B3b*), Aroona-*Glu-B3c*, *B3d*, *B3h* and *B3i* contain the unique B3-688 proteins (Fig. 4). The B3-688 protein spot in Aroona-*Glu-B3c* was much weaker than those in the other three NILs. Except for LMW-GS proteins, only one or two small gliadin spot differences were identified among the four NILs (Fig. 4). The other three Aroona NILs (i. e., Aroona-*Glu-B3a*, *B3f* and *B3g*) had similar protein spot patterns with Aroona (*Glu-B3b*; Fig. 4). Among these four NILs, a unique LMW-GS protein B3-544 was detected in Aroona-*Glu-B3g*, whereas the other three alleles shared the same LMW-GS protein spot patterns (Fig. 4). Thus, the 2D-PAGE patterns of Aroona and seven *Glu-B3* NILs suggested that their different proteins mainly were unique LMW-GS proteins rather than gliadins.

For the *Glu-D3* NILs, the protein spots encoded by the *D3-441* gene in Aroona-*Glu-D3a*, *D3b* and *D3d* had larger molecular weights and pIs than its allelic variant *D3-432* in Aroona (Fig. 5). No other different LMW-GS spots were detected among these *Glu-D3* NILs, and Aroona-*Glu-D3a* and *D3d*, each contained only one unique gliadin spot (Fig. 5). The Aroona-*Glu-D3f* allele was unique, lacking two LMW-GS proteins (i. e., D3-385 and D3-441) and at least three medium gliadin spots (Fig. 5). The absence of four LMW-GS genes was also observed using the LMW-GS gene marker system (Table 1). Thus, except for Aroona-*Glu-D3f*, only one or two different protein spots were detected among *Glu-D3* NILs, and the proteins encoded by genes at the *Glu-D3* locus were highly conserved.

4. Quality properties of the LMW-GS NILs

Dough properties such as Zeleny-sedimentation value

(ZSV), Farinograph and Extensograph parameters, GMP parameters, and pan bread quality parameters, were measured on the Aroona NILs (Table 2). The 16 genotypes showed significant differences ($P<0.05$ or $P<0.01$) in most quality parameters, suggesting that genotype had an important influence on variation in wheat quality properties. Significant differences in some parameters (e.g., kernel protein, ZSV, Farinograph water absorption, glutenin and gliadin contents, and external bread color and structure) were also observed between the two locations. However, no significant interaction effects between genotypes and locations were detected for most quality parameters, except for external color of bread and Farinograph development time. Analysis of variance (ANOVA) and multiple comparisons of all quality parameters were performed among all NILs within each *Glu-3* group (Table 2).

Table 2 F values of one way ANOVA of wheat quality parameters within each *Glu-3* group[a]

Group	LV	LVS	EC	Sha	IC	Smo	Spr	Str	TF	PBTS
Aroona-*Glu-A3*	5.31*	2.33	0.7	3.35	1.14	12.14**	0.67	6.45*	2	4.75*
Aroona-*Glu-B3*	2.85*	2.07	0.8	1.57	1.28	2.09	0.75	5.09**	2	5.78**
Aroona-*Glu-D3*	3.74	2.29	1.71	2.57	1	1	0.33	0.52	0.12	0.68

Group	KP	ZSV	ST	DT	Wab	Rmax	EA	Ext
Aroona-*Glu-A3*	3.3	59.35**	6.55*	11.57**	27.98**	10.34**	13.46**	4.51*
Aroona-*Glu-B3*	2.54	43.89**	6.19**	8.47**	5.4**	9.97**	10.06**	2.34
Aroona-*Glu-D3*	3.01	0.85	2.47	3.05	4.04*	2.58	1.18	2.33

Group	Glutenin	Gliadin	Gli/Glu	EPP	UPP	%UPP
Aroona-*Glu-A3*	0.28	3.13	4.62*	2.87	7.1**	9.58**
Aroona-*Glu-B3*	0.55	8.81**	4.65**	2.14	3.15*	3.63*
Aroona-*Glu-D3*	3.64	0.32	6.55*	1.51	1.53	1.48

* Significant at $P<0.05$; ** Significant at $P<0.01$.

a. LV, Loaf volume (cm^3); LVS, Loaf volume score; EC, External color; Sha, Shape; IC, Inner color; Smo, Smoothness; Spr, Springiness; Str, Structure; TF, Taste flavor; PBTS, Pan bread total score; KP, Kernel protein (%, 14% m. b.); ZSV, Zeleny sedimentation value (ml); ST, Farinograph stability time (min); DT, Farinograph development time (min); Wab, Farinograph water absorption (%); Rmax, Extensograph maximum resistance (B. U.); EA, Energy area (cm^2); Ext, Extensograph extensibility (mm); EPP, Extractable glutenin polymeric protein; UPP, Unextractable glutenin polymeric protein; Gli/Glu, Ratio of gliadin to glutenin; %UPP=UPP/ (UPP+EPP) ×100.

ANOVA showed that the *Glu-A3* NILs were significantly different in most quality parameters (Table 2). This suggested that *Glu-A3* alleles affect the breadmaking quality of bread wheat. Subsequent multiple comparisons indicated that Aroona-*Glu-A3e* always produced the worst quality parameters, e.g., ZSV, Farinograph stability time (ST), Extensograph maximum resistance (Rmax), Extensograph extensibility (Ext), percentage of SDS-unextractable fraction in total polymeric protein (%UPP), and pan bread total score (PBTS), whereas the other five *Glu-A3* NILs had similar values for Rmax, ST, %UPP and Ext (Fig. 6). Although Aroona-*Glu-A3d* had the largest ZSV, it showed only moderate PBTS (Fig. 6).

Glu-B3 NILs showed large differences in the most important quality parameters (Table 2), indicating that *Glu-B3* alleles were also significantly associated with breadmaking quality. Among eight *Glu-B3* NILs, Aroona-*Glu-B3g* produced the highest PBTS, ZSV, ST, Rmax, %UPP and some other parameters (Fig. 6). Similarly, Aroona (*Glu-B3b*) and Aroona-*Glu-B3i* had higher quality parameters than the other *Glu-B3* NILs, excluding Aroona-*Glu-B3g*. In contrast, Aroona-*Glu-B3c* performed the lowest in ZSV, ST, Rmax, %UPP, PBTS, and some other parameters (Fig. 6), suggesting

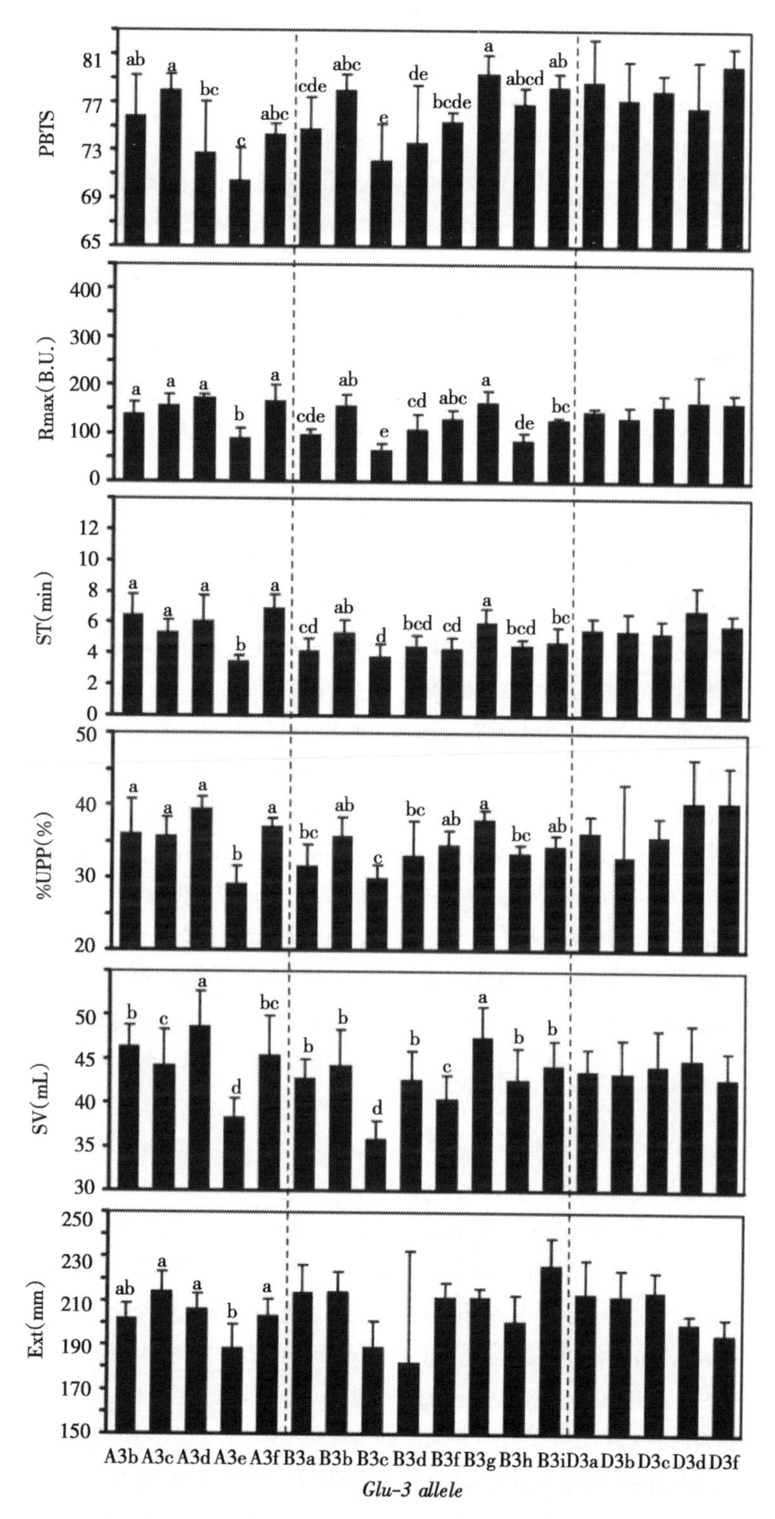

Fig. 6 Comparison of wheat quality properties among *Glu-A3* NILs, *Glu-B3* NILs and *Glu-D3* NILs.
Statistical analysis of the quality parameters of Aroona NILs for each glutenin locus by ANOVA and Duncan LSR ($P<0.05$). The bars are labeled by different letters or letter combinations based on multiple statistical comparisons. No statistical significance exists between alleles labeled by one or more identical letters. PBTS, Pan bread total score; Rmax, Extensograph maximum resistance; ST, Farinograph stability time; %UPP, Percent of SDS-unextractable fraction in total polymeric protein; ZSV, Zeleny-sedimentation value; Ext, Extensograph extensibility.

that *Glu-B3c* was an undesirable allele for bread-making quality. The other *Glu-B3* NILs (Aroona-*Glu-B3a*, *B3d*, *B3f* and *B3h*) possessed similar moderate quality parameters, without significant differences.

Five *Glu-D3* NILs produced similar quality parameters, without significant differences, except for Farinograph water absorption and the ratio of gliadin to glutenin (Table 2). This was consistent with the similar composition of LMW-GS genes and proteins among *Glu-D3* NILs.

Discussion

In the present study, SDS-PAGE, 2D-PAGE and LMW-GS gene molecular marker system were used for characterization of LMW-GS genes and proteins in 16 Aroona NILs. Each NIL contained a unique LMW-GS allele. The genetic and protein compositions of each LMW-GS allele were dissected. In addition, we analyzed the functional differences among the LMW-GS alleles in bread-making quality using these NILs. The molecular mechanisms behind the functional differences are discussed based on the characterization of LMW-GS genes and proteins in the NILs.

1. Characterization of LMW-GS genes and proteins in the NILs

SDS-PAGE is a useful method to accurately identify HMW-GS in bread wheat, but is inefficient for separating LMW-GS alleles, because of the presence of large numbers of LMW-GS proteins with similar mobilities with each other and with gliadins (Fig. 1b). In the present study, the LMW-GS gene marker system was successfully used to separate 14-20 LMW-GS from each NIL, and also distinguished 16 *Glu-3* NILs, except for Aroona-*Glu-B3c* and *B3d* without the length polymorphisms among *B3-688-1* and *B3-688-2* (Table 1; Fig. 2). These results indicated that the LMW-GS gene marker system was efficient and accurate not only in separating members of the LMW-GS gene family[21], but also in distinguishing allelic variants of individual LMW-GS genes. Recently, allele-specific markers were widely used in distinguishing *Glu-A3* and *Glu-B3* alleles[12-14]. However, the high conservation among *Glu-D3* allelic variants made it difficult to develop allele-specific markers for discriminating different allelic variants. Because these allelic variants showed DNA sequence polymorphisms in length, the LMW-GS gene marker system worked well in dissecting the complex genes and allelic variants at the *Glu-D3* locus (Table 1). On the other hand, compared with only one gene in individual alleles identified with gene-specific primers[13, 14], almost all the genes in *Glu-3* alleles were displayed using the LMW-GS gene marker system (Table 1). Identification of all genes in each allele will greatly contribute to an understanding of the molecular mechanism determining functional differences among *Glu-3* alleles in bread wheat.

In the present study, Aroona and its NILs contained most of the *Glu-3* allelic variations identified from the worldwide wheat germplasm in a cooperative program[12]. Each LMW-GS allele was encoded by several linked genes or haplotypes[13, 14, 16, 17, 21, 22]. The dissection of LMW-GS genes in Aroona NILs would facilitate the haplotype analysis of LMW-GS genes in common wheat. At the *Glu-A3* locus, besides two m-type genes, 2-4 i-type genes were identified from individual *Glu-A3* alleles. Unlike the conserved m-type genes, i-type genes were completely different among five alleles. These i-type genes might be tightly linked, forming unique haplotype in each *Glu-A3* allele, e. g. *A3-484*/*A3-565*/*A3-568*/*A3-662* in *Glu-A3d* (Table 1)[13, 18, 19, 21]. At the *Glu-B3* locus, 1-2m-type gene (s) and 1-3 s-type gene (s) were characterized in individual alleles. Generally, these s-type genes could be divided into two groups, one containing *B3-578*/*B3-544*/*593*/*601*/*607*/*B3-621*/*624*, the other having *B3-688*/*B3-691*/*813*/*N* (Table 1, Fig. 2). These s-type genes in each group might cosegregate and form unique haplotypes in bread wheat[14]. Among four allelic variants of m-type gene *B3-530*, *B3-530-2* was present in both Aroona (*Glu-B3b*) and Aroona-*Glu-B3h*, containing difference s-type haplotypes *B3-578*/*B3-607*/*B3-621-1* and *B3-688*/*B3-813*, respectively. Thus, at the *Glu-B3* locus, the m- and s-type genes appeared not to be tightly linked (Table 1; Fig. 2)[6, 18]. All the *Glu-D3* alleles except *Glu-D3f* contained eight LMW-GS genes (Table 1). Only one or two allelic variants (*D3-432/441*, *D3-525/528* and *D3-578-1/-2* were identified for each LMW-GS gene and only a few SNPs were detected among allelic variants. Thus, *Glu-D3* genes were highly conserved among wheat varieti es[16, 17], wherease *Glu-A3* and *Glu-B3* genes showed high diversities, which were derived from multiple allelic variations of i- and s-type genes, respectively.

To analyze differences in seed storage proteins among LMW-GS NILs, flour proteins were displayed using 2D-PAGE (Figs. 3, 4 and 5). LMW-GS proteins shared some similar characteristics, such as mostly located in the B-group region and larger pIs than gliadins. Although it is difficult to detect small variation in molecular weight and quantity of each protein spot in 2D-PAGE, the composition and allelic variations of seed storage proteins for each *Glu-3* allele were successfully characterized. These results were in agreement with the 2D protein patterns of Xiaoyan 54, Jing 411, Norin 61, Butte 86 and some other varieties identified with LC-MS or N-terminal sequencing[18, 19, 36].

For example, the *Glu-A3d* and *Glu-B3d* alleles showed similar electrophoretic patterns to Xiaoyan 54 and Norin 61 in i-type (A3-568 and A3-662) and s-type (B3-688) proteins, respectively[18, 19], and *Glu-B3g* shared the same spot pattern of B3-544 with Glenlea[19]. Moreover, comparison of 2D-PAGE patterns showed that, except for the different LMW-GS proteins, only 1-3 unique small gliadin spots were present in individual Aroona NILs (Figs. 3, 4 and 5). These results suggested that the differences among the Aroona NILs were mainly derived from allelic variation of LMW-GS genes.

2. LMW-GS alleles and bread-making quality

The NIL population was used to study the effects of LMW-GS on dough properties, GMP parameters, and pan bread quality parameters. The results confirmed that the LMW-GS played important roles in determining variation in wheat quality properties. Among the three *Glu-3* loci, alleles *Glu-A3* and *Glu-B3* were of major importance in determining differences in processing qualities among the NILs (Table 2).

Glu-A3 alleles

Among five *Glu-A3* alleles, *Glu-A3e* (*A3-391*, *A3-400*, *A3-502-3* and *A3-646*) performed the poorest in almost all quality properties (Table 1, Fig. 6; Supplementary Tables 2, 3 and 4). This was consistent with previous reports in which *Glu-A3e* was associated with lower extensibility and Rmax than *Glu-A3d*, *A3b* and *A3c*[33]. The negative effect of *Glu-A3e* on dough rheological properties was reported in several studies previously[24, 33, 37, 38]. No unique i-type protein band was detected from the *Glu-A3e* allele using SDS-PAGE (Fig. 1a)[39]. The protein product of the *A3-646* gene was also not identified in 2D-PAGE although the i-type gene *A3-646* in Aroona-*Glu-A3e* contained the intact ORF (Fig. 3b, Supplementary Fig. 2). Less i-type proteins and more gliadins in *Glu-A3e* increased the ratio of gliadin and glutenin and greatly reduced %UPP (Fig. 6; Supplementary Table 4), resulting in the worst performance of the *Glu-A3e* genotype in dough resistance and extensibility and pan bread total score[39]. The other *Glu-A3* NILs produced similar quality parameters, including Rmax, ST, Ext and % UPP. These data indicated that *Glu-A3a*, *A3c*, *A3d* and *A3f* all had equivalent positive effects on UPP content and dough resistance and extensibility (Fig. 6). Among them, *Glu-A3d* had a significant effect on high ZSV, which was consistent with results from Xiaoyan 54 and Jing 411 RILs[18]. Some other studies also reported that the *Glu-A3d* allele had a superior effect on dough strength[40-42]. Compared with only one active gene at the *Glu-A3* locus in the other alleles, the *Glu-A3d* allele possessed three active LMW-GS genes and produced the highest ZSV, Rmax, and %UPP (Fig. 6). The large number of active genes in *Glu-A3d* might be the basis of the superior performance in wheat quality properties[18].

Glu-B3 alleles

Our study confirmed the important contribution of the *Glu-B3* locus to quality variation of bread wheat varieties. Among eight *Glu-B3* alleles, *Glu-B3c* (*B3-530-1*, *B3-548* and *B3-688-1*) produced the lowest quality parameters, including ZSV, ST, Rmax, %UPP, and PBTS (Fig. 6). The inferior effects of *Glu-B3c* allele were also reported in previous studies[24, 33, 43, 44]. *Glu-B3c* and *Glu-B3d* (*B3-530-1*, *B3-548* and *B3-688-2*) shared an identical LMW-GS composition, except for five amino acid differences between B3-688-1 and B3-688-2 (Table 1; Fig. 2), but produced significantly different Rmax and ZSV (Fig. 6). *B3-688-2* appeared to produce better quality properties than *B3-688-1*. On the other hand, *Glu-B3c* and *Glu-B3h* (*B3-530-2*, *B3-548*, *B3-688-3* and *B3-813*) also contained similar active LMW-GS genes and had only a few SNPs in the *B3-530* and *B3-688* allelic variants, but *Glu-B3h* had significantly more positive effects on PBTS and ZSV than *Glu-B3c* (Fig. 6). The 2D-PAGE showed that *Glu-B3c* had a smaller quantity of B3-688-1 protein than *Glu-B3d* and *B3h* (Fig. 4). This suggested that the unique SNPs in *B3-688-1* might result in lower expression or difficulty in translation and polymerization. The small amount of B3-688-1 might directly cause the higher ratio of gliadins and glutenin and the lower %UPP for the *Glu-B3c* allele, finally resulting in its inferior effects on wheat quality. Compared to two active genes in the three *Glu-B3* alleles above, *Glu-B3i* contained four active genes, m-type haplotype *B3-510*/*B3-570* and s-type haplotype *B3-688*/*B3-691*. Both haplotypes provided the highest number of active genes and lead to the best quality performance among the four alleles[18].

The *Glu-B3a*, *B3b*, *B3f* and *B3g* alleles belonged to the same group of s-type haplotypes. They shared similar electrophoretic patterns and each possessed three active LMW-GS genes, but had significantly different effects on quality properties (Fig. 6). *Glu-B3g* (*B3-530-3*, *B3-548*, *B3-544* and *B3-621-2*) produced the best quality parameters, including ZSV, Rmax, ST, %UPP, LV and PBTS (Fig. 6), in agreement with previous reports[40, 42, 45]. *Glu-B3b* (*B3-530-2*, *B3-548*, *B3-607* and *B3-621-1*) also had positive effects on most quality parameters (Fig. 6), again consistent with the previous studies[46, 47]. The SE-HPLC parameters indicated that the low content of gliadin and the high content of UPP might be the main reasons for the superior performance of *Glu-B3b* and *B3g* (Fig. 6). And the high content of UPP might be derived from the SNPs or InDels in the nucleotide sequences of three active genes (*B3-530-2*/*3*, *B3-544*/607 and *B3-621-1*/*-2*), which enhanced the ability of the protein products to form large glutenin macropolymers. More evidence was obtained by comparing *Glu-B3f* and *Glu-B3g*. Both alleles contained the same m-type gene, *B3-530-3*, but had different s-type haplotypes and significant differences in ZSV, ST, UPP, and PBTS. The s-type haplotype *B3-578*/*B3-544*/*B3-621-2* formed more UPP and produced better bread-making quality than *B3-578*/*B3-601*/*B3-621-1*.

Glu-D3 alleles

The five *Glu-D3* alleles produced similar values for almost all quality properties in the present study (Fig. 6; Table 2), which confirmed previous reports that *Glu-D3* alleles produced similar Rmax and extensibilities among large collections of wheat varieties[24, 33, 48], although some studies indicated different effects of *Glu-D3* alleles on dough strength or mixing properties[27, 29, 38, 49]. *Glu-D3a*, *D3b*, *D3c* and *D3d* contained similar 2D-PAGE spot patterns, and all six active genes were highly conserved among the alleles (> 99% identities). Their similarity in LMW-GS genes and whole proteins was consistent with their equivalent effects on all quality properties. Although *Glu-D3f* allele lacks two LMW-GS proteins and three gliadin spots, it produced similar quality properties to the other *Glu-D3* alleles (Fig. 5; Table 2). These results suggested that *D3-385*, *D3-432* and the three missing gliadins were not related to quality improvement. However, the lack of active genes *D3-394* and *D3-528* at the *Glu-D3* locus in Jing 411 (*Glu-D3l*) showed significant negative effects on ZSV[18]. Thus, except for *Glu-D3l*, *Glu-D3* alleles appeared to play only minor roles in determining quality variation among bread wheat varieties, and they should be given the lowest priority among LMW-GS alleles in selecting for improved bread-making quality in wheat.

Conclusion

In the present study, we dissected the genetic and protein composition of 16 LMW-GS NILs, measured the dough property and bread making quality properties of individual NILs, and performed functional analyses for each allele. Among five *Glu-A3* alleles, *Glu-A3e* (i-type haplotype *A3-502-3/A3-646*) was inferior with negative effects on all quality properties. Among eight *Glu-B3* alleles, *Glu-B3b* (m-type gene *B3-530-2* and s-type haplotype *B3-578/B3-607/B3-621-1*), *Glu-B3g* (m-type gene *B3-530-3* and s-type haplotype *B3-578/B3-544/B3-621-2*) and *Glu-B3i* (m-type haplotype *B3-510/B3-570* and s-type haplotype *B3-688-4/B3-691*) were correlated with superior bread-making quality, whereas *Glu-B3c* (m-type gene *B3-530-1* and s-type haplotype *B3-688-1/N*) produced inferior quality properties. Among five *Glu-D3* alleles, there were no significant differences in all quality parameters measured in the present study. Moreover, all alleles with superior dough properties and pan bread quality also possessed high contents of UPP and %UPP. Thus, it is possible that LMW-GS alleles determine dough viscoelasticity by modifying the size distribution of glutenin polymers and aggregative properties of glutenins[50]. These results significantly enhance our understanding of the composition of LMW-GS, confirm the strong effects of LMW-GS on not only dough extensity but dough strength, and provide useful information for quality improvement in bread wheat.

Acknowledgments

The authors are grateful to Prof. Robert McIntosh, University of Sydney, for reviewing this manuscript. We thank Drs Marie Appelbee and Ken Shepherd, SARDI Grain Quality Research Laboratory, Adelaide, South Australia, for seed of the spring wheat variety Aroona and near-isogenic derivatives. This study was supported by the National Basic Research Program (2009CB118300), the National Science Foundation of China (30830072), International Collaboration Project from Ministry of Agriculture (2011-G3), and China Agriculture Research System (CARS-3-1-3).

References

[1] Osborne TB: *The vegetable proteins*. London: Longmans Green and Co.; 1924: 154.

[2] Wasik RJ, Bushuk W: Sodium dodecyl sulfate-polyacrylamide gel-electrophoresis of reduced glutenin of durum wheats of different spaghetti-making quality. *Cereal Chem.* 1975, 52 (3): 328-334.

[3] Payne PI: Genetics of wheat storage proteins and the effect of allelic variation on bread-making quality. *Ann Rev Plant Physiol.* 1987, 38: 141-153.

[4] D'Ovidio R, Masci S: The low-molecular-weight glutenin subunits of wheat gluten. *J Cereal Sci.* 2004, 39 (3): 321-339.

[5] Shewry PR, Tatham AS, Barro F, Barcelo P, Lazzeri P: Biotechnology of breadmaking: unraveling and manipulating the multi-protein gluten complex. *Nat Biotechnol.* 1995, 13 (11): 1185-1190.

[6] Gao S, Gu YQ, Wu J, Coleman-Derr D, Huo N, Crossman C, Jia J, Zuo Q, Ren Z, Anderson OD *et al*: Rapid evolution and complex structural organization in genomic regions harboring multiple prolamin genes in the polyploid wheat genome. *Plant Mol Biol.* 2007, 65 (1-2): 189-203.

[7] Pogna NE, Autran JC, Mellini F, Lafiandra D, Feillet P: Chromosome *1B* encoded gliadins and glutenin subunits in durum wheat: genetics and relationship to gluten strength. *J Cereal Sci.* 1990, 11 (1): 15-34.

[8] Ruiz M, Carrillo JM: Linkage relationships between prolamin genes on chromosome *1A* and chromosome *1B* of durum wheat. *Theor Appl Genet.* 1993, 87 (3): 353-360.

[9] Cassidy BG, Dvorak J, Anderson OD: The wheat low-molecular-weight glutenin genes: characterization of six new genes and progress in understanding gene family structure. *Theor Appl Genet.* 1998, 96 (6-7): 743-750.

[10] Huang XQ, Cloutier S: Molecular characterization and genomic organization of low molecular weight glutenin subunit genes at the *Glu-3* loci in hexaploid wheat (*Triticum aestivum* L.). *Theor Appl Genet.* 2008, 116 (7): 953-966.

[11] Singh NK, Shepherd KW, Cornish GB: A simplified SDS-PAGE procedure for separating LMW subunits of

glutenin. *J Cereal Sci.* 1991, 14 (3): 203-208.

[12] Liu L, Ikeda TM, Branlard G, Pena RJ, Rogers WJ, Lerner SE, Kolman MA, Xia X, Wang L, Ma W *et al*: Comparison of low molecular weight glutenin subunits identified by SDS-PAGE, 2-DE, MALDI-TOF-MS and PCR in common wheat. *BMC Plant Biol.* 2010, 10: 124.

[13] Wang LH, Li GY, Pena RJ, Xia XC, He ZH: Development of STS markers and establishment of multiplex PCR for *Glu-A3* alleles in common wheat (*Triticum aestivum* L.). *J Cereal Sci.* 2010, 51 (3): 305-312.

[14] Wang LH, Zhao XL, He ZH, Ma W, Appels R, Peña RJ, Xia XC: Characterization of low-molecular-weight glutenin subunit *Glu-B3* genes and development of STS markers in common wheat (*Triticum aestivum* L.). *Theor Appl Genet.* 2009, 118 (3): 525-539.

[15] Zhang W, Gianibelli MC, Rampling LR, Gale KR: Characterisation and marker development for low molecular weight glutenin genes from *Glu-A3* alleles of bread wheat (*Triticum aestivum*. L). *Theor Appl Genet.* 2004, 108 (7): 1409-1419.

[16] Zhao XL, Xia XC, He ZH, Gale KR, Lei ZS, Appels R, Ma W: Characterization of three low-molecular-weight *Glu-D3* subunit genes in common wheat. *Theor Appl Genet.* 2006, 113 (7): 1247-1259.

[17] Zhao XL, Xia XC, He ZH, Lei ZS, Appels R, Yang Y, Sun QX, Ma W: Novel DNA variations to characterize low molecular weight glutenin *Glu-D3* genes and develop STS markers in Common Wheat. *Theor Appl Genet.* 2007, 114 (3): 451-460.

[18] Dong LL, Zhang XF, Liu DC, Fan HJ, Sun JZ, Zhang ZJ, Qin HJ, Li B, Hao ST, Li ZS *et al*: New insights into the organization, recombination, expression and functional mechanism of low molecular weight glutenin subunit genes in bread wheat. *PLoS ONE.* 2010, 5 (10): e13548. doi: 10.1371/journal.pone.0013548.

[19] Ikeda TM, Araki E, Fujita Y, Yano H: Characterization of low-molecular-weight glutenin subunit genes and their protein products in common wheats. *Theor Appl Genet.* 2006, 112 (2): 327-334.

[20] Ikeda TM, Nagamine T, Fukuoka H, Yano H: Identification of new low-molecular-weight glutenin subunit genes in wheat. *Theor Appl Genet.* 2002, 104 (4): 680-687.

[21] Zhang XF, Liu DC, Jiang W, Guo XL, Yang WL, Sun JZ, Ling HQ, Zhang AM: PCR-based isolation and identification of full-length low-molecular-weight glutenin subunit genes in bread wheat (*Triticum aestivum* L.). *Theor Appl Genet.* 2011, 123 (8): 1293-1305.

[22] Zhang XF, Liu DC, Yang WL, Liu KF, Sun JZ, Guo XL, Li YW, Wang DW, Ling HQ, Zhang AM: Development of a new marker system for identifying the complex members of the low-molecular-weight glutenin subunit gene family in bread wheat (*Triticum aestivum* L.). *Theor Appl Genet.* 2011, 122 (8): 1503-1516.

[23] Payne PI, Nightingale MA, Krattiger AF, Holt LM: The relationship between HMW glutenin subunit composition and the bread-making quality of British grown wheat varieties. *J Sci Food Agr.* 1987, 40 (1): 51-65.

[24] Branlard G, Dardevet M, Saccomano R, Lagoutte F, Gourdon J: Genetic diversity of wheat storage proteins and bread wheat quality. *Euphytica*. 2001, 119 (1-2): 59-67.

[25] Brett GM, Mills ENC, Tatham AS, Fido RJ, Shewry PR, Morgan MRA: Immunochemical identification of LMW subunits of glutenin associated with bread-making quality of wheat flours. *Theor Appl Genet.* 1993, 86 (4): 442-448.

[26] Flaete NES, Uhlen AK: Association between allelic variation at the combined *Gli-1*, *Glu-3* loci and protein quality in common wheat (*Triticum aestivum* L.). *J Cereal Sci.* 2003, 37 (2): 129-137.

[27] Gupta RB, Bekes F, Wrigley CW: Prediction of physical dough properties from glutenin subunit composition in bread wheats - correlation studies. *Cereal Chem.* 1991, 68 (4): 328-333.

[28] Gupta RB, Paul JG, Cornish GB, Palmer GA, Bekes F, Rathjen AJ: Allelic variation at glutenin subunit and gliadin loci, *Glu-1*, *Glu-3* and *Gli-1*, of Common Wheat. 1. Its additive and interaction effects on dough properties. *J Cereal Sci.* 1994, 19 (1): 9-17.

[29] Luo C, Griffin WB, Branlard G, McNeil DL: Comparison of low- and high molecular-weight wheat glutenin allele effects on flour quality. *Theor Appl Genet.* 2001, 102 (6-7): 1088-1098.

[30] Nietotaladriz MT, Perretant MR, Rousset M: Effect of gliadins and HMW and LMW subunits of glutenin on dough properties in the F-6 recombinant inbred lines from a bread wheat cross. *Theor Appl Genet.* 1994, 88 (1): 81-88.

[31] Nagamine T, Kai Y, Takayama T, Yanagisawa T,

Taya S: Allelic variation at the *Glu-1* and *Glu-3* loci in southern Japanese wheats, and its effects on gluten properties. *J Cereal Sci.* 2000, 32 (2): 129-135.

[32] Bekes F, Kemeny S, Morell M: An integrated approach to predicting end-product quality of wheat. *Eur J Agron.* 2006, 25 (2): 155-162.

[33] Eagles HA, Hollamby GJ, Gororo NN, Eastwood RF: Estimation and utilisation of glutenin gene effects from the analysis of unbalanced data from wheat breeding programs. *Aust J Agr Res.* 2002, 53 (4): 367-377.

[34] Cane K, Sharp PJ, Eagles HA, Eastwood RF, Hollamby GJ, Kuchel H, Lu M, Martin PJ: The effects on grain quality traits of a grain serpin protein and the VPM1 segment in southern Australian wheat breeding. *Aust J Agr Res.* 2008, 59 (10): 883-890.

[35] Zheng SS, Byrne PF, Bai GH, Shan XY, Reid SD, Haley SD, Seabourn BW: Association analysis reveals effects of wheat glutenin alleles and rye translocations on dough-mixing properties. *J Cereal Sci.* 2009, 50 (2): 283-290.

[36] Dupont FM, Vensel WH, Tanaka CK, Hurkman WJ, Altenbach SB: Deciphering the complexities of the wheat flour proteome using quantitative two-dimensional electrophoresis, three proteases and tandem mass spectrometry. *Proteome Sci.* 2011, 9: 10 doi: 10.1186/1477-5956-9-10.

[37] Gupta RB, Singh NK, Shepherd KW: The cumulative effect of allelic variation in LMW and HMW glutenin subunits on dough properties in the progeny of two bread wheats. *Theor Appl Genet.* 1989, 77 (1): 57-64.

[38] Park CS, Kang C-S, Jeung J-U, Woo S-H: Influence of allelic variations in glutenin on the quality of pan bread and white salted noodles made from Korean wheat cultivars. *Euphytica.* 2011, 180 (2): 235-250.

[39] Gupta RB, Macritchie F: Allelic variation at glutenin subunit and gliadin loci, *Glu-1*, *Glu-3* and *Gli-1* of Common Wheats. 2. Biochemical basis of the allelic effects on dough properties. *J Cereal Sci.* 1994, 19 (1): 19-29.

[40] Ito M, Fushie S, Maruyama-Funatsuki W, Ikeda TM, Nishio Z, Nagasawa K, Tabiki T, Yamauchi H: Effect of allelic variation in three glutenin loci on dough properties and bread-making qualities of winter wheat. *Breeding Sci.* 2011, 61 (3): 281-287.

[41] Liu L, He ZH, Yan J, Zhang Y, Xia XC, Peña RJ: Allelic variation at the *Glu-1* and *Glu-3* loci, presence of the 1B. 1R translocation, and their effects on mixographic properties in Chinese bread wheats. *Euphytica.* 2005, 142 (3): 197-204.

[42] Meng XG, Xie F, Shang XW, An LZ: Association between allelic variations at the *Glu-3* loci and wheat quality traits with Lanzhou Alkaline Stretched Noodles quality in northwest China spring wheats. *Cereal Res Commun.* 2007, 35 (1): 109-118.

[43] Bekes F, Gras PW, Anderssen RS, Appels R: Quality traits of wheat determined by small-scale dough testing methods. *Aust J Agr Res.* 2001, 52 (11-12): 1325-1338.

[44] Ma W, Appels R, Bekes F, Larroque O, Morell MK, Gale KR: Genetic characterisation of dough rheological properties in a wheat doubled haploid population: additive genetic effects and epistatic interactions. *Theor Appl Genet.* 2005, 111 (3): 410-422.

[45] Martinez-Cruz E, Espitia-Rangel E, Villasenor-Mir HE, Molina-Galan JD, Benitez-Riquelme I, Santacruz-Varela A, Pena-Bautista RJ: Dough rheology of wheat recombinant lines in relation to allelic variants of *Glu-1* and *Glu-3* loci. *Cereal Res Commun.* 2011, 39 (3): 386-393.

[46] Li Y, Zhou R, Branlard G, Jia J: Development of introgression lines with 18 alleles of glutenin subunits and evaluation of the effects of various alleles on quality related traits in wheat (*Triticum aestivum* L.). *J Cereal Sci.* 2010, 51 (1): 127-133.

[47] Tabiki T, Ikeguchi S, Ikeda TM: Effects of high-molecular-weight and low-molecular-weight glutenin subunit alleles on common wheat flour quality. *Breeding Sci.* 2006, 56 (2): 131-136.

[48] Liu L, He ZH, Ma WJ, Liu JJ, Xia XC, Pena RJ: Allelic variation at the *Glu-D3* locus in Chinese bread wheat and effects on dough properties, pan bread and noodle qualities. *Cereal Res Commun.* 2009, 37 (1): 57-64.

[49] Gobaa S, Bancel E, Branlard G, Kleijer G, Stamp P: Proteomic analysis of wheat recombinant inbred lines: Variations in prolamin and dough rheology. *J Cereal Sci.* 2008, 47 (3): 610-619.

[50] Popineau Y, Cornec M, Lefebvre J, Marchylo B: Influence of high M (r) glutenin subunits on glutenin polymers and rheological properties of glutens and gluten subfractions of near-isogenic lines of wheat Sicco. *J Cereal Sci.* 1994, 19 (3): 231-241.

[51] Saghai-Maroof MA, Soliman KM, Jorgensen RA, Allard RW: Ribosomal DNA spacer-length polymor-

phisms in barley: mendelian inheritance, chromosomal location, and population dynamics. *Proc Nat Acad Sci USA*. 1984, 81 (24): 8014-8018.

[52] Kumar S, Nei M, Dudley J, Tamura K: MEGA: a biologist-centric software for evolutionary analysis of DNA and protein sequences. *Brief bioinform* 2008, 9 (4): 299-306.

[53] Sambrook J, Russell DW: *Molecular cloning: a laboratory manual*. New York: Cold Spring Harbor Laboratory Press; 2001.

[54] Wang W, Vignani R, Scali M, Cresti M: A universal and rapid protocol for protein extraction from recalcitrant plant tissues for proteomic analysis. *Electrophoresis*. 2006, 27 (13): 2782-2786.

[55] He ZH, Liu L, Xia XC, Liu JJ, Pena RJ: Composition of HMW and LMW glutenin subunits and their effects on dough properties, pan bread, and noodle quality of Chinese bread wheats. *Cereal Chem*. 2005, 82 (4): 345-350.

[56] Zhang PP, He ZH, Zhang Y, Xia XC, Chen DS, Zhang Y: Association between % SDS-unextractable polymeric protein (%UPP) and end-use quality in Chinese bread wheat cultivars. *Cereal Chem*. 2008, 85 (5): 696-700.

Allelic variants of phytoene synthase 1 (*Psy1*) genes in Chinese and CIMMYT wheat cultivars and development of functional markers for flour colour

X. Y. He[1], Z. H. He[1,2], W.J. Ma[3], R. Appels[3], and X. C. Xia[1]

[1] *Institute of Crop Science, National Wheat Improvement Centre/The National Key Facility for Crop Gene Resources and Genetic Improvement, Chinese Academy of Agricultural Sciences (CAAS), 12 Zhongguancun South Street, Beijing 100081, China*; [2] *International Maize and Wheat Improvement Centre (CIMMYT) China Office, c/o CAAS, 12 Zhongguancun South Street, Beijing 100081, China*; [3] *Western Australia Department of Agriculture and Food, State Agriculture Biotechnology Center, Murdoch University, Murdoch, WA 1650, Australia*

Abstract: Phytoene synthase genes influence yellow pigment (YP) content in wheat grain, and are associated with the quality of end-use products. In the present study, a suite of 217 Chinese winter wheat cultivars and 342 CIMMYT spring wheat cultivars were used to search for phytoene synthase 1 gene variations and to detect and compare their genetic effects in different genetic backgrounds. An initial focus on the Chinese winter wheat cultivars revealed four allelic variants of this gene on chromosome 7B (*Psy-B1*), designated as *Psy-B1a*, *Psy-B1b*, *Psy-B1c* and *Psy-B1d*. The frequencies of these four alleles were 39.6%, 43.8%, 15.7% and 0.9%, respectively. A co-dominant marker *YP7B-1* based on a 5-bp InDel of poly C in the fifth intron of *Psy-B1* amplified a 151-bp PCR fragment in accessions with the medium YP content allele *Psy-B1a*, and a 156-bp fragment in lower YP content accessions with *Psy-B1b*. Two dominant markers *YP7B-2* (428bp) and *YP7B-3* (884bp) were designed for accessions with *Psy-B1c* and *Psy-B1d*, respectively. Allele *Psy-B1c* was associated with high YP content, but the phenotypic effect of *Psy-B1d* was not determined due to the limited number of accessions. In CIMMYT spring wheat cultivars, *Psy-B1a*, *Psy-B1b*, *Psy-B1d* and a further allelic variant, *Psy-B1e*, were detected with frequencies of 50.6%, 29.2%, 19.6% and 0.6%, respectively. *Psy-B1c* was not found in the CIMMYT germplasm. However, no significant differences were detected for mean YP content among CIMMYT wheat lines with different *Psy-B1* genotypes. A new allelic variant of *Psy-A1*, designated *Psy-A1c*, was identified in three CIMMYT wheat lines, and this allele was associated with higher YP content.

Key words: Common wheat, Yellow pigment content, Phytoene synthase 1 gene, 1B・1R translocation, Allelic variation

Introduction

Grain yellow pigment (YP) content is an important quality parameter in both common and durum wheats with different end-use products made from these wheats having different preferences for specific levels of yellow pigment. Pasta, a staple food made from durum and consumed

Published in Molecular Breeding, 2009, 23 (4): 553-563

mainly in Europe, America and the Middle East, requires a bright yellow colour. This colour is also desired for yellow alkaline noodles that are made from common wheat and consumed in Japan and southeastern Asia, and while the pH dependent colour shift of flavonoids is the primary factor, YP also contributes to the yellow colour (Mares and Campbell, 2001; Fu, 2008). Wheat lines for these products need higher YP content (Kruger et al., 1992). In contrast, a bright white to creamy colour, requiring a lower YP content in common wheat cultivars, is preferred for Chinese style foods such as steamed bread and Chinese noodles (He et al., 2004; Fu, 2008).

Many reports indicate that loci on the long arms of chromosomes 7A and 7B are associated with grain YP content. Among 150 SSD lines derived from Schomburgk/Yarralinka, Parker et al. (1998) detected major QTLs on chromosomes 7A and 3A, accounting for 60% and 13% of the phenotypic variance. Mares and Campbell (2001) located two major QTLs on chromosomes 7A and 3B in a DH population derived from Sunco/Tasman, explaining 27% and 20% of the phenotypic variance, respectively. Ma et al. (1999) mapped QTLs for flour colour (L^* and b^*) on homoeologous group 1 chromosomes and chromosome 7B. Kuchel et al. (2006) found a major QTL associated with flour yellowness b^* on chromosome 7B, explaining 48% and 61% of phenotypic variance in different cropping seasons. Similarly, QTLs for YP content were also mapped on chromosomes 7A and 7B in durum wheat (Elouafi et al., 2001; Pozniak et al., 2007; Patil et al., 2008; Zhang and Dubcovsky, 2008), demonstrating the importance of these two loci in conditioning grain YP content. In addition, the important role of the 1B·1R chromosome in grain YP content has also been reported (Yang et al., 2004; Zarco-Hernandez et al., 2005; Zhang et al., 2005, Zhang et al., 2008).

Carotenoids are the main components of flour yellow pigment (Miskelly, 1984; Adom et al., 2003), synthesized through a complex pathway, involving more than 10 enzymatic steps (Hirschberg, 2001). Phytoene synthase (PSY), catalyzing the condensation of two geranylgeranyl pyrophosphate molecules into phytoene, is generally accepted as the rate-limiting enzyme in carotenoid biosynthesis (Lindgren et al., 2003). In the grass family, duplicated PSY genes were identified (Gallagher et al., 2004) and designated as *Psy1* and *Psy2*, respectively. In maize, Palaisa et al. (2003) and Gallagher et al. (2004) demonstrated that *Psy1*, but not *Psy2*, exhibited a strong association with YP content of endosperm. Pozniak et al. (2007) localized the durum wheat *Psy1* and *Psy2* genes to homoeologous group 7 and 5 chromosomes, respectively, and demonstrated that *Psy1*, rather than *Psy2*, was associated with grain YP content. A similar conclusion was made by Zhang and Dubcovsky (2008). These reports led to a conclusion that the genes responsible for the QTLs detected on chromosomes 7A and 7B were orthologues of *Psy1*.

In our previous study (He et al., 2008), the *Psy1* gene on chromosome 7A was cloned and designated *Psy-A1*, and two allelic variants, *Psy-A1a* and *Psy-A1b*, were detected in Chinese winter wheat cultivars. A co-dominant functional marker, *YP7A*, was developed and validated. Subsequently, Zhang et al. (2008) demonstrated a significant influence of *Psy-A1* on flour YP content in a RIL population from PH82-2/Neixiang 188, accounting for 33.9% of the phenotypic variance. However, no functional markers are available for *Psy-B1*, the *Psy1* gene on chromosome 7B, in common wheat. The objectives of this study were to identify allelic variants at the *Psy-B1* locus, to develop functional markers for the different alleles, and to determine the associations of the allelic variants with grain YP content, in Chinese and CIMMYT common wheat cultivars and advanced lines.

Materials and methods

Plant materials

A total of 217 Chinese winter wheat cultivars and 342 CIMMYT spring wheat lines were used to identify allelic variants of *Psy-B1* and to investigate their association with YP content. During the 2001-2002 and 2002-2003 cropping seasons, the 217 Chinese wheat cultivars were sown in a randomized complete block with three replicates at the Anyang Experimental Station, Chinese Academy of Agricultural Sciences, located in Henan Province. Each plot consisted of two 2-m rows spaced 25 cm apart, with 100 plants in each row. The 342 CIMMYT spring wheat lines were from 37th International Bread Wheat Screening Nursery (37th IBWSN) and planted at Linhe, Inner Mongolia Autonomous Region, in 2007. Each plot comprised two 1.5-m rows, spaced 20 cm apart. The field trials were kept free of weeds and diseases, with two applications of broad-range herbicides and fungicides. A set of Chinese Spring nulli-tetrasomic lines (except nullisomic 2A and nullisomic 4B) and ditelosomic line 7BL, were used to confirm the location of the newly developed marker *YP7B-1*. A nullisomic 2A-tetrasomic 2B seed for DNA extraction was selected from the self-pollinated progeny of a monosomic 2A-tetrasomic 2B (M2A-T2B) individual.

Grain yellow pigment assay

The procedure for evaluating YP content in wheat grains followed the recommended AACC method 14-50. Flour samples of 3.0g were extracted with water-saturated n-butyl alcohol for 1h at room temperature on a gentle rocker table, and then stood for 10min. After centrifuging at 4, 000rpm for 10min, the supernatant was used to measure the absorbance at 436.5 nm. YP content was expressed as mg/kg using a correction coefficient 30.1 (AACC 1995).

Cloning and sequence analyses of the *Psy-B1* gene

The common wheat *Psy-B1* gene was obtained through a PCR-based cloning approach. In our previous study (He et al., 2008), a contig① designated as contig 3 was assembled and localized on chromosome 7B, comprising five ESTs CA598664, CD862515, CD862556, CD875290 and CD898848. This contig, representing the downstream sequence of the *Psy-B1* gene, was used as a seed sequence to clone the full coding sequence of *Psy-B1* in common wheat cv. Chinese Spring. Firstly, a primer set *P7B1* (Table 1) was designed based on the contig and its PCR product was sequenced. Then, another primer set *P7B2* (Table 1) was designed to amplify the upstream sequence of *Psy-B1*, with its forward primer generated according to the upstream sequence of *Psy-A1a* (He et al., 2008) and the reverse primer based on the PCR product of *P7B1*. Subsequently, a third primer set *P7B3* (Table 1) was generated to clone the 5' end sequence of *Psy-B1*, following the method for designing *P7B2*. The intron positions of *Psy-B1* were determined by alignment of the gene with the contig and the cDNA sequence of *Psy-A1*, using the software DNAMAN (http://www.lynnon.com). Primers were designed using the software Premier Primer 5 (http://www.premierbiosoft.com) and synthesized by Beijing Augct Biological Technology Co., Ltd (http://www.augct.com). Transposable elements in *Psy-B1* were detected by BLAST searches in the Triticeae Repeat Sequence Database (TREP, http://wheat.pw.usda.gov/ITMI/Repeats/index.shtml).

Genomic DNA was isolated from seeds following a method modified from Lagudah et al. (1991). PCR reactions were performed in an MJ Research PTC-200 thermal cycler in a total volume of 20μl, including 20 mM of Tris-HCl (pH 8.4), 20mM of KCl, 100μM of each dNTP, 1.5mM of $MgCl_2$, 5pmol of each primer, 1unit of *Taq* DNA polymerase (TIANGEN Biotech Co., Ltd., Beijing, http://www.tiangen.com) and 50ng of genomic DNA. Cycling conditions were 95℃ for 5min, followed by 40 cycles of 95℃ for 30s, 56℃ to 63℃ for 30s (according to the annealing temperatures of different primer sets), and 72℃ for 1.5min, with a final extension of 72℃ for 5min. The PCR products were cloned into the pMD18-T vector (TaKaRa Biotechnology Co., Ltd., Dalian, China) following the manufacturer's instructions and sequenced by Shanghai Sangon Biological Engineering & Technology and Service Co., Ltd. (http://www.sangon.com). To eliminate errors in sequencing, the PCR reaction and DNA sequencing were repeated 2-4 times for each

① An assembled gene sequence generated by screening a nucleotide sequence database or a gene library, consisted of consecutive and overlapping ESTs or genetic DNA sequences.

primer set.

Identification of allelic variants of *Psy-B1* and development of functional markers

Ten Chinese wheat cultivars with various YP contents and each possessing the *Psy-A1b* allele, were used for the identification of sequence polymorphisms at *Psy-B1* locus. All of the cultivars were amplified with the primer set *P7B1*, and their PCR products were sequenced and aligned for the detection of *Psy-B1* alleles. The full-length coding sequences of the newly identified allelic variants were cloned as described above. Functional markers were developed based on the sequence polymorphisms among the allelic variants of *Psy-B1*, and subsequently validated in 217 Chinese winter wheat cultivars and 342 CIMMYT spring wheat lines. The PCR products of the newly developed marker, *YP7B-1*, were separated on 4% polyacrylamide gels and subjected to silver staining (Bassam et al., 1991), whereas those of other markers were separated on 1.5% agarose gels, stained with ethidium bromide and visualized using UV light.

Detection of the 1B・1R translocation

The procedure for detecting the 1B・1R translocation followed Francis et al. (1995). The 1RS specific SCAR marker AF1/AF4 was used to detect wheat lines with the 1B・1R translocation, with forward primer (AF1) 5'-GGAGACATCATGAAACATTTG-3' and reverse primer (AF4) 5'-CTGTTGTTGGGCAGAAAG-3'. PCR reaction included 20mM of Tris-HCl (pH 8.4), 20mM of KCl, 250μM of each dNTP, 1.5mM of $MgCl_2$, 4pmol of each primer, 1unit of *Taq* DNA polymerase and 80ng of genomic DNA, with a total volume of 20μl. Cycling conditions were 94℃ for 5min, followed by 35 cycles of 94℃ for 1min, 60℃ for 2min, and 72℃ for 30s, with a final extension of 72℃ for 5min. PCR products were detected on 1.5% agarose gels, and a band of 1.5kb can be observed in the wheat lines with 1B・1R translocation, whereas no PCR products were amplified in those without 1B・1R translocation.

Statistical analyses

For the 217 Chinese wheat lines, the YP content of each accession was evaluated ineach of two cropping seasons and averaged for the test of the association between YP content and newly developed markers. The YP content of each of the 342 CIMMYT wheat lines was measured with two replications and averaged for statistical analyses. Analysis of variance was performed by PROC MIXED in the Statistical Analysis System (SAS Institute, 2000) with genotype clusters as categorical variables to calculate the mean YP content for each cluster and to test differences among clusters. The genotype cluster, indicated by different *Psy1* alleles or the 1B・1R translocation, was treated as fixed effects, with genotypes nested within clusters assumed to be random.

Results

Sequence characteristics of *Psy-B1*

The cloned genomic DNA sequence of the *Psy-B1* allele from Chinese Spring, designated *Psy-B1a* (GenBank accession EU650392), contained 3313 base pairs, including the 5' and 3' flanking regions. The exon/intron structure of *Psy-B1a* was very similar to that of *Psy-A1a*, containing six exons and five introns, and the sizes of exons were conserved between the two genes except for the first exon, which was seven codons shorter in *Psy-B1a* than in *Psy-A1a* (Fig. 1). Furthermore, the exons between the two genes also showed high sequence identities, ranging from 93.6% to 98.8%. The introns differed in size between the two genes, with sequence identities of 84.9% to 91.4%.

The deduced cDNA sequence of the cloned *Psy-B1a* was 1644bp, containing an open reading frame (ORF) of 1263bp, a 222-bp 5' UTR and a 159-bp 3' UTR. The deduced protein sequence of *Psy-B1a* consisted of 421 amino-acid residues, with a calculated molecular mass of ~47.0kDa. The protein sequence of *Psy-B1a* was very similar to that of *Psy-A1a*, with a sequence identity of 96.4%.

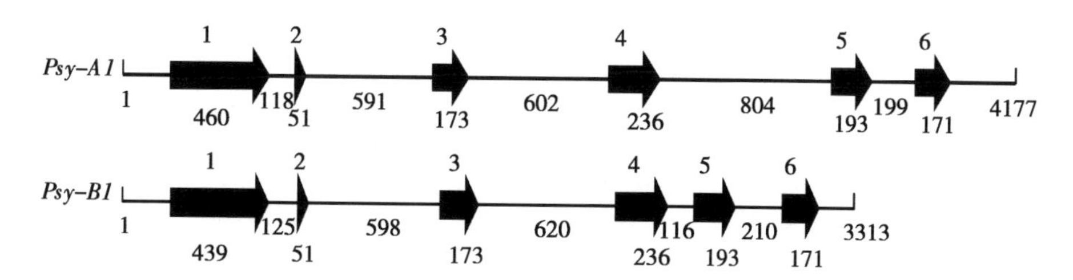

Fig. 1 Gene structures of *Psy-A1 and Psy-B1*

The numbered solid arrows denote exons, and the lines between exons represent introns. The numbers under exons and introns indicate their size (bp)

Allelic variants at the *Psy-B1* locus in Chinese wheat cultivars and development of functional markers

The 10 Chinese wheat cultivars with different YP contents used for the detection of sequence polymorphisms of *Psy-B1* were all genotype *Psy-A1b*, and the phenotypic values were assumed to be associated with variation at the *Psy-B1* locus. The primer set *P7B1* amplified PCR fragments of approximately 1000bp from the 10 cultivars, and four allelic variants of *Psy-B1* locus were detected after sequencing the PCR products. One of the four alleles was identical to that of Chinese Spring, *Psy-B1a*, and the other three were designated *Psy-B1b*, *Psy-B1c* and *Psy-B1d* (GenBank accessions EU650393, EU650394 and EU650395, respectively). Utilizing the primer sets *P7B1* through *P7B6* (Table 1), the full-length genomic DNA sequences of *Psy-B1b* and *Psy-B1d* were obtained, but the sequence of *Psy-B1c* was not completely determined because the 5' UTR and the first exon were missing.

Table 1 Primer sets for cloning *Psy-B1* alleles and the STS markers developed in this study

	Primer sequences (5'-3')	Tm (℃)	Targeted allele
P7B1	F: GGACCTCAAGAAGGCAAGAT R: CGGGACCGACAACGAGTATA	63	*Psy-B1a*, *Psy-B1b*, *Psy-B1c*, *Psy-B1d*
P7B2	F: GTCCCAACGCGTAGCACATC R: CTGGTTCGCCAACCCCGAGA	62	*Psy-B1a*, *Psy-B1b*
P7B3	F: GGCAGGCTAGTGGTCGGTA R: GGGGAACTTGGTGATGGTGTC	62	*Psy-B1a*, *Psy-B1b*, *Psy-B1e*
P7B4	F: CGAGATCTGCGAGGAGTACGCC R: CCAAGGTGAGGGTCTTCAAC	60	*Psy-B1c*
P7B5	F: TATGGTGCAGGAGGACAGAC R: CTCTTCGCGCACTGGAAAAG	62	*Psy-B1d*
P7B6	F: GGCAGGCTAGTGGTCGGTA R: TCCCAGTTTCCCACTATACG	56	*Psy-B1d*
P7B7	F: CAACGCGTCGCACATCACA R: GCAACATTGATCGCTGAGGA	60	*Psy-B1e*
YP7A-2	F: GCCAGCCCTTCAAGGACATG R: CAGATGTCGCCACACTGCCA	60	*Psy-A1a*, *Psy-A1b*, *Psy-A1c*
YP7B-1	F: GCCACAACTTGAATGTGAAAC R: ACTTCTTCCATTTGAACCCC	60	*Psy-B1a*, *Psy-B1b*
YP7B-2	F: GCCACCCACTGATTACCACTA R: CCAAGGTGAGGGTCTTCAAC	60	*Psy-B1c*

（续）

	Primer sequences (5'-3')	Tm (℃)	Targeted allele
YP7B-3	F: GAGTAAGCCACCCACTGATT R: TCGCTGAGGAATGTACTGAC	62	*Psy-B1d*
YP7B-4	F: AGGTACCAGCCAGCCCATA R: CTCGTCAAAGTTCGTGTACC	58	*Psy-B1e*

Among the four *Psy-B1* alleles identified in Chinese wheat cultivars, *Psy-B1a* and *Psy-B1b* shared the highest sequence identity, with the only difference being a 5-bp InDel in the Poly C region of the fifth intron (Fig. 2). Both alleles exhibited lower sequence identities with *Psy-B1c* and *Psy-B1d*, ranging from 76.8% to 90.0%. Differences in intron regions contributed mainly to the low sequence identities among the alleles. In the second and third introns of *Psy-B1c* and in the third intron of *Psy-B1d*, three insertion sequences of 93bp, 193bp and 193bp, respectively, were found. A BLAST search against the TREP database showed that three insertion sequences were members of the *Stowaway* family, with typical terminal inverted repeats and TA recognition target sites, characteristic of *Stowaway* elements. The four alleles had higher sequence identities among the exons. The sizes from the second through sixth exons were conserved among the alleles, whereas the first exon of *Psy-B1d* was 30bp longer than those of *Psy-B1a* and *Psy-B1b*. They also showed high sequence identities in deduced amino acid sequences, ranging from 97.9 to 100%. The deduced amino acid sequences of *Psy-B1a* and *Psy-B1b* were identical, whereas those of *Psy-B1c* and *Psy-B1d* differed only in two residues.

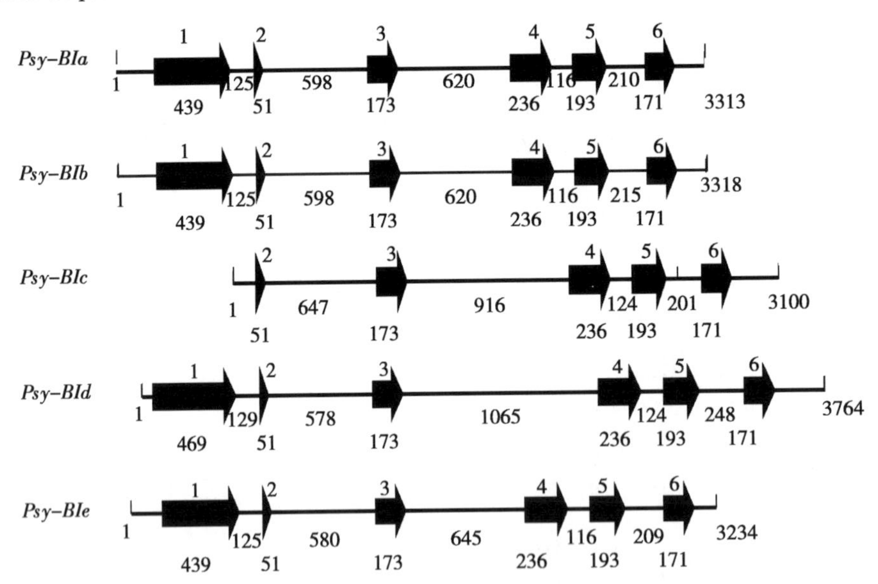

Fig. 2 Gene structures of allelic variants at the *Psy-B1* locus

The numbered solid arrows denote exons, and the lines between exons represent introns. The numbers under exons and introns indicate their size (bp)

Note: the 5' UTR and the first exon of *Psy-B1c* were not able to be cloned in this study.

A co-dominant marker, *YP7B-1*, was developed based on the 5-bp polymorphic sequence difference between *Psy-B1a* and *Psy-B1b* (Table 1). This marker amplified 151-bp and 156-bp fragments from the genotypes with *Psy-B1a* and *Psy-B1b*, respectively (Fig. 3). Based on sequences specific to *Psy-B1c* and *Psy-B1d*, two dominant markers, *YP7B-2* and *YP7B-3*, were designed for these alleles (Table 1), amplifying 428-bp and 884-bp fragments, respectively (Fig. 3).

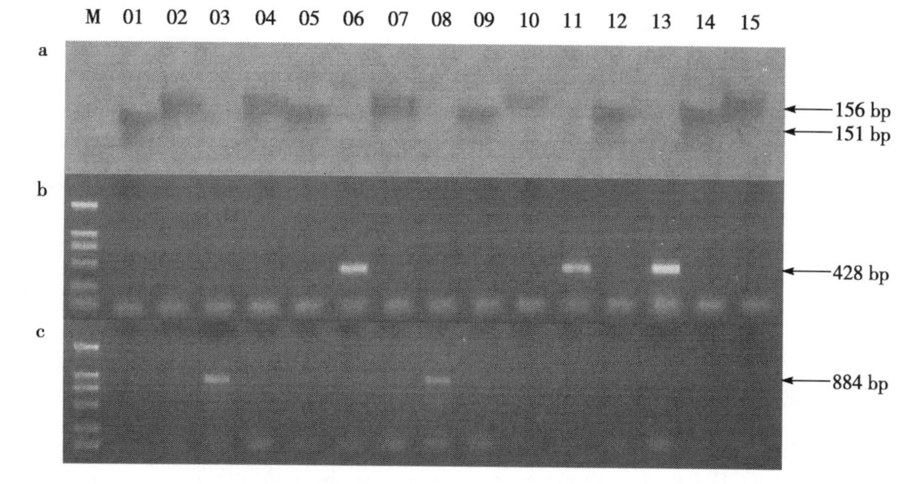

Fig. 3 PCR amplification with markers *YP7B-1* (a), *YP7B-2* (b) and *YP7B-3* (c) in 15 Chinese winter wheat lines

M DNA ladder DL2000; 01 Jingdong 8 (*Psy-B1a*, yellow pigment content 2.14 mg/kg); 02 Jimai 38 (*Psy-B1b*, 1.11); 03 Ning 98084 (*Psy-B1d*, 1.57); 04 Yunmai 42 (*Psy-B1b*, 1.22); 05 Wanmai 38 (*Psy-B1a*, 1.67); 06 Yannong 18 (*Psy-B1c*, 2.71); 07 Shan 150 (*Psy-B1b*, 0.86); 08 Ning 99415-8 (*Psy-B1d*, 1.39); 09 Shannong 1355 (*Psy-B1a*, 3.36); 10 Yumai 49 (*Psy-B1b*, 1.63); 11 Zhengzhou 992 (*Psy-B1c*, 2.84); 12 Henong 2552 (*Psy-B1a*, 2.20); 13 Ji 3475 (*Psy-B1c*, 2.02); 14 Jinmai 50 (*Psy-B1a*, 1.97); 15 Zhongyou 9701 (*Psy-B1b*, 1.19)

Chromosomal localization of *YP7B-1*

The location of *YP7B-1* on chromosome 7B was determined by amplifying genomic DNA from a set of Chinese Spring nulli-tetrasomic lines and ditelosomic line 7BL. No PCR product was detected in N7B-T7D (lane 21), whereas a 151-bp fragment was amplified in DT7BL (lane 23), confirming the location of *YP7B-1* on the long arm of chromosome 7B.

Association between the allelic variants of *Psy-B1* and YP content in Chinese wheat lines

The 217 Chinese wheat lines were genotyped with the markers *YP7B-1*, *YP7B-2* and *YP7B-3*, and the frequencies of *Psy-B1a*, *Psy-B1b* and *Psy-B1c* were 39.6%, 43.8% and 15.7%, respectively, with two lines, Ning 98084 and Ning 99415-8, having the *Psy-B1d* allele. Statistical analyses indicated that cultivars with *Psy-B1c* had the highest YP content, followed by those with *Psy-B1a*, whereas those with *Psy-B1b* had the lowest mean value for YP content (Table 2). The mean YP contents of the three groups were significantly different (Table 2). The phenotypic effect of *Psy-B1d* was not evaluated due to the limited number of accessions with this allele.

Table 2 Association between the allelic variants of *Psy-B1* and YP content in Chinese winter wheat cultivars

Genotype[1)]	Number of accessions	Mean YP content (mg/kg)[2)]	Standard deviation	Range
Psy-B1a	86	1.71 b	0.57	0.57-3.36
Psy-B1b	95	1.40 c	0.48	0.48-2.69
Psy-B1c	34	2.01 a	0.35	0.35-3.42

1) The allelic variant *Psy-B1d* was not included due to the limited number of accessions with this allele.

2) Different letters following the mean YP content indicate significant differences between genotypic groups ($P<0.05$).

Allelic variation at the *Psy-A1* and *Psy-B1* loci in CIMMYT wheat lines

Molecular detection with the marker *YP7A* showed that genotypes *Psy-A1a* and *Psy-A1b* were the main allelic variants of *Psy-A1* in the 342 CIMMYT spring wheat lines, accounting for 48.0% and 51.2% of total accessions, respectively. In addition, a new variant of

Psy-A1, designated as *Psy-A1c* (GenBank accession EU650391), was identified in three CIMMYT wheat lines, viz. 37th IBWSN 241, 37th IBWSN 245 and 37th IBWSN 246. Compared with *Psy-A1a*, *Psy-A1c* had a 688-bp deletion in the fourth intron, a single nucleotide insertion in the third intron and a trinucleotide insertion in the fifth intron. Moreover, 28 SNPs were found between the two alleles with 22 of them in introns and six in exons. However, only one of the six SNPs found in exons resulted in a transition from Serine (in *Psy-A1a*) to Threonine (in *Psy-A1c*) in the deduced polypeptides, whereas other five SNPs were synonymous mutations. All of the sites polymorphic between *Psy-A1a* and *Psy-A1c* were also between *Psy-A1b* and *Psy-A1c*, but an additional synonymous SNP and a 37-bp insertion were present in the first exon and in the 5' end of the second intron, respectively. A co-dominant STS marker, *YP7A-2* (Table 1), based on the polymorphic sequence differences among the three allelic variants of *Psy-A1*, amplified a 1, 001-bp fragment in genotypes with *Psy-A1c* and a 1686-bp fragment in those with *Psy-A1a* and *Psy-A1b*.

Allelic variants at *Psy-B1* in CIMMYT wheat lines were detected with markers *YP7B-1*, *YP7B-2* and *YP7B-3*, indicating frequencies of lines with *Psy-B1a*, *Psy-B1b* and *Psy-B1d* at 50.6%, 29.2% and 19.6%, respectively. No line with *Psy-B1c* was detected, but a new *Psy-B1* allele, *Psy-B1e* (GenBank accession EU263021), was identified in sister lines 37th IBWSN 166 and 37th IBWSN 167. This allele showed an unusual sequence characteristic; its upstream sequence before the third exon exhibited 99.5% identity with *Psy-B1a*, but the similarity dropped to 92.0% in the remaining sequence. Overall, *Psy-B1e* showed 94.8% sequence identity with *Psy-B1a*, higher than those of *Psy-B1a*/*Psy-B1c* and *Psy-B1a*/*Psy-B1d*; but its deduced polypeptide shared lower similarities with the other alleles. According to the sequence specific to *Psy-B1e*, a dominant marker *YP7B-4* was developed (Table 1), and it yielded a 717-bp fragment in *Psy-B1e* and no PCR product in the other four genotypes.

Association between the allelic variants of *Psy-A1* and *Psy-B1* and YP content in CIMMYT wheat lines

The allelic variants of *Psy-A1* gene were significantly associated with YP content in CIMMYT spring wheat lines, similar to the results found with Chinese wheat cultivars. Lines with *Psy-A1a* had a significantly higher mean YP content than those with *Psy-A1b* (Table 3); lines with *Psy-A1c* were not included in the statistical analyses because of limited numbers. However, phenotypic differences between different *Psy-B1* genotypes of the CIMMYT materials were not significant, even though comparisons were made among lines with the same *Psy-A1* alleles (data not shown).

Table 3 Association between the allelic variants of *Psy-A1* and YP content in CIMMYT spring wheat accessions

Genotype	Number of accessions	Mean YP content1)	Standard deviation	Range
Psy-A1a	164	2.56 a	0.81	1.32-6.11
Psy-A1b	175	2.12 b	0.76	0.94-7.50

1) Different letters following the mean YP content indicate significant differences between groups ($P<0.01$).

The combined phenotypic effects of *Psy1* and 1B·1R translocation

Combining the marker detections of *Psy1* alleles and the presence or absence of the 1B·1R translocation, the relation between genotypes and phenotypes became more significant. Among Chinese wheat lines, accessions with all the alleles favorable for high YP content, i. e. *Psy-A1a*, *Psy-B1c* and 1B·1R, showed the highest mean value, whereas those with alleles for low YP content, i. e. *Psy-A1b*, *Psy-B1b* and non-1B·1R, exhibited the lowest mean value (Table 4). Generally, the more of the alleles for high YP content at the three loci were present in a cultivar, the higher YP content was obtained, and *vice versa* (Table 4). A similar situation also applied to CIMMYT wheat lines, where the loci conditioning grain YP content were *Psy-A1* and 1B·1R (Table 5).

Table 4 Association between *Psy-A1*/*Psy-B1*/1B·1R genotypes and YP content in Chinese winter wheat accessions

Genotype[1)]	Number of accessions	Mean YP content (mg/kg)[2)]	Standard deviation	Range
Psy-A1a/*Psy-B1c*/1B·1R	18	2.42 a	0.65	1.04-3.42
Psy-A1a/*Psy-B1a*/1B·1R	35	2.08 b	0.51	1.22-3.36
Psy-A1a/*Psy-B1b*/1B·1R	26	1.80 c	0.46	0.93-2.69
Psy-A1a/*Psy-B1a*/non-1B·1R	23	1.60 c	0.46	1.08-2.91
Psy-A1a/*Psy-B1b*/non-1B·1R	27	1.30 d	0.44	0.62-2.27
Psy-A1b/*Psy-B1a*/non-1B·1R	19	1.16 d	0.34	0.57-1.84
Psy-A1b/*Psy-B1b*/non-1B·1R	30	1.15 d	0.37	0.48-1.78

1) Allelic combinations with low numbers of accessions were not included.

2) Different letters following the mean YP content indicate significant differences between genotypic groups ($P<0.05$).

Table 5 Association between *Psy-A1*/1B·1R genotypes and YP content in CIMMYT spring wheat accessions

Genotype	Number of accessions	Mean YP content (mg/kg)[1)]	Standard deviation	Range
Psy-A1a/1B·1R	56	3.10 a	0.61	2.00-4.44
Psy-A1b/1B·1R	101	2.46 b	0.70	1.39-7.50
Psy-A1a/non-1B·1R	108	2.28 b	0.77	1.32-6.11
Psy-A1b/non-1B·1R	74	1.65 c	0.57	0.95-5.34

1) Different letters following the mean YP content indicate significant differences between genotypic groups ($P<0.05$).

Discussion

Identification and characterization of *Psy-B1* alleles

In this study, the allele-specific primer sets were adapted for cloning allelic variants of *Psy-B1*, other than *Psy-A1 or Psy-D1*. It was found that PCR products needed to be carefully compared to the targeted and homoeologous alleles to ascertain their origin to identify chimeric products. For sequence assembly, overlaps longer than 200bp for adjacent PCR products were required.

Psy-B1 alleles shared highly conserved exon/intron structures, but exhibited divergent sequences, with significant numbers of SNPs and InDels. Although *Psy-B1a* and *Psy-B1b* shared almost identical sequences, differing only in a 5-bp InDel of poly C, other alleles exhibited many sequence differences from these two alleles. *Psy-B1c* and *Psy-B1d* shared a higher sequence identity than those they shared with other alleles, and as inferred from the common presence of a *Stowaway* element in the third introns they probably evolved from a common ancestor early in the divergence of the *Psy-B1* alleles. The unusual sequence characteristics of *Psy-B1e* were very similar to the durum Kofa *PSY-B1* allele (Zhang and Dubcovsky, 2008; EU096092); the two sequences differed only in a SNP and an 8-bp InDel in the fifth intron. According to Zhang and Dubcovsky (2008), the Kofa *PSY-B1* allele was derived from the *ph1c* mutant of the Italian durum cultivar Cappelli, and was present in about 30% of durum accessions. Pedigree analyses indicated that the two CIMMYT spring wheat lines with *Psy-B1e* had a common synthetic wheat ancestor. Synthetic wheats are produced from crosses between tetraploid wheat (including durum) and *Ae. tauschii*①.

Allelic variants of *Psy1* and grain YP content

In our previous study (He et al. 2008), two allelic variants *Psy-A1a* and *Psy-A1b* were cloned and their association with YP content was attributed to alternative splicing of the second intron. In the present study, two

① The pedigrees of the two lines were identical, i.e. CHEN/*Aegilops tauschii*//BCN/3/BAV92 of the five allelic variants at the *Psy-B1* locus, *Psy-B1a and Psy-B1b*, were associated with YP content in Chinese wheat lines. However, the two sequences differed only in a 5-bp InDel of poly C in the fifth intron and they actually shared the same ORF, encoding identical polypeptides. The 5-bp InDel may not be the causative factor for different YP contents because of its location in an intron and the nature of simple sequence repeat. Thus there may be polymorphic sequences in the promoter region, or in other regulatory sequences of *Psy-B1* that influence the expression of the gene. Further studies on these sequences may reveal the molecular mechanisms for the different YP levels between lines with these genotypes.

Although an association between allelic variation of *Psy-B1* and YP content was established in Chinese wheat lines, the relationship did not hold among CIMMYT lines, even though the different *Psy-B1* alleles were compared across the same allelic backgrounds of *Psy-A1* and 1B·1R. Thus it appeared that *Psy-B1* exerted little, if any, influence on grain YP content in CIMMYT wheat lines. This difference might be due to the different ecotypes of winter and spring wheats, since all of the Chinese cultivars were winter wheats, whereas all of the CIMMYT lines were spring wheats. Konopka et al. (2006) determined the contents of carotenoids, the major component of YP, in spring and winter wheat lines, and found that spring wheats had significantly higher mean levels of carotenoids than winter wheats. The difference may be caused by different molecular mechanisms, in which *Psy-B1* alleles were involved.

Psy-A1c, a new allelic variant of *Psy-A1*, was only found in three CIMMYT wheat lines, and thus its phenotypic effect could not be evaluated through statistical analysis. However, the deduced polypeptides of *Psy-A1a and Psy-A1c* are nearly identical, with only a single serine/threonine transition. Since the two amino acids have similar chemical properties, the polypeptides encoded by the two alleles probably exert similar enzymatic activities, and thus *Psy-A1c* may also be associated with higher YP content in common with *Psy-A1a*. In terms of genomic DNA sequence, *Psy-A1c* differed from *Psy-A1a* and *Psy A1b* mainly in a 688-bp InDel in the fourth intron. Zhang and Dubcovsky (2008) characterized this sequence as a transposable element, and found it occurring in durum and common wheat lines at frequencies of 12% and 94%, respectively. Based on the fact that durum wheats generally have YP contents higher than those of common wheats, they proposed that absence of the 688-bp sequence was associated with higher YP content, in agreement with the conclusions drawn above.

The influences of 1B·1R translocation on grain YP content

Previous studies implied that the 1B·1R translocation exerts a significant influence on grain YP content (Yang et al., 2004; Zarco-Hernandez et al., 2005; Zhang et al., 2005). A major QTL for YP content, recently detected on the 1B·1R chromosome in a RIL population derived from PH82-2/Neixiang 188, explained 31.9% of the phenotypic variation (Zhang et al., 2008). The increased effect came from the Neixiang 188 parent which has the 1B·1R translocation. In the present study, the 1B·1R translocation exerted a significant influence on grain YP content in both Chinese and CIMMYT wheat lines, and the inclusion of 1B·1R detection into the marker system for YP content greatly improved the predictability of phenotypic outcomes (Tables 4 and 5). Zhou et al. (2007) pointed out that the frequency of 1B·1R was 42.6% among cultivars grown in the Northern China Winter Wheat Region, the most important wheat grown area in China. In this study, the percentages of 1B·1R lines were 47.9% and 53.8% in Chinese and CIMMYT wheat lines, respectively, showing the extensive prevalence of 1B·1R in common wheat lines.

Implementation of the markers for YP content in wheat breeding programs

For the five allelic variants of *Psy-B1*, four STS markers were developed in this study. Two of them, *YP7B-1* and *YP7B-2*, were validated in Chinese winter wheat cultivars and could be used in wheat breeding programs. Since lower YP content is beneficial for most Chinese wheat products, *Psy-B1b* should be preferred over *Psy-B1a* and *Psy-B1c*. In this circumstance, breeding materials with the 156-bp fragment amplified by *YP7B-1* (*Psy-B1b*), instead of those with the 151-bp fragment amplified by *YP7B-1* (*Psy-B1a*) and those with the 428-bp fragment yielded by *YP7B-2* (*Psy-B1c*), should be selected. For the purpose of breeding wheat cultivars with low YP content, the genotype *Psy-A1b*/*Psy-B1b*/non-1B・1R would be preferred, whereas for those with high YP content, the genotype *Psy-A1a*/*Psy-B1c*/1B・1R could be emphasized.

Nevertheless, there were still relatively wide ranges of variation within genotype groups, classified by *Psy1* alleles and 1B・1R translocation (Tables 4 and 5). The reason for this situation was due to the multigenic control of YP content in wheat grain. In addition to the major QTLs at *Psy1* loci and 1B・1R translocation, QTLs with smaller phenotypic effects have also been detected on chromosomes 3A (Parker et al., 1998), 3B (Mares and Campbell, 2001), 4A and 5A (Hessler et al., 2002), 2D and 4D (Zhang et al., 2006), and 2A, 4B and 6B (Pozniak et al., 2007). Thus further works are needed to clone the genes responsible for these QTLs, and to develop more functional markers for grain YP content.

Acknowledgments

The authors are grateful to Prof. RA McIntosh and Prof. PD Chen for providing the Chinese Spring nullitetrasomic, ditelosomic, and monosomic 2A - tetrasomic 2B lines. This study was supported by the National Science Foundation of China (30771335), National Basic Research Program (2002CB11300), National 863 Program (2006AA10Z1A7 and 2006AA100102), and International Collaboration Project from the Ministry of Agriculture (2006-G2).

References

AACC (1995) Approved Methods of the American Association of Cereal Chemists. 9th ed. MN, St. Paul, USA.

Adom KK, Sorrells ME, Liu RH (2003) Phytochemical profiles and antioxidant activity of wheat varieties. J Agric Food Chem. 51: 7825-7834.

Bassam BJ, Caetano-Anolles G, Gresshoff PM (1991) Fast and sensitive silver staining of DNA in polyacrylamide gels. Anal Biochem. 196: 80-83.

Elouafi I, Nachit MM, Martin LM (2001) Identification of a microsatellite on chromosome 7B showing a strong linkage with yellow pigment in durum wheat (*Triticum turgidum* L. var. *durum*). Hereditas. 135: 255-261.

Francis HA, Leitch AR, Koebner RMD (1995) Conversion of a RAPD-generated PCR product, containing a novel dispersed repetitive element, into a fast and robust assay for the presence of rye chromatin in wheat. Theor Appl Genet. 90: 636-642.

Fu BX (2008) Asian noodles: History, classification, raw materials, and processing. Food Res Int. 41: 888-902.

Gallagher CE, Matthews PD, Li F, Wurtzel ET (2004) Gene duplication in the carotenoid biosynthetic pathway preceded evolution of the grasses. Plant Physiol. 135: 1776-1783.

He XY, Zhang YL, He ZH, Wu YP, Xiao YG, Ma CX, Xia XC (2008) Characterization of phytoene synthase 1 gene (*Psy1*) located on common wheat chromosome 7A and development of a functional marker. Theoretical and Applied Genetics, 116: 213-221.

He ZH, Yang J, Zhang Y, Quail K, Pena, RJ (2004) Pan bread and dry white Chinese noodle quality in Chinese winter wheats. Euphytica. 139: 257-267.

Hessler TG, Thomson MJ, Benscher D, Nachit MM, Sorrells ME (2002) Association of a lipoxygenase locus, *Lpx-B1*, with variation in lipoxygenase activity in durum wheat seeds. Crop Sci. 42: 1695-1700.

Hirschberg J (2001) Carotenoid biosynthesis in flowering plants. Curr Opin Plant Biol. 4: 210-218.

Konopka I, Czaplicki S, Rotkiewicz D (2006) Differences in content and composition of free lipids and carotenoids in flour of spring and winter wheat cultivated in Poland. Food Chem. 95: 290-300.

Kruger JE, Matsuo RR, Preston K (1992) A comparison of methods for the prediction of Cantonese noodle colour. Can J Plant Sci. 72: 1021-1029.

Kuchel H, Langridge P, Mosionek L, Williams K, Jefferies SP (2006) The genetic control of milling yield, dough rheology and baking quality of wheat. Theor Appl Genet. 112: 1487-1495.

Lagudah ES, Appels R, McNeil D (1991) The *Nor-D3* locus of

Triticum tauschii: natural variation and genetic linkage to markers in chromosome 5. Genome. 34: 387-395.

Lindgren LO, Stalberg KG, Hoglund AS (2003) Seed-specific overexpression of an endogenous *Arabidopsis* phytoene synthase gene results in delayed germination and increased levels of carotenoids, chlorophyll, and abscisic acid. Plant Physiol. 132: 779-785.

Ma W, Daggard G, Sutherland M, Brennan P (1999) Molecular markers for quality attributes in wheat. In: Williamson P, Banks P, Haak I, Thompson J, Campbell A (eds.) Proc 9th Assembly of the Wheat Breeding Society of Australia, Toowoomba, vol. 1, pp 115-117.

Mares DJ, Campbell AW (2001) Mapping components of flour and noodle colour in Australian wheat. Aust J Agric Res. 52: 1297-1309.

Miskelly DM (1984) Flour components affecting paste and noodle colour. J Sci Food Agric. 35: 463-471.

Palaisa KA, Morgante M, Williams M, Rafalski A (2003) Contrasting effects of selection on sequence diversity and linkage disequilibrium at two phytoene synthase loci. Plant Cell. 15: 1795-1806.

Parker GD, Chalmers KJ, Rathjen AJ, Langridge P (1998) Mapping loci associated with flour colour in wheat. Theor Appl Genet. 97: 238-245.

Patil RM, Oak MD, Tamhankar SA, Sourdille P, Rao VS (2008) Mapping and validation of a major QTL for yellow pigment content on 7AL in durum wheat (*Triticum turgidum* L. ssp. *durum*). Mol Breed. 21: 485-496.

Pozniak CJ, Knox RE, Clarke FR, Clarke JM (2007) Identification of QTL and association of a phytoene synthase gene with endosperm colour in durum wheat. Theor Appl Genet. 114: 525-537.

Yang J, Zhang Y, He ZH, Yan J, Wang DS, Liu JJ, Wang MF (2004) Association between wheat quality traits and performance of pan bread and dry white Chinese noodle. Acta Agron Sin. 30: 739-744.

Zarco-Hernandez JA, Santiveri F, Michelena A, Peña RJ (2005) Durum wheat (*Triticum turgidum* L.) carrying the 1BL·1RS chromosomal translocation: agronomic performance and quality characteristics under Mediterranean conditions. Eur J Agron. 22: 33-43.

Zhang LP, Yan J, Xia XC, He ZH, Sutherland MW (2006) QTL mapping for kernel yellow pigment content in common wheat. Acta Agron Sin. 32: 41-45.

Zhang W, Dubcovsky J (2008) Association between allelic variation at the *Phytoene synthase 1* gene and yellow pigment content in the wheat grain. Theor Appl Genet. 116: 635-645.

Zhang Y, Quail K, Mugford DC, He ZH (2005) Milling quality and white salt noodle color of Chinese winter wheat cultivars. Cereal Chem. 82: 633-638.

Zhang YL, Wu YP, Xiao YG, He ZH, Zhang Y, Yan J, Zhang Y, Ma CX, Xia XC (2008) QTL mapping for flour colour components, yellow pigment content and polyphenol oxidase activity in common wheat (*Triticum aestivum* L.). Euphytica, DOI 10.1007/s10681-008-9744-z.

Zhou Y, He ZH, Sui XX, Xia XC, Zhang XK, Zhang GS (2007) Genetic improvement of grain yield and associated traits in the Northern China winter wheat region from 1960 to 2000. Crop Sci. 47: 245-253.

Allelic variants at the *Psy-A1* and *Psy-B1* loci in durum wheat and their associations with grain yellowness

X. Y. He[1], J. W. Wang[1,2], K. Ammar[3], R. J. Peña[3], X. C. Xia[1], and Z. H. He[1,4]

[1] *Institute of Crop Science, National Wheat Improvement Center/The National Key Facility for Crop Gene Resources and Genetic Improvement, Chinese Academy of Agricultural Sciences (CAAS), 12 Zhongguancun South Street, Beijing* 100081, *China*; [2] *College of Agronomy, Northwest Sci-Tech University of Agriculture and Forestry, Yangling* 712100, *Shaanxi, China*; [3] *International Maize and Wheat Improvement Center (CIMMYT), Apdo Postal* 6-641, 06600 *Mexico, D. F., Mexico*; [4] *International Maize and Wheat Improvement Center (CIMMYT) China Office, c/o CAAS*, 12 *Zhongguancun South Street, Beijing* 100081, *China*.

Abstract: Phytoene synthase (PSY) genes are involved in the biosynthesis of carotenoid pigments in durum wheat, significantly influencing grain yellowness. This study was conducted to identify new allelic variants at the *Psy-A1* and *Psy-B1* loci in durum wheat, and to evaluate the applicability of functional markers developed from common wheat for durum wheat breeding. Two new allelic variants, *Psy-A1d* and *Psy-A1e*, were identified at the *Psy-A1* locus, and both the co-dominant markers *YP7A* and *YP7A-2* can be used to discriminate the two haplotypes, yielding 194-bp and 231-bp PCR products with *YP7A* and 1, 001-bp and 1, 684-bp fragments with *YP7A-2*, respectively. At the *Psy-B1* locus, three allelic variants were identified. *Psy-B1e* was also found in common wheat, whereas *Psy-B1f* and *Psy-B1g* were detected only in durum wheat. The co-dominant marker *YP7B-1* can be used to distinguish *Psy-B1f* and *Psy-B1g*, generating 151-bp and 153-bp PCR fragments, respectively, and the dominant marker *YP7B-4* was specific to haplotype *Psy-B1e*, producing a 717-bp PCR product. In a set of 100 CIMMYT durum wheat lines with widely variable grain yellowness, the frequencies of *Psy-A1d*, *Psy-A1e*, *Psy-B1e*, *Psy-B1f* and *Psy-B1g* were 99%, 1%, 0%, 67% and 33%, respectively, and the genotype *Psy-B1f* showed a significant association with higher grain yellowness, whereas the presence of *Psy-B1g* led to lower yellowness. A phylogenetic tree generated from the gene sequences of the allelic variants at *Psy-A1* and *Psy-B1* loci indicated two parallel lineages of durum/common wheats, suggesting that more than one tetraploid *T. dicoccum* genotypes were involved in the origin of common wheat. Our results suggested that *Psy-B1f* should be paid more attention in durum breeding programs for its association with elevated grain yellowness.

Abbreviations: CAPS, cleaved amplified polymorphic sequence; DH, doubled haploid; InDel, insertion/deletion; MAS, marker assisted selection; PCR, polymerase chain reaction; PSY, phytoene synthase; QTL, quantitative trait locus; RIL, Recombinant inbred line; SNP, single nucleotide polymorphism; YP, yellow pigment.

Published in Crop Science, 2009, 49 (6): 2058-2064

Introduction

Durum wheat (*Triticum turgidum* L. var. *durum* Desf) is widely used to make pasta, a staple food consumed extensively in Europe, America, North Africa and the Middle East. A bright yellow color is particularly preferred for pasta, which is due to the presence of carotenoid pigments, predominantly lutein, with a small proportion of zeaxanthin and β-cryptoxanthin (Hentschel et al., 2002; Adom et al., 2003).

Many studies have been conducted to localize loci for durum grain yellowness, and the results indicated that the trait was influenced by one or two major loci plus several minor genes. On a region near the end of chromosome 7BL, Elouafi et al. (2001) and Pozniak et al. (2007) found major QTLs, both of which linked to an SSR marker *Xgwm344*. Zhang and Dubcovsky (2008) also reported a QTL in this region, linking to an SSR marker *Xgwml46*, in agreement with the result reported by Pozniak et al. (2007), implying that the three QTLs identified in different populations were from the same locus. Similarly, in the studies of Elouafi et al. (2001), Patil et al. (2008) and Zhang and Dubcovsky (2008), QTLs near the end of chromosome 7A have also been identified, suggesting a homoeologous locus with that on chromosome 7B. In addition, major QTLs on chromosomes 7A and 7B were also detected in common wheat (Parker et al., 1998; Ma et al., 1999; Mares and Campbell 2001; Kuchel et al., 2006; He et al. 2008; Zhang et al., 2009), demonstrating the importance of the two chromosome regions for grain yellowness.

Phytoene synthase (PSY), which catalyzes the dimerization of two geranylgeranyl pyrophosphate molecules to produce phytoene, is generally considered the rate-limiting enzyme for the accumulation of carotenoid in endosperm (Lindgren et al., 2003). Three *PSY* genes, *PSY1*, *PSY 2* and *PSY3*, have been reported in the grass family so far (Gallagher et al., 2004; Li et al., 2008). However, only *PSY1* showed a strong association with endosperm yellowness in maize (Palaisa et al., 2003; Gallagher et al., 2004; Li et al., 2008). In durum wheat, Pozniak et al. (2007) mapped the *PSY1* and *PSY2* genes to homoeologous group 7 and 5 chromosomes, respectively, and demonstrated that *PSY1*, rather than *PSY2*, was associated with grain yellow pigment (YP) content. Zhang and Dubcovsky (2008) confirmed the association of *PSY1* with YP content, demonstrating that the genes responsible for the QTLs detected on chromosomes 7A and 7B were orthologs of *PSY1*.

Using primers designed fromthe rice *PSY1* gene (AY445521.1), Pozniak et al. (2007) cloned partial sequences of the *PSY1* genes located on chromosomes 7A and 7B, i.e. *Psy-A1* and *Psy-B1* (designated as *Psy1-2* and *Psy1-1* in the paper), respectively. Furthermore, they identified two alleles at the *Psy-B1* locus and developed a CAPS marker to discriminate them. In the DH population derived from Kofa/W9262-260D3, the CAPS marker co-segregated with the QTL on chromosome 7B. Recently, Zhang and Dubcovsky (2008) cloned full-length sequences of *Psy-A1* and *Psy-B1* from durum cultivars U1113 and Kofa, and found that the two cultivars shared the same *Psy-A1* allele, but had different *Psy-B1* alleles.

In our previous study (He et al., 2008; 2009a), *Psy-A1* and *Psy-B1* and their allelic variants were identified in common wheat cultivars, and functional markers were developed for the alleles. The objectives of the present study were to identify new allelic variants of *Psy-A1* and *Psy-B1* in durum wheat, validate the available functional markers in a set of reference genetic resources and a group of elite advanced lines from the CIMMYT program, and analyze the phylogenetic relationships of the allelic variants at the two loci in durum and common wheats.

Materials and methods

Plant materials and yellow color evaluation

The durum wheat materials used in this study included 16 lines from the National Key Facilities for Crop Genetic Resources and Improvement (NFCRI), Institute of Crop Science, CAAS, China, and 100 advanced breeding lines from the CIMMYT program.

One hundred advanced breeding lines (F_7 to F_{10} generations) were selected to cover a wide range of yellow color values based on 2 or 3 years data, obtained from replicated yield trials conducted from 2005 to 2007 at the Ciudad Obregon experimental station in Northern Mexico, under full irrigation, optimum nitrogen fertilization and weed control for maximum yield potential (5.5-9.0t/ha). Experimental plots consisted of four rows of 2.1m, sown at a constant seeding density of 250 live plants/m^2, corresponding to an average commercial seeding rate of 90-100kg/ha. Trials were arranged in 10×11 rectangular lattice designs with two or three replications in different years, and the commercial check JUPARE C2001 (a high yielding, rust resistant cultivar with low YP content) was present in each column. Plots were combine-harvested at commercial maturity and 40g of sub-samples from each plot (all replications including checks) were used for yellow color evaluations. Yellow color was evaluated on ground whole wheat (cyclone mill equipped with a 0.5mm screen) by determining the b*-value or yellow index using a Minolta colorimeter. A reference value was used here to evaluate yellow color, calculated from the difference between the b*-value of a plot and that of the check included in the same sub-block of eight plots (i.e. b*-value_entry - b*-value_check). The 100 genotypes were grouped into five classes based on the mean reference values, i.e. Very High (reference values higher than 3.0), High (from 2.0 to 3.0), Medium (from 1.0 to 2.0), Low (from 0 to 1.0) and Very Low (lower than 0), with each group consisting of 20 genotypes.

Identification of *Psy-A1* and *Psy-B1* alleles in durum wheat

Allelic variants of *Psy-A1* and *Psy-B1* were detected with markers *YP7A* (He et al., 2008), *YP7A-2*, and *YP7B-1*, *YP7B-2*, *YP7B-3*, and *YP7B-4* (He et al., 2009a) in the 16 durum wheat lines from NFCRI, and the PCR products were sequenced to detect new allelic variants at the *Psy-A1* and *Psy-B1* loci. For the newly identified allelic variants, full-length genomic DNA sequences were amplified using the primers developed previously for cloning the corresponding orthologous alleles in common wheat (He et al., 2008; 2009a).

Genomic DNA was extracted from seeds following a method from Lagudah et al. (1991), with minor modifications. PCR were performed in an MJ Research PTC-200 thermal cycler. The PCR condition was 20mM of Tris-HCl (pH 8.4), 20mM of KCl, 100μM of each dNTPs, 1.5mM of $MgCl_2$, 5pmol of each primer, 1unit of *Taq* DNA polymerase (TIANGEN Biotech Co., Ltd., Beijing, http://www.tiangen.com) and 50ng of genomic DNA, in a total volume of 20μl. Thermo-cycling conditions were 95℃ for 5min, followed by 40 cycles of 95℃ for 30s, 56℃ to 63℃ for 30s (according to the annealing temperatures of different primer sets), and 72℃ for 1.5min, with a final extension of 72℃ for 5min. The PCR products were cloned into a pMD18-T vector (TaKaRa Biotechnology Co., Ltd., Dalian, China) following the manufacturer's instructions and sequenced by Shanghai Sangon Biological Engineering & Technology and Service Co., Ltd. (http://www.sangon.com). The PCR and DNA sequencing were repeated 2-4 times for each primer set to eliminate errors in sequencing.

Statistical analysis

Functional markers showing polymorphisms in the 16 durum wheat lines from NFCRI were further verified for their association with grain yellowness in the 100 CIMMYT durum lines. An analysis of variance was performed by PROC MIXED in the Statistical Analysis System (SAS Institute, 2000), in which genotype clusters, determined by different *Psy-B1* alleles, served as categorical variables to calculate the mean reference value for each cluster and to test differences among the clusters. Genotypic clusters of different *Psy-B1* alleles were treated as fixed effects, and genotypes nested within clusters as random.

Phylogenetic analysis

The genomic DNA sequences of *Psy-A1* and *Psy-B1* alleles identified in this study, together with those of common wheat cloned in previous studies (He et al., 2008; 2009a) and six durum *PSY1* genes deposited in GenBank (accession numbers DQ642439, DQ642440, DQ642443, DQ642444, EU096090 and EU096093; Pozniak et al., 2007; Zhang and Dubcovsky, 2008), were used to construct a phylogenetic tree. Gene sequences were aligned with the software ClustalW 1.83 (Thompson et al., 1997) and a neighbor joining tree was generated by the program MEGA version 3.1 (Kumar et al., 2004) with default parameters, in which bootstrap tests were performed with 1, 000 replicates.

Results

Characterization of allelic variants at the *Psy-A1* locus

Among the 16 durum wheat lines from NFCRI, DR4 and DR8 amplified a 231-bp fragment with *YP7A*, whereas a 194-bp product was produced in the other 14 accessions (Fig. 1a). Similarly, when *YP7A-2* was used, a 1,686-bp fragment was detected in DR4 and DR8, and a 1,001-bp fragment was amplified in the remaining accessions (Fig. 1b). Sequence analyses indicated the presence of two new *Psy-A1* alleles in these accessions, designated *Psy-A1d* and *Psy-A1e* (GenBank accessions EU263018 and EU263019, respectively).

Psy-A1d and *Psy-A1e* showed sequence identity of 98.9% (Table 1). The sequence differences between them were mainly present in two InDels of 37bp and 688bp, located in the second and fourth introns, respectively, being identical to the two InDels between *Psy-A1b* and *Psy-A1c*. In addition, there were two InDels, with a 1-bp and a 3-bp InDels in the third and the fifth introns, respectively. Of the 28 SNPs between the two alleles, 22 were in introns and six in exons. Compared to *Psy-A1* alleles from common wheat, *Psy-A1d* showed only one 1-bp InDels and one SNP difference relative to *Psy-A1c*, all found in introns, and there was only one synonymous SNP difference between *Psy-A1e* and *Psy-A1b*, found in the first exon. Among the five allelic variants at the *Psy-A1* locus, *Psy-A1a*, *Psy-A1b* and *Psy-A1e* shared the same encoded polypeptide, whereas *Psy-A1c* and *Psy-A1d* shared the second, and the two polypeptide sequences differed only in one residue.

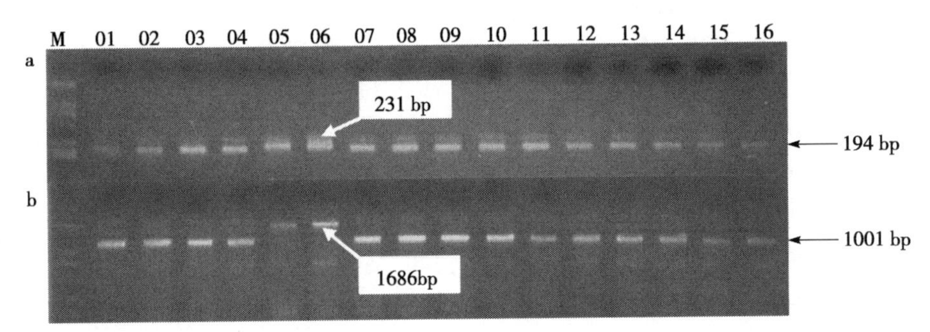

Fig. 1 PCR amplification of 16 NFCRI durum lines with markers *YP7A* (a) and *YP7A-2* (b). M, DNA Ladder DL2000; 01, Langdon (with allele *Psy-A1d*); 02, DR1 (*Psy-A1d*); 03, DR2 (*Psy-A1d*); 04, DR3 (*Psy-A1d*); 05, DR4 (*Psy-A1e*); 06, DR8 (*Psy-A1e*); 07, DR10 (*Psy-A1d*); 08, DR11 (*Psy-A1d*); 09, DR12 (*Psy-A1d*); 10, DR13 (*Psy-A1d*); 11, DR14 (*Psy-A1d*); 12, DR17 (*Psy-A1d*); 13, DR18 (*Psy-A1d*); 14, DR22 (*Psy-A1d*); 15, DR24 (*Psy-A1d*); 16, DR28 (*Psy-A1d*).

Table 1 Sequence comparison of five allelic variants at the *Psy-A1* locus (below diagonal) **and their deduced amino-acid sequences** (above diagonal, %)

Allele	*Psy-A1a*	*Psy-A1b*	*Psy-A1c*	*Psy-A1d*	*Psy-A1e*
Psy-A1a	**100**	100	99.8	99.8	100
Psy-A1b	99.9	**100**	99.8	99.8	100
Psy-A1c	98.8	98.8	**100**	100	99.8
Psy-A1d	98.9	98.9	99.9	**100**	99.8
Psy-A1e	99.9	99.9	98.8	98.9	**100**

Characterization of allelic variants at the *Psy-B1* locus

In the detection of allelic variants of *Psy-B1* among the 16 NFCRI durums, markers *YP7B-2* and *YP7B-3* failed to amplify, whereas *YP7B-1* and *YP7B-4* exhibited polymorphisms (Fig. 2). The marker *YP7B-1* yielded a 151-bp fragment in Langdon, DR2, DR14, DR17, DR24 and DR28, and a 153-bp fragment in DR1, DR10, DR11, DR18 and DR22, but did not amplify any products from the other five lines (Fig. 2a). *YP7B-4* amplified a 717-bp fragment in lines DR3, DR4, DR8, DR12 and DR13, and no PCR product in the other 11 accessions (Fig. 2b). Sequencing of the PCR products of *YP7B-1* and *YP7B-4* indicated three allelic variants at the *Psy-B1* locus. The haplotype in the lines with the 717-bp fragment amplified by *YP7B-4* was *Psy-B1e*, which was also found in common wheat (He et al., 2009a), whereas the other two variants were novel alleles, designated *Psy-B1f* (GenBank accession EU263020) and *Psy-B1g* (GenBank accession EU650396), with 171-bp and 173-bp PCR fragments amplified by *YP7B-1*, respectively.

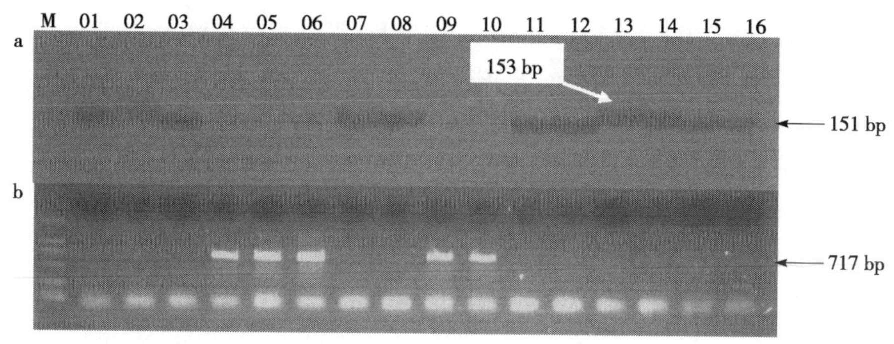

Fig. 2 PCR amplification of 16 NFCRI durum lines with markers *YP7B-1* (a) and *YP7B-4* (b). M, DNA Ladder DL2000; 01, Langdon (with allele *Psy-B1f*); 02, DR1 (*Psy-B1g*); 03, DR2 (*Psy-B1f*); 04, DR3 (*Psy-B1e*); 05, DR4 (*Psy-B1e*); 06, DR8 (*Psy-B1e*); 07, DR10 (*Psy-B1g*); 08, DR11 (*Psy-B1g*); 09, DR12 (*Psy-B1e*); 10, DR13 (*Psy-B1e*); 11, DR14 (*Psy-B1f*); 12, DR17 (*Psy-B1f*); 13, DR18 (*Psy-B1g*); 14, DR22 (*Psy-B1g*); 15, DR24 (*Psy-B1f*); 16, DR28 (*Psy-B1f*).

Psy-B1f and *Psy-B1g* shared a very high sequence identity of 99.7% (Table 2), differing by only eight SNPs and one 2-bp InDel. However, they showed lower sequence identities with *Psy-B1e* (Table 2). Compared with alleles from the common wheat *Psy-B1* locus, *Psy-B1f* showed only one 1-bp InDel in the fifth intron and two SNPs in the 3' untranslated region with *Psy-B1a*. In addition, *Psy-B1f* shared the same polypeptide sequence with *Psy-B1a* and *Psy-B1b*, differing from *Psy-B1g* by only two residues. However, *Psy-B1c*, *Psy-B1d* and *Psy-B1e* showed lower sequence identities with the other four alleles at the *Psy-B1* locus at both the nucleotide and protein levels (Table 2).

Table 2 Sequence comparison of seven allelic variants at the *Psy-B1* locus (below diagonal) **and their deduced amino-acid sequences** (above diagonal, %)

Allele	*Psy-B1a*	*Psy-B1b*	*Psy-B1c*	*Psy-B1d*	*Psy-B1e*	*Psy-B1f*	*Psy-B1g*
Psy-B1a	**100**	100	98.2	97.9	98.3	100	99.3
Psy-B1b	99.9	**100**	98.2	97.9	98.3	100	99.3
Psy-B1c	75.8	76.9	**100**	99.3	99.3	98.2	98.5
Psy-B1d	90.0	89.8	92.5	**100**	98.6	97.9	98.6
Psy-B1e	94.8	94.7	77.6	89.8	**100**	98.3	99.0
Psy-B1f	99.9	99.9	77.2	90.1	94.8	**100**	99.3
Psy-B1g	99.7	99.7	77.3	90.2	94.9	99.7	**100**

Associations of allelic variants of *Psy-A1* and *Psy-B1* with grain yellowness

The 100 CIMMYT durum lines with a wide range of yellowness were genotyped with markers *YP7A*, *YP7A-2*, *YP7B-1* and *YP7B-4*. Almost all lines possessed *Psy-A1d*, with only one line having *Psy-A1e*. At the *Psy-B1* locus, 67 lines carried *Psy-B1f* and 33 had *Psy-B1g*. Statistical analysis demonstrated that lines with *Psy-B1f* had a significantly higher mean reference value than those with *Psy-B1g* (Table 3). Groups of lines with high yellowness compared to the check showed a higher frequency of *Psy-B1f*, and as yellowness decreased, the frequency of lines carrying *Psy-B1g* increased (Fig. 3). Therefore, *Psy-B1f* was considered more favorable than *Psy-B1g* for expression of high yellowness. The two alleles can be distinguished with the co-dominant marker *YP7B-1*, and its 151-bp and 153-bp PCR products were associated with higher and lower yellowness, respectively.

Table 3 Associations between *Psy-B1* alleles and YP content in 100 durum breeding lines tested with *YP7B-1*

Allele†	No. of accessions	Mean YP content‡	Standard deviation	Range§
Psy-B1f	67	1.83a	1.55	−1.3-4.1
Psy-B1g	33	0.59b	1.75	−2.3-4.1

† *Psy-B1f* and *Psy-B1g* were detected by *YP7B-1*, with 151-bp and 153-bp PCR fragments, respectively.

‡ Different letters following the mean YP content indicate highly significant differences between the two groups ($P<0.01$).

§ The YP content used here was the difference between the bvalues of an entry and the check.

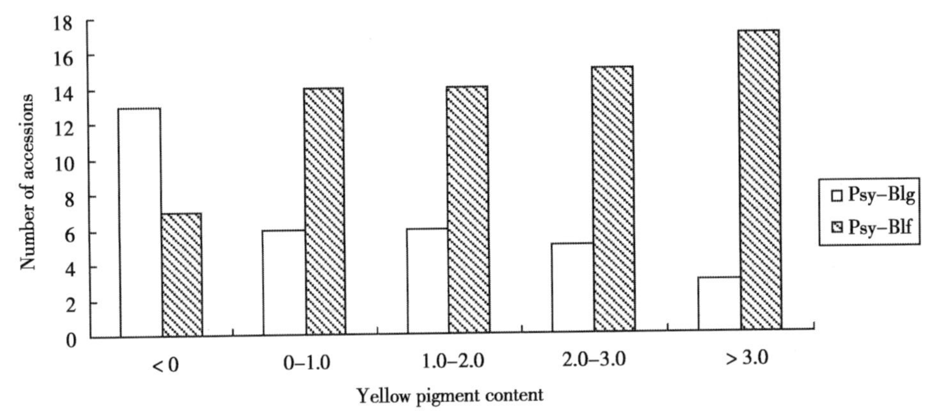

Fig. 3 Association between two *Psy-B1* alleles and grain YP content in 100 durum wheat breeding lines. Yellow pigment content was represented by the difference between the b-value of an entry and that of the check.

Phylogenetic analysis of *Psy-A1* and *Psy-B1* alleles in durum and common wheats

The phylogenetic tree consisted of two subtrees (Fig. 4). Subtree I comprised eight alleles at the *Psy-A1* locus and eight of ten alleles at *Psy-B1*, whereas subtree II included the other two alleles of *Psy-B1*, i.e. *Psy-B1c* and *Psy-B1d*. Within subtree I, there were two groups, the *Psy-A1* group and the *Psy-B1* group. The *Psy-A1* group was further divided into two clusters, α1 and α2, and the former included *Psy-A1c* from common wheat, *Psy-A1d* from durum wheat, and other three *Psy-A1* alleles reported previously, whereas the latter contained *Psy-A1a* and *Psy-A1b* from common wheat and *Psy-A1e* from durum (Fig. 4). The *Psy-B1* group also comprised two clusters, β1 and β2; the former included *Psy-B1e* found in both durum and common wheat and a *Psy-B1* allele identified in Kofa, and the latter contained *Psy-B1a* and *Psy-B1b* from common wheat, *Psy-B1f* and *Psy-B1g* from durum, and two *Psy-B1* alleles found in durum cultivars W9262-260D3 and U1113 (Fig. 4). In consideration of the sequence comparisons mentioned above, alleles from a common cluster showed very high sequence identities, whereas those from different clusters showed lower sequence identities (Fig. 4). It is notable that apart from subtree II with alleles only from common wheat, all four clusters within subtree I comprised alleles both from durum and common wheat, indicating two genetic lineages in both the *Psy-A1* and *Psy-B1* groups (Fig. 4).

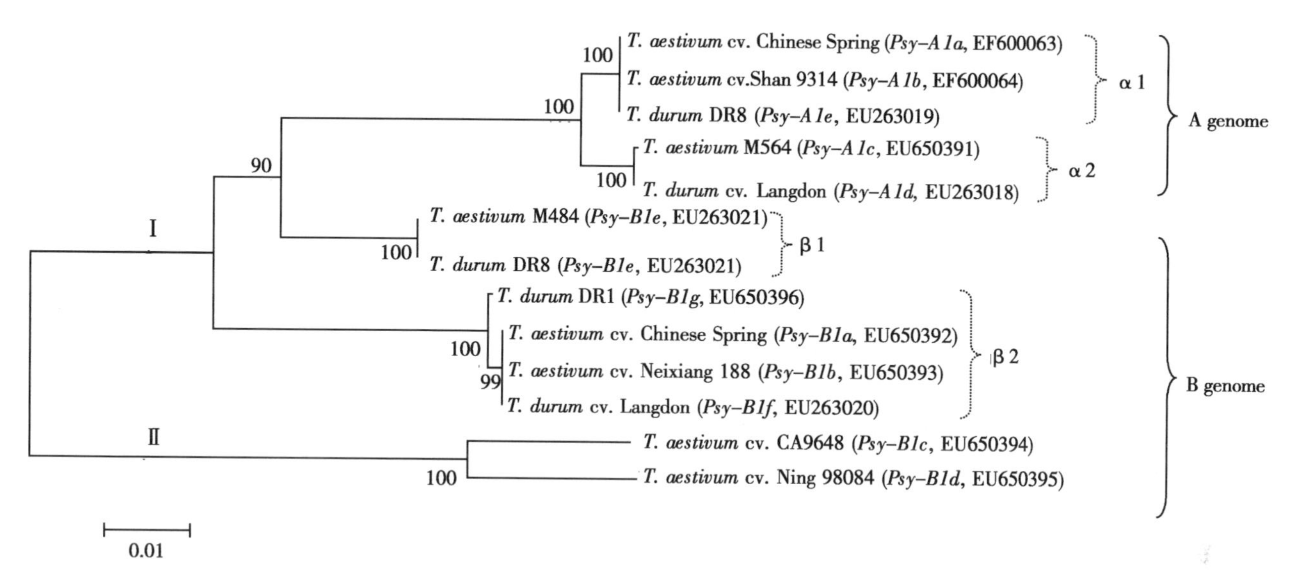

Fig. 4 Phylogenetic tree of the alleles at *Psy-A1* and *Psy-B1* loci in common and durum wheats. The tree was constructed by the software MEGA version 3. 1 with the neighbor joining algorithm, including five alleles at the *Psy-A1* locus and seven at the *Psy-B1* locus. Genes are labeled by species name, accession identifier, allele name and GenBank accession number. Bootstrap values are shown and the scale bar indicates the number of nucleotide substitutions per site.

Discussion

Allelic variants at the *Psy-B1* locus and grain yellowness

Three allelic variants, *Psy-B1e*, *Psy-B1f* and *Psy-B1g*, were identified at *Psy-B1* locus, and associations of *Psy-B1f* and *Psy-B1g* with higher and lower grain yellowness, respectively, were validated in 100 durum breeding lines varying widely in grain yellowness (Table 3). However, no information on the phenotypic effect of *Psy-B1e* could be obtained from this set of lines due to absence of the allele. In our study, 177 RIL lines from a cross of Mohawk (with *Psy-B1e* allele) and Cocorit (with *Psy-B1f* allele) was used to analyze the association of *Psy-B1* alleles with grain YP content. However, no significant difference was detected between the mean phenotypic values of the two alleles (data not shown). There were two possible reasons for this situation. Firstly, the QTL on *Psy-B1* locus was not significant in the two environments in which field trials were conducted for the population, as reported previously that the QTL on chromosome 7B could not be detected in some environments (Pozniak et al., 2007); secondly, it is likely that both *Psy-B1e* and *Psy-B1f* were associated with higher grain yellowness, showing similar phenotypic effects. The latter hypothesis was also supported by previous works. In the 155 DH lines derived from W9262-260D3/Kofa, Pozniak et al. (2007) found a major QTL on chromosome 7B, for which the Kofa allele contributed positively to YP content. In the study of Zhang and Dubcovsky (2008), a major QTL on chromosome 7B was localized in the RIL population from UC1113/Kofa, and the allele associated with increased YP content was also from Kofa. Based on the phylogenetic tree constructed in this study, Kofa had an allele highly similar to *Psy-B1e*, whereas W9262-260D3 and UC1113 had alleles similar to *Psy-B1g* (Fig. 4), and thus *Psy-B1e* may confer a higher YP content than *Psy-B1g*.

Implementation of functional markers for grain yellowness in durum wheat breeding programs

YP7B-1, the co-dominant marker designed for the discrimination of *Psy-B1a* and *Psy-B1b* in common wheat, can also be used to detect *Psy-B1f* and *Psy-B1g* in durum. As for *Psy-B1e*, an allele shared by common and durum wheats, can be detected with *YP7B-4*. In durum breeding programs, lines with the 151-bp fragment amplified by *YP7B-1* (*Psy-B1f* allele) should be given preference because of their association with higher grain yellowness. In addition, the lines with the 717-bp product yielded by *YP7B-4* (*Psy-B1e* allele) may also be paid some attention, for their possible association with grain yellowness discussed in the previous section, but this needs further validation. However, there were still wide ranges of variation within each *Psy-B1* genotype (Fig. 3), indicating that *Psy-B1* locus alone can not explain all the phenotypic variation of grain yellowness. This was due to the nature of multigenic control of grain yellowness. In addition to the major QTLs on chromosomes 7A and 7B at *Psy1* loci, QTLs with minor effects have also been detected on chromosomes 1A (Patil et al., 2008), 3A (Parker et al. 1998), 3B (Mares and Campbell, 2001; Patil et al., 2008), 4A and 5A (Hessler et al. 2002), 5B (Patil et al., 2008) and 2A, 4B and 6B (Pozniak et al., 2007). Recently, Reimer et al. (2008) published results based on an association mapping, in which markers associated with YP were detected on all chromosomes of the durum genome. Therefore, cautions should be taken when applying *YP7B-1* and *YP7B-4* in marker assisted selection (MAS); In combination with other YP-related markers in MAS, they may more precisely predict the grain yellowness of a given cultivar. Nevertheless, *YP7B-1* and *YP7B-4* could be very useful and phenotypically predictable in breeding populations segregating at *Psy-B1* locus, in which the genetic background was more uniform than a collection of durum cultivars.

Phylogenetic analysis of *Psy-1* genes and the origin of common wheat

Common wheat originated from the hybridization of cultivated emmer wheat (*T. turgidum* ssp. *dicoccum*, AABB) and *Aegilops tauschii* (DD) approximately 8 000 years ago (Huang et al., 2002). Several reports suggested that more than one hybridization event was involved in the origin of hexaploid wheat (Gu et al., 2004; Isidore et al., 2005; Ragupathy et al., 2008), and this conclusion was also supported by our previous work on the phylogenetics of polyphenol oxidase genes (He et al., 2009b). In this study, two parallel lineages, α1 and α2, were found in the *Psy-A1* group, with each lineage comprising *Psy-A1* alleles from both common wheat and durum. It is notable that durum and emmer wheats share the same ancestor (*T. dicoccoides*) and they are closely related to each other (Salamini et al., 2002; Ozkan et al., 2005; Jauhar, 2008); thus our results imply that at least two emmer wheat genotypes participated in the origin of common wheat. Similarly, there were parallel lineages, β1 and β2, for the *Psy-B1* group within subtree I. However, the allele *Psy-B1e* found in common wheat was actually introduced from durum wheat by a recent introgression event (He et al., 2009a). Nevertheless, the conclusion of polyphyletic origin of common wheat can still be inferred from the phylogenetic analysis of *Psy-B1* locus. Subtree II included *Psy-B1* alleles, *Psy-B1c* and *Psy-B1d*, from common wheat, showing much greater genetic distances from other *Psy-B1* alleles in subtree I (Fig. 4). From our previous results, the two alleles were derived from a common ancestor early in the divergence of *Psy-B1* alleles (He et al., 2009a). Considering the large genetic distance of *Psy-B1c* and *Psy-B1d* from other alleles, their divergence must have preceded the origin of common wheat. Thus, including the ancestory of cluster β2, there might be likely three tetraploid ancestors involved in the origin of common wheat. This is in accordance with the theory of recurrent formation of polyploid plants proposed by Soltis and Soltis (1999).

Acknowledgments

The authors are very grateful to the critical review of this manuscript by Prof. R. A. McIntosh, Plant Breeding Institute, University of Sydney, and to Dr. Lihui Li, Institute of Crop Science, CAAS, for kindly providing the NFCIR durum lines. This study was supported by the National Science Foundation of China (30871522 and 30830072), National Basic Research Program (2009CB118300), National 863 Program (2006AA10Z 1A7and 2006AA100102), and International Collaboration Project from the Ministry of Agriculture (2006-G2).

References

Adom, K. K., M. E. Sorrells, and R. H. Liu. 2003. Phytochemical profiles and antioxidant activity of wheat varieties. J. Agric. Food. Chem., 51: 7825-7834.

Elouafi, I., M. M. Nachit, and L. M. Martin. 2001. Identification of a microsatellite on chromosome 7B showing a strong linkage with yellow pigment in durum wheat (*Triticum turgidum* L. var. *durum*). Hereditas, 135: 255-261.

Gallagher, C. E., P. D. Matthews, F. Li, and E. T. Wurtzel. 2004. Gene duplication in the carotenoid biosynthetic pathway preceded evolution of the grasses. Plant Physiol., 135: 1776-1783.

Gu, Y. Q., D. Coleman-Derr, X. Y. Kong, and O. D. Anderson. 2004. Rapid genome evolution revealed by comparative sequence analysis of orthologous regions from four Triticeae genomes. Plant Physiol., 135: 459-470.

He, X. Y., Y. L. Zhang, Z. H. He, Y. P. Wu, Y. G. Xiao, C. X. Ma, and X. C. Xia. 2008. Characterization of phytoene synthase 1 gene (*Psy1*) located on common wheat chromosome 7A and development of a functional marker. Theor. Appl. Genet., 116: 213-221.

He, X. Y., Z. H. He, W. Ma, R. Appels, and X. C. Xia. 2009a. Allelic variants of phytoene synthase 1 (*Psy1*) genes in Chinese and CIMMYT wheat cultivars and development of functional markers for flour colour. Mol. Breeding, DOI 10.1007/s11032-009-9255-1

He, X. Y., Z. H. He, C. F. Morris, and X. C. Xia. 2009b. Cloning and phylogenetic analysis of polyphenol oxidase genes in common wheat and related species. Genet. Resour. Crop Evol., DOI 10.1007/s10722-008-9365-3

Hentschel, V., K. Kranl, J. Hollmann, M. G. Lindhauer, V. Bohm, and R. Bitsch. 2002. Spectrophotometric determination of yellow pigment content and evaluation of carotenoids by high-performance liquid chromatography in durum wheat grain. J. Agric. Food Chem., 50: 6663-6668.

Hessler TG, Thomson MJ, Benscher D, Nachit MM, Sorrells ME (2002) Association of a lipoxygenase locus, *Lpx-B1*, with variation in lipoxygenase activity in durum wheat seeds. Crop Sci., 42: 1695-1700.

Huang, S., A. Sirikhachornkit, X. Su, J. Faris, B. Gill, R. Haselkorn, and P. Gornicki. 2002. Genes encoding plastid acetyl-CoA carboxylase and 3-phosphoglycerate kinase of the *Triticum/Aegilops* complex and the evolutionary history of polyploid wheat. Proc. Natl. Acad. Sci. USA. 99: 8133-8138.

Isidore, E., B. Scherrer, B. Chalhoub, C. Feuillet, and B. Keller. 2005. Ancient haplotypes resulting from extensive molecular rearrangements in the wheat A genome have been maintained in species of three different ploidy levels. Genome Res., 15: 526-536.

Jauhar, P. P. 2007. Meiotic restitution in wheat polyhaploids (amphihaploids): a potent evolutionary force. J. Hered, 98: 188-193.

Kuchel, H., P. Langridge, L. Mosionek, K. Williams, and S. P. Jefferies. 2006. The genetic control of milling yield, dough rheology and baking quality of wheat. Theor. Appl. Genet., 112: 1487-1495.

Kumar, S., K. Tamura, and M. Nei. 2004. MEGA3: integrated software for molecular evolutionary genetics analysis and sequence alignment. Brief. Bioinform, 5: 150-163.

Lagudah, E. S., R. Appels., and D. McNeil. 1991. The *Nor-D3* locus of *Triticum tauschii*: natural variation and genetic linkage to markers in chromosome 5. Genome, 34: 387-395.

Li, F., R. Vallabhaneni, J. Yu, T. Rocheford, E. T. Wurtzel. 2008. The maize phytoene synthase gene family: Overlapping roles for carotenogenesis in endosperm, photomorphogenesis, and thermal stress tolerance. Plant Physiol., 147: 1334-1346.

Lindgren, L. O., K. G. Stalberg, and A. S. Hoglund. 2003. Seed-specific overexpression of an endogenous *Arabidopsis* phytoene synthase gene results in delayed germination and increased levels of carotenoids, chlorophyll, and abscisic acid. Plant Physiol., 132: 779-785.

Ma, W., G. Daggard, M. Sutherland, and P. Brennan. 1999. Molecular markers for quality attributes in wheat. *In* P. Williamson, P. Banks, I. Haak, J. Thompson, A. Campbell (eds.) Proc. 9th Assembly, Wheat Breeding Society of Australia, Toowoomba, vol. 1, pp 115-117.

Mares, D. J., and A. W. Campbell. 2001. Mapping components of flour and noodle colour in Australian wheat. Aust. J. Agric. Res., 52: 1297-1309.

Ozkan, H., A. Brandolini, C. Pozzi, S. Effgen, J. Wunder, and F. Salamini. 2005. A reconsideration of the domestication geography of tetraploid wheats. Theor. Appl. Genet., 110: 1052-1060.

Palaisa, K. A., M. Morgante, M. Williams, and A. Rafalski. 2003. Contrasting effects of selection on sequence diversity and linkage disequilibrium at two phytoene synthase loci. Plant Cell, 15: 1795-1806.

Parker, G. D., K. J. Chalmers, A. J. Rathjen, and P. Langridge. 1998. Mapping loci associated with flour color in wheat. Theor. Appl. Genet., 97: 238-245.

Patil, R. M., M. D. Oak, S. A. Tamhankar, P. Sourdille, and V. S. Rao. 2008. Mapping and validation of a major QTL for yellow pigment content on 7AL in durum wheat (*Triticum turgidum* L. ssp. *durum*). Mol. Breed, 21: 485-496.

Pozniak, C. J., R. E. Knox, F. R. Clarke, and J. M. Clarke. 2007. Identification of QTL and association of a phytoene synthase gene with endosperm colour in durum wheat. Theor. Appl. Genet., 114: 525-537.

Ragupathy, R., H. A. Naeem, E. Reimer, O. M. Lukow, H. D. Sapirstein, and H. Cloutier. 2008. Evolutionary origin of the segmental duplication encompassing the wheat *GLU-B1* locus encoding the overexpressed Bx7 (Bx7OE) high molecular weight glutenin subunit. Theor. Appl. Genet., 116: 283-296.

Reimer, S., C. J. Pozniak, F. R. Clarke, J. M. Clarke, D. J. Somers, R. E. Knox, and A. K. Singh. 2008. Association mapping of yellow pigment in an elite collection of durum wheat cultivars and breeding lines. Genome., 51: 1016-1025.

Salamini, F., H. Ozkan, A. Brandolini, R. Schäfer-Pregl, and W. Martin. 2002. Genetics and geography of wild cereal domestication in the Near East. Nat. Rev. Genet., 3: 429-441.

Soltis, D. E., and P. S. Soltis. 1999. Polyploidy: recurrent formation and genome evolution. Trends Ecol. Evol., 14: 348-352.

Thompson, J. D., T. J. Gibson, F. Plewniak, F. Jeanmougin, and D. G. Higgins. 1997. The CLUSTAL _ X Windows interface: flexible strategies for multiple sequence alignment aided by quality analysis tools. Nucleic Acids Res., 25: 4876-4882.

Zhang, W., and J. Dubcovsky. 2008. Association between allelic variation at the *Phytoene synthase 1* gene and yellow pigment content in the wheat grain. Theor. Appl. Genet., 116: 635-645.

Zhang, Y. L., Y. P. Wu, Y. G. Xiao, Z. H. He, Y. Zhang, J. Yan, Y. Zhang, X. C. Xia, and C. X. Ma. 2009. QTL mapping for flour and noodle colour components and yellow pigment content in common wheat. Euphytica, 165: 435-444.

CIMMYT 普通小麦品系 Waxy 蛋白类型及淀粉糊化特性研究

穆培源[1,2]，何中虎[1,3,*]，徐兆华[1]，王德森[1]，张艳[1]，夏先春[1]

[1]中国农业科学院作物科学研究所/国家小麦改良中心，北京 100081；[2]新疆农垦科学院作物研究所，石河子 832000；[3]国际玉米小麦改良中心（CIMMYT）中国办事处，北京 100081

摘要：将来自 CIMMYT 第 35 届 IBWSN 和第 20 届 SAWSN 国际鉴定圃的 488 份品系于 2003 年种植在内蒙古临河，对其 Wx 蛋白亚基组成和淀粉糊化特性进行了研究。利用单向 SDS-PAGE 鉴定出 191 份 Wx-B1 蛋白亚基缺失和 28 份和 Wx-D1 蛋白亚基缺失品系。除峰值时间外，其余淀粉糊化特性参数在品系间都有较大变异，与中国春小麦相比，CIMMYT 小麦表现峰值黏度、低谷黏度、最终黏度高，稀澥值和反弹值大，峰值时间长而稳定。Wx-B1 和 Wx-D1 蛋白亚基缺失品系的最终黏度和反弹值显著高于和大于 Wx 蛋白亚基正常品系，Wx-D1 蛋白亚基缺失品系的峰值黏度和稀澥值显著低于和小于 Wx 蛋白亚基正常品系，Wx-B1 蛋白亚基缺失品系的峰值黏度和稀澥值显著高于和大于 Wx-D1 蛋白亚基缺失品系。第 35 届 IBWSN 品系的淀粉糊化特性优于第 20 届 SAWSN 品系，两者低谷黏度、最终黏度和反弹值间差异显著。

关键词：普通小麦；Wx 蛋白亚基；RVA；淀粉糊化特性

Waxy Protein Identification and Starch Pasting Properties of CIMMYT Wheat lines

Mu Peiyuan[1,2], He Zhonghu[1,3], Xu Zhaohua[1], Wang Desen[1], Zhang Yan[1] Xia Xianchun[1]

[1]*Institute of Crop Sciences/National Wheat Improvement Center, Chinese Academy of Agricultural Sciences, Beijing* 100081; [2]*Crop Research Institute, Xinjiang Academy of Agri-recalamation Sciences, Shihezi* 832000, *Xinjiang*; [3]*CIMMYT-China Office, Beijing* 100081, *China*

Abstract: Starch pasting properties, one of the major factors influencing the quality of Dry White Chinese Noodle, is great affected by Wx-A1, Wx-B1 and Wx-D1 protein subunits, among which Wx-B1 null makes the biggest contribution. In order to detect new gemplasm to improve the quality of Chinese noodles, four hundred and eighty-eight spring wheat lines from CIMMYT 35th International Bread Wheat Screening Nursery (35th IBWSN) and 20th Semiarid Wheat Screening Nursery (20th SAWSN) were collected and planted in Linhe, Inner Mongolia in 2003. Samples were harvested to identify their Waxy protein subunits and the relationship with RVA (rapid viscosity analyzer) parameters. 191 lines with Wx-B1 null subunit and 28 lines with Wx-D1 null subunit were identified by SDS-PAGE. The results indicated that final viscosity and set back of lines with subunits of Wx-B1 null and Wx-D1 null were significantly higher than those with normal Wx protein subunit, while peak viscosity and break down of Wx-D1 null

原文发表在《作物学报》，2006，32 (7)：1071-1075

lines were significantly lower than those of normal Wx lines, peak viscosity and break down of Wx-B1 null lines were significantly higher than those of Wx-D1 null lines. Significant variation among the lines for RVA parameters except for peak time was also observed. Compared with China's spring wheat, the CIMMYT spring wheat lines expressed higher peak viscosity, though viscosity, final viscosity, larger break down and set back, as well as longer and stable peak time. RVA parameters for lines from 35th IB-WSN were much higher than those from 20th SAWSN.

Key words: *Triticum aestivum* L.; Waxy protein subunit; Rapid viscosity analyser; Starch pasting characteristic

淀粉是小麦（*Triticum aestivum* L.）胚乳的主要成分，约占籽粒干重的65%～70%，其中直链淀粉为22%～35%，支链淀粉为78%～65%[1]。颗粒结合型淀粉合成酶（granule-bound starch synthase，GBSS）是淀粉合成过程中的关键酶，亦称Wx蛋白[2]。普通小麦籽粒中含有Wx-A1、Wx-B1和Wx-D1 3种Wx蛋白，其编码基因*Wx-7A*、*Wx-4A*和*Wx-7D*分别位于7AS、4AL和7DS[3,4]。在六倍体小麦中已发现的缺失类型中，Wx-B1缺失型最常见，其次是Wx-A1缺失型，而Wx-D1缺失型非常罕见，双缺失型中以Wx-A1和Wx-B1同时缺失较常见，其余双缺失型及三缺失型（糯小麦）在自然界尚未发现。Wx蛋白亚基缺失会导致直链淀粉合成酶活性下降，直链淀粉含量降低和支链淀粉含量增加，从而改变小麦的淀粉构成、面粉品质和食用价值。但3种Wx蛋白亚基对直链淀粉含量的作用不同，Wx-B1亚基缺失的影响最大，而Wx-A1和Wx-D1亚基缺失的影响较小，并且缺失Wx-B1亚基的小麦品种的直链淀粉含量低、高峰黏度大、膨胀势高，具有良好的面条品质[5-8]。

鉴于淀粉特性与面条品质间的密切关系，令人们对小麦淀粉性状的研究日益重视。研究表明，具有较高峰值黏度和较大稀澥值、较小反弹值和较高最终黏度等特性的面粉能够加工出优质口感的日本白盐面条和朝鲜面条，澳大利亚的小麦育种项目已把淀粉糊化峰值黏度作为改良日本面条品质的选择指标[9-12]。笔者已有的研究表明，品种蛋白质数量和质量、淀粉糊化特性、面粉颜色、籽粒硬度是影响中国面条品质的主要因素，其中峰值黏度和稀澥值与中国干白面条品质关系密切，对质地、外观品质和口感均有较大正向作用[13-16]，这说明改良淀粉糊化特性有助于提高我国小麦的面条品质。

国际玉米小麦改良中心（CIMMYT）育成的小麦种质对全世界的小麦育种和生产做出了重要贡献，CIMMYT小麦于20世纪70年代初引入我国，在春麦育种中发挥了重要作用，丰富了我国春小麦的遗传基础，特别是其广适性、矮秆和抗病性，极大促进了当地品种的更新换代[17]。由于CIMMYT小麦资源在我国春麦育种的重要作用，以及春麦区面条品质改良的迫切性，本文对来自CIMMYT的488份普通春小麦品系进行了Wx蛋白及淀粉糊化特性的研究。

1 材料与方法

1.1 试验材料

488份CIMMYT普通春小麦品系来自第35届国际春小麦筛选圃（35th Internatioal Bread Wheat Screening Nursery，简称35th IBWSN）和第20届旱地小麦筛选圃（20th Semiarid Wheat Screening Nursery，简称20th SAWSN），前者由灌溉地区育种课题组育成，后者由干旱地区育种课题组育成。以上试验材料于2003年统一种植在内蒙古临河巴盟农科所，2行区，行长1.5m，行距20cm，按常规管理，正常成熟，及时收获。这些材料是CIMMYT小麦项目育成的最新品系，代表目前的育种现状和最新进展，已将这些材料分发给国内多家育种单位，供进一步筛选和利用。

1.2 淀粉糊化特性测定

全麦粉制备以及淀粉糊化特性的测定参照Yasui等的方法，利用澳大利亚Newport快速黏度分析仪（Rapid Viscosity Analyser Super 3）测定，记录峰值黏度（peak viscosity，PV）和稀澥值（breakdown，BD）等参数[18]。每个材料样品测定2次，取其平均值。测定时用1mmol/L的硝酸银溶液代替蒸馏水，以抑制α-淀粉酶活性，消除其对淀粉糊化特性的影响[19]。

1.3 Waxy蛋白的电泳检测

Waxy蛋白亚基的电泳检测参照 Zhao 等[20]和徐兆华等[26]的方法，并做适当改进。在提取淀粉时洗液保持低温（18℃），180V 电压预电泳 2h 后，换 230V 电压至结束，时间控制在 16～18h。

1.4 统计分析

用 SAS Ver. 6.12 统计软件计算各参数平均值和变异系数，并进行差异显著性比较。

2 结果与分析

2.1 Wx蛋白缺失类型的分布

利用 SDS-PAGE 方法分析了 488 份参试品系的 Wx蛋白亚基组成，筛选出 191 份 Wx-B1 蛋白亚基缺失品系和 28 份 Wx-D1 蛋白亚基缺失品系，分别占参试品系的 39.1% 和 5.7%，没有出现 Wx-A1 蛋白亚基缺失及其他缺失类型的品系。将 Wx-D1 蛋白亚基缺失品系名称和组合名称列于表 1。

表 1 Wx-D1 蛋白亚基缺失的品系汇总

Table 1 The list of Wx-D1 null lines from CIMMYT in 35th IBWSN and 20th SAWSN

品系 Line	组合名称 Cross
35th IBWSN-91	JUN/BORL95
35th IBWSN-94	VEE/TRAP#1//ANGRA/3/PASTOR
35th IBWSN-113	LUAN/MILAN
35th IBWSN-114	CNDO/R143//ENTE/MEXI_2/3/*Aegilops squarrosa*（TAUS）/4/WEAVER/5/PICUS
35th IBWSN-116	CNDO/R143//ENTE/MEXI_2/3/ *Aegilops squarrosa*（TAUS）/4/WEAVER/5/PICUS
35th IBWSN-188	WEAVER/4/NAC/TH. AC//3*PVN/3/MIRLO/BUC
35th IBWSN-189	WEAVER/4/NAC/TH. AC//3*PVN/3/MIRLO/BUC
35th IBWSN-190	WEAVER/4/NAC/TH. AC//3*PVN/3/MIRLO/BUC
35th IBWSN-289	PASTOR//TRAP#1/BOW/3/CHEN/ *Aegilops squarrosa*（TAUS）//BCN
35th IBWSN-296	MUNIA/CHTO/3/PFAU/BOW//VEE#9/4/CHEN/ *Aegilops squarrosa*（TAUS）//BCN
35th IBWSN-491	SKAUZ/KS93U76//SKAUZ
35th IBWSN-496	SKAUZ/KS93U76//SKAUZ
35th IBWSN-497	SKAUZ/KS94U215//SKAUZ
35th IBWSN-542	MON/ALD//BOW
35th IBWSN-543	MON/ALD//BOW
20th SAWSN-15	JUP/BJY//URES/3/HD2206/HORK//BUC/BUL
20th SAWSN-19	PIOS/DUCULA
20th SAWSN-24	VEE#5//DOVE/BUC/3/BCN/4/RL6043/3*GEN
20th SAWSN-31	PFAU/VEE#9//URES/3/PSN/BOW//SERI/4/BUC/CHRC//PRL/VEE#6
20th SAWSN-37	BJY/COC//PRL/BOW/5/ND/VG9144//KAL/BB/3/YACO/4/CHIL
20th SAWSN-130	CHEN/ *Aegilops squarrosa*（TAUS）//2*OPATA
20th SAWSN-133	CHEN/ *Aegilops squarrosa*（TAUS）//WEAVER
20th SAWSN-135	BCN//CETA/ *Aegilops searsii*（34D）
20th SAWSN-138	TSI/VEE#5//KAUZ
20th SAWSN-139	TSI/VEE#5//KAUZ
20th SAWSN-141	C182.24/C168.3/3/CNO67*2/7C//CC/TOB/4/CHAM6
20th SAWSN-142	C182.24/C168.3/3/CNO67*2/7C//CC/TOB/4/CHAM6
20th SAWSN-147	OPATA//SORA/ *Aegilops squarrosa*（TAUS）（323）

上述结果表明，CIMMYT小麦中Wx-B1蛋白亚基缺失品系的频率相当高，而且在第35届IBWSN和第20届SAWSN中的分布不同，前者为144份，后者为47份，这些材料对面条品质改良有重要利用价值。28份Wx-D1蛋白亚基缺失品系是非常独特的育种资源，共涉及21个组合，其中有6个组合的2～3个系均出现Wx-D1蛋白亚基缺失品系，说明Wx-D1蛋白亚基出现或缺失与某些特定的亲本有关，又如第35届IBWSN的第114号品系（简称35thIBWSN-114，下同）、35th IBWSN-116、35th IBWSN-289、35th IBWSN-296和第20届SAWSN的第130号品系（简称20th SAWSN-130，下同）、20th SAWSN-133、20th SAWSN-147等7个品系的亲本组合中都含有人工合成小麦的材料*Aegilops squarrosa*（TAUS），而35th IBWSN-94、20th SAWSN-24、20th SAWSN-31、20th SAWSN-138和20thSAWSN-139等5个品系的亲本组合中都有材料VEE。由于亲本难以追踪，故不能确定其最初来源。

2.2 淀粉糊化特性分析

488份CIMMYT普通春小麦品系的淀粉糊化特性结果列于表2。总体来说，CIMMYT普通春小麦品系的RVA参数中除峰值时间外，峰值黏度、低谷黏度、稀澥值、反弹值和最终黏度的变异范围都较大，说明淀粉糊化特性的遗传变异广泛，可以筛选出淀粉品质优良的品系。

在供试品系中，有200份材料的峰值黏度值大于300RVU，占总数的41.0%，其中35th IBWSN-287、35th IBWSN-518、35th IBWSN-276和20th SAWSN-128、20th SAWSN-129的峰值黏度值较高，在340RVU上。有235份品系的低谷黏度值高于平均水平，占总数的36.2%，其中35thIBWSN-18、35thIBWSN-119、35thIBWSN-518、35thIBWSN-542、35thIBWSN-28和20thSAWSN-128的值较高，在220RUV上，仅35thIBWSN-442的值低于100RUV，为97.3RUV。稀澥值较大（>115RVU）的品系有180份，占品系总数的36.9%，其中35thIBWSN-207、35thIBWSN-5、35thIBWSN-244和20thSAWSN-122、20thSAWSN-99、20thSAWSN-7、20thSAWSN-18稀澥值较大，大于150RVU，而35th IBWSN-468、35th IBWSN-322、35th IBWSN-147和20th SAWSN-138、20th SAWSN-139、20thSAWSN-136、20th SAWSN-42的稀澥值低于80RUV。最终黏度值高于400RVU的品系有244份，占品系总数的50.2%，其中35th IBWSN-127、35th IBWSN-399、35th IBWSN-114、35th IBWSN-17、35thIBWSN-18和20th SAWSN-128的值高于500RUV，仅20th SAWSN-196、20th SAWSN-84和20th SAWSN-81的最终黏度值小于300RVU。反弹值大于平均值的品系有228份，占品系总数的46.7%，其中35th IBWSN-18、35th IBWSN-116、35th IBWSN-399、35th IBWSN-17、35th IBWSN-25、35th IBWSN-118和35th IBWSN-114的反弹值较大，在290RVU以上，仅20th SAWSN-84和20th SAWSN-81的反弹值小于160RVU。品系间峰值时间差异不大，在6.0～7.7min之间，仅35th IBWSN-287的高于7min，达到7.7min。

表2 CIMMYT普通小麦品系淀粉特性

Table 2 Starch properties of common wheat lines from CIMMYT

参数 Parameter	35thIBWSN			20thSAWSN		
	平均值 Mean	变幅 Range	变异系数 *CV*（%）	平均值 Mean	变幅 Range	变异系数 *CV*（%）
峰值黏度PV（RVU）	295.0a	157.3～346.4	7.5	292.0a	142.3～351.0	8.5
低谷黏度HT（RVU）	183.3a	97.3～225.5	8.6	179.5b	139.5～225.8	7.8
稀澥值BD（RVU）	111.5a	69.2～157.9	12.7	113.4a	72.4～152.8	13.5
最终黏度FV（RVU）	409.6a	305.5～516.3	9.2	387.5b	222.4～506.1	8.7
反弹值SB（RVU）	226.4a	162.2～298.1	12.0	209.0b	151.3～280.3	10.6
峰值时间PT（min）	6.3a	6.0～7.7	2.1	6.3a	6.0～6.5	1.7

不同字母表示差异达5%显著水平。下同。

Values followed by a different letter are significantly different at 5% probability level.

PV=peak viscosity; HT=hold trough; BD=breakdown; FV=final viscosity; SB=setback; PT=peak time; *CV*=coefficience of variation; RVU= the unit of rapid visco analyzer (RVA). The same below.

从表 2 可以看出，第 35 届 IBWSN 和第 20 届 SAWSN 品系的淀粉糊化特性中低谷黏度、最终黏度和反弹值间存在显著差异，这可能是由于灌溉地区育种课题和抗旱育种课题的育种目标不同所致。第 35 届 IBWSN 中的品系为水地材料，在亲本选配和后代选择中更重视品质性状，而第 20 届 SAWSN 品系对抗旱性选择更为重视。总体来说，第 35 届 IBWSN 品系的淀粉糊化特性优于第 20 届 SAWSN 品系。

Wx 蛋白亚基正常与 Wx 蛋白亚基缺失品系淀粉糊化特性的比较结果列于表 3。

表 3 Wx 蛋白亚基正常与缺失型品系的淀粉糊化特性比较

Table 3 Comparison of starch pasting properties between normal and null types

类型 Type	品系数 No. of line	参数 Parameter	峰值黏度 PV (RVU)	低谷黏度 HT (RVU)	稀澥值 BD (RVU)	最终黏度 FV (RVU)	反弹值 SB (RVU)	峰值时间 PT (min)
正常小麦 Normal	269	平均 Mean	294.2a	181.6a	112.3a	398.4b	217.1b	6.3a
		范围 Range	224.0～351.0	97.3～225.8	79.4～145.1	222.4～506.1	152.8～289.0	6.0～7.7
缺 B 小麦 Wx-B1 null	191	平均 Mean	294.4a	182.3a	113.1a	406.5a	224.7a	6.3a
		范围 Range	142.3～338.5	145.1～225.5	69.2～157.9	296.8～516.3	151.3～296.8	6.0～6.5
缺 D 小麦 Wx-D1 null	28	平均 Mean	288.9b	184.2a	102.9b	411.9a	227.7a	6.3a
		范围 Range	237.7～331.3	160.5～220.7	72.4～126.5	337.7～502.0	177.1～289.1	6.1～6.4

从表 3 可看出，Wx 蛋白亚基正常品系分别与 Wx-B1 蛋白亚基缺失品系的最终黏度、反弹值间，与 Wx-D1 蛋白亚基缺失品系的峰值黏度、稀澥值、最终黏度、反弹值间存在显著差异。Wx-B1 蛋白亚基缺失品系和 Wx-D1 蛋白亚基缺失品系的峰值黏度、稀澥值间也存在显著差异。与 Wx 蛋白亚基正常品系相比，Wx-B1 蛋白亚基缺失品系的淀粉糊化特性各参数（除峰值时间外）值均略高，Wx-D1 蛋白亚基缺失品系的低谷黏度、最终黏度和反弹值略高，Wx-B1 蛋白亚基缺失品系仅有峰值黏度、稀澥值比 Wx-D1 蛋白亚基缺失品系高。

3 讨论

各国对 Wx 蛋白亚基缺失类型鉴定做了许多工作，Yamamori 等[21]发现朝鲜、日本和土耳其品种中 Wx-A1 蛋白亚基缺失类型频率较高，48%的澳大利亚和印度小麦为 Wx-B1 蛋白亚基缺失类型，仅有 1 个中国农家品种“白火麦”为 Wx-D1 蛋白亚基缺失类型，9 个日本品种同时缺失 Wx-A1 和 Wx-B1 蛋白亚基，没有发现其他缺失类型。美国和加拿大小麦品种中 Wx 蛋白亚基缺失类型数量不多，其中在美国硬红冬品种中以 Wx-A1 和 Wx-B1 蛋白亚基缺失为主要类型[22,23]。姚大年等[24]分析了国内外 429 份小麦品种（系）材料，发现 7 份材料是 Wx-A1 缺失体，29 份是 Wx-B1 缺失体，3 份是 Wx-D1 缺失体（江苏白火麦、内乡白火麦和来自 CIMMYT 的人工小麦 Syntheticas-elite96-97 No. 33），同时缺失 Wx-A1 和 Wx-B1 的有 4 个。王子宁等[25]分析了 900 份河北省地方品种以及引进地方品种，发现 6 份是 Wx-B1 缺失型，仅有 1 份地方品种“白芒白”是 Wx-D1 缺失型。对 260 份中国冬小麦品种的研究表明，Wx-B1 蛋白亚基缺失类型的频率为 14.2%[26]。本研究中 CIMMYT 春小麦 Wx-B1 缺失型的频率很高，与澳大利亚和印度小麦品种相似，与其他国家小麦不同的是 Wx-D1 缺失型出现相对较多，显示了其春小麦种质的独特性。随着中国与 CIMMYT 合作的进一步加强，穿梭育种规模的不断扩大，CIMMYT 种质 Wx 蛋白亚基的特点将为我国小麦育种利用，有助于我国小麦面条品质的改良工作。

已有研究表明[13-16,27,28]，面条品质与小麦淀粉糊化特性有关，应将淀粉品质改良列为育种的重要目标。鉴于我国小麦的淀粉品质较差，与之相比 CIMMYT 春小麦表现为峰值黏度高、低谷黏度和最终黏度高，稀澥值和反弹值大，峰值时间较高且比较稳定，可以用于我国春小麦淀粉品质的改良。此外，本研究发现与 Wx 蛋白亚基正常品系相比，Wx-B1 蛋白亚基缺失品系的淀粉糊化特性各参数值均略高；Wx-D1 蛋白亚基缺失品系的低谷黏度、最终黏度和反弹值略高，其峰值黏度和稀澥值则略低，这与前人的研究结果有所差异[5-8]，其原因可能在于本研究中部分 Wx-D1 蛋白亚基缺失品系的亲本中含有人工合成小麦（如 35th IBWSN-114、35th IBWSN-116、35th IBWSN-289、35th IBWSN-296、20th SAWSN-130、20th

SAWSN-133和20th SAWSN-147)。与正常Wx-D1蛋白亚基缺失品系相比，这部分材料的淀粉糊化特性各参数值均较低（资料未列出），从而间接影响了Wx-D1蛋白亚基缺失品系淀粉糊化特性的表现；同时这也可能与所用材料的数量不同以及Wx-D1蛋白亚基缺失品系不易鉴定有关，相关研究正在进行之中。

参考文献

[1] Igrejas G, Faucher B, Bertrand D, Guibert D, Leroy P, Branlard G. Genetic analysis of the size of endosperm starch granules in a mapped segregating wheat population. *J Cereal Sci*, 2002, 35: 103-107.

[2] Zhao X C, Sharp P J. Production of all eight genotype of null alleles at 'waxy' loci in bread wheat (*Triticum aestivum* L.). *Plant Breed*, 1998, 117: 488-490.

[3] Chao S, Sharp P J, Worland A J, Warham E J, Koebner R M D, Dale M D. RFLP-based genetic maps of wheat homoeologous group 7 chromosomes. *Theor Appl Genet*, 1989, 78: 495-504.

[4] Nakamura T, Yamamori M, Hirano H, Hidaka S. Identification of three Wx proteins in wheat (*Triticum aestivum* L.). *Biochem Genet*, 1993, 31: 75-86.

[5] Yamamori M, Endo T R. Variation of starch granule proteins and chromosome mapping of their coding genes in common wheat. *Theor Appl Genet*, 1996, 93: 275-281.

[6] Miura H, Sugawara A. Dosage effects of the three Wx genes on amylose synthesis in wheat endosperm. *Theor Appl Genet*, 1996, 93: 1066-1070.

[7] Crosbie G. B. The relationship between starch swelling properties, paste viscosity and boiled noodle quality in wheat flours. *J Cereal Sci*, 1991, 131: 45-150.

[8] Yun S H, Quail K. Physicochemical properties of Australian wheat flours for white salted noodles. *J Cereal Sci*, 1996, 23: 181-189.

[9] Miura H, Tanii S. Endosperm starch properties in several wheat vanties preferred for Jappanese noodles. *Euphytica*, 1994, 72: 171-175.

[10] Lee C H, Gore P J, Lee H O. Utilization of Australian wheat for Korean style dried noodle making. *J Cereal Sci*, 1987, 5: 283-297.

[11] Crosbie G B, Lambe W J, Tsutsui H, Gilmour R F. Further evaluation of the flour swelling volume test for identifying wheats potentially suitable for Japanese noodles. *J Cereal Sci*, 1992, 15: 271-280.

[12] Panozzo J F, McCormick K M. The Rapid Viscoanalyser as a method of testing for noodle quality in a wheat breeding programme. *J Cereal Sci*, 1993, 17: 25-32.

[13] Liu J J, He Z H, Zhao Z D, Peña R J, Rajaram S. Wheat quality traits and quality parameters of cooked dry white Chinese noodles. *Euphytica*, 2003, 131: 147-154.

[14] Liu J-J (刘建军), He Z-H (何中虎), Zhao Z-D (赵振东), Liu A-F (刘爱峰), Song J-M (宋建民), Peña R J. Investigation on relationship between wheat quality traits and quality parameters of dry white Chinese noodles (小麦品质性状与干面条品质参数关系的研究). *Acta Agron Sin* (作物学报), 2002, 28 (6): 738-742. (in Chinese with English abstract)

[15] Liu J J (刘建军), He Z-H (何中虎), Yan J (阎俊), Xu Z-H (徐兆华), Liu A-F (刘爱峰), Zhao Z-D (赵振东). Variation of starch properties in wheat cultivars and their relationship with dry white Chinese noodle quality. *Sci Agric Sin* (中国农业科学), 2003, 36 (1): 7-12. (in Chinese with English abstract)

[16] Yan J (杨金), Zhang Y (张艳), He Z-H (何中虎), Yan J (阎俊), Wang D-S (王德森), Liu J-J (刘建军), Wang M-F (王美芳). Association between wheat quality traits and performance of pan bread and dry white Chinese noodle. *Acta Agron Sin* (作物学报), 2004, 30 (8): 739-744. (in Chinese with English abstract)

[17] Wu Z-L (吴振录), Zhang Y (张勇), He Z-H (何中虎), Fan Z-R (樊哲儒), Xin W-L (辛文利), Shao L-G (邵立刚), Li Y-Q (李元清), Yang W-X (杨文雄), Wei Y-Q (魏亦勤), Ma X-G (马晓刚), Pan C (潘超), Liu Y-P (刘艳萍). Performance on yield and quality of CIMMYT wheat in China. *J Triticeae Crops* (麦类作物学报), 2004, 24 (3): 34-39. (in Chinese with English abstract)

[18] Yasui T, Sasaki T, Matsuki J. Starch properties of a bread wheat (*Triticum aestivum* L.) mutant with an altered flour-pasting profile. *J Cereal Sci*, 2002, 35: 11-16.

[19] Crosbie G B, Ross A S, Moro T, Chiu P C. Starch and protein quality requirements of Japanese Alkaline noodles (ramen). *Cereal Chem*, 1999, 3: 328-334.

[20] Zhao X C, Sharp P J. An improved 1D-SDS-PAGE method for the identification of three bread wheat Waxy proteins. *J Cereal Sci*, 1996, 23: 191-193.

[21] Yamamori M, Nakamura T, Endo T R, Nagamine

T. Waxy protein deficiency and chromosomal location of coding genes in common wheat. *Theor Appl Genet*, 1994, 89: 179-184.

[22] Demeke T, Hucl P, Nair R B, Nakamuru T, Chibbar R N. Evaluation of Canadian and other wheats for waxy proteins. *Cereal Chem*, 1997, 4: 442-444.

[23] Graybosch R A, Peterson C J, Hansen L E, Rahman S, Hill A, Skerritt J N. Identification and characterization of US wheats carrying null alleles at the Wx loci. *Cereal Chem*, 1998, 1: 162-165.

[24] Yao D-N（姚大年）, Wang X-W（王新望）, Liu Z-Y（刘志勇）, Liu G-T（刘广田）. Identification and screen of waxy protein in wheat. *J Agric Biotech*（农业生物技术学报）, 1999, 7 (1): 1-9. (in Chinese with English abstract)

[25] Wang Z-N（王子宁）, Guo B-H（郭北海）, Zhang Y-M（张艳敏）, Wen Z-Y（温之雨）, Li H-J（李洪杰）, Shi Y-S（石云素）. Discovery and analysis of wheat cultivar (*T. aestivum*) with *Wx* genes. *Acta Agric Boreali-Sin*（华北农学报）, 1999, 14 (3): 5-9. (in Chinese with English abstract)

[26] Xu Z-H（徐兆华）, Zhan Y（张艳）, Xia L-Q（夏兰芹）, Xia X-C（夏先春）, He Z-H（何中虎）. Genetic variation of starch properties in Chinese winter wheats. *Acta Agron Sin*（作物学报）, 2005, 31 (5): 587-591. (in Chinese with English abstract)

[27] Zhang Y（张勇）, He Z-H（何中虎）. Investigation on paste property of spring-sown Chinese wheat. *Sci Agric Sin*（中国农业科学）, 2002, 35 (5): 471-475. (in Chinese with English abstract)

[28] Yan J（阎俊）, Zhang Y（张勇）, He Z-H（何中虎）. Investigation on starch pasting properties of wheat varities. *Sci Agric Sin*（中国农业科学）, 2001, 34 (1): 9-13. (in Chinese with English abstract)

成株抗性种质筛选鉴定

Seedling and adult-plant resistance to powdery mildew in Chinese bread wheat cultivars and lines

Z. L. Wang[1,3], L. H. Li[1], Z. H. He[1,2], X. Y. Duan[4], Y. L. Zhou[4], X. M. Chen[1], M. Lillemo[5], R. P. Singh[5], H. Wang[3], and X. C. Xia[1]

[1] *Institute of Crop Breeding and Cultivation/National Wheat Improvement Center, Chinese Academy of Agricultural Sciences (CAAS), No.* 12 *Zhongguancun South Street*, 100081, *Beijing, China*; [2] *CIMMYT China Office, C/O CAAS, No.* 12 *Zhongguancun South Street*, 100081, *Beijing, China*; [3] *Northwest Sci-Tech University of Agriculture and Forestry, Yangling*, 712100, *Shaanxi, China*; [4] *Institute of Plant Protection, CAAS, No.* 2 *Yuanmingyuan West Road*, 100094, *Beijing, China*; [5] *International Maize and Wheat Improvement Center (CIMMYT), Apdo. Postal* 6-641, 06600, *Mexico D. F., Mexico.*

Abstract: Powdery mildew, caused by *Blumeria graminis* f. sp. *tritici*, is a widespread wheat disease in China. Identification of race-specific genes and adult-plant resistance (APR) is of major importance in breeding for an efficient genetic control strategy. The objectives of this study were to (i) identify genes that confer seedling resistance to powdery mildew in Chinese bread wheat cultivars and introductions used by breeding programs in China, and (ii) evaluate their APR in the field. The results showed that: (i) 98 out of 192 tested wheat cultivars and lines were postulated to have one or more resistance genes to powdery mildew; (ii) *Pm8* and *Pm4b* are the most common resistance genes in Chinese wheat cultivars, while *Pm8* and *Pm3d* are present most frequently in wheat cultivars introduced from CIMMYT, the US, and European countries; (iii) genotypes carrying *Pm1*, *Pm3e*, *Pm5*, and *Pm7* were susceptible, whereas those carrying *Pm12*, *Pm16*, and *Pm20* were highly resistant to almost all powdery mildew isolates of *B. graminis* f. sp. *tritici* tested; and (iv) 22 genotypes expressed APR. Our data showed that the area under the disease progress curve (AUDPC), maximum disease severity (MDS) on penultimate leaves, and the disease index are good indicators of the degree of APR in the field. It may be a good choice to combine major resistance genes and APR genes in wheat breeding to obtain effective resistance to powdery mildew.

Key words: common wheat, powdery mildew, *Blumeria graminis* f. sp. *tritici*, gene-for-gene specificity, adult-plant resistance

Introduction

Powdery mildew, caused by *Blumeria graminis* f. sp. *tritici*, is an important disease of wheat (*Triticum aestivum*) worldwide, especially in highly productive areas with a maritime or semi-continental climate[1]. Losses in grain yield may be as high as 12 to 45% in the US[8,18,19,21]. In China, powdery mildew on wheat occurred occasionally in the southwestern plateau and coastal area of Shandong Province before the 1970's[23,31]. However, its importance has increased notably from the early 1980's, mainly due to the high rates of nitrogenous fertilizer, production of semidwarf wheat cultivars, and expansion of irrigated areas. Wheat area affected by powdery mildew was 2.9 million ha in 1981

Published in Plant Disease, 2005, 89: 457-463

and increased rapidly to 12.0 million ha, causing grain yield losses of 14.4 million tons in 1990[33].

Utilization of resistant cultivars is the most economical and environmentally safe means of controlling powdery mildew[1]. To date, 48 *Pm* genes/alleles at 32 loci are known to confer race-specific resistance in wheat[14,16]. Selection pressure exerted by cultivars that carry single race-specific genes results in a rapid build-up of the pathogen population possessing matching virulence genes and contributes to reducing resistance durability. Knowledge of the presence and frequency of virulence genes in the pathogen population is, therefore, useful for replacing ineffective genes with effective ones. Information on resistance genes present in important wheat cultivars and knowledge of avirulence/virulence in pathogen populations is valuable for planning crosses in a wheat breeding program, and for developing management strategies including cultivar recommendations and deployment schemes[15,19,34]. Such information is also valuable to predict the breakdown of resistance genes caused by shifting pathogen populations.

Resistance to powdery mildew that retards infection, growth, and reproduction of the pathogen in adult plants, but not in seedlings, has been defined as slow mildewing[28] or adult-plant resistance (APR)[5]. This type of resistance can be identified in cultivars with defeated race-specific genes or cultivars that lack race-specific resistance genes[1]. APR to powdery mildew is more durable than race-specific resistance. For example, cv. Knox and its derivatives (e.g., cv. Massey) have remained resistant to powdery mildew infection for over 20 years, despite being grown commercially on large areas[9,28]. Some wheat breeders are interested in using APR genes, and many genotypes with APR have been identified and used in wheat improvement[4,9,36].

China is the largest wheat producer in the world, with about 25 million ha sown annually. The Chinese Academy of Agricultural Sciences has divided the country's wheat area into 10 agro-ecological zones based on wheat types and the response to biotic and abiotic stresses. Winter, facultative, and spring wheat types sown in both autumn and spring are grown, and Chinese wheat germplasm is, therefore, very diverse in many aspects. However, little attention has been paid to investigating APR to powdery mildew and achieving durable resistance. The objectives of our study were: (i) to identify probable *Pm* genes that confer seedling resistance in Chinese wheat cultivars and various introductions frequently used by breeding programs in China, and (ii) to identify the presence of adult-plant resistance in these cultivars based on field trials.

Materials and methods

Isolates of *B. graminis* f. sp. *tritici*

Twenty isolates of *B. graminis* f. sp. *tritici* (Table 1) with different virulence patterns[7], collected from various regions of China, were used for identifying genes conferring resistance at the seedling stage. Isolate E20, with a broad spectrum of virulence, was used for evaluating APR to powdery mildew in field trials.

Table 1 Reaction patterns of 20 differential wheat cultivars and lines with known resistance genes to 20 isolates of *Blumeria graminis* f. sp. *tritici* collected from various regions of China

Cultivar or line	Resistance gene[a]	Infection types to *B. graminis* f. sp. *tritici* isolates[b]																			
		E01	E02	E03	E05	E06	E07	E10	E13	E15	E16	E17	E18	E20	E21	E23	E26	E30	E31	E32	E42
Chancellor	None	4	4	4	4	4	4	4	4	4	4	4	4	4	4	4	4	4	4	4	4
Axminster/8 * CC	*Pm1*	4	4	4	4	4	4	4	4	4	4	4	4	4	4	4	4	4	4	4	4
Ulka/8 * CC	*Pm2*	0;	4	0;	4	0;	0;	0;	3	0;	0;	4	4	4	4	1	0;	0;	3	4	4
Maris Huntsman	*Pm2+6*	0;	0;	0;	0;	0;	0;	0;	4	0;	—	0;	4	4	3	0;	0;	0;	0;	4	0;
Maris Dove	*Pm2+Mld*	0;	0;	0;	0;	0;	0;	0;	4	0;	0	0;	0;	3	4	0	0;	0;	0;	4	0;
Brock	*Pm2+Ta*	0;	4	0;	4	0;	0;	0;	0;	2	2	4	0;	0;	4	0;	0;	0;	0;	0;	4

（续）

Cultivar or line	Resistance gene[a]	Infection types to *B. graminis* f. sp. *tritici* isolates[b]																			
		E01	E02	E03	E05	E06	E07	E10	E13	E15	E16	E17	E18	E20	E21	E23	E26	E30	E31	E32	E42
Asosan/8 * CC	*Pm3a*	0;	3	0;	4	3	4	4	4	4	3	0;	3	4	4	4	4	4	4	4	4
Kolibrí	*Pm3d*	4	4	4	4	4	4	4	4	4	4	0;	3	4	4	4	2	4	4	4	4
W150	*Pm3e*	4	4	4	4	4	4	4	4	4	4	4	4	4	4	4	4	4	4	4	4
Michigen Amber / 8 * CC	*Pm3f*	4	4	3	0;	4	4	4	0;	4	3	4	4	4	4	0;	4	4	4	4	4
Khapli/8 * CC	*Pm4a*	0;	4	0	0	1	0;	4	0	4	4	4	4	4	4	1	0;	1	0	0;	0
Armada	*Pm4b*	0;	0;	0;	0;	0;	0;	2	1	0;	4	0;	4	4	4	0;	3	0;	0;	0;	0;
Baimian3	*Pm4+8*	0;	4	0;	0;	0;	0;	4	0;	0;	4	3	4	4	4	0;	0;	0;	0;	0;	0;
Hope	*Pm5*	4	4	4	4	4	4	4	4	4	4	4	4	4	4	4	4	4	4	4	4
Timgalen	*Pm6*	4	3	4	4	4	4	3	4	0;	3	3	4	4	3	4	4	4	3	3	3
CI 14189	*Pm7*	4	4	4	4	4	4	4	4	4	4	4	4	4	4	4	4	4	4	4	4
Kavkaz	*Pm8*	4	4	4	4	4	4	4	4	4	4	4	4	4	4	0;	4	4	4	0;	4
Wembley	*Pm12*	0;	0;	0;	0;	0;	0;	0;	0;	0;	0;	0;	0;	0;	0;	0;	0;	0;	0;	0;	0;
Brigrand	*Pm16*	0;	0;	0;	0;	1	0;	0;	0;	0;	0	0;	0;	2	0;	0;	0;	0;	0	0;	0
Amigo	*Pm17*	4	4	4	4	3	4	3	4	0;	4	0;	4	4	4	0;	4	4	3	0;	4
Pm20	*Pm20*	0;	0;	0;	0;	0;	0;	0;	3	0;	0;	0;	0;	2	4	0;	2	0;	0	0;	0;

a. Information obtained from Huang and Röder[16] and McIntosh et al.[24].

b. Infection types are: 0=no visible symptoms, 0; =hypersensitive necrotic flecks, 1=tiny colonies with little sporulations, 2=colonies with moderately developed hyphae and little conidia, 3=colonies with well developed hyphae and conidia, but colonies not joined together, and 4=colonies with well developed hyphae and conidia, and colonies mostly joined together.

Wheat germplasm

Included in this study were 56 widely grown Chinese cultivars and 28 advanced breeding lines, as well as 108 cultivars and lines introduced from Europe, USA, Brazil, and the International Maize and Wheat Improvement Center (CIMMYT) in Mexico (Tables 2 and 3). The cultivars and lines are available at the National Wheat Improvement Center of China upon request. Various wheat breeding programs in China currently use the introduced germplasm. Twenty differential lines (Table 1) with known *Pm* genes[16,24] were included in each test, and the low and high infection types displayed by them toward *B. graminis* f. sp. *tritici* isolates were used for postulating resistance genes in the wheat cultivars tested.

Table 2 Origin, growth habit, and 1B/1R translocation status of 94 cultivars and lines that were susceptible to all 20 *Blumeria graminis* f. sp. *tritici* isolates

Cultivar or line	Origin	Growth habit[a]	1B/1R status[b]
Bainong 64	China	W	1B/1B
Baofeng 7228	China	W	1B/1R
CA 9532	China	W	1B/1R
Hongjuanmang	China	W	1B/1B
Hobbit/CA837	China	W	/
Ji 966185	China	W	/
Jimai 30	China	W	1B/1R
Jimai 36	China	W	1B/1R
Jimai 38	China	W	1B/1R

（续）

Cultivar or line	Origin	Growth habit[a]	1B/1R status[b]
Jin 411	China	W	1B/1B
Jinan 17	China	W	1B/1B
Jingdong 6	China	W	1B/1R
Jingdong 8	China	W	1B/1R
Jinmai 45	China	W	1B/1R
Jinmai 47	China	W	/
Linfen 6010	China	W	/
Lumai 21	China	W	1B/1B
Lumai 22	China	W	1B/1B
Lumai 23	China	W	1B/1B
Lunhui 201	China	W	1B/1R
Nongda123	China	W	1B/1B
Pm 20491	China	W	1B/1B
Pm 499-1	China	W	1B/1B
Pm94081	China	W	/
Pm94082	China	W	/
Shan 213	China	W	1B/1R
Shan 354	China	W	1B/1R
Shangdong 935031	China	W	1B/1B
Shannong413863	China	W	1B/1R
Tiegangmai	China	W	1B/1B
Wenmai 6	China	W	1B/1B
Xiaoyan 6	China	W	1B/1B
Xingmai 99	China	W	1B/1R
Yumai 13	China	W	1B/1R
Yumai 18	China	W	1B/1B
Yumai 21	China	W	1B/1R
Yumai 25	China	W	1B/1B
Yumai 47	China	W	1B/1B
Yuzhan 9705	China	W	/
Zhengmai 9405	China	W	1B/1B
Zhongyou 9507	China	W	1B/1B
Zhongyou 9701	China	W	1B/1B
Zhongyou 9843	China	W	1B/1B
Zhongyou 9844	China	W	1B/1B
Zhoumai 11	China	W	1B/1R
Zhoumai 16	China	W	1B/1R
Zhoumai 17	China	W	1B/1R
24Ibwsn#278/Soissons	CIMMYT	W	1B/1B
Bacanora T88	CIMMYT	S	1B/1R
Batera	CIMMYT	W	/
Bow/Nkt//Ducula/3/Ducula	CIMMYT	S	1B/1B
Emb27/Cep8825//Milan-1	CIMMYT	S	1B/1B
Emb27/Cep8825//Milan-2	CIMMYT	S	1B/1B

（续）

Cultivar or line	Origin	Growth habit[a]	1B/1R status[b]
F12. 71/Bez//Tjb368. 251/Buc/4/D6301/Heine VII//ERA/3/BUC	CIMMYT	W	1B/1B
Filin	CIMMYT	S	1B/1R
Filin/Milan-1	CIMMYT	S	1B/1B
Filin/Milan-2	CIMMYT	S	1B/1B
Gen/Kauz	CIMMYT	S	1B/1R
Hxl 8088/Ducula	CIMMYT	S	/
Inqalab 91	CIMMYT	S	1B/1B
Milan	CIMMYT	S	1B/1B
Milan/Pastor	CIMMYT	S	1B/1B
Saar	CIMMYT	S	1B/1B
Sha3/Seri//Sha4/Lira	CIMMYT	S	1B/1R
Thb//Maya/Nac/3/Rabe/4/Milan	CIMMYT	S	1B/1B
Ures/Kauz	CIMMYT	S	1B/1R
Dulus	CIMMYT	S	1B/1B
Heilo	CIMMYT	S	1B/1B
Prointa Guazu	Argentina	S	1B/1R
Prointa Cauquen	Argentina	S	1B/1B
110254-1. 77W	France	W	/
Cp9106-1-2-1	France	W	/
Cp9111-3-2	France	W	/
Pistov-2	France	W	/
Alidos	Germany	W	/
Andros	Germany	W	1B/1B
Mv Emma	Hungary	W	1B/1R[c]
Mv Magdalena	Hungary	W	1B/1R
Mv Magvas	Hungary	W	1B/1B
Mv Palma	Hungary	W	1B/1R[c]
Mv Vilma	Hungary	W	/
Dropia	Romania	W	/
Arina	Switzerland	W	1B/1B
Kariega	South Africa	S	1B/1B
Manning/SDV1//DOGU88	Turkey	W	1B/1B
Abn/Jun	USA	W	1B/1B
Ga841465-2-1-1-4	USA	W	/
Ga85164-3-2-4-2	USA	W	/
Houser	USA	W	1B/1B
Ks90175-1-2	USA	W	1B/1R
Redcoat	USA	W	1B/1B
Mercia	UK	W	1B/1B
FAP74161	/	W	1B/1B
FAP77517	/	W	1B/1R

a. W=winter wheat; S=spring wheat.

b. Identified according to the method described by Gupta and Shepherd[12], except for those with superscripted c.

c. Information obtained from Schlegel[27].

/=unknown.

Table 3 Reactions of 98 wheat cultivars or lines to 20 isolates of *Blumeria graminis* f. sp. *tritici* in seedling tests

Cultivar or line	Origin	Growth habit[a]	1B/1R status[b]	Resistance gene	Infection types to *B. graminis* f. sp. *tritici* isolates[d]																			
					E01	E02	E03	E05	E06	E07	E10	E13	E15	E16	E17	E18	E20	E21	E23	E26	E30	E31	E32	E42
98301	China	W	1B/1R	*Pm8*+?	4	4	4	4	4	4	4	4	4	4	4	4	4	3	1	4	0	4	0	4
975297	China	W	1B/1R	?	4	4	4	4	4	4	4	4	4	4	4	4	4	4	3	4	4	4	0	4
98Zhong 33	China	W	1B/1B	?	0;	0	4	4	0;	0;	4	0	0	4	0;	4	4	4	0	0	0	0	0	4
Aiyuandong 3	China	W	1B/1R	*Pm8*	4	4	4	4	4	4	4	4	4	4	4	4	4	4	0	4	3	4	0	4
Aizao 4110	China	W	1B/1B	*Pm2*+*4b*	0;	0	0	0;	0;	0;	0;	1	0	2	0;	4	4	4	0;	0;	0;	0	0	0
BPM 14	China	W	1B/1B	*Pm4b*	0;	0;	0	0;	2	0;	0	0	0	4	0	4	3	4	0;	4	2	0;	0	0
BPM 15	China	W	1B/1B	*Pm2*	0	4	0	4	0;	0;	1	3	0	0	4	4	4	4	0	0	0	3	4	4
BPM 28	China	W	1B/1B	*Pm4b*	0	0	0	0	2	0;	0	0	0	4	0;	3	4	4	1	4	2	0	0	0
CA 8646	China	W	1B/1R	*Pm8*	4	4	4	4	4	4	4	4	4	4	4	4	4	4	0	4	3	4	0	4
CA 9550	China	W	1B/1B	*Pm2*+*4b*	0	0	0;	0	0	0	0;	0	0	0	0	4	4	4	0;	0	0;	0	0;	0
CA 9553	China	W	1B/1B	*Pm4b*	0;	0;	0;	0	0;	0	1	1	0	3	0	4	3	4	0	4	0	0	0	0
CA 9632	China	W	1B/1R	*Pm8*	4	4	4	4	4	4	4	4	4	4	4	4	4	4	1	4	3	4	0	4
CA 9641	China	W	1B/1B	*Pm3a*+?	0	0	0	0;	4	0	0	4	0	0	0	4	4	4	0;	1	0	0	4	0
CA 8686	China	W	1B/1R	?	4	3	4	4	4	4	4	4	4	4	4	4	4	4	2	4	4	4	4	4
CA 9722	China	W	1B/1B	?	0	4	0	0;	3	0	4	4	4	4	4	4	4	4	4	3	2	0	0	0
Jimai 3	China	W	1B/1B	?	4	4	4	4	4	0	4	4	4	4	4	4	4	4	4	4	0	4	0	4
Jinnong 8112	China	W	1B/1R	*Pm8*	4	4	4	4	4	4	4	4	4	4	4	4	4	4	1	4	3	4	0	4
Lankao 906	China	W	1B/1B	*Pm4b*	0	0	0;	0	2	0;	0;	0;	0;	4	0;	4	4	4	0	4	2	0;	0;	0;
Lin 9303	China	W	1B/1R	?	4	4	4	4	4	4	4	4	4	4	4	4	4	4	4	4	2	4	0	4
Lu7you/2/C39	China	W	1B/1B	*Pm4b*	0;	0;	0;	2	0;	0;	0	0;	0;	3	0	4	4	4	0;	3	0;	0;	0;	0;
Lumai 13	China	W	1B/1R	*Pm8*	4	4	4	4	4	4	4	4	4	4	4	4	4	4	1	4	3	4	0	4
Lumai 14	China	W	1B/1R	*Pm8*+?	4	4	4	4	4	4	4	4	4	4	4	4	4	4	1	4	0	4	0	4
Neixiang 991	China	W	1B/1R	*Pm2*+*6*+?	—	0;	0;	0	0;	0;	0;	0;	0;	4	0;	0;	4	4	4	0;	0;	0;	0;	0;
Neixiang 188	China	W	1B/1R	?	4	4	4	4	4	4	4	4	4	0	4	4	4	4	4	4	4	4	4	4
Nongda 280218	China	W	1B/1B	*Pm4a*	0;	4	0;	0;	1	0;	3	0;	4	4	3	3	4	4	0;	0;	1	1	0;	0
Nongda 523	China	W	1B/1R	*Pm8*	4	4	4	4	4	4	4	4	4	4	4	4	4	4	0;	4	3	4	0	4
Nongda 883	China	W	/	*Pm4a*	0;	4	0;	0;	2	0;	3	0;	4	4	4	4	3	4	0;	0;	1	2	0;	0
Nongdayoufan 5	China	W	/	*Pm4b*	0;	0;	0;	2	0;	0;	0	2	0;	3	0	4	4	4	0;	3	0;	0;	0;	0;

（续）

Cultivar or line	Origin	Growth habit[a]	1B/1R status[b]	Resistance gene	Infection types to *B. graminis* f. sp. *tritici* isolates[d]																			
					E01	E02	E03	E05	E06	E07	E10	E13	E15	E16	E17	E18	E20	E21	E23	E26	E30	E31	E32	E42
Qiandong 6	China	W	/	*Pm2*+Ta	2	4	0;	4	0;	0;	2	0	0	2	4	0;	0	4	0	0	2	0;	0;	4
Pm93-625-4	China	W	1B/1B	*Pm30*	0	0;	0	0;	0;	0;	0;	0;	0	0;	0;	0;	0	0	0	0;	0;	0;	0;	0;
Shan 229	China	W	1B/1R	*Pm3d*	4	4	4	4	4	4	4	4	4	4	0;	3	4	4	4	0;	4	4	4	4
Shangdong 20015537	China	W	1B/1R	*Pm8*+?	4	4	4	4	4	4	4	4	4	4	4	4	4	4	1	4	0;	4	0	4
Shangdong 418	China	W	1B/1R	?	4	4	0	4	4	4	4	4	4	3	4	4	4	4	4	4	4	4	4	4
Xingmai 883	China	W	/	*Pm6*	4	4	4	4	4	4	4	4	0;	4	4	4	4	4	4	4	4	4	4	4
YE-2416-7a	China	W	1B/1B	?	4	0	4	4	4	3	4	4	0	4	0;		4	4	0;	3	1	0	0	0
Zheng 86115	China	W	1B/1B	*Pm3d*+?	4	3	0	3	4	4	4	4	4	4	2	3	4	4	4	3	4	4	4	4
Zhoumai 13	China	W	1B/1R	*Pm8*	4	4	4	4	4	4	4	4	4	4	4	4	4	4	0;	4	3	4	0;	4
Zhoumai 88	China	W	1B/1B	?	4	4	4	4	4	4	4	4	4	4	4	4	4	4	4	4	3	4	0;	4
494J6. 11	CIMMYT	W	1B/1B	?	4	0;	4	4	4	4	4	4	4	4	4	4	4	4	4	4	4	0;	4	4
8030 Versailles/Edch//Cd	CIMMYT	W	1B/1R	*Pm3d*+?	4	0	4	4	3	2	4	4	4	4	0	3	4	4	4	2	4	3	4	4
Altar 84/Ae. squarrosa(219)//2 * Seri	CIMMYT	S	1B/1R	*Pm8*+?	4	0;	0;	4	4	4	4	4	0;	4	0;	0;	0	4	0;	4	4	0	0	4
Au/Up301//Gll/Sx/3/Pew/4/Mai/Maya//Pew/5/Kea/6/2*R37/Ghl121//Kal/Bb/3/Buc/Bul	CIMMYT	S	1B/1B	?	4	0;	4	4	4	4	4	4	4	4	4	4	4	4	4	4	4	4	4	4
Catbird	CIMMYT	S	1B/1R	?	4	0;	4	4	4	4	4	4	4	4	4	4	4	4	4	4	4	3	4	4
Croc_1/Ae. squarrosa(205)//Kauz	CIMMYT	S	1B/1R	?	4	4	4	4	2	4	4	4	4	4	4	4	4	4	4	4	4	4	4	4
Choix M95	CIMMYT	S	1B/1R	*Pm8*+?	4	4	4	4	4	4	4	4	4	4	4	4	4	4	1	4	0;	4	0	4
Corydon	CIMMYT	S	1B/1R	*Pm8*	4	4	4	4	4	4	4	4	4	4	4	4	4	4	1	4	3	4	0;	4
Desconocido	CIMMYT	S	1B/1R	?	4	4	0;	4	4	4	4	4	4	4	4	4	4	4	4	4	4	4	4	4
Ducula	CIMMYT	S	1B/1B	?	4	4	4	4	4	4	4	4	4	4	4	0;	4	4	4	4	4	4	4	4
Hahn/2 * Weaver	CIMMYT	S	1B/1R	*Pm8*+?	4	4	4	4	4	4	4	4	4	4	4	4	4	4	0;	4	0;	4	0	4
Irena/Weaver	CIMMYT	S	1B/1R	*Pm8*+?	4	4	4	4	4	4	4	4	4	4	4	4	4	4	2	4	0	4	0;	4
Ketupa	CIMMYT	S	1B/1R	?	4	0	0;	4	4	4	4	4	0	4	0;	4	4	3	4	4	4	0	4	4
Kukuna	CIMMYT	S	1B/1R	*Pm8*	4	3	4	4	4	4	4	4	4	4	4	4	4	4	1	4	3	4	0	4
Lfn/Ii58. 57//Prl/3/Hahn/4/ 2 * Mo88	CIMMYT	S	1B/1R	*Pm8*+?	4	4	4	4	4	4	4	4	4	4	4	4	4	4	0	4	0	4	0	4

（续）

Cultivar or line	Origin	Growth habit[a]	1B/1R status[b]	Resistance gene	Infection types to *B. graminis* f. sp. *tritici* isolates[d]																			
					E01	E02	E03	E05	E06	E07	E10	E13	E15	E16	E17	E18	E20	E21	E23	E26	E30	E31	E32	E42
Milan/Ducula	CIMMYT	S	1B/1B	*Pm3d*	4	4	4	4	4	4	4	4	4	4	0	4	4	4	4	2	4	4	4	4
Mon/Tmu//Ald/Pvn	CIMMYT	S	1B/1R	*Pm8*+?	4	4	4	4	4	4	4	4	4	4	4	4	4	4	1	4	0;	4	0	4
Nadadores M63	CIMMYT	S	1B/1B	?	4	4	0	4	4	4	4	4	4	4	4	0;	4	4	4	4	4	4	0	4
Otus	CIMMYT	S	1B/1R	?	4	3	4	4	4	4	4	4	4	4	4	4	4	4	0	4	4	4	4	4
Pastor	CIMMYT	S	1B/1B	?	4	0	2	4	4	4	4	4	4	4	4	4	4	4	4	4	3	3	4	4
Pbw343	CIMMYT	S	1B/1R	?	4	0	0	4	4	4	4	4	4	4	4	4	4	4	4	4	4	0	4	4
Prinia	CIMMYT	S	1B/1R	*Pm8*+6+?	4	0	0	4	4	4	4	4	0;	4	3	0	3	4	0;	4	0;	0	0;	4
Seri/Attila	CIMMYT	S	1B/1R	?	4	0;	0	4	4	4	4	4	4	4	4	4	4	4	4	4	4	3	4	4
Seri/Rayon	CIMMYT	S	1B/1B	*Pm3d*+?	4	0;	0;	4	4	4	4	4	4	4	0	0	4	4	4	1	4	3	4	4
Tinamou	CIMMYT	S	1B/1R	*Pm8*+6+?	4	0;	0;	4	4	4	4	4	0	4	4	4	0	4	0;	4	0;	0	0;	4
Tnmu/Milan-1	CIMMYT	S	1B/1R	*Pm8*+?	4	0;	4	4	4	4	4	4	0	4	0;	0;	4	4	0	4	0	0	0	4
Tnmu/Milan-2	CIMMYT	S	1B/1R	*Pm3d*+?	4	0;	0	4	4	4	4	4	4	4	0;	4	4	4	1	2	1	3	4	4
Tnmu/Milan-3	CIMMYT	S	1B/1R	*Pm8*+?	4	3	0	4	4	4	4	4	4	4	4	4	4	4	0;	4	0;	3	0;	4
Tnmu/Milan-4	CIMMYT	S	1B/1R	*Pm8*+?	4	4	4	4	4	4	4	4	4	4	4	4	4	4	0	4	0	4	0	4
Tx71A1039. V1 * 3/Ami//Trap #1	CIMMYT	W	1B/1R	*Pm3d*+6+?	4	0;	0	4	4	4	4	4	0;	4	0	3	3	3	0	2	4	0	4	4
Tx71A1039. V1 * 3/Ami//Turaco	CIMMYT	W	1B/1B	?	4	0;	4	4	4	4	0	4	0	4	2	3	4	4	0;	4	3	0	0	0
Vee/Koel//Hel/3 * Cno79/3/Kauz	CIMMYT	S	1B/1R	*Pm8*+*3d*+?	4	0;	0;	4	4	4	4	4	4	4	0;	3	0;	4	0;	2	4	3	0	4
Yang87-158/Ducula	CIMMYT	S	1B/1B	?	4	0;	4	4	4	4	4	4	4	4	4	0;	4	4	4	4	4	3	4	4
Br23/Emb40	Brazil	S	1B/1R	*Pm8*+?	4	0;	0;	4	4	4	4	4	4	4	4	0;	3	4	0;	4	0;	0;	0;	4
Chat/Cep7780//Prl/Bow	Brazil	S	1B/1R	?	4	4	4	4	4	4	4	4	4	4	4	2	4	4	4	4	4	4	0;	4
Embrapa 119	Brazil	S	/	?	4	4	0	4	4	4	4	4	4	0	4	3	4	4	4	4	4	3	4	4
Jur/2 * Iapar29//Pf83144	Brazil	S	1B/1R	*Pm8*	4	4	4	4	4	4	4	4	4	4	4	4	4	4	0	4	3	4	0	4
Nd610/Kauz//Tui	Brazil	S	1B/1R	*Pm8*+?	0;	0	0;	0	0	0	0;	4	0;	3	0	4	4	4	0;	0	0	3	0	0
Stozher	Bulgaria	W	1B/1B	?	4	0;	4	3	4	4	4	4	0	4	0;	3	4	4	2	4	4	3	0	3
Soissons-1	France	W	1B/1B[c]	?	4	0;	4	3	4	3	4	4	3	4	4	4	4	4	4	4	3	3	4	4
Soissons-2	France	W	1B/1B[c]	*Pm3d*+?	4	0;	4	4	4	3	0	4	3	4	0	4	4	3	4	2	4	3	4	4
Soissons-3	France	W	1B/1B[c]	*Pm3d*+?	4	0	4	3	4	3	4	4	3	4	0;	4	4	4	4	2	4	3	4	4

（续）

Cultivar or line	Origin	Growth habit[a]	1B/1R status[b]	Resistance gene	Infection types to *B. graminis* f. sp. *tritici* isolates[d]																			
					E01	E02	E03	E05	E06	E07	E10	E13	E15	E16	E17	E18	E20	E21	E23	E26	E30	E31	E32	E42
Contra	Germany	W	1B/1B[c]	*Pm4b*+?	0	0;	0	—	0;	0;	0	0;	0;	0	0;	4	4	4	0;	0;	0;	0;	0	0
German 8661	Germany	W	/	*Pm4b*	1	0;	0;	2	0;	0;	0	0;	1	3	0	4	4	4	0;	3	0;	0;	0;	0;
Mv Madrigal	Hungary	W	1B/1R	*Pm8*+?	4	4	4	4	4	4	4	4	4	4	4	4	4	4	2	4	1	4	0	4
Mv Magdalena	Hungary	W	1B/1R	*Pm8*+?	4	4	4	4	4	4	4	4	4	4	4	4	4	4	1	4	1	4	0	4
MvOptima	Hungary	W	1B/1R	*Pm8*	4	4	4	4	4	4	4	4	4	4	4	4	4	4	1	4	4	4	0	4
Mv-17	Hungary	W	/	?	0;	0;	0;	0;	0;	0;	0;	0;	0	0;	0;	0;	0	0	0	0;	0;	0;	0;	0;
Mv-23	Hungary	W	1B/1R	?	4	4	4	4	4	4	4	4	4	4	3	4	4	4	1	4	4	4	3	4
Campion	Romania	W	1B/1B	*Pm3a*+6+?	0;	0;	0;	0;	4	4	3	4	0;	3	0;	0;	0;	3	2	4	4	0;	3	4
Rah122/94	Romania	W	/	*Pm4a*	0;	4	0;	0;	0;	0;	3	0;	4	4	4	4	3	4	0;	0;	1	2	0;	0
Lona	Switzerland	S	1B/1B[c]	?	4	4	4	4	4	0	4	4	4	3	4	4	4	4	0	4	4	4	4	4
Hussar	UK	S	1B/1R[c]	*Pm8*+?	0;	0;	4	0	0	0	0;	0;	0;	0;	0	4	4	4	0;	0;	0;	0;	0	0;
Cty * 3/Ta2450	USA	W	1B/1R	*Pm8*+?	4	0;	3	3	4	4	0	4	0;	4	2	3	3	3	0;	1	2	0;	0;	0;
Massey	USA	W	1B/1B	?	4	0;	4	3	4	4	4	4	4	0	4	4	4	3	4	3	4	3	4	0;
Va91-51-26	USA	W	/	?	0	0	0	0;	0;	0;	0;	0;	0	0;	0;	0;	0	0	0	0;	0;	0;	0;	0;
Va91-52-65	USA	W	/	?	0;	0;	0;	0	0	0;	0	0;	0	0;	0;	0;	0	0	0	0;	0;	0;	0;	0;
90-I145	/	W	/	?	4	0	4	4	4	4	4	4	4	4	3	4	2	4	0	4	3	0	3	4
90-I89	/	W	/	*Pm3d*	4	4	4	4	4	4	4	4	4	4	0;	4	4	4	4	0;	4	4	4	4
Sxafa4-3	/	W	1B/1B	*Pm3f*+?	4	4	0;	0	0;	0;	4	0;	4	4	3	4	4	4	0	0	0	0	4	0;

a. W = winter wheat, S = spring wheat.

b. Identified according to the method described by Gupta and Shepherd[12] except for those with superscripted c.

c. Information obtained from Kazman and Lein[17].

d. Infection types are: 0 = no visible symptom, 0; = hypersensitive necrotic flecks, 1 = tiny colonies with little sporulations, 2 = colonies with moderately developed hyphae and little conidia, 3 = colonies with well developed hyphae and conidia, but colonies not joined together, and 4 = colonies with well developed hyphae and conidia, and colonies mostly joined together.

/ and ? = unknown.

Seedling studies

A total of 192 wheat cultivars and advanced lines were evaluated using 20 *B. graminis* f. sp. *tritici* isolates in the greenhouse at the Institute of Plant Protection, Chinese Academy of Agricultural Sciences, in the spring of 1999 and 2003 according to the method described by Xiang et al.[35]. Between five and seven plants of each cultivar and differential line were grown in flat plastic trays 70cm×45cm×18cm in size. Highly susceptible check cultivar (Jingshuang 16) was randomly planted four times in each tray to evaluate uniformity of inoculum distribution. Inoculation with each of the 20 isolates was performed in separate trays, when the first leaves were fully expanded. Host reactions were recorded 7 to 10 days after inoculation, when the susceptible checks were heavily infected. The 0 to 4 infection type scale[29] was used for recording the host response to infection, where 0=no visible symptoms; 0; = hypersensitive necrotic flecks; 1 = tiny colonies with little sporulations; 2 = colonies with moderately developed hyphae and little conidia; 3=colonies with well developed hyphae and conidia, but colonies not joined together; and 4 = colonies with well developed hyphae and conidia, and colonies mostly joined together.

Field trials

During the 2000 to 2001 and 2002 to 2003 crop seasons, 52 and 76 winter wheat cultivars, respectively, were evaluated for APR to powdery mildew at the Institute of Crop Breeding and Cultivation, Chinese Academy of Agricultural Sciences, and 30 cultivars were common to both experiments. Field trials were planted in a randomized complete block design with three replicates; each plot consisted of two 2-m rows spaced 25cm apart. Approximately 150 seeds were sown in each plot. Spreader rows of the susceptible cultivar Jingshuang 16 surrounded the tested cultivars and were also planted every 15 plots. The nursery was surrounded by rye (*Secale cereale*) about 2-m high to prevent contamination by other powdery mildew isolates. Inoculation was carried out with the isolate E20 of *B. graminis* f. sp. *tritici* before stem elongation. The cultivar response to powdery mildew was scored for the first time 2 weeks after inoculation and then at weekly intervals until leaves were physiologically mature. The infection severity on 10 arbitrarily selected plants from each plot was recorded based on the 0 to 9 scale described by Saari and Prescott[26]. Powdery mildew severity on penultimate leaves (e. g., F-1 leaves) was rated 2 weeks after flowering, based on the actual percentage (1-100%) of leaf area covered by powdery mildew, and then every 6 to 8 days. A disease index (I) was calculated using the formula

$$I=\frac{(0\times n_0+1\times n_1+\cdots+9\times n_9)}{9\times(n_0+n_1+\cdots+n_9)}$$

where n_i = numbers of plants at i-th scale of infection severity[10]. The area under the disease progress curve (AUDPC) was calculated based on the severity on penultimate leaves according to the formula described by Bjarko and Line[2]. Six wheat cultivars (Alidos, Andros, FAP77517, Massey, Mercia, and Redcoat) with known APR to powdery mildew were used as checks.

Statistical analysis

Relative AUDPC was calculated by dividingthe AUDPC of each cultivar/line by the AUDPC of susceptible check Jingshuang 16. The percentage of maximum disease severity (MDS) on penultimate leaves was transferred into inverse sine by the formula $x = \mathrm{Sin}^{-1}\sqrt{MDS}$ for subsequent analysis of variance (ANOVA) and LSD test. SAS software was used to compute ANOVA as well as the LSD test of differences between the relative AUDPC and MDS of each cultivar or line and the mean value of six check cultivars with known APR.

Results

Identification of resistance genes effective at the seedling stage

It is shown in Table 1 that resistance genes *Pm1*, *Pm3e*, *Pm5*, and *Pm7* were susceptible to all isolates of *B. graminis* f. sp. *tritici* tested. Genes *Pm12* and *Pm16* conferred resistance to all isolates, and *Pm20* gave a susceptible reaction to isolates E13 and E21 only. Genes *Pm2* and *Pm6* conferred resistance to 10 and one isolates, respectively. However, Maris Huntsman, which carries both the *Pm2* and *Pm6* genes, was resistant to 14 isolates of *B. graminis* f. sp. *tritici*.

Of the 192 wheat cultivars, 94 were susceptible to all isolates (Table 2). Ninety-eight wheat cultivars and lines were postulated to carry one or more resistance genes (Table 3). Of them, 11 cultivars and lines (Aiyuandong 3, CA8646, CA9632, Jinnong 8112, Lumai 13, Nongda 523, Zhoumai 13, Corydon, Jur/2 * Iapar29//Pf83144, Kukuna, and Mv Optima) had the same reaction pattern as Kavkaz, which carries *Pm8*. Seven cultivars (BPM14, BPM28, CA9553, Lankao 906, Lu7you/C39, Nongdayoufan 5, and German 8661) showed the same reaction patterns as Armada, which carries *Pm4b*. Aizao 4110 and CA9550 displayed a combination of reaction patterns of Ulka/8 * CC and Armada, which carry *Pm2* and *Pm4b*, respectively. Three cultivars (90-I89, Shan 229, and Milan/Ducula) showed the same reaction pattern as Kolibri that carries *Pm3d*, while 7 cultivars or lines (Zheng86115, 8030Versailles/Edch//Cd, Tnmu/Milan-2, Tx71A1039. V1 * 3/Ami//Trap # 1, Vee/Koel//Hel/3 * Cn079/3/Kauz, Soissons-2, and Soissons-3) were postulated to have unidentified *Pm* genes in addition to *Pm3d* (Table 3). Three cultivars (Nongda 280218, Nongda 883, and Rah122/94) showed the same resistance pattern as Khapli/8 * CC, which carries *Pm4a*. BPM15 and Xingmai 883 displayed the same reaction patterns as Ulka/8 * CC and Timgalen, respectively (Tables 2 and 3).

Twenty-one cultivars and lines (98301, Lumai 14, Shangdong200015537, Altar 84/Ae. squarrosa (219) // 2 * Seri, Br23/Emb40, Choix M95, Cty * 3/Ta2450, Hahn/2 * Weaver, Hussar, Irena/Weaver, Lfn/Ii58. 57//Prl/3/Hahn/4/2 * Mo88, Mon/Tmu//Ald/Pvn, Mv Madrigal, Mv Magdalena, Nd610/Kauz//Tui, Prinia, Tinamou, Tnmu/Milan-1, Tnmu/Milan-3, Tnmu/Milan-4, and Vee/Koel//Hel/3 * Cn079/3/Kauz) were postulated to have unidentified resistance genes in addition to *Pm8*.

Thirty-six cultivars were resistant to one or several isolates; however, their reaction patterns did not match any of the known resistance genes or their combinations. Of these, Mv-17, Va91-51-26, Va91-52-65, and Pm93-625-4 were resistant to all isolates and showed the same reaction pattern as those of *Pm12* and *Pm16*.

Adult-plant resistance to powdery mildew

Cultivars and lines differed significantly for relative AUDPC and MDS in the field trials (Table 4). Twenty-two cultivars and lines showed APR to powdery mildew according to the LSD test between them and the mean value of the six APR checks (Table 5). As shown in Fig. 1, the disease developed quickly on the susceptible check, Jingshuang 16, indicating a favorable environment for disease development. Compared to susceptible cultivars, disease developed much more slowly on adult-plant resistant cultivars (Fig. 1).

Table 4 Analysis of variance of relative AUDPC and maximum disease severity (MDS) on penultimate leaves in 52 and 76 wheat cultivars or lines tested in the 2000 to 01 and 2002 to 03 crop seasons, respectively

Season	Parameter	Source ofvariation	D. F.	Mean of squares	F-value
2000 to 01	AUDPC	Genotypes	51	0. 264 3	528. 6**
		Replicates	2	0. 000 3	<1. 0
		Error	102	0. 000 5	
	MDS	Genotypes	51	315. 673	242. 6**
		Replicates	2	0. 427	<1. 0
		Error	102	1. 301	
2002 to 03	AUDPC	Genotypes	75	0. 123 3	137. 0**
		Replicates	2	0. 000 2	<1. 0
		Error	150	0. 000 9	
	MDS	Genotypes	75	513. 479	385. 2**
		Replicates	2	0. 311	<1. 0
		Error	150	1. 333	

** indicates significant difference at $P=0.01$.

Table 5 Twenty-two wheat cultivars and lines with adult-plant resistance (APR) to powdery mildew identified in the 2000 to 2001 and/or 2002 to 2003 crop seasons

Genotype	2000 to 2001		2002 to 2003		Seedling reaction to E20[d]
	Relative[c] AUDPC	MDS (%)	Relative[c] AUDPC	MDS (%)	
98301	/	/	0.04	2.2	4
8030 Versailles/Edch//Cd	/	/	0.04	2.5	4
Arina	/	/	0.02	1.0	4
Bainong 64	0.08	3.0	0.05	3.4	4
CA8686	/	/	0.02	1.4	4
CA9550	0.03	4.0	0.04	2.0	4
CA9641	0.03	4.0	0.02	1.0	4
Hussar	0.03	2.0	0.02	1.0	4
Lona	/	/	0.03	1.3	4
MvMadrigal	/	/	0.03	1.7	4
MvMagdalena	/	/	0.04	2.3	4
Mv-23	0.05	6.8	0.01	0.8	4
Shandong 418	/	/	0.02	1.4	4
Soissons-1	/	/	0.04	2.8	4
Soissons-2	/	/	0.03	1.7	4
Soissons-3	/	/	0.03	1.5	4
Sxafa4-3	0.04	4.0	0.00	0.0	4
Xingmai 99	0.03	3.0	0.02	1.5	4
YE-2416-7A	0.03	4.0	0.00	0.0	4
Yumai47	/	/	0.04	2.5	4
Yuzhan9705	/	/	0.03	1.7	4
Zhoumai 17	/	/	0.03	1.9	4
Alidos[a]	0.05	5.0	0.07	3.1	4
Andros[a]	0.06	6.0	0.05	3.0	4
FAP77517[a]	0.06	5.0	0.04	2.0	4
Massey[a]	0.04	6.0	0.00	0.6	4
Mercia[a]	0.06	7.0	0.02	1.4	4
Redcoat[a]	0.06	5.0	0.02	1.0	4
Jingshuang 16[b]	1.00	92.0	1.00	95.0	4

a. Resistant check cultivar with known APR to powdery mildew.

b. Susceptible check cultivar.

c. AUDPC of cultivar/line divided by AUDPC of susceptible check Jingshuang 16.

/= not tested.

d. 4=colonies with well developed hyphae and conidia, and colonies mostly joined together.

Among the cultivars and lines, 11 (including 5 APR checks) were susceptible to all isolates of *B. graminis* f. sp. *tritici* at the seedling stage, but showed slow disease progress at the adult stage in the field. Seventeen cultivars (including one APR check) with resistance genes overcome by isolate E20 at the seedling stage showed moderate resistance in the field (Tables 3 and 5). In contrast, Aizao4110, which carries *Pm2*+*4b*, was highly susceptible in the field, with a relative AUDPC and MDS of 0.73 and 42.7, respectively. A highly significant correlation was found between AUDPC and MDS on penultimate leaves ($r=0.96$, $P<0.001$).

Discussion

In the present study, known powdery mildew resistance genes were detected in 62 wheat cultivars and lines. Of these resistance genes, *Pm12* and *Pm16* showed resistant to all isolates of *B. graminis* f. sp. *tritici* tested, indicating that they are effective resistance genes against powdery mildew in China. *Pm20* were resistant to most of the isolates and could confer effective resistance at least in areas where virulence may be absent. Genes *Pm2* and *Pm6* conferred resistance to 10 and one isolates, respectively. However, Maris Huntsman, which carries both the *Pm2* and *Pm6* genes, was resistant to 14 isolates of *B. graminis* f. sp. *tritici*, indicating the presence of a previously unidentified resistance gene (s) in this cultivar. The resistance gene in the line Pm93-625-4 was derived from wild emmer (accession C20), which should be *Pm30* according to its origin.

Pm8 was the most frequently identified resistance gene, followed by *Pm4b* in Chinese wheat cultivars. *Pm8* in these cultivars is mainly derived from the 1B/1R cultivars Lovrin 13, Predgornaia 2, Kavkaz, Neuzucht, and their derivatives, which have been widely employed in wheat breeding programs in China since the early 1970's[13]. At present, 46% of Chinese wheat cultivars possess the 1B/1R translocation (data not shown). The high frequency of virulence to *Pm8* in

Chinese *B. graminis* f. sp. *tritici* populations has remained stable at about 94%[6]. This has led to increased losses in China due to powdery mildew.

All Chinese wheat cultivars and lines carrying the *Pm8* gene possess the 1B/1R translocation, as shown in Table 3. However, 71.5% of 1B/1R translocation lines did not show the reaction pattern expected for *Pm8* (Table 3), which might be due to the presence of suppressor genes for *Pm8*[25]. This was consistent with the results of a previous study by Ren et al.[25], who found that on average only one-third of all CIMMYT wheats with 1B/1R tested actually expressed the *Pm8* gene.

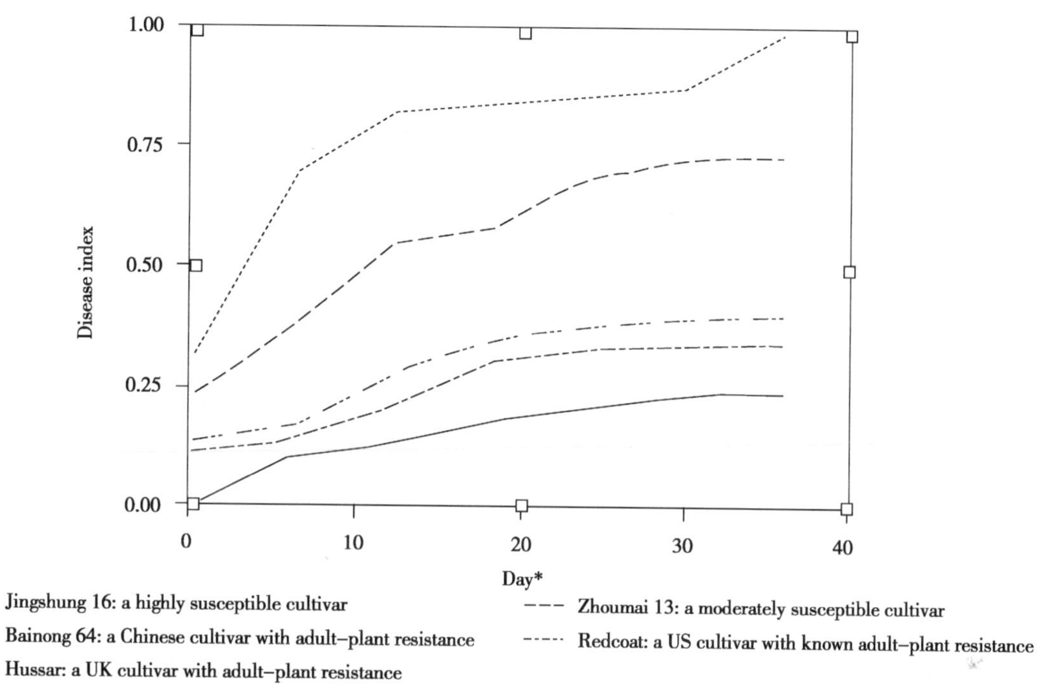

Fig. 1 Progress curve of powdery mildew development in wheat based on the index of disease at seven scoring dates (indicated as cumulative days: day 0=4 weeks after inoculation). Jingshuang 16 is a highly susceptible cultivar. Zhoumai 13 is a moderately susceptible cultivar, Bainong 64 is a Chinese cultivar with adult plant resistance (APR), Redcoat is a US cultivar with known APR, and Hussar is a UK cultivar with APR.

* 0=four weeks after inoculation

We could not postulate resistance genes in 36 cultivars and linesdue to the lack of *B. graminis* f. sp. *tritici* isolates or differential lines (Table 3). In addition, 94 genotypes were susceptible to all *B. graminis* f. sp. *tritici* isolates (Table 2), which indicates that they either carry no resistance genes or possess resistance genes that were overcome by matching virulence in all 20 *B. graminis* f. sp. *tritici* isolates used in this study.

There is an urgent need for a better genetic strategy for powdery mildew control. Several strategies have been proposed for developing durable resistance to powdery mildew, including the pyramiding of effective major resistance genes[32], the use of multilines with different resistance genes[3], and breeding cultivars with APR or slow mildewing resistance[28]. The use of molecular markers to pyramid major genes seemed promising[22], but may cause the rapid formation of isolates with multiple matching virulence[10]. APR has, therefore, been suggested as the most sustainable strategy for breeding wheat cultivars with durable powdery mildew resistance[1,28].

A series of studieson APR to powdery mildew were conducted during the last decades, and both qualitative and quantitative genes were detected in various wheat cultivars[4,5,9]. Chae and Fischbeck[4] reported the effect of 14 chromosomes on APR to powdery mildew in cv. Diplomat, while in cv's Knox 62, Massey, Redcoat, and Houser, APR is governed by 2 to 3 genes with moderate to high heritability[5,9]. CIMMYT has conducted several studies on slow rusting, and found that combinations of 4 to 5 minor genes for APR to leaf rust and stripe rust give adequate protection in the field[30].

Bainong 64, a leading cultivar in China until 2000, was grown on about 700, 000 ha annually for 8 years. Our study confirmed that it carries adult plant resistance (Fig. 1). The cultivar was widely used as a parent, and several derived lines are being tested for possible release in the Huang-Huai winter wheat zone of China.

Description of APR can be carried outbased on different parameters such as a disease index, AUDPC, and MDS on penultimate leaves[11,36]. In the present study, we used AUDPC, MDS, and a disease index to evaluate APR in the field. All cultivars or lines with adult-plant resistance showed lower values for AUDPC, MDS, and the disease index than susceptible cultivars (Table 5). In practical breeding, it is good to use a simple parameter to evaluate APR in the field. Investigations of AUDPC and the disease index are time-consuming, and because AUDPC and MDS are highly correlated, it is more practical to use the latter parameter for powdery mildew screening, with a single scoring at an appropriate time.

Generally, APR to powdery mildew is detected in cultivars that either have no identified major resistance genes or whose major-gene resistance has been overcome[1]. As an example, most powdery mildew populations worldwide show high virulence frequencies for *Pm5* at the seedling stage[14], but the gene serves as a source of APR[14,20]. In our study 17 cultivars and lines whose race-specific resistance genes were overcome in the seedling stage by isolate E20 showed APR in the field, indicating that they might be carrying minor APR genes besides the major genes (Tables 3 and 5). Results indicate that it is possible to combine major resistance genes and APR genes to achieve durable resistance to powdery mildew in wheat cultivars.

Acknowledgments

The authors are grateful to Dr. Xiuqiang Huang for his critical review of this paper, and to Alma McNab for editing it. This research was supported by the international collaboration program of the National Natural Science Foundation of China (NSFC).

References

[1] Bennett, F. G. A. 1984. Resistance to powdery mildew in wheat: a review of its use in agriculture and breeding programmes. Plant Pathol. 33: 279-300.

[2] Bjarko, M. E., and Line, R. F. 1988. Heritability and number of genes controlling leaf rust resistance in four cultivars of wheat. Phytopathology. 78: 457-461.

[3] Browning, J. A., and Frey, K. J. 1969. Multiline cultivars as a means of disease control. Annu. Rev. Phytopathol. 7: 355-382.

[4] Chae, Y. A., and Fischbeck, G. W. 1979. Genetic analysis of powdery mildew resistance in wheat cultivar "Diplomat". Z. Pflanzenzucht. 83: 272-280.

[5] Das, M. K., and Griffey, C. A. 1994. Heritability and number of genes governing adult-plant resistance to powdery mildew in Houser and Redcoat winter wheats. Phytopathology. 84: 406-409.

[6] Duan, X. Y., and Sheng, B. Q. 1998. Identification of isolates of *Blumeria graminis* f. sp. *tritici* and the monitoring of their virulence frequencies. Acta Phytopathol. Sinica. 25: 31-36.

[7] Duan, X. Y., Xiang, Q. J., Zhou, Y. L., and Sheng, B. Q. 1993. Establishment of wheat powdery mildew isolates with putative virulence genotype. Plant Protection. 19 (5): 27-28.

[8] Fried, P. M., Mackenzie, D. R., and Nelson, R. R. 1981. Yield loss caused by *Erysiphe graminis* f. sp. *tritici* on single culms of "Chancellor" wheat and four multilines Z. Pflanzenkrankh. Pflanzenschutz. 88: 256-264.

[9] Griffey, C. A., and Das, M. K. 1994. Inheritance of adult-plant resistance to powdery mildew in Knox 62 and Massey winter wheats. Crop Sci. 34: 641-646.

[10] Groth, J. V. 1976. Multilines and "super-races": a

simple model. Phytopathology. 66：937-939.

[11] Guo，Y. Z.，Duan，S. K.，and Zhang，T. 1999. A study of classifying resistant types to scab in wheat with the index of disease severity. Acta Agriculturae Universitatis Henanensis. 33：336-339.

[12] Gupta，R. B.，and Shepherd，K. W. 1992. Identification of rye chromosome 1R translocation and subunits in hexaploid wheat using storage proteins as genetic markers. Plant Breeding. 109：130-140.

[13] He，Z. H.，Rajaram，S.，Xin，Z. Y.，and Huang，G. Z. (eds). 2001. A History of Wheat Breeding in China. Mexico，D. F.：CIMMYT.

[14] Hsam，S. L. K.，and Zeller，F. J. 2002. Breeding for powdery mildew resistance in common wheat (*Triticum aestivum* L.). Pages 219-238 In：The powdery mildews：a comprehensive treatise. (eds. R. Belanger，B. Bushnell，A. Dik，and T. Carver). The American Phytopathological Society，ST Paul，MN.

[15] Huang，X. Q.，Hsam，S. L. K.，and Zeller，F. J. 1997. Identification of powdery mildew resistance genes in common wheat (*Triticum aestivum* L. em Thell). 9. Cultivars，land races and breeding lines grown in China. Plant Breeding. 116：233-238.

[16] Huang，X. Q.，and Röder，M. S. 2004. Molecular mapping of powdery mildew resistance genes in wheat：A review. Euphytica. 137：203-223.

[17] Kazman，M. E.，and Lein，V. 1996. Cytological and SDS-PAGE characterization of 1994-95-grown European wheat cultivars. Annu. Wheat Newsletter. 42：86-92.

[18] Leath，S.，and Bowen，K. L. 1989. Effects of powdery mildew，triadimenol seed treatment and triadimefon foliar sprays on yield of winter wheat in North Carolina. Phytopathology. 79：152-155.

[19] Leath，S.，and Heun，M. 1990. Identification of powdery mildew resistance genes in cultivars of soft winter wheat. Plant Dis. 74：747-752.

[20] Lebsock，K. L.，and Briggle，L. W. 1974. Gene *Pm5* for resistance to *Erysiphe graminis* f. sp. *tritici* in Hope wheat. Crop Sci. 14：561-563.

[21] Lipps，P. E.，and Madden，L. V. 1988. Effect of triadimenol seed treatment and triadimefon foliar treatment on powdery mildew epidemics and grain yield of winter wheat cultivars. Plant Dis. 72：887-892.

[22] Liu，J.，Liu，D.，Tao，W.，Li，W.，Wang，S.，Chen，P.，Cheng，S.，and Gao，D. 2000. Molecular marker-facilitated pyramiding of different genes for powdery mildew resistance in wheat. Plant Breeding. 119：21-24.

[23] Liu，X. K. 1989. Prospects of wheat powdery mildew research in China. Inf. Agric. Husbandry August. 1989：1-10.

[24] McIntosh，R. A.，Yamazaki，Y.，Devos，K. M.，Dubcovsky，J.，Rogers，W. J. and Appels，R.，2003：Catalogue of gene symbols for wheat，in Proceedings of 10th International Wheat Genetics Symposium，Vol. 4：1-34.

[25] Ren，S. X.，McIntosh，R. A.，and Lu，Z. J. 1997. Genetic suppression of the cereal rye-derived gene *Pm8* in wheat. Euphytica. 93：353-360.

[26] Saari，E. E.，and Prescott，J. M. 1975. A scale for appraising the foliar intensity of wheat diseases. Plant Dis. Report. 59：377-381.

[27] Schlegel，R. 1997. Current list of wheats with rye introgressions of homoeologous group 1. 2nd update. Wheat Information Service. 84：64-69.

[28] Shaner，G. 1973. Evaluation of slow-mildewing resistance of Knox wheat in the field. Phytopathology. 63：867-872.

[29] Shi，Q. M.，Zhang，X. X.，and Duan，X. Y. 1987. Identification of isolates of *Blumeria graminis* f. sp. *tritici*. Scientia Agricultura Sinica. 20：64-70.

[30] Singh，R. P.，Huerta-Espino，J.，and William，M. 2001. Slow rusting genes based resistance to leaf and yellow rusts in wheat. Pages 103-108 in：R. Eastwood，G. Hollamby，T. Rathjen and N. Gororo (eds.) Proc. 10th Assembly of Wheat Breeding Society of Australia Inc.，2001，Mildura，Australia.

[31] Tao，J. F.，Shen，X. Z.，Qin，J. Z.，and Qin，Y. 1982. Studies on the resistance to powdery mildew in wheat species and varieties. Acta Phytopathol. Sinica. 12：7-14.

[32] Vanderplank，J. E. 1984. Disease Resistance in Plant. Page 194 In：(2nd ed.) Academic Press，New York.

[33] Wu，Z. S. 1990. Breeding for Wheat Disease Resistance. Pages 235-272 In：Wu，Z. S. ed.，Wheat Breeding. Agric. Pub. Press of China，Beijing.

[34] Xia，X. C.，Hasm，S. L. K.，Stephan，U.，Yang，T. M.，and Zeller，F. J. 1995. Identification of powdery mildew resistance genes in common wheat (*Triticum aestivum* L). VI. Wheat cultivars grown in China. Plant Breeding. 114：174-175.

[35] Xiang，Q. J.，Sheng，B. Q.，Zhong，Y. L.，Duan，

X. Y., and Zhang, K. C. 1994. Analyses of resistance genes of three differential varieties to the isolates of *Blumeria graminis* f. sp. *tritici* in wheat. Race. Acta Agriculturae Boreali-Sinica. 9: 94-97.

[36] Yu, D. Z., Yang, X. J., Yang, L. J., Jeger, M. J., and Brown, J. K. M. 2001. Assessment of partial resistance to powdery mildew in Chinese wheat varieties. Plant Breeding. 120: 279-284.

Seedling and slow rusting resistance to stripe rust in Chinese common wheats

Z. F. Li[1], X. C. Xia[1*], X. C. Zhou[2], Y. C. Niu[3], Z. H. He[1,4], Y. Zhang[1], G. Q. Li[1], A. M. Wan[3], D. S. Wang[1], X. M. Chen[1], Q. L. Lu[2], and R. P. Singh[5]

[1] *Institute of Crop Science/National Wheat Improvement Center, Chinese Academy of Agricultural Sciences, Zhongguancun South Street* 12, 100081, *Beijing, China*; [2] *Gansu Winter Wheat Research Institute, Duan Jia Tan* 418, 730020, *Lanzhou, China*; [3] *Plant Protection Institute, Chinese Academy of Agricultural Science, Yuanmingyuan West Road* 2, 100094, *Beijing, China*; [4] *CIMMYT China Office, C/O CAAS, Zhongguancun South Street* 12, 100081, *Beijing, China*; [5] *International Maize and Wheat Improvement Center (CIMMYT), Apdo. Postal* 6-641, 06600, *Mexico D. F.*, *Mexico*

Abstract: Identification of seedling and slow stripe rust resistance genes is important for gene pyramiding, gene deployment and developing slow-rusting wheat cultivars to control the disease. A total of 98 Chinese lines were inoculated with 26 pathotypes of *Puccinia striiformis* f. sp. *tritici* (PST) for postulation of stripe rust resistance genes effective at the seedling stage. One hundred thirty five wheat lines were planted at two locations to characterize their slow rusting responses to stripe rust in the 2003-2004 and 2004-2005 cropping seasons. Genes *Yr2*, *Yr3a*, *Yr4a*, *Yr6*, *Yr7*, *Yr9*, *Yr26*, *Yr27*, and *YrSD*, either singly or in combinations, were postulated in 72 lines, whereas known resistance genes were not identified in the other 26 accessions. The resistance genes *Yr9 and Yr26* were found in 42 and 19 accessions, respectively. *Yr3a* and *Yr4a* were detected in two lines, and four lines may contain *Yr6*. Three lines were postulated to possess *YrSD*, one carried *Yr27*, and one may possess *Yr7*. Thirty-three lines showed slow stripe rusting resistance at two locations in both seasons.

Key words: bread wheat, yellow rust, *Puccinia striiformis* f. sp. *tritici*, gene-for-gene specificity, slow rusting resistance

Introduction

Stripe rust, caused by *Puccinia striiformis* f. sp. *tritici* (PST), is an important disease in many wheat-growing regions of the world. Epidemics of stripe rust have occurred periodically in China, and in some years, caused destructive yield losses in the major wheat production regions, particularly in northwestern and southwestern China[35]. The most recent widespread epidemic in 2002, was caused by Chinese PST pathotype CYR32, which is virulent to almost all Chinese wheat cultivars except those with resistance genes *Yr5*, *Yr10*, *Yr15*, *Yr24* or *Yr26*[11, 16, 35, 44].

Following the gene-for-gene hypothesis proposed by Flor[7], Zadoks[45] found a gene-for-gene interaction between wheat cultivars and PST pathotypes. Based on the gene-for-gene concept, Loegering and Burton[14] reported that race-specific resistance genes in wheat cultivars could be identified by comparing their reactions to different pathotypes of *Puccinia graminis*

Published in Plant Disease, 2006, 90: 1302-1312

tritici with those of lines with known resistance genes. Perwaiz and Johnson[24] tested seedlings of 26 wheat cultivars from Pakistan with 18 British PST pathotypes and postulated *Yr2*, *Yr6*, *Yr7*, and/or *Yr9* to be present in the cultivars. Bartos et al.[2] postulated *Yr1*, *Yr2*, *Yr3a* + *Yr4a*, *Yr9*, and *Yr32* (*YrCV*) in 17 Czechoslovakian and two Russian wheat cultivars following tests with 18 PST pathotypes. Singh et al.[28] found *Yr2* and *Yr7* in a seedling test of 11 Indian wheat cultivars using 17 PST pathotypes, and confirmed the presence of *Yr2* in three cultivars by testing the F_1 and F_2 generations of crosses between these cultivars and Heines VII, a cultivar that contains *Yr2*. Sharma et al.[26] inoculated seedlings of 38 wild emmer derivatives and 53 advanced wheat lines from Nepal with 18 PST pathotypes, and found 28 wild emmer derivatives were resistant to all pathotypes with unidentified resistance genes, and five resistance genes *Yr2*, *Yr6*, *Yr7*, *Yr9* and *YrA* were present in Nepalese wheat cultivars and advanced lines tested. In China, Wang et al.[36, 37] identified eight resistance genes (*Yr1*, *Yr2*, *Yr3*, *Yr7*, *Yr9*, *Yr10*, *YrSu* and *YrSD*) in 39 Chinese wheat cultivars and 20 important resistant germplasm lines from the provinces of Shaanxi, Gansu, and Sichuan, using 20 PST pathotypes. Niu et al.[22] screened 50 Chinese wheat cultivars from the provinces of Henan, Shandong and Anhui with 26 PST pathotypes, and found that *Yr9* was the most frequent resistance gene. However, no information is available about the identity of stripe rust resistance genes in current leading cultivars, newly released cultivars, and advanced breeding lines in China. Therefore, characterization of stripe rust resistance genes in widely grown Chinese wheats is extremely important for breeding new resistant cultivars and for gene deployment schemes[35].

In general, seedling resistance is race-specific and thus short-lived due to frequent changes of virulence in the pathogen population[13, 32]. In contrast, resistance at the adult plant stage, such as high-temperature adult-plant (HTAP) resistance or slow rusting, is considered more durable than seedling resistance[5, 6, 13, 18, 31]. Wheat cultivars with slow rusting genes are often susceptible at the seedling stage, but may be moderately to highly resistant to all pathotypes at the adult plant stage in the field[4,32]. The components of slow rusting include a longer latent period, low infection frequency, smaller uredial size, and reduced duration and quantity of spore production[4]. Three currently named stripe rust resistance genes, *Yr18*, *Yr29* and *Yr30*, confer slow rusting[29, 33, 40] and many more are suspected. Singh[29, 30] and McIntosh[18] found the slow rusting resistance gene *Yr18* in Anza and Bezostaja, and this gene confers a moderate level of resistance when present alone. However, combinations of *Yr18* with two to four additional slow rusting genes resulted in high resistance in most environments[32]. William et al.[40] identified slow rusting gene *Yr29*, on chromosome 1B. This gene was closely linked with slow leaf rusting gene *Lr46*. Suenaga et al.[33] reported resistance gene *Yr30* closely linked with durable adult plant stem rust resistance gene *Sr2*.

From the early 1970's, the 1B/1R translocation with *Yr9* was introduced to Chinese wheat breeding programs through introductions such as Lovrin 10, Lovrin 13, Predgornaia 2, Kavkaz, and Neuzucht[9, 21]. The widespread growing of cultivars with *Yr9* resulted in major epidemics in 1990 and 2002, since the stripe rust resistance was overcome by the new race of the pathogen CYR 32, which has forced Chinese wheat breeders to focus on the development of cultivars with durable resistance. Although breeders are using slow rusting resistance and major genes for durable control of stripe rust, very little information is available on sources and efficacy of resistance in Chinese wheat cultivars.

A total of 145 widely grown Chinese wheat cultivars, newly released cultivars, and advanced lines from 12 provinces in the major wheat production regions were investigated in this study. The objectives were to identify probable genes conferring resistance to stripe rust at the seedling stage and to investigate the likelihood of slow rusting resistance from adult plant tests in the field.

Materials and methods

Wheat germplasm

The wheat lines comprised 145 Chinese leading cultivars, newly released cultivars, and advanced lines from major breeding programs across the autumn-sown wheat regions of China, including the provinces of Hebei, Henan, Shandong, Shaanxi, Gansu, Anhui, Hubei, Jiangsu, Chongqing, Guizhou, Sichuan, and Yunnan. Ninety-eight lines were tested for seedling response to 26 PST pathotypes in the greenhouse, and 135 were tested for slow stripe rusting resistance in the field (Table 1).

Table 1 Pedigrees of 145 Chinese wheat cultivars and lines tested for stripe rust response and arranged according to province of origin

Line number	Line	Pedigree	Origin	1B/1R status[a]	Stripe rust test[b]
1	Han 3475	Peixian30421/Han 4162	Hebei	−	S and A
2	Han 4564	88-6012/Shi 5144	Hebei	+	S* and A
3	Han 4599	Han 4032/85Zhong47	Hebei	+	S and A
4	Han 5316	Han 7808/CA 8059//85Zhong47	Hebei	+	S* and A
5	Han 6172	Han 4032/Zhongyin 1	Hebei	−	S* and A
6	Lankao52-24	Triticale 84-184/Yumai 2//90 Xuan	Henan	+	S* and A
7	Lankao 411	893B25/Lankao 906	Henan	+	S* and A
8	Lankao 8	84 (184) /90 Selection	Henan	+	S* and A
9	Lankao 906-4	84 (184) /Yumai 2//90 Selection	Henan	+	S* and A
10	Luohan 2	Luoyang 78 (111) /Jinmai 33	Henan	−	S* and A
11	Neixiang 188	Mianyang 84-27/Neixiang82C6//Yumai 17	Henan	+	S* and A
12	Neixiang 991	87C27/Zhengzhou 84115// Yumai 13	Henan	+	S* and A
13	Xinmai 11	Zhou 8826/Xinxiang 3577	Henan	−	S* and A
14	Xinmai 9	Baiquan 3047-3/Neixiang82C6	Henan	−	S* and A
15	Yanzhan 4110	C39/Xibei 78 (6) 9-2//FR 81-3/Aizao 781-4/3/Aizao 781-4	Henan	−	S* and A
16	Yumai 18	Zhengzhou 761/Yanshi 4	Henan	−	S* and A
17	Yumai 34	Aifeng 3//Meng 201/Neuzucht/3/Yumai 2	Henan	−	S* and A
18	Yumai 49	Bainong 791//Wenxuan 1/Zhengzhou 761/3/Yumai 2	Henan	−	S* and A
19	Yumai 62	Zhou 8425B/SWP73295	Henan	+	S and A
20	Zhengmai 9023	(Xiaoyan 6/Xinong 65//8323-3/8443) /Shaan 213	Henan	+	S* and A
21	Zhou 8425B	Guangmai 74/Lianfeng 1//Predgornayi 2/3/Annong 7959	Henan	+	S
22	Zhou 88114	Zhou 8425B/ Yumai 17	Henan	+	S
23	Zhoumai 11	Zhou 8425B/Yumai 17	Henan	+	S* and A
24	Zhoumai 16	Zhou 9/Zhou 8425B	Henan	+	S* and A
25	Jinan 16	Tal Shannongfu 63/AimengniuV	Shandong	+	S* and A
26	Lumai 21	Lumai 13/Yumai 2	Shandong	−	S* and A
27	Yannong 19	Yan 1933/Shan 82-29	Shandong	−	S* and A
28	Yanyou 361	Yan 1933/Shan 82-29	Shandong	−	S* and A
29	Zimai 12	917065/910292	Shandong	−	S* and A

（续）

Line number	Line	Pedigree	Origin	1B/1R status[a]	Stripe rust test[b]
30	(95).18	?	Shaanxi	+	S* and A
31	1718	Xinong 242//84G6/Shaan 167	Shaanxi	+	S and A
32	2208	Shaan 229//Shaan 213/Xinong 8623	Shaanxi	+	S and A
33	2611	Xinong 881/Shaan 229	Shaanxi	−	S and A
34	953	Xinong 918/Zhi 87-135	Shaanxi	−	S and A
35	N139	Xiaoyan 22//94156/N9134	Shaanxi	−	S
36	99104	9434/9527	Shaanxi	−	S
37	H-46	?	Shaanxi	?	S* and A
38	N95175	92R/Yuancai 142//Xiaoyan 6	Shaanxi	−	S
39	Nonglin 9823	Xiaoyan 4/73 (36) //Baofeng 7228	Shaanxi	−	S and A
40	Qinnong 151	Zhengzi R84019-0-7-5/Xinong 1376	Shaanxi	+	S* and A
41	Shaan 512	Shaan 150/Shaan 354	Shaanxi	−	S and A
42	Shaanzi 1869	Shaan 354/Shaanmai 898	Shaanxi	+	S and A
43	Shaan 715	Shaan 354/Yanmai 8911	Shaanxi	+	S* and A
44	Shaannong 981	Derived from Shaanyou 225	Shaanxi	−	S* and A
45	Xiaoyan 143	87C332/Xiaoyan 135	Shaanxi	−	S* and A
46	Xiaoyan 22	Xiaoyan 6/775-1//Xiaoyan 107	Shaanxi	−	S* and A
47	Xiaoyan 921	?	Shaanxi	−	S and A
48	Xinong 1163-4	84Jia79/Xinong 1376	Shaanxi	+	S and A
49	Xinong 1376	Xinong84G6/Bi 16	Shaanxi	+	S
50	Xinong 291	?	Shaanxi	+	S and A
51	Xinong 8925-13	Xiaoyan 168/Zhi 763	Shaanxi	−	S and A
52	Xinong 914	Xinong 1376/2611	Shaanxi	?	S
53	Yunfeng 988	?	Shaanxi	−	S and A
54	90-99-5	?	Gansu	+	S* and A
55	93-23-13	83-44-29/Guinong 90-21-7	Gansu	−	S* and A
56	94t-133	Hybrid 46/Xian 4//Tao 157	Gansu	−	S* and A
57	95-108	92R137/Lantian 6	Gansu	−	S* and A
58	95-111-6-3-10	92R137/87-121-2	Gansu	+	S and A
59	953111-13	Mega/Lantian10	Gansu	−	S and A
60	95-3-5-1	Mega/Lantian 10	Gansu	−	S* and A
61	9562-1	Ibis/Lantian 10	Gansu	+	S and A
62	96-22-2-3	Cappelle Desprez/935	Gansu	−	S* and A
63	Lantian 9	Xifeng 16/76-89-13	Gansu	+	S* and A
64	Annong 98005	Yumai 18/Wanmai 19	Anhui	−	S and A
65	Een 1	Lovrin 10/761//Sumai 3	Hubei	+	S* and A
66	Emai 14	Fan 6/Yanda 72-629	Hubei	−	S and A
67	Huaimai 18	Zhengzhou 891/Yan 1604	Jiangsu	+	S* and A
68	Yangfu 2	Yangmai 158/101-90	Jiangsu	−	S and A
69	Yangmai 10	Yuma/8*Cc//Yangmai 5/3/4*Yang 85-85/4/2*Yangmai 158	Jiangsu	−	S and A

（续）

Line number	Line	Pedigree	Origin	1B/1R status[a]	Stripe rust test[b]
70	Yangmai 158	Yangmai 4/ST 1472/506	Jiangsu	—	S and A
71	Yu 94-7	Chuanyu 84-2/91-125	Chongqing	+	S and A
72	An 96-8	L9288022-21/Xingnong 5	Guizhou	—	S and A
73	Bi 2002-2	8513-1624/Ji 1002	Guizhou	+	S and A
74	Guizhou 98-18	Sumai 3/C39//Qing 30	Guizhou	+	S and A
75	01-018	Chuannong 21/89-8//89-107	Sichuan	+	S and A
76	01-10251	73A/RI	Sichuan	—	S* and A
77	01-3570	96Xia440/Guinong 21	Sichuan	—	S* and A
78	01-DH689	94-D. H-375/Mianyang 26	Sichuan	+	S and A
79	298-1	Mianyang 4/45590	Sichuan	—	S and A
80	323-2	Zhongzhi 3586/50609	Sichuan	—	S and A
81	3472	YanzhanA11/Yiyuan 2	Sichuan	—	S and A
82	351-15	Mian 99-310/GansuM180	Sichuan	—	S and A
83	58512-2	Mian 90-309/92R171	Sichuan	—	S and A
84	58749	SW 3243/35050//21530	Sichuan	+	S and A
85	58769-6	30220/8619-10	Sichuan	—	S* and A
86	93Xia032-1	93-38-1/93-35	Sichuan	+	S and A
87	98-1266	Chuanyu 12/87-429	Sichuan	+	S and A
88	99-0502	911/MY86-51	Sichuan	+	S and A
89	99422	Mianyang 87-19/R1301	Sichuan	—	S and A
90	99-607	Syn-CD769/SW 3243//Chuan 6415	Sichuan	—	S and A
91	C2001-2	P92/P88	Sichuan	+	S and A
92	CD 1485-6	98718/Chuanyu 12	Sichuan	—	S and A
93	Chen 98-18	Mianyang 26/Zixuan 1	Sichuan	—	S and A
94	Chuanmai 42	Syn-CD768/SW3243//Chuan6415	Sichuan	—	S and A
95	Chuanmai 43	Syn-CD768/SW 3243//Chuan 6415	Sichuan	—	S* and A
96	Chuanmai 44	96Xia440/Guinong 21	Sichuan	+	S
97	Chuannong 16	Chuanyu 12/87-429	Sichuan	+	S and A
98	Chuannong 17	91S-23/A302	Sichuan	+	S and A
99	Chuannong 21	91S-23/Qianhui 3	Sichuan	—	S and A
100	Chuan 5436	Cereta/*Ae. tauschii*//Mianyang 26	Sichuan	?	S
101	Chuanyu 18	Chuanyu 5 /Mo 460//94F2-4	Sichuan	—	S and A
102	Chuanyu 19	Chuanyu 5/Mo 460//Mianyang 26	Sichuan	—	S and A
103	GY 96-35	?	Sichuan	+	S and A
104	Jinfeng 626	9481/93N1001	Sichuan	—	S and A
105	Kefeng 14	8420//Neixiang82C6F1/Yumai 17	Sichuan	—	S and A
106	LB 0288	Mianyang 90-310/M180	Sichuan	—	S and A
107	LB 0438	Mian 95-339/92R178	Sichuan	—	S and A
108	LB 0458	Mianyang 90-310/M180	Sichuan	—	S and A
109	Lemai 3	Mianyang 89-224/7705	Sichuan	—	S and A

（续）

Line number	Line	Pedigree	Origin	1B/1R status[a]	Stripe rust test[b]
110	Liangmai 2	94-2/Miannong 4	Sichuan	+	S and A
111	Mian 2000-13	Mianyang 01821/90Zhong165//Guinong 19-4	Sichuan	+	S and A
112	Mian 2000-18	Mianyang 96-5/Liaochun 10	Sichuan	−	S and A
113	Mian 2000-34	Mianyang 01821/903//95-14426	Sichuan	−	S and A
114	Mian 2000-36	Mian92-8/Mian88-304//Guinong 19-4	Sichuan	+	S and A
115	Mian 2000-8	Mianyang86-78/Guinong21-1	Sichuan	−	S and A
116	Mian 99-3	Mianyang 87-24/81026-0-1-2//053638-1	Sichuan	+	S and A
117	Mianmai 37	96EW37/Mianyang 90-100	Sichuan	−	S and A
118	Mianyang 4	75-21-4/76-19//Miannong 1	Sichuan	+	S* and A
119	Mianyang 11	70-5858/Fan 6	Sichuan	−	S* and A
120	Mianyang 26	Chuanyu 9/Mianyang 20	Sichuan	−	S* and A
121	Mianyang 28	T79350-1-4-1-2/Mianyang 11	Sichuan	+	S and A
122	Mianyang 29	Mianyang 11/Jiangyou 83-5	Sichuan	−	S and A
123	N1621-19-4	Mianyang 26/Yiyuan 2	Sichuan	+	S and A
124	N711	92-4/Mianyang 4	Sichuan	+	S and A
125	Nei 2938	Mianyang 26/92R178	Sichuan	−	S and A
126	Neimai 9	Mianyang 26/92R178	Sichuan	−	S and A
127	R111	961-225/91S-5-4	Sichuan	−	S and A
128	R122	96I-22/133-3//Mianyang 26	Sichuan	−	S and A
129	R25	91S-23/A302	Sichuan	+	S and A
130	R88	1104A/R935	Sichuan	−	S and A
131	R97	Mianyang 92-8/91S-5-7	Sichuan	+	S and A
132	SW8588	Milan's'/SW 5193	Sichuan	−	S and A
133	W7	Durum/Xianyang large spike//W7268	Sichuan	−	S and A
134	Xike 01015	Mianyang 26/Zhi 39060-4	Sichuan	−	S and A
135	Xike0106-1	Chuanyu 11/9310-6	Sichuan	+	S and A
136	Xingmai 2	Mianyang 26/92R178	Sichuan	−	S and A
137	Yi 99-49	Chuanyu 12/85-5	Sichuan	+	S and A
138	Yi 99-81	90036-5-1/89-311	Sichuan	+	S and A
139	9021	Fengmai 10/Mo 120//110-75/3/Zhongkangai2 /4/04835/5/882-182	Yunnan	+	S and A
140	Fengmai 24	Yunmai 36/Mo 965	Yunnan	−	S and A
141	Jing 035-1	?	Yunnan	+	S and A
142	Jing 9308	TB78-212/Yin 48/092-1	Yunnan	+	S and A
143	Yin 11-12	NG8319//SHA4/LIRA	Yunnan	−	S and A
144	Yumai 3	81-2/8334	Yunnan	+	S and A
145	3S857	Bai//108/Flaminio-IDO2229/3/BT 881	Beijing	?	S

a. +=presence of 1B/1R translocation, -=non-presence of 1B/1R translocation, identified according to the method described by Fransis et al.[8]

b. S, Seedling tests with 26 PST pathotypes including CYR32, S*, Seeding tests only with the pathotype CYR32, A, Adult-plant tests ? =Unknown

Seedling testing

A set of 25 differential lines with known stripe rust resistance genes[20] were included in the seedling tests, and the infection types displayed by them were used as a basis for postulating resistance genes in the Chinese lines. The seedling tests for the 98 cultivars and lines were conducted in 2004 and 2005, respectively, using 26 pathotypes with the method as described by Bariana and McIntosh[1] and Niu et al. [22] (Table 2). Six to eight plants of each line tested were grown in a growth chamber. Seedling inoculations were performed by brushing urediniospores of pathotypes from a fully infected susceptible cultivar on to the test seedlings, when the first leaf was fully expanded. Inoculated seedlings were subsequently placed in plastic-covered cages and incubated at 9℃ and 100% RH for 24 hr. Seedlings were then transferred to growth chambers with the day/night regime of 14 hr of light (22, 000 lux) at 17℃ and 10 hr of darkness at 12℃, with 70% RH. Infection types (ITs) were scored 15-16 days after inoculation based on a 0-4 scale[1] when rust was fully developed on the susceptible check, Mingxian 169. Plants with IT 0 to 2^{+} were considered to be resistant and those with IT of 3^{-} to 4, susceptible.

Identification of 1B /1R status

1B/1R translocation lines were identified according to the method described by Fransis et al. [8].

Field testing

Of the 145 cultivars and lines employed in this study, 135 lines and slow rusting check Lantian 12 and highly susceptible check Tiaogan 601 were planted in a randomized complete block design with three replicates in Beijing and Tianshui inthe 2003-2004 and 2004-2005 cropping seasons for the test of adult-plant resistance (Table 1). Each plot consisted of a 2-m row spaced 25 cm apart with 150 seeds. Beijing is the location of main winter wheat breeding program. Tianshui, located in the Gansu province, is considered a 'hot-spot' for stripe rust[42, 47]. Cultivar (cv.) Lantian 12 with typical slow rusting[47] and highly susceptible line Tiaogan 601 were used as slow rusting and susceptible checks, respectively. Spreader rows of Tiaogan 601 were planted perpendicular and adjacent to the rows of tested cultivars. Stripe rust epidemics were initiated by spraying aqueous suspensions of urediospores of PST pathotype CYR32, to which a few drops of Tween 20 (0.03%) had been added, on to the spreader rows at tillering. The stripe rust severity was assessed for the first time 4 weeks after inoculation and then at weekly intervals with four times in total using the modified Cobb scale[25], and infection types were scored using a 0～4 scale[1]. The disease severity data were used to calculate the area under disease progress curve (AUDPC) based on the method described by Broers et al. [3]. AUDPC was calculated as

$$AUDPC = \sum_{i=1}^{n}[(X_i + X_{i+1})/2][T_{i+1} - T_i]$$

Where T_i = number of days after inoculation, X_i = disease severity. Relative AUDPC (rAUDPC) was calculated using the actual AUDPC divided by the AUDPC of susceptible check Tiaogan 601[15,17], and final disease severity (FDS) when the susceptible check (Tiaogan 601) displayed maximum disease severity, were used for statistical analyses. The field data obtained in Beijing during the 2004-2005 cropping season was excluded from analysis due to inadequate infection.

Statistical analysis

The Statistical Analysis System (SAS) procedural PROC GLM was used to conduct analysis of variance with lines, environments that were the combination of location and year, and their interaction effects as fixed, and replicates nested in environments as random. Fisher's F-protected least significant difference (LSD) was used to separate the means for AUDPC and FDS among the lines tested. Lines that were seedling susceptible to CYR32 and showing lower or not significantly higher values of FDS and AUDPC than those of the slow rusting check in field trials were considered to be slow rusting lines of potential use for crop protection and breeding.

Table 2 Seedling infection types produced by 25 differentials (bold) with known resistance genes and 98 wheat cultivars and lines when infected with 26 pathotypes of *Puccinia striiformis* f. sp *tritici* collected from China and other countries. Lines are arranged in groups according to postulated *Yr* genes

Line number	Yr gene	Infection types to *Puccinia striiformis* f. sp. *tritici* pathotypes																									
		61009 (P1)	78028 (P2)	58893 (P3)	75078 (P4)	60105 (P5)	59791 (P6)	82517 (P7)	78080 (P8)	86094 (P9)	72107 (P10)	74187 (P11)	86036 (P12)	76088 (P13)	76093 (P14)	86106 (P15)	68009 (P16)	86107 (P17)	82061 (P18)	80551 (P19)	85019 (P20)	PE92 (P21)	CYR26 (P22)	CYR27 (P23)	CYR29 (P24)	CYRSu-1 (P25)	CYR32 (P26)
Clement	***Yr9*,*YrCle***	**0**	**0**	**0**	**0**	**0**	**0**	**0**	**0**	**3**	**0**	**0**	**0**	**0**	**0**	**0**	**0**	**4**	**4**	**4**	**0;**	**0**	**0**	**0**	**4**	**0;**	**4**
19	*Yr9*	0	0	0	0	0;	0	0	0	4	0	0	0;	0,0;	0	0	0	4	4	4	0;	0	0	0	4	0	4
42	*Yr9*	0	0	0	0	0	0	0	0;	2	0;	0	0	0	0	0	0;	3	2	3	0;	0	0	0;	4	0	4
71	*Yr9*	0	0	0	0	0	0	0	0;	4	0	0	0	0	0,0;	0	0;	4	3	4	0;	0	0	0	4	0	4
78	*Yr9*	0,0;	0	0	0	0;	0;	0,0;	0;	3	0	0;	0,0;	0;	0	0	0;	3	4	4	0,0;	0;	0	0	2,2+	0	3
84	*Yr9*	0	0	0	0;	0;	0	0	0	3,4	0;	0	0	0;	0;	0	0;	4	3	3	0;	0	0	0	4	0	4
91	*Yr9*	0	0	0,0;	0;	0	0;	0;	0;	3	0;	0	0,0;	0	0;	0	0;	3	3	3	0;	0;	0,0;	0;	3	0	4
110	*Yr9*	0	0;	0;	0;	0;	0;	0	0;	4	0;	0	0;	0	0	0	0;	3	3	4	0;	0;	0,0;	0;	4	0	3
116	*Yr9*	0	0	0;	0;	0;	0	0	0;	4	0	0	0;	0;	0,0;	0	0;	4	4	4	0	0	0	0;	4	0	4
121	*Yr9*	0	0	0	0;	0;	0	0	0;	3	0;	0	0,0;	0	0,0;	0	0	4	3	4	0	0;	0	0	4	0	4
123	*Yr9*	0	0;	0	0;	0;	0;	0	0	3	0	0;	0;	0;	0	0;	0;	3	4	2	0,0;	0;	0	0;	3	0	3
135	*Yr9*	0,0;	—	0	0	0	0	0	0	4	0	0	0	0	0;	0	0;	4	4	4	0,0;	0;	0	0	4	0	4
137	*Yr9*	0	0	0	0	0	0	0	0	4	0	0	0	0	0	0	0	3	4	4	0	0	0	0;	4	0	4
138	*Yr9*	0	0;	0	0;	0;	0;	0;	0	4	0	0	0;	0;	0;	0	0	3	3	3	0;	0	0	0;	3+	0	4
139	*Yr9*	0;	0	0	0	0;	0	0;	0	3	0;	0	0	0	0	0	0;	3	4	4	0;	0	0,0;	0	4	0	3
141	*Yr9*	0	0	0,0;	0	0	0	0	0	3	0	0	0;	0,0;	0,0;	0	0	3	4	4	0,0;	0	0	0	3	0	4
142	*Yr9*	0	0	0	0;	0,0;	0	0;	0	4	0;	0	0;	0;	0;	0,0;	0	3	3	4	0,0;	0	0	0	4	0	4
143	*Yr9*	0	0	0	0;	0	0	0;	0;	3	0	0	0	0,0;	0	0	0;	3	3	4	0	0	0,0;	0;	4	0	4
144	*Yr9*	0,0;	0	0	0	0	0	0	0	4	0;	0	0	0	0	0;	0;	4	4	4	0;	0;	0,0;	0	4	0	3
3	*Yr9*,+	0	0	0	0	0	0	0	0	0	0	0	0	0	0	0	0;	3	0;	3	0;	0	0,0;	0;	4	0,0;	4
48	*Yr9*,+	0	0	0	0	0	0	0	0;	0	0	0	0	0	0	0	0	0;,1	3	3	0	0	0	0;	0	0	4
50	*Yr9*,+	0	0	0	0	0	0	0	0	0	0	0	0,0;	0	0	0	0	3	0	0;,1	0	0	0	0;	4	0	3+
73	*Yr9*,+	0,0;	0	0	0	0	0,0;	0	0	0	0	0	0	0,0;	0	0	0	1,2	0;	0;,2	0	0	0	0	3+	0	3
75	*Yr9*,+	0	0	0	0,0;	0	0;	0	0	0	0	0	0	0	0;	0	0;	0	3	0,0;	0	0	0	0	0	0	3
97	*Yr9*,+	0	0	0	0	0	0,0;	0	0;	0,0;	0	0	0	0	0	0	0;	0;,1	3	0;	0;	0,0;	0	0	1+,2+	0	3,4
98	*Yr9*,+	0	0	0	0	0	0;	0	0;	0,0;	0	0	0	0	0	0	0;	0;,1	3	0	0	0	0	0	0;	0	3
131	*Yr9*,+	0	0	0,0;	0	0	0	0	0	0,0;	0	0	0;	0	0	0	0;	0	3	0	0;	0,1+	0	0;	0	0	3

（续）

Line number	Yr gene	Infection types to *Puccinia striiformis* f. sp. *tritici* pathotypes																									
		61009 (P1)	78028 (P2)	58893 (P3)	75078 (P4)	60105 (P5)	59791 (P6)	82517 (P7)	78080 (P8)	86094 (P9)	72107 (P10)	74187 (P11)	86036 (P12)	76088 (P13)	76093 (P14)	86106 (P15)	68009 (P16)	86107 (P17)	82061 (P18)	80551 (P19)	85019 (P20)	PE92 (P21)	CYR26 (P22)	CYR27 (P23)	CYR29 (P24)	CYRSu-1 (P25)	CYR32 (P26)
129	*Yr9*,+	0	0	0	0	0	0	0,0;	0;	0,0;	0	0	0	0	0	0	0	0;,1	3	0,0;	0	0	0	0	0	0	2,2+
74	*Yr9*,+	0	0	0	0	0	0;	0	0	0;	0	0,0;	0	0	0	0	0;	3	2	3+	0	0	0,0;	0	4	0	3
88	*Yr9*,+	0,0;	0	0	0	0	0;	0	0	0;	0	0	0	0	0	0	0;	0;	3	0	0	0	0	0	0	0	4
111	*Yr9*,+	0	0,0;	0	0	0	0	0	0	0;	0	0,0;	0	0	0	0	0	1,2	0;,2	0;,1	0	0	0	0;	0	0	4
124	*Yr9*,+	0	0,0;	0	0	0	0	0	0;	0;	0	0	0	0	0	0	0	0;	0;	0;,1	0	0	0	0	4	0	4
61	*Yr9*,+	0	0;	0	0	0	0	0	0;	0;,1	0	0	0	0	0	0	0	2	0;	3,3+	0;	0,0;	0	0	0;	0	2
87	*Yr9*,+	0;	0	0	0	0	0;	0	0;	0;,1	0	0;	0	0	0	0	0;	1	3	0,0;	0;	0	0	0;	0;,2+	0	2+
21	*Yr9*,+	0	0,0;	0	0	0	0	0	0	2	0	0	0	0	0	0	0	2	3	3+,4	0;	0	0	0;	1+,2	0	1+,2
49	*Yr9*,+	0	0;	0	0	0	0,0;	0	0	2	0;	0	0	0	0	0	0	2+	0,0;	0;	0,0;	0;	0,0;	0	4	0	4
22	*Yr9*,+	0	0,0;	0	0	0	0	0	0	3	0	0	0	0	0	0	0	2	3	3+,4	0;	0	0	0;	1+,2	0	1+,2
31	*Yr9*,+	0	0	0;	0;	0	0	0;	0;	3	0	0	0;	0	0	0	0	4	0,0;	0;	0	0	0	0;	4	0	4
114	*Yr9*,+	0	0	0	0	0	0	0	0;	3	0	0	0;	0	0,0;	0	0;	4	0;,2	0;,1	0	0	0	0	4	0	4
32	*Yr9*,+	0	0;	0;	0	0	0;	0;	0;	4	0;	0	0;	0	0	0	0;	4	0	4	0	0	0;	0;	3	0	4
59	*Yr9*,+	0	0	0	0;	0;	0	0	0	4	0;	0,0;	0;	0;	0	0	0;	4	0;	1,2	0	0	0	0;	4	0	4
86	*Yr9*,+	0	0;	0	0;	0;	0	0	0	4	0	0;	0;	0	0;	0,0;	0	0;	3	4	0;	0	0	0;	2,2+	0	3
103	*Yr9*,+	0,0;	0	0	0;	0;	0	0	0	4	0;	0	0;	0,0;	0;	0	0	3	0	4	0;	0;	0	0	3	0	3,4
***Yr24*/3 * Avocet**	***Yr24***	**0;**	**0;**	**0;**	**2+**	**0;**	**0;,1**	**0;**	**0;**	**0;,1**	**0;,1+**	**0;**	**0;**	**0**	**0;,1+**	**0;**	**0;**	**0;**	**0;**	**0;**	**0;,1**	**0;**	**0;**	**0;**	**0;**	**0;**	**0;**
***Yr26*/3 * Avocet**	***Yr26***	**0;**	**0;**	**0;**	**2+,3**	**0;**	**0;,1**	**0;**	**0;**	**0;**	**0;,1**	**0;**	**0;**	**0;**	**0;**	**0;,2**	**0;**	**0;**	**0;**	**0;**	**0;,1**	**0;**	**0;**	**0;**	**0;**	**0;**	**0;**
36	*Yr26*	0;	0	0;	2+,3	0,0;	0,0;	0;,1+	0;	0	2	0	0;	0	0,0;	0;	0;	0,0;	0	0;	0;	0;	0	0	0	0	0;
38	*Yr26*	0,0;	0;	0;	3	0;	0,0;	0;	0;	0,0;	2+	0;	0;	0	0;	0;	0	0	0	0;	0;	0;	0	0;	0;	0,0;	0;
58	*Yr26*,	0;	0	0;	3	0;	0	0;	0;	0;	0;	0;	0;	0,0;	0,0;	0;	0;	0;	0;	0;	0;	0;	0	0;	0;	0	0,0;
83	*Yr26*	0;	0;	0;	3,4	0;	0;	0;	0;	0;	0;,1	0;	0;	0,0;	0;	0;,1+	0;	0;	0;	0;	0;,1	0;	0,0;	0;	0,0;	0	0;
90	*Yr26*	0;	0,0;	0;	2+,3	0,0;	0;	0;	0;	0;	2	0;	0;	0;	0;	0;	0;	0;	0;	0;	0;	0;	0	0	0;	0;	0;
94	*Yr26*	0,0;	0,0;	0;	2+,3	0;	0	0;	0;	0;	0;,1	0;	0;	0;	0;	0;	0;	0;	0;	0;	0;	0;	0	0,0;	0;	0;	0;
96	*Yr26*	0;	0;	0;	2	0	0;	0;	0;	0;	0;,1	0;	0;	0;	0;	0	0;	0;	0;	0;	0;	0;	0	0;	0;	0,0;	0;
100	*Yr26*	0,0;	0,0;	0;	2+,3	0;	0;	0;	0;	0;	0;,1	0;	0;	0;	0;	0;	0;	0;	0;	0;	0;	0;	0	0	0;	0;	0;
104	*Yr26*	0	0	0;	3	0	0;	0,0;	0;	0	0;	0,0;	0	0,0;	0;	0;	0;	0	0;	0	0;	0	0	0;	0;	0,0;	0;

（续）

Line number	Yr gene	Infection types to *Puccinia striiformis* f. sp. *tritici* pathotypes																									
		61009 (P1)	78028 (P2)	58893 (P3)	75078 (P4)	60105 (P5)	59791 (P6)	82517 (P7)	78080 (P8)	86094 (P9)	72107 (P10)	74187 (P11)	86036 (P12)	76088 (P13)	76093 (P14)	86106 (P15)	68009 (P16)	86107 (P17)	82061 (P18)	80551 (P19)	85019 (P20)	PE92 (P21)	CYR26 (P22)	CYR27 (P23)	CYR29 (P24)	CYRSu-1 (P25)	CYR32 (P26)
113	*Yr26*	0;	0	0,0;	2	0;	0;	0;	0;	0;	0;	0;	0;	0	0;	0,0;	0;,1	0;	0;	0;	0;	0;	0	0;	0	0;	0;
117	*Yr26*	0	0	0;	3	0;	0	0,0;	0;	0,0;	0;	0	0	0	0,0;	0;	0	0;	0;	0;	0;	0;	0	0	0,0;	0	0,0;
126	*Yr26*	0;	0	0;	2,3	0;	0;	0;	0;	0;	0;	0	0;	0;	0;	0;	0;	0,0;	0	0;	0;,1	0;	0	0	0;	0;	0;
127	*Yr26*	0,0;	0	0;	3	0,0;	0;	0;	0;	0,0;	0;,1	0	0,0;	0,0;	0;	0,0;	0;	0;	0;	0;	0;	0;	0	0;	0	0	0;,1+
128	*Yr26*	0,0;	0,0;	0;	3	0;	0;	0;	0;	0,0;	0;,1	0	0;	0;	0,0;	0;,1+	0;	0;	0;	0	0;	0;	0	0;	0;	0	0;
136	*Yr26*	0,0;	0	0;	3	0;	0;	0;	0;	0;	1	0,0;	0;	0;	0;	0;,2	0	0,0;	0;	0;	0;	0;	0	0	0;	0	0;
107	*Yr26*,+	0;	0	0;	0;	0;	0;	0;	0;,1	0;	0;	0	0	0;	0;	0;	0;	0;	0;	0;	0;	0,0;	0	0;	0;	0;	0;
115	*Yr26*,+	0;	0,0;	0;	0;	0;	0	0;	0;	0,0;	0;	0,0;	0;	0,0;	0;	0;,1	0;	0,0;	0;,1	0;	0	0;	0	0	0;	0,0;	0;
125	*Yr26*,+	0;	0;	0;	0+	0;	0;	0;	0;	0;	0;	0	0;	0,0;	0;	0;	0;	0;	0;	0	0;	0	0	0	0;	0	0;
134	*Yr26*,+	0,0;	0	0;	0;,1,2	0;	0;	0;	0;	0;	1,1+	0,0;	0,0;	0,0;	0,0;	0;	0;	0;	0;	0;	0;	0;	0	0;	0	0	0;
Maris Huntsman	***Yr2*,*Yr3a*,*Yr4a*,*Yr13***	**0;**	**0;**	**0;,1+**	**3**	**0;,1+**	**3**	**4**	**4**	**0;**	**0;**	**3**	**0;**	**0**	**0;**	**0;,1+**	**4**	**0;**	**4**	**3**	**4**	**1+,2+**	**3+**	**3**	**4**	**0;**	**3+,4**
82	*Yr2*,*Yr3a*,*Yr4a*,+	0,0;	0	0;	0;	0	0	0;	0;,2	0,0;	1	0;	0;	0	0;	0;	0;	0	2	0;	3	0;	3	3	0,1	0;	4
Heines VII	***Yr2***	**0;,1**	**1,2+**	**0;,1+**	**3,3+**	**0;,2+**	**4**	**4**	**3**	**0**	**0;,1**	**0**	**0;**	**0;**	**1+**	**4**	**4**	**3**	**4**	**3**	**2**	**2**	**3**	**3**	**4**	**1**	**4**
Yr6/6*Avocet S	***Yr6***	**3**	**4**	**1,1+**	**3**	**0;**	**0;**	**1+,2**	**4**	**4**	**2+**	**4**	**0;**	**3**	**3+,4**	**4**	**4**	**4**	**3**	**2+**	**4**	**3+,4**	**3+**	**4**	**4**	**3**	**4**
133	*Yr2*,*Yr6*,+	0;	0	0;	0,0;	0;	0;	0;	2,0;	0	1+	0;,1	0;	0;	0;	3	0;,1	0;,1	2	0;	2	0;	4	3	4	0;	4
41	*Yr6*,+	3	0;	0;,2	4	0;	0	0;,1+	3	3	3	0	0;,1+	3	3+,4	0;,1	0;	3	4	2+	3	0;	4	4	0;	4	4
52	*Yr6*,+	0;	2	0;,1	0;	0;	0;	0;	0;	4	0;	0;	0;	0;	2,2+	4	4	4	0;,1	2+	0;	0;	0;	4	4	2	4
106	*Yr6*,+	0;	3	0,0;	0;,1,2	0,0;	0;	0;	0;,1	0,0;	1	0	0	0;,2	0;,2	0;,2	0;	3	0;,1	2+	2	3	2+	0;	1+,2	3	4
Vilmorin 23	***Yr3a*,*Yr4a*,**	**4**	**0;**	**0;**	**3**	**0;,2+**	**1**	**4**	**4**	**1**	**0;**	**3**	**0;,2+**	**0;**	**0;**	**0;**	**2**	**0,1**	**4**	**3**	**3**	**0;**	**1,2**	**4**	**4**	**0**	**2+**
Nord Desprez	***Yr3a*,*Yr4a*,**	**4**	**1,2**	**0;,1+**	**3**	**0;,1**	**2+**	**4**	**4**	**2**	**0;**	**3**	**0;**	**0;**	**1,1+**	**0;**	**2**	**2**	**4**	**3**	**4**	**0;**	**4**	**4**	**4**	**1**	**4**
130	*Yr3a*,*Yr4a*,+	2	0;,1	1,1+	2+	1,2+	0;	3+	0;	2	0;,1	3	1,2	0;	2,2+	2+	0;	0;,2	0;	2,2+	0;,1	0;,1	3	4	3+,4	0;	4
Lee*	***Yr7*,*Yr22***	**0;**	**4**	**0;**	**2+,3**	**0;,1**	**0;**	**0;,1**	**0;**	**4**	**4**	**0;**	**0;**	**4**	**4**	**4**	**3+**	**4**	**0;**	**2+**	**0;,1**	**3**	**4**	**0;**	**4**	**4**	**4**

（续）

Line number	Yr gene	Infection types to *Puccinia striiformis* f. sp. *tritici* pathotypes																									
		61009 (P1)	78028 (P2)	58893 (P3)	75078 (P4)	60105 (P5)	59791 (P6)	82517 (P7)	78080 (P8)	86094 (P9)	72107 (P10)	74187 (P11)	86036 (P12)	76088 (P13)	76093 (P14)	86106 (P15)	68009 (P16)	86107 (P17)	82061 (P18)	80551 (P19)	85019 (P20)	PE92 (P21)	CYR26 (P22)	CYR27 (P23)	CYR29 (P24)	CYRSu-1 (P25)	CYR32 (P26)
Reichersberg 42	***Yr7***	0;	4	0;	1+	0;,1	0;	0;,1	0;	2+	0;	0;,1	0;	1+,2	2+	4	4	4	0;	1+	0;	2,2+	2+	0;,1	4	3	2+
92	*Yr7*,+	2	2+,3	0;,1+	3+	2	0;	2	0;	1,2	3+	0;,1	0;	3	3	0;,2	0;	3	0;	2+	0;,1	1+,2	3	0;	4	4	4
Yr15/6 * Avocet S	***Yr15***	0;	0	0	0	0	0	0	0	0;	0	0	0	0	0	0	0;	0;	0;	0	0;	0	0	0;	0	0	0,0;
35	*Yr15* or other genes	0;	0	0;	0	0,0;	0	0;	0;	0	0;	0	0	0	0;	0;	0	0,0;	0	0;	0;	0;	0	0;	0,0;	0	0;
80	*Yr15* or other genes	0	0,0;	0	0	0	0;	0	0;	0	0	0	0,0;	0	0	0	0	0	0	0	0;	0	0	0;	0	0	0,0;
145	*Yr15* or other genes	0	0	0;	0;	0;	0;	0;	0;	0	0	0;	0;	0	0	0;	0;	0	0;	0	0;,1	0;	0	0;	0	0	0;
Selkirk	***Yr27***	1	2,2+	3	3	3	0;	2	2	2	4	3	0;	3	3	0;	0;	2	0;	3	0;	3	3	2	3+	2+	4
112	*Yr27*,+	0	0,0;	0	3	0,0;	0;	0;	0;	0,0;	4	0	0,0;	0,2+	1+	0;,1	0;,1	0	0;	2+	0;	0;,1	3	0	0	1,2	3
Strubes Dickkopf	***YrSD***	4	0;,1+	4	3	4	4	4	4	3	0;,1	0	4	0;	1+	4	4	2	4	3	3	0;	0;	3	4	2	3
64	*YrSD*,+	3	0;,1+	1+,2	3	1,2	2	4	0;	3	0;	0	3	0	2	4	2	0;,2	4	3	3	0;	1,2	4	4	0;	2+,3
72	*YrSD*,+	4	0;,1	0;,2	3	1,2	0;	3	3	4	0;	0;	2+	0;,1	0;	2	2	0;,1	3	0;,1+	3	0;,2	0	3	3+,4	0;	0;,1+
109	*YrSD*,+	3	0,0;	0;	0,2	1,2	0;,1,2	3	3	0,0;	1,1+	0,0;	2+	0;,1	0;,1+	0,0;	3	0;,1	3	2+	3+	0;,1	0;	0;	0	0;,1	1+,2
Chinese 166	***Yr1***	0;	0,0;	0;	0;	3+	0;	4	0	0	3	0	0;	0	4	3+	4	0	0;	4	4	0,0;	4	4	4	4	4
Hybrid 46	***Yr3b*, *Yr4b*,**	0;	0;	0;,1	3,3+	0;	0;	2+,3	4	0;	0;	0; 1	0;	0;	0;	0;	0;	1	4	1,2	4	0;	0;	0;	3	0;	3
Compair	***Yr8***	0;	4	0,0;	4	0	0	0,0;	0	3	4	0;	0;	3+	4	3	0	3	0;	4	0	4	0;	3	0;	3	0;,1
Moro	***Yr10***	0	0	0,0;	3+,4	0	0	0	0	0	4	0	0,0;	0	0	0	0	0	0	0	0	0;	0	0;	0	0	0
VPM	***Yr17***	0;	2+,3	0;	1,2	1,2	0;,1	1+,2	2	0;	0	0	0,0;	0	2+	2	0;	0;	0;	3	0;	0;,1+	0;	0;	4	2	4
Funo	***YrA***	3	3+,4	0;	3,3+	0;,1	3	2+	3	4	1,1+	3	0;,1	2+	1+	0;,2	2+	4	2	2+	2	4	4	4	4	0;	4
Alba*	***YrAlba***	3	0;	0;,1+	3,3+	3	3,4	3+	4	2	0;	0;,1	0;,1+	0,0;	0;	3,4	0;	2	0;	2	3	2+	4	4	4	1	3+
Carstens V	***YrCV***	3	0	2+	3	4	0;	3	3	4	2+	4	0;	0;	0;	0;	3	1	4	0;,1	4	0;	0;	2	4	0;	3
Gaby	***YrGaby***	0;	2+	0;	0;	0;	0;	0;	4	3	0;	0;	0;	0;	3	4	4	2+	3	0;,1+	4	3	0;	4	4	2	2+

（续）

Line number	Yr gene	Infection types to *Puccinia striiformis* f. sp. *tritici* pathotypes																									
		61009 (P1)	78028 (P2)	58893 (P3)	75078 (P4)	60105 (P5)	59791 (P6)	82517 (P7)	78080 (P8)	86094 (P9)	72107 (P10)	74187 (P11)	86036 (P12)	76088 (P13)	76093 (P14)	86106 (P15)	68009 (P16)	86107 (P17)	32061 (P18)	80551 (P19)	85019 (P20)	PE92 (P21)	CYR26 (P22)	CYR27 (P23)	CYR29 (P24)	CYRSu-1 (P25)	CYR32 (P26)
Resulka	***YrRes***	**0**	**0**	**0**	**0;**	**0;**	**0**	**0**	**0**	**0**	**0**	**0**	**—**	**0**	**—**	**0**	**0**	**0;**	**—**	**3**	**0**	**0;**	**0**	**—**	**4**	**0**	**3**
Spaldings Prolific	***YrSpp***	**0**	**0;**	**0;,1**	**2+,3**	**2,3**	**0**	**0;**	**1**	**0**	**0**	**3**	**0;**	**0,0;**	**0**	**2+**	**0;**	**0;**	**3**	**0;**	**3**	**0,0;**	**0**	**4**	**0**	**0**	**4**
Suwon 92/Omar	***YrSu***	**0,0;**	**0;**	**0;**	**4**	**0;**	**0**	**0;,1+**	**4**	**0**	**4**	**4**	**0**	**0;**	**2+**	**0;**	**0**	**0**	**3**	**3-**	**0**	**0**	**0,2**	**0**	**0;**	**4**	**4**
1	+	3	3	4	3	1,2	3	4	4	3	4	3	3	0;	0	3	4	3	3	4	0;	3	4	4	0;	1,2	4
33	+	3	3	3,3+	3	3	0;	3+	0;	0	2+	3	3,4	0;	2+	3	0;	3	4	3	0;	3,3	0	3	4	4	4
34	+	0	0	2,2+	2	3	0,0;	3	0;	4	3	3	0;	0,0;	0;,2	0	0	0	4	0;	3	0;	0	4	4	0	4
39	+	0;,2	0	3	3	2+	0,0;	3	3	3	3	3	0;,1	0;,1	0;,1	1,2	0;	3	0	4	3	3	4	3	2+	1,2	3
47	+	4	3	4	4	4	4	4	3	4	3	4	3	4	3	4	4	4	4	4	4	4	4	4	3	3	4
51	+	3	0;	3	4	3	2,3	4	3	2+	3	1	3	3	2	3	3	3	4	3	3	3	4	3	4	3	4
53	+	4	4	4	3	3	4	3	0;	3	4	3	4	2+	3,4	4	3	3	3	4	4	4	3	4	3	3	4
66	+	3	3	4	3	4	3	3	3	3	3	3	0;	0,0;	0,0;	3	2	4	4	4	4	4	4	4	3,4	1,2	4
68	+	4	3	4	4	4	3	4	4	4	4	4	4	4	4	4	4	4	4	4	4	4	4	4	4	4	4
69	+	4	3	4	4	4	4	4	3	4	4	4	4	3	4	4	4	4	4	4	4	4	4	4	4	4	4
70	+	4	3	4	4	4	4	4	4	4	4	4	4	3	3	4	3	4	4	4	4	4	3	4	4	3	4
79	+	4	0;,1	+3,3+	3+	3+,4	0;	3	3	3	1+,2	4	0,0;	3	0	0	0;	3	1	3,3+	4	0;,1	3	3	3+	3	4
81	+	1,2	3	3	3	0;,1	0	3	3	0;,2	0;	3	3	0	3	0	1,2	3	4	4	4	0	0	3	3,4	4	4
89	+	4	4	4	4	4	4	3	3	3	4	4	4	3	4	4	3	3	3	4	4	4	4	4	3	4	3
93	+	0;,2	3	3	2	1,2	0	3	3	2	0;	2	3	3	2+	0,1	3	3	4	4	4	4	4	4	4	3	4
99	+	3,4	0	4	4	3	0	0,2	1,2	3	3	3	4	3	0;	1,2	1,2	0;,1	1	4	3	4	0;	3	4	4	3
101	+	4	4	3	4	3	1,2	3	3	4	3	3	3	3	0;	0;,1	2+	3	0;,1	4	3	4	0;	2	4	3	4
102	+	4	3	3	4	3	3	3	3	4	3	4	4	4	4	3	3+	4	3	4	4	4	4	3	4	3	4
105	+	4	4	4	4	4	4	4	4	4	4	4	4	4	4	4	4	3	4	4	4	0	0;,1	4	4	4	4
108	+	3	4	4	3+	0;	0;	0;,1	0;,1	3	3	4	3	3	4	0;,1	3	2	0;	3	3	4	2	3	4	0;,2	3
122	+	4	3	3	4	3	3	4	3+	2+	4	4	3	1,2	4	0;,2	3	4	3	4	4	3	4	4	3	3	3
132	+	0,0;	0;	0,0;	0,0;	0	0;	0;	0	0,0;	4	0	0,0;	0;	0;	0;	0	0,0;	0;	0;,1+	0;	0;	0;	0;	3+	0;	4
140	+	3	0	4	3+	0;,2	3	3	4	2	3	0;,2	3	2	3	0;,1	0;	3	3	4	4	3	2	4	3	4	4

+ = unknown gene (s)

— = untested

Results

Seedling reaction

The reactions of the standard differential lines with known resistance genes to the 26 PST pathotypes are included in Table 2. Line *Yr15*/6 * Avocet S displayed high resistance to all pathotypes tested. Lines *Yr24*/3 * Avocet S and *Yr26*/3 * Avocet S showed a moderately susceptible reaction only to pathotype P4 (75078). Moro with *Yr10* gave high infection types to pathotypes P4 (75078) and P10 (72107). Thus *Yr10*, *Yr15*, *Yr24*, and *Yr26* confer effective resistance against Chinese prevailing PST pathotypes, whereas lines with *Yr1 and Yr6* were susceptible to them.

The resistance genes *Yr2*, *Yr3a*, *Yr4a*, *Yr6*, *Yr7*, *Yr9*, *Yr26*, *Yr27*, and *YrSD*, either independently or in certain combinations, were postulated in 72 lines, whereas 26 accessions did not contain any known stripe rust resistance genes.

Eighteen lines possessed only *Yr9*, whereas another 24 lines possessed *Yr9* plus one or more unknown resistance genes according to their reaction patterns (Table 2). All lines postulated to have *Yr9* also amplified a SCAR marker for rye chromatin[8] (Tables 1 and 2).

Fifteen lines were postulated to have *Yr26* on the basis that only P4 (75078) was virulent. Pedigree information (Table 1) indicated that four additional lines (L107=LB 0438, L115=Mian 2000-8, L125=Nei 2938, and L134=Xike 01015), highly resistant to all pathotypes, probably carry *Yr26* and one more unknown gene.

Four lines (L133=W7, L41=Shaan 512, L52=Xinong 914, and L106=LB 0288) carried *Yr6* based on their responses to the pathotypes avirulent to *Yr6*/6 * Avocet S (Table 2). L133 conferred resistance to pathotypes lacking pathogenicity for *Yr2*, indicating that it contained *Yr2* and *Yr6*. Two lines (L82=351-15, L130 = R88) displayed low infection types to pathotypes avirulent to differentials Vilmorin 23 and Nord Desprez, indicating they may contain *Yr3a* and *Yr4a*. In addition, line L82 also carried *Yr2*, being resistant to pathotypes lacking virulence for *Yr2*. Three lines (L64 = Annong 98005, L72 = An 96-8, and L109=Lemai 3) carried *YrSD*, displaying low ITs to the pathotypes avirulent to *YrSD*. L92 (CD 1485-6) conferred resistance to all pathotypes avirulent to *Yr7* and some pathotypes virulent for *Yr7*, indicating it carried *Yr7* and unknown resistance genes.

Low infection typesproduced by L112 (Mian 2000-18) to pathotypes avirulent to Selkirk (*Yr27*) and some pathotypes virulent to *Yr27* indicated the presence of *Yr27* and other unidentified resistance genes.

Six lines (L47=Xiaoyan 921, L68=Yangfu 2, L69=Yangmai 10, L70=Yangmai 158, L89=99422, and L102=Chuanyu 19) showed high infection types to all pathotypes tested, indicating they did not have any detectable stripe rust resistance genes present in the reference differentials. The reactions of 17 lines (L1=Han 3475, L33=2611, L34=953, L39=Nonglin 9823, L51=Xinong 8925-13, L53=Yunfeng 988, L66=Emai 14, L79=298-1, L81=3472, L93=Chen 98-18, L99=Chuannong 21, L101=Chuanyu 18, L105=Kefeng 14, L108=LB 0458, L122=Mianyang 29, L132=SW8588, and L140=Fengmai 24) were not consistent with those of any known resistance genes. Hence more work is needed to identify the resistance genes in these lines.

Three lines L35 (N139), L80 (323-2), and L145 (3S857) confered resistance to all pathotypes tested, indicating they possess either *Yr15*, a gene combination, or one or more unidentified resistance genes.

Slow rusting resistance in field tests

Highly significant differences were foundfor genotypes and environments for relative AUDPC and FDS in the field trials (Table 3). The genotype-environment interaction was also significant, but its effect on variation was much lower than that for genotype differences.

FDS of susceptible check, Tiaogan 601, was 90-100 in Beijing and Tianshui (Table 4), indicating disease developed well at both locations. The FDS of the slow rusting check, Lantian 12, was 13 and 19, respectively, while its relative AUDPC was 0.17 and 0.16, respectively, at the two locations (Table 4).

Table 3 Analysis of variance of relative AUDPC and final disease severity (FDS) in 137 wheat lines including slow rusting and susceptible checks tested in Beijing in 2003-2004, and in Tianshui in 2003-2004 and 2004-2005 cropping seasons

Parameter	Source of variation	DF	MS	F-value	Pr>F
AUDPC	line	136	0.92	106.89	<0.000 1
	env	2	1.36	157.51	<0.000 1
	replicate	6	0.009 4	1.10	0.36
	lines×env	272	0.062	7.15	<0.000 1
	Error	816	0.008 6		
FDS	line	136	9 972.99	207.46	<0.000 1
	env	2	13 181.77	274.21	<0.000 1
	replicate	6	35.90	0.75	0.61
	lines×env	272	416.60	8.67	<0.000 1
	Error	816	48.07		

Table 4 Wheat lines with slow rusting resistance to stripe rust identified by using Fisher's F-protected LSD (LSD alpha=0.05) **test in Beijing in 2003-2004, and in Tianshui in 2003-2004 and 2004-2005 cropping seasons**

Line number	Seedling IT to CYR32	Beijing (2003-2004)			Tianshui (2003-2004)			Tianshui (2004-2005)		
		IT	FDS	rAUDPC	IT	FDS	rAUDPC	IT	FDS	rAUDPC
Lantian12[a]	4	1, 2	13	0.17	1, 2	17	0.14	3	21	0.18
Tiaogan 601[b]	4	4	90	1	4	95	1	4	100	1
Lankao 906-4 (L9)	4	3, 4	25	0.24	3, 4	12	0.12	4	12	0.1
Xinong 291 (L50)	4	0;	1	0.01	2	1	0.01	0;, 2	1	0.01
Shaannong 981 (L44)	4	2, 3	12	0.19	2, 3	15	0.1	2, 3	25	0.21
90-99-5 (L54)	4	3, 4	3	0.11	3, 4	7	0.09	4	17	0.21
94t-133 (L56)	3+	0;	1	0.01	1, 2	3	0.06	1, 2	2	0.07
96-22-2-3 (L62)	4	1, 2	6	0.07	2, 3	10	0.07	2, 3	6	0.09
Lantian 9 (L63)	4	2, 3	6	0.06	0;, 1	1	0.02	1, 2	5	0.1
Guizhou 98-18 (L74)	3	2, 3	6	0.15	3, 4	24	0.15	3, 4	24	0.25
01-018 (L78)	3	1, 2	4	0.1	3, 3+	15	0.14	3, 4	1	0.05
01-10251 (L76)	4	1, 2	19	0.25	2, 3	16	0.13	2, 3	8	0.1
351-15 (L82)	4	1	2	0.02	2	2	0.03	2	11	0.12
58749 (L84)	4	3, 4	9	0.11	3, 4	4	0.04	3, 4	4	0.1
93Xia032-1 (L86)	3	3, 3+	18	0.21	3, 3+	8	0.08	3, 3+	7	0.11
99422 (L89)	3	2, 3	16	0.2	2, 3	4	0.07	2, 3	4	0.07
Bi 2002-2 (L73)	3	2, 3	14	0.1	3	8	0.07	3	12	0.11
Chen 98-18 (L93)	4	2, 3	16	0.26	2, 3	22	0.23	2	24	0.23
Chuannong 17 (L98)	3	1, 2	11	0.11	2, 3	12	0.12	2, 3	11	0.14
Chuannong 21 (L99)	3	1, 2	5	0.09	2, 3	12	0.11	3, 4	15	0.14
Chuanyu 18 (L101)	4	1, 2	1	0.1	3	2	0.03	3+	2	0.02

（续）

Line number	Seedling IT to CYR32	Beijing (2003-2004)			Tianshui (2003-2004)			Tianshui (2004-2005)		
		IT	FDS	rAUDPC	IT	FDS	rAUDPC	IT	FDS	rAUDPC
Chuanyu 19 (L102)	4	3, 4	8	0.14	3, 4	5	0.06	4	4	0.11
LB 0288 (L106)	4	1, 2	15	0.07	2, 3	3	0.04	2, 3	14	0.11
LB 0458 (L108)	3	1, 2	7	0.05	1, 2	4	0.05	2	12	0.1
Mian 2000-13 (L111)	3	1, 2	9	0.18	3	12	0.08	3, 4	7	0.1
Mian 2000-18 (L112)	3	1	1	0.01	2	1	0.02	2, 3	2	0.06
Mian 2000-36 (L114)	4	1, 2	2	0.04	3, 4	9	0.09	3, 4	21	0.2
N1621-19-4 (L123)	3	1, 2	2	0.16	2	1	0.03	3, 4	5	0.11
N711 (L124)	4	0;	5	0.01	1, 2	1	0.03	2, 3	2	0.07
R88 (L130)	4	0;	1	0.01	3	2	0.02	3, 4	1	0.03
R97 (L131)	4	3	1	0.09	3	9	0.08	1, 2	2	0.06
SW 8588 (L132)	4	0;	1	0.01	2+	1	0.01	3	3	0.03
9021 (L139)	3	0;, 1	5	0.02	2, 3	22	0.19	2, 3	11	0.15
Jing 035-1 (L141)	4	0;, 1	11	0.19	0;, 2	21	0.21	2, 3	21	0.15
Yin 11-12 (L143)	4	3	3	0.04	2, 3	6	0.09	3	12	0.15
LSD (0.05)			13.01	0.21		10.73	0.12		9.40	0.10

a. Slow rusting check

b. Susceptible check

Twenty-six lines, including L23 (Zhoumai 11), L55 (93-23-13), L57 (95-108), L58 (95-111-6-3-10), L60 (95-3-5-1), L61 (9562-1), L72 (An 96-8), L77 (01-3570), L80 (323-2), L83 (58512-2), L90 (99-607), L94 (Chuanmai 42), L95 (Chuanmai 43), L104 (Jinfeng 626), L107 (LB 0438), L109 (Lemai 3), L113 (Mian 2000-34), L115 (Mian 2000-8), L117 (Mianmai 37), L125 (Nei 2938), L126 (Neimai 9), L127 (R111), L128 (R122), L129 (R25), L134 (Xike 01015), and L136 (Xingmai 2), were identified as highly resistant cultivars with high resistance to the pathotype CYR32 at both seedling and adult stages. Thirty-three lines (Table 4) showed slow rusting resistance at two locations, exhibiting high infection types at seedling stage, but low FDS and AUDPC in the field tests.

The correlations of FDS and relative AUDPC between Beijing and Tianshui were 0.88 and 0.80, respectively, indicating a high consistency of the test in two locations for slow rusting resistance. The correlation between relative AUDPC and FDS at the two locations varied from 0.91 to 0.98 indicating that FDS could be used as a reliable indicator of AUDPC.

Discussion

Seedling test for stripe rust

In the present study, *Yr9* was identified in 42 of 98 (43%) lines, which is consistent with the results reported by Niu et al.[22] and Wu et al.[43]. *Yr26* was postulated to be present in 19 lines. This gene was derived from Chinese line γ80-1 (*Triticum turgidum* L.) and its derivatives such as 92R137, 92R171, and 92R178 from Nanjing Agricultural University (Table 1)[16]. The lines possessing *Yr26* are grown over a large area in Gansu and Sichuan provinces where climatic conditions are favorable to stripe rust development. If a new PST race virulent to *Yr26* appears, a new epidemic could again cause severe losses as the case in 1990 and with the Fan 6-derived resistance in 2002. The danger of using limited sources of resistance has become a great concern to wheat breeders and pathologists. It is imperative that new stripe rust resistance genes be identified and utilized in breeding programs. Although *Yr26* currently provides effective resistance in China, breeding programs should be anticipating its downfall by improving the genetic diversity of breeding materials.

The *Yr26*-virulent race (P4 = 75078) originated from Egypt[21]. Apart from its *Yr26* virulence it is probably not a threatening pathotype. This culture was also virulent on the line *Yr24*/3 * Avocet S, indicating that *Yr26* and *Yr24* are likely the same gene, which is in agreement with our previous study[10]. Although *Yr24* was also derived from *T. turgidum* [19], its origin and naming was independent of that for *Yr26*.

Near-isogenic lines (NIL) with one resistance gene were often utilized in the postulation of resistance genes to wheat leaf rust[23,34]. Nevertheless, in the postulation of stripe rust resistance genes, differential lines with one or more genes were often employed[24, 26]. In the present study, 8 of 25 differential lines contain more than one resistance gene, leading to some difficulties in gene postulation in genetically unknown materials. Some gene combinations may be identified, but others cannot be postulated due to a lack of testers with confirmed single resistance genes. The near-isogenic lines developed in Avocet S may be a good option for reference lines[39], but the series including limited known genes needs to be extended to cover a wider range of genes. The differential lines used in this study are all with universally known genes[20]. Most of them contain only one gene, assuring the postulation of resistance genes. In addition, the most frequent resistance gene *Yr9* identified in the seedling test was confirmed by a SCAR marker of 1B/1R (Table1), and the pedigree information was also used in the postulation of seedling resistance genes.

In the seedling test to stripe rust, it causesvery likely confusion with IT 0 confusing true immune response (low) and escape (missing data) with the other response that infection has occurred but variability between tests can still be a problem. In the present study, the seedling tests were conducted for two times and the results were consistent between two seedling tests with very few exceptions. In case disease scores with some pathotypes tested were not consistent between two repeats, the data with disease developed better were used for the postulation of seedling resistance genes, because the low infection types might be due to inadequate inoculation.

Field resistance

Of 33 lines with slow rusting resistance to CYR32, 26 were tested at seedlings with 26 PST pathotypes; 15 lines carried *Yr9* and 6 lines gave high ITs to most of the pathotypes (Table 2). In these lines the major seedling resistance genes were overcome by pathotype CYR32 and the slow rusting resistances probably resulted from minor genes. The resistance in 96-22-2-3 (L62) may be conferred by *Yr16* derived from Cappelle Desprez (Table 1)[41]. Slow rusting in 94t-133 (L56) may be derived from Xiannong 4 with durable resistance to stripe rust (Table 1)[46]. Slow rusting resistance in Chuanyu 18 (L101), Chuanyu 19 (L102), LB 0288 (L106), LB 0458 (L108), 9021 (L139), and Yin 11-12 (L143) may be derived from CIMMYT wheat germplasm (Table 1). All lines except for Lankao 906-4 (L9) with slow rusting resistance identified in this study were from northwestern and southwestern China where stripe rust resistance is a major breeding objective, indicating that more attention was paid to wheat stripe rust resistance in western China than other wheat grown regions.

Breeding for durable resistance with seedling and slow rusting genes

In China, urediospores of PST lead over summer in southeastern Gansu and northeastern Sichuan, where climatic conditions, geographic characteristics, and cropping systems are favorable for stripe rust every year[11]. The spores spread from west to east in the autumn, winter and early spring, and result in epidemics of stripe rust in the main wheat grown region of China. Therefore, reducing the amount of initial inoculum in the over-summer regions is a key strategy to achieve sustainable control of stripe rust in China[35]. Several strategies such as pyramiding of major resistance genes, deployment of different genes in mountainous region (where urediospores can live in summer) and plain (over-winter region), and reducing wheat acreage by planting other crops can largely reduce inoculum in the over-summer regions[35,42]. In the present study, 26 lines with good agronomic performance were found to be highly resistant to stripe rust both in the seedling and adult plant stage with different resistance genes, which may be used as parents in the wheat breeding program for gene deployment and gene pyramiding in southeastern Gansu and northeastern Sichuan to reduce the amount of urediospores of *Puccinia striiformis* f. sp. *tritici*. This will certainly benefit for the control of stripe rust in other wheat grown regions.

The eastern and northern epidemic region covers major wheat production areas with less frequent occurrence of favorable climatic conditions, especially in Hebei, Henan, Shandong, and Shanxi provinces[35]. Strategies for controlling stripe rust in these regions are slightly different from those in the over-summer regions of western China. In addition to pyramiding and deploying major resistance genes, utilization of minor additive genes was strongly proposed for controlling stripe rust[22]. During the past 5 years, a shuttle-breeding program was conducted between CIMMYT and China for developing slow rusting cultivars with a limited backcrossing method[32]. CIMMYT wheat cultivars Chapio, Tukuru, Kukuna, and Vivitsi with 4 to 5 minor additive genes were used as parents in wheat breeding. In this study, six lines including Chuanyu 18 (L101), Chuanyu 19 (L102), LB 0288 (L106), LB 0458 (L108), 9021 (L139), and Yin 11-12 (L143) were identified to have slow rusting resistance, which were derived from the CIMMYT wheat germplasm, indicating that it is possible to combine slow rusting resistance with good agronomic performance through shuttle breeding between Mexico and China. Because multiple genes may confer durable resistance[27,32], the use of slow rusting resistance should be a more sustainable strategy for developing wheat cultivars in eastern and northern China.

In the evaluation of adult-plant resistance (APR) to wheat rusts, AUDPC is certainly a more reliable indicator than FDS. However, in the wheat breeding program, breeders need toevaluate the disease severity for hundreds of thousands of individuals and thousands of lines. They may have no time to score the disease for several times to calculate AUDPC. A simple method is, therefore, needed for evaluating such a great number of materials. In the present study, a high correlation was found between AUDPC and FDS, which is consistent with our previous studies in the characterization of APR to powdery mildew in wheat[12,38]. Hence, we suggested using FDS as an indicator for the evaluation of APR to stripe rust in wheat breeding with a single scoring at an appropriate time. The high correlations of FDS and AUDPC between Beijing and Tianshui indicated a high degree of repeatability between sites and seasons. The Tianshui environment with low temperature and much more rain is very favorable to the development of stripe rust and for selection of slow rusting resistance in wheat breeding populations.

Acknowledgments

The authors are very grateful to Prof. R. A. McIntosh for reviewing the manuscript. This project was funded by the National 863 program (2003AA207090) and National Natural Science Foundation of China (30220140636 and 30471083).

References

[1] Bariana, H. S., and McIntosh, R. A. 1993. Cytogenetic studies in wheat XV. Location of rust resistance genes in VPM1 and their genetic linkage with other disease resistance genes in chromosome 2A. Genome. 36: 476-482.

[2] Bartos, P., Johnson, R., and Stubbs, R. W. 1987. Postulated genes for resistance to yellow rust in Czechoslavakian wheat cultivars. Cereal Rusts Bull. 15: 79-84.

[3] Broers, L. H. M., Cuesta Subias, X., and Lopez Atilano, R. M. 1996. Field assessment of quantitative resistance to yellow rust in ten spring bread wheat cultivars. Euphytica. 90: 9-16.

[4] Caldwell, R. M. 1968. Breeding for general and/or

specific plant disease resistance. In: K. W. Findlay and K. W. Shepherd (eds.), Proc. 3rd Int. Wheat Genetics Symp. Australian Acad Sci Canberra, Australia. p. 263-272.

[5] Chen, X. M., and Line, R. F. 1995. Gene action in wheat cultivars for durable, high-temperature, adult-plant resistance and interaction with race-specific, seedling resistance to *Puccinia striiformis*. Phytopathology. 85: 567-572.

[6] Chen, X. M., and Line, R. F. 1995. Gene number and heritability of wheat cultivars with durable, high-temperature, adult-plant (HTAP) resistance and interaction of HTAP and race-specific, seedling resistance to *Puccinia striiformis*. Phytopathology. 85: 573-578.

[7] Flor, H. H. 1955. Host parasite interaction in flax rust: Its genetics and implication. Phytopathology. 45: 680-685.

[8] Fransis, H. A., Leitch, A. R., and Koebner, R. M. D. 1995. Conversion of a RAPD-generated PCR product, containing a novel dispersed repetitive element, into a fast robust assay for the presence of rye chromatin in wheat. Theor. Appl. Genet. 90: 636-642.

[9] He, Z. H., Rajaram, S, Xin, Z. Y., and Huang, G. Z. (eds.) 2001. A History of Wheat Breeding in China. Mexico, D. F.: CIMMYT.

[10] Li, G. Q., Li, Z. F., Yang, W. Y., Zhang, Y., He, Z. H., Xu, S. C., Singh, R. P., Qu, Y. Y., and Xia, X. C. 2006. Molecular mapping of stripe rust resistance gene *YrCH42* in Chinese wheat cultivar Chuanmai 42 and its allelism with *Yr24* and *Yr26*. Theor. Appl. Genet. DOI 10.1007/s00122-006-0245-y. (in press)

[11] Li, Z. Q., and Zeng, S. M. 2000. Wheat Rusts in China. China Agriculture Press, Beijing.

[12] Liang, S. S., Suenaga, K., He, Z. H., Wang, Z. L., Liu, H. Y., Wang, D. S., Singh, R. P., Sourdille, P., and Xia, X. C. 2006. Quantitative trait loci mapping for adult-plant resistance to powdery mildew in bread wheat. Phytopathology. (Accepted for publication)

[13] Line, R. F., and Chen, X. M. 1995. Successes in breeding for and managing durable resistance to wheat rusts. Plant Dis. 79: 1254-1255.

[14] Loegering, W. Q., and Burton, C. H. 1974. Computer-generated hypothetical genotypes for reaction and pathogenicity of wheat cultivars and cultures of *Puccinia graminis tritici*. Phytopathology. 64: 1380-1384.

[15] Ma, H., and Singh, R. P. 1996. Expression of adult resistance of stripe rust at different growth stages of wheat. Plant Dis. 80: 375-379.

[16] Ma, J. X., Zhou, R. H., Dong, Y. S., Wang, L. F., Wang, X. M., and Jia, J. Z. 2001. Molecular mapping and detection of the yellow rust resistance gene *Yr26* in wheat transferred from *Triticum turgidum* L. using microsatellite markers. Euphytica. 120: 219-226.

[17] Martinez, F., Niks, R. E., and Rubiales, D. 2001. Partial resistance to leaf rust in a collection of ancient Spanish barleys. Hereditas. 135: 199-203.

[18] McIntosh, R. A. 1992. Close genetic linkage of genes confering adult-plant resistance to leaf rust and stripe rust in wheat. Plant Pathol. 41: 523-527.

[19] McIntosh, R. A., and Lagudah, E. S. 2000. Cytogenetical studies in wheat XVIII. Gene *Yr24* for resistance to stripe rust. Plant Breed. 119: 81-83.

[20] McIntosh, R. A., Yamazaki, Y., Devos, K. M., Dubcovsky, J., Rogers, J., and Appels, R. 2003. Catalogue of gene symbols for wheat. Proceedings of 10th International Wheat Genetics Symposium, Vol 1, Instituto Sperimentale per la Cerealcoltura, Rome Italy.

[21] Niu, Y. C., and Wu, L. R. 1997. The breakdown of resistance to stripe rust in Fan 6 and Mianyang wheat cultivars and strategies for its control. *Acta Phtopathologica Sinica*. 27: 5-8.

[22] Niu, Y. C., Qiao, Q., and Wu, L. R. 2000. Postulation of resistance genes to stripe rust in commercial wheat cultivars from Henan, Shandong, and Anhui provinces. *Acta Phtopathologica Sinica*. 30: 122-128.

[23] Oelke, L. M., and Kolmer, J. A. Characterization of leaf rust resistance in hard red spring wheat cultivars. Plant Dis. 88: 1127-1133.

[24] Perwaiz, M. S., and Johnson, R. 1986. Genes for resistance to yellow rust in seedlings of wheat cultivars from Pakistan tested with British isolates of *Puccinia striiformis*. Plant Breed. 97: 289-296.

[25] Peterson, R. F., Campbell, A. B., and Hannah, A. E. 1948. A diagrammatic scale for estimating rust intensity of leaves and stems of cereals. Can. J. Res. 26: 496-500.

[26] Sharma, S., Louwers, J. M., Karki, C. B., and Snijders, C. H. A. 1995. Postulation of resistance genes to yellow rust in wild emmer wheat derivatives and advanced wheat lines from Nepal. Euphytica. 81:

271-277.

[27] Sharp, E. L., and Volin, R. B. 1970. Additive genes in wheat conditioning resistance to stripe rust. Phytopathology. 60: 1146-1147.

[28] Singh, H., Johnson, R., and Seth, D. 1990. Genes for race-specific resistance to yellow rust (*Puccinia striiformis*) in Indian wheat cultivars. Plant Pathol. 39: 424-433.

[29] Singh, R. P. 1992. Genetic association of leaf rust resistance gene *Lr34* with adult plant resistance to stripe rust in bread wheat. Phytopathology. 82: 835-838.

[30] Singh, R. P. 1993. Genetic association of gene *Bdvl* for tolerance to barley yellow dwarf virus with genes *Lr34* and *Yr18* for adult plant resistance to rusts in bread wheat. Plant Dis. 77: 1103-1106.

[31] Singh, R. P., and Rajaram, S. 1994. Genetics of adult plant resistance to stripe rust in ten bread wheats. Euphytica. 72: 1-7.

[32] Singh, R. P., Huerta-Espino, J., and Rajaram, S. 2000. Achieving near-immunity to leaf and stripe rusts in wheat by combining slow rusting resistance genes. *Acta Phytopathologica Hungarica*. 35: 133-139.

[33] Suenaga, K., Singh, R. P., Huerta-Espino, J., and William, H. M. 2003. Microsatellite markers for genes *Lr34/Yr18* and other quantitative trait loci for leaf rust and stripe rust resistance in bread wheat. Phytopathology. 93: 881-890.

[34] Wamishe, Y. A., and Milus, E. A. 2004. Seedling resistance genes to leaf rust in soft red winter wheat. Plant Dis. 88: 136-146.

[35] Wan, A. M., Zhao, Z. H., Chen, X. M., He, Z. H., Jin, S. L., Jia, Q. Z., Yao, G., Yang, J. X., Wang, B. T., Li, G. B., Bi, Y. Q., and Yuan, Z. Y. 2004. Wheat stripe rust epidemic and virulence of *Puccinia striiformis* f. sp. *tritici*. Plant Dis. 88: 896-904.

[36] Wang, F. L., Wu, L. R., Wan, A. M., Song, W. Z., Yuan, W. H., and Yang, J. X. 1994. Postulated genes for resistance to stripe rust in seedlings of wheat cultivars from Shaanxi, Gansu and Sichuan provinces. *Acta Agronomica Sinica*. 20 : 589-594.

[37] Wang, F. L., Wu, L. R., Xie, S. X., and Wan, A. M. 1994. Postulation of genes and adult resistance to stripe rust of Chinese important wheat resistance resources. *Acta Phtopathologica Sinica*. 24 : 175-180.

[38] Wang, Z. L., Li, L. H., He, Z. H., Duan, X. Y., Zhou, Y. L., Chen, X. M., Lillemo, M., Singh, R. P., Wang, H., and Xia, X. C. 2005. Seedling and adult plant resistance to powdery mildew in Chinese bread wheat cultivars and lines. Plant Dis. 89: 457-463.

[39] Wellings, C. R., Singh, R. P., McIntosh, R. A., and Pretorius, Z. A. 2004. The development and application of near isogenic lines for the wheat stripe (yellow) rust pathosystem. 11th International Cereal Rusts and Powdery Mildews Conference, John Innes Centre, Norwich, England, 22nd to 27th August. 2004. A1. 39.

[40] William, M., Singh, R. P., Huerta-Espino, J., Ortiz Islas, S., and Hoisington, D. 2003. Molecular marker mapping of leaf rust resistance gene *Lr46* and its association with stripe rust resistance gene *Yr29* in wheat. Phytopathology. 93: 153-159.

[41] Worland, A. J., and Law, C. N. 1986. Genetic analysis of chromosome 2D of wheat. I. The location of genes affecting height, day-length insensitivity, hybrid dwarfism and yellow-rust resistance. Z. Pflanzenzüchtg. 96: 331-345.

[42] Wu, L. R., Wan, A. M., Niu, Y. C., Jin, S. L., and Li, J. P. 1999. Integrated pest management on wheat in the Longnan region. In: Research Progress in Plant Protection and Plant Nutrition. China Association of Agricultural Science Societies, China Agricultural Press, Beijing, China. p. 34-38.

[43] Wu, L. R., Yang, H. A., Yuan, W. H., Song, W. Z., Yang, J. X., Li, Y. F., and Bi, Y. Q. 1993. On the physiological specialization of stripe rust of wheat in China during 1985-1990. *Acta Phtopathologica Sinica*. 23: 269-274.

[44] Yang, Z. M., Xie, C. J., and Sun, Q. X. 2003. Situation of the sources of stripe rust resistance of wheat in the Post-CY32 Era in China. *Acta Agronomica Sinica*. 29: 161-168.

[45] Zadoks, J. C. 1961. Yellow rust on wheat: studies on epidemiology and physiological specialization. Tijdschr. Plantenziekten T. Pl. Ziekten. 67: 69-256.

[46] Zhang, Z. J., Yang, G. H., Li, G. H., Jin, S. L., and Yang, X. B. 2001. Transgressive segregation, heritability, and number of genes controlling durable resistance to stripe rust in one Chinese and two Italian wheat cultivars. Phytopathology. 91: 680-686.

[47] Zhou, X. C. 2003. Strategies and research methodologies for a sustainable genetic control of wheat stripe rust in China. Petria. 13: 125-138.

Seedling and slow rusting resistance to leaf rust in Chinese wheat cultivars

Z. F. Li[1], X. C. Xia[2], Z. H. He[2,3], X. Li[1], L. J. Zhang[1], H. Y. Wang[1], Q. F. Meng[1], W. X. Yang[1], G. Q. Li[4], and D. Q. Liu[1]

[1] *Department of Plant Pathology, College of Plant Protection, Agricultural University of Hebei, Biological Control Center for Plant Diseases and Plant Pests of Hebei*, 289 *Lingyusi Street*, *Baoding*071001, *Hebei*, *China*; [2] *Institute of Crop Science*, *National Wheat Improvement Center*, *Chinese Academy of Agricultural Sciences*, 12 *Zhongguancun South Street*, *Beijing* 100081, *China*; [3] *Institute of Crop Science*, *National Wheat Improvement Center*, *Chinese Academy of Agricultural Sciences and CIMMYT China Office*, 12 *Zhongguancun South Street*, *Beijing* 100081, *China*; [4] *Plant Protection Institute*, *Chinese Academy of Agricultural Sciences*, 2 *Yuanmingyuan West Road*, *Beijing* 100091, *China*.

Abstract: Identification of resistance genes is important for developing leaf rust resistant wheat cultivars. A total of 102 Chinese winter wheat cultivars and advanced lines were inoculated with 24 pathotypes of *Puccinia triticina* for postulation of leaf rust resistance genes effective at the seedling stage. These genotypes were also planted in the field for characterization of slow rusting responses to leaf rust in the 2006-2007 and 2007-2008 cropping seasons. Fourteen leaf rust resistance genes *Lr1*, *Lr2a*, *Lr3bg*, *Lr3ka*, *Lr14a*, *Lr16*, *Lr17a*, *Lr18*, *Lr20*, *Lr23*, *Lr24*, *Lr26*, *Lr34*, and *LrZH84*, either singly or in combinations, were postulated in 65 genotypes, whereas known resistance genes were not identified in the other 37 accessions. Resistance gene *Lr26* was present in 44 accessions. Genes *Lr14a* and *Lr34* were each detected in seven entries. *Lr1* and *Lr3ka* were each found in six cultivars, and five lines possessed *Lr16*. *Lr17a*, and *Lr18* were each identified in four lines. Three cultivars were postulated to possess *Lr3bg*. Genes *Lr20*, *Lr24*, and *LrZH84* were each present in two cultivars. Each of the genes *Lr2a* and *Lr23* may exist in one line. Fourteen genotypes showed slow leaf rusting resistance in two cropping seasons.

Key words: bread wheat, leaf rust, gene-for-gene specificity, slow rusting resistance

Introduction

Leaf rust, caused by *Puccinia triticina* Eriks., is one of the most important and widespread diseases of common wheat (*Triticum aestivum* L.) worldwide. It is adapted to a wide range of environments, occurs wherever wheat is grown, and can cause significant yield and economic losses[38]. In China, destructive epidemics of leaf rust occurred in 1969, 1973, 1975, and 1979[4]. During the past decade, leaf rust has periodically caused destructive yield losses in the major wheat production regions, particularly in North China and the Yellow and Huai Valleys[42]. Resistant cultivars are the most efficient and environmentally friendly method for reducing damage caused by leaf rust[25].

Genetic interactions between wheat and *P. triticina* involve gene-for-gene relationships[27]. Gene postulations assume gene-for-gene specificities in hypothesizing

Published in Plant Disease, 2010, 94 (1): 45-53

probable resistance genes present in host genotypes[23]. The presence of a specific resistance gene in a host line can be postulated from response arrays of pathogen cultures with known avirulence and virulence characteristics. Many researchers have used this approach to identify leaf rust resistance genes (*Lr*) in various sets of wheat lines. For example, Statler[37] postulated *Lr1*, *Lr2a*, *Lr2c*, *Lr10*, *Lr17a*, and *Lr18* in a group of 25 hard red spring wheat genotypes. McVey[19] identified genes *Lr1*, *Lr3*, *Lr10*, *Lr16*, *Lr24*, and *Lr26* in 86 winter wheat cultivars from 26 countries using 15 *P. triticina* races. Unidentified resistance genes were also present in some cultivars. McVey and Long[20] postulated 14 resistance genes, *Lr1*, *Lr2a*, *Lr3*, *Lr3ka*, *Lr9*, *Lr10*, *Lr11*, *Lr14a*, *Lr16*, *Lr17a*, *Lr18*, *Lr24*, *Lr26*, and *Lr30*, in 30 cultivars, and 56 hard red winter wheat lines from 9 breeding programs using 17 isolates of *P. triticina* with known avirulence or virulence to 16 *Lr* genes. Fourteen genes, *Lr1*, Lr3, *Lr3bg*, *Lrl0*, *Lr13*, *Lrl4a*, *Lr16*, *Lr17a*, *Lrl9*, *Lr23*, *Lr26*, Lr27+*Lr31*, and *Lr34*, were identified among 76 Mexican wheat cultivars[30,32]. Resistance genes were also postulated in selected soft red winter wheat[11,38], and hard red spring wheat cultivars[22] from the USA, and 36 wheat cultivars from Ethiopia and Germany[21]. Singh et al.[34] screened 61 spring, and 102 facultative and winter wheat cultivars from China with 14 *P. triticina* races, and postulated 9 seedling resistance genes, *Lr1*, *Lr3*, *Lr3bg*, *Lr10*, *Lr13*, *Lr14a*, *Lr16*, *Lr23*, and *Lr26*, in some of the cultivars. However, limited information is available in respect of possible leaf rust resistance genes in current commercial wheat cultivars and advanced lines in China. Therefore, characterization of leaf rust resistance genes in current widely grown Chinese wheat cultivars would be extremely useful for breeding new resistant cultivars and for gene deployment schemes using Chinese wheat germplasm.

To date, 61 leaf rust resistance genes have been catalogued in wheat[18]. Most of them confer hypersensitive reactions and interact with the pathogen in a gene-for-gene fashion[7]. The effectiveness of such genes is often absent or short-lived due to the appearance of pathotypes with corresponding alleles for virulence[10]. The pyramiding of several resistance genes in a single cultivar is considered important since their combined effects confer a wider spectrum of resistance[26] and perhaps extends the period of effectiveness. Currently, only a few designated leaf rust resistance genes, such as *Lr9*, *Lr19*, *Lr24*, and *Lr38*, are effective against prevalent Chinese *P. triticina* races[39]. Hence, it is very important to search for new resistance genes to cope with the rapidly evolving pathogen population.

Adult-plant resistance (APR) of the slow rusting type is considered more durable than seedling resistance[33], or major gene APR. Slow rusting is characterized by slow disease development in the field despite a high infection type[35], and involves a longer latent period, a low infection frequency, smaller uredial size and reduced duration of sporulation, and less spore production[2]. Of the 61 leaf rust resistance genes catalogued, *Lr34* and *Lr46* are assumed to be slow rusting genes[31,36]. Dyck and Samborski[6] found slow rusting resistance gene *Lr34* in the Brazilian wheat cultivar Frontana. Dyck[5] located *Lr34* on chromosome 7D. Singh[29] and McIntosh[17] reported that *Lr34* is genetically associated with gene *Yr18*, conferring slow rusting resistance to stripe rust. Both genes are also linked with gene *Ltn* for leaf tip necrosis in adult plants, a gene that can often be used as a morphological marker for *Lr34* and *Yr18*[28]. Krattinger et al.[12] showed the *Lr34* protein resembled ABC (ATP-binding cassette) transporters of a pleiotropic drug resistance subfamily. *Lr34* confers a moderate level of leaf rust resistance when present alone, however, combinations of *Lr34* with two to four additional slow rusting genes resulted in high resistance in most environments[35]. In addition, Singh et al.[36] identified the gene *Lr46* for slow leaf rusting and located it on chromosome 1B of cultivar Pavon 76. This gene has remained effective since 1976. Lillemo et al.[15] reported that *Lr46* was closely linked to the slow stripe rusting gene *Yr29* and powdery mildew resistance gene *Pm39*. The frequent breakdown of leaf rust resistant cultivars has forced wheat breeders to focus on the development of cultivars with durable resistance based on slow rusting genes. Chen et al.[3] screened 7 wheat cultivars with slow rusting resistance to leaf rust in 37 Chinese wheat cultivars based on latent period, epidemic rate, area under disease progress curve (AUDPC), infection type (IT), final disease rating, and accumulation temperature of latent period by the time of the pustule eruption up to 50%. Although slow rusting resistance and major genes have been used for durable control of leaf rust, very little information is available on either type of resistance in most of current Chinese wheat cultivars.

One hundred and two widely grown Chinese wheat cultivars and advanced lines from 12 provinces in the major wheat production regions were investigated in this study. The objectives were to postulate genes conferring seedling resistance and to assess the likelihood of slow rusting resistance in adult plants grown in the field.

Materials and methods

Wheat germplasm and *P. triticina* pathotypes

The 102 winter wheat cultivars and advanced lines were from major breeding programs across the autumn-sown wheat regions of China, including the provinces of Beijing, Hebei, Henan, Shandong, Shanxi, Shaanxi, Gansu, Anhui, Hubei, Jiangsu, Guizhou, Sichuan, and Yunnan (Table 1). The lines were tested for seedling reaction to 24 *P. triticina* pathotypes in the greenhouse and for slow leaf rusting resistance in the field. CIMMYT line Saar with typical slow leaf rusting resistance[15,40] and highly susceptible line Zhengzhou 5389 were used as slow rusting and susceptible controls, respectively. A set of 35 differential lines, mostly near-isogenic lines in the background of Thatcher with known leaf rust resistance genes, were included in the seedling tests, and the infection types displayed by them were used as a basis for postulating resistance genes in the Chinese lines (Tables 2 and 3). The differential lines were kindly provided by the USDA-ARS Cereal Disease Laboratory, University of Minnesota, St Paul, USA. The 24 pathotypes were designated following the coding system of Long and Kolmer[16], with addition of the fourth letter for the reactions to the fourth set of differentials (http://www.ars.usda.gov/SP2UserFiles/ad_hoc/36400500 Cerealrusts/pt_nomen.pdf). Four Thatcher isogenic lines with genes *Lr1*, *Lr2a*, *Lr2c*, and *Lr3* were the first set of differentials; with genes *Lr9*, *Lr16*, *Lr24*, and *Lr26* the second set of differentials; with genes *Lr3ka*, *Lr11*, *Lr17a*, and *Lr30* the third set of differentials; and with genes *LrB*, *Lr10*, *Lr14a*, and *Lr18* the fourth set of differentials.

Table 1 Pedigrees, origins, and likely presence or absence of the 1B.1R wheat-rye chromosomal translocation and the *Lr34* leaf rust resistance gene based on molecular markers of 102 Chinese wheat genotypes

Line number	Genotype	Origin	Pedigree	1B.1R[a]	*Lr34*[b]
1	Han 4599	Hebei	Han 4032/85Zhong47	+	−
2	Han 5316	Hebei	Han 7808/CA 8059//85Zhong47	+	−
3	Aikang 58	Henan	Zhoumai 11//Wenmai 6/Zhengzhou 8960	+	−
4	Xinmai 208	Henan	Yumai 18/Jimai 5418	+	−
5	Zhoumai 16	Henan	Zhou 9/Zhou 8425B	+	−
6	Zhoumai 18	Henan	Zhou 9/Neixiang 185	+	−
7	(95) 18	Shaanxi	843380/81170-1-2-1	+	−
8	Lantian 14	Gansu	Qingshan 895/Zhongliang 17	+	−
9	Lantian 9	Gansu	Xifeng 16/76-89-13	+	−
10	Chuanyu 19	Sichuan	Chuanyu 5/Mo 460//Mianyang 26	+	−
11	N1621-19-4	Sichuan	Mianyang 26/Yiyuan 2	+	−
12	Jing 035-1	Yunnan	/	+	−

（续）

Line number	Genotype	Origin	Pedigree	1B. 1R[a]	*Lr34*[b]
13	Zhoumai 12	Henan	Zhou8425A/SWD73295	+	—
14	5R608	Beijing	R1//Xuezao/D164	+	—
15	5R610	Beijing	6521-89630278/K3—ABE//A121	+	—
16	Henong326	Hebei	84 (252) /86 (298)	+	—
17	Shi 4185	Hebei	Tal/Zhi 8094//Yumai 2/3/Jimai 26	+	—
18	Lankao 906-4	Henan	84-184/ Yumai 2//90 Selection	+	—
19	Jinan 16	Shandong	Tal Shannongfu 63/AimengniuV	+	—
20	Shaan 715	Shaanxi	Shaan 354/Yanmai 8911	+	—
21	Tian 95HF2	Gansu	Slected from*Tal* wheat F_2 recurrent selection population	+	—
22	Huaimai 18	Jiangsu	Zhengzhou 891/Yan 1604	+	—
23	Bi 2002-2	Guizhou	8513-1624/Ji 1002	+	—
24	99-0502	Sichuan	911/MY86-51	+	—
25	Mian 2000-36	Sichuan	Mian92-8/Mian88-304//Guinong 19-4	+	—
26	N711	Sichuan	92-4/Mianyang 4	+	—
27	Yumai 47	Henan	Yumai 2/Baiquan 3199	—	—
28	Yannong 19	Shandong	Yan 1933/Shaan 82-29	—	—
29	Lumai 21	Shandong	Lumai 13/Yumai 2	—	—
30	Emai 14	Hubei	Fan 6/Yanda 72-629	—	—
31	Guizhou 98-18	Guizhou	Sumai 3/C39//Qing 30	+	—
32	Zhengmai 9023	Henan	(Xiaoyan 6/Xinong 65//8323-3/8443) /Shaan 213	—	—
33	5R615	Beijing	D221/891162 (Agent)	+	—
34	5R616	Beijing	BL 7You * 2/Agent	—	—
35	Shaannong 981	Shaanxi	Derived from Shaanyou 225 (Xiaoyan 6/NS2761)	—	—
36	Een 1	Hubei	Lovrin 10/761//Sumai 3	+	—
37	Xinmai 19	Henan	C5/Xinxiang 3577//Xinmai 9	—	—
38	Xinong 1043	Shaanxi	Changwu 131/Y 8402	—	—
39	An 96-8	Anhui	L9288022-21/Xingnong 5	—	—
40	58749	Sichuan	SW 3243/35050//21530	+	—
41	Xinong 1163-4	Shaanxi	84Jia79/Xinong 1376	+	—
42	93Xia032-1	Sichuan	93-38-1/93-35	+	—
43	Y 1496-15	Sichuan	94-2/Mianyang 4	+	—
44	Tian 9470	Gansu	Tian 817/Lantian 6	+	—
45	W7	Sichuan	Durum/Xianyang large spike//W7268	—	—
46	LB 0458	Sichuan	Mianyang 90-310/M180	—	—
47	SW 8588	Sichuan	Milan's/SW 5193	—	—
48	Lankaoaizao 8	Henan	84 (184) /90 Selection	+	—
49	01-DH689	Sichuan	94-D. H-375/Mianyang 26	+	—
50	Chuannong 17	Sichuan	91S-23/A302	+	—
51	Mian 2000-13	Sichuan	Mianyang 01821/90Zhong165//Guinong 19-4	+	—
52	Chuannong 19	Sichuan	Qian1104A/R935	—	—
53	Qinnong 151	Shaanxi	Zhengzi R84019-0-7-5/Xinong 1376	+	—
54	351-15	Sichuan	Mian 99-310/GansuM180	—	—
55	Neixiang 991	Henan	87C27/Zhengzhou 84115// Yumai 13	+	—
56	LB 0288	Sichuan	Mianyang 90-310/M180	—	—
57	Zhou 8425B	Henan	Guangmai 74/Lianfeng 1//Predgornayi 2/3/Annong 7959	+	—
58	Zhoumai 11	Henan	Zhou 8425B/Yumai 17	+	—
59	01-018	Sichuan	Chuannong 21/89-8//89-107	+	+
60	R 97	Sichuan	Mianyang 92-8/91S-5-7	+	+

（续）

Line number	Genotype	Origin	Pedigree	1B. 1R[a]	*Lr34*[b]
61	Lantian 12	Gansu	Qingnong 4/Xiannong 4	—	+
62	libellula	Gansu	Tevere/Giuliani//San Pastore (Italy)	—	+
63	Strampelli	Gansu	Introduced from Italy	—	+
64	99422	Sichuan	Mian 87-19/R1301	—	+
65	Chinese Spring	Sichuan	Landrace	—	+
66	3S587	Beijing	Bai//108/Flaminio-IDO2229/3/BT 881	—	—
67	5R631	Beijing	93/Yantsr	—	—
68	Zhongyou 9507	Beijing	Reselection of Zhongyou 8	—	—
69	Henong 822	Hebei	88S522/92	—	—
70	Luohan 2	Henan	Luoyang 78 (111) ai/Jinmai 33	—	—
71	Pingyuan 50	Henan	Landrace	—	—
72	Xinmai 16	Henan	Yumai 13/Neixiang82C6//Yumai 2	—	—
73	Xinmai 9	Henan	Baiquan 3047-3/Neixiang82C6	—	—
74	Yumai 49	Henan	Bainong 791//Wenxuan 1/Zhengzhou 761/3/Yumai 2	—	—
75	Yannong 9	Shandong	Derived from Bima 1	—	—
76	Zimai 12	Shandong	917065/910292	—	—
77	N95175	Shaanxi	92R/Yuancai 142//Xiaoyan 6	—	—
78	Shaan 512	Shaanxi	Shaanmai 150/Shaan 354	—	—
79	Xiaoyan 926	Shaanxi	Derived from Xiaoyan 6	—	—
80	Xinong 8925-13	Shaanxi	Xiaoyan 168/Zhi 763	—	—
81	96-22-2-3	Gansu	Cappelle Desprez/935	—	—
82	Lantian 17	Gansu	92R137/Lantian 6	—	—
83	ZhongliangX9610	Gansu	92R178/92R178//Chuanjian 134	—	—
84	Yangmai 158	Jiangsu	Yangmai 4/ST1472/506	—	—
85	Fengmai 24	Yunnan	Yunmai 36/Mo 965	—	—
86	4S589	Beijing	Yanda1817/wel//Tuyou/3/Jing 411 * 3	—	—
87	4S615	Beijing	T601/Jing 411 * 3	—	—
88	Nongda 189	Beijing	Lv7you/Yanshi 9//Changfeng 1/H. w y1775/3/Ji 82-4255	—	—
89	Henong 215	Hebei	Anyang 10/Aifeng 1//Beijing 8/3/Lovrin 10/4/HH/5/Orofen	—	—
90	Xinmai 11	Henan	Zhou 8826/Xinxiang 3577	—	—
91	Xinmai 13	Henan	Yuanyuanchangbai//C5/Xinxiang 3577	—	—
92	Yanzhan 1	Henan	89 (35) -14/Yumai 18	—	—
93	Yanzhan 4110	Henan	C39/Xibei 78 (6) 9-2//FR 81-3/Aizao 781-4/3/Aizao 781-4	—	—
94	Yumai 18	Henan	Zhengzhou 761/Yanshi 4	—	—
95	Zhoumai 19	Henan	Zhoumai 13/Shaanyou 225//PH82-2	—	—
96	Mingxian 169	Shanxi	Landrace	—	—
97	Shaanmai 139	Shaanxi	Xiaoyan 22/6/ [*Triticum dicoccoides* AS846// (Shaanmai 8003/ Shaanmai 8007) F_4/3/Shaan 229/4/Aizaofeng] F_3/5/N9134	—	—
98	94t-133	Gansu	Hybrid 46/Xiannong 4//Tao 157	—	—
99	Zhongliang 98702	Gansu	Zhongliang 21/92R137	—	—
100	Annong 98005	Anhui	Aizao 781/Wansu 8802	—	—
101	Yangfu 2	Jiangsu	Yangmai 158/101-90	—	—
102	Chuannong 21	Sichuan	91S-23/Qianhui 3	—	—

a. +=presence of 1BL. 1RS, —=presence of normal 1BL. 1BS, determined by the method of Zhang et al. [40].

b. +=presence of *Lr34*, —=absence of *Lr34*, according to STS marker csLV34[12].

/=Unknown.

Table 2 Seedling infection types on 35 wheat lines with known leaf rust resistance genes when tested with 24 pathotypes of *Puccinia triticina* collected from China

Tester	*Lr* gene	Infection types to *Puccinia triticina* pathotypes																							
		PH	FC	FC	PC	TC	PG	FG	FK	PC	FB	FH	PH	FH	FH	PG	PC	TC	PH	FH	FH	PH	TH	FC	PC
		QT	QR	ST	BT	GT	SN	SQ	QT	HS	HT	ST	SS	GS	TS	TT	JT	HT	ST	HQ	TR	JT	TT	TT	GR
RL 6003	*Lr1*	3	; 0	0;	3	3	4	;	2	3	; 2	; 2	3	0	0;	3	3	3	3	0;	0;	3	4	;	3
Rl 6016	*Lr2a*	1	1	;	; 1	3	1+	1+	; 2	; 1	; 2	; 1	;	0	1	1	; 2	3	;	0;	0; 2	;	4	;	;
RL 6019	*Lr2b*	3	4	3	4	3	4	3	3	3	3	3	3	3	3	3	3	3	3	3	3	3	4	4	3
RL 6047	*Lr2c*	3	3	3	4	3	3C	4	3	3	3	3	3	3	4	3	3	3	3	3	3	3	4	4	3
RL 6002	*Lr3*	3	4	3	3	3	4	4	3	3	3	4	3	3	3	3	3	3	3	3	3	4	4	4−	3
RL 6042	*Lr3bg*	3	3	3	3	3	4	4	3	3	3	3	3	3	3	3	3	2	2	3	2	3	3	4	3
RL 6007	*Lr3ka*	3	3	3	; 1	;	4	4	3	2	; 2	3	3	2	3	3	0;	0;	3	0;	3	;, 2	3	3	; 2
RL 6010	*Lr9*	0	;	0	0	0	0	0	0	0	;	0	0	0	0	0	0	0	0	0	0	0	0	;	0
RL 6004	*Lr10*	3	3	3	3	3	2+, 3−	4	3	3	3	3	3	3	3	3	3	3	3	3	3	3	3	4	3
RL 6053	*Lr11*	3	3	3	; 12	3	3	3+	3	3	3	3	3	3	3	3	3	3	3	3	3	3	3	4	3
RL 6013	*Lr14a*	3	3	3	4	3	3C	2	3	3	3	3	3	3	3	3	3	3	3	2	2	3	3	4	2
RL 6052	*Lr15*	0;	0	;	; 1	; 2	4	1+	; 2	3	; 1	1	3	;	1	3	;	1, 2	1	2	; 2	; 1	3	; 1	3
RL 6005	*Lr16*	3	1	1	1+	2;	3	4	3	2	2	3	3	3	3	3	2	2	3	3	4	3	4	; 2	2
RL 6008	*Lr17a*	2	2	3	;	2	3	3+	2	1	1	3	3	2	3	3	3	; 1	3	1	3	3	3	3	2
RL 6009	*Lr18*	3	3	3	3	3	2	1	3	2	3	3	2	2	2	3	3	3	3	2	3	3	3	4	3
Rl 6040	*Lr19*	0	;	;	0	;	0	;	0	0;	0	0;	0	0	;	1	0;	2	0	0;	0	0;	;	0;	;
RL 6092	*Lr20*	3	3	; 1	0; 1	3	4	;	;	3	0;	3	0	0;	0	0;	3	3	; 2	3	3	3	3	3	3
RL 6043	*Lr21*	0;	; 1	3	; 2	; 2	3	3	2	2	3	3	12	2	3	2	2	3	3	2	3	3	3	3	2
RL 6012	*Lr23*	;	;	3	; 1	;, 2	2+	4	; 1	1	;	3	;	12	; 1	4	4	4	3	; 2	0;	;	3	3−	; 1
RL 6064	*Lr24*	0	0	;	;	0;	;	;	;	;	;	;	0;	0	0	;	;	0	0;	;	0	;	;	;	;
RL 6084	*Lr25*	3	4	3	4	3	4	4	3	3	3	3	3	3	3	3	3	3	3	3	3	3	3	4	3
Rl 6078	*Lr26*	3	3	3	3	4	1+	0;	3	3	;	3	3	3	3	2+	3	3	3	3	3	3−	3	4	3
RL 6079	*Lr28*	0	0;	;	0;	0;	0;	;	3	;	2	0	0;	3	0	0	;	0	;	;	3	0	0	;	0
RL 6080	*Lr29*	3	3	3	; 2	3	;	4	;	;	;	3	3	3	3	3	1	; 2	3	3	3	; 2	3	3	3
RL 6049	*Lr30*	0	;	;	; 1	; 1	1	1	2	3	3	;	2	2	3	3	0; 2	3	2	3	3	; 2	3	3	2
RL 5497	*Lr32*	;	;	;	; 1	;	12+, 3−	4	12	; 1	;	0; 12	1	; 1	1+	1	; 1	; 2	;	3	0	; 2	3	4	2
RL 6057	*Lr33*	3	3	3	3	3	4	4	3	4	3	4	3	4	3	3	4	4	4	3	3	3	3	3−	3

（续）

Tester	*Lr* gene	Infection types to *Puccinia triticina* pathotypes																							
		PH	FC	FC	PC	TC	PG	FG	FK	PC	FB	FH	PH	FH	FH	PG	PC	TC	PH	FH	FH	PH	TH	FC	PC
		QT	QR	ST	BT	GT	SN	SQ	QT	HS	HT	ST	SS	GS	TS	TT	JT	HT	ST	HQ	TR	JT	TT	TT	GR
E84018	*Lr36*	3	; 1	; 1	;	;	2	2+	0;	2	; 1	;	;	1	0;	;	2	1	; 1	2	0	;	3	;	;
RL 6097	*Lr38*	0	0	;	0	0	; 1	; 1	;	0	0;	;	;	0;	0	0	0	0;	;	0	0	0;	;	;	;
KS86NGRC02	*Lr39*	3	3	;	; 2	2	1	1	2	2	2	; 1	; 1	1	2	2	; 2	; 2	3	3	3	; 2	;	1	2
KS91WGRC11	*Lr42*	0;	0;	;	0; 12	; 1	; 1	;	12	2	; 2	2	2	2	2	2	3	; 2	2	;	3	0	0; 1	0; 2	;
RL 6147	*Lr44*	3C	2	3	3	3	1+	3	12	2	2	2	2	2	2	2	1+	1	2	3	1	3	3	3	2
RL 6144	*Lr45*	3	3	;	2;	1	;	1	3	;	3	3	3	3	3	1	0;	2	2	3	3	0	;	0;	3
TcLr50	*Lr50*	0;	0;	;	;	;	4	4	;	; 2	;	; 1	;	0;	;	3	;	0;	;	0;	0	; 1	3	;	2
RL 6051	*LrB*	4	3	3	3	3	4	4	3	3	4	3	3	3	3	3	3	3	3	3	3	3	3	4	3

Table 3 Seedling infection types of 102 wheat cultivars and lines when tested with 24 pathotypes of *Puccinia triticina* collected from China

Line number[a]	*Lr* gene	Infection types to *Puccinia triticina* pathotypes																							
		PH	FC	FC	PC	TC	PG	FG	FK	PC	FB	FH	PH	FH	FH	PG	PC	TC	PH	FH	FH	PH	TH	FC	PC
		QT	QR	ST	BT	GT	SN	SQ	QT	HS	HT	ST	SS	GS	TS	TT	JT	HT	ST	HQ	TR	JT	TT	TT	GR
1	*Lr26*	3C	4	4	4	4	1+	; 1	4	4	; 1	3	3	4	4	1+	4	4	4	4	4	4	4	4	4
2	*Lr26*	3C	4	4	3C	4	1+	; 1	4	4	; 1	3	3	4	4	1+	4	4	4	4	4	4	4	4	4
3	*Lr26*	3C	3+	4	3+	4	2	;	4	3	1	4	3	4	4	1	4	3	4	4	4	4	4	4	4
4	*Lr26*	3	4	3	3	4	2	; 1	4	3+	0	4	4	4	3	1+,2	4	4	4	3	3	4	4	4	4
5	*Lr26*	3	4	4	3	4	1+	; 1	4	3	;	3	4	4	4	12	4	4	4	4	4	4	4	4	4
6	*Lr26*	3C	4	4	3C	4	1+	; 1	4	3	0	4	3	4	3	1+	4	4	4	3+	3	4	4	4	4
7	*Lr26*	3	3+	3+	3	3+	1+	;	4	3	;	3	3	4	3	2	4	4	3	4	4	4	4	4	4
8	*Lr26*	3	3+	4	4	4	1+	;	4	3	; 1	3	3	4	4	12	4	4	4	4	4	4	4	4	4
9	*Lr26*	3	4	4	4	4	; 1	0	3	3	0	4	3	4	4	2	4	4	4	4	4	4	4	4	4
10	*Lr26*	3	3	4	3	4	1	;	3	3	;	3	3+	3	3	1	4	4	3	3+	4	4	3−	4	4
11	*Lr26*	3	4	4	3	4	1+	;	4	3	2	3	4	4	4	2	4	4	4	4	4	4	4	4	4
12	*Lr26*	3	4	4	3+	4	1+	;	4	3	0;	3	3	4	3	2	3	4	4	4	4	4	4	4	4
13	*Lr26*[b]	4	4	4	4	4	4	4	4	4	4	4	4	4	4	4	4	4	4	4	4	4	4	4	4

（续）

Line number[a]	Lr gene	Infection types to *Puccinia triticina* pathotypes																							
		PH	FC	FC	PC	TC	PG	FG	FK	PC	FB	FH	PH	FH	FH	PG	PC	TC	PH	FH	FH	PH	TH	FC	PC
		QT	QR	ST	BT	GT	SN	SQ	QT	HS	HT	ST	SS	GS	TS	TT	JT	HT	ST	HQ	TR	JT	TT	TT	GR
14	*Lr26*[c]+	0	0	0	0	0	;	0	0	0	;	0	0	0	0	0	0	0	0	0	0	0	0	;	0
15	*Lr26*[c]+	;	;	;	;	;	0	0	1	0	0	0	0	;	0	0	;	;	;	;	;	0;	0;	;	;
16	*Lr26*+	2+	3+	2+,3C	22+	2	1+	;	4	2+,3−	1	3	3	3	4	;1+	4	4	3	4	4	3	3	4	4
17	*Lr26*+	2	0	0	0	0	;	0;	0	3	0;	3	3	0	3	0	3	3	;	2	0	3	;	3	3
18	*Lr26*+	3	3C	2	2	2	;1	0	4	2	0	3	3	3	4	;1	3	4	2	4	4	2+,3−	3	3	4
19	*Lr26*+	3	3	3	3	4	1	;1	;2	2	0;	2	3	;0	2	2	2	3	3	3	4	4	3	3	0
20	*Lr26*+	2	3	3	3	3	1+	1	3	3	0	3	3	2	3	2	3	3	3	3	3	3	3	3	3
21	*Lr26*[c]+	2	0	;	0;	;1	;	;	;	0;	;	;	0	0	0	0	0;	;2	;	0;	0;	0;	;	;	0;
22	*Lr26*+	0;	0	3	2	3−	1+	;1	3	3	0;	3	3	2	3	2	3	3	3	3	3	3	3	3	3
23	*Lr26*[c]+	0;	;	;	0;	0	;	;	0	0;	;1	;	;	0	0	0	0;	0;	;	0	3	0	;	0	;2
24	*Lr26*+	2	3	3	2+	3	1+	1	3	3	2	3	3	2	3	0	3	3	3	3	3	3	3	3	2
25	*Lr26*+	2	;2	3	;1	;2	1+	1+	;2	2	2	2	2	1	3	2	3	3	3	2	2	;2	3	3	3
26	*Lr26*[b]+	1	3	2	2	3	34	3	;1	12	0;	2	3	1	;1	3	;	3	3	2	2	2	3	4	3
27	*Lr1*	3	;	1+	3	3	4	;	1	3	;	;	4	;	1	4	3	4	3	;	1+	4	4	1	3
28	*Lr1*	3	0;	0	3	3	3	1	0;	3	;	0	3	0	;	3	3	3	3	0;	;	3	4	;	3
29	*Lr1*+	;	0	0	3	3	4	1	0;	;	0	0	3	0	;	3	3	3	;	;	0	0	4	;	3
30	*Lr1*+	3	;	;	0	3	1+	;1	;	0	0	0	3	0	0	;	3	3	;	0;	0	3	3	0	3
31	*Lr1*, *Lr26*+	;2	;	;	;	;1	1	;1	;1	;	;1	;	1;	;	1	0	0;	2	;	;	0;	0,;	;1	;	3
32	*Lr16*+	2	0	2	;1	1+	4	3	2	2	2	2	2	3	2	1	2	2	2;	12	3	2	3−	2	1
33	*Lr26*[c], *Lr24*	0	0	0	0	0	0	0	0	0	0	0	0	0	0	0	0	0	0	0	0	0	0	0	0
34	*Lr24*	0	;	0	0	0	;	0	0	0	0		0	0	0	0;	0	0	0	0	;	0	0;	0	0
35	*Lr3ka*, *Lr16*+	3	2	1	2	;2	3	3	2	;1	;2	2	2	2	1	0	;1	2	2;	;1	3	2+	4	2	2
36	*Lr26*, *Lr16*, *Lr3ka*+	3	1	2	2	1+	1+	1	2	2	;	2	2	0	3	1	0;	1+	2	2	3	2	4	2	2
37	*Lr17a*+	;2	1	1	1	1+,2	3	3C	2	2	2	3	3	2	2	1;	3	2	3	2	2	3	3	3	2

(续)

Line number[a]	Lr gene	PH	FC	FC	PC	TC	PG	FG	FK	PC	FB	FH	PH	FH	FH	PG	PC	TC	PH	FH	FH	PH	TH	FC	PC
		QT	QR	ST	BT	GT	SN	SQ	QT	HS	HT	ST	SS	GS	TS	TT	JT	HT	ST	HQ	TR	JT	TT	TT	GR
38	*Lr17a*+	2	0	1	2	2	4	4	2	2	2	3	3	2	2	3	3	;	2	1	3	2	3	2	1
39	*Lr3ka*, *Lr17a*	2	2+	2	2+	2	3+	3+	2	2	2	3	3	2	1	2	2	2	3	1	2	1+	3	3	2
40	*Lr26*, *Lr3ka*, *Lr16*, *Lr17a*+	2	2	2	2	2	;	1	2	2	0;	2	3	2	2	;	2	2	3	2	3	2	3	2	1+
41	*Lr26*, *Lr14a*+	;	0;	3	0;	0	1	; 1	0;	;	0	;	;	0	;	0	; 1	;	;	0	2	;	;	;	2
42	*Lr26*, *Lr14a*+	2	3	3	2	2	1	1	2	2	;	3	2	1	2	0	2	3	; 2	; 1	2	3	3	3−	1
43	*Lr26*, *Lr14a*+	; 2	0;	3	; 2	; 2	;	;	2	0	0;	3	1	;	1	0	2	3	;	0;	0	3	3	3	0;
44	*Lr26*, *Lr14a*+	;	0 ;	;	; 1	; 2	1	1	2	1	; 2	3	1	1	;	;	; 2	3	3	; 2	0	; 2	3	3−	2
45	*Lr3ka*, *Lr14a*+	; 2	2	3	; 1	; 1	;	1	2	0	;	;	1	0	;	;	0;	;	;	0	2	0; 2	4	;	1
46	*Lr14a*+	3	0; 2	0	3	3	;	;	0	0	3	0	; 1	0	0	3	0	2+	3	0	0	0	3	;	;
47	*Lr14a*+	3	; 0	;	0	3	;	;	0	0	0	;	1	2	3	0	0	3	;	0;	0	3	4	0	;
48	*Lr26*, *Lr18*+	2	3	3	2+,3−	3−	1+	;	2	1	2	3	2	2	2	2	3	3	2	2	2	3	3	3	3
49	*Lr26*, *Lr18*+	2	3	3	; 2	3−	1	0;	2	2	0;	3	2	0	1	2+	0	3	2	; 2	3	3	3	3	3
50	*Lr26*, *Lr18*+	2	0	3	; 1	2	1	1	2	2	0	3	1	0	1	2+	2+	3	3	2	3	3−	3	3−	2
51	*Lr26*, *Lr18*+	;	3	3	3	3	1	0;	; 2	; 1	0;	3	2	0	;	0	0;	3	2	; 2	3	3	3	3	; 1
52	*Lr3bg*+	; 2	0	1	; 2	; 1	3	3	; 2	2	;	2;	3	1	1	1	3	2	; 1	;	1	3	3	;	3
53	*Lr26*, *Lr3bg*+	3	; 2	3	3	2+	1+	1+	3	3	0;	3	2	; 2	3	0	3	2	2	1	2	3	3	3	3
54	*Lr2a*, *Lr3bg*	;	0	0	0;	3	1	;	;	0	;	0	;	0	0	1	;	2	;	0	0	0	4	0;	0;
55	*Lr26*, *Lr1*, *Lr20*+	3	0	;	; 2	3	1	; 1	0	0	0	;	0	0	0	0	0;	2	0	0;	0;	0	3	;	3
56	*Lr16*, *Lr20*, *Lr23*+	;	0	0;	0; 1	; 2	;	;	;	0	0	3	;	0	0	0	0	;	0	0	0	0	0;	; 0	1
57	*Lr26*[c], *LrZH84*	0	0;	; 1	; 1	2+	1+	; 1	12	1	1	2	2	0;	1	0; 2	1	3	;	0;	2	1	2	2	2
58	*Lr26*[c], *LrZH84*	;	0	;	; 2	2+	; 1	1	; 1	0	0;	; 2	; 2	0	;	0	;	; 2	;	;	0;	;	; 1	;	1
59	*Lr26*, *Lr34*[d]	; 1	;	2	;	2	;	;	4	2	; 1	3	3	3	3	;	3	4	3	3	4	4	4	4	4
60	*Lr26*[b], *Lr34*[d]	3	3	3	3	3	;	;	3	3	;	2	3	2	3	3	3	3	3	3	3	3	3	3	3
61	*Lr34*[d]	3	3	3	4	3	1	3	3	3	3	3	3	3	3	3	3	3	3	; 2	3	3	3	3	3

Infection types to *Puccinia triticina* pathotypes (columns PH QT … PC GR)

(续)

Line number[a]	Lr gene	Infection types to *Puccinia triticina* pathotypes																							
		PH	FC	FC	PC	TC	PG	FG	FK	PC	FB	FH	PH	FH	FH	PG	PC	TC	PH	FH	FH	PH	TH	FC	PC
		QT	QR	ST	BT	GT	SN	SQ	QT	HS	HT	ST	SS	GS	TS	TT	JT	HT	ST	HQ	TR	JT	TT	TT	GR
62	*Lr34*[d]	3	2	2	1+ 2	2	4	4	3	3	3	3	3	3	3	3	3	2	3	2	3	1+,2	4	2	2
63	*Lr34*[d]	4	4	4	4	4	4	3	4	4	4	4	4	4	3+	4	4	4	4	4	4	4	4	4	4
64	*Lr34*[d]	;	0; 2	;	3	3	3−	; 1	3	3	3	;	3	0	3	0	3	3	;	0	0	3	4	0	3
65	*Lr34*[d]	3+	4	3+	3+	;	4	4	4	4	3	4	4	4	4	4	4	4	3+	4	3	4	4	4	4
66	+	2	0	3	2	2	3	2	2	3	3	3	3	3	3	2	3	3	3	2	3	2+	3	2	3
67	+	0	0	0	0	;	0	0	0	0	0	0	0	0	0	0	0	0	;	0	0	0	0	0	0
68	+	3	4	3	4	3	3	3	3	3	3	3	3	3	3	3	3	3	3	2	3	3	4	3	3
69	+	3	2	2	1+ 2	2	3+	4	2	3	3	3	2	2	3	3	2	2+	3	2	3	1+,2	3	2	2
70	+	2	3	3	3	3−	3	3+	2	2	3	3	3	2	2	3	3	3	3	2	3	3	3	3	3
71	+	3	3	3	4	3	3	3	3	3	3	3	3	2	3	3	3	2+	3	3	3	3	4	3	3
72	+	4	4	4	4	4	4	4	4	4	; 1	3	3	4	4	4	4	4	3	3	4	4	4	4	4
73	+	1+	4	2+,3−	3	3	4	4	3	4	0	4	3	4	4	4	4	4	3	4	3	4	4	3	4
74	+	3	3	3	3	3	4	; 1	3	2	3	3	3	3	3	3	3	3	3	3	3	3	3	4	3
75	+	;	0	0	3	2	3+	; 1	; 2	;	3	;	3	0	3	1	2	3	2	;	2	2+	3	3	3
76	+	0;	3	3	3	3	4	4	2	3	3	3	3	3	2	3	2	3	3	2	3	3	3	3	3
77	+	3	0	3	2	3−	4	3+	3	3	3	3	3	2	3	0	3	3	2	3	3	2+,3−	3	3−	2
78	+	4	4	3	4	3	4	4	4	4	;	3	4	4	4	4	4	4	3	4	4	4	4	4	4
79	+	3	3	3	4	3	4	4	2	2	3	3	3	2	3	3	3	3	3	3	2	3	4	4	3
80	+	3	3	3	3	3	3+	4	3	0	3	3	3	0	3	3	3	3	3	3	3	2	3	3	3
81	+	;	3	2	2	3	1+	2+	; 2	; 2	;	2	3	1	3	3	; 2	3	3	; 2	;	; 2	3	4	3
82	+	2	2+	3	2	2	3	2	3	3	2	2	3	2	3	1	2	2	3	3	2	2	3	3	2
83	+	3	2	3	1+,2	1+	4	4	2	3	3	3	3	3	3	1	3	3	3	3	3	3	3	2	3
84	+	2	0	3	12	1	4	4	2	2	3	3	2+	3	3	3	3	2	3	2	3	3	3	2	2
85	+	0	2	3	; 2+	1	4	4	; 2	3	2	2	2	2	1+	2	2	2+	3	2	2	2	4	3	2

（续）

Line number[a]	*Lr* gene	Infection types to *Puccinia triticina* pathotypes																							
		PH	FC	FC	PC	TC	PG	FG	FK	PC	FB	FH	PH	FH	FH	PG	PC	TC	PH	FH	FH	PH	TH	FC	PC
		QT	QR	ST	BT	GT	SN	SQ	QT	HS	HT	ST	SS	GS	TS	TT	JT	HT	ST	HQ	TR	JT	TT	TT	GR
86	+	3+	3+	3	3+	4	4	4	4	4	3	4	4	3+	3	4	4	3	3	4	3	4	4	3	4
87	+	4	4	4	4	4	4	4	4	4	4	4	4	4	4	4	4	4	4	4	4	4	4	4	4
88	+	4	4	4	4	4	4	4	4	4	3	4	4	4	4	4	4	4	4	4	4	4	4	4	4
89	+	4	4	4	4	4	4	4	4	4	4	4	4	4	4	4	4	4	4	4	4	4	4	4	4
90	+	4	4	4	4	4	4	4	3	4	4	4	3	4	4	4	4	3	3	3+	4	4	4	3	4
91	+	4	4	4	4	4	4	4	4	4	4	3	3	4	3	4	4	4	4	4	4	4	4	4	4
92	+	4	4	3+	4	4	3+	4	3	4	4	4	4	4	4	4	4	4	3	4	3	4	4	4	4
93	+	4	4	4	4	4	4	4	4	4	4	4	3	4	4	4	4	4	4	4	4	4	4	4	4
94	+	4	4	4	4	4	4	4	4	4	4	4	4	4	4	4	4	4	4	4	4	4	4	4	4
95	+	4	4	4	4	4	4	4	4	4	4	3	3	4	4	4	4	4	4	4	4	4	4	4	4
96	+	4	4	4	4	4	4	4	3	4	4	3	4	4	4	4	4	4	4	4	4	4	4	4	4
97	+	4	4	4	4	4	4	4	3+	4	3	4	4	4	3	4	4	4	3	3	4	4	4	4	4
98	+	4	4	4	4	4	4	4	4	4	4	4	4	4	4	4	4	4	4	4	4	4	4	4	4
99	+	3+	4	4	4	4	4	4	3	4	3	4	4	4	3	4	4	3	3	4	3	4	4	4	4
100	+	3	4	4	3	4	4	4	4	4	4	4	4	4	4	4	4	4	4	4	4	4	4	4	4
101	+	4	4	4	4	4	4	4	4	4	3	4	4	4	4	4	4	4	4	4	4	4	4	4	4
102	+	3+	4	4	4	4	4	4	4	4	4	4	4	4	4	4	4	4	4	4	4	4	4	4	4

a. The line numbers are corresponding to those in Table 1 and are arranged in groups according to postulated *Lr* genes.

b. Postulation of 1BL. 1RS was based on molecular markers. These lines were susceptible to all or some of four *Lr26*-avirulent pathotypes.

c. Postulation of 1BL. 1RS was based on molecular markers. These lines are resistant to all or most of pathotypes.

d. Postulation of *Lr34* was based on molecular markers. These lines also showed symptoms of leaf tip necrosis (*Ltn*) and lower disease severity at the adult-plant stage.

Seedling testing

The seedling tests for the 102 cultivars and lines were conducted in 2007 and repeated in 2008, using 24 pathotypes (Table 3) with the method described by Singh et al.[34] with minor modifications. Forty cultivars or lines, with 6-8 seeds each, were sown in a plastic growth chamber (35cm × 24cm). Zhengzhou 5389 was included as the susceptible control. Seedling inoculations were performed by brushing urediniospores from sporulating susceptible seedlings on to test seedlings when the first leaves were fully expanded. Inoculated seedlings were subsequently placed in plastic-covered cages and incubated at 18℃ and 100% relative humidity (RH) for 24h. They were then transferred to a growth chamber maintained with 12h light/12h darkness at 18 to 20℃ with 70% RH. Infection types were scored 10 to 14 days after inoculation according to the 0-4 Stakman scale as modified by Roelfs et al.[26]. In addition to the genes detected based on reaction patterns to different phenotypes, *Lr26*, *Lr34*, and *LrZH84* were also confirmed by molecular markers.

Identification of *Lr26* and *Lr34*

To determine thepresence of the 1BL.1RS translocation carryintg *Lr26*, two molecular markers, one with a specific sequence from the *ω-secalin* rye gene generating a 1076-bp fragment in genotypes with the 1BL.1RS translocation, and the other from the wheat *Glu-B3* gene generating a 636-bp fragment amplified in genotypes without the 1BL.1RS translocation, were used to assay the 102 cultivars and lines following the procedure developed by Zhang et al.[41] with minor modifications. The slow rusting resistance gene *Lr34* was identified using the STS marker csLV34 as described by Lagudah et al.[13] (Table 1).

Field testing

The 104 cultivars and lines, including SAAR and Zhengzhou 5389, were planted in a randomized complete block design with three replicates in Baoding, Hebei province, in the 2006-2007 and 2007-2008 cropping seasons. Fifty seeds of each line were sown in a single-row plot with 1.5-m length and 30 cm between rows. Spreader rows of Zhengzhou 5389 were planted perpendicular and adjacent to the test rows. Leaf rust epidemics were initiated by spraying aqueous suspensions of urediniospores of *P. triticina* pathotype THTT, to which a few drops of Tween 20 (0.03%) were added, on to the spreader rows at the tillering stage. Disease severities were assessed for three times at weekly intervals with the first scoring four weeks after inoculation, using the modified Cobb scale[24], and infection types were scored using a 0-4 scale[26]. The disease severity data were used to calculate AUDPC based on the method described by Broers et al.[1]. AUDPC was calculated as

$$AUDPC = \sum_{i=1}^{n}[(X_i + X_{i+1})/2][T_{i+1} - T_i]$$

Where T_i = number of days after inoculation, X_i = disease severity. The relative AUDPC (rAUDPC) value for each wheat genotype in each cropping season was calculated as the percentage of the AUDPC value of the susceptible check (Zhengzhou 5389). The maximum disease severity (MDS, %) was recorded for each genotype in each cropping season as percentage of leaf area infected when the susceptible check displayed its MDS.

Statistical analysis

The SAS software was used for analysis of variance (ANOVA) and for determining least standard deviations (LSDs) for comparing the rAUDPC and MDS values among the wheat genotypes. Cultivars and lines that were seedling-susceptible to THTT and had lower or non-significantly higher values of MDS and AUDPC than those of the slow rusting check in field trials were considered as slow rusting genotypes.

Results

Resistance genes identified from seedling reactions or molecular markers

Variation in infection types (ITs) conferred by known *Lr* genes in differential lines, inoculated with 24 *P. triticina* pathotypes (Table 2), provided an ability to postulate 24 resistance genes (*Lr1*, *Lr2a*, *Lr3*, *Lr3bg*, *Lr3ka*, *Lr11*, *Lr14a*, *Lr15*, *Lr16*, *Lr17a*, *Lr18*, *Lr20*, *Lr21*, *Lr23*, *Lr26*, *Lr28*, *Lr29*, *Lr30*, *Lr32*, *Lr36*, *Lr39*, *Lr42*, *Lr44*, and *Lr50*). Resistance genes *Lr9*, *Lr19*, *Lr24*, and *Lr38* conferred low ITs with all pathotypes. Postulation of genes *Lr2b*, *Lr2c*, *Lr3*, *Lr10*, *Lr25*, *Lr33*, and *LrB* was not possible because high ITs were recorded with all pathotypes.

Fourteen resistance genes (*Lr1*, *Lr2a*, *Lr3bg*, *Lr3ka*, *Lr14a*, *Lr16*, *Lr17a*, *Lr18*, *Lr20*, *Lr23*, *Lr24*, *Lr26*, *Lr34*, and *LrZH84*) either singly or in combinations, were postulated to be present in 65 genotypes, whereas 37 accessions carried undetectable or unknown resistance genes (Table 3).

Forty-four entries contained *Lr26*; 12 displayed low infection types with four *Lr26*-avirulent pathotypes (P6, P7, P10, and P15) and high reactions with the remaining 20 pathotypes, showing the same response pattern as that of RL 6078 carrying *Lr26*. Thirty-one lines were postulated to possess *Lr26* plus other known or unknown *Lr* genes according to their low reactions with certain *Lr26*-virulent pathotypes (Tables 2 and 3). All entries postulated to have *Lr26* were also confirmed using the markers for the presence or absence of the 1BL.1RS translocation (Tables 1 and 3). Three lines (L13 = Zhoumai 12, L26 = N711, and L60 = R97) displayed high ITs to all or some of the four *Lr26*-avirulent pathotypes, but were positive for the presence of 1BL.1RS translocation using the markers. The high ITs with *Lr26*-avirulent pathotypes might be due to an inhibitor of *Lr26*, or the presence of a 1RS translocated chromosome not possessing a leaf rust resistance gene. Seven lines might carry *Lr1*, showing low ITs to 10 pathotypes avirulent to *Lr1*. Two cultivars (L27=Yumai 47 and L28=Yannong 19) are likely to carry gene *Lr1* alone because they displayed only low ITs with ten pathotypes avirulent to *Lr1*. Guizhou 98-18 (L31) carried *Lr1*, *Lr26*, and the other unidentified genes based on its reaction patterns. Another three cultivars (L29=Lumai 21, L30=Emai 14, and L55=Neixiang 991) were postulated to have *Lr1* and other resistance genes.

Seven lines were postulated to contain *Lr14a*; they gave low or intermediate reactions to five *Lr14a* avirulent pathotypes. *Lr14a* was present in combination with *Lr26* in four lines (L41=Xinong 1163-4, L42=93Xia032-1, L43=Y 1496-15, and L44=Tian 9470). The other three lines (L45=W7, L46=LB 0458, and L47=SW 8588) might carry *Lr14a* and other unknown or known genes.

Lr34 was postulated in seven lines with the molecular marker *csLv34* (13) and these lines also showed symptoms of leaf tip necrosis (*Ltn*) at the adult stage. Two lines (L59=01-018 and L60=R97) carried *Lr34* in addition to *Lr26*.

Lr3ka was postulated in combination with other known *Lr* genes in six lines according to their low reactions to nine *Lr3ka*-avirulent pathotypes and some *Lr3ka*-virulent pathotypes.

Gene*Lr16* may be present in five lines, showing low ITs with the 10 pathotypes that were avirulent on the *Lr16* NIL (Tables 2 and 3). *Lr16* was present in combination with *Lr3ka* in Shaannong 981 (L35) because low ITs were observed in seedlings inoculated with 10 *Lr16*-avirulent pathotypes and three *Lr3ka*-avirulent pathotypes. Een 1 (L36) carried *Lr16* in addition to *Lr3ka* and *Lr26* according to its low reactions to all pathotypes avirulent to the three genes. Another three lines (L32=Zhengmai 9023, L40=58749, and L56=LB 0288) may carry *Lr16* and other known or unidentified genes.

Lr17a and *Lr18* were each found in four lines. The four lines carrying *Lr17a* displayed low ITs to eleven *Lr17a*-avirulent pathotypes. *Lr17a* and *Lr3ka* were present in An 96-8 (L39) and *Lr17a* were postulated in combination with *Lr3ka*, *Lr16*, and *Lr26* in line 58749 (L40). Xinmai 19 (L37) and Xinong 1043 (L38) may contain *Lr17a* and other unknown genes. *Lr18* was postulated in combination with *Lr26* in four lines (L48=Lankaoaizao 8, L49 = 01-DH689, L50 = Chuannong 17, and L51=Mian 2000-13) based on low ITs with four *Lr26*-avirulent pathotypes and intermediate ITs with five *Lr18*-avirulent pathotypes.

Three lines displayed intermediate or low ITs to three pathotypes (P17, P18, and P20) lacking virulence for *Lr3bg*, indicating that they contain *Lr3bg*. *Lr3bg* was postulated in combination with *Lr26* in Qinnong 151 (L53) and in combination with *Lr2a* in 351-15 (L54). Chuannong 19 (L52) was postulated to have *Lr3bg* and other unidentified genes.

Lr20 was detected in Neixiang 991 (L55) in combination with *Lr1* and LB 0288 (L56) carried known *Lr* genes additional to *Lr20*. 5R615 (L33) and 5R616 (L34) showed high levels of resistance to all 24 pathotypes and may contain *Lr24* because the two lines were derived from Agent, a known *Lr24* carrier (Table 1). 5R615 also possessed *Lr26* as it had the 1BL. 1RS translocation detected with molecular markers. *LrZH84* was present in combination with *Lr26* in Zhou 8425B (L57). According to Zhao et al. [42], Zhou 8425B contains gene *LrZH84*. Zhoumai 11 (L58) is a derivative of Zhou 8425B, and had a reaction pattern similar to Zhou 8425B, indicating that it may also carry *LrZH84*.

Line 351-15 (L54) displayed low infection types with all 21 pathotypes avirulent to *Lr2a* and an intermediate IT to P17 (TCHT) which is avirulent for *Lr3bg*, indicating that *Lr2a* is present in combination with *Lr3bg* in this line. *Lr23* was present in combination with *Lr16* and *Lr20* in LB 0288 (L91), displaying low ITs with pathotypes lacking virulence for *Lr16*, *Lr20* and *Lr23*.

Eighteen genotypes (L85-L102) showed high infection types with all pathotypes, indicating that they do not have any resistance gene corresponding to an avirulence gene present in any of the pathotypes. The reactions of 18 cultivars and lines (L66 and L68-L84) were not consistent with the response of any known resistance gene or predicted response of any gene combination.

5R631 (L67) conferred resistance to all pathotypes tested, indicating that it carries one or more unidentified resistance genes.

Slow rusting resistance in field tests

Highly significant differences were found for wheat genotypes and environments (years) for rAUDPC and MDS in the field trials (Table 4). The genotype-environment interaction was also significant, but its effect on variation was much less than the genotypic differences.

Table 4 Analysis of variance of relative area under the disease progress curve (rAUDPC) **and maximum disease severity** (MDS) **in 104 wheat cultivars and lines including slow rusting and susceptible checks tested in Baoding in the 2006-2007 and 2007-2008 cropping seasons**

Parameter	Source of variation	DF	MS	F value	P
rAUPDC	Cultivar	103	5 710. 81	72. 00	<0. 000 1
	Season	1	1 745. 76	21. 94	<0. 000 1
	Replication	2	58. 71	0. 74	0. 48
	Cultivar * Season	103	1 005. 96	12. 68	<0. 000 1
	Cultivar * Replication	206	78. 24	0. 99	0. 54
	Error	208	79. 34		
MDS	Cultivar	103	5 595. 44	112. 53	<0. 000 1
	Season	1	21 443. 85	431. 27	<0. 000 1
	Replication	2	52. 15	1. 05	0. 35
	Cultivar * Season	103	930. 84	18. 72	<0. 000 1
	Cultivar * Replication	206	51. 73	1. 04	0. 39
	Error	208	49. 72		

The MDS value of the susceptible check, Zhengzhou 5389, was 100% and 85%, respectively, in the 2006-2007 and 2007-2008 cropping seasons (Table 5), indicating that the disease developed well in both seasons. The MDS value of the slow rusting check, SAAR, was 18% and 12%, respectively, and its rAUDPC was 15% in both seasons (Table 5).

Table 5 Infection types (IT) in the seedling test with *Puccinia triticina* pathotype THTT, presence and absence of the 1B.1R wheat-rye chromosomal translocation and *Lr34* based on molecular markers, and IT, mean maximum disease severity (MDS), and mean relative area under the disease progress curve (rAUDPC) in the field experiments with the same pathotype in Baoding, Hebei in the 2006-2007 and 2007-2008 growing seasons for wheat genotypes with slow rusting resistance to leaf rust

Genotype (line number)	Seedling IT to THTT	1B.1R and *Lr34*[c]	2006-2007			2007-2008		
			IT	MDS (%)	rAUDPC (%)	IT	MDS (%)	rAUDPC (%)
Saar[a]	4	*Lr34*	4	18	15.11	4	12	15.43
Zhengzhou 5389[b]	4	—	4	100	100.00	4	85	100.00
(95) 18 (L7)	4	1B.1R	3	4	2.22	4	18	20.31
Lantian 9 (L9)	4	1B.1R	3	7	5.89	4	22	25.14
Jing 035-1 (L12)	4	1B.1R	3	7	7.45	3, 4	12	9.71
Lankao 906-4 (L18)	3	1B.1R	3	1	0.66	3, 4	4	3.81
Huaimai 18 (L22)	3	1B.1R	3	7	5.78	4	13	15.43
58749 (L40)	3	1B.1R	3	1	1.29	3	1	1.53
Y 1496-15 (L43)	3	1B.1R	3	1	0.85	3	7	5.67
LB0458 (L46)	3	—	3	1	0.55	2, 3	1	0.95
SW 8588 (L47)	4	—	3	1	0.66	4	1	1.21
Mian 2000-13 (L51)	3	1B.1R	3	3	4.44	4	5	5.21
351-15 (L54)	4	—	3	1	0.67	3	1	0.87
Neixiang 991 (L55)	3	1B.1R	2, 3	1	1.22	2, 3	4	4.37
01-018 (L59)	4	1B.1R, *Lr34*	4	5	3.64	4	12	11.43
R97 (L60)	3	1B.1R, *Lr34*	3	2	2.11	4	13	14.29
LSD (P=0.05)		—		13	15.25		15	14.36

a. Slow rusting check.

b. Susceptible check.

c. 1B.1R=presence of 1B.1R, *Lr34* = presence of *Lr34*, —=absence of 1B.1R and *Lr34*.

Twelve entries, including 5R608 (L14), 5R610 (L15), Tian 95HF2 (L21), Bi 2002-2 (L23), Guizhou 98-18 (L31), 5R615 (L33), 5R616 (L34), Xinong 1163-4 (L41), LB0288 (L56) Zhou 8425B (L57), Zhoumai 11 (L58), and 5R631 (L67), were highly resistant to pathotype THTT at both the seedling and adult stages. Fifteen genotypes exhibiting high infection types at the seedling stage showed slow leaf rusting resistance in the field to the same pathotype in both seasons (Table 5). The correlations of MDS and relative AUDPC between the two cropping seasons were 0.77 and 0.74 (N=104 and P=0.01), respectively, indicating highly consistent results for slow rusting resistance. The correlation between relative AUDPC and MDS at the two locations varied from 0.96 to 0.99 (N=104 and P=0.01), indicating that MDS was a reliable indicator of AUDPC (14).

Discussion

Seedling tests for leaf rust response

Yuan et al.[39] postulated 10 known genes, *Lr1*, *Lr3*, *Lr3bg*, *Lr9*, *Lr10*, *Lr13*, *Lr16*, *Lr23*, *Lr26*, and *Lr34* in 47 new wheat cultivars from China using 17 pathotypes of *P. triticina*. Six of them, *Lr1*, *Lr3bg*, *Lr16*, *Lr23*, *Lr26*, and *Lr34*, were also identified in our seedling tests. The postulation of genes *Lr3*, *Lr9*, and *Lr10* was not possible in the present study because high ITs or low ITs were recorded with all pathotypes. *Lr13* was not tested in the seedling tests because the gene is an APR gene.

During the early 1970s, the 1BL. 1RS translocation with *Yr9* and *Lr26* for resistance to stripe rust (*P. striiformis* f. sp. *tritici*), respectively was introduced to Chinese wheat breeding programs through the introductions of wheat germplasms such as Lovrin 10, Lovrin 13, Predgornaia 2, Kavkaz, and Neuzucht[9], leading to a high frequency of *Lr26* in Chinese wheat cultivars and lines. Among 102 wheat lines tested in the present study, *Lr26* was identified in 44 (43.1%). Three lines (L13 = Zhoumai 12, L26 = N711, and L60=R97) were susceptible to all or some of four *Lr26*-avirulent pathotypes, but had a 1BL. 1RS translocation based on the test with molecular markers. This indicated that an inhibitor gene might suppress the expression of *Lr26* in these two lines, or that a 1BL. 1RS translocation lacking *Lr26* might be present. Hanušová et al. [8] reported an inhibitor of *Pm8* in certain 1BL. 1RS wheat genotypes, but suppression of leaf rust resistance gene *Lr26* has not been reported and therefore needs to be verified.

Lr1 was postulated to be present in seven cultivars and lines. A study by Singh et al. [34] indicated that *Lr1* was present in many Chinese cultivars. *Lr1* in Yumai 47 might be derived from Baiquan 3199, because Bainong 3217 with *Lr1*[34] had the same parent Funo as Baiquan 3199, indicating that *Lr1* in Yumai 47 and Bainong 3217 was likely derived from Funo. *Lr1* in Lumai 21 might be derived from Lumai 13 (Table 1), because Baiyoubao carrying *Lr1*[34] was one of parents of Lumai 13. *Lr1* in Emai 14 was derived from Yanda 72-629, as Emai 14 and Bainong 64 carrying *Lr1* had the same parent Yanda 72-629[34]. *Lr1* in Neixiang 991 might be derived from Bainong 3217 based on its pedigree (Table 1). *Lr16*, detected in five lines, was postulated in En 1, in agreement with the result reported by Singh et al. [34]. *Lr16* in Zhengmai 9023 and Shaannong 981 might be derived from Shaan 213 and Xiaoyan 6[34], respectively, according to their pedigrees (Table 1). Most lines with *Lr1* and *Lr16* were highly susceptible to pathotype THTT at the adult plant stage. Genes *Lr3ka*, *Lr14a*, *Lr17a*, and *Lr18* could be present in some lines (Tables 2 and 3); however, these postulation will require further verification because few avirulent pathotypes were available for testing, and the resistant reactions displayed by the four genes were vulnerable to genetic background of host and environment.

Various lines were widely resistant to many of the pathotypes used in the study, but the resistance could not be attributed to known leaf rust resistance genes. This indicates that new leaf rust resistance genes may be present in some lines. Genetic analyses are now being undertaken on lines with superior leaf rust resistance, and the results of those studies will permit an assessment of the postulations made herein. *Lr1* and an unknown *Lr* gene were identified in Tian 95HF2, and *LrZH84* was detected in Xinong 1163-4 and Guizhou 98-18 using SSR markers (data not shown).

Slow rusting resistance

The 14 genotypes with slow leaf rusting resistance to THTT were tested as seedlings with all 24 *P. triticina* pathotypes. Twelve of them contained *Lr26*; two of those (L59 = 01-018 and L60 = R97) carried *Lr34* based on molecular marker detection. Four genotypes (L40=58749, L43=Y1496-15, L51=Mian 2000-13, and L55=Neixiang 991) had other known leaf rust resistance genes (*Lr1*, *Lr3ka*, *Lr14a*, *Lr16*, *Lr17*, *Lr18*, and *Lr20*), and three (L18=Lankao 906-4 and L22=Huaimai 18) likely possessed unidentified genes (Tables 3 and 4). In these genotypes, the seedling resistance gene (s) were overcome by pathotype THTT and the slow rusting resistances probably resulted from *Lr34* and other minor APR genes. Slow rusting resistance in LB 0458 (L46) may be derived from CIMMYT wheat germplasm (Table 1). Of all the slow leaf rusting cultivars and lines, 10 genotypes displayed slow rusting resistance to stripe rust[14]. Slow rusting in these lines to both diseases might be conferred by common genes, as in the examples of *Lr34/Yr18* and *Lr46/Yr29* [15].

Singh et al. [35] reported that acceptable levels of slow rusting are conferred by combinations of minor additive genes. In the present study we identified six lines with *Lr34*, but only two lines displayed a significant level of slow rusting resistance in the two-years of field trials, indicating that *Lr34* alone does not provide adequate resistance. However, *Lr34* combined with other minor resistance genes can provide near-immune levels of leaf rust resistance.

Acknowledgments

The authors are very grateful to Prof. R. A. McIntosh for reviewing this manuscript. This study was supported by the National Natural Science Foundation of China (30700505) and National Scientific and Technical Supporting Programs funded by the Ministry of Science & Technology of China (2006BAD08A05).

References

[1] Broers, L. H. M., Cuesta Subias, X., and Lopez Atilano, R. M. 1996. Field assessment of quantitative resistance to yellow rust in ten spring bread wheat cultivars. Euphytica. 90: 9-16.

[2] Caldwell, R. M. 1968. Breeding for general and/or specific plant disease resistance. Pages 263-272 in: Proc. 3rd Int. Wheat Genetics Sympos. K. W. Findlay and K. W. Shepherd, eds. Australian Acad Sci., Canberra, Australia.

[3] Chen, W. Q., Qin, Q. M., Chen, Y. L., and Wang, H. W. 1998. Preliminarily screening of slow leaf-rusting in the adult plant period in Chinese wheat cultivars. Southwest China J. Agric. Sci. 11: 54-61.

[4] Dong, J. G.. 2001. Agricultural Plant Pathology. China Agriculture Press, Beijing.

[5] Dyck, P. L. 1987. The association of a gene for leaf rust resistance with the chromosome 7D suppressor of stem rust resistance in common wheat. Genome. 29: 467-469.

[6] Dyck, P. L., and Samborski, D. J. 1982. The inheritance of resistance to *Puccinia recondita* in a group of common wheat cultivars. Can. J. Genet. Cytol. 24: 273-283.

[7] Flor, H. H. 1942. Inheritance of pathogenicity in *Melampsora lini*. Phytopathology. 32: 653-669.

[8] Hanušová, R., Hsam, S. L. K., Bartoš, P., and Zeller, F. J. 1996. Suppression of powdery mildew resistance gene *Pm8* in *Triticum aestivum* L. (common wheat) cultivars carrying wheat-rye tranlocation T1BL·1RS. Heredity. 77: 383-387.

[9] He, Z. H., Rajaram S., Xin Z. Y., and Huang, G. Z. 2001. A history of wheat breeding in China. CIMMYT, Mexico, D. F.

[10] Kilpatrick, R. A. 1975. New wheat cultivars and longevity of rust resistance, 1971-1975. Beltsville, MD: US Department of Agriculture, Agricultural Research Service.

[11] Kolmer, J. A. 2003. Postulation of leaf rust resistance genes in selected soft red winter wheats. Crop Sci. 43: 1266-1274.

[12] Krattinger, S. G., Lagudah, E. S., Spielmeyer, W., Singh, R. P., Huerta-Espino, J., McFadden, H., Bossolini, E., Selter, L. L., and Keller, B. 2009. A putative ABC transporter confers durable resistance to multiple fungal pathogens in wheat. Science. 323: 1360-1363.

[13] Lagudah, E. S., McFadden, H., Singh, R. P., Huerta-Espino, J., Bariana, H. S., and Spielmeyer, W. 2006. Molecular genetic characterization of the *Lr34/Yr18* slow rusting resistance gene region in wheat. Theor. Appl. Genet. 114: 21-30.

[14] Li, Z. F., Xia, X. C., Zhou, X. C., Niu, Y. C., He, Z. H., Zhang, Y., Li, G. Q., Wan, A. M., Wang, D. S., Chen, X. M., Lu, Q. L., and Singh, R. P. 2006. Seedling and slow rusting resistance to stripe rust in Chinese common wheats. Plant Dis. 90: 1302-1312.

[15] Lillemo, M., Asalf, B., Singh, R. P., Huerta-Espino, J., Chen, X. M., He, Z. H., and Bjornstad, A. 2008. The adult plant rust resistance loci *Lr34/Yr18* and *Lr46/Yr29* are important determinants of partial resistance to powdery mildew in bread wheat line Saar. Theor. Appl. Genet. 116: 1155-1166.

[16] Long, D. L., and Kolmer, J. A. 1989. A North

American system of nomenclature for *Puccinia recondita f.* sp. *tritici*. Phytopathology. 79：525-529.

[17] McIntosh, R. A. 1992. Close genetic linkage of genes conferring adult-plant resistance to leaf rust and stripe rust in wheat. Plant Pathol. 41：523-527.

[18] McIntosh, R. A., Devos, K. M., Dubcovsky, J., Rogers, W. J., Morris, C. F., Appels, R., Somers, D. J., and Anderson, O. A. 2008. Catalogue of gene symbols for wheat：2008 supplement. Annu. Wheat Newsl. Vol. 54.

[19] McVey, D. V. 1992. Genes for rust resistance in International Winter Wheat Nurseries XII through VXII. Crop Sci. 32：891-895.

[20] McVey, D. V., and Long, D. L. 1993. Genes for leaf rust resistance in hard red winter wheat cultivars and parental lines. Crop Sci. 33：1373-1381.

[21] Mebrate, S. A., Dehne, H. W., Pillen, K., and Oerke, E. C. 2008. Postulation of seedling leaf rust resistance genes in selected Ethiopian and German bread wheat cultivars. Crop Sci. 48：507-516.

[22] Oelke, L. M., and Kolmer, J. A. 2004. Characterization of leaf rust resistance in hard red spring wheat cultivars. Plant Dis. 88：1127-1133.

[23] Person, C. 1959. Gene-for-gene relationships in host：parasite systems. Can. J. Bot. 37：1101-1130.

[24] Peterson, R. F., Campbell, A. B., and Hannah, A. E. 1948. A diagrammatic scale for estimating rust intensity of leaves and stems of cereals. Can. J. Res. 26：496-500.

[25] Pink, D. A. C. 2002. Strategies using genes for non-durable disease resistance. Euphytica . 124：227-236.

[26] Roelfs, A. P., Singh, R. P., and Saari, E. E. 1992. Rust diseases of wheat：concepts and methods of disease management. CIMMYT, Mexico, D. F.

[27] Samborski, D. J. 1963. A mutation in *Puccinia recondita* Rob. ex Desm. f. sp. *tritici* to virulence on transfer, Chinese Spring × *Aegilops umbellulata* Zhuk. Can. J. Bot. 41：475-479.

[28] Singh, R. P. 1992. Association between gene *Lr34* for leaf rust resistance and leaf tip necrosis in wheat. Crop Sci. 32：874-878.

[29] Singh, R. P. 1992. Genetic association of leaf rust resistance gene *Lr34* with adult plant resistance to stripe rust in bread wheat. Phytopathology. 82：835-838.

[30] Singh, R. P. 1993. Resistance to leaf rust in 26 Mexican wheat cultivars. Crop Sci. 33：633-637.

[31] Singh, R. P., and Gupta, A. K. 1991. Genes for leaf rust resistance in Indian and Pakistani wheats tested with Mexican pathotypes of *Puccinia recondita* f. sp. *tritici*. Euphytica. 57：27-36.

[32] Singh, R. P., and Rajaram, S. 1991. Resistance to *Puccinia recondita* f. sp. *tritici* in 50 Mexican bread wheat cultivars. Crop Sci. 31：1472-1479.

[33] Singh, R. P., and Rajaram, S. 1992. Genetics of adult-plant resistance to leaf rust in 'Frontana' and three CIMMYT wheats. Genome. 35：24-31.

[34] Singh, R. P., Chen, W. Q., and He, Z. H. 1999. Leaf rust resistance of spring, facultative, and winter wheat cultivars from China. Plant Dis. 83：644-651.

[35] Singh, R. P., Huerta-Espino, J., and Rajaram, S. 2000. Achieving near-immunity to leaf and stripe rusts in wheat by combining slow rusting resistance genes. Acta Phytopathol. Hung. 35：133-139.

[36] Singh, R. P., Mujeeb-Kazi, A., and Huerta-Espino, J. 1998. *Lr46*：A gene conferring slow-rusting resistance to leaf rust in wheat. Phytopathology. 88：890-894.

[37] Statler, G. D. 1984. Probable genes for leaf rust resistance in several hard red spring wheats. Crop Sci. 24：883-886.

[38] Wamishe, Y. A., and Milus, E. A. 2003. Seedling resistance genes to leaf rust in soft red winter wheat. Plant Dis. 88：136-146.

[39] Yuan, J. H., Liu, T. G., and Chen, W. Q. 2007. Postulation of leaf rust resistance genes in 47 new wheat cultivars at seedling stage. Sci. Agric. Sin. 40：1925-1935.

[40] Zhang, L. J., Li, Z. F., Lillemo, M., Xia, X. C., Liu D. Q., Yang W. X., Luo, J. C., and Wang, H. Y. 2009. QTL Mapping for adult-plant resistance to leaf rust in CIMMYT wheat cultivar Saar. Sci. Agric. Sinica. 42：388-397.

[41] Zhang, X. K., Liu, L., He, Z. H., Sun, D. J., He, X. Y., Xu, Z. H., Zhang, P. P., Chen, F., and Xia, X. C. 2008. Development of two multiplex PCR assays targeting improvement of bread-making and noodle qualities in common wheat. Plant Breed. 127：109-115.

[42] Zhao, X. L., Zheng, T. C., Xia, X. C., He, Z. H., Liu, D. Q., Yang, W. X., Yin, G. H., and Li, Z. F. 2008. Molecular mapping of leaf rust resistance gene *LrZH84* in Chinese wheat line Zhou 8425B. Theor. Appl. Genet. 117：1069-1075.

Effective resistance to wheat stripe rust in a region with high disease pressure

B. Bai[1], J. Y. Du[2], Q. L. Lu[2], C. Y. He[2],
L. J. Zhang[2], G. Zhou[2], X. C. Xia[3], Z. H. He[3,4], and C. S. Wang[1]

[1] *State Key Laboratory of Crop Stress Biology in Arid Areas/College of Agronomy, Northwest A & F University, Yangling* 712100, *Shaanxi, China; and Wheat Research Institute, Gansu Academy of Agricultural Sciences*, 1 *Nongkeyuanxincun, Lanzhou* 730070, *China*; [2] *Wheat Research Institute, Gansu Academy of Agricultural Sciences*, 1 *Nongkeyuanxincun, Lanzhou* 730070, *China*; [3] *Institute of Crop Science, National Wheat Improvement Center, Chinese Academy of Agricultural Sciences (CAAS)*, 12 *Zhongguancun South Street, Beijing* 100081, *China*; [4] *Institute of Crop Science, National Wheat Improvement Center, Chinese Academy of Agricultural Sciences (CAAS), Zhongguancun South Street* 12, *Beijing* 100081, *China, and International Maize and Wheat Improvement Center (CIMMYT), CIMMYT China Office, c/o CAAS*, 12 *Zhongguancun South Street, Beijing* 100081, *China.*

Abstract: Stripe rust is a major fungal disease of wheat. It frequently becomes epidemic in southeastern Gansu province, a stripe rust hot spot in China. Evaluations of wheat germplasm response are crucial for developing cultivars to control the disease. A total of 57 wheat cultivars and lines from Europe and other countries, comprising 36 cultivars with documented stripe rust resistance genes and 21 with unknown genes were tested annually with multiple races of *Puccinia striiformis* f. sp. *tritici* in the field at Tianshui in Gansu province from 1993 to 2013. Seven wheat lines were highly resistant with IT 0 during the entire period; 16 were moderately resistant (IT 0; -2); and 26 were moderately susceptible (IT 0; -4) with low maximum disease severity compared to the susceptible control Huixianhong. Cultivars Strampelli and Libellula, with three and five quantitative trait loci for stripe rust resistance, respectively, have displayed durable resistance in this region for four decades. Ten cultivars, including Lantian 15, Lantian 26, and Lantian 31 with stripe rust resistance derived from European lines, were developed in our breeding program, and have made a significant impact on controlling stripe rust in southeastern Gansu. Breeding resistant cultivars with multiple adult-plant resistance genes seems to be a promising strategy in wheat breeding for managing stripe rust in this region and other hot spots.

Key words: *Triticum aestivum*, resistance evaluation, stripe rust

Introduction

Wheat stripe rust (or yellow rust, YR), caused by *Puccinia striiformis* Westend. f. sp. *tritici* Erikss. (*Pst*), is one of the most destructive foliar diseases of wheat globally. China has one of the largest epidemic regions for YR in the world, and epidemics of the disease occur frequently in the autumn planted wheat growing regions of northwestern and southwestern China. From 2004 to 2009, YR occurred on about 4.2 million ha each year and caused serious damage to wheat production in China[13]. In recent years, YR was particularly prevalent in Sichuan, Chongqing, Shaanxi and Gansu provinces, Shiyan and Xiangfan regions of Hubei province, and the Xinyang and Nanyang

Published in Plant Disease, 2014, 98 (7): 891-897

regions of Henan province.

Resistance breeding is an effective, economic and environmentally friendly approach to control YR[4]. However, many wheat cultivars have become susceptible to YR due to changes in the pathogen population, and even some new breeding lines became susceptible before being officially released into production. Many Chinese wheat cultivars with *Yr9* (1B.1R translocation) became susceptible after the spread of *Pst* race CYR29 during the 1990s, and then Fan 6 and its derivatives became susceptible in 2002[6]. More than 50 genes for resistance to YR are formally named in wheat and most are assigned to various chromosomal regions[25]. However, the majority of these genes are either not effective in China or were overcome by the pathogen populations at different times. Only a few named major genes (such as *Yr5* and *Yr15*, and some temporarily designated genes, such as *YrGaby*) and several adult-plant resistance (APR) genes (such as *Yr18*, *Yr29*, and *Yr46*) are effective in limiting the size of the Chinese *Pst* population. Thus, there is an ongoing need to identify and develop new resistant materials, and to understand the resistance genes present in wheat cultivars and lines. Considerable work has been done in a number of countries, including China, on the basis of YR resistance in spring wheat germplasm, but relatively little has been reported on durable resistance in winter wheat from European countries.

Southeastern Gansu, especially Tianshui and Longnan in China, is a well documented hot spot for YR[6,19,33,36]. In these areas *Pst* can complete a year-round uredinial cycle, because about 250, 000 ha of wheat are grown in areas that extend from lowland valleys at 800m asl to highland terraces at 2, 400m asl, providing a "green bridge" for *Pst* that migrates throughout the year from late-maturing highland areas to early-sown wheat fields in the lowlands[6,19,36]. Several *Berberis* species (Barberry) occur in the region and following the recent discovery that barberry is a potential alternate host for *Pst*[14,38], the possibility of sexual reproduction and genetic recombination contributing to the apparently high level of pathogen variability in the region is only now being appreciated and studied[38]. This continuous epidemic recycling maintains large pathogen populations and high pathogenic variability that leads to frequent 'break-downs' in resistance and thus an urgency to broaden the genetic basis of resistance in wheat cultivars used in the region.

The objective of this study was to report on long term assessments of European wheat germplasm in southeastern Gansu since 1993, and to identify germplasm with effective resistance to YR in a region with high disease pressure, several of which have been used in our breeding program.

Materials and methods

Wheat germplasm

Fifty-seven wheat cultivars and lines from Europe and other countries were evaluated annually for YR response in the field at Tianshui of southeastern Gansu from 1993 to 2013. All cultivars and lines were originally provided by the International Maize and Wheat Improvement Center (CIMMYT), and Chinese Academy of Agricultural Science (CAAS). The resistance genes to YR, origin, and pedigrees of cultivars or lines are listed in Table 1. Although many of the cultivars and lines were known, or postulated, to possess well documented resistance genes based on classical seedling tests, it was not known if those genes or other genes for seedling (all-stage) resistance and adult plant resistance (APR) contributed to the observed responses at Tianshui.

Pst races

The trial comprised both inoculated and naturally infected nurseries. The Chinese *Pst* races used in artificial inoculations differed over time with changes in the local populations. The key races included CYR29 (the first race that became epidemic on cultivars with *Yr9* in the 1990s), CYR30 (virulent on Hybrid 46 with *Yr3b* and *Yr4b*, and Fan 6 and its derivatives such as Miannong 4, Mianyang and several lines in the Chuanmai series in 1991), CYR31 (with similar virulences to CYR30, appearing in 1993), CYR32 (with the widest virulence spectrum combining the virulences of CYR31 with virulence for *Yr9* and Suwon 11 (*YrSu*) first detected in 1994 and becoming predominant in 2000-2002), CYR33 (virulent on Suwon 11, first detected in 1997 and formally named in 2008), and Gui pathotype (the first race virulent on Guinong 22, Chuanmai 42 and other cultivars with *Yr24*, *Yr26*, and *YrCH42*, as well as *Yr10* in 2010). The main races identified from samples taken from the trial sites in different years are listed in Table 2.

Table 1 Pedigrees and origin of wheat cultivars and lines used in this study[a]

Cultivar/line	*Yr* gene	Pedigree	Origin
Aquila I	/[b]	San Giovanni/Damiano	ITA
Atou	*Yr3a*, *Yr4a*, *Yr16*	Cappelle Desprez/Garnet	FRA
Bouquet	*Yr3a*, *Yr4a*, *Yr14*, *Yr16*	2-7, Versailles / Cappelle Desprez//Cappelle Desprez	FRA
Cappelle Desprez	*Yr3a*, *Yr4a*, *Yr16*	Vilmorin 27/Hybride Du Joncquois	FRA
CarstensV	*Yr32* (*YrCV*)	CarstensIII/Dickkopf//Dickkopf/Criewener 104	DEU
Champlein	*Yr3a*, *Yr4a*, *Yr16*	Yga Blondeau/Tadepi	FRA
Cik Vee	/	/	/
Compair	*Yr8*, *Yr19*	Chinese Spring/Ae. Co	UK
C591	*YrC591*	PBType 8B/PB Type 9	PK
Dippes Triumph	*Yr15*	Derenburger Silber/Erbachshofer Braun	DEU
Elite Le Peuple	*Yr2*, DR[c]	Bellevue/Hybride De Bersee	FRA
Fenman	*Yr1*, *Yr2*	TJB 268/175/Hobbit 'Sib'	UK
Flanders	*Yr1*, *Yr3a*, *Yr4a*, *Yr16*	Champlein/FD 2816-348	FRA
Flinor	DR	Elite Le-Peuple/Poncheau	FRA
FR 81-1	/	/	FRA
Gaby	*YrGaby*	Koga/Fylby	BEL
German 2	/	/	DEU
German 8661	/	/	DEU
HeinesVII	*Yr2*, *Yr25*, *YrHVII*	Hybrid A Courte Palle/Kronen, Swe	DEU
Heines Kolben	*Yr2*, *Yr6*	/	DEU
Hobbit	*Yr3*, *Yr4*, *Yr14*, *YrHVII*	Professeur Marchal//Marne-Desprez/VG9144/3 /// TJB 16 *	UK
Holdfast	DR	Yeoman/White Fife	UK
Hybrid 46	*Yr3b*, *Yr4b*, *YrH46*	Benoist 40/Yeoman	UK
Ibis	*Yr1*, *Yr2*, APR QTL	Merlin//HeinesVII/Heines 2167	DEU
Joss Cambier	*Yr2*, *Yr11*	HeinesVII/Tadepi//Cappelle Desprez	FRA
Jubilar	/	Pfeuffers Schernauer//Taca/Derenburger Silber	DEU
Libellula	*Yr18* + APR QTL	Tevere/Giuliari (1482-54-3) //San Pastore	ITA
Little Joss	DR	Squareheads Master/Ghirka	UK
Longbow	*Yr1*, *Yr2*, *Yr6*, *Yr9*, *Yr13*	TJB 268-175/Hobbit	UK
Lperfer	/	/	/
Maris Huntsman	*Yr2*, *Yr3a*, *Yr4a*, *Yr13*	CI 12633/5 * Cappelle Desprez//Hybrid 46/Cappelle Desprez/3/2 * Professeur Marchal	UK

（续）

Cultivar/line	Yr gene	Pedigree	Origin
Maris Widgeon	*Yr3a*, *Yr4a*, *Yr16*	Holdfast/Cappelle Desprez	UK
Mega	*Yr3a*, *Yr4a*, *Yr12*	Cappelle/H 2596//6003	UK
Mercia	/	Talent/Virtue //Flanders	UK
Moulin	*Yr6*, *Yr14*	CB 306-Y-70/Maris Widgeon//Maris Hobbit	FRA
Moro	*Yr10*, *YrMor*	PI 178383/2 * Omar	USA
Mo (s) 490	/	/	/
Mo (w) 406	/	/	/
Mo (w) 470	/	/	/
Mo (w) 499	/	/	/
Norman	*Yr2*, *Yr6*	TJB 268-175/ (Sib) Hobbit	UK
Pagode	/	(S) Composite Cross of 36 Cultivars	NLD
Paiimiot 113 wi	/	/	FRA
PI 178383	*Yr10*+ minor genes	/	TR
Rendezvous	*Yr17*	VPM 1/ (Sib) Hobbit//Virtue	UK
Resulka	*YrRes*	/	/
Selkirk	*Yr27*	Mcmurachy/Exchange//3 * Redman	CAN
Strampelli	*Yr18* + APR QTL	Libero//San Pastore 14/Jacomtti 49	ITA
Strubes Dickkopf	*Yr25*, *YrSD*	/	DEU
SXAF 4-7	/	/	FRA
T. spelta album	*Yr5*	/	/
Vilmorin 23	*Yr4a*, *YrV23*	Melbor/Grosse Tete//Japhet/Parsel	FRA
Vilmorin 27	*Yr3a*, *Yr4a*, *Yr16*	Dattel//Japhet/Parsel/3/Hatif Inversable/Bon Fermier	FRA
VM 82-24-20-44	/	VM 628-17-2/Rescler//VM480-4-2-5	FRA
VPM 1	*Yr17*	Ae. Ve/Tr. Ca//3 * Marne	FRA
VPM 1/4 * Cook	*Yr17*, *Yr18* +	/	AU
83-508-42	/	/	FRA

a. pedigrees and origin of cultivars and lines are from http: //genbank. vurv. cz/wheat/pedigree/pedigree. asp and http: //www. ars-grin. gov.

b. /, unknown.

c. DR, durable resistance to YR.

Table 2 Predominant *Puccinia striiformis* f. sp. *tritici* races or pathotypes identified from naturally infected and inoculated nurseries in southeastern Gansu province during 1993-2013

Year	Races or pathotypes in decreasing order by frequency[a]
1993	CYR29, CYR30, CYR31, Lovrin13II, Lovrin13 VII, Su11, HY pathotypes
1994	CYR29, CYR30, CYR31, Lovrin13 VII, Su11, HY, Su pathotypes
1995	CYR31, CYR29, CYR30, Su11, HY, Lovrin pathotypes, Su pathotypes
1996	CYR31, HY pathotypes, Su pathotypes, CYR29, CYR30
1997	CYR31, Su pathotypes, HY pathotypes
1998	CYR31, Su7, Su10, Su14, CYR29
1999	Su7, Su14, Su10, CYR31, CYR29

（续）

Year	Races or pathotypes in decreasing order by frequency[a]
2001	CYR32, Su14, Su3, Su7, Su4, Su10, Su11, CYR29
2002	Su14, CYR32, Su4, Su7, Su5, CYR31
2003	Su14, CYR32, Su4, Su7, Su5, CYR31
2004	Su14, CYR32, Su7, Su4, Su5, CYR31
2005	CYR33, CYR32, Su7, Su4, CYR31
2006	CYR33, CYR32, Su4, Su13, Su7, Su5, CYR31
2007	CYR33, CYR32, Su7, Su4, Su13, Su5, CYR31
2008	CYR33, CYR32, Su7, Su4, HY8, CYR31
2009	CYR33, CYR32, Su7, Su4, HY8, CYR29
2010	CYR33, CYR32, Su7, Su4, HY8, CYR29
2012	CYR33, CYR32, Su4, Su7, Gui2-9, Gui2-14
2013	CYR33, CYR32, Su4, Su7, Gui2-9, Gui2-14

a. Data provided by Dr. Q. Z. Jia and S. Q. Cao, Institute of Plant Protection, Gansu Academy of Agricultural Science, Lanzhou.

The races and pathotypes were identified by a set of wheat differentials for *Puccinia striiformis* f. sp. *tritici* in China, using 17 wheat genotypes (Trigo Eureka, Fulhard, Letescens 128, Mentana, Virgilio, Abbondanza, Early Premium, Funo, Danish 1, Jubilejina 2, Fengchan 3, Lovrin 13, Kangyin 655, Suwon 11, Zhong 4, Lovrin 10, Hybrid 46), and supplemental differential wheat lines Guinong 22 etc. Lovrin 13, Su, HY, and Gui indicate that the pathotypes were virulent on cultivars Lovrin 13, Suwon, Hybrid 46, and Guinong 22, respectively. Lovrin 13 Ⅱ and Lovrin 13 Ⅶ mean pathotypes of Lovrin 13 type 2 and Lovrin 13 type 7, respectively; HY8 is the Hybrid 46 type 8; Su3, SY4, Su5, Su7, Su10, Su11, Su13, and Su14 are the pathotypes of Suwon type 3, Suwon type 4, Suwon type 5, Suwon type 7, Suwon type 10, Suwon type 11, Suwon type 13, and Suwon type 14, respectively.

Evaluation of YR response

YR responses at Tianshui (altitude 1, 373m asl, annual rainfall 450-560mm) were evaluated from 1993 to 2013. Cultivars and lines were grown in three-row 1.5 m plots with a 0.2m row spacing. The highly susceptible cultivar Huixianhong was sown perpendicular and adjacent to the test rows as a spreader to ensure ample inoculum and also to serve as a susceptible control. Artificial inoculations were performed with contemporary and prevalent pathotypes at the three-leaf stage in spring (Table 2). The scoring of YR response followed the standard 0-4 infection type (IT) scale described by McIntosh et al.[26]. Maximum disease severities (MDS), percentages of leaf area infected when disease severity on the control reached its maximum (90-100%) around June 8 each year was based on the modified Cobb scale[28]. The data in 2000 and 2011 were not included in this paper due to the poor disease development in the dry springs of these two years.

Results

Cultivars and lines with documented single major resistance gene or major gene combinations effective against YR at the adult-plant stage

The adult-plant reactions of 19 cultivars and lines with one or more major genes to YR during 1993-2013 are shown in Table 3. Six cultivars were highly resistant (HR) or moderately resistant (MR) with IT 0-2 during 1993-2013, and the genes underlying the resistance in the cultivars were identified as *Yr5*, *Yr10*, *Yr15*, and other unknown genes. For example, PI 178383 (*Yr10* + minor genes) (34), Rendezvous, and *T. spelta album* (*Yr5*)[23] displayed IT 0 in all 19 years. Gaby (*YrGaby*) exhibited HR to MR with IT 0-2 in all 19 years, Dippes Triumph[3] and Hybrid 46 (*Yr3a*, *Yr4b*, *H46*)[7,22] showed IT 0 in most years, and IT 0; -2 with lower MDS (less than 5%) in a few years. Compair (*Yr8*, *Yr19*)[9,29], Fenman (*Yr1*, *Yr2*)[9,29],

Heines Kolben (*Yr2*, *Yr6*)[5,23], Norman (*Yr2*, *Yr6*)[30], Resulka (*YrRes*)[11], and Strubes Dickkopf (*YrSD*) (37) were HR or MR (IT 0; —2) to MS (IT 3-4) with low MDS (less than 25%) in all 19 years. Although the predominant races in the region changed from CYR29 to CYR33 and Gui pathotype during 1993-2013, these cultivars remained resistant. For example, *Yr8* in Compair was overcome by race CYR29 in 1993, and resistance in Resulka and Strubes Dickkopf was compromised by CYR32 in 1999-2001, but the three cultivars had low MDS over a long period, indicating that unknown APR genes commonly provided effective YR resistance in various cultivars and lines.

Table 3 Adult-plant reactions of 19 wheat cultivars with major genes to mixed prevalent races of Chinese *Puccinia striiformis* f. sp. *tritici* in the field during 1993-2013

Cultivar/line	Infection type (0-4)																			MDS[a]
	1993	1994	1995	1996	1997	1998	1999	2001	2002	2003	2004	2005	2006	2007	2008	2009	2010	2012	2013	(%)
CarstensV	0;-2	1-2	0;-2	0;-1	0;-2	2-3	2-3	0	3-4	0	2-4	3-4	3	3-4	3-4	3-4	3-4	3-4	3-4	0-50
Compair	2-3	2-3	2-3	2-4	3-4	3-4	2-3	3-4	1-2	0	0	1-4	1-2	3	0	2-3	2-3	0	0	0-25
C591	0	0	0;	1	1	1	2-3	3-4	3-4	1-2	0;-1	2-4	0	0	0	2-3	2-3	0	0;-1	0-60
Dippes Triumph	0	0	0	0	1	0	0	0	1-2	0	0	0	0	0;-1	0	0	0	0	0	0-5
Fenman	0	0	0	0	0	0	0	0	0	0	0;	3-4	0	0	0	0;-1	0;	0	0	0-25
Gaby	0;-1	0;	0;	0;	0;-2	1	0;	0;-1	0	0;-1	0;-2	0;-2	0	0;-1	0	0;-2	0;-1	0;-2	0;-2	0-40
Heines VII	0;-3	0;-3	0;-3	0;-3	0;-3	2-3	3-4	3	3	1-3	3-4	3	2-3	2-3	3	3-4	2-3	2-3	3-4	5-60
Heines Kolben	0;	0;	0;	2-3	0	0;	0	0	0	0	0;-1	0;	0	0	0	3	2	0	0	0-5
Hybrid 46	0	0	0	0	0	0	0	0	0	0	0	0;	0	0	0	0	0	0	0	0-1
Moro	0	0;	0	0	0	1	0	0	0	0	0	0	0	0	0	0	0	2-3	2-4	0-15
Norman	1	1-2	0;-2	1-2	0;-1	0	2-3	1	1	0	0;-2	0;-2	0	0	0;-2	1-2	1	1	0	0-5
PI 178383	0	0	0	0	0	0	0	0	0	0	0	0	0	0	0	0	0	0	0	0
Rendezvous	0	0	0	0	0	0	0	0	0	0	0	0	0	0	0	0	0	0	0	0
Resulka	0;-1	1-2	0;-1	1-2	0;-1	2-3	1-2	3-4	2-3	2-3	2-3	3	2-3	2-3	2-3	3	3	2-3	2-3	1-15
Selkirk	0;-1	1	0;	0;-1	0;	1	3-4	3-4	3-4	3-4	3-4	4	2-3	4	4	3-4	3	1-4	2-4	1-60
Strubes Dickkopf	0;-1	0;-1	0;-2	0;-2	0	0;	2-3	3-4	2-3	0	0	3-4	0	0	3	2-3	2-3	3	3	0-25
T. spelta album	0	0	0	0	0	0	0	0	0	0	0	0	0	0	0	0	0	0	0	0
Vilmorin 23	0;-1	0;-2	0;-2	0;-2	1-2	1-2	1-2	3-4	3-4	0	0;-1	3-4	0;-2	0;-1	0;-2	3-4	2-3	2-4	2-4	0-60
VPM 1	0;-1	0;-2	0;	0;-2	0	0;	0;-2	3-4	3-4	0	1-2	3-4	1-2	0;-2	3	1-2	1-2	0	4	0-25
Huixianhong (CK)[b]	4	4	4	4	4	4	4	4	4	4	4	4	4	4	4	4	4	4	4	90-100

a. MDS, Maximum disease severity (%), the data range across 19 years.

b. Huixianhong, a winter wheat landrace from Henan province of China, is highly susceptible to most Chinese races of *Puccinia striiformis* f. sp. *tritici* in both the seedling and adult-plant stages, as a susceptible check in the field.

The other seven cultivars displayed HR to MR, MS or very susceptible (VS) responses during 1993-2013. For example, Heines Ⅶ (*Yr2*, *Yr25*, *YrHVII*)[4,5,11] was susceptible to YR from 1993 to 2013, indicating these genes were not effective against Chinese *Pst* races and that there were no additional APR genes. Carstens V (*Yr32*)[11], Selkirk (*Yr27*)[27], Vilmorin 23 (*Yr4a*, *YrV23*)[7], and VPM 1 (*Yr17*)[1], were resistant with IT 0; —1 during 1993-1997, but became susceptible at different times after 1998. C591 (*YrC591*)[17] had IT 0 during 1993-1994, from 1995 to 1998, but showed MR to YR, and MS responses after 1999, and in 2005 it was HS with a high MDS (60%). The resistance became ineffective in these cultivars as new races emerged; CYR29 with virulence to Heines Ⅶ in the 1990s, CYR31 with virulence to Carstens V in 1998, and CYR32 with virulence to C591, Selkirk, Vilmorin 23, and VPM 1 in 2001 (Table 2 and Table 3). Moro (*Yr10*, *YrMor*)[10,24] had IT 0-0; from 1993 to 2010, and then became susceptible to YR (IT 2, 3,

4) after 2012. This change was associated with the appearance of a new race overcoming resistance conferred by *Yr26/YrCH42*, because it is now evident from worldwide data that virulence for these genes is associated with virulence to *Yr10*[20] even though *Yr10* is a clearly different gene located at a different locus on chromosome 1B.

Reaction of cultivars and lines with race non-specific resistance to YR at the adult-plant stage

Based on classical resistance testing, the documented resistance genes in the cultivars and lines such as Cappelle Desprez (*Yr3a*, *Yr4a*, *Yr16*)[35], Flanders (*Yr1*, *Yr3a*, *Yr4a*, *Yr16*)[15, 30], Hobbit (Yr3, *Yr4*, *Yr14*, *YrHVII*)[12], Ibis (*Yr1*, *Yr2*, *QYr. caas-2BS. 1*, *QYr. caas-6BS. 1*)[2], Joss Cambier (*Yr2*, *Yr11*)[22], Longbow (*Yr1*, *Yr2*, *Yr6*, *Yr9*, *Yr13*), Maris Huntsman (*Yr2*, *Yr3a*, *Yr4a*, *Yr13*), Maris Widgeon (*Yr3a*, *Yr4a*, *Yr16*), Mega (Yr3a, *Yr4a*, *Yr12*), Moulin (*Yr6*, *Yr14*), and Vilmorin 27 (Yr3a, *Yr4a*, *Yr16*) showed race non-specific resistance which is often referred to as APR or slow-rusting resistance[4]. In the field trials (Table 4), three cultivars, Ibis, Mega, Strampelli[21], VPM 1/4 * Cook (Probably at least *Yr17* and *Yr18*) had IT 0 in most years, and 0; -2 in a few years; Flanders, Hobbit, Joss Cambier, Libellula[21], Longbow, Maris Huntsman, Maris Widgeon, and Vilmorin 27 exhibited MR to MS with lower MDS (less than 25%) during 1993-2013. However, the MDS in cultivar Cappelle Desprez increased gradually during 1993-2013. In fact, it was not known if these genes or other genes for seedling (all-stage) resistance and APR contributed to the observed responses at Tianshui in the period.

Table 4 Adult-plant reactions of wheat cultivars with race non-specific resistance to mixed prevalent races of Chinese *Puccinia striiformis* f. sp. *tritici* in the field during

Cultivar/line	Infection type (0-4)																			MDS (%)
	1993	1994	1995	1996	1997	1998	1999	2001	2002	2003	2004	2005	2006	2007	2008	2009	2010	2012	2013	
Atou	1-2	2	1-2	1-2	0;-1	2-3	1-2	3-4	1-2	1-2	2	1-4	2	1-2	1-2	2	1-2	2	2-3	5-10
Bouquet	1-2	2	1-2	2	0;	2-3	1-2	3-4	3-4	2-3	2-3	1-3	2-3	2-3	2-3	3-4	2-3	2-3	0;	5-15
Cappelle Desprez	0;-2	1-2	0;-2	0;-2	0;-2	0;	1-3	3-4	2-3	0	2-4	2-4	1	0;-2	0	2-4	3-4	2-3	3	0-30
Champlein	0;-1	1	1	1	0	1	0	2-3	2-3	1-2	1-2	3	2-3	1-2	1-2	2-3	1-2	1-2	0;	0-15
Elite Le Peuple	0	0;	0	0;	0	0;	0	0	0	0;	0;	0;-1	0;	0;-1	0;-1	0;	0;	0;	0;-1	0-15
Flanders	0	0	0	0	0	0;	0	3	0	0	0;	0;	0	0	0	0	0	0	0	0-5
Flinor	0;-1	0;-1	0;-1	0;-1	0;-1	0;-1	0	3	0	0	0;-1	0;	0	0;	0	0;-1	0	0	0	0-5
Hobbit	0;-2	0;-2	0;-3	2	0;-2	0;-1	1-2	3-4	2-3	1-2	2-4	2-4	1-2	0;-2	0;-1	1-3	1-2	2-4	2-4	5-15
Holdfast	0;-1	0;	0	0	0	0;	3-4	3-4	3-4	0;-1	0;-1	3-4	0	0;-1	4	3-4	3-4	2-3	2-3	0-45
Ibis	0	0	0	0	0	0;	0	0	0	0;-1	0	0	0	0	0	0	0	0;	0;	0-1
Joss Cambier	0;-2	0;-2	0;-2	0	0;-2	1	2-3	3-4	3-4	0	1-3	0;-3	1-2	0;-2	2	2-3	2	2-3	0;-3	0-25
Libellula	0;	2	2-3	2-3	2	2-3	2-3	2-3	2-3	0;-2	1-2	3-4	2-3	3-4	2-3	3-4	2-3	0;-2	3-4	5-10
Little Joss	1-2	1-2	1	1	0;-2	0;	0;-1	3	2-3	0	0;	0;-2	0	0;-1	0	0;-1	0;	0;	0	0-5
Longbow	0;-1	0	0	0	0	0	0	2-3	0	0	0;-1	0;-2	0	0	0;-1	2-4	0	0;-2	0	0-5
Maris Huntsman	1-2	0;-1	0;-2	1-2	1-2	1	0	3	2-3	0	0;-2	1-4	0	3	3	2-3	3	3-4	4	0-15
Maris Widgeon	1-2	1-2	1-2	0;-1	0	0;	1-2	3-4	3-4	0	3-4	1-3	0	0;-1	3-4	1-3	3-4	0;-2	0	0-15
Mega	0	0;-1	0	0	0;-1	0	0	0	0	0	0	0;-1	0	0;-1	0	0	0	0	0	0-10
Moulin	0;-1	0;-1	1	0;-1	0;-1	0	0;	3	1	0	0	0;-1	0	0;	0	1-2	0	0;-1	0	0-5
Pagode	0;-1	0;	0;	0;	0;	0;	0	3-4	2-3	0	0;-1	0;-1	0	0;	0;-1	2-3	2-3	1-2	2	0-8

(续)

Cultivar/line	Infection type (0-4)																			MDS
	1993	1994	1995	1996	1997	1998	1999	2001	2002	2003	2004	2005	2006	2007	2008	2009	2010	2012	2013	(%)
Strampelli	0;-1	0;-2	0	0;-3	0;-3	0	1-2	0	0	0	1-2	0;-1	0;-1	1-2	0;-1	0;-1	0;-1	0;-1	0;-1	0-8
Vilmorin 27	0;-2	0;-2	0	0	0;-2	0	0	3-4	2-3	0	1	1-3	1	0	0;-1	0;-1	1	0	0	0-8
VPM 1/4 * Cook	0	0	0	0	0	0	0	0	0	0;	0	0;-2	0	1-2	0	0	0;	1-2	0;-2	0-15
Huixianhong (CK)	4	4	4	4	4	4	4	4	4	4	4	4	4	4	4	4	4	4	4	90-100

The other cultivars and lines with unknown resistance genes, such as Atou, had IT 0-2 in most years and IT 3-4 in only 1998, 2001, 2005, and 2013. The susceptible response in Atou could be caused by different races becoming predominant in the period, for example, Su7 pathotype, CYR32, CYR33, and Gui pathotype become the predominant race in 1999, 2001, 2005, and 2013, respectively. Elite Le Peuple, Flinor, and Moulin had IT ranging from 0; to 2 in most years. Champlein and Little Joss were MR to MS with IT 0-3. Bouquet and Pagode were MR to MS from 1993 to 2013, and particularly, Holdfast had IT 4 and MDS of 45% in 2005, but was heavily rusted in later years, perhaps caused by the epidemic of CYR33 in the region in 2005.

Adult-plant reactions of cultivars and lines with unknown resistance genes

Reactions of 17 cultivars with unknown resistance genes to YR are shown in Table 5. Four cultivars, FR81-1, Mo (w) 499, Mo (w) 406, and German 8661 had IT 0 from 1993 to 2013. Lperfer was very resistant with IT 0-0; in most years. Paiimiot113wi, 83-508-42, Aquila I, Cik Vee, and Mo (w) 470 showed HR to MR with IT 0-2. Other cultivars, such as Jubilar and SXAF4-7, exhibited MR to MS with low MDS (less than 20%) in most years.

Table 5 Adult-plant reactions of wheat cultivars with unknown resistance genes to mixed prevalent races of Chinese *Puccinia striiformis* f. sp. *tritici* in the field during 1993-2013

Cultivar/line	Infection type (0-4)																			MDS
	1993	1994	1995	1996	1997	1998	1999	2001	2002	2003	2004	2005	2006	2007	2008	2009	2010	2012	2013	(%)
83-508-42	0	0	0	0	0	0	1-2	1-2	0	0	0	0	0	0	0	0	0	0	0	0-5
Aquila I	0;-1	0;-1	1	0;	0;-1	0	0;-2	1-2	1-2	0	0	0;	0	0	0	0	0	0	0	0-5
Cik Vee	0;	0;	0;	1	0;	0;	0	1-2	0	0;-1	1	1	0;-1	0;-1	0;-1	1-2	1	0;-1	0;-1	0-5
FR 81-1	0	0	0	0	0	0	0	0	0	0	0	0	0	0	0	0	0	0	0	0
German 2	0	0	0	0	0;	0	0	0	1-2	0	0	0	0	0	0	0	0	0	0	0-5
German 8661	0	0	0	0	0	0	0	0	0	0	0	0	0	0	0	0	0	0	0	0
Jubilar	0;-1	0;-1	0;-1	0;-1	0;-2	0;-1	1	3-4	0;-2	1-2	1-2	1-4	2	1-2	1-2	2-3	1-2	1-4	2	5-12
Lperfer	0	0	0	0	0	0	0	0;	0	0	0	0	0	0	0	0	0	0	0	0-5
Mercia	0	0	0	0	0	0	0	2	0	0	0;	0	0	0	0;	0	0	0	0	0-5
Mo (s) 490	0	0	0	0	0	0;	0;	1-2	1	0;-1	1	0;-1	1	1	1	1-2	1	0;-1	1-2	0-5
Mo (w) 406	0	0	0	0	0	0	0	0	0	0	0	0	0	0	0	0	0	0	0	0
Mo (w) 470	0;-1	0;-1	0;-1	0;-1	0;-1	0	1-2	0	0	0;-1	0;-2	0;-1	0;-1	0;-1	0;-1	1-2	0;-2	0;-1	1-2	0-5
Mo (w) 499	0	0	0	0	0	0	0	0	0	0	0	0	0	0	0	0	0	0	0	0
Paiimiot113wi	1	1	1	1	1	1	1-2	1-2	2	2	1	1-2	2	1-2	1-2	2	2	1	1-2	1-5
SXAF 4-7	0;-1	0;-1	0;-1	0;-1	0;-1	4	2-3	2-3	0	1-3	1-2	1-2	1-2	1-2	1-2	2-3	2-3	1-2	2-3	0-8
VM 82-24-20-44	0	0	0	0	0	0	0	0	0;	0	0	0	0	0	0	0	0	0	0	0-5
Huixianhong (CK)	4	4	4	4	4	4	4	4	4	4	4	4	4	4	4	4	4	4	4	90-100

Use of European cultivars and lines in wheat breeding

Ten commercial winter wheat cultivars were released in this region for highland terrace or lowland valley cultivation during 2004-2013 (Table 6). These included Lantian 15, Lantian 18, Lantian 19, Lantian 20, Lantian 22, Lantian 23, Lantian 25, Lantian 26, Lantian 27, and Lantian 31. The YR resistances in these cultivars were derived from Ibis, Flinor, Mega, Cappelle Desprez, German 2, SXAF 4-7, Mo (s) 311, Flanders, Norman, and Longbow, respectively, and they have shown effective YR resistance since they were released for commercial production from 2004 to 2013. Lantian 15, Lantian 22, Lantian 23, Lantian 25, Lantian 26, and Lantian 31 have been grown in southeastern Gansu for many years and still exhibited highly effective APR to YR, indicating that the use of APR could be a promising strategy for sustainable control of YR in this area. Lantian 18, Lantian 19, Lantian 20, and Lantian 27 showed moderate to high resistance to YR at the seedling stage, and high resistance at the adult-plant stage, indicating that they had all-stage resistance for YR. Lantian 18 and Lantian 20 have also shown potentially durable resistance to powdery mildew (data not shown), indicating that they may possess APR genes besides major genes to YR.

Table 6 Commercial winter wheat cultivars with YR resistance genes derived from European cultivars and lines

Commercial cultivar	Pedigree	Resistance to YR[a]		Growing Region	Year released
		Seedling	Adult		
Lantian 15	Ibis/Lantian 10[s]	MS	HR	Highland terrace	2004
Lantian 18	Flinor/Lovrin 13[s]	HR	HR	Highland terrace	2007
Lantian 19	Mega/Lantian 10[s]	HR	HR	Highland terrace	2007
Lantian 20	Cappelle Desprez/Lantian 10[s]	HR	HR	Highland terrace	2007
Lantian 22	German2/Lantian 11[s]	MS	HR	Highland terrace	2007
Lantian 23	SXAF 4-7/87-121[s]	MS	HR	Lowland valley	2007
Lantian 25	92-72-3-3/ Mo (s) 311	MS	HR	Lowland valley	2009
Lantian 26	Lantian 10[s]/ Flanders	MS	HR	Highland terrace	2010
Lantian 27	Norman/Lantian 10[s]	HR	HR	Highland terrace	2010
Lantian 31	Longbow/Lantian 10[s]	MS	HR	Highland terrace	2013

a. The test of seedling-stage resistance to YR in each cultivar was inoculated by races of CYR29, CYR31, CYR32, CYR33, Lovrin 13 Ⅲ, and Su pathotypes, respectively. The test of adult-plant stage resistance to YR in these cultivars was inoculated by mixed races same as its seedling-stage plant.

s. Susceptible to YR.

Discussion

Climatic conditions, geographic characteristics, and cropping systems in southeastern Gansu are ideal for year-round cycling of *Pst*[19]. Continuing epidemics with high levels of inoculum are associated with the generation and maintenance of high levels of pathogenic variability in the pathogen population, which possibly also includes sexual reproduction on barberry[38]. Over the last three decades from 1990 to 2013, severe YR epidemics occurred in three years and moderate epidemics occurred in 16 years, and the predominant *Pst* races sampled from wheat in this region have been changed from CYR29 to CYR33 and Gui (Table 2). The change in the pathogen population resulted in ineffectiveness of resistance in many cultivars for three to five years after release[13]. Presence of CYR29 resulted in the ineffectiveness of resistance in Lovrin 10, Lovrin 13, and other cultivars with *Yr9* (1B. 1R translocation); CYR31 was virulent on Hybrid 46, CYR32 combining virulences on Hybrid 46 and Fan 6 and its derivatives and virulent to *Yr1*, *Yr2*, *Yr3*, *Yr4*, *Yr6*, *Yr7*, *Yr9*, *Yr17*, *Yr22*, *Yr23*, *Yr27*, *YrA*, *YrCV1*,

YrCV2, *YrCV3*, *YrG*, *YrSD*, and *YrSO* (33); CYR33 was virulent on Suwon 11, and the Gui pathotype was virulent on Chuanmai 42, Moro (*Yr10*), Guinong 22, Lantian 17 (probably *Yr26*), and Lantian 30 (probably *Yr26*) during 1993-2013. Based on our field evaluation, C591, Carstens V, Resulka, Selkirk, Strubes Dickkopf, Vilmorin 23, VPM 1, Bouquet, Cappelle Desprez, Champlein, Holdfast, Maris Huntsman, Maris Widgeon, Jubilar, and SXAF 4-7 were susceptible to YR during the epidemics caused by CYR32.

In southeastern Gansu, many cultivars with seedling resistance or single major genes became susceptible to YR within three to five years after release into commercial production. From the 1950s to 1960s, many foreign cultivars with YR resistance were released directly into production in southeastern Gansu. For example, Lovrin 10, Lovrin 13, Abbondanza, Funo, and Jubilejna II grew in the region, but many of them had a short life in commercial production. From the 1970s to 1980s, many foreign cultivars and lines with YR resistance were used in wheat breeding program. Cultivars Tianxuan 15 and Tianxuan 16 with YR resistance derived from Abbondanza, Zhongliang 1, Zhongliang 2 and Zhongliang 3 with YR resistance from Funo, Zhongliang 13, Zhongliang 14 and Lantian 1 with YR resistance from Lovrin 13, also had short lives in production because they have only seedling resistance genes. However, two Italian cultivars, Strampelli and Libellula, with APR genes have remained resistant to YR in this region over the past four decades. In the period since 1993, deployment of different resistance genes in the highland terrace regions (over-summering area for *Pst*) and lowland valleys (over-wintering area)[36], and using a diversity of resistance including seedling resistance, APR was employed in wheat breeding programs, which could be a sustainable way to control YR in this area[39]. Many foreign cultivars and lines with different YR resistance genes were evaluated in field trials and some were used in wheat breeding programs. A series of commercial cultivars were developed by use of different effective resistance sources such as *Yr12* and *Yr16*, as well as some unknown genes, and these commercial cultivars were grown in the over-summering or over-wintering regions (Table 6). Lantian 15, Lantian 18, Lantian 19 (possibly *Yr12*), Lantian 20 (possibly *Yr16*), Lantian 21, Lantian 22, Lantian 26, Lantian 27, Lantian 31, Strampelli (*Yr18* +), and Libellula (*Yr18* +) were grown in the over-summering region, whereas Lantian 23, Lantian 24 (probably *Yr26* +), and Lantian 25 were grown in over-wintering regions. These cultivars are still resistant and provide effective control of YR in this region. Many major genes were overcome by predominant races in different periods, and only *Yr5* and *Yr15* are still effective against YR in southeastern Gansu. However, APR to YR in wheat, which is generally race non-specific and quantitatively inherited, is providing durable protection against the disease[4]. Pyramiding four to five minor resistance genes can achieve near-immune resistance and long-lasting protection from the rusts[31]. Therefore, the use of APR or slow-rusting genes in wheat breeding programs is a promising strategy for sustainable control of YR in southeastern Gansu. For example, European cultivars Ibis, Flinor, Mega, Cappelle Desprez, Flanders, Longbow, and Holdfast with documented adult-plant resistance for YR have been successfully used in wheat breeding program, and commercial cultivars Lantian 15, Lantian 18, Lantian 19, Lantian 20, Lantian 26, Lantian 31, and Zhongliang 30 with YR resistance derived from these cultivars still showed high APR for YR in the field.

Based on the YR responses of 57 cultivars in Tianshui during 1993-2013 (Tables 3, 4, 5, and Sub-Table 1), Dippes Triumph, Flanders, Flinor, FR 81-1, German 2, German 8661, Hybrid 46 (*Yr4* +), Ibis, Mega, Mo (w) 499, Mo (w) 406, PI 178383 (*Yr10* +), Rendezvous (*Yr17* +), *T. spelta album* (*Yr5*), and VPM 1/4 * Cook (*Yr17* +) were highly resistant (IT 0, 0;) to YR in all 19 years. They carried major genes, seedling and APR and undoubtedly other unknown genes, that can be utilized in future breeding. Cultivars Cappelle Desprez, Compair, Heines Kolben,

Hobbit, Joss Cambier, and Maris Huntsman exhibited MR to MS with low MDS (less than 30%) in all 19 years; these materials can be used in future wheat breeding programs for gene deployment and gene pyramiding. Cultivar Gaby with the single major gene *YrGaby* was HR to MR with lower MDS (0-40%); pyramiding *YrGaby* and other genes in an elite genotype could be an effective way for improving resistance in wheat breeding. Whereas C591 (*YrC591*), Carstens V (*YrCV*), Heines VII (*Yr2*, *Yr25*, *YrH2VII*), Moro (*Yr10*), Strubes Dickkopf (*YrSD*), Vilmorin 23 (*Yr4a*, *Yr23*), and VPM 1 (*Yr17*) became susceptible to YR in recent years, indicating that *Yr2*, *Yr4a*, *Yr10*, *Yr17*, *Yr23*, *Yr25*, *YrSD*, *YrC591* were not effective against current Chinese *Pst* races.

In this study, the cultivars and lines shown in Table 4 were identified as having race non-specific resistance to YR, based on the YR resistance genes contained in the cultivars and lines, and several cultivars had durable resistance for the disease (Table 1). Cultivar Ibis, which possesses multiple resistance genes to YR[2], displayed HR in all 19 years. Atou, Bouquet, Champlein, Flanders, Flinor, Hobbit, Joss Cambier, Little Joss, Longbow, Maris Huntsman, Maris Widgeon, Moulin, Pagode, and Vilmorin 27 were susceptible to one or more races among CYR29, CYR31, and CYR32, but these cultivars and lines had low adult-plant stage MDS (0-25%) during 19 years of testing, Cappelle Desprez had lower MDS (0-10%) in 18 years, and only exhibited a higher MDS of 30% in 2005; Holdfast had a lower MDS of 0-25% in 18 years, and a higher MDS of 45% in 2005. These two cultivars still showed APR in the field, indicating that they probably continue to have value if combined with other resistance genes. In contrast, C591 with a single resistance gene was resistant to YR before 1998, but became susceptible in 1999. Carstens V was resistant to YR with IT 1-2 during 1993-1998, and then became susceptible in 1999. Vilmorin 23 was resistant to YR during 1993-1999, and became susceptible in 2001. The resistances of three cultivars were overcome after spread of the race CYR32, and these cultivars also had high MDS (5-60%).

In order to be confident in the effectiveness and durability of YR resistance, wheat genotypes should be tested under naturally infected and inoculated conditions in the field for a long period. In this study, several cultivars showed HR to MR in the first four to five years, but then gradually became susceptible. For example, C591 was very resistant (IT, 0) in 1993 to 1994; we made crosses of C591 and many elite lines were bred, but C591 and its derivatives became susceptible after 1999 with the spread of race CYR32.

Acknowledgments

The authors are grateful to Prof. R. A. McIntosh, Plant Breeding Institute, University of Sydney, for critical review of this manuscript. This study was supported by the National Key Basic Research Program of China (2013CB127700), National Natural Science Foundation of China (31261140370), International Collaboration Projects from the Ministries of Science and Technology (2011DFG32990) and Agriculture (2011-G3), National 863 Project (2012AA101105), and the China Agriculture Research System (CARS-3-1-3).

References

[1] Bariana, H. S., and McIntosh, R. A. 1993. Cytogenetic studies in wheat. XIV. Location of rust resistance genes in VPM 1 and their genetic linkage with other disease resistance genes in chromosome 2A. Genome. 36: 476-482.

[2] Bai, B., Ren, Y., Xia, X. C., Du, J. Y., Zou, G., Wu, L., Zhu, H. Z, He, Z. H., and Wang C. S. 2012. Mapping of quantitative trait loci for adult plant resistance to stripe rust in German wheat cultivar Ibis. J. Integr. Agric. 11: 528-536.

[3] Bai, Y. L., Sun, Q., Zhang, C. Y., Cui, N., Lin, F., Xu, S. C., Zhang, Z. Y., Gao, Y., and Xu, X. D. 2010. Molecular detection and resistance evaluation of 59 cultivars from the Northwest of the United States to Chinese stripe rust races. Sci. Agric. Sinica. 43:

1147-1155.

[4] Chen, X. M. 2005. Epidemiology and control of stripe rust (*Puccinia striiformis* f. sp. *tritici*) on wheat. Can. J. Plant Pathol. 27: 314-337.

[5] Calonnec, A., Johnson, R., and De Vallavieille-Pope, C. 1997. Identification and expression of the gene *Yr2* for resistance to *Puccinia striiformis* in the wheat differential cultivars Heines Kolben, Heines Peko and Heines VII. Plant Pathol. 46: 387-396.

[6] Chen, W. Q., Wu, L. R., Liu, T. G., Xu, S. C., Jin, S. L., Peng, Y. L., and Wang, B. T. 2009. Race dynamics, diversity, and virulence revolution in *Puccinia striiformis* f. sp. *tritici*, the causal agent of wheat stripe rust in China from 2003 to 2007. Plant Dis. 93: 1093-1101.

[7] Chen, X. M., and Line, R. F. 1993. Inheritance of stripe rust resistance in wheat cultivars postulated to have resistance genes at *Yr3* and *Yr4* loci. Phytopathology. 83: 382-388.

[8] Chen, X. M., and Line R. F. 1993. Inheritance of stripe rust (yellow rust) resistance in the wheat cultivar Carstens V. Euphytica. 71: 107-113.

[9] Chen, X. M., Line, R. F., and Jones, S. S. 1995. Chromosomal location of genes for stripe rust in spring wheat cultivars Compair, Fielder, Lee, and Lemhi and interactions of aneuploid wheats with races of *Puccinia striiformis*. Phytopathology. 85: 375-381.

[10] Chen, X. M., Line, R. F., Shi, Z. X., and Leung, H. 1998. Genetics of wheat resistance to stripe rust. Pages 237-239 in: Proc. 9th Int. Wheat Genetics Sympos. A. E. Slinkard, eds.. University Extension Press, University of Saskatchewan, Saskatoon, Sask.

[11] Eriksen, L., Afshari, F., Christiansen, M. J., McIntosh, R. A., Jahoor, A., and Wellings, C. R. 2004. *Yr32* for resistance to stripe (yellow) rust present in the wheat cultivar Carstens V. Theor. Appl. Genet. 108: 567-575.

[12] Feng, J., Zhang, Z. Y., Lin, R. M., and Xu, S. C. 2009. Postulation of seedling resistance genes in 20 wheat cultivars to yellow rust (*Puccinia striiformis* f. sp. *tritici*). Agric. Sci. China. 8: 1429-1439.

[13] He, Z. H., Lan, C. X., Chen, X. M., Zou, Y. C., Zhuang, Q. S. and Xia, X. C. 2011. Progress and perspective in research of adult-plant resistance to stripe rust and powdery mildew in wheat. Sci. Agric. Sinica. 44: 2193-2215.

[14] Jin, Y. 2011. Role of *Berberis* spp. as alternate hosts in generating new races of *Puccinia graminis* and *P. striiformis*. Euphytica. 179: 105-108.

[15] Keshavarzi, M., Hallajian, T. M., Bagheri, A., and Afshari, F. 2004. Identification of resistance gene (s) to yellow rust in wheat bulked genomic DNAs using RGAP and RAPD marker. Int. Cereal Rusts and Powdery Mildews Conference, John Innes Centre, Norwich, UK.

[16] Law, C. N., and Worland, A. J. 1997. The control of adult plant resistance to yellow rust by the translocated chromosome 5BS-7BS of bread wheat. Plant Breed. 116: 59-63.

[17] Li, Y., Niu, Y. C., and Chen X. M. 2009. Mapping a stripe rust resistance gene *YrC591* in wheat variety C591 with SSR and AFLP markers. Theor. Appl. Genet. 118: 339-346.

[18] Li, Z. F., Xia, X. C., Zhou, X. C., Niu, Y. C., He, Z. H., Zhang, Y., Li, G. Q., Wan, A. M., Wang, D. S., Chen, X. M., Lu, Q. L., and Singh, R. P. 2006. Seedling and slow rusting resistance to stripe rust in Chinese common wheats. Plant Dis. 90: 1302-1312.

[19] Li, Z. Q., and Zeng, S. M. 2002. Wheat rusts in China. China Agriculture Press, Beijing.

[20] Liu, T. G., Peng, Y. L., Chen, W. Q., and Zhang Z. Y. 2010. First detection of virulence in *Puccinia striiformis* f. sp. *tritici* in China to resistance genes *Yr24* (=*Yr26*) present in wheat cultivar Chuanmai 42. Plant Dis. 94: 1163.

[21] Lu, Y. M., Lan, C. X., Liang, S. S., Zhou, X. C., Liu, D., Zhou, G., Lu, Q. L., Jing, J. X., Wang, M. N., and Xia, X. C. 2009. QTL mapping for adult-plant resistance to stripe rust in Italian common wheat cultivars Libellula and Strampelli. Theor. Appl. Genet. 119: 1349-1359.

[22] Lupton, F. G. H., and Macer, R. C. F. 1962. Inheritance of resistance to yellow rust (*Puccinia glumarum* Erikss. and Henn.) in seven varieties of wheat. Trans. Br. Mycol. Soc. 45: 21-45.

[23] Macer, R. C. F. 1966. The formal and monosomic genetic analysis of stripe rust (*Puccinia striiformis*) resistance in wheat. Pages 127-142 in Proc. 2nd Int. Wheat Genetics Symp. J. MacKey (ed.) Hereditas, Lund, Sweden.

[24] Macer, R. C. F. 1975. Plant pathology in a changing world. Trans. Br. Mycol. Soc. 65: 351-374.

[25] McIntosh, R. A., Dubcovsky, J., Rogers, W. J.,

Morris, C. F., Appels, R., and Xia, X. C. 2012. Catalogue of gene symbols for wheat: 2012 supplement. Online: http://www.shigen.nig.ac.jp/wheat/komugi/genes/macgene/supplement2012.pdf

[26] McIntosh, R. A., Wellings, C. R., and Park, R. F. 1995. Wheat rusts: an atlas of resistance genes. Commonwealth Scientific and Industrial Research Organisation (CSIRO), Australia.

[27] McIntosh, R. A., Yamazaki. Y., Devos, K. M., Dubcovsky. J., Rogers, J., and Appels, R. 2003. Catalogue of gene symbols for wheat. Pages 31-40 in: 10th Int. Wheat Genetics Symp., Paestum, Italy.

[28] Peterson, R. F., Campbell, A. B., and Hannah, A. E. 1948. A diagrammatic scale for estimating rust intensity on leaves and stems of cereals. Can. J. Res. Sect. C. 26: 496-500.

[29] Riley, R., Chapman, V., and Johnson, R. 1968. The incorporation of alien disease resistance in wheat by genetic interference with the regulation of meiotic chromosome synapsis. Genet. Res., Camb. 12: 713-715.

[30] Singh, D., Park, R. F., McIntosh, R. A., and Bariana, H. S. 2008. Characterisation of stem rust and stripe rust seedling resistance genes in selected wheat cultivars from the United Kingdom. J. Plant Pathol. 90: 553-562.

[31] Singh, R. P., Huerta-Espino, J., and Rajaram, S. 2000. Achieving near-immunity to leaf and stripe rusts in wheat by combining slow rusting resistance genes. Acta Phytopathol. et Entomol. Hungarica. 35: 133-139.

[32] Wan, A. M., Chen, X. M., and He, Z. H. 2007. Wheat stripe rust in China. Aust. J. Agri. Res. 58: 605-619.

[33] Wan, A. M., Zhao, Z. H., Chen, X. M., He, Z. H., Jin, S. L., Jia, Q. Z., Yao, G., Yang, J. X., Wang, B. T., Li, G. B., Bi, Y. Q., and Yuan, Z. Y. 2004. Wheat stripe rust epidemic and virulence of *Puccinia striiformis* f. sp. *tritici*. Plant Dis. 88: 896-904.

[34] Wang, L. F., Ma, J. X., Zhou, R. H., Wang, X. M., and Jia, J. Z. 2002. Molecular tagging of the yellow rust resistance gene *Yr10* in common wheat, P. I. 178383 (*Triticum aestivum* L.). Euphytica. 124: 71-73.

[35] Worland, A. J., and Law, C. N. 1986. Genetic analysis of chromosome 2D of wheat. Z. Pflanzenzuecht. 96: 331-345.

[36] Zeng, S. M., and Luo, Y. 2006. Long-distance spread and interregional epidemics of wheat stripe rust in China. Plant Dis. 90: 980-988.

[37] Zhang, J. Z., Lin, R. M., Cao, L. H., He, Y. Q., and Xu, S. C. 2007. Monosomic analysis of the wheat stripe rust resistant genes in Taichung29 * 6/Strubes Dickkopf. Acta Phytophyl. Sinica. 34: 343-346.

[38] Zhao, J., Wang, L., Wang, Z. Y., Chen, X. M., Zhang, H. C., Yao, J. N., Zhan, G. M., Chen, W., Huang, L. L., and Kang Z. S. 2013. Identification of eighteen *Berberis* species as alternate hosts of *Puccinia striiformis* f. sp. *tritici* and virulence variation in the pathogen isolates from natural infection of Barberry plants in China. Phytopathology. 103: 927-934.

[39] Zhou, X. C., Wu, L. R., Song, J. R., and Jin, S. L. 2008. Control of wheat stripe rust based on genetic diversity of cultivars in Longnan. Acta Phytophyl. Sinica. 35: 97-101.

兼抗型成株抗性基因定位与分子检测

小麦条锈病和白粉病成株抗性研究进展与展望

何中虎[1,2]，兰彩霞[1,3]，陈新民[1]，庄巧生[1]，邹裕春[4]，夏先春[1]

[1]中国农业科学院作物科学研究所/国家小麦改良中心，北京 100081；[2]国际玉米小麦改良中心（CIMMYT）中国办事处，北京 100081；[3]华中农业大学植物科学技术学院，武汉 430070；[4]四川省农业科学院作物研究所，成都 610066

摘要：利用成株抗性已成为小麦抗病育种的重要方向，本文综述了小麦条锈病和白粉病成株抗性鉴定方法、基因定位和克隆及其在育种中的应用。将报道的72个条锈病成株抗性数量性状遗传位点（quantitative trait loci，QTL）和82个白粉病成株抗性QTL整合到一张连锁图谱，控制两种病害的基因簇（≥5个QTL）有8个，其中位于7DS的*Yr18*/*Lr34*/*Pm38*和1BS的*Yr29*/*Lr46*/*Pm39*对条锈病、叶锈病和白粉病均表现成株抗性，位于4DL的*Yr46*/*Lr67*位点可能也对白粉病表现成株抗性，*Yr18*/*Lr34*/*Pm38*和*Yr36*已被克隆，*Yr29*/*Lr46*/*Pm39*的克隆已取得良好进展，为培育兼抗和成株抗性相结合的品种提供了可用基因。总结了成株抗性在中国小麦育种中的应用现状，并用实例证实了培育成株抗性品种的可行性，建议对兼抗条锈病和白粉病成株抗性的咸农4号和小偃6号等进行遗传分析，育种工作者和品种审定部门需要转变观念，将成株抗性利用作为国内条锈病和白粉病抗性育种的重要内容。

关键词：小麦；条锈病；白粉病；成株抗性；分子定位；基因克隆

Adult-Plant Resistance to Stripe Rust and Powdery Mildew in Wheat, Progress and Perspective

He Zhonghu[1,2], Lan Caixia[1,3], Chen Xinmin[1], Zhuang Qiaosheng[1], Zou Yuchun[4], Xia Xianchun[1]

[1]*Institute of Crop Sciences, National Wheat Improvement Center, Chinese Academy of Agricultural Sciences (CAAS), Beijing* 100081;[2]*International Maize and Wheat Improvement Center (CIMMYT) China Office, c/o CAAS, Beijing* 100081;[3]*College of Plant Science and Technology, Huazhong Agricultural University, Wuhan* 430070;[4]*Crop Research Institute, Sichuan Academy of Agricultural Sciences, Chengdu* 610066

Abstract: Stripe rust, caused by *Puccinia striiformis* f. sp. *tritici*, and powdery mildew, caused by *Blumeria graminis* f. sp. *tritici*, are the devastating diseases in common wheat (*Triticum aestivum* L.) worldwide. Use of adult-plant resistance (APR) genes is an important method for the development of durable resistant cultivars. A total of 72 quantitative trait loci (QTLs) for APR to stripe rust hand 82 QTLs for APR to powdery mildew were integrated into a linkage map based on the information of DNA markers linked to individual QTL. Eight gene clusters (≥5 QTLs) conferred resistance to both stripe rust and powdery powdery, among them, *Yr18*/*Lr34*/*Pm38*, *Yr29*/*Lr46*/*Pm39*, and *Yr46*/*Lr67* showed

原文发表在《中国农业科学》，2011，44（11）：2193-2215

resistance to stripe rust, leaf rust and powdery mildew. *Yr18/Lr34/Pm38* and *Yr36* have been cloned. Xiannong 4 and Xiaoyan 6 were very important resistant germplasm for APR to stripe rust and powdery midew. We summarized the application of APR to wheat breeding in China. Use of APR genes will become the major method for improving stripe rust and powdery mildew resistance in wheat breeding. The strategies for APR on wheat breeding were also discussed.

Key words: common wheat (*Triticum aestivum* L.); stripe rust; powdery mildew; adult-plant resistance; molecular mapping; gene cloning

中国小麦生产中的主要病害包括小麦条锈病、白粉病、赤霉病、纹枯病和全蚀病等。由小麦条锈病菌（*Puccinia striiformis* f. sp. *tritici*）引起的小麦条锈病曾是全国第一大病害，由于粉锈宁的普遍使用等原因，其相对重要性较以前有所下降，2004—2009年中国条锈病每年平均发生面积约420万 hm^2[1-6]，对四川省、重庆市、云南省、陕西省、甘肃省、湖北省十堰和襄樊地区及河南省信阳和南阳地区来说，条锈病仍是当地小麦生产第一大病害，因此抗条锈病是上述地区除产量外的最基本育种目标。20世纪70年代以前，由小麦白粉病菌（*Blumeria graminis* f. sp. *tritici*）引起的小麦白粉病主要在中国湿润多雨的西南地区及山东沿海地区流行；20世纪80年代以后，该病在中国主要冬麦区逐渐从次要病害上升为主要病害，且常年发生[7]。2004—2009年白粉病在全国发生平均面积为685万 hm^2[1-6]。对北部冬麦区和黄淮冬麦区而言，白粉病是最重要的小麦病害；在长江中下游和西南麦区，白粉病是仅次于赤霉病或条锈病的第二大病害，因此抗白粉病也是上述地区最基本的育种目标。

由于小麦条锈病和白粉病分布广泛、病原菌生理小种复杂多变等特点，常常导致品种抗性频繁丧失，因此防治和控制这两种病害不仅十分重要，而且是一项长期任务。尽管通过化学药剂等措施可以有效防治病害，但应用抗病品种是防治病害最经济有效、安全的途径[8]，利用分子标记辅助选择技术聚合多个抗病基因是培育持久抗性小麦品种的重要手段[9-11]。小麦的抗病性主要有两类：一类是垂直抗性，又称生理小种专化抗性[12]、苗期抗性[13]、全生育期抗性[14]或主效基因抗性[15]，它由一个或少数几个主效基因控制，对病原菌的侵染产生过敏性坏死反应从而表现出高抗或免疫，具有病原菌生理小种专化性，即随着生理小种的变化常导致抗性丧失，致使抗病性不持久、不稳定[12]；另一类是水平抗性，又称非小种专化抗性[12]、成株抗性[14]、高温成株抗性[16]、慢病性[12]或部分抗性[17]，这些名称从不同侧面反映了同一遗传现象，因此对于遗传育种工作者而言，其实质是相同的，故本文统称为成株抗性。该类抗性基因对病原菌无小种专化性或专化性弱，减少了品种对病原菌生理小种的选择压力，从而表现为潜育期延长14%～49%、孢子堆缩小34%～78%、产孢量下降42%～98%、病程曲线下面积（area under the disease progress curve, AUDPC）缩小50%～99%[18]、抗病性持久并稳定等特点[19]。选用单个生理小种或混合小种接种时，苗期表现为感病，成株期的严重度或病害发展速率则较低，而非免疫或坏死反应。以往的研究认为成株抗性由若干个微效基因所控制[15,19]，而现有研究表明聚合4～5个效应相对较大的微效基因即可培育出接近免疫的成株抗性品种，也就是说成株抗性并非像以前报道的那样复杂[9,11]。国际上多数国家已将成株抗性的利用作为小麦抗病育种的主要方向[20]。20世纪60年代至今，国际玉米小麦改良中心（CIMMYT）对成株抗性的应用进行了深入系统的研究，形成了一套行之有效的育种方法，育成一大批成株抗性品种，并在生产上大面积推广，如含有*Yr18*或*Yr29*及2～3个微效基因的小麦品种Jupateco 73R、Parula、Trap、Cook、Tonichi 81、Sonoita 81、Yaco、Chapio、Tukuru、Kukuna、Vivitsi、Pavon 76、Attila和Amadina[21,22]，其抗性已保持30多年[23-25]。目前，CIMMYT约60%的小麦品系均携带成株抗性基因，其中最典型的例子是具有多种病害抗性位点*Yr18/Lr34/Pm38*，当其单独存在时产量损失可达31%～52%，与3～4个微效基因并存时，产量损失则小于10%[18]。因此，通过聚合*Yr18/Lr34/Pm38*、*Yr29/Lr46/Pm39*和*Yr46/Lr67*等几个成株抗性基因，获得兼抗几种病害的持久抗性是CIMMYT小麦抗病育种的主要策略。美国对高温成株抗性进行了深入研究并成功用于生产实践[16]，澳大利亚的研究重点近10年已从垂直抗性转向成株抗性[26]，欧洲也有类似趋势[27,28]。

鉴于研究成株抗性对中国小麦生产和育种的重要性和紧迫性，本文综述了小麦条锈病和白粉病成株抗性鉴定方法、基因定位和克隆及其在育种中的应用，并利用各个 QTL 间连锁信息，将已定位的条锈病和白粉病成株抗性 QTL 整合到同一张遗传连锁图谱中，目的是利用分子标记培育成株抗性品种，为实现抗病育种思路转变提供方法和信息。

1 成株抗性鉴定

小麦成株抗性基因与主效基因的研究策略差别很大，传统的成株抗性鉴定方法是在苗期选用尽可能多的条锈菌或白粉菌生理小种进行鉴定，在成株期则选用 1 个或几个优势小种接种鉴定，只有在苗期对所有生理小种均表现为感病，成株期表现为中抗或高抗的小麦品种（系）才被视为成株抗性材料。常用的成株抗性鉴定指标是反应型（infection type，IT)[18]、AUDPC[29]和最大严重度（maximum disease severities，MDS)[30]，其中 IT 参数可以反映病原菌对寄主的致病性强弱，但由于成株抗性通常在苗期表现为感病而成株期抗病，此类抗性小麦品种 IT 有时可达到高感级，但普遍率较低，即叶片上病原菌孢子堆较小，显现出特有的局限性。AUDPC 是学术研究中常用的鉴定指标，通过几次田间病害（至少 3 次）调查计算而来，能准确反映病害在发病过程中的变化趋势，但对于成株抗性的田间表现型数据通常需要多年多点调查，因此田间调查工作量很大。然而，MDS 的利用使成株抗性在多环境下进行研究成为现实，MDS 和 AUDPC 在多个遗传背景下的相关系数变化在 0.89～0.96[31-33]，表明在病害发生最严重时选用 MDS 可以有效衡量小麦成株抗性，对于学术研究其工作量仅为 AUDPC 的 1/4～1/3，但由于需在对照发病充分时进行调查，所以要求调查者对该时期的把握及病害评价标准具有较丰富的经验。

2 成株抗性基因定位

迄今，正式命名的小麦条锈病抗性基因有 48 个，分布于 43 个染色体位点，暂时命名的有 33 个基因[34]，其中 *Yr11*、*Yr12*、*Yr13*、*Yr14*、*Yr16*、*Yr18*、*Yr29*、*Yr30*、*Yr36*、*Yr39* 和 *Yr46* 为成株抗性基因[32,35-37]。60 个小麦白粉病抗性基因定位在 40 个染色体位点，即 *Pm1* ～*Pm43*，其中 *Pm38* 和 *Pm39* 为成株抗性基因[34,38]，其余均为具有小种专化性的主效基因。由于病原菌生理小种易变，大部分基因已丧失抗性[39,40]。因此，成株抗性基因的利用对培育持久抗性小麦品种至关重要[19]。

QTL 定位能有效确定抗病基因的数目、效应及染色体位置，与其紧密连锁的分子标记可用于小麦抗病基因聚合[41]。目前，已定位了 72 个条锈病成株抗性 QTL，除 1A、1D、3D 和 7A 染色体外，其余各染色体均有分布（表 1）。82 个白粉病成株抗性 QTL 分布于小麦基因组的 21 条染色体上（表 2）。

表 1 已定位的条锈病成株抗性 QTL/基因

Table 1 Summary of QTLs for adult-plant resistance to stripe rust in wheat

基因位点 QTL	染色体位置 Chromosome	供体亲本 Donor	标记区间 Marker interval	贡献率 R^2（%）	参考文献 Reference
QYr. jirc-1B	1BS	Fukuho-komugi	*Xwmc320*	10.4	[42]
QYrex. wgp-1BL	1BL	Express	*Xwgp78—Xwmc631*	6.8～9.4	[43]
QYr. csiro-1BL	1BL	Attila	*LTN—XP35/M55*	/	[44]
QPst. jic-1B	1BL	Guardian	*Xgwm818*	18.0～22.0	[45]
QYr. cimmyt-1BL	1BL	Pavon 76	*Xgwm259*	33.0～33.9	[46]
QYrtm. pau-2A	2AS	*T. monococcum*	*Xwmc407—Xwmc170*	14.0～27.0	[10]
QTL 2AS	2AS	PAU14087/Kris	*Xwmc407—Xgwm071d*	27.0	[47]
QYR2	2AL	Camp Remy	*Xgwm356—Xgwm382*	10.7～15.4	[28]
QYr. inra-2AL	2AL	Camp Remy	*Xgwm282a—Xgwm359*	20.0～40.0	[48]

（续）

基因位点 QTL	染色体位置 Chromosome	供体亲本 Donor	标记区间 Marker interval	贡献率 R^2（%）	参考文献 Reference
QTL 2AL	2AL	Deben/Kris/Soloist	*Xwmc198a*—*Xwmc170B*	40.0～41.0[a]	[47]
QYR3	2BS	Opata 85	*Xcdo405*—*Xbcd152*	30.7	[28]
QYr. inra-2BS	2BS	Camp Remy	*Xgpw3032*—*Xcfd50a*	22.0～70.0	[48]
QYr. csiro-2BS	2BS	Attila	*XP32/M62*—*XP88/M64*（*Yr27*）	/	[44]
QYr. sgi-2B. 1	2BS	Kariega	*Xgwm148*—*s12m60A*	17.0～46.0	[49]
QYrlu. cau-2BS. 1	2BS	Luke or Aquileja	*Xwmc154*—*Xgwm148*	36.6	[50]
QYr. caas-2BS	2BS	Pingyuan 50	*Xbarc13*—*Xbarc230*	5.1～9.5	[51]
QYr. csiro-2BL	2BL	Avocet-S	*Xgwm1027*—*Xgwm619*	/	[44]
QYR1	2BL	Camp Remy	*Xgwm47*—*Xgwm501*	45.8～46.0[b]	[28]
QYr. inra-2BL	2BL	Camp Remy	*Xbarc101*—*Xgwm120*	42.0～61.0[c]	[48]
QTL 2BL	2BL	Deben	*Xwmc149*—*Xwmc317a*	21.0	[47]
QYraq. cau-2BL	2BL	Luke or Aquileja	*Xwmc175*—*Xwmc332*	61.5	[50]
QYr. inra-2DS	2DS	Camp Remy	*Xgwm102*—*Xgwm539*	24.0～69.0[d]	[48]
QPst. jic-2D	2DS	Guardian	*Xgwm539*	8.0～11.0	[45]
QYr. jirc-2DS	2DS	Oligoculm	*Xcfd51*—*Xgwm261*	5.4～5.7	[42]
QYr. caas-2DS	2DS	Libellula	*Xcfd51*—*Xgwm261*	8.1～12.4	[11]
QYr. jirc-2DL	2DL	Fukuho-komugi	*Xgwm349*	10.1～11.4	[42]
Yrns-B1	3BS	Lgst. 79-74	*Xgwm493*—*Xgwm1329*	/	[52]
Yrns-B1	3BS	Lgst. 79-74	*Xgwm533*	/	[53]
QYr. cimmyt-3BS	3BS	Opata 85	*Xfba190*—*XksuG53*	16.0～28.0	[54]
QYr. jirc-3BS	3BS	Oligoculm	*Xgwm389*	0.2～3.1	[42]
QYr. cimmyt-3BS	3BS	Pavon76	*XPstAATMseCAC2*	3.3～5.9	[46]
QYrex. wgp-3BL	3BL	Express	*Xgwm299*—*Xwgp66*	22.1～27.4	[43]
QYr. cimmyt-3D	3DS	Opata 85	*Xfba241*—*Xfba91*	14.0	[54]
QYR6	3DS	Opata 85	*Xcdo407*—*XksuA6*	11.7	[28]
QYr. sgi-4A. 1	4AL	Kariega	*Xs21m40A*—*Xs22m55A*	6.0～15.0	[49]
QYr. jirc-4B	4BS	Oligoculm	*Xgwm538*	1.8～12.3	[42]
QPst. jic-4B	4BS	Guardian	*Xwmc652*—*Xwmc692*	8.0～12.0	[45]
QYr. cimmyt-4BL	4BL	Avocet S	*Xgwm495*	7.4～12.7	[46]
QYr. caas-4BL	4BL	Libellula Strampelli	*Xgwm165*—*Xgwm149*	3.6～5.1	[11]
QYr. cimmyt-4DS	4DS	Synthetic	*Xbcd265*—*Xmwg634*	9.0～31.0	[54]
QYr. jirc-4DL	4DL	Oligoculm	*Xwmc399*	2.5～8.0	[42]
QYR5	5AL	Opata 85	*Xfbb209*—*Xabg391*	15.0	[28]
QYrtm. pau-5A	5AL	*T. boeoticum* PAU5088	*Xbarc151*—*Xcfd12*	24.0	[10]
QYr. caas-5AL	5AL	Pingyuan 50	*Xwmc410*—*Xbarc261*	5.0～19.9	[51]
QYr. inra-5BL. 1	5BL	Camp Remy	*Xgwm639a*—*Xgwm639c*	18.0～26.0	[48]
QYr. inra-5BL. 2	5BL	Camp Remy	*Xgwm234*—*XDuPW115a*	29.0～34.0	[48]
QYr. jirc-5BL	5BL	Oligoculm	*Xwmc415*	2.4～16.1	[42]
QYr. caas-5BL. 1	5BL	Libellula 和 Strampelli	*Xwmc415*—*Xwmc537*	3.4～8.6	[11]
QYr. caas-5BL. 2	5BL	Libellula	*Xbarc142*—*Xgwm604*	2.6	[11]

（续）

基因位点 QTL	染色体位置 Chromosome	供体亲本 Donor	标记区间 Marker interval	贡献率 R^2（%）	参考文献 Reference
QYr. cimmyt-5DS	5DS	Opata	*Xfbb238—Xfba114*	8.0	[54]
QYr. jirc-5DL	5DL	Oligoculm	*Xwmc215*	3.9	[42]
QYr. nsw-5DL	5DL	Otane	*Xgwm583a*	6.0	[53]
QYrex. wgp-6AS	6AS	Express	*Xgwm344—Xwgp56*	24.5～30.9	[43]
QYr. cimmyt-6AL	6AL	Avocet S	*Xgwm617*	5.9～8.1	[46]
QYr. jirc-6B	6BS	Oligoculm	*Xgwm935. 1*	0.5～3.8	[42]
Yr36	6BS	Langdon	*Xbarc101—Xbarc136*	/	[56]
QYrst. wgp-6BS. 1	6BS	Stephens	*Xbarc101—Xbarc136*	32.0～45.0	[57]
QYr. caas-6BS	6BS	Pingyuan 50	*Xgwm361—Xbarc136*	4.5～7.7	[51]
QYrst. wgp-6BS. 2	6BS	Stephens	*Xgwm132—Xgdm113*	25.0～43.0	[57]
QTL 6BL	6BL	Soloist/Kris	*Xwmc397—Xwmc105b*	25.0～29.0	[47]
QYr. cimmyt-6BL	6BL	Pavon76	*XPstAGGMseCGA1*	10.1～17.6	[46]
QYR7	6DL	W-7984	*Xbcd1510—XksuD27*	13.1	[28]
QYr. jirc-7BS	7BS	Oligoculm	*Xgwm935. 3*	1.0～5.2	[42]
QYr. nsw-7B	7BL	Tiritea	*Xgwm611*	26.0	[55]
Yr39	7BL	Alpowa	*Xwgp36—Xwgp45*	59.1～64.2[e]	[36]
QYr. csiro-7BL	7BL	Attila	*XP32/M59—Xgwm344*	/	[44]
YrCK	7DS	Cook	*M59/P41-215—M49/P33-280*	18.0～30.0	[58]
QYr. jirc-7DS	7DS	Fukuho-komugi	*Xgwm295*	10.7～23.7	[42]
QYr. sgi-7D	7DS	Kariega	*Xgwm295—Ltn*	9.0～29.0	[49]
QYr. nsw-7DS	7DS	Otane	*Xgwm44*	13.0	[55]
QYr. caas-7DS	7DS	Libellula and Strampelli	*csLV34—Xgwm295*	14.6～39.1	[11]
QYr. cimmyt-7D	7DS	Opata 85	*Xbcd1438—Xwg834*	12.0～36.0	[54]
QYR4	7DS	Opata 85	*XWg834—Xbcd1438*	13.9	[28]

/：贡献率不清；a：推测为主效基因 *Yr17* 或者 *Yr32* 的残留效应；b：选用 Mapmaker-QTL 分析（SIM）；c：推测与主效基因 *Yr7* 有关；d：可能与条锈病抗性基因 *Yr16* 有关；e：成株抗性基因 *Yr39* 遗传效应。

"/", R^2 of this QTL was unknown; a, it was speculted that it is the residual effect of major resistance genes *Yr 17* or *Yr32*; b, QTL was detected by the software Mapmaker QTL; c, it was speculted that it may be associated with major resistance gene *Yr7*; d, it might have relationship with stripe rust resistance gene *Yr16*; e, it was the effect of adult-plant resistance gene *Yr39*.

表 2　已定位的白粉病成株抗性 QTL/基因

Table 2　Summary of QTLs for adult-plant resistance to powdery mildew in wheat

基因位点 QTL	染色体位置 Chromosome	供体亲本 Donor	标记区间 Marker interval	贡献率 R^2（%）	参考文献 Reference
QPm. osu-1A	1AS	2174	*Pm3a*	63.0[a]	[59]
QPm. caas-1AS	1AS	Fukuho-komugi	*Xgdm33—Xpsp2999*	19.9～26.6	[32]
QPm. sfr-1A	1AL	Oberkulmer	*Xpsr1201b—Xpsr941*	7.7	[27]
QPm. crag-1A	1AL	RE714	*Xcdo572—Xbad442*	39.3～43.0[b]	[60]
QPm. caas-1AL	1AL	Bainong 64	*Xbarc148—Xwmc550*	7.4～9.9	[33]
QPm. sfr-1B	1BS	Forno	*CD9b—Xpsr593a*	11.6	[27]
QPm. ttu-1B	1BS	*Triticum militinae*	*Xgwm3000*	4.0～5.0	[61]

（续）

基因位点 QTL	染色体位置 Chromosome	供体亲本 Donor	标记区间 Marker interval	贡献率 R^2（%）	参考文献 Reference
QPm. vt-1BL	1BL	Massey	*Xgwm259—Xbarc80*	15.0—17.0	[62]
QPm. vt-1B	1BL	USG3209	*WG241*	17.0	[35]
Yr29/Lr46/Pm39	1BL	Saar	*Xwmc719—Xhbe248*	7.3~35.9	[38]
QPm. osu-1B	1BL	2174	*Xwmc134*	14.0	[59]
QPm. inra-1D. 1	1DS	RE9001	*Xgwm106*	12.6	[63]
QPm. sfr-1D	1DL	Forno	*Xpsr168—Xglk558b*	9.5	[27]
QPm. sfr-2A	2AS	Oberkulmer	*Xpsr380—Xglk293b*	7.7	[27]
QPm. inra-2A	2AS	Courtot	*Xgwm275*	7.4	[63]
QPm. crag-2A	2AL	RE714	*Pm4b—gbxG303*	22.7~33.6	[60]
QPm. ttu-2A	2AL	*Triticum militinae*	*Xgwm311—Xgwm382*	5.0	[61]
QPm. vt-2AL	2AL	Massey	*Xgwm304—Xgwm312*	29.0	[35]
QPm. vt-2°	2AL	USG3209	*Xgwm304—Xgwm294*	26.0~29.0	[62]
QPm. crag-2B	2BS	Festin	*Xgwm148—gbxG553*	23.6~71.5	[60]
QPm. caas-2BS	2BS	Lumai 21	*Xbarc98—Xbarc1147*	10.6~20.6	[64]
QPm. vt-2B	2BL	Massey	*WG338—Xgwm526a*	11.0	[35]
QPm. caas-2B	2BL	Fukuho-komugi	*Xgwm877—Xgwm47*	5.7~8.0	[32]
QPm. inra-2B	2BL	RE9001	*Xrtp114R—Xcfd267b*	10.3~36.3	[63]
QPm. vt-2BL	2BL	USG3209	*Xgwm501—Xgwm191*	11.0~15.0	[62]
QPm. caas-2BL	2BL	Lumai 21	*Xbarc1139—Xgwm47*	5.2~10.1	[64]
QPm. inra-2D-a	2DS	RE9001	*Xgwm102*	19.0	[63]
QPm. inra-2D-b	2DS	RE9001	*Xcfd2e*	16.5	[63]
QPm. sfr-2D	2DL	Oberkulmer	*Xpsr932—Xpsr331a*	10.0	[27]
Qpm. ipk-2D	2DL	W7984	*Xglk558—XksuD23*	/	[65]
QPm. caas-2DL	2DL	Lumai 21	*Xwmc18—Xcfd233*	5.7~11.6	[64]
QPm. sfr-3A	3AS	Forno	*Xpsr598—Xpsr570*	10.4	[27]
QPm. crag-3[a]	3AS	Festin	*Xpsr598—Xgwm5*	21.4~25.9	[60]
QPm. nuls-3AS	3AS	Saar	*Xstm844tcac—Xbarc310*	8.1~20.7	[38]
QPm. inra-3B	3BS	Courtot	*Xgwm389*	22.7	[63]
QPm. osu-3B	3BS	2174	*Xwmc533*	10.0	[59]
QPm. sfr-3D	3DS	Oberkulmer	*Xpsr1196a—Lrk10-6*	15.7	[27]
QPm. inra-3D	3DS	RE9001	*Xcfd152，Xgwm707*	9.3~15.2	[63]
QPm. sfr-4A. 1	4AL	Forno	*Xgwm111c—Xpsr934a*	14.7	[27]
QPm. sfr-4A. 2	4AL	Forno	*Xmwg710b—Xglk128*	14.3	[27]
QPm. ttu-4A	4AL	*Triticum militinae*	*Xgwm232—Xgwm160*	35.0~54.0[c]	[61]
QPm. inra-4A	4AL	RE714	*XgbxG036*	4.9~6.9	[66]
QPm. crag-4A	4AL	RE714	*XgbxG036—XgbxG542*	22.3	[60]
QPm. inra-4[a]	4AL	Courtot	*Xcfd71b*	8.9	[63]
QPm. osu-4A	4AL	2174	*Xwms160*	12	[59]
QPm. sfr-4B	4BL	Forno	*Xpsr593b—Xpsr1112*	7.5	[27]
QPm. ipk-4B	4BL	W7984	*Xcdo795—Xbcd1262*	/	[65]

（续）

基因位点 QTL	染色体位置 Chromosome	供体亲本 Donor	标记区间 Marker interval	贡献率 R^2（%）	参考文献 Reference
QPm. caas-4BL	4BL	Oligoculm	*Xgwm375—Xgwm251*	5.9	[32]
QPm. nuls-4BL	4BL	Avocet	*XwPt1505—Xgwm149*	21.0～40.2	[38]
QPm. sfr-4D	4DL	Forno	*Xglk302b—Xpsr1101a*	14.4	[27]
QPm. caas-4DL	4DL	Bainong 64	*Xbarc200—Xwmc33*	15.2～22.7	[33]
QPm. sfr-5A. 1	5AS	Oberkulmer	*Xpsr644a—Xpsr945a*	22.9	[27]
QPm. ttu-5A	5AS	*Triticum militinae*	*Xgwm186—Xgwm415*	4.0～6.0	[61]
QPm. sfr-5A. 2	5AL	Oberkulmer	*Xpsr1194—Xpsr918b*	16.6	[27]
QPm. sfr-5A. 3	5AL	Oberkulmer	*Xpsr911—Xpsr120a*	10.5	[27]
QPm. nuls-5A	5AL	Saar	*Xgwm617b—Xwmc327*	4.2～15.2	[38]
QPm. ttu-5B	5BS	Tahti	*Xgwm133. mi6—Xgwm205. mi1*	4.0～6.0	[61]
QPm. nuls-5B	5BS	Saar	*Xbarc4—Xgwm274b*	9.7	[38]
QPm. sfr-5B	5BL	Oberkulmer	*Xpsr580b—Xpsr143*	12.6	[27]
QPm. inra-5B. 2	5BL	Courtot	*Xgwm790b*	11.1	[63]
QPm. inra-5D	5DS	RE9001	*cfd189*	9.0	[63]
QPm. crag-5D. 1	5DL	RE714	*Xgwm639a—Xgwm174*	30.2～38.9[b]	[60]
QPm. crag-5D. 2	5DL	RE714	*Xcfd8B9—Xcfd4A6*	24.0～37.8	[60]
QpmVpn. inra-5D	5DL	Courtot	*Xcfd8*	11.0	[63]
QPm. inra-5D. 1	5DL	RE714	*Xcfd26*	28.1～37.7	[66]
QPm. inra-5D. 2	5DL	RE714	*XgbxG083c*	37.7	[66]
QPm. inra-6A	6AL	RE714	*MIRE*（*Xgwm427*）	8.8～13.4	[66]
QPm. crag-6[a]	6AL	RE714	*MIRE*	19.8—53.9[d]	[60]
QPm. sfr-6B	6BS	Forno	*Xpsr167b—Xpsr964*	8.7	[27]
QPm. caas-6BS	6BS	Bainong 64	*Xbarc79—Xgwm518*	10.3～16.0	[33]
QPm. osu-6D	6DS	2174	*Xbarc196*	5.0	[59]
QPm. inra-7A	7AS	RE714	*Xfba069—Xgwm344*	2.9～6.4	[66]
QPm. caas-7A	7AS	Bainong 64	*Xbarc127—Xbarc174*	6.3～7.1	[33]
QPm. sfr-7B. 1	7BL	Forno	*Xpsr593c—Xpsr129c*	11.3	[27]
QPm. sfr-7B. 2	7BL	Forno	*Xglk750—Xmwg710a*	31.8	[27]
QPm. crag-7B	7BL	RE714	*XpdaC01—XgbxR035b*	22.8～33.5	[60]
QPm. inra-7B	7BL	RE714	*Xgwm577*	1.7	[66]
QPm. nuls-7BL	7BL	Saar	*Xwmc581—XwPt8007*	4.9	[38]
QPm. ipk-7D	7DS	Optata	*Xwg834—Xbcd1872*	/	[65]
QPm. caas-7DS	7DS	Fukuho-komugi	*Ltn—Xgwm295. 1*	12	[32]
Yr18/Lr34/Pm38	7DS	Saar	*Xgwm1220—Xswm10*	19.0～56.5	[38]
QPm. inra-7D. 1	7DS	Courtot	*Xgpw1106*	10.6	[63]

/：贡献率不清；a：*Pm3a* 残留效应；b：选用 Mapmaker-QTL 分析（SIM）；c：选用 Map Manager QTX Version b16 分析；d：推测 *MIRE* 的残留效应。

"/", R^2 of this QTL was unknown; a, it was the residual effect of major resistance genes *Pm3a*; b, QTL was detected by the software Mapmaker QTL; c, QTL was detected by the software Map Manager QTX Version b16; d, it was speculted that it was the residual effect of *MIRE*.

植物抗病基因在基因组中有两种不同的分布模式：一种是抗病基因位于单一的基因座，但具有编码不同专化抗性的等位基因，如小麦白粉病抗性基因 *Pm3* 位点的不同等位基因[67]；另一种是抗病基因成簇分布于特定的染色体区域，共同组成紧密连锁的复合抗性基因座[68-70]。依据 Somers 等[71] 和 http://wheat.pw.usda.gov/GG2/index.shtml 网址所公布的分子标记图谱，笔者将已定位的条锈病和白粉病成株抗性 QTL 整合于同一张连锁图谱（图 1）。可以看出，含有条锈病成株抗性 QTL 的基因簇有 4 个，分别位于 2AS、2DS、5BL 和 6BL 染色体上，说明小麦各个基因组均有分布。含有白粉病成株抗性 QTL 基因

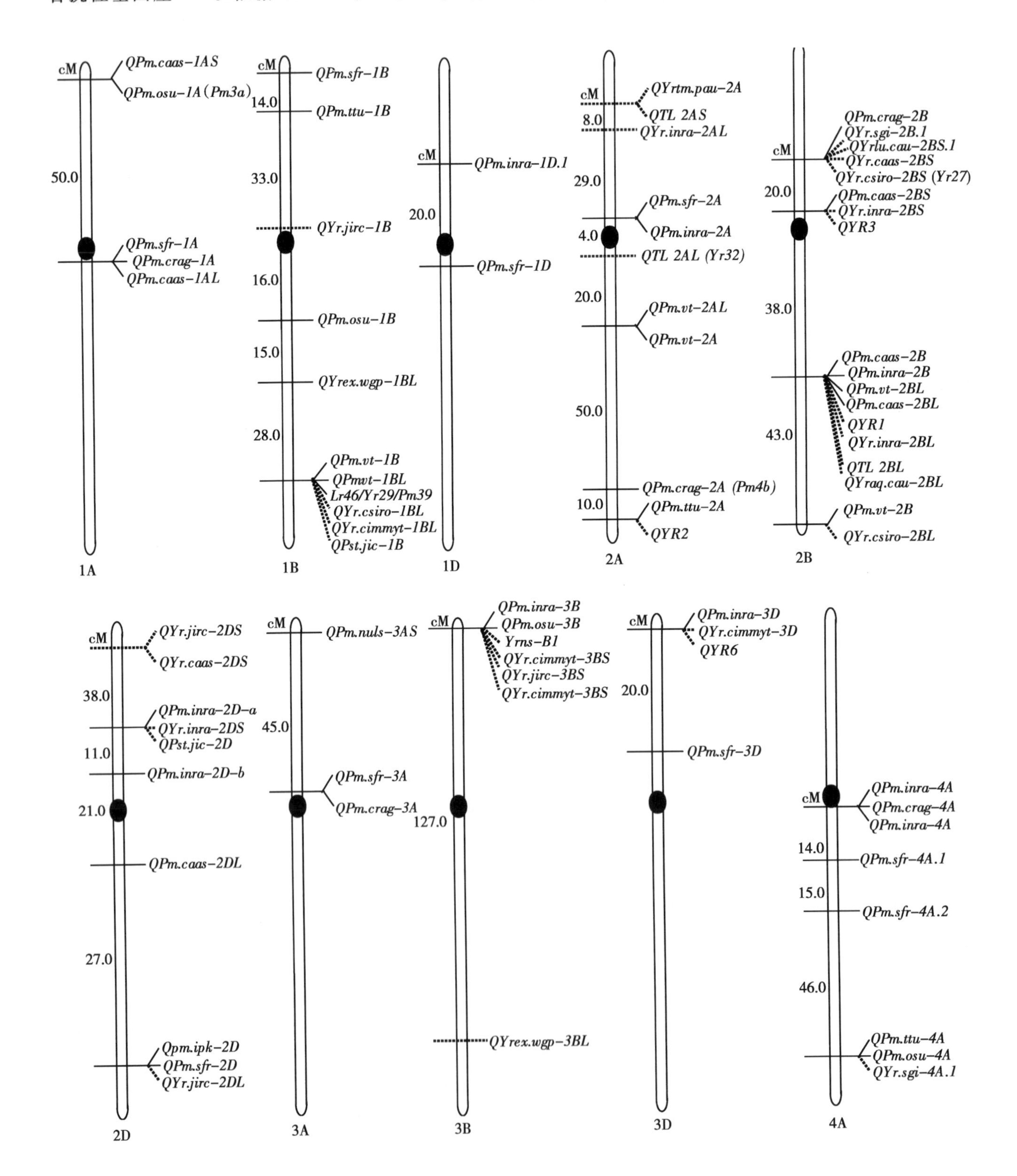

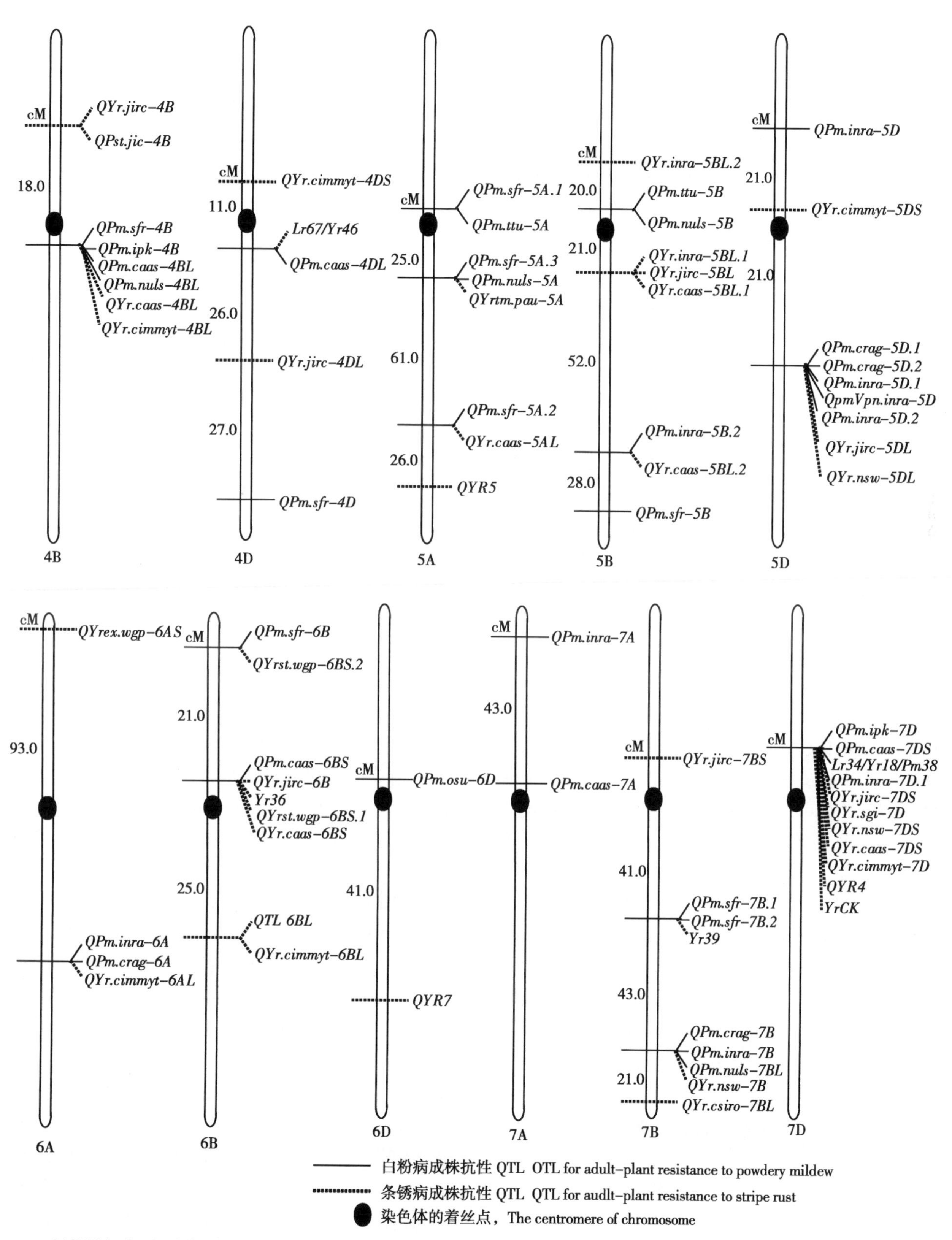

QTL 间的距离（cM）是分子标记图谱中标记间遗传距离的累加值，The distances (cM)between QTLs in a linkage group are the accumulated genetic distances between moleaular markers in wheat consensus linkaye map (somers 等[71]；http://wheat.pw.usda.gov/GG2/index.shtml)

图 1　整合的小麦条锈病和白粉病成株抗性 QTL/基因连锁图谱

Fig. 1　Integrated linkage map of QTLs/genes for adult-plant resistance to stripe rust and powdery mildew in wheat

簇有8个，分别位于1AS、1AL、2AS、2AL、3AS、4AL、5AS和5BS染色体上，表明控制白粉病成株抗性QTL大多集中在小麦A基因组，B和D基因组相对较少。同时含有条锈病和白粉病成株抗性QTL基因簇有22个，除1D、6D和7A染色体外，其余各个染色体均含有1～2个兼抗2种病害的成株抗性基因簇，其中含有5（包括5）个以上QTL的大基因簇有8个，分别位于小麦1BL、2BS、2BL、3BS、4BL、5DL、6BS和7DS染色体，约占总QTL的34%（图1）。下文着重对兼抗条锈病和白粉病成株抗性基因簇进行论述。

2.1 已证实兼抗多种病害的成株抗性基因簇

目前公认的兼抗多种病害成株抗性基因簇分别位于1BL和7DS染色体上。Lillemo等[38]利用成株抗性品种Saar与感病品种Avocet构建的113个重组自交系，检测到6个成株抗性QTL，其中位于1BL和7DS染色体上的2个QTL解释的表现型变异最大，并将1BL染色体上的条锈病成株抗性基因*Yr29*、叶锈病成株抗性基因*Lr46*和白粉病成株抗性基因*Pm39*和7DS染色体上条锈病成株抗性基因*Yr18*、叶锈病成株抗性基因*Lr34*和白粉病成株抗性基因*Pm38*分别定义为同一基因，即以往认为3种病害分别由3个基因所控制，实际上是同一基因。说明这两个位点均对条锈病、叶锈病和白粉病表现成株抗性，*Yr18/Lr34/Pm38*已被克隆[72]，DNA序列分析显示这3种病害由同一位点所控制，进一步证实了其定义为同一基因的正确性，*Yr29/Lr46/Pm39*的克隆工作正在进行中。此外，*Yr29/Lr46/Pm39*和*Yr18/Lr34/Pm38*除对以上3种病害表现成株抗性外，常伴随叶尖坏死现象，可作为田间选择的形态标记[38,73]。Singh[74]在7DS染色体区域检测到一个在成株期耐大麦黄矮病基因*Bdv1*，并对秆锈病具有一定成株抗性[75]，它们可能与*Yr18/Lr34/Pm38*是同一基因。Herrera-Foessel等[37]在成株抗性小麦品系RL6077与感病品种Avocet构建的136个F_4家系中检测到1个兼抗叶锈病和条锈病的成株抗性基因，位于小麦基因组的4DL染色体上，与SSR标记*Xcfd71*、*Xbarc98*和*Xcfd23*紧密连锁，命名为*Yr46/Lr67*，依据小麦分子标记图谱[71]，该位点与Lan等[33]在中国小麦品种百农64中检测到的白粉病成株抗性QTL*Qpm.caas-4DL*位于同一位点。由于它们的抗性来源均为普通小麦，所以推测该位点很可能是第三个兼抗条锈病、叶锈病和白粉病的多种病害成株抗性位点。虽然该基因的抗性效应不及*Yr18/Lr34/Pm38*，但由于百农64的农艺性状相对较好，该位点将是小麦持久抗病育种的另一个重要基因资源。

2.2 有待进一步发掘的兼抗多种病害基因簇

2.2.1 2BS和2BL染色体的抗病基因成簇分布

小麦条锈病主效抗性基因*Yr27*[76]、*Yr31*[77]和*Yr41*[78]及叶锈病主效抗性基因*Lr23*[77]紧密连锁并位于2BS染色体上的基因簇（如*Qpm.crag-2B*等5个QTL），其中*Yr41*存在于小麦品系Ch377/Cn19中，苗期对条锈菌强毒性生理小种CYR32表现为抗病，*Yr27*和*Yr31*分别来自小麦品种Opata 85和Pastor，但苗期对CYR32表现感病[78]。此外，Rosewarne等[44]在抗病品种Attila中检测到的条锈病成株抗性QTL *QYr.csiro-2BS*对病原菌生理小种的反应型与*Yr27*相同，因此推测该QTL效应可能源于*Yr27*抗性基因。由此可知，位于2BS染色体上的基因簇可能是*Yr27*或*Yr31*的残留效应，当然这还有待于进一步研究证实，该位点对白粉病、秆锈病和叶锈病同样具有一定的抗性。

位于2BL染色体基因簇（如*Qpm.caas-2B*等8个QTL）的苗期抗性基因有白粉病主效抗性基因*Pm6*[79]和*Pm33*[80]、条锈病主效抗性基因*Yr5*[81]和*Yr7*[81]及秆锈病主效抗性基因*Sr9g*[81]，其中*Pm6*、*Pm33*和*Yr5*分别来源于小麦近缘种*T. timopheevi*、*T. carthlicum*和*T. spelta album*，而*Yr7*源于普通小麦，该基因在苗期对CYR32表现为感病[82]，Mallard等[48]认为*QYr.inra-2BL*可能位于一个含有*Yr7*的基因簇中，从而表达抗性。目前，该基因簇研究相对较少，至于这些QTL是*Yr7*的残留效应还是位于同一位点的不同抗性基因，有待于进一步证实。

2.2.2 3BS和6BS染色体的抗病基因成簇分布

位于3BS染色体基因簇（如*Qpm.inra-3B*等6个QTL）的小麦条锈病成株抗性基因*Yrns-B1*与*Yr30*[54]、秆锈病成株抗性基因*Sr2*[26,83]、叶锈病抗性基因*Lr27*[84]、叶锈病成株抗性QTL[65]及赤霉病抗性QTL[85]紧密连锁，显示该位点同时具有抗条锈病、白粉病、秆锈病、叶锈病和赤霉病的特点。Crossa等[86]在60多个国际环境下，利用连锁不平衡对170份CIMMYT春小麦品种进行秆锈病、叶锈病、条锈病、白粉病和产量相关分析，进一步证实*Sr2*、*Yr30*为非小种专化抗性基因，并在不同环境下均表达中度抗性，目前认为*Sr2*和*Yr30*可能是同一基因，或者是紧密连锁的两个基因，其作用方式类似一个基因（Singh，个人交流）。由此推测，3BS染色体上的基因可能存在一因多效现象，这也有待于进一步研究利用。

在位于6BS染色体上基因簇（如*Qpm. caas-6BS*等5个QTL）中，高温成株抗性基因*Yr36*及籽粒蛋白基因*Gpc-B1*[87]紧密连锁[88]，条锈病抗性基因*Yr35*[89]与叶锈病抗性基因*Lr53*[89]紧密连锁，不过*Yr36*来源于野生二粒小麦[56]，而许多成株抗性QTL存在于普通小麦中。由此看出，该基因簇中至少存在两个条锈病成株抗性基因，其中一个基因可能同时发挥着白粉病成株抗性，并且该基因簇对小麦抗病性及品质改良有一定影响。

2.2.3 4BL和5DL染色体的抗病基因成簇分布

依据Crossa等[86]分析结果，位于4BL染色体上的基因簇（如*Qpm. sfr-4B*等6个QTL）与叶锈病主效抗性基因*Lr30*[90]、白粉病主效抗性基因*Pm7*[91]和两个与高产相关的QTL[92,93]紧密连锁，但由于*Pm7*来源于黑麦，因此该基因簇的白粉病抗性效应并非*Pm7*的残留效应。由此可知，4BL染色体上的基因簇不仅对条锈病、叶锈病和白粉病发挥抗性作用，同时对提高产量也有一定效应。此外，位于4BL染色体基因簇的QTL *QPm. caas-4BL*、*QYr. cimmyt-4BL*和*QPm. nuls-4BL*分别来源于感病亲本Oligoculm和Avocet S[32,38,46]，从分子水平上进一步证实了在育种实践中利用中度感病品种进行抗病性改良的有效性。

由表1和表2还可以看出，小麦5DL染色体上存在4个效应较大的白粉病成株抗性QTL，即*QPm. crag-5D. 1*、*QPm. crag-5D. 2*、*QPm. inra-5D. 1*和*QPm. inra-5D. 2*，该位点对条锈病成株抗性同样发挥一定作用，并且与叶锈病主效抗性基因*Lr1*[94]和春化基因*Vrn-A3*[95]紧密连锁，但该基因簇目前研究尚少，有待于进一步发掘利用。

综上所述，条锈病和白粉病成株抗性QTL及二者所构成的基因簇广泛分布于小麦基因组的21条染色体。由于大量的QTL只是粗略地定位于染色体的特定区域，并且所用的作图群体和不同环境下田间发病情况均不尽相同，加之QTL间等位性测定的困难，因此实际存在的QTL个数可能少于已命名的基因位点。为了进一步澄清基因簇内各个QTL间的关系，对这些QTL进行精细定位及克隆显得尤为重要。

3 成株抗性基因克隆及分子标记发掘

近年来，随着分子标记技术的广泛应用，一些效应较大的成株抗性基因在染色体上的位置逐步明确[25,56,96,97]，为进一步基因克隆奠定了基础。Krattinger等[72]和Fu等[88]利用图位克隆法，分别克隆了成株抗性基因*Yr18/Lr34/Pm38*和*Yr36*。序列分析显示，在*Yr18/Lr34/Pm38* BAC跨叠群中含有8个编码蛋白的开放阅读框，其中包括1个ATP结合位点转移酶（ABC）、2个细胞色素P450、2个血凝素激酶、1个半胱氨酸蛋白酶和1个糖基转移酶，功能验证显示，ABC区域即为*Yr18/Lr34/Pm38*[75]。*Yr36*分析表明，该区域含有3个基因，即*WKS1*、*WKS2*和*IBR1*；突变体分析显示，*WKS1*即为*Yr36*，包括1个激酶和1个脂质结合区域（START），二者缺一不可[88]。由此可知，成株抗性基因分子结构与主效抗性基因所具有的保守结构域NBS-LRR[98-100]完全不同，并且*Yr18/Lr34/Pm38*和*Yr36*之间没有相同的结构域。

3.1 *Yr18/Lr34/Pm38*

20世纪早期，*Yr18/Lr34/Pm38*在中国、意大利、北美和南美及欧洲的小麦品种中广泛分布[101]，特别是在中国小麦地方品种中的频率较高，已保持了70多年的抗性[11,102,103]。目前在CIMMYT种质中分布频率较高[18]。据报道，在发展中国家，含有*Yr18/Lr34/Pm38*的小麦品种约有2 600万hm^2，在病害流行年份对稳产发挥着重要作用[104]。在分子水平上，该位点的初步定位始于21世纪初，检测到SSR标记*Xgwm295. 1*和*Xgwm1220*与*Yr18/Lr34/Pm38*紧密连锁[42,97]。随后，Lagudah等[96]依据小麦与水稻基因序列的共线性原理，选用小麦表达序列标签（wEST）进行分析，开发的STS标记*csLV34*与*Yr18/Lr34/Pm 38*间连锁距离缩短到0.4cM。Spielmeyer等[97]通过诱导突变产生更多变异，再次利用wEST标记进行筛选，从而获得与*Yr18/Lr34/Pm 38*连锁更为紧密的分子标记*csLVMS1*（0.13cM）。Krattinger等[72]利用同样的方法，筛选到与*Yr18/Lr34/Pm38*的连锁距离仅为0.03cM的分子标记，随后利用3个不同来源的品种群体对该基因进行图位克隆，结果显示该基因含有24个外显子，编码1 401个氨基酸。携带*Yr18/Lr34/Pm38*和不携带

Yr18/*Lr34*/*Pm38* 的小麦品种在基因序列上存在 3 个易变区，即 2 个 SNP（第 4 内含子和第 12 外显子）和 1 个位于第 11 外显子上的 3bp 缺失；最近在 Jagger 中发现第 22 外显子的 1 个 SNP 变异导致品种感病[105]。通过 RT-PCR 以及基因突变等方法，证实 *Lr34*、*Yr18*、*Pm38* 和 *Ltn1* 是同一基因所控制的多种持久病害。目前该基因的功能标记已经开发完成[101]，可以鉴定大部分国外材料，85%的中国小麦农家品种含有该基因特异性扩增片段[103]，但 25.8%品种在田间的条锈病最终严重度达到 100%，说明中国农家品种在该基因位点含有新的等位变异或存在抑制基因。

3.2 *Yr36*

条锈病高温成株抗性基因 *Yr36* 来源于野生二粒小麦[56]，高温条件下（25～35℃）在田间表现为广谱抗性，低温下（15℃）则感病，并且与高产显著相关[56]。该基因的克隆思路基于已检测到与 *Yr36* 紧密连锁的分子标记 *Xucw71* 和 *Xbarc136*，对 4 500 个 F_2 单株进行检测，发现 121 个单株发生重组，同样利用小麦与水稻基因组序列的共线性，进行精细定位，获得与其紧密连锁的分子标记（0.14cM），最后利用图位克隆法得到 *Yr36*。Alpy 等[106]报道，在人类中，START 区域蛋白在脂质运输、新陈代谢和人的感观中起重要作用。所以，Fu 等[88]推测，在高温条件下，*WKS1* 结合脂质使其远离病原物，同时激活激酶，最终导致病原菌细胞死亡实现抗病[88]。此外，将 *Yr36* 导入含有 *Yr18* 的小麦品种中，其抗性有所提高[88]，所以将成株抗性基因进行聚合对持久抗病育种极其重要。

综上所述，相对于主效基因固有的保守区域如 NBS-LRR，成株抗性基因结构较为复杂，没有特定结构域，致使成株抗性基因作用机理和通用功能标记的开发等研究受到限制。随着分子标记在基因定位中的广泛应用和 QTL 精细定位的不断发展，估计在不久的将来会有更多的成株抗性基因被分离、克隆，将有助于揭示成株抗性基因的功能、信号传导途径及其进化的分子机制，为病害防治提供新的育种策略[107,108]，特别是通过分析已克隆抗病基因在抗、感材料中的等位变异，依据基因序列设计特异性的功能标记，将为育种实践提供更有效的分子标记，大大提高分子标记辅助选择的准确性和实用性。

4 中国小麦成株抗性研究与应用

中国小麦条锈病成株抗性研究始于 20 世纪 70 年代，针对小麦品种抗锈性丧失现象，曾士迈率先介绍植物水平抗性的概念[109]，指出在育种过程中由于过分强调垂直抗性而忽略了水平抗性，人为造成植物抗性丧失，并提出小麦条锈病水平抗性综合鉴定方法。Shaner[110]报道了白粉病成株抗性小麦品种 Knox 抗性较持久稳定，引起小麦育种界的高度重视。20 世纪 80 年代中期到 21 世纪初，中国学者对条锈病和白粉病成株抗性进行了鉴定，现将主要结果分别列于表 3 和表 4。

表 3 国内鉴定的条锈病成株抗性品种

Table 3 Presence of the adult-plant resistance to stripe rust in Chinese wheat cultivars

品种（系）名 Cultivar	系谱 Pedigree	参考文献 Reference
阿桑 San pastore	/	[111]
保加利亚 10 号 Bulgarian 10	/	[111]
保加利亚 14 号 Bulgarian 14	/	[111]
东方红 3 号 Dongfanghong 3	农大 45 选系	[111]
邯郸蚰子麦 Handan Youzi Mai	地方品种	[111]
农大 168 Nongda 168	/	[111]
农大 198 Nongda 198	/	[111]
农大 4356 Nongda 4356	/	[111]
农大 6085 Nongda 6085	/	[111]
平原 50 Pingyuan 50	地方品种	[111]
徐州 8 号 Xuzhou 8	碧蚂 1 号/苏联早熟 1 号	[111]
陕西蚂蚱麦 Shaanxi Mazha Mai	地方品种	[111]
草帽黄选系 Line of Caomao Huangxuan	地方品种	[112]
岐山蚂蚱麦 Qishan Mazha Mai	地方品种	[112]
望水白 Wangshuibai	地方品种	[112]
无名 7 号 Wuming 7	/	[112]
咸农 4 号 Xiannong 4	/	[112]
小偃 6 号 Xiaoyan 6	郑引 4 号/小偃 96	[112]
里勃留拉 Libellula	Tevere/Giuliari（1482-54-3）//San-Pastore	[113]
斯特拉姆佩列 Strampelli	Libero//San-Pastore-14/Jacometti-19	[113]
清农 1 号 Qingnong 1	/	[113]
清农 3 号 Qingnong 3	山前麦/6922	[113]
中梁 5 号 Zhongliang 5	/	[113]

/：系谱不详。Pedigrees of the lines/cultivars were not available.

表4 国内鉴定的白粉病成株抗性品种

Table 4 Presence of the adult-plant resistance to powdery mildew in Chinese wheat cultivars

品种（系）名 Cultivar	系谱 Pedigree	参考文献 Reference
豫麦2号 Yumai 2	65 (14)$_3$/抗锈辉县红	[114]
周8846 Zhou 8846	偃师4号/盘江3号	[114]
里勃留拉 Libellula	Evere/Giuliari （1482-54-3）//San-Pastore	[115]
阿勃 Abbondanza	Autonomia/Fontartonco	[115]
咸农4号 Xiannong 4	/	[115]
小偃6号 Xiaoyan 6	郑引4号/小偃96	[115]
25413-23-4-5	百农3217/CA8065	[116]
521021	矮孟牛/辐663//矮孟牛2//高38/726294//矮孟牛高38//辐63	[116]
435209	矮孟牛/辐66	[116]
435775	矮孟牛/辐66	[116]
临远85-7053 Linyuan 85-7053	米西切尔//太古286/矮丰四号	[116]
临远85-60495 Linyuan 85-60495	73 (29) 2/郑州7308	[116]
临远85-6015 Linyuan 85-6015	20301/72-629-26/伏2845	[116]
86-5144	石81-4474/90354	[116]
豫麦16 Yumai 16	郑州761/无芒77	[116]
鲁麦13 Lumai 13	[洛夫林13/3/蚰选57//小罂粟/欧柔] 莱阳584	[116]
平阳27 Pingyang 27	平阳79391/平阳76262	[116]
唐麦4号 Tangmai 4	7201-1/74060-0-2-2-3-1//北京5123	[116]
陕213 Shaan 213	山前/郑洲721//TJB259-29/6521-2/3/小偃6号	[116]
98301	/	[117]
CA8686	丰抗4号//百泉5号-有芒白4号/有芒红7号-洛夫林10	[117]
CA9550	CA8695/C39//京411	[117]
CA9641	CA8695/C39//京411	[117]
YE-2416-7A	/	[117]
百农64 Bainong 64	百农8717/[百农84-4046-1//偃大72-629-52/石82-5594]	[117]
山东418 Shandong 418	/	[117]
兴麦99 Xingmai 99	C39/庆30	[117]
豫麦47 Yumai 47	豫麦2号/百泉3199	[117]
周麦17 Zhoumai 17	矮早781/周8425B//周麦9号	[117]
花培3号 Huapei 3	花953350-1-2/花962437-1-1	[118]
豫麦57 Yumai 57	矮早781/80 (6) -3-3-10	[118]
鲁麦21 Lumai 21	鲁麦13/豫麦2号	[119]

/：系谱不详。Pedigrees of the lines/cultivars were not available.

由表3和表4可知，同时对条锈病和白粉病具有成株抗性的小麦品种有咸农4号、小偃6号和Libellula，生产上大面积推广的成株抗性小麦品种有豫麦2号、小偃6号、豫麦47、百农64和鲁麦21[120-122]。表3和表4还显示多个小麦成株抗性品种（系）含有共同亲本，说明其成株抗性基因来源可能相同。

4.1 豫麦2号成株抗性

豫麦2号为豫麦47和鲁麦21的共同亲本，多年生产应用和抗性鉴定皆表明，豫麦2号为白粉病成株抗性小麦品种，由此可以初步推测，豫麦47和鲁麦21的白粉病成株抗性可能来源于豫麦2号。倪小文等[120]利用数量遗传学方法对鲁麦21/京双16进行白粉病成株抗性遗传分析，显示该品种至少含有3个白粉病成株抗性基因；Lan等[64]利用分子标记对鲁麦21/京双16的F_2衍生混合群体进行白粉病成株抗性QTL分析，发现3个白粉病成株抗性QTL，位于小麦基因组2B和2D染色体，因此，选用已获得的紧密连锁分子标记对豫麦47和豫麦2号进行检测，即可明确3个白粉病成株抗性小麦品种的抗性来源。豫麦2号的组合为65 (14)$_3$/抗锈辉县红，由于2个亲本对条锈和白粉都有一定的抗性，因此其成株抗性来源还值得进一步研究。

4.2 San Pastore成株抗性

Libellula和Strampelli含有共同亲本San Pastore。殷学贵等[123,124]分别对Strampelli和Libellula进行持久抗条锈病遗传机制研究，表明这两个品种的抗性至少由两个基因所控制。Lu等[11]分别对252个Libellula/辉县红$F_{2:3}$家系和255个Strampelli/辉县红$F_{2:3}$家系进行条锈病成株抗性QTL分析，在Libellula/辉县红群体中检测到5个条锈病成株抗性QTL，均由抗病亲本Libellula所提供，其中*QYr.caas-4BL*、*QYr.caas-5BL.1*和*QYr.caas-7DS*同样稳定地存在于Strampelli中，进一步证实了含有共同亲本小麦品种的成株抗性基因在一定程度上是一致的。此外，在Libellula、Strampelli中分别检测到QTL间上位性效应，但在作用方式上两个品种存在明显差异。如在Libellula中，条锈病成株抗性QTL对表现型主要以加性效应为主，QTL间上位性效应较小，而Strampelli品种中，QTL间加性效应和上位性效应对表现型起着同等重要的作用[11]。

4.3 小偃6号和矮早781成株抗性

表4显示，陕213的一个亲本为小偃6号，而小偃6号为典型的白粉病成株抗性，由此可以初步推测，陕213的成株抗性可能源于小偃6号。周麦17和豫麦57的共同亲本为矮早781，而矮早781即为豫麦18，来源于郑州761/偃师4号，而周8846的亲本之一为偃师4号，这可能暗示周麦17、周8846和豫麦18的共同抗源为偃师4号。由于小偃6号和偃师4号的共同亲本为St2422/464，也许上述品种的成株抗性来自St2422/464。San Pastore和St2422/464皆来自意大利，它们是否有共同的亲本和抗源也值得进一步研究。

4.4 其他成株抗性品种遗传分析

平原50曾是中国黄淮海麦区推广的一个重要地方品种，在生产上保持50多年条锈病抗性。Lan等[51]对平原50/铭贤169的137个DH系进行条锈病成株抗性QTL分析，检测到3个条锈病成株抗性QTL，位于2BS、5AL和6BS染色体上，分别命名为*QYr. caas-2BS*、*QYr. caas-5AL*和*QYr. caas-6BS*，这三个QTL均来源于抗病亲本平原50。王竹林等[125]利用数量遗传学方法对生产上保持抗性15年的小麦品种百农64进行白粉病成株抗性遗传分析，该品种含有3个白粉病成株抗性基因。随后，Lan等[33]在百农64/京双16的DH群体中共检测到4个白粉病成株抗性QTL，分别位于1A、4DL、6BS和7A染色体，命名为*QPm. caas-1A*、*QPm. caas-4DL*、*QPm. caas-6BS*和*QPm. caas-7A*，单个QTL可解释表型变异的6.3%～22.7%，这4个QTL均由抗病亲本百农64所提供，与这些QTL紧密连锁的分子标记在一定程度上可应用于分子辅助选择。Liu等[126]通过对*Yr10*的近等基因系和感病品种Avocet S间的多态性分析，发现一个homeobox类基因*TaHLRG*，该基因不仅对条锈病苗期有一定抗性，并且与条锈病和白粉病成株期抗性有关。此外，Liang等[32]对Fukuho-komugi/Oligoculm的DH群体进行分析，检测到4个白粉病成株抗性QTL，其中位于1AS、2BL和7DS染色体上QTL的抗性均由Fukuho-komugi所提供，4BL染色体上的QTL来源于Oligoculm，单个QTL可解释表现型变异变化在5.7%～26.6%。张坤普等[118]利用SSR标记在花培3号/豫麦57的DH群体中检测到一个白粉病成株抗性QTL，位于小麦基因组4DS染色体上。

此外，小麦白粉病成株抗性品系521021、435209和435775均含有共同亲本矮孟牛，CA9550、CA9641和兴麦99含有共同亲本C39，由于相关遗传分析研究尚少，多个成株抗性品系虽含有共同亲本，但其抗性是否均来自同一亲本，还有待于进一步研究证实。

4.5 成株抗性的育种实践

CIMMYT的成株抗性育种已有40多年的历史，方法比较成熟，具体做法见图2（Ravi Singh，个人交流）。与选择全生育期抗性不同的是，在BC_1至F_4选择农艺性状的同时，选择中感到中抗类型，淘汰高感和高抗（主基因抗性）类型，到F_4对抗性的要求提高到中抗到高抗。但国内有目的地进行成株抗性育种尚未见报道，为此我们在抗条锈病和抗白粉病育种中进行了尝试，目的是积累经验，为大范围应用起示范作用。

中国与CIMMYT合作进行小麦穿梭育种已有25年的历史，从2000年开始将目标转向培育具有成株抗性的小麦品种，四川省农业科学院的进展相对较快。具体做法是选用CIMMYT成株抗性种质与四川小麦品种杂交，并用四川小麦品种回交，将成株抗性基因导入农艺性状优良、丰产性高及广适性好的四川小麦品种（图2）。目前，表现突出的3份成株抗性品系已参加长江上游区试和四川省区试，均优于对照品种川麦42（Syn-CD769/SW89-3243/川6415）和绵麦37（SW2148/绵阳90-100）（表5）。

由表5可知，08RC2525（川麦32 * 2/Chapio）在预试中增产15.1%，苗期表现为感病，田间则对条锈病表现为高抗（5HR）；07RC3941（SW119 * 2/Tukuru）和07RC3929（SW119 * 2/Tukuru）苗期同样表现为高感，但成株期则为中抗（20MR）；08RC2525和07RC3941对白粉病和赤霉病分别表现为中感，而07RC3929为高抗。此外，云南省农业科学院粮食作物所于2001年从CIMMYT引入的F_4（四川品种与CIMMYT材料杂交后代）中选育出1份小麦条锈病成株抗性新品种云选3号，在多个区域试验点均表现为高抗条锈病、白粉病、耐寒、耐旱、抗倒伏，目前已进入生产试验。上述工作进一步证实了培育成株抗性与高产相结合的品种是可行的。

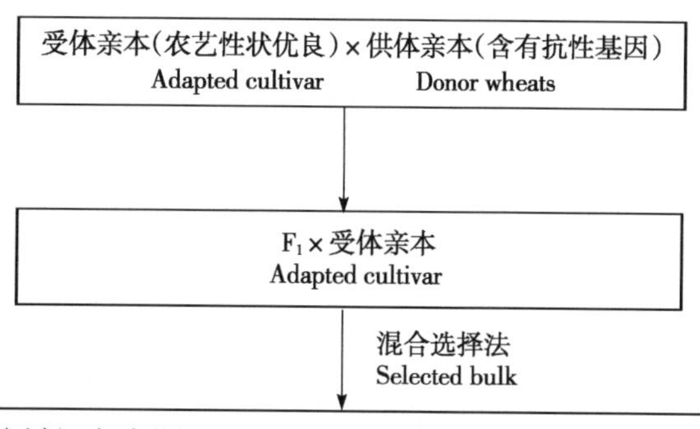

图2 单粒回交混合选择育种策略

Fig. 2 Single-backcross selected bulk breeding strategy

表5 2010年四川省3个成株抗性品系的产量和抗性表现[a, b]

Table 5 Performance of 3 lines on yield and disease in 2010 in Sichuan

品种名 Cultivar	长江上游预试 PYTYR[c]		四川省区试A组 PYTS[d]（A）		四川省区试B组 PYTS[d]（B）	
	08RC2525	川麦42（CK）	07RC3941	绵麦37（CK）	07RC3929	绵麦37（CK）
平均产量（kg/hm^2） Averaged yield	6 114	5 450	5 757	5 654	5 787	5 652
增产 Increase（%）	+15.1		+1.8		+2.4	
田间条锈病[a] Stripe rust under natural infection	高抗（5HR）	高感（95HS）	中-高抗（20MR）	高抗（1HR）	中-高抗（20MR）	高抗（1HR）
条锈病接种鉴定 Stripe rust under artificial inoculation	—	—	高抗 HR	高抗 HR	高抗 HR	高抗 HR

（续）

品种名 Cultivar	长江上游预试 PYTYR[c]		四川省区试A组 PYTS[d]（A）		四川省区试B组 PYTS[d]（B）	
	08RC2525	川麦42（CK）	07RC3941	绵麦37（CK）	07RC3929	绵麦37（CK）
田间白粉病 Powdery mildew under natural infection	中感MS	中抗MR	中感MS	高抗HR	高抗HR	高抗HR
白粉病接种鉴定 Powdey mildew under artificial inoculation	—	—	中感MS	高抗HR	高抗HR	高抗HR
田间赤霉病 Scab under natural infection	中感MS	中抗MR	中感MS	中抗MR	中抗MR	中抗MR
赤霉接种鉴定 Scab under artificial inoculation	—	—	中感MS	中抗MR	中抗MR	中抗MR

a. 2010年因干旱总体发病轻，但成都点发病重，因此田间条锈病为成都点的资料，其他为长江上游预试和四川省区试汇总，Disease data was only from Chengdu.

b. 资料来源：四川省农业科学院作物研究所朱华忠研究员提供，Data wasprovided by Prof. Zhu Huazhong from Crop Research Institute of Sichuan Academy of Agricultural Sciences.

c. 扬子区区试试验点，PYTYR，Prelimiary yield trial in Yangtze Region.

d. 四川省区试试验点，PYTS，Procincial yield trial in Sichuan.

在前期对鲁麦21和百农64遗传分析的基础上[119,125]，中国农业科学院国家小麦改良中心对这两个品种进行白粉病成株抗性基因聚合，在21个F_6家系中，C181、C219和C263分别聚合5个白粉病成株抗性QTL，含有4个白粉病成株抗性QTL的家系有11个（表6）。这3个系的田间农艺性状和抗病性均优于亲本，为小麦白粉病成株抗性品种选育提供了重要的材料，证实了利用连锁分子标记进行成株抗性QTL聚合的可行性及其有效性，进一步说明4～5个白粉病成株抗性QTL足以在田间表达高水平抗性。

表6　7个白粉病成株抗性QTL紧密连锁分子标记对24份试验材料的检测结果及2010年北京田间表现型

Table 6　Performance of 24 lines with SSR markers closely linked to 7 adult-plant resistance QTLs and their maximum disease severities (MDS) in Beijing 2010

株系 Line	*QPm caas-1A*	*QPm caas-4DL*	*QPm caas-6BS*	*QPm caas-7A*	*QPm caas-2BS*	*QPm caas-2BL*	*QPm caas-2DL*	QTL汇总	MDS
京双16 Jingshuang16	—	—	—	—	—	—	—	0	80
百农64 Bainong64	+	+	+	+	—	—	—	4	3
鲁麦21 Lumai21	—	—/?	—/?	—/?	+	+	+	3	3
C181	+	+	+/?	+	+	—	—	5	2
C191	+	+	+/?	—	—	—	—	3	6.5
C197	+	+	+/?	—	—	—	—	3	6.5
C209	+	+/?	?	—/?	+	+	—	4	3
C219	+	+/?	?	—/?	+	+	+	5	2
C233	+	+/?	?	—	+	+	—	4	2
C237	+	+/?	?	—	+	—	—	3	7.5
C249	—	+/?	—	—/?	+	—	—	2	15
C251	+	+	?	—	—	—	+	3	5.5
C255	+	+	+	—	—	—	+	4	1
C259	+	+	+	—	—	—	—	3	6.5
C263	+	+/?	?	+	+	—	+	5	1
C265	+	+	?	+	—	—	+	4	4
C269	+	+	?	+	—	—	+	4	4

（续）

株系 Line	*QPm caas-1A*	*QPm caas-4DL*	*QPm caas-6BS*	*QPm caas-7A*	*QPm caas-2BS*	*QPm caas-2BL*	*QPm caas-2DL*	QTL 汇总	MDS
C271	+	+/?	?	+	−	−	+	4	2
C275	+	+/?	?	+	−	−	+	4	3
C281	+	+/?	?	+	−	−	+	4	5.5
C289	+	+	?	+	−	−	+	4	1
C291	+	+	?	+	−	−	+	4	8
C293	+	+	?	+	−	−	+	4	5
C301	+	+	?	+	−	−	+	4	6.5

+：可能含有对应的白粉病成株抗性 QTL；With resistance QTL.

−：可能不含有对应的白粉病成株抗性 QTL；Without resistance QT.

?：该位点含有未知带型，with unknown band.

5 结论与展望

由于气候变化等原因，在过去 10 年，无论在美国、澳大利亚、欧洲还是中国，均出现了毒性更强、毒谱更广的新小麦条锈菌生理小种。2010 年，英国、中东和中国再次出现可以克服已有抗性基因的新条锈菌生理小种，导致一些垂直抗性基因丧失抗性，如前几年表现高抗条锈病的川麦 42 于 2010 年在成都表现为高感，验证了 McIntosh 教授几年前的预测。与此同时，2009 年河南省及北部冬麦区的一些白粉病抗性品种也变为感病品种，虽然尚未见到国内相关白粉菌新生理小种的报道，但育种家需启用新的抗源。反思国内外条锈病和白粉病抗性育种的历史经验，预计育种家与新生理小种的博弈还将继续，但更重要的是现在必须改变观念，人类与病原菌必须和平相处，实现和谐发展。

中国近几十年抗条锈育种的实践已清楚表明，育种工作总是在被动地追赶病原菌小种的变化，新品种（系）的抗性在出圃前后就开始变样，大面积推广3～5 年后抗性就明显丧失，有的品种还未审定抗性就已丧失。另外，由于所用国外抗源的农艺性状明显较差（很晚熟、成穗少、抗逆性差、红皮等），杂交后早期世代材料综合性状明显下降，通过连续几年的定向选择才能勉强恢复到本地推广品种的产量水平，因此新品种的增产幅度往往有限。若能培育和推广抗性更持久一些的成株抗性品种，育种家则可把更多的精力用于产量和品质等性状的改良，全面提高新品种的水平。我们建议两条腿走路，两类抗性同时并举，在病害的偶发区要加强成株抗性育种研究，小种专化抗性育种已有较多经验，仍要发挥其应用作用，将来转向以“持久”抗性为主的轨道，这对生产没有不良影响，而对提高育种效率、节约人力及各方面资源都十分有利；病害常发区则仍以小种专业抗性为主，但也要注意“持久抗性”的利用，对于甘肃西南部的一些地区则以不种或少种小麦作为将来的奋斗目标。

本文对国外的总结分析已清楚表明，利用成株抗性是未来实现品种兼抗和持久抗性的最佳选择。当然育种中还可继续发掘和利用垂直抗性基因，通过分子标记辅助选择可有目的地进行垂直抗性基因累加，但其前提是明确亲本中抗病基因的分布及其等位关系，抗病基因的精细定位是基础。没有与目的基因紧密连锁的分子标记很难进行基因累加，因此必须加强抗病基因精细定位这一基础性工作。除此之外，普及分子标记知识，分子标记发掘与主流育种项目的密切合作也很关键。

我们更主张利用成株抗性来实现持久抗病性，对条锈病和白粉病的抗性育种更是如此，主要原因有三：第一，成株抗性的理论和应用问题已基本解决。现有的理论研究和育种实践都表明，持久抗性或成株抗性比原来估计要简单的多，聚合 3～5 个效应相对较大的微效基因，即可培育出接近免疫的持久抗性品种，尽管还不能证实所有成株抗性基因都是持久抗性基因，但至少相当一部分成株抗性品种已保持 50 年以上抗性，足以满足育种家对持久抗性的需求，目前成株抗性的育种方法已经成熟。第二，现有成株抗性基因容易做到兼抗与持久抗性的结合。育种家的目标是多方面的，改良产量、品质和适应性比抗病性更难，有时还要兼抗几种病害，对我国大部分麦区来说，兼抗条锈病和白粉病至关重要。而本文分析表明

Yr18、*Yr29* 和 *Yr46* 等对条锈病、叶锈病和白粉病等皆表现成株抗性，利用这些基因可取得事半功倍的成效。第三，只要具备成株抗性亲本，所有育种单位均可进行品种选育。上述基因的分子标记已具备，没有太多经验的育种家可利用标记聚合这些基因；对于经验丰富的育种家，即便标记使用条件不具备，只要稍微降低一下选择标准（F_2～F_4 选择中感到中抗类型），不再盲目追求高抗或免疫，用常规方法也能培育出具成株抗性的多抗品种。

总体来说，国内成株抗性的研究与应用还处于起步阶段，为了进一步加强条锈病和白粉病成株抗性育种工作，建议加强以下三方面的工作：首先，要从国内外（尤其是国内农艺性状优良的）品种和材料中发掘更多的成株抗性基因和亲本资源。据观察，中国小麦品种和 CIMMYT 品种中的成株抗性基因较为丰富，由于工作量较大，过去对成株抗性基因的鉴定和定位较少。其次，不断改进和利用新的分子标记技术，将其广泛用于抗病育种。随着分子标记类型的增加和测序技术的改进，小麦全基因组测序也已取得很大进展，今后 3～5 年定会有更多的成株抗性基因被克隆，从而使基于 SNP 标记的基因芯片开发成为可能。由图 1 可知，同时拥有两种病害抗性的基因簇颇多，通过遗传分析显示，大多数成株抗性基因具有加性效应，即随着微效基因数目的增加，其抗性水平有所提高[127]。因此，将所有的条锈病、白粉病和叶锈病等病害的 SNP 标记集成到一个芯片上，商品化生产，将显现出基因芯片的快捷与高通量优势。实现只需提取 DNA，即可快速检测其含有或欠缺哪些基因的设想，从而提高分子育种效率。第三，有针对性地培育成株抗性品种。特别是将含有 *Yr18*、*Yr29* 和 *Yr46* 的品种及鲁麦 21 和百农 64 分别与条锈病或白粉病易发区的主栽感病品种杂交，并用主栽感病品种回交 1～2 次，在回交后代结合农艺性状选择，利用已获得的紧密连锁分子标记进行条锈病或白粉病成株抗性基因聚合，可在最短时间内获得农艺性状优良并具有持久抗性的小麦新品种。

参考文献

[1] 赵中华．2003 年全国小麦条锈病的流行特点及治理策略．中国植保导刊，2004，2：16-18.
Zhao Z H. The character and control strategy for stripe rust epidemic in China 2003. *China Plant Protection*, 2004，2：16-18.（in Chinese）

[2] 张跃进，王建强，姜玉英，夏冰．2005 年全国农作物重大病虫发生趋势预报．中国植保导刊，2005，4：28-30.
Zhang Y J，Wang J Q，Jiang Y Y，Xia B. Forecast the trend of big crop diseases in China 2005，*China Plant Protection*，2005，4：28-30.（in Chinese）

[3] 张跃进，王建强，姜玉英，冯晓东，夏冰．2006 年全国农作物重大病虫发生趋势预报．中国植保导刊，2006，4：5-8.
Zhang Y J，Wang J Q，Jiang Y Y，Feng X D，Xia B. Forecast the trend of big crop diseases in China 2006，*China Plant Protection*，2006，4：5-8.（in Chinese）

[4] 张跃进，王建强，姜玉英，冯晓东，夏冰，刘宇．2007 年全国农作物重大病虫发生趋势预报．中国植保导刊，2007，2：32-35.
Zhang Y J，Wang J Q，Jiang Y Y，Feng X D，Xia B，Liu Y. Forecast the trend of big crop diseases in China 2007，*China Plant Protection*，2007，2：32-35.（in Chinese）

[5] 张跃进，王建强，姜玉英，冯晓东，夏冰，刘宇．2008 年全国农作物重大病虫发生趋势预报．中国植保导刊，2008，3：38-40.
Zhang Y J，Wang J Q，Jiang Y Y，Feng X D，Xia B，Liu Y. Forecast the trend of big crop diseases in China 2008，*China Plant Protection*，2008，3：38-40.（in Chinese）

[6] 张跃进，姜玉英，冯晓东，夏冰，曾娟，刘宇．2009 年全国农作物重大病虫害发生趋势．中国植保导刊，2009，3：33-35.
Zhang Y J，Jiang Y Y，Feng X D，Xia B，Zeng J，Liu Y. Forecast the trend of big crop diseases in China 2009，*China Plant Protection*，2009，3：33-35.（in Chinese）

[7] 李振岐，曾士迈．中国小麦锈病．北京：中国农业出版社，2002：180-190.
Li Z Q，Zeng S M. *Stripe rust in Chinese wheat*. Beijing：China Agriculture Press，2002：180-190.（in Chinese）

[8] Line R F，Chen X M. Success in breeding for and managing durable resistance to wheat rusts. *Plant Disease*，1995，79：1254-1255.

[9] Singh R P，Huerta-Espino J，Rajaram S. Achieving near-immunity to leaf and stripe rusts in wheat by combining slow rusting resistance genes. *Acta. Phytopathologica. et Entomologica Hungarica Hungary*，2000，35：133-139.

[10] Chhuneja P, Kaur S, Garg T, Ghai M, Kaur S, Prashar M, Bains N S, Goel R K, Keller B, Dhaliwal H S, Singh K. Mapping of adult plant stripe rust resistance genes in diploid A genome wheat species and their transfer to bread wheat. *Theoretical and Applied Genetics*, 2008, 116: 313-324.

[11] Lu Y M, Lan C X, Liang S S, Zhou X C, Liu D, Zhou G, Lu L Q, Jing J X, Wang M N, Xia X C, He Z H. QTL mapping for adult-plant resistance to stripe rust in Italian common wheat cultivarsLibellula and Strampelli. *Theoretical and Applied Genetics*, 2009, 119: 1349-1359.

[12] Roberts J, Caldwell R. General resistance (slow mildewing) to *Erysiphe graminis* f. sp. *tritici* in Knox wheat. *Molecular Genetics*, 1970, 60: 1310.

[13] Qayoum A, Line R F. High-temperature, adult-plant resistance to stripe rust of wheat. *Phytopathology*, 1985, 75: 1121-1125.

[14] Gustafson G, Shaner G. Influence of plant age on the expression of slow-mildewing resistance in wheat. *Phytopathology*, 1982, 72: 746-749.

[15] Chen X M, Line R F. Inheritance of stripe rust resistance in wheat cultivars used to differentiate races of *Puccinia striiformis* in North America. *Euphytica*, 1993, 71: 107-113.

[16] Chen X M, Line R F. Gene action in wheat cultivars for durable high-temperature adult-plant resistance and interactions with race-specific, seedling resistance to stripe rust caused by *Puccinia striiformis*. *Phytopathology*, 1995, 85: 567-572.

[17] Hautea R, Coffman W, Sorrels M, Bergstrom G. Inheritance of partial resistance to powdery mildew in spring wheat. *Theoretical and Applied Genetics*, 1987, 73: 609-615.

[18] Singh R P, Huerta-Espino J, William H M. Genetics and breeding for durable resistance to leaf and stripe rusts in wheat. *Turkish Journal of Agriculture and Forestry*, 2005, 29: 1-7.

[19] Bennett F. Resistance to powdery mildew in wheat: a review of its use in agriculture and breeding programmes. *Plant Pathology*, 1984, 33: 297-300.

[20] 何中虎，夏先春，罗晶，辛志勇，孔秀英，景蕊莲，吴振录，李杏普．国际小麦育种研究趋势分析．麦类作物学报，2006, 26 (2): 154-156.

He Z H, Xia X C, Luo J, Xin Z Y, Kong X Y, Jing R L, Wu Z L, Li X P. Trend analysis of international wheat breeding. *Journal of Triticeae Crops*, 2006, 26 (2): 154-156. (in Chinese)

[21] Marasas C N, Smale M, Singh R P. The impact of agricultural maintenance research: the case of leaf rust resistance breeding in CIMMYT-related spring bread wheat//*CD-ROM Proceeding Internal Congress on Impacts of Agricultural Research and Development*. San Jose, Costa Rica, 4-7 February, 2002. (CIMMYT, Mexico)

[22] Singh R P, William H M, Huerta-Espino J, Rosewarne G. Wheat rust in Asia: Meeting the challenges with old and new technologies//*New Directions for a Diverse Planet. Proceedings of the 4th International Crop Science Congress*, 2004, 26, Brisbane, Australia.

[23] Singh R P, Rajaram S. Genetics of adult-plant resistance to leaf rust in 'Frontana' and three CIMMYT wheats. *Genome*, 1992, 35: 24-31.

[24] Singh R P, Kazi-Mujeeb A, Huerta-Espino J. *Lr46*: A gene conferring slow rusting resistance to leaf rust in wheat. *Phytopathology*, 1998, 88: 890-894.

[25] William H M, Singh R P, Huerta-Espino J, Ortiz-Islas S, Hoisington D. Molecular marker mapping of leaf rust resistance gene *Lr46* and its association with stripe rust resistance gene *Yr29* in wheat. *Phytopathology*, 2003, 93: 153-159.

[26] Bariana H S, Kailasapillai S, Brown G N, Sharp P J. Marker assisted identification of *Sr2* in the National Cereal Rust Control Program in Australia//Slinkard A E. *Proceedings of 9th International Wheat and Genetic Symposium*. University of Extension Press, University of Saskatchewan, Saskatoon, 1998, 3: 83-91.

[27] Keller M, Keller B, Schachermayr G, Winzeler M, Schmid J E, Stamp P, Messmer M M. Quantitative trait loci for resistance against powdery mildew in a segregating wheat×spelt population. *Theoretical and Applied Genetics*, 1999, 98: 903-912.

[28] Boukhatem N, Baret P V, Mingeot D, Jacquemin J M. Quantitative trait loci for resistance against yellow rust in two wheat-derived recombinant inbred line populations. *Theoretical and Applied Genetics*, 2002, 104: 111-118.

[29] Bjarko M E, Line R F. Heritability and number of genes controlling leaf rust resistance in four cultivars of wheat. *Phytopathology*, 1988, 78: 457-461.

[30] Donini P, Koebner R M D, Ceoloni C. Cytogenetic and molecular mapping of the wheat-*Aegilops longis-*

sima chromatin breakpoints in powdery mildew resistant introgression lines. *Theoretical and Applied Genetics*, 1995, 91: 738-743.

[31] Wang Z L, Li L H, He Z H, Duan X Y, Zhou Y L, Chen X M, Lillemo M, Singh R P, Wang H, Xia X C. Seeding and adult-plant resistance to powdery mildew in Chinese bread wheat cultivars and lines. *Plant Disease*, 2005, 89: 457-463.

[32] Liang S S, Suenaga K, He Z H, Wang Z L, Liu H Y, Wang D S, Singh R P, Sourdile P, Xia X C. Quantitative trait loci mapping for adult-plant resistance to powdery mildew in bread wheat. *Phytopathology*, 2006, 96: 784-789.

[33] Lan C X, Liang S S, Wang Z L, Yan J, Zhang Y, Xia X C, He Z H. Quantitative trait loci mapping for adult-plant resistance against powdery mildew in Chinese wheat cultivar Bainong 64. *Phytopathology*, 2009, 99: 1121-1126.

[34] McIntosh R A, Dubcovsky J, Rogers W J, Morris C F, Appels R, XiaX C. Catalogue of gene symbols for wheat: 2010 (suppl). 2010, http: //www. wheat. pw. usda. gov/GG2/pubs. shtml.

[35] Liu S X, Griffey C A, Saghai-Maroof M A. Identification of molecular markers associated with adult plant resistance to powdery mildew in common wheat cultivar Massey. *Crop Science*, 2001, 41: 1268-1275.

[36] Lin F, Chen X M. Genetics and molecular mapping of genes for race specific and all-stage resistance and non-specific high temperature adult-plant resistance to stripe rust in spring wheat cultivar Alpowa. *Theoretical and Applied Genetics*, 2007, 114: 1277-1287.

[37] Herrera-Foessel S A, Lagudah E S, Huerta-Espino J, Hayden M, Bariana H S, Singh R P. New slow-rusting leaf rust and stripe rust resistance gene *Lr67* and *Yr46* in wheat are pleiotropic or closely linked. *Theoretical and Applied Genetics*, 2010, Doi 10.1007/s00122-010-1439-x.

[38] Lillemo M, Asalf B, Singh R P, Huerta-Espino J, Chen X M, He Z H, Bjørnstad Å. The adult plant rust resistance loci *Lr34/Yr18 and Lr46/Yr29* are important determinants of partial resistance to powdery mildew in bread wheat line Saar. *Theoretical and Applied Genetics*, 2008, 116: 1155-1166.

[39] McDonald B A, Linde C. The population genetics of plant pathogens and breeding strategies for durable resistance. *Euphytica*, 2002, 124: 163-180.

[40] Skinnes H. Breakdown of race specific resistance to powdery mildew in Norwegian wheat. *Cereal Rusts and Powdery Mildew Bulletin* vol. 30 online, publication [http: //www. crpmb. org] 2002/1201skinnes. 2002.

[41] Young N D. QTL mapping and quantitative disease resistance in plants. *Annual Review Phytopathology*, 1996, 34: 479-501.

[42] Suenaga K, Singh R P, Huerta-Espino J, William H M. Microsatellite markers for genes *Lr34/Yr18* and other quantitative trait loci for leaf rust and stripe rust resistance in bread wheat. *Phytopathology*, 2003, 93: 881-890.

[43] Lin F, Chen X M. Quantitative trait loci for non-race-specific, high-temperature adult-plant resistance to stripe rust in wheat cultivar express. *Theoretical and Applied Genetics*, 2009, 118: 631-642.

[44] Rosewarne G M, Singh R P, Huerta-Espino J, Rebetzke G J. Quantitative trait loci for slow-rusting resistance in wheat to leaf rust and stripe rust identified with multi-environment analysis. *Theoretical and Applied Genetics*, 2008, 116: 1027-1034.

[45] Melichar J P E, Berry S, Newell C, MacCormack R, Boyd L A. QTL identification and microphenotype characterization of the developmentally regulated yellow rust resistance in the UK wheat cultivar Guardian. *Theoretical and Applied Genetics*, 2008, 117: 391-399.

[46] William H M, Singh R P, Huerta-Espino J, Palacios G, Suenaga K. Characterization of genetic loci conferring adult plant resistance to leaf rust and stripe rust in spring wheat. *Genome*, 2006, 49: 977-990.

[47] Christiansen M J, Feenstra B, Skovgaard I M, Andersen S B. Genetic analysis of resistance to yellow rust in hexaploid wheat using a mixture model for multiple crosses. *Theoretical and Applied Genetics*, 2006, 112: 581-591.

[48] Mallard S, Gaudet D, Aldeia A, Abelard C, Besnard A L, Sourdille P, Dedryver F. Genetic analysis of durable resistance to yellow rust in bread wheat. *Theoretical and Applied Genetics*, 2005, 110: 1401-1409.

[49] Ramburan V P, Pretorius Z A, Louw J H, Boyd L A, Smith P H, Boshoff W H P, Prins R. A genetic analysis of adult plant resistance to stripe rust in the wheat cultivar Kariega. *Theoretical and Applied Genetics*, 2004, 108: 1426-1433.

[50] Guo Q, Zhang Z J, Xu Y B, Li G H, Feng J, Zhou Y. Quantitative trait loci for high-temperature adult-

plant and slow-rusting resistance to*Puccinia striiformis f.* sp. *tritici* in wheat cultivars. *Phytopathology*, 2008, 98: 803-809.

[51] Lan C X, Liang S S, Zhou X C, Liu D, Zhou G, Lu Q L, Xia X C, He Z H. Identification of genomic regions controlling adult plant stripe rust resistance to in Chinese wheat landrace Pingyuan 50 through bulked segregant analysis. *Phytopathology*, 2010, 100: 313-318.

[52] Börner A, Röder M S, Unger O, Meinel A. The detection and molecular mapping of a major gene for non-specific adult-plant disease resistance against stripe rust (*Puccinia striiformis*) in wheat. *Theoretical and Applied Genetics*, 2000, 100: 1095-1099.

[53] Khlestkina E K, Röder M S, Unger O, Meinel A, Börner A. More precise map position and origin of a durable non-specific adult plant disease resistance against stripe rust (*Puccinia striiformis*) in wheat. *Euphytica*, 2007, 153: 1-10.

[54] Singh R P, Nelson J C, Sorrells M E. Mapping *Yr28* and other genes for resistance to stripe rust in wheat. *Crop Science*, 2000, 40: 1148-1155.

[55] Imtiaz M, Ahmad M, Cromey M G, Griffin W B, Hampton J G. Detection of molecular markers linked to the durable adult plant stripe rust resistance gene *Yr18* in bread wheat (*Triticum aestivum* L.). *Plant Breeding*, 2004, 123: 401-404.

[56] Uauy C, Brevis J C, Chen X M, Khan I, Jackson L, Chicaiza O, Distelfeld A, Fahima T, Dubcovsky J. High-temperature adult-plant (HTAP) stripe rust resistance gene *Yr36* from *Triticum turgidum* ssp. *diccoccoides* is closely linked to the grain protein content locus *Gpc-B1*. *Theoretical and Applied Genetics*, 2005, 112: 97-105.

[57] Santra D K, Chen X M, Santra M, Campbell K G, Kidwell K K. Identification and mapping QTL for high-temperature adult-plant resistance to stripe rust in winter wheat (*Triticum aestivum* L.) cultivar Stephens. *Theoretical and Applied Genetics*, 2008, 117: 793-802.

[58] Navabi A, Tewari J P, Singh R P, McCallum B, Laroche A, Briggs K G. Inheritance and QTL analysis of durable resistance to stripe and leaf rusts in an Australian cultivar. *Triticum aestivum* Cook. *Genome*, 2005, 48: 97-107.

[59] Chen Y H, Hunger R M, Carver B , Zhang H L, Yan L L. Genetic characterization of powdery mildew resistance in U. S. hard winter wheat. *Molecular Breeding*, 2009, 24: 141-152.

[60] Mingeot D, Chantret N, Baret P V, Dekeyser A, Boukhatem N, Sourdille P, Doussinault G, Jacquemin J M. Mapping QTL involved in adult plant resistance to powdery mildew in the winter wheat line RE714 in two susceptible genetic backgrounds. *Plant Breeding*, 2002, 121: 133-140.

[61] Jakobson I, Peusha H, Timofejeva L, Jarve K. Adult plant and seedling resistance to powdery mildew in a *Triticum aestivum* × *Triticum militinae* hybrid line. *Theoretical and Applied Genetics*, 2006, 112: 760-769.

[62] Tucker D M, Griffey C A, Liu S, Brown-Guedira G, Marshall D S, Saghai Maroof M A. Confirmation of three quantitative trait loci conferring adult plant resistance to powdery mildew in two winter wheat populations. *Euphytica*, 2007, 155: 1-13.

[63] Bougot Y, Lemoine J, Pavoine M T, Guyomarc'h H, Gautier V, Muranty H, Barloy D. A major QTL effect controlling resistance to powdery mildew in winter wheat at the adult plant stage. *Plant Breeding*, 2006, 125: 550-556.

[64] Lan C X, Ni X W, Yan J, Zhang Y, Xia X C, Chen X M, He Z H. Quantitative trait loci mapping of adult-plant resistance to powdery mildew in Chinese wheat cultivar Lumai 21. *Plant Breeding*, 2010, 25: 615-622.

[65] Börner A, Schumann E, Fürste A, Cöster H, Leithold B, Röder M S, Weber W E. Mapping of quantitative trait loci determining agronomic important characters in hexaploid wheat (*Triticum aestivum* L.). *Theoretical and Applied Genetics*, 2002, 105: 921-936.

[66] Chantret N, Mingeot D, Sourdille P, Bernard M, Jacquemin J M, Doussinault G. A major QTL for powdery mildew resistance is stable over time and at two development stages in winter wheat. *Theoretical and Applied Genetics*, 2001, 103: 962-971.

[67] Yahiaoui N, Srichumpa P, Dudler R, Keller B, Genome analysis at different ploidy levels allows cloning of the powdery mildew resistance gene *Pm3b* from hexaploid wheat. *The Plant Journal*, 2004, 37: 528-538.

[68] Islam M R, Shepherd K W. Present status of genetics of rust resistance in flax. *Euphytica*, 1991, 55: 255-267.

[69] Ellis J G, Lawrence G J, Finnegan E J, Anderson P

A. Contrasting complexity of two rust resistance loci in flax. *Proceedings of the National Academy of Sciences, USA*, 1995, 92: 4185-4188.

[70] Yang Z, Sun X, Wang S, Zhang Q. Genetic and physical mapping of a new gene for bacterial blight resistance in rice. *Theoretical and Applied Genetics*, 2003, 106: 1467-1472.

[71] Somers D J, Isaac P, Edwards K. A high-density microsatellite consensus map for bread wheat (*Triticum aestivum* L.). *Theoretical and Applied Genetics*, 2004, 109: 1105-1114.

[72] Krattinger S G, Lagudah E S, Spielmeyer W, Singh R P, Huerta-Espino J, McFadden H, Bossolini E, Selter L L, Keller B. A putative ABC transporter confers durable resistance to multiple fungal pathogens in wheat. *Science*, 2009, 323: 1360-1363.

[73] Rosowarne G M, Singh R P, Huerta-Espino J, William H M, Bouchet S, Cloutier S, McFadden H, Lagudah E S. Leaf tip necrosis, molecular markers andβ-proteasome subunits associated with the slow rusting resistance gene *Lr46/Yr29*. *Theoretical and Applied Genetics*, 2006, 112: 500-508.

[74] Singh R P. Genetic association of leaf rust resistance gene *Lr34* with adult plant resistance to stripe rust in bread wheat. *Phytopathology*, 1992, 82: 835-838.

[75] Dyck P L, Kerber E R, Aung T. An interchromosomal reciprocal translocation in wheat involving leaf rust resistance gene *Lr34*. *Genome*, 1994, 37: 556-559.

[76] McDonald D B, McIntosh R A, Wellings C R, Sing R P, Nelson J C. Cytogenetical studies in wheat XIX. Location and linkage studies on gene *Yr27* for resistance to stripe rust. *Euphytica*, 2004, 136: 239-248.

[77] Singh R P, William H M, Huerta-Espino J, Crosby M. Identification and mapping of gene *Yr31* for resistance to stripe rust in *Triticum aestivum* cultivar Pastor. *Proceedings 10th International Wheat Genetics Symposium*, Instituto Sperimentale per la Cerealcoltura, Rome, Italy, 2003, 1: 411-413.

[78] Luo P G, Hu X Y, Ren Z L, Zhang H Y, Shu K, Yang Z J. Allelic analysis of stripe rust resistance genes on wheat chromosome 2BS. *Genome*, 2008, 51: 922-927.

[79] Tao W, Liu D, Liu J, Feng Y, Chen P. Genetic mapping of the powdery mildew resistance gene *Pm6* in wheat by RFLP analysis. *Theoretical and Applied Genetics*, 2000, 100: 564-568.

[80] Zhu Z D, Zhou R H, Kong X Y, Dong Y C, Jia J Z. Microsatellite markers linked to 2 powdery mildew resistance genes introgressed from *Triticum carthlicum* accession *PS5* into common wheat. *Genome*, 2005, 48: 585-590.

[81] McIntosh R A, Luig N H, Johnson R, Hare R A. Cytogenetical studies in wheat. XI. *Sr9g* for reaction to *Puccinia graminis tritici*. *Zeitschrift für Pflanzenzuchtung*, 1981, 87: 274-289.

[82] Li Z F, Xia X C, Zhou X C, Niu Y C, He Z H, Zhang Y, Li G Q, Wan A M, Wang D S, Chen X M, Lu Q L, Singh R P. Seedling and slow rusting resistance to stripe rust in Chinese common wheats. *Plant Disease*, 2006, 90: 1302-1312.

[83] Spielmeyer W, Sharp P J, Lagudah E S. Identification and validation of markers linked to broad-spectrum stem rust resistance gene *Sr2* in wheat (*Triticum aestivum* L.). *Crop Science*, 2003, 43: 333-336.

[84] Nelson J C, Singh R P, Autrique J E, Sorrells M E. Mapping genes conferring and suspecting leaf rust resistance in wheat. *Crop Science*, 1997, 37: 1928-1935.

[85] Zhou W C, Kolb F L, Bai G H, Domier L L, Boze L K, Smith N J. Validation of a major QTL for scab resistance with SSR markers and use of marker assisted selection in wheat. *Plant Breeding*, 2003, 122: 40-46.

[86] Crossa J, Burgueño J, Dreisigacker S, Vargas M, Herrera-Foessel SA, Lillemo M, Singh RP, Trethowan R, Warburton M, Franco J, Reynolds M, Crouch JH, Ortiz R. Association analysis of historical bread wheat germplasm using additive genetic covariance of relatives and population structure. *Genetics*, 2007, 177: 1889-1913.

[87] Chen X M, Luo Y H, Xia X C, Xia L Q, Chen X, Ren Z L, He Z H, Jia J Z. Chromosomal location of powdery mildew resistance gene *Pm16* in wheat using SSR marker analysis. *Plant Breeding*, 2005, 124: 225-228.

[88] Fu D L, Uauy C, Distelfeld A, Blechl A, Epstein L, Chen X M, Sela H, Fahima T, Dubcovsky J. A kinase-start gene confers temperature-dependent resistance to wheat stripe rust. *Science*, 2009, 323: 1357-1360.

[89] Marais G. F, Pretorius Z A, Wellings C R, McCallum B, Marais A F. Leaf and stripe rust resistance genes transferred to common wheat from *Triticum dicoccoides*. *Euphytica*, 2005, 143: 115-123.

[90] Dyck PL. The inheritance of leaf rust resistance in the

wheat cultivar Pasqua. *Canada Journal Plant Science*, 1993, 73: 903-906.

[91] Friebe B, Jiang J, Raupp W J, McIntosh R A, Gill B S. Characterization of wheat-alien translocations conferring resistance to disease and pests: current status. *Euphytica*, 1996, 91: 59-87.

[92] Kirigwi F M M, Van Ginkel G, Brown-Gedira B S, Gill G M, Paulsen. Markers associated with a QTL for grain yield in wheat under drought. *Molecula Breeding*, 2007, 20: 401-413.

[93] Kumar N, Kulwal P L, Balyan H S, Gupta P K. QTL mapping for yield and yield contribution traits in two mapping populations of bread wheat. *Molecular Breeding*, 2007, 19: 163-177.

[94] Ling H Q, Qiu J W, Singh R P, Keller B. Identification and characterization of an *Aegilops tauschii* ortholog of the wheat leaf rust disease resistance gene *Lr1*. *Theoretical and Applied Genetics*, 2004, 109: 1230-1236.

[95] Bonnin I, Rousset M, Madur D, Sourdille P, Dupuits C, Brunel D, Goldringer I. FT genome A and D polymorphisms are associated with the variation of earliness components in hexaploid wheat. *Theoretical and Applied Genetics*, 2008, 116: 383-394.

[96] Lagudah E S, McFadden H, Singh R P, Huerta-Espino J, Bariana H S, Spielmeyer W. Molecular genetic characterization of the *Lr34/Yr18* slow rusting resistance gene region in wheat. *Theoretical and Applied Genetics*, 2006, 114: 21-30.

[97] Spielmeyer W, Singh R P, McFadden H, Wellings C R, Huerta-Espino J, Kong X, Appels R, Lagudah E S. Fine scale genetic and physical mapping using interstitial deletion mutants of *Lr34/Yr18*: A disease resistance locus effective against multiple pathogens in wheat. *Theoretical and Applied Genetics*, 2008, 116: 481-490.

[98] Feuillet C, Travella S, Stein N, Albar L, Nublat A, Keller B. Map-based isolation of the leaf rust disease resistance gene *Lr10* from the hexaploid wheat (*Triticum aestivum* L.) genome. *Proceedings of the National Academy of Sciences*, 2003, 100: 15253-15258.

[99] Li H, Brooks S A, Li W L, Fellers J P, Trick H N, Gill B S. Map-based cloning of leaf rust resistance gene *Lr21* from the large and polyploid genome of bread wheat. *Genetics*, 2003, 164: 655-664.

[100] Yahiaoui N, Srichumpa P, Dudler R, Keller B. Genome analysis at different ploidy levels allows cloning of the powdery mildew resistance gene *Pm3b* from hexaploid wheat. *The Plant Journal*, 2004, 37: 528-538.

[101] Lagudah E S, Krattinger S G, Herrera-Foessel S, Singh R P, Huerta-Espino J, Spielmeyer W, Brown-Guedira G, Selter L L, Keller B. Gene-specific markers for the wheat gene *Lr34/Yr18/Pm38* which confers resistance to multiple fungal pathogens. *Theoretical and Applied Genetics*, 2009, 119: 889-898.

[102] Kolmer J A, Singh R P, Garvin D F, Viccars L, William H M, Huerta-Espino J, Ogbonnaya F C, Raman H, Orford S, Bariana H S, Lagudah E S. Analysis of the *Lr34/Yr18* rust resistance region in wheat germplasm. *Crop Science*, 2008, 48: 1841-1852.

[103] 杨文雄，杨芳萍，梁丹，何中虎，尚勋武，夏先春．中国小麦育成品种和农家种中慢锈基因 *Lr34/Yr18* 的分析检测．作物学报，2008，34：1109-1113.
Yang W X, Yang F P, Liang D, He Z H, Shang X W, Xia X C. Molecular characterization of slow-rusting genes *Lr34/Yr18* in Chinese wheat cultivars. *Acta Agronomica Sinica*, 2008, 34: 1109-1113. (in Chinese)

[104] Marasas C N, Smale M, Singh R P. The economic impact of productivity maintenance research: Breeding for leaf rust resistance in modern wheat. *Agricultural Economics*, Blackwell, 2003, 29: 253-263.

[105] Cao S H, Carver B F, Zhu X K, Fang T L, Chen Y H, Hunger R M, Yan L L. A single-nucleotide polymorphism that accouts for allelic variation in the *Lr34* gene and leaf rust reaction in hard winter wheat. *Theoretical and Applied Genetics*, 2010, 121: 385-392.

[106] Alpy F, Tomasetto C. Give lipids a start: the StAR-related lipid transfer (START) domain in mammals. *Journal Cell Science*, 2005, 118: 2791-2801.

[107] Hammond-Kosack K E, Jones J D G. Plant disease resistance genes. *Annual Review of Plant Physiology and Plant Molecular Biology*, 1997, 48: 575-607.

[108] Friedman A R, Baker B J. The evolution of resistance genes in multi-protein plant resistance systems. *Current Opinion in Genetics and Development*, 2007, 17: 493-499.

[109] 曾士迈．关于植物的水平抗病性．中国学术期刊电子出版社，1994，1-6.
Zeng S M. The horizontal resistance to diseases in

crop. *China Academic Journal Electronic Publishing House*, 1994, 1-6 (in Chinese)

[110] Shaner G.. Reduced infectability and inoculum production as factors of slow-mildewing in Knox wheat. *Phytopathology*, 1973, 69: 1307-1311.

[111] 曾士迈，王沛有，武修英，张万，王吉庆，宋位中．小麦对条锈病的水平抗病性研究初报．植物保护学报，1979，6 (1)：1-10.
Zeng S M, Wang F Y, Wu X Y, Zhang W Q, Wang J Q, Song W Z, Wang S Y. Analysis of horizontal resistance against stripe rust in wheat. *Plant Protection*, 1979, 6 (1): 1-10. (in Chinese)

[112] 汪可宁，谢水仙，刘孝坤，吴立人，王剑雄，陈杨林．我国小麦条锈病防治研究的进展．中国农业科学，1988，21 (2)：1-8.
Wang K N, Xie S X, Liu X K, Wu L R, Wang J X, Cheng Y L. Progress in studies on control of wheat stripe rust in China. *Scientia Agricultura Sinica*, 1988, 21 (2): 1-8. (in Chinese)

[113] 周祥椿，杜久元．陇南小麦生产品种抗条锈病持久性研究．麦类作物学报，2006，26 (1)：108-112.
Zhou X C, Du J Y. Study on resistance durability of commercial wheat cultivars grown in southern region of Gansu province to stripe rust. *Journal of Triticeae Crops*, 2006, 26 (1): 108-112. (in Chinese)

[114] 王锡锋，何文兰，何家泌．小麦品种的慢白粉性田间鉴定．植物保护学报，1991，18 (3)：230.
Wang X F, He W L, He J M. The identification of slow-mildewing resistance in common wheat. *Acta Phytophylacica Sinica*, 1991, 18 (3): 230. (in Chinese)

[115] 张志德，李振岐，刘卿．四个小麦品种的慢白粉病抗性研究．植物病理学报，1994，21 (3)：197-201.
Zhang Z D, Li Z Q, Liu Q. Slow- mildewing resistance in four wheat cultivars. *Acta Phytopatho _ Ogica Sinica*, 1994, 21 (3): 197-201. (in Chinese)

[116] 李立华．北方冬麦区小麦慢白粉病品种的筛选与鉴定．北京：中国农业科学院，2001：10.
Li L H. Identification of slow-mildewing varieties in common wheat from north China. Beijing: Chinese Academy of Agricultural Sciences, 2001: 10. (in Chinese)

[117] Wang Z L, Li L H, He Z H, Duan X Y, Zhou Y L, Chen X M, Lillemo M, Singh R P, Wang H, Xia X C. Seeding and adult-plant resistance to powdery mildew in Chinese bread wheat cultivars and lines. *Plant Disease*, 2005, 89: 457-463.

[118] 张坤普，赵亮，海燕，陈广凤，田纪春．小麦白粉病成株抗性和抗倒伏性及穗下节长度的 QTL 定位．作物学报，2008，34 (8)：1350-1357.
Zhang K P, Zhao L, Hai Y, Chen G F, Tian J C. QTL mapping for adult-plant resistance to powdery mildew, lodging resistance and internode length below spike in wheat. *Acta Agronomica Sinica*, 2008, 34 (8): 1350-1357. (in Chinese)

[119] 倪小文，阎俊，陈新民，夏先春，何中虎，张勇，王德森，Lillemo M. 鲁麦 21 慢白粉病抗性基因数目和遗传力分析．作物学报，2008，34 (8)：1317-1322.
Ni X W, Yan J, Chen X M, Xia X C, He Z H, Zhang Y, Wang D S, Lillemo M. Heritability and number of genes controlling slow-mildewing resistance in wheat cultivar Lumai 21. *Acta Agronomica Sinica*, 2008, 34: 1317-1322. (in Chinese)

[120] 李万隆，李振声，穆素梅．小偃 6 号小麦旗叶直立基因的染色体定位．遗传学报，1992，19 (1)：71-73.
Li W L, Li Z S, Mu S M. Chromoaomal location of genes for erect flag leavea of commom wheat variety Xiaoyan No. 6. *Acta Genetica Sinica*, 1992 (1): 71-73. (in Chinese)

[121] 吴政卿，雷振生，杨会民，章家长，刘媛媛．豫麦 47 号示范推广及其产业化开发．河南农业科学，2001，6：9-10.
Wu Z Q, Lei Z S, Yang H M, Zhang J Z, Liu Y Y. Demonstration release and industrial development of Yumai 47. *Journal of Henan Agricultural Sciences*, 2001, 6: 9-10. (in Chinese)

[122] 王竹林，刘曙东，王辉，何中虎，夏先春，陈新民，段霞瑜，周益林．小麦慢病性的遗传育种研究进展．麦类作物学报，2006，26 (1)：129-134.
Wang Z L, Liu S D, Wang H, He Z H, Xia X C, Chen X M, Duan X Y, Zhou Y L. Advances of study on adult-plant resistance in bread wheat. *Journal of Triticeae Crops*, 2006, 26 (1): 129-134. (In Chinese)

[123] 殷学贵，张莹花，阎秋洁，尚勋武．小麦持久抗条锈品种斯汤佩利的遗传机制研究．植物遗传资源学报，2005，6 (4)：390-393.
Yin X G, Zhang Y H, Yan Q J, Shang X W. Resistant characteristics to stripe rust and genetic analysis of durable resistance on wheat cultivar N. strampelli. *Journal of Plant Genetics Resources*, 2005, 6 (4): 390-393. (In Chinese)

[124] 殷学贵，尚勋武，宋建华，张莹花，阎秋洁．小麦品

种里勃留拉的持久抗条锈病遗传机制．麦类作物学报，2006，26（2）：147-150.
Yin X G.，Shang X W，Song J R，Zhang Y H，Yan Q J. Genetic mechanism of durable resistance to stripe rust of wheat cultivar Libellula. *Journal of Triticeae Crops*，2006，26（2）：147-150.（In Chinese）
[125] 王竹林，刘曙东，王辉，何中虎，夏先春，陈新民．百农 64 慢白粉性的遗传分析．西北植物学报，2006，26：0332-0336.
Wang Z L，Liu S D，Wang H，He Z H，Xia X C，Chen X M，Duan X Y. Genetic analysis of the resistance of the wheat cultivar Bainong 64 to powdery mildew. *Acta Botanica*. *Boreali*. -*Occidentalia*. *Sinica*，2006，26：0332-0336.（In Chinese）
[126] Liu D，Xia X C，He Z H，Xu S X. A novel homeobox-like gene associated with reaction to stripe rust and powdery mildew in common wheat. *Phytopathology*，2008，98：1291-1296.
[127] Singh R P，Rajaram S. Genetics of adult plant resistance to stripe rust in ten spring bread wheats. *Euphytica*，1994，72：1-7.

Quantitative trait loci of stripe rust resistance in wheat

G. M. Rosewarne[1,2], S. A. Herrera-Foessel[2], R. P. Singh[2], J. Huerta-Espino[3], C. X. Lan[2], and Z. H. He[2,4]

[1] *Crop Research institute, Key Laboratory of Biology and Genetic Breeding in Wheat* (*Southwest*), *Sichuan Academy of Agricultural Science*, #4 *Shizishan Rd*, *Jinjiang*, *Chengdu*, *Sichuan Province*, 610066, *P.R. of China*, [2] *International Maize and Wheat Improvement Centre*, (*CIMMYT*) *Apdo. Postal* 6-6-41, 06600 *Mexico, D. F.*, *Mexico*, [3] *Campo Experimental Valle de Mexico-INIFAP*, *Apartado Postal* 10, 56230 *Chapingo*, *Edo. de Mexico*, *Mexico*, [4] *Crop Science Institute*, *Chinese Academy of Agricultural Sciences*. 12 *Zhongguancun South St*, *Beijing*, 100081, *China*.

Abstract: Over thirty publications during the last ten years have identified more than 140 QTLs for stripe rust resistance in wheat. It is likely that many of these QTLs are identical genes that have been spread through plant breeding into diverse backgrounds through phenotypic selection under stripe rust epidemics. Allelism testing can be used to differentiate genes in similar locations but in different genetic backgrounds, however this is problematic for QTL studies where multiple loci segregate from any one parent. This review utilizes consensus maps to illustrate important genomic regions that have had effects against stripe rust in wheat, and although this methodology cannot distinguish alleles from closely linked genes, it does highlight the extent of genetic diversity for this trait and identifies the most valuable loci and the parents possessing them for utilization in breeding programs. With the advent of cheaper, high throughput genotyping technologies, it is envisioned that there will be many more publications in the near future describing ever more QTLs. This review sets the scene for the coming influx of data and will quickly enable researchers to identify new loci in their given populations.

Abbreviations: Amplified Fragment Length Polymorphism, AFLP; Area Under Disease Progress Curve, AUDPC; Cleaved Amplified Polymorphic Sequences, CAPS; Centimorgans, cM; Diversity Array Technology, DArT; Expressed Sequence Tag, EST; High Temperature Adult Plant, HTAP; Infection Type, IT; Logarithm (base 10) of Odds, LOD; Pleiotropic Adult Plant Resistance, PAPR; Polymerase Chain Reaction, PCR; Phenotypic Explained Variance, PEV; Quantitative Trait Locus, QTL; Quantitative Trait Loci, QTLs; Restriction Fragment Length Polymorphism, RFLP; Resistance Gene Analogue Polymorphism, RGAP; Random Amplified Polymorphic DNA, RAPD; Single Nucleotide Polymorphism, SNP; Sequence-Tagged Site, STS

Author Contribution Statement: GMR wrote the manuscript, developed maps and summary tables; SAH was consulted on accuracy of information and edited manuscript; RPS was consulted on accuracy of information and edited manuscript; JH was consulted on accuracy of information and edited manuscript; CXL was consulted on accuracy of information and edited manuscript; ZHH was consulted on accuracy of information and edited manuscript.

Published in TAG, 2013, 126: 2427-2449

Key Message: Over 140 QTLs for resistance to stripe rust have been published and through mapping flanking markers on consensus maps, 49 chromosomal regions are identified.

Stripe Rust Resistance in Wheat

Stripe (or yellow) rust, caused by *Puccinia striiformis* Westend. f. sp. *tritici* Erikss., is an important biotic constraint of wheat production globally, with regular epidemics occurring in almost all areas where wheat is grown. Control of this disease can be achieved through the timely use of fungicides although this can be expensive to resource poor farmers, and ineffective if not completed in a timely fashion. Furthermore, unprotected neighboring crops that are heavily infected act as reservoirs of inoculums, continually bombarding fungicide treated crops that can ultimately succumb to the disease. Genetic resistance is a more effective way to control the disease as once the resistant variety is sown; no further effort is required by the farmer in relation to disease control.

Resistance genesto fungal diseases in plants can be broadly categorized into two main classes, namely, major and minor resistance genes. Major genes were used by Flor (1956) to describe the complementary gene-for-gene interaction between flax and flax rust. This work highlights the race-specificity of these major genes and implies the non-durable nature of such resistance mechanisms. Generally these genes are involved in a host response to the invading pathogen very early in the infection process and elicit hypersensitivity in which plant host cells that are in close proximity to the invading fungus, or are being attacked by the fungus, undergo programmed cell death. This stops the fungus from establishing feeding structures within the plant, ultimately leading to fungal death. Effective major resistance genes eliminate, or significantly reduce, the ability of the fungal pathogen to reproduce, placing a strong selection pressure in the fungus to evolve and overcome the resistance gene. History shows that when a single major gene protects a large area of a wheat growing region, the fungus can overcome this resistance in a relatively short period of time. Some advanced lines even lose their resistance before or just after release. These types of genes have also been termed seedling genes as they are effective in both seedlings and adult plants, however a more accurate description of these is all-stage resistance (Lin and Chen, 2007).

Minor resistance genes generally have a different mode of action and do not provide the immunity, or high level of resistance, that a single major gene does. These genes have variously been called horizontal, partial, non-race specific, slow-rusting, durable or adult plant resistances (Caldwell, 1968; Johnson, 1988; Parlevliet, 1975; Van der Plank, 1963). The mechanisms by which fungal disease is inhibited by minor resistances include an increase in the latency period, reduced uredinia size, reduced infection frequency and reduced spore production (Caldwell, 1968; Ohm and Shaner, 1976; Parlevliet, 1975). The additive nature of these genes has long been known with transgressive segregation of resistance in progeny of certain crosses being observed by Farrer (1898). More recently, Singh et al. (2000a) developed wheat lines with near-immunity to stripe rust based on four to five minor resistance loci.

With the advent of modern molecular mapping techniques, our understanding of the location and numbers of adult plant resistance loci has steadily increased. Quantitative trait mappingin wheat was first applied to stripe rust at the turn of the 21st century (Börner et al., 2000; Singh et al., 2000b) and since then there has been over 30 publications describing QTL mapping for stripe rust resistance (Table 1). As each paper generally describes multiple QTLs, there have been over 140 loci described. There is a large amount of redundancy in these loci as many of the more useful ones are common amongst the different studies.

Table 1 Summary of stripe rust regions on group 1 chromosomes associated with stripe rust resistance QTLs. Markers in parentheses were not described in the original publication but were used as closely associated substitutes on consensus maps where the original markers were not available. References (Ref) are listed in Supplementary Table 1. Frequency (Freq) describes to number of environments that the locus had a significant QTL, out of the total number of environments tested.

Chromosome Region	Source	Markers	Field			Infection Type		Freq	Ref
			LOD	PEV	Freq	LOD	PEV		
QRYr1A. 1	Janz	Xgwm164	3. 0-3. 3	6. 5-7. 0	4/6				3
QRYr1A. 1	Renan	Xfba118b	6. 5	9. 4	1/2				8
QRYr1A. 1	Pastor	wPt-6005 (Xgwm497)	4. 9	3. 6-4. 1	2/4				26
QRYr1A. 2	Naxos	Xwmc59 Xbarc213	3. 8	8. 2	1/3				23
QRYr1B. 1	Pastor	wPt-6240 (Xgwm11, Xgwm273)	6. 6	5. 0-5. 1	2/3				26
QRYr1B. 2	Express	Xwmc631 Xgwm268	3. 1-4. 1	4. 5-5. 2	2/3			0/3F	17
QRYr1B. 3	Kukri	Xbarc80	3. 1	6	2/6				3
QRYr1B. 3	Brigadier	Xwmc735	3. 3-4. 5	7. 9-13. 1	2/2	3. 2	9. 9	1/2F	13
QRYr1B. 3	Guardian	Xgwm259 Xgwm818	5. 7-11. 7	15-45	2/2	3. 8-6. 5	10-22	2/2F	20
QRYr1B. 3	Pavon 76	Xgwm259	11-17	48-56	3/3	6. 3	18. 5	1/1F	32
QRYr1B. 3	CPI133872	wPt-1313 (Xwmc44, Xgwm259, Xgwm140)	2. 8-4. 8	10-17	3/4				34
QRYr1B. 3	CD87	Xpsr305	14. 3	9	2/2				2
QRYr1B. 3	Saar	Xwmc719 Xhbe248	3. 8	17. 4	1/1				15
QRYr1B. 3	Attila	LTN	13-24	33-64	3/3				25
QRYr1B. 3	Pastor	XcsLV46	23	15. 1-22. 1	3/3				26
QRYr1D. 1	CPI133872	Xwmc147	3	13	1/4				34
QRYr1D. 1	Stephens	379227 Xbcd1434	3. 3	11	1/2				31
QRYr1D. 2	Naxos	Xwmc432	3. 9	5. 8	1/3				23

F. Infection type scored in field on adult plants

Marker Technologies

Concurrent to the detailed rust QTL work, marker systems have also developed rapidly, with the earlier maps described with RFLP markers. This evolved to include the much easier to apply SSR markers with their genetic location being well documented through many different mapping studies. Other PCR based marker systems have been implemented and include RGAP, CAPS, EST, STS, AFLP and RAPD markers. More recently, multiplex platforms of SNP, DArT and whole genome sequencing markers make the production of detailed maps commonplace. A major issue with all of these marker systems is that they are not yet well integrated. SSR markers, being the best categorized, are often used to link maps that contain different marker types and build consensus between different maps. Although this has limitations, mainly around the micro-order of specific markers between maps, it does help to locate useful traits within the genome as well as describing chromosomal reordering through translocations.

Consensus Mapping

The incorporation of quantitative resistances to facilitate near-immunity is seen as providing durable resistance and has been the breeding strategy in the CIMMYT wheat breeding program since the 1960s. Breeding for durability is becoming a priority in many breeding program globally as they move away from major gene resistances. It is therefore critical to gain a better understanding of the location and breeding values of these loci. In this review, all known stripe rust QTLs are located on consensus maps by using the published information on flanking markers in an effort to locate their position. Consensus mapping of flanking markers has the ability to differentiate important regions on individual chromosomes as well as highlighting important loci that have been identified in many studies. It does have the limitation that loci falling in the same chromosomal region cannot be differentiated; however it does identify the minimum number of regions contributing to resistance and gives insights into future studies that could lead to gene discoveries.

The consensus maps used were located on the cmap website (http://ccg.murdoch.edu.au/cmap/ccg-live/) with the main maps being "Consensus Map August 11 2003" "Somers Consensus April 04" and "Consensus 2010-2011", the latter being a series of single chromosome consensus maps developed over the years 2010 and 2011. Consensus 2010-2011 maps generally had the best coverage of markers and contained the most markers. These consensus maps were used as the basis for all figures, although on occasion, map positions of some markers were inferred from the earlier mentioned consensus maps. Some of the DArT markers were not in any of the consensus maps, and their position was also inferred from flanking SSR markers in other mapping populations. The identification of specific chromosomal regions were determined by either by a position of a single reported QTL, or more commonly by the clustering of flanking markers from two or more studies. Often, the flanking markers from different studies identified overlapping segments and the limits of each region were set by the outermost flanking markers from these segments. Regions were labeled according to the following nomenclature: QRYrChromosome. position, where QR stands for QTL Region, Yr for stripe (yellow) rust, Chromosome conventional name (group number 1-7 followed by genome A, B or D) and position, where regions were labeled numerically with the first position being the region closest to the telomere of the short arm of that respective chromosome.

QTL studies have used different software to describe aspects of QTLs, of particular note is the variance explained by the relevant QTL. Some packages have used R^2 to describe this term but for consistency we use the term Phenotypic Explained Variance (PEV) throughout the text.

QTL regions associated with stripe rust resistance

There were 47 regions identified that had an effect against stripe rust severity and these were found on all chromosomes with the exception of 5D. Many of the regions are known to contain more than one gene but limitations with consensus mapping did not permit further delineation of this region. For example, chromosome 2B had a region that contained *Yr27* and *Yr31*, two linked race specific seedling genes, as well as a number of QTLs that were effective in the adult plant stage. This suggests a minimum number of regions have been identified. Described below are seven sections relating to each of the different chromosome groups of wheat, sequentially highlighting regions where stripe rust QTLs have been found and indicating which regions are most important.

Group 1

The QTLs associated with group 1 chromosomes are listed in Table 1 and shown diagrammatically in Fig. 1. Chromosome 1A had four QTLs described (Bariana et al., 2010; Dedryver et al., 2009; Ren et al., 2012a;

Rosewarne et al., 2012). The flanking markers of these form two distinct groups although they are both on the long arm of chromosome 1A. The QTLs from Janz, Renan and Pastor are proximal on 1AL, while flanking markers from Naxos appear to be somewhat more distal on 1AL. All four QTLs resulted in similar LOD and PEV scores (Table 1), but they are relatively inconsistent in that they are picked up in approximately half of the environments tested. Given the closeness of the Naxos QTL with the other three, there is a possibility that they are at the same locus. The PEV scores are moderate and this is a locus of intermediate value. Application of markers across the populations would help to clarify if there are two regions of importance for stripe rust on this chromosome.

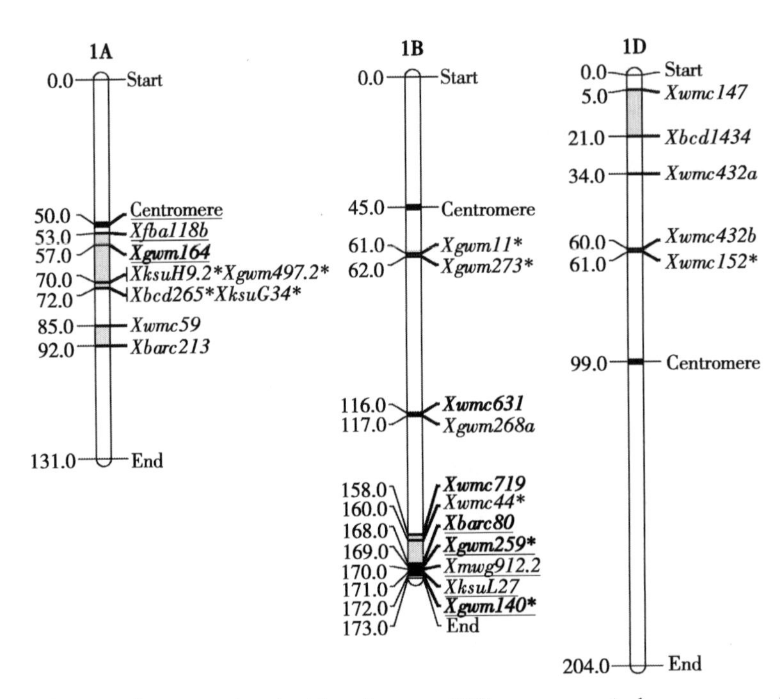

Fig. 1 Location of flanking markers associated with stripe rust QTLs on group 1 chromosomes. Marker locations are taken from consensus maps from the CMap website (http://ccg.murdoch.edu.au/cmap/ccg-live/). Map distances for markers in plain text are from "Consensus Maps 2010-11", underlined markers have map positions estimated from "Consensus 2003" and markers in bold have map positions estimated from "Sommers 2004" consensus maps. * identified markers that were used as associated (linked) substitutes for QTL markers that were not available on any consensus maps.

Chromosome 1B contains the important locus of *Lr46*/*Yr29*, hereafter referred to as *Yr29*. This had been identified in numerous studies and is described as group 3 on 1BL (QRYr1B.3) in Table 1 (Bariana et al., 2001; Bariana et al., 2010; Jagger et al., 2011; Lillemo et al., 2008; Melichar et al., 2008; Rosewarne et al., 2008; Rosewarne et al., 2012; William et al., 2006; Zwart et al., 2010). There is a wide range in the LOD (2.8-23) and PEV (4.5-65) scores indicating the variability that this locus can contribute. However, *Yr29* is important as is demonstrated by the number of significant environments tested. Nearly every study finds this locus significant in every environment tested with the exception of Bariana et al. (2010). Interestingly, that study also had *Lr34*/*Yr18* (hereafter termed *Yr18*) segregating in that population and it is now becoming apparent that when those two loci are in the same genetic background, the *Yr29* locus has lesser effect (Lillemo et al., 2008; Suenaga et al., 2003; Yang et al., 2013). This is clearly a useful locus on which to build a basis for multi-genic resistance. Chromosome 1B also had two other regions that contributed to stripe rust resistance. These are both also likely to be on 1BL with the proximal locus identified by

Rosewarne et al. (2012) in Pastor. This was effective in two of three environments tested and had intermediate LOD and PEV scores. A QTL identified by Lin and Chen (2009) from the cultivar Express was clearly located between the proximal Pastor QTL and the *Yr29* locus. In their paper, they could not conclude whether their locus was different from *Yr29* although clearly it falls in a different position in the consensus maps and is likely to be a different gene.

There were three QTLs identified on Chromosome 1DS (Ren et al., 2012a; Vazquez et al., 2012, Zwart et al., 2010), all with relatively minor and inconsistent effects against the pathogen. These loci probably form two separate regions with the Stephens (Vazquez et al., 2012) and CPI133872 (Zwart et al., 2010) QTLs being in the telomeric region of 1DS. The location of the Naxos locus was ill-defined as *Xwmc432* has two forms in the consensus maps that are 26cM apart and Ren et al. (2012a) did not differentiate which form was associated with the Naxos resistance. However it would appear that *Xwmc432b* was the allele identified as this marker was mapped close to *Xwmc152*, a marker associated with their flanking DArT marker *wPt-6979*. Either way, the identification of two minor loci on Chromosome 1DS is still supported.

Group 2

The group 2 chromosomes have several important regions for stripe rust resistance and these are outlined inTable 2 and Fig. 2. Chromosome 2A has one region associated with resistance on the short arm and another region on the long arm. Both Récital (Dedryver et al., 2009) and Camp Remy (Boukhatem et al., 2002; Mallard et al., 2005) have QTLs in both of these regions. The region around the 2AS QTLs are also associated with the major, race-specific gene *Yr17*, introgressed from *Aegilops ventricosa* (Bariana and McIntosh, 1993). Both Pioneer 26R61 (Hao et al., 2011) and Y16DH70 (Agenbag et al., 2012) have QTLs with LOD scores consistent with major genes in this region although markers confirmed that the alien introgression containing *Yr17* was not present. It seems likely that these lines contain potentially new major genes for resistance. The other QTLs in the 2AS region appeared to be more consistent with minor resistance genes where the LOD scores were lower and effect across environments was generally inconsistent. Both the 2AS and 2AL minor resistance genes are likely to be useful in contributing to stable resistance.

Table 2 Summary of stripe rust regions on group 2 chromosomes associated with stripe rust resistance QTLs. Markers in parentheses were not described in the original publication but were used as closely associated substitutes on consensus maps where the original markers were not available. References (Ref) are listed in Supplementary Table 1. Frequency (Freq) describes to number of environments that the locus had a significant QTL, out of the total number of environments tested.

Chromosome Region	Source	Markers	Field			Infection Type		Freq	Ref
			LOD	PEV	Freq	LOD	PEV		
QRYr2A. 1	Y16DH70	Xgwm636	14-19	37-49	2/2	9-16	31-53	2/2^F	1
QRYr2A. 1	Pioneer26R61	Xbarc124 Xgwm359	15-25	23-24	3/3				11
QRYr2A. 1	Recital	Xcfd36	5. 0-6. 8	5. 5-7. 7	2/4				8
QRYr2A. 1	Camp Remy	Xgwm356 Xgwm122, Xgpw2111	5. 4-6. 1	20-40	4/4				19
QRYr2A. 1	Stephens	wPt-0003 (Xgwm359)	3. 8-6. 7	9-20	4/6				31
QRYr2A. 1	Kukri	Xbarc5	3. 7-4. 1	13-15	2/6				3
QRYr2A. 1	*T. monococcum*	Xwmc170 (Xbarc5)	7. 8-15	7-12	3/3	13-15	11-13	3/3^F	7

（续）

Chromosome Region	Source	Markers	Field			Infection Type		Freq	Ref
			LOD	PEV	Freq	LOD	PEV		
QRYr2A. 2	Recital	Xgwm382 Xbarc122	4. 4-5. 4	4. 5-8. 1	2/4				8
QRYr2A. 2	Camp Remy	Xgwm359 Xgwm382	1. 7	10. 7	2/2	3	15	2/2F	5
QRYr2B. 1	Renan	Xgwm210a Xfbb67c	12-13	9-16	4/4				8
QRYr2B. 1	Stephens	wPt-5738	3-3. 6	10	2/6				31
QRYr2B. 2	Attila	Xwmc257 Xwmc154		4. 7-9. 3	3/3				25
QRYr2B. 2	Luke	Xwmc154 Xgwm148				5. 1-7. 7	32-37	3/3F	10
QRYr2B. 2	Opata 85	Xcdo405，Xbcd152				7. 4	30. 7	2/2F	5
QRYr2B. 2	Chapio	wPt-0079 (wPt-8583，Xgwm410)	3. 3-7. 8	4. 9-13. 6	3/8				33
QRYr2B. 2	Luke	Xgwm148 Xbarc167				5. 5-8. 8	33-42	3/3F	10
QRYr2B. 2	Kareiga	Xgwm148	55	30	1/1	95	46	1/1F	22
QRYr2B. 2	Stephens	wPt-0408	2. 8-4. 3	8-13	5/6				31
QRYr2B. 3	Pingyuan 50	Xbarc13 Xbarc55	1. 7-3. 2	5. 1-9. 5	3/4				14
QRYr2B. 3	Louise	Xwmc474 Xbarc230	5. 5-30	11-58	7/8				6
QRYr2B. 3	Camp Remy	Xgwm47 Xgwm501	11. 8	46	2/2	12	46	2/2F	5
QRYr2B. 3	Naxos	wPt-8460	4. 9	12. 2	1/3				23
QRYr2B. 3	Pastor	Yr31 (Lr23，Lr13)	10-17	32-66	3/4				26
QRYr2B. 4	Cranbrook	Xwmc339 Yr7							2
QRYr2B. 4	Camp Remy	Xbarc101 Xgwm120，Xwmc175	22-36	42-61	4/4				19
QRYr2B. 4	Aqulia	Xwmc175 Xwmc332				7. 8-13. 9	49-62	3/3SN,F	10
QRYr2B. 4	Avocet	Xgwm619	3. 6	6. 3	1/3				25
QRYr2D. 1	Libellula	Xcfd51 Xgwm261	5-10	8-10	2/4				18
QRYr2D. 2	Camp Remy	Xgwm102 Xgwm539	5. 7-11	24-69	4/4				19
QRYr2D. 2	Yr16DH70	Xgwm102	5. 4	8. 4	1/2	6. 1-6. 6	10	2/2F	1
QRYr2D. 2	Sunco	Xgdm005 Xwmc190	15. 8		1/1				2
QRYr2D. 2	Guardian	Xgwm539 Xgwm349	4. 4	11	1/2	4. 3	14	1/2F	20

（续）

Chromosome Region	Source	Markers	Field			Infection Type		Freq	Ref
			LOD	PEV	Freq	LOD	PEV		
QRYr2D. 2	Naxos	Xgwm539 Xcfd62	5. 6	10. 2	1/3				23
QRYr2D. 2	Fukuho-komugi	Xgwm349	14. 5	9. 6	1/3				29
QRYr2D. 3	Alcedo	Xgwm320 Xgwm301	14-18	32-36	2/2	18-28	36-53	2/2^{F}	13

Likelihood ratio

Also scored in seedling test but was not significant

SN. Stripe number per 10 cm^2 leaf area

F. Infection type scored in field on adult plants

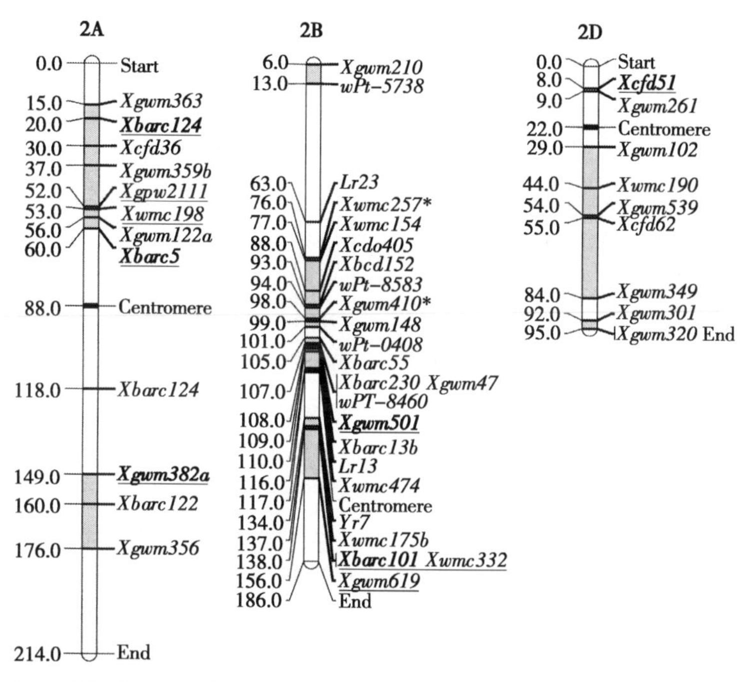

Fig. 2 Location of flanking markers associated with stripe rust QTLs on group 2 chromosomes. Marker locations are taken from consensus maps from the CMap website (http://ccg.murdoch.edu.au/cmap/ccg-live/). Map distances for markers in plain text are from "Consensus Maps 2010-11", underlined markers have map positions estimated from "Consensus 2003" and markers in bold have map positions estimated from "Sommers 2004" consensus maps. * identified markers that were used as associated (linked) substitutes for QTL markers that were not available on any consensus maps.

There are at least four regions associated with rust resistance on chromosome 2B. The QRYr2B. 1 region was identified by Dedryver et al. (2009) and Vazquez et al. (2012) with the marker *Xgwm210* and *wPt-5738* respectively, placing this region at the telomere of 2BS.

The QRYr2B. 2 region was identified in six studies and is the location of several seedling resistance genes. Rosewarne et al. (2008) indicated that Attila had *Yr27* and Yang et al. (2013) identified *Yr31* in Chapio, both falling in this region. Pastor (Rosewarne et al., 2012) also contained *Yr31*, with this gene being flanked by the leaf rust seedling resistance genes *Lr23* and *Lr13*. The cultivar Luke (Guo et al., 2008) contained two loci, 23cM apart, in this region that were observed with infection type data and were shown to be high temperature adult plant (HTAP) resistances. Opata 85 (Boukhatem et al., 2002) also contained a major

QTL identified through IT data in this region and Ramburan et al. (2004) used IT, final disease severity and seedling data to show an adult plant resistance with major effect in Kariega. Finally, Vazquez et al. (2012) used final disease severity to identify a highly significant QTL in the majority of environments tested. The combination of these data show that QRYr2B. 2 is a gene rich region which contains a number of seedling and HTAP resistance genes, as shown by relatively high LOD and PEV scores. Furthermore, these types of genes were effective across most environments, with only *Yr31* not being significant in environments where a virulent pathotype was used (Rosewarne et al., 2012; Yang et al., 2013).

The next region also contained a seedling resistance gene derived from Camp Remy (Boukhatem et al., 2002), HTAP resistance in Louise (Carter et al., 2009) and minor QTLs in Pingyuan50 (Lan et al., 2010) and Naxos (Ren et al., 2012a). This region was located close to QRYr2B. 2 and although the flanking markers from the loci within these two regions did not overlap, limitations in consensus mapping do not rule out the possibility that the same genes may contribute to some resistances in both regions. Although further studies are required to separate individual genes, it is clear that the QRYr2B. 2 and QRYr2B. 3 regions have been used extensively in resistance breeding.

The fourth region on 2BS was also associated with seedling resistances with Bariana et al. (2001) mapping *Yr7* from Cranbrook. Mallard et al. (2005) asserted that the 2BL QTL from Thatcher was most likely this same gene. Our study supports this as the Mallard et al. (2005) markers fall very closely to *Yr7* in the consensus maps. A locus from Aquileja (Guo et al., 2008) also clusters in this region and showed seedling resistance. The high LOD and PEV scores and its effectiveness in all three environments support this being a major gene, however it is not *Yr7* as the pathotype used (CYR32) is virulent on *Yr7*. Mallard et al. (2005) point out that *Yr7* and *Yr5* are probably allelic so it seems the Aquileja 2BL locus is likely to contain the latter, with CYR32 being avirulent on this gene. The final QTL in this region was identified in Avocet-*YrA*, hereafter termed Avocet (Rosewarne et al., 2008), a line often used as a susceptible parent. Other minor QTLs from Avocet have been identified on 3A, 4B, 6A and 7A (Lillemo et al., 2008; Rosewarne et al. 2012; William et al., 2006). These loci invariably have relatively low LOD and PEV scores and are often inconsistent across environments. These very minor QTLs still provide some level of resistance as transgressive segregation can sometimes be observed in Avocet crosses where a progeny line is more susceptible than Avocet (Melichar et al. 2008; Rosewarne et al. 2008; Rosewarne et al., 2012).

There are three regions associated with resistance on chromosome 2D. One region on 2DS was identified in a single study with Libellula (Luet al., 2009). This locus had intermediate LOD and PEV scores and was significant in two of the four environments tested. This characteristic minor gene can contribute to durable resistance when combined with a number of other loci.

There was a large region in the proximal area of 2DL that contained QTLs from 6 parents. The flanking markers for the Guardian QTL were*Xgwm539* and *Xgwm349*. There is conjecture in the linkage maps as to how close these markers are, with Somers Grain Genes Consensus having them 2. 3cM apart, yet Consensus 2003 and Consensus 2010 maps had them about 30cM apart. QTLs from Camp Remy (Mallard et al., 2005) and Naxos (Ren et al., 2012a) also had *Xgwm 539* as a flanking marker and Fukuho-komugi had *Xgwm 349* as a flanking marker. Flanking markers from YrDH70 (Agenbag et al., 2012) and Sunco (Bariana et al., 2001) also fall within this region as does the adult plant resistance gene *Yr16* (Worland and Law, 1986). The loci outlined here had inconsistent effects across environments as has been observed with other minor loci. Further work would be required to differentiate all of these loci from *Yr16* however it still appears to be a valuable locus in combination with other genes.

A final locus, QRYr2D. 3 was identified close to the telomere of 2DL in Alcedo and was effective in both seedling and adult plant stages. The high LOD and PEV scores support a major gene for resistance in this region.

Group 3

There were 14 studies that identified QTLs on the group 3 chromosomes. Lillemo et al. (2008) identified a 3AS QTL in Saar but this was only investigated in one stripe rust environment so its breeding value is yet to be determined. The 3B chromosome appears to have at least three regions associated with stripe rust resistance. Consensus mapping in this region was difficult as a group of SSR markers are multiallelic and produced up to three bands each that were located in different positions of the chromosome. Most of the publications do not discriminate which band was identified. For example, *Xbcd907*, *Xgwm389*, *Xbarc147*, *Xbarc133*, *Xgwm533* and *Xgwm493* have at least an "a" and a "b" locus, with the "a" loci of these markers clustering near the telomere of 3BS, whereas the "b" loci cluster more towards the middle of 3BS. We have presented data based on the "b" clustering loci however we cannot be sure if some reported QTLs that make up this cluster are from the "a" locus region. Nonetheless, the majority of QTLs identified on 3B are on the short arm. The consensus mapping approach can only give a broad picture of chromosomal regions that are involved in resistance. For example, Hao et al. (2011) identified three QTLs that all fall within QRYr3B. 1, suggesting that there is more than one locus contributing to resistance in this region. However, QTL analyses often report multiple peaks within a region that can be bought together by reordering the markers. The QRYr3B. 1 region is known to be extremely important as it is the location of *Yr30*, a partial resistance gene that is very tightly linked, or pleiotropic, to the slow-stem rusting locus *Sr2* (shown in Fig. 3) and the phenotypic marker of pseudo-black chaff (*Pbc*). *Yr30* is a valuable gene that has shown to work well in combination with other genes such as *Yr18* (Yang et al., 2013) and gives an intermediate effect in most environments. It is expressed only in the adult plant stage as William et al. (2006) did not find any evidence of resistance at this locus in a seedling assay. Singh et al. (2000b) and William et al. (2006) did find a significant QTL at this site with IT data on adult plants, and this chlorotic effect can be seen in advanced stripe rust infections on lines protected solely by adult plant resistances (Singh et al., 2005). The LOD and PEV scores within the studies listed in Table 3 indicate that QRYr3B. 1 has a consistent intermediate effect on stripe rust and is fairly consistent across environments. Bariana et al. (2010) described a QTL from Kukri on 3BS and assumed it was *Yr30*. We have included the main DArT marker *wPt-6802* used by Bariana et al. (2010) in consensus mapping and this extends Group QRYr3B. 1 by 23cM. However it should be noted that the Bariana et al. (2010) study describes a significant region of over 82cM, and this region also includes *Xgwm533*. This SSR marker has been associated with *Yr30* in four other studies (Börner et al., 2000; Dedryver et al. 2009; Spielmeyer et al., 2005; William et al., 2006).

Table 3 Summary of stripe rust regions on group 3 chromosomes associated with stripe rust resistance QTLs. Markers in parentheses were not described in the original publication but were used as closely associated substitutes on consensus maps where the original markers were not available. References (Ref) are listed in Supplementary Table 1. Frequency (Freq) describes to number of environments that the locus had a significant QTL, out of the total number of environments tested.

Chromosome Region	Source	Markers	Field			Infection Type		Freq	Ref
			LOD	PEV	Freq	LOD	PEV		
QRYr3A. 1	Saar	Xstm844tcac Xbarc310		8. 7	1/1				15
QRYr3B. 1	AGS2000	wPt-2557 Xbarc133	3. 8	5	1/3				11
QRYr3B. 1	Chapio	Xbarc147 wPt-3038	3. 8-18	4. 9-15	7/8				33

（续）

Chromosome Region	Source	Markers	Field			Infection Type		Freq	Ref
			LOD	PEV	Freq	LOD	PEV		
QRYr3B. 1	Oligoculm	Xgwm389	23. 1	2. 9-4. 9	3/4				29
QRYr3B. 1	Opata	Xfba190 Xbcd907	3. 9	16	1/3	7. 2	28	1/1F	28
QRYr3B. 1	AGS2000	wPt-730063 wPt-9579	3. 6	7	1/3				11
QRYr3B. 1	AGS2000	wPt-1612 wPt-7486	3. 1-3. 2	4-5	2/3				11
QRYr3B. 1	Lgst. 79-74	Xgwm533 Xgwm493			2/2				4
QRYr3B. 1	Pavon 76	Xgwm533	2	4. 6-6	0/3	2. 7	8. 3	1/1F	32
QRYr3B. 1	Renan	Xgwm533	4. 1-7. 7	3. 3-11	2/2				8
QRYr3B. 1	Kukri	wPt-6802	2. 6	5	1/6				3
QRYr3B. 2	Pastor	wPt-2458 wPt-0036	16. 2	3. 8-5. 8	3/4				26
QRYr3B. 2	Renan	Xgwm121b Xbcd131	4. 4	6. 3	1/2				8
QRYr3B. 3	Express	Xgwm340 Xgwm299	5. 2-8. 4	9. 1-13	3/3	2. 6-5	6-11	3/3F	17
QRYr3D. 1	Opata 85	Xcdo407 XksuA5				2. 8	1	1/1F	5
QRYr3D. 1	Opata 85	Xfba241 Xfba91	3. 5	14	1/4				28
QRYr3D. 2	Recital	Xbarc125 Xgwm456a	5. 2-6. 2	4. 7-7. 5	2/2				8
QRYr3D. 2	Chapio	Xgdm8 Xgdm128	3. 2-8	3. 2-11	3/8				33

Likelihood ratio
F. Infection type scored in field on adult plants

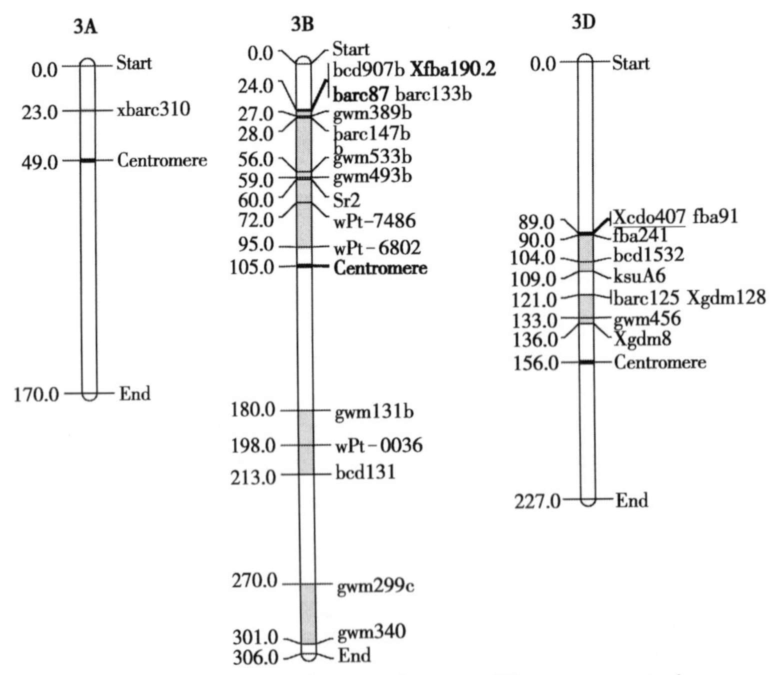

Fig. 3 Location of flanking markers associated with stripe rust QTLs on group 3 chromosomes. Marker locations are taken from consensus maps from the CMap website (http: //ccg. murdoch. edu. au/cmap/ccg-live/). Map distances for markers in plain text are from "Consensus Maps 2010-11", underlined markers have map positions estimated from "Consensus 2003" and markers in bold have map positions estimated from "Sommers 2004" consensus maps. *identified markers that were used as associated (linked) substitutes for QTL markers that were not available on any consensus maps.

The second region associated with stripe rust resistance on chromosome 3B was centrally located on the long arm. Renan (Dedryver et al., 2009) and Pastor (Rosewarne et al., 2012) contributed with intermediate to small effect QTLs and it was not consistently detected across environments. The third region associated with resistance was a HTAP gene (Lin and Chen, 2009) located near the telomere of 3BL. This locus is clearly different from those mentioned above as it was consistently identified across environments, had an intermediate level of effect and contributed significantly to lowering IT in adult plants.

There were two regions associated with resistance onchromosome 3D. Fig 3 shows that the nearest flanking markers from these two groups are only 12cM apart and the possibility exists that all of the described QTLs are in the same region. Region QRYr3D. 1 was identified from Opata 85 by Boukhatem et al. (2002) in a seedling assay on IT and later confirmed by Singh et al. (2000b) in a field based assessment of adult plants. The QRYr3D. 2 region was identified in Recital (Dedryver et al., 2009) and Chapio (Yang et al., 2013). The intermediate LOD and PEV scores were of the same order as those observed by Singh et al. (2000b) and finer mapping would be required to clearly differentiate QRYr3D. 1 and QRYr3D. 2.

Group 4

The group 4 chromosomes had limited QTLs and these are outlined in Table 4 and Fig. 4. Two studies have identified a region contributing to resistance on the long arm of chromosome 4A. Kariega has a locus that conferred all stage resistance and is therefore likely a major gene (Ramburan et al., 2004). Vazquez et al. (2012) identified a region that had a small effect in one of the two environments tested. They identified this QTL with the DArT marker *wPt-9901*, which was not present in any consensus map, however it was mapped within 1. 6cM of *Xbarc70* in Arina/NK93604 population (Howes pers. comm.). *Xbarc70* is located very close to the Kariega QTL and we have described it as a single region although it may contain two genes.

Table 4 Summary of stripe rust regions on group 4 chromosomes associated with stripe rust resistance QTLs. Markers in parentheses were not described in the original publication but were used as closely associated substitutes on consensus maps where the original markers were not available. References (Ref) are listed in Supplementary Table 1. Frequency (Freq) describes to number of environments that the locus had a significant QTL, out of the total number of environments tested.

Chromosome Region	Source	Markers	Field			Infection Type		Freq	Ref
			LOD	PEV	Freq	LOD	PEV		
QRYr4A. 1	Kariega	Xgwm160	23	15	1/1	8	6	1/1F	22
		Xgwm742, Xgwm832			40	24	1/1S		
QRYr4A. 1	Stephens	wPt-9901 (Xbarc70)	3	7	1/2				31
QRYr4B. 1	Palmiet	Xgwm165 Xgwm495			0/2	7. 6	12	1/2F	1
QRYr4B. 1	Alcedo	Xwmc692 Xstm535	13-15	24-29	2/2	8-18	22-37	3/3F	13
QRYr4B. 1	Oligoculm	Xgwm538		4. 4-12. 3	3/4				29
QRYr4B. 1	Avocet	Xgwm495 Xgwm368	2. 6-4. 4	8-13	3/3	4. 7	13	1/1F	32
QRYr4B. 1	Janz	wPt-8543 Xwmc238, Xgwm368	3. 8-5. 2	9-17	3/4				34
QRYr4B. 1	Libellula	Xgwm165 Xgwm149	2. 9-3. 6	4-5	2/4				18
QRYr4B. 1	Strampelli	Xgwm165 Xgwm149	3	5	1/5				18

(续)

Chromosome Region	Source	Markers	Field			Infection Type		Freq	Ref
			LOD	PEV	Freq	LOD	PEV		
QRYr4B. 2	Guardian	Xwmc652 Xwmc692	3	5	1/2	2.5	7	1/2F	20
QRYr4D. 1	Bainong 64	Xgwm165 Xwmc331	3	8	1/3				24
QRYr4D. 1	Pastor	wPt-6880 wPt-4572 (Xmwg634)	11.8	2.8-4.9	2/4				26
QRYr4D. 1	W-219	Xmwg634	2.3-4.4	9-17	2/4	8.6	31	1/1F	28
QRYr4D. 1	RL6077	Xgwm165 Xgwm192							12
QRYr4D. 2	Oligoculm	Xwmc399	32	3-8	4/4				29

Likelihood ratio

F. Infection type scored in field on adult plants

S. Infection type scored at seedling stage in glasshouse

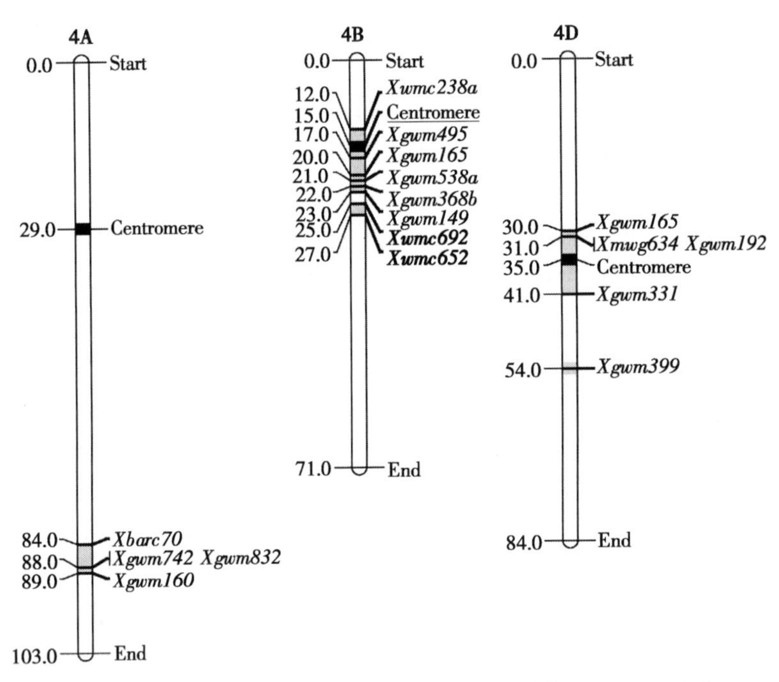

Fig. 4 Location of flanking markers associated with stripe rust QTLs on group 4 chromosomes. Marker locations are taken from consensus maps from the CMap website (http://ccg.murdoch.edu.au/cmap/ccg-live/). Map distances for markers in plain text are from "Consensus Maps 2010-11", underlined markers have map positions estimated from "Consensus 2003" and markers in bold have map positions estimated from "Sommers 2004" consensus maps. * identified markers that were used as associated (linked) substitutes for QTL markers that were not available on any consensus maps.

Chromosome 4B had two regions associated with stripe rust resistance and these were only separated by 2cM on the consensus maps. Six lines, Alcedo (Jagger et al., 2011), Avocet (William et al., 2006), Janz (Zwart et al., 2010), Libellula, Strampelli (Lu et al., 2009) and Palmiet (Agenbag et al., 2012) were identified with QTLs on the long arm of chromosome 4B. Alcedo had relatively high LOD and PEV scores whilst all the others had low to moderate scores. Several of the studies used both disease severity and IT data from field studies to show these mostly consistent QTLs (Agenbag et al., 2012; Jagger et al., 2011; William et al., 2006). Both Avocet and Palmiet were the susceptible parent in the cross and QTLs from the susceptible parent have been discussed previously. The lines Libellula and Strampelli (Lu et al., 2009) both have San Pastore in their pedigree and it seems likely that they share the same gene. This locus was not effective in all environments and although it is in a similar position to the other QRYr4B.1 QTLs, the inconsistent nature of this QTL suggests that it may be a different gene. Melichar et al. (2008) identified a QTL (QRYr4B.2) in Guardian with flanking markers *Xwmc692* and *Xwmc652*. In the consensus map, these were 2cM and 4cM distal to the last marker in QRYr4B.1 cluster and could easily have been classified as belonging to the QRYr4B.1 group. If so, its comparatively lower LOD and PEV scores make it more similar to the Libellula and Strampelli QTLs rather that the other QTLs observed in this region.

Relatively few QTLs were identified on chromosome 4D, yet it has turned out to contain what is an important and under-utilized locus. Hiebert et al. (2010) and Herrera-Foessel et al. (2011) characterized *Lr67/Yr46/Sr55/Pm46* (hereafter termed *Yr46*) in this region and have shown that it confers partial resistance to multiple pathogens including leaf rust, stripe rust, stem rust and powdery mildew. Furthermore it also confers the phenotypic marker "leaf tip necrosis" (*Ltn3*), making it very similar to the *Yr18 and Yr29* loci. These types of loci, when combined with two to three other minor loci, can provide near-immunity to stripe rust. Bainong 64 (Ren et al., 2012b), Pastor (Rosewarne et al., 2012) and W-219 (Singh et al. 2000b) also contained QTLs with intermediate LOD scores in this region and these were effective in approximately half of the environments tested. It is possible that some of these QTLs correspond to *Yr46* with Ren et al. (2012b) reporting both stripe rust and powdery mildew QTL at this locus. The location of the Pastor QTL was problematic as the DArT markers flanking this QTL were present on very few maps, however *wPt-4572* was located in the same linkage group as *Xmwg634* in an Opata/synthetic map and it is assumed that the Pastor QTL falls in this region. The oligoculm QTL (Suenaga et al., 2003) appears to be distinct from the *Yr46* locus.

Group 5

Group 5 chromosomes have only had QTLs identified on the A and B genomes (Table 5). The 5AL region likely contains two regions associated with resistance. The first was identified in the diploid *Triticum boeticum* and displays seedling susceptibility. LOD, PEV values and the effectiveness in all environments indicate this could be a major gene similar to the HTAP types of resistance. The QRYr5A.2 chromosomal region (Fig. 5) contained QTLs from four parents, Pastor (Rosewarne et al. 2012), Pingyuan 50 (Lan et al., 2010), SHA3/CBRD (Ren et al., 2012a) and Opata 85 (Boukhatem et al., 2002). The first three QTLs were identified in field based studies and had low to moderate LOD and PEV scores, as well as being inconsistently identified across multiple environments. The Opata 85 QTL was observed in seedling tests however it had a low LOD score (suggested but not significant) which fits well with the other loci identified in this region in field based scores. Hence, all named parents may contain the same locus. Two of the studies (Ren et al., 2012a; Rosewarne et al., 2012) identified QTLs with DArT markers that were not present in any consensus maps. However, *wPt-5231*, *wPt-1903 and wPt-3334* had previously been associated within 2.9cM of each other and within 5.2cM of *Xgwm179* in linkage maps generated using the Cadoux/Reeves population (Francki et al., 2009). In a Berkut/Krichauff population, Huynh et al. (2008) also identified *wPt-5231* within 2cM of the *Vrn-A1* locus, which also falls within this genomic region. This is interesting as later maturing lines in a segregating population are often scored lower for stripe rust as the younger leaves look greener. As maturity scores were not disclosed in any of the above populations, it cannot be ruled out that the QTL observed in this region

(QRYr5A. 2) are related to maturity rather than rust resistance *per se*.

Table 5 Summary of stripe rust regions on group 5 chromosomes associated with stripe rust resistance QTLs. Markers in parentheses were not described in the original publication but were used as closely associated substitutes on consensus maps where the original markers were not available. References (Ref) are listed in Supplementary Table 1. Frequency (Freq) describes to number of environments that the locus had a significant QTL, out of the total number of environments tested.

Chromosome Region	Source	Markers	Field			Infection Type		Freq	Ref
			LOD	PEV	Freq	LOD	PEV		
QRYr5A. 1	*T. boeticum*	Xbarc151 Xcfd12	19-21	17-19	3/3	21-23	18-20	3/3[F]	7
QRYr5A. 2	Opata 85	Xfbb209 Xabg391				2. 8	15	1/1[F]	5
QRYr5A. 2	Pastor	wPt-5231 wPt-0837 (Xgwm179, Vrn-1A)	13. 2	4-7	2/4				26
QRYr5A. 2	Pingyuan 50	Xwmc410 Xbarc261	1. 7-5. 6	5-20	3/5				14
QRYr5A. 2	SHA3/CBRD	wPt-1903 wPt-3334, Xwmc727	2. 1-2. 4	3-4	0/3				23
QRYr5B. 1	AGS2000	Xgdm152 Xwmc740	3. 3	5	1/3				11
QRYr5B. 1	Chapio	Xbarc267 Xbarc74	3. 6-10. 3	6-16	4/8				33
QRYr5B. 1	Flinor	Xgwm67 Xbarc89				4. 4	37	1/1[HTS]	9
QRYr5B. 1	Libellula	Xwmc415 Xwmc537	4-9	2-10	3/4				18
QRYr5B. 1	Strampelli	Xwmc415 Xwmc537	3-4. 3	3-6	2/3				18
QRYr5B. 1	Oligoculm	Xwmc415	47. 8	3-16	4/4				29
QRYr5B. 1	Yr16DH70	wPt-7114 Xbarc74	2. 8	4. 3	1/2	3. 5	5. 3	2/2[F]	1
QRYr5B. 1	Camp Remy	Xgwm639 Xgwm499, Xgwm544	6. 6-9. 7	18-26	4/4				19
QRYr5B. 2	Janz	wPt-3030 wPt-2707	2. 5-3. 5	6-8	5/6				3
QRYr5B. 2	SHA3/CBRD	wPt-2707 Xwmc75	2. 9-5. 2	5-8	2/3				23
QRYr5B. 3	Camp Remy	Xgwm234 Xgwm604	9. 7-11. 2	29-34	4/4				19
QRYr5B. 3	Flinor	Xwmc235 Xgwm604				5	33	1/1[HTS]	9
QRYr5B. 3	Libellula	Xbarc142 Xgwm604	2. 5	2. 6	1/4				18

Likelihood ratio
F. Infection type scored in field on adult plants
HTS. Infection type scored at high temperatures in glasshouse grown seedlings

The 5B chromosome had three clusters of markers associated with stripe rust resistance. Flinor (Feng et al., 2011), Camp Remy (Mallard et al., 2005) and Libellula (Lu et al., 2009) had loci in both QRYr5B. 1 and QRYr5B. 3. It is likely that these were inherited together. Pedigree information outlined by Feng et al. (2011) showed that these three lines, along with Strampelli, have a common parent in Hatif Inversable, which may have been the donor of these two QTLs. However, Strampelli only contains one of these QTLs (QRYr5B. 1). AGS2000 (Hao et al., 2011), Chapio (Yang et al., 2013), Yr16DH70 (Agenbag et al.,

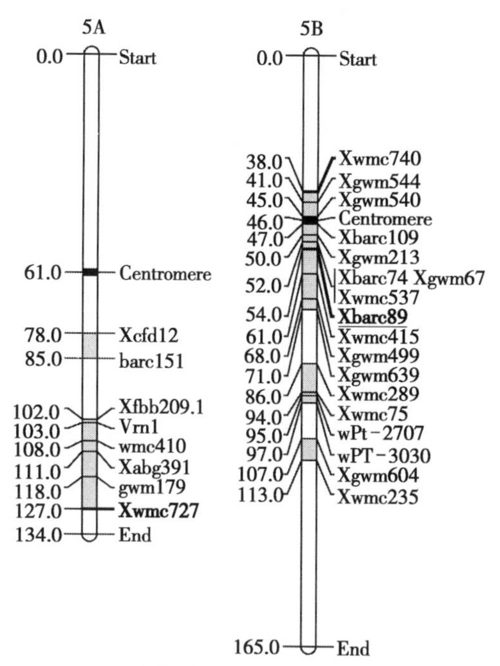

Fig. 5 Location of flanking markers associated with stripe rust QTLs on group 5 chromosomes. Marker locations are taken from consensus maps from the CMap website (http://ccg.murdoch.edu.au/cmap/ccg-live/). Map distances for markers in plain text are from "Consensus Maps 2010-11", underlined markers have map positions estimated from "Consensus 2003" and markers in bold have map positions estimated from "Sommers 2004" consensus maps. * identified markers that were used as associated (linked) substitutes for QTL markers that were not available on any consensus maps.

2012), Libelulla and Strampelli (Lu et al., 2009) contain QTLs in the QRYr5B.1 cluster of a low to intermediate effect that were inconsistent across environments, raising the possibility that all these lines contain the same gene. Flinor was described to contain a temperature sensitive seedling resistance gene that is more effective in higher temperatures. Yang et al. (2013) showed that the Chapio locus on QRYr5B.1 was only effective in Chinese environments and not in the cooler highland environments in Mexico. If this is the same gene, this difference in effectiveness could be due to environmental factors. Alternately there could be different race virulence in Mexico that renders this gene ineffective. Further studies are needed to determine whether these lines carry the same gene.

The second region on 5B contained QTLs from Janz (Zwart et al., 2010) and SHA3/CBRD (Ren et al., 2012a). These shared a common DArT marker and had similar LOD and PEV values. It seems most likely that these are the same locus if not the same gene.

Group 6

Chromosome 6A hadthree clearly defined regions associated with stripe rust resistance and these are likely to be conferred by three distinct genes (Table 6, Fig. 6). The first region is at the telomere of 6AS and was first described by Lin and Chen (2009) as a HTAP gene from the cultivar Express. This locus had high LOD and PEV scores for both final disease severity and lowered IT in field conditions. Hao et al. (2011) identified a locus in Pioneer 26R61 that shared a flanking marker, *Xgwm334*, with the Express locus. Both of these QTLs were effective in all tested environments and these similarities suggest they are associated with the same gene.

Table 6 Summary of stripe rust regions on group 6 chromosomes associated with stripe rust resistance QTLs. Markers in parentheses were not described in the original publication but were used as closely associated substitutes on consensus maps where the original markers were not available. References (Ref) are listed in Supplementary Table 1. Frequency (Freq) describes to number of environments that the locus had a significant QTL, out of the total number of environments tested.

Chromosome Region	Source	Markers	Field			Infection Type		Freq	Ref
			LOD	PEV	Freq	LOD	PEV		
QRYr6A.1	Express	Xgwm459 Xgwm334	5.8-9.8	11-16	3/3	3.6-5.6	8-13	3/3^F	17
QRYr6A.1	Pioneer 26R61	Xgwm334 wPt-7840	3.1-4.6	6-7	3/3				11
QRYr6A.2	Avocet	Xbarc3 xPT-7063		14	1/1				15

（续）

Chromosome Region	Source	Markers	Field			Infection Type		Freq	Ref
			LOD	PEV	Freq	LOD	PEV		
QRYr6A. 2	Avocet	Xgwm427 Xwmc256	1. 9-2. 8	6-8	2/3			0/1F	32
QRYr6A. 2	Avocet	wPt-0959	7. 4	2-7	2/4				26
QRYr6A. 3	Platte	378849 wPt-1642 (Xgwm617, Xcdo836, mwg2053)	3	6	1/2				31
QRYr6B. 1	Bainong 64	Xwmc487 Xcfd13	1. 9-2. 4	4-6	3/3				24
QRYr6B. 1	Naxos	Xwmc104 wPt-0259, wPt-7906	2. 9	6	1/3				23
QRYr6B. 1	Stephens	Xgwm132 Xgdm113	3. 5-8. 3	30-43	3/4				27
QRYr6B. 1	Oligoculm	Xgwm935. 1 Xwmc398	23	4	2/4				29
QRYr6B. 2	Janz	wPt-8183 wPt-1700	2. 5-2. 9	4-8	4/6				3
QRYr6B. 2	Pingyuan 50	Xgwm361 Xgwm136	1. 6～2. 5	5～8	2/4				14
QRYr6B. 2	Recital	Xcdo270 Xgwm193	3. 7-3. 8	2-4	2/2				8
QRYr6B. 2	Stephens	Xbarc101 Xbarc136	3. 9-6. 8	32-45	4/4				27
QRYr6B. 2	*T. turgidum* ssp.	*dicoccoides* FA15-3 Xbarc101, Xgwm193							30
QRYr6B. 3	Pastor	wPt-6329 wPt-5176	5. 6	2-4	3/4				26
QRYr6B. 3	Pavon 76	Xgwm58 Xgwm626	3-6. 5	9-19	3/3	2. 5	7. 9	1/1F	32
QRYr6D. 1	Yr16DH70	Xgwm32 Xbarc175	4. 2	6. 2	1/1			0/1F	15
QRYr6D. 2	W-7984	Xbcd1510 XksuD27				2. 4	13. 1	1/1F	5

Likelihood ratio

F. Infection type scored in field on adult plants

The second 6A region associated with resistance was close to the centromere but on the long arm of this chromosome. It was identified in three different studies and always derived from the parent Avocet (Lillemo et al., 2008; William et al. 2006; Rosewarne et al., 2012). This QTL had intermediate LOD and PEV scores and although it was not effective in all environments, it is likely to be useful in combination with other genes. Previously we have discussed the presence of minor QTLs in susceptible parents and here we have clear evidence from three different studies that such a QTL is important although its effect in isolation is very minor. The consensus map around the QRYr6A. 2 region covers 34cM, a value that is quite large and does not give confidence in defining this region. This is likely due to the use of some DArT markers in two of the Avocet QTLs and the relative paucity of mapping data surrounding these markers. As better consensus maps of the DArT markers are produced, it is expected that this region would narrow and probably be more focused

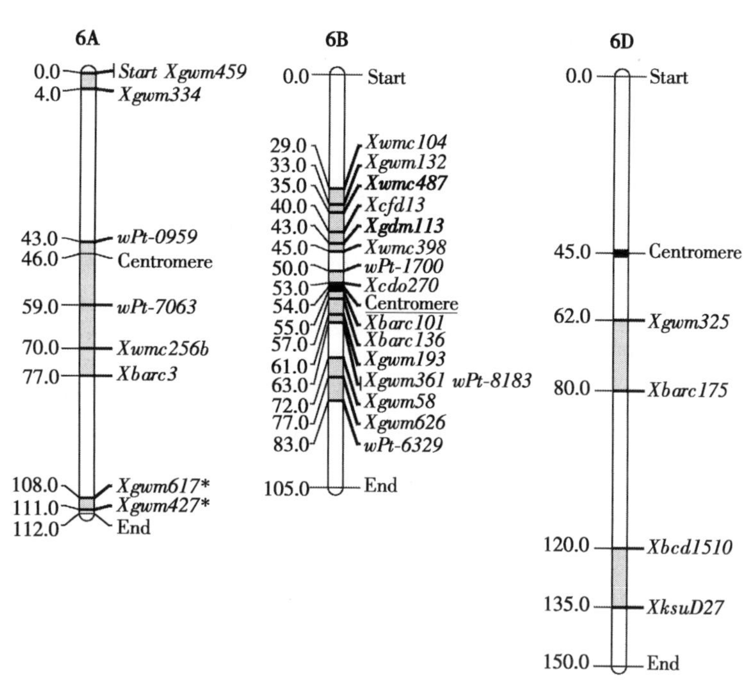

Fig. 6 Location of flanking markers associated with stripe rust QTLs on group 6 chromosomes. Marker locations are taken from consensus maps from the CMap website (http://ccg.murdoch.edu.au/cmap/ccg-live/). Map distances for markers in plain text are from "Consensus Maps 2010-11", underlined markers have map positions estimated from "Consensus 2003" and markers in bold have map positions estimated from "Sommers 2004" consensus maps. * identified markers that were used as associated (linked) substitutes for QTL markers that were not available on any consensus maps.

around the SSR markers identified. Avocet contains the *Thinopyrum elongatum* translocation on 6AL which confers stem rust resistance with *Sr26* (McIntosh et al., 1995). Translocations often carry multiple resistance genes and the positioning of a near-centromeric location for QRYr6A.2 is consistent with it being on the *Sr26* translocation (Rao, 1996).

The third region was near the teleomere of 6AL and was identified in Stephens (Vazquez et al., 2012). The closest marker to this QTL was *wPt-1642* and although not present on any consensus maps, it had been mapped to this telomeric region in two studies. The marker was mapped to within 4.5cM of *Xgwm617* in a Cranbrook/Halberd population (Akbari et al., 2006) and to 4.7cM of *Xgwm427* in a durum population of Colosseo/Lloyd (Mantovani et al., 2008). The close association of these latter two SSR markers and their established position near the telomere of 6AL in consensus maps provides good support for the positioning the stripe rust QTL.

There were three regions of importance in chromosome 6B. The short arm contained QRYr6B.1 which was determined by flanking markers from Bainong 64 (Ren et al., 2012b), Naxos (Ren et al., 2012a), Stephens (Santra et al., 2008) and Oligoculm (Suenaga et al., 2003). Judging by PEV scores, there appears to be two types of loci involved in this region, with the Stephens locus being conferred by a HTAP gene. Stephens also had another HTAP locus in group QRYr6B.2. These were the only two loci in the Santra et al. (2008) study and combined to contribute between 62%-79% of variance with both loci having significant effects on IT. The other loci on QRYr6B.1 had much lower PEVs and LOD scores and were inconsistently significant across environments indicating the likely involvement of a different gene in this region.

The QRYr6B.2 region contained flanking markers

from Janz (Bariana et al., 2010), Pingyuan 50 (Lan et al., 2010), Recital (Dedryver et al., 2009) and the aforementioned Stephens (Santra et al., 2008). Again it seems likely that there are two distinct loci based on the extremely high PEV scores associated with the Stephens QTL, and the relative inconsistency and much lower PEVs of the QTLs associated with the other three lines. This region also contained the HTAP gene *Yr36* (Uauy et al., 2005) derived from *T. turgidum* ssp. *dicoccoides* (Körn.) Thell. accession FA15-3. This gene has been cloned and the predicted protein has similarities to a kinase domain followed by a steroidogenic acute regulatory protein-lipid transfer domain and may play a role in recognizing and binding to lipids in the invading fungus and initiating a programmed cell death signaling cascade (Fu et al., 2009). It seems likely that the Santra et al. (2008) study may have identified the HTAP *Yr36* but it is unknown if the other QTLs in this region contain a similar gene.

The final region on 6B was identifiedin Pastor (Rosewarne et al., 2012) and Pavon (William et al., 2006) and had moderate to high LOD and PEV scores. The Pavon QTL was also identified with IT data from field studies. These two lines share a common parent, Kalyansona/Bluebird, a line that is known to contribute minor rust resistance genes in CIMMYT germplasm (Rosewarne et al., 2012). From this data it seems likely that a single gene in this region contributes with resistance.

Only two studies have identified resistance loci on chromosome 6D and in both cases on the long arm. Agenbag et al. (2012) identified a minor QTL proximal in 6DL in the line Yr16DH70. Boukhatem et al. (2002) found a seedling resistance gene more distal on 6DL although their QTL analysis stated that this was only suggestive and not significant. These studies were conducted in relatively few environments and more work needs to be done to determine their breeding value.

Group 7

Chromosome 7A had up to five regions where QTLs were described (Table 7, Fig. 7). All of these were of relative minor effect. Avocet (Rosewarne et al., 2012) had two apparent QTLs, here designated as QRYr7A. 1 and QRYr7A. 5. The first was located near the telomere of 7AS and the flanking DArT markers had been placed on consensus maps. QRYr7A. 2 was derived from Recital and was defined on the consensus map by the marker *Xfbal27c* that appears to be centrally located on 7AS (Dedryver et al., 2009). Ren et al. (2012b) identified a fairly minor QTL in a more proximal region of the short arm of 7A (QRYr7A. 3) from Jingshuang 16. The CPI133972 (Zwart et al., 2010) QTL in QRYr7A. 4 was defined by several DArT markers however only *wPt-4345* could be found on a linkage map derived from the cross P92201D5-2/P91193D1-10 (Francki et al., 2009) and was flanked by *Xbarcl74* (2. 2cM) and *Xcfa2174a* (0. 8cM) tentatively placing this locus near the centromere of 7AS. The second Avocet QTL (Rosewarne et al., 2012) was in a similar location to a QTL identified from Stephens (Vazquez et al., 2012), centrally located on 7AL. Both of these QTLs were described with DArT markers that were not available on consensus maps with the position of the Avocet QTL (*wPt-2600*) being inferred by a map generated from a Cadoux/Reeves population (Francki et al., 2009) within 2. 8cM of the SSR marker *Xcfa2257*. The Stephens QTL (wPt-1023) co-segregates with XksuH9c in the Cranbrook/Halberd population (Akbari et al., 2006) and this RFLP marker is well defined on many maps. Both markers defined in the consensus maps are within close proximity to each other. The high LOD scores from the QRYr7A. 5 QTLs indicate this as a particularly useful region on chromosome 7AL.

Chromosome 7B had three regions associated with resistance. Oligoculm (Suenaga et al., 2003), Stephens (Vazquez et al., 2012) and SHA3/CBRD (Ren et al., 2012a) had QTLs in the centromeric region (QRYr7B. 1) and showed intermediate effects against stripe rust. The DArT marker from Stephens (*wPt-7653*) was not present on consensus maps and its position was inferred by its close association (0. 9cM) to

Xwmc76 in a Spark/Rialto population (Howes pers. comm.). Region QRYr7B. 2, proximal on the long arm of 7B, was identified in Alpowa (Lin and Chen, 2007) with a HTAP QTL that had strong effects when lines were scored for final disease severity and IT. They identified two SSR markers that were on the consensus map with *Xgwml31* being within 7cM of the QTL peak and *Xgwm43* being more than 30cM from the peak. Kukri (Bariana et al., 2010) also had a QTL in this region identified by the marker *wPt-8921*. This QTL was described over a long interval with the aforementioned DArT marker being centrally located, and other markers *wPt-6372* and *wPt-8106* flanking it by 20. 8 and 13cM respectively. Alpowa and Kukri appear to contain similar loci in regard to the location and effectiveness across all environments. The final region associated with stripe rust on the long arm of 7B was identified in two CIMMYT lines, Attila (Rosewarne et al., 2008) and Pastor (Rosewarne et al., 2012). Both of these loci are linked to a leaf rust QTL by 11-19cM and as they share a common parent (Seri M82), they are likely to be the same locus.

Table 7 Summary of stripe rust regions on group 7 chromosomes associated with stripe rust resistance QTLs. Markers in parentheses were not described in the original publication but were used as closely associated substitutes on consensus maps where the original markers were not available. References (Ref) are listed in Supplementary Table 1. Frequency (Freq) describes to number of environments that the locus had a significant QTL, out of the total number of environments tested.

Chromosome Region	Source	Markers	Field			Infection Type			Ref
			LOD	PEV	Freq	LOD	PEV	Freq	
QRYr7A. 1	Avocet	wPt-8149 wPt-4172	5. 9	3	2/4				26
QRYr7A. 2	Recital	Xfba127c Xbcd129b	3. 8	9	1/1				8
QRYr7A. 3	Jingshuang16	Xbarc127	2. 3-2. 4	6	2/3				24
QRYr7A. 4	CPI133972	wPt-4345 (Xcfa2174, Xbarc108)	4	11	1/4				34
QRYr7A. 5	Avocet	wPt-2260 (Xcfa2257)	9. 1	3-6	3/4				26
QRYr7A. 5	Stephens	wPt-1023 (XksuH9c)	5. 2-10	12-20	5/6				31
QRYr7B. 1	Oligoculm	Xgwm935. 3 Xgwm46	30. 3	9-17	3/4				29
QRYr7B. 1	Stephens	wPt-7653 Xwmc76	3. 1	6	1/2				31
QRYr7B. 1	SHA3/CBRD	Xbarc176 wPt-8106, wPt-9467	2. 6	8. 2	1/3				23
QRYr7B. 2	Alpowa	Xggp36 Xgwm131, Xgwm43	22-29	52-63	4/4	18-25	45-57	3/3F	16
QRYr7B. 2	Kukri	wPt-3723 wPt-8921	2. 6-3	8-9	6/6				3
QRYr7B. 3	Attila	Xgwm344	Not Des	3. 1	3/3				25
QRYr7B. 3	Pastor	wPt-3190 (Xgwm577, Xpsr680b)	25. 8	4-8	3/3				26
QRYr7B. 3	SHA3/CBRD	Xgwm577 wPt-4300, wPt-5309	6. 4	14. 7	1/3				23
QRYr7D. 1	CD87	Xwmc405b	27. 4	15	2/2				2
QRYr7D. 1	Chapio	Xgwm295 XcsLV34	6. 7-51	10-52	8/8				33
QRYr7D. 1	Cook	Xgwm295	ND	5-11	3/3				21
QRYr7D. 1	Fukuho-komugi	Xgwm295	25. 3	11-24	4/4				29

（续）

Chromosome Region	Source	Markers	Field			Infection Type		Freq	Ref
			LOD	PEV	Freq	LOD	PEV		
QRYr7D. 1	Janz	wPt-3328	2. 5-6. 8	7-19	6/6				3
QRYr7D. 1	Janz	Xwmc405 wPt-3727，(Xbcd1438)	4. 6-12	15-42	4/4				34
QRYr7D. 1	Kariega	Xgwm295 LTN	53	29	1/1	14. 9 22	9 16	1/1F 1/1S	22
QRYr7D. 1	Libellula	XcsLV34	9. 1-15	15-35	4/4				18
QRYr7D. 1	Opata	Xwg834 Xbcd1438	3. 4	13. 9	2/2				5
QRYr7D. 1	Saar	wPt-3328 XcsLV34	ND	40	1/1				15
QRYr7D. 1	Strampel-li	XcsLV34	6. 8-17	17-39	5/5				5
QRYr7D. 1	Opata 85	Xwg834 Xbcd1438，Xbcd1872	3-9. 7	12-36	3/3	3 5. 4	12 22	1/1F 1/1F	28

Likelihood ratio

F. Infection type scored in field on adult plants

S. Infection type scored at seedling stage in glasshouse

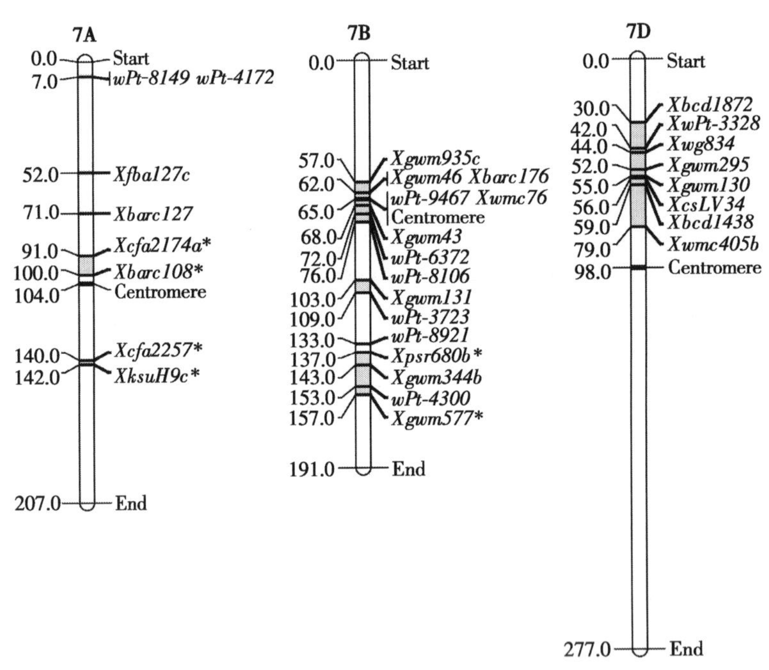

Fig. 7 Location of flanking markers associated with stripe rust QTLs on group 7 chromosomes. Marker locations are taken from consensus maps from the CMap website (http：//ccg. murdoch. edu. au/cmap/ccg-live/). Map distances for markers in plain text are from “Consensus Maps 2010-11”, underlined markers have map positions estimated from “Consensus 2003” and markers in bold have map positions estimated from “Sommers 2004” consensus maps. ＊ identified markers that were used as associated (linked) substitutes for QTL markers that were not available on any consensus maps.

The 7D chromosome containsonly one region associated with stripe rust resistance and that is the *Yr18* locus. Numerous studies have identified this region (Table 7) on 7DS and it provides strong, stable resistance over every environment tested. Some studies (Ramburan et al. 2004; Singh et al. , 2000b) have linked this gene with lowering infection type in the field. These factors would generally suggest a seedling resistance major gene, yet the *Yr18* locus is associated with seedling susceptibility, making it a very important adult plant resistance gene. *Yr18* shares common features with *Yr29* and *Yr46* in that they all confer non-race specific and presumable durable resistance, as well as being effective against multiple pathogens (discussed below).

Seedling resistance genes

The strongest effect QTLs against rust resistance are generally associated with major, seedling resistance genes. In this review we have summarized several studies where this has been the case. 2AS contains the race-specific seedling resistance gene *Yr17* derived from an *Aegilops ventricosa* translocation (Bariana and McIntosh, 1993). This gene was identified in a QTL study by Dedryver et al. (2009) and had high LOD and PEV scores (up to 40 and 43 respectively) which are typical of a major genes. However, this locus was only significant in one of the two environments tested, as the second environment (10 years later) used a rust race that was virulent to *Yr17*. Both Agenbag et al. (2012) and Hao et al. (2011) identified QTLs in this region that had major effects, however they eliminated the presence of the *Yr17* with the use of molecular markers and by showing that the parents had seedling susceptibility to the rust pathotypes. Chhuneja et al. (2008) also identified an adult plant resistance gene in this region and it was derived from *T. moncoccum*. This gene also had moderate to high effects against the disease (PEV 7-11) that were consistent across all environments tested.

Chromosome 2B also contains a number of seedling resistance genes with *Yr27* from Attila (Rosewarne et al., 2008), *Yr31* from Pastor and Chapio (Rosewarne et al., 2012; Yang et al., 2013) and *Yr7* from Cranbrook (Bariana et al., 2001) having been identified in QTL studies. The seedling resistance gene *Yr31* on 2BS has an intermediate effect on stripe rust infection type and was defeated in Mexico in 2008 with the aforementioned studies showing genotype x environment interactions when the virulent pathotype was used in inoculated field trials. In trials with the avirulent pathotype, PEV scores of between 33-66% in Pastor and 5-14% in Chapio were obtained. This locus was significant in all environments in which the avirulent pathotype was used. The QTL associated with *Yr27*, a gene closely linked to *Yr31* with a recombination value of 0.148 (http://www.wheat.pw.usda.gov) was in the Rosewarne et al. (2008) study and was significant in all three environments tested, with PEVs in the range of 4.7-9.3. *Yr27* has been used extensively in germplasm globally. However virulence to this gene was first identified in South America in mid 1990s and in south Asia in 2004 and is no longer effective in most parts of the world.

There is virulence for *Yr7* and although this was identified in Cranbrook (Bariana et al., 2001) and is present in Australian germplasm, it has a limited role in controlling stripe rust epidemics.

High temperature adult plant (HTAP) resistance

Chenand Line (1995) described HTAP resistance as conferred by a class of genes that show susceptibility in seedlings under normal conditions, yet provide generally high levels of resistance in infected adult plants when grown under higher temperatures. There is also evidence of effectiveness at earlier growth stages at higher temperatures with the genes *YrCK* on chromosome 2DS of Cook (Bariana et al., 2001), two loci on chromosome 5B of Flinor (Feng et al., 2009) and in the cloned HTAP gene *Yr36* (Fu et al., 2009). It is assumed that this type of resistance is durable as race-specificity has yet to be shown on any of the characterized loci. Susceptibility under low temperatures is of little consequence as this generally occurs when the wheat plants are young. As plants mature, the temperature usually increases and these loci become effective. Accurate characterization of these loci as HTAP resistance genes is somewhat problematic as a specific low and high temperature screens should be completed to show susceptibility; however most QTL studies only investigate field infections under prevailing temperature regimes, along with seedling tests to confirm susceptibility of parental stocks.

Due to the requirements of a specialized screen to identify HTAP loci, relatively few studies have investigated these loci. Lin and Chen (2007) were the first to localize a HTAP locus when they identified a 7BL locus in "Alpowa" which was designated as *Yr39*. This locus was one of two resistance genes found in "Alpowa" and gave extremely high LOD and PEV scores for both AUDPC and IT data. This highly effective locus was significant across all environments. Santra et al. (2008) identified two HTAP loci on chromosome 6BS from Stephens. Both of these loci were very effective across most environments with AUDPC LOD scores ranging from 3.5～8.3 and PEV scores between 30% and 45%. Lin and Chen (2009) identified two consistent HTAP QTLs from Express on 3BL and 6AS. The 3BL was located near the telomere, indicating it is a unique locus on this chromosome. The 6AS locus was in the same region as a QTL identified by Hao et al. (2011) that had similar LOD and PEV scores, although this later study did not specifically test for HTAP resistance. Both of the QTLs from Express were effective across three environments with high LOD and PEV scores for both AUDPC and IT. A third locus on 1BL was only effective in two of the three environments for AUDPC and did not affect IT. The location of this locus indicates that it is likely different from the well characterized *Yr29* locus.

Carteret al. (2009) and Guo et al. (2008) identified HTAP loci on 2BS from Louise and Luke which group in similar regions along with numerous other resistance loci that include seedling resistances and other small effect QTLs. Guo et al. (2008) identified two loci that shared a common flanking marker and these could possibly belong to a single locus. The loci from Luke and Louise all have high LOD and PEV scores for both AUDPC and IT and were effective in nearly all environments tested. The pedigree data for Luke and Louise was not given so we are unable to comment on whether they contain the same gene, however as these were the only HTAP genes identified in these populations, an allelism test would be possible to clarify this issue. Other cultivars that had QTLs in this region include Kariega (Ramburan et al., 2004), Naxos (Ren et al., 2012a), Camp Remy (Boukhatem et al., 2002; Mallard et al., 2005) and Pingyuan 50 (Lan et al., 2010) however these were not specifically tested for HTAP resistance. Along with the seedling resistance gene *Yr31*, (Rosewarne et al., 2012; Yang et al., 2013), this highlights a very complicated genetic region with multiple resistance loci.

Pleiotropic adult plant resistances (PAPR)

With the cloning of *Lr34/Yr18/Sr57/Pm38/Ltn1* (Kolmer et al., 2011; Krattinger et al., 2009) came the definitive proof that this very important gene conferred resistance to major biotrophic pathogens of leaf rust, stripe rust, and powdery mildew as well as conferring the phenotypic marker of leaf tip necrosis. Furthermore, QTL mapping indicates that this locus also confers resistance to spot blotch with the *Sb1* designation (Lillemo et al., 2012). Additional studies confirmed its effect on stem rust resistance with the corresponding designation of *Sr57* (RP Singh pers. comm.). There are a number of other loci that are candidates for PAP resistance genes although these need to be confirmed empirically through gene cloning. *Lr46/Yr29/Sr58/Pm39/Ltn2* (Lillemo et al., 2008; Rosewarne et al., 2006) and *Lr67/Yr46/Sr55/Pm46/Ltn3* (Herrera-Foessel et al., 2011; Hiebert et al., 2010) appear to have very similar functional roles in their pleiotropic responses although their absolute values tend to vary across environments.

Evidence of mapping populations containing both *Yr18* and *Yr29* suggest that although these genes generally act additively with most other minor QTLs, they are not completely additive in the presence of each other (Lillemo et al., 2008) and indeed *Yr29* may be less effective against stripe in the presence of *Yr18* (Suenaga et al., 2003; Yang et al., 2013). This suggests that these loci may work on the same molecular pathway in inhibiting fungal growth. Sequence data suggests that *Yr18* is an ABC transporter (Krattinger et al., 2009) and is presumably involved in the export of antifungal compounds out of the cytosol into the apoplast. Despite extensive molecular investigations, Lagudah (2011) has not identified any sequence similarities between the *Yr18* locus and regions surrounding the *Yr29* locus on chromosome 1BL. This suggests a different gene family may be involved in *Yr29* resistance, but this locus somehow taps into the same defensive pathway as *Yr18*. A possible theory of their interaction is that *Yr29* could be a different type of transporter for the same antifungal substrate as *Yr18*, yet may have a higher K_m (Michaelis constant that measures affinity for a particular substrate), resulting in *Yr29* being a less efficient transporter at a given substrate concentration. This is probably the simplest of a number of scenarios that could explain the observed interaction of these two loci.

A fourth candidate for PAP resistance is the*Sr2/Lr27/Yr30/Pbc* locus. This adult plant resistance locus is effective against a variety of pathogens including stem rust, leaf rust, stripe rust and powdery mildew (Mago et al., 2011). Furthermore, a morphological marker of pseudo black chaff, characterized by a darkening of the glumes and internodes is thought to be pleiotropic or tightly linked to the resistances (Kota et al., 2006). This locus has been effective against stem rust for over 80 years since its original transference to hexaploid wheat from emmer wheat (*T. dicoccum*) by McFadden (1930). The current wide deployment in stem rust prone areas is testament to the durability of *Sr2*. The tight linkage between *Sr2* and *Pbc* was reportedly broken (Mishra et al., 2005), although *Pbc* is known to be a quantitative trait. Again, gene cloning will reveal the true nature of potential pleiotropism of this locus. None-the-less, *Yr30*, associated with *Sr2*, is an important APR locus in wheat germplasm worldwide and has been identified in several QTL studies, generally having a significant effect across the majority of environments tested.

Other QTLs

The aforementioned classes of resistance genes have relatively few members as identified in QTL studies with the vast majority of QTLs falling into a class that can be characterized as seedling susceptible, not dependent on temperature regimes (as far as known) and are not Pleiotropic in nature. These QTLs have quite minor effects as shown by typically low LOD and PEV scores, and are often inconsistently observed across environments. Several studies have identified QTLs being derived from a susceptible parent suggesting that they may be hard to detect when in a genetic background that does not contain other additive resistance loci. However these loci are critical in the development of durably resistant lines. Placed in combination with HTAP, PAP and other QTLs, they can provide near immunity against stripe rust. Indeed, four to five loci have been shown to confer near immunity (Singh et al., 2000a) and breeding for such resistance, although more complicated than using single, seedling resistance genes, is not overly difficult. This can be greatly enhanced with the improved use of markers or phenotypic selection under severely diseased nurseries.

Future Directions

This review identifies 49 chromosomal regions that contain QTLs that lower stripe rust disease severity and it is likely that many of these regions contain more than one locus. It is expected that with the advent of cheaper genotyping, there will be many more studies identifying regions of importance. Clearly there is an abundance of partial resistance loci that can be used in combating stripe rust epidemics. For the effective deployment of these loci there are still a lot of unanswered questions. Of the loci that appear to be more sensitive to the environment, we need to know in which environments they are effective and what leads to their effectiveness. No doubt there are complicated interactions with prevailing environmental conditions interacting with the timing and severity of an epidemic.

The chromosomal regions identified in numerous studies are likely to contain important loci that are effective across multiple environments and may warrant a greater focus for future research. The PAPR genes feature extensively throughout the research with *Yr18*, *Yr29* and *Yr30* being identified independently in 12, ninc and ten studies respectively. Regions QRYr2A. 1 and QRYr2B. 2 were both identified in seven studies and appear to be gene rich regions containing several seedling resistances and well as some minor QTLs. The QRYr2A. 1 also corresponds to a region where several translocations have been incorporated. QRYr2D. 2 is a region that contains one of the first identified adult plant resistance genes for stripe rust, *Yr16* with six QTL studies showing this region to have a moderate effect. The regions QRYr4B. 1 and QRYr5B. 1, identified in seven and eight parents respectively, also tended to show moderate effects that were not identified in all environments tested. These results highlight the importance of these minor QTLs and even though they may not have a significant effect in all environments, these regions do contribute to stable resistance when combined with other loci.

It would also be beneficial to understand how different genetic combinations of the most influential loci interact, such that marker assisted breeding packages can be developed for breeders and ensure diverse deployment of resistance loci is achieved to further enhance durability. A reductionist approachwould be to obtain gene designations through the development of single gene lines. Such work was carried out for the leaf rust QTL now designated as *Lr68* (Herrera-Foessel et al., 2012). This approach was to identify lines with a single locus from within a mapping population based on the presence/absence of appropriate QTL flanking markers along with intermediate rust scores. The single gene lines were used as parents in a cross with a susceptible parent and rust scores showed that an intermediate final disease severity score was inherited as a single gene. Problems can arise when isolating some of the minor effect QTLs in that it may be difficult under certain environmental conditions to determine the intermediate resistance phenotype, and that these loci in isolation have a very minor effect that is difficult to detect. Despite these difficulties, single gene populations do provide the opportunity to dissect the role of these loci as well as aid in marker development and ultimate cloning of the resistance gene. The pyramiding of two or more loci will also be facilitated through the use of closely linked markers and further questions can be answered surrounding genetic interactions and effective gene combinations.

References

Agenbag GM, Pretorius ZA, Boyd LA, Bender CM, Prins R (2012) Identification of adult plant resistance to stripe rust in the cultivar Cappelle-Desprez. Theor Apppl Genet 125: 109-120.

Akbari M, Wenzl P, Caig V, Carling J, Xia L, Yang Suszynski G, Mohler V, Lehmensiek A, Kuchel H, Hayden M, Howes N, Sharp P, Vaughan P, Rathmell W, Huttner E, Killian A (2006) Diversity arrays technology (DArT) for high throughput profiling of the hexaploid wheat genome. Theor Appl Genet 113: 1409-1420.

Bariana HS, Bansal UK, Schmidt A, Lehmensiek A, Kaur J, Miah H, Howes N, McIntyre CL (2010) Molecular mapping of adult plant stripe rust resistance in wheat and identification of pyramided QTL genotypes. Euphytica 176: 251-260.

Bariana HS, Hayden MJ, Ahmed NU, Bell JA, Sharp PJ, McIntosh RA (2001) Mapping of durable adult plant and seedling resistances to stripe rust and stem rust diseases in wheat. Australian Journal of Agricultural Research 52: 1247-1255.

Bariana HS, McIntosh RA (1993) Cytogenetic studies in wheat. XV. Location of rust resistance gene in VPM1 and their genetic linkage with other disease resistance genes in chromosome 2A. Genome 36: 476-482.

Börner A, Röder MS, Unger O, Meinel A (2000) The detection and molecular mapping of a major resistance gene for non-specific adult-plant disease resistance against stripe rust (*Puccinia striiformis*) in wheat. Theor Appl Genet 100: 1095-1099.

Boukhatem N, Baret PV, Mingeot D, Jacquemin JM (2002) Quantitative trait loci for resistance against yellow rust in two wheat-derived recombinant inbred line populations. Theor Appl Genet 104: 111-118.

Carter, AH, Chen, XM, Garland-Campbell, K, Kidwell, KK (2009) Identifying QTL for high-temperature adult-plant resistance to stripe rust (*Puccinia striiformis* f. sp. *Tritici*) in the spring wheat (*Triticum aestivum* L.) cultivar 'Louise'. Theor Appl Genet 119: 1119-1128.

Caldwell, R. M., 1968: Breeding for general and/or specific plant disease resistance. In: K. W. Finlay and K. W. Shepherd (eds.), Proc. 3rd Int. Wheat Genetics Symp., 263-272. Australian Academy of Science, Canberra.

Chen XM, Line RF (1995) Gene number and heritability of wheat cultivars with durable, high-tempreature, adult plant (HTAP) resistance and interaction of HTAP and race-specific seedling resistance to Puccinia striiformis Phytopathology 85: 573-578.

Chhuneja P, Kaur S, Garg T, Ghai M, Kaur S, Prashar M, Bains NS, Goel RK, Keller B, Dhaliwal HS, Singh K (2008) Mapping of adult plant stripe rust resistance genes in diploid A genome wheat species and their transfer to bread wheat. Theor Appl Genet 116: 313-324.

Dedryver F, Paillard S, Mallard S, Robert O, Trottet M, Nègre S, Verplancke G, Jahier J (2009) Characterization of genetic components involved in durable resistance to stripe rust in the bread wheat 'Renan'. Phytopathology 99: 968-973.

Farrer W (1898) The making and improvement of wheats for Australian conditions. Agricultural Gazette of New South Wales 9: 131-168.

Feng J, Zuo LL, Zhang ZY, Lin R M, Cao YY, Xu SC (2011) Quantitative trait loci for tempertature-sensitive resistance to *Puccinia striiformis* f. sp. *Tritici* in wheat cultivar Flinor. Euphytica 178: 321-329.

Flor HH (1956) The complementary genic systems in flax and flax rust. Advances in Genetics 8: 29-54.

Francki GM, Walker E, Crawford AC, Broughton S, Ohm HW, Barclay I, Wilson RE, McLean R (2009) Comparison of genetic and cytogenetic maps of hexaploid wheat (*Triticum aestivum* L.) using SSR and DArT markers. Mol Genet Genomics 281: 181-191.

Fu D, Uauy C, Distelfeld A, Blechl A, Epstein L, Chen XM, Sela H, Fahima T, Dubcovsky J (2009) A kinase-START gene confers temperature-dependent resistance to wheat stripe rust. Science 323: 1357.

Guo Q, Zhang ZJ, Xu YB, Li GH, Feng J, Zhou Y (2008) Quantitative trait loci for high-temperature adult-plant and slow-rusting resistance to *Puccinia striiformis* f. sp. *Tritici* in wheat cultivars. Phytopathology 98: 803-809.

Hao YF, Chan ZB, Wang YY, Bland D, Buck J, Brown-Guedira G, Johnson J (2011) Characterization of a major QTL for adult plant resistance to stripe rust in US soft red winter wheat. Theor Appl Genet 123: 1401-1411.

Hiebert CW, Thomas JB, McCallum BD, Humphreys DG, DePauw RM, Hayden MJ, Mago R, Schnippenkoetter W, Spielmeyer W (2010) An introgression on wheat chromosome 4DL in RL6077 (Thatcher6/PI250413) confers adult plant resistance to stripe rust and leaf rust. Theor Appl Genet 121: 1083-1091.

Herrera-Fossel SA, Lagudah ES, Huerta-Espino J, Hayden MJ, Bariana HS, Singh D, Singh RP (2011) New slow-rusting leaf rust and stripe rust resistance genes Lr67 and Yr46 in wheat are pleiotropic or closely linked. Theor Appl Genet 122: 239-249.

Herrera-Foessel SA, Singh RP, Huerta-Espino J, Rosewarne GM, Periyannan SK, Viccars L, Calvo-Salazar V, Lan C, Lagudah ES (2012) Lr68: a new gene conferring slow rusting resistance to leaf rust in wheat. Theor Appl Genet 124: 1475-1486.

Huynh BL, Wallwork H, Stangoulis JCR, Graham RD, Willsmore KL, Olson S, Mather DE (2008) Quantitative trait analysis for grain fructan concentration in wheat (*Triticum aestivum* L.). Theor Appl Genet 117: 701-709.

Jagger LJ, Newell C, Berry ST, MacCormack R, Boyd LA (2011) The genetic characterisation of stripe rust resistance in the German wheat cultivar Alcedo. Theor Appl Genet 122: 723-733.

Johnson R (1988) Durable resistance to yellow (stripe) rust in wheat and its implications in plant breeding. In: Simmonds NW, Rajaram S (eds) Breeding Strategies for Resistance to the Rusts of Wheat. CIMMYT, Mexico D. F., Mexico, pp 63-75.

Kolmer JA, Garvin DF, Jin Y (2011) Expression of a Thatcher wheat adult plant stem rust resistance QTL on chromosome arm 2BL is enhanced by Lr34. Crop Sci 51: 526-533.

Kota R, Spielmeyer W, McIntosh RA, Lagudah ES (2006) Fine genetic mapping fails to dissociate durable stem rust resistance gene *Sr2* from pseudo-black chaff in common wheat (*Triticum aestivum* L.). Theor Appl Genet 112: 492-499.

Krattinger SG, Lagudah ES, Spielmeyer W, Singh RP, Huerta-Espino J, McFadden H, Bossolini E, Selter LL, Keller B (2009) A putative ABC transporter confers durable resistance to multiple fungal pathogens in wheat. Science 323: 1360-1363.

Lagudah ES (2011) Molecular genetics of race non-specific

rust resistance in wheat. Euphytica 179：81-91.

Lan CX, Liang SS, Zhou XC, Zhou G, Lu QL, Xia XC, He ZH (2010) Identification of genomic regions controlling adult-plant stripe rust resistance in Chinese landrace Pingyuan 50 through bulked segregant analysis. Phytopathology 100：313-318.

Lillemo M, Asalf B, Singh RP, Huerta-Espino J, Chen XM, He ZH, Bjørnstad Å (2008) The adult plant rust resistance loci *Lr34/Yr18* and *Lr46/Yr29* are important determinants of partial resistance to powdery mildew in bread wheat line Saar. Theor Appl Genet 116：1155-1166.

Lillemo M, Joshi AK, Prasad R, Chand R, Singh RP (2013) QTL for spot blotch resistance in bread wheat line Saar co-locate to the biotropic disease resistance loci Lr34 and Lr46. Theor Appl Genet doi：10.1007/s00122-012-2012-6 126：711-719.

Lin F, Chen XM (2007) Genetics and molecular mapping of genes for race-specific all-stage resistance and non-race-specific high-temperature adult-plant resistance to stripe rust in spring wheat cultivar Alpowa. Theor Appl Genet 114：1277-1287.

Lin F, Chen XM (2009) Quantitative trait loci for non-race-specific, high-temperature adult-plant resistance to stripe rust in wheat cultivar Express. Theor Appl Genet 118：631-642.

Lu YM, Lan CX, Linag SS, Zhou XC, Liu D, Zhou G, Lu QL, Jing JX, Wang MN, Xia XC, He ZH (2009) QTL mapping for adult-plant resistance to stripe rust in Italian common wheat cultivars Libellula and Strampelli. Theor Appl Genet 119：1349-1359.

Mago R, Tabe L., McIntosh RA, Pretorius Z, Kota R, Paux E, Wicker T, Breen J, Lagudah ES, Ellis JG, (2011) A multiple resistance locus on chromosome arm 3BS in wheat confers resistance to stem rust (*Sr2*), leaf rust (*Lr27*) and powdery mildew. Theoretical and Applied Genetics 123：615-623 .

Mallard S, Gaudet D, Aldeia A, Abelard C, Besnard AL, Sourdille P, Dedryver F (2005) Genetic analysis of durable resistance to yellow rust in bread wheat. Theor Appl Genet 110：1401-1409.

McFadden ES (1930) A successful transfer of emmer characters to *vulgare* wheat. J Am Soc Agron 22：1020-1034.

McIntosh RA, Wellings CR, Park RF (1995) Wheat rusts：an atlas of resistance genes. CSIRO, Australia.

Mantovani P, Maccaferri M, Sanguineti MC, Tuberosa R, Catizone I, Wenzl P, Thompson B, Carling J, Huttner E, Deambrogio E, Kilian A (2008) An integrated DArT-SSR linkage map of durum wheat. Mol Breeding 22：629-648.

Melichar JPE, Berry S, Newell C, MacCormack R, Boyd LA (2008) QTL identification and microphenotype characterisation of the developmentally regulated yellow rust resistance in the UK wheat cultivar Guardian. Theor Appl Genet 117：391-399.

Mishra AN, Kaushal K, Yadav SR, Shirsekar GS, Pandey HN (2005) The linkage between stem rust resistance gene *Sr2* and pseudo-black chaff in wheat can be broken. Plant Breeding 124：520-522.

Ohm HW, Shaner GE, (1976) Three components of slow-leaf rusting at different growth stages in wheat. Phytopathology 66：1356-1360.

Parlevliet JE (1975) Components of resistance that reduce the rate of epidemic development. Annual Review of Phytopathology 17：203-222.

Ramburan VP, Pretorius ZA, Louw JH, Boyd LA, Smith PH, Boshoff WHP, Prins R (2004) A genetic analysis of adult plant resistance to stripe rust in wheat cultivar Kariega. Theor Appl Genet 108：1426-1433.

Rao MVP (1996) Close linkage of the *Agropyron elongatum* gene *Sr26* for stem rust resistance to the centromere of wheat chromosome 6A. Wheat Information Service 82：8-10.

Ren Y, He ZH, Li J, Lillemo M, Wu L, Bai B, Lu QX, Zhu HZ, Zhou G, Du JY, Lu QL, Xia XC (2012a) QTL mapping of adult-plant resistance to stripe rust in a population derived from common wheat cultivars Naxos and Shanghai 3/Catbird. Theor Appl Genet 125：1211-1221.

Ren Y, Li ZF, He ZH, Wu L, Bai B, Lan CX, Wang CF, Zhou G, Zhu HZ, Xia XC (2012b) QTL mapping of adult-plant resistances to stripe rust and leaf rust in a Chinese wheat cultivar Bainong 64. Theor Appl Genet 125：1253-1262.

Rosewarne GM, Singh RP, Huerta-Espino J, Herrera-Foessel SA, Forrest KL, Hayden MJ, Rebetzke GJ (2012) Analysis of leaf and stripe rust severities reveals pathotype changes and multiple minor QTLs associated with resistance in an Avocet × Pastor wheat population. Theor Appl Genet 124：1283-1294.

Rosewarne GM, Singh RP, Huerta-Espino J, Rebetzke GJ (2008) Quantitative trait loci for slow-rusting resistance in wheat to leaf rust and stripe rust identified with multi-environment analysis. Theor Appl Genet 116：1027-1034.

Rosewarne GM, Singh RP, Huerta-Espino J, William HM, Bouchet S, Cloutier S, McFadden H, Lagudah ES (2006) Leaf tip necrosis, molecular markers and β1-protease subunits associated with the slow rusting resistance genes

Lr46/Yr29. Theor Appl Genet 112: 500-508.

Santra DK, Chen XM, Santra M, Campbell KG, Kidwell KK (2008) Identification and mapping QTL for high-temperature adult-plant resistance to stripe rust in winter wheat (*Triticum aestivum* L.) cultivar 'Stephens'. Theor Appl Genet 117: 793-802.

Singh RP, Huerta-Espino J, Rajaram S (2000a) Achieving near-immunity to leaf and stripe rusts in wheat by combining slow rusting resistance genes. Acta Phytopathol Hun 35: 133-139.

Singh RP, Nelson JC, Sorrells ME (2000b) Mapping *Yr28* and other genes for resistance to stripe rust in wheat. Crop Sci 40: 1148-1155.

Singh RP, Huerta-Espino J, William HM (2005) Genetics and breeding for durable resistance to leaf and stripe rusts in wheat. Turk J Agric For 29: 121-127.

Spielmeyer W, McIntosh RA, Kolmer J, Lagudah ES (2005) Powdery mildew resistance and *Lr34/Yr18* genes for durable resistance to leaf and stripe rust cosegregate at a locus on the short arm of chromosome 7D of wheat. Theor Appl Genet 111: 731-735.

Suenaga K, Singh RP, Huerta-Espino J, William HM (2003) Microsatellite Markers for genes *Lr34/Yr18* and other quantitative trait loci for leaf rust and stripe rust resistance in bread wheat. Phytopathology 93: 881-890.

Uauy C, Brevis JC, Chen XM, Khan I, Jackson L, Chicaiza O, Distelfeld A, Fahima T, Dubcovsky J (2005) High-temperature adult-plant (HTAP) stripe rust resistance gene *Yr36* from *Triticum turgidum* ssp. *dicoccoides* is closely linked to grain protien content locus *Gpc-B1*. Theor Appli Genet 112: 97-105.

Van der Plank JE (1963) Plant diseases; epidemics and control. Academic Press, New York and London.

Vazquez MD, Peterson CJ, Riera-Lizarazu O, Chen X, Heesacker A, Ammar K, Crossa J, Mundt CC (2012) Genetic analysis of adult plant, quantitiative resistance to stripe rust in wheat cultivar 'Stephens' in multi-environment trials. Theor Appl Genet 124: 1-11.

William HM, Singh RP, Huerta-Espino J, Palacios G, Suenaga K (2006) Characterization of genetic loci conferring adult plant resistance to leaf rust and stripe rust in spring wheat. Genome 49: 977-990.

Worland AJ, Law CN (1986) Genetic analysis of chromosome 2D of wheat 1. The location of genes affecting height, day-length insensitivity, hybrid dwarfism and yellow-rust resistance. Zeitschrift für Pflanzenzüchtung 96: 331-345.

Yang EN, Rosewarne GM, Herrera-Foessel SA, Huerta-Espino J, Tang ZX, Sun CF, Ren ZL, Singh RP (2013) QTL analysis of the spring wheat "Chapio" identifies stable stripe rust resistance despite inter-continental genotype × environment interactions. Theor Appl Genet 126: 1721-1732.

Zwart RS, Thompson JP, Milgate AW, Bansal UK, Williamson PM, Raman H, Bariana HS (2010) QTL mapping of multiple foliar disease and root-lesion nematode resistances in wheat. Mol Breeding 26: 107-124.

Supplementary Table 1 References that relate to stripe rust QTL studies and reference numbers as referred to Tables 1 - 7.

Reference Number	Authors	Reference Number	Authors
1	Agenbag et al. (2012)	18	Lu et al. (2009)
2	Bariana et al. (2001)	19	Mallard et al. (2005)
3	Bariana et al. (2010)	20	Melichar et al. (2008)
4	Börner et al. (2000)	21	Navabi et al. (2005)
5	Boukhatem et al. (2002)	22	Ramburan et al. (2004)
6	Carter et al. (2009)	23	Ren et al. (2012a)
7	Chhuneja et al. (2008)	24	Ren et al. (2012b)
8	Dedryver et al. (2009)	25	Rosewarne et al. (2008)
9	Feng et al. (2011)	26	Rosewarne et al. (2012)
10	Guo et al. (2008)	27	Santra et al. (2008)
11	Hao et al. (2011)	28	Singh et al. (2000)
12	Herrera-Foessel et al. (2011)	29	Suenaga et al. (2003)
13	Jagger et al. (2011)	30	Uauy et al. (2005)
14	Lan et al. (2010)	31	Vazquez et al. (2012)
15	Lillemo et al. (2008)	32	William et al. (2006)
16	Lin and Chen (2007)	33	Yang et al. (2013)
17	Lin and Chen (2009)	34	Zwart et al. (2010)

Overview and application of QTL for adult plant resistance to leaf rust and powdery mildew in wheat

Z. F. Li[1], C. X. Lan[2], Z. H. He[2,3], R. P. Singh[2], G. M. Rosewarne[2], X. M. Chen[3], and X. C. Xia[3]

[1]*College of Plant Protection, Hebei Agricultural University, Biological Control Center for Plant Diseases and Plant Pests of Hebei*, 289 *Lingyusi Street*, *Baoding* 071001, *Hebei*, *China*; [2]*International Maize and Wheat Improvement Center* (*CIMMYT*), *Apdo. Postal* 6-641, 06600 *México D. F.*, *México*; [3]*Institute of Crop Science*, *National Wheat Improvement Center*, *Chinese Academy of Agricultural Sciences* (*CAAS*), 12 *Zhongguancun South Street*, *Beijing* 100081, *China*.

Abstract: Leaf rust and powdery mildew, caused by *Puccinia triticina* and *Blumeria graminis* f. sp. *tritici*, respectively, are widespread fungal diseases of wheat (*Triticum aestivum* L.). Development of cultivars with durable resistance is crucially important for global wheat production. This paper reviews the progress of genetic study and application of adult plant resistance (APR) to wheat leaf rust and powdery mildew. Eighty leaf rust and 119 powdery mildew APR QTL (quantitative trait locus/loci) have been reported on 16 and 21 chromosomes, respectively, based on over 50 publications during the last 15 years. More importantly, we found 11 loci located on chromosomes 1BS, 1BL, 2AL, 2BS[2], 2DL, 4DL, 5BL, 6AL, 7BL, and 7DS showing pleiotropic effects on resistance to leaf rust, stripe rust, and powdery mildew. Among these, QTL on chromosomes 1BL, 4DL and 7DS also correlate with leaf tip necrosis. Fine mapping and cloning of these QTL will be achieved with the advent of cheaper high throughput genotyping technologies. Germplasm carrying these potential resistance genes will be useful for developing cultivars with durable multi-disease resistance. In addition to its non-NBS-LRR (Nucleotide Binding Site-Leucine Rich Repeat) structure, the senescence-like processes induced by *Lr34* could be the reason for durability of resistance; however, more information is needed for a full understanding of the molecular mechanism related to durability. APR genes have been utilized by CIMMYT for more than 30 years and have also been transferred to many Chinese wheat varieties through shuttle breeding.

Key words: *Triticum aestivum*, APR, durable resistance, plant breeding

Introduction

Wheat leaf (brown) rust, caused by *Puccinia triticina* (*Pt*), is one of the most widely distributed wheat (*Triticum aestivum* L.) diseases in the world. It can cause yield loss up to 40% in susceptible cultivars by decreasing kernel number per spike and reducing kernel weight (Khan et al., 2013). A particularly severe leaf rust epidemic in northwestern Mexico during 1976-1977 caused estimated yield losses of up to 70% (Dubin and Torres, 1981), and four significant leaf rust epidemics were documented in China in 1969, 1973, 1975 and 1979 (Dong, 2001). Leaf rust has become increasingly important in the major Chinese wheat production regions, particularly in Northern China and the southern part of the Yellow and Huai Valleys (Zhao et al., 2008) possibly due to climate change. In 2012,

Published in Crop Science, 2014, 54: 1907-1925

significant yield losses were recorded in some regions of Gansu, Sichuan, Shaanxi, Henan and Anhui (Zhou et al., 2013).

Powdery mildew, caused by *Blumeria graminis* f. sp. *tritici* (*Bgt*), is an important wheat disease in regions with temperate and maritime climates, such as Europe, North and South America, Africa, and major parts of China (Hsam and Zeller, 2002; He et al., 2011). The use of semi-dwarf, high-yielding cultivars in combination with high levels of nitrogen fertilization led to rapid increases in frequencies and severities of powdery mildew epidemics. Powdery mildew causes estimated yield losses of 5 to 34% (Conner et al., 2003). It has been widespread across large areas of China since the 1980s and is now a major disease in Chinese winter wheat areas (Li and Zeng, 2002; He et al., 2011). Most of the Chinese wheat cultivars are susceptible to the currently widespread *Bgt* isolate E20. An average of 6.9 million ha of wheat was reported to be affected by powdery mildew during the period 2005-2008 (He et al., 2011).

Stripe (yellow) rust, caused by *P. striiformis* f. sp. *tritici* (*Pst*), affects up to 40% of the wheat production area in countries such as Mexico, India, Pakistan, Bangladesh, and China (Dubin and Brennan, 2009), and more than 20 stripe rust epidemics have been documented worldwide (Wellings, 2011). It was the most important wheat disease in China before the 1990s, and is still the major disease in the northwestern and southwestern areas, including the provinces of Shaanxi, Gansu, Sichuan, and Yunnan.

Stem rust, caused by *P. graminis* f. sp. *tritici* (*Pgt*), has historically been a menace to wheat production worldwide (Saari and Prescott, 1985). A new *Pgt* race, TTKSK (commonly referred to as Ug99), detected in Uganda in 1998, had virulence to most of the widely deployed specific resistance genes and was seen as threat to global food security. Since then, variants of this race with virulence to additional stem rust resistance genes, and races belonging to the same lineage, have appeared in Kenya, Ethiopia, Zimbabwe, South Africa, Sudan, Yemen, Iran (Singh et al., 2011) and Tanzania (Hale et al., 2013). Yield losses caused by the Ug99 stem rust lineage in Uganda, Kenya, Ethiopia, Yemen, the Middle East and South Asia were estimated to be approximately 3 billion USD since 1998 (Khan et al., 2013).

Resistance to stripe rust and stem rust will not be reviewed in detail in this study since stripe rust QTL were summarized recently (Rosewarne et al., 2013) and current information on APR to stem rust is limited. However, some genes which give resistance to one fungal disease are known to have pleiotropic effects against other fungal pathogens as well. Thus, the objectives of this review are to: 1) summarize APR QTL for leaf rust and powdery mildew; 2) identify potential pleiotropic resistance loci for leaf rust, stripe rust and powdery mildew; 3) explore the potential mechanism of APR genes; and 4) report progress on the application of APR genes in developing durable multiple disease resistant cultivars in CIMMYT and Chinese wheat breeding programs.

Resistance to leaf rust and powdery mildew in wheat

There are two types of resistance to wheat rusts and powdery mildew. The first is race-specific or major gene resistance that involves hypersensitive reactions, and interacts with the pathogen in a gene-for-gene manner (Flor, 1942). This kind of interaction imposes a strong selection pressure on the pathogen to drive the increase of new virulent mutants or of virulent races already present at low frequency. Most of the formally designated rust resistance genes are race-specific. Gene combinations, strategic gene deployment and multi-line cultivars are suggested strategies to prolong race-specific resistance, and abundant numbers of effective genes are necessary for optimal deployment. Forty six designated race-specific leaf rust resistance loci, viz. *Lr1*, *Lr3*, *Lr9*, *Lr10*, *Lr13*, *Lr14*, *Lr17*, *Lr19*, *Lr20*, *Lr21*, *Lr24*, *Lr25*, *Lr26*, *Lr27*, *Lr28*, *Lr29*, *Lr31*, *Lr32*, *Lr34*, *Lr35*, *Lr37*, *Lr38*, *Lr39*, *Lr42*, *Lr44*, *Lr45*, *Lr46*, *Lr47*, *Lr48*, *Lr49*, *Lr50*,

Lr51, *Lr52*, *Lr53*, *Lr57*, *Lr58*, *Lr60*, *Lr61*, *Lr63*, *Lr64*, *Lr65*, *Lr67*, *Lr68*, *Lr70*, *Lr71*, and *Lr72*, and 31 loci for powdery mildew resistance, viz. *Pm1*, *Pm2*, *Pm3*, *Pm4*, *Pm5*, *Pm6*, *Pm8*, *Pm12*, *Pm13*, *Pm16*, *Pm17*, *Pm21*, *Pm24*, *Pm25*, *Pm26*, *Pm27*, *Pm29*, *Pm30*, *Pm31*, *Pm33*, *Pm34*, *Pm35*, *Pm36*, *Pm37*, *Pm40*, *Pm41*, *Pm42*, *Pm43*, *Pm44*, *Pm45*, and *Pm47*, are mapped to wheat chromosomes and some linked molecular markers with variable potential for application in marker assisted selection are available (McIntosh et al., 2013).

The second type is race non-specific resistance, which is usually effective in the post-seedling growth stage. It is associated with longer latent period, lower infection frequency, smaller uredinial size, reduced duration of sporulation, and less spore production per infection site (Caldwell, 1968). This type of resistance is also called APR or slow rusting (or slow mildewing) resistance, and most of APR genes showed race non-specific characteristics; however, some sources of specific resistance may also fulfil these criteria. Although the time of onset of non-specific resistance and level of effectiveness vary with growth stage and environment, there is general agreement that high levels of resistance can be achieved by combining the so-called minor genes underlying such resistance. Adequate levels of resistance are usually achieved when 3 to 5 minor genes are combined. Historically, when accessions or varieties displaying adequate levels of this kind of resistance were analysed genetically there were indications of quasi-quantitative inheritance (Johnson and Law, 1973; Das et al., 1992), and subsequent studies estimated that three or more often interactive genes were involved. Features or suggested principles emerging from this are that genotypes with quantitative resistance display resistance to all or most pathotypes, and as the resistance does not display immunity, there is greatly reduced selection pressure on the pathogen, thus decreased likelihood of increase of new virulent mutants if they occur, resulting in longevity of resistance. The outcome, therefore, is that slow rusting (mildewing) resistance is more durable than race-specific resistance (Bjarko and Line, 1988). For example, Chinese landrace Pingyuan 50, US variety Anza selected from CIMMYT germplasm and Mexican cultivar Pavon 76 have shown moderate and stable resistance for extended time periods.

Among the 72 leaf rust and 47 powdery mildew resistance genes formally catalogued (McIntosh et al., 2013), only 12 genes, viz. *Lr12*, *Lr13*, *Lr22* (alleles *a* and *b*), *Lr34*, *Lr35*, *Lr37*, *Lr46*, *Lr48*, *Lr49*, *Lr67*, and *Lr68*, confer APR to leaf rust and three (*Pm38*, *Pm39* and *Pm46*) confer APR to powdery mildew (McIntosh et al., 2013). Alleles at three slow rusting APR loci, *Lr34/Yr18/Pm38/Sr57* (Spielmeyer et al., 2005), *Lr46/Yr29/Pm39/Sr58* (William et al., 2003), and *Lr67/Yr46/Pm46/Sr55* (Herrera-Foessel et al., 2014), are believed to be race non-specific and confer pleiotropic effects on other disease responses, including leaf rust, stripe rust, powdery mildew, stem rust, and spot blotch.

Use of molecular markers to identify APR genes for leaf rust and powdery mildew

During the past two decades, molecular markers have been widely used in identification of APR genesfor leaf rust and powdery mildew (Tables 1 and 2). Marker types have changed with advancing technologies and have included restriction fragment length polymorphisms (RFLP), random amplified polymorphic DNA (RAPD), amplified fragment length polymorphisms (AFLP), simple sequence repeats (SSR), diversity arrays technology (DArT), single nucleotide polymorphisms (SNP), and genotyping-by-sequencing (GBS).

Among these systems, RFLP markers were mostly codominant and restricted to regions with low-copy sequences and four seedlingresistance genes *Lr10*, *Lr23*, *Lr27*, *Lr31*, and one APR gene *Lr34* were mapped by this method (Nelson et al., 1997). RAPD markers are usually dominant, coupled with low levels of polymorphism, and have reproducibility problems in wheat (Devos and Gale, 1992). They were used on a Parula/Siete Cerros RIL populaion to identify leaf rust APR

QTL on chromosomes 1B, 1D and 7BL (William et al., 1997). Compared with the above two systems, AFLP markers give higher reproducibility and resolution at the whole genome level; however, the procedure of AFLP analysis is complex and costly (Mueller et al., 1999). Once available, SSR markers became the preferred system in the late 1990s due to co-dominance, accuracy, high repeatability, high levels of polymorphism, chromosome specificity, and ease of manipulation (Röder et al., 1998).

Table 1 Summary of QTL or genes for adult-plant resistance to leaf rust in wheat

QTL	Chr.	Donor	Marker interval	R^2 (%)	Reference
QLr. sfrs-1BS	1BS	Forno	*Xpsr949-Xgwm18*	10. 6	Messmer et al., 2000
QLr. sfr-1BS	1BS	Forno	*Xgwm604-OA93*	28-31. 5	Schnurbusch et al., 2004
QLr. cimmyt-1BS. 1	1BS	Parula	*Xcmtr03-500*	7-10	William et al., 1997
QLr. cimmyt-1BS. 2	1BS	Pastor	*wPT5580-wPT3179*	4. 1-6. 1	Rosewarne et al., 2012
QLr. pbi-1BS	1BS	Beaver	1BL/1RS	17. 3	Singh et al., 2009
Lr46/Yr29/Pm39	1BL	Saar	*Xwmc719-Xhbe248*	49. 1	Lillemo et al., 2008
Lr46	1BL	Pavon76	*Xksul27-PAAGMCTA-1*	46. 1-53. 9	William et al., 2003
Lr46	1BL	Pastor	*cslv46-Xgwm818*	16. 7-25. 4	Rosewarne et al., 2012
Lr46	1BL	Oligo	*Xwmc44-Xgwm793*	12. 9-17. 4	Suenaga et al., 2003
Lr46	1BL	Pavon	*Xgwm140*, *Xgwm259*	55. 8	William et al., 2006
QLr. caas-1BL	1BL	Bainong 64	*Xgwm153. 2-Xwmc44*	21. 1-28. 5	Ren et al., 2012b
QLr. pser-1BL	1BL	Ning7840	lm, *Xscm9-Xwmc85. 1*	60. 8	Li and Bai, 2009
QLr. csiro-1BL	1BL	Attila	*XP84/M78-LTN*	/	Rosewarne et al., 2008
Lr37	2AS	Madsen	*Xcmwg682*	/	Helguera et al., 2003
QLr. cimmyt-2AL	2AL	Avocet	*wPT4419-wPT8226*	5. 8-7. 2	Rosewarne et al., 2012
QLr. sfr-2AL	2AL	Forno	*cfa2263c-sfr. BE590525*	9. 5-12	Schnurbusch et al., 2004
QLr. ubo - 2A	2AL	Lloyd	*wPt-386-310911*	18. 6-30	Maccaferri et al., 2008
Lr48	2BS	CSP44	*Xgwm429b-Xbarc07*	/	Bansal et al., 2008
Lr13	2BS	ThLr13	*Xgwm630*	/	Seyfarth et al., 2000
QLr. csiro-2BS	2BS	Attila	*Xgwm682-XP32/M62*	/	Rosewarne et al., 2008
QLr. cimmyt-2BS	2BS	Pastor	*wPT6278*, *Yr31*	5. 2-9. 6	Rosewarne et al., 2012
QLr. ksu-2BS	2BS	W-7984	*Per2* (*Lr23*?)	15. 7	Faris et al., 1999
Lr35	2BS	ThatcherLr35	*Xwg996*, *Xpsr540*, *Xbcd260*	/	Seyfarth et al., 1999
QLr. osu-2B	2BL	*CI 13227*	*Xagc. tgc135-Xcatg. atgc60*	18. 8	Xu et al., 2005a
QLrlp. osu-2B	2BL	CI 13227	Xcag. cgat70-Xcatg. atgc60	16. 2	Xu et al., 2005b
QLr. sfrs-2BL	2BL	Oberkulmer	*Xpsr924-Xglk699a*	7. 2	Messmer et al., 2000
QLr. cimmyt-2DS	2DS	Avocet	*wPT8319-wPT3728*	3. 8-9. 8	Rosewarne et al., 2012
QLrlp. osu-2DS	2DS	CI 13227	*Xactg. gtg185-Xbarc124*	42. 8	Xu et al., 2005b
QLr. hbau-2DS	2DS	Saar	*Xbarc124-Xgwm296a*	12. 2-12. 5	Zhang et al., 2009
Lr22a	2DS	RL6044	*Xgwm296*	/	Hiebert et al., 2007
QLrid. osu-2DS	2DS	CI 13227	*Xgwm261*, *XGCTG. CGCT118*	21. 5-26. 4	Xu et al., 2005a
QLr. sfr-2DS	2DS	Forno	gdm35-cfd53	10. 3-14. 8	Schnurbusch et al., 2004
QLr. sfr-2DL	2DL	Arina	glk302-gwm539	11. 4-12. 7	Schnurbusch et al., 2004

（续）

QTL	Chr.	Donor	Marker interval	R^2 (%)	Reference
QLr. ubo-3A	3AS	Lloyd	*311707-Xwmc664*	24. 8	Maccaferri et al. , 2008
QLr. sfrs-3A	3AL	Forno	*Xpsr570-Xpsr543*	13. 5	Messmer et al. , 2000
QLr. fcu-3AL	3AL	TA4152-60	*Xcfa2183 - Xgwm666*	10-18	Chu et al. , 2009
QLr. sfrs-3B	3B	Oberkulmer	*Xpsr919a-Xpsr1101b*	9. 3	Messmer et al. , 2000
QLr. fcu-3BL	3BL	TA4152-60	*Xbarc164 - Xfcp544*	19-20	Chu et al. , 2009
Lr12	4A	Exchange	nd	/	Dyck and Kerber, 1971
QLr. sfrs-4B	4B	Forno	*Xpsr921-Xpsr953b*	10	Messmer et al. , 2000
Lr49	4BL	VL404	*Xbarc163-Xwmc349*	/	Bansal et al. , 2008
QLr. pbi-4BL	4BL	Beaver	wPt-1708	12. 4	Singh et al. , 2009
QLr. cimmyt-4BL	4BL	Avocet	*Xgwm495*, *Xgwm368*	6. 4-8. 9	William et al. , 2006
QLr. sfr-4BS	4BS	Forno	gwm368-gwm540a	10. 7	Schnurbusch et al. , 2004
Lr67	4D	RL6077	*Xgwm165/Xgwm192*	nd	Herrera-Foessel et al. , 2011
QLr. fcu-4DL	4DL	ND495	*Xgdm61 - Xcfa2173*	7-13	Chu et al. , 2009
QLr. sfrs-4DL	4DL	Forno	*Xglk302b-Xpsr1101a*	8. 7-19. 8	Messmer et al. , 2000
QLr. pbi-5AS	5AS	Beaver	wPt1931-wPt8756	11. 2	Singh et al. , 2009
QLr. sfrs-5AS	5AS	Forno	*Xpsr945a-Xglk424*	7. 7	Messmer et al. , 2000
QLr. cimmyt-5AL	5AL	Avocet	wPT0373-wPT0837	5. 2-7. 4	Rosewarne et al. , 2012
QLr. hbau-5BL	5BL	SAAR	*XDuPw395-Xgwm777*	4. 9-11. 2	Zhang et al. , 2009
QLr. sfrs-5BL	5B	Forno	*Xpsr580b-Xpsr143*	14. 6	Messmer et al. , 2000
QLr. fcu-5BL	5BL	TA4152-60	*Xgdm116 - Xbarc59*	7-10	Chu et al. , 2009
QLr. sfrs-5DL	5DL	Oberkulmer	*Xpsr906a-Xpsr580a*	9. 1	Messmer et al. , 2000
QLr. cimmyt-6AL	6AL	Avocet	*Xgwm617*	4. 8-6. 3	William et al. , 2006
QLr. hbau-6AL	6AL	Avocet	*Xgwm617a-Xgwm169*	7	Zhang et al. , 2009
QLr. caas-6BS. 1	6BS	Bainong 64	*Xwmc487-Xcfd13*	10-10. 8	Ren et al. , 2012b
QLr. caas-6BS. 2	6BS	Jingshuang 16	*Xgwm518-Xwmc398*	8. 2-9	Ren et al. , 2012b
QLr. fcu-6BL	6BL	TA4152-60	*Xbarc5 - Xgwm469. 2*	12	Chu et al. , 2009
QLr. cimmyt-6BL. 1	6BL	Pastor	wPT6329-wPT5176	5. 4-10. 8	Rosewarne et al. , 2012
QLr. cimmyt-6BL. 2	6BL	Pavon 76	*XpAGGmCGA1*	4. 37	William et al. , 2006
QLr. ubo-7B. 1	7BS	Colosseo	*Xwmc405. 1-Xgwm573*	9. 7-19. 4	Maccaferri et al. , 2008
QLr. sfrs-7B. 1	7BS	Oberkulmer	*Xpsr952-Xgwm46*	7. 6	Messmer et al. , 2000
QLr. sfr-7BS	7BS	Arina	sfr. BE427461-gwm573b	8. 8	Schnurbusch et al. , 2004
QLr. osu-7BL	7BL	CI 13227	*Xaca. cacg126/Xbarc50*	12. 9-20. 8	Xu et al. , 2005a
QLrlp. osu-7BL	7BL	CI 13227	*XBarc182-Xcatg. atgc125*	13. 8	Xu et al. , 2005b
Lr68	7B	Parula	Psy1-1-gwm146	/	Herrera-Foessel et al. , 2012
QLr. ubo-7B. 2	7BL	Colosseo	*Xbarc340. 2-Xgwm146*	49. 8-76. 9	Maccaferri et al. , 2008
QLr. sfrs-7B. 2	7B	Forno	*Xpsr593c-Xpsr129c*	35. 8	Messmer et al. , 2000
QLr. sfrs-7B. 3	7B	Forno	*Xglk750-Xmwg710a*	12. 8	Messmer et al. , 2000
QLr. cimmyt-7BL. 1	7BL	Parula	*Xcmtg05-50*, *Xcmti16-1500*	18-30	William et al. , 1997
QLr. csiro-7BL. 2	7BL	Attila	*Xgwm146*, *Xwmc273*	/	Rosewarne et al. , 2008
QLr. cimmyt-7BL	7BL	Pastor	wPT4342-wPT8921	3. 8-11. 5	Rosewarne et al. , 2012

（续）

QTL	Chr.	Donor	Marker interval	R^2 (%)	Reference
QLr. sfr-7BL (*Lr14a*)	7BL	Forno	ksuD2-gbxG218b	15. 9	Schnurbusch et al. , 2004
QLr. ksu-7BL	7BL	Opata	Cht1b/Tha1/Cat	11-42. 2	Faris et al. , 1999
Lr34/Yr18/Pm38	7DS	SAAR	*XwPt3328-XcsLV34*	73. 1	Lillemo et al. , 2008
Lr34	7DS	Cook	*Xgwm37*, *Xgwm295*	/	Navabi et al. , 2005
Lr34	7DS	Forno	cfd66-gwm1002	32. 6-42. 9	Schnurbusch et al. , 2004
Lr34	7DS	Fukuho	*Xgwm295. 1-Xgwm130*	36. 4-45. 2	Suenaga et al. , 2003
Lr34	7DS	Opata	*Xwg834*	8-24. 1	Faris et al. , 1999

R^2 (%), percentage of variance explained by the QTL (PVE);

/, R^2 of this QTL is unknown.

Table 2 Summary of QTL for adult-plant resistance to powdery mildew in wheat

QTL	Chr.	Donor	Marker interval	R^2 (%)	Reference
QPm. osu-1A	1AS	2174	*Pm3a*	63. 0[a]	Chen et al. , 2009
QPm. caas-1AS	1AS	Fukuho-komugi	*Xgdm33-Xpsp2999*	19. 9-26. 6	Liang et al. , 2006
QPm. sfr-1A	1AL	Oberkulmer	*Xpsr1201b-Xpsr941*	7. 7	Keller et al. , 1999
QPm. crag-1A	1AL	RE714	*Xcdo572-Xbad442*	39. 3-43. 0[b]	Mingeot et al. , 2002
QPm. caas-1AL	1AL	Bainong 64	*Xbarc148-Xwmc550*	7. 4-9. 9	Lan et al. , 2009
QPm. sfr-1B	1BS	Forno	*CD9b-Xpsr593a*	11. 6	Keller et al. , 1999
QPm. ttu-1B	1BS	*Triticum militinae*	*Xgwm3000*	4. 0-5. 0	Jakobson et al. , 2006
QPm. vt-1BL	1BL	Massey	*Xgwm259-Xbarc80*	15. 0-17. 0	Tucker et al. , 2007
QPm. vt-1B	1BL	USG3209	*WG241*	17. 0	Liu et al. , 2001
Yr29/Lr46/Pm39	1BL	Saar	*Xwmc719-Xhbe248*	7. 3-35. 9	Lillemo et al. , 2008
QPm. osu-1B	1BL	2174	WMC134	14. 0	Chen et al. , 2009
Qaprpm. cgb-1B	1B	Hanxuan 10	WMC269. 2-CWM90	4. 8-20. 3	Huang et al. , 2008
QPm. inra-1D. 1	1DS	RE9001	*Xgwm106*	12. 6	Bougot et al. , 2006
QPm. sfr-1D	1DL	Forno	*Xpsr168-Xglk558b*	9. 5	Keller et al. , 1999
QPm. sfr-2A	2AS	Oberkulmer	*Xpsr380-Xglk293b*	7. 7	Keller et al. , 1999
QPm. inra-2A	2AS	Courtot	*Xgwm275*	7. 4	Bougot et al. , 2006
QPm. crag-2A	2AL	RE714	*Pm4b-gbxG303*	22. 7-33. 6	Mingeot et al. , 2002
QPm. ttu-2A	2AL	*Triticum militinae*	*Xgwm311-Xgwm382*	5. 0	Jakobson et al. , 2006
QPm. vt-2AL	2AL	Massey	*Xgwm304- Xgwm312*	29. 0	Liu et al. , 2001
QPm. vt-2A	2AL	USG3209	*Xgwm304-Xgwm294*	26. 0-29. 0	Tucker et al. , 2007
QPm. crag-2B	2BS	Festin	*Xgwm148-gbxG553*	23. 6-71. 5	Mingeot et al. , 2002
QPm. caas-2BS	2BS	Lumai 21	*Xbarc98-Xbarc1147*	10. 6-20. 6	Lan et al. , 2010b
QPm. umb-2BS	2BS	Folke	*wPt-9402*	3. 9-13. 0	Lillemo et al. , 2012
QPm. umb-2BS	2BS	Folke	*Xgwm410b-Xgwm148*	8. 0-10. 2	Lillemo et al. , 2012
CP5	2BS	Pedroso	*wPt-5513*	12. 3	Marone et al. , 2013
QPm. vt-2B	2BL	Massey	*WG338-Xgwm526a*	11. 0	Liu et al. , 2001
QPm. caas-2B	2BL	Fukuho-komugi	*Xgwm877-Xgwm47*	5. 7-8. 0	Liang et al. , 2006

（续）

QTL	Chr.	Donor	Marker interval	R^2（%）	Reference
QPm. inra-2B	2BL	RE9001	*Xrtpl14R-Xcfd267b*	10. 3-36. 3	Bougot et al.，2006
QPm. vt-2BL	2BL	USG3209	*Xgwm501-Xgwm191*	11. 0-15. 0	Tucker et al.，2007
QPm. caas-2BL	2BL	Lumai 21	*Xbarc1139-Xgwm47*	5. 2-10. 1	Lan et al.，2010b
Qaprpm. cgb-2B	2B	Hanxuan 10	*Xwmc477-Xwmc272*	5. 4	Huang et al.，2008
Qpm. caas-2DS	2DS	Libellula	*Xcfd51-Xcfd56*	2. 3-3. 4	Asad et al.，2012
QPm. inra-2D-a	2DS	RE9001	*Xgwm102*	19. 0	Bougot et al.，2006
QPm. inra-2D-b	2DS	RE9001	*Xcfd2e*	16. 5	Bougot et al.，2006
QPm. sfr-2D	2DL	Oberkulmer	*Xpsr932-Xpsr331a*	10. 0	Keller et al.，1999
Qpm. ipk-2D	2DL	W7984	*Xglk558-XksuD23*	/	Börner et al.，2002
QPm. caas-2DL	2DL	Lumai 21	*Xwmc18-Xcfd233*	5. 7-11. 6	Lan et al.，2010b
QPm. umb-2DL	2DL	Folke	*Xwmc167-Xgwm301*	4. 3-9. 5	Lillemo et al.，2012
QPm. sfr-3A	3AS	Forno	*Xpsr598-Xpsr570*	10. 4	Keller et al.，1999
QPm. crag-3A	3AS	Festin	*Xpsr598-Xgwm5*	21. 4-25. 9	Mingeot et al.，2002
QPm. nuls-3AS	3AS	Saar	*Xstm844tcac-Xbarc310*	8. 1-20. 7	Lillemo et al.，2008
Qaprpm. cgb-3A	3A	Hanxuan 10	*Xwmc21-Xwmc505. 2*	9. 8	Huang et al.，2008
QPm. inra-3B	3BS	Courtot	*Xgwm389*	22. 7	Bougot et al.，2006
QPm. osu-3B	3BS	2174	WMS533	10. 0	Chen et al.，2009
QPm. caas-3B	3BS	Opata 85	*XksuG53-Xfba190*	7. 3	Huo et al.，2005
CP2	3BS	Creso	*F103*	10. 6	Marone et al.，2013
Qaprpm. cgb-3BL	3B	Hanxuan 10	*Xgwm181-Xgwm340*	13. 3	Huang et al.，2008
QPm. sfr-3D	3DS	Oberkulmer	*Xpsr1196a-Lrk10-6*	15. 7	Keller et al.，1999
QPm. inra-3D	3DS	RE9001	*Xcfd152，Xgwm707*	9. 3-15. 2	Bougot et al.，2006
QPm. sfr-4A. 1	4AL	Forno	*Xgwm111c-Xpsr934a*	14. 7	Keller et al.，1999
QPm. sfr-4A. 2	4AL	Forno	*Xmwg710b-Xglk128*	14. 3	Keller et al.，1999
QPm. ttu-4A	4AL	*Triticum militinae*	*Xgwm232-Xgwm160*	35. 0-54. 0[c]	Jakobson et al.，2006
QPm. inra-4A	4AL	RE714	*XgbxG036*	4. 9-6. 9	Chantret et al.，2001
QPm. crag-4A	4AL	RE714	*XgbxG036-XgbxG542*	22. 3	Mingeot et al.，2002
QPm. inra-4A	4AL	Courtot	*Xcfd71b*	8. 9	Bougot et al.，2006
QPm. osu-4A	4AL	2174	WMS160	12. 0	Chen et al.，2009
QPm. tut-4A	4A	8. 1	*Xwmc232-Xrga3. 1*	24-46	Jakobson et al.，2012
QPm. sfr-4B	4BL	Forno	*Xpsr593b-Xpsr1112*	7. 5	Keller et al.，1999
QPm. ipk-4B	4BL	W7984	*Xcdo795-Xbcd1262*	/	Börner et al.，2002
QPm. caas-4BL	4BL	Oligoculm	*Xgwm375-Xgwm251*	5. 9	Liang et al.，2006
QPm. nuls-4BL	4BL	Avocet	*XwPt1505-Xgwm149*	21. 0-40. 2	Lillemo et al.，2008
QPm. Caas-4BL. 1	4BL	Libellula	*Xgwm149-Xgwm495*	9. 1-14. 7	Asad et al.，2012
QPm. sfr-4D	4DL	Forno	*Xglk302b-Xpsr1101a*	14. 4	Keller et al.，1999
qApr4D	4D	Yumai 57	*Xgwm194-Xcfa2173*	20	Zhang et al.，2008

（续）

QTL	Chr.	Donor	Marker interval	R^2 (%)	Reference
QPm. caas-4DL	4DL	Bainong 64	*Xbarc200-Xwmc33*	15. 2-22. 7	Lan et al., 2009
QPm. sfr-5A. 1	5AS	Oberkulmer	*Xpsr644a-Xpsr945a*	22. 9	Keller et al., 1999
QPm. ttu-5A	5AS	*Triticum militinae*	*Xgwm186 - Xgwm415*	4. 0-6. 0	Jakobson et al., 2006
QPm. sfr-5A. 2	5AL	Oberkulmer	*Xpsr1194-Xpsr918b*	16. 6	Keller et al., 1999
QPm. sfr-5A. 3	5AL	Oberkulmer	*Xpsr911-Xpsr120a*	10. 5	Keller et al., 1999
QPm. nuls-5A	5AL	Saar	*Xgwm617b-Xwmc327*	4. 2-15. 2	Lillemo et al., 2008
Qaprpm. cgb-5A	5A	Hanxuan 10	*P3616-185-P3616-195*	13. 2	Huang et al., 2008
QPm. tut-5A	5A	8. 1	*Xgwm666-Xcfd30-Xbarc319*	14-16	Jakobson et al., 2012
QPm. nau-5AL	5AL	TA2027	*Xcfd39/Xmag1491-Xmag1493*	59	Jia et al., 2009
QPm. umb-5AL	5AL	Folke	*wPt-2426*	4. 0-9. 7	Lillemo et al., 2012
QPm. umb-5BS	5BS	T2038	*wPt-1261*	3. 1	Lillemo et al., 2012
QPm. umb-5BS	5BS	Folke	*Xbarc128a-Xgwm213*	8. 1-12. 9	Lillemo et al., 2012
QPm. ttu-5B	5BS	Tahti	*Xgwm133. mi6-Xgwm205. mi1*	4. 0-6. 0	Jakobson et al., 2006
QPm. nuls-5B	5BS	Saar	*Xbarc4-Xgwm274b*	9. 7	Lillemo et al., 2008
QPm. sfr-5B	5BL	Oberkulmer	*Xpsr580b-Xpsr143*	12. 6	Keller et al., 1999
QPm. inra-5B. 2	5BL	Courtot	*Xgwm790b*	11. 1	Bougot et al., 2006
Qaprpm. cgb-5B	5B	Lumai 14	*Xgwm213-Xgwm499*	19. 8	Huang et al., 2008
QPm. inra-5D	5DS	RE9001	*cfd189*	9. 0	Bougot et al., 2006
QPm. crag-5D. 1	5DL	RE714	*Xgwm639a-Xgwm174*	30. 2-38. 9[b]	Mingeot et al., 2002
QPm. crag-5D. 2	5DL	RE714	*Xcfd8B9-Xcfd4A6*	24. 0-37. 8	Mingeot et al., 2002
QpmVpn. inra-5D	5DL	Courtot	*Xcfd8*	11. 0	Bougot et al., 2006
QPm. inra-5D. 1	5DL	RE714	*Xcfd26*	28. 1-37. 7	Chantret et al., 2001
QPm. inra-5D. 2	5DL	RE714	*XgbxG083c*	37. 7	Chantret et al., 2001
QPm. caas-5D	5D	W7984	*Xmwg922-Xbcd1103*	5. 9	Huo et al., 2005
qApr5D	5D	Yumai 57	*Xwmc215-Xgdm63*	1. 3	Zhang et al., 2008
CP1	6AS	Pedroso	*MAG1200b*	12. 6	Marone et al., 2013
QPm. inra-6A	6AL	RE714	*MIRE* (*Xgwm427*)	8. 8-13. 4	Chantret et al., 2001
QPm. crag-6A	6AL	RE714	*MIRE*	19. 8-53. 9[d]	Mingeot et al., 2002
QPm. sfr-6B	6BS	Forno	*Xpsr167b-Xpsr964*	8. 7	Keller et al., 1999
QPm. umb-6BS	6BS	Folke	*wPt-6437-Xwmc494*	6. 5-10. 3	Lillemo et al., 2012
QPm. caas-6BS	6BS	Bainong 64	*Xbarc79-Xgwm518*	10. 3-16. 0	Lan et al., 2009
Qaprpm. cgb-6B	6B	Hanxuan 10	*Xgwm193-P3470-210*	21. 0	Huang et al., 2008
QPm. caas-6BL. 1	6BL	Huixianhong	*Xgwm219-Xbarc24*	2. 5-5. 2	Asad et al., 2012
QPm. caas-6BL. 2	6BL	Huixianhong	*Xbarc24-Xbarc345*	0. 5-1. 9	Asad et al., 2012
CP3	6BL	Pedroso	*Xgwm219-Xgwm889*	14. 8-18. 5	Marone et al., 2013
CP4	6BL	Pedroso	*wPt-5270*	13. 4	Marone et al., 2013
QPm. osu-6D	6DS	2174	BARC196	5. 0	Chen et al., 2009

（续）

QTL	Chr.	Donor	Marker interval	R^2（%）	Reference
QPm. inra-7A	7AS	RE714	*Xfba069-Xgwm344*	2. 9-6. 4	Chantret et al. , 2001
QPm. caas-7A	7AS	Bainong 64	*Xbarc127-Xbarc174*	6. 3-7. 1	Lan et al. , 2009
Qaprpm. cgb-7A	7A	Hanxuan 10	*CWM462. 2-Xgwm635. 2*	8. 0	Huang et al. , 2008
QPm. tut-7A	7A	8. 1	*Xgwm635-Xbarc70-Waxy*	9-28	Jakobson et al. , 2012
QPm. umb-7AL	7AL	T2038	*Xgwm428-Xcfa2040*	6. 4-13. 0	Lillemo et al. , 2012
QPm. sfr-7B. 1	7BL	Forno	*Xpsr593c-Xpsr129c*	11. 3	Keller et al. , 1999
QPm. sfr-7B. 2	7BL	Forno	*Xglk750-Xmwg710a*	31. 8	Keller et al. , 1999
QPm. crag-7B	7BL	RE714	*XpdaC01-XgbxR035b*	22. 8-33. 5	Mingeot et al. , 2002
QPm. inra-7B	7BL	RE714	*Xgwm577*	1. 7	Chantret et al. , 2001
QPm. nuls-7BL	7BL	Saar	*Xwmc581- XwPt8007*	4. 9	Lillemo et al. , 2008
Qaprpm. cgb-7B	7B	Lumai 14	*Xwmc273-Xwmc276*	12. 6	Huang et al. , 2008
QPm. caas-7DS	7DS	Libellula	*XcsLV34-Xgwm295*	7. 6-13. 8	Asad et al. , 2012
QPm. ipk-7D	7DS	Optata	*Xwg834-Xbcd1872*	/	Börner et al. , 2002
QPm. caas-7DS	7DS	Fukuho-komugi	*Ltn-Xgwm295. 1*	12. 0	Liang et al. , 2006
Yr18/Lr34/Pm38	7DS	Saar	*Xgwm1220-Xswm10*	19. 0-56. 5	Lillemo et al. , 2008
QPm. inra-7D. 1	7DS	Courtot	*Xgpw1106*	10. 6	Bougot et al. , 2006
Qaprpm. cgb-7D	7D	Hanxuan 10	*Xwmc436-Xgwm44*	3. 8-4. 6	Huang et al. , 2008
QPm. Caas-7D	7D	Opata 85	*Xwg834-Xbcd1438*	29. 6	Huo et al. , 2005

R^2（%），percentage of variance explained by the QTL（PVE）；

/，R^2 of this QTL is unknown.

a. Residual effect of major resistance gene *Pm3a*；

b. QTL detected by the software Mapmaker QTL；

c. QTL detected by the software Map Manager QTX Version b16；

d. QTL was attributed to the residual effect of MlRE.

More recently，high-throughput technologies，DArT，SNP and GBS，have become the major genotyping platforms. DArT was developed as a hybridisation-based system capable of generating whole-genome fingerprints by scoring presence *versus* absence of DNA fragments in genomic representations generated from samples of genomic DNA. However，all DArT markers are dominant and need SSR markers to confirm their exact chromosomal locations. Some DArT markers have been located on the physical map（http：//www. cerealsdb. uk. net/CerealsDB/Documents/FORM _ DArT _ 1A. php）. SNP assays directly interrogate sequence variation，reducing genotyping errors compared to assays based on size discrimination. SNPs are ideally suited for construction of high-resolution genetic maps，investigations of population evolutionary history，and discovery of marker-trait associations（Aranzana et al. , 2005；Zhao et al. , 2007）. However，limited D genome markers，too many co-segregating markers at the same position，and non-specific chromosome locations from some SNP markers in the Illumina iSelect 90K SNP assays（Wang et al. , 2014）may affect its widespread use in genetic analysis（J. Z. Jia，personal comm.）. To address these problems，the Institute of Crop Science，Chinese Academy of Agricultural Sciences and Affymetrix in USA are developing the 670K and 820K wheat SNP assays（http：//www. cerealsdb. uk. net/cerealgenomics/CerealsDB/SNPs/Documents/Axiom _download . php），respectively，with markers specific to each of the A，B and D genomes. It is expected that this will lead to major benefits for both mapping and cloning of wheat genes（J. Z. Jia，personal comm.）. With the development of sequencing technologies，the approaches applied to genetic variation analysis will shift from SNP-based genotyping to direct sequencing of all individuals in populations (Hamilton and Buell, 2012). The GBS approach was recently applied to construct genetic maps of crops with large genomes (Elshire et al., 2011; Poland et al., 2012), allowing direct analysis of genetic variation and reducing the effect of ascertainment bias caused by the SNP discovery process.

Although these early studies demonstrate the utility of complexity-reduced sequencing for genome analysis, detailed investigations are needed to understand the limitations of low-coverage, next-generation sequence data for variant calling in the large wheat genome and developing approaches for mapping genes of interest, ordering physical contigs, and performing association mapping (Saintenac et al., 2013a). Thus, combining more marker systems together is the best option for genetic studies in wheat; one such example is the Wheat GBS 1.0 marker set, with approximately 50,000 loci evaluated for SNP and DArT polymorphisms through sequencing, developed by the DArT Company in Australia (https://www.diversityarrays.com/cgi-bin/order/order.pl).

Linkage mapping and genome-wide association analysis (GWAS) were used to map QTL for leaf rust resistance in wheat (Maccaferri et al., 2010; Ando and Pumphrey, 2013; Basnet et al., 2013). They are primary tools for gene discovery, localization and functional analysis. Conventional QTL linkage mapping is an effective tool for the identification of genetic loci conferring APR to leaf rust and powdery mildew. The number, genetic effects and locations of QTL can be estimated by linkage mapping to molecular markers in segregating populations. Among segregating populations RIL and DH are better than $F_{2:3}$, as most or all of the QTL are homozygous leading to improved accuracy of phenotyping and clearer dissection of interaction between QTL and environments. Genes identified by the linkage mapping are restricted to those segregating in the particular cross. GWAS studies have recently received increasing attention for identification of QTL in plants, as an alternative to or in combination with linkage mapping approaches (Lu et al., 2010; Huang et al., 2010; Huang et al., 2012; Li et al., 2013a). Association analysis, also known as linkage disequilibrium (LD) mapping or association mapping, allows correlation analyses between traits and molecular markers based on LD (Flint-Garcia et al., 2003). Although GWAS provides much higher mapping resolution than linkage mapping, GWAS mapping has limitations caused by false positives due to population structure, variable marker densities and presence of rare alleles (Shriner et al., 2007; Wang et al., 2005a). Therefore, combined GWAS mapping and classical linkage mapping has been successfully applied in analysis of traits such as time of flowering (Thornsberry et al., 2001), pro-vitamin A (Harjes et al., 2008), drought resistance (Lu et al., 2010) and oil biosythesis of kernels (Li et al., 2013a) in maize, flowering time, grain yield and agronomic traits in rice (Huang et al., 2010, 2012), and seed size and flour quality properties in wheat (Breseghello and Sorrells, 2006).

Eighty leaf rust APR QTL and 119 powdery mildew APR QTL have been mapped on 16 (Table 1) and 21 wheat chromosomes (Table 2), respectively. Many of these QTL are identical or closely linked genes that may have been transferred into diverse backgrounds by phenotypic selection in disease nurseries during plant breeding. We summarized these QTL into an integrated linkage map (Fig. 1) based on wheat consensus maps (Somers et al., 2004, http://wheat.pw.usda.gov/GG2/index.shtml) in order to identify identical resistance genes and potential pleiotropic APR loci. Five leaf rust APR QTL clusters are located on chromosomes 1BS, 2BS, 2DS, 6BL, and 7BL. Overall, the B genome contains more leaf rust APR QTL than the A or D genomes. Eight powdery mildew APR QTL clusters were mapped on chromosomes 1AS, 1AL, 2AS, 2AL, 3AS, 4AL, 5AL and 5BS; interestingly most of them are in the A genome. Sixteen APR QTL conferred resistance to both leaf rust and powdery mildew, including 11 loci conferring resistance to leaf rust, stripe rust and powdery mildew.

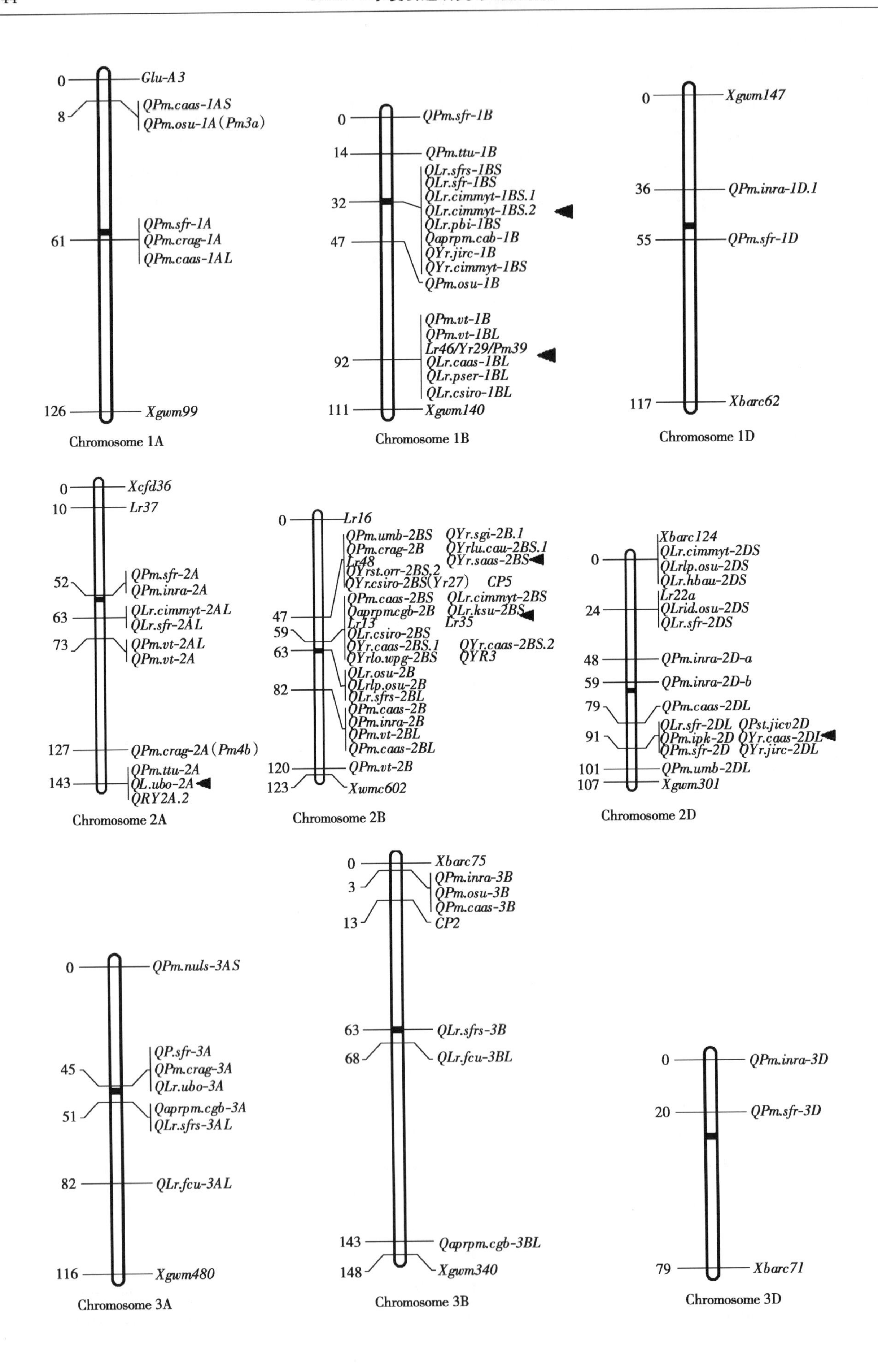
0 Glu-A3
8 QPm.caas-1AS
QPm.osu-1A(Pm3a)
61 QPm.sfr-1A
QPm.crag-1A
QPm.caas-1AL
126 Xgwm99
Chromosome 1A
0 QPm.sfr-1B
14 QPm.ttu-1B
32 QLr.sfrs-1BS
QLr.sfr-1BS
QLr.cimmyt-1BS.1
QLr.cimmyt-1BS.2
QLr.pbi-1BS
47 Qaprpm.cab-1B
QYr.jirc-1B
QYr.cimmyt-1BS
QPm.osu-1B
QPm.vt-1B
QPm.vt-1BL
Lr46/Yr29/Pm39
92 QLr.caas-1BL
QLr.pser-1BL
QLr.csiro-1BL
111 Xgwm140
Chromosome 1B
0 Xgwm147
36 QPm.inra-1D.1
55 QPm.sfr-1D
117 Xbarc62
Chromosome 1D
0 Xcfd36
10 Lr37
52 QPm.sfr-2A
QPm.inra-2A
63 QLr.cimmyt-2AL
QLr.sfr-2AL
73 QPm.vt-2AL
QPm.vt-2A
127 QPm.crag-2A(Pm4b)
143 QPm.ttu-2A
QL.ubo-2A
QRY2A.2
Chromosome 2A
0 Lr16
QPm.umb-2BS QYr.sgi-2B.1
QPm.crag-2B QYrlu.cau-2BS.1
Lr48 QYr.saas-2BS
QYrst.orr-2BS.2
QYr.csiro-2BS(Yr27) CP5
47 QPm.caas-2BS QLr.cimmyt-2BS
Qaprpmcgb-2B QLr.ksu-2BS
59 Lr13 Lr35
63 QLr.csiro-2BS
QYr.caas-2BS.1 QYr.caas-2BS.2
QYrlo.wpg-2BS QYR3
82 QLr.osu-2B
QLrlp.osu-2B
QLr.sfrs-2BL
QPm.caas-2B
QPm.inra-2B
QPm.vt-2BL
QPm.caas-2BL
120 QPm.vt-2B
123 Xwmc602
Chromosome 2B
Xbarc124
0 QLr.cimmyt-2DS
QLrlp.osu-2DS
QLr.hbau-2DS
Lr22a
24 QLrid.osu-2DS
QLr.sfr-2DS
48 QPm.inra-2D-a
59 QPm.inra-2D-b
79 QPm.caas-2DL
QLr.sfr-2DL QPst.jicv2D
91 QPm.ipk-2D QYr.caas-2DL
QPm.sfr-2D QYr.jirc-2DL
101 QPm.umb-2DL
107 Xgwm301
Chromosome 2D
0 QPm.nuls-3AS
45 QP.sfr-3A
QPm.crag-3A
QLr.ubo-3A
51 Qaprpm.cgb-3A
QLr.sfrs-3AL
82 QLr.fcu-3AL
116 Xgwm480
Chromosome 3A
0 Xbarc75
3 QPm.inra-3B
QPm.osu-3B
QPm.caas-3B
13 CP2
63 QLr.sfrs-3B
68 QLr.fcu-3BL
143 Qaprpm.cgb-3BL
148 Xgwm340
Chromosome 3B
0 QPm.inra-3D
20 QPm.sfr-3D
79 Xbarc71
Chromosome 3D

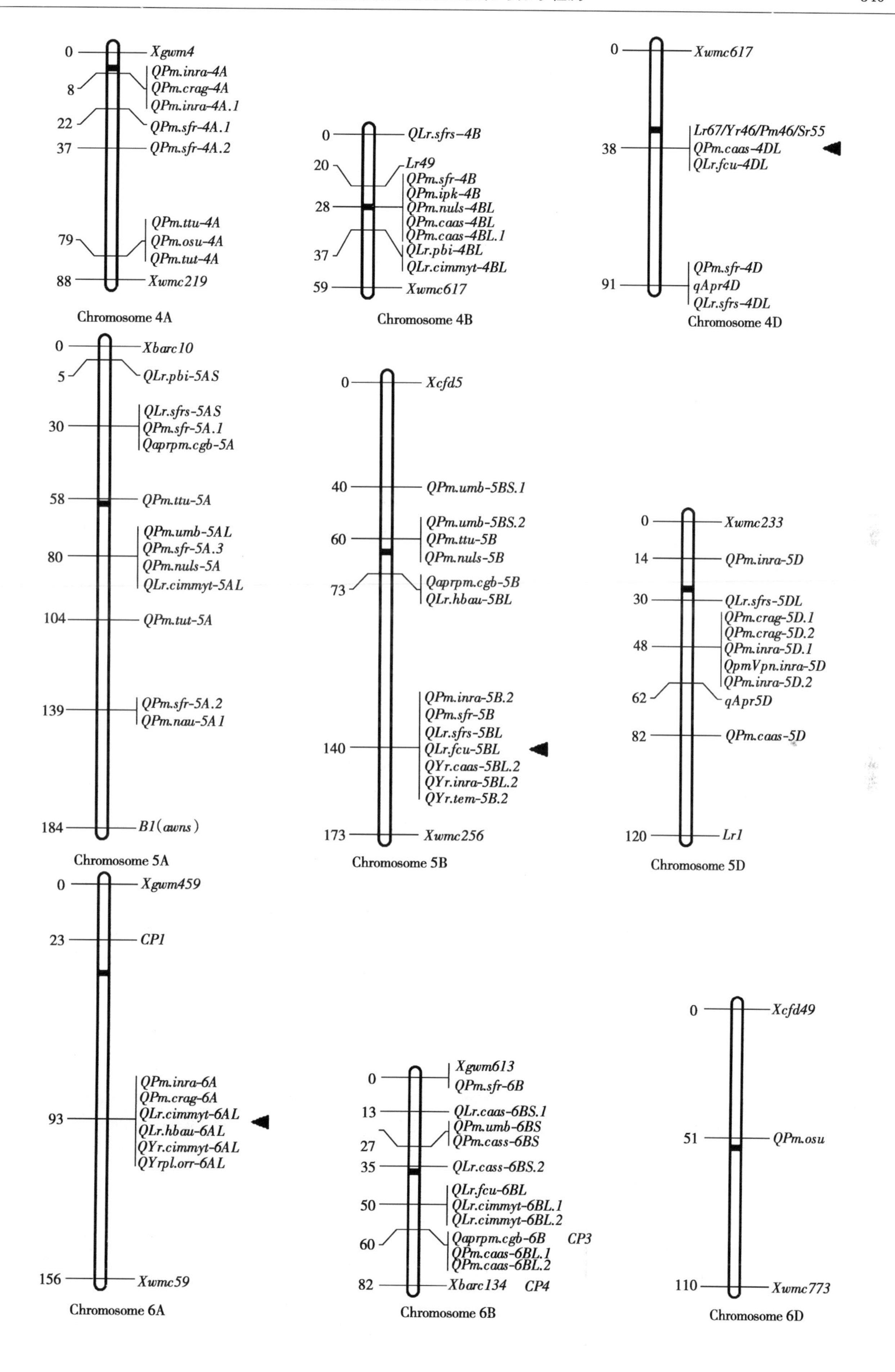
0 Xgwm4
8 QPm.inra-4A
QPm.crag-4A
QPm.inra-4A.1
22 QPm.sfr-4A.1
37 QPm.sfr-4A.2
79 QPm.ttu-4A
QPm.osu-4A
QPm.tut-4A
88 Xwmc219
Chromosome 4A
0 QLr.sfrs-4B
20 Lr49
QPm.sfr-4B
QPm.ipk-4B
28 QPm.nuls-4BL
QPm.caas-4BL
QPm.caas-4BL.1
37 QLr.pbi-4BL
QLr.cimmyt-4BL
59 Xwmc617
Chromosome 4B
0 Xwmc617
38 Lr67/Yr46/Pm46/Sr55
QPm.caas-4DL
QLr.fcu-4DL
91 QPm.sfr-4D
qApr4D
QLr.sfrs-4DL
Chromosome 4D
0 Xbarc10
5 QLr.pbi-5AS
30 QLr.sfrs-5AS
QPm.sfr-5A.1
Qaprpm.cgb-5A
58 QPm.ttu-5A
80 QPm.umb-5AL
QPm.sfr-5A.3
QPm.nuls-5A
QLr.cimmyt-5AL
104 QPm.tut-5A
139 QPm.sfr-5A.2
QPm.nau-5A1
184 B1(awns)
Chromosome 5A
0 Xcfd5
40 QPm.umb-5BS.1
60 QPm.umb-5BS.2
QPm.ttu-5B
QPm.nuls-5B
73 Qaprpm.cgb-5B
QLr.hbau-5BL
140 QPm.inra-5B.2
QPm.sfr-5B
QLr.sfrs-5BL
QLr.fcu-5BL
QYr.caas-5BL.2
QYr.inra-5BL.2
QYr.tem-5B.2
173 Xwmc256
Chromosome 5B
0 Xwmc233
14 QPm.inra-5D
30 QLr.sfrs-5DL
48 QPm.crag-5D.1
QPm.crag-5D.2
QPm.inra-5D.1
QpmVpn.inra-5D
QPm.inra-5D.2
62 qApr5D
82 QPm.caas-5D
120 Lr1
Chromosome 5D
0 Xgwm459
23 CP1
93 QPm.inra-6A
QPm.crag-6A
QLr.cimmyt-6AL
QLr.hbau-6AL
QYr.cimmyt-6AL
QYrpl.orr-6AL
156 Xwmc59
Chromosome 6A
0 Xgwm613
QPm.sfr-6B
13 QLr.caas-6BS.1
27 QPm.umb-6BS
QPm.cass-6BS
35 QLr.cass-6BS.2
50 QLr.fcu-6BL
QLr.cimmyt-6BL.1
QLr.cimmyt-6BL.2
60 Qaprpm.cgb-6B CP3
QPm.caas-6BL.1
QPm.caas-6BL.2
82 Xbarc134 CP4
Chromosome 6B
0 Xcfd49
51 QPm.osu
110 Xwmc773
Chromosome 6D

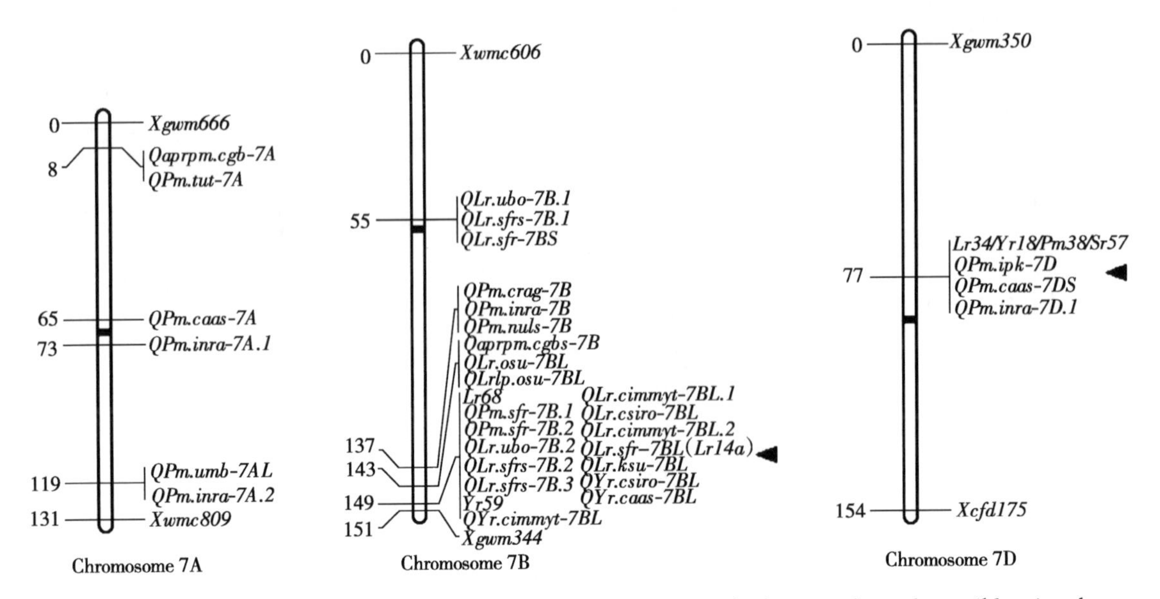

Fig. 1 Integrated linkage map of QTL for adult-plant resistance to leaf rust and powdery mildew in wheat

The positions of QTL in linkage groups are estimated from molecular markers in wheat consensus linkage maps (Somers et al., 2004; http://wheat.pw.usda.gov/GG2/index.shtml). Black triangles show resistance QTL clusters for leaf rust, powdery mildew and stripe rust.

Known pleiotropic APR genes in wheat

Among APR genes, as stated previously, three slow rusting genes are well charcaterized. *Lr34/Yr18/Pm38/Sr57* (Spielmeyer et al., 2005), *Lr46/Yr29/Pm39/Sr58* (William et al., 2003), and *Lr67/Yr46/Pm46/Sr55* (Herrera-Foessel et al., 2011) confer resistance to leaf rust, stripe rust, powdery mildew, and stem rust. These pleiotropic APR genes can be used for breeding resistant cultivars with durable resistance to multiple wheat diseases. *Lr34/Yr18* likely originates from Chinese and Italian wheat germplasm (Keller et al., 2013). Located on chromosome 7DS, this gene has provided APR to leaf rust and stripe rust for almost 100 years, with resistance to additional diseases such as stem rust becoming evident more recently, and finally confirmed by phenotypic analysis of loss-of-function mutants of the same gene under the name *Sr57* (R. P. Singh, pers. comm.). It also co-segregates with leaf tip necrosis (*Ltn1*) (Singh, 1992; Rosewarne et al., 2006), and APR to powdery mildew (*Pm38*) (Lillemo et al., 2008). *Lr34* confers a moderate level of leaf rust resistance when present alone; however, combinations of *Lr34* with two to four additional slow rusting genes provide high levels of resistance in many environments (Singh et al., 2000). Nelson et al. (1997) identified two QTL for leaf rust resistance, one on chromosome 7DS (*Lr34*) and the other on chromosome 2BS, and together they explained 45% of the phenotypic variation in leaf rust response. *Lr34* also enhances the resistant responses of most specific leaf rust resistance genes (German and Kolmer, 1992; Dakouri et al., 2013). Krattinger et al. (2009) cloned *Lr34* and found that the translated protein encoded an ABC (ATP-binding cassette) transporter belonging to the pleiotropic drug resistance subfamily.

The second pleiotropic APR gene *Lr46/Yr29*, first detected in CIMMYT wheat line Pavon 76 (Singh et al., 1998) and located on chromosome 1BL (William et al., 2003), has provided APR to leaf rust and stripe rust for more than 40 years. It co-segregates with the leaf tip necrosis gene *Ltn2* (Rosewarne et al., 2006), *Pm39* for APR to powdery mildew (Lillemo et al., 2008) and *Sr58* for APR to stem rust (Singh et al., 2013). *Lr46/Yr29* reportedly confers weaker levels of resistance to leaf rust and powdery mildew than *Lr34/Yr18* (Lagudah, 2011), but displayed larger effects in the Avocet/Pastor and Avocet/Francolin # 1 genetic backgrounds (Rosewarne et al., 2012; Lan et al., 2014). However, this gene contributed a significant negative effect when combined with *Lr34/Yr18* in Avocet/Saar based on a simple additive model of analysis (Lillemo et al., 2008). Rosewarne et al. (2008) also reported that *Lr46/Yr29* had a negative additive effect on stripe rust response in an Avocet/Attila RIL population, and a similar result was obtained with an Avocet/Francolin#1 population (Lan et al., 2014). In South America *Lr68* had a larger effect than *Lr34* on leaf rust resistance (Silva et al., 2013).

The third pleiotropic APR gene, *Lr67/Yr46*, was originally identified in a Pakistani accession and transferred to Thatcher (near-isogenic line RL6077). This was mapped on chromosome 4DL, with closely linked SSR loci of *Xgwm165* and *Xgwm192* (Hiebert et al., 2010; Herrera-Foessel et al., 2011). It also confers resistance to powdery mildew (*Pm46*) and stem rust (*Sr55*) (Herrera-Foessel et al., 2014). It has a similar effect to *Lr34/Yr18* on leaf rust and stripe rust responses. Stripe rust resistance QTL have been mapped on this chromosome arm in Bainong 64 (Ren et al., 2012b), Pastor (Rosewarne et al., 2012), W-219 (Singh et al., 2000) and Oligoculm (Suenaga et al., 2003), and QTL were detected in ND495 (Chu et al., 2009) and Forno (Messmer et al., 2000) for leaf rust resistance. The relationship between these QTL and *Lr67/Yr46* will eventually be confirmed through gene sequencing.

Potential APR QTL for multiple disease resistance in wheat

Chromosome 1BS

Five leaf rust QTL (William et al., 1997; Messmer et al., 2000; Schnurbusch et al., 2004; Singh et al., 2009; Rosewarne et al., 2012) (Table 1) and three powdery mildew QTL (Keller et al., 1999; Jakobson et al., 2006; Huang et al., 2008) (Table 2) were mapped on chromosome 1BS. Flanking markers for these loci form one cluster with the exceptions of *QPm. sfr-1B* and *QPm. ttu-1B* (Fig. 1). Among the multiple disease resistance cluster, five leaf rust APR QTL were located near the centromere on 1BS (Fig. 1), *QLr. sfr-1BS* was closely linked to SSR locus *Xgwm11* and the phenotypic variance explained (PVE) was 28% in the Arina × Forno population (Schnurbusch et al., 2004). Powdery mildew APR QTL *Qaprpm. cgb-1B* in Chinese wheat line Hanxuan 10 (Huang et al., 2008) and stripe rust APR QTL *QYr. jirc-1B* in wheat line Fukuho-komugi (Suenaga et al., 2003) were respectively linked to SSR loci *Xwmc269* and *Xwmc320* close to *QLr. sfr-1BS* based on the wheat consensus map (Somers et al., 2004). Further tests are necessary to confirm this potentially pleiotropic APR QTL on 1BS, and its effect on diseases other than leaf rust and powdery mildew in a single segregating population.

Chromosome 2AL

Three leaf rust QTL (Rosewarne et al., 2008; Schnurbusch et al., 2004; Maccaferri et al., 2008) (Table 1), four powdery mildew QTL (Mingeot et al., 2002; Jakobson et al., 2006; Liu et al., 2001; Tucker et al., 2007) (Table 2) and two related stripe rust APR QTL (Rosewarne et al., 2013) were identified on chromosome 2AL (Fig. 1). These loci may form one leaf rust resistance gene cluster, one powdery mildew resistance gene cluster, and one cluster of genes for resistance to leaf rust, powdery mildew and stripe rust in the distal region of chromosome 2AL based on the integrated linkage map (Fig. 1). *QRYr2A. 2* were derived from common wheat lines Camp Remy and Recital (Rosewarne et al., 2013), and *QLr. ubo-2A* was derived from durum wheat Lloyd (Maccaferri et al., 2008), whereas *QPm. ttu-2A* was derived from *Triticum militinae* (Jakobson et al., 2006). The effect of this locus on both leaf rust and stripe rust responses might be from common wheat or durum wheat, but the effect on powdery mildew response is probably from *Triticum militinae* (Jakobson et al., 2006).

Chromosome 2B

Chromosome 2B carries many disease resistance genes (Fig. 1). There are at least two clusters of QTL for leaf rust, powdery mildew and stripe rust resistance on 2BS, as well as one leaf rust cluster and one powdery mildew resistance cluster on 2BL (Fig. 1).

Two potential multiple disease APR QTL were mapped on 2BS. The first was mapped at 47cM in the integrated linkage map (Fig. 1); the race-specific resistance genes *Lr48* and *Yr27* and three powdery mildew APR QTL *QPm. umb-2BS*, *QPm. crag-2B* and *CP5* (Lillemo et al., 2012; Mingeot et al., 2002; Marone et al., 2013), and four stripe rust APR QTL (Guo et al., 2008; Ramburan et al., 2004; Vazquez et al., 2012; Yang et al., 2013) were also mapped in this region. *Lr48* is a recessive hypersensitive resistance gene in CSP44, a selection from Australian cultivar Condor. *Yr27* originated from Selkirk (pedigree: McMurachy (*Yr27*) /Exchange//3 * Redman) and is present in a number of CIMMYT lines such as Ciano 79 (Wellings, 1992). However, it became ineffective to a Mexican *P. striiformis* race typified by isolate Mex96.11 (Lan et al., 2014). This region of chromosome 2BS has closely linked race-specific loci for leaf rust and stripe rust resistances and also possesses multiple minor QTL.

The second potential pleiotropic APR QTL is at position 59cM (Fig. 1). Leaf rust APR genes *Lr13* (Seyfarth et al., 2000) and *Lr35* (Seyfarth et al., 1999), and stripe rust race-specific resistance genes *Yr31* (Rosewarne et al., 2012), *Yr41* (Luo et al., 2008) and *YrKK* (Li et al., 2013b), were located in this chromosome region. *Lr35* occurs in a chromosome segment translocated from the wild wheat relative *Aegilops speltoides* to common wheat and confers a high level of resistance to mixed Chinese *P. triticina* races at the adult plant stage (Z. F. Li, unpublished data). Stripe rust race-specific resistance gene *Yr31* was ineffective in China in 2009 and 2010 (Yang et al., 2013) and in Mexico since 2008 (Rosewarne et al., 2012) whereas virulent races for *Yr41* in Chuannong 19 and *YrKK* in Kenya Kudu have not been detected in China or Mexico. In addition to the above genes, three leaf rust APR QTL (Faris et al., 1999; Rosewarne et al., 2008, 2012), two powdery mildew APR QTL (Huang et al., 2008; Lan et al., 2010b), and four stripe rust APR QTL (Boukhatem et al., 2002; Carter et al., 2009; Lan et al., 2010a; Ren et al., 2012a) were also mapped to this region (Fig. 1, He et al., 2011). The multiple disease resistances in this region could be attributed to the pleiotropic APR QTL, but the presence of a cluster of separate resistance genes cannot be ruled out.

Chromosome 2DL

Two regions were associated with powdery mildew resistance (Lan et al., 2010b; Lillemo et al., 2012) and one region showed effects on leaf rust, powdery mildew and stripe rust responses at about 91cM (Fig. 1). The PVE by the latter region were 11.4-12.7, 10 and 10.1-11.4% for leaf rust (Table 1), powdery mildew (Table 2) and stripe rust (He et al., 2011), respectively. SSR locus *Xgwm539* was linked to two stripe rust APR QTL *QYr. caas-2DL* (Ren et al., 2012a) and *QPst. jic-2D* (Melichar et al., 2008), and leaf rust APR QTL *QLr. sfr-2DL* (Schnurbusch et al., 2004) within a genetic distance of 4 cM from the locus *Xglk558*, flanking powdery mildew APR QTL *QPm. ipk-2D* (Sourdille et al., 2004) and 2 cM from the SSR locus *Xgwm349* near the stripe rust APR QTL *QYr. jirc-2DL* (Suenaga et al., 2003). The latter region gave inconsistent effects across environments. However, it still appears to be an interesting region for multiple disease resistance.

Chromosome 5AL

APR QTL to three diseases were mapped at a position near80 cM on chromosome 5AL (Fig. 1). Leaf rust APR QTL *QLr. cimmyt-5AL* was located between *wPT0373* and *wPT0837* and about 2cM from *Vrn-A1* in wheat cultivar Avocet (Rosewarne et al., 2012). Three powdery mildew APR QTL *QPm. umb-5AL* (Lillemo et al., 2012), *QPm. sfr-5A.3* (Keller et al., 1999), and *QPm. nuls-5A* (Lillemo et al., 2008) in wheat lines Folke, Berkulmer and Saar, respectively, were also located in this region. *QYrtm.*

pau-5A identified in the diploid *Triticum boeoticum* displays APR (Chhuneja et al., 2008). Relatively high LOD scores and high PVE values and stable effects across environments suggest that this locus could be a major gene exhibiting high temperature APR to stripe rust. Further work is required to determine any pleiotropic effects of this locus.

Chromosome 5BL

One multiple disease resistance QTL was located at 140cM (Fig. 1). Powdery mildew APR QTL *QPm. sfr-5B* was mapped to the *psr580b-psr143* interval in a Forno × Oberkulmer spelt RIL population (Keller et al., 1999). Messmer et al. (2000) detected leaf rust APR QTL *QLr. sfrs-5BL* in the same region using the same population, but the QTL was derived from the other parent (Forno), indicating that *QPm. sfr-5B* and *QLr. sfrs-5BL* were different genes. Three minor effect stripe rust APR QTL *QYr. caas-5BL. 2* (Lu et al., 2009), *QYr-tem-5B. 2* (Feng et al., 2011) and *QYr. inra-5BL* (Mallard et al., 2005) were also mapped in this region (Fig. 1). Evaluation of the Forno × Oberkulmer population in a stripe rust situation may confirm pleiotropic effects at this locus.

Chromosome 6AL

Chromosome 6AL had only one clearly defined region associated with leaf rust, powdery mildew and stripe rust resistance at 93cM (Fig. 1). William et al. (2006) identified a locus or closely linked loci on 6AL for resistance to stripe rust (*QYr. cimmyt-6AL*) and leaf rust (*QLr. cimmyt-6AL*) in Avocet with flanking marker *Xgwm617* in a Avocet × Pavon76 RIL population; Zhang et al. (2009) also detected leaf rust APR QTL *QLr. hbau-6AL* in Avocet with flanking markers *Xgwm617* and *Xgwm169* in the same region. However, Lillemo et al. (2008) did not find an effect of 6AL on powdery mildew response in Avocet/Saar, whereas this locus showed minor effects on both leaf rust and stripe rust responses at the adult plant stage. This combined resistance is likely contributed by a translocation from *Thinopyrum ponticum* that carries the stem rust resistance gene *Sr26* (William et al., 2006). The Avocet APR QTL associated with the translocation had intermediate LOD and PVE scores and was detected in several segregating populations (William et al., 2006; Lillemo et al., 2008; Rosewarne et al., 2012). This translocation has a linkage drag for low yield and is not widely used in breeding. Other *Sr26* lines with shortened translocations will have to be tested for stripe rust resistance. Stripe rust APR QTL *QYrpl. orr-6AL* with flanking marker *Xgwm617* (Vazquez et al., 2012) and a powdery mildew APR QTL derived from wheat line RE714 (Chantret et al., 2001; Mingeot et al., 2002) with flanking marker *Xgwm427* about 2cM from *Xgwm617* were also mapped in this region. The powdery mildew APR QTL might be due to a residual effect of powdery mildew resistance gene *MlRE* (Chantret et al., 2001; Mingeot et al., 2002).

Chromosome 7BL

Chromosome 7BL has up to three resistance QTL clusters (Tables 1 and 2, Fig. 1), including a powdery mildew resistance QTL cluster at 137cM, a QTL cluster for leaf rust and powdery mildew resistances at 143cM, and a cluster for leaf rust, powdery mildew and stripe rust resistances at 149cM in the region of seedling resistance genes *Lr14*, *LrBi16* (Zhang et al., 2011), APR gene *Lr68*, and stripe rust high temperature adult-plant (HTAP) resistance gene *Yr59* (X. M. Chen, pers. comm.). Seven additional leaf rust APR QTL, two powdery mildew APR QTL and three stripe rust APR QTL were also mapped in this region (Fig. 1). All of these loci conferred low to moderate APR. However, a pleiotropic effect of *Lr68* on stripe rust and powdery mildew response was not observed (Lillemo et al., 2011), suggesting there is no pleiotropic effect on stripe rust and powdery mildew response or that there is a temperature effect similar to that experienced with *Lr46/Yr29* (Lagudah, 2011). Hence, this region might carry potentially important pleiotropic APR loci.

Fine mapping and cloning of APR QTL

Nearly every cloned rust and powdery mildew resistance gene has been cloned via map-based approaches; these genes include *Lr1*, *Lr10*, *Lr21*, *Lr34*, *Yr36*, *Sr33*, *Sr35*, *Pm3*, *Pm8* and *Pm21* (*Lr34* and *Yr36* are APR genes) (Cao et al., 2011; Cloutier et al., 2007; Feuillet et al., 2003; Fu et al., 2009; Huang et al., 2003; Hurni et al., 2013; Krattinger et al., 2009; Periyannan et al., 2013; Saintenac et al., 2013b; Yahiaoui et al., 2004). The single exception was the homology-based cloning of *Pm8*. In order to clone APR genes, accurate phenotyping and fine mapping is essential. SSR markers have often been used for primary molecular mapping, with SNP markers and comparative genomics approaches used for fine mapping (Zhang et al., 2013). As a final step, Fu et al. (2009) used chromosome walking combined with bacterial artificial chromosome (BAC) library screening to clone resistance gene *Yr36*. Recently, selective genotyping was used in fine mapping and cloning of a minor QTL in maize (Sun et al., 2010). This cost-efficient method was used to genotype individuals from the extremes of a segregating population instead of the entire population. When phenotyped segregating populations are available, genotyping of the tails is often more efficient than genotyping entire linkage or GWAS populations. In the future, RNA sequencing techniques combined with candidate gene association analysis and high-density wheat SNP arrays, such as the 670K chip, will be used for fine mapping and cloning of APR genes in wheat. Currently, there is an urgent need for further analyses of several APR genes in wheat. Functional analyses should then reveal whether there is a clear distinction between APR genes that protect against single pathogens and the pleiotropic multiple disease resistance genes that are the main focus of this paper.

Molecular mechanisms of APR in wheat

The proposed mechanisms of race-specificity and the actions of major genes, including R-Avr gene interactions and the guard hypothesis, were discussed in several previous reviews (Hulbert et al., 2001; Jones and Dangl, 2006; Bent and Mackey, 2007). Few major genes confer resistance to prevalent *Pt* and *Bgt* pathotypes in China (Li et al., 2010; Wang et al., 2005b). Minor resistance genes often appear to be race non-specific and therefore confer durable resistance. However, there is little information regarding functional relatedness, combinability, or potential durability of minor resistance genes that can be explained by currently known molecular mechanisms of disease resistance. So far, the two cloned APR genes, *Lr34*/*Yr18*/*Pm38*/*Sr57* (Krattinger et al., 2009) and *Yr36* (Fu et al., 2009), suggest a heterogeneous group of functionally diverse genes.

Lr34/*Yr18*/*Pm38*/*Sr57* encodes a protein with homology to ATP-binding cassette (ABC) transporters, a trans-kingdom superfamily of trans-membrane proteins involved in the transport of a wide variety of substrates across cellular membranes (Krattinger et al., 2009). In addition to its non-NBS-LRR structure and long-standing durability, other evidence suggests that the partial resistance conferred by *Lr34*/*Yr18*/*Pm38*/*Sr57* is truly race non-specific (Lowe et al., 2011). Risk et al. (2012) reported that *Lr34* did not increase pathogenesis-related (PR) gene induction in flag leaves to a significantly higher level after pathogen infection, and the findings are in accordance with Rubiales and Niks (1995), who concluded that *Lr34*-based resistance is not conferred by a typical hypersensitive response. Therefore, a distinct type of resistance must underlie *Lr34*-based resistance. Several intriguing observations were made concerning the molecular basis of leaf tip necrosis (LTN). The senescence-associated gene HvS40 was induced in flag leaves of *Lr34*-containing lines (Krattinger et al., 2009; Risk et al., 2012). Thus, the senescence-like processes induced by *Ltn1*, might simply make the tissue less conducive for biotrophic activity and result in a quantitative limitation of fungal growth, hence explaining the partial resistance effect. Such an effect might not be overcome easily by the pathogen, and therefore might explain the durability and multiple pathogen protection conferred by this gene (Keller et al., 2013). The difference between the encoded proteins of the resistant and susceptible *Lr34* alleles is two amino acid changes (Dakouri et al., 2010). These two changes are apparently sufficient to change the biochemical properties of the resistant *Lr34* transporter in such a way that the plant becomes resistant (Keller et al., 2013). These authors speculated that there was a slight conformational change in the resistant form of the protein, resulting in either modified specificity, or altered kinetics of the transported molecule, or different binding properties to an unknown second protein interacting with the *Lr34* protein. Krattinger et al. (2011) provided evidence suggesting that the critical mutations resulting in the gain of resistance function in *Lr34* occurred after formation and domestication of common wheat.

Chen (2013) proposed that race-specific resistance controlled by R genes of the NBS-LRR type regulate a relatively small number of genes involved in narrow based defense mechanisms. In contrast, race non-specific HTAP resistance is controlled by non-NBS-LRR genes, and especially genes without LRR domains as such domains are thought to be involved in specific recognition of avirulence factors. HTAP resistance was suggested to be more diverse in its molecular basis, with each HTAP resistance gene interacting with genes having diverse functions in defense. Broadly based molecular mechanisms may explain the durability of HTAP resistance (Chen, 2013). Faris et al. (1999) reported on the application of the candidate-gene approach to the mapping of QTL for disease resistance. A cluster of closely linked DR (defense resistance) genes on the long arm of chromosome 7B, including genes for catalase, chitinase, thaumatins, and an ion channel regulator, had major effects on response to leaf rust in adult plants under conditions of natural infestation. They suggested that many race non-specific resistance QTL may involve the action of DR genes, and that a candidate-gene approach would be an efficient way to study mechanisms underlying durable resistance.

Lowe et al. (2011) presented four steps to study the molecular mechanisms of durable resistance: 1) Cloning APR genes is the first step for further biology-oriented research of functional classification of APR genes; 2) Once the resistance gene is cloned, characterization of host-pathogen interactions using new and powerful methods will become available; 3) Disclose the defense-related gene networks; and 4) Identify clues from the pathogen. Molecules within the pathogen secretome that interact with host resistance proteins can be identified using yeast-two-hybrid libraries of rust-infected susceptible wheat leaves. Since resistance response pathways which initiate upon recognition of highly conserved regions of pathogen molecules are more likely to lead to durable resistance than those that initiate upon recognition of more variable regions; it offers a potential means of assessing the potential durability of any given resistance genes (Lowe et al., 2011).

Utilization of APR genes in wheat breeding and future perspectives

The CIMMYT wheat breeding program in Mexico has successfully used rust APR genes for almost 40 years, from the 1970s when wheat breeder Dr. Sanjay Rajaram and pathologist Dr. Jesse Dubin were inspired by Dr. Ralph Caldwell from Purdue University, U. S. A. The milestone genotype Pavon 76 developed by CIMMYT has maintained moderate levels of durable resistance to leaf rust, stripe rust, stem rust, and powdery mildew for nearly 40 years. However, the genetic mapping of APR QTL and systematic utilization of these QTL in strategic wheat breeding programs to develop high yielding lines with near-immunity became possible only after the 1980s. Selection for resistance based on additive minor genes in the field can be difficult, but experienced breeding staff with a good knowledge of variation and interaction of minor genes can achieve adequate APR by phenotypic selection in appropriately designed and administered nurseries.

The CIMMYT approach is to use a limited (single) backcross to incorporate additive minor genes into elite wheat cultivars/germplasm (Singh et al., 2005). In this strategy, a leading or selected cultivar is crossed with a group of resistance donors and then 20 spikes from F_1 plants in each cross are backcrossed or top crossed with a high-yielding parent to obtain 400-500 BC_1 or TC seeds. Selection for resistance and agronomic features is practiced from BC_1 or TC generations onwards under high rust pressure. BC_1/TC plants carrying most of the genes show intermediate resistance levels and can be selected; the low extremes of resistance are avoided. Plants with low to moderate disease severities in early generations (BC_1, TC, F_2 and F_3), and plants with low severities in later generations (F_4 and F_5) are retained. During this period, a selected-bulk scheme is practiced where one spike from each selected plant is harvested and threshed in bulk and resown until the F_5 generation, when plants are harvested individually. Because repeated evidence has shown that high resistance levels require the presence of four or more additive genes, the level of homozygosity from the F_4 generation onwards is sufficient to identify plants that combine adequate resistance with good agronomic features. Selection for seed characteristics is carried out on grain obtained from individually harvested F_5 plants. Small plots of the F_6 lines are then evaluated for agronomic traits and uniformity of resistance prior to yield trialing. Some lines developed through this strategy not only carry high levels of resistance to leaf rust and/or stripe rust, but have exhibited 5-15% higher yield potential than the original elite lines. This implies that through limited backcrossing the outstanding agronomic characteristics of the original cultivar can be maintained or improved upon while achieving disease resistance.

Cultivars with resistance to both stripe rust and powdery mildew are major objectives of many breeding programs in China. Following recurrent failures in protection using race-specific resistance genes, there has been a recent change by the Chinese Academy of Agricultural Sciences (CAAS) and Sichuan Academy of Agricultural Sciences (SAAS) to use shuttle breeding with CIMMYT for spring wheat and introgression of race non-specific resistance from CIMMYT sources into winter wheat germplasm. Similar methods of bulk selection in the early generations and use of disease hot-spot sites for powdery mildew in winter wheat and stripe rust in spring wheat are employed.

A shuttle breeding program similar to the one at CIMMYT has been ongoing at SAAS for more than 25 years. Sichuan elite cultivars were crossed to CIMMYT lines with APR and the F_1 plants were backcrossed with elite Sichuan cultivars to introgress the APR genes, while maintaining high yield potential and local adaptation characteristics of Sichuan germplasm. Several advanced lines such as 08RC2525 (Chuanmai 32 * 2/Chapio), 07RC3941 (SW119 * 2/Tukuru), and 07RC3929 (SW119 * 2/Tukuru) were developed by this methodology. All the lines are susceptible at the seedling stage, but are highly resistant to stripe rust at the adult plant stage in field trials and also display higher yield potential than the original cultivars (He et al., 2011). Since most of the APR genes, such as *Lr34/Yr18/Pm38/Sr57* and *Lr46/Yr29/Pm39/Sr58*, used in the program conferred resistance to stripe rust, leaf rust and powdery mildew; these lines have potential resistance to multiple wheat diseases, for example the line SW 8588 derived from the shuttle breeding program carries high levels of APR to stripe rust and leaf rust (Li et al., 2006, 2010). This program has established a model for using minor gene-based APR resistance in Chinese wheat breeding programs and the derived lines have been distributed for use in other breeding programs.

A new breeding program usinga combination of conventional breeding and molecular marker application has been established to develop germplasm with multiple APR in the CAAS-CIMMYT wheat program in China. Firstly, large numbers of elite cultivars and advanced lines were screened in the field and gene postulation methods were used to identify genotypes with APR to powdery mildew, stripe rust, and leaf rust. Secondly, elite winter wheat cultivars and landraces that had retained durable APR for over 50 years were used in QTL analyses. Thirdly, backcrosses were made with elite cultivars to transfer identified QTL using molecular markers. For example, Lumai 21 was crossed with Bainong 64 to combine QTL identified in both parents by QTL mapping of genes for APR to powdery mildew in Lumai 21 (Lan et al., 2010b) and Bainong 64 (Lan et al., 2009), which were competitive cultivars in China for many years. Among 21 selected F_6 lines, 12 had significantly lower area under disease progress curve (AUDPC) values, better agronomic traits than their parents, and carried 3-5 minor QTL determined by testing with closely linked molecular markers (Bai et al., 2012). Two APR QTL *QPm. caas-1AL* and *QPm. caas-4DL* from Bainong 64 played more significant roles in reducing disease severity, and pyramiding these two QTL with other QTL from Lumai 21 would be an ideal strategy to obtain high to near-immune levels of resistance (Bai et al., 2012). The lines with pyramided minor genes showed good APR to powdery mildew based on field evaluations (Bai et al., 2012). Thus, the current result confirmed that pyramiding QTL can be a suitable strategy for breeding highly and durably resistant wheat cultivars. This is in agreement with a previous report that four or five minor genes for resistance to leaf rust and stripe rust were sufficient to confer near immunity (Singh et al., 2011). In addition, a high-disease-pressure nursery is still needed for at least two generations of disease responses, and selection by molecular markers will still require uniform field testing for disease phenotype to validate the resistance.

Conclusions

While it can be assumed that APR genes are more durable than seedling resistance genes, breeders must be aware that some APR genes are race specific and more such genes will likely be confirmed in the future. For example, APR gene *Lr13* was overcome within a few years of release in the United States and Canada (Kolmer, 1989), South Africa, Europe, Mexico, and South America (Singh and Rajaram, 1992). Nevertheless, resistance conferred by many APR genes, such as *Yr34/Yr18/Pm38/Sr57* and *Lr46/Yr29/Pm39/Sr58* have conferred resistance for a long time. Improved understanding and identification of minor and race non-specific resistance genes and their interactions allows their use as an alternative to single or combined major genes is increasing. So far, only 11 leaf rust and three powdery mildew APR genes have been named, and while most of them may have closely linked molecular markers, those markers are often unsuitable for use in breeding programs because the specific marker alleles in donor resistance sources may be present in the elite lines chosen for combining useful traits. Published genomic sequences of the A (Jia et al., 2013) and D (Ling et al., 2013) genomes will now allow the development of high throughput genome-specific SNP arrays, which will enable the efficient identification, fine mapping, and map-based cloning of APR genes for leaf rust, powdery mildew and stripe rust. Resistance gene-specific markers (functional or perfect markers) developed from the cloned resistance alleles will allow accurate molecular marker assisted selection of those resistance genes.

Acknowledgments

The authors are grateful to Prof. R. A. McIntosh, Plant Breeding Institute, University of Sydney, for review of this manuscript. This study was supported by the National Key Basic Research Program of China (2013CB127700), National Natural Science Foundation of China (31361140367 and 31261140370), and International Collaboration Project from the Ministry of Science and Technology (2011DFG32990).

References

Aranzana, M. J., S. Kim, K. Zhao, E. Bakker, M. Horton, K. Jakob, C. Lister, J. Molitor, C. Shindo, C. Tang, C. Toomajian, B. Traw, H. Zheng, J. Bergelson, C. Dean, P. Marjoram, and M. Nordborg. 2005. Genomewide association mapping in *Arabidopsis* identifies previously known flowering time and pathogen resistance genes. PLoS Genet. 1: e60.

Ando, K., and M. O. Pumphrey. 2013. Genome-wide association analysis on seedling and adult plant resistance of stripe rust in elite Pacific Northwest spring wheat lines. Proceedings of Borlaug Global Rust Initiative, 2013 Technical Workshop, 19-22 August, New Dehli, India. p 27.1.

Asad, M. A., B. Bai, C. X. Lan, J. Yan, X. C. Xia, Y. Zhang, and Z. H. He. 2012. Molecular mapping of quantitative trait loci for adult-plant resistance to powdery mildew in Italian wheat cultivar Libellula. Crop Pasture Sci. 63: 539-546.

Bai, B., Z. H. He, M. A. Asad, C. X. Lan, Y. Zhang, X. C. Xia, J. Yan, X. M. Chen, and C. S. Wang. 2012. Pyramiding adult-plant powdery mildew resistance QTLs in bread wheat. Crop Pasture Sci. 63: 606-611.

Bansal, U. K., M. J. Hayden, B. P. Venkata, R. Khanna, R. G. Saini, and H. S. Bariana. 2008. Genetic mapping of adult plant leaf rust resistance genes *Lr48* and *Lr49* in common wheat. Theor. Appl. Genet. 117: 307-312.

Basnet, B. R., R. P. Singh, A. M. H. Ibrahim, S. A. Herrera-Foessel, J. Huerta-Espino, C. Lan, and J. C. Rudd. 2013. Characterization of *Yr54* and other genes associated with adult plant resistance to yellow rust and leaf rust in common wheat Quaiu 3. Mol. Breed. DOI 10.1007/s11032-013-9957-2.

Bent, A. F., and D. Mackey. 2007. Elicitors, effectors, and R genes: the new paradigm and a lifetime supply of questions. Annu. Rev. Phytopathol. 45: 399-436.

Bjarko, M. E., and R. F. Line. 1988. Heritability and number of genes controlling leaf rust resistance in four cultivars of wheat. Phytopathology. 78: 457-461.

Bougot, Y., J. Lemoine, M. T. Pavoine, H. Guyomarc'h, V. Gautier, H. Muranty, and D. Barloy. 2006. A major QTL effect controlling resistance to powdery mildew in winter wheat at the adult plant stage. Plant Breed. 125: 550-556.

Boukhatem, N., P. V. Baret, D. Mingeot, and J. M. Jacquemin. 2002. Quantitative trait loci for resistance against yellow rust in two wheat-derived recombinant inbred line populations. Theor. Appl. Genet. 104: 111-118.

Breseghello, F., and M. E. Sorrells. 2006. Association mapping of kernel size and milling quality in wheat (*Triticum aestivum* L.) cultivars. Genetics. 172: 1165-1177.

Börner, A., E. Schumann, A. Fürste, H. Cöster, B. Leithold, M. S. Röder, and W. E. Weber. 2002. Mapping of quantitative trait loci determining agronomic important characters in hexaploid wheat (*Triticum aestivum* L.). Theor. Appl. Genet. 105: 921-936.

Caldwell, R. M. 1968. Breeding for general and/or specific plant disease resistance. *In*: Finlay KW, Shepherd KW (eds). Proceedings of the Third International Wheat Genetics Symposium. Canberra, Australia: Australian Academy of Sciences, pp. 263-272.

Cao, A., L. Xing, X. Wang, X. Yang, W. Wang, Y. Sun, C. Qian, J. Ni, Y. Chen, D. Liu, X. Wang, and P. Chen. 2011. Serine/threonine kinase gene Stpk-V, a key member of powdery mildew resistance gene *Pm21*, confers powdery mildew resistance in wheat. Proc. Natl. Acad. Sci. USA. 108: 7727-7732.

Carter, A. H., X. M. Chen, K. Garland-Campbell, and K. K. Kidwell. 2009. Identifying QTL for high-temperature adult-plant resistance to stripe rust (*Puccinia striiformis* f. sp. *tritici*) in the spring wheat (*Triticum aestivum* L.) cultivar 'Louise'. Theor. Appl. Genet. 119: 1119-1128.

Chantret, N., D. Mingeot, P. Sourdille, M. Bernard, J. M. Jacquemin, and G. Doussinault. 2001. A major QTL for powdery mildew resistance is stable over time and at two development stages in winter wheat. Theor. Appl. Genet. 103: 962-971.

Chen, X. M. 2013. Review article: High-temperature adult-plant resistance, key for sustainable control of stripe rust. Amer. J. Plant Sci. 4: 608-627.

Chen, Y. H., R. M. Hunger, B. F. Carver, H. L. Zhang, and L. L. Yan. 2009. Genetic characterization of powdery mildew resistance in U. S. hard winter wheat. Mol. Breed. 24: 141-152.

Chhuneja, P., S. Kaur, T. Garg, M. Ghai, S. Kaur, M. Prashar, N. S. Bains, R. K. Goel, B. Keller, H. S. Dhaliwal, and K. Singh. 2008. Mapping of adult plant stripe rust resistance genes in diploid A genome wheat species and their transfer to bread wheat. Theor. Appl. Genet. 116: 313-324.

Chu, C. G., T. L. Friesen, S. S. Xu, J. D. Faris, and J. A. Kolmer. 2009. Identification of novel QTLs for seedling and adult plant leaf rust resistance in a wheat doubled haploid population. Theor. Appl. Genet. 119: 263-269.

Cloutier, S., B. D. McCallum, C. Loutre, T. W. Banks, T. Wicker, C. Feuillet, B. Keller, and M. C. Jordan. 2007. Leaf rust resistance gene *Lr1*, isolated from bread wheat (*Triticum aestivum* L.) is a member of the large psr567 gene family. Plant Mol. Biol. 65: 93-106.

Conner, R. L., A. D. Kuzyk, and H. Su. 2003. Impact of powdery mildew on the yield of soft white spring wheat cultivars. Can. J. Plant Sci. 83: 725-728.

Dakouri, A., B. D. McCallum, A. Z. Walichnowski, and

S. Cloutier. 2010. Fine-mapping of the leaf rust *Lr34* locus in *Triticum aestivum* (L.) and characterization of large germplasm collections support the ABC transporter as essential for gene function. Theor. Appl. Genet. 121: 373-384.

Dakouri, A., B. D. McCallum, N. Radovanovic, and S. Cloutier. 2013. Molecular and phenotypic characterization of seedling and adult plant leaf rust resistance in a world wheat collection. Mol. Breed. 32: 663-677.

Das M. K., S. Rajaram, C. C. Mundt, W. E. Kronstad, and R. P. Singh, 1992. Inheritance of slow rusting resistance in wheat. Crop Sci. 32: 1452-1456.

Devos, K. M., and M. D. Gale. 1992. The use of random amplified polymorphic DNA markers in wheat. Theor. Appl. Genet. 84: 567-572.

Dong, J. G. (ed) 2001. Agricultural Plant Pathology. China Agriculture Press, Beijing.

Dubin, H. J., and J. P. Brennan. 2009. Combating stem and leaf rust of wheat: Historical perspective, impacts, and lessons learned. IFPRI Discussion Paper 910, Washington, D. C. Dubin, H. J., and E. Torres. 1981. Causes and consequences of the 1976-1977 wheat leaf rust epidemic in Northwest Mexico. Annu. Rev. Phytopathol. 19: 41-49.

Dyck, P. L., and E. R. Kerber. 1971. Chromosome location of three genes for leaf rust resistance in common wheat. Can. J. Genet. Cytol. 13: 480-483.

Elshire, R. J., J. C. Glaubitz, Q. Sun, J. A. Poland, K. Kawamoto, E. S. Buckler, and S. E. Mitchel. 2011. A robust, simple genotyping-by-sequencing (GBS) approach for high diversity species. PLoS One 6: e19379.

Faris, J. D., W. L. Li, D. J. Liu, P. D. Chen, and B. S. Gill. 1999. Candidate gene analysis of quantitative disease resistance in wheat. Theor. Appl. Genet. 98: 219225.

Feng, J., L. L. Zuo, Z. Y. Zhang, R. M. Lin, Y. Y. Cao, and S. C. Xu. 2011. Quantitative trait loci for tempertature-sensitive resistance to *Puccinia striiformis* f. sp. *tritici* in wheat cultivar Flinor. Euphytica 178: 321-329.

Feuillet, C., S. Travella, N. Stein, L. Albar, A. Nublat, and B. Keller. 2003. Map-based isolation of the leaf rust disease resistance gene *Lr10* from the hexaploid wheat (*Triticum aestivum* L.) genome. Proc. Natl. Acad. Sci. USA. 100: 15253-15258.

Flint-Garcia, S. A., J. M. Thornsberry, and E. S. Buckler. 2003. Structure of linkage disequilibrium in plants. Annu. Rev. Plant Biol. 54: 357-374.

Flor, H. H. 1942. Inheritance of pathogenicity in *Melampsora lini*. Phytopathology 32: 653-669.

Fu, D. L., C. Uauy, A. Distelfeld, A. Blechl, L. Epstein, X. M. Chen, H. Sela, T. Fahima, and J. Dubcovsky. 2009. A kinase-start gene confers temperature dependent resistance to wheat stripe rust. Science. 323: 1357-1360.

German, S. E., and J. A. Kolmer. 1992. Effect of *Lr34* in the enhancement of resistance to leaf rust of wheat. Theor. Appl. Genet. 84: 97-105.

Guo, Q., Z. J. Zhang, Y. B. Xu, G. H. Li, J. Feng, and Y. Zhou. 2008. Quantitative trait loci for high-temperature adult-plant and slow-rusting resistance to *Puccinia striiformis* f. sp. *tritici* in wheat cultivars. Phytopathology. 98: 803-809.

Hale, I. L., I. Mamuya, and D. Singh. 2013. *Sr31*-virulent races (TTKSK, TTKST, and TTTSK) of the wheat stem rust pathogen *Puccinia graminis* f. sp. *tritici* are present in Tanzania. Plant Dis. 97: 557.

Hamilton, J. P., and C. R. Buell. 2012. Advances in plant genome sequencing. Plant J. 70: 177-190.

Harjes, C. E. T. R. Rocheford, L. Bai, T. P. Brutnell, C. B. Kandianis, S. G. Sowinski, A. E. Stapleton, R. Vallabhaneni, M. Williams, E. T. Wurtzel, J. Yan, and E. S. Buckler. 2008. Natural genetic variation in lycopene epsilon cyclase tapped for maize biofortification. Science. 319: 330-333.

He, Z. H., C. X. Lan, X. M. Chen, Y. C. Zou, Q. S. Zhuang, and X. C. Xia. 2011. Progress and perspective in research of adult-plant resistance to stripe rust and powdery mildew in wheat. Sci. Agric. Sin. 44: 2193-2215.

Helguera, M., I. A. Khan, J. Kolmer, D. Lijavetzky, L. Zhong-qi, and J. Dubcovsky. 2003. PCR assays for the *Lr37-Yr17-Sr38* cluster of rust resistance genes and their use to develop isogenic hard red spring wheat lines. Crop Sci. 43: 1839-1847.

Herrera-Foessel, S. A., E. S. Lagudah, J. Huerta-Espino, M. J. Hayden, H. S. Bariana, D. Singh, and R. P. Singh. 2011. New slow-rusting leaf rust and stripe rust resistance genes *Lr67* and *Yr46* in wheat are pleiotropic or closely linked. Theor. Appl. Genet. 122: 239-249.

Herrera-Foessel, S. A., R. P. Singh, J. Huerta-Espino, G. M. Rosewarne, S. K. Periyannan, L. Viccars, V. Calvo-Salazar, C. X. Lan, and E. S. Lagudah. 2012. *Lr68*: a new gene conferring slow rusting resistance to leaf rust in wheat. Theor. Appl. Genet. 124: 1475-1486.

Herrera-Foessel, S. A., R. P. Singh, M. Lillemo, J. Huerta-Espino, S. Bhavani, S. Singh, C. X. Lan, V. Calvo-Salazar, and E. S. Lagudah 2014. *Lr67/Yr46* confers adult plant resistance to stem rust and powdery mildew in wheat. Theor. Appl. Genet., DOI 10.1007/s00122-013-2256-9.

Hiebert, C. W., J. B. Thomas, B. D. McCallum, D. G. Humphreys, R. M. DePauw, M. J. Hayden, R. Mago, W. Schnippenkoetter, and W. Spielmeyer. 2010. An introgression on wheat chromosome 4DL in RL6077 (Thatcher * 6/PI 250413) confers adult plant resistance to stripe rust and leaf rust (*Lr67*). Theor. Appl. Genet. 21: 1083-1091.

Hiebert, C. W., J. B. Thomas, D. J. Somers, B. D. McCallum, and S. L. Fox. 2007. Microsatellite mapping of adult-plant leaf rust resistance gene *Lr22a* in wheat. Theor. Appl. Genet. 115: 877-884.

Hsam, S. L. K., and F. J. Zeller. 2002. Breeding for powdery mildew resistance in common wheat (*Triticum aestivum* L.). *In*: Belanger R. R., W. R. Bushnell, A. J. Dik, and T. L. W. Carver (eds) The powdery mildews: a comprehensive treatise. The American Phytopathological Society, St Paul, MN, pp. 219-238.

Huang, L., S. A. Brooks, W. Li, J. P. Fellers, H. N. Trick, and B. S. Gill. 2003. Map-based cloning of leaf rust resistance gene *Lr21* from the large and polyploid genome of bread wheat. Genetics 164: 655-664.

Huang, Q. H., R. L. Jing, X. Y. Wu, L. P. Cao, X. P. Chang, X. Z. Zhang, and T. R. Huang. 2008. QTL mapping for adult-plant resistance to powdery mildew in common wheat. Sci. Agric. Sin. 41: 2528-2536.

Huang, X., X. Wei, T. Sang, Q. Zhao, Q. Feng, Y. Zhao, C. Li, C. Zhu, T. Lu, Z. Zhang, M. Li, D. Fan, Y. Guo, A. Wang, L. Wang, L. Deng, W. Li, Y. Lu, Q. Weng, K. Liu, T. Huang, T. Zhou, Y. Jing, W. Li, Z. Lin, E. S. Buckler, Q. Qian, Q. F. Zhang, J. Li, and B. Han. 2010. Genome-wide association studies of 14 agronomic traits in rice landraces. Nat. Genet. 42: 961-967.

Huang, X., Y. Zhao, X. Wei, C. Li, A. Wang, Q. Zhao, W. Li, Y. Guo, L. Deng, C. Zhu, D. Fan, Y. Lu, Q. Weng, K. Liu, T. Zhou, Y. Jing, L. Si, G. Dong, T. Huang, T. Lu, Q. Feng, Q. Qian, J. Li, and B. Han. 2012. Genome-wide association study of flowering time and grain yield traits in a worldwide collection of rice germplasm. Nat. Genet. 44: 32-39.

Hulbert, S. H., C. A. Webb, S. M. Smith, and Q. Sun. 2001. Resistance gene complexes: evolution and utilization. Annu. Rev. Phytopathol. 39: 285-312.

Huo, N. X., R. H. Zhou, L. F. Zhang, and J. Z. Jia. 2005. Mapping quantitative trait loci for powdery mildew resistance in wheat. Acta Agron. Sin. 6: 692-296.

Hurni, S., S. Brunner, G. Buchmann, G. Herren, T. Jordan, P. Krukowski, T. Wicker, N. Yahiaoui, R. Mago, and B. Keller. 2013. Rye *Pm8* and wheat *Pm3* are orthologous genes and show evolutionary conservation of resistance function against powdery mildew. Plant J. 76: 957-969.

Jakobson, I., D. Reis, A. Tiidema, H. Peusha, L. Timofejeva, M. Valárik, M. Kladivová, H. Šimková, and J. Doležel. 2012. Fine mapping, phenotypic characterization and validation of race non-specific resistance to powdery mildew in a wheat-*Triticum militinae* introgression line. Theor. Appl. Genet. 125: 609-623.

Jakobson, I., H. Peusha, L. Timofejeva, and K. Jarve. 2006. Adult plant and seedling resistance to powdery mildew in a *Triticum aestivum* × *Triticum militinae* hybrid line, Theor. Appl. Genet. 112: 760-769.

Jia, H. Y., G. Q. Yao, Z. Z. Zhang, H. X. Xu, B. S. Fu, Z. X. Kong, and Z. Q. Ma. 2009. Mapping of a major QTL for powdery mildew resistance in a *Triticum monococcum* accession. Mol. Plant Breed. 4: 646-652.

Jia, J., S. Zhao, X. Kong, Y. Li, G. Zhao, W. He, R. Appels, M. Pfeifer, Y. Tao, X. Zhang, R. Jing, C. Zhang, Y. Ma, L. Gao, C. Gao, M. Spannagl, K. F. X. Mayer, D. Li, S. Pan, F. Zheng, Q. Hu, X. Xia, J. Li, Q. Liang, J. Chen, T. Wicker, C. Gou, H. Kuang, G. He, Y. Luo, B. Keller, Q. Xia, P. Lu, J. Wang, H. Zou, R. Zhang, J. Xu, J. Gao, C. Middleton, Z. Quan, G. Liu, J. Wang, IWGSC, H. Yang, X. Liu, Z. He, and J. Wang. 2013. *Aegilops tauschii* draft genome sequence reveals a gene repertoire for wheat adaptation. Nature 496: 91-95.

Johnson, R., and C. N. Law. 1973. Cytogenetic studies in the resistance of the wheat variety Bersée to *Puccinia striiformis*. Cereal Rusts Bul. 1: 38-43.

Jones, J. D., and J. L. Dangl. 2006. The plant immune system. Nature 444: 323-329.

Keller, B., E. S. Lagudah, L. L. Selter, J. M. Risk, C. Harsh, and S. G. Krattinger. 2013. How has *Lr34/Yr18* conferred effective rust resistance in wheat for so long? www. globalrust. org/db/attachments/bgriworkshop/13/1/keller_web. pdf

Keller, M., B. Keller, G. Schachermayr, M. Winzeler, J.

E. Schmid, P. Stamp, and M. M. Messmer. 1999. Quantitative trait loci for resistance against powdery mildew in a segregating wheat × spelt population. Theor. Appl. Genet. 98: 903-912.

Khan, M., A. Bukhari, Z. Dar, and S. Rizvi. 2013. Status and strategies in breeding for rust resistance in wheat. Agr. Sci. 4: 292-301.

Kolmer, J. A. 1989. Virulence and race dynamics of *Puccinia recondita* f. sp. *tritici* in Canada during 1956 - 1987. Phytopathology. 79: 349-356.

Krattinger, S. G., E. S. Lagudah, W. Spielmeyer, R. P. Singh, J. Huerta-Espino, H. McFadden, E. Bossolini, L. L. Selter, and B. Keller. 2009. A putative ABC transporter confers durable resistance to multiple fungal pathogens in wheat. Science. 323: 13601363.

Krattinger, S. G., E. S. Lagudah, T. Wicker, J. M. Risk, A. R. Ashton, L. L. Selter, T. Matsumoto, and B. Keller. 2011. *Lr34* multi-pathogen resistance ABC transporter: molecular analysis of homoeologous and orthologous genes in hexaploid wheat and other grass species. Plant J. 65: 392-403.

Lagudah, E. S. 2011. Molecular genetics of race non-specific rust resistance in wheat. Euphytica. 179: 81-91.

Lan, C. X., R. P. Singh., J. Huerta-Espino., V. Calvo-Salazar., and S. A. Herrera-Foessel. 2014. Genetic analysis of resistance to leaf rust and stripe rust in wheat cultivar Francolin#1. Plant Dis. PDIS-07-13-0707.

Lan, C. X., S. S. Liang, X. C. Zhou, G. Zhou, Q. L. Lu, X. C. Xia, and Z. H. He. 2010a. Identifcation of genomic regions controlling adult-plant stripe rust resistance in Chinese landrace Pingyuan 50 through bulked segregant analysis. Phytopathology. 100: 313-318.

Lan, C. X., S. S. Liang, Z. L. Wang, J. Yan, Y. Zhang, X. C. Xia, and Z. H. He. 2009. Quantitative trait loci mapping for adult-plant resistance against powdery mildew in Chinese wheat cultivar Bainong 64. Phytopathology. 99: 1121-1126.

Lan, C. X., X. W. Ni, J. Yan, Y. Zhang, X. C. Xia, X. M. Chen, and Z. H. He. 2010b. Quantitative trait loci mapping of slow powdery mildewing resistance in Chinese wheat cultivar Lumai 21. Mol. Breed. 25: 615-622.

Li, H., Z. Peng, X. Yang, W. Wang, J. Fu, J. Wang, Y. Han, Y. Chai, T. Guo, N. Yang, J. Liu, M. L. Warburton, Y. Cheng, X. Hao, P. Zhang, J. Zhao, Y. Liu, G. Wang, J. Li, and J. Yan. 2013a. Genome-wide association study dissects the genetic architecture of oil biosynthesis in maize kernels. Nat. Genet. 45: 43-50.

Li, T., and G. H. Bai. 2009. Lesion mimic associates with adult plant resistance to leaf rust infection in wheat. Theor. Appl. Genet. 119: 13-21.

Li, Z. F., S. Singh, R. P. Singh, E. E. Lopez-Vera, and J. Huerta-Espino. 2013b. Genetics of resistance to yellow rust in PBW343 × Kenya Kudu recombinant inbred line population and mapping of a new resistance gene *YrKK*. Mol. Breed. 32: 821-829.

Li, Z. F., X. C. Xia, X. C. Zhou, Y. C. Niu, Z. H. He, Y. Zhang, G. Q. Li, A. M. Wan, D. S. Wang, X. M. Chen, Q. L. Lu, and R. P. Singh. 2006. Seedling and slow rusting resistance to stripe rust in Chinese common wheats. Plant Dis. 90: 1302-1312.

Li, Z. F., X. C. Xia, Z. H. He, X. Li, L. J. Zhang, H. Y. Wang, Q. F. Meng, W. X. Yang, G. Q. Li, and D. Q. Liu. 2010. Seedling and slow rusting resistance to leaf rust in Chinese wheat cultivars. Plant Dis. 94: 45-53.

Li, Z. Q., and S. M. Zeng (eds) 2002. WheatRusts in China. China Agriculture Press, Beijing.

Liang, S. S., K. Suenaga, Z. H. He, Z. L. Wang, H. Y. Liu, D. S. Wang, R. P. Singh, P. Sourdille, and X. C. Xia. 2006. Quantitative trait loci mapping for adult-plant resistance to powdery mildew in bread wheat. Phytopathology. 96: 784-789.

Lillemo, M., B. Asalf, R. P. Singh, J. Huerta-Espino, X. M. Chen, Z. H. He, and Å. Bjørnstad. 2008. The adult plant rust resistance loci *Lr34/Yr18* and *Lr46/Yr29* are important determinants of partial resistance to powdery mildew in bread wheat line Saar. Theor. Appl. Genet. 116: 1155-1166.

Lillemo, M., Å. Bjørnstad, and H. Skinnes. 2012. Molecular mapping of partial resistance to powdery mildew in winter wheat cultivar Folke. Euphytica. 185: 47-59.

Ling, H. Q., S. Zhao, D. Liu, J. Wang, H. Sun, C. Zhang, H. Fan, D. Li, L. Dong, Y. Tao, C. Gao, H. Wu., Y. Li, Y. Cui, X. Guo, S. Zheng, B. Wang, K. Yu, Q. Liang, W. Yang, X. Lou, J. Chen, M. Feng, J. Jian, X. Zhang, G. Luo, Y. Jiang, J. Liu, Z. Wang, Y. Sha, B. Zhang, H. Wu, D. Tang, Q. Shen, P. Xue, S. Zou, X. Wang, X. Liu, F. Wang, Y. Yang, X. An, Z. Dong, K. Zhang, X. Zhang, M. -C. Luo, J. Dvorak, Y. Tong, J. Wang, H. Yang, Z. Li, D. Wang, A. Zhang, and J. Wang. 2013. Draft genome of the wheat A-genome progenitor *Triticum urartu*. Nature. 496: 87-90.

Liu, S. X., C. A. Griffey, and M. A. S. Maroof. 2001. Identification of molecular markers associated with adult plant resistance to powdery mildew in common wheat cultivar Massey. Crop Sci. 41：1268-1275.

Lowe I.，D. Cantu，and J. Dubcovsky. 2011. Durable resistance to the wheat rusts：integrating systems biology and traditional phenotype-based research methods to guide the deployment of resistance genes. Euphytica. 179：69-79.

Luo，P. G.，X. Y. Hu，Z. L. Ren，H. Y. Zhang，K. Shu，and Z. J. Yang. 2008. Allelic analysis of yellow rust resistance genes on wheat chromosome 2BS. Genome. 51：922-927.

Lu，Y. M.，C. X. Lan，S. S. Linag，X. C. Zhou，D. Liu，G. Zhou，Q. L. Lu，J. X. Jing，M. N. Wang，X. C. Xia，and Z. H. He. 2009. QTL mapping for adult-plant resistance to stripe rust in Italian common wheat cultivars Libellula and Strampelli. Theor. Appl. Genet. 119：1349-1359.

Lu，Y.，S. Zhang，T. Shah，C. Xie，Z. Hao，X. Li，M. Farkhari，J. M. Ribaute，M. Cao，T. Rong，and Y. Xu. 2010. Joint linkage － linkage disequilibrium mapping is a powerful approach to detecting quantitative trait loci underlying drought tolerance in maize. Proc. Natl. Acad. Sci. USA. 107：19585-19590.

Maccaferri，M.，M. C. Sanguineti，P. Mantovani，A. Demontis，A. Massi，K. Ammar，J. A. Kolmer，J. H. Czembor，S. Ezrati，and R. Tuberosa. 2010. Association mapping of leaf rust response in durum wheat. Mol. Breed. 26：189-228.

Maccaferri，M.，P. Mantovani，R. Tuberosa，E. DeAmbrogio，S. Giuliani，A. Demontis，A. Massi，and M. C. Sanguineti. 2008. A major QTL for durable leaf rust resistance widely exploited in durum wheat breeding programs maps on the distal region of chromosome arm 7BL. Theor. Appl. Genet. 117：1225-1240.

Mallard，S.，D. Gaudet，A. Aldeia，C. Abelard，A. L. Besnard，P. Sourdille，and F. Dedryver. 2005. Genetic analysis of durable resistance to yellow rust in bread wheat. Theor. Appl. Genet. 110：1401-1409.

Marone，D.，M. A. Russo，G. Laidò，P. D. Vita，R. Papa，A. Blanco，A. Gadaleta，D. Rubiales，and A. M. Mastrangelo. 2013. Genetic basis of qualitative and quantitative resistance to powdery mildew in wheat：from consensus regions to candidate genes. BMC Genomics. 14：562.

McIntosh，R. A.，Y. Yamazaki，J. Dubcovsky，W. J. Rogers，C. Morris，R. Appels，and X. C. Xia. 2013. Catalogue of gene symbols for wheat：2013 supplement. http：//www. shigen. nig. ac. jp/wheat/komugi/genes/macgene/2013/GeneCatalogueIntroduction. pdf

Melichar，J. P. E.，S. Berry，C. Newell，R. MacCormack，and L. A. Boyd. 2008. QTL identification and microphenotype characterisation of the developmentally regulated yellow rust resistance in the UK wheat cultivar Guardian. Theor. Appl. Genet. 117：391-399.

Messmer，M. M.，R. Seyfarth，M. Keller，G. Schachermayr，M. Winzeler，S. Zanetti，C. Feuillet，and B. Keller. 2000. Genetic analysis of durable leaf rust resistance in winter wheat. Theor. Appl. Genet. 100：419-431.

Mingeot，D.，N. Chantret，P. V. Baret，A. Dekeyser，N. Boukhatem，P. Sourdille，G. Doussinault，and J. M. Jacquemin. 2002. Mapping QTL involved in adult plant resistance to powdery mildew in the winter wheat line RE714 in two susceptible genetic backgrounds. Plant Breed. 121：133-140.

Mueller，U. G.，and L. L. Wolfenbarger. 1999. AFLP genotyping and fingerprinting. Trends Ecol. Evol. 14：389-394.

Navabi，A，J. P. Tewari，R. P. Singh，B. McCallum，A. Laroche，and K. G. Briggs. 2005. Inheritance and QTL analysis of durable resistance to stripe and leaf rusts in an Australian cultivar，*Triticum aestivum* ‘Cook’. Genome. 48：97-107.

Nelson，J. C.，R. P. Singh，J. E. Autrique，and M. E. Sorrells. 1997. Mapping genes conferring and suppressing leaf rust resistance in wheat. Crop Sci. 37：1928-1935.

Periyannan，S.，J. Moore，M. Ayliffe，U. Bansal，X. Wang，L Huang，K. Deal，M. Luo，X. Kong，H. Bariana，R. Mago，R. McIntosh，P. Dodds，J. Dvorak，and E. Lagudah. 2013. The gene *Sr33*，an ortholog of barley *Mla* genes，encodes resistance to wheat stem rust race Ug99. Science. 341：786-768.

Poland，J. A.，P. J. Brown，M. E. Sorrells，and J. L. Jannink. 2012. Development of high-density genetic maps for barley and wheat using a novel two-enzyme Genotyping-by-Sequencing approach. PLoS One 7：e32253.

Ramburan，V. P.，Z. A. Pretorius，J. H. Louw，L. A. Boyd，P. H. Smith，W. H. P. Boshoff，and R. Prins. 2004. A genetic analysis of adult plant resistance to stripe rust in wheat cultivar Kariega. Theor. Appl. Genet. 108：1426-1433.

Ren，Y，Z. H. He，J. Li，M. Lillemo，L. Wu，B. Bai，Q. X. Lu，H. Z. Zhu，G. Zhou，J. Y. Du，Q. L. Lu，and X. C. Xia. 2012a. QTL mapping of adult-plant resistance to stripe rust in a population derived from common wheat cultivars Naxos and Shanghai 3/Catbird. Theor. Appl. Genet. 125: 1211-1221.

Ren, Y., Z. F. Li, Z. H. He, L. Wu, B. Bai, C. X. Lan, C. F. Wang, G. Zhou, H. Z. Zhu, and X. C. Xia. 2012b. QTL mapping of adult-plant resistances to stripe rust and leaf rust in Chinese wheat cultivar Bainong 64. Theor. Appl. Genet. 125: 1253-1262.

Risk, J. M., L. L. Selter, S. G. Krattinger, L. A. Viccars, T. M. Richardson, G. Buesing, G. Herren, E. S. Lagudah, and B. Keller. 2012. Functional variability of the *Lr34* durable resistance gene in transgenic wheat. Plant Biotech. J. 10: 477-487.

Rosewarne, G. M., R. P. Singh, J. Huerta-Espino, and G. J. Rebetzke. 2008. Quantitative trait loci for slow-rusting resistance in wheat to leaf rust and stripe rust identified with multi-environment analysis. Theor. Appl. Genet. 116: 1027-1034.

Rosewarne, G. M., R. P. Singh, J. Huerta-Espino, H. M. William, S. Bouchet, S. Cloutier, H. McFadden, and E. S. Lagudah. 2006. Leaf tip necrosis, molecular markers and β1-proteasome subunits associated with the slow rusting resistance genes *Lr46/Yr29*. Theor. Appl. Genet. 112: 500-508.

Rosewarne, G. M., R. P. Singh, J. Huerta-Espino, S. A. Herrera-Foessel, K. L. Forrest, M. J. Hayden, and G. J. Rebetzke. 2012. Analysis of leaf and stripe rust severities reveals pathotype changes and multiple minor QTLs associated with resistance in an Avocet × Pastor wheat population. Theor. Appl. Genet. 124: 1283-1294.

Rosewarne, G. M., S. A. Herrera-Foessel, R. P. Singh, J. Huerta-Espino, C. X. Lan, and Z. H. He. 2013. Quantitative trait loci of stripe rust resistance in wheat. Theor. Appl. Genet. 126: 2427-2449.

Röder, M. S., V. Korzun, K. Wendehake, J. Plaschke, M. H. Tixier, P. Leroy, and M. W. Ganal. 1998. A microsatellite map of wheat. Genetics. 149: 2007-2023.

Rubiales, D., and R. E. Niks. 1995. Characterization of *Lr34*, a major gene conferring non-hypersensitive resistance to wheat leaf rust. Plant Dis. 79: 1208-1212.

Saari, E. E., and J. M. Prescott. 1985. World distribution in relation to economic losses. In: Roelfs A. P. and W. R. Bushnell eds. The Cereal Rusts Vol. Ⅱ: Diseases, distribution, epidemiology and control, Academic Press, Orlando, Fl. pp: 259-298.

Saintenac, C., D. Jiang, S. Wang, and E. Akhunov. 2013a. Sequence-based mapping of the polyploid wheat genome. G3 (Bethesda). 3: 1105-1114.

Saintenac, C., W. Zhang, A. Salcedo, M. N. Rouse, H. N. Trick, E. Akhunov, and J. Dubcovsky. 2013b. Identification of wheat gene *Sr35* that confers resistance to Ug99 stem rust race group. Science. 341: 783-786.

Schnurbusch, T., S. Paillard, A. Schori, M. Messmer, G. Schachermayr, M. Winzeler, and B. Keller. 2004. Dissection of quantitative and durable leaf rust resistance in Swiss winter wheat reveals a major resistance QTL in the *Lr34* chromosome region. Theor. Appl. Genet. 108: 477-484.

Seyfarth, R., C. Feuillet, G. Schachermayr, M. Messmer, M. Winzeler, and B. Keller. 2000. Molecular mapping of the adult-plant rust resistance gene *Lr13* in wheat (*Triticum aestivum* L.). J. Genet. Breed. 54: 193-198.

Seyfarth, R., C. Feuillet, G. Schachermayr, M. Winzeler, and B. Keller. 1999. Development of a molecular marker for the adult plant leaf rust resistance gene *Lr35* in wheat. Theor. Appl. Genet. 99: 554-560.

Shriner, D., L. K. Vaughan, M. A. Padilla, and H. K. Tiwari. 2007. Problems with genome-wide association studies. Science. 316: 1840-1842.

Silva, P., V. Calvo-Salazar, F. Condon, M. Quinke, L. Gutierrez, A. Castro, S. Herrera-Foessel, J. von Zitzewitz, and S. Germán. 2013. Effect and interaction of wheat leaf rust adult plant resistance genes in Uruguay. BGRI 2013 Technical Workshop, 19-22. August, New Delhi, India.

Singh, D., J. Simmonds, R. F. Park, H. S. Bariana, and J. W. Snape. 2009. Inheritance and QTL mapping of leaf rust resistance in the European winter wheat cultivar 'Beaver'. Euphytica. 169: 253-261.

Singh, R. P. 1992. Association between gene *Lr34* for leaf rust resistance and leaf tip necrosis in wheat. Crop Sci. 32: 874-878.

Singh, R. P., A. Mujeeb-Kazi, and J. Huerta-Espino. 1998. *Lr46*: A gene conferring slow-rusting resistance to leaf rust in wheat. Phytopathology. 88: 890-894.

Singh, R. P., and S. Rajaram. 1992. Genetics of adult-plant resistance of leaf rust in 'Frontana' and three CIMMYT wheats. Genome. 35: 24-31.

Singh, R. P., S. A. Herrera-Foessel., J. Huerta-Espino., C. X. Lan., B. R. Basnet., S. Bhavani., and E. S. Lagudah. 2013. Pleiotropic gene *Lr46/Yr29/Pm39/Ltn2* confers slow rusting, adult plant resistance to wheat stem rust fungus. Proceedings Borlaug Global Rust Initiative, 2013 Technical Workshop, 19-22. August, New Dehli, India. p 17. 1.

Singh, R. P., J. Huerta-espino, and H. M. William. 2005. Genetics and breeding for durable resistance to leaf and stripe rusts in wheat. Turk. J. Agric. For. 29: 121-127.

Singh, R. P., J. Huerta-Espino, and S. Rajaram. 2000. Achieving near-immunity to leaf and stripe rusts in wheat by combining slow rusting resistance genes. Acta Phytopathol. Hung. 35: 133-139.

Singh, R. P., J. Huerta-Espino, S. Bhavani, S. A. Herrera-Foessel, D. Singh, P. K. Singh, G. Velu, R. E. Mason, Y. Jin, P. Njau, and J Crossa. 2011. Race non-specific resistance to rust diseases in CIMMYT spring wheats. Euphytica. 179: 175-186.

Somers, D. J., P. Isaac, and K. Edwards. 2004. A high-density microsatellite consensus map for bread wheat (*Triticum aestivum* L.). Theor. Appl. Genet. 109: 1105-1114.

Sourdille, P., B. Gandon, V. Chiquet, N. Nicot, D. Somers, A. Murigneux, and M. Bernard. 2004. Wheat Genoplante SSR mapping data release: a new set of markers and comprehensive Genetic and physical mapping data. http: //wheat. pw. usda. gov/ggpages/SSRclub/GeneticPhysical/

Spielmeyer W., R. A. McIntosh, J. Kolmer, and E. S. Lagudah. 2005. Powdery mildew resistance and *Lr34*/*Yr18* genes for durable resistance to leaf and stripe rust cosegregate at a locus on the short arm of chromosome 7D of wheat. Theor. Appl. Genet. 111: 731-735.

Suenaga, K., R. P. Singh, J. Huerta-Espino, and H. M. William. 2003. Microsatellite markers for genes *Lr34*/*Yr18* and other quantitative trait loci for leaf rust and stripe rust resistance in bread wheat. Phytopathology. 93: 881-890.

Sun, Y., J. Wang, J. H. Crouch, and Y. Xu. 2010. Efficiency of selective genotyping for genetic analysis of complex traits and potential applications in crop improvement. Mol. Breed. 26: 493-511.

Thornsberry, J. M., M. M. Goodman, J. Doebley, S. Kresovich, D Nielsen, and E. S. Buckler. 2001. Dwarf 8 polymorphisms associate with variation in flowering time. Nat. Genet. 28: 286-289.

Tucker, D. M., C. A. Griffey, S. Liu, G. Brown-Guedira, D. S. Marshall, and M. A. Saghai Maroof. 2007. Confirmation of three quantitative trait loci conferring adult plant resistance to powdery mildew in two winter wheat populations. Euphytica. 155: 1-13.

Vazquez, M. D., C. J. Peterson, O. Riera-Lizarazu, X. Chen, A. Heesacker, K. Ammar, J. Crossa, and C. C. Mundt. 2012. Genetic analysis of adult plant, quantitative resistance to stripe rust in wheat cultivar 'Stephens' in multi-environment trials. Theor. Appl. Genet. 124: 1-11.

Wang, S., D. Wong, K. Forrest, A. Allen, S. Chao, B. E. Huang, M. Maccaferri, S. Salvi, S. G. Milner, L. Cattivelli, A. M. Mastrangelo, A. Whan, S. Stephen, G. Barker, R. Wieseke, J. Plieske; International Wheat Genome Sequencing Consortium, M. Lillemo, D. Mather, R. Appels, R. Dolferus, G. Brown-Guedira, A. Korol, A. R. Akhunova, C. Feuillet, J. Salse, M. Morgante, C. Pozniak, M. C. Luo, J. Dvorak, M. Morell, J. Dubcovsky, M. Ganal, R. Tuberosa, C. Lawley, I. Mikoulitch, C. Cavanagh, K. J. Edwards, M. Hayden, and E. Akhunov. 2014. Characterization of polyploid wheat genomic diversity using a high-density 90, 000 single nucleotide polymorphism array. Plant Biotechnol. J., Doi: 10. 1111/pbi. 12183.

Wang, W. Y. S., B. J. Barratt, D. G. Clayton, and J. A. Todd. 2005a. Genome-wide association studies: theoretical and practical concerns. Nat. Rev. Genet. 6: 109-118.

Wang, Z. L., L. H. Li, Z. H. He, X. Y. Duan, Y. L. Zhou, X. M. Chen, M. Lillemo, R. P. Singh, H. Wang, and X. C. Xia. 2005b. Seedling and adult plant resistance to powdery mildew in Chinese bread wheat cultivars and lines. Plant Dis. 89: 457-463.

Wellings, C. R. 1992. Resistance to stripe (yellow) rust in selected spring wheats. Vortrage für Pflanzenzuchtung. 24: 273-275.

Wellings, C. R. 2011. Global status of stripe rust: a review of historical and current threats. Euphytica. 179: 129-141.

William, H. M., D. Hoisington, R. P. Singh, and D. Gonzalez-de-Leon. 1997. Detection of quantitative trait loci associated with leaf rust resistance in bread wheat. Genome. 40: 253-260.

William, H. M., R. P. Singh, J. Huerta-Espino, G. Palacios, and K. Suenaga. 2006. Characterization of genetic loci conferring adult plant resistance to leaf rust and stripe rust in spring wheat. Genome. 49: 977-990.

William, M., R. P. Singh, J. Huerta-Espino, S. Ortiz Islas, and D. Hoisington. 2003. Molecular marker mapping of leaf rust resistance gene *Lr46* and its association with stripe rust resistance gene *Yr29* in wheat. Phytopathology. 93：153-159.

Xu, X. Y, G. H. Bai, B. F. Carver, G. E. Shaner, and R. M. Hunger. 2005a. Molecular characterization of slow leaf-rusting resistance in wheat. Crop Sci. 45：758-765.

Xu, X. Y., G. H. Bai, B. F. Carver, G. E. Shaner, and R. M. Hunger. 2005b. Mapping of QTLs prolonging the latent period of *Puccinia triticina* infection in wheat. Theor. Appl. Genet. 110：244-251.

Yahiaoui, N., P. Srichumpa, R. Dudler, and B. Keller, 2004 Genome analysis at different ploidy levels allows cloning of the powdery mildew resistance gene *Pm3b* from hexaploid wheat. Plant J. 37：528-538.

Yang, E. N., G. M. Rosewarne, S. A. Herrera-Foessel, J. Huerta-Espino, Z. X. Tang, C. F. Sun, Z. L. Ren, and R. P. Singh. 2013. QTL analysis of the spring wheat "Chapio" identifies stable stripe rust resistance despite inter-continental genotype × environment interactions. Theor. Appl. Genet. 126：1721-1732.

Zhang, H., X. C. Xia, Z. H. He, X. Li, Z. F. Li, and D. Q. Liu. 2011. Molecular mapping of leaf rust resistance gene *LrBi16* in Chinese wheat cultivar Bimai 16. Mol. Breed. 28：527-534.

Zhang, K. P., L. Zhao, Y. Hai, G. F. Chen, and J. C. Tian. 2008. QTL mapping for adult-plant resistance to powdery mildew, lodging resistance, and internode length below spike in wheat. Acta Agron. Sin. 34：1350-1357.

Zhang, L., Z. Li, M. Lillemo, X. Xia, D. Liu, W. Yang, J. Luo, and H. Wang. 2009. QTL mapping for adult-plant resistance to leaf rust in CIMMYT wheat cultivar Saar. Sci. Agri. Sin. 42：388-397.

Zhang, X., D. Han., Q. Zeng, Y. Duan, F. Yuan, J. Shi, Q. Wang, J. Wu, L. Huang, and Z. Kang. 2013. Fine mapping of wheat stripe rust resistance gene *Yr26* based on collinearity of wheat with *Brachypodium distachyon* and rice. PLoS One 8：e57885.

Zhao, K., M. J. Aranzana, S. Kim, C. Lister, C. Shindo, C. Tang, C. Toomajian, H. Zheng, C. Dean, P. Marjoram, and M. Nordborg. 2007. An Arabidopsis example of association mapping in structured samples. PLoS Genet. 3：e4.

Zhao, X. L., T. C. Zheng, X. C. Xia, Z. H. He, D. Q. Liu, W. X. Yang, G. H. Yin, and Z. F. Li. 2008. Molecular mapping of leaf rust resistance gene *LrZH84* in Chinese wheat line Zhou 8425B. Theor. Appl. Genet. 117：1069-1075.

Zhou, H. X., X. C. Xia, Z. H. He, X. Li, C. F. Wang, Z. F. Li, and D. Q. Liu. 2013. Molecular mapping of leaf rust resistance gene *LrNJ97* in Chinese wheat line Neijiang 977671. Theor. Appl. Genet. 126：2141-2147.

Quantitative trait loci mapping for adult-plant resistance to powdery mildew in bread wheat

S. S. Liang[1], K. Suenaga[2], Z. H. He[1,3], Z. L. Wang[4], H. Y. Liu[1], D. S. Wang[1], R. P. Singh[5], P. Sourdile[6], and X. C. Xia[1]

[1] *Institute of Crop Science/National Wheat Improvement Center, Chinese Academy of Agricultural Sciences (CAAS), Zhongguancun South Street* 12, 100081, *Beijing, China*; [2] *Japan International Research Center for Agricultural Sciences (JIRCAS)*, 1-1 *Ohwashi, Tsukuba, Ibaraki* 305-8686, *Japan*; [3] *CIMMYT China Office, C/O CAAS, Zhongguancun South Street* 12, 100081, *Beijing, China*; [4] *Northwest Sci-Tech University of Agriculture and Forestry, Yangling* 712100, *Shaanxi, China*; [5] *International Maize and Wheat Improvement Center (CIMMYT), Apdo, Postal*-6-641, 06600 *Mexico, D. F., Mexico*; [6] *INRA Amélioration et Santé des Plantes, Domaine de Crouelle*, 234, *Avenue du Brézet*, 63100, *Clermont-Ferrand Cedex* 2, *France.*

Abstract: Powdery mildew, caused by *Blumeria graminis* f. sp. *tritici*, is a major wheat disease worldwide. Use of adult-plant resistance (APR) is an effective method to develop wheat cultivars with durable resistance to powdery mildew. In the present study, a total of 432 molecular markers were used to map QTL for adult-plant resistance to powdery mildew in a doubled haploid (DH) population with 107 lines derived from the cross Fukuho-komugi × Oligoculm. Field trials were conducted in Beijing and Anyang of China during 2003-2004 and 2004-2005 cropping seasons, respectively. The DH lines were planted in a randomized complete block design with three replicates. Artificial inoculation was carried out in Beijing with highly virulent isolate E20 of *Blumeria graminis* f. sp. *tritici* and the powdery mildew severity on penultimate leaf was evaluated four times, and the maximum disease severity (MDS) on penultimate leaf was investigated in Anyang under natural inoculation in May 2004 and 2005. The heritability of resistance to powdery mildew for maximum disease severity (MDS) in two years and two locations ranged from 0. 82 to 0. 93, while the heritability for AUDPC was between 0. 84 and 0. 91. With the method of composite interval mapping (CIM), four quantitative trait loci (QTL) for APR to powdery mildew were detected on chromosomes 1AS, 2BL, 4BL, and 7DS, explaining 5. 7-26. 6% of the phenotypic variance. Three QTL on chromosomes 1AS, 2BL, and 7DS were derived from the female, Fukuho-komugi, while the one on chromosome 4BL was from the male, Oligoculm. The QTL on chromosome 1AS showed high genetic effect on powdery mildew resistance, accounting for 19. 5-26. 6% of phenotypic variance across two environments. The QTL on 7DS associated with the locus *Lr34/Yr18*, flanked by microsatellite *Xgwm295. 1* and *Ltn* (leaf tip necrosis). These results will benefit for improving powdery mildew resistance in wheat breeding programs.

Published in Phytopathology, 2006, 96 (7): 784-789

Introduction

Powdery mildew, caused by *Blumeria graminis* f. sp. *tritici*, is a very destructive leaf disease of common wheat (*Triticum aestivum* L.), which causes great yield losses in many wheat production areas of the world, especially in the regions with high rainfall and with a maritime or semi-continental climate [1]. In China, the vulnerability to powdery mildew has increased significantly since the early 1980s due to the introduction of semi-dwarf varieties and subsequent use of large amounts of fertilizers [40].

Although fungicide application is effective in controlling powdery mildew, the most economically and environmentally sound way of control is to use resistant cultivars [1]. Therefore, breeding for resistance to powdery mildew is a major objective in the main wheat producing regions, such as Yellow and Huai valley and Yangtze region in China. During the past decades, the race-specific resistance genes (*Pm* genes), conferring complete resistance caused by a hypersensitive reaction, have been used extensively. However, this type of resistance was often short lived due to the emergence of new pathogen races with matching virulence[24,26]. To prolong and enhance the effectiveness of race-specific resistance, gene pyramiding [36], multilines [4] and cultivar mixtures[39] were proposed and used in wheat breeding programs.

Alternatively, partial resistance [9] was proposed for durable resistance to powdery mildew, which was characterized by a compatible interaction in all growth stages, but a lower frequency of infection, a longer latent period, a lower rate or a shorter period of spore production at adult plant stage [22]. This type of resistance is also called adult-plant resistance (APR) or slow mildewing that can be identified in cultivars with defeated race-specific genes or lacking known race-specific resistance genes [8,24]. APR to powdery mildew is more durable than race-specific resistance, which has provided durable control of powdery mildew in wheat[26], barley[11] and oat [12].

Since APR is conditioned by quantitative resistance genes[8], molecular markers associated with the genes should be a useful tool for breeding. During the past decade, a series of studies on adult-plant resistance to powdery mildew were carried out. Griffey and Das[7] reported two to three genes conferred APR to powdery mildew in Massey. Keller et al. [13] detected 18 QTL for APR to powdery mildew in a segregating wheat/spelt (*Triticum spelt* L.) population, which explained 77% of the phenotypic variance. Liu et al. [16] identified three QTL for APR to powdery mildew in an $F_{2:3}$ mapping population derived from Becker/Massey, which were located on the chromosomes 1B, 2A, and 2B, respectively, explaining 17%, 29%, and 11% of the total variation for powdery mildew resistance. Chantret et al. [5] found three QTL on chromosomes 4A, 5D, and 6A, and five QTL on chromosomes 5D, 6A, 7A and 7B in a DH and $F_{2:3}$ population derived from the cross RE714/Hardi, of which the QTL on chromosome 5D was a major QTL, explaining 28.1-37.9% of phenotypic variance. Mingeot et al. [19] detected one to seven QTL for adult-plant resistance to powdery mildew in two DH populations from the cross between RE714 and the susceptible parents 'Festin' and 'Hardi' in different environments. They found two major QTL on chromosome 5D and at the *MlRE* locus that displayed stable resistance across different environments.

Fukuho-komugi (i. e. Norin 124) is a Japanese wheat cultivar with good agronomic traits and resistance to stripe rust, leaf rust and powdery mildew [23,33,34]. The adult-plant resistance genes *Yr18*, *Lr34* and other quantitative trait loci against stripe rust and leaf rust were identified in a previous study [34]. The aim of this study was to identify QTL for APR to powdery mildew and their associated molecular markers in a DH population from the cross between Fukuho-komugi and Israeli wheat Oligoculm.

Materials and methods

Plant Materials

A doubled haploid (DH) population with 107 lines used in this study was developed from a cross between Japanese wheat Fukuho-komugi and Israeli wheat Oligoculm by a wheat × maize cross technique [32]. Fukuho-komugi is moderately susceptible to the Chinese predominant isolate E20 of *B. graminis* f. sp. *tritici* at seeding stage, yet is highly resistant at the adult plant stage. Oligoculm is highly susceptible to the isolate E20 of *B. graminis* f. sp. *tritici* at seedling stage and moderately resistant at adult plant stage.

Field trials

During the 2003-2004 and 2004-2005 cropping seasons, the DH lines and their parents were evaluated for APR to powdery mildew, respectively, at the experimental station of the Institute of Crop Science, Chinese Academy of Agricultural Sciences (CAAS), Beijing, and the Cotton Research Institute, CAAS, Anyang, Henan province. Field trials were conducted in a randomized complete block design with three replicates. Plots consisted of single row with a row spacing of 25 cm and 2 m in length. Approximately 150 seeds were sown in each row.

In Beijing, the DH lines and their parents were sown in the spring. The spring wheat cultivar, Morocco, highly susceptible to powdery mildew, was planted every 10 rows as susceptible check, and planted around the test lines to ensure ample powdery mildew inoculum. Artificial inoculation with the highly virulent isolate E20 of *B. graminis* f. sp. *tritici* was performed prior to plants reaching stem elongation. The disease severity on penultimate leaf (F-1 leaf) on 10 randomly selected plants from each line was scored based on the actual percentage of leaf area covered by powdery mildew for the first time 2 weeks after inoculation and then at weekly intervals until leaves were physiologically mature when the leaves turns yellow. Disease severity of the 10 selected plants was averaged to obtain mean powdery mildew severity for each line.

In Anyang, the seeds were sown in the autumn. Jingshuang 16, a highly susceptible cultivar to powdery mildew, was planted every 10 rows and around the test lines as susceptible check and spreader. Powdery mildew severity on the penultimate leaf (F-1 leaf) of 10 randomly selected plants from each line was rated based on the actual percentage of leaf area covered by powdery mildew, when the disease severity of susceptible check cv. Jingshuang 16 reached maximum level around May 18, in 2004 and 2005. Disease severity of the 10 randomly selected plants from each line was averaged to obtain mean powdery mildew severity for each line.

Statistical Analysis

Data Analysis

Relative maximum disease severity (MDS) on the penultimate leaf was calculated by dividing the MDS of each line by the MDS of the susceptible check, Morocco or Jingshuang16. Because of the skewed distribution of the percentage of maximum disease severity (MDS) on penultimate leaf among the DH lines, the MDS value was transformed into inverse sine by the formula $x=\mathrm{Sin}^{-1}\sqrt{MDS}$ for subsequent analysis of variance (ANOVA) and QTL analysis. The area under the disease progress curve (AUDPC) was calculated for each line using the following formula described by Bjarko and Line [2]

$$\mathrm{AUDPC}=\sum_{i=1}^{n}[(X_i+X_{i+1})/2][T_{i+1}-T_i]$$

Where X_i is the disease severity on assessment date i, T_i = number of days after inoculation on assessment date i, n is the total times of disease assessments. SAS software (SAS Institute. Inc., Cary. NC) was used to compute ANOVA. Heritabilities (h^2) were estimated from the analysis of variance by the following formula [5]

$$h^2=\sigma_g^2/[\sigma_g^2+(\sigma_e^2/n)]$$

Where σ_g^2 is the genetic variance [$\sigma_g^2=1/n$ (MS$_e$-

$MS_g)]$, σ_e^2 is the environmental variance ($\sigma_{e2}=MS_e$), and n is the number of replicates.

Map Construction and QTL Detection

To construct the framework map for the QTL analysis, we chose 432 loci, including 367 SSR, 43 RFLP, 10 RAPD, seven RAPD-ISSR, two STS, one grain protein, one glume hair, and one *Ltn* loci [33,34]. The linkage groups were established with the software Map Manager QTXb20 [17]. Genetic distances between markers were estimated using the mapping function Kosambi [15]. QTL were detected by CIM (composite interval mapping) using the software Cartographer 2.5[37]. A logarithm of odds (LOD) of 2.5 was set to declare QTL as significant. The phenotypic variance (R^2) explained by a QTL was obtained by the square of the partial correlation coefficient.

Results

Distribution of MDS and AUDPC, and their correlation

The susceptible check cv. Morocco was amply infected with 80-90% of MDS on penultimate leaf in Beijing, and the MDS on penultimate leaf of Jingshuang 16 reached 50-60% in Anyang. The frequency distribution of powdery mildew severity parameters (MDS and AUDPC) in the DH lines at two locations is shown in Fig. 1. The mean relative MDS of Fukuho-komugi and Oligoculm was 1.4% and 4.8% in Anyang, and 0.5% and 3.2% in Beijing, respectively. The average of relative MDS of the DH lines in Beijing over two years was 8.2%, ranging from 0 to 70.4%, while the mean relative MDS in Anyang for two years was 7.1%

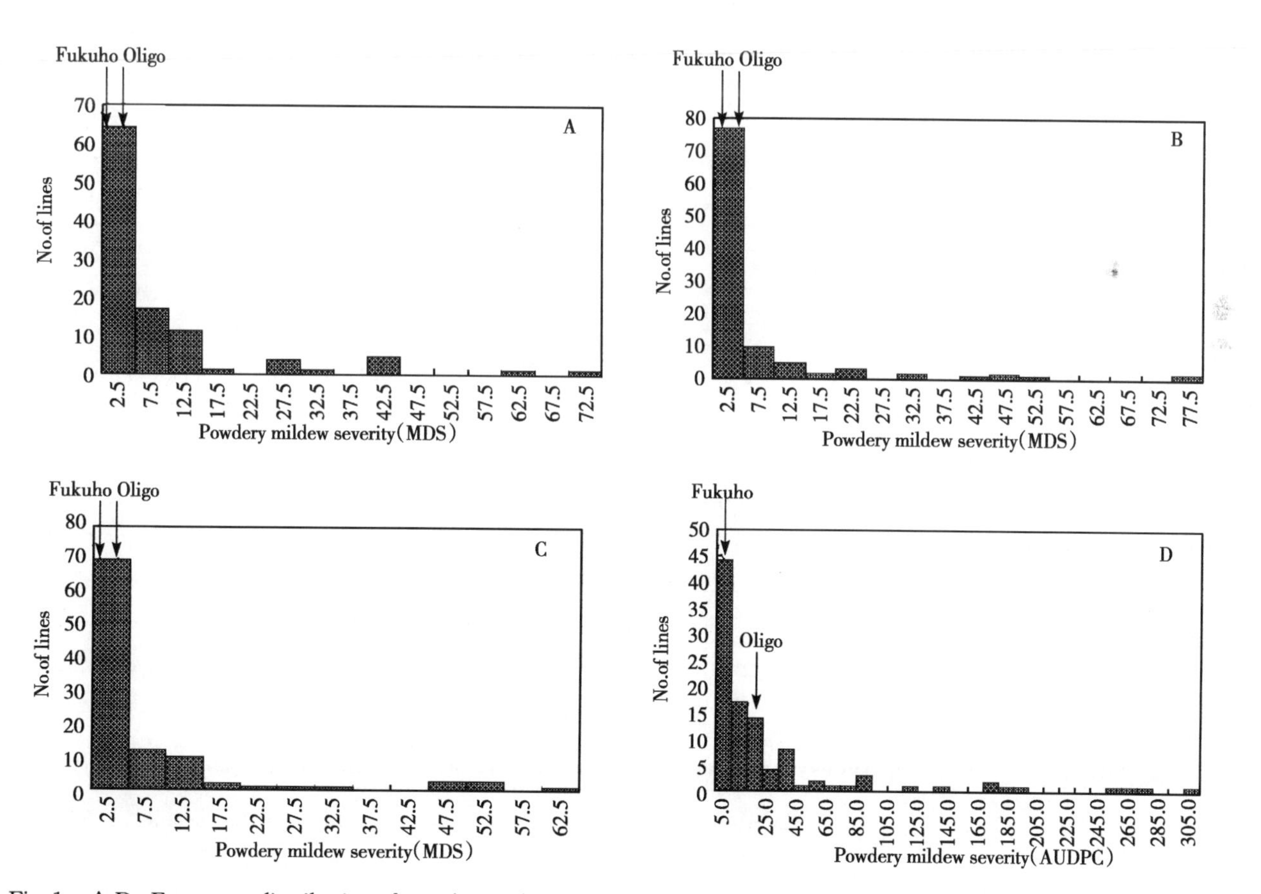

Fig. 1 A-D, Frequency distribution of powdery mildew maximum disease severity (MDS) and area under the disease progress curve (AUDPC) in the DH lines derived from the cross Fukuho-komugi/Oligoculm

A, average of relative MDS in Beijing for 2 years, B, average of relative MDS in Anyang for 2 years, C, average of relative MDS in two locations for 2 years, and D, average of AUDPC in Beijing for 2 years. Mean values for the parents Fukuho-komugi (Fukuho) and Oligoculm (Oligo) are indicated by arrows.

ranging from 0 to 79.0%. In Beijing, the average of AUDPC over two years was 38.1, ranging from 0 to 307.7. ANOVA revealed a significant variation among the DH lines (Table 1). The MDS and AUDPC were significantly correlated with each other for the test in Beijing over two years ($r=0.90$, $P<0.01$).

Table 1 Analysis of variance of relative maximum disease severity (MDS) on penultimate leaf and the area under disease progress curve (AUDPC) for powdery mildew index in the DH population derived from the cross Fukuho-komugi /Oligoculm

Parameter	Source of variance	d. f.	Mean of squares	F values	P
MDS	Lines	106	1 848.57	15.58**	<0.000 1
	Location	1	1 139.82	9.60**	0.002
	Year	1	9 656.77	81.37**	<0.000 1
	Replicate	2	273.09	2.30	0.100 6
	Error	1 173	118.68		
AUDPC	Lines	106	24 258.79	7.1**	<0.000 1
	Year	1	395 195.51	115.65**	<0.000 1
	Replicate	2	17 542.82	5.13**	0.006 2
	Error	532	3 417.17		

Heritabilities

In Anyang, the heritabilities for MDS were 0.84 and 0.82 in two years, respectively, while the heritabilities for MDS in Beijing over two years were 0.85 and 0.95, respectively. The values of heritabilities for AUDPC in Beijing were 0.84 and 0.91 in two years, respectively.

QTL analysis for APR to powdery mildew

Four QTL for APR were detected in the DH population in two environments over two years (Table 2 and Fig. 2). Based on the mean MDS of the 2003-2004 and 2004-2005 cropping seasons in Anyang, two QTL for powdery mildew resistance were found on chromosomes 1AS, and 2BL, explaining 26.6% and 5.7% of the phenotypic variance, respectively. The additive effects of the QTL were −8.06 and −3.46, respectively. According to the mean of MDS in Beijing in the 2003-2004 and 2004-2005 cropping seasons, four QTL were detected on chromosomes 1AS, 2BL, 4BL, and 7DS, accounting for 19.5%, 7.4%, 5.9%, and 12.0% of the phenotypic variance, respectively. The additive effects of these QTL were −6.19, −3.64, 3.31, and −4.61, respectively. With the average value of MDS from two locations during two years, three QTL were mapped on chromosomes 1AS, 2BL, and 7DS, explaining 8.0-22.3% of the phenotypic variance. The additive effects of these QTL were −6.52, −3.71, and −4.11, respectively. Using the mean of AUDPC in Beijing in two cropping seasons, three QTL were mapped on chromosomes 1AS, 2BL, and 7DS, explaining 7.5-24.3% of the phenotypic variance. The additive effects of these QTL were −32.29, −17.20, and −18.89, respectively. Among the QTL identified in this study, those on chromosomes 1AS, 2BL, and 7DS were from the female parent, Fukuho-komugi, while the one on chromosomes 4BL was derived from the male parent, Oligoculm.

Table 2 QTL detected for APR to powdery mildew in the DH population derived from the cross Fukuho-komugi/Oligoculm

Parameter	Chro. [a]	Interval	Length[b]	Position[c]	LOD score[d]	Additive[e]	R^2[f] (%)
Anyang MDS	1AS	*Xgdm33-Xpsp2999*	3.9	3.9	9.17	−8.06	26.6
	2BL	*Xgwm877.1-Xwmc435.1*	11.8	0	2.29*	−3.46	5.7
Beijing MDS	1AS	*Xgdm33-Xpsp2999*	3.9	3.9	8.41	−6.19	19.5
	2BL	*Xwmc877.1-Xwmc435.1*	11.8	4.0	3.57	−3.64	7.4
	4BL	*Xgwm375-Xgwm251*	1.1	1.1	2.89	3.31	5.9
	7DS	*Ltn-Xgwm295.1*	5.7	0	5.29	−4.61	12.0

（续）

Parameter	Chro. [a]	Interval	Length[b]	Position[c]	LOD score[d]	Additive[e]	$R^{2\,f}$ (%)
Average of MDS at two locations	1AS	*Xgdm33-Xpsp2999*	3.9	3.9	9.20	−6.52	22.3
	2BL	*Xwmc877.1-Xwmc435.1*	11.8	4.0	3.75	−3.71	8.0
	7DS	*Ltn-Xgwm295.1*	5.7	0	4.27	−4.11	9.8
Beijing AUDPC	1AS	*Xgdm33-Xpsp2999*	3.9	3.9	9.66	−32.29	24.3
	2BL	*Xwmc877.1-Xwmc435.1*	11.8	4.0	3.44	−17.20	7.5
	7DS	*Ltn-Xgwm295.1*	5.7	2.0	3.90	−18.89	9.0

a. Chromosome involved.

b. Interval length in cM between the two markers flanking the peak position.

c. Peak position in cM from the first interval marker.

d. Logarithm of odds (LOD) score, thresholds equivalent to likelihood ratio (LR) =11.7.

e. Additive effects

f. R^2 is the proportion of the phenotypic variance explained by the QTL.

* LOD value of the QTL detected in Anyang is lower than the threshold 2.5.

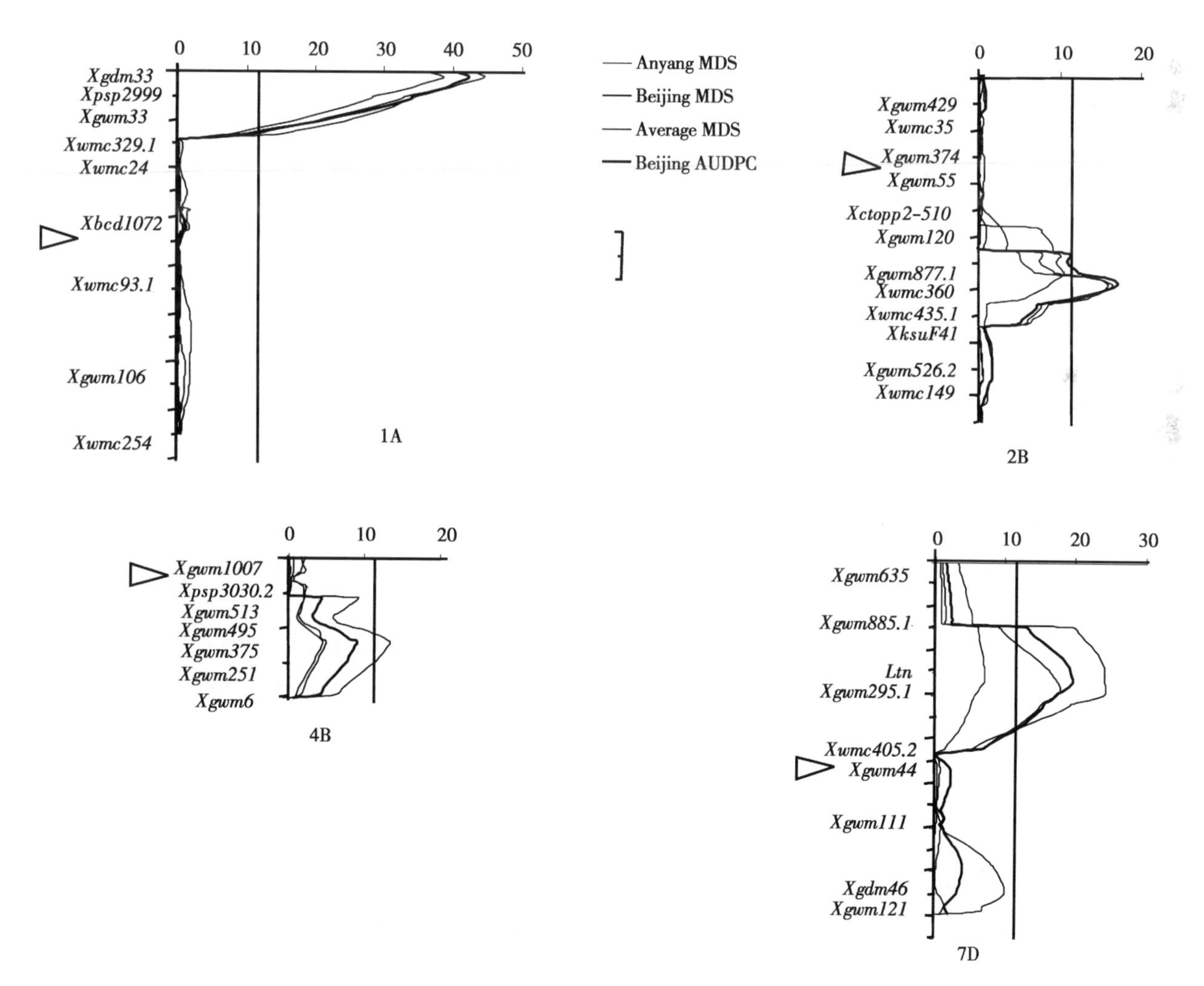

Fig. 2 A-D, likelihood ration (LR) contours obtained by composite interval mapping for four quantitative trait loci mapped on chromosomes 1AS, 2BL, 4BL, and 7DS that reduce powdery mildew severity in Fukuho-komugi/Oligoculm doubled haploid population. Bold contours indicate LR for the AUDPC in Beijing. LR thresholds, equivalent to LOD=2.5, are 11.7. Short arms are toward to the top and the open arrow indicates centromere.

Discussion

Adult plant resistance genes with minor or major but additive effects on powdery mildew were reported in the previous studies [5,7,13,16]. Sourdille et al. [30] found a gene *Mlar* conferring resistance to powdery mildew in Courtot, located on the short arm of the chromosome 1AS with a genetic distance of 5.2cM from the locus *XGli-A5* coding for storage proteins. *Mlar* was an allele of the *Pm3* locus (*Pm3g*) involved in the resistance to powdery mildew. Bougot et al. [3] mapped *Pm3g* in the recombinant inbred lines from the cross RE9001/Courtot and found *PSP2999* cosegregated with the allele. The location of the QTL on chromosome 1AS detected in the present study is almost the same as that of the *Pm3* locus, which is consistent with previous reports [3,30]. Originally, the powdery mildew resistance of Fukuho-komugi was derived from wheat cultivar Norin 29 that possessed the resistance gene *Pm3a* [23]. Nevertheless, the wheat line Asosan/8 * CC with *Pm3a* was highly susceptible to the predominant Chinese isolate E20 of *Blumeria graminis* f. sp. *tritici* at seedling stage [38]. Therefore, this QTL from Fukuho-komugi is likely the residual effect of *Pm3a* conferring resistance to powdery mildew at adult-plant stage. Keller et al. [13] found a QTL on chromosome 7B for APR to powdery mildew in a segregating wheat/spelt (*Triticum spelta* L.) population, and proved to be the residual effects of *Pm5*. Mingeot et al. [19] reported the residual effect of the race-specific resistance gene *MlRE* (on 6AL) and *Pm4b* (on 2A) in a DH population from the cross RE714 × Festin. The residual effects of *Pm3c* and *Pm4a* in reducing disease severity were also reported, although the isolates were virulent to these genes at seedling stage[21]. Martin and Ellingboe [18], Nagassa[20] and Royer et al. [25] also reported that *Pm* genes overcome by virulent isolates still contribute to partial resistance. The QTL on chromosome 2BL identified in this study was flanked by the microsatellite *Xgwm877.1* and *Xwmc435.1*, Which is different from the QTL at the marker interval *WG338* and *Xgwm526.1* detected by Liu et al. [16] with a distance of 40cM. Tao et al. [35] found that the gene *Pm6* is located on chromosome 2BL and flanked by the loci *Xpsr934-Xbcd135*, which is close to the QTL identified in our study. Keller et al. [13] detected a QTL for powdery mildew resistance on chromosome 4BL between the RFLP markers *Xpsr593b* and *Xpsr1112*, in a segregation population from a cross of wheat with spelt. Its location is very close to that of the QTL detected in the present study. Huo et al. [10] found a major QTL conferring resistance to powdery mildew located on chromosome 7DS, flanked by the loci *Xwg834-Xbcd1438*, which is also close to the QTL on 7DS identified in the present study.

Fukuho-komugi displayed leaf tip necrosis (LTN) in field trials, and hence, was considered to possess *Lr34*/*Yr18* [6,27]. Using the same DH population as that employed in the present study, Suenaga et al. [34] detected a QTL on chromosome 7DS for resistance to both leaf rust and stripe rust, possibly due to the resistance genes *Lr34*/*Yr18*. The microsatellite locus *Xgwm295.1*, located almost at the peak of the likelihood ratio contours for both leaf and stripe rust severity, was closely linked to *Lr34*/*Yr18*. *Lr34* and *Yr18* were previously shown to be associated with enhanced tolerance to stem rust and barley yellow dwarf virus infection [14,29]. In the present study, a QTL for powdery mildew resistance was found in the same region. This chromosomal region in wheat has now been found to be associated with resistance to five different pathogens. It indicates that the resistance genes often tend to cluster together. Association of the resistance genes is likely due to the close linkage of them, or pleitrophic effects of a same locus for resistance to different diseases. In a most recent report, Spielmeyer et al. [31] also found cosegregation of powdery mildew resistance with the durable leaf and stripe rust resistance conferred by *Lr34* and *Yr18*, respectively. Lillemo et al. (unpublished data, personal communication) tested the near-isogenic lines for *Lr34* and *Lr46* in the genetic background of Avocet-*YrA* and *YrLrPrl1* in the background of Lalbahadur and found that all three genes were associated with significantly reduced levels of leaf rust, stripe rust and powdery mildew in comparison to their susceptible genetic backgrounds. They concluded that resistance to both rust and powdery mildew is not only confined to *Lr34*, but could be a general phenomenon of LTN-associated resistance genes (Lillemo et al., unpublished data).

Of thefour QTL identified in this study, two QTL located on chromosomes 1AS and 2BL were stably detected across two environments. The QTL on chromosome 2BL detected in Anyang had a lower LOD score value, which may be due to the inadequate disease infection under natural inoculation. This indicated that the environment affected the infection of powdery mildew and the action of the resistance gene, which was also reported in the previous studies [16,19].

The resistance to powdery mildew was estimated by different disease parameters. The correlation coefficient between mean MDS and mean AUDPC is 0.90 over two years in Beijing ($P<0.01$), which is consistent with the result in our previous study in the characterization of Chinese wheat cultivars [38]. It indicates that the maximum disease severity (MDS) on penultimate leaf is also a good indicator of APR in the field, which is suitable to be used for the characterization of APR to stripe rust in wheat breeding programs with a single scoring at an appropriate time.

Acknowledgments

This project was funded by National 863 program (2003AA207090) and National Natural Science Foundation of China (30220140636).

References

[1] Bennett, F. G. A. 1984. Resistance to powdery mildew in wheat: A review of its use in agriculture and breeding programmes. Plant Pathol. 33: 279-300.

[2] Bjarko, M. E. and Line, R. F. 1988. Heritability and number of genes controlling leaf rust resistance on four cultivars of wheat. Phytopathology 78: 457-461.

[3] Bougot, Y., Lemoine, J., Pavoine, M. T., Barloy, D., and Doussinault, G. 2002. Identification of a microsatellite marker associated with *Pm3* resistance alleles to powdery mildew in wheat. Plant Breed. 121: 325-329.

[4] Browning, J. A. and Frey, K. J. 1969. Multiline cultivars as a means of disease control. Annu. Rev. Phytopathol. 7: 355-382.

[5] Chantret, N., Mingeot, D., Sourdillle, P., Bernard, M., Jacquemin, J. M., and Dousdinault, G. 2001. A major QTL for powdery mildew resistance is stable over time and at two development stages in winter wheat. Theor. Appl. Genet. 103: 962-971.

[6] Dyck, P. L. 1991. Genetics of adult-plant leaf rust resistance in 'Chinese Spring' and 'Sturdy' wheats. Crop Sci. 31: 309-311.

[7] Griffey, C. A., and Das. M. K. 1994. Inheritance of adult-plant resistance to powdery mildew in Knox62 and Massey winter wheats. Crop Sci. 34: 641-646.

[8] Gustafson. G., and Shaner, G. 1982. Influence of plant age on the expression of slow-mildewing resistance in wheat. Phytopathology 72: 746-749.

[9] Hautea, R. A., Coffman, W. R., Sorrells, M. E., and Bergstrom, G. C. 1987. Inheritance of partial resistance to powdery mildew in spring wheat. Theor. Appl. Genet. 73: 609-615.

[10] Huo, N. X., Zhou, R. H., Zhang, L. F., and Jia, J. Z. 2005. Mapping quantitative trait loci for powdery mildew resistance in wheat. *Acta Agronomica Sinica* 31: 692-696. (in Chinese with English abstract)

[11] Jones, I. T., and Davies, I. J. E. R. 1985. Partial resistance to *Erysiphe graminis* f. sp. *hordei* in old European barley varieties. Euphytica 34: 499-507.

[12] Jones, I. T., and Hayes, J. D. 1971. The effect of sowing date on adult plant resistance to *Erysiphe graminis* f. sp. *avenae* in oats. Annu. Appl. Biol. 68: 31-39.

[13] Keller, M., Keller, B., Schachermayr, G., Winzeler, M., Schmid, J. E., Stamp, P., and Messmer, M. M. 1999. Quantitative trait loci for resistance against powdery mildew in a segregating wheat×spelt population. Theor. Appl. Genet. 98: 903-912.

[14] Kerber, E. R. and Aung, T. 1999. Leaf rust resistance gene *Lr34* associated with non-suppression of stem rust resistance in the wheat cultivar Canthatch.

Phytopathology 89：518-521.

[15] Kosambi, D. D. 1944. The estimation of map distance from recombination values. Annu Eugen 12：172-175.

[16] Liu, S. X., Griffey, C. A., and Maroof, M. A. S. 2001. Identification of molecular markers associated with adult plant resistance to powdery mildew in common wheat cultivar Massey. Crop Sci. 41：1268-1275.

[17] Manly, K. F., Cudmore, Jr. R. H., and Meer, J. M. 2001. Map Manager QTX, cross-platform software for genetic mapping. Mammal. Genome 12：930-932.

[18] Martin, T. J., and Ellingboe, A. H. 1976. Differences between compatible parasite/host genotypes involving the *Pm4* locus of wheat and the corresponding genes in *Erysiphe graminis* f. sp. *tritici*. Phytopathology 66：1435-1438.

[19] Mingeot, D., Chantret, N., Baret, P. V., Dekeyser, A., Boukhatem, N., Sourdille, P., Doussinault, G., and Jacquemin, J. M. 2002. Mapping QTL involved in adult plant resistance to powdery mildew in the winter wheat line RE714 in two susceptible genetic backgrounds. Plant Breed. 121：133-140.

[20] Nagassa, M. 1987. Possible new genes for resistance to powdery mildew, *Septoria* glume blotch and leaf rust of wheat. Plant Breed. 98：37-46.

[21] Nass, H. A., Pedersen, W. L., MacKenzie, D. R., and Nelson, R. R. 1981. The residual effects of some "defeated" powdery mildew resistance loci in cereals. Theor. Appl. Genet. 93：1078-1082.

[22] Parlevliet, J. E., and van Ommeren, 1975. Partial resistance of barley to leaf rust, *Puccinia hordei*. Ⅱ. Relationship between field trials, micro plot tests and latent period. Euphytica 24：293-303.

[23] Peusha, H., Hsam, S. L. K., and Zeller, F. J. 1996. Chromosomal location of powdery mildew resistance genes in common wheat (*Triticum aestivum* L. em. Thell.) 3. Gene *Pm22* in cultivar Virest. Euphytica 91：149-152.

[24] Roberts. J., and Caldwell, R. 1970. General resistance (slow mildewing) to *Erysiphe graminis* f. sp. *tritici* 'Knox' Wheat (Abs.). Phytopathology 60：1310.

[25] Royer, M. H., Nelson, R. R., Mackenzie, D. R., and Diehle, D. A. 1984. Partial resistance of near-isogenic wheat lines compatible with *Erysiphe graminis*. f. sp. *tritici*. Phytopathology 74：1001-1006.

[26] Shaner, G., 1973. Evaluation of slow-mildewing resistance of Knox wheat in the field. Phytopathology 63：867-872.

[27] Singh, R. P. 1992. Association between gene *Lr34* for leaf rust resistance and leaf tip necrosis in wheat. Crop Sci. 32：874-878.

[28] Singh, R. P. 1992. Genetic association of leaf rust resistance gene *Lr34* with adult plant resistance to stripe rust in bread wheat. Phytopathology 82：835-838.

[29] Singh, R. P. 1993. Genetic association of gene *Bdv1* for tolerance to Barley Yellow Dwarf Virus with genes *Lr34* and *Yr18* for adult plant resistance to rusts in bread wheat. Plant Dis. 77：1103-1106.

[30] Sourdille, P., Röbe, P., Tixier, M. H., Doussinault, G., Pavoine, M. T., and Bernard, M. 1999. Location of *Pm3g*, a powdery mildew resistance allele in wheat, by using a monosomic analysis and by identifying associated molecular markers. Euphytica 110：193-198.

[31] Spielmeyer, W., McIntosh, R. A., Kolmer, J., and Lagudah, E. S. 2005. Powdery mildew resistance and *Lr34/Yr18* genes for durable resistance to leaf and stripe rust cosegregate at a locus on the short arm of chromosome 7D of wheat. Theor. Appl. Genet. 111：731-735.

[32] Suenaga, K., and Nakajima, K. 1993. Segregation of genetic markers among wheat doubled haploid lines derived from wheat × maize crosses. Euphytica 65：145-152.

[33] Suenaga, K., Khairallah, M., William, H. M., and Hoisington, D. A. 2005. A new intervarietal linkage map and its application for quantitative trait locus analysis of "gigas" features in bread wheat. Genome 48：65-75.

[34] Suenaga, K., Singh, R. P., Huerta-Espino, J., and William, H. M. 2003. Microsatellite markers for genes *Lr34/Yr18* and other quantitative trait loci for leaf rust and stripe rust resistance in bread wheat. Phytopathology 93：881-890.

[35] Tao W. J., Liu D. J., Liu J. Y., and Chen P. D. 2000. Genetic mapping of the powdery mildew resistance gene *Pm6* in wheat by RFLP analysis. Theor. Appl. Genet. 100：564-568.

[36] Vanderplank, J. E. 1968. Disease Resistance in Plants. Academic Press, New York.

[37] Wang, S., Basten, C. J., and Zeng, Z. B. 2005. Windows QTL Cartographer v2.5. Statistical Genetics. North Carolina State Univ.

[38] Wang, Z. L., Li, L. H., He, Z. H., Duan, X. Y., Zhou, Y. L., Chen, X. M., Lillemo. M., Singh R. P., Wang, H., and Xia, X. C. 2005. Seeding and adult-plant resistance to powdery mildew in Chinese bread wheat cultivars and lines. Plant Dis. 89: 457-463.

[39] Wolfe, M. S., and Barrett, J. A. 1977. Population genetics of powdery mildew epidemics. Annu. New York Acad. Sci. 287: 151-163.

[40] Wu, Z. S. 1990. Breeding for Wheat Disease Resistance. Pages 235-272. In: Wu, Z. S. ed., Wheat Breeding. Agric. Pub. Press of China, Beijing.

The adult plant rust resistance loci *Lr34/Yr18* and *Lr46/Yr29* are important determinants of partial resistance to powdery mildew in bread wheat line Saar

M. Lillemo[1], B. Asalf[1], R. P. Singh[2], J. Huerta-Espino, X. M. Chen[3], Z. H. He[3,4] and Å. Bjørnstad[1]

[1]*Department of Plant and Environmental Sciences, Norwegian University of Life Sciences, P. O. Box* 5003, *N*-1432 *Ås, Norway*; [2]*CIMMYT, Apdo. Postal* 6-641, 06600, *México, D. F., México*; [3]*Institute of Crop Sciences/National Wheat Improvement Center, Chinese Academy of Agricultural Sciences, Zhongguancun South Street* 12, 100081, *Beijing, P. R. China*; [4]*CIMMYT China Office, c/o Chinese Academy of Agricultural Sciences, Zhongguancun South Street* 12, *Beijing* 100081, *P. R. China.*

Abstract: Powdery mildew, caused by *Blumeria graminis* f. sp. *tritici* is a major disease on wheat (*Triticum aestivum* L.) that can be controlled by resistance breeding. The CIMMYT bread wheat line Saar is known for its good level of partial and race non-specific resistance, and the aim of the present study was to map QTLs for resistance to powdery mildew in a population of 113 recombinant inbred lines from a cross between Saar and the susceptible line Avocet. The population was tested over two years in field trials at two locations in south-eastern Norway and once in Beijing, China. SSR markers were screened for association with powdery mildew resistance in a Bulked Segregant Analysis, and linkage maps were created based on selected SSR markers and supplemented with DArT genotyping. The most important QTLs for powdery mildew resistance derived from Saar were located on chromosomes 7D and 1B and corresponded to the adult plant rust resistance loci *Lr34/Yr18* and *Lr46/Yr29*. A major QTL was also located on 4B with resistance contributed by Avocet. Additional QTLs were detected at 3A and 5A in the Norwegian testing environments and at 5B in Beijing. The population was also tested for leaf rust (caused by *Puccinia triticina*) and stripe rust (caused by *P. striiformis* f. sp. *tritici*) resistance and leaf tip necrosis in Mexico. QTLs for these traits were detected on 7D and 1B at the same positions as the QTLs for powdery mildew resistance, and confirmed the presence of *Lr34/Yr18* and *Lr46/Yr29* in Saar. The powdery mildew resistance gene at the *Lr34/Yr18* locus has been named *Pm38*. We propose here that the corresponding resistance gene at the *Lr46/Yr29* locus is named *Pm39*.

Introduction

Powdery mildew, caused by *Blumeria graminis* f. sp. *tritici* is an important disease on wheat (*Triticum aestivum* L.) in regions with temperate and maritime climates, like Europe, North and South America, Africa and parts of China (Bennett, 1984; Hsam and Zeller, 2002). The disease is favoured by intensive cultivation methods associated with modern agriculture such as the use of semi-dwarf and high-yielding cultivars in combination with high levels of nitrogen fertili-

Published in TAG, 2008, 116 (8): 1155-1166

zation, and yield losses have been reported in the range from 5 to 34% (Conner et al., 2003; Griffey et al., 1993; Lipps and Madden, 1988).

Host resistance is considered to be a cost effective and environmentally friendly way of controlling the disease (Bennett, 1984; Hsam and Zeller, 2002). Resistance breeding is often based on the incorporation of race-specific resistance genes that give complete protection of the crop, but such genes are usually associated with a very short durability since they can be overcome by simple genetic changes in the pathogen (McDonald and Linde, 2002; Skinnes, 2002). As a consequence, there is increased interest in the development of germplasm with partial or race non-specific resistance which allows the plants to be infected with the pathogen, but significantly retards the development of disease in adult plants (Hautea et al., 1987; Shaner, 1973). Such resistance has also been termed slow mildewing (Roberts and Caldwell, 1970) and adult plant resistance (Gustafson and Shaner, 1982). Partial resistance is inherited as a quantitative trait, and has shown to be durable. Examples include the winter wheat cultivar Knox (Shaner, 1973) and the derived cultivar Massey (Liu et al., 2001) which have provided effective resistance against powdery mildew in south-eastern United States for half a century. Also many European wheat cultivars are well known for their good level of partial resistance.

The quantitative nature of partial resistance to powdery mildew makes it more complicated to handle in a breeding program compared to race-specific resistance, especially if the symptoms in the field are confounded with the effects of race-specific resistance genes with a low frequency of matching virulence genes in the pathogen population (Yu et al., 2001). The selection for partial resistance in a breeding program could therefore be more efficient with the aid of molecular markers. Molecular markers have recently been used to map Quantitative Trait Loci (QTL) for partial resistance to powdery mildew in several wheat cultivars, including the Swiss winter wheat Forno (Keller et al., 1999), the French winter wheats RE714 (Chantret et al., 2001; Chantret et al., 2000; Mingeot et al., 2002) and RE9001 (Bougot et al., 2006), the North American winter wheats Massey (Liu et al., 2001) and USG3209 (Tucker et al., 2007) and the Japanese cultivar Fukuho-komugi (Liang et al., 2006).

In addition to complementing tradition selection for resistance in the field, molecular markers linked to genes for resistance can also facilitate resistance breeding in environments where the disease does not occur. One example is Mexico, where the main breeding operations of CIMMYT (The International Maize and Wheat Improvement Center) are located, and the absence of natural epidemics of powdery mildew makes resistance breeding difficult although powdery mildew is considered and important disease in many areas of the world where CIMMYT germplasm is grown. International testing has recently identified several high-yielding and widely adapted spring wheat lines with good partial resistance to powdery mildew. One of them is the spring wheat line Saar, which also has good partial resistance to leaf rust (caused by *Puccinia triticina*) and stripe rust (caused by *P. striiformis* f. sp. *tritici*). Quantitative genetic analysis based on a segregating population from a cross with the susceptible cultivar Avocet showed that the powdery mildew resistance in Saar is governed by at least three genes with additive effects (Lillemo et al., 2006). High correlations among the disease scores of powdery mildew, leaf rust and stripe rust further indicated that the resistance to these biotrophic fungal pathogens could be under some common genetic control (Lillemo et al., 2007).

The objectives of the present study were to identify and map the main genetic factors behind the powdery mildew resistance in Saar with molecular markers and to test whether some of the detected genes also are involved in the resistance to leaf rust and stripe rust.

Materials and methods

Plant material

The present study was based on a segregating population of 113 recombinant inbred F_6 lines from the cross between Saar and Avocet-*YrA*. The pedigree of Saar is Sonoita F81/Trap#1//Baviacora M92 and the population was based on a pure inbred line with selection history CG25-099Y-099M-4Y-2M-3Y-0B. Avocet-*YrA* is a line selected from the heterogeneous Australian cultivar Avocet for lacking yellow rust resistance gene *YrA*, and will throughout the rest of this paper simply be referred to as Avocet. The population was developed from randomly selected F_5 lines that were kindly provided by A. Navabi, the University of Alberta, Edmonton, Canada (Navabi et al., 2003, 2004) and grown in a net house at El Batan, Mexico in 2003/2004. Bulk harvest of the seed from each plot was used for the quantitative genetic analysis previously reported (Lillemo et al., 2006), while one single head was selected from each plot and gave rise to the F_6 lines used in the present study.

Field evaluation

Powdery mildew (PM) was evaluated over two years at two locations in south-eastern Norway: Vollebekk research farm at Ås, about 30 km south of Oslo (59°N, 90 m above sea level) and Staur research farm close to Hamar (60°N, 153 m above sea level). Both locations experience severe natural epidemics of PM every year, but are characterized by a different virulence composition (Skinnes, 2002). Field trials were conducted with a randomized complete block design with two reps at each location in 2005 and 2006. The lines were planted in hillplots to provide favourable conditions for mildew development, and the planting was delayed 3-4 weeks compared to the normal planting time for spring wheat to ensure an ample source of natural inoculum. The percentage of leaf area covered with PM was recorded on penultimate leaves at weekly intervals based on a modified Cobb scale (Peterson et al., 1948) commencing at the time of heading (GS 50-59) and ending when Avocet had reached maximum severity (around GS 69-71). The area under the disease progress curve was calculated according to Bjarko and Line (1988). PM was also assessed in Beijng, China, in 2005 based on a 0-9 scale (Saari and Prescott, 1975). The trial was planted on 2 m rows with 30 cm row spacing and followed a randomized complete block design with 3 reps.

The population was evaluated for leaf rust (LR) resistance at CIANO, near Ciudad Obregon in north-western Mexico during the 2004-2005 cropping season and stripe rust (YR) at Toluca in the Mexican highlands during the 2005 cropping season. The plot size at both locations was 2 rows of 1 m length with two reps. Artificial disease epidemics were created by inoculating spreader rows with selected races of the respective pathogens to which seedlings of the two parents were susceptible. Disease severity on flag leaves was scored according to the modified Cobb scale (Peterson et al., 1948) at the time when Avocet had just reached maximum severity. In addition, leaf tip necrosis (LTN) was scored at CIANO as absence (0) or presence (1).

Molecular marker genotyping

Genomic DNA was extracted from young leaves of the parents and recombinant inbred lines using the DNeasy Plant DNA extraction Kit (QIAGEN). Simple Sequence Repeat (SSR) markers were screened for polymorphism between Saar and Avocet. SSR analysis was performed with fluorescently labelled primers and PCR products were either separated by polyacrylamid gel electrophoresis on an ABI PRISM 377 DNA Sequencer or subjected to capillary electrophoresis on an ABI 3730 Gene Analyzer. PCR was conducted as described elsewhere (Semagn et al., 2006). Resistant and susceptible bulks were created by mixing equal amounts of DNA from the five most susceptible and the five most resistant lines, respectively, based on the PM field data from the two locations in Norway in 2005. SSR markers that showed a similar pattern of polymorphism between the bulks as between the parents were used to genotype individual lines of the population. Additional markers in the vicinity of those showing association with resistance were selected based on published linkage maps of wheat, and used to genotype the population. In addition, 111 recombinant inbred lines of the population were genotyped with DArT (Diversity Array Technologies) markers by Triticarte Pty. Ltd. (Canberra, Australia).

Statistical analysis

Analysis of variance was performed with the PROC GLM procedure in SAS (SAS Institute Inc., v 9.1.). The information in the ANOVA table was used to calculate the heritability (h^2) of phenotypic traits based on the formula $h^2 = \sigma_g^2/\sigma_p^2$, where $\sigma_g^2 = (\sigma_L^2 - \sigma_E^2)/r$, and $\sigma_p^2 = \sigma_g^2 + \sigma_E^2$; in this formula, σ_p^2 = phenotypic variance, σ_g^2 = genetic variance, σ_L^2 = variance of the F_6 lines, σ_E^2 = error variance, and r = number of replications (Singh et al., 1995). Pearson correlation coefficients among traits were calculated by the PROC CORR procedure in SAS. PROC CORR was also used initially to determine statistically significant associations between single markers and powdery mildew resistance after the bulked segregant analysis (BSA).

Map construction and QTL analysis

The genotypic data of SSR and DArT markers was used to construct genetic linkage maps with the software JoinMap v. 3.0 (van Oijen and Voorrips, 2001). Markers were assigned to linkage groups with a minimum LOD threshold of 4.0, and map distances were calculated based on the Kosambi mapping function (Kosambi, 1944). QTL analysis was done with PLABQTL v. 1.2 (Utz and Melchinger, 2003). Simple Interval Mapping (SIM) was conducted first to detect the major QTLs for powdery mildew resistance. The most closely linked marker to each of the QTLs that consistently showed effects in all environments were then selected as cofactors in Composite Interval Mapping (CIM). The LOD threshold for declaring a significant QTL in one single environment was set to 3.2 after 1000 permutation tests (type 1 error rate of $\alpha = 0.05$). QTLs reaching this level in one environment were also reported for other environments if their LOD scores reached an arbitrary LOD level of 2.0 and showed significant effects in multiple regression. Genetic maps and LOD curves were drawn using the program MapChart, v. 2.1 (Voorrips, 2002).

Results

Phenotypic evaluation

The disease development was good in alltesting environments. Histograms of the average of the last PM score in the four testing environments in Norway, PM in Beijing and YR in Toluca revealed continuous distributions close to normality (Fig. 1.), which is typical of traits with quantitative inheritance. The LR data from CIANO showed a more bimodal distribution which indicated that this trait is likely under control of a major genetic factor. Transgressive segregation with a substantial number of lines with higher susceptibility than Avocet was apparent for PM, which indicated that the susceptible parent most likely carried a gene for resistance. The same was also observed for the F_5 RILs of the same population (Lillemo et al., 2006).

Phenotypic correlations among traits and heritabilities are presented in Table 1. It shows that the PM scores across the Norwegian environments were highly correlated and the heritability estimates were also very high except for the Hamar 2006 trial which was affected by poor germination of some plots resulting in less favourable conditions for disease development. The lower heritability for the PM data in Beijing and weaker correlation with the Norwegian test environments is likely reflecting the different scale used for disease rating (0-9), which gives less discrimination among the lines. The PM data was significantly correlated with the disease scores for LR and YR, indicating that the resistance to these three biotrophic pathogens might be under some common genetic control in this population. Moreover, LTN, the phenotypic marker for the adult plant rust resistance genes *Lr34/Yr18* was significantly correlated not only with LR and YR but also with PM (Table 1).

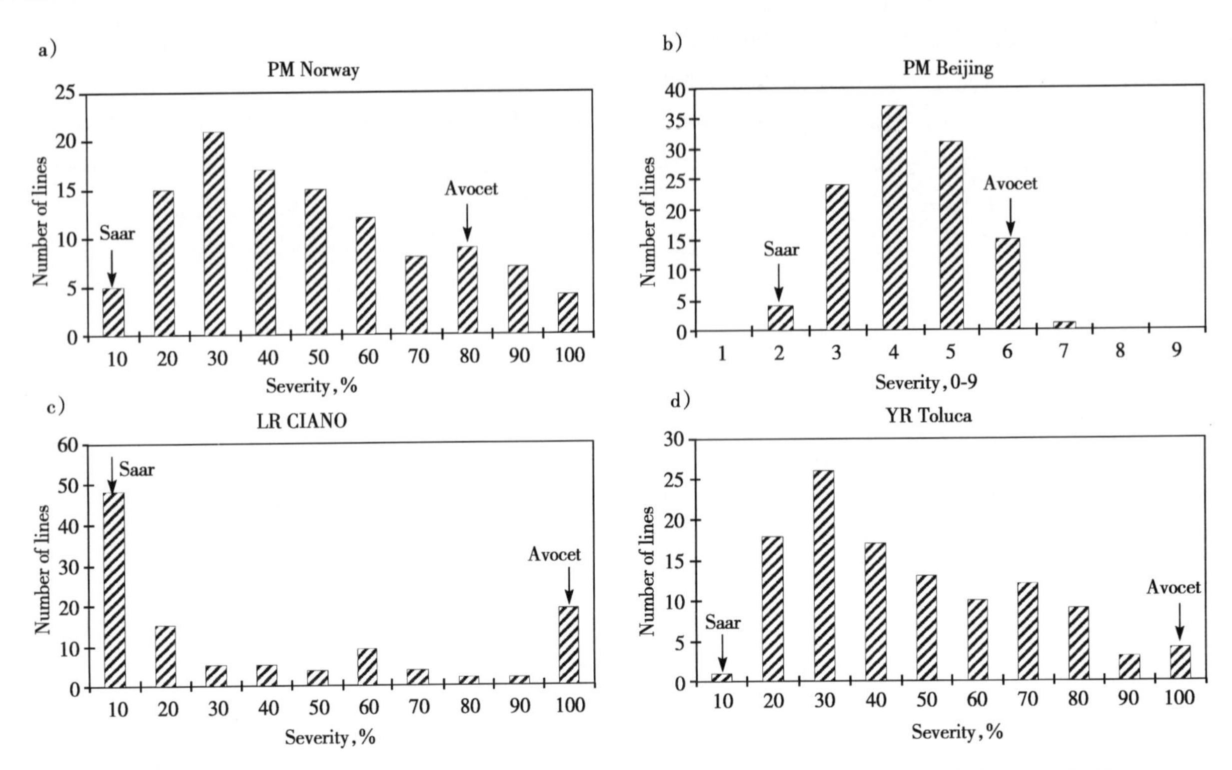

Fig. 1 Frequency distributions of disease severities for 113 RILs of the cross between Saar and Avocet. a) The average of the last severity score for PM across four testing locations in Norway (Vollebekk and Staur, 2005 and 2006), b) PM in Beijing, China in 2005, c) LR at CIANO, Mexico, 2005, d) YR in Toluca, Mexico, 2005.

Construction of linkage maps

A total of 572 SSR markers were selected from the wheat consensus map (Somers et al., 2004), based on coverage of all chromosome arms and approximately even distribution along the genome. Of these, 334 markers (58%) were polymorphic between the parents, and 56 marker loci showed differentiation when tested on the resistant and susceptible bulks for PM. These markers were genotyped on individual RILs from the population and used to develop linkage maps for QTL analysis. The population was in addition genotyped with DArT markers, and 209 polymorphic DArT loci were included in the linkage maps. A total of 34 linkage groups were created, representing chromosomal areas from 16 chromosomes. Interestingly, one of the linkage groups revealed a translocation consisting of the centromeric part of 5B and the distal part of the long arm of 7B (Fig. 2), which is commonly found in many West European wheat cultivars (Riley et al., 1967) and associated with adult plant resistance to YR (Law and Worland, 1997).

Detection of QTLs for PM resistance

Simple Interval Mapping (SIM) based on the preliminary linkage maps detected QTLs on 1B (close to *gwm44*), 4B (close to *wPt-6209*) and 7D (close to *gwm1220*) in all testing environments for PM, with the resistance on 1B and 7D being contributed by the resistant parent Saar and the resistance on 4B contributed by Avocet. In addition, putative QTLs were detected in some environments on 2D (close to *gwm296a*), 3A (close to *barc310*) and 5A (close to *gwm617b*) with the resistance contributed by Saar (data not shown). The linkage maps around these loci were refined with more SSR markers based on other published maps (Bougot et al., 2006; Hayden et al., 2006; Parida et al., 2006; Somers et al., 2004; Torada et al., 2006) and two markers tightly linked to *Lr34* on 7D; swm10 (Bossolini et al., 2006) and csLV34 (Lagudah et al., 2006).

Table 1 Pearson correlation coefficients among phenotypic traits and broad sense heritability estimates. All correlations were highly significant at the level α=0. 0001.

	Correlation coefficients											Heritability h^2
	PM Ås 2005, AUDPC	PM Ås 2006, Severity	PM Ås 2006, AUDPC	PM Hamar 2005, Severity	PM Hamar 2005, AUDPC	PM Hamar 2006, Severity	PM Hamar 2006, AUDPC	PM Beijing 2005, Severity	LR CIANO 2005, Severity	YR Toluca 2005, Severity	Leaf Tip Necrosis	
PM Ås 2005, Severity	0. 989	0. 867	0. 848	0. 903	0. 905	0. 864	0. 810	0. 672	0. 797	0. 768	−0. 713	*0. 95*
PM Ås 2005, AUDPC		0. 875	0. 860	0. 910	0. 917	0. 866	0. 814	0. 661	0. 785	0. 779	−0. 688	*0. 97*
PM Ås 2006, Severity			0. 983	0. 860	0. 872	0. 896	0. 876	0. 630	0. 577	0. 692	−0. 438	*0. 92*
PM Ås 2006, AUDPC				0. 840	0. 854	0. 882	0. 888	0. 612	0. 552	0. 678	−0. 432	*0. 91*
PM Hamar 2005, Severity					0. 989	0. 878	0. 821	0. 695	0. 764	0. 718	−0. 674	*0. 93*
PM Hamar 2005, AUDPC						0. 877	0. 820	0. 688	0. 752	0. 721	−0. 663	*0. 93*
PM Hamar 2006, Severity							0. 960	0. 663	0. 645	0. 702	−0. 523	*0. 79*
PM Hamar 2006, AUDPC								0. 584	0. 554	0. 660	−0. 481	*0. 76*
PM Beijing 2005, Severity									0. 547	0. 552	−0. 575	*0. 51*
LR CIANO 2005, Severity										0. 762	−0. 849	*0. 98*
YR Toluca 2005, Severity											−0. 640	*0. 96*

Table 2 Results of Composite Interval Mapping(CIM) of PM resistance with the markers wmc719(1B), gwm149(4B) and swm10(7D) as cofactors. The table shows R^2 values obtained from multiple regression in PlabQTL. QTLs that were detected with a LOD score above the threshold of 3. 2 determined by permutation tests are highlighted in bold. Other putative QTL are also listed if they showed significant association with the trait in the multiple regression.

Chromosome	Position (cM)	marker interval	Source of resistance	Ås 2005 severity	Ås 2005 AUDPC	Ås 2006 severity	Ås 2006 AUDPC	Hamar 2005 severity	Hamar 2005 AUDPC	Hamar 2006 severity	Hamar 2006 AUDPC	Mean severity Norway	Mean AUDPC Norway	Beijing 2005 severity
1B	23	wmc719-hbe248	S	**32. 0**	**31. 6**	**18. 8**	9. 5	**35. 7**	**35. 8**	**18. 2**	7. 7	**31. 4**	**24. 2**	9. 6
3A	1	stm844-barc310	S	**19. 1**	**20. 3**	**18. 0**	**19. 6**	8. 5	**11. 8**	**18. 6**	**18. 2**	**20. 0**	**21. 1**	
4B	28	wPt-1505 -gwm149	A	**30. 8**	**30. 2**	**34. 0**	**31. 0**	**33. 8**	**34. 2**	**37. 2**	**28. 7**	**40. 6**	**35. 6**	**21. 2**
5A	54	gwm617b- wmc327	S	4. 1	4. 0	7. 9	9. 6	**12. 7**	**14. 8**	5. 7	6. 9	**10. 0**	**9. 9**	
5B/7B	22	barc4 -gwm274b	S	7. 9	7. 4	8. 0		8. 9	8. 5	4. 4		8. 7	5. 3	**9. 6**
5B/7B	70	wmc581-wPt-8007												**4. 8**
7D	31	gwm1220-swm10	S	**56. 1**	**53. 2**	**20. 4**	**18. 9**	**49. 5**	**48. 7**	**31. 8**	**24. 3**	**46. 0**	**40. 1**	**27. 8**
% of the phenotypic variation explained				72. 0	70. 8	59. 7	54. 3	71. 7	72. 3	61. 9	52. 0	72. 1	67. 5	48. 9
% of the genotypic variation explained				75. 5	73. 3	64. 7	59. 3	77. 2	77. 3	78. 2	68. 5	80. 5	77. 9	96. 1

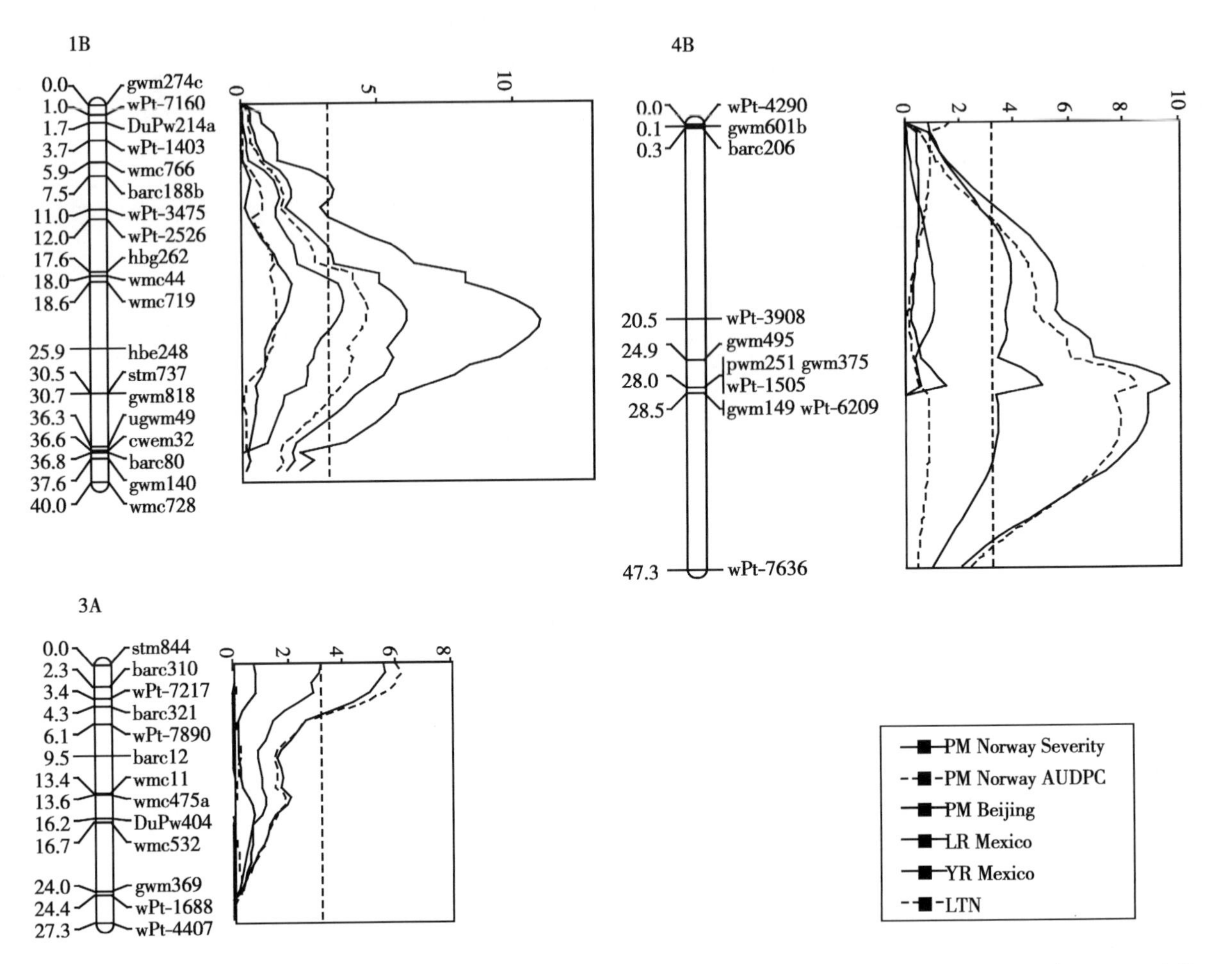

Fig. 2 Linkage groups showing significant association with resistance to PM, LR, YR and LTN, with corresponding LOD curves obtained from either CIM or SIM as described in the text. Genetic distances are shown in centimorgans to the left of each linkage group. The LOD significance threshold of 3.2 is indicated by a dashed line.

Results of the final Composite Interval Mapping (CIM) with the consistent QTL factors on 1B, 4B and 7D as cofactors are shown in Table 2 and Fig. 2. The most important QTL for PM resistance was located on chromosome 7D, with the LOD curves peaking at the *Lr34* locus between the closely linked markers gwm1220 and swm10 (Fig. 2). This QTL explained from 18.9 to 56.1% of the phenotypic variation of the trait, and the resistance was contributed by Saar. The second largest QTL was located on 4B, around the SSR markers gwm251 and gwm375 and the DArT marker wPt-1505. The resistance at this locus was derived from the susceptible parent Avocet, and explained from 21.2 to 40.6% of the phenotypic variation. A third QTL with significant effects in all environments was located on 1B, between the SSR markers wmc719 and hbe248, corresponding to the location of *Lr46*. The resistance at this locus was contributed by Saar and explained from 7.7 to 35.8% of the phenotypic variation. Two additional QTLs for PM resistance were detected in the Norwegian environments, on 3A and 5A with resistance contributed by Saar. The QTL on 3A was located between stm844 and barc310 and explained from 8.5 to 21.1% of the phenotypic variation. The QTL on 5A was located close to gwm617b and was detected above the LOD threshold of 3.2 only in Hamar 2005 and for the mean data from Norway. The contribution of this QTL in multiple regression was, however, significant in all Norwegian environments and ranged from 4.0 to 14.8%. The QTLs on 3A and 5A were not detected in Beijing, but the data from this location revealed a QTL on 5B explaining 9.6% of the phenotypic variation with marginal effects also in the Norwegian environments. The

LOD curve for PM in Beijing also revealed a peak at the presumed location for *Pm 5a*, close to wmc581 (Nematollahi et al., 2007), but the contribution of this QTL in multiple regression was only 4.8% and barely significant at the $\alpha=0.05$ level. The putative QTL at 2D, although showing a significant, but small effect on PM resistance when tested individually, was not detected in the CIM analysis and showed no significant contribution in multiple regression when tested together with the other detected QTLs. No significant epistatic interactions were found among the QTLs for PM resistance.

Effects of detected QTLs on LR, YR and LTN

SIM withthe LR and YR data from Mexico detected two consistent QTLs for both diseases, corresponding to the *Lr46/Yr29* and *Lr34/Yr18* loci on 1B and 7D with resistance contributed by Saar. CIM with these two QTLs as cofactors did not detect any other significant QTL for LR (Table 3, Fig. 2). The effects of 1B and 7D on LR showed a significant and negative deviation from a simple additive model, and the final fitted model with the interaction term accounted for 82.3% of the genetic variation for LR, with the biggest contribution coming from the *Lr34* locus on 7D (Table 3). The possibly duplicate interaction between *Lr46* and *Lr34* is illustrated in Fig. 3, showing that *Lr46* only has a marginal effect on LR in the presence of *Lr34*.

Table 3 Results of Composite Interval Mapping (CIM) of LR and YR resistance with the markers wmc719 (1B) and swm10 (7D) as cofactors, and Simple Interval Mapping (SIM) for Leaf Tip Necrosis. The table shows R^2 values obtained from multiple regression in PlabQTL. QTLs that were detected with a LOD score above the threshold of 3.2 determined by permutation tests are highlighted in bold.

Chromosome	Position (cM)	marker interval	Source of resistance	LR Cd. Obregon, Mexico 2005	YR Toluca, Mexico 2005	Leaf Tip Necrosis
1B	23	wmc719-hbe248	S	**48.9**	**17.4**	12.3
3A	0	stm844-barc310	S		**8.9**	
6A	13	barc3 - wPt-7063	A		**14.6**	
7D	31	wPt-3328 -csLV34	S	**72.5**	**36.7**	**76.6**
1B x 7D				34.0		
% of the phenotypic variation explained				81.0	52.9	77.0
% of the genotypic variation explained				82.3	55.3	—

CIM detected two further QTLs for YR resistance with a LOD score above the significance threshold of 3.2. One was located at 3AS, at the same position as the QTL for PM resistance, while the other was mapped on 6A between barc3 and wPt-7063, and had resistance contributed by Avocet (Table 3, Fig. 2). In contrast to the situation for LR, no significant interactions were found among the detected QTLs for YR resistance. Altogether, the four QTLs explained about half of the genetic variation for the trait (Table 3).

The phenotypic marker LTN deviated slightly from the expected 1∶1 ratio of a single gene with 61 lines with and 45 lines without LTN ($\chi^2=2.42$, $p=0.12$). Instead of trying to place it on the linkage map as a marker, SIM was used to study the genetic basis for the trait. As expected, the LOD curve peaked at the *Lr34/Yr18* locus on 7D (Fig. 2), which accounted for most of the variation, but there was also a small, but significant contribution from the *Lr46/Yr29* locus on 1B.

Discussion

In this study we have conducted QTL mapping for resistance to three different biotrophic pathogens in the same population, which was feasible due to the high levels of partial and race non-specific resistance to PM,

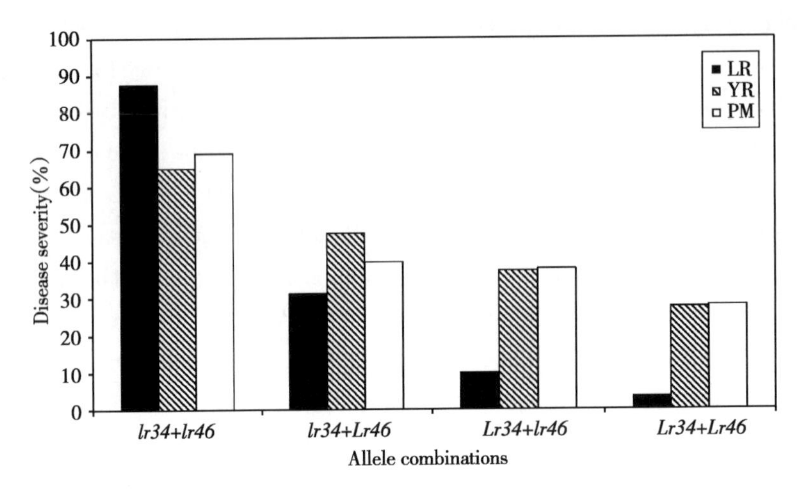

Fig. 3 Effects of the *Lr34/Yr18* and *Lr46/Yr29* loci on LR, YR and PM of 113 RILs derived from the cross between Saar and Avocet. The effects are calculated as the average disease severity of lines with different allelic combinations as determined by the flanking markers gwm1220 and swm10 for *Lr34/Yr18* and wmc719 and hbe248 for *Lr46/Yr29*.

LR and YR in Saar and the corresponding susceptibility to all three diseases by Avocet. The high correlations among the disease data suggested that resistance to these three diseases might be under some common genetic control, and this was indeed confirmed by the QTL analysis. Although the bulked segregant analysis (BSA) and subsequent QTL mapping was only based on the PM data, two of the detected QTLs for PM resistance also showed to have major effects on LR and YR. The genetic aspects and potential impact of this important finding will be discussed at the end of this section.

Genetic control of PM resistance in Saar

The PM resistance of Saar in the Norwegian testing environments was shown to be mainly controlled by three major QTLs on 7D, 1B and 3A as well as a minor QTL on 5A and a barely detectable locus on 5B. This is largely in agreement with the quantitative genetic study of the same population (Lillemo et al., 2006) indicating at least three genes for partial resistance from Saar. The other indication of a resistance gene from Avocet was also confirmed by a major QTL mapped on 4B. Altogether, these QTLs explained from 59 to 81% of the genetic variation for PM resistance in the Norwegian environments (Table 2), which indicates that the BSA strategy was successful in detecting the most important genetic factors behind the resistance.

That partial resistance to PM is under control of a few loci with relatively large effects has also been confirmed in otherrecent QTL mapping studies. About half of the phenotypic variation for resistance to PM in the winter wheat Massey (Liu et al., 2001) and the derived cultivar USG3209 (Tucker et al., 2007) was shown to be controlled by three genes. Liang et al. (2006) detected four QTLs for PM resistance in the Fukuhokomugi by Oligoculm population, each explaining from 6 to 27% of the phenotypic variance. Several QTLs were detected in the winter wheat line RE714, though only two major loci showed stable expression across environments and genetic backgrounds (Mingeot et al., 2002). Likewise, although Keller et al. (1999) reported 18 QTLs for PM resistance in a segregating wheat by spelt population, only two QTLs with major effects were consistent across environments.

As demonstrated in the present study, the two most important genetic factors behind the PM resistance in Saar were located at the *Lr34/Yr18* and *Lr46/Yr29* loci on chromosomes 7D and 1B, respectively. Previous studies have also detected important QTLs for PM resistance in chromosomal areas corresponding to these loci, although no direct genetic relationship with the rust resistance was suggested in these studies. The QTL on 7D corresponds to similar QTLs detected in the Synthetic × Opata (Börner et al., 2002) and Fukuho-komugi × Oligoculm (Liang et al., 2006) mapping populations. Likewise, the QTL on 1B corresponds with a QTL detected at the same position in the winter wheat Massey (Liu et al., 2001) and the derived cultivar USG3209 (Tucker et al., 2007).

To our best knowledge, this is the first report of a QTL for powdery mildew resistance on the short arm of chromosome 3A. Although this locus was only significant in the Norwegian testing environments, the LOD curves indicate that this QTL also had effects on PM in Beijing and on YR in Toluca (Fig. 2). No gene for race-specific resistance to PM is known at this location (Huang and Röder, 2004), and the race non-specific nature of the PM resistance in Saar would suggest that the QTL detected is a gene for partial resistance to PM.

The minor QTL for PM resistance 5A close to the marker gwm617b could correspond to similar loci for partial resistance to PM detected in the Swiss winter wheat Forno (Keller et al., 1999) and *T. militinae* (Jakobson et al., 2006).

A major genetic factor for PM resistance was found to be inherited from the susceptible parent Avocet. This QTL was located on chromosome 4B at the same position as similar QTLs for PM resistance detected in the Forno×Oberkulmer (Keller et al., 1999), Synthetic ×Opata (Börner et al., 2002) and Fukuho-komugi× Oligoculm (Liang et al., 2006) populations. The same QTL from Avocet showed significant effects on LR and YR in the Avocet×Pavon population (William et al., 2006), but although the LOD curves for LR and YR also peaked at the same location in the present study (Fig. 2), the effects were far from being significant.

Although the three most important QTLs on 1B, 4B and 7D in the present study showed significant and relatively similar effects across all the testing environments for PM, there were some discrepancies among the QTLs detected in Norway and China. The lack of significance of the 3A and 5A QTLs in Beijing and a correspondingly higher contribution by the 5B QTL at that location could be due to the different scales used for scoring the disease. While PM severity in Norway was scored as the percentage of leaf area of penultimate leaves covered with the disease, the whole canopy was assessed on a 0-9 scale in Beijing. These methods are likely to emphasize different aspects of the PM resistance, and the slightly different results obtained in the QTL mapping were therefore not unexpected. Environmental differences could also account for some of the variation in the QTL effects between Norway and China, while the different virulence composition of *Blumeria graminis* f. sp. *tritici* in the two countries was not expected to influence the results due to the partial and race non-specific nature of the PM resistance in Saar (Lillemo et al., 2006). Saar has been tested against a collection of 18 differential isolates at the seedling stage and only found to have a reaction pattern corresponding to *Pm5a* (M. Lillemo, unpublished). The presence of *Pm5a* was also confirmed by the QTL analysis, but this gene did not affect the PM scores in Norway and had only negligible effects on the PM severity in Beijing, which is in agreement with the widespread virulence against this defeated resistance gene in both in Norway (Skinnes, 2002) and China (Yu, 2000). Since *Pm 5a* was the only race-specific resistance detected at the seedling stage, and all the other QTLs showed relatively consistent effects across all environments, one might conclude that all the detected QTLs for PM resistance in Saar are likely to represent genes for partial and race non-specific resistance.

One interesting finding of the map construction was the detection ofa 5B/7B translocated chromosome in Saar, which is known from Capelle Desprez and many other European wheat cultivars (Riley et al., 1967), and associated with increased adult plant resistance to YR (Law and Worland, 1997). The YR resistance was attributed to the 5BS arm of the translocation (Law and Worland, 1997), and in the present study we found a significant QTL for PM resistance on 5BS, close to the SSR marker barc4, which showed a much bigger effect in Beijing than the Norwegian testing environments. However, we did not find any significant QTL for YR resistance associated with the 5B/7B translocation, which may indicate that the translocation in Saar is different from the one in Capelle Desprez or that the YR resistance was not effective in the Toluca testing environment.

Genetic control of LR and YR resistance in Saar

Since the primary objective of the present QTL study was to detect the main genetic factors behind the PM resistance in Saar, the linkage maps obtained from the BSA approach have on purpose excluded most of the wheat genome that did not represent chromosomal areas involved in the PM resistance. Thus, the QTL mapping of LR and YR resistance presented in this study does not constitute any complete molecular genetic analysis of these traits, and was included just for the purpose of testing whether some of the detected QTLs for PM resistance could also be involved in resistance to LR and YR.

This analysis confirmed the presence of *Lr34/Yr18* in Saar, while the presence of *Lr46/Yr29* had not been inferred by previous studies (Navabi et al., 2003, 2004). While the analysis of the LR data indicated that these two loci account for most of the genetic variation for this trait at Cd. Obregon, their magnitude of effect on YR is more similar to the effect on PM (Tables 2 and 3). The QTL on 6A with YR resistance contributed by Avocet was also detected in the Avocet × Pavon population and is likely due to a translocation from *Agropyron elongatum* in Avocet that is known to carry the stem rust resistance gene *Sr26* (William et al., 2006). The Avocet × Saar population is possibly segregating for other important, but yet undetected genetic factors for YR resistance, as the four QTLs on 1B, 3A, 6A and 7D did not explain much more than half of the genetic variation for the trait (Table 3).

The significantly negative interaction between *Lr34* and *Lr46* for LR (Table 3, Fig. 3) indicates a partially duplicate gene action of these two loci, which can probably be explained by the similar resistance mechanisms of these genes. They both confer partial resistance with increased latent period and decreased infection frequency and uredium size, and both are more clearly expressed in adult plants than in seedlings (Martinez et al., 2001; Rubiales and Niks, 1995).

General disease resistance loci on 7D and 1B

The co-location of genes for partial and race non-specific resistance to three biotrophic pathogens at the *Lr34/Yr18* and *Lr46/Yr29* loci on 7D and 1B is the most important finding of this study. The tight linkage or potential pleiotropy of genes for LR and YR resistance at these two loci has been widely known and utilized in wheat breeding (Singh, 1992b; William et al., 2003), but their involvement in resistance to PM has not been suggested until recently. The first direct evidence for the association of *Lr34/Yr18* with resistance to PM came from a field experiment in Australia where a population of recombinant inbred lines that segregated for *Lr34/Yr18* in the near-isogenic background of the susceptible cultivar Thatcher became infected with PM (Spielmeyer et al., 2005). The uniform genetic background of the population allowed for the unambiguous scoring of lines as either homozygous resistant or homozygous susceptible to all three diseases, and no recombination was detected among the resistance genes to these three diseases in the 107 lines tested (Spielmeyer et al., 2005). Independently from this study, Lillemo et al. (2007) tested near-isogenic lines of *Lr34/Yr18* and *Lr46/Yr29* in the genetic background of Avocet and found significant effects of both loci on PM. Another common feature of these two disease resistance loci is the expression of LTN, a premature senescence of the leaf tips about 1-2 weeks after flowering, which is widely used as a phenotypic marker for the rust resistance at these loci (Rosewarne et al., 2006; Singh, 1992a). The two loci also share another similarity which might indicate that the resistances to LR, YR and PM are not just caused by genetic linkage. Studies with near-isogenic lines have shown that the resistance effect of the *Lr34/Yr18* locus is always bigger than that of *Lr46/Yr29* (Lillemo et al., 2007; Martinez et al., 2001), which is also confirmed by the present study. The partial resistance to PM at the *Lr34/Yr18* locus has been given the gene designation *Pm38* (R. McIntosh, pers. comm.), and based on the analogous effect of the *Lr46/Yr29* locus on PM as shown in this study and previously on the Avocet NILs (Lillemo et al., 2007) we propose that the corresponding PM resistance gene should be named *Pm 39*.

Interestingly, Saar was never exposed to PM and selected for resistance to this disease during its breeding in Mexico, but nevertheless obtained a good partial resistance to PM through the intense selection for race non-specific resistance to LR and YR based on *Lr34/Yr18* and *Lr46/Yr29*. It is a major advantage to resistance breeding that race non-specific and potentially durable resistance to three of the most important biotrophic pathogens in wheat can be selected for simultaneously by these two loci by field testing for only one of the diseases. Moreover, the resistance of these loci can also be selected for based on the expression of LTN or facilitated by the closely linked molecular markers presented in this study.

Acknowledgments

This work was supported by a grant from the Norwegian Research Council and International Collaboration Project from Ministry of Agriculture of China (2006-G2). The helpful assistance of Anne Guri Marøy with molecular marker genotyping and Yalew Tarkegne for the field trials is greatly acknowledged. We also thank Alireza Navabi for providing the seed of the Avocet×Saar population that was used to produce the F_6 lines used in the study, and Martin Ganal at TraitGenetics Inc., Gatersleben, Germany and Atushi Torada at Hokuren Agricultural Reserach Institute, Hokkaido, Japan for providing unpublished primer sequences.

References

Bennett FGA (1984) Resistance to powdery mildew in wheat: a review of its use in agriculture and breeding programmes. Plant Pathol 33: 279-300.

Bjarko ME, Line RF (1988) Heritability and number of genes controlling leaf rust resistance in four cultivars of wheat. Phytopathology 78: 457-461.

Bossolini E, Krattinger SG, Keller B (2006) Development of simple sequence repeat markers specific for the *Lr34* resistance region of wheat using sequence information from rice and *Aegilops tauschii*. Theor Appl Genet 113: 1049-1062.

Bougot Y, Lemoine J, Pavoine MT, Guyomar'ch H, Gautier V, Muranty H, Barloy D (2006) A major QTL effect controlling resistance to powdery mildew in winter wheat at the adult plant stage. Plant Breeding 125: 550-556.

Börner A, Schumann E, Fürste A, Cöster H, Leithold B, Röder MS, Weber WE (2002) Mapping of quantitative trait loci determining agronomic important characters in hexaploid wheat (*Triticum aestivum* L.). Theor Appl Genet 105: 921-936.

Chantret N, Mingeot D, Sourdille P, Bernard M, Jacquemin JM, Doussinault G (2001) A major QTL for powdery mildew resistance is stable over time and at two development stages in winter wheat. Theor Appl Genet 103: 962-971.

Chantret N, Sourdille P, Roder M, Tavaud M, Bernard M, Doussinault G (2000) Location and mapping of the powdery mildew resistance gene *MLRE* and detection of a resistance QTL by bulked segregant analysis (BSA) with microsatellites in wheat. Theor Appl Genet 100: 1217-1224.

Conner RL, Kuzyk AD, Su H (2003) Impact of powdery mildew on the yield of soft white spring wheat cultivars. Can J Plant Sci 83: 725-728.

Griffey CA, Das MK, Stromberg EL (1993) Effectiveness of adult-plant resistance in reducing grain yield loss to powdery mildew in winter wheat. Plant Dis 77: 618-622.

Gustafson GD, Shaner G (1982) Influence of plant age on the expression of slow-mildewing resistance in wheat. Phytopathology 72: 746-749.

Hautea RA, Coffman WR, Sorrells ME, Bergstrom GC (1987) Inheritance of partial resistance to powdery mildew in spring wheat. Theor Appl Genet 73: 609-615.

Hayden MJ, Stephenson P, Logojan AM, Khatkar D, Rogers C, Elsden J, Koebner RMD, Snape JW, Sharp PJ

(2006) Development and genetic mapping of sequence-tagged microsatellites (STMs) in bread wheat (*Triticum aestivum* L.). Theor Appl Genet 113: 1271-1281.

Hsam SLK, Zeller FJ (2002) Breeding for powdery mildew resistance in common wheat (*Triticum aestivum* L.). In: Belanger RR, Bushnell WR, Dik AJ, Carver TLW (eds) The Powdery mildews A comprehensive treatise. The American Phytopathological Society, St. Paul, MN., pp 219-238.

Huang XQ, Röder MS (2004) Molecular mapping of powdery mildew resistance genes in wheat: A review. Euphytica 137: 203-223.

Jakobson I, Peusha H, Timofejeva L, Jarve K (2006) Adult plant and seedling resistance to powdery mildew in a *Triticum aestivum* × *Triticum militinae* hybrid line. Theor Appl Genet 112: 760-769.

Keller M, Keller B, Schachermayr G, Winzeler M, Schmid JE, Stamp P, Messmer MM (1999) Quantitative trait loci for resistance against powdery mildew in a segregating wheat x spelt population. Theor Appl Genet 98: 903-912.

Kosambi DD (1944) The estimation of map distance from recombination values. Annu Eugen 12: 172-175.

Lagudah ES, McFadden H, Singh RP, Huerta-Espino J, Bariana HS, Spielmeyer W (2006) Molecular genetic characterization of the *Lr34/Yr18* slow rusting resistance gene region in wheat. Theor Appl Genet 114: 21-30.

Law CN, Worland AJ (1997) The control of adult-plant resistance to yellow rust by the translocated chromosome 5BS-7BS of bread wheat. Plant Breeding 116: 59-63.

Liang SS, Suenaga K, He ZH, Wang ZL, Liu HY, Wang DS, Singh RP, Sourdille P, Xia XC (2006) Quantitative trait loci mapping for adult-plant resistance to powdery mildew in bread wheat. Phytopathology 96: 784-789.

Lillemo M, Singh RP, Huerta-Espino J, Chen XM, He ZH, Brown JKM (2007) Leaf rust resistance gene *Lr34* is involved in powdery mildew resistance of CIMMYT bread wheat line Saar. In: Buck HT, Nisi JE, Salomon N (eds) Developments in Plant Breeding Vol 12 Wheat productions in stressed environments Proceedings of the 7th International Wheat Conference Mar del Plata, Argentina Nov 27-Dec 2, 2005, pp 97-102.

Lillemo M, Skinnes H, Singh RP, van Ginkel M (2006) Genetic analysis of partial resistance to powdery mildew in bread wheat line Saar. Plant Dis 90: 225-228.

Lipps PE, Madden LV (1988) Effect of triadimenol seed treatment and triadimefon foliar treatment on powdery mildew epidemics and grain yield of winter wheat cultivars. Plant Dis 72: 887-892.

Liu SX, Griffey CA, Maroof MAS (2001) Identification of molecular markers associated with adult plant resistance to powdery mildew in common wheat cultivar Massey. Crop Sci 41: 1268-1275.

Martinez F, Niks RE, Singh RP, Rubiales D (2001) Characterization of *Lr46*, a gene conferring partial resistance to wheat leaf rust. Hereditas 135: 111-114.

McDonald BA, Linde C (2002) The population genetics of plant pathogens and breeding strategies for durable resistance. Euphytica 124: 163-180.

Mingeot D, Chantret N, Baret PV, Dekeyser A, Boukhatem N, Sourdille P, Doussinault G, Jacquemin JM (2002) Mapping QTL involved in adult plant resistance to powdery mildew in the winter wheat line RE714 in two susceptible genetic backgrounds. Plant Breeding 121: 133-140.

Navabi A, Singh RP, Tewari JP, Briggs KG (2003) Genetic analysis of adult-plant resistance to leaf rust in five spring wheat genotypes. Plant Dis 87: 1522-1529.

Navabi A, Singh RP, Tewari JP, Briggs KG (2004) Inheritance of high levels of adult-plant resistance to stripe rust in five spring wheat genotypes. Crop Sci 44: 1156-1162.

Nematollahi G, Mohler V, Wenzel G, Zeller F, Hsam S (2007) Microsatellite mapping of powdery mildew resistance allele *Pm5d* from common wheat line IGV1-455. Euphytica online first.

Parida SK, Kumar KAR, Dalal V, Singh NK, Mohapatra T (2006) Unigene derived microsatellite markers for the cereal genomes. Theor Appl Genet 112: 808-817.

Peterson RF, Campbell AB, Hannah AE (1948) A diagrammatic scale for estimating rust intensity on leaves and stems of cereals. Can J Res C 26: 496-500.

Riley R, Coucoli H, Chapman V (1967) Chromosomal interchanges and phylogeny of wheat. Heredity 22: 233-248.

Roberts JJ, Caldwell RM (1970) General resistance (slow mildewing) to *Erysiphe graminis* f. sp. *tritici* in Knox wheat. Phytopathology 60: 1310.

Rosewarne GM, Singh RP, Huerta-Espino J, William HM, Bouchet S, Cloutier S, McFadden H, Lagudah ES (2006) Leaf tip necrosis, molecular markers and β1-proteasome subunits associated with the slow rusting resistance genes *Lr46/Yr29*. Theor Appl Genet 112: 500-508.

Rubiales D, Niks RE (1995) Characterization of *Lr34*, a major gene conferring nonhypersensitive resistance to wheat leaf rust. Plant Dis 79：1208-1212.

Semagn K，Bjornstad A，Skinnes H，Maroy AG，Tarkegne Y，William M（2006）Distribution of DArT，AFLP，and SSR markers in a genetic linkage map of a doubled-haploid hexaploid wheat population. Genome 49：545-555.

Shaner G（1973）Evaluation of slow-mildewing resistance of Knox wheat in the field. Phytopathology 63：867-872.

Singh RP（1992a）Association between gene *Lr34* for leaf rust resistance and leaf tip necrosis in wheat. Crop Sci 32：874-878.

Singh RP（1992b）Genetic association of leaf rust resistance gene *Lr34* with adult plant resistance to stripe rust in bread wheat. Phytopathology 82：835-838.

Singh RP，Ma H，Rajaram S（1995）Genetic analysis of resistance to scab in spring wheat cultivar Frontana. Plant Dis 79：238-240.

Skinnes H（2002）Breakdown of race specific resistance to powdery mildew in Norwegian wheat. Cereal Rusts and Powdery Mildew Bulletin vol. 30 [http：//www. crpmb. org] 2002/1201skinnes.

Somers DJ，Isaac P，Edwards K（2004）A high-density microsatellite consensus map for bread wheat（*Triticum aestivum* L.）. Theor Appl Genet 109：1105-1114.

Spielmeyer W，McIntosh RA，Kolmer J，Lagudah ES（2005）Powdery mildew resistance and *Lr34/Yr18* genes for durable resistance to leaf and stripe rust cosegregate at a locus on the short arm of chromosome 7D of wheat. Theor Appl Genet 111：731-735.

Saari EE，Prescott JM（1975）A scale for appraising the foliar intensity of wheat diseases. Plant Disease Reporter 59：377-380.

Torada A，Koike M，Mochida K，Ogihara Y（2006）SSR-based linkage map with new markers using an intraspecific population of common wheat. Theor Appl Genet 112：1042-1051.

Tucker DM，Griffey CA，Liu S，Brown-Guedira G，Marshall DS，Maroof MAS（2007）Confirmation of three quantitative trait loci conferring adult plant resistance to powdery mildew in two winter wheat populations. Euphytica 155：1-13.

Utz HF，Melchinger AE（2003）PLABQTL：A computer program to map QTL，Version 1.2. Institute of plant breeding，seed science and population genetics，University of Hohenheim，Stuttgart，Germany.

van Oijen JW，Voorrips RE（2001）Joinmap 3.0 software for the calculation of genetic linkage maps. Plant Research International，Wageningen，The Netherlands.

Voorrips RE（2002）MapChart：Software for the graphical presentation of linkage maps and QTLs. Journal of Heredity 93：77-78.

William HM，Singh RP，Huerta-Espino J，Palacios G，Suenaga K（2006）Characterization of genetic loci conferring adult plant resistance to leaf rust and stripe rust in spring wheat. Genome 49：977-990.

William M，Singh RP，Huerta-Espino J，Islas SO，Hoisington D（2003）Molecular marker mapping of leaf rust resistance gene *Lr46* and its association with stripe rust resistance gene *Yr29* in wheat. Phytopathology 93：153-159.

Yu DZ（2000）Wheat powdery mildew in Central China：Pathogen population structure and host resistance. Ph. D. Thesis. Wageningen University and Research Centre Wageningen，The Netherlands，p 134.

Yu DZ，Yang XJ，Yang LJ，Jeger MJ，Brown JKM（2001）Assessment of partial resistance to powdery mildew in Chinese wheat varieties. Plant Breeding 120：279-284.

QTL mapping for adult-plant resistance to stripe rust in Italian common wheat cultivars Libellula and Strampelli

Y. M. Lu[1,2], C. X. Lan[1], S. S. Liang[1], X. C. Zhou[3], D. Liu[1], G. Zhou[3], Q. L. Lu[3], J. X. Jing[2], M. N. Wang[2], X. C. Xia[1], and Z. H. He[1,4]

[1] *Institute of Crop Science, National Wheat Improvement Centre/The National Key Facility for Crop Gene Resources and Genetic Improvement, Chinese Academy of Agricultural Sciences* (*CAAS*), 12 *Zhongguancun South Street, Beijing* 100081, *China*; [2] *College of Plant Protection, Northwest A&F University, Yangling* 712100, *Shaanxi, China*; [3] *Gansu Wheat Research Institute, Gansu Academy of Agricultural Sciences, Lanzhou* 730070, *Gansu Province, China*; [4] *International Maize and Wheat Improvement Centre* (*CIMMYT*), *China Office, c/o CAAS*, 12 *Zhongguancun South Street, Beijing* 100081, *China.*

Abstract: Italian common wheat cultivars Libellula and Strampelli, grown for over three decades in Gansu province of China, have shown effective resistance to stripe rust. To elucidate the genetic basis of the resistance, F_3 populations were developed from crosses between the two cultivars and susceptible Chinese wheat cultivar Huixianhong. The F_3 lines were evaluated for disease severity in Beijing, Gansu and Sichuan from 2005 to 2008. Joint and single-environment analyses by composite interval mapping identified five quantitative trait loci (QTLs) in Libellula for reduced stripe rust severity, designated *QYr. caas-2DS*, *QYr. caas-4BL*, *QYr. caas-5BL. 1*, *QYr. caas-5BL. 2* and *QYr. caas-7DS*, and explained 8.1-12.4%, 3.6-5.1%, 3.4-8.6%, 2.6%, and 14.6-35.0%, respectively, of the phenotypic variance across four environments. Six interactions between different pairs of QTLs explained 3.2-7.1% of the phenotypic variance. The QTLs *QYr. caas-4BL*, *QYr. caas-5BL. 1* and *QYr. caas-7DS* were also detected in Strampelli, explaining 4.5%, 2.9-5.5% and 17.1-39.1% of phenotypic variance, respectively, across five environments. Three interactions between different pairs of QTLs accounted for 6.1-35.0% of the phenotypic variance. The QTL *QYr. caas-7DS* flanked by markers *csLV34* and *Xgwm295* showed the largest effect for resistance to stripe rust. Sequence analyses confirmed that the lines with the *QYr. caas-7DS* allele for resistance carried the resistance allele of the *Yr18/Lr34* gene. Our results indicated that the adult-plant resistance gene *Yr18* and several minor genes confer effective durable resistance to stripe rust in Libellula and Strampelli.

Introduction

Stripe rust, caused by *Puccinia striiformis* f. sp. *tritici* (*Pst*), is an important disease of common wheat (*Triticum aestivum* L.) worldwide (Stubbs, 1985; Chen, 2005). The most effective approach to control the disease is growing resistant wheat cultivars (Line,

Published in TAG, 2009, 119 (8): 1349-1359

2002; Chen, 2005). However, most of the resistance genes used in the past few decades were race-specific, eliciting hypersensitive responses in host plants and easily overcome when new virulent strains increased in the pathogen population. In China, the vulnerability of this type of stripe rust resistance has occurred repetitively since the 1950s with consequent high yield losses (Wan et al., 2004).

The development of wheat cultivars with adult-plant resistance (APR) has been given increasing emphasis in recent years because of its higher durability (Line, 2002). Cultivars conferring APR often show susceptible responses at the seedling stage, but have low disease severities at the adult plant stage in the field. APR is characterized by a lower frequency of infections, a longer latent period, less urediniospore production, a smaller uredinial size and its polygenic nature (Chen and Line, 1995; Liang et al., 2006). It may be identified in cultivars with defeated race-specific genes, or in those lacking known race-specific resistance genes (Singh and Rajaram, 1994; Santra et al., 2008).

Genetic analysis of cultivars with APR indicated that this type of resistance was conferred by the additive effects of several minor genes (Singh and Rajaram, 1994; Navabi et al., 2004; Singh et al., 2005). This conclusion was further confirmed by studies of an increasing number of cultivars with adult-plant resistance, such as Kariega (Ramburan et al., 2004), Camp Rémy (Mallard et al., 2005), Pavon 76 (William et al., 2006), Attila (Rosewarne et al., 2008), and Express (Lin and Chen, 2009). Singh et al. (2000a) demonstrated that the combination of 4-5 slow rusting genes with small to intermediate effects, but acting additively provides up to near-immune levels of adult plant resistance.

Since APR is quantitatively inherited, molecular markers can be employed to determine the number, genomic location and effect of the individual resistance genes (Young, 1996). Molecular markers closely linked to those resistance loci can then be used in gene pyramiding in facilitating wheat breeding programs (Young, 1996). Many quantitative trait loci (QTLs) in wheat for reducing stripe rust severity at adult plant stage have been identified with molecular markers (Bariana et al., 2001; Boukhatem et al., 2002; Mallard et al., 2005; Santra et al., 2008; Lin and Chen, 2009). Of them, the most important slow stripe rusting and designated loci are *Yr18* (Suenaga et al., 2003), *Yr29* (William et al., 2003, 2006), and *Yr30* (Singh et al., 2000b, 2005). These three genes are widely distributed in CIMMYT wheat germplasm (Singh et al., 2005). More recently, *Yr36* (Uauy et al., 2005) and *Yr39* (Lin and Chen, 2007) were identified. Due to the significant contributions of these genes for stripe rust resistance, several studies were conducted to map them more precisely (William et al., 2003; Uauy et al., 2005; Lagudah et al., 2006; Spielmeyer et al., 2008), and this resulted in the recent cloning of *Yr18* and *Yr36* (Fu et al., 2009; Krattinger et al., 2009). Both genes differed from the NBS-LRR structures that characterize most of the specific resistance genes cloned to date.

The Italian wheat cultivars Libellula and Strampelli, introduced into China in 1973 (Zheng, 1993), have been grown in Gansu province, a hot spot for stripe rust, for over 30 years. In spite of the occurrence of many new pathogen races, they continued to confer effective adult plant resistance to stripe rust, justifying their classification as durable resistance (Zhou et al., 2003a). Inheritance of the APR in these two cultivars was previously reported based on the reaction patterns of F_1 and F_2 progenies derived from crosses Libellula/Huixianhong and Strampelli/Huixianhong using conventional quantitative genetic analysis (Yin et al., 2005, 2006). However, the precise positions of the resistance loci in the two cultivars remained unknown. Accordingly, the objective of the present study was to identify and locate the QTLs for adult-plant resistance to stripe rust in the two cultivars using molecular markers.

Materials and methods

Plant materials

F_3 populations used for QTL mapping were derived from the crosses Libellula/Huixianhong and Strampelli/Huixianhong, totaling 244 and 252 lines, respectively. The two cultivars have been grown in Gansu province, a hot spot for stripe rust, for over 30 years, exhibiting high APR expressed as longer latent period, lower disease severity and lower damage to kernel weight (Zhou et al., 2003a), whereas Huixianhong is highly susceptible to almost all isolates of *Pst* at both the seedling and adult-plant stages. The F_3 lines generated from individual F_2 plants were planted and harvested as bulks with over 50 plants of each line to produce F_3 populations that can be maintained as bulk populations, each deriving from a single F_2 plant.

Field trials

The two populations were evaluated for disease severity to stripe rust inBeijing, Gansu and Sichuan provinces from 2005 to 2008, providing data for the populations of Libellula/Huixianhong and Strampelli/Huixianhong for 4 and 5 environments, respectively. Field trials were conducted in randomized complete blocks design with three replicates. Each plot consisted of two 1.5 m rows spaced 25 cm apart. Approximately 100 seeds were sown in each plot. The highly susceptible line, Tiaogan 601, was used as a susceptible check in Gansu and Beijing and was planted after every 10 plots, and Mingxian 169 was used as a susceptible check in Sichuan. Infection rows of Tiaogan 601 or Mingxian 169 were planted perpendicular and adjacent to the test rows to ensure adequate inocula. Artificial inoculation was performed with the prevalent *Pst* race CYR32 at the three-leaf stage in the spring. Stripe rust severities were assessed for the first time 4 weeks after inoculation, and then at weekly intervals for two further weeks using the modified Cobb scale (Peterson et al., 1948) in Beijing and Gansu. In Sichuan, stripe rust severities were visually rated, when the disease severities on Mingxian 169 reached a maximum level around April 20, 2008.

Statistical analysis of variance was conducted by PROC GLM in the Statistical Analysis System (SAS Institute, 1997), with genotype as a fixed effect, and environments, a combination of locations and years, and replicates as random effects. Broad-sense heritability (h^2) for stripe rust reaction was calculated using the formula $h^2 = \sigma_g^2 / (\sigma_g^2 + \sigma_{ge}^2/e + \sigma_\varepsilon^2/re)$, where σ_g^2, σ_{ge}^2, andσ_ε^2 were estimates of genotypic, genotype × environment interaction and error variances, respectively, and *e* and *r* were the numbers of environments and replicates per environment, respectively. Phenotypic correlation coefficients between maximum disease severities (MDS) in different environments were calculated on a mean basis using the Microsoft Excel analytical tool.

Microsatellite markeranalysis and gene sequencing

Genomic DNA was extracted from young leaves of the parents and F_3 lines (40-50 plants per F_3 line as a bulk) using the CTAB method (Sharp et al., 1988). Simple sequence repeat (SSR) markers were screened for polymorphism between the two parents by polyacrylamide gel electrophoresis. PCR and gel staining were conducted as described by Li et al. (2006) and Bassam et al. (1991). The SSR primers were from the GWM (Röder et al., 1998), BARC (developed by P. Cregan, Q. Song and associates at the USDA-ARS Beltsville Agriculture Research Station), WMC (developed by a team led by P. Isaac, IDnagenetics, Norwich, UK), and CFD (Guyomarc'h et al., 2002) marker series and one STS marker, *csLV34* (Lagudah et al., 2006). Resistant and susceptible bulks were established by mixing equal amounts of DNA from the five most resistant and the five most susceptible lines, respectively, based on the averaged stripe rust severity across environments. SSR markers showing polymorphisms between the resistant and susceptible bulks, as well as between the parents, were used to genotype 15-20 most resistant and most susceptible lines, respectively. Subsequently, the SSRs showing linkage with stripe rust resistance were used to genotype the entire population. Additional markers for enriching the chromosome regions linked to resistance genes were selected from published wheat consensus maps (http://www.shigen.nig.ac.jp/wheat; http://wheat.pw.usda.gov; Somers et al., 2004) and tested for polymorphisms between the parents and bulks. Those showing polymorphism were also used to genotype the population for linkage analysis. The PCR primers used for sequencing the *Yr18* gene in Libellula and Strampelli were kindly provided by Dr. Evans Lagudah, at CSIRO Plant Industry, Canberra, Australia. All the sequencings were performed by Beijing Augct Biological Technology Co., Ltd. (http://www.augct.com) and Shanghai Sangon Biological Engineering Technology & Service Co., Ltd. (http://www.sangon.com). Sequence alignments were performed using the software DNAMAN (http://www.lynnon.com).

QTL analysis

QTL mapping was based on the averaged MDS of three replicates in each environment, and also the averaged data across all environments. Linkage groups were established with the software Map Manager QTXb20 (Manly et al., 2001). Recombination values were converted to genetic distances using the Kosambi mapping function (Kosambi, 1944). The positions of the detected QTLs were determined by composite interval mapping (CIM) using the software Cartographer 2.5 (Wang et al., 2005). A logarithm of odds (LOD) of 2.5 was set to declare QTL as significant. Each QTL was represented by a 20 cM interval with the local LOD maximum at its center. QTL with overlapping 20 cM intervals among different environments were considered as being in common. QTL effects were estimated as the proportion of phenotypic variance (R^2) explained by the QTL. Digenic interactions between non-allelic QTLs were analyzed by inclusive composite interval mapping (ICIM) method, using the software IciMapping 2.2 (Li et al., 2007, 2008). The chromosomal assignments of the linkage groups were based on published wheat maps (Somers et al., 2004), and the Graingenes (http://wheat.pw.usda.gov) and Komugi integrated wheat consensus maps (http://www.shigen.nig.ac.jp/wheat).

Results

Distribution of MDS and correlation analysis

Libellula and Strampelli were susceptible to the prevalent races CYR31 and CYR32 of *Pst* at the seedling stage, but were highly resistant at the adult-plant stage, and they displayed a MDS of less than 10% in all environments, whereas the susceptible parent, Huixianhong, showed an MDS ranging from 30 to 80% across different environments (Fig. 1). This indicates typical APR to stripe rust in these two cultivars. Both populations exhibited a continuous distribution of stripe rust severities, indicating the polygenic characteristics of the slow rusting resistance.

Correlations for MDS in the Libellula/Huixianhong population ranged from 0.58 to 0.70 ($P<0.0001$) among different environments, and the heritability of MDS was 0.95. Significant correlations ($r=0.58$ to 0.74, $P<0.0001$) for MDS were also detected in the Strampelli/Huixianhong population among different environments, and the heritability was 0.97. ANOVA of the two populations revealed significant differences ($P<0.0001$) in MDS among lines in the two populations. Highly significant differences ($P<0.0001$) were also observed among different environments and for genotype×environment interactions (Table 1).

QTL analyses for APR to stripe rust

Libellula /Huixianhong population

A total of 943 SSR markers were screened for polymorphism between Libellula and Huixianhong. Of them, 133 showing polymorphisms between the two parents were used to test the resistant and susceptible bulks. Subsequently, 39 markers producing polymorphic bands between the two bulks were used to genotype the entire population. Based on the mean MDS in each environment and that averaged from all environments, five QTLs were detected on chromosomes 2DS, 4BL, 5BL (2 QTLs) and 7DS (Table 2 and Fig. 2).

Table 1 Analysis of variance of MDS scores for F_3 lines of crosses Libellula /Huixianhong and Strampelli /Huixianhong populations

Population	Source of variation	df	MS	F value
Libellula /Huixianhong	Line	243	2 415	10.71**
	Environment	3	236 950	1 050.99**
	Replicate	2	3 043	13.50**
	Line×environment	729	398	1.76**
	Error	1 947	225	
Strampelli/Huixianhong	Line	251	1 222	15.65**
	Environment	4	80 728	1 033.53**
	Replicate	2	1 504	19.26**
	Line×environment	990	182	2.33**
	Error	2 444	78	

** Significant at $P<0.0001$

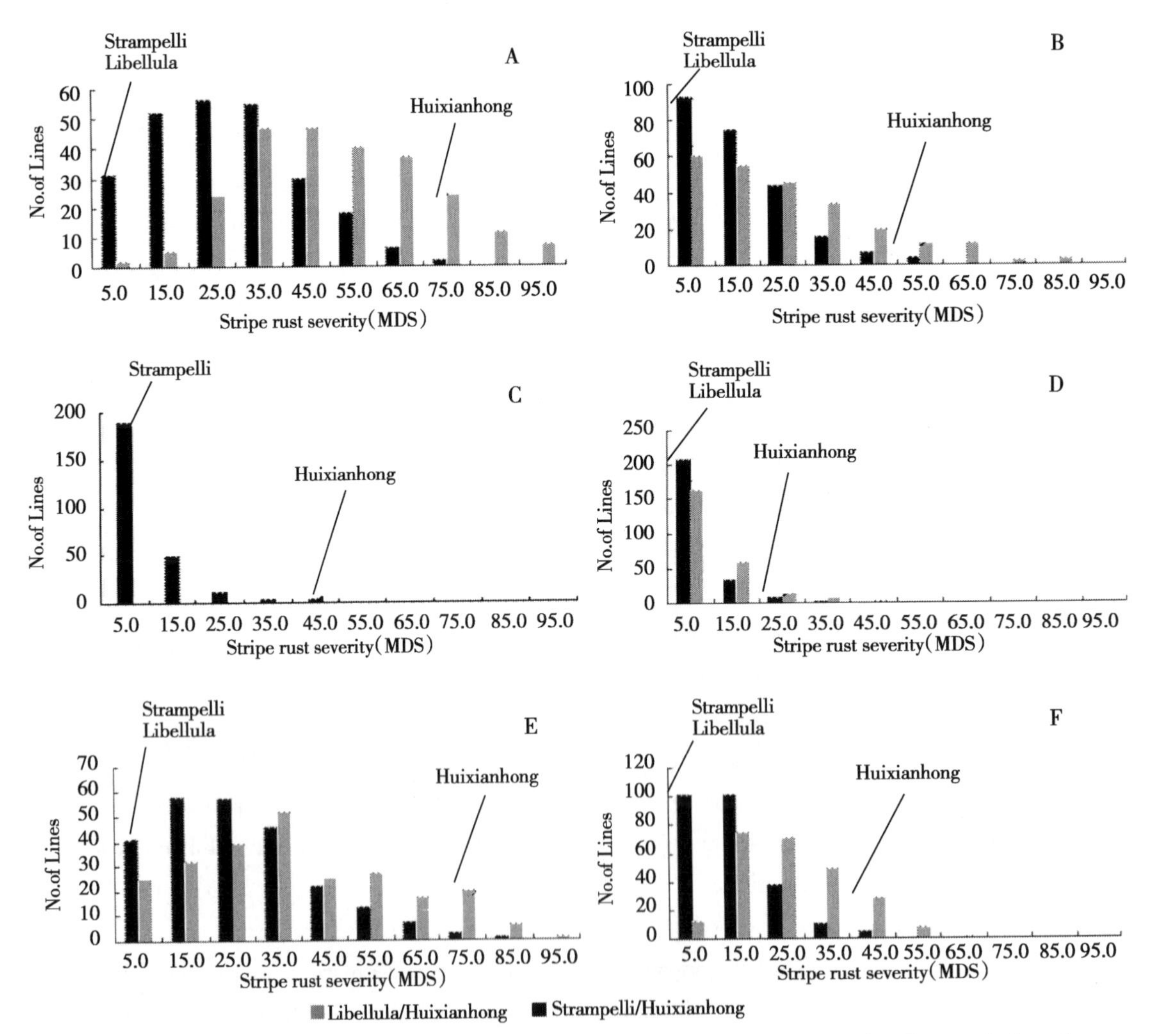

Fig. 1 Frequency distribution of stripe rust maximum disease severities (MDS) for F_3 populations from two wheat crosses in four or five environments. A, Gansu 2005; B, Gansu 2006; C, Beijing 2006; D, Gansu 2007; E, Sichuan 2008; F, Average MDS in four or five environments; Grey and black columns indicate the Strampelli/Huixianhong and Libellula/Huixianhong populations, respectively.

Table 2 Summary of QTLs for MDS to stripe rust detected by CIM in Libellula /Huixianhong population across four environments. For each QTL the corresponding marker interval, individual explained phenotypic variances R^2 (%), additive effect and LOD value are given

Environment	QTL[a]	Marker Interval	LOD	AE[b]	R^2 (%)	TotalR^2 (%)
Gansu 2005	*QYr. caas-4BL*	*Xgwm165-Xgwm149*	2.92	4.96	3.6	31.8
	QYr. caas-7DS	*csLV34-Xgwm295*	11.33	14.15	28.2	
Gansu 2006	*QYr. caas-2DS*	*Xcfd51-Xgwm261*	4.94	6.81	8.1	46.8
	QYr. caas-5BL. 1	*Xwmc415-Xwmc537*	4.00	5.08	3.7	
	QYr. caas-7DS	*csLV34-Xgwm295*	15.17	15.36	35.0	
Gansu 2007	*QYr. caas-5BL. 1*	*Xwmc415-Xwmc537*	3.95	1.96	3.4	30.2
	QYr. caas-7DS	*csLV34-Xgwm295*	9.06	5.41	26.8	
Sichuan 2008	*QYr. caas-2DS*	*Xcfd51-Xgwm261*	10.01	10.08	12.4	43.3
	QYr. caas-4BL	*Xgwm165-Xgwm149*	3.57	7.22	5.1	
	QYr. caas-5BL. 1	*Xwmc415-Xwmc537*	9.35	9.54	8.6	
	QYr. caas-5BL. 2	*Xbarc142-Xgwm604*	2.53	5.11	2.6	
	QYr. caas-7DS	*csLV34-Xgwm295*	8.20	11.64	14.6	
Average in four environments	*QYr. caas-2DS*	*Xcfd51-Xgwm261*	8.00	5.92	9.9	52.4
	QYr. caas-4BL	*Xgwm165-Xgwm149*	3.19	3.61	3.1	
	QYr. caas-5BL. 1	*Xwmc415-Xwmc537*	8.19	5.53	7.2	
	QYr. caas-7DS	*csLV34-Xgwm295*	16.16	11.37	32.2	

a. Only QTL with LOD > 2.5 are shown

b. AE = additive effect of resistance allele

The most consistent locus with the largest effect found in all environments was *QYr. caas-7DS*, located between *csLV 34* and *Xgwm 295* on the short arm of chromosome 7D. This QTL explained 28.2%, 35.0%, 26.8% and 14.6% of the phenotypic variances (R^2) in Gansu 2005, Gansu 2006, Gansu 2007 and Sichuan 2008, respectively (Table 2 and Fig. 2E). R^2 was as high as 32.2% for the QTL computed by the averaged MDS of all environments. The locus with the second largest effect, *QYr. caas-2DS*, was located in the marker interval *Xcfd51-Xgwm261* on chromosome 2DS. This gene was detected in two environments as well as for the averaged MDS, explaining from 8.1 to 12.4% of the phenotypic variance (Table 2 and Fig. 2A). One QTL, *QYr. caas-4BL*, in the interval *Xgwm165-Xgwm149* on chromosome 4BL, explained from 3.1 to 5.1% of the phenotypic variances in two environments as well as the averaged data over four environments (Table 2 and Fig. 2B). Two QTLs, *QYr. caas-5BL. 1* and *QYr. caas-5BL. 2*, identified on chromosome 5BL, explained 3.4 to 8.6%, and 2.6% of the phenotypic variance, respectively (Table 2, Fig. 2C and Fig. 2D). All five QTLs for APR to stripe rust came from the resistant parent Libellula (Table 2). These QTLs accounted for 30.2-46.8% of the total phenotypic variance in a simultaneous fit across four environments (Table 2), suggesting a significant effect of the QTLs in reducing disease severity.

Strampelli /Huixianhong population

A total of 1136 SSR markers were screened for polymorphism between Strampelli and Huixianhong. Thirty-four markers showing polymorphisms between the resistant and susceptible bulks were used to genotype the entire population. Based on the mean MDS in each

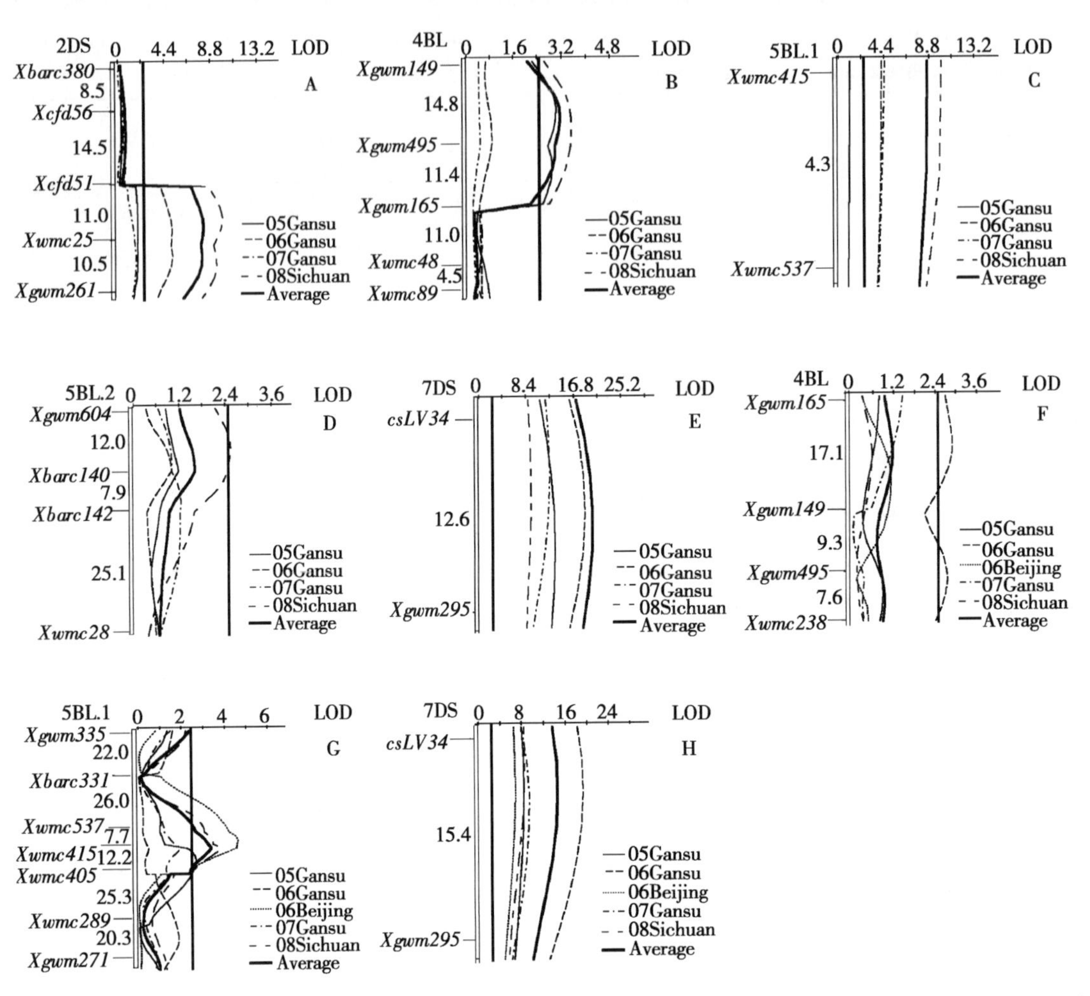

Fig. 2 Likelihood plots of QTL for stripe rust APR on chromosomes 2DS, 4BL, 5BL and 7DS identified by composite interval mapping in cross Libellula/Huixianhong (A, B, C, D, E), and on chromosomes 4BL, 5BL and 7DS in cross Strampelli/Huixianhong (F, G, H). The LOD threshold for significance is 2.5. Positions (in cM) of the molecular markers along chromosomes are shown on the vertical axes; genetic distances between markers are shown.

environment and that averaged from five environments, three QTLs were identified on chromosomes 4BL, 5BL and 7DS (Table 3 and Fig. 2). All three were located at similar chromosomal positions to those detected in the Libellula/Huixianhong population. Similarly, the largest and most consistent resistance locus was mapped on chromosome 7DS, again designated *QYr. caas-7DS*, explaining from 17.1 to 39.1% of the phenotypic variance across five environments (Table 3 and Fig. 2H). The second QTL, *QYr. caas-5BL. 1*, on chromosome 5BL in the interval *Xwmc415-Xwmc537* detected in two individual environments as well as the overall mean, explained 2.2 to 5.5% of the phenotypic variances (Table 3 and Fig. 2G). The third QTL, *QYr. caas-4BL*, located on chromosome 4BL was detected only in Gansu 2006, explaining 4.5% of the phenotypic variance (Table 3 and Fig. 2F). No QTL was detected on chromosome 2DS in this population, due either to the lack of polymorphic markers between the two parents in this region, or because of the absence of a QTL in this chromosome region in Strampelli. The total phenotypic variance explained by the three QTLs ranged from 17.1 to 43.6% in a simultaneous fit across five environments (Table 3).

Table 3 Summary of the QTLs for MDS to stripe rust detected by CIM in Strampelli/Huixianhong population across five environments. For each QTL the corresponding marker interval, individual explained phenotypic variances R^2 (%), additive effect and LOD value are given

Environment	QTL[a]	Marker Interval	LOD	AE[b]	R^2 (%)	Total R^2 (%)
Gansu 2005	*QYr. caas-7DS*	*csLV34-Xgwm295*	8.46	8.68	17.1	17.1
Gansu 2006	*QYr. caas-4BL*	*Xgwm165-Xgwm149*	2.82	2.92	4.5	43.6
	QYr. caas-7DS	*csLV34-Xgwm295*	17.03	9.42	39.1	
Beijing 2006	*QYr. caas-5BL. 1*	*Xwmc415-Xwmc537*	4.34	2.37	5.5	30.4
	QYr. caas-7DS	*csLV34-Xgwm295*	6.77	3.60	24.9	
Gansu 2007	*QYr. caas-7DS*	*csLV34-Xgwm295*	6.78	3.29	19.8	19.8
Sichuan 2008	*QYr. caas-5BL. 1*	*Xwmc415-Xwmc537*	3.00	5.23	2.9	21.5
	QYr. caas-7DS	*csLV34-Xgwm295*	7.30	9.92	18.6	
Average in	*QYr. caas-5BL. 1*	*Xwmc415-Xwmc537*	2.54	2.12	2.2	30.5
five environments	*QYr. caas-7DS*	*csLV34-Xgwm295*	13.07	6.91	28.3	

a. Only QTL with LOD > 2.5 are shown

b. AE = additive effect of resistance allele

Epistasis between non-allelic QTLs

Among the five QTLs for APR to stripe rust in Libellula/Huixianhong population, six significant interactions between different pairs of QTLs were detected in four environments, explaining from 3.2 to 7.1% of the phenotypic variance (Table 4). In Strampelli/Huixianhong population, three interactions between different pairs of QTLs were stably identified across five environments, explaining from 6.1 to 35.0% of the phenotypic variance.

Table 4 Summary of significant (LOD>2.5) epistatic interactions between pairs of QTLs in two F_3 populations of Libellula /Huixianhong and Strampelli /Huixianhong across different environments

Population	Environment	$QTL_1 \times QTL_2$	LOD	R^2 (%)[a]
Libellula/Huixianhong	Gansu 2005	*QYr. caas-2DS×QYr. caas-4BL*	2.81	4.6
	Gansu 2006	*QYr. caas-2DS×QYr. caas-4BL*	3.39	7.1
		QYr. caas-2DS×QYr. caas-7DS	4.17	6.5
		QYr. caas-4BL×QYr. caas-7DS	2.99	3.7
	Gansu 2007	*QYr. caas-2DS×QYr. caas-4BL*	3.07	4.5
		QYr. caas-2DS×QYr. caas-5BL. 2	2.62	6.4
		QYr. caas-5BL. 1×QYr. caas-7DS	3.49	6.6
	Sichuan 2008	*QYr. caas-4BL×QYr. caas-5BL. 2*	3.67	6.6
	Average in four environments	*QYr. caas-2DS×QYr. caas-4BL*	3.02	3.2
Strampelli/Huixianhong	Gansu 2005	*QYr. caas-4BL×QYr. caas-7DS*	2.64	6.1
		QYr. caas-5BL. 1×QYr. caas-7DS	3.20	6.6
	Gansu 2006	*QYr. caas-4BL×QYr. caas-5BL. 1*	4.86	25.5
		QYr. caas-4BL×QYr. caas-7DS	13.15	35.0
		QYr. caas-5BL. 1×QYr. caas-7DS	11.55	31.6
	Beijing 2006	*QYr. caas-4BL×QYr. caas-5BL. 1*	5.28	16.8
		QYr. caas-4BL×QYr. caas-7DS	4.70	8.5

(续)

Population	Environment	$QTL_1 \times QTL_2$	LOD	R^2 (%)[a]
		QYr. caas-5BL. 1×*QYr. caas-7DS*	6.41	10.0
	Gansu 2007	*QYr. caas-4BL*×*QYr. caas-5BL. 1*	6.89	28.2
		QYr. caas-4BL×*QYr. caas-7DS*	11.50	30.5
		QYr. caas-5BL. 1×*QYr. caas-7DS*	17.43	32.9
	Sichuan 2008	*QYr. caas-4BL*×*QYr. caas-7DS*	5.85	10.9
	Average in five environments	*QYr. caas-4BL*×*QYr. caas-5BL. 1*	4.69	18.1
		QYr. caas-4BL×*QYr. caas-7DS*	6.13	11.8
		QYr. caas-5BL. 1×*QYr. caas-7DS*	6.07	19.5

a. R^2 is the percentage of phenotypic variance explained by the QTLs.

Discussion

During the past decades, Italian wheat cultivars contributed greatly to Chinese wheat improvement and production (Zheng, 1993). Libellula and Strampelli have been planted in China for over 30 years and showed effective APR to stripe rust. In the present study, we detected five QTLs for stripe rust resistance in Libellula and three in Strampelli. These results are in agreement with previous reports on the inheritance of APR, indicating that a few additive genes often conferred APR to stripe rust (Bariana et al., 2001; Boukhatem et al., 2002; Suenaga et al., 2003; Mallard et al., 2005; Rosewarne et al., 2008). Using conventional quantitative genetic analysis, Zhang et al. (2001) inferred that 2 to 3 stripe rust resistance genes were present in Libellula based on the frequency distribution of the area under the disease progress curve (AUDPC) of F_2 and F_3 lines in the cross of Libellula/Mingxian 169. Yin et al. (2005, 2006) concluded that at least two genes were involved in resistance to stripe rust in the same two cultivars, but their study was based on F_2 plant phenotypes. The greater number of resistance genes found in our study illustrates the advantage of using later generation materials and the greater power of QTL analysis in resolving individual gene (or gene region) effects.

In this study, three QTLs were common to both populations. This was likely because both resistant cultivars had a common ancestor. Libellula has the pedigree Tevere/Giuliari//San Pastore, and Strampelli came from Libero//San Pastore-14/Jacometti-49 (http://genbank.vurv.cz/wheat/pedigree). The common parental cultivar San Pastore was derived from Villa Glori/Balilla. Villa Glori was one of four cultivars selected from the cross Riete/Wilhelmina Tare//Akakomugi made by the Italian breeder Nazareno Strampelli in 1913. The other three cultivars were Ardito, Mentana and Damiano Chiesa (Borojevic and Borojevic, 2005). Using diagnostic STS marker *csLV34*, Kolmer et al. (2008) traced the origin of the *Lr34*/*Yr18* rust resistance region in many current wheat cultivars to the Italian wheat cultivars Ardito and Mentana. In addition, DNA sequencing revealed the presence of *Lr34*/*Yr18* gene in Libellula and Strampelli but not in Huixianhong, indicating that *QYr. caas-7DS* identified in Libellula and Strampelli is most likely *Yr18*. The *Lr34*/*Yr18* gene is an important slow rusting gene and can confer high levels of resistance when combined with other minor genes (Singh and Rajaram, 1994; Navabi et al., 2004). Cultivars possessing *Yr18* have been widely used in CIMMYT germplasm (Singh et al., 2005). Chinese landraces may have a relatively high frequency of *Yr18* (Kolmer et al., 2008). Yang et al. (2008) screened 422 Chinese landraces with the marker *csLV34* and found 85.1% of them contained the specific allele for *Yr18*. Field test of the landraces indicated that most of the genotypes with the specific allele for *Yr18* showed moderate to high resistance to stripe rust (data not shown). Therefore, both Italian wheat cultivars and Chinese landraces can be important wheat germplasm with durable resistance gene *Yr18*.

Mallard et al. (2005) identified a QTL, *QYr. inra-5BL. 1*, on chromosome 5BL in the French cultivar Camp Rémy within the marker interval *Xgwm499-Xgwm639*. This QTL explained 18-26% of the phenotypic variance for AUDPC. It is approximately 3 cM from the *QYr. caas-5BL. 1* found in the present study based on the wheat consensus map (Somers et al. 2004). Suenaga et al. (2003) also reported a QTL for stripe rust severity on 5BL in the cultivar Oligoculm near marker locus *Xwmc415*, which falls within the interval carrying *QYr. caas-5BL. 1* in our study.

We identified a second QTL, *QYr. caas-5BL. 2*, on the telomeric region of chromosome 5BL. This gene was more than 40cM from *QYr. caas-5BL. 1* based on the wheat consensus map (Somers et al., 2004). Mallard et al. (2005) reported *QYr. inra-5BL. 2* flanked by *Xgwm234* and *DuPw115a* in this region in the cross Camp Rémy/Récital. Because these two markers were consistently mapped on 5BS in several populations, those authors proposed that *QYr. inra-5BL. 2* might be in a translocated region from chromosome 5BS of cultivar Cappelle-Desprez. In our study the markers flanking the *QYr. caas-5BL. 2* were located on 5BL based on several consensus wheat maps (http: //www. shigen. nig. ac. jp/wheat/komugi/maps/markerMap. jsp; Somers et al., 2004), indicating no translocation happened in this chromosomal region in our population.

QYr. caas-4BL identified in both crosses examined in this study was in the marker interval *Xgwm165-Xgwm149*. William et al. (2006) identified a QTL on 4BL near marker *Xgwm495* in the cross Avocet S/Pavon 76. It was derived from Avocet S and reduced the stripe rust response by 7.4-12.7% over three years. This QTL coincided with the position of *QYr. caas-4BL* in Libellula and Strampelli. Suenaga et al. (2003) also reported a QTL for stripe rust severity in the cultivar Oligoculm. The LOD peak for this QTL, near *Xgwm538*, and was more than 15cM away from *QYr. caas-4BL* identified in this study based on the Somers et al. (2004) consensus map.

Bariana et al. (2001) identified a QTL flanked by the loci *Xwmc111* and *Xwmc25* on chromosome 2DS for disease severity from the cultivar Katepwa. This QTL was also detected by Suenaga et al. (2003) in a Fukuho-Komugi/Oligoculm population, and was possibly mapped as *QYr. caas-2DS* in the present study. Mallard et al. (2005) identified *QYr. inra-2DS* from cultivar Camp Rémy on chromosome 2DS. However, according to the Somers et al. (2004) map, the distance between the peaks for this QTL and *QYr. caas-2DS* is more than 30 cM. Therefore, the QTL of Camp Rémy is likely to be different from *QYr. caas-2DS*.

It has been well known that disease resistance genes in plant genomes frequently occur in clusters on particular chromosomes (McIntosh et al., 2003; Islam et al., 1989). For example, the resistance gene *Yrns-B1* was found to be at a similar position as the adult plant resistance genes *Yr30* against stripe rust (Singh et al., 2000b) and *Sr2* against stem rust (Bariana et al., 1998; Spielmeyer et al., 2003). Similarly, the leaf rust resistance gene *Lr27* (Nelson et al., 1997), a QTL for Fusarium head blight resistance (Zhou et al., 2003b) and a QTL for leaf rust resistance (Börner et al., 2002) were also mapped on the same location. Therefore, although a few QTLs identified in this study were mapped on the similar chromosome regions to those of QTLs reported previously, the allelism among them still needs to be investigated. The similar chromosome locations of these QTLs indicate that they are either at one locus or closely linked loci.

Seedling resistance in Libellula was reported by Li et al. (2007) in the cross Libellula/Mingxian 169 following inoculation with certain *Pst* races. Based on the infection types of F_1 and the segregation ratios of F_2 and BC_1 populations, they found one recessive gene in Libellula conferring resistance to races CYR22 and CYR25, and two genes giving resistance to races

CYR30 and Su4. Thus Libellula also has race-specific seedling resistance genes in addition to APR genes. The combination of APR and seedling resistance in a same genotype is not uncommon. Cultivars Stephens and Druchamp have both high-temperature APR and race-specific seedling resistances (Chen and Line, 1995). Camp Rémy contained a major seedling resistance factor, *QYr. inra-2BL* (probably *Yr7*), together with five other QTLs responsible for APR (Mallard et al., 2005). Cultivar Express possessed two seedling resistance genes *YrExp1* and *YrExp2* as well as three QTLs effective at the adult plant stage (Lin and Chen, 2009). In order to avoid seedling resistance genes in the identification of APR, we used a race (CYR32) of *Pst* that was virulent on seedlings of both Libellula and Strampelli. CYR32 is a current predominant race in China. Clearly, the APR genes identified in Libellula and Strampelli could be a valuable resource to control the disease, but may not be unique.

APR to stripe rust are conferred by the combined effects of several resistance loci (Singh and Rajaram, 1994; Navabi et al., 2004; Singh et al., 2005), and the effects of some resistance gene may not be stable across different environments (Singh et al., 2000b; Suenaga et al., 2003; Rosewarne et al., 2008; Lin and Chen, 2009). In our study, the effect of *QYr. caas-7DS* was very stable, suggesting that QTLs with major effect are most likely to be detected across environments (Tables 2 and 3). Boukhatem et al. (2002) reported that QTLs that contributed less than 10% of the phenotypic variance were difficult to detect across years and environments. However, in the present study both *QYr. caas-4BL* and *QYr. caas-5BL. 1* were detected across different environments, in spite of the fact that they simply explained around 5% of phenotypic variance. Therefore, molecular markers *Xwmc165* and *Xwmc415* closely linked to *QYr. caas-4BL* and *QYr. caas-5BL. 1*, are likely to be useful for marker-assisted selection in wheat breeding programs.

Additive effects among APR genes have been reported in many studies (Singh and Rajaram, 1994; Singh et al., 2000a, b; Suenaga et al., 2003; Navabi et al., 2004; Singh et al., 2005). However, epistasis among them was less stressed. In the present study, several significant epistasis interactions among pairs of APR epistatic QTLs were detected in both the Libellula/Huixianhong and Strampelli/Huixianhong populations, with relatively large effects across different environments (Table 4), this indicates that epistatic interactions among APR genes contribute to the overall resistance of these wheat lines to stripe rust. This is in agreement with a previous report indicating that epistasis often occurred among genes affecting complex traits (Carlborg and Haley, 2004).

Previously, Bulk Segregant Analysis (BSA) strategy used widely for characterizing loci with qualitative effects (Bassam et al., 1991; Li et al., 2006), also be used to characterize and identify markers for stripe rust by loci with a continuous pattern of distribution (Lin et al., 2007, 2009). In wheat, which full linkage map construction is both time consuming and resource intensive, BSA compared with full linkage mapping is a desirable alternative. Although there might be additional minor QTLs not detected in our BSA method, we made the bulks with 5 lines in order to reducing this possibility and detected 3 and 5 QTLs for stripe rust in our two populations, respectively. This was in agreement with Singh et al. (2000a), who reported that the adult plant resistance cultivar often carries 4 to 5 slow rusting genes with small to intermediate effects. Furthermore, the practice told us that the more lines were composed for bulks, the less polymorphisms were detected, 4 to 5 lines for each bulk usually will be fine.

Acknowledgments

The authors are very grateful to Prof. R. A. McIntosh, Plant Breeding Institute, University of Sydney for the critical review of this manuscript. This study was supported by the National Science Foundation of China (30671294 and 30810214).

References

Bariana HS, Kailasapillai S, Brown GN, Sharp PJ (1998) Marker assisted identification of *Sr2* in the National Cereal Rust Control Program in Australia. In: Slinkard AE (ed) Proceedings of 9th international wheat and genetic symposium, University of Extension Press, University of Saskatchewan, Saskatoon, 3: 38-91.

Bariana HS, Hayden MJ, Ahmed NU, Bell JA, Sharp PJ, McIntosh RA (2001) Mapping of durable adult plant and seedling resistances to stripe rust and stem rust diseases in wheat. Aust J Agric Res 52: 1247-1255.

Bassam BJ, Caetano-Anollés G, Gresshoff PM (1991) Fast and sensitive silver staining of DNA in polyacrylamide gels. Anal Biochem 196: 80-83.

Börner A, Schumann E, Fü rste A, Cöster H, Leithold B, Röder MS, Weber WE (2002) Mapping of quantitative trait loci determining agronomic important characters in hexaploid wheat (*Triticum aestivum* L.). Theor Appl Genet 105: 921-936.

Borojevic K, Borojevic K (2005) Historic role of the wheat variety Akakomugi in southern and central European wheat breeding programs. Breeding Sci 55: 253-256.

Boukhatem N, Baret PV, Mingeot D, Jacquemin JM (2002) Quantitative trait loci for resistance against yellow rust in two wheat-derived recombinant inbred line populations. Theor Appl Genet 104: 111-118.

Carlborgö, Haley C (2004) Epistasis: too often neglected in complex trait studies? Nat Rev Genet 5: 618-625.

Chen XM, Line RF (1995) Gene action in wheat cultivars for durable, high-temperature, adult-plant resistance and interaction with race-specific, seedling resistance to *Puccinia striiformis*. Phytopathology 85: 567-572.

Chen XM (2005) Epidemiology and control of stripe rust (*Puccinia striiformis* f. sp. *tritici*) on wheat. Can J Plant Pathol 27: 314-337.

Fu DL, Uauy C, Distelfeld A, Blechl A, Epstein L, Chen XM, Sela H, Fahima T, Dubcovsky J (2009) A kinase-start gene confers temperature-dependent resistance to wheat stripe rust. Science 323: 1357-1360.

Guyomarc'h H, Sourdille P, Charmet G, Edwards KJ, Bernard M (2002) Characterisation of polymorphic microsatellite markers from *Aegilops tauschii* and transferability to the D-genome of bread wheat. Theor Appl Genet 104: 1164-1172.

Islam MR, Shepherd KW, Mayo GME (1989) Recombination among genes at the L group in flax conferring resistant to rust. Theor Appl Genet 77: 540-546.

Kolmer JA, Singh RP, Garvin DF, Viccars L, William HM, Huerta-Espino J, Ogbonnaya FC, Raman H, Orford S, Bariana HS, Lagudah ES (2008) Analysis of the *Lr34/Yr18* rust resistance region in wheat germplasm. Crop Sci 48: 1841-1852.

Kosambi DD (1944) The estimation of map distance from recombination values. Annu Eugen 12: 172-175.

Krattinger SG, Lagudah ES, Spielmeyer W, Singh RP, Huerta-Espino J, McFadden H, Bossolini E, Selter LL, Keller B (2009) A putative ABC transporter confers durable resistance to multiple fungal pathogens in wheat. Science 323: 1360-1363.

Lagudah ES, McFadden H, Singh RP, Huerta-Espino J, Bariana HS, Spielmeyer W (2006) Molecular genetic characterization of the *Lr34/Yr18* slow rusting resistance gene region in wheat. Theor Appl Genet 114: 21-30.

Li HH, Li Z, Wang JK (2008) Inclusive composite interval mapping (ICIM) for digenic epistasis of quantitative traits in biparental population. Theor Appl Genet 116: 243-260.

Li HH, Ye GY, Wang JK (2007) A modified algorithm for the improvement of composite interval mapping. Genetics 175: 361-374.

Li GQ , Li ZF, Yang WY, Zhang Y , He ZH, Xu SC , Singh RP , Qu YY, and Xia XC (2006) Molecular mapping of stripe rust resistance gene *YrCH42* in Chinese wheat cultivar Chuanmai 42 and its allelism with *Yr24* and *Yr26*. Theor Appl Genet 112: 1434-1440.

Li Q, Jing JX, Wang BT, Zhou XC, Du JY (2007) Genetic analysis of resistance to stripe rust in durable resistance wheat variety Libellula. Acta Phytophylacica Sinica 34: 432-433.

Liang SS, Suenaga K, He ZH, Wang ZL, Liu HY, Wang DS, Singh RP, Sourdille P, Xia XC (2006) Quantitative trait loci mapping for adult-plant resistance to powdery mildew in bread wheat. Phytopathology 96: 784-789.

Lin F, Chen XM (2007) Genetics and molecular mapping of genes for race-specific all-stage resistance and non-race-specific high-temperature adult-plant resistance to stripe rust in spring wheat cultivar Alpowa. Theor Appl Genet 114: 277-287.

Lin F, Chen XM (2009) Quantitative trait loci for non-race-specific, high-temperature adult-plant resistance to stripe rust in wheat cultivar Express. Theor Appl Genet118: 631-642.

Line RF (2002) Stripe rust of wheat and barley in North

America: a retrospective historical review. Annu Rev Phytopathol 40: 75-118.

Mallard S, Gaudet D, Aldeia, Abelard C, Besnard AL, Sourdille P, Dedryver F (2005) Genetic analysis of durable resistance to yellow rust in bread wheat. Theor Appl Genet 110: 1401-1409.

Manly KF, Cudmore RH, Jr, Meer JM (2001) Map Manager QTX, cross-platform software for genetic mapping. Genome12: 930-932.

McIntosh RA, Yamazaki Y, Devos KM, Dubcovsky J, Rogers J, Appels R (2003) Catalogue of gene symbols for wheat http: //www. grs. nig. ac. jp/wheat/komugi/ genes.

Navabi A, Singh RP, Tewari JP, Briggs KG (2004) Inheritance of high levels of adult-plant resistance to stripe rust in five spring wheat genotypes. Crop Sci 44: 1156-1162.

Nelson JC, Singh RP, Autrique JE, Sorrells ME (1997) Mapping genes conferring and suspecting leaf rust resistance in wheat. Crop Sci 37: 1928-1935.

Peterson RF, Campbell AB, Hannah, AE (1948) A diagrammatic scale for estimating rust intensity on leaves and stems of cereals. Can J Res Sect C 26: 496-500.

Ramburan VP, Pretorius ZA, Louw JH, Boyd LA, Smith PH, Boshoff WHP, Prins R (2004) A genetic analysis of adult plant resistance to stripe rust in the wheat cultivar Kariega. Theor Appl Genet 108: 1426-1433.

Rosewarne GM, Singh RP, Huerta-Espino J, Rebetzke GJ (2008) Quantitative trait loci for slow-rusting resistance in wheat to leaf rust and stripe rust identified with multi-environment analysis. Theor Appl Genet 116: 1027-1034.

Röder MS, Korzun V, Wendehake K, Plaschke J, Tixier M-H, Leroy P, Ganal MW (1998) A microsatellite map of wheat. Genetics 149: 2007-2023.

Santra DK, Chen XM, Santra M, Campbell KG, Kidwell KK (2008) Identification and mapping QTL for high-temperature adult-plant resistance to stripe rust in winter wheat (*Triticum aestivum* L.) cultivar Stephens. Theor Appl Genet 117: 793-802.

Sharp PJ, Kreis M, Shewry PR, Gale MD (1988) Location of β-amylase sequence in wheat and its relatives. Theor Appl Genet 75: 286-290.

Singh RP, Rajaram S (1994) Genetics of adult plant resistance to stripe rust in ten spring bread wheats. Euphytica 72: 1-7.

Singh RP, Huerta-Espino J, Rajaram S (2000a) Achieving near-immunity to leaf and stripe rusts in wheat by combining slow rusting resistance genes. Acta Phytopathol Entomological Hungarica 35: 133-139.

Singh RP, Nelson JC, Sorrells ME (2000b) Mapping *Yr28* and other genes for resistance to stripe rust in wheat. Crop Sci 40: 1148-1155.

Singh RP, Huerta-Espino J, William HM (2005) Genetics and breeding for durable resistance to leaf and stripe rusts in wheat. Turk J Agric For 29: 121-127.

Somers DJ, Isaac P, Edwards K (2004) A high-density microsatellite consensus map for bread wheat (*Triticum aestivum* L.). Theor Appl Genet 109: 1105-1114.

Spielmeyer W, Sharp PJ, Lagudah ES (2003) Identification and validation of markers linked to broad-spectrum stem rust resistance gene *Sr2* in wheat (*Triticum aestivum* L.) Crop Sci 43: 333-336.

Spielmeyer W, Singh RP, McFadden H, Wellings CR, Huerta-Espino J, Kong X, Appels R, Lagudah ES (2008) Fine scale genetic and physical mapping using interstitial deletion mutants of *Lr34/Yr18*: A disease resistance locus effective against multiple pathogens in wheat. Theor Appl Genet 116: 481-490.

Stubbs RW (1985) Stripe rust. In: Roelfs AP, Bushnell WR (Eds) The cereal rusts Ⅱ. Academic Press, Orlando, Florida, USA, pp 61-101.

Suenaga K, Singh RP, Huerta-Espino J, William HM (2003) Microsatellite markers for genes *Lr34/Yr18* and other quantitative trait loci for leaf rust and stripe rust resistance in bread wheat. Phytopathology 93: 881-890.

Uauy C, Brevis JC, Chen XM, Khan I, Jackson L, Chicaiza O, Distelfeld A, Fahima T, Dubcovsky J (2005) High-temperature adult-plant (HTAP) stripe rust resistance gene *Yr36* from *Triticum turgidum* ssp. *dicoccoides* is closely linked to the grain protein content locus *Gpc-B1*. Theor Appl Genet 112: 97-105.

Wan AM, Zhao ZH, Chen XM, He ZH, Jin SL, Jia QZ, Yao G, Yang JX, Wang BT, Li GB , Bi YQ, Yuan ZY (2004) Wheat stripe rust epidemic and virulence of *Puccinia striiformis* f. sp. *tritici*. Plant Dis 88: 896-904.

Wang S, Basten CJ, Zeng ZB (2005) Windows QTL Cartographer v2. 5. Statistical Genetics. North Carolina State University.

William HM, Singh RP, Huerta-Espino J, Ortiz-Islas S, Hoisington D (2003) Molecular marker mapping of leaf rust resistance gene *Lr46* and its association with stripe rust resistance gene *Yr29* in wheat. Phytopathology 93: 153-159.

William HM, Singh RP, Huerta-Espino J, Palacios G,

Suenaga K (2006) Characterization of genetic loci conferring adult plant resistance to leaf rust and stripe rust in spring wheat. Genome 49：977-990.

Yang WX, Yang FP, Liang D, He ZH, Shang XW, Xia XC (2008) Molecular characterization of slow-rusting genes *Lr34/Yr18* in Chinese wheat cultivars. Acta Agronomica Sinica 34：1109-1113.

Yin XG, Zhang YH, Yan QJ, Shang XW (2005) Resistant characteristics to stripe rust and genetic analysis of durable resistance on wheat cultivar Strampelli. J. Plant Genet Res 6：390-393.

Yin XG, Shang XW, Song JR, Zhang YH, Yan QJ (2006) Genetic mechanism of durable resistance to stripe rust of wheat cultivar Libellula. J *Triticeae* Crops 26：147-150.

Young ND (1996) QTL mapping and quantitative disease resistance in plants. Annu Rev Phytopathol 34：479-501.

Zhang ZJ, Yang GH, Li GH, Jin SL, Yang XB (2001) Transgressive segregation, heritability, and number of genes controlling durable resistance to stripe rust in one Chinese and two Italian wheat cultivars. Phytopathology 91：680-686.

Zheng DS (1993) Use of Italian wheat varieties in China. Genet Resour Crop Evol 40：137-142.

Zhou XC, Du JY, Yang JH (2003a) A 30 successive years' observation on the performance of several wheat cultivars in resistance to stripe rust (*Puccinia striiformis* West.) in the southern region of Gansu province of China. Acta Phytopath. Sin. 33：550-554.

Zhou WC, Kolb FL, Bai GH, Domier LL, Boze LK, Smith NJ (2003b) Validation of a major QTL for scab resistance with SSR markers and use of marker assisted selection in wheat. Plant Breed 122：40-46.

Molecular mapping of quantitative trait loci for adult-plant resistance to powdery mildew in Italian wheat cultivar Libellula

M. A. Asad[1], B. Bai[1], C. X. Lan[1], J. Yan[2], X. C. Xia[1], Y. Zhang[1], and Z. H. He[2,3]

[1] *Institute of Crop Science, National Wheat Improvement Center/The National Key Facility for Crop Gene Resources and Genetic Improvement, Chinese Academy of Agricultural Sciences* (*CAAS*), 12 *Zhongguancun South Street, Beijing* 100081, *China*; [2] *Cotton Research Institute, Chinese Academy of Agricultural Sciences* (*CAAS*), *Huanghedadao, Anyang, Henan* 455000, *China*; [3] *International Maize and Wheat Improvement Center* (*CIMMYT*) *China Office, c/o CAAS*, 12 *Zhongguancun South Street, Beijing* 100081, *China.*

Abstract: Powdery mildew, caused by *Blumeria graminis* f. sp. *tritici* (*Bgt*), is a fungal disease that causes significant yield losses in many wheat growing regions of the world. Previously, five QTLs for adult-plant resistance (APR) to stripe rust resistance were identified in Libellula. The objectives of this study were to map QTLs for APR to powdery mildew in 244 $F_{2:3}$ lines of Libellula/Huixianhong, to analyze the stability of detected QTLs across environments, and to assess the association of these QTLs with stripe rust resistance. Powdery mildew response was evaluated for two years in Beijing and for one year in Anyang. In Beijing artificial inoculations were used, whereas in Anyang there was natural infection. The correlation between averaged MDS (Maximum disease severity) and averaged AUDPC (Area under disease progress curve) over two years in Beijing was 0.98, and heritabilities of MDS and AUDPC were 0.65 and 0.81, respectively, based on the mean values averaged across environments. SSR markers were used to screen the parents and mapping population. Five QTLs were identified by inclusive composite interval mapping, designated as *QPm.caas-2DS*, *QPm.caas-4BL.1*, *QPm.caas-6BL.1*, *QPm.caas-6BL.2* and *QPm.caas-7DS*. Three QTLs *QPm.caas-2DS* and *QPm.caas-6BL.1* and *QPm.caas-6BL.2* seem to be new resistance loci for powdery mildew. QTLs *QPm.caas-2DS* and *QPm.caas-4BL.1* were identified at the same position as previously mapped QTLs for stripe rust resistance in Libellula. *QPm.caas-7DS*, derived from Libellula, coincided with the slow rusting and slow mildewing locus *Lr34*/*Yr18*/*Pm38*. These results and the identified markers could be useful for wheat breeders aiming for durable resistance to both powdery mildew and stripe rust.

Key words: *Triticum aestivum* L., *Blumeria graminis* f. sp. *tritici*, disease resistance, SSR markers, quantitative trait loci.

Published in Crop & Pasture Science, 2012, 63 (6): 539-546

Introduction

Wheat is an important food crop across many countries. Powdery mildew, caused by *Blumeria graminis* f. sp. *tritici*, is an important disease of wheat causing serious yield losses, not only in China but also in other countries with cool maritime weather conditions (Bennett, 1984). The best way to control powdery mildew is disease resistant cultivars because chemical control can be costly and hazardous to human health and environment.

Forty five powdery mildew resistance genes (*Pm1*-*Pm45*) have been designated on various wheat chromosomes. Some of these genes were transferred from relative species such as *Secale cereale* and *Dasypyrum villosum* (Lillemo et al., 2008; Luo et al., 2009; He et al., 2009; Hua et al., 2009; Ma et al., 2011). Resistance conferred by major genes is race specific, and it has not been durable due to frequent changes in the pathogen population (McDonald and Linde, 2002; Skinnes, 2002). Quantitative resistance on the other hand is conferred by minor genes with small to intermediate effects with each gene, usually contributing additively to the overall level of resistance. This type of resistance seems to be race non-specific and therefore its effectiveness is potentially durable. Rather than displaying powdery mildew immunity phenotype, adult plant resistance (APR) genes do not confer adequate resistance, however combinations of 4-5 such genes usually result in "near-immunity" or high level of resistance (Gustafson and Shaner, 1982; Liu et al., 2000). Pyramiding of minor genes is considered a valuable method to control disease (Chen et al., 2002). Gene pyramiding has been successfully applied in several crops such as bacterial blight resistance in rice (Hittalmani et al., 2000) and powdery mildew resistance in wheat (Liu et al., 2000; Wang et al., 2001), and recently QTL pyramiding for powdery mildew resistance in wheat (Bin Bai, pers. comm. in our own laboratory).

With the development of molecular markers, it became much easier to identify and pyramid resistance genes (Liu et al., 2000). Molecular markers were used to map genes for APR to powdery mildew in various wheat genotypes including Forno (Keller et al., 1999), RE714 (Chantret et al., 2001), Festin (Mingeot et al., 2002), USG3209 (Tucker et al., 2007), and Lumai 21 (Lan et al., 2010).

Libellula, an Italian wheat introduced into China in 1973 (Zheng, 1993) has been performing well in the field against stripe rust for the last four decades, particularly in Gansu province, a hotspot for stripe rust. Libellula also showed good resistance to powdery mildew in our field experiments. Kolmer et al. (2008) identified the *Lr34/Yr18* locus, known to confer durable leaf rust and stripe rust resistances, in wheat cultivars of Italian origin using the diagnostic marker *csLV34* (Lagudah et al., 2006) and Lu et al. (2009) reported the same locus for APR to stripe rust in Libellula. The purpose of the present work was to determine if Libellula possesses other valuable loci contributing to durable powdery mildew resistance.

Materials and methods

Plant materials

A mapping population of 244 $F_{2:3}$ lines, derived from cross Libellula/Huixianhong were investigated for QTL identification. Libellula is an Itallian introduction to Northwestern China made in 1973 (Hu and Wang, 1989). Libellula has durable resistance to stripe rust (Zhang, 1995), and also showed good APR to powdery mildew in our field. Huixianhong, a landrace of Henan, is susceptible to powdery mildew both at seedling and adult-plant stages. The F_3 lines, developed from individual F_2 plants, were planted and harvested as bulks of more than 50 plants of each line, and thereafter maintained as bulk populations.

Field disease assessment

The 244 $F_{2:3}$ lines were grown for assessment of powdery mildew response at the CAAS Experimental Station in Beijing (39°N, 43.5 m above sea level), and the CAAS Cotton Research Institute in Anyang (36°N, 70-80 m above sea level), Henan province, in 2009-2010 and 2010-2011 cropping seasons (Four environments in this study are mentioned as Beijing 2010, Beijing 2011, Anyang 2010 and Anyang 2011). The field trials were conducted in randomized complete block design with three replications. Fifty seeds of each line were sown in a circular plot (diameter, 8 cm) as hill plot, and plots were separated by 30 cm apart (Frey, 1965). Irrigation, fertilization and other agronomic practices followed standard production practices. Cv. Jingshuang 16 was planted as a susceptible check every tenth row and surrounding the experiment to ensure adequate inoculum in spring season. In Beijing plots were inoculated with *Blumeria graminis* f. sp. *tritici* (*Bgt*) isolate E20 prior to stem elongation. Disease ratings were first recorded five weeks after inoculation based on percentage of leaf area covered by powdery mildew on penultimate leaves (F-1 leaf), and again one week later when the disease had reached its maximum level (around May 20). In Anyang, powdery mildew severities resulting from natural infection were evaluated only when infection levels on Jingshuang 16 check had reached its maximum level around the third week of May. Due to dry weather conditions disease levels reached a maximum of only 5% on the check at Anyang in 2011, and field data from that trial could not be included in the analysis.

Statistical analysis

Maximum disease severities (MDS), phenotypic correlation coefficients (r) and a frequency distribution of powdery mildew response based on MDS in different environments were calculated in Microsoft Excel 2007. The area under disease progress curve (AUDPC) was calculated according to Bjarko and Line (1988). The MDS and AUDPC were used in subsequent analysis of variance (ANOVA) and quantitative trait loci (QTL) analysis. ANOVA was conducted by PROC GLM in the statistical analysis system (SAS Institute, 1997). Broad-sense heritability (h^2) of powdery mildew resistance was calculated as the ratio of genotypic variance to phenotypic variance: $h^2=\sigma_g^2/(\sigma_g^2+\sigma_{ge}^2/e+\sigma_\varepsilon^2/re)$, where σ_g^2, σ_{ge}^2 and σ_ε^2 are estimates of genotypic, genotype × environment interaction and error variances, respectively, and e and r are the numbers of environments and replications per environment, respectively (Allard, 1960).

Molecular marker analysis

Genomic DNA was extracted from leaf tissues of the parents and $F_{2:3}$ lines (30-40 plants per line as a bulk) following the procedure described in Sharp et al. (1988). A total of 1, 528 pairs of simple sequence repeats (SSR) primers consisting of 240 pairs of GWM (Röder et al., 1998), 382 pairs of BARC (Song et al., 2002), 640 pairs of WMC (IDnagenetics, Norwich, UK), 144 pairs of CFD (Guyomarc'h et al., 2002), 64 pairs of GDM (Pestsova et al., 2000), 58 pairs of CFA (Sourdille et al., 2004), and one STS (Sequence Tagged Site) marker *csLV34* (Lagudah et al., 2006) were screened for polymorphism between the parents. PCR protocols, electrophoresis of PCR products and gel staining were conducted as described by Bryan et al. (1997) and Bassam et al. (1991). For bulked segregant analysis, equal amounts of DNA from the five highly resistant and five highly susceptible lines, based on the average disease severities, were used to constitute resistant and susceptible bulks (Michelmore et al., 1991). SSR markers that showed polymorphism between the parents and between the resistant and susceptible bulks were then used to genotype the whole population. Previously reported SSR markers were also used to enrich the chromosome region linked to resistance genes (Somers et al., (2004); http://www.shigen.nig.ac.jp/wheat; http://wheat.pw.usda.gov). Data of polymorphic markers were used for mapping linkage and QTL analysis.

Quantitative trait loci (QTL) analysis

Linkage groups were established with the software Map Manager QTXb20 (Manly et al., 2001). Genetic distances between markers were determined from recombination values using the Kosambi mapping function (Kosambi, 1944). Chromosomal alignments to linkage groups were deduced from published wheat consensus maps (http://www.shigen.nig.ac.jp/wheat; http: //wheat. pw. usda. gov; Somers et al., 2004). QTL analysis was performed with the software IciMapping v3.1 (Wang et al., 2011) by inclusive composite interval mapping (ICIM). A logarithm of odds (LOD) of 2.0 was set to declare QTL as significant. The percentages of phenotypic variance explained (R^2) by individual QTL and additive effects at LOD peaks were obtained through stepwise regression (Wang et al., 2011). Adjacent QTLs on the same chromosome were considered different when the distances between the curve peaks were greater than 20cM. QTLs overlapping within 20cM intervals on the same chromosome across different environments were considered to be the same.

Results

Phenotypic analysis of powdery mildew response

Over the three environments (Anyang 2010, Beijing 2010 and Beijing 2011), Libellula expressed maximum disease severities (MDS) of less than 5%, whereas Huixianhong displayed 15 to 27%. The MDS scores were significantly correlated between environments (r = 0.60 to 0.74). The powdery mildew severities of the $F_{2:3}$ lines showed a unimodal segregation suggesting the presence of quantitative resistance genes (Fig. 1). In Beijing, the average AUDPC over two years was 90.2, ranging from 20.0 to 234.7. The MDS and AUDPC in Beijing over two years were significantly correlated with each other for disease scoring (r = 0.98, P < 0.000 1). Analyses of variance of MDS and AUDPC confirmed significant variation (P < 0.000 1) among $F_{2:3}$ lines (Table 1). The broad-sense heritabilities of MDS and AUDPC across environments were 0.65 and 0.81, respectively.

Table 1 Analyses of variance of maximum disease severities (MDS) on penultimate leaves and area under the disease progress curve (AUDPC) values for powdery mildew responses on F_3 lines from Libellula /Huixianhong

Parameter	Source of variation	*df*	Sum of squares	Mean of squares	*F* values
MDS	Replicate	2	3 604	1 802	38.80*
	Environment	2	62 564	31 282	673.50*
	Line	243	117 959	485	10.45*
	Line×Environment	486	48 906	100	2.17*
	Error	1 423	66 094	46	
AUDPC	Replicate	2	88 779	44 389	36.79*
	Environment	1	461 269	461 269	382.33*
	Line	243	2 210 369	9 096	7.54*
	Line×Environment	243	418 178	1 720	1.43*
	Error	974	1 175 089	1 206	

* indicates P < 0.000 1.

QTL analysis for adult-plant resistance (APR) to powdery mildew

One hundred and thirty three of 1, 528 SSR markers were polymorphic for the parents and bulks of resistant and susceptible lines. These markers were used to genotype the mapping population. Five QTLs were consistently detected by Icimapping v3.1, using mean MDS in each environment, and the average AUDPC over two years in Beijing (Table 2; Fig. 2). QTLs designated as *QPm. caas-2DS*, *QPm. caas-4BL. 1*, and *QPm. caas-7DS* were contributed by Libellula whereas *QPm. caas-6BL. 1* and *QPm. caas-6BL. 2* from Huixianhong. The *QPm. caas-2DS* detected in marker interval *Xcfd51-Xcfd56*, explained 2.3, 3.4, 2.4, 2.3, and 3.5% of the phenotypic variance in Anyang 2010, Beijing 2010, Beijing 2011, averaged MDS over three environments, and averaged AUDPC over two years in Beijing, respectively. The *QPm. caas-4BL. 1* flanked by markers *Xgwm149* and *Xgwm495*, ex-

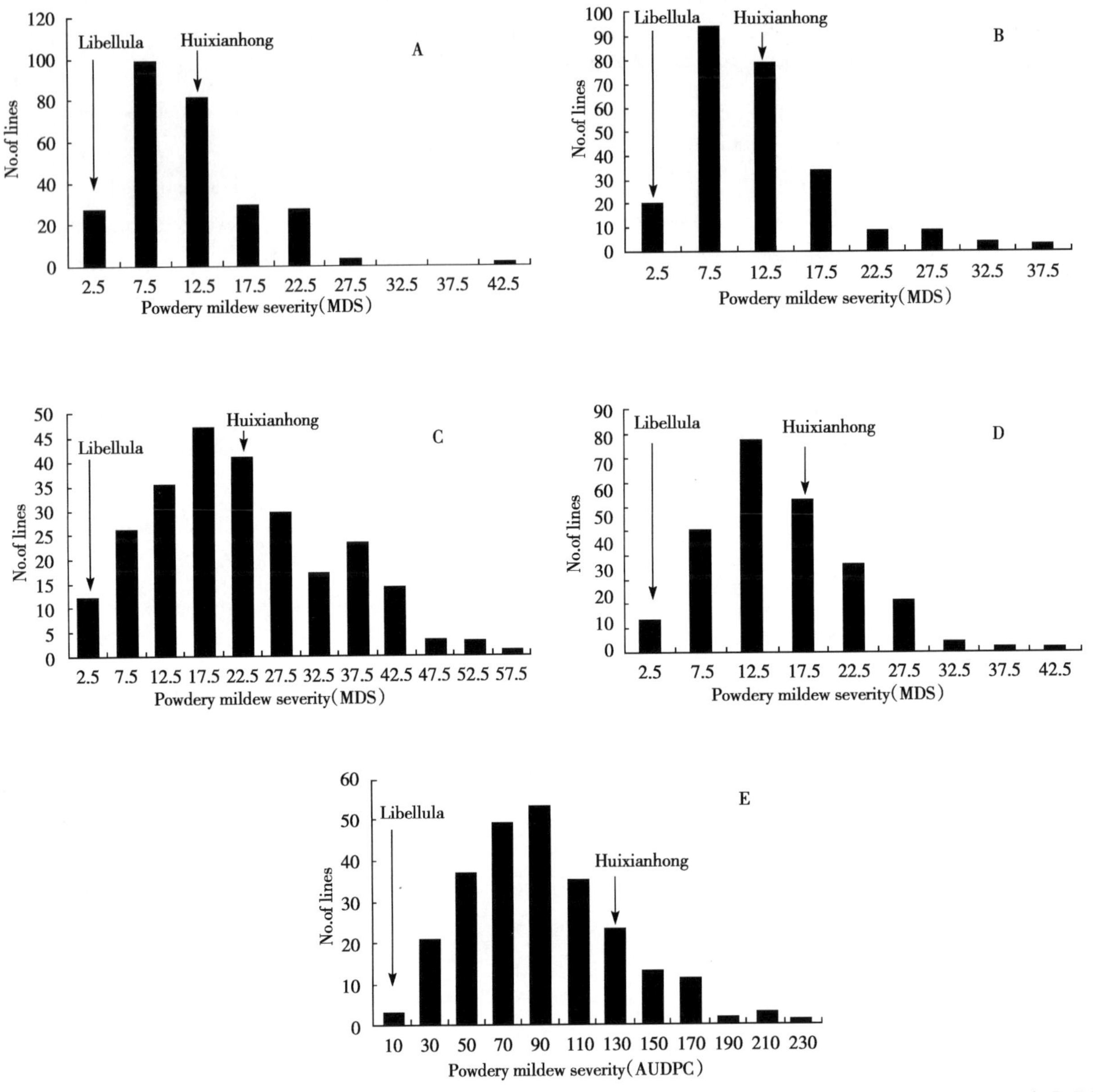

Fig. 1 Frequency distributions of maximum disease severities (MDS) and area under the disease progress curve (AUDPC) values of powdery mildew in F_3 population of Libellula/Huixianhong. A, Anyang 2010; B, Beijing 2010; C, Beijing 2011; D, Averaged MDS across all three environments; E, Averaged AUDPC in Beijing over 2 years. Mean values for the parents are indicated by arrows.

plained phenotypic variances of 12.1, 12.4, 9.1, 14.7, and 15.4%, with additive effects of −2.88, −3.52, −5.47, −3.85 and −22.45 in Anyang 2010, Beijing 2010, Beijing 2011, averaged MDS, and averaged AUDPC, respectively. Phenotypic variances ranged from 2.5 to 5.5% and 0.5 to 9.3% were explained by *QPm. caas-6BL. 1* and *QPm. caas-6BL. 2* in the marker interval *Xgwm 219-Xbarc24*, and *Xbarc24-Xbarc354* in Anyang 2010, Beijing 2010, Beijing 2011, averaged MDS and averaged AUDPC, respectively. These two QTLs expressed additive effects ranged from 0.55 to 10.66 and 0.51 to 3.29 across environments. As distance between the peaks of these two QTLs *QPm. caas-6BL. 1* and *QPm. caas-6BL. 2* was more than 20cM, considered them two different QTLs. The fifth QTL, *QPm. caas-7DS* was mapped between the SSR markers *XcsLV34* and *Xgwm 295*. The phenotypic variances and additive effects ranged from 7.6 to 18.5% and from −2.35 to −22.22, respectively, across environments.

Table 2 Quantitative trait loci (QTL) for adult-plant resistance (APR) to powdery mildew in $F_{2:3}$ lines derived from cross Libellula/Huixianhong

Location and year	QTL[a]	Marker interval	LOD[b]	AE[c]	R^2 (%)[d]	TotalR^2 (%)
MDS[e]						
Anyang 2010	*QPm. caas-2DS*	*Xcfd51-Xcfd56*	1.9	−0.47	2.3	26.9
	QPm. caas-4BL. 1	*Xgwm149-Xgwm495*	7.8	−2.88	12.1	
	QPm. caas-6BL. 1	*Xgwm219-Xbarc24*	3.0	0.55	2.5	
	QPm. caas-6BL. 2	*Xbarc24-Xbarc354*	1.7	0.52	0.5	
	QPm. caas-7DS	*XcsLV34-Xgwm295*	6.4	−2.35	9.5	
Beijing 2010	*QPm. caas-2DS*	*Xcfd51-Xcfd56*	2.6	−1.63	3.4	33.1
	QPm. caas-4BL. 1	*Xgwm149-Xgwm495*	6.7	−3.52	12.4	
	QPm. caas-6BL. 1	*Xgwm219-Xbarc24*	3.0	1.20	2.6	
	QPm. caas-6BL. 2	*Xbarc24-Xbarc354*	2.2	1.48	0.9	
	QPm. caas-7DS	*XcsLV34-Xgwm295*	8.1	−3.41	13.8	
Beijing 2011	*QPm. caas-2DS*	*Xcfd51-Xcfd56*	1.9	−2.38	2.4	28.5
	QPm. caas-4BL. 1	*Xgwm149-Xgwm495*	4.8	−5.47	9.1	
	QPm. caas-6BL. 1	*Xgwm219-Xbarc24*	3.8	3.24	5.2	
	QPm. caas-6BL. 2	*Xbarc24-Xbarc354*	2.1	1.52	1.9	
	QPm. caas-7DS	*XcsLV34-Xgwm295*	4.5	−4.28	7.6	
Averaged MDS in three environments	*QPm. caas-2DS*	*Xcfd51-Xcfd56*	2.1	−1.04	2.3	33.3
	QPm. caas-4BL. 1	*Xgwm149-Xgwm495*	9.2	−3.85	14.7	
	QPm. caas-6BL. 1	*Xgwm219-Xbarc24*	4.2	1.70	4.7	
	QPm. caas-6BL. 2	*Xbarc24-Xbarc354*	2.5	1.81	1.3	
	QPm. caas-7DS	*XcsLV34-Xgwm295*	7.2	−2.83	10.3	
AUDPC[f]						
Averaged AUDPC in Beijing	*QPm. caas-2DS*	*Xcfd51-Xcfd56*	2.6	−7.24	3.5	52.2
	QPm. caas-4BL. 1	*Xgwm149-Xgwm495*	8.4	−22.45	15.4	
	QPm. caas-6BL. 1	*Xgwm219-Xbarc24*	4.5	10.66	5.5	
	QPm. caas-6BL. 2	*Xbarc24-Xbarc354*	3.2	3.29	9.3	
	QPm. caas-7DS	*XcsLV34-Xgwm295*	10.6	−22.22	18.5	

a. QTL that extend across single one-log support confidence intervals were assigned the same symbol, b. Logarithm of odds (LOD) score, c. Additive effects, d. R^2 is the proportion of phenotypic variance explained by the QTL, e. Maximum disease severity, f. Area under the disease progress curve.

Discussion

A better understanding of quantitative traits enhances genetic improvement of crop plants (Gupta et al., 2010), and molecular markers can be used to enhance our understanding of quantitative resistance in wheat (Priyamvada et al., 2011). In the present study, significant variation among genotypes and high broad sense heritabilities of MDS and AUDPC (0.65 and 0.81) for powdery mildew resistance indicated that great genetic variation for powdery mildew resistance exists in the investigated population. Unimodal distributions of MDS and AUDPC across environments revealed that powdery mildew resistance was quantitatively inherited. Five QTLs for powdery mildew resist-

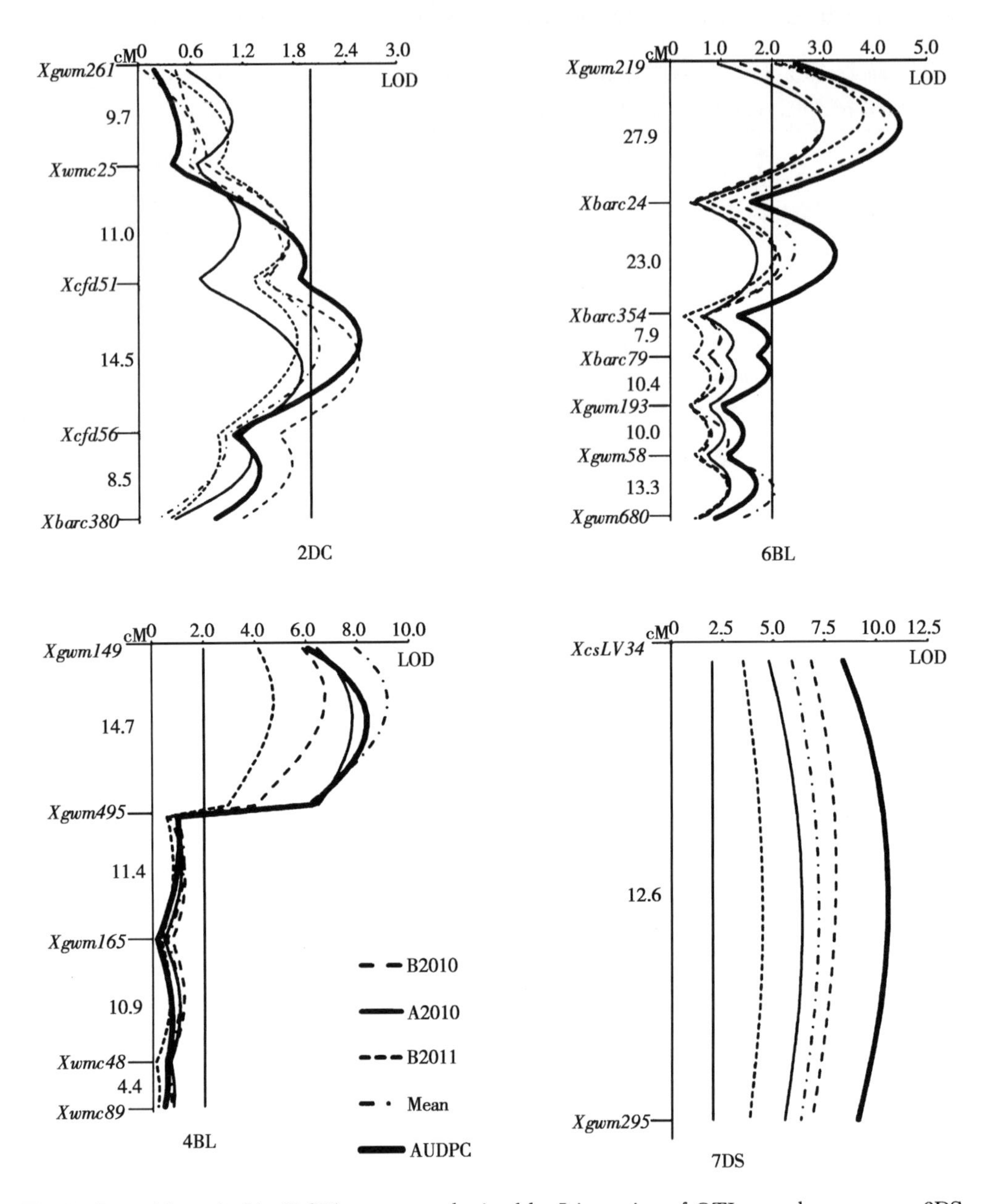

Fig. 2 Logarithm of odds (LOD) contours obtained by Icimapping of QTLs on chromosomes 2DS, 4BL, 6BL, and 7DS determining powdery mildew severities in Libellula/Huixianhong F_3 lines. A2010, maximum disease severities (MDS) in Anyang 2010; B2010 and B2011, maximum disease severities in Beijing 2010 and 2011, respectively; Mean, mean MDS across three environments; AUDPC, average area under the disease progress curve in Beijing over two years. Logarithm of odds (LOD) thresholds = 2.0.

ance were detected on chromosomes 2D, 4B, 6B, and 7D, explaining 33.3% of the phenotypic variation in averaged MDS in a simultaneous fit across three environments.

QPm.caas-2DS was detected in marker interval *Xcfd51-Xcfd56*. Previously, a number of QTLs have been identified on chromosome 2D. Bougot et al. (2006) detected QTLs, *QPm.inra-2D-a* and *QPm.inra-2D-b*, on chromosome 2DS closely linked with *Xgwm102* and *Xcfd2* located about 38 cM and 49cM away from *QPm.caas-2DS*, respectively based on linkage map (Somers et al., 2004). In Chinese wheat cultivar Lumai 21, Lan et al. (2010) mapped a QTL,

QPm. caas-2DL in marker interval *Xwmc18-Xcfd233*, about 70cM from *QPm. caas-2DS*. Börner et al. (2002) identified *QPm. ipk-2D* in the synthetic hexaploid wheat line W7984 in the markers interval *Xglk 558-XksuD23*, about 90cM from *QPm. caas-2DS*. This appears to be new adult-plant resistance (APR) loci for powdery mildew located on short arm of chromosome 2D. Lu et al. (2009) identified QTL *QYr. caas-2DS* for APR to stripe rust in the same genomic region on chromosome 2DS in the same population. Bariana et al., 2001 identified a temperature sensitive gene, *YrCK*, flanked by *Xgdm005* and *Xwmc190* on the short arm of chromosome 2D, and Navabi et al. (2005) confirmed the involvement of *YrCK* in adult plant stripe rust resistance in cv. Cook. The *QPm. caas-2DS* for powdery resistance had lower LOD value than *QYr. caas-2DS* for stripe rust resistance in the same population (Lu et al., 2009), which may be due to low disease pressure or weaker effect to powdery mildew than stripe rust pathogen, but it was constant across all three environments. The similar locations of these two loci indicate that *QPm. caas-2DS* may also confer resistance to stripe rust.

A stable QTL designated *QPm. caas-4BL. 1* was flanked by SSR markers *Xgwm149* and *Xgwm495*, and explained 14.7 and 15.4% of the phenotypic variance across all three environments and AUDPC over two years in Beijing, respectively. Lillemo et al. (2008) detected *QPm. muls-4BL* flanked by DArT (Diversity Arrays Technology) marker *XwPt1505* and SSR marker *Xgwm149*, in the same chromosome region of cv. Avocet in both Beijing and Norway. *QPm. caas-4BL* in the marker interval *Xgwm375-Xgwm251* contributed by cv. Oligoculm with 5.9% of phenotypic variance was identified at the same position on chromosome 4BL (Liang et al., 2006). Borner et al. (2002) identified *QPm. ipk-4B* in the chromosome 4B centromere region (*Xcdo795-Xbcd1262*) in ITMI mapping population Opata 85/W7984. *QPm. sfr-4B* in the marker interval *Xpsr593b-Xpsr1112* from cv. Forno was only identified at one location (Keller et al., 1999). *QPm. caas-4BL. 1* appears to be more consistent across the environments than previously identified QTLs for powdery mildew resistance in wheat. William et al. (2006) and Lu et al. (2009) identified QTLs for stripe rust resistance on chromosome 4BL at the similar position to *QPm. caas-4BL. 1* in the population Avocet S /Pavon 76 and Libellula/Huixianhong, respectively. This locus appears to have potential to be used for multiple disease resistance.

QPm. caas-6BL. 1 and *QPm. caas-6BL. 2* were identified in the marker intervals *Xgwm219-Xbarc24*, and *Xbarc24-Xbarc354* on the long arm of chromosome 6B whereas, Keller et al. (1999) and Lan et al. (2009) identified *QPm. sfr-6BS* and *QPm. caas-6BS* on the short arm of chromosome 6B. Both *QPm. caas-6BL. 1* and *QPm. caas-6BL. 2* explained 2.5-5.2% and 0.5-9.3% of the phenotypic variance, and conferred consistent resistance across environments This locus has not been reported before on long arm of chromosome 6B for adult-plant resistance to powdery mildew.

The most effective powdery mildew resistance QTL, *QPm. caas-7DS*, contributed by Libellula was closely linked to SSR marker *Xgwm295* and STS marker *Xcs-LV34*, and explained 7.6-18.5% of phenotypic variance across environments. In mapping powdery mildew resistance genes in Saar/Avocet RILs population, Lillemo et al. (2008) found a QTL closely linked with SSR markers *Xgwm1220* and *Xswm10* in the same genomic region on chromosome 7DS. The *QPm. caas-7DS* for adult-plant resistance to powdery mildew in Libellula might be the same gene as reported by Liang et al. 2006 and Borner et al. 2002 in cv. Fukuho-Komugi and W7984, respectively.

The QTLs, *QPm. caas-7DS* for powdery mildew resistance and *QYr. caas-7DS* for stripe rust resistance (Lu et al., 2009) mapped in the same region of chromosome 7DS. This locus is well documented for durable resistance to leaf rust (*Lr34*) and stripe rust (*Yr18*) (Lagudah et al., 2009), and is also associated with APR to powdery mildew (*Pm38*) (Spielmeyer et al., 2005; 2008). The QTL, *QPm. caas-7DS*, is probably the *Lr34/Yr18/Pm38* locus which has become prime breeding target for adult-plant race non-specific resistance to multiple pathogens. Using diagnostic STS marker *csLV34*, the *Lr34/Yr18* locus was reported to be originated from Italian wheat cultivars Ardito and Mentana, and *Lr34/Yr18* was common in Chinese landraces (Kolmer et al., 2008). Pedigree of Libellula also showed the presence of cultivars Ardito and Mentana in its parentage (Borojevic and Borojevic, 2005). Based on these relationships it is likely that Libellula possesses *Lr34/Yr18/Pm38*. This locus in Libellula can be worthwhile in improving adult plant resistance to powdery mildew and stripe rust simultaneously.

In this study five QTLs for powdery mildew were identified with SSR markers. The resistance alleles of three QTLs *QPm. caas-2DS*, *QPm. caas-6BL. 1* and *QPm. caas-6BL. 2* identified on chromosomes 2DS and 6BL were contributed by Libellula and Huixianhong, respectively, and they might be new resistance loci for powdery mildew. QTLs *QPm. caas-2DS*, *QPm. caas-4BL. 1* and *QPm. caas-7DS* derived from Libellula, were identified in the same position as previously mapped adult plant resistance loci for stripe rust resistance in the same Libellula /Huixianhong population (Lu et al., 2009). Previously, we have identified 3 to 4 adult plant resistance loci for powdery mildew in cv. Bainong 64 (Lan et al., 2009) and Lumai 21 (Lan et al., 2010) and pyramided them successfully in elite lines with closely linked molecular markers (Bin et al. pers. comm.) to enhance the durable resistance to powdery mildew. These identified quantitative trait loci and linked molecular markers in this study could be useful for wheat breeders aiming for durable resistance to both powdery mildew and stripe rust through marker assisted breeding.

Acknowledgments

We are grateful to the critical review of this manuscript by Prof. R. A. McIntosh, Plant Breeding Institute, University of Sydney. This study was supported by the National Science Foundation of China (30821140351) and China Agriculture Research System.

References

Allard RW (1960) Principles of Plant Breeding. John Wiley and Sons, Inc., New York, London.

Bariana HS, Hayden MJ, Ahmed NU, Bell JA, Sharp PJ, McIntosh RA (2001) Mapping of durable adult plant and seedling resistances to stripe rust and stem rust diseases in wheat. *Australian Journal of Agricultural Research* 52, 1247-1255.

Bassam BJ, Caetano-Anolles G, Gresshoff PM (1991) Fast and sensitive silver staining of DNA in polyacrylamide gels. *Analytical Biochemistry* 196, 80-83.

Bennett F (1984) Resistance to powdery mildew in wheat: a review of its use in agriculture and breeding programmes. *Plant Patholo1gy* 33, 297-300.

Bjarko ME, Line RF (1988) Heritability and number of genes controlling leaf rust resistance on four cultivars of wheat. *Phytopathology* 78, 457-461.

Borner A, Schumann E, Furste A, Coster H, Leithold B, Roder S, Weber E (2002) Mapping of quantitative trait loci determining agronomic important characters in hexaploid wheat (*Triticum aestivum* L.). *Theoretical and Applied Genetics* 105, 921-936.

Borojevic K, Borojevic K (2005) Historic role of the wheat variety Akakomugi in southern and central European wheat breeding programs. *Breeding Science* 55, 253-256.

Bougot Y, Lemoine J, Pavoine MT, Guyomar'ch H, Gautier V, Muranty H, Barloy D (2006) A major QTL effect controlling resistance to powdery mildew in winter wheat at the adult plant stage. *Plant Breeding* 125, 550-556.

Bryan GJ, Collins AJ, Stephenson P, Orry A, Smith JB, Gale MD (1997) Isolation and characterization of microsatellites from hexaploid bread wheat. *Theoretical and Applied Genetics* 94, 557-563.

Chantret N, Mingeot D, Sourdille P, Bernard M, Jacquemin JM, Doussinault G (2001) A major QTL for powdery mildew resistance is stable over time and at two development stages in winter wheat. *Theoretical and Applied Genetics* 103, 962-971.

Chen XM, Cui SL, Zhang WX, Chen X (2002) Identification of powdery mildew resistance gene in four new winter wheat lines. *Acta Phytopathologia Sinica* 32, 138-141.

Frey KJ (1965) The utility of hill plots in oat research. *Euphytica* 14, 196-208.

Gustafson GD, Shaner G (1982) Influence of plant age on the expression of slow mildewing resistance in wheat. *Phytopathology* 72, 746-749.

Gupta PK, Balyan HS, Varshney RK (2010) Quantitative genetics and plant genomics: an overview. *Molecular Breeding* 26, 133-134.

Guyomarc'h H, Sourdille P, Edwards KJ, Bernard M (2002) Studies of the transferability of microsatellites derived from Triticum tauschii to hexaploid wheat and to diploid related species using amplification, hybridization and sequence comparisons. *Theoretical and Applied Genetics* 105, 736-744.

He R, Chang Z, Yang Z, Yuan Z, Zhan H, Zhang X, Liu J (2009) Inheritance and mapping of powdery mildew resistance gene *Pm43* introgressed from *Thinopyrum intermedium* into wheat. *Theoretical and Applied Genetics* 118, 1173-1180.

Hittalmani S, Parco A, Mew TV, Zeigler RS, Huang N (2000) Fine mapping and DNA marker-assisted pyramiding of the three major genes for blast resistance in rice. *Theoretical and Applied Genetics* 100, 1121-1128.

Hu Y, Wang Y (1989) Records of wheat varieties of Gansu province. Gansu Academy of Agricutural Science, Lanzhou.

Hua W, Liu Z, Zhu J, Xie C, Yang T, Zhou Y, Duan X, Sun Q (2009) Identification and genetic mapping of *pm42*, a new recessive wheat powdery mildew resistance gene derived from wild emmer (*Triticum turgidum* var. *dicoccoides*). *Theoretical and Applied Genetics* 119, 223-230.

Keller M, Keller B, Schachermayr G, Winzeler M, Schmid JE, Stamp P, Messmer MM (1999) Quantitative trait loci for resistance against powdery mildew in a segregating wheat x spelt population. *Theoretical and Applied Genetics* 98, 903-912.

Kolmer JA, Singh RP, Garvin DF, Viccars L, William HM, Huerta-Espino J, Ogbonnaya FC, Raman H, Orford S, Bariana HS, Lagudah ES (2008) Analysis of the *Lr34/Yr18* rust resistance region in wheat germplasm. *Crop Science* 48, 1841-1852.

Kosambi DD (1944) The estimation of map distance from recombination values. *Annual Eugen* 12, 172-175.

Lagudah ES, Krattinger SG, Herrera-Foessel S, Singh RP, Huerta-Espino J, Spielmeyer W, Brown-Guedira G, Selter LL, Keller B (2009) Gene-specific markers for the wheat gene *Lr34/Yr18/Pm38* which confers resistance to multiple fungal pathogens. *Theoretical and Applied Genetics* 119, 889-898.

Lagudah ES, McFadden H, Singh RP, Huerta-Espino J, Bariana HS, Spielmeyer W (2006) Molecualr genetic characterization of the *Lr34/Yr18* slow rusting resistance gene region in wheat. *Theoretical and Applied Genetics* 114, 21-30.

Lan CX, Liang SS, Wang Z, Yan J, Zhang Y, Xia XC, He ZH (2009) Quantitative trait loci mapping for adult-plant resistance to powdery mildew in Chinese wheat cultivar Bainong 64. *Phytopathology* 99, 1121-1126.

Lan CX, Ni XW, Yan J, Zhang Y, Xia XC, Chen XM, He ZH (2010) Quantitative trait loci mapping for adult-plant resistance to powdery mildew in Chinese wheat cultivar Lumai21. *Molecular Breeding* 25, 615-622.

Liang SS, Suenaga K, He ZH, Wang ZL, Liu HY, Wang DS, Singh RP, Sourdille P, Xia XC, (2006) Quantitative trait Loci mapping for adult-plant resistance to powdery mildew in bread wheat. *Phytopathology* 96, 784-789.

Lillemo M, Asalf B, Singh RP, Huerta-Espino J, Chen XM, He ZH, Bjornstad A (2008) The adult plant rust resistance loci *Lr34/Yr18* and *Lr46/Yr29* are important determinants of partial resistance to powdery mildew in bread wheat line Saar. *Theoretical and Applied Genetics* 116, 1155-1166.

Liu JY, Tao WJ, Liu DJ, Chen PD (2000) Screening and study of RAPD markers tightly linked to wheat powdery mildew resistance gene *Pm2*. *Acta Genetica Sinica* 27, 139-145.

Lu YM, Lan CX, Liang SS, Zhou X, Liu D, Zhou G, Lu Q, Jing J, Wang M, Xia XC, He ZH (2009) QTL mapping for adult-plant resistance to stripe rust in Italian common wheat cultivars Libellula and Strampelli. *Theoretical and Applied Genetics* 119, 1349-1359.

Luo PG, Luo HY, Chang ZJ, Zhang HY, Zhang M, Ren ZL (2009) Characterization and chromosomal location of *Pm40* in common wheat: a new gene for resistance to powdery mildew derived from *Elytrigia intermedium*. *Theoretical and Applied Genetics* 118, 1059-1064.

Ma HQ, Kong ZX, Fu BS, Li N, Zhang LX, Jia HY, Ma ZQ (2011) Identification and mapping of a new powdery mildew resistance gene on chromosome 6D of common wheat. *Theoretical and Applied Genetics* 123, 1099-1106.

Manly KF, Cudmore Jr. RH, Meer JM (2001) Map Manager QTX, cross-platform software for genetic mapping. *Mammalian Genome* 12, 930-932.

McDonald BA, Linde C (2002) The population genetics of

plant pathogens and breeding strategies for durable resistance. *Euphytica* 124, 163-180.

Michelmore RW, Paran I, Kesseli RV (1991) Identification of markers linked to disease resistance genes by bulked segregant analysis: A rapid method to detect markers in specific genomic regions by using segregating populations. *Proceedings Natitonal Academy of Sciences* USA 88, 9828-9832.

Mingeot D, Chantret N, Baret PV, Dekeyser A, Boukhatem N, Sourdille P, Doussinault G, Jacquemin JM (2002) Mapping QTL involved in adult plant resistance to powdery mildew in the winter wheat line RE714 in two susceptible genetic backgrounds. *Plant Breeding* 121, 133-140.

Navabi A, Tewari JP, Singh RP, McCallum B, Laroche A, Briggs KG (2005) Inheritance and QTL analysis of durable resistance to stripe and leaf rusts in an Australian cultivar, *Triticum aestivum* 'Cook'. *Genome* 48, 97-107.

Pestsova E, Ganal MW, Röder MS (2000) Isolation and mapping of microsatellite markers specific for the D genome of bread wheat. *Genome* 43, 689-697.

Priyamvada A, Saharan MS, Tiwari R (2011) A review: durable resistance in wheat. *International Journal of Genetics and Molecular Biology* 3, 108-114.

Roder MS, Korzun V, Wendehake K, Plaschke J, Tixier MH, Leroy P, Ganal MW (1998) A microsatellite map of wheat. *Genetics* 149, 2007-2023.

Sharp PJ, Kreis M, Shewry PR, Gale MD (1988) Location of β-amylase sequence in wheat and its relatives. *Theoretical and Applied Genetics* 75, 286-290.

Skinnes H (2002) Breakdown of race specific resistance to powdery mildew in Norwegian wheat. *Cereal Rusts and Powdery Mildew Bulletin vol. 30 online publication http://www.crpmb.org*.

Somers DJ, Isaac P, Edwards K (2004) A high-density microsatellite consensus map for bread wheat (*Triticum aestivum* L.). *Theoretical and Applied Genetics* 109, 1105-1114.

Song QJ, Fickus EW, Cregan, PB (2002) Characterization of trinucleotide SSR motifs in wheat. *Theoretical and Applied Genetics* 104, 286-293.

Sourdille P, Singh S, Cadalen T, Brown-Guedira GL, Gay G, Qi L, Gill BS, Dufour P, Murigneux A, Bernard M (2004) Microsatellite-based deletion bin system for the establishment of genetic-physical map relationships in wheat (*Triticum aestivum* L.). *Functional Integrative Genomics* 4, 12-25.

Spielmeyer W, McIntosh RA, Kolmer J, Lagudah ES (2005) Powdery mildew resistance and *Lr34/Yr18* genes for durable resistance to leaf and stripe rust cosegregate at a locus on the short arm of chromosome 7D of wheat. *Theoretical and Applied Genetics* 111, 731-735.

Spielmeyer W, Singh RP, McFadden H, Wellings CR, Huerta-Espino J, Kong X, Appels R, Lagudah ES (2008) Fine scale genetic and physical mapping using interstitial deletion mutants of *Lr34/Yr18*: a disease resistance locus effective against multiple pathogens in wheat. *Theoretical and Applied Genetics* 116, 481-490.

Tucker DM, Griffey CA, Liu S, Brown-Guedira G, Marshall DS, Maroof MAS (2007) Confirmation of three quantitative trait loci conferring adult plant resistance to powdery mildew in two winter wheat populations. *Euphytica* 155, 1-13.

Wang JK, Li HH, Zhang LY, Li CH, Meng L (2011) QTL IciMapping v3.1. Institute of Crop Sciences, CAAS, Beijing 100081, China and Crop Research Informatics Lab, CIMMYT, Apdo. Postal 6-641, 06600 Mexico, D. F., Mexico.

Wang XY, Chen PD, Zhang SZ (2001) Pyramiding and marker-assisted selection for powdery mildew resistance genes in common wheat. *Acta genetica Sinica* 28, 640-646.

William HM, Singh RP, Huerta-Espino J, Palacios G, Suenaga K (2006) Characterization of genetic loci conferring adult plant resistance to leaf rust and stripe rust in spring wheat. *Genome* 49, 977-990.

Zhang ZJ (1995) Evidence of durable resistance in nine Chinese land races and one Italian cultivar of Triticum aestivum to *Puccinia striiformis*. *European Journal of Plant Patholology* 101, 405-409.

Zheng DS (1993) Use of Italian wheat varieties in China. *Genetic Resources and Crop Evolution* 40, 137-142.

QTL mapping for adult plant resistance to powdery mildew in Italian wheat cv. Strampelli

M. A. Asad[1], B. Bai[1], C. X. Lan[1], J. Yan[2], X. C. Xia[1], Y. Zhang[1] and Z. H. He[1,3]

[1] *Institute of Crop Science, National Wheat Improvement Center/The National Key Facility for Crop Gene Resources and Genetic Improvement, Chinese Academy of Agricultural Sciences* (*CAAS*), 12 *Zhongguancun South Street, Beijing* 100081, *China*; [2] *Cotton Research Institute, Chinese Academy of Agricultural Sciences* (*CAAS*), *Huanghe Dadao, Anyang, Henan* 455000, *China*; [3] *International Maize and Wheat Improvement Center* (*CIMMYT*) *China Office, c/o CAAS*, 12 *Zhongguancun South Street, Beijing* 100081, *China*

Abstract: The Italian wheat cv. Strampelli displays high resistance to powdery mildew caused by *Blumeria graminis* f. sp. *tritici*. The objective of this study was to map quantitative trait loci (QTLs) for resistance to powdery mildew in a population of 249 $F_{2:3}$ line bulks from Strampelli/Huixianhong. Adult plant powdery mildew tests were conducted over two years in Beijing and one year in Anyang and simple sequence repeat (SSR) markers were used for genotyping. QTLs *Qpm. caas-3BS*, *Qpm. caas-5BL. 1* and *Qpm. caas-7DS* were consistent across environments whereas, *Qpm. caas-2BS. 1* found in two environments, explained 0. 4-1. 6%, 5. 5-6. 9%, 27. 1-34. 5%, and 1. 0-3. 5% of the phenotypic variation respectively. *Qpm. caas-7DS* corresponded to the genomic location of *Pm38/Lr34/Yr18*. *Qpm. caas-4BL* was identified in Anyang 2010 and Beijing 2011, accounting for 1. 9-3. 5% of phenotypic variation. *Qpm. caas-2BS. 1* and *Qpm. caas-5BL. 1* contributed by Strampelli, and *Qpm. caas-3BS* by Huixianhong, seem to be new QTL for powdery mildew resistance. *Qpm. caas-4BL*, *Qpm. caas-5BL. 3* and *Qpm. caas-7DS* contributed by Strampelli appear to be in the same genomic regions as those mapped previously for stripe rust resistance in the same population, indicating that these loci conferred resistance to both stripe rust and powdery mildew. Strampelli could be a valuable genetic resource for improving durable resistance to both powdery mildew and stripe rust in wheat.

Key words: QTL analysis, SSR markers, *Blumeria graminis*, Durable resistance, *Triticum aestivum* L.

Introduction

Wheat is the most widely used food crop for humankind. Powdery mildew caused by *Blumeria graminis* f. sp. *tritici* is a major wheat disease worldwide and causes significant yield losses every year in many wheat growing regions in China (Zeng et al., 2010). Host plant resistance, being cost effective and environmentally friendly compared to the use of fungicides, is the best option to control the disease (Gurung et al., 2009). Qualitative and quantitative forms of resistance to powdery mildew have been reported in wheat and its wild relatives. Qualitative resistance is usually highly protective throughout the growth cycle of the host, but when exploited in agriculture usually becomes ineffective with-

Published in Journal of Integrate Agriculture, 2013, 12 (5): 756-764 in a relatively short period of adoption. This lack of durability is the consequence of increased frequencies of previously rare races of the pathogen, mutation from avirulence to virulence, or in some cases, if resistance is based on combinations of such genes, by genetic recombination in the pathogen. *Blumeria graminis* f. sp. *tritici* is a haploid pathogen and although it usually propagates asexually by means of conidiospores it can also undergo sexual recombination, particularly at the beginning of the epidemic cycle. Most powdery mildew resistance genes commonly present in wheat are not effective in China (Zhou et al., 2002).

The contrasting type of resistance, known as quantitative resistance, tends to confer partial resistance at post seedling growth stages, and although less effective, is considered to be durable. It is also called as adult plant resistance (APR). Although individual adult plant resistance (APR) genes or QTLs often confer inadequate resistance, combinations of a few such genes usually result in "near-immunity" or high levels of resistance (Gustafson and Shaner, 1982, Liu et al., 2001).

Studies of adult plant partial resistance to powdery mildew in wheat have generally found that it is an oligogenic trait and individual genes/QTLs were mapped on many different chromosomes (Keller et al., 1999; Bougot et al., 2002; Lillemo et al., 2008; Lan et al., 2009, 2010b; Muranty et al., 2010). Generally the highest levels of resistance occur in lines with the most resistance genes/QTLs. The best characterized gene of this type is *Pm38* (also known as *Lr34* and *Yr18*) which is involved in multi-pathogen durable resistance. This gene has been cloned and characterized as a putative ABC transporter (Krattinger et al., 2009). Whether multiple disease protection is a general characteristic of other partial resistance genes is yet to be established.

The Italian common wheat cv. Strampelli, introduced to Gansu province in 1973 has adult plant resistance to stripe rust controlled by three QTLs, including *Yr18* (Lu et al., 2009). It has also shown good resistance to powdery mildew in the field. The objective of this study was to determine the genetic basis of durable powdery mildew resistance in Strampelli.

Materials and methods

Plant materials

A total of 249 $F_{2:3}$ lines developed from Strampelli/Huixianhong were available for QTL mapping. Strampelli has shown powdery mildew resistance at the adult plant stage in our nurseries over many years. Huixianhong, a landrace from Henan, is very susceptible at the seedling stage and moderately resistant at the adult stage. The $F_{2:3}$ lines, derived from individual F_2 plants are maintained as perpetual bulks of more than 50 plants.

Powdery mildew disease assessment

$F_{2:3}$ lines were sown in randomized complete blocks with three replications over two years (2009-2010 and 2010-2011) at the CAAS Experimental Station, Beijing, and CAAS Cotton Research Institute, Anyang, Henan, (herein referred to as Beijing 2010, Beijing 2011, Anyang 2010). Fifty seeds of each line were sown in circular plots (8 cm diameter), and plots were spaced 30 cm apart to facilitate disease evaluation. Jingshuang16 was planted at each tenth plot and around the experiment as a susceptible check and spreader to ensure maximum inoculum for disease development. In Beijing plots were inoculated prior to stem elongation with *Blumeria graminis* f. sp. *tritici* (*Bgt*) isolate E20. All entries were scored for the first time five weeks after inoculation based on the actual percentage of leaf area covered by powdery mildew on five randomly selected penultimate (F-1) leaves and again one week later when disease was at its maximum level around May 20. In Anyang, powdery mildew severities resulting from natural infection were evaluated once when infection levels on Jingshuang 16 had reached maximum levels during the third week of May. Phenotypic data from Anyang in 2011 could not be used because the disease did not develop sufficiently due to dry spring conditions.

Statistical analysis

The area under the disease progress curve (AUDPC) was calculated for each line using the formula: AUDPC $=\sum_{i=1}^{n}[(X_i+X_{i+1})/2](T_{i+1}-T_i)$ (Bjarko and Line, 1988), where X_i is the severity value on date i, T_i the number of days after inoculation on scoring date i, and n is the number of times of disease scoring. The correlation coefficients (r) and frequency distributions of powdery mildew response based on maximum disease severities (MDS) in different environments were calculated with Microsoft Excel 2007. The statistical analyses of maximum disease severity (MDS) and area under disease progress curve (AUDPC) were performed using the statistical package SAS v. 9.1 by PROC GLM (SAS Institute 1997). Broad-sense heritability (h^2) was calculated using the formula $h^2=$ Var (G) /Var (P), where, Var (G) and Var (P) are genotypic and phenotypic variances, respectively (Allard, 1960).

Simple sequence repeat (SSR) analysis

Fresh young leaf tissue of the parents and each $F_{2:3}$ line (30-40 plants per $F_{2:3}$ line as a bulk) was used to extract genomic DNA using the procedure described by Sharp et al. (1988). A total of 943 sequences of available SSR markers (http://www.wheat.pw.usda.gov) and one STS marker *csLV34* closely linked with the locus *Lr34/Yr18* (Lagudah et al., 2006) were used. Based on maximum disease severities, resistant and susceptible bulks were constructed by mixing equal amounts of DNA from the five most resistant and five most susceptible $F_{2:3}$ lines, respectively (Michelmore et al., 1991; Lan et al., 2010a). SSRs showing polymorphisms between the two parents and between the two bulks were used to genotype all 249 lines to determine their associations with powdery mildew responses. PCR reaction protocols, electrophoresis of PCR products and gel staining were conducted as described by Bryan et al. (1997) and Bassam et al. (1991). Polymorphic markers were used to genotype the population for linkage and QTL analysis.

Quantitative trait loci (QTL) analysis

Linkage groups were established with the software Map Manager QTXb20 (Manly et al., 2001) and recombination values were converted to centiMorgans using the Kosambi mapping function (Kosambi, 1944). Each linkage group was assigned to a chromosome based on previously published wheat genome maps (http://www.shigen.nig.ac.jp/wheat, http://wheat.pw.usda.gov, Somers et al., 2004). QTL detection was performed using inclusive composite interval mapping (ICIM) with the software IciMapping v3.1 (Wang et al., 2011). A LOD threshold value of 2.0 was used for declaration of a QTL. The phenotypic variance explained (*PVE* or R^2) by individual QTL and additive effects were obtained through stepwise regression (Wang et al., 2011). The *PVE* or R^2 was calculated by the formula,

$$PVE=\frac{V_G}{V_P}\times 100\%$$

where V_G is the genotypic variance of a QTL, and V_P is the phenotypic variance of the trait in QTL mapping (Wang et al., 2011). QTLs overlapping within 20 cM intervals on the same chromosome across different environments were considered to be the same.

Results

Field evaluation of mapping population

Strampelli and Huixianhong expressed maximum disease severities of 3.5 and 15%, 2.33 and 16.7%, and 4.0 and 33.3%, respectively, whereas the susceptible control Jingshuang 16 displayed 80 to 100%, 60 to 90%, and 90 to 100% in Anyang 2010, Beijing 2010, and Beijing 2011, respectively. Strampelli and Huixianhong expressed mean MDS of less than 5, and 15 to 25%, respectively, over three environments. Powdery mildew MDS across the three environments were significantly correlated ($r=0.67$ to 0.69). Frequency distributions of MDS and AUDPC showed continuous variation in all environments, suggesting a quantitative basis of resistance to powdery mildew (Fig. 1). Broad-sense heritabilities of MDS and AUDPC were 0.78 and

0.73, respectively. Analyses of variance showed significant variation among $F_{2:3}$ lines for both MDS and AUDPC (Table 1).

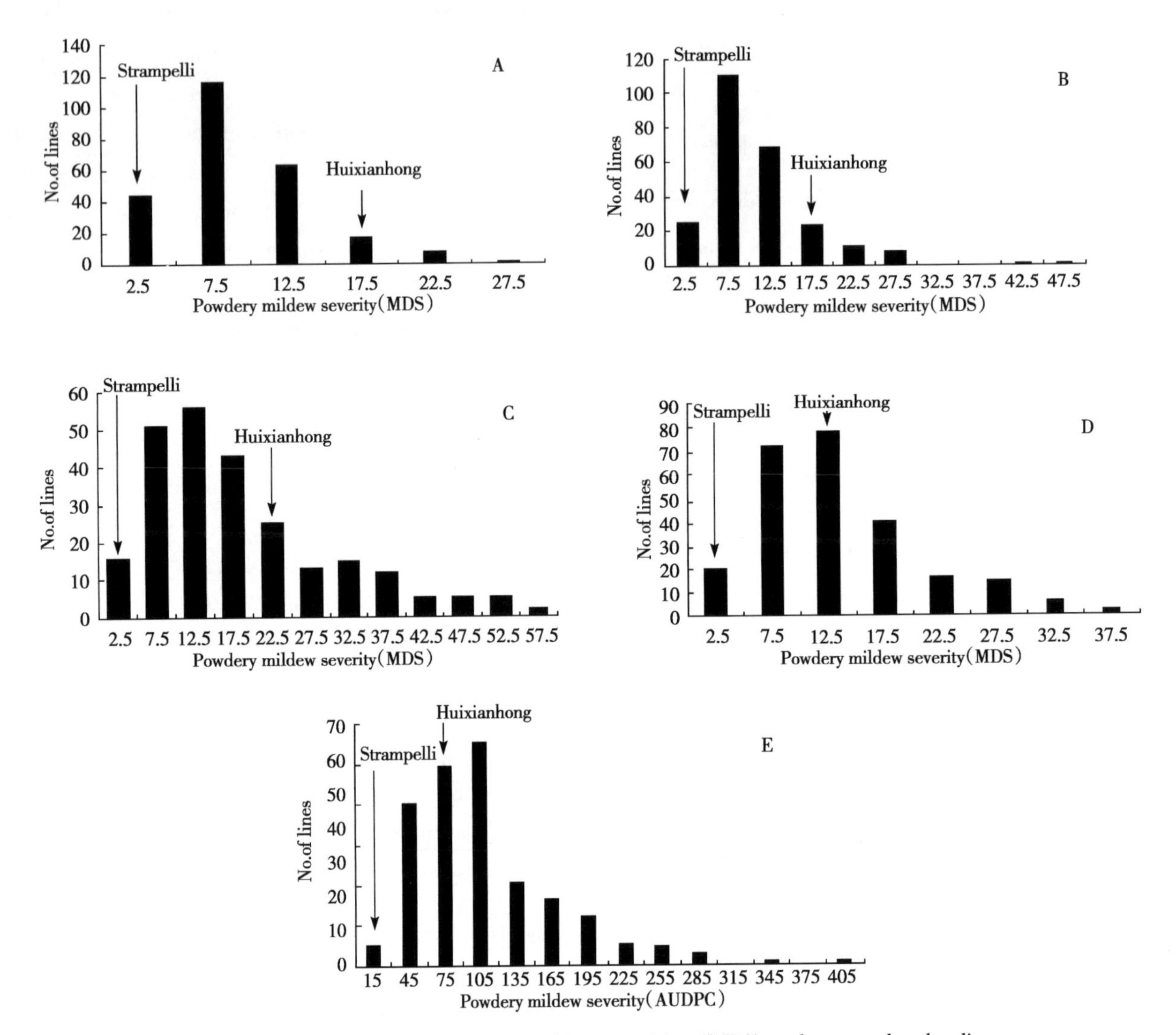

Fig. 1 Frequency distributions of maximum powdery mildew severities (MDS) and area under the disease progress curve (AUDPC) values in $F_{2:3}$ lines derived from Strampelli/Huixianhong. A, MDS in Anyang in 2010; B, MDS in Beijing in 2010; C, MDS in Beijing in 2011; D, Average MDS across three environments; E, average AUDPC in Beijing over 2 years. Mean values for the parents, Strampelli and Huixianhong are indicated by arrows.

Table 1 Analysis of variance of maximum disease severity (MDS) on penultimate leaves and area under the disease progress curve (AUDPC) for powdery mildew responses on $F_{2:3}$ lines derived from Strampelli /Huixianhong

Parameter	Source of variation	*df*	Sum of squares	Mean square	*F* value
MDS	Replicates	2	224	112	2.88
	Environments	2	36 232	18 116	464.51**
	Lines	248	109 050	439	11.27**
	Lines ×Environments	494	47 076	95	2.44**
	Error	1 379	53 782	39	

（续）

Parameter	Source of variation	df	Sum of squares	Mean square	F value
AUDPC	Replicates	2	30 477	15 238	10.77**
	Environments	1	216 566	216 566	153.05**
	Lines	248	2 393 427	9 650	6.82**
	Lines ×Environments	248	638 137	2 573	1.82**
	Error	944	1 406 507	1 414	

** $P < 0.0001$

QTLs for adult plant resistance (APR) to powdery mildew

Seven QTLs based on MDS, five QTLs in averaged MDS over three environments, and three QTLs in averaged AUDPC over two years in Beijing were mapped on chromosomes 2BS, 3BS, 4BL, 5BL and 7DS (Fig. 2, Table 2). QTLs on 3BS and 7DS were stable across all

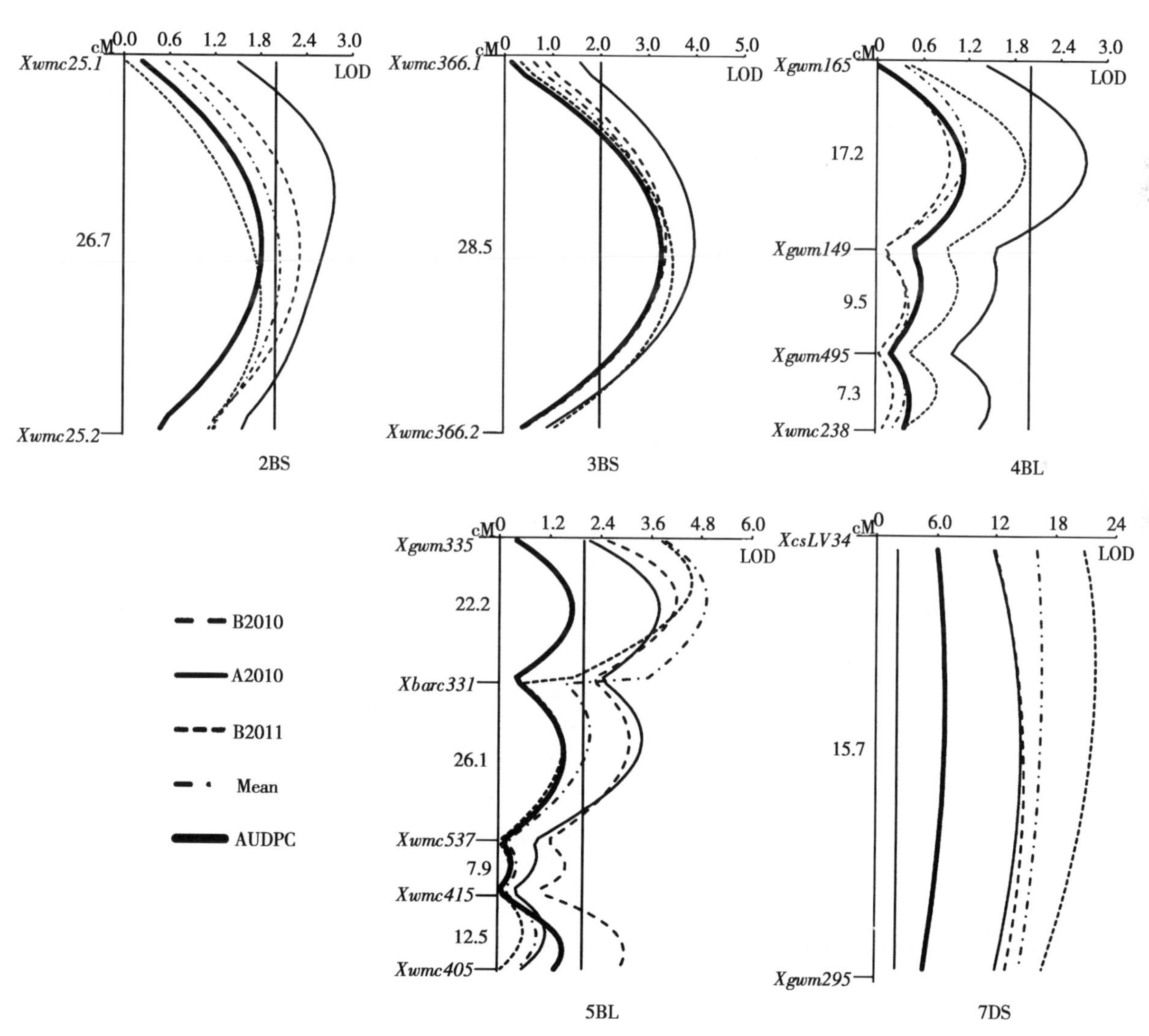

Fig. 2 Logarithm of odds (LOD) contours obtained by inclusive composite interval mapping (ICIM) of QTLs for adult plant powdery mildew resistance in $F_{2:3}$ lines of Strampelli/Huixianhong. A2010, maximum disease severities (MDS) in Anyang 2010; B2010 and B2011, maximum disease severities in Beijing 2010 and 2011, respectively; Mean, mean MDS across three environments; AUDPC, average of area under the disease progress curve in Beijing over two years. Logarithm of odds (LOD) threshold, 2.0.

environments, average MDS in three environments, and averaged AUDPC in Beijing over two years. Except in Beijing 2011, QTL on 2BS was consistently identified in other environments, averaged MDS of three environments, and averaged AUDPC in Beijing over two years, whereas *Qpm. caas-5BL. 1* was detected in almost all environments except averaged AUDPC in Beijing over two years. *Qpm. caas-4BL*, *Qpm. caas-5BL. 2* and *Qpm. caas-5BL. 3* were detected in two, three and one environment, respectively. QTLs on 2BS, 4BL, 5BL and 7DS, and one on 3BS were contributed by Strampelli and Huixianhong, respectively.

Table 2 Quantitative trait loci (QTLs) **for adult-plant resistance** (APR) **to powdery mildew in $F_{2:3}$ lines derived from Strampelli /Huixianhong**

Location/ year	QTL[a]	Marker interval	LOD[b]	AE[c]	R^2 (%)[d]	TotalR^2 (%)
MDS[e]						
Anyang 2010	*QPm. caas-2BS. 1*	*Xwmc25. 1-Xwmc25. 2*	2. 7	1. 36	3. 5	46. 1
	QPm. caas-3BS	*Xwmc366. 1-Xwmc366. 2*	4. 0	0. 56	1. 6	
	QPm. caas-4BL. 1	*Xgwm165-Xgwm149*	2. 7	0. 74	3. 5	
	QPm. caas-5BL. 1	*Xgwm335-Xbarc331*	3. 8	1. 42	5. 5	
	QPm. caas-5BL. 2	*Xbarc331-Xwmc537*	3. 4	1. 26	4. 8	
	QPm. caas-7DS	*XcsLV34-Xgwm295*	14. 5	3. 34	27. 2	
Beijing 2010	*QPm. caas-2BS. 1*	*Xwmc25. 1-Xwmc25. 2*	2. 3	1. 00	1. 7	48. 5
	QPm. caas-3BS	*Xwmc366. 1-Xwmc366. 2*	3. 4	0. 01	0. 4	
	QPm. caas-5BL. 1	*Xgwm335-Xbarc331*	4. 2	1. 30	6. 9	
	QPm. caas-5BL. 2	*Xbarc331-Xwmc537*	3. 1	1. 06	4. 0	
	QPm. caas-5BL. 3	*Xwmc415-Xwmc405*	3. 0	1. 32	4. 2	
	QPm. caas-7DS	*XcsLV34-Xgwm295*	14. 9	4. 74	27. 1	
Beijing 2011	*QPm. caas-3BS*	*Xwmc366. 1-Xwmc366. 2*	3. 5	0. 97	0. 5	43. 8
	QPm. caas-4BL. 1	*Xgwm165-Xgwm149*	1. 9	1. 09	1. 9	
	QPm. caas-5BL. 1	*Xgwm335-Xbarc331*	4. 6	3. 39	5. 9	
	QPm. caas-7DS	*XcsLV34-Xgwm295*	22. 0	10. 0	34. 5	
Averaged MDS in three environments	*QPm. caas-2BS. 1*	*Xwmc25. 1-Xwmc25. 2*	2. 1	0. 72	1. 0	34. 9
	QPm. caas-3BS	*Xwmc366. 1-Xwmc366. 2*	3. 3	0. 06	0. 1	
	QPm. caas-5BL. 1	*Xgwm335-Xbarc331*	4. 9	2. 36	6. 1	
	QPm. caas-5BL. 2	*Xbarc331-Xwmc537*	2. 1	0. 38	1. 8	
	QPm. caas-7DS	*XcsLV34-Xgwm295*	16. 6	5. 90	25. 9	
AUDPC[f]						
Averaged AUDPC in Beijing	*QPm. caas-2BS. 1*	*Xwmc25. 1-Xwmc25. 2*	1. 8	4. 74	0. 8	13. 5
	QPm. caas-3BS	*Xwmc366. 1-Xwmc366. 2*	3. 3	0. 24	0. 1	
	QPm. caas-7DS	*XcsLV34-Xgwm295*	6. 8	19. 3	12. 6	

a. QTL that extend across single one-log support confidence intervals were assigned the same symbol,
b. Logarithm of odds (LOD) score ,
c. Additive effects ,
d. Proportion of phenotypic variance explained by the QTL,
e. Maximum disease severity,
f. Area under the disease progress curve.

Qpm. caas-2BS mapped in the marker interval *Xwmc25. 1-Xwmc25. 2*, explained 1. 0-3. 5% of the phenotypic variances and was stable across two environments, in average MDS, and average AUDPC in

Beijing. A consistently detected QTL, *Qpm. caas-3BS*, was identified on the short arm of chromosome 3B between *Xwmc366. 1* and *Xwmc366. 2* across all environments. *Qpm. caas-4BL* explaining 1. 9-3. 5% of the phenotypic variance was located in the marker interval *Xgwm165-Xgwm149* in two environments. *Qpm. caas-5BL. 1*, identified in the marker interval *Xgwm335-Xbarc331* across all the environments, accounted for 5. 5-6. 9% of the phenotypic variance with additive effects of 1. 42-3. 39. *Qpm. caas-5BL. 2* in marker interval *Xbarc331-Xwmc537* explained phenotypic variation 1. 8-4. 8% whereas *Qpm. caas-5BL. 3*, in marker interval *Xwmc415 -Xwmc405* on the same chromosome explained 4. 2% of the phenotypic variance in one environment. Use of STS marker *csLV34* indicated the major stable QTL explaining 12. 6-34. 5% of phenotypic variances across all environments was *Pm38/Yr18/Lr34*. Seven QTLs together accounted for 43. 8-48. 5% of the total phenotypic variance across three environments.

Discussion

In this mapping population, maximum disease severity (MDS) and area under disease progress curve (AUDPC) showed continuous variation (Fig. 1A to E). The apparent lack of full susceptibility may be because the parents are homozygous at common loci conferring partial resistance, which could result in decreased variance and absence of transgressive segregation. Disease severity range of $F_{2:3}$ lines was quite below the susceptibility levels of cv. Jingshuang 16 suggesting that both parents possess some resistance loci in common that did not segregate in the cross, and there may be more minor resistance loci that could not be detected because of less number of simple sequence repeat markers used in this study. However, molecular mapping of the Strampelli/Huixianhong population indicated the presence of seven QTLs for powdery mildew resistance on chromosomes 2BS, 4BL, 5BL and 7DS in Strampelli and one on 3BS in Huixianhong.

Previously, QTLs, *QPm. crag-2BS* (Mingeot et al., 2002) and *QPm. caas-2BS* (Lan et al., 2010b) were detected on chromosome 2BS in Festin and Lumai 21, respectively. These genes are estimated to be about 22 and 43. 9cM, respectively, from *QPm. caas-2BS. 1* based on wheat consensus maps (Somers et al., 2004; http: //wheat. pw. usda. gov/GG2/ index. shtml). The *Pm26* (Rong et al., 2000) and *Pm42* (Hua et al., 2009) resistance genes derived from wild emmer (*Triticum turgidum* var. *dicoccoides*) were estimated to be 33 and 22cM from the *QPm. caas-2BS. 1* peak. This less effective QTL detected in two environments, average MDS in three environments, and averaged AUDPC in Beijing over two years, seems to be a new locus for powdery mildew resistance on chromosome 2BS.

A stable QTL, *QPm. caas-3BS*, was detected across all environments on chromosome 3BS. Chen et al. (2009) mapped a QTL on chromosome 3BS in hard winter wheat line 2,174. It was centered on *Xwms533* and 56cM from *QPm. caas-3BS*. *Pm13* derived from *Ae. longissima* is also located on 3BS (Donini et al., 1995). *QPm. caas-3BS* appears to be a new gene conferring adult plant resistance to powdery mildew.

QPm. caas-4BL. 1 was flanked by SSR markers *Xgwm149* and *Xgwm495*. *QPm. sfr-4B* and *QPm. ipk-4B* were mapped in cv. Forno (Keller et al., 1999) and the ITMI (Opata 85/W7984) population (Borner et al., 2002), flanked by markers *Xpsr593b* and *Xpsr1112*, and *Xcdo795* and *Xbcd1262*, respectively, a similar position to *QPm. caas-4BL. 1*. Liang et al. (2006) detected *QPm. caas-4BL* flanked by *Xgwm375* and *Xgwm251* in cv. Oligoculm in a similar position based on data from one test location. *QPm. nuls-4BL* flanked by *XwPt1505* and *Xgwm149*, was mapped on the same chromosomal region of cv. Avocet in both Beijing and Norway (Lillemo et al., 2008). *QYr. caas-4BL* detected at one site and explaining 4. 5% of the phenotypic variance in stripe rust response was mapped to a similar position (Lu et al., 2009). William et al., (2006) and Zwart et al. (2010) mapped stripe rust resistance QTL in crosses Avocet S/Pavon 76 and synthetic hexaploid CPI133872/Janz and these also coincided with the position of *QPm. caas-4BL. 1.* Thus results from a number of studies located QTLs for powdery mildew resistance in similar positions to those for stripe rust resistance.

Three QTLs, *QPm. ttu-5B*, *QPm. inra-5B. 2* and *QPm. nuls-5B* were previously mapped on chromosome 5BS in cv. Tahti, Courtot and Saar, respectively (Bougot et al., 2006; Jakobson et al., 2006; Lillemo et al., 2008). Keller et al. (1999) mapped *QPm. sfr-5B* flanked by RFLP markers *Xpsr580b-Xpsr143* on chromosome 5B. Our study has not determined a relationship of *QPm. sfr-5B* with *QPm. caas-5BL. 1* or *QPm. caas-5BL. 2* or *QPm. caas-5BL. 3* due to the different types of markers being used, and the positions of these markers are not comparable. QTLs, *QPm. caas-5BL. 1* and *QPm. caas-5BL. 2* flanked by *Xgwm335 -Xbarc331* and Xbarc331 -Xwmc537 were located nearby with less than 20cM distance between the resistance genes. We considered them same QTL as *QPm. caas-5BL. 1*. QTL, *QPm. caas-5BL. 3* flanked by *Xwmc405* and *Xwmc415*, identified at distance more than 20cM away from the peak of *QPm. caas-5BL. 2*, considered as independent QTL. *QPm. caas-5BL. 1* conferred stable and consistent resistance across all environments, explaining 5.5-6.9% of the phenotypic variance. *QYr. inra-5BL. 1* accounting for 25% of the phenotypic variation in stripe rust response was flanked by *Xgwm499* and *Xgwm639* in the same region as *QPm. caas-5BL. 3* (Mallard et al., 2005). Lu et al. (2009) reported stripe rust APR QTLs on chromosome 5BL in cultivars Libellula and Strampelli located at a similar position to *QPm. caas-5BL. 3*, indicating a common locus might confer resistance to both stripe rust and powdery mildew. *QPm. caas-5BL. 1* identified in the present study showed a stronger and more consistent effect across environments than *QPm. caas-5BL. 3* which explained only 4.2% of the phenotypic variance in one environment. *QPm. caas-5BL. 1* appears to be a new locus for APR to powdery mildew.

The most effective QTL was *QPm. caas-7DS*, inherited from Strampelli and explaining 27.1-34.5% of phenotypic variance across different environments. QTLs for slow mildewing in cv. W7984, Fukuho-Komugi and Saar were previously mapped in the same genomic region as *QPm. caas-7DS* on chromosome 7DS (Borner et al., 2002; Liang et al., 2006; Lillemo et al., 2008). Lu et al. (2009) detected *QYr. caas-7DS* for stripe rust resistance in a similar position to *QPm. caas-7DS* for powdery mildew resistance in Strampelli. The only presumed powdery mildew resistance gene in Strampelli was *Pm38.*

The *Pm38/Yr18/Lr34* gene is currently a prime target for durable resistance to multiple diseases because it is amenable to accurate molecular selection (Morgounov et al., 2012). Genes for resistance to powdery mildew in Strampelli seem to be more consistent and stable across environments than previously identified in Chinese wheat cultivars such as Yumai 2, Xiaoyang 6, Yumai 47, Lumai 21 and Bainong 64 (He et al., 2011, and per. comm.). So far, *Pm38/Yr18/Lr34* has been identified in powdery mildew studies of Mexican (W7984 and Saar) and Japanese (Fukuho-Komugi) cultivars, and its identification in Strampelli, well adapted in Chinese environment seems to play a large role and should be useful for developing cultivars with potentially durable powdery mildew, stripe rust and leaf rust resistance simultaneously. During the past years, Strampelli was crossed with several widely grown Chinese wheat cultivars Zhongmai 175, Lunxuan 987, Jimai 22 and Zhoumai 18 in our breeding program. The molecular markers csLV34, wmc537, gwm149, gwm165 and barc331 closely linked to *Pm38/Yr18/Lr34* and other QTLs (Lu et al., 2009) were used for marker-assisted selection in BC_1 or BC_2 progenies, and over 20 lines with good yield potential and resistance to powdery mildew, stripe rust and leaf rust were developed. This will provide a good example for marker-assisted selection targeting for durable resistance to powdery mildew, stripe rust and leaf rust in Chinese wheat breeding program.

Acknowledgments

We are grateful to the critical review of this manuscript by Prof. R. A. McIntosh, Plant Breeding Institute, University of Sydney. This study was supported by the National Key Basic Research Program of China (2013CB 127700), International Collaboration Project from the Ministry of Agriculture (2011-G3), the National Natural Science Foundation of China (30821140351) and China Agriculture Research System (CARS-3-1-3).

References

Allard R W. 1960. *Principles of Plant Breeding*. John Wiley and Sons, Inc. , New York, London.

Bassam B J, Caetano-Anolles G, Gresshoff P M. 1991. Fast and sensitive silver staining of DNA in polyacrylamide gels. *Analytical Biochemistry*, 196, 80-83.

Bjarko M E, Line R F. 1988. Heritability and number of genes controlling leaf rust resistance on four cultivars of wheat. *Phytopathology*, 78, 457-461.

Börner A, Schumann E, Furste A, Cöster H, Leithold B, Röder S, Weber E. 2002. Mapping of quantitative trait loci determining agronomic important characters in hexaploid wheat (*Triticum aestivum* L.). *Theoretical and Appllied Genetics*, 105, 921-936.

Bougot Y, Lemoine J, Pavoine M T, Guyomar'ch H, Gautier V, Muranty H, Borley D. 2006. A major QTL effect controlling resistance to powdery mildew in winter wheat at the adult plant stage. *Plant Breeding*, 125, 550-556.

Bryan G J, Collins A J, Stephenson P, Orry A, Smith J B, Gale M D. 1997. Isolation and characterization of microsatellites from hexaploid bread wheat. *Theoretical and Appllied Genetics*, 94, 557-563.

Chen Y H, Hunger R M, Carver B F, Zhang H L, Yan L L. 2009. Genetic characterization of powdery mildew resistance in U. S. hard winter wheat. *Molecular Breeding*, 24, 141-152.

Donini P, Koebner R M D, Ceoloni C. 1995. Cytogenetic and molecular mapping of the wheat *Aegilops longissima* chromatin breakpoints in powdery mildew resistant introgression lines. *Theoretical and Appllied Genetics*, 91, 738-743.

Gurung S, Bonman J M, Ali S, Patel J, Myrfield M, Mergoum M, Singh P K, Adhikari T B. 2009. New and diverse sources of multiple disease resistance in wheat. *Crop Science*, 49, 1655-1666.

Gustafson G D, Shaner G. 1982. Influence of plant age on the expression of slow mildewing resistance in wheat. *Phytopathology*, 72, 746-749.

He Z H, Lan C X, Chen X M, Zou Y C, Zhuang Q S, Xia X C. 2011. Progress and perspective in research of adult-plant resistance to stripe rust and powdery mildew in wheat. *Scientia Agricultura Sinica*, 44, 2193-2215.

Hua W, Liu Z J, Zhu J, Xie C J, Yang T M, Zhou Y L, Duan X Y, Sun Q X, Liu Z Y. 2009. Identification and genetic mapping of *pm42*, a new recessive wheat powdery mildew resistance gene derived from wild emmer (*Triticum turgidum* var. *dicoccoides*). *Theoretical and Appllied Genetics*, 119, 223-230.

Jakobson I, Peusha H, Timofejeva L, Jarve K. 2006. Adult plant and seedling resistance to powdery mildew in a *Triticum aestivum* x *Triticum militinae* hybrid line. *Theoretical and Appllied Genetics*, 112, 760-769.

Keller M, Keller B, Schachermayr G, Winzeler M, Schmid J E, Stamp P, Messmer M M. 1999. Quantitative trait loci for resistance against powdery mildew in a segregating wheat × spelt population. *Theoretical and Appllied Genetics*, 98, 903-912.

Kosambi D D. 1944. The estimation of map distance from recombination values. *Annals of Eugenics*, 12, 172-175.

Krattinger S G, Lagudah E S, Spielmeyer W, Singh R P, Huerta-Espino J, McFadden H, Bossolini E, Selter L L, Keller B. 2009. A putative ABC transporter confers durable resistance to multiple fungal pathogens in wheat. *Science*, 323, 1360-1363.

Lagudah E S, McFadden H, Singh R P, Huerta-Espino J, Bariana H S, Spielmeyer W. 2006. Molecualr genetic characterization of the *Lr34/Yr18* slow rusting resistance gene region in wheat. *Theoretical and Appllied Genetics*, 114, 21-30.

Lan C X, Liang S S, Wang Z, Yan J, Zhang Y, Xia X C, He Z H. 2009. Quantitative trait loci mapping for adult-plant resistance to powdery mildew in Chinese wheat cultivar Bainong 64. *Phytopathology*, 99, 1121-1126.

Lan C X, Liang S S, Zhou X C, Zhou G, Lu Q L, Xia X C, He Z H. 2010a. Identification of genomic regions controlling adult-plant stripe rust resistance in Chinese landrace Pingyuan 50 through bulked segregant analysis. Phytopathology 100: 313-318.

Lan C X, Ni X W, Yan J, Zhang Y, Xia X C, Chen X M, He Z H. 2010b. Quantitative trait loci mapping for adult-plant resistance to powdery mildew in Chinese wheat cultivar Lumai 21. *Molecular Breeding*, 25, 615-622.

Liang S S, Suenaga K, He Z H, Wang Z L, Liu H Y, Wang D S, Singh R P, Sourdille P, Xia X C. 2006. Quantitative trait loci mapping for adult-plant resistance to powdery mildew in bread wheat. *Phytopathology*, 96, 784-789.

Lillemo M, Asalf B, Singh R P, Huerta-Espino J, Chen X M, He Z H, Bjornstad A. 2008. The adult plant rust resistance loci *Lr34*/*Yr18* and *Lr46*/*Yr29* are important determinants of partial resistance to powdery mildew in bread wheat line Saar. *Theoretical and Appllied Genetics*, 116, 1155-1166.

Liu S X, Griffey C A, Maroof M A S. 2001. Identification of molecular markers associated with adult plant resistance to powdery mildew in common wheat cultivar Massey. *Crop Science*, 41, 1268-1275.

Lu Y M, Lan C X, Liang S S, Zhou X C, Liu D, Zhou G, Lu Q L, Jing J X, Wang M N, Xia X C, He Z H. 2009. QTL mapping for adult-plant resistance to stripe rust in Italian common wheat cultivars Libellula and Strampelli. *Theoretical and Appllied Genetics*, 119, 1349-1359.

Mallard S, Gaudet D, Aldeia A, Abelard C, Besnard A L, Sourdille P, Dedryver F. 2005. Genetic analysis of durable resistance to yellow rust in bread wheat. *Theoretical and Appllied Genetics*, 110, 1401-1409.

Manly K F, Cudmore R H Jr., Meer J M. 2001. Map Manager QTX, cross-platform software for genetic mapping. *Mammalian Genome*, 12, 930-932.

Michelmore R W, Paran I, Kesseli R V. 1991. Identification of markers linked to disease resistance genes by bulked segregant analysis: A rapid method to detect markers in specific genomic regions by using segregating populations. *Proceedings of National Academy of Sciences USA*, 88, 9828-9832.

Mingeot D, Chantret N, Baret P V, Dekeyser A, Boukhatem N, Sourdille P, Doussinault G, Jacquemin J M. 2002. Mapping QTL involved in adult plant resistance to powdery mildew in the winter wheat line RE714 in two susceptible genetic backgrounds. *Plant Breeding*, 121, 133-140.

Morgounov A, Tufan H A, Sharma R, Akin B, Bagci A, Braun H-J, Kaya Y, Keser M, Payne T S, Sonder K, McIntosh R. 2012. Global incidence of wheat rusts and powdery mildew during 1969 - 2010 and durability of resistance of winter wheat variety Bezostaya 1. *European Journal of Plant Patholology*, 132, 323-340.

Muranty H, Pavoine M T, Doussinault G, Barloy D. 2010. Origin of powdery mildew resistance factors in RE714, a wheat breeding line obtained from two interspecific crosses. *Plant Breeding*, 129, 465-471.

Rong J K, Millet E, Manisterski J, Feldman M. 2000. A new powdery mildew resistance gene: Introgression from wild emmer into common wheat and RFLP-based mapping. *Euphytica*, 115, 121-126.

Sharp P J, Kreis M, Shewry P R, Gale M D. 1988. Location of β-amylase sequence in wheat and its relatives. *Theoretical and Appllied Genetics*, 75, 286-290.

Somers D J, Isaac P, Edwards K. 2004. A high-density microsatellite consensus map for bread wheat (*Triticum aestivum* L.). *Theoretical and Appllied Genetics*, 109, 1105-1114.

Wang, J K, Li H H, Zhang L Y, Li C H, Meng L. 2011. QTL IciMapping v3.1. Institute of Crop Sciences, CAAS, Beijing 100081, China, and Crop Research Informatics Lab, CIMMYT, Apdo. Postal 6-641, 06600 Mexico, D. F., Mexico.

William H M, Singh R P, Huerta-Espino J, Palacios G, Suenaga K. 2006. Characterization of genetic loci conferring adult plant resistance to leaf rust and stripe rust in spring wheat. *Genome*, 49, 977-990.

Zeng X W, Luo Y, Zheng Y M, Duan X Y, Zhou Y L. 2010. Detection of latent infection of wheat leaves caused by *Blumeria graminis* f. sp *tritici* using nested PCR. *Journal of Phytopathology*, 158, 227-235.

Zhou Y L, Duan X Y, Chen G, Sheng B Q, Zhang Y. 2002. Analyses of resistance genes of 40 wheat cultivars or lines to wheat powdery mildew. *Acta Phytopathology Sinica*, 32, 301-305.

Zwart R S, Thompson J P, Milgate A W, Bansal U K, Williamson P M, Raman H, Bariana H S. 2010. QTL mapping of multiple foliar disease and root-lesion nematode resistances in wheat. *Molecular Breeding*, 26, 107-124.

QTL mapping of adult - plant resistance to stripe rust in a population derived from common wheat cultivars Naxos and Shanghai 3/Catbird

Y. Ren[1], Z. H. He[1,5], J. Li[1], M. Lillemo[2], L. Wu[3], B. Bai[4], Q. X. Lu[2], H. Z. Zhu[3], G. Zhou[4], J. Y. Du[4], Q. L. Lu[4], and X. C. Xia[1]

[1] *Institute of Crop Science, National Wheat Improvement Center/The National Key Facility for Crop Gene Resources and Genetic Improvement, Chinese Academy of Agricultural Sciences (CAAS), 12 Zhongguancun South Street, Beijing* 100081, *China*; [2] *Dept. of Plant and Environmental Sciences, Norwegian University of Life Sciences, P.O. Box* 5003, *NO*-1432 *Ås, Norway*; [3] *Crop Research Institute, Sichuan Academy of Agricultural Sciences, Chengdu* 610066, *Sichuan, China*; [4] *Wheat Research Institute, Gansu Academy of Agricultural Sciences, Lanzhou* 730070, *Gansu, China*; [5] *International Maize and Wheat Improvement Center (CIMMYT) China Office, c/o CAAS, 12 Zhongguancun South Street, Beijing* 100081, *China.*

Abstract: Stripe rust, caused by *Puccinia striiformis* Westend. f. sp. *tritici* Erikss., is a severe foliar disease of common wheat (*Triticum aestivum* L.) worldwide. Use of adult-plant resistance (APR) is an efficient approach to provide long-term protection of crops from the disease. The German spring wheat cultivar Naxos showed a high level of APR to stripe rust in the field. To identify the APR genes in this cultivar, a mapping population of 166 recombinant inbred lines (RILs) was developed from a cross between Naxos and Shanghai 3/Catbird (SHA3/CBRD), a moderately susceptible line developed by CIMMYT. The RILs were evaluated for maximum disease severity (MDS) in Sichuan and Gansu in the 2009-2010 and 2010-2011 cropping seasons. Composite interval mapping (CIM) identified four QTL, *QYr.caas-1BL.1RS*, *QYr.caas-1DS*, *QYr.caas-5BL.3* and *QYr.caas-7BL.1*, conferring stable resistance to stripe rust across all environments, each explaining 1.9-27.6%, 2.1-5.8%, 2.5-7.8% and 3.7-9.1% of the phenotypic variance, respectively. *QYr.caas-1DS* flanked by molecular markers *XUgwm353-Xgdm33b* was likely a new QTL for APR to stripe rust. Because the interval between flanking markers for each QTL was less than 6.5cM, these QTL and their closely linked markers are potentially useful for improving resistance to stripe rust in wheat breeding.

Introduction

Stripe rust or yellow rust (YR), caused by *Puccinia striiformis* Westend. f. sp. *tritici* Erikss. (*Pst*), is a worldwide devastating disease of common wheat (*Triticum aestivum* L.), and is particularly prevalent in temperate and maritime wheat growing regions (Stubbs, 1985; Boyd, 2005; Chen, 2005). In China, the disease is most destructive in the northwestern and southwestern regions, such as Gansu, Sichuan, Chongqing, Shaanxi and Yunnan, due to the cool and moist spring conditions (He et al., 2011). Since the 1950s wheat YR epidemics have occurred in China for more than 15 times, most notably the four large scales of pandemics in 1950, 1964, 1990 and 2002 (Wan et al., 2004). In recent years,

Published in TAG, 2012, 125: 1211-1221

many wheat cultivars have lost their resistance to YR due to the occurrence and spread of Chinese *Pst* races CYR32 and CYR33 with broad virulence patterns. From 2003 to 2011, YR occurred in about 4.1 million hectares each year posing serious risks to wheat production in China (Tang, 2004; Zhao, 2004; Zhang et al., 2005, 2006, 2007, 2008, 2009; NAESC, 2010, 2011).

Although YR can be controlled through the application of fungicides, resistance breeding is a more economic, effective, and environmentally friendly approach to control the disease (Line, 2002; Chen, 2005). Resistance to YR can be categorized as race specific resistance or race non-specific resistance (Johnson, 1988), but the distinction is not always clear. Race specific resistance is normally highly effective both in seedlings and adult plant stages. However, it is readily overcome by new races that are then selected on the previously resistant cultivars (Chen and Line, 1995; Carter et al., 2009). For example, due to the wide cultivation of cultivars with *Yr9* (1B.1R translocation) *Pst* races CYR29 and CYR32 became prevalent. Also, cultivars with resistance from Fan 6 succumbed to race CYR32 in 2002 (Chen et al., 2009). In contrast, race non-specific resistance, often associated with APR, partial resistance or slow-rusting is expressed at later stages of plant development, generally qualitatively inherited and often durable (Johnson, 1984; Chen, 2005). Examples include the slow-rusting resistance genes *Yr18* and *Yr29*, which have provided effective resistance to YR since the early 20th century (Krattinger et al., 2009). These two genes also show partial and race non-specific resistance to leaf rust (LR), powdery mildew (PM) and stem rust (Suenaga et al., 2003; Rosewarne et al., 2006; Lillemo et al., 2008; Bhavani et al., 2011).

To date, 52 genes (*Yr1*-*Yr49*) for YR resistance have been catalogued, and these are assigned to 49 loci (McIntosh et al., 2011). Most of them are race-specific resistance genes that are no longer effective or likely to have short duration of effectiveness if widely deployed in cultivars (McDonald and Linde, 2002; Lu et al., 2009). As a consequence, APR is being increasingly emphasized in breeding for resistance (Line, 2002; Lu et al., 2009). During the last decade, many APR quantitative trait loci (locus) (QTL) for reducing YR severity have been identified with the help of molecular markers (Singh et al., 2000b; Bariana et al., 2001; Boukhatem et al., 2002; Suenaga et al., 2003; Imtiaz et al., 2004; Ramburan et al., 2004; Mallard et al., 2005; Uauy et al., 2005; Christiansen et al., 2006; Rosewarne et al., 2006; William et al., 2006; Lin and Chen, 2007, 2009; Chhuneja et al., 2008; Guo et al., 2008; Melichar et al., 2008; Rosewarne et al., 2008; Santra et al., 2008; Carter et al., 2009; Dedryver et al., 2009; Lu et al., 2009; Herrera-Foessel et al., 2010; Lan et al., 2010; Hao et al., 2011; Jagger et al., 2011; Lowe et al., 2011; Chen et al., 2012). They cover almost all chromosomes except for 1A, 1D, 3A and 7A, and the QTL on 1BL, 2BS, 2BL, 3BS, 4BL, 5DL, 6BS and 7DS were frequently coincided with PM resistance QTL (He et al., 2011). Since APR is generally quantitatively inherited, it is more complicated for selection in wheat breeding than race-specific resistance (Yu et al., 2001; Lillemo et al., 2008). Nevertheless, molecular markers that are closely linked to these QTL will greatly facilitate selection for APR in breeding programs.

The German spring wheat cultivar Naxos exhibits a high level of resistance to YR and PM in the field although it is susceptible to Chinese *Pst* races CYR29, CYR32 and CYR33 and *Blumeria graminis* f. sp. *tritici* isolate E20 at the seedling stage, indicating typical APR to both diseases. Recently, QTL mapping for APR to PM in Naxos was performed by Lu et al. (2012); however, the inheritance of APR to YR in this cultivar has not been reported. The objectives of this study were to identify and locate APR QTL to YR in a RIL population from cross SHA3/CBRD × Naxos with molecular markers, and to assess the stability of detected QTL across environments.

Materials and methods

Plant materials

One hundred and sixty-six F_6 RILs were developed by

single-seed descent from the cross Shanghai 3/Catbird (SHA3/CBRD) × Naxos, which was provided by Dr. Morten Lillemo, Norwegian University of Life Sciences. Naxos (Pedigree: Tordo/St. Mir808-Bastion// Minaret), an adapted spring wheat cultivar from Germany, is highly susceptible to currently prevalent *Pst* races CYR32 and CYR33 at the seedling stage (IT=4), but shows a high level of resistance to YR in the field. SHA3/CBRD (Pedigree: Shanghai 3//Chuanmai 18/Bagula), developed by the International Maize and Wheat Improvement Center (CIMMYT), is highly susceptible to *Pst* race CYR32 at the seedling stage (IT=4) and is moderately susceptible to YR at the adult plant stage. SHA3/CBRD have the 1B. 1R translocation based on AF1/AF4, SECA2/SECA3, SCM9, IB-267, iag95 and Bmac0213 markers (Francis et al., 1995; Froidmont, 1998; Saal and Wricke, 1999; Mago et al., 2002; Nagy et al., 2003) and genomic in situhybridization (GISH) analysis (Sup-Figs. 1 and 2). Both parents are adapted to growing conditions in China.

Genotyping

Nine hundred and fifty two simple sequence repeat (SSR) markers were screened on SHA3/CBRD and Naxos, including BARC (Song et al., 2002), CFA, CFD and GPW (Sourdille et al., 2004), CNL and KSUM (Yu et al., 2004), GDM (Pestsova et al., 2000), GWM (Röder et al., 1998), MAG (Xue et al., 2008), SWM (Bossolini et al., 2006; Krattinger et al., 2009), UGWM (Parida et al., 2006) and WMC (Gupta et al., 2002) markers. Polymorphic SSR markers were then used to genotype all 166 RILs from the cross SHA3/CBRD × Naxos by polyacrylamide gel electrophoresis. The population was also screened with diversity array technology (DArT) markers. In addition, six rye (*Secale cereale* L.) 1RS specific markers (Tang et al., 2009; Yu et al., 2011) were also included for detection of 1B. 1R translocation.

Genetic linkage map construction

The genotypic data for markers were used to construct genetic linkage maps with the software Map Manager QTX20 (Manly et al., 2001). Genetic distances between markers were calculated based on the Kosambi mapping function (Kosambi, 1944). The assignment of linkage groups on chromosomes were checked against previously published wheat consensus maps (Somers et al., 2004; http: //wheat. pw. usda. gov).

Field trials

The F_6 RILs and their parents were evaluated in Sichuan and Gansu provinces in the 2009-2010 and 2010-2011 cropping seasons. The trial in Sichuan was conducted in Pi county (103°E, 30°N, 555m a. s. l.), close to Chengdu, and the trial in Gansu was conducted in Qingshui county (106°E, 34°N, 1572m a. s. l.), close to Tianshui. Both locations are hotspots for YR in China with ideal conditions for rust infection and spread. Field trials were conducted in randomized complete blocks with three replicates at each location. Each plot consisted of a single row 1. 5min length with 25cm between rows. Approximately 50 seeds were sown in each row. Every tenth row was planted with the highly susceptible Mingxian169 to aid the spread of the pathogen within the trial. To ensure ample field inoculum, infection rows of Mingxian169 were also planted perpendicularly and adjacent to the test rows. Inoculations at both sites were performed at the three-leaf stage with a mixture of prevalent Chinese *Pst* races. Maximum disease severities (MDS) were scored, when YR severities on Mingxian169 reached a maximum level around the 15th of April in Sichuan and 10th of June in Gansu, respectively. Field data from Gansu 2010-2011 was excluded from the statistical analysis and QTL detection due to the low YR development caused by the dry weather condition in the spring.

Statistical analysis

Analysis of variance was performed with the PROC GLM in the statistical analysis system (SAS) software package (SAS institute, V8). The information in the ANOVA table was used to calculate the broad sense heritability (h_b^2) for YR reaction based on the formula $h_b^2=\sigma_g^2/(\sigma_g^2+\sigma_{ge}^2/e+\sigma_\varepsilon^2/re)$, where $\sigma_g^2=(MS_f-MS_{fe})/re$, $\sigma_{ge}^2=(MS_{fe}-MS_e)/r$ and $\sigma_\varepsilon^2=MS_e$ (Allard 1960); in this formula, σ_g^2=genetic variance, σ_{ge}^2=genotype×environment interaction variance, σ_ε^2 = error variance, MS_f = mean square of genotype, MS_{fe} = mean square of genotype × environment interaction, MS_e = mean square of error, r = number of replicates and e = number of environments.

QTL analysis

Stripe rust resistance QTL were detected in the population based on the mean MDS of three replicates in each environment and also on the averaged data from three environments. Composite interval mapping (CIM) analysis was performed using the software QTL Cartographer 2.5 (Wang et al., 2005). After performing a 1, 000 permutation test, a LOD threshold of 2.5 was set to declare QTL as significant. A walk speed of 2.0cM was chosen for all QTL detections. QTL effects were estimated as the proportion of phenotypic variance (R^2) explained by the QTL. The data were also analyzed with the software QTL IciMapping V3.1 (Li et al., 2007, 2008; Wang, 2009) by inclusive composite interval mapping (ICIM) for detecting interactions between two QTL.

Results

Phenotypic evaluation

Stripe rust developed well in Sichuan in 2010 and 2011 and in Gansu in 2010. The MDS of the susceptible Mingxian169 control ranged from 90 to 95%, from 95 to 100%, from 60 to 100% in Sichuan 2010, Sichuan 2011 and Gansu 2010, respectively. Naxos had a mean MDS of 4.7-11.7% across the three environments, whereas SHA3/CBRD was rated with a mean MDS of 31.7 to 66.7%. The frequency distributions of MDS for the 166 RILs ranging over 1-80%, 4-95%, 3-80% and 3-78% in Sichuan 2010, Sichuan 2011, Gansu 2010 and the averaged MDS for all three environments, respectively, showed continuous distributions (Sup-Fig. 3), indicative of polygenic inheritance. The broad-sense heritability of MDS was 0.65. The ANOVA showed significant differences (P = 0.01) in MDS among RILs, environments, replicates within environments and line × environment interactions (Table 1).

Table 1 Analysis of variance of MDS scores for RILs generated from the cross SHA3/CBRD × Naxos

Source of variance	*df*	Mean square	*F* value
Line	165	2, 317	2.8**
Environment	2	266, 735	326.9**
Replicate	2	6, 026	49.4**
Line×environment	330	816	6.7**
Error	992	122	

** Significant at P = 0.01.

Construction of linkage maps

A total of 952 SSR markers were tested for polymorphism between SHA3/CBRD and Naxos; 279 markers (29.3%) were polymorphic. The latter were used to genotype individual RILs from the population for construction of genetic linkage maps. Additionally, 283 polymorphic DArT markers covering all chromosomes and six molecular markers specific for 1B.1R translocation were used for mapping the population. Among the 568 polymorphic marker loci, 373 loci were assigned into 26 linkage groups.

QTL for APR to YR

Using CIM analysis, 10 QTL for APR to YR were identified on chromosomes 1AL, 1BL.1RS, 1DS, 2BL, 2DL, 5AL, 5BL, 6BS and 7BL (2 QTL) based on the mean MDS in each environment and the averaged values from all three environments (Fig. 1; Table 2). The resistance alleles of the QTL on 1AL, 1DS, 2BL, 2DL and 6BS were contributed by Naxos, whereas those on 1BL.1RS, 5AL, 5BL, and 7BL (2 QTL) were from SHA3/CBRD.

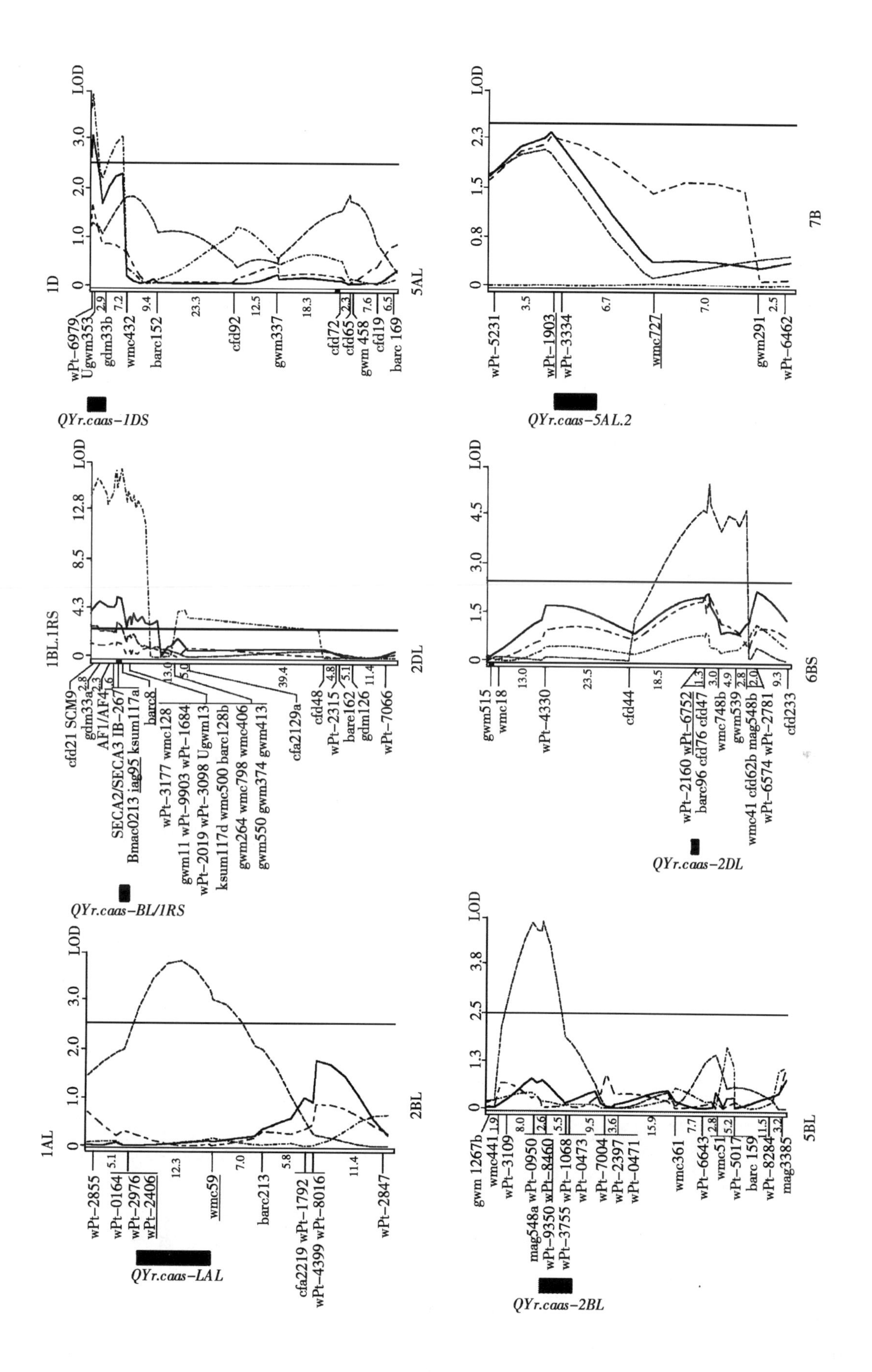
QYr.caas-1DS
QYr.caas-5AL.2
QYr.caas-BL/1RS
QYr.caas-2DL
QYr.caas-LAL
QYr.caas-2BL

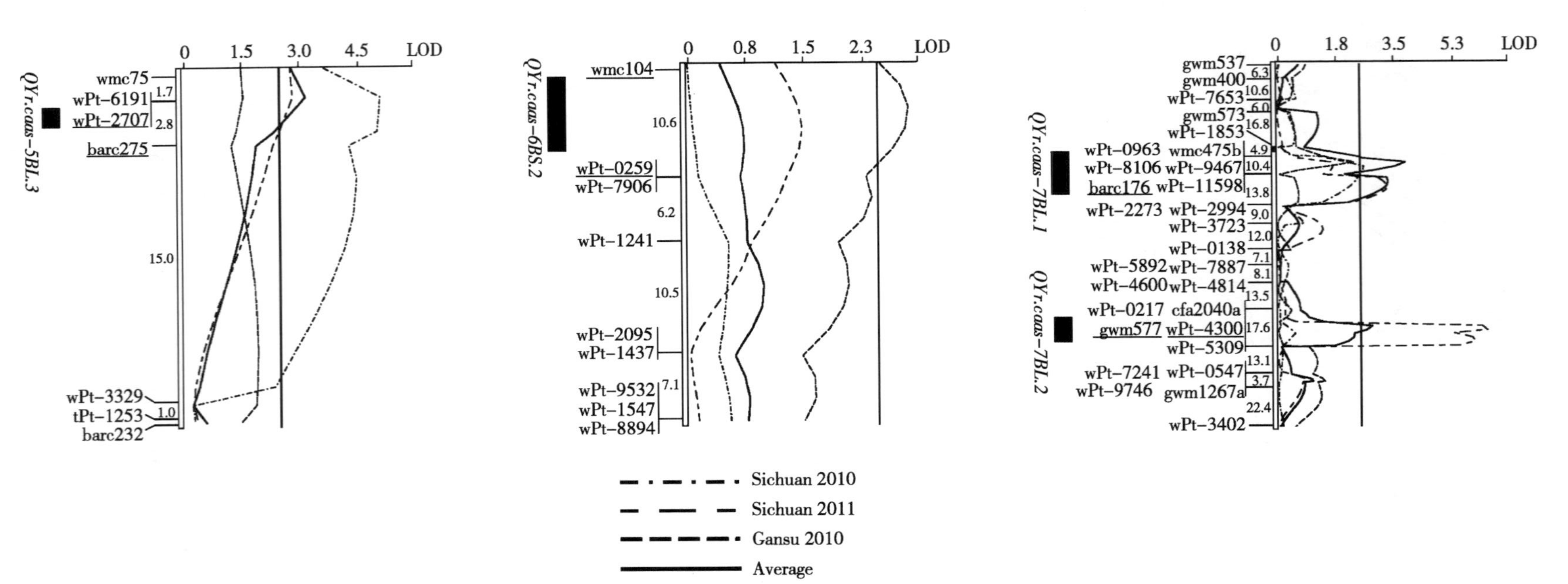

Fig. 1 LOD contours for QTL on chromosomes 1AL, 1BL.1RS, 1DS, 2BL, 2DL, 5AL, 5BL, 6BS, 7BL (2 QTL) that reduce stripe rust severity in the RIL population from SHA3/CBRD × Naxos. The short arms are toward the top and the approximate positions of centromeres are indicated by solid squares in the vertical axis. Genetic distances are shown in centiMorgans (cM) to the left of the vertical axis. Markers closely linked to each QTL are underlined. The approximate positions of QTL are indicated by oblong black squares to the left of markers and their heights represent the interval distances of flanking markers. LOD thresholds of 2.5 are indicated by dashed vertical lines in the graphs.

Table 2 Quantitative trait loci for stripe rust resistance detected by CIM in the SHA3 /CBRD × Naxos RIL population across three environments and averaged MDS over three environments

QTL[a]	Sichuan 2010			Sichuan 2011			Gansu 2010			Average			Source of resistance
	LOD[b]	Add[c]	R^2(%)[d]	LOD	Add	R^2(%)	LOD	Add	R^2(%)	LOD	Add	R^2(%)	
QYr. caas-1AL							3.8	−5.1	8.2				Naxos
QYr. caas-1BL. 1RS	16.3	9.5	27.6	1.3	4.1	1.9	3.0	4.2	5.3	5.3	4.8	7.9	SHA3/CBRD
QYr. caas-1DS	3.9	−4.1	5.8	1.6	−4.3	2.4	1.3	−2.6	2.1	3.1	−3.5	4.5	Naxos
QYr. caas-2BL							4.9	−6.0	12.2				Naxos
QYr. caas-2DL				2.1	−5.0	3.1	5.6	−5.7	10.2	2.1	−2.9	2.9	Naxos
QYr. caas-5AL. 2				2.3	5.4	3.7	2.1	3.3	3.4	2.4	3.0	3.2	SHA3/CBRD
QYr. caas-5BL. 3	5.2	4.8	7.8	2.9	6.2	4.5	1.6	2.8	2.5	3.2	3.5	4.7	SHA3/CBRD
QYr. caas-6BS. 2				1.5	−4.8	2.9	2.9	−4.2	5.7				Naxos
QYr. caas-7BL. 1	2.6	5.7	8.2	1.7	5.7	3.7	2.3	4.8	6.3	3.9	5.2	9.1	SHA3/CBRD
QYr. caas-7BL. 2				6.4	11.9	14.7				2.9	4.5	6.3	SHA3/CBRD

a. QTL were detected with a minimum LOD score of 2.5 in at least one environment,

b. Logarithm of odds (LOD) score,

c. Add, additive effect of resistance allele,

d. R^2, percentages of phenotypic variance explained by individual QTL.

Using the six rye specific molecular markers, a major QTL for YR resistance was located on chromosome 1BL. 1RS, designated *QYr. caas-1BL. 1RS*, most closely associated with *Xiag95*. This QTL explained 27.6%, 1.9%, 5.3% and 7.9% of the phenotypic variance in Sichuan 2010, Sichuan 2011, Gansu 2010 and averaged MDS, respectively.

The second consistently detected QTL with larger effect, *QYr. caas-5BL. 3*, was located on chromosome 5BL between *XwPt-2707* and *Xbarc275*, and explained 2.5 to 7.8% of the phenotypic variance in three environments and averaged MDS. The third QTL, *QYr. caas-7BL. 1*, was flanked by *XwPt-8106* and *Xbarc176*, and explained 8.2%, 3.7%, 6.3% and 9.1% of the phenotypic variance in Sichuan 2010, Sichuan 2011, Gansu 2010, and averaged MDS, respectively. The fourth QTL, *QYr. caas-1DS*, was identified in the marker interval *XUgwm353-Xgdm33b* on chromosome 1DS. This QTL was detected in three environments as well as the overall mean, explaining the phenotypic variance of 2.1 to 5.8%.

Two QTL, *QYr. caas-2DL* and *QYr. caas-5AL. 2*, were detected in two environments and averaged MDS, respectively, and explained 2.9-10.2% and 3.2-3.7% of the phenotypic variance. *QYr. caas-2DL*, flanked by *XwPt-6752* and *Xcfd47*, was located on chromosome 2DL whereas *QYr. caas-5AL. 2* was located in marker interval *XwPt-1903-5AL* and *Xwmc727-5AL*. One QTL, *QYr. caas-6BS. 2*, in the interval *Xwmc104-XwPt-0259* on chromosome 6BS, explained 2.9-5.7% of the phenotypic variance across two environments.

The remaining three QTL were detected in single environment. *QYr. caas-1AL* and *QYr. caas-2BL* were in marker intervals *XwPt-2406-Xwmc59* and *XwPt-8460-XwPt-3755*, and explained 8.2% and 12.2%, respectively, of the phenotypic variance in Gansu 2010. *QYr. caas-7BL. 2* was located between *Xgwm577* and *XwPt-4300* on chromosome 7BL, explaining 14.7% of the phenotypic variance in Sichuan 2011.

The total phenotypic variance explained by detected QTL ranged from 36.9 to 55.9% in a simultaneous fit across the three environments, suggesting a significant effect of these QTL in reducing YR severity.

Interactions among different QTL could not be stably identified across the three environments using IciMapping V3.1. All these 10 QTL associated with YR resistance showed additive effects (Table 2). To obtain the combined effects of these QTL, the flanking markers for these QTL regions were used to select RILs possessing the corresponding QTL, for example, RILs possessing the 1AL resistance allele were selected to present the effect of the 1AL QTL using the closely linked marker *XwPt-2406*. The results shown in Fig. 2 indicated that the more resistance genes a subset of RILs possess, the lower the disease severity. In addition, when four to five genes were combined in a RIL, the disease severity was less than 35% on average (Fig. 2; Sup-Table 1). This is consistent with a report by Singh et al. (2000a), who demonstrated that four to five APR genes, each with small to intermediate effects, may provide a high level of resistance to YR when combined in a cultivar.

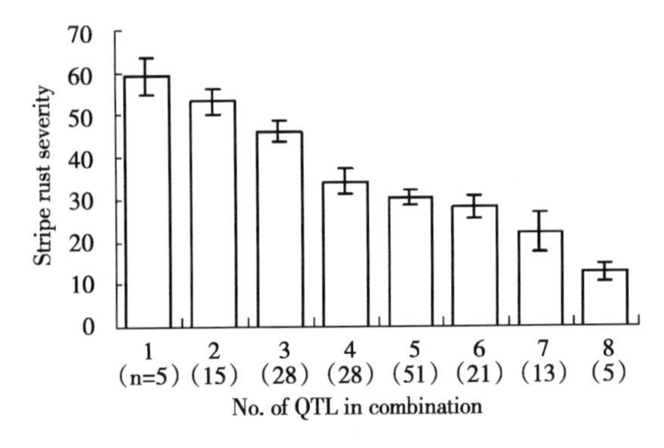

Fig. 2 The effect of different QTL combinations on stripe rust severity in different classes. The number in abscissa indicate the number of QTL combined in each subset of RILs, for example, 1 means the class of RILs possessing one QTL. The numbers in *parentheses* indicate the number of RILs in each class. The error bars indicate the standard error of sample means. A detailed list of the RILs in each class is shown in Sup-Table 1.

Discussion

Although the German cultivar Naxos was susceptible to currently prevalent *Pst* races CYR32 and CYR33 at the seedling stage, it showed a high level of resistance when inoculated with these races in the field. Five QTL for YR resistance were detected in Naxos. Although SHA3/CBRD was moderately susceptible to YR, it actually carried five QTL for APR.

Comparison with previous reports

QYr. caas-1AL

QYr. caas-1AL, with a resistance allele contributed by Naxos, was not stable across environments, but could be related to known resistance QTL against YR. Ramburan et al. (2004) identified an APR QTL on chromosome 1A, designated *QYr. sgi-1A*, which was also inconsistently detected across environments. Furthermore, Dedryver et al. (2009) identified a QTL on 1A in cultivar Renan, only at one scoring. Because of different kinds of flanking markers used in these studies, the relationships among these QTL are uncertain.

QYr. caas-1DS

This is the first report of QTL for YR resistance on chromosome 1D. Considering that Naxos was highly susceptible to currently prevalent *Pst* races CYR32 and CYR33 at seedling stages, *QYr. caas-1DS* contributed by Naxos is likely a new QTL for APR to YR.

QYr. caas-1BL. 1RS

The short arm of rye chromosome 1 (1RS) carries a variety of disease resistance genes, viz., *Yr9*, *Pm8*, *Lr26* and *Sr31* from 'Petkus' and *SrR* from 'Imperial' rye (Mago et al., 2002). Based on GISH analysis, SHA3/CBRD carries a translocated chromosome in which the entire 1RS has replaced the entire 1BS. Theoretically, there should not be recombination between 1BS and 1RS. However, we still found a cluster of markers with low recombination frequency on 1RS (Fig. 1). This is most likely due to inaccurate genotypic data as most of these markers are dominant markers. Although *Xcfd21* and *Xgdm33a* were mapped on the short arm of wheat homologous group 1 in several consensus maps (Somers et al., 2004; http: //wheat. pw. usda. gov), both of them gave specific amplification in SHA3/CBRD while no corresponding PCR products were detected in Naxos. The two SSR markers are therefore specific to 1B. 1R translocation in the present population. *QYr. caas-1BL. 1RS* was close to centromere on chromosome 1BL. 1RS, most closely associated with *Xiag95-1RS*. This QTL was possibly located on chromosome 1RS in a 1BL. 1RS translocation line SHA3/CBRD. The stripe rust resistance gene *Yr9* was also located on 1RS, which in turn was located 26.1±4.3cM from the centromere (Singh et al., 1990). However, this gene has succumbed to the *Pst* races used in the present study both at seedling and adult-plant stages (Yang and Ren, 2001; Ren et al., 2009). Therefore, *QYr. caas-1BL. 1RS* is likely different from *Yr9*. Recently, Luo et al. (2008) reported two genes (*YrCN17* and *YrR212* in cultivars CN17 and R212) for stripe rust resistance located on 1RS. Both genes areresistant to *Pst* races CYR31 and CYR32. Ren et al. (2009) also identified the gene (*YrCn17*) in the line R14, and concluded that this gene is likely to be an allele of *Yr9* based on allelism tests. But, these are major resistance genes rather than a QTL.

QYr. caas-2BL

Several QTL for YR resistance were reported on wheat chromosome 2BL, including *QYR1*, *QYr. inra-2BL*, *QTL 2BL*, *QYraq. cau-2BL* and *QYr. csiro-2BL* (Boukhatem et al., 2002; Mallard et al., 2005; Christiansen et al., 2006; Guo et al., 2008; Rosewarne et al., 2008). *QTL 2BL* was located between *Xwmc149* and *Xwmc317a* with a map distance of about 15cM from *QYr. caas-2BL* based on the consensus map (Somers et al., 2004), indicating that this QTL was unlikely to be the same locus as *QYr. caas-2BL*. *QYr. csiro-2BL* flanked by *Xgwm1027-Xgwm619* is also different from *QYr. caas-2BL* because it is more than 20cM from *QYr. caas-2BL* (Somers et al., 2004). In contrast, *QYR1*, *QYr. inra-2BL* and *QYraq. cau-2BL* were located in the marker interval *Xgwm47-Xgwm501*, *Xbarc101-Xgwm120* and *Xwmc175-Xwmc332*, respectively, which are very closely linked to *QYr. caas-2BL* (Somers et al., 2004). *QYR1* and *QYr. inra-2BL* were derived from cultivar Camp Remy. They accounted for 46.0% and 61% of the phenotypic variance, respectively. Another one, *QYraq. cau-2BL*, was contributed by cultivar Aquileja and explained up to 61.5% of the phenotypic variance. However, *QYr. caas-2BL* detected in the present study explained 12.2% of the phenotypic variance and the resistance gene came from Naxos. Pedigree analysis did not show any common ancestors between Naxos and the two cultivars. It is possible that *QYr. caas-2BL* is different from these QTL. However, this needs to be confirmed by allelism test in the future.

QYr. caas-2DL

The APR gene *Yr16* was located on the centromeric region of chromosome 2D (Worland and Law, 1986; Mallard et al., 2005), with a map distance of about 23cM from *QYr. caas-2DL* (Hart et al., 1993; Somers et al., 2004). This gene is therefore different from *QYr. caas-2DL*. Recently, Jagger et al. (2011) identified a QTL flanked by SSR markers *Xgwm320* and *Xgwm301* on 2DL in the German cultivar Alcedo. It was approximately 18cM from *QYr. caas-2DL* (Somers et al., 2004), suggesting that this locus is also likely different from *QYr. caas-2DL*. Suenaga et al. (2003) reported a QTL on 2DL in the cultivar Fukuho-Komugi, closely linked to the marker *Xgwm349*, which corresponds to the interval for *QYr. caas-2DL* (Somers et al., 2004). It explained 10.1 to 11.4% of the phenotypic variance in YR severity across environments, which is similar to *QYr. caas-2DL* in the present study. Melichar et al. (2008) also identified a QTL at the same chromosomal region in the UK wheat cultivar Guardian, which explained 8.0% to 11.0% of the phenotypic variance. Thus, these two QTL and *QYr. caas-2DL* might be same or closely linked.

QYr. caas-5AL. 2

Boukhatem et al. (2002) mapped a QTL, flanked by RFLP markers *Xfbb209* and *Xabg391*, and linked to the SSR marker *Xgwm126*. This QTL should be different from *QYr. caas-5AL. 2* based on a map distance of about 25cM between them (Somers et al., 2004). Chhuneja et al. (2008) identified a QTL on 5AL between *Xbarc151* and *Xcfd12*, with a map distance of 58cM from *QYr. caas-5AL. 2* (Somers et al., 2004). In addition, it was derived from the diploid A genome wheat species *T. boeoticum* (acc. Pau5088), indicating that this QTL was also different from *QYr. caas-5AL. 2*. Lan et al. (2010) located a QTL on 5AL in the Chinese landrace wheat Pingyuan 50, flanked by

Xwmc410 and *Xbarc261*, in the same region as *QYr. caas-5AL. 2*. Because pedigree analyses showed no relationship between SHA3/CBRD and Pingyuan 50, this QTL is also likely different from *QYr. caas-5AL. 2*. Recently, a QTL (*Yr48*) with large effect, flanked by *Xwmc727* and *Xgwm291* on 5AL, was validated in PI610750 (Lowe et al., 2011), at a similar position to *QYr. caas-5AL. 2* (Somers et al., 2004). However, PI610750 [Pedigree: Crocl/*Aegilops tauschii* (Synthetic 205) // Kauz] is a synthetic wheat derivative from CIMMYT's Wide Cross Program, and it has no common ancestor with SHA3/CBRD, suggesting that *Yr48* may be different from *QYr. caas-5AL. 2*. Therefore, *QYr. caas-5AL. 2* is possibly a new QTL for APR to YR.

QYr. caas-5BL. 3

Mallard et al. (2005) reported a QTL for YR resistance, flanked by *Xgwm234* and *XDuPw115a* on 5BL, in the same region as *QYr. caas-5BL. 3*. It might be in a translocated region from chromosome 5BS of cultivar Cappelle Desprez (Mallard et al., 2005; Lu et al., 2009). Lu et al. (2009) reported two APR QTL to YR on 5BL in cultivar Libellula. One of two QTL, *QYr. caas-5BL. 2*, flanked by *Xbarc142* and *Xgwm604*, was located to a similar position as *QYr. caas-5BL. 3* (Somers et al., 2004). It explained 2.6% of the phenotypic variance in one environment, which showed lower effect and stability than *QYr. caas-5BL. 3*.

QYr. caas-6BS. 2

A number of genes for YR resistance have been reported on chromosome 6BS. Among them, *QYr. jirc-6B* (Suenaga et al., 2003), *Yr36* (Uauy et al., 2005), *QYrst. wgp-6BS. 1* (Santra et al., 2008) and *QYr. caas-6BS* (Lan et al., 2010) were close to the centromere on 6BS, more than 20cM away from *QYr. caas-6BS. 2*. These QTL were therefore different from *QYr. caas-6BS. 2*. All-stage resistance gene *Yr35* (Dadkhodaie et al., 2010) on chromosome 6BS is linked to SSR marker *Xgwm508*, with a map distance of about 18cM from *QYr. caas-6BS. 2* (Somers et al., 2004). Dedryver et al. (2009) reported *QYr. inra-6B* flanked by SSR markers *Xgwm518* and *Xgwm608* with a map distance of about 13cM from *QYr. caas-6BS. 2* (Somers et al., 2004). Santra et al. (2008) detected a QTL on bin 6BS-satellite in cultivar Stephens, flanked by SSR markers *Xgwm132* and *Xgdm113* that coincided with *QYr. caas-6BS. 2* (Somers et al., 2004). This QTL explained a large proportion of the phenotypic variance (R^2=25-43%) across environments, whereas *QYr. caas-6BS. 2* only explained 2.9-5.7% across environments.

QYr. caas-7BL. 1 and *QYr. caas-7BL. 2*

Lin and Chen (2007) identified *Yr39* from cultivar Alpowa on chromosome 7BL, flanked by *Xgwm43* and *Xgwm131*. In the present study, the QTL *QYr. caas-7BL. 1* was located between *XwPt-8106* and *Xbarc176*. The map distance between that QTL and *QYr. caas-7BL. 1* is approximately 6cM (Somers et al., 2004). Thus, these two genes might be either at the same position or closely linked with each other. According to significant differences in map distance between *QYr. caas-7BL. 1* and the other QTL on 7BL (Imtiaz et al., 2004; Somers et al., 2004; Rosewarne et al., 2008), *QYr. caas-7BL. 1* described here might be different.

We detected a second QTL, *QYr. caas-7BL. 2*, which was in the marker interval *Xgwm577-XwPt-4300*, at the end of chromosome 7BL. As shown in Fig. 1, this QTL was approximately 63.5cM from *QYr. caas-7BL. 1*. Rosewarne et al. (2008) also identified a QTL in the telomere region of chromosome 7BL, flanked by *XP32/M59* and *Xgwm344*, but it was approximately 20cM away from *QYr. caas-7BL. 2* (Somers et al., 2004). Nevertheless, Imtiaz et al. (2004) identified *QYr. nsw-7B* closely linked to *Xgwm611*, with a map distance of only 1cM from *QYr. caas-7BL. 2* (Somers et al., 2004). It was contributed by the susceptible variety "Tiritea" and accounted for 42.0 and 37.0% of variation in adult plant resistance under greenhouse and field conditions. Consequently, *QYr. caas-7BL. 2* may be the same as *QYr. nsw-7B*, or they are closely linked genes.

Pleiotropic effects of detected genes

Previous studies indicated that APR genes involved in

resistance to rusts are clustered on wheat chromosomes or showed pleiotropic effects, such as, *Yr18/Lr34* (Suenaga et al., 2003; Spielmeyer et al., 2007), *Yr29/Lr46* (Rosewarne et al., 2006), *Yr30/Sr2* (Singh et al., 2000b), and *Yr46/Lr67* (Herrera-Foessel et al., 2010). Many studies confirmed that the same principle also applies to rusts and PM. For example, Lillemo et al. (2008) conducted QTL mapping for resistance to PM, LR and YR in a population from the cross between Saar and Avocet S, and found that QTL for PM resistance coincided with *Yr18/Lr34* and *Yr29/Lr46*, and designated as *Pm38* and *Pm39*, respectively. He et al. (2011) integrated all APR QTL for YR and PM resistance into a linkage map based on DNA marker information, and found eight gene clusters (≥5 QTL) conferring resistance to both YR and PM simultaneously. In addition, the durable stem rust resistance gene *Sr2* was also associated with LR (*Lr27*) and PM resistance (Lagudah, 2011; Mago et al., 2011).

The APR QTL to PM in the same population of SHA3/CBRD × Naxos was characterized by Lu et al. (2012). Compared with the study, *QYr. caas-1BL. 1RS* contributed by SHA3/CBRD was in the same region as a PM resistance QTL from the same parent, and *QYr. caas-2BL* corresponded well with a PM resistance QTL which was from Naxos, indicating pleiotropic effects of the resistance loci. Besides, there are coincident QTL for YR and PM resistance close to the centromere on 2DL, but the resistance is contributed by opposite parents. *QYr. caas-1BL. 1RS* and *QYr. caas-2BL* may play an important role for resistance breeding to both YR and PM.

The same populationwas also used for QTL mapping of Fusarium head blight (FHB) resistance (Lu et al., manuscript to be submitted). Comparison with this study shows that both *QYr. caas-1DS* and *QYr. caas-2BL* are co-located with FHB resistance QTL with the alleles for resistance contributed by Naxos. *QYr. caas-1DS* is therefore a valuable resistance resource for breeding to both YR and FHB while *QYr. caas-2BL* could be used to simultaneously improve resistance levels to YR, PM and FHB. Based on the same study, *QYr. caas-5AL* with YR resistance contributed by SHA3/CBRD coincided with a FHB resistance QTL from the same parent, while *QYr. caas-2DL* and *QYr. caas-5BL. 3* located to similar positions as FHB resistance QTL, but with resistance contributed by the opposite parents. Further studies are necessary to elucidate these effects, which can more likely be explained by close linkage of resistance genes than pleiotropy as both these FHB resistance QTL were associated with anther extrusion, which does not play any role as a resistance mechanism against YR.

Breeding and application

In this study, we detected four QTL conferring stable resistance to YR across environments, including *QYr. caas-1BL. 1RS*, *QYr. caas-1DS*, *QYr. caas-5BL. 3* and *QYr. caas-7BL. 1*, and they were closely linked to the SSR markers *Xbarc8*, *XUgwm353*, *Xbarc275* and *Xbarc176*, respectively. As the interval of flanking markers for each QTL was less than 6.5cM, these closely linked SSR markers could be used for improving resistance to YR resistance in wheat breeding.

As the research on race non-specific genes becomes more and more extensively, especially following the cloning of APR genes *Yr18* and *Yr36* (Krattinger et al., 2009; Fu et al., 2009), the mechanisms of APR are becoming better understood. The durability of any gene or gene combination continues to elude prediction and remains a matter of "time will tell" (Lowe et al., 2011). Nevertheless, the QTL reported in the present study, particularly *QYr. caas-1DS* was a new QTL for APR to YR, which should enable diversification of the genetic basis of partial and durable resistance to YR. Their closely linked molecular markers can be used in marker-assisted selection and pyramiding of APR genes to YR in wheat breeding.

Acknowledgments

We are grateful to the critical review of this manuscript

by Prof. R. A. McIntosh, Plant Breeding Institute, University of Sydney, Australia. We also thank Dr. Fangpu Han for his help in making Genomic in situ hybridization. This study was supported by the National Science Foundation of China (30821140351) and China Agriculture Research System (CARS-3-1-3).

References

Allard RW (1960) Princilpes of Plant Breeding. John Wiley and Sons, Inc. , New York, London.

Bariana HS, Hayden MJ, Ahmed NU, Bell JA, Sharp PJ, McIntosh RA (2001) Mapping of durable adult plant and seedling resistances to stripe rust and stem rust diseases in wheat. Aust J Agric Res 52: 1247-1255.

Bhavani S, Singh RP, Argillier O, Huerta-Espino J, Singh S, Njau P, Brun S, Lacam S, Desmouceaux N (2011) Mapping durable adult plant stem rust resistance to the race Ug99 group in six CIMMYT wheats. 2011 BGRI Technical Workshop, June 13-16, University of Minnesota and USDA Cereal Disease Lab, St. Paul, Minnesota, http: //www. globalrust. org/db/attachments/knowledge/122/2/Bhavani-revised. pdf

Boukhatem N, Baret PV, Mingeot D, Jacquemin JM (2002) Quantitative trait loci for resistance against yellow rust in two wheat derived recombinant inbred line populations. Theor Appl Genet 104: 111-118.

Bossolini E, Krattinger SG, Keller B (2006) Development of simple sequence repeat markers specific for the *Lr34* resistance region of wheat using sequence information from rice and *Aegilops tauschii*. Theor Appl Genet 113: 1049-1062.

Boyd LA (2005) Can Robigus defeat an old enemy? — Yellow rust of wheat. J Agric Sci 143: 233-243.

Carter AH, Chen XM, Garland-Campbell K, Kidwell KK (2009) Identifying QTL for high-temperature adult-plant resistance to stripe rust (*Puccinia striiformis* f. sp. *tritici*) in the spring wheat (*Triticum aestivum* L.) cultivar 'Louise' . Theor Appl Genet 119: 1119-1128.

Chen JL, Chu CG, Souza EJ, Guttieri MJ, Chen XM, Xu S, Hole D, Zemetra R (2012) Genome-wide identification of QTL conferring high-temperature adult-plant (HTAP) resistance to stripe rust (*Puccinia striiformis* f. sp. *tritici*) in wheat. Mol Breeding 29: 791-800.

Chen WQ, Wu LR, Liu TG, Xu SC, Jin SL, Peng YL, Wang BT (2009) Race dynamics, diversity, and virulence evolution in *Puccinia striiformis* f. sp. *tritici*, the causal agent of wheat stripe rust in China from 2003 to 2007. Plant Dis 93: 1093-1101.

Chen XM (2005) Epidemiology and control of stripe rust [*Puccinia striiformis* f. sp. *tritici*] on wheat. Can J Plant Pathol 27: 314-337.

Chen XM, Line RF (1995) Gene action in wheat cultivars for durable, high-temperature, adult-plant resistance and interaction with race-specific, seedling resistance to *Puccinia striiformis*. Phytopathology 85: 567-572.

Chhuneja P, Kaur S, Garg T, Ghai M, Kaur S, Prashar M, Bains NS, Goel RK, Keller B, Dhaliwal HS, Singh K (2008) Mapping of adult plant stripe rust resistance genes in diploid A genome wheat species and their transfer to bread wheat. Theor Appl Genet 116: 313-324.

Christiansen MJ, Feenstra B, Skovgaard IM, Andersen SB (2006) Genetic analysis of resistance to yellow rust in hexaploid wheat using a mixture model for multiple crosses. Theor Appl Genet 112: 581-591.

Dadkhodaie NA, Karaoglou H, Wellings CR, Park RF (2010) Mapping genes *Lr53* and *Yr35* on the short arm of chromosome 6B of common wheat with microsatellite markers and studies of their association with *Lr36*. Theor Appl Genet 122: 479-487.

Dedryver F, Paillard S, Mallard S, Robert O, Trottet M, Nègre S, Verplancke G, Jahier J (2009) Characterization of genetic components involved in durable resistance to stripe rust in the bread wheat 'Renan' . Phytopathology 99: 968-973.

Francis HA, Leitch AR, Koebner RMD (1995) Conversion of a RAPD-generated PCR product, containing a novel dispersed repetitive element, into a fast and robust assay for the presence of rye chromatin in wheat. Theor Appl Genet 90: 636-642.

Froidmont DD (1998) A co-dominant marker for the 1BL/1RS wheat-rye translocation via multiplex PCR. J Cereal Sci 27: 229-232.

Fu DL, Uauy C, Distelfeld A, Blechl A, Epstein L, Chen XM, Sela H, Fahima T, Dubcovsky J (2009) A kinase-start gene confers temperature-dependent resistance to wheat stripe rust. Science 323: 1357-1360.

Guo Q, Zhang ZJ, Xu YB, Li GH, Feng J, Zhou Y (2008) Quantitative trait loci for high-temperature adult-plant and slow-rusting resistance to *Puccinia striiformis* f. sp. *tritici* in wheat cultivars. Phytopathology 98: 803-809.

Gupta PK, Balyan HS, Edwards KJ, Isaac P, Korzun V, Röder M, Gautier MF, Joudrier P, Schlatter AR, Dubcovsky J, De la Pena RC, Khairallah M, Penner G, Hayden MJ, Sharp P, Keller B, Wang RCC, Hardouin J, Jack P, Leroy P (2002) Genetic mapping of 66 new microsatellite (SSR) loci in bread wheat. Theor Appl Genet 105: 413-422.

Hao YF, Chen ZB, Wang YY, Bland D, Buck J, Brown-Guedira G, Johnson J (2011) Characterization of a major QTL for adult plant resistance to stripe rust in US soft red winter wheat. Theor Appl Genet 123: 1401-1411.

Hart GE, Gale MD, McIntosh RA (1993) Linkage maps of *Triticum aestivum* (hexaploid wheat, 2n = 42, genomes A, B, and D) and *T. tauschii* (2n = 14, genome D). In: O' Brien SJ (ed) Genetic Maps: Locus Maps of Complex Genomes. Cold Spring Harbor Laboratory Press, Cold Spring Harbor, pp 6204-6219.

He ZH, Lan CX, Chen XM, Zou YC, Zhuang QS, Xia XC (2011) Progress and perspective in research of adult-plant resistance to stripe rust and powdery mildew in wheat. Scientia Agricultura Sinica 44: 2193-2215.

Herrera-Foessel SA, Lagudah ES, Huerta-Espino J, Hayden MJ, Bariana HS, Singh D, Singh RP (2010) New slow-rusting leaf rust and stripe rust resistance genes *Lr67* and *Yr46* in wheat are pleiotropic or closely linked. Theor Appl Genet 122: 239-249.

Imtiaz K, Ahmad M, Cromey MG, Griffin WB, Hampton JG (2004) Detection of molecular markers linked to the durable adult plant stripe rust resistance gene *Yr18* in bread wheat (*Triticum aestivum* L.). Plant Breeding 123: 401-404.

Jagger LJ, Newell C, Berry ST, MacCormack R, Boyd LA (2011) The genetic characterisation of stripe rust resistance in the German wheat cultivar Alcedo. Theor Appl Genet 122: 723-733.

Johnson R (1984) A critical analysis of durable resistance. Annu Rev Phytopathol 22: 309-330.

Johnson R (1988) Durable resistance to yellow (stripe) rust in wheat and its implications in plant breeding. In: Simmonds NW, Rajaram S (eds) Breeding Strategies for Resistance to the Rusts of Wheat. CIMMYT, Mexico D. F., pp 63-75.

Kosambi DD (1944) The estimation of map distance from recombination values. Annu Eugen 12: 172-175.

Krattinger SG, Lagudah ES, Spielmeyer W, Singh RP, Huerta-Espino J, McFadden H, Bossolini E, Selter LL, Keller B (2009) A putative ABC transporter confers durable resistance to multiple fungal pathogens in wheat. Science 323: 1360-1363.

Lagudah ES (2011) Molecular genetics of race non-specific rust resistance in wheat. Euphytica 179: 81-91.

Lan CX, Liang SS, Zhou XC, Zhou G, Lu QL, Xia XC, He ZH (2010) Identification of genomic regions controlling adult-plant stripe rust resistance in Chinese landrace Pingyuan 50 through bulked segregant analysis. Phytopathology 100: 313-318.

Li HH, Ribaut JM, Li ZL, Wang JK (2008) Inclusive composite interval mapping (ICIM) for digenic epistasis of quantitative traits in biparental populations. Theor Appl Genet 116: 243-260.

Li HH, Ye GY, Wang JK (2007) A modified algorithm for the improvement of composite interval mapping. Genetics 175: 361-374.

Lillemo M, Asalf B, Singh RP, Huerta-Espino J, Chen XM, He ZH, Bjørnstad Å (2008) The adult plant rust resistance loci *Lr34/Yr18* and *Lr46/Yr29* are important determinants of partial resistance to powdery mildew in bread wheat line Saar. Theor Appl Genet 116: 1155-1166.

Lin F, Chen XM (2007) Genetics and molecular mapping of genes for race-specific all-stage resistance and non-race-specific high-temperature adult-plant resistance to stripe rust in spring wheat cultivar Alpowa. Theor Appl Genet 114: 1277-1287.

Lin F, Chen XM (2009) Quantitative trait loci for non-race-specific, high-temperature adult-plant resistance to stripe rust in wheat cultivar Express. Theor Appl Genet 118: 631-642.

Line RF (2002) Stripe rust of wheat and barley in North America: a retrospective historical review. Annu Rev Phytopathol 40: 75-118.

Lowe I, Jankuloski LC, Chao SM, Chen XM, See D, Dubcovsky J (2011) Mapping and validation of QTL which confer partial resistance to broadly virulent post-2000 North American races of stripe rust in hexaploid wheat. Theor ApplGenet 123: 143-157.

Lu QX, Bjørnstad Å, Ren Y, Asad MA, Xia XC, Chen XM, Ji F, Shi JR, Lillemo M (2012) Partial resistance to powdery mildew in German spring wheat 'Naxos' is based on multiple genes with stable effects in diverse environments. Theor Appl Genet, DOI 10.1007/s00122-012-1834-6.

Lu YM, Lan CX, Liang SS, Zhou XC, Liu D, Zhou G, Lu QL, Jing JX, Wang MN, Xia XC, He ZH (2009) QTL mapping for adult-plant resistance to stripe rust in Italian common wheat cultivars Libellula and Strampelli. Theor

Appl Genet 119：1349-1359.

Luo PG，Zhang HY，Shu K，Zhang HQ，Luo HY，Ren ZL（2008）Stripe rust（*Puccinia striformis* f. sp. *tritici*）resistance in wheat with the wheat-rye 1BL/1RS chromosomal translocation. Can J Plant Pathol 30：254-259.

Mago R，Spielmeyer W，Lawrence G，Lagudah E，Ellis J，Pryor A（2002）Identification and mapping of molecular markers linked to rust resistance genes located on chromosome 1RS of rye using wheat-rye translocation lines. Theor Appl Genet 104：1317-1324.

Mago R，Tabe L，McIntosh RA，Pretorius Z，Kota R，Paux E，Wicker T，Breen J，Lagudah ES，Ellis JG，Spielmeyer W（2011）A multiple resistance locus on chromosome arm 3BS in wheat confers resistance to stem rust（*Sr2*），leaf rust（*Lr27*）and powdery mildew. Theor Appl Genet 123：615-623.

Mallard S，Gaudet D，Aldeia A，Abelard C，Besnard AL，Sourdille P，Dedryver F（2005）Genetic analysis of durable resistance to yellow rust in bread wheat. Theor Appl Genet 110：1401-1409.

Manly KF，Cudmore JRH，Meer JM（2001）Map Manager QTX，cross-platform software for genetic mapping. Genome 12：930-932.

McDonald B，Linde C（2002）The population genetics of plant pathogens and breeding strategies for durable resistance. Euphytica 124：163-180.

McIntosh RA，Dubcovsky J，Rogers WJ，Morris C，Appels R，Xia XC（2011）Catalogue of gene symbols for wheat：2011 supplement，http：//www. shigen. nig. ac. jp/wheat/komugi/genes/macgene/supplement2011. pdf

Melichar JPE，Berry S，Newell C，MacCormack R，Boyd LA（2008）QTL identification and microphenotype characterization of the developmentally regulated yellow rust resistance in the UK wheat cultivar Guardian. Theor Appl Genet 117：391-399.

Nagy ED，Eder C，Molnar-Lang M，Lelley T（2003）Genetic mapping of sequence-specific PCR-based markers on the short arm of the 1BL. 1RS wheat-rye translocation. Euphytica 132：243-250.

National Agro-technical Extension and Service Center（NAESC）（2010）Epidemic characteristics of wheat stripe rust in China in 2010 and its control strategies. China Plant Protection 3：32-35.

National Agro-technical Extension and Service Center（NAESC）（2011）Epidemic characteristics of wheat stripe rust in China in 2011 and its control strategies. China Plant Protection 2：28-32.

Parida SK，Kumar KAR，Singh VDNK，Mohapatra T（2006）Unigene derived microsatellite markers for the cereal genomes. Theor Appl Genet 112：808-817

Pestsova E，Ganal MW，Röder MS（2000）Isolation and mapping of microsatellite markers specific for the D genome of bread wheat. Genome 43：689-697.

Ramburan VP，Pretorius ZA，Louw JH，Boyd LA，Smith PH，Boshoff WHP，Prins R（2004）A genetic analysis of adult plant resistance to stripe rust in the wheat cultivar Kariega. Theor Appl Genet 108：1426-1433.

Ren TH，Yang ZJ，Yan BJ，Zhang HQ，Fu SL，Ren ZL（2009）Development and characterization of a new 1BL. 1RS translocation line with resistance to stripe rust and powdery mildew of wheat. Euphytica 169：207-213.

Röder MS，Korzun V，Wendehake K，Plaschke J，Tixier MH，Leroy P，Ganal MW（1998）A microsatellite map of wheat. Genetics 149：2007-2023.

Rosewarne GM，Singh RP，Huerta-Espino J，Rebetzke GJ（2008）Quantitative trait loci for slow-rusting resistance in wheat to leaf rust and stripe rust identified with multi-environment analysis. Theor Appl Genet 116：1027-1034.

Rosewarne GM，Singh RP，Huerta-Espino J，William HM，Bouchet S，Cloutier S，McFadden H，Lagudah ES（2006）Leaf tip necrosis，molecular markers and β1-proteasome subunits associated with the slow rusting resistance genes *Lr46/Yr29*. Theor Appl Genet 112：500-508.

Saal B，Wricke G（1999）Development of simple sequence repeat markers in rye（*Secale cereale* L.）. Genome 42：964-972.

Santra DK，Chen XM，Santra M，Campbell KG，Kidwell KK（2008）Identification and mapping QTL for high-temperature adult-plant resistance to stripe rust in winter wheat（*Triticum aestivum* L.）cultivar 'Stephens'. Theor Appl Genet 117：793-802.

Singh NK，Shepherd KW，McIntosh RA（1990）Linkage mapping of genes for resistance to leaf，stem and stripe rusts and ω-secalins on the short arm of rye chromosome 1R. Theor Appl Genet 80：609-616.

Singh RP，Huerta-Espino J，Rajaram S（2000a）Achieving near-immunity to leaf and stripe rusts in wheat by combining slow rusting resistance genes. Acta Phytopathol Hun 35：133-139.

Singh RP，Nelson JC，Sorrells ME（2000b）Mapping *Yr28* and other genes for resistance to stripe rust in wheat. Crop Sci 40：1148-1155.

Somers DJ，Isaac P，Edwards K（2004）A high-density microsatellite consensus map for bread wheat（*Triticum aesti-*

vum L.). Theor Appl Genet 109: 1105-1114.

Song QJ, Fickus EW, Cregan PB (2002) Characterization of trinucleotide SSR motifs in wheat. Theor Appl Genet 104: 286-293.

Sourdille P, Singh S, Cadalen T, Brown-Guedira GL, Gay G, Qi L, Gill BS, Dufour P, Murigneux A, Bernard M (2004) Microsatellite-based deletion bin system for the establishment of genetic-physical map relationships in wheat (*Triticum aestivum* L.). Funct Integr Genomics 4: 12-25.

Spielmeyer W, Singh RP, McFadden H, Wellings CR, Huerta-Espino J, Kong X, Appels R, Lagudah ES (2007) Fine scale genetic and physical mapping using interstitial deletion mutants of *Lr34* /*Yr18*: a disease resistance locus effective against multiple pathogens in wheat. Theor Appl Genet 116: 481-490.

Stubbs RW (1985) Striperust. In: Roelfs AP, Bushnell WR (eds) The Cereal Rusts Vol. II: Diseases, Distribution, Epidemiology, and Control. Academic Press, Orlando F. L., pp 61-101.

Suenaga K, Singh RP, Huerta-Espino J, William HM (2003) Microsatellite markers for genes *Lr34*/*Yr18* and other quantitative trait loci for leaf rust and stripe rust resistance in bread wheat. Phytopathology 93: 881-890.

Tang HJ, Yin GH, Xia XC, Feng JJ, Qu YY, He ZH (2009) Evaluation of molecnlar markers specific for 1BL・1RS translocation and characterization of 1RS chromosome in wheat varieties from different origins. Acta Agronomica Sinica 35: 2107-2115.

Tang JY (2004) Epidemic characteristics of wheat stripe rust in China in 2004 and its control strategies. China Rural Science 4: 28-29.

Uauy C, Brevis JC, Chen XM, Khan I, Jackson L, Chicaiza O, Distelfeld A, Fahima T, Dubcovsky J (2005) High-temperature adult-plant (HTAP) stripe rust resistance gene *Yr36* from *Triticum turgidum* ssp. *dicoccoides* is closely linked to the grain protein content locus *Gpc-B1*. Theor Appl Genet 112: 97-105.

Wan AM, Zhao ZH, Chen XM, He ZH, Jin SL, Jia QZ, Yao G, Yang JX, Wang BT, Li GB, Bi YQ, Yuan ZY (2004) Wheat stripe rust epidemic and virulence of *Puccinia striiformis* f. sp. *tritici* in China in 2002. Plant Dis 88: 896-904.

Wang JK (2009) Inclusive composite interval mapping of quantitative trait genes. Acta Agronomica Sinica 35: 239-245.

Wang S, Basten CJ, Zeng ZB (2005) Windows QTL cartographer v2.5 statistical genetics. North Carolina State University, Raleigh, NC.

William HM, Singh RP, Huerta-Espino J, Palacios G, Suenaga K (2006) Characterization of genetic loci conferring adult plant resistance to leaf rust and stripe rust in spring wheat. Genome 49: 977-990.

Worland AJ, Law CN (1986) Genetic analysis of chromosome 2D of wheat I. The location of genes affecting height, day-length insensitivity, hybrid dwarfism and yellow-rust resistance. Z. Pflanzenzuchtung 96: 331-345.

Xue SL, Zhang ZZ, Lin F, Kong ZX, Cao Y, Li CJ, Yi HY, Mei MF, Zhu HL, Wu JZ, Xu HB, Zhao DM, Tian DG, Zhang CQ, Ma ZQ (2008) A high-density intervarietal map of the wheat genome enriched with markers derived from expressed sequence tags. Theor Appl Genet 117: 181-189.

Yang ZJ, Ren ZL (2001) Chromosomal distribution and genetic expression of *Lophopyrum elongatum* (Host) A. Löve genes for adult plant resistance to stripe rust in wheat background. Genet Resour Crop Evol 48: 183-187.

Yu DZ, Yang XJ, Yang LJ, Jeger MJ, Brown JKM (2001) Assessment of partial resistance to powdery mildew in Chinese wheat varieties. Plant Breeding 120: 279-284.

Yu JK, Dake TM, Singh S, Benscher D, Li WL, Gill B, Sorrells ME (2004) Development and mapping of EST-derived simple sequence repeat markers for hexaploid wheat. Genome 47: 805-818.

Yu L, He F, Chen GL, Cui F, Qi XL, Wang HG, Li XF (2011) Identification of 1BL・1RS wheat-rye chromosome translocations via 1RS specific molecular markers and genomic in situ hybridization. Acta Agronomica Sinica 37: 563-569.

Zhang YJ, Jiang YY, Feng XD, Xia B, Zeng J, Liu Y (2009) Epidemic characteristics of wheat stripe rust in China in 2009 and its control strategies. China Plant Protection 3: 33-35.

Zhang YJ, Wang JQ, Jiang YY, Feng XD, Xia B, Liu Y, Zeng J (2007) Epidemic characteristics of wheat stripe rust in China in 2007 and its control strategies. China Plant Protection 2: 32-35.

Zhang YJ, Wang JQ, Jiang YY, Feng XD, Xia B, Liu Y, Zeng J (2008) Epidemic characteristics of wheat stripe rust in China in 2008 and its control strategies. China Plant Protection 3: 38-40.

Zhang YJ, Wang JQ, Jang YY, Xia B (2005) Epidemic characteristics of wheat stripe rust in China in 2005 and its control strategies. China Plant Protection 25: 28-30.

Zhang YJ, Wang JQ, Jang YY, Xia B (2006) Epidemic characteristics of wheat stripe rust in China in 2006 and its control strategies. China Plant Protection 4: 5-8.

Zhao ZH (2004) Epidemic characteristics of wheat stripe rust in China in 2003 and its control strategies. China Plant Protection 24: 16-18.

QTL mapping of adult-plant resistance to leaf rust in a RIL population derived from a cross of wheat cultivars Shanghai 3/Catbird and Naxos

Y. Zhou[1,2], Y. Ren[3], M. Lillemo[4], Z. J. Yao[1], P. P. Zhang[1], X. C. Xia[5], Z. H. He[5,6], Z. F. Li[1], and D. Q. Liu[1]

[1]*Department of Plant Pathology, College of Plant Protection, Hebei Agricultural University, Biological Control Center for Plant Diseases and Plant Pests of Hebei*, 289 *Lingyusi Street, Baoding* 071001, *Hebei, China*; [2]*Baoding University*, 3027 *Qiyi Donglu Street, Baoding* 071001, *Hebei, China*; [3]*College of Agronomy, Henan Agricultural University*, 63 *Nongye Road, Zhengzhou* 450002, *Henan, China*; [4]*Department of Plant Science, Norwegian University of Life Sciences, P. O. Box* 5003, *NO*-1432 *Ås, Norway*; [5]*Institute of Crop Science, National Wheat Improvement Center/The National Key Facility for Crop Gene Resources and Genetic Improvement, Chinese Academy of Agricultural Sciences (CAAS)*, 12 *Zhongguancun South Street, Beijing* 100081, *China*; [6]*International Maize and Wheat Improvement Center (CIMMYT) China Office, c/o CAAS*, 12 *Zhongguancun South Street, Beijing* 100081, *China*.

Abstract: Leaf rust is an important wheat disease worldwide and utilization of adult-plant resistance (APR) may be the best approach to achieve long-term protection from the disease. The CIMMYT spring wheat line Shanghai 3/Catbird showed a high level of APR to Chinese *P. triticina* pathotypes in the field. To identify the APR genes in the cultivar, a mapping population of 164 recombinant inbred lines (RILs) was developed from a cross between Shanghai 3/Catbird (SHA3/CBRD) and spring wheat Naxos, a moderately susceptible line from Germany. The RILs were evaluated for final disease severity (FDS) in Baoding, Hebei province, and Zhoukou, Henan province, in the 2010-2011 and 2011-2012 cropping seasons. Six QTL for resistance were detected on chromosomes 1AL, 2BS, 2DL, 5B, 7BS, and 7DS, explaining 5.3-5.5%, 15.3-37.4%, 5.7%, 6.0%, 4.2-5.3%, and 4.4% of the phenotypic variance, respectively. SHA3/CBRD also possessed seedling resistance gene *Lr26*, and Naxos contained *Lr1* based on gene postulation from tests with an array of *P. triticina* pathotypes and molecular marker assays. These seedling resistance and APR genes and their closely linked molecular markers are potentially useful for improving leaf rust resistance in wheat breeding programs.

Introduction

Leaf rust (LR), caused by *Puccinia triticina*, is one of the most important wheat diseases worldwide. It is adapted to a wide range of environments, occurs wherever wheat is grown, and can cause significant yield losses (Knott, 1989). During the past decades, leaf rust has periodically caused destructive yield losses in various Chinese wheat regions, particularly northern China and the Yellow and Huai valleys (Zhao et al., 2008). In 2012, significant yield losses were recorded in the provinces of Gansu, Sichuan, Shaanxi, Henan and Anhui (Zhou et al., 2013). Resistant cultivars are

Published in TAG, 2014, 127: 1873-1883

the most efficient, economic and environmentally friendly means to reduce losses caused by leaf rust.

Much progress has been made in searching for resistance to leaf rust in wheat. To date, 72 leaf rust resistance genes have been catalogued (McIntosh et al., 2013). Most of known genes confer race-specific resistance that generally lacks durability. Therefore, it is important to identify and use slow rusting types of adult-plant resistance (APR) in breeding programs as current evidence suggest that such sources are more likely to be durable. Cultivars with slow rusting non-hypersensitivity types of APR produce rust symptoms at later growth stages, but at levels that do not lead to significant losses, or at levels that lead to losses significantly lower than susceptible controls. Whereas this type of resistance tends to be more durable (Bjarko and Line, 1988), that is, less likely to be overcome by changes in the pathogen population, it tends to be more interactive with the environment. In epidemiological terms slow rusting APR is envisaged as slow disease development in the field despite high infection types, longer latent periods, low infection frequencies, smaller uredinial size and reduced duration of sporulation, and less spore production (Caldwell, 1968). Chinese landrace Pingyuan 50, Californian CIMMYT-derived variety Anza, and Mexican line Pavon 76 with these characteristics have shown high, stable, and durable resistance to leaf rust (Zhang et al., 2009). Of the 72 cataloged leaf rust resistance genes, four, viz. *Lr34* (Dyck, 1977), *Lr46* (Singh et al., 1998), *Lr67* (Herrera-Foessel et al., 2011), and *Lr68* (Herrera-Foessel et al., 2012), are APR or slow leaf rusting genes that also confer resistance to stripe rust and powdery mildew. Dyck and Samborski (1982) found first reported slow rusting resistance gene *Lr34* in the Brazilian wheat cultivar Frontana. Dyck (1987) located *Lr34* on chromosome 7D. McIntosh (1992) and Singh (1992) independently discovered that *Lr34* is genetically associated with gene *Yr18*, conferring slow rusting resistance to stripe rust. *Lr34* confers a moderate level of leaf rust resistance when present alone; however, combinations with additional slow rusting genes result in high resistance levels in most environments (Singh et al., 2000a). Nelson et al. (1997) identified two QTL for leaf rust resistance, one located on chromosome 7DS (*Lr34*), the other was located on 2BS, and together they explained 45% of the phenotype variation. Krattinger et al. (2009) showed the *Lr34* protein resembled ABC (ATP-binding cassette) transporters. Singh et al. (1998) identified the gene *Lr46* for slow leaf rusting on chromosome 1B in cultivar Pavon 76. William et al. (2003) mapped *Lr46* to the terminal bin of 1BL, closely linked to AFLP marker PstAAgMseCTA-1, and co-located with a slow stripe rusting gene later named *Yr29*. Lillemo et al. (2008) reported that *Lr46/Yr29* co-located with a powdery mildew resistance gene later named *Pm39*, and that *Lr34/Yr18* co-located with *Pm38*. Mutation studies have now confirmed the pleiotropy of these genes (Krattinger et al., 2011). Herrera-Foessel et al. (2011) mapped a third slow rusting resistance gene *Lr67/Yr46* on chromosome 4DL, closely linked to SSR markers *gwm165* and *gwm192*. This gene is also associated with partial resistance to powdery mildew (Herrera-Foessel et al., 2014). Herrera-Foessel et al. (2012) found at least three APR QTL for leaf rust in wheat line Parula, including those on chromosomes 7DS and 1BL controlled by *Lr34* and *Lr46*, respectively, and a third gene on chromosome 7BL designated as *Lr68*. At present, around 80 leaf rust APR QTL were located on 16 wheat chromosomes.

Although quantitative resistance is more durable than major gene resistance, there is an opinion that it will be overcome by slow evolution of the pathogen (McDonald and Linde, 2002). It is therefore very important to identify further sources of slow rusting genes for breeding cultivars with durable resistance. Ren et al. (2012b) detected three QTL for resistance to leaf rust on chromosomes 1BL and 6BS (2 genes) in Chinese wheat cultivar Bainong 64; those QTL explained 14.9-21.2%, 10.0-11.2% and 9.0-9.7% of the phenotypic variance, respectively.

The CIMMYT spring wheat cultivar Shanghai 3/Catbird has a high level of APR to Chinese *P. triticina* pathotypes in the field, but is susceptible to some Chinese *P. triticina* pathotypes at the seedling stage. QTL mapping of APR to powdery mildew and stripe rust in a SHA3/CBRD× Naxos RIL population recently reported by Lu et al. (2012) and Ren et al. (2012a), respectively. However, similar studies on APR to LR in this population have not been reported. The objectives of this study were to identify the QTL for APR to LR in the SHA3/CBRD × Naxos population, and to assess the stability of the QTL detected across environments.

Materials and methods

Plant materials and *P. triticina* pathotypes

One hundred and sixty-four F_6 RILs developed by single-seed descent from the cross of SHA3/CBRD × Naxos were used for mapping of QTL for APR to leaf rust. Naxos (Tordo/St. Mir808-Bastion//Minaret), a spring wheat cultivar from Germany, is susceptible at the seedling stage to some Chinese *P. triticina* pathotypes, and usually displays susceptibility in the field. SHA3/CBRD (Shanghai 3//Chuanmai 18/Bagula), developed by the International Maize and Wheat Improvement Center (CIMMYT), is susceptible to some Chinese *P. triticina* pathotypes at the seedling stage, but is highly resistant in the field. SHA3/CBRD has the 1B. 1R translocation (Ren et al., 2012a). Both parents are adapted to growing conditions in China. A set of 36 differential lines, mostly near-isogenic lines in the background of Thatcher with known leaf rust resistance genes, and 14 *P. triticina* pathotypes collected from China were included in the seedling tests (Table 1). The 14 pathotypes were designated following the coding system of Long and Kolmer (1989), with addition of a fourth letter for the reactions to a fourth quartet differentials (http: //www. ars. usda. gov/SP2 UserFiles/ad _ hoc/36400500Cerealrusts/pt _ nomen. pdf).

Seedling tests in the greenhouse

Naxos, SHA3/CBRD and 36 wheat lines with known leaf rust resistance genes were tested with 14 Chinese *P. triticina* pathotypes (Table 1). Seedlings were grown in a growth chamber. When the first leaf was fully expanded, inoculations were performed by brushing urediniospores from fully infected susceptible genotype Zhengzhou 5389 onto the seedlings to be tested. Inoculated seedlings were placed in plastic-covered cages and incubated at 15℃ and 100% relative humidity (RH) for 24h in darkness. They were then transferred to a growth chamber programmed with 12h light/12h darkness at 18 to 22℃ and 70% RH. Infection types (IT) were scored 10 to 14days after inoculation according to the Stakman scale as modified by Roelfs et al. (1992). Plants with IT 0 to 2+ were considered to be resistant and those with IT 3 to 4 were susceptible. Genes in the lines were postulated following Dubin et al. (1989).

Field trials

The 164 F_6 RILs and their parents were evaluated for leaf rust reaction in Baoding in Hebei and Zhoukou in Henan in the 2010-2011 and 2011-2012 cropping seasons. Both locations are hotspots for leaf rust with ideal conditions for rust infection and spread. Field trials were conducted in randomized complete blocks with three replicates at each location. Each plot consisted of a single 1.5m row with 30cm between rows. Approximately 100 seeds were sown in each row. Every tenth row was planted with the highly susceptible line Zhengzhou 5389 as a control and to aid the spread of the spores within the trial. Additional rows of Zhengzhou 5389 were planted perpendicularly and adjacent to the test rows. LR epidemics were initiated by spraying aqueous suspensions of urediniospores containing equal amounts of *P. triticina* pathotypes MHJS, THJL, and PHGP to which a few drops of Tween 20 (0.03%) were added, onto the spreader rows at tillering. These pathotypes were virulent on SHA3/CBRD and Naxos at the seedling test (Table 1). Disease severities were scored according to the modified Cobb scale (Peterson et al., 1948) two or three times at weekly intervals with the first scoring 4 weeks after inoculation. The final disease severity (FDS, %) was recorded for each line

Table 1 Seedling infection types[a] on 36 wheat lines with known leaf rust resistance genes when tested with 14 pathotypes of *Puccinia triticina* collected from China

Tester	*Lr* gene	Infection types to *Puccinia triticina* pathotypes[a]													
		PHKS	MHJS	FHDQ①	FGBQ	FHBQ	FHDQ②	THJL	FHDR	FGDQ	FHDS	THJP	TGTT	PHGP	THJC
RL6003	*Lr1*	4	4	;	;	;	;	4	0	;	0	4	4	4	4
Naxos	*Lr1*	4	4	;	0	0;	0	3	;	0	0	3	4	3	1
RL6016	*Lr2a*	;	;	1+	;	;	1	3	;	;	2	3	3	;	4
RL6047	*Lr2c*	4	1	4	4	4	4	4	4	4	4	4	4	4	4
RL6002	*Lr3*	4	4	4	4	4	4	4	4	4	4	4	4	4	4
RL6010	*Lr9*	;	;	;	;	0;	0	;	0	0	;	;	;	0	;
RL6005	*Lr16*	4	4	4	4	4	4	3+	4	4	4	4	4	3	4
RL6064	*Lr24*	; 1	;	;	;	;	;	;	;	;	;	;	;	;	;
RL6078	*Lr26*	4	4	4	1	4	4	4	4	1	4	4	; 1	4	4
SHA3/CBRD	*Lr26+*	1	4	3	0	3	3	4	3	0	;	1	;	3+	3+
RL6007	*Lr3ka*	X	X	;	;	;	;	1	;	;	;	1	4	;	X
RL6053	*Lr11*	4	4	1	;	;	1+	3+	1	1	2	4	3+	4	4
RL6008	*Lr17*	4	3+	3+	2	2	3+	4	3+	4	4	4	4	2+	4
RL6049	*Lr30*	3C	1	1	;	;	;	1	;	;	;	;	4	;	1
RL6051	*LrB*	3+	4	4	4	3+	4	3+	4	4	4	4	4	4	X
RL6004	*Lr10*	3	3	4	4	4	4	2	4	4	4	2+	4	1	X
RL6013	*Lr14a*	4	4	X	X	X	X	X	X	2	4	4	4	3+	X
RL6009	*Lr18*	1	1+	2	2	2	2	1+	4	2+	2	4	3+	3C	3
RL6019	*Lr2b*	1	0;	4	;	3	3+	4	3	3+	3+	2	4	3C	4
RL6042	*Lr3bg*	4	4	4	4	4	3+	4	4	4	4	4	4	4	4
Manitou	*Lr13*	3	4	4	3	3	4	3	3	2	3+	4	4	4	4
RL6006	*Lr14b*	4	4	4	4	4	4	4	4	4	4	X	4	X	4
RL6052	*Lr15*	1	;	;	;	;	;	4	1	;	;	4	3+	4	4
RL6040	*Lr19*	0	0	;	0	0	;	0	0	0	;	0	0	0	;
RL6092	*Lr20*	4	4	;	;	;	;	;	;	;	;	4	1	4	;
RL6043	*Lr21*	4	2	2	;	2+	3	;	1	;	1+	;	3	1	1
RL6012	*Lr23*	4	4	4	3+	3+	4	1	4	4	4	4	4	3+	4
RL6079	*Lr28*	0	0	0	0	0	0	0	0	0	0	0	0	0	0
RL6080	*Lr29*	0	0	0	0	;	0	;	;	0	3+	4	;	0	0
RL6057	*Lr33*	3	4	3+	3+	3+	4	3C	3+	3	4	4	4	3+	3+
E84018	*Lr36*	4	2	1+	;	2	2	1	2	2+	3	2+	3+	2+	3+
KS86 NGRC02	*Lr39*	;	; 1	;	;	;	;	;	;	;	;	;	;	;	;
KS91 WGRC11	*Lr42*	;	;	;	0	0	;	;	1	;	;	0	;	0	1
RL6147	*Lr44*	1	;	4	4	4	4	1	4	4	4	;	1+	; 1	1
RL6144	*Lr45*	4	4	4	4	4	4	;	4	4	4	4	;	;	;
PAVON + Lr47	*Lr47*	0	0	0	0	0	0	0	0	0	0	0	0	0	0
C78. 5	*Lr51*	;	;	;	;	1	;	;	;	;	;	0	;	;	;
98M71	*Lr53*	;	0	0	0	0	0	;	0	0	0	0	0	0	0

a. According to the 0-4 Stakeman scale as modified by Roelfs et al. (1992).

+Uredinia somewhat larger than normal for the IT.

in each environment as a percentage leaf area infected when the susceptible check Zhengzhou 5389 displayed its FDS around the 25th of May in Zhoukou and 10th of June in Baoding.

Genotyping

Nine hundred and fifty-two simple sequence repeat (SSR) markers were screened on SHA3/CBRD and Naxos, including BARC (Songet al., 2002), CFA, CFD and GPW (Sourdille et al., 2004), CNL and KSUM (Yu et al., 2004), GDM (Pestsova et al., 2000), GWM (Röder et al., 1998), MAG (Xue et al., 2008), SWM (Bossolini et al., 2006; Krattinger et al., 2009), UGWM (Parida et al., 2006) and WMC (Gupta et al., 2002) markers. Polymorphic SSR markers were then used to genotype all 164 RILs from the cross of SHA3/CBRD × Naxos by polyacrylamide gel electrophoresis. The population was also screened with diversity array technology (DArT) markers. In addition, the rye (*Secale cereale* L.) 1RS-specific marker (Chai et al., 2006) was included for detection of the 1B.1R translocation.

Statistical analysis

Analysis of variance was performed with PROC GLM in the Statistical Analysis System (SAS Institute, V8), with genotype as a fixed effect, and environments, locations × years, and replicates as random effects. The information in the ANOVA (Table 2) was used to calculate broad sense heritabilities (h_b^2) based on: $h_b^2 = \sigma_g^2 / (\sigma_g^2 + \sigma_{ge}^2/e + \sigma_\varepsilon^2/re)$ (Allard 1960), where $\sigma_g^2 = (MS_f - MS_{fe})/re$, $\sigma_{ge}^2 = (MS_{fe} - MS_e)/r$ and $\sigma_\varepsilon^2 = MS_e$; σ_g^2 is genetic variance, σ_{ge}^2 is genotype × environment interaction variance, σ_ε^2 is error variance, MS_f is mean square of genotype, MS_{fe} is mean square of genotype × environment interaction, MS_e is mean square of error, r is number of replicates and e is number of environments.

Map construction and QTL analysis

The genotypic data for markers were used to construct genetic linkage maps with the software MapManager QTX20 (Manly et al., 2001). Genetic distances between markers were estimated using the Kosambi mapping function (Kosambi, 1944). The assignment of linkage groups on chromosomes was checked according to previously published wheat consensus maps (Somers et al., 2004; http://www.wheat.pw.usda.gov).

QTL for leaf rust resistance were detected using the mean FDS of three replicates in each environment. QTL analysis was conducted using the ICIM-ADD function with the software QTL IciMapping 3.1 (Li et al., 2007). A logarithm of odds (LOD) threshold of 2.5, calculated from 1,000 permutations at a probability of 0.05, was used for declaring definitive QTL. A walk speed of 1.0cM was chosen for all QTL estimations. QTL effects were estimated as the proportion of phenotypic variance (R^2) explained (PVE) by the QTL.

Results

Phenotypic evaluation

Leaf rust developed well in all trials. The frequency distribution of leaf rust FDS for the 164 RILs in each environment revealed a continuous distribution skewed towards resistance (Fig. 1), indicating polygenic inheritance. The FDS of the susceptible control Zhengzhou 5389 ranged from 70 to 100%, from 90 to 100%, from 80 to 90%, and from 80 to 90% in Baoding 2011, Baoding 2012, Zhoukou 2011 and Zhoukou 2012, respectively. SHA3/CBRD had a mean FDS of 0-5% across the four environments, whereas Naxos was rated with a mean FDS of 15-70%. Similar results were obtained across locations and years (Fig. 1). The broad-sense heritability of FDS was 0.88. ANOVA showed significant differences ($P = 0.01$) in FDS among RILs, environments, and line × environment interactions (Table 2).

Construction of linkage maps

A total of 952 SSR markerswas tested for polymorphism between SHA3/CBRD and Naxos, and 279 (29.3%) were polymorphic. The latter were used to genotype individual RILs from the population for construction of genetic linkage maps. Additionally, 283 polymorphic DArT markers covering all chromosomes and six molecular markers

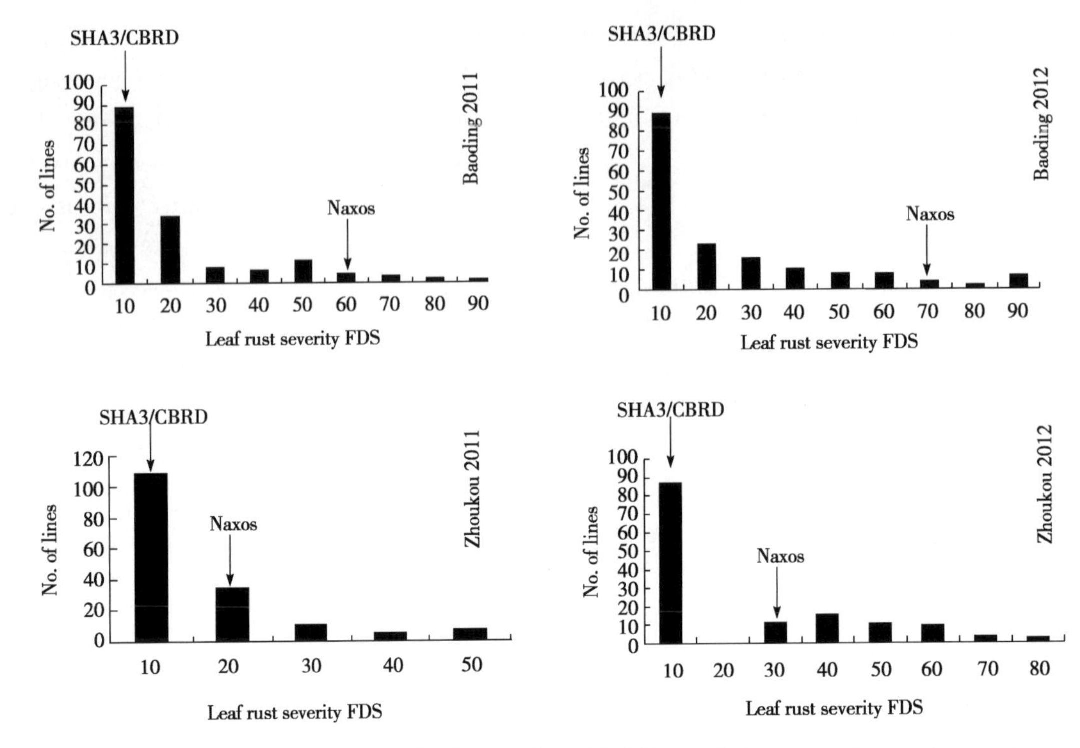

Fig. 1 Distribution for FDS in the SHA3/CBRD × Naxos F_6 RIL population across four environments

Table 2 Analysis of variance of FDS scores for RILs generated from the cross SHA3/CBRD × Naxos

Source of variance	*df*	Mean square	*F* value	*P*
Lines	163	2, 853	110. 1**	<0. 0001
Environments	3	15, 160	585. 2**	<0. 0001
Replicates	2	14	0. 5	0. 5848
Line × environment	489	344	13. 3**	<0. 0001
Error	984	26		

specific for 1B. 1R translocation were used for mapping the population.

Resistance genes identified from seedling reactions or molecular markers

Variation in ITand IT arrays conferred by 36 known leaf rust genes in differential lines, inoculated with 14 *P. triticina* pathotypes (Table 1), provided an ability to postulate 17 resistance genes (*Lr1*, *Lr2a*, *Lr2b*, *Lr3ka*, *Lr10*, *Lr11*, *Lr14a*, *Lr15*, *Lr17a*, *Lr18*, *Lr20*, *Lr21*, *Lr26*, *Lr30*, *Lr36*, *Lr44*, and *Lr45*). Resistance genes *Lr9*, *Lr19*, *Lr24*, *Lr28*, *Lr29*, *Lr39*, *Lr42*, *Lr47*, *Lr51* and *Lr53* conferred low ITs to all pathotypes. Postulation of genes *Lr2c*, *Lr3*, *Lr3bg*, *Lr13*, *Lr14b*, *Lr16*, *Lr23*, *Lr33*, and *LrB* was not possible because high infection types (ITs) were recorded with most pathotypes. SHA3/CBRD displayed low ITs with three *Lr26*-avirulent pathotypes (FGBQ, FGDQ, and TGTT) and three *Lr26*-virulent pathotypes (PHKS, FHDS, and THJP) (Table 1), indicating SHA3/CBRD possessed *Lr26*, which was confirmed using the 1BL. 1RS specific markers (Chai et al., 2006), plus other unknown genes. Naxos was postulated to contain *Lr1* or *Lr11* because it gives low ITs to seven *Lr1*-avirulent or *Lr11*-avirulent pathotypes. The presence of *Lr1* in Naxos was confirmed using molecular marker WR003 (Qiu et al., 2007).

QTL for LR resistance

Usingin clusive composite interval mapping (ICIM) analysis, 6 QTL for APR were identified on chromosomes 1AL, 2BS, 2DL, 5B, 7BS, and 7DS based on the mean FDS in Baoding 2011, Baoding 2012, Zhoukou 2011 and Zhoukou 2012 (Fig. 2, Table 3). The resistance

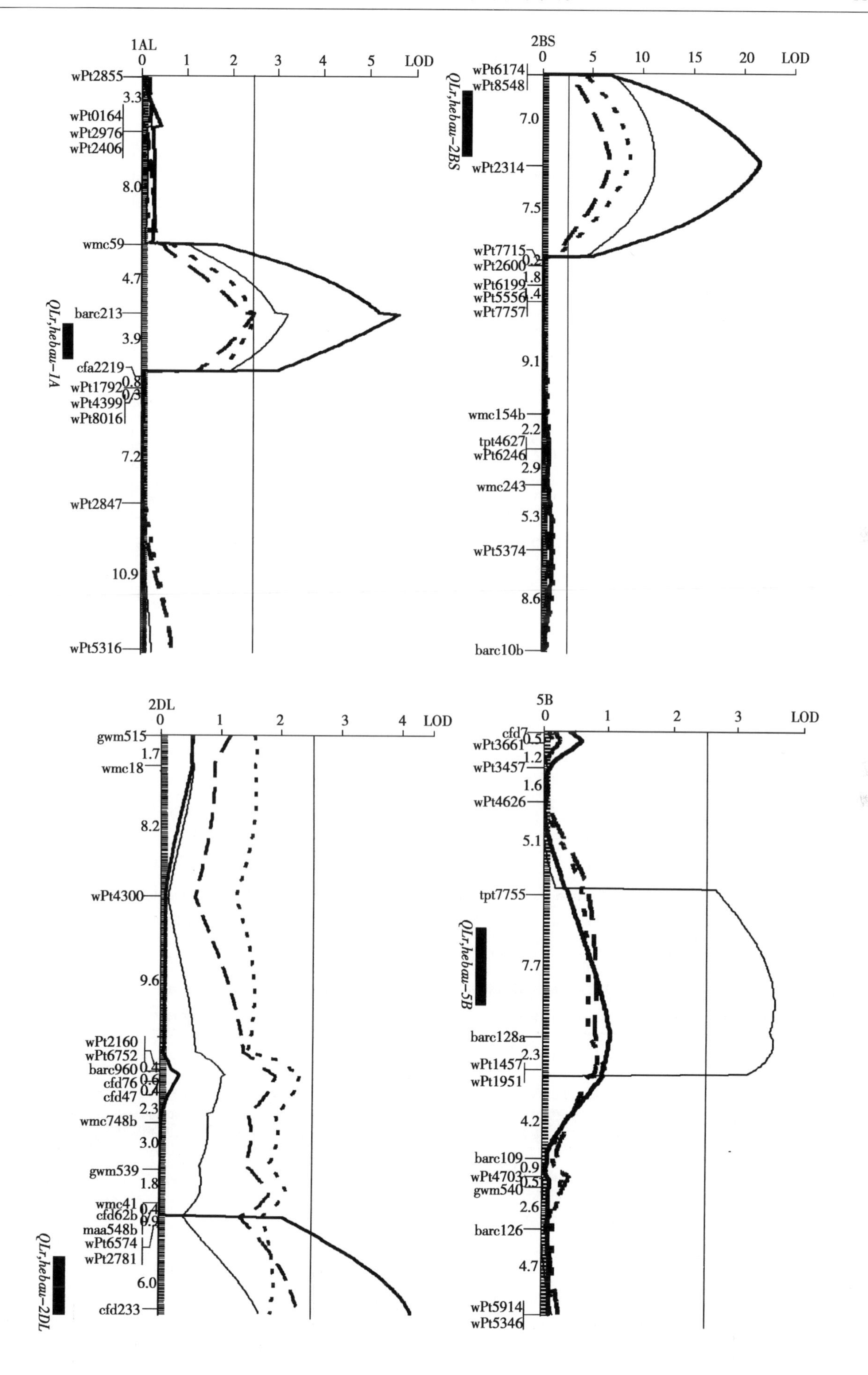
1AL
0 1 2 3 4 5 LOD
wPt2855
3.3
wPt0164
wPt2976
wPt2406
8.0
wmc59
4.7
barc213
3.9
cfa2219
wPt1792
wPt4399
wPt8016
7.2
wPt2847
10.9
wPt5316
QLr,hebau-1A
2BS
0 5 10 15 20 LOD
wPt6174
wPt8548
7.0
wPt2314
7.5
wPt7715
wPt2600
wPt6199
wPt5556
wPt7757
9.1
wmc154b
2.2
tpt4627
wPt6246
2.9
wmc243
5.3
wPt5374
8.6
barc10b
QLr,hebau-2BS
2DL
0 1 2 3 4 LOD
gwm515
1.7
wmc18
8.2
wPt4300
9.6
wPt2160
wPt6752
barc960
cfd76
cfd47
2.3
wmc748b
3.0
gwm539
1.8
wmc41
cfd62b
maa548b
wPt6574
wPt2781
6.0
cfd233
QLr,hebau-2DL
5B
0 1 2 3 LOD
cfd7
wPt3661
1.2
wPt3457
1.6
wPt4626
5.1
tpt7755
7.7
barc128a
2.3
wPt1457
wPt1951
4.2
barc109
wPt4703
gwm540
2.6
barc126
4.7
wPt5914
wPt5346
QLr,hebau-5B

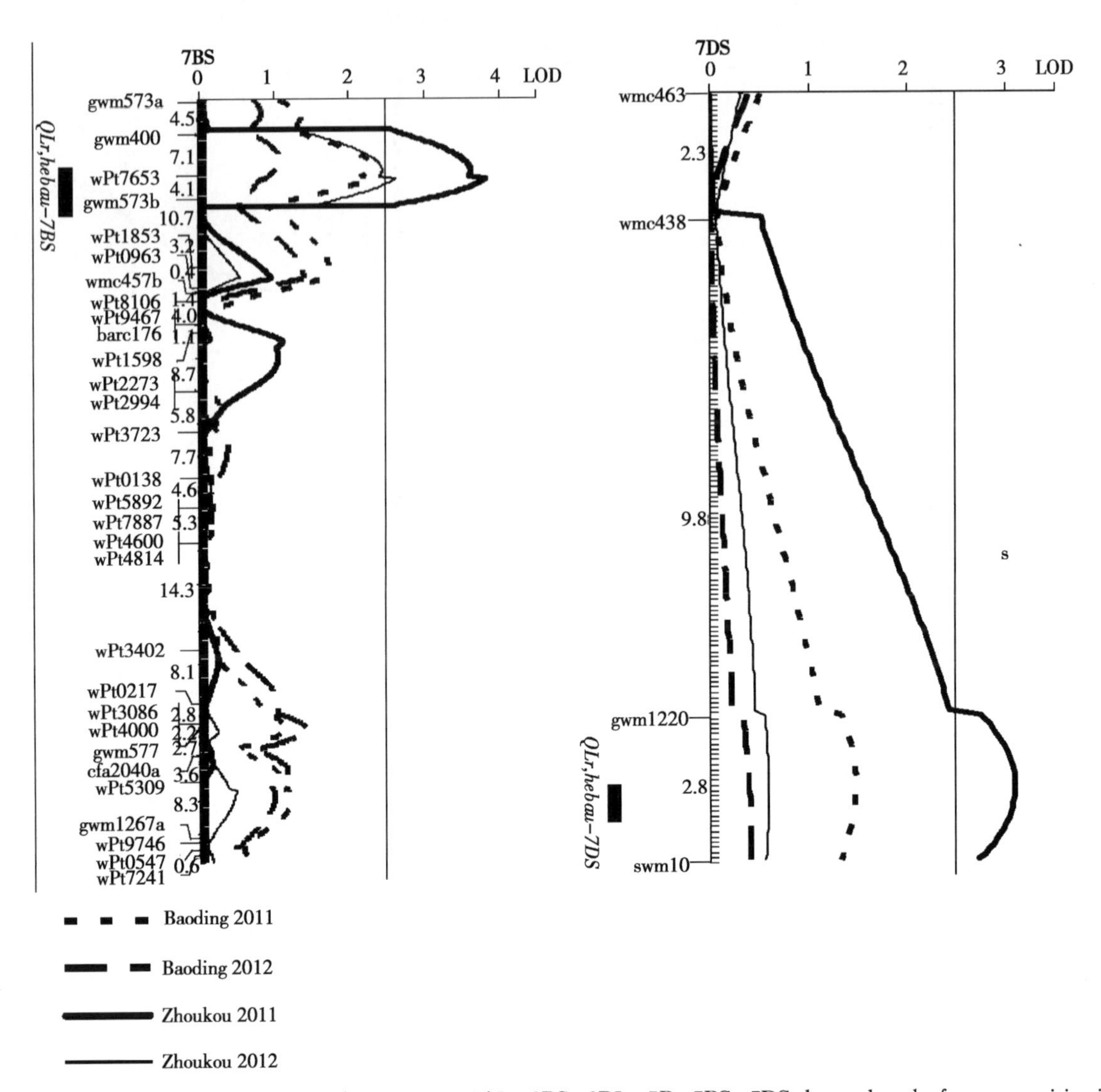

Fig. 2 LOD contours for QTL on chromosomes 1AL, 2BS, 2DL, 5B, 7BS, 7DS that reduce leaf rust severities in the RIL population from SHA3/CBRD × Naxos. Genetic distances are shown in centiMorgans (cM) to the left of the vertical axis. The approximate positions of QTL are indicated by oblong black squares to the left of markers and their lengths represent the interval distances of flanking markers. LOD threshold of 2.5 is indicated by a dashed vertical line in each graph

Table 3 Quantitative trait loci (QTL) for adult-plant resistance (APR) to LR detected by inclusive composite interval mapping (ICIM) in the SHA3 /CBRD × Naxos RIL population across four environments

QTL[a]	Marker interval	Location and year	LOD[b]	Add[c]	R^2 (%)[d]
QLr. hebau-1AL	*Xbarc213-Xcfa2219*	Baoding 2011	2.5	−4.4	5.5
		Zhoukou 2011	5.6	−3.6	8.2
		Zhoukou 2012	3.2	−4.6	5.3
QLr. hebau-2BS	*XwPt8548-XwPt2314*	Baoding 2011	8.7	−8.6	19.3
		Baoding 2012	6.7	−9.6	15.3
		Zhoukou 2011	21.6	−8	37.4
		Zhoukou 2012	11.2	−9.2	19.6
QLr. hebau-2DL	*XwPt2781-Xcfd233*	Zhoukou 2011	4.1	3	5.7
QLr. hebau-5B	*Xtpt7755-Xbarc128a*	Zhoukou 2012	3.6	4.9	6

（续）

QTL[a]	Marker interval	Location and year	LOD[b]	Add[c]	R^2 (%)[d]
QLr. hebau-7BS	*XwPt7653-Xgwm573*	Zhoukou 2011	3.8	2.9	5.3
		Zhoukou 2012	2.6	4.2	4.2
QLr. hebau-7DS	*Xgwm1220-Xswm10*	Zhoukou 2011	3	2.7	4.4

a. QTL were detected with a minimum LOD score of 2.5 in at least one environment,

b. LOD, logarithm of odds score,

c. Add, additive effect of resistance allele,

d. R^2, percentages of the phenotypic variance explained by individual QTL.

alleles of the QTL on 1AL and 2BS were contributed by SHA3/CBRD, whereas those on 2DL, 5B, 7BS, and 7DS were derived from Naxos.

A major QTL for LRlocated on chromosome 2BS, designated *QLr. hebau-2BS*, was detected in all the four environments. The QTL *QLr. hebau-2BS*, flanked by *XwPt8548* and *XwPt2314*, explained 19.3%, 15.3%, 37.4% and 19.6% of the phenotypic variance in Baoding 2011, Baoding 2012, Zhoukou 2011 and Zhoukou 2012, respectively.

The consistently detected QTL with larger effect, *QLr. hebau-1AL*, was located on chromosome 1AL, flanked by *Xbarc213* and *Xcfa2219*, and explained 5.5%, 8.2% and 5.3% of the phenotypic variance in Baoding 2011, Zhoukou 2011, and Zhoukou 2012, respectively.

QTL *QLr. hebau-7BS*, flanked by *XwPt7653* and *Xgwm573*, were detected in two environments and explained 5.3% and 4.2% of the phenotypic variance in Zhoukou 2011 and Zhoukou 2012, respectively.

Three QTL, *QLr. hebau-2DL*, *QLr. hebau-5B* and *QLr. hebau-7DS* were detected in single environment. *QLr. hebau-5B* and *QLr. hebau-7DS* were in marker intervals *Xtpt7755-Xbarc128a* and *Xgwm1220-Xswm10*, and explained 6.0% and 4.4%, respectively, of the phenotypic variance in Zhoukou 2012. Another QTL *QLr. hebau-2DL* was located between *XwPt2781* and *Xcfd233* on chromosome 2DL, explaining 5.7% of the phenotypic variance in Zhoukou 2011.

The total phenotypic variance explained by detected QTL ranged from 15.3 to 61% across the four environments in a simultaneous fit, suggesting a significant effect of these QTL in reducing LR severity. Interactions among different QTL were not identified across the four environments using IciMapping V3.1, indicating all these six QTL had additive effects (Table 3).

Discussion

Although the wheat line SHA3/CBRD was susceptible to *P. triticina* pathotypes MHJS, THJL, and PHGP at the seedling stage (Table 1), it showed high resistance with the mean FDS from 0 to 5% when inoculated with these pathotypes in the field, and two APR QTL for leaf rust were detected in SHA3/CBRD. The other parent Naxos was also susceptible to the three *P. triticina* pathotypes at the seedling stage. Although it showed moderately resistant to moderately susceptible reaction with FDS from 15 to 70% in the field test, four APR QTL for leaf rust detected in the population were derived from this line.

Environmental influence on APR QTL

Expression of minor genes is often affected by environments, including temperature, light and moisture (Roelfs et al., 1992). It is very necessary to evalute the populationamong different environments for identifing stable QTLs. For example, two leaf rust APR QTLs on 6BS were identified in the RIL population from the Bainong 64 × Jingshuang 16 cross; they explained 11.2% and 9.7% of phenotype variance, respectively, in Baoding 2011, but they were not found in Zhoukou 2011 (Ren et al., 2012b). In the present study, ANOVA showed significant differences ($P = 0.01$) in

FDS among line × environment interactions (Table 2), indicating environments had significant effect for expression of APR genes. Three APR QTL, viz. *QLr. hebau-2DL*, *QLr. hebau-7BS*, *QLr. hebau-7DS*, and two APR QTL, viz. *QLr. hebau-5B*, *QLr. hebau-7BS*, from Naxos were detected in Zhoukou 2011 and Zhoukou 2012, respectively, but they were not identified in Baoding 2011 and Baoding 2012 (Fig. 1).

Comparisons with previous reports

QLr. hebau-1AL

This is the first QTL for LR resistance identified on chromosome 1AL and is therefore likely to be a new APR QTL for leaf rust. A QTL for stripe rust resistance, *QYr. caas-1AL*, was reported by Ren et al. (2012a) using the same RIL population from SHA3/CBRD × Naxos, but *QYr. caas-1AL* was contributed by Naxos whereas *QLr. hebau-1AL* is from SHA3/CBRD.

A major leaf rust APR QTL on chromosome 2BS

The QTL on chromosome 2BS was derived from SHA3/CBRD. *QLr. hebau-2BS* was stably detected across all four environments, and explained 19.3-37.4% of the phenotypic variance. Marker *wPt2314* near *QLr. hebau-2BS* is not present on the consensus map (Somers et al., 2004), but its position was inferred to be 15cM from the distal end of the chromosome based on its linkage (about 15cM distal) to *wmc154* (Somers et al., 2004). There are five *Lr* genes, *Lr13* (Seyfarth et al., 2000), *Lr16* (McCartney et al., 2005), *Lr23* (McIntosh and Dyck 1975), *Lr35* (Seyfarth et al., 1999) and *Lr48* (Bansal et al., 2008) on 2BS. *Lr13*, *Lr35* and *Lr48* are APR genes. *Lr13* and *Lr35* are estimated to be 44cM, and *Lr48* are 32cM proximal from *QLr. hebau-2BS* based on the wheat consensus map (Somers et al., 2004), indicating that *QLr. hebau-2BS* is different from *Lr13*, *Lr35 and Lr48*. Based on the wheat consensus map *Lr16* is at the terminal of chromosome 2BS (Somers et al., 2004) and linked to *QLr. hebau-2BS* by a genetic distance of 15cM. Moreover, the reference line with *Lr16* was highly susceptible to all *P. triticina* pathotypes used seedling tests in this study and was also highly susceptible (80S) in the field trial (data not shown), thus *QLr. hebau-2BS* is likely different from *Lr16*. It was concluded that *QLr. hebau-2BS* might be a new major APR QTL for leaf rust resistance and could be a potentially important APR gene for wheat breeding.

QLr. hebau-2DL

Schnurbusch et al. (2004) identified a QTL flanked by *Xglk302* and *Xgwm539* on 2DL in Swiss winter wheat population from Arina × Forno. It was approximately 12cM from *QLr. hebau-2DL* based on the wheat consensus map (Somers et al., 2004), indicating that this locus is likely the same as *QLr. hebau-2DL*, and the relationship between these two loci should be further assessed in the future. A QTL for stripe rust resistance on chromosome 2DL, *QYr. caas-2DL*, was reported by Ren et al. (2012a) using the present population of SHA3/CBRD × Naxos, with both of the resistance alleles contributed by Naxos. They might be the same QTL based on their chromosomal locations. Lu et al. (2012) mapped a major QTL for powdery mildew to the distal *Xwmc817-Xcfd50* region of 2DL derived from Naxos. Although the reported maps suggested about 20cM between the PM QTL and *QLr. hebau-2DL*, when we re-analyzed the data using a combined model for leaf rust and powdery mildew (data not shown), the two QTL were much more closely associated and could be the same allele. In conclusion APR QTL for leaf rust *QLr. hebau-2DL* was likely to associate with QTL for stripe rust and powdery mildew.

QLr. hebau-5B

Messmer et al. (2000) identified a QTL for leaf rust resistance on chromosome 5B in winter wheat Forno, closely linked to the markers *Xpsr580b* and *Xpsr143*. This QTL should be different from *QLr. hebau-5B* based on a map distance of about 100cM separating them on the wheat consensus map (Somers et al., 2004).

QLr. hebau-7BS

Schnurbusch et al. (2004) identified a QTL *QLr. sfr-7BS* flanked by *Xsfr. BE427461* and *Xgwm573b* on chromosome 7BS in the winter wheat population from

Arina × Forno. It explained 8.8% of the phenotypic variance in LR severity across environments. In the present study *QLr. hebau-7BS*, flanked by *XwPt7653* and *Xgwm573*, explained 5.3% and 4.2% of the phenotypic variance in Zhoukou 2011 and Zhoukou 2012, respectively. *QLr. hebau-7BS* was located on the same chromosome region as *QLr. sfr-7BS*.

QLr. hebau-7DS

So far only the pleiotropic APR gene *Lr34/Yr18/Pm38/Sr57* maps to chromosome 7DS, conferring resistance to all three rusts and powdery mildew (Singh, 1992; Lillemo et al., 2008; Spielmeyer et al., 2005). In the present study *QLr. hebau-7DS* also mapped on 7DS, and flanked by *Xgwm1220* and *Xswm10* which are linked to *Lr34* (Somers et al., 2004). However, genotyping with the diagnostic marker *cssfr5* (Lagudah et al., 2009) showed that both parents carry the 523bp allele associated with susceptibility (results not shown), indicating that *QLr. hebau-7DS* is different from *Lr34*. Generally *Lr34* provids a strong effect in decreasing leaf rust severity (Nelson et al., 1997; Singh et al., 2000b), whereas *QLr. hebau-7DS* showed a relatively weak effect and was detected only in one environment. One APR QTL for powdery mildew from Naxos mapped to the same chromosome region using this population. Therefore *QLr. hebau-7DS* is likely a new minor APR QTL for leaf rust and a potential pleiotropic APR QTL conferring resistance for leaf rust and powdery mildew simultaneously.

Breeding and application

Six QTL for resistance to leaf rust were detected on chromosomes 1AL, 2BS, 2DL, 5B, 7BS, and 7DS (Table 3). All the closely linked markers can be used for improving resistance to leaf rust in wheat breeding. The APR QTL *QLr. hebau-2BS* provided a significant level of stable resistance to leaf rust in all environments, and may be an important APR gene for use in wheat breeding.

Two potentially pleiotropic APR QTL were detected in the population and both were derived from Naxos. The first was on 2DL, and evidence from different studies indicated effects on leaf rust, stripe rust and powdery mildew responses. The second was mapped on 7DS at a similar position to, but evidently different from the pleiotropic *Lr34/Yr18/Pm38/Sr57*. It conferred resistance to leaf rust and powdery mildew, but no effect on stripe rust response found in a previous study. Known pleiotropic APR genes, such as *Lr34/Yr18/Pm38/Sr57* and *Lr46/Yr29/Pm39/Sr58* (William et al., 2003), can be combined with potentially pleiotropic APR QTL identified in the present study for breeding durably resistant wheat cultivars to multiple diseases.

Acknowledgments

We are grateful to the critical review of this manuscript by Prof. R. A. McIntosh, Plant Breeding Institute, University of Sydney, Australia. The project was supported by National Natural Science Foundation of China (31361140367 and 31300562), the National Key Basic Research Program of China (2013CB127700), International Science & Technology Cooperation Program of China, and the China Agriculture Research System (CARS-3-1-3).

References

Allard RW (1960) Principles of Plant Breeding. John Wiley and Sons, New York.

Bansal UK, Hayden MJ, Venkata BP, Khanna R, Saini RG, Bariana HS (2008) Genetic mapping of adult plant leaf rust resistance genes *Lr48* and *Lr49* in common wheat. Theor Appl Genet 117: 307-312.

Bjarko ME, Line RF (1988) Heritability and number of genes controlling leaf rust resistance in four cultivars of wheat. Phytopathology 78: 457-461.

Bossolini E, Krattinger SG, Keller B (2006) Development of simple sequence repeat markers specific for the *Lr34* resistance region of wheat using sequence information from rice and *Aegilops tauschii*. Theor Appl Genet 113: 1049-1062.

Caldwell RM (1968) Breeding for general and/or specific plant disease resistance. *In*: Finlay KW, Shepherd KW (eds). Proceedings of the third international wheat genetics symposium. Canberra, Australia: Australian Academy of Sciences, pp. 263-272.

Chai JF, Zhou RH, Jia JZ, Liu X (2006) Development and application of a new codominant PCR marker for detecting 1BL · 1RS wheat - rye chromosome translocations. Plant Breed 125: 302-304.

Dyck PL (1977) Genetics of leaf rust reaction in three introductions of common wheat. Can J Genet Cytol 19: 711-716.

Dyck PL (1987) The association of a gene for leaf rust resistance with the chromosome 7D suppressor of stem rust resistance in common wheat. Genome 29: 467 - 469.

Dyck PL, Samborski DJ (1982) The inheritance of resistance to *Puccinia recondita* in a group of common wheat cultivars. Can J Genet Cytol 24: 273-283.

Dubin HJ, Johnson R, Stubbs RW (1989) Postulated genes to stripe rust in selected CIMMYT and related wheats. Plant Dis 73: 472-475.

Gupta PK, Balyan HS, Edwards KJ, Isaac P, Korzun V, Röder M, Gautier M-F, Joudrier P, Schlatter AR, Dubcovsky J, De la Pena RC, Khairallah M, Penner G, Hayden MJ, Sharp P, Keller B, Wang RCC, Hardouin JP, Jack P, Leroy P (2002) Genetic mapping of 66 new microsatellite (SSR) loci in bread wheat. Theor Appl Genet 105: 413-422.

Herrera-Foessel SA, Lagudah ES, Huerta-Espino J, Hayden MJ, Bariana HS, Singh D, Singh RP (2011) New slow-rusting leaf rust and stripe rust resistance genes *Lr67* and *Yr46* in wheat are pleiotropic or closely linked. Theor Appl Genet 122: 239-249.

Herrera-Foessel SA, Singh RP, Huerta-Espino J, Rosewarne GM, Periyannan SK, Viccars L, Calvo-Salazar V, Lan C, Lagudah ES (2012) *Lr68*: a new gene conferring slow rusting resistance to leaf rust in wheat. Theor Appl Genet 124: 1475-1486.

Herrera-Foessel SA, Singh RP, Lillemo M, Huerta-Espino J, Bhavani S, Singh S, Lan C, Calvo-Salazar V, Lagudah ES (2014). *Lr67/Yr46* confers adult plant resistance to stem rust and powdery mildew in wheat. Theor Appl Genet. (online) DOI: 10.1007/s00122-013-2256-9

Knott DR (1989) The wheat-rust breeding for resistance. Monographs on Theoretical and Applied Genetics 12. Springer-Verlag, Berlin, Germany.

Kosambi DD (1944) The estimation of map distance from recombination values. Annu Eugen 12: 172-175.

Krattinger SG, Lagudah ES, Spielmeyer W, Singh RP, Huerta-Espino J, McFadden H, Bossolini E, Selter LL, Keller B (2009) A putative ABC transporter confers durable resistance to multiple fungal pathogens in wheat. Science 323: 1360-1363.

Krattinger SG., Lagudah ES, Wicker T, Risk JM, Ashton AR, Selter LL, Matsumoto T, Keller B (2011) *Lr34* multi-pathogen resistance ABC transporter: molecular analysis of homoeologous and orthologous genes in hexaploid wheat and other grass species. Plant J 65: 392-403.

Lagudah ES, Krattinger SG, Herrera-Foessel S, Singh RP, Huerta-Espino J, Spielmeyer W, Brown-Guedira G, Selter LL, Keller B (2009) Gene-specific markers for the wheat gene *Lr34/Yr18/Pm38* which confers resistance to multiple fungal pathogens. Theor Appl Genet 119: 889-898.

Li HH, Ye GY, Wang JK (2007) A modified algorithm for the improvement of composite interval mapping. Genetics 175: 361-374.

Lillemo M, Asalf B, Singh RP, Huerta-Espino J, Chen XM, He ZH, Bjørnstad Å (2008) The adult plant rust resistance loci *Lr34/Yr18* and *Lr46/Yr29* are important determinants of partial resistance to powdery mildew in bread wheat line Saar. Theor Appl Genet 116: 1155-1166.

Long DL, Kolmer JA (1989) A North American system of nomenclature for *Puccinia recondita* f. sp. *tritici*. Phytopathology 79: 525-529.

Lu QX, Bjørnstad Å, Ren Y, Asad MA, Xia XC, Chen XM, Ji F, Shi JR, Lillemo M (2012) Partial resistance to powdery mildew in German spring wheat 'Naxos' is based on multiple genes with stable effects in diverse environments. Theor Appl Genet 125: 297-309.

Manly KF, Cudmore RH Jr, Meer JM (2001) Map manager QTX, cross-platform software for genetic mapping. Genome 12: 930-932.

McCartney CA, Somers DJ, McCallum BD, Thomas J, Humphreys DG, Menzies JG, Brown PD (2005) Microsatellite tagging of the leaf rust resistance gene*Lr16* on wheat chromosome 2BSc. Mol Breed 15: 329-337.

McDonald BA, Linde C (2002) Pathogen population genetics, evolutionary potential, and durable resistance. Annu Rev Phytopathol 40: 349-379.

McIntosh RA (1992) Close genetic linkage of genes conferring adult plant resistance to leaf rust and stripe rust in wheat. Plant Pathol 41: 523-527.

McIntosh RA, Dyck PL (1975) Cytogenetical studies in wheat VII. Gene *Lr23* for reaction to *Puccinia recondita* in Gabo and related cultivars. Aust J Biol Sci 28: 201-211.

McIntosh RA, Yamazaki Y, Dubcovsky J, Rogers WJ, Morris

C, Appels R, Xia XC (2013) Catalogue of gene symbols for wheat: 2013. http: //www. shigen. nig. ac. jp/wheat/komugi/genes/macgene/2013/GeneCatalogueIntroduction. pdf

Messmer MM, Seyfarth S, Keller M, Schachermayr G, Winzeler M, Zanetti S, Feuillet C, Keller B (2000) Genetic analysis of durable leaf rust resistance in winter wheat. Theor Appl Genet 100: 419-431.

Nelson JC, Singh RP, Autrique JE, Sorrells ME (1997) Mapping genes conferring and suppressing leaf rust resistance in wheat. Crop Sci 37: 1928-1935.

Parida SK, Anand Raj Kumar K, Dalal V, Singh NK, Mohapatra T (2006) Unigene derived microsatellite markers for the cereal genomes. Theor Appl Genet 112: 808-817.

Pestsova E, Ganal MW, Röder MS (2000) Isolation and mapping of microsatellite markers specific for the D genome of bread wheat. Genome 43: 689-697.

Peterson RF, Campbell AB, Hannah AE (1948) A diagrammatic scale for estimating rust intensity of leaves and stems of cereals. Can J Res 26: 496-500.

Qiu JW, Schürch AC, Yahiaoui N, Dong LL, Fan HJ, Zhang ZJ, Keller B, Ling HQ (2007) Physical mapping and identification of a candidate for the leaf rust resistance gene *Lr1* of wheat. Theor Appl Genet, 2007, 115: 159-168.

Ren Y, He, ZH, Li J, Lillemo M, Wu L, Bai B, Lu QG, Zhu HZ, Zhou G, Du JY, Lu QL, Xia XC (2012a) QTL mapping of adult-plant resistance to stripe rust in a population derived from common wheat cultivars Naxos and Shanghai 3/Catbird. Theor Appl Genet 125: 1211-1221.

Ren Y, Li ZF, He ZH, Wu L, Bai B, Lan CX, Wang CF, Zhou G., Zhu HZ, Xia XC (2012b) QTL mapping of adult-plant resistances to stripe rust and leaf rust in Chinese wheat cultivar Bainong 64. Theor Appl Genet 125: 1253-1262.

Röder MS, Korzun V, Wendehake K, Plaschke J, Tixier MH, Leroy P, Ganal MW (1998) A microsatellite map of wheat. Genetics 149: 2007-2023.

Roelfs AP, Singh RP, Saari EE (1992) Rust diseases of wheat: concepts and methods of disease management. CIMMYT, Mexico.

Schnurbusch T, Paillard S, Schori A, Messmer M, Schachermayr G, Winzeler M, Keller B (2004) Dissection of quantitative and durable leaf rust resistance in Swiss winter wheat reveals a major resistance QTL in the *Lr34* chromosome region. Theor Appl Genet 108: 477-484.

Seyfarth R, Feuillet C, Schachermayr G, Messmer M, Winzeler M, Keller B (2000) Molecular mapping of the adult-plant rust resistance gene *Lr13* in wheat (*Triticum aestivum* L.). J Genet Breed 54: 193-198.

Seyfarth R, Feuillet C, Schachermayr G, Winzeler M, Keller B (1999) Development of a molecular marker for the adult plant leaf rust resistance gene *Lr35* in wheat. Theor Appl Genet 99: 554-560.

Singh RP (1992) Genetic association of leaf rust resistance gene *Lr34* with adult plant resistance to stripe rust in bread wheat. Phytopathology 82: 835-838.

Singh RP, Mujeeb-Kazi A, Huerta-Espino J (1998) *Lr46*: A gene conferring slow-rusting resistance to leaf rust in wheat. Phytopathology 88: 890-894.

Singh RP, Huerta-Espino J, Rajaram S (2000a) Achieving near-immunity to leaf and stripe rusts in wheat by combining slow rusting resistance genes. Acta Phytopathol Entomol Hung 35: 133-139.

Singh RP, Nelson JC, Sorrells ME (2000b) Mapping *Yr28* and other genes for resistance to stripe rust in wheat. Crop Sci 40: 1148-1155.

Somers DJ, Isaac P, Edwards K (2004) A high-density microsatellite consensus map for bread wheat (*Triticum aestivum* L.). Theor Appl Genet 109: 1105-1114.

Song QJ, Fickus EW, Cregan PB (2002) Characterization of trinucleotide SSR motifs in wheat. Theor Appl Genet 104: 286-293.

Sourdille P, Singh S, Cadalen T, Brown-Guedira GL, Gay G, Qi L, Gill BS, Dufour P, Murigneux A, Bernard M (2004) Microsatellite-based deletion bin system for the establishment of genetic-physical map relationships in wheat (*Triticum aestivum* L.). Funct Integr Genomics 4: 12-25.

Spielmeyer W, McIntosh RA, Kolmer J, Lagudah ES (2005) Powdery mildew resistance and *Lr34*/*Yr18* genes for durable resistance to leaf and stripe rust cosegregate at a locus on the short arm of chromosome 7D of wheat. Theor Appl Genet 111: 731-735.

William M, Singh RP, Huerta-Espino J, Ortiz Islas S, Hoisington D (2003) Molecular marker mapping of leaf rust resistance gene *Lr46* and its association with stripe rust resistance gene *Yr29* in wheat. Phytopathology 93: 153-159.

Xue SL, Zhang ZZ, Lin F, Kong ZX, Cao Y, Li CJ, Yi HY, Mei MF, Zhu HL, Wu JZ, Xu HB, Zhao DM, Tian DG, Zhang CQ, Ma ZQ (2008) A high-density intervarietal map of the wheat genome enriched with markers derived from expressed sequence tags. Theor Appl Genet 117: 181-189.

Yu JK, Dake TM, Singh S, Benscher D, Li WL, Gill B, Sorrells ME (2004) Development and mapping of EST-derived simple sequence repeat markers for hexaploid wheat. Genome 47：805-818.

Zhang LJ，Li ZF，Lillemo M，Xia XC，Liu DQ，Yang WX，Luo JC，Wang HY（2009）QTL mapping for adult-plant resistance to leaf rust in CIMMYT wheat cultivar Saar. Sci Agric Sin 42：388-397.

Zhao XL，Zheng TC，Xia XC，He ZH，Liu DQ，Yang WX，Yin GH，Li ZF（2008）Molecular mapping of leaf rust resistance gene *LrZH84* in Chinese wheat line Zhou 8425B. Theor Appl Genet 117：1069-1075.

Zhou HX，Xia XC，He ZH，Li X，Wang CF，Li ZF，Liu DQ（2013）Molecular mapping of leaf rust resistance gene *LrNJ97* in Chinese wheat line Neijiang 977671. Theor Appl Genet 126：2141-2147.

Quantitative trait loci mapping for adult-plant resistance to powdery mildew in Chinese wheat cultivar Bainong 64

C. X. Lan[1], S. S. Liang[1,2], Z. L. Wang[3], J. Yan[4], Y. Zhang[1], X. C. Xia[1], and Z. H. He[1,5]

[1] *Institute of Crop Science, National Wheat Improvement Center/The National Key Facility for Crop Gene Resources and Genetic Improvement, Chinese Academy of Agricultural Sciences (CAAS), 12 Zhongguancun South Street, Beijing* 100081, *China*; [2] *Department of Primary Industries, Victorian AgriBiosciences Center, La Trobe R&D Park*, 1 *Park Drive, Bundoora, Vic* 3083, *Australia*; [3] *College of Agronomy, Northwest Sci-Tech University of Agriculture and Forestry, Yangling, Shaanxi* 712100, *China*; [4] *Cotton Research Institute, Chinese Academy of Agricultural Sciences (CAAS), Huanghedadao, Anyang, Henan* 455000, *China*; [5] *International Maize and Wheat Improvement Center (CIMMYT) China Office, c/o CAAS*, 12 *Zhongguancun South Street, Beijing* 100081, *China*

Abstract: Adult-plant resistance (APR) is an effective means of controlling powdery mildew in wheat. In the present study, 406 simple sequence repeat (SSR) markers were used to map quantitative trait loci (QTLs) for APR to powdery mildew in a doubled haploid (DH) population of 181 lines derived from the cross Bainong 64 × Jingshuang 16. The DH lines were planted in a randomized complete block design with three replicates in Beijing and Anyang during the 2005 to 2006 and 2007 to 2008 cropping seasons. Artificial inoculations were carried out in Beijing using the highly virulent *Blumeria graminis* f. sp. *tritici* isolate E20. Disease severities on penultimate leaves were scored twice in Beijing, whereas at Anyang, maximum disease severities (MDS) were recorded following natural infection. Broad-sense heritabilities of MDS and areas under the disease progress curve (AUDPC) were 0. 89 and 0. 77, respectively, based on the mean values averaged across environments. Composite interval mapping detected four QTLs for APR on chromosomes 1A, 4DL, 6BS, and 7A; these were designated ***QPm. caas-1A***, ***QPm. caas-4DL***, ***QPm. caas-6BS***, and ***QPm. caas-7A***, respectively, and explained from 6. 3 to 22. 7% of the phenotypic variance. QTLs ***QPm. caas-4DL*** and ***QPm. caas-6BS*** were stable across environments with high genetic effects on powdery mildew response, accounting for 15. 2 to 22. 7% and 9. 0 to 13. 2% of the phenotypic variance, respectively. These results should be useful for the future improvement of powdery mildew resistance in wheat.

Introduction

Powdery mildew, caused by *Blumeria graminis* f. sp. *tritici* (*Bgt*), is a very destructive foliar disease of wheat (*Triticum aestivum* L.), especially in regions with high rainfall and with a maritime or semi-continental climate[1]. Race-specific major resistance genes (*Pm* genes) confer complete resistance to powdery mildew resulting from hypersensitive responses (HR) to infection[9]. However, this kind of resistance is often not durable, being overcome by new pathogen races

published in Phytopathology, 2009, 99 (10): 1121-1126

possessing the corresponding virulence genes[26]. Some wheat cultivars exhibit resistance that delays infection and reduces growth and reproduction of the pathogen on post-seedling plants. Such resistance has been termed slow mildewing[26], adult-plant resistance (APR)[9] or partial resistance[11]. So far, 58 powdery mildew resistance genes at 39 loci have been characterized and assigned to specific wheat chromosomes (http://www.shigen.nig.ac.jp/wheat/komugi/genes/symbolClassList.jsp), and some of them were introduced from wheat-related species. Most of these genes are race-specific, and many of them have been overcome by new pathogen variants.

In contrast, APR genes are likely to be more useful for long-term disease resistance strategies, and thus have greater potential for durable resistance[1]. APR genes are expressed quantitatively and typically have small genetic effects, thereby requiring multiple genes to provide adequate levels of protection against pathogens[9,20]. Investigations on the chromosomal locations of such genes and on their biological effects are important to ensure their appropriate deployment in elite germplasm and commercial wheat cultivars.

APR to powdery mildew is easy to identify in cultivars possessing no seedling-effective resistance genes or when natural populations of *Bgt* overcome any resistance gene that may be present[1,26]. Since it is time-consuming and difficult to assess APR phenotypes in breeding populations in the field[9], molecular markers will have many advantages in the selection of genotypes with combinations of APR genes. Molecular markers have been used to map APR genes for powdery mildew in wheat. For example, quantitative trait loci (QTLs) for APR were mapped in the Swiss winter wheat Forno[15], the French winter wheat lines RE714[4,23] and RE9001[3], the North American winter wheats Massey[20] and USG3209[37], the Japanese cultivar Fukuho-Komugi (a QTL on 7DS identified as *Pm38*)[18], the Korean wheat Suwon 92[42], and International Maize and Wheat Improvement Center (CIMMYT) line Saar (two QTLs identified as *Pm38* and *Pm39*)[19].

Bainong 64 was a leading wheat cultivar in the Yellow-Huai wheat region of China from 1993 to 2000, occupying about 700,000ha annually. It showed highly susceptible responses to many isolates of *Bgt* at the seedling stage, but is highly resistant at the adult stage[39]. This cultivar had good quality, high yield and broad adaptability, and was used widely as a parent in Chinese wheat breeding programs. The aim of the present study was to identify QTLs for APR to powdery mildew and associated molecular markers in a doubled haploid (DH) population from a cross between Bainong 64 and Jingshuang 16, and to assess the stability of the effects of the detected QTLs across different environments. It was expected that such knowledge could lead to a more efficient strategy in breeding for durable resistance.

Materials and methods

Plant materials

The 181 DH lines used in the study were developed from a cross between Chinese winter wheat cultivars Bainong 64 and Jingshuang 16, and were generated by the wheat × maize crossing method[17]. Bainong 64 (Bainong 8717/3/Yeda 72-629-52/Shi 82-5594//Bainong 84-4046-1) is susceptible to Chinese isolates of *Bgt* at the seedling stage, but highly resistant at the adult-plant stage, whereas Jingshuang 16 is highly susceptible at both the seedling and adult-plant stages[39]. These isolates were collected in Beijing, Henan, Guizhou, Yunnan and Sichuan provinces from 1990 through 1998, and maintained on wheat seedlings in growth chambers, with 24h of light at 4℃ to 8℃, in the Institute of Plant Protection, CAAS.

Field trials

The DH lines and parents were evaluated for APR to powdery mildew during the 2005 to 2006 and 2007 to 2008 cropping seasons at the Chinese Academy of Agricultural Sciences (CAAS) experimental station of the Institute of Crop Science in Beijing, and at the CAAS Cotton Research Institute, Anyang. Field trials were planted in randomized complete blocks with three replicates. Plots consisted of single 1.5m rows with 25cm between rows. Approximately 75 seeds were sown in each row. The susceptible parent cv. Jingshuang 16 was planted every 10 rows as a susceptible check, and planted around the test lines to ensure ample powdery mildew inoculum in spring.

Artificial inoculations in Beijing were carried out using the highly virulent *Bgt* isolate E20 prior to plants reaching stem elongation. The disease severity on penultimate leaf (F-1 leaf) on 10 randomly selected plants from each line was scored based on the percentage of leaf area covered by powdery mildew for the first time in 6 weeks after inoculation, and then after one week for the second time when the disease severity reached a maximum level around May 20. Disease severity of the 10 selected plants was averaged to obtain mean powdery mildew severity for each line.

Powdery mildew severities on each line in Anyang were rated on the percentages of leaf area covered by powdery mildew under natural infection conditions when the disease severities on Jingshuang 16 reached a maximum level around 18 May 2006. Field data from the 2008 trial were not used for analysis because disease severities failed to reach satisfactory levels, for instance, the averaged maximum disease severities (MDS) of susceptible check cv. Jingshuang 16 was less than 5%.

Statistical analysis

The area under the disease AUDPC $= \sum_{i=1}^{n}[(X_i + X_{i+1})/2](T_{i+1} - T_i)$ calculated for each line using the following formula from Bjarko and Line[2]: Where X_i is the disease severity on assessment date i, T_i is the number of days after inoculation on assessment date i, and n is the total times of disease assessments. MDS and AUDPC were used for the subsequent analysis of variance (ANOVA) and QTL analysis. SAS software (SAS Institute, Cary, NC) was used to compute ANOVA. Broad-sense heritability (h^2) for powdery mildew resistance was calculated from the ANOVA, on an across-environment genotype mean basis, by the formula: $h^2 = \sigma_g^2/(\sigma_g^2 + \sigma_{ge}^2/e + \sigma_\varepsilon^2/re)$, where σ_g^2, σ_{ge}^2, *and* σ_ε^2 are estimates of genotypic, genotype × environment interaction and error variances, respectively, and e and r are the numbers of environments and replications per environment, respectively. Each QTL was represented by a 20-cM interval with the local Logarithm of odds (LOD) maximum as center. QTL with overlapping 20-cM intervals among different environments were considered as being in common.

SSR analysis

To construct the framework map for the QTL analysis, we chose 406 simple sequence repeats (SSR), including 140 pairs of BARC (Beltsville Agriculture Research Station) primers[33], 134 pairs of WMC (Wheat Microsatellite Consortium) primers[8], 81 pairs of GWM (Gatersleben Wheat Microsatellite) primers[27], 30 pairs of CFD (Clermont Ferrand D-genome) primers[10], 14 pairs of GDM (Gatersleben D-genome Microsatellite) primers[25], and 7 pairs of CFA (Clermont Ferrand A-genome) primers[34].

Bulkedsegregant analysis

Two DNA bulks were constructed by mixing equal amounts of DNA from the five most resistant and the five most susceptible lines, respectively, based on severity data from all three environments. SSRs showing the same patterns of polymorphism between the parents and between the bulks were used to genotype all 181 DH lines. Additional SSRs on the linkage maps (http://wheat.pw.usda.gov/GG2/index.shtml, http://www.shigen.nig.ac.jp/wheat/komugi/maps/markerMap[32] around the initially indicated loci associated with APR were chosen to genotype the population for linkage and QTL analyses as described by Michelmore et al.[22].

Map construction and QTL detection

Linkage groups were generated with the software Map Manager QTXb20[21]. Genetic distances between markers were estimated using the Kosambi[16] mapping function. QTL were detected by composite interval mapping using the software Cartographer 2.5[38]. A logarithm of odds (LOD) of 2.0 was set to declare

QTL as significant. For each QTL, estimates of phenotypic variance (R^2) and additive effects at the LOD peaks were obtained from QTL Cartographer 2.5.

Results

Phenotypic analysis of MDS and AUDPC, and their correlations and heritabilities

The MDS scoresfrom the three environments were significantly correlated ($P < 0.0001$), with correlation coefficients ranging from 0.61 to 0.76. The frequency distributions of the powdery mildew severity parameters (MDS and AUDPC) of the DH lines over the three environments showed continuous variation, confirming quantitative inheritance (Fig. 1). The mean MDS of Bainong 64 and Jingshuang 16 were 2.3% and 28.3% in Anyang in 2006, whereas the means were 1.0% and 30.0%, 0.7% and 86.7% in Beijing in 2006 and 2008, respectively. The average of MDS of the DH lines in Beijing over two years was 10.4%, ranging from 0.3 to 86.7%, and 7.3% in Anyang, ranging from 0 to 60.0%. In Beijing, the average of AUDPC over two years was 51.4, ranging from 1.2 to 298.3.

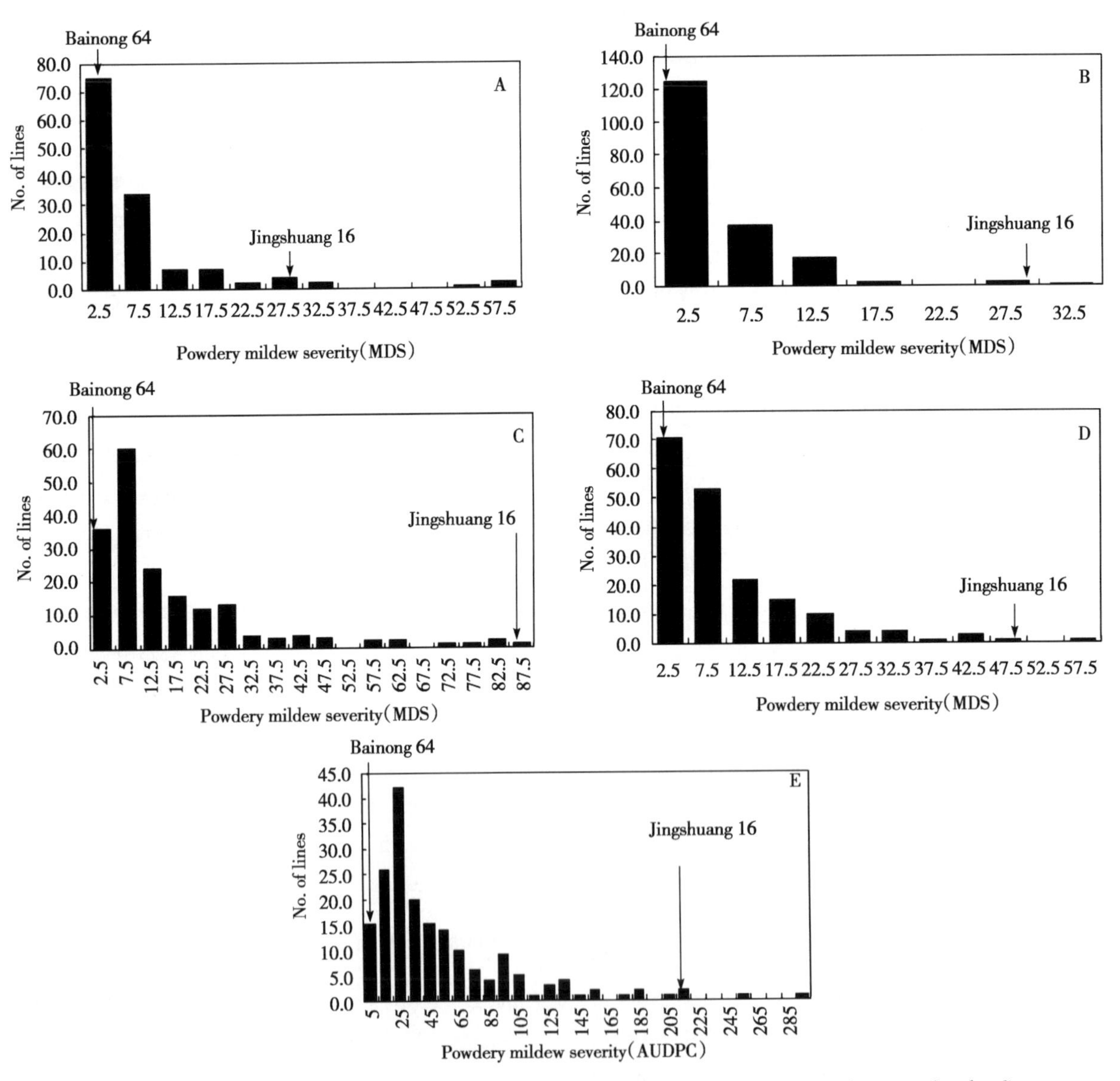

Fig. 1 A-E, Frequency distributions of powdery mildew maximum disease severities (MDS) and area under the disease progress curve (AUDPC) values in doubled haploid (DH) lines derived from Bainong 64 × Jingshuang 16. A, MDS, Anyang 2006; B, MDS, Beijing 2006; C, MDS, Beijing 2008; D, average MDS across three environments; and E, average AUDPC, Beijing over two years. Mean values for the parents, Bainong 64 and Jingshuang 16, are indicated by arrows.

The MDS and AUDPC were significantly correlated for the test inBeijing over two years (r =0.89, P <0.0001). The broad-sense heritabilities of MDS and AUDPC across environments were 0.89 and 0.77, respectively. The ANOVA confirmed a significant variation among DH lines (Table 1).

Table 1 Analysis of variance of maximum disease severities (MDS) values for penultimate leaves and area under the disease progress curve (AUDPC) values for powdery mildew responses on doubled haploid (DH) lines derived from the cross Bainong 64 × Jingshuang 16.

Parameter	Source of variation	df	Sum of squares	Mean of squares	F values
MDS	Replicate	2	77.11	38.55	1.37
	Environment	2	27 571.50	13 785.75	491.14**
	Line	180	77 918.81	432.88	15.42**
	Line×Environment	360	47 941.89	157.70	5.62**
	Error	815	24 448.03	28.07	
AUDPC	Replicate	2	3 477.93	1 738.97	1.95
	Environment	1	1 081 028.60	1 081 028.60	1 209.77**
	Line	180	2 266 148.83	12 589.72	14.09**
	Line×Environment	180	1 293 457.93	7 266.62	8.13**
	Error	649	581 719.75	893.58	

** Significant at P < 0.0001.

QTL for APR to powdery mildew

Four QTLs for APR were detected in the DH population in three environments (Table 2, Fig. 2). Using the MDS from the 2005-2006 cropping season in Anyang, three QTLs were detected on chromosomes 1A, 4DL, and 6BS, with additive effects of 4.00, 6.95, and 4.98, explaining 7.4%, 20.9%, and 13.2% of the phenotypic variance, respectively. Four MDS QTLs were identified in Beijing in the 2005-2006 cropping season, accounting for 9.9%, 22.7%, 11.5%, and 6.3% of the phenotypic variance, respectively. The additive effects of these QTLs were 1.68, 2.82, 1.93, and 1.52, respectively. In Beijing 2007-2008, three QTLs were detected on chromosomes 4DL, 6BS, and 7A, explaining 16.7%, 10.3%, and 7.1% of the phenotypic variance, respectively. The additive effects of these QTLs were 6.33, 6.79, and 5.03, respectively. Using the average values of MDS from three environments over two years, four QTLs were mapped on chromosomes 1A, 4DL, 6BS, and 7A with additive effects of 3.10, 3.42, 4.36, and 2.90, respectively, explaining from 6.7 to 15.2% of the phenotypic variance. Based on averaged AUDPC values in Beijing over two cropping seasons, three QTLs were found on the chromosomes 4DL, 6BS, and 7A, explaining from 6.7 to 19.0% of the phenotypic variance, and the additive effects of these QTLs were 16.14, 19.63, and 17.05, respectively. All QTLs identified in this population were from the resistant parent Bainong 64.

Table 2 Quantitative trait loci (QTLs) detected for adult-plant resistance (APR) to powdery mildew in the doubled haploid (DH) population derived from the cross Bainong 64 × Jingshuang 16

Parameter	Year and location	QTL[b]	Interval	Position[c]	LOD score[d]	Addi. effect	R^{2e} (%)	Total R^2 (%)
MDS[a]	06Anyang	*QPm.caas-1A*	*Xbarc148-Xwmc550*	22.5	1.95	4.00	7.4	41.5
		QPm.caas-4DL	*Xbarc200-Xwmc331*	18.0	1.96	6.95	20.9	
		QPm.caas-6BS	*Xbarc79-Xgwm518*	0.0	4.51	4.98	13.2	

（续）

Parameter	Year and location	QTL[b]	Interval	Position[c]	LOD score[d]	Addi. effect	R^{2e} (%)	Total R^2 (%)
	06Beijing	*QPm. caas-1A*	*Xbarc148-Xwmc550*	22.0	3.83	1.68	9.9	50.4
		QPm. caas-4DL	*Xbarc200-Xwmc331*	14.0	3.41	2.82	22.7	
		QPm. caas-6BS	*Xbarc79-Xgwm518*	11.9	3.99	1.93	11.5	
		QPm. caas-7A	*Xbarc127-Xbarc174*	26.0	1.95	1.52	6.3	
	08Beijing	*QPm. caas-4DL*	*Xwmc331-Xgwm165*	0.8	4.70	6.33	16.7	34.1
		QPm. caas-6BS	*Xbarc79-Xgwm518*	10.0	3.03	6.79	10.3	
		QPm. caas-7A	*Xbarc127-Xbarc174*	28.0	2.52	5.03	7.1	
	Average MDS in three environments	*QPm. caas-1A*	*Xbarc148-Xwmc550*	20.0	2.88	3.10	8.1	41.7
		QPm. caas-4DL	*Xwmc331-Xgwm165*	0.8	3.94	3.42	15.2	
		QPm. caas-6BS	*Xbarc79-Xgwm518*	8.0	3.59	4.36	11.7	
		QPm. caas-7A	*Xbarc127-Xbarc174*	28.0	2.43	2.90	6.7	
AUDPC[f]	Average AUDPC in Beijing	*QPm. caas-4DL*	*Xwmc331-Xgwm165*	0.8	3.24	16.14	19.0	34.7
		QPm. caas-6BS	*Xbarc79-Xgwm518*	10.0	2.73	19.63	9.0	
		QPm. caas-7A	*Xbarc127-Xbarc174*	26.0	2.04	17.05	6.7	

a. Maximum disease severity.

b. QTL that extend across single one-log support confidence intervals were assigned the same symbol.

c. Peak position in centimorgans from the first interval marker.

d. Logarithm of odds (LOD) score.

e. R^2 is the proportion of phenotypic variance explained by the QTL.

f. Area under the disease progress curve.

Discussion

In a previous study, it was estimated that Bainong 64 possessed 3 to 4 genes for APR to powdery mildew, based on phenotypic data from $F_{2:3}$ populations of Bainong 64×Jingshuang 16[40]. This was in agreement with the present study where four QTLs were identified in Bainong 64 using DH lines from the same cross.

The QTL in the centromeric region of chromosome 1A was detected in three environments in the SSR interval *Xbarc148-Xwmc550*. This was in a similar position to a QTL found by Keller et al.[15], but flanked by AFLP markers *Xpsr1201b* and *Xpsr941M* in a segregating wheat×*T. spelta* population. Mingeot et al.[23] detected a QTL in the marker interval *Xcdo572b-Xbcd442* in the French line RE714, and this location is close to *QPm. caas-1A* based on the consensus map of Somers et al.[32]. Therefore, these QTLs could be at the same locus or linked closely. In contrast, Liang et al.[18] identified a QTL in the marker interval *Xgdm33-Xpsp2999* in a DH population from Fukuho-Komugi×Oligoculm at a similar location to the *Pm3* locus; this position is clearly different from that of *QPm. caas-1A*. Two seedling-effective race-specific resistance genes have been identified on chromosome 1A, viz. *Pm25* derived from *T. aegilopoides*[30], and *Pm17* from rye[12], but *QPm. caas-1A* is unlikely to be related to those genes. Although *QPm. caas-1A* had a small effect (Table 2), it was detected in two environments in 2006 and the average MDS in three environments, and was close to wheat SSR marker *Xwmc 550*. Its relatively stable contribution to resistance in two environments and its detection in a previous study indicates a significant role in APR.

The *QPm. caas-4DL* was proximal to the wheat microsatellite marker *Xwmc331* on the long arm of chromosome 4D. Keller et al.[15] previously detected a QTL at the end of chromosome 4DL in marker interval of *Xglk302b-Xpsr1101a* in a segregating wheat × *T.*

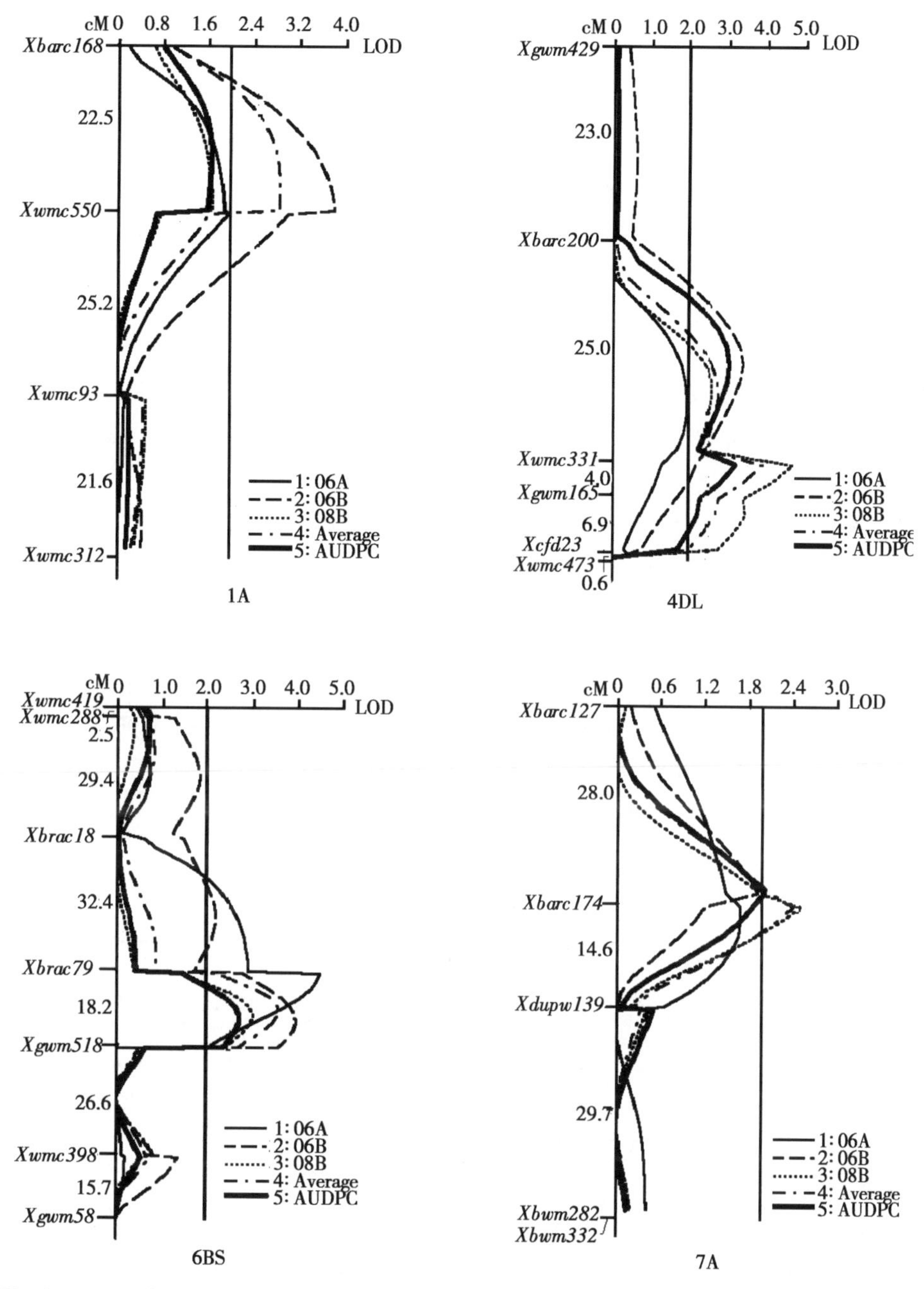

Fig. 2 Logarithm of odds (LOD) contours obtained by composite interval mapping for QTLs on chromosomes 1A, 4DL, 6BS, and 7A affecting powdery mildew severity in Bainong 64 × Jingshuang 16 doubled haploid (DH) lines. **06A**, maximum disease severities (MDS) in Anyang, 2006; **06B** and **08B**, maximum disease severities in Beijing, 2006 and 2008, respectively. **Average**, average maximum disease severities across three environments; **AUDPC**, average of area under the disease progress curve in Beijing over two years. Logarithm of odds (LOD) thresholds, 2.0.

spelta population. This location was different from that of *QPm. caas-4DL* by a map distance of nearly 50cM (http: //wheat. pw. usda. gov/cgi-bin/graingenes/browse. cgi? class=marker). Therefore, *QPm. caas-4DL* is likely to be a new gene conferring APR to powdery mildew. It had a high LOD score, a large genetic effect on resistance, and was stably detected across three environments. The SSR marker *Xwmc331*, closely linked to *QPm. caas-4DL*, could be used for marker-assisted selection in wheat breeding.

QTL *QPm. caas-6BS* was proximal to the molecular marker *Xbarc79*, at a genetic distance of 0.0 to 11.9cM. Keller et al.[15] detected a QTL in the AFLP marker interval *Xpsr167b-Xpsr964* on chromosome 6BL in a segregating wheat × *T. spelta* population. Based on the consensus map that gene should be different from *QPm. caas-6BS*. Five race-specific seedling-effective resistance genes have been identified on the chromosome 6B, viz. *Pm11*[36], *Pm12*[14], *Pm14*[35], *Pm20*[6], and *Pm27*[13]. *Pm11* and *Pm14* are effective only against certain wheat grass *Bgt* cultures, and *Pm12*, *Pm20* and *Pm27* were derived from alien species, viz. *Aegilops speltoides*[14], rye[6], and *T. timopheevii*[13], respectively. Thus, *QPm. caas-6BS* is unlikely to correspond to a known *Pm* gene or QTL on chromosome 6B. It showed a stable genetic effect against powdery mildew across three environments, explaining from 9.0 to 13.2% of the phenotypic variance. SSR marker *Xbarc79*, closely linked to *QPm. caas-6BS*, could be used for improving wheat powdery mildew resistance in molecular breeding programs.

The QTLdetected at the centromeric region of chromosome 7A was proximal to the wheat microsatellite marker *Xbarc174*. Chantret et al.[5] identified a QTL in the interval *Xfba069- Xgwm344* in an RE714 × Hardi population, but that region is quite distal from the location of *QPm. caas-7A* based on the consensus map. Several seedling-effective major gene loci have been located on chromosome 7A, including *Pm1*[29], *Pm9*[28], *Pm37*[24], and *mlRD30*[31]. *QPm. caas-7A* is unlikely to be any of these genes, or is unlikely to represent the effects of defeated alleles at any of these loci. It was detected in Beijing in two environments, and also identified in the mean MDS and AUDPC analyses. This QTL was proximal to marker *Xbarc174* within genetic distance of 0.0 to 2.0cM region. That marker could be used for marker-assisted selection.

In the crosses of the 11 cultivars and advanced lines, including the released cultivars 04 Zhong 36, Huaimai 0320, Xu 9908, and Yuanyu 3, and advanced lines 13397, 03 Zhong 35, Bainong 69, Junmai 35, Ruifeng 97-6, Tianmin 668, and Xumai 4060. All of them have showed moderate resistance to powdery mildew at adult plant stage in the field, except for Huaimai 0320[7,43]. In the crosses for the 11 cultivars and advanced lines, all the other parents were susceptible to powdery mildew, except for Zhoumai 16, another resistant parent of Junmai 35[41,44]. This indicates that Bainong 64 is a good resistant parent in Chinese wheat breeding programs.

In this study, we detected four QTLs in Bainong 64, which in total explained 34.1% to 50.4% of the phenotypic variance for powdery mildew response across three environments (Table 2). *QPm. caas-4DL* and *QPm. caas-6BS* had large effects on resistance to powdery mildew, and the molecular markers *Xwmc331* and *Xbarc79*, closely linked to two QTLs, could be used in molecular breeding for APR to powdery mildew in wheat.

Acknowledgments

We are grateful to the critical review of an earlier draft of this manuscript by Prof. R. A. McIntosh, Plant Breeding Institute, University of Sydney. This study was supported by the National Science Foundation of China (30810214 and 30671294).

References

[1] Bennett, F. 1984. Resistance to powdery mildew in wheat: a review of its use in agriculture and breeding programmes. Plant Pathol. 33: 297-300.

[2] Bjarko, M. E., and Line, R. F. 1988. Heritability and number of genes controlling leaf rust resistance on four cultivars of wheat. Phytopathology 78: 457-461.

[3] Bougot, Y., Lemoine, J., Pavoine, M. T., Guyomarc' h, H., Gautier, V., Muranty, H., and Barloy, D. 2006. A major QTL effect controlling resistance to powdery mildew in winter wheat at the adult plant stage. Plant Breed. 125: 550-556.

[4] Chantret, N., Sourdille, P., Röder, M., Tavaud, M., Bernard, M., and Doussinault, G. 2000. Location and mapping of the powdery mildew resistance

gene *MlRE* and detection of a resistance QTL by bulked segregant analysis (BSA) with microsatellites in wheat. Theor. Appl. Genet. 100: 1217-1224.

[5] Chantret, N., Mingeot, D., Sourdille, P., Bernard, M., Jacquemin, J. M., and Doussinault, G. 2001. A major QTL for powdery mildew resistance is stable over time and at two development stages in winter wheat. Theor. Appl. Genet. 103: 962-971.

[6] Friebe, B., and Larter, E. N. 1988. Identification of a complete set of isogenic wheat/rye D-genome substitution lines by means of Giemsa C-banding. Theor. Appl. Genet. 76: 473-479.

[7] Gu, Z. Z., Zhou, Y. M., Wang, A. B., Su, S. Y., Xia, Z. H., and Wang, Y. J. 2007. Characteristic of a new cultivar Huimai 21 (Huimai 0302) with high yield, good quality and broad diseases resistance. Jiangsu Agri. Sci. 2: 50-51. (In Chinese)

[8] Gupta, P. K., Rustgi, S., Sharma, S., Singh, R., Kumar, N., and Balyan, H. S. 2003. Transferable EST-SSR markers for the study of polymorphism and genetic diversity in bread wheat. Mol. Gen. Genet. 270: 315-323.

[9] Gustafson, G., and Shaner, G. 1982. Influence of plant age on the expression of slow-mildewing resistance in wheat. Phytopathology 72: 746-749.

[10] Guyomarc' h, H., Sourdille, P., Edwards, K. J., and Bernard, M. 2002. Studies of the transferability of microsatellites derived from *Triticum tauschii* to hexaploid wheat and to diploid related species using amplification, hybridization and sequence comparisons. Theor. Appl. Genet. 105: 736-744.

[11] Hautea, R., Coffman, W., Sorrels, M., and Bergstrom, G. 1987. Inheritance of partial resistance to powdery mildew in spring wheat. Theor. Appl. Genet. 73: 609-615.

[12] Heun, M., Friebe, B., and Bushuk, W. 1990. Chromosomal location of the powdery mildew resistance gene of Amigo wheat. Phytopathology 80: 1129-1133.

[13] Järve, K., Peusha, H. O., Tsymbalova, J., Tamm, S., Devos, K. M., and Enno, T. M. 2000. Chromosomal location of a *Triticum timopheevii* derived powdery mildew resistance gene transferred to common wheat. Genome 43: 377-381.

[14] Jia, J., Devos, K. M., Chao, S., Miller, T. E., Reader, S. M., and Gale, M. D. 1996. AFLP-based maps of the homoeologous group-6 chromosomes of wheat and their application in the tagging of *Pm12*, a powdery mildew resistance gene transferred from *Aegilops speltoides* to wheat. Theor. Appl. Genet. 92: 559-565.

[15] Keller, M., Keller, B., Schachermayr, G., Winzeler, M., Schmid, J. E., Stamp, P., and Messmer, M. M. 1999. Quantitative trait loci for resistance against powdery mildew in a segregating wheat×*spelt* population. Theor. Appl. Genet. 98: 903-912.

[16] Kosambi, D. D. 1944. The estimation of map distance from recombination values. Annu. Eugen. 12: 172-175.

[17] Laurie, D. A., and Bennett, M. D. 1988. The production of haploid wheat plants from wheat × maize crosses. Theor. Appl. Genet. 76: 393-397.

[18] Liang, S. S., Suenaga, K., He, Z. H., Wang, Z. L., Liu, H. Y., Wang, D. S., Singh, R. P., Sourdille, P., and Xia, X. C. 2006. Quantitative trait loci mapping for adult-plant resistance to powdery mildew in bread wheat. Phytopathology 96: 784-789.

[19] Lillemo, M., Asalf, B., Singh, R. P., Huerta-Espino, J., Chen, X. M., He, Z. H., and Bjørnstad, Å. 2008. The adult plant rust resistance loci *Lr34/Yr18* and *Lr46/Yr29* are important determinants of partial resistance to powdery mildew in bread wheat line Saar. Theor. Appl. Genet. 116: 1155-1166.

[20] Liu, S. X., Griffey, C. A., and Maroof, M. A. S. 2001. Identification of molecular markers associated with adult plant resistance to powdery mildew in common wheat cultivar Massey. Crop Sci. 41: 1268-1275.

[21] Manly, K. F., Cudmore, R. H. J., and Meer, J. M. 2001. Map Manager QTX, cross-platform software for genetic mapping. Genome 12: 930-932.

[22] Michelmore, R. W., Paran, I., and Kesseli, R. V. 1991. Identification of markers linked to disease-resistance genes by bulked segregant analysis: A rapid method to detect markers in specific genomic regions by using segregating populations. Proc. Natl. Acad. Sci. USA 88: 9828-9832.

[23] Mingeot, D., Chantret, N., Baret, P. V., Dekeyser, A., Boukhatem, N., Sourdille, P., Doussinault, G., and Jacquemin, J. M. 2002. Mapping QTL involved in adult plant resistance to powdery mildew in the winter wheat line RE714 in two susceptible genetic backgrounds. Plant Breed. 121: 133-140.

[24] Perugini, L. D., Murphy, J. P., Marshall, D. S., and Brown-Guedira, G. 2008. *Pm37*, a new broadly effective powdery mildew resistance gene from *Triticum timophee-*

vii. Theor. Appl. Genet. 116: 417-425.

[25] Pestsova, E., Ganal, M. W., and Röder, M. S. 2000. Isolation and mapping of microsatellite markers specific for the D genome of bread wheat. Genome 43: 689-697.

[26] Roberts, J., and Caldwell, R. 1970. General resistance (slow mildewing) to *Erysiphe graminis* f. sp. *tritici* in Knox Wheat. (Abstr.) Mol. Gen. Genet. 60: 1310.

[27] Röder, M. S., Korzun, V., Wendehake, K., Plaschke, J., Tixier, M. H., Leroy, P., and Ganal, M. W. 1998. A microsatellite map of wheat. Genetics 149: 2007-2023.

[28] Schneider, D., Heun, M., and Fischbeck, G. 1991. Inheritance of the powdery mildew resistance gene *Pm9* in relation to *Pm1* and *Pm2* of wheat. Plant Breed. 107: 161-164.

[29] Sears, E. R., and Briggle, L. W. 1969. Mapping the *Pm1* gene for resistance to *Erysiphe graminis* f. sp. *tritici* on chromosome 7A of wheat. Crop Sci. 9: 96-97.

[30] Shi, A. N., Leath, S., and Murphy, J. P. 1998. A major gene for powdery mildew resistance transferred to common wheat from wild einkorn wheat. Phytopathology 88: 144-147.

[31] Singrün, C. H., Hsam, S. L., Zeller, F. J., Wenzel, G., and Mohler, V. 2004. Localization of a novel recessive powdery mildew resistance gene from common wheat line RD30 in the terminal region of chromosome 7AL. Theor. Appl. Genet. 109: 210-214.

[32] Somers, D. J., Isaac, P., and Edwards, K. 2004. A high-density microsatellite consensus map for bread wheat (*Triticum aestivum* L.). Theor. Appl. Genet. 109: 1105-1114.

[33] Song, Q. J., Fickus, E. W., and Cregan, P. B. 2002. Characterization of trinucleotide SSR motifs in wheat. Theor. Appl. Genet. 104: 286-293.

[34] Sourdille, P., Singh, S., Cadalen, T., Brown-Guedira, G. L., Gay, G., Qi, L., Gill, B. S., Dufour, P., Murigneux, A., and Bernard, M. 2004. Microsatellite-based deletion bin system for the establishment of genetic-physical map relationships in wheat (*Triticum aestivum* L.). Funct. Int. Gen. 4: 12-25.

[35] Tosa, Y., and Sakai, K. 1990. The genetics of resistance of hexaploid wheat to the wheat grass powdery mildew fungus. Genome 33: 225-230.

[36] Tosa, Y., Tokunaga, H., and Ogura, H. 1988. A gene involved in the resistance to wheat to wheatgrass powdery mildew fungus in common wheat cultivar Chinese Spring. Genome 30: 612-614.

[37] Tucker, D. M., Griffey, C. A., Liu, S., Brown-Guedira, G., Marshall, D. S., and Saghai Maroof, M. A. 2007. Confirmation of three quantitative trait loci conferring adult plant resistance to powdery mildew in two winter wheat populations. Euphytica 155: 1-13.

[38] Wang, S., Basten, C. J., and Zeng, Z. B. 2005. Windows QTL Cartographer v2.5. Statistical Genetics. North Carolina State University.

[39] Wang, Z. L., Li, L. H., He, Z. H., Duan, X. Y., Zhou, Y. L., Chen, X. M., Lillemo, M., Singh, R. P., Wang, H., and Xia, X. C. 2005. Seeding and adult-plant resistance to powdery mildew in Chinese bread wheat cultivars and lines. Plant Dis. 89: 457-463.

[40] Wang, Z. L., Liu, S. D., Wang, H., He, Z. H., Xia, X. C., Chen, X. M., and Duan, X. Y. 2006. Genetic analysis of the resistance of the wheat cultivar Bainong 64 to powdery mildew. Acta Bot. Boreal. -Occident. Sin. 26: 0332-0336. (In Chinese with English abstract)

[41] Xia, Z. H., Gu, Z. Z., Sun, S. Y., Liu, Y. H., and Zhang, Y. F. 2002. Characteristic of a new cultivar Huimai 17 with high yield and good quality. J. Anhui Agri. Sci. 30: 225. (In Chinese)

[42] Xu, X. Y., Bai, G. H., Carver, B. F., Shaner, G. E., and Hunger, R. M. 2006. Molecular characterization of a powdery mildew resistance gene in wheat cultivar Suwon 92. Phytopathology 96: 496-500.

[43] Yun, Q. F., Zhao, S. L., Zhao, D. W., Liu, C. H., Yang, C. Y., and Yi, Z. H. 2007. Breeding a new cultivar Yuanyu 3 with high protein and broad diseases resistance. Bull. Agri. Sci. Tech. 7: 28-29. (In Chinese)

[44] Zhao, Z. W., Ma, H. P., Jiang, Z. K., Dong, Y., and Zhou, Q. Z. 2000. Breeding a new cultivar Xinmai 9 with high yield and good quality. Henan Agri. Sci. 4: 3-5. (In Chinese)

QTL mapping of adult-plant resistances to stripe rust and leaf rust in Chinese wheat cultivar Bainong 64

Y. Ren[1], Z. F. Li[2], Z. H. He[1,5], L. Wu[3], B. Bai[1,4], C. X. Lan[1], C. F. Wang[2], G. Zhou[4], H. Z. Zhu[3], and X. C. Xia[1]

[1] *Institute of Crop Science, National Wheat Improvement Center/The National Key Facility for Crop Gene Resources and Genetic Improvement, Chinese Academy of Agricultural Sciences (CAAS), 12 Zhongguancun South Street, Beijing* 100081, *China*; [2] *Department of Plant Pathology, College of Plant Protection, Agricultural University of Hebei, Biological Control Center for Plant Diseases and Plant Pests of Hebei*, 289 *Lingyusi Street, Baoding* 071001, *Hebei, China*; [3] *Crop Research Institute, Sichuan Academy of Agricultural Sciences, Chengdu* 610066, *Sichuan, China*; [4] *Gansu Winter Wheat Research Institute, Gansu Academy of Agricultural Sciences, Lanzhou* 730070, *Gansu, China*; [5] *International Maize and Wheat Improvement Center (CIMMYT) China Office, c/o CAAS*, 12 *Zhongguancun South Street, Beijing* 100081, *China*

Abstract: Stripe rust and leaf rust, caused by *Puccinia striiformis* Westend. f. sp. *tritici* Erikss. and *P. triticina*, respectively, are devastating fungal diseases of common wheat (*Triticum aestivum* L.). Chinese wheat cultivar Bainong 64 has maintained acceptable adult-plant resistance (APR) to stripe rust, leaf rust and powdery mildew for more than ten years. The aim of this study was to identify quantitative trait loci/locus (QTL) for resistance to the two rusts in a population of 179 doubled haploid (DH) lines derived from Bainong 64 × Jingshuang 16. The DH lines were planted in randomized complete blocks with three replicates at four locations. Stripe rust tests were conducted using a mixture of currently prevalent *P. striiformis* races, and leaf rust tests were performed with *P. triticina* race THTT. Leaf rust severities were scored two or three times, whereas maximum disease severities (MDS) were recorded for stripe rust. Using bulked segregant analysis (BSA) and simple sequence repeat (SSR) markers, five independent loci for APR to two rusts were detected. The QTL on chromosomes 1BL and 6BS contributed by Bainong 64 conferred resistance to both diseases. The loci identified on chromosomes 7AS and 4DL had minor effects on stripe rust response, whereas another locus, close to the centromere on chromosome 6BS, had a significant effect only on leaf rust response. The loci located on chromosomes 1BL and 4DL also had significant effects on powdery mildew response. These were located at the same positions as the *Yr29/Lr46* and *Yr46/Lr67* genes, respectively. The multiple disease resistance locus for APR on chromosome 6BS appears to be new. All three genes and their closely linked molecular markers could be used in breeding wheat cultivars with durable resistance to multiple diseases.

Introduction

Stripe rust (or yellow rust, YR) and leaf rust (LR), caused by *Puccinia striiformis* Westend. f. sp. *tritici* Erikss. and *P. triticina*, respectively, are major foliar diseases of common wheat (*Triticum aestivum* L.) in many wheat-growing regions of the world. Rust epidemics are recurrent events that cause significant grain yield losses and reduced quality (Samborski, 1985;

Published in TAG, 2012, 125 (6): 1253-1262

Line and Chen, 1995). Yield losses caused by YR range from 10-70% depending upon the cultivar, earliness of the initial infection, rate of disease development and duration of the disease (Chen, 2005; Afzal et al., 2007), whereas, LR can cause yield losses of up to 40% in favorable conditions (Knott, 1989; Zhao et al., 2008). Although fungicides can provide adequate control of rusts, resistant cultivars are a more economic and effective approach to control these diseases, as it has no cost to growers and is environmentally friendly (Line, 2002; Chen, 2005).

Currently, at least49 YR and 68 LR resistance loci are cataloged in wheat and assigned to specific chromosomes or chromosome arms (McIntosh et al., 2011; Herrera-Foessel et al., 2012). Most of these resistance genes are race-specific and are conferred by single or a few major genes (Kilpatrick et al., 1975; Zhao et al., 2008; Lu et al., 2009). Race-specific resistance has been often used by wheat breeders because of its high level of effectiveness throughout the entire growth cycle of the crop. Unfortunately, it is readily overcome by mutation and/or selection in the pathogen population (Chen and Line, 1995a, b; Carter et al., 2009). Currently, only a few named YR and LR resistance genes (including *Yr5*, *Yr10*, *Yr15*, and *Yr24*/*Yr26*; *Lr9*, *Lr19*, *Lr24* and *Lr38*) are effective against prevalent Chinese *P. striiformis* and *P. triticina* races, respectively (Yang et al., 2003; Yuan et al., 2007). In contrast, non race-specific or adult-plant resistance (APR) is generally quantitatively inherited. This type of resistance is often characterized by lower frequencies of infections, longer latent periods, smaller uredinial size and less urediniospore production (Caldwell, 1968; Chen and Line, 1995a; Liang et al., 2006; Lu et al., 2009; Li et al., 2010). Although individual genes of this type do not confer adequate levels of resistance, combinations of four or five slow-rusting genes may confer near immunity (Singh et al., 2000; Herrera-Foessel et al., 2011).

To date, several important genes for slow rusting have been identified in wheat (Herrera-Foessel et al., 2011, 2012; Hiebert et al., 2010; Singh et al., 2011). Evidence suggests that *Yr18*/*Lr34* on chromosome 7DS (Suenaga et al., 2003; Lagudah et al., 2006) and *Yr29*/*Lr46* on 1BL (William et al., 2003) have been effective since the early 20th century (Krattinger et al., 2009). Both loci also confer partial resistance to powdery mildew (PM) (Lillemo et al., 2008) and stem rust (SR) (Bhavani et al., 2011) and are associated with leaf tip necrosis (LTN) (Singh 1992; Rosewarne et al., 2006). *Yr18*/*Lr34* was cloned and shown to encode a putative ATP-binding cassette (ABC) transporter (Krattinger et al., 2009). Gene-based DNA markers derived from the sequence enable precise marker-assisted breeding (Lagudah et al., 2009). A third locus on 4DL contains *Yr46* (Herrera-Foessel et al., 2011), *Lr67* (Hiebert et al., 2010) and the recently named *Pm46* (McIntosh et al., 2012). These genes can be utilized in combination with other slow rusting or slow mildewing genes to develop high levels of durable APR to YR, LR and PM. A fourth slow-rusting resistance gene, *Lr68*, located on chromosome 7BL of CIMMYT cv. Parula, is also available for breeding durable and stable APR to LR (Herrera-Foessel et al., 2012). This gene is likely to be widely distributed in CIMMYT spring bread wheat germplasm (Singh et al., 2011).

Bainong 64 was a leading winter wheat cultivar in the Yellow-Huai wheat region of China at the end of 1990s and beginning of 2000s, and it occupied about 700,000ha on average annually for eight years since its release in Henan Province in 1998 (Wang et al., 2005b, 2006). This cultivar continues to exhibit resistance to YR, LR and PM in the field. Because it is susceptible to Chinese *P. striiformis* race CYR32, *P. triticina* race THTT and *Blumeria graminis* f. sp. *tritici* isolate E20 at the seedling stage, it carries APR to these three diseases. Although Lan et al. (2009) conducted an analysis of APR to PM in Bainong 64, little is known about the genetics of resistance to YR and LR in this cultivar. The aim of the present study was to detect genetic loci for resistance to YR and LR in a doubled haploid (DH) population derived from a

Bainong 64 × Jingshuang 16 cross.

Materials and methods

Plant materials

A DH population of 179 lines was developed from Bainong 64×Jingshuang 16 by the wheat×maize method. Bainong 64 was derived from the cross Bainong 8717/3/Yeda 72-629-52/Shi 82-5594//Bainong 84-4046-1. Jingshuang 16 (Lovrin 10×Youmanghong 7), released in Beijing in 1985 (Wang et al., 2006), is susceptible to both YR and LR at seedlings.

Field trials

Bainong 64, Jingshuang 16 and 179 DH lines were evaluated for YR response in Tianshui, Gansu Province, and Chengdu, Sichuan Province, during the 2009-2010 and 2010-2011 cropping seasons. They were also evaluated for LR in Baoding, Hebei Province, and Zhoukou, Henan Province, in 2010-2011. The population was planted in randomized complete blocks with three replicates at each location. Trials were managed according to local practices in the respective regions.

YR tests

BothTianshui and Chengdu are hotspots for YR in China and experience severe epidemics almost every year. Plots consisted of single rows in 1.5 m length and 25 cm between rows. Approximately 50 seeds were sown in each row. The highly susceptible control Mingxian 169 was planted every 10th row, and also perpendicular and adjacent to the test rows to ensure ample inoculum. Inoculations were carried out using mixtures of *P. striiformis* races CYR31, CYR32, CYR33, Shui4, Shui6, Hy6 and Hy7 in Chengdu around January 3, and mixtures of *P. striiformis* races CYR29, CYR31, CYR32, CYR33, Shui4, Shui5 and Hy8 were used for inoculation in Tianshui around April 20. The maximum disease severities (MDS) were assessed (Peterson et al., 1948) when the disease severities on Mingxian 169 reached a maximum level around the 15th of April in Chengdu and the 10th of June in Tianshui, respectively.

LR tests

Leaf rust was tested in Baoding and Zhoukou with ideal conditions for rust infection and spread. Fifty seeds of each line were sown in single-row plots of 1.5 m length and 30 cm between rows. Spreader rows of Zhengzhou 5389 were planted perpendicular and adjacent to the test rows. LR epidemics were initiated by spraying aqueous suspensions of urediniospores of *P. triticina* pathotype THTT, to which a few drops of Tween 20 (0.03%) were added, onto the spreader rows at the tillering stage. Disease severities were assessed two or three times at weekly intervals with the first scoring four weeks after inoculation based on the modified Cobb scale (Peterson et al., 1948). Areas under the disease progress curve (AUDPC) were calculated according to Bjarko and Line (1988).

Statistical analysis

Analysis of variance was performed with PROC GLM in the Statistical Analysis System (SAS Institute, V8), with genotype as a fixed effect, and environments, a combination of locations and years, and replicates as random effects. The information in the ANOVA table was used to calculate the broad sense heritabilities (h_b^2) of resistance to the two diseases reactions based on the formula $h_b^2 = \sigma_g^2 / (\sigma_g^2 + \sigma_{ge}^2/e + \sigma_\varepsilon^2/re)$ (Allard 1960), where $\sigma_g^2 = (MS_f - MS_{fe})/re$, $\sigma_{ge}^2 = (MS_{fe} - MS_e)/r$ and $\sigma_\varepsilon^2 = MS_e$; in this formula, σ_g^2 = genetic variance, σ_{ge}^2 = genotype × environment interaction variance, σ_ε^2 = error variance, MS_f = mean square of genotype, MS_{fe} = mean square of genotype × environment interaction, MS_e = mean square of error, r = number of replicates and e = number of environments. Field data from Tianshui in 2011 were excluded from the statistical analysis and QTL detection due to the low YR development caused by the dry weather condition in the spring.

SSR marker assay and bulked segregant analysis

The parental lines Bainong 64 and Jingshuang 16 and the contrasting bulk for powdery mildew were screened for polymorphism with 406 simple sequence repeat

(SSR) markers by Lan et al. (2009). Then 375 more SSR markers were used for screening the two parents to enable more saturated linkage maps in the present study; these included BARC (Song et al., 2002), CFA and CFD (Sourdille et al., 2004), GDM (Pestsova et al., 2000), GWM (Röder et al., 1998) and WMC (Gupta et al., 2002) markers. Based on results of two years of field data for YR, equal amounts of DNA from the five most resistant and five most susceptible lines, respectively, were mixed to form resistant and susceptible bulks. SSR markers that showed similar patterns of polymorphism between the bulks and parents were used to genotype individual lines in the population. Additional SSRs around the QTL for resistance to YR or LR were also selected to genotype the DH lines based on several wheat consensus maps (Somers et al., 2004; http://wheat.pw.usda.gov).

Map construction and QTL analysis

Linkage groups were constructed using Map Manager QTX20 (Manly et al., 2001). Genetic distances between markers were calculated based on the Kosambi mapping function (Kosambi, 1944). The information on a publicly available wheat consensus map (Somers et al., 2004) was used to assign linkage groups to chromosomes. Cartographer 2.5 was used to detect QTL by composite interval mapping (CIM) (Wang et al., 2005a). A logarithm of odds (LOD) threshold of 2.0 was set to declare QTL as significant. A walk speed of 2.0cM was chosen for all QTL detections. QTL effects were estimated as the proportion of phenotypic variance (R^2) explained by the QTL.

Results

Phenotypic evaluation

Stripe rust and LR developed well across environments, except for YR at Tianshui in 2011. The frequency distributions of YR MDS for the 179 DH lines in three environments and LR response (MDS and AUDPC) of the DH lines in two environments revealed continuous distributions, indicating polygenic inheritance. For YR, the averaged MDS of the DH lines across three environments over two years was 35.4%, ranging from 3.7 to 86.1%. Bainong 64 was rated with a mean MDS of 8.3%, 35.0% and 8.7% in Chengdu 2010, Chengdu 2011 and Tianshui 2010, respectively, whereas Jingshuang 16 had mean MDS of 12.7%, 57.5% and 27.5% in three environments, respectively. For LR, the mean MDS of Bainong 64 and Jingshuang 16 were 56.7% and 58.3% in Baoding 2011, respectively, whereas their mean MDS were 13.3% and 18.3% in Zhoukou 2011, respectively. The averaged MDS of the DH lines in Baoding 2011 was 41.0% ranging from 2.3 to 85%, and 18.3% in Zhoukou ranging from 2.3 to 53.3%. In addition, the averaged AUDPC of the DH lines across two environments were 302.6, ranging from 42.3 to 675.0.

Stripe rust MDS were significantly correlated among three environments, with correlation coefficients of 0.50 to 0.57 ($P<0.0001$), and the heritability of YR MDS was 0.76. Significant correlations for LR MDS or AUDPC were also detected in two environments ($r=0.57$ and 0.64, respectively, $P<0.0001$), and the heritabilities of LR MDS and AUDPC were 0.63 and 0.52, respectively. Furthermore, the LR MDS and LR AUDPC were significantly correlated in Baoding 2011 and Zhoukou 2011 ($r=0.95$ and 0.97, respectively, $P<0.0001$). ANOVA of the two traits revealed significant differences ($P=0.01$) in MDS and AUDPC among RILs, environments, replicates within environments and line × environment interactions (Table 1).

QTL for YR resistance

Three QTL for resistance to YR were identified on chromosomes 4DL, 6BS and 7AS based on CIM using the MDS in Chengdu 2010, Chengdu 2011 and Tianshui 2010 and averaged MDS from all three environments (Fig. 1; Table 2). They were designated *QYr.caas-4DL*, *QYr.caas-6BS.3* and *QYr.caas-7AS*, respectively. *QYr.caas-6BS.3*, flanked by *Xwmc487*-*Xcfd13* and located in the telomeric region, explained from 3.8 to 6.2% of the phenotypic variance. The resistance allele of this QTL was contributed by Bainong

64. *QYr. caas-4DL* was located between *Xwmc331* and *Xgwm165*. This QTL was detected in Chengdu 2011, and explained 8.0% of the phenotypic variance. It also came from Bainong 64. The third QTL, *QYr. caas-7AS*, was identified between *Xbarc127* and *Xbarc174*, and explained 6.0% and 6.1% of the phenotypic variances in Chengdu 2011 and the averaged MDS, respectively. This gene came from Jingshuang 16.

Table 1 Analysis of variance of maximum disease severities (MDS) and the area under the disease progress curve (AUDPC) values for YR and LR responses on doubled-haploid (DH) lines derived from Bainong 64 × Jingshuang 16

Phenotype	Source of variance	*df*	Mean square	*F* values
YR	**MDS**			
	Lines	178	3,101	4.2**
	Environments	2	222,737	304.3**
	Replicates	2	5,578	31.3**
	Lines × environments	356	732	4.1**
	Error	1071	178	
LR	**MDS**			
	Lines	178	1,065	2.7**
	Environments	1	138,835	348.8**
	Replicates	2	870	5.5**
	Lines × environments	178	398	2.5**
	Error	714	158	
	AUDPC			
	Lines	178	112,680	2.1**
	Environments	1	34,221,842	629.1**
	Replicates	2	412,481	37.0**
	Lines × environments	178	54,396	4.9**
	Error	714	11,142	

** Significant at $P = 0.01$.

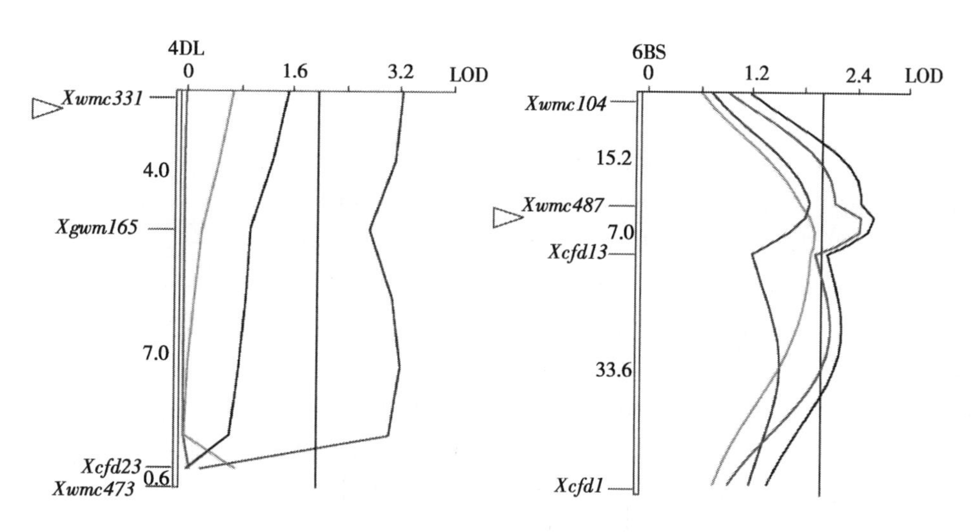

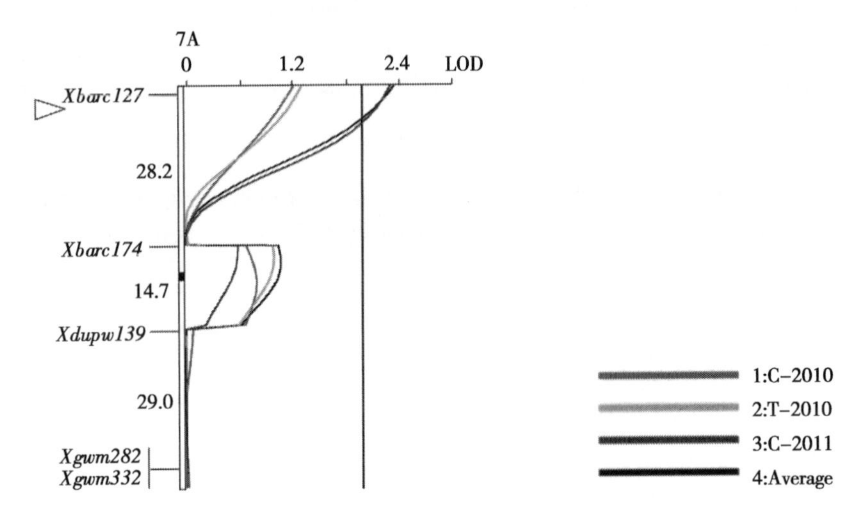

Fig. 1 LOD contours obtained by CIM analysis for QTL on chromosomes 4DL, 6BS and 7AS affecting YR MDS in Bainong 64 × Jingshuang 16 DH lines. The approximate positions of centromeres are indicated by solid squares in the vertical axis. Genetic distances are shown in centiMorgans to the left of vertical axis. The approximate positions of the QTL are indicated by arrowheads to the left of markers. LOD thresholds of 2.0 are indicated by a dashed vertical line in graphs. C-2010 and C-2011 = MDS in Chengdu 2010 and 2011, respectively, and T2010= MDS in Tianshui 2010. Average=average of MDS across three environments.

Table 2 Quantitative trait loci (QTL) for adult plant resistance (APR) to YR, LR and PM detected by composite interval mapping (CIM) in the Bainong 64 × Jingshuang 16 DH population across environments

	Location and year	QTL[a]	Marker interval	LOD[b]	Add[c]	R^2 (%)[d]
YR	**MDS**					
	Chengdu 2010	*QYr. caas-6BS. 3*	*Xwmc487-Xcfd13*	2.4	4.8	6.2
	Chengdu 2011	*QYr. caas-4DL*	*Xwmc331-Xgwm165*	3.3	8.3	8.0
		QYr. caas-6BS. 3	*Xwmc487-Xcfd13*	1.9	5.8	3.8
		QYr. caas-7AS	*Xbarc127-Xbarc174*	2.3	−7.0	6.0
	Tianshui 2010	*QYr. caas-6BS. 3*	*Xwmc487-Xcfd13*	1.9	5.6	4.5
	Averaged MDS	*QYr. caas-6BS. 3*	*Xwmc487-Xcfd13*	2.6	5.2	6.1
	in 2 environments	*QYr. caas-7AS*	*Xbarc127-Xbarc174*	2.4	−5.0	6.1
LR	**MDS**					
	Baoding 2011	*QLr. caas-1BL*	*Xgwm153. 2-Xwmc44*	5.3	8.7	15.6
		QLr. caas-6BS. 1	*Xwmc487-Xcfd13*	5.0	7.9	11.2
		QLr. caas-6BS. 2	*Xgwm518-Xwmc398*	3.7	−6.7	9.7
	Zhoukou 2011	*QLr. caas-1BL*	*Xgwm153. 2-Xwmc44*	5.4	4.2	14.9
	Averaged MDS	*QLr. caas-1BL*	*Xgwm153. 2-Xwmc44*	7.1	6.8	21.1
	in 2 environments	*QLr. caas-6BS. 1*	*Xwmc487-Xcfd13*	4.6	5.0	10.0
		QLr. caas-6BS. 2	*Xgwm518-Xwmc398*	3.3	−4.4	9.0
	AUDPC					
	Baoding 2011	*QLr. caas-1BL*	*Xgwm153. 2-Xwmc44*	8.1	130.7	26.4
		QLr. caas-6BS. 1	*Xwmc487-Xcfd13*	5.5	91.6	11.4
		QLr. caas-6BS. 2	*Xgwm518-Xwmc398*	3.2	−72.6	8.7

（续）

	Location and year	QTL[a]	Marker interval	LOD[b]	Add[c]	R^2 (%)[d]
	Zhoukou 2011	*QLr. caas-1BL*	*Xgwm153. 2-Xwmc44*	4. 6	27. 7	14. 4
	Averaged AUDPC	*QLr. caas-1BL*	*Xgwm153. 2-Xwmc44*	8. 8	82. 0	28. 5
	in 2 environments	*QLr. caas-6BS. 1*	*Xwmc487-Xcfd13*	5. 2	54. 2	10. 8
		QLr. caas-6BS. 2	*Xgwm518-Xwmc398*	3. 1	−42. 7	8. 2
PM[e]	Averaged MDS	*QPM. caas-1BL*	*Xgwm153. 2-Xwmc44*	2. 4	2. 4	5. 7

a. QTL were detected with a minimum LOD score of 2. 0 in at least one environment.

b. LOD, logarithm of odds score.

c. Add, additive effect of resistance allele.

d. R^2, percentages of the phenotypic variance explained by individual QTL.

e. Four QTL for APR to PM were identified in Bainong 64 by Lan et al. , (2009); in addition, *QPm. caas-1BL* contributed by Bainong 64 was detected in the present study.

QTL for LR resistance

Based on the mean MDS in Baoding 2011, Zhoukou 2011 and MDS averaged from the two environments, three QTL for resistance to LR were detected on chromosomes 1BL and 6BS (2 QTL) (Fig. 2; Table 2). The most stable locus with the largest effect across environments was *QLr. caas-1BL*, located on 1BL between *Xgwm153. 2* and *Xwmc44*. This QTL explained 15. 6%, 14. 9% and 21. 1% of the phenotypic variance in Baoding 2011, Zhoukou 2011 and averaged MDS, respectively. *QLr. caas-6BS. 1*, flanked by *Xwmc487* and *Xcfd13*, explained 11. 2 and 10. 0% of the phenotypic variance in Baoding 2011 and averaged MDS, respectively. The third QTL, *QLr. caas-6BS. 2*, in the interval *Xgwm518-Xwmc398* close to the centromere, was approximately 22cM from *QLr. caas-6BS. 1* based on the wheat consensus map (Somers et al. , 2004). This QTL explained from 9. 0 to 9. 7% of the phenotypic variance across environments. *QLr. caas-1BL* and *QLr. caas-6BS. 1* were contributed by Bainong 64, whereas *QLr. caas-6BS. 2* was from Jingshuang 16.

The three loci for LR resistance were also detected using the mean AUDPC in each environment and when averaged from both environments (Fig. 2, Table 2). *QLr. caas-1BL*, derived from Bainong 64, explained 26. 4%, 14. 4%, and 28. 5% of the phenotypic variance in Baoding 2011, Zhoukou 2011 and the averaged AUDPC. *QLr. caas-6BS. 1* from Bainong 64 explained 11. 4% and 10. 8% of the phenotypic variance in Baoding 2011 and the overall mean, respectively. The third locus, *QLr. caas-6BS. 2*, was contributed by Jingshuang 16 and explained 8. 2-8. 7% of the phenotypic variance.

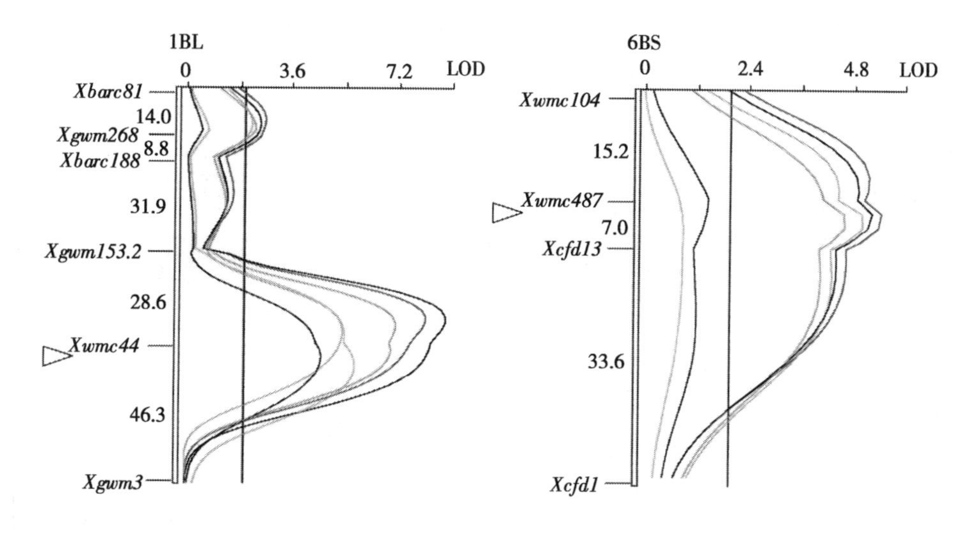

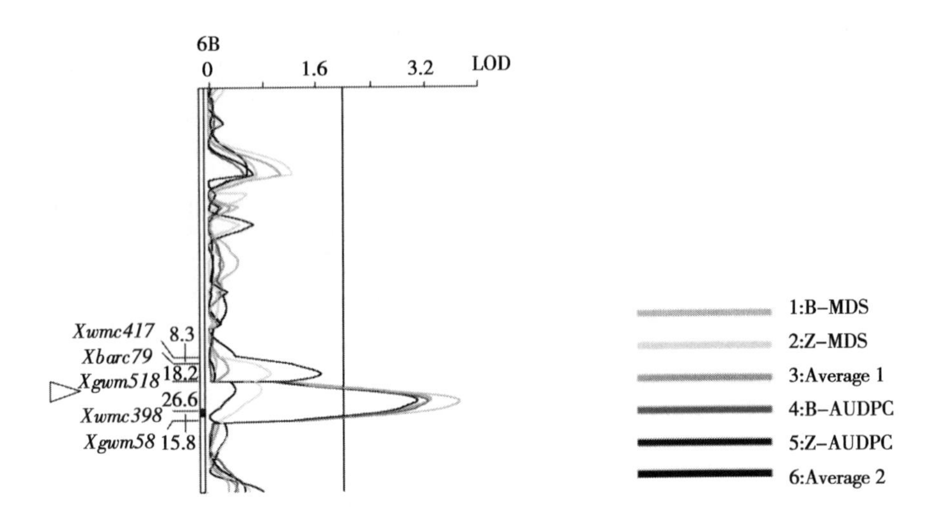

Fig. 2 LOD contours obtained by CIM analysis for QTL on chromosomes 1BL and 6BS (2 QTL) affecting LR MDS and AUDPC in Bainong 64 × Jingshuang 16 DH lines. The short arms are toward the top and the approximate positions of centromeres are indicated by solid squares in the vertical axis. Genetic distances are shown in centiMorgans to the left of vertical axis. The arrowheads indicate the likely positions of the QTL based on LR MDS and AUDPC. LOD thresholds of 2.0 are indicated by a dashed vertical line in graphs. B-MDS = MDS in Baoding 2011, Z-MDS = MDS in Zhoukou 2011 and Average 1=averaged MDS across two environments; B-AUDPC=AUDPC in Baoding 2011, Z-AUDPC=AUDPC in Zhoukou 2011 and Average 2=averaged AUDPC across two environments.

Pleiotropic effects of detected QTL

The same population of Bainong 64 × Jingshuang 16 was earlier used for QTL mapping of PM resistance and four QTL for APR to PM were identified on chromosomes 1A, 4DL, 6BS and 7A (Lan et al., 2009). In comparison with that study, *QYr. caas-4DL* co-located with a QTL for PM resistance with the allele for resistance contributed by Bainong 64. Therefore, *QYr. caas-4DL* may be used in breeding for resistance to both YR and PM. Based on the same study *QYr. caas-7AS* and *QLr. caas-6BS. 2* also co-located to similar positions to PM resistance QTL, but with resistance contributed by opposite parents. The parental lines and two bulks had been screened with 406 SSR markers by Lan et al. (2009), and in subsequent QTL mapping 73 SSR markers were used to genotype all DH lines. However, in the present study, a total of 99 SSR markers were used to genotype individual lines after detection of polymorphisms between the two parents and contrasting bulks. Because more markers were selected to genotype the DH lines in the present study, we identified a new PM gene (*QPm. caas-1BL*) in Bainong 64 based on CIM using the PM data from Lan et al. (2009). This QTL was located on chromosome 1BL between *Xgwm153. 2* and *Xwmc44* and corresponded with a LR resistance QTL from the same parent (Table 2; Fig. 3). Thus, the gene located on chromosome 1BL could be used to simultaneously improve resistance to LR and PM. Furthermore, *QYr. caas-6BS. 3* with YR resistance contributed by Bainong 64 coincided with a LR resistance QTL from the same parent. *QYr. caas-6BS. 3/QLr. caas-6BS. 1* is therefore a valuable gene for resistance to both YR and LR.

Discussion

Leaf rust MDS was significantly associated with LR AUDPC across environments in this study (r =0.95−0.97, P<0.000 1). This is in agreement with previous reports (Wang et al., 2005b; Liang et al., 2006; Lan et

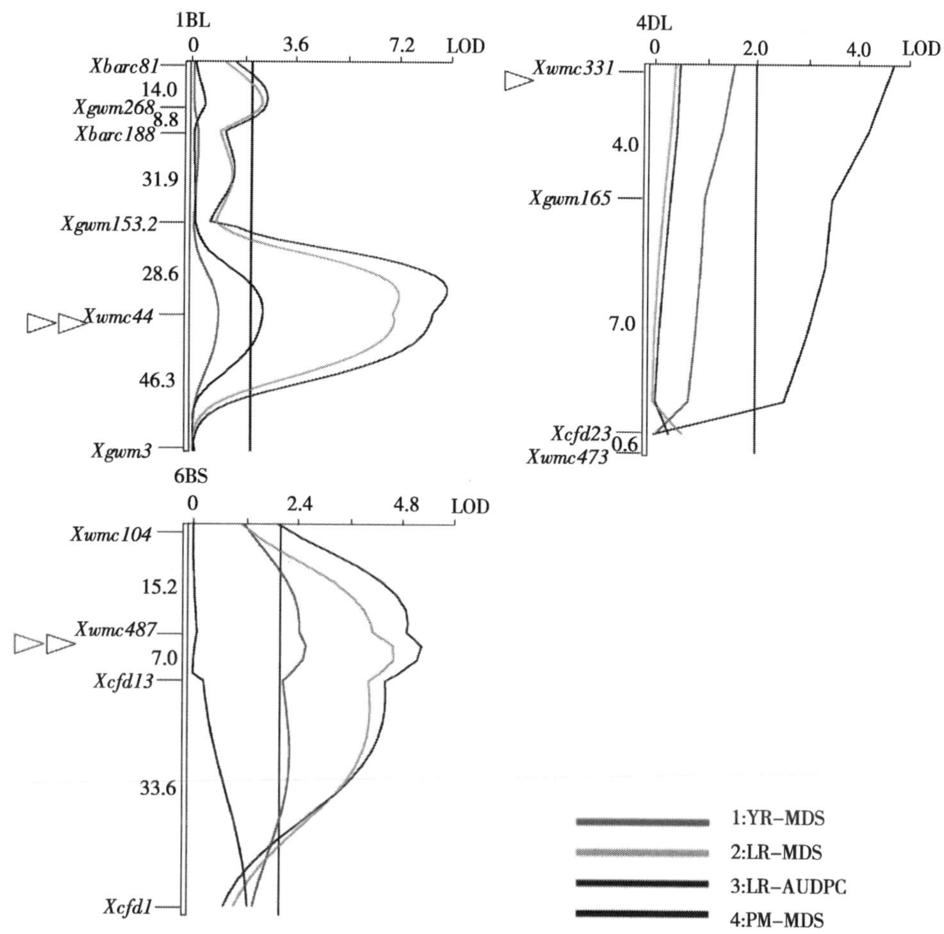

Fig. 3 LOD contours obtained by CIM analysis for QTL associated with averaged YR MDS, LR MDS, LR AUDPC and PM MDS in the Bainong 64 × Jingshuang 16 population. Genetic distances are shown in centiMorgans to the left of vertical axis. The red, blue and black arrowheads indicate the likely positions of the QTL based on YR MDS, LR MDS and AUDPC, and PM MDS, respectively. LOD thresholds of 2.0 are indicated by a dashed vertical line in graphs. YR-MDS=averaged YR MDS across three environments (Tianshui 2010, Chengdu 2010 and 2011), LR-MDS and LR-AUDPC=averaged LR MDS and AUDPC across two environments (Baoding 2011 and Zhoukou 2011), respectively, and PM-MDS=averaged PM MDS across three environments (Anyang 2006, Beijing 2006 and 2008, Lan et al., 2009).

al., 2009) indicating that it is feasible to replace AUDPC with MDS. In addition, the use of MDS reduces the labor and time for field investigations, as it is only assessed once when the disease severities on susceptible controls reach maximum levels.

In the present study, all QTL for resistance to YR, LR and PM in the Bainong 64×Jingshuang 16 population were integrated into linkage maps. As shown in Fig. 3, linkage groups on chromosomes 1BL, 4DL and 6BS showed significant associations of resistance to two or three diseases. Lillemo et al. (2008) concluded that resistance to YR, LR and PM might be under common genetic control as in the present study. Two loci located on chromosomes 1BL and 4DL were in the same position as the *Yr29/Lr46* and *Yr46/Lr67* loci, respectively, and we report here a new resistance locus conferring APR to YR and LR on chromosome 6BS.

Yr29/Lr46 located at the distal region of chromosome 1BL with significant effects on response to YR and LR was detected in several mapping populations (Suenaga et al., 2003; Rosewarne et al., 2006; William et al., 2003, 2006). However, Zhang et al. (2009) suggested that *Lr46* was significantly affected by environment. Lillemo et al. (2008) conducted QTL mapping for resistance to PM, LR and YR in the same population, and found that *Yr29/Lr46* was linked with a major PM resistance QTL (designated as *Pm39*) in wheat line Saar. In the present study, the allele on 1BL derived from Bainong 64 significantly reduced LR and PM severities, but the effect on YR response was very weak. This was likely due to the different expression levels of QTL in different genetic backgrounds and environments. Considering that the LOD peak for *QLr. caas-1BL*, near *Xwmc44*, was in the same position as *Yr29/Lr46*, these two genes might be either at the same locus or very closely linked. The slow-rusting gene *Yr29/Lr46* has provided APR to YR and LR for almost 30 years (William et al., 2006), and is widely distributed in CIMMYT germplasm (Singh et al., 2005). Because the linked markers for *Yr29/Lr46* are often population-specific or parent related, the molecular detection of this gene in some wheat genotypes might be difficult. The QTL in this report and its closely linked marker may be helpful for selecting *Yr29/Lr46* in Chinese wheat germplasm.

Suenaga et al. (2003) reported a QTL for YR resistance closely linked to *Xwmc399* on chromosome 4DL in Israeli wheat Oligoculm. It accounted for low levels of phenotypic variance ranging from 2.5 to 8.0%. This QTL was approximately 26cM from *QYr. caas-4DL* (Suenaga et al., 2003; He et al., 2011) indicating that it is probably different from *QYr. caas-4DL*. A recently reported multiple disease resistance locus on chromosome 4DL confers APR to YR (*Yr46*), LR (*Lr67*) and PM (*Pm46*) (Hiebert et al., 2010; Herrera-Foessel et al., 2011; McIntosh et al., 2012). The APR gene *Yr46/Lr67/Pm46*, closely linked to *Xgwm165* and *Xgwm192*, was located in a similar position to *QYr. caas-4DL* based on the consensus map of Somers et al. (2004). However, *QYr. caas-4DL* in Bainong 64 was co-located only with a PM resistance QTL *QPm. caas-4DL* (Lan et al., 2009) and had no effect on LR response. This is probably due to the relatively lower heritability of the LR data compared to YR. Further study is needed to test the allelism between *QYr. caas-4DL* and *Yr46/Lr67* to confirm whether they are at the same locus.

To date, a total of six YR APR QTL on chromosome 6BS have been reported, including *QYr. jirc-6B* in Oligoculm (Suenaga et al., 2003), *QYrst. wgp-6BS. 1* and *QYrst. wgp-6BS. 2* in Stephens (Santra et al., 2008), *Yr36* in wild emmer (*T. turgidum* ssp. *dicoccoides* accession FA15-3) (Fu et al., 2009), *QYr. inra-6B* in Renan (Dedryver et al., 2009) and *QYr. caas-6BS* in Pingyuan 50 (Lan et al., 2010). One of these QTL, *QYrst. wgp-6BS. 2* (Santra et al., 2008), flanked by *Xgwm132* and *Xgdm113*, was located in the same region as *QYr. caas-6BS. 3* based on the consensus map (Somers et al., 2004). Pedigree analyses showed no relationship between Bainong 64 and Stephens (http://genbank.vurv.cz/wheat/pedigree/pedigree.asp). To date, no LR APR genes were located on chromosome 6BS. In the present study, we detected two LR APR QTL on chromosome 6BS, with one being contributed by each parent. Both QTL are possibly new. *QLr. caas-6BS. 1* contributed by Bainong 64 was also associated with *QYr. caas-6BS. 3* from the same parent.

QYr. caas-7AS in Jingshuang 16 was close to centromere. Although this QTL was significant only in Chengdu 2011 and averaged MDS, the LOD curves indicate it also had effects on YR response in other environments (Fig. 1). Dedryver et al. (2009) identified an APR QTL on chromosome 7A in wheat cultivar Récital, designated *QYr. inra-7A*. This QTL was located between AFLP markers *Xbcd129b* and *Xfba127c*. Because of different kinds of flanking markers used in these studies, the relationship between *QYr. inra-7A* and *QYr. caas-7AS* can't be determined.

Previous studies indicate that slow rusting resistance is controlled by genetic factors with moderateheritability, and generally with additive gene action or interactions (Bjarko and Line, 1988; William et al., 2006). In the present study all five loci for APR to YR or LR showed additive effects (Table 2), whereas interactions among different additive QTL were not identified across environments using IciMapping V3. 1 (Li et al., 2008).

Several loci that have effects on multiple disease responses were detected in the present study. The first locus located on chromosome 1BL, at a same chromosome region to *Yr29/Lr46*, showed significant effects on response to LR and PM. The second locus located on 4DL and conferring significant effects on both PM and YR responses was possibly at the same locus as *Yr46/Lr67*. A new multiple resistance locus was detected on chromosome 6BS with significant effects on both YR and LR. These three multiple resistance QTL in Bainong 64, and their corresponding closely linked molecular markers, *Xwmc44*, *Xwmc331* and *Xwmc487*, will be useful for marker-assisted selection in breeding for resistance to YR, LR and PM. *QLr. caas-6BS. 2* is likely a new APR gene for resistance to leaf rust. This QTL and its flanking markers might serve to diversify the genetic basis of APR to LR and to accelerate the breeding process.

Acknowledgements

We are grateful to the critical review of this manuscript by Prof. R. A. McIntosh, Plant Breeding Institute, University of Sydney, Australia. This study was supported by the China Agriculture Research System (CARS-3-1-3), International Collaboration Project from the Ministry of Agriculture (2011-G3), National 863 project (2012AA101105) and National Natural Science Foundation of China (30821140351).

References

Afzal SN, Haque MI, Ahmedani MS, Bashir S, Rattu AUR (2007) Assessment of yield losses caused by *Puccinia striiformis* triggering stripe rust in the most common wheat varieties. Pak J Bot 39: 2127-2134.

Allard RW (1960) Princilpes ofPlant Breeding. John Wiley and Sons, Inc., New York, London.

Bhavani S, Singh RP, Argillier O, Huerta-Espino J, Singh S, Njau P, Brun S, Lacam S, Desmouceaux N (2011) Mapping durable adult plant stem rust resistance to the race Ug99 group in six CIMMYT wheats. http://www.globalrust.org/db/ attachments/knowledge/122/2/Bhavani-revised.pdf

Bjarko ME, Line RF (1988) Heritability and number of genes controlling leaf rust resistance on four cultivars of wheat. Phytopathology 78: 457-461.

Caldwell RM (1968) Breeding for general and/or specific plant disease resistance. *In*: Findlay KW, Shepherd KW (eds) Proc 3rd Int. Wheat Genetics Symp Australian Academy of Science, Canberra, Australia, pp 263-272.

Carter AH, Chen XM, Garland-Campbell K, Kidwell KK (2009) Identifying QTL for high-temperature adult-plant resistance to stripe rust (*Puccinia striiformis* f. sp. *tritici*) in the spring wheat (*Triticum aestivum* L.) cultivar 'Louise'. Theor Appl Genet 119: 1119-1128.

Chen XM (2005) Epidemiology and control of stripe rust [*Puccinia striiformis* f. sp. *tritici*] on wheat. Can J Plant Pathol 27: 314-337.

Chen XM, Line RF (1995a) Gene action in wheat cultivars for durable high-temperature adult-plant resistance and interactions with race-specific, seedling resistance to stripe rust caused by *Puccinia striiformis*. Phytopathology 85: 567-572.

Chen XM, Line RF (1995b) Gene number and heritability of wheat cultivars with durable, high-temperature, adult-plant resistance and race-specific resistance to *Puccinia striiformis*. Phytopathology 85: 573-578.

Dedryver F, Paillard S, Mallard S, Robert O, Trottet M, Nègre S, Verplancke G, Jahier J (2009) Characterization of genetic components involved in durable resistance to stripe rust in the bread wheat 'Renan'. Phytopathology 99: 968-973.

Fu DL, Uauy C, Distelfeld A, Blechl A, Epstein L, Chen XM, Sela H, Fahima T, Dubcovsky J (2009) A kinase-start gene confers temperature-dependent resistance to wheat stripe rust. Science 323: 1357-1360.

Gupta PK, Balyan HS, Edwards KJ, Isaac P, Korzun V, Röder M, Gautier M-F, Joudrier P, Schlatter AR, Dubcovsky J, De la Peña RC, Khairallah M, Penner G, Hayden MJ, Sharp P, Keller B, Wang RCC, Hardouin JP, Jack P, Leroy P (2002) Genetic mapping of 66 new microsatellite (SSR) loci in bread wheat. Theor Appl Genet 105: 413-422.

He ZH, Lan CX, Chen XM, Zou YC, Zhuang QS, Xia XC (2011) Progress and perspective in research of adult-plant resistance to stripe rust and powdery mildew in wheat. Sci

Agric Sin 44：2193-2215.

Herrera-Foessel SA，Lagudah ES，Huerta-Espino J，Hayden MJ，Bariana HS，Singh D，Singh RP（2011）New slow-rusting leaf rust and stripe rust resistance genes *Lr67* and *Yr46* in wheat are pleiotropic or closely linked. Theor Appl Genet 122：239-249.

Herrera-Foessel SA，Singh RP，Huerta-Espino J，Rosewarne GM，Periyannan SK，Viccars L，Calvo-Salazar V，Lan CX，Lagudah ES（2012）*Lr68*：a new gene conferring slow rusting resistance to leaf rust in wheat. Theor Appl Genet 124：1475-1486.

Hiebert CW，Thomas JB，McCallum BD，Humphreys DG，DePauw RM，Hayden MJ，Mago R，Schnippenkoetter W，Spielmeyer W（2010）An introgression on wheat chromosome 4DL in RL6077（Thatcher * 6/PI 250413）confers adult plant resistance to stripe rust and leaf rust（*Lr67*）. Theor Appl Genet 121：1083-1091.

Kilpatrick RA（1975）New wheat cultivars and longevity of rust resistance，1971-1975. US Department of Agriculture，Agricultural Research Service，Beltsville.

Knott DR（1989）The Wheat Rusts - Breeding for Resistance. *In*：Monographs on Theoretical and Applied Genetics. Vol. 12. Springer Verlag，Berlin，pp. 201.

Kosambi DD（1944）The estimation of map distance from recombination values. Annu Eugen 12：172-175.

Krattinger SG，Lagudah ES，Spielmeyer W，Singh RP，Huerta-Espino J，McFadden H，Bossolini E，Selter LL，Keller B（2009）A putative ABC transporter confers durable resistance to multiple fungal pathogens in wheat. Science 323：1360-1363.

Lagudah ES，Krattinger SG，Herrera-Foessel S，Singh RP，Huerta-Espino J，Spielmeyer W，Brown-Guedira G，Selter LL，Keller B（2009）Gene-specific markers for the wheat gene *Lr34/Yr18/Pm38* which confers resistance to multiple fungal pathogens. Theor Appl Genet 119：889-898.

Lagudah ES，McFadden H，Singh RP，Huerta-Espino J，Bariana HS，Spielmeyer W（2006）Molecular genetic characterization of the*Lr34/Yr18* slow rusting resistance gene region in wheat. Theor Appl Genet 114：21-30.

Lan CX，Liang SS，Wang ZL，Yan J，Zhang Y，Xia XC，He ZH（2009）Quantitative trait loci mapping for adult-plant resistance to powdery mildew in Chinese wheat cultivar Bainong 64. Phytopathology 99：1121-1126.

Lan CX，Liang SS，Zhou XC，Zhou G，Lu QL，Xia XC，He ZH（2010）Identification of genomic regions controlling adult-plant stripe rust resistance in Chinese landrace Pingyuan 50 through bulked segregant analysis. Phytopathology 100：313-318.

Li HH，Ribaut J-M，Li ZL，Wang JK（2008）Inclusive composite interval mapping（ICIM）for digenic epistasis of quantitative traits in biparental populations. Theor Appl Genet 116：243-260.

Li ZF，Xia XC，He ZH，Li X，Zhang LJ，Wang HY，Meng QF，Yang WX，Li GQ，Liu DQ（2010）Seedling and slow rusting resistance to leaf rust in Chinese wheat cultivars. Plant Dis 94：45-53.

Liang SS，Suenaga K，He ZH，Wang ZL，Liu HY，Wang DS，Singh RP，Sourdille P，Xia XC（2006）Quantitative trait loci mapping for adult-plant resistance to powdery mildew in bread wheat. Phytopathology 96：784-789.

Lillemo M，Asalf B，Singh RP，Huerta-Espino J，Chen XM，He ZH，Bjørnstad Å（2008）The adult plant rust resistance loci *Lr34/Yr18* and *Lr46/Yr29* are important determinants of partial resistance to powdery mildew in bread wheat line Saar. Theor Appl Genet 116：1155-1166.

Line RF（2002）Stripe rust of wheat and barley inNorth America：a retrospective historical review. Annu Rev Phytopathol 40：75-118.

Line RF，Chen XM（1995）Success in breeding for and managing durable resistance to wheat rusts. Plant Dis 79：1254-1255.

Lu YM，Lan CX，Liang SS，Zhou XC，Liu D，Zhou G，Lu QL，Jing JX，Wang MN，Xia XC，He ZH（2009）QTL mapping for adult-plant resistance to stripe rust in Italian common wheat cultivars Libellula and Strampelli. Theor Appl Genet 119：1349-1359.

Manly KF，Cudmore RH Jr，Meer JM（2001）Map Manager QTX，cross-platform software for genetic mapping. Genome 12：930-932.

McIntosh RA，Dubcovsky J，Rogers WJ，Morris C，Appels R，Xia XC（2011）Catalogue of gene symbols for wheat：2011 supplement. http：//www. shigen. nig. ac. jp/wheat/komugi/genes/macgene/supplement 2011. pdf

McIntosh RA，Dubcovsky J，Rogers WJ，Morris C，Appels R，Xia XC（2012）Catalogue of gene symbols for wheat：2012 supplement. http：//www. shigen. nig. ac. jp/wheat/komugi/genes/macgene/ supplement 2012. pdf

Peterson RF，Campbell AB，Hannah AE（1948）A diagrammatic scale for estimating rust intensity of leaves and stems of cereals. Can J Res 26：496-500.

Pestsova E，Ganal MW，Röder MS（2000）Isolation and mapping of microsatellite markers specific for the D genome of bread wheat. Genome 43：689-697.

Röder MS，Korzun V，Wendehake K，Plaschke J，Tixier

M-H, Leroy P, Ganal MW (1998) A microsatellite map of wheat. Genetics 149: 2007-2023.

Rosewarne GM, Singh RP, Huerta-Espino J, William HM, Bouchet S, Cloutier S, McFadden H, Lagudah ES (2006) Leaf tip necrosis, molecular markers and β1-proteasome subunits associated with the slow rusting resistance genes*Lr46*/*Yr29*. Theor Appl Genet 112: 500-508.

Samborski DJ (1985) Wheat Leaf Rust. *In*: Roelfs AP, Bushnell WR (eds) The cereal rusts. Vol. 2. Academic Press, Orlando, Fla., pp. 39-59.

Santra DK, Chen XM, Santra M, Campbell KG, Kidwell KK (2008) Identification and mapping QTL for high-temperature adult-plant resistance to stripe rust in winter wheat (*Triticum aestivum* L.) cultivar 'Stephens'. Theor Appl Genet 117: 793-802.

Singh RP (1992) Association between gene *Lr34* for leaf rust resistance and leaf tip necrosis in wheat. Crop Sci 32: 874-878.

Singh RP, Huerta-Espino J, Bhavani S, Herrera-Foessel SA, Singh D, Singh PK, Velu G, Mason RE, Jin Y, Njau P, Crossa J (2011) Race non-specific resistance to rust diseases in CIMMYT spring wheats. Euphytica 179: 175-186.

Singh RP, Huerta-Espino J, Rajaram S (2000) Achieving near immunity to leaf and stripe rusts in wheat by combining slow rusting resistance genes. Acta Phytopathologica et Entomologica Hungarica 35: 133-139.

Singh RP, Huerta-Espino J, William HM (2005) Genetics and breeding for durable resistance to leaf and stripe rusts in wheat. Turk J Agric For 29: 121-127.

Somers DJ, Isaac P, Edwards K (2004) A high-density microsatellite consensus map for bread wheat (*Triticum aestivum* L.). Theor Appl Genet 109: 1105-1114.

Song QJ, Fickus EW, Cregan PB (2002) Characterization of trinucleotide SSR motifs in wheat. Theor Appl Genet 104: 286-293.

Sourdille P, Singh S, Cadalen T, Brown-Guedira GL, Gay G, Qi L, Gill BS, Dufour P, Murigneux A, Bernard M (2004) Microsatellite-based deletion bin system for the establishment of genetic-physical map relationships in wheat (*Triticum aestivum* L.). Funct Integr Genomics 4: 12-25.

Suenaga K, Singh RP, Huerta-Espino J, William HM (2003) Microsatellite markers for genes*Lr34*/*Yr18* and other quantitative trait loci for leaf rust and stripe rust resistance in bread wheat. Phytopathology 93: 881-890.

Wang S, Basten CJ, Zeng ZB (2005a) Windows QTL cartographer v2.5 statistical genetics. North Carolina State University, Raleigh, NC.

Wang ZL, Li LH, He ZH, Duan XY, Zhou YL, Chen XM, Lillemo M, Singh RP, Wang H, Xia XC (2005b) Seeding and adult-plant resistance to powdery mildew in Chinese bread wheat cultivars and lines. Plant Dis 89: 457-463.

Wang ZL, Liu SD, Liu HY, He ZH, Xia XC, Chen XM (2006) Genetic linkage map in Bainong 64 × Jingshuang 16 of wheat. Acta Bot Boreal Occident Sin 26: 886-892.

William M, Singh RP, Huerta-Espino J, Ortiz Islas S, Hoisington D (2003) Molecular marker mapping of leaf rust resistance gene *Lr46* and its association with stripe rust resistance gene *Yr29* in wheat. Phytopathology 93: 153-159.

William HM, Singh RP, Huerta-Espino J, Palacios G, Suenaga K (2006) Characterization of genetic loci conferring adult plant resistance to leaf rust and stripe rust in spring wheat. Genome 49: 977-990.

Yang ZM, Xie CJ, Sun QX (2003) Situation of the sources of stripe rust resistance of wheat in the post-CYR32 era in China. Acta Agron Sin 29: 161-168.

Yuan JH, Liu TG, Chen WQ (2007) Postulation of leaf rust resistance genes in 47 new wheat cultivars at seedling stage. Sci Agric Sin 40: 1925-1935.

Zhang LJ, Li ZF, Lillemo M, Xia XC, Liu DQ, Yang WX, Luo JC, Wang HY (2009) QTL mapping for adult-plant resistance to leaf rust in CIMMYT wheat cultivar Saar. Sci Agric Sin 42: 388-397.

Zhao XL, Zheng TC, Xia XC, He ZH, Liu DQ, Yang WX, Yin GH, Li ZF (2008) Molecular mapping of leaf rust resistance gene *LrZH84* in Chinese wheat line Zhou 8425B. Theor Appl Genet 117: 1069-1075.

鲁麦21慢白粉病抗性基因数目和遗传力分析

倪小文[1]，阎俊[2]，陈新民[1]，夏先春[1]，何中虎[1]，张勇[1]，王德森[1]，Morten Lillemo[3]

[1]中国农业科学院作物科学研究所/国家小麦改良中心/国家农作物基因资源与基因改良重大科学工程，北京 100081；[2]中国农业科学院棉花研究所，安阳 455000；[3]Department of Plant and Environmental Sciences，Norwegian University of Life Sciences，5003，Ås，Norway

摘要：以多年鉴定具有慢白粉抗性的小麦品种鲁麦21和感白粉病品种京双16及其杂交组合$F_{2:3}$和$F_{2:4}$代株系200个为材料，于2005—2007年连续两个生长季，在北京和安阳两地分别进行田间病害鉴定，并采用质量性状和数量性状两种分析方法估算鲁麦21的慢病基因数目和遗传力。结果表明，在这两个群体中至少存在4对抗性基因，其广义遗传力为0.53～0.78。由于出现超亲分离，因此推测京双16可能贡献1对微效抗病基因，而鲁麦21至少含有3对慢白粉病抗性基因。

关键词：鲁麦21；白粉病慢抗性；抗病基因数目；遗传力

Heritability and Number of Genes Controlling Slow-Mildewing Resistance in Wheat Cultivar Lumai 21

Ni Xiaowen[1]，Yan Jun[2]，Chen Xinmin[1]，Xia Xianchun[1]，He Zhonghu[1]，Zhang Yong[1]，Wang Desen[1]，Morten Lillemo[3]

[1] *National Wheat Improvement Center / National Key Facility for Crop Gene Resource and Genetic Improvement*，*Institute of Crop Sciences*，*Chinese Academy of Agricultural Sciences*，*Beijing* 100081；[2] *Institute of Cotton*，*Chinese Academy of Agricultural Sciences*，*Anyang* 455000，*Henan*，*China*；[3]*Department of Plant and Environmental Sciences*，*Norwegian University of Life Sciences*，5003，*Ås*，*Norway*

Abstract：It is very important to apply slow-mildewing resistance in wheat（*Triticum aestivum* L.）breeding，because slow-mildewing resistance is more durable than hypersensitive resistance. However，little information is available about the genetics of slow-mildewing resistance in Chinese wheat cultivars. Lumai 21 is identified as a slow-mildewing resistant wheat cultivar. To estimate the number of genes and its heritability，200 lines of $F_{2:3}$ and $F_{2:4}$ populations derived from the cross between Lumai 21 and Jingshuang 16（susceptible to mildew）and their parents were planted at Beijing and Anyang，Henan for disease evaluation in 2005-2007 growing seasons. The resistance was analyzed on the bases of both quantitative and qualitative genetic models. At least 4 resistance genes were detected in the 2 populations. The broad-sense heritability of the resistance was 0.53-0.78. Transgressive segregation result indicated that

原文发表在《作物学报》，2008，34（8）：1317-1322

Jingshuang 16 might hold 1 minor gene for the resistance, and Lumai 21 involves at least 3 genes, accordingly.

Key words: Lumai 21; Adult-plant resistance to powdery mildew; Number of resistance genes; Heritability

小麦白粉病是由白粉菌（*Blumeria graminis* f. sp. *tritici*）引起的小麦主要病害之一。虽然可以通过药剂防治病害，但是培育抗病品种是最经济有效和环保的方法。自 Flor 1955 年提出基因对基因学说以来，质量抗性基因的筛选、鉴定和利用工作取得了很大进展，在小麦抗病育种中发挥重要作用。但由于病原菌小种变异快，造成质量抗病基因很快丧失，抗病品种寿命短，生产上存在极大隐患[1]。慢病性亦称成株抗性[2-4]或部分抗性[5]，为非小种专化性，其苗期感病，成株期抗病，主要是通过成株期延迟病菌的侵入和繁殖而表现出抗性[6]。慢病性相对质量抗病性更持久[7]，在小麦[1]、大麦[8]、燕麦[9]中都得到证实。因此，慢白粉抗性的利用对于培育持久抗病品种具有重要意义，而慢白粉病的遗传研究是利用慢抗性的基础。

国内对小麦慢白粉病的遗传研究很少。王竹林[10]等对百农 64 的慢白粉抗性遗传分析表明其由三对基因控制，其中两对基因的显性作用较强，另一对基因的显性作用较弱，广义遗传力为 0.66～0.69，狭义遗传力 0.17～0.43。国外多数研究表明，小麦慢白粉抗性是由 2～3 对基因控制。Das 等[11]认为小麦慢白粉病品种 Houser 和 Redcoat 是由 2～3 对基因控制，基因作用方式为部分显性和加性作用，广义遗传力为 0.57～0.94。Griffey 等[12]对 Massey 和 Knox62 的慢病性分析表明，其慢病性受 2～3 对基因控制，广义遗传力为 0.79～0.95。Lillemo 等[13]对 CIMMYT 小麦品种 Saar 的研究认为其慢病性至少由 3 对基因控制，基因为加性作用，遗传力为0.83～0.92。

鲁麦 21 是山东省烟台市农业科学研究所于 1991 年育成，具有高产、抗病和适应性广等特点，是黄淮麦区主推品种之一。从 1994 年至 2006 年累计推广6 069万亩*，1997 年最大面积 1 604 万亩，现在仍然具有较好的白粉病抗性。经过我们多年的抗病鉴定，鲁麦 21 具有慢白粉抗性[14]，且在百农 64/鲁麦 21 杂交组合的 F_4代出现了 8 个抗病性超过双亲的株系，但至今对鲁麦 21 的慢抗白粉病遗传还不清楚。本研究运用数量性状与质量性状分析方法，以确定鲁麦 21 抗性基因数目及遗传力，为抗病育种提供依据。

1 材料与方法

1.1 试验材料

2005 年秋在北京和河南安阳种植慢白粉病抗性亲本鲁麦 21、感病亲本京双 16（也作为感病对照）及其杂种后代 $F_{2:3}$群体，北京共 200 个株系，由于 $F_{2:3}$代种子数量有限，安阳试验点只种植了 195 个株系。采用完全随机区组设计，两次重复，单行区，行长 1.5m，每行 50 粒，在每个重复开始种植亲本各两行。北京试验点人工接种白粉菌 E20 菌种，由中国农业科学院植物保护研究所段霞瑜研究员提供；安阳试验点采用自然发病，每隔 10 行种一行感病对照京双 16，试验材料周围种植接种行。每个 F2：3 株系选择 30 个穗混合组成 F2：4 株系。

2006 年秋采用相同田间设计，在两个试验点种植双亲和 F2：4 株系 200 个，三次重复。

1.2 病害调查及统计方法

2006 年 5～6 月，在北京和安阳分别进行田间病害调查。调查前每行随机选取 10 株挂牌编号，小麦刚抽穗时进行第一次调查，以目测估计白粉孢子堆面积所占倒二叶总面积的百分率，每行调查挂牌的 10 株，取平均值为该株行的病害严重度，每 7d 调查一次，直到发病高峰过后，叶片变黄为止，最后一次调查的结果即为倒二叶最大病害严重度（maximum disease severity，MDS）。北京试验点共调查 3 次，安阳试验点共调查 2 次。病程曲线下面积（area under the disease progress curve，AUDPC）按如下公式[15]计算，

$$AUDPC = \sum_{i=1}^{n}(X_i + X_{i+1})(T_{i+1} - T_i)/2$$

式中，i 为调查次数，n 为总调查次数，X 为倒二叶病害严重度，T 为调查时间。

对 2006 年调查数据进行相关分析，发现 MDS·

---

* 亩为非法定计量单位。15 亩=1 公顷——编者注

与 AUDPC 之间存在极显著相关，相关系数在北京和安阳试验点分别是 0.91 和 0.93。所以 2007 年只调查了 MDS，北京试验点由于发病比 2006 年早 10 d 左右，故 MDS 调查时间为 5 月 13 日，安阳试验点为 5 月 18 日。

1.3　抗病基因数目和遗传力计算

参考前人对小麦慢白粉抗性[11,13]和慢锈抗性[16,17]的质量性状和数量性状的遗传分析报道的方法来估计鲁麦 21 的抗性基因数目。

在分析质量性状时，因后代群体中不容易区分感病类型与部分中间偏感病亲本型，所以将 $F_{2:3}$ 和 $F_{2:4}$ 的株系病害反应分为两类，即纯合慢抗病型（R）和包括感病型和中间型的其他类型（S+I）。当某一株系的平均倒二叶病害严重度和所有 10 个单株的病害严重度小于或等于抗病亲本鲁麦 21 平均值加 1 个标准差时，则为纯合慢抗病类型。R 和 S+I 型比例进行卡方检验，由于自由度为 1，用校正公式 $\chi 2=\sum[(|O-E|-0.5)^2/E]$，显著水平为 $\alpha=0.05$。

数量性状分别用 MDS 和 AUDPC 值进行分析，根据公式 $n=D^2/[8\sigma_g^2/(2-1/2^{g-2})]$[18]计算基因数目，式中 n 为基因数目，D 为双亲均值之差，σ_g^2 为遗传方差，g 为世代数。对于 $F_{2:3}$ 世代，$n=D^2/5.33\sigma_g^2$；对于 $F_{2:4}$ 世代，$n=D^2/4.57\sigma_g^2$。用 SAS 9.0 软件中 ANOVA 求 σ_g^2，其前提条件是假设没有连锁、上位性和显性作用，且各基因位点作用相等，抗性基因均来自同一个亲本，没有超亲分离。当出现上述任何假设情况时，n 会变小[19]；如果抗性基因来自两个亲本，则 D 的值是 F_g 世代中的极差[16,20]。为了排除环境影响，并给出更准确的基因数目估值，用 F_g 世代中的极差值乘以遗传力来估计 D 值[21]。

广义遗传力 $h^2=\sigma_g^2/\sigma_p^2$，式中 σ_p^2 为表型方差，σ_g^2 为遗传方差，均可用 ANOVA 方法估值，$\sigma_g^2=(\sigma_L^2-\sigma_E^2)/n$，$\sigma_p^2=\sigma_g^2+\sigma_E^2$，$\sigma_L^2$ 是第 L 世代方差，σ_E^2 是环境方差，n 是重复数。

2　结果与分析

2.1　病情分布

在两年两点的试验中，感病亲本（也是感病对照）发病良好（表 1），保证了数据的准确性。两年田间病害 MDS 和 AUDPC 分布见图 1。虽然安阳试验点是自然发病，北京点为人工接种，但两地点总的发病趋势基本一致。两个世代两点病情呈连续性、接近正态分布，属于数量性状，并且都出现超亲分离现象（表 1），即有一小部分株系病害抗性比抗病亲本鲁麦 21 还强，表明感病亲本京双 16 也为其后代提供了抗性基因。

MDS 和 AUDPC 的遗传力为 0.51～0.78（表 1），其中 AUDPC 的遗传力高于 MDS，二者均表现为安阳自然发病条件下的遗传力低于人工接种条件下的北京，其中北京 $F_{2:3}$ 代通过 AUDPC 计算得到的遗传力是 0.78，为最大值，其余在 0.5 以上，表明鲁麦 21 的慢抗病性可以较稳定遗传。

2.2　基因数目

采用数量性状方法估算结果显示，两个世代在北京和安阳两点的基因数（n 值）为 3.49～4.35，而且 MDS 和 AUDPC 两种参数估值基本一致（表 2）。表明该群体的慢白粉病抗性由 3～4 对基因控制。由于假设所有抗病基因的作用是等效的，而这种情况实际较难存在，因此该群体实际含有抗性基因的数目比估值要多。

采用质量性状方法的计算结果也显示，群体慢抗病性也由 3～4 对基因控制（表 2）。在北京试验点根据 MDS 和 AUDPC 判断的 $F_{2:3}$ 群体纯合抗病株系均是 5 个，其余 195 个株系为感病和中间型类型。自由度为 1，$\chi_{0.05}{}^2=3.84$。按照含有 3 对抗性基因的期望比率 0.053∶0.947，得到的 $\chi^2=2.59<3.84$，说明符合 3 对基因的遗传规律；当按照含有 4 对抗性基因时，期望比率为 0.020∶0.980，$\chi^2=0.06<3.84$，表明也符合 4 对基因遗传，但较 3 对基因概率更大；当 5 对基因时，期望比率为 0.007∶0.993，$\chi^2=6.95>3.84$，说明不符合 5 对基因遗传。安阳试验点 $F_{2:3}$ 群体的结果与北京点类似，但在 $F_{2:4}$ 群体，无论是北京试点还是安阳试点均不符合 3 对抗病基因的遗传，而符合 4 对或 5 对基因遗传，同时 4 对基因的概率（0.50～0.75）大于 5 对基因（0.05～0.10）。

结合质量性状和数量性状分析结果，在鲁麦 21×京双 16 组合群体中约有 4 对基因控制慢白粉病抗性。由于出现超亲遗传，感病亲本京双 16 也提供一些抗性。

表 1 亲本、$F_{2:3}$、$F_{2:4}$的发病范围、均值和广义遗传力

Table 1 Range and mean power mildew score for parents and $F_{2:3}$, $F_{2:4}$ lines, and broad-sense heritability

世代 Generation	地点 Location	亲本 Parents 母本 P_1	父本 P_2	均值 Mean	后代 Offspring 品系数 No. of lines	范围 Range	均值 Mean	广义遗传力 Broad-sense heritability
$F_{2:3}$	北京 MDS	6.0	85.0	45.5	200	1.7～90.0	53.6	0.68
$F_{2:3}$	北京 AUDPC	42.5	950.5	496.3	200	31.1～1115.5	625.9	0.78
$F_{2:3}$	安阳 MDS	3.0	78.0	40.5	195	1.0～85.0	48.2	0.53
$F_{2:3}$	安阳 AUDPC	10.5	385.0	194.3	195	8.5～455.0	261.3	0.61
$F_{2:4}$	北京 MDS	1.5	80.0	40.8	200	0.4～86.0	42.0	0.72
$F_{2:4}$	安阳 MDS	1.0	71.0	37.5	200	0.3～76.3	39.0	0.51

P_1为鲁麦 21，P_2为京双 16。

P_1 is Lumai21, P_2 is Jingshuang16.

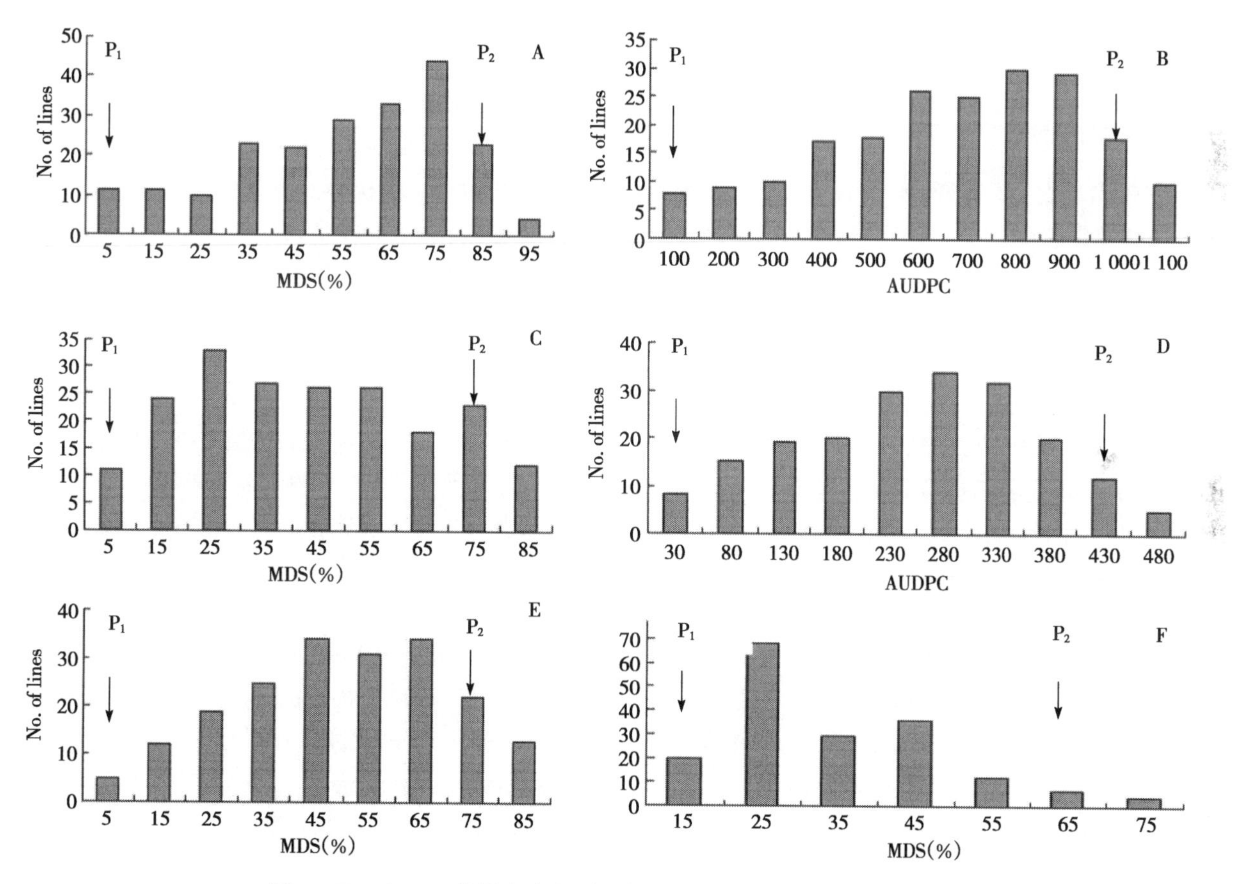

图 1 $F_{2:3}$和$F_{2:4}$代最大病害严重度及病情曲线下面积的频率分布

Fig. 1 Distribution of frequencies for MDS and AUDPC of $F_{2:3}$ and $F_{2:4}$ lines

A and B: $F_{2:3}$ in Beijing site; C and D: $F_{2:3}$ in Anyang site, Henan; E: $F_{2:4}$in Beijing site; F: $F_{2:4}$ in Anyang site, Henan.
P_1: Lumai 21; P_2: Jingshuang16

表 2 根据 Wright 公式和卡平方分析估算的抗病基因数目

Table 2 Gene number estimates for powdery mildew resistance based on Wright's formula and $\chi2$ analysis

世代 Generation	地点 Location	基因数[a] No. of genes[a]	品系数 No. of lines R	S+I	卡平方和 P 值[b] $\chi2$ and P values [b] 3 对基因	3genes	4 对基因	4genes	5 对基因	5genes
$F_{2:3}$	北京 MDS	4.21	5	195	2.59	0.05～0.1	0.06	0.75～0.9	6.91	0.005～0.01

（续）

世代 Generation	地点 Location	基因数[a] No. of genes[a]	品系数 No. of lines		卡平方和P值[b] χ2 and P values [b]					
			R	S+I	3对基因	3genes	4对基因	4genes	5对基因	5genes
$F_{2:3}$	北京 AUDPC	4.35	5	195	2.59	0.05～0.1	0.06	0.75～0.9	6.91	0.005～0.01
$F_{2:3}$	安阳 MDS	3.78	5	190	2.88	0.05～0.1	0.32	0.25～0.5	15.20	<0.005
$F_{2:3}$	安阳 AUDPC	4.03	6	189	1.48	0.1～0.25	0.67	0.25～0.5	12.61	<0.005
$F_{2:4}$	北京 MDS	4.13	6	194	6.89	0.005～0.01	0.11	0.5～0.75	1.68	0.1～0.25
$F_{2:4}$	安阳 MDS	3.49	6	194	6.89	0.005～0.01	0.11	0.5～0.75	1.68	0.1～0.25

a. 用 Wright 公式估算的基因数目。a. Number of genes estimated based on Wright' s formula.

b. $F_{2:3}$代对于 3、4、5 对基因的期望分离比率分别为 0.053∶0.947、0.020∶0.980 、0.007∶0.993，$F_{2:4}$代对于 3、4、5 对基因的期望分离比率分别为 0.084∶0.916、0.037∶0.963、0.016∶0.984。b. Expected segregation values for chi-square analysis were 0.053∶0.947，0.020∶0.980，0.007∶0.993，respectively for 3、4、5 independent genes in $F_{2:3}$ Expected segregation values for chi-square analysis were 0.084∶0.916，0.037∶0.963，0.016∶0.984，respectively for 3、4、5 independent genes in $F_{2:4}$.

3 讨论

3.1 慢白粉病的鉴定方法

由于慢病性表现为数量性状特征，数值呈连续性分布，病害鉴定较质量抗性困难。白粉病的田间慢病性鉴定一般采用基于倒二叶病害严重度的 AUDPC、MDS 以及病情指数法和 0～9 级法[22]。国外大多采用 AUDPC，需要调查多次，费工费时。Wang 等[23]和 Yu 等[24]在慢病性的研究中发现 MDS 和 AUDPC 呈极显著相关，与本文研究结果相同，且 AUDPC 和 MDS 两种方法不论是病情分布图还是计算基因的数目十分接近。因此，可以用 MDS 代替 AUDPC。同时，MDS 指标仅需要田间调查一次，减少了工作量，特别适合于对大量品种或群体的慢白粉抗性鉴定。但由于 MDS 必须在发病最严重时调查，需要调查者有足够的经验，否则容易延迟调查，常常出现叶片干枯，影响数据准确性。根据我们的经验，在感病对照发病达到 80%～90%时，进行调查即可。

3.2 抗和感基因型的区分

Singh[25]在用质量性状方法分析小麦 F_3代群体成株抗锈性时，将田间抗病性分为纯合抗病型、纯合感病型和抗性分离型 3 种，以及抗病亲本纯合型、感病亲本纯合型、偏抗性亲本分离型和偏感病亲本分离型 4 种[17]。Das 和 Griffey[11]在进行小麦成株抗白粉病遗传时，将 F_3代群体病害分为抗病、感病和中间型 3 种。当某个品系的病害严重度小于或等于抗病亲本严重度平均值加 1 个标准差时，为抗病型；而大于或等于感病亲本严重度平均值减 1 个标准差时，为感病类型；居两者之间的则是中间类型。Lillemo 和 Skinnes[13]将 F_5代群体分为抗病型和其他类型共两类，当某个品系的病害严重度等于抗病亲本严重度平均值±1 个标准差时，为抗病型。由于大田试验环境条件（如边行、倒伏等）对植株发病程度有一定影响，以及实际调查时较难区分感病类型与部分偏感病亲本类型，而纯合抗病品系很容易观察，因此本试验采用与 Lillemo 和 Skinnes 相似的方法，将后代群体抗病性分为纯合抗病型和其他两种类型。纯合抗病型的判断标准是不仅当某个株系的病害严重度平均值小于或等于抗病亲本严重度平均值加 1 个标准差，而且该品系调查的所有 10 个单株均达此标准，才定为纯合抗病类型。实际上对纯合抗病类型所有单株都进行了观察，这样使质量性状分析结果更可靠，并且 $F_{2:3}$代的 5 个抗病株系在 $F_{2:4}$代仍表现抗病，由此可以确定 $F_{2:3}$代纯合抗病株系的准确性。但 $F_{2:4}$ 群体中比 $F_{2:3}$代多出 1 个纯合抗病株系，该株系在 $F_{2:3}$代的 10 株中除了 1 株发病较高（10%），不符合纯合抗病类型的标准外，其余都符合。表明该株系绝大部分抗病基因已纯合。经过一代自交，使纯合基因频率增加，另外 $F_{2:4}$代群体还不足够大，所以全部表现为纯合抗病类型。

3.3 控制慢白粉病的基因数目

本研究在鲁麦 21×京双 16 组合群体中约有 4 对基因控制慢白粉抗性。由于出现超亲遗传，感病亲本京双 16 也提供一些抗性。假定京双 16 提供 1 对抗病基因，那么鲁麦 21 至少提供 3 对抗病基因，这与以前研究者认为慢白粉病是由 2～3 对基因控制的结果一致。小麦慢白粉病品种 Houser、Redcoat 和 Knox62

都由2～3对基因控制[11,12]，Massey由3对基因控制[12]，百农64的慢白粉抗性由3对基因控制[10]，Saar的慢病性至少由3对基因控制[13]。鲁麦21的慢抗病性广义遗传力为0.51～0.78，与其他的研究广义遗传力范围0.57～0.92[10-13]接近。虽然遗传力较高，但鲁麦21至少含有3对抗病基因，只有随着世代的进展，稳定抗病植株（纯合抗病基因型）频率才能增加。因此，在育种实践中利用慢白粉病抗性品种（如鲁麦21、百农64等）作为抗病亲本杂交育种或进行慢抗基因聚合时，要与质量性状有所区别，后代群体要大，早代（F_2或F_3）病害选择标准应放宽，而在较高世代（F_4或F_5），待基因型相对纯合后，病害选择标准要严。

4 结论

鲁麦21的慢白粉病抗性至少由3对基因控制；由于出现超亲分离，推测感病亲本京双16可能贡献1对微效抗病基因。其杂交后代$F_{2:3}$和$F_{2:4}$群体的抗病基因广义遗传力为0.53～0.78。

参考文献

[1] McDonald B A, Linde C. The population genetics of plant pathogens and breeding strategies for durable resistance. *Euphytica*, 2002, 124: 163-180.

[2] Bennett F G A. Resistance to powdery mildew in wheat: A review of its use in agriculture and breeding programmes. *Plant Pathol*, 1984, 33: 279-300.

[3] Griffey C A, Das M K, Stromberg E L. Effectiveness of adult-plant resistance in reducing grain yield loss to powdery mildew in winter wheat. *Plant Dis*, 1993, 77: 619-622.

[4] Griffey C A, Das M K. Inheritance of adult-plant resistance to powdery mildew in Knox62 and Massey winter wheats. *Crop Sci*, 1994, 34: 641-646.

[5] Hautea R A, Corrman W R, Sorrells M E, Bergstrom G C. Inheritance of partial resistance to powdery mildew in spring wheat. *Theor Appl Genet*, 1987, 73: 609-615.

[6] Das M K, Griffey C A. Gene action for adult-plant resistance to powdery mildew in wheat. *Genome*, 1995, 38: 277-282.

[7] Shaner G. Evaluation of slow-mildewing resistance of Knox wheat in the filed. *Phytopathology*, 1973, 63: 867-872.

[8] Jones I T, Davies I J E R. Partial resistance to *Erysiphe graminis hordei* in old European barley varieties. *Euphytica*, 1985, 34: 499-507.

[9] Jones I T, Hayes J D. The effect of sowing date on adult plant resistance to *Erysiphe graminis* sp. *avenae* in oats. *Ann Appl Biol*, 1971, 68: 31-39.

[10] Wang Z-L（王竹林），Liu-S-D（刘曙东），Wang H（王辉），He Z-H（何中虎）. Genetic analysis of slow resistance of the wheat variety Bainong 64 to slow powdery mildew. *Acta Bot Boreal-Occident Sin*（西北植物学报），2006，26（2）：332-336.（in Chinese with English abstract）

[11] Das M K, Griffey C A. Heritability and number of genes governing adult-plant resistance to powdery mildew in Houser and Redcoat winter wheats. *Phytopathology*, 1994, 84: 406-409.

[12] Griffey C A, Das M K. Inheritance of adult plant resistant to powdery mildew in Knox 62 and Massey winter wheats. *Plant Dis*, 1996, 83: 424-428.

[13] Lillemo M, Skinnes H. Genetic Analysis of a partial resistance to powdery mildew in bread wheat line Saar. *Plant Dis*, 2005, 90: 225-228.

[14] Wang Z-L（王竹林），Liu-S-D（刘曙东），Wang H（王辉），He Z-H（何中虎）. Advances of study on adult-plant resistance in bread wheat. *J Triticeae Crops*（麦类作物学报），2006，26（1）：129-134.（in Chinese with English abstract）

[15] Jeger M J, Viljanen-Rollinson S L H. The use of the area under the disease-progress curve (AUDPC) to assess quantitative disease resistance in crop cultivars. *Theor Appl Genet*, 2001, 102: 32-40.

[16] Bjarko M E, Line R F. Heritability and number of genes controlling leaf rust resistance in for cultivars of wheat. *Phytopathology*, 1988, 78: 457-461.

[17] Singh R P, Rajaram S. Genetics to adult-plant resistance to leaf rust in Frontana and three CIMMYT wheat. *Genome*, 1992, 35: 24-31.

[18] Wright S. Evolution and the genetics of populations. In: Genetic and Biometric Foundations, Vol. 1. Chicago, University of Chicago Press, 1968. p. 469.

[19] Burton G W. Quantitative inheritance in pearl millet (*Pennisetum glaucum*). *Agron J*, 1951, 43: 409-417.

[20] Mylutze D K, Baker R J. Genotype assay and method of moments analyses of qualititative traits in a spring wheat cross. Crop Sci. 1985, 25: 162-167.

[21] Singh R P, Ma H, Rajaram S. Genetic analysis of re-

sistance to scab in spring wheat cultivar Frontana. *Plant Dis*, 1995, 79: 238-240.

[22] Saarie E, Prescott J M. A scale for appraising the foliar intensity of wheat diseases. *Plant Dis*, 1975, 595: 337-380.

[23] Wang Z L, Li L H, He Z H, Duan X Y, Zhou Y L, Chen X M, Lillemo M, Singh R P, Wang H, Xia X C. Seedling and adult plant resistance to powder mildew in Chinese bread wheat cultivars and lines. *Plant Dis*, 2005, 89: 457-463.

[24] Yu D Z, Yang X J, Yang L J, Jeger M J. Assessment of partial resistance to powdery mildew in Chinese wheat varieties. *Plant Breed*, 2001, 120: 279-284.

[25] Singh R P. Genetic association of leaf rust resistance gene*Lr34* with adult plant resistance to stripe rust in bread wheat. *Phytopathology*, 1992, 82: 835-838.

Quantitative trait loci mapping of adult-plant resistance to powdery mildewing in Chinese wheat cultivar Lumai 21

C. X. Lan[1], X. W. Ni[1], J. Yan[2], Y. Zhang[1], X. C. Xia[1], X. M. Chen[1], and Z. H. He[1,3]

[1] *Institute of Crop Science, National Wheat Improvement Center/The National Key Facility for Crop Gene Resources and Genetic Improvement, Chinese Academy of Agricultural Sciences* (*CAAS*), 12 *Zhongguancun South Street, Beijing* 100081, *China*; [2] *Cotton Research Institute, Chinese Academy of Agricultural Sciences* (*CAAS*), *Huanghedadao, Anyang, Henan* 455000, *China*; [3] *International Maize and Wheat Improvement Center* (*CIMMYT*) *China Office, c/o CAAS*, 12 *Zhongguancun South Street, Beijing* 100081, *China*

Abstract: Powdery mildew, caused by *Blumeria graminis* f. sp. *tritici*, is a major fungal disease in common wheat (*Triticum aestivum* L.) worldwide. The Chinese winter wheat cultivar Lumai 21 has shown good and stable slow mildewing resistance for 19 years. The aim of this study was to map quantitative trait loci (QTLs) for resistance to powdery mildew in a population of 200 F_3 lines from the cross Lumai 21/Jingshuang 16. The population was tested for powdery mildew reaction in Beijing and Anyang in the 2005-2006 and 2006-2007 cropping seasons, providing data for 4 environments. A total of 1, 375 simple sequence repeat (SSR) markers were screened for associations with powdery mildew reactions, initially in bulked segregant analysis (BSA). Based on the mean disease values averaged across environments, broad-sense heritabilities of maximum disease severity (MDS) and area under the disease progress curve (AUDPC) were 0.96 and 0.77, respectively. Three QTLs for slow mildewing resistance were detected by inclusive composite interval mapping (ICIM). These were designated *QPm caas-2BS*, *QPm caas-2BL* and *QPm caas-2DL*, respectively, and explained from 5.4 to 20.6% of the phenotypic variance across environments. *QPm caas-2BS* and *QPm caas-2DL* were likely new slow mildewing resistance QTLs flanked by SSR markers *Xbarc98* - *Xbarc1147* and *Xwmc18* - *Xcfd233*, respectively. These markers could be useful for improving wheat powdery mildew resistance in breeding programs.

Key words: Common wheat, Durable resistance, Microsatellites, Powdery mildew, QTL

Introduction

Powdery mildew, caused by *Blumeria graminis* f. sp. *tritici*, is an important wheat (*Triticum aestivum* L.) disease in the regions with temperate and maritime climates, such as Europe, North and South America, Africa and parts of China (Bennett, 1984; Hsam and Zeller, 2002). Deployment of resistant cultivars is the most environmentally friendly and economical way of controlling the disease (Bennett, 1984). To date, about 60 genes for powdery mildew resistance have been located at 43 loci (*Pm1-Pm43*). Most of them are major resistance genes derived from common wheat and its relatives (McIntosh et al., 2009; Hua et al., 2009). Major race-specific resistance genes usually do not pro-

Published in Molecular Breeding, 2010, 25 (40): 615-622

vide long-term protection from the disease, because they are usually overcome by simple genetic changes in the pathogen population (McDonald and Linde, 2002; Skinnes, 2002). In contrast, slow mildewing resistance that is usually race non-specific, delays the infection, growth and reproduction of the pathogen at post-seedling stages, and confers a durable type of resistance to the disease. For example, the North American winter wheat cultivar Massey (Liu et al., 2001), French winter wheat lines RE9001 (Bougot et al., 2006) and RE714 (Mingeot et al., 2002), and Chinese winter wheat cultivar Bainong 64 (Wang 2005; Lan et al., 2009) with slow mildewing characteristics have provided effective powdery mildew resistance in their respective countries for extended time periods over 20 years.

Quantitative traitloci (QTLs) for slow mildewing resistance were reported in the Swiss winter wheat cultivar Forno (Keller et al., 1999), the French winter lines RE714, Festin, Courtot and RE9001 (Chantret et al., 2001; Mingeot et al., 2002; Bougot et al., 2006), the North American winter wheats Massey and USG3209 (Liu et al., 2001; Tucker et al., 2007), the CIMMYT lines Opata 85, W7984 and Saar (Börner et al., 2002; Lillemo et al., 2008), the Japanese cultivar Fukuho-Komugi (Liang et al., 2006), the Israeli cultivar Oligoculm (Liang et al., 2006), the Australian cultivar Avocet (Lillemo et al., 2008) and the Chinese cultivar Bainong 64 (Lan et al., 2009). However, many of these cultivars or lines do not have good agronomic traits in China and the number of QTLs for slow mildewing resistance is still limited compared with race-specific resistance genes. Therefore, it would be advantageous for breeding to identify more QTLs for slow mildewing in locally adapted commercial wheat cultivars with good agronomic traits than to depend on introduced materials.

Lumai 21 was a leading cultivar with an average annual area of 300,000ha in the Yellow and Huai valleys winter wheat region of China from 1994 to 2007, largely due to its high yield potential, outstanding performance under reduced irrigation and good resistance to powdery mildew. It was highly susceptible to all prevalent races of *B. graminis.* f. sp. *tritici* at the seedling stage, but was resistant at the adult stage (Wang, 2005). It has been widely used in wheat breeding programs. The objective of this study was to identify QTLs for slow mildewing resistance in Lumai 21 and their closely linked molecular markers in an F_2-derived bulk populations from a cross between the winter wheat cultivars Lumai 21 and Jingshuang 16.

Materials and methods

Plant materials

Two hundred F_2-derived bulk lines (called F_3 lines below) were produced from Lumai 21/Jingshuang 16. Lumai 21 was susceptible to 20 isolates of *B. graminis* f. sp. *tritici* at the seedling stage, but was highly resistant at the adult plant stage, whereas Jingshuang 16 was highly susceptible at both the seedling and adult stages (Wang, 2005). These isolates, collected in Beijing, Henan, Guizhou, Yunnan and Sichuan provinces from 1990 through 1998, are maintained on wheat seedlings in growth chambers at 4℃ to 8℃ and 24h of light, at the Institute of Plant Protection, Chinese Academy of Agricultural Sciences (CAAS). The F_3 lines generated from individual F_2 plants were planted and harvested as bulks of over 50 plants to produce F_3 bulk populations maintained as bulks in each generation.

Field trials

The F_3 bulk populations were evaluated for disease severity in Beijing and Anyang in the 2005-2006 and 2006-2007 cropping seasons. Field trials were conducted in randomized complete blocks with 2 replicates. Plots consisted of single rows with 1.5m length and 30 cm between rows. Approximately 80 seeds were sown in each row. The susceptible parent Jingshuang 16 was planted after every 10 rows as a susceptible check, and planted around the tested lines to ensure ample inoculum in spring.

Inoculations were carried out in Beijing using the highly

virulent isolate E20 prior to stem elongation. The percentages of leaf area covered by powdery mildew on penultimate leaves (F-1) of 10 randomly selected plants from per plot were recorded for the first time 5 weeks after inoculation, and then at weekly intervals on two further occasions until disease severities reached maximum levels around May 20, 2006. Powdery mildew severities were evaluated only once when disease severities on Jingshuang 16 had reached maximum levels around May 20 in Beijing 2007. In Anyang 2006, powdery mildew severities were rated for the first time around May 10, and for a second time around May 18 when severities on Jingshuang 16 had reached a maximum level. In Anyang 2007, severities were evaluated once when disease levels on Jingshuang 16 had reached maximum levels around May 18. The severity scores from 10 plants were averaged to obtain the mean severity for per plot in each site-year test.

Statistical analysis

The area under the disease progress curve (AUDPC) were calculated according to Bjarko and Line (1988). The MDS and AUDPC were used for the subsequent analysis of variance (ANOVA) and QTL analysis. ANOVA and estimates of correlation coefficients were conducted by PROC GLM in the Statistical Analysis System (SAS) software. Broad-sense heritabilities (h^2) for powdery mildew resistance were calculated by the formula: $h^2=\sigma_g^2/(\sigma_g^2+\sigma_{ge}^2/e+\sigma_\varepsilon^2/re)$, where σ_g^2, σ_{ge}^2 *and* σ_ε^2 are estimates of genotypic, genotype × environment interaction and error variances, respectively, and e and r are the numbers of environments and replications per environment, respectively.

SSR analysis

A total of 1, 375 SSRs were used to screen the parents and bulks and to construct the framework map for QTL analysis. These SSRs were from the series of Beltsville Agriculture Research Center (BARC) (Song et al., 2000), Wheat Microsatellite Consortium (WMC) (Gupta et al., 2003), Gatersleben Wheat Microsatellite (GWM) (Röder et al., 1998), Gatersleben D-genome Microsatellite (GDM) (Pestsova et al., 2000), Clermont Ferrand D-genome (CFD) (Guyomarc'h et al., 2002) and Clermont Ferrand A-genome (CFA) (Sourdille et al., 2004).

Bulked segregant analysis

Based on disease severities data from the four environments, two DNA bulks were constructed by mixing equal amounts of DNA from the six most resistant and six most susceptible lines, respectively. SSRs showing the same patterns of polymorphism between the parents and between the bulks were used to genotype all 200 F_3 lines. Additional SSRs on the linkage maps (Somers et al., 2004; http://wheat.pw.usda.gov/GG2/index.shtml, http://www.shigen.nig.ac.jp/wheat/komugi) around the loci associated with slow mildewing resistance were chosen to genotype the population for linkage and QTL analysis as described by Michelmore et al. (1991).

Map construction and QTL detection

Linkage groups were established with the software Mapmaker 3.0b (Lincoln et al., 1992). The map distances between markers were calculated using the Kosambi (1944) mapping function. QTL analysis was performed with the software IciMapping 2.2 by inclusive composite interval mapping (ICIM) method (Li et al., 2007). A logarithm of odds (LOD) of 2.5 was set to declare significance of QTLs. The percentages of phenotypic variance explained (PVE) by individual QTL and additive effects at LOD peaks were obtained through stepwise regression (Li et al., 2007). Each QTL was represented by a 20-centimorgan (cM) interval with the local LOD maximum as central point. Adjacent QTLs on the same chromosome were considered different when the distances between the curve peaks were greater than 20cM, or when the support intervals were non-overlapping.

Results

Phenotypic evaluations

The MDS scores were significantly correlated ($P <$

0. 000 1) across all four environments, with correlation coefficients ranging from 0. 53 to 0. 73. The frequency distributions of powdery mildew severity parameters (MDS and AUDPC) showed continuous distribution across the F_3 populations over different environments, indicating typical quantitative inheritance (Fig. 1). The mean MDS of Lumai 21 and Jingshuang 16 were 3. 0 and 78. 0%, 1. 0 and 65. 0% in Anyang 2006 and 2007, respectively, whereas the means were 6. 0 and 85. 0%, 1. 5 and 80. 0% in Beijing 2006 and 2007, respectively. The average of MDS of the F_3 lines across four environments was 39. 5%, ranging from 0. 0 to 86. 0%. In 2006, the average of AUDPC of the F_3 lines over two locations was 421. 0, ranging from 21. 7 to 737. 1.

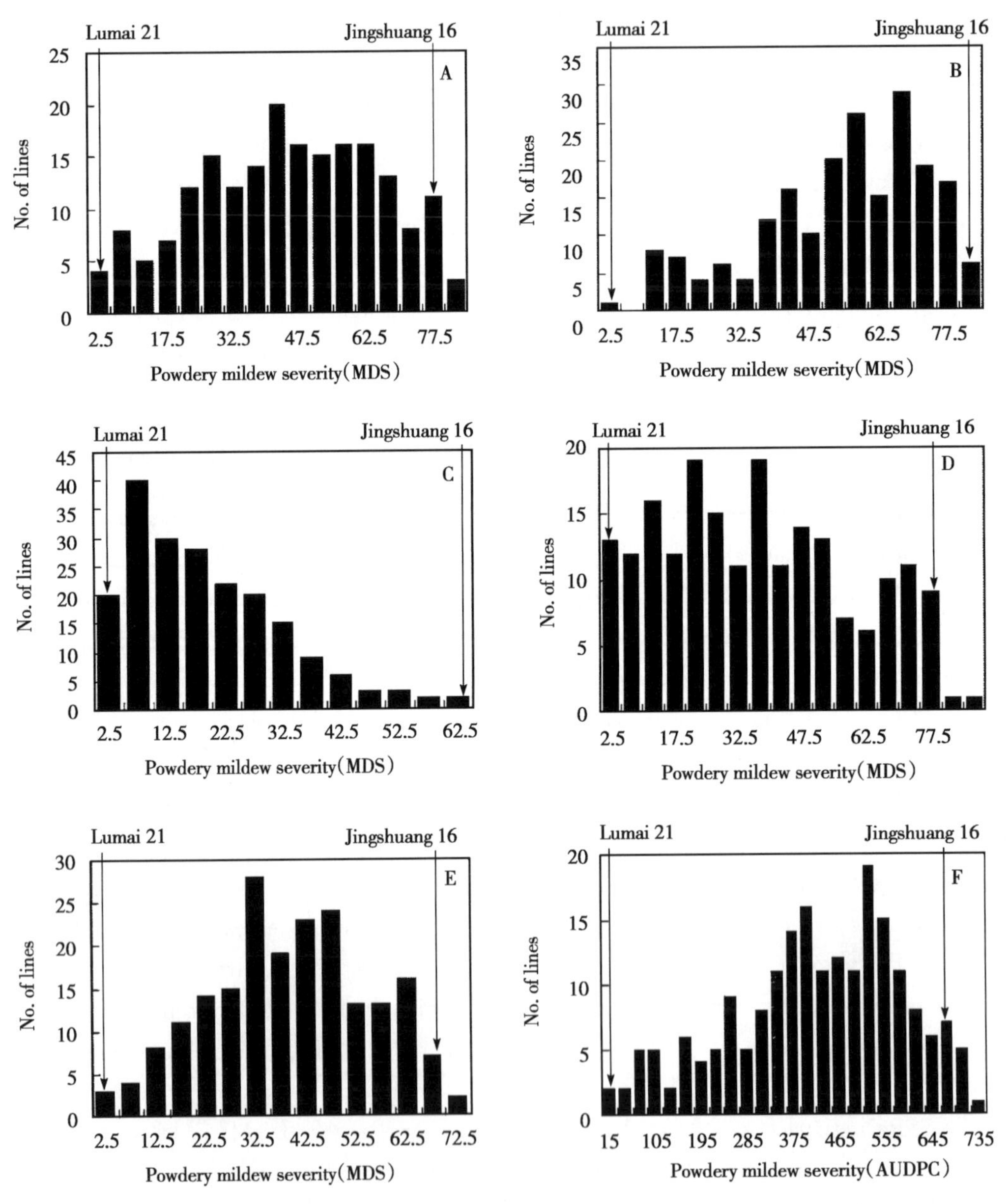

Fig. 1 A-F, Frequency distributions of powdery mildew maximum disease severities (MDS) and area under the disease progress curve (AUDPC) values in the F_3 bulk population derived from the cross Lumai 21/Jingshuang 16. A, MDS, Anyang 2006; B, MDS, Beijing 2006; C, MDS, Anyang 2007; D, MDS, Beijing 2007; E, average MDS in four environments; and F, average AUDPC, 2006 over two locations. Mean values for the parents, Lumai 21 and Jingshuang 16, are indicated by arrows.

Table 1 Analysis of variance of maximum disease severities (MDS) on penultimate leaves and area under the disease progress curve (AUDPC) values for powdery mildew responses on F_3 lines derived from the cross Lumai 21/Jingshuang 16

Parameter	Source of variation	df	Sum of squares	Mean square	*F* value
MDS	Replicates	1	2.85	2.85	0.02
	Environments	3	244 010.23	81 336.74	450.20**
	Lines	199	407 596.29	2 048.22	11.34**
	Lines×environments	597	172 244.96	290.95	1.61**
	Error	762	138 570.88	180.67	
AUDPC	Replicates	1	217 483.81	217 483.81	13.51
	Environments	1	44 933 873.77	44 933 873.77	2 792.24**
	Lines	199	18 544 783.22	93 189.87	5.79**
	Lines×environments	199	9 175 813.11	47 298.01	2.94**
	Error	361	5 889 819.74	16 092.40	

** $P < 0.0001$.

The MDS and AUDPC were significantly correlated for the test in 2006 over two locations ($r = 0.92$, $P < 0.0001$). The broad-sense heritabilities of MDS and AUDPC across four environments were 0.96 and 0.77, respectively. The ANOVA confirmed significant variation among F_3 populations (Table1).

QTL for slow mildewing resistance to powdery mildew

Based on the MDSand AUDPC data, three QTLs for slow mildewing resistance were detected by ICIM across four environments (Table 2; Fig. 2). They were designated *QPm caas-2BS*, *QPm caas-2BL* and *QPm caas-2DL*, and all came from the resistant parent Lumai 21.

TheQTL *QPm caas-2BS*, detected on chromosome 2BS in the SSR interval *Xbarc98* - *Xbarc1147*, and explained 10.9, 20.6, 10.6, 14.1 and 18.9% of the phenotypic variance in Anyang 2006, Beijing 2006 and 2007, and the averaged MDS and AUDPC, respectively (Table 2). The additive effects of this QTL across different environments were −9.01, −2.40, −10.28, −8.26 and −136.75 and dominance effects were −4.64, −17.08, −4.38, −3.93 and 12.13, respectively. Another QTL on the long arm of this chromosome, *QPm caas-2BL*, was mapped between markers *Xbarc1139* and *Xgwm47*, and identified in Beijing 2006 and 2007, and Anyang 2007, as well as the averaged MDS and AUDPC, accounting for 5.4, 10.1, 6.7, 5.7 and 5.2% of the phenotypic variance, respectively, with the additive effects ranging from −72.77 to −1.87, and dominance effects from −9.37 to 22.35. The third QTL, *QPm caas-2DL*, on the long arm of chromosome 2D, was flanked by *Xwmc18* and *Xcfd233*, and explained from 5.7 to 11.6% of the phenotypic variance in three environments and the averaged MDS and AUDPC. The additive effects of this QTL were −3.12, −1.54, −5.95, −2.33 and −46.76 and dominance effects were −11.31, −5.49, −8.21, −8.20 and −65.59, respectively.

Table 2 QTLs for slow powdery mildewing resistance in F_3 lines derived from Lumai 21/Jingshuang 16

Parameter	Location and year	QTL[b]	Marker interval	Position[c]	AE[d]	DE[e]	LOD[f]	PVE (%)[g]	Total PVE (%)
MDSa	Anyang 2006	*QPm caas-2BS*	*Xbarc98-Xbarc1147*	11.1	−9.01	−4.64	3.9	10.9	10.9
	Beijing 2006	*QPm caas-2BS*	*Xbarc98-Xbarc1147*	10.1	−2.40	−17.08	6.2	20.6	37.6
		QPm caas-2BL	*Xbarc1139-Xgwm47*	14.2	−8.53	2.84	2.8	5.4	
		QPm caas-2DL	*Xwmc18-Xcfd233*	11.0	−3.12	−11.31	5.7	11.6	
	Anyang 2007	*QPm caas-2BL*	*Xbarc1139-Xgwm47*	13.2	−3.1	−5.64	5.2	10.1	15.8
		QPm caas-2DL	*Xwmc18-Xcfd233*	11.0	−1.54	−5.49	2.8	5.7	

（续）

Parameter	Location and year	QTL[b]	Marker interval	Position[c]	AE[d]	DE[e]	LOD[f]	PVE (%)[g]	Total PVE (%)
	Beijing 2007	*QPm. caas-2BS*	*Xbarc98-Xbarc1147*	10. 1	−10. 28	−4. 38	4. 7	10. 6	26. 0
		QPm. caas-2BL	*Xbarc1139-Xgwm47*	9. 2	−3. 22	−9. 37	3. 9	6. 7	
		QPm. caas-2DL	*Xwmc18-Xcfd233*	9. 0	−5. 95	−8. 21	4. 5	8. 7	
	Average MDS in 4 environments	*QPm. caas-2BS*	*Xbarc98-Xbarc1147*	10. 1	−8. 26	−3. 93	5. 3	14. 1	28. 4
		QPm. caas-2BL	*Xbarc1139-Xgwm47*	12. 2	−1. 87	−6. 31	3. 6	5. 7	
		QPm. caas-2DL	*Xwmc18-Xcfd233*	11. 0	−2. 33	−8. 20	4. 3	8. 6	
AUDPCh	Average AUDPC in 2006	*QPm. caas-2BS*	*Xbarc98-Xbarc1147*	8. 1	-136. 75	12. 13	6. 0	18. 9	33. 8
		QPm. caas-2BL	*Xbarc1139-Xgwm47*	15. 2	−72. 77	22. 35	2. 7	5. 2	
		QPm. caas-2DL	*Xwmc18-Xcfd233*	8. 0	−46. 76	−65. 59	4. 9	9. 7	

a. Maximum disease severity,

b. QTL that extend across single one-log support confidence intervals were assigned the same symbol,

c. Peak position in centimorgans from the first interval marker,

d. Additive effect of resistance allele,

e. Dominance effect of resistance allele,

f. Logarithm of odds (LOD) score,

g. PVE is the percentages of phenotypic variance explained by individual QTL,

h. Area under the disease progress curve.

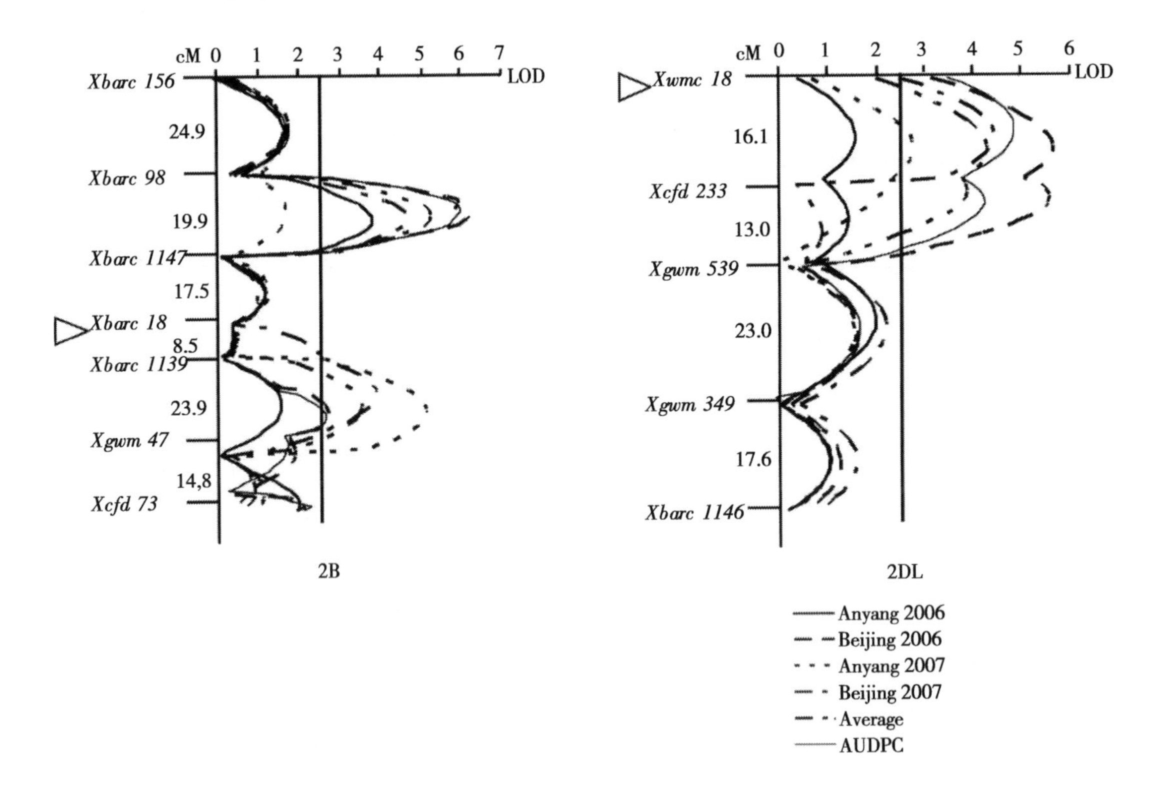

Fig. 2 Logarithm of odds (LOD) contors obtained by inclusive composite interval mapping (ICIM) for quantitative traits loci (QTLs) on slow mildewing resistance in the F_3 bulk population of Lumai 21/Jingshuang 16. **Anyang 2006** and **Anyang 2007**, maximum disease severities (MDS) in Anyang, 2006 and 2007, respectively; **Beijing 2006** and **Beijing 2007**, MDS in Beijing, 2006 and 2007, respectively. **Average**, average MDS across four environments; **AUDPC**, average of area under the disease progress curve in 2006 over two locations. Logarithm of odds (LOD) thresholds, 2. 5. Short arms are toward the top and the open arrow indicates the centromere.

Discussion

Previously, we reported that at least three genes for slow mildewing resistance were present in the population of Lumai 21/Jingshuang 16 based on the analysis of quantitative and qualitative genetic models (Ni et al., 2008). This was in agreement with the present results from QTL analysis, with a greater power in resolving individual gene location and effects.

Mingeot et al. (2002) detected a QTL for adult plant resistance to powdery mildew on the short arm of chromosome 2B, in the marker interval *Xgwml48-XgbxG553* from the parent Festin, with a map distance about 21.9cM from *QPm. caas-2BS* based on the consensus map (Somers et al., 2004; http://wheat.pw.usda.gov/GG2/index.shtml). This QTL might be different from *QPm. caas-2BS* because they were mapped to different locations. In addition, two genes for effective all-stage resistance on chromosome 2BS were derived from wild emmer (*T. turgidum* var. *dicoccoides*), viz. *Pm26* co-segregating with the RFLP marker *Xwg516* (Rong et al., 2000), and *Pm42* co-segregating with the RFLP-derived STS marker *BF146221* (Hua et al., 2009), with map distances of about 49cM and 29cM, respectively, from *QPm. caas-2BS* based on the Somers et al. (2004) map. Thus *QPm. caas-2BS* is unlikely to be a residual effect of one of these genes based on both their origins and locations. We conclude that *QPm. caas-2BS* is likely be a new gene conferring slow mildewing resistance to powdery mildew. It showed a large genetic effect on powdery mildew resistance, and SSR markers *Xbarc98* and *Xbarc1147* flanked to this QTL could be useful for improving wheat powdery mildew resistance in molecular breeding programs.

QPm. caas-2BL was located in a similar position to a QTL found by Liang et al. (2006), flanked by microsatellites *Xgwm887.1* and *Xwmc435.1* in the Fukuho-Komugi/Oligoculm population (Liang et al., 2006). This locus was also likely detected by Bougot et al. (2006) in French winter wheat cultivar FE9001, where it was closely linked to *Xgwm877*, *Xgwm47*, *Xrtpl14R* and *Xcfd267b*. Liu et al. (2001) found *QPm. vt-2B* in winter wheat cultivar Massey at a similar position to *QPm. caas-2BL* based on wheat consensus map (http://wheat.pw.usda.gov/GG2/index.shtml; Somers et al., 2004), and Tucker et al. (2007) again identified *QPm. vt-2B* in USG 3209, the resistant parent of Massey, where it was flanked by *Xgwm501* and *Xgwm191*. Therefore, these QTLs could be at the same or closely linked loci. *QPm. caas-2BL* was detected in two cropping seasons, and also identified in the mean MDS and AUDPC analyses. Thus it showed a relatively stable genetic effect against powdery mildew across different environments.

QPm. caas-2DL was closely linked to SSR marker *Xcfd233*. Keller et al. (1999) detected a QTL for adult plant resistance to powdery mildew in the Swiss spelt cultivar Oberkulmer on the chromosome 2DL. It was flanked by *Xpsr932* and *Xpsr331a* with a map distance about 46cM from *QPm. caas-2DL* (based on Somers et al., 2004; http://wheat.pw.usda.gov/GG2/index.shtml). Börner et al. (2002) identified a QTL in marker interval *Xglk558-2DL - XksuD23-2DL* in the W7984 synthetic. This QTL was about 28cM from *QPm. caas-2DL* (http://wheat.pw.usda.gov/GG2/index.shtml; Somers et al., 2004). Therefore, *QPm. caas-2DL* is probably different from these two QTLs. Recently, He et al. (2009) documented gene *Pm43* on chromosome 2DL and flanked by SSR markers *Xwmc41* and *Xgwm539*, that is, at a similar location to *QPm. caas-2DL*. However, *Pm43* was introgressed from *Thinopyrum intermedium*, and thus could be different from *QPm. caas-2DL* in common wheat. Therefore, *QPm. caas-2DL* is likely to be a new gene for slow mildewing resistance.

In conclusion, *QPm. caas-2BS*, *QPm. caas-2BL* and *QPm. caas-2DL* for slow mildewing resistance were detected in Lumai 21, explaining from 10.9 to 37.6% of phenotypic variance across four environments in a simultaneous fit (Table 2). QTLs *QPm. caas-2BS* and *QPm. caas- 2DL*, detected across a range of environments appear to be new resistance genes. The flanking markers of these three QTLs were *Xbarc98*-*Xbarc1147*, *Xbacr1139*-*Xgwm47* and *Xwmc18*-*Xcfd233*, respectively. These markers could be used in marker-assisted breeding for slow mildewing resistance in wheat.

Acknowledgments

The authors are very grateful to Prof. R. A. McIntosh, Plant Breeding Institute, University of Sydney, for the critical review of this manuscript. This study was supported by the National Science Foundation of China (30810214 and 30671294), International Collaboration Project from the Ministry of Agriculture (2006-G2) and the core research budget of CAAS, a nonprofit governmental research institution.

References

Bennett F (1984) Resistance to powdery mildew in wheat: a review of its use in agriculture and breeding programmes. Plant Pathol 33: 297-300.

Bjarko ME, Line RF (1988) Heritability and number of genes controlling leaf rust resistance on four cultivars of wheat. Phytopathology 78: 457-461.

Börner A, Schumann E, Fürste A, Cöster H, Leithold B, Röder MS, Weber WE (2002) Mapping of quantitative trait loci determining agronomic important characters in hexaploid wheat (*Triticum aestivum* L.). Theor Appl Genet 105: 921-936.

Bougot Y, Lemoine J, Pavoine MT, Guyomarc' h H, Gautier V, Muranty H, Barloy D (2006) A major QTL effect controlling resistance to powdery mildew in winter wheat at the adult plant stage. Plant Breed 125: 550-556.

Chantret N, Mingeot D, Sourdille P, Bernard M, Jacquemin JM, Doussinault G (2001) A major QTL for powdery mildew resistance is stable over time and at two development stages in winter wheat. Theor Appl Genet 103: 962-971.

Gupta PK, Rustgi S, Sharma S, Singh R, Kumar N, Balyan HS (2003) Transferable EST-SSR markers for the study of polymorphism and genetic diversity in bread wheat. Mol Gen Genet 270: 315-323.

Guyomarc' h H, Sourdille P, Edwards KJ, Bernard M (2002) Studies of the transferability of microsatellites derived from *Triticum tauschii* to hexaploid wheat and to diploid related species using amplification, hybridization and sequence comparisons. Theor Appl Genet 105: 736-744.

He RL, Chang ZJ, Yang ZJ, Yuan ZJ, Zhan HX, Zhang XJ, Liu JX (2009) Inheritance and mapping of powdery mildew resistance gene*Pm43* introgressed from *Thinopyrum intermedium* into wheat. Theor Appl Genet 118: 1173-1180.

Hsam SLK, Zeller FJ (2002) Breeding for powdery mildew resistance in common wheat (*Triticum aestivum* L.). In: Belanger RR, Bushnell WR, Dik AJ, Carver TLW (eds) The powdery mildews a comprehensive treatise. The American Phytopathological Society, St Paul, pp. 219-238.

Hua W, Liu ZJ, Zhu J, Xie CJ, Yang TM, Zhou YL, Duan XY, Sun QX, Liu ZY (2009) Identification and genetic mapping of*Pm42*, a new recessive wheat powdery mildew resistance gene derived from wild emmer (*Triticum turgidum* var. *dicoccoides*). Theor Appl Genet 119: 223-230.

Keller M, Keller B, Schachermayr G, Winzeler M, Schmid JE, Stamp P, Messmer MM (1999) Quantitative trait loci for resistance against powdery mildew in a segregating wheat×*spelt* population. Theor Appl Genet 98: 903-912.

Kosambi DD (1944) The estimation of map distance from recombination values. Annu Eugen 12: 172-175.

Lan CX, Liang SS, Wang ZL, Yan J, Zhang Y, Xia XC, He ZH (2009) Quantitative trait loci mapping for adult-plant resistance to powdery mildew in Chinese wheat cultivar Bainong 64. Phytopathology 99: 1121-1126.

Li HH, Ye GY, Wang JK (2007) A modified algorithm for the improvement of composite interval mapping. Genetics 175: 361-374.

Liang SS, Suenaga K, He ZH, Wang ZL, Liu HY, Wang DS, Singh RP, Sourdille P, Xia XC (2006) Quantitative trait loci mapping for adult-plant resistance to powdery mildew in bread wheat. Phytopathology 96: 784-789.

Lillemo M, Asalf B, Singh RP, Huerta-Espino J, Chen XM, He ZH, Bjørnstad Å (2008) The adult plant rust resistance loci *Lr34*/*Yr18* and *Lr46*/*Yr29* are important determinants of partial resistance to powdery mildew in bread wheat line Saar. Theor Appl Genet 116: 1155-1166.

Lincoln S, Daly M, Lander E (1992) Constructing genetic maps with Mapmaker/EXP3.0. Whitehead Institute Techn Rep, 3rd edn. Whitehead Institute, Cambridge

Liu SX, Griffey CA, Maroof MAS (2001) Identification of molecular markers associated with adult plant resistance to powdery mildew in common wheat cultivar Massey. Crop

Sci 41：1268-1275.

McDonald BA，Linde C（2002）The population genetics of plant pathogens and breeding strategies for durable resistance. Euphytica 124：163-180.

McIntosh RA，Dubcovsky J，Rogers WJ，Morris CF，Appels R，Xia CX（2009）Catalogue of gene symbols for wheat：2009（suppl）http：//www.wheat.pw.usda.gov/GG2/pubs.shtml

Michelmore RW，Paran I，Kesseli RV（1991）Identification of markers linked to disease-resistance genes by bulked segregant analysis：A rapid method to detect markers in specific genomic regions by using segregating populations. Proc Natl Acad Sci USA 88：9828-9832.

Mingeot D，Chantret N，Baret PV，Dekeyser A，Boukhatem N，Sourdille P，Doussinault G，Jacquemin JM（2002）Mapping QTL involved in adult plant resistance to powdery mildew in the winter wheat line RE714 in two susceptible genetic backgrounds. Plant Breed 121：133-140.

Ni XW，Yan J，Chen XM，Xia XC，He ZH，Zhang Y，Wang DS，Lillemo M（2008）Heritability and number of genes controlling slow-mildewing resistance in wheat cultivar Lumai 21. Acta Agronomica Sinica 34：1317-1322.［In Chinese with English abstract.］

Pestsova E，Ganal MW，Röder MS（2000）Isolation and mapping of microsatellite markers specific for the D genome of bread wheat. Genome 43：689-697.

Röder MS，Korzun V，Wendehake K，Plaschke J，Tixier MH，Leroy P，Ganal MW（1998）A microsatellite map of wheat. Genetics 149：2007-2023.

Rong JK，Millet E，Manisterski J，Feldman M（2000）A new powdery mildew resistance gene：Introgression from wild emmer into common wheat and RFLP-based mapping. Euphytica 115：121-126.

Skinnes H（2002）Breakdown of race specific resistance to powdery mildew in Norwegian wheat. Cereal Rusts and Powdery Mildew Bulletin vol. 30 online，publication［http：//www.crpmb.org］2002/1201skinnes.

Somers DJ，Isaac P，Edwards K（2004）A high-density microsatellite consensus map for bread wheat（*Triticum aestivum* L.）. Theor Appl Genet 109：1105-1114.

Song QJ，Fickus EW，Cregan PB（2002）Characterization of trinucleotide SSR motifs in wheat. Theor Appl Genet 104：286-293.

Sourdille P，Singh S，Cadalen T，Brown-Guedira GL，Gay G，Qi L，Gill BS，Dufour P，Murigneux A，Bernard M（2004）Microsatellite-based deletion bin system for the establishment of genetic-physical map relationships in wheat（*Triticum aestivum* L.）. Funct Integr Genomics 4：12-25.

Tucker DM，Griffey CA，Liu S，Brown-Guedira G，Marshall DS，Saghai Maroof MA（2007）Confirmation of three quantitative trait loci conferring adult plant resistance to powdery mildew in two winter wheat populations. Euphytica 155：1-13.

Wang ZL（2005）Seeding and adult-plant resistance to powdery mildew in bread wheat and QTL analysis of powdery mildew resistance in Bainong 64. Dissertation for Doctoral Degree（Graduation）of Northwest Sci-Tech University of Agriculture and Forestry in 2005，China，pp. 38-45.

QTL mapping of adult-plant resistance to stripe rust in a Lumai 21 × Jingshuang 16 wheat population

Y. Ren[1,2], L. S. Liu[2], Z. H. He[1,5], L. Wu[3], B. Bai[4], and X. C. Xia[1]

[1] *Institute of Crop Science, Chinese Academy of Agricultural Sciences (CAAS), 12 Zhongguancun South Street, Beijing* 100081, *China*; [2] *College of Agronomy, Henan Agricultural University, Zhengzhou, Henan* 450002, *China*; [3] *Crop Research Institute, Sichuan Academy of Agricultural Sciences, Chengdu, Sichuan* 610066, *China*; [4] *Wheat Research Institute, Gansu Academy of Agricultural Sciences, Lanzhou, Gansu* 730070, *China*; [5] *International Maize and Wheat Improvement Center (CIMMYT) China Office, c/o CAAS*, 12 *Zhongguancun South Street, Beijing* 100081, *China.*

Abstract: Stripe rust, caused by *Puccinia striiformis* f. sp. *tritici*, is a devastating fungal disease in common wheat (*Triticum aestivum* L.) worldwide. Chinese wheat cultivars Lumai 21 and Jingshuang 16 show moderate levels of adult-plant resistance (APR) to stripe rust in the field, and they showed a mean maximum disease severity (MDS) ranging from 24 to 56.7% and 26 to 59%, respectively, across different environments. The aim of this study was to identify quantitative trait loci (QTL) for resistance to stripe rust in an F_3 population of 199 lines derived from Lumai 21×Jingshuang 16. The F_3 lines were evaluated for MDS in Qingshui, Gansu province, and Chengdu, Sichuan province, in the 2009-2010 and 2010-2011 cropping seasons. Five QTL for APR were detected on chromosomes 2B (2 QTL), 2DS, 4DL and 5DS based on mean MDS in each environment and averaged values from all three environments. These QTL were designated *QYr.caas-2BS.2*, *QYr.caas-2BL.2*, *QYr.caas-2DS.2*, *QYr.caas-4DL.2* and *QYr.caas-5DS*, respectively. *QYr.caas-2DS.2* and *QYr.caas-5DS* were detected in all three environments, explaining 2.3-18.2% and 5.1-18.0% of the phenotypic variance, respectively. In addition, *QYr.caas-2BS.2* and *QYr.caas-2BL.2* co-located with QTL for powdery mildew resistance reported in a previous study. These APR genes and their linked molecular markers are potentially useful for improving stripe rust and powdery mildew resistances in wheat breeding.

Key words: Common wheat, *Triticum aestivum*, *Puccinia striiformis* f. sp. *Tritici*, non-specific resistance, bulked segregant analysis, molecular markers, pleiotropic effects, quantitative trait locus

Introduction

Stripe (yellow) rust (YR), caused by *Puccinia striiformis* f. sp. *tritici* (*Pst*), is a worldwide disease of common wheat (*Triticum aestivum* L.) (Stubbs, 1985; Boyd, 2005; Chen, 2005). Since the 1950s, about 23 large-scale stripe rust epidemics have occurred in major wheat-production areas of the world, including China, USA, the United Kingdom, Australia, and India (Wellings, 2011). In China, the disease is very destructive in the northwestern and southwestern regions due to the cool and moist spring conditions (Wan et al., 2004). The most destructive epidemics of stripe rust in China occurred in 1950, 1964, 1990, and 2002, resulting in yield loss estimates of up to

Published in Plant Breeding, 2015, 134 (5): 501-507

6.0, 3.2, 1.8, and 1.3 million tonnes, respectively (Wan et al., 2004). Over the last ten years, the importance of stripe rust in China may have fallen due to the application of fungicides and climate change. However, local stripe rust epidemics are still serious in Sichuan, Chongqing, Yunnan, Shanxi and Gansu, as well as the Shiyan and Xiangfan prefectures of Hubei, and Xinyang and Nanyang prefectures of Henan (He et al., 2011).

Although fungicides can provide adequate control of rusts, growing resistant wheat cultivars are a more economic and effective approach to control the disease, having no additional costs to farmers and no environmental pollution (Line and Chen, 1995; Line, 2002; Chen, 2005). Of the 67 permanently designated *Yr* loci (*Yr1-Yr67*) and numerous temporarily designated resistance genes identified in wheat (Basnet et al., 2013; McIntosh et al., 2013) most are race-specific in effect (Chen, 2005, McIntosh et al., 2013). This kind of resistance gene has been extensively used in wheat breeding because of the high levels of resistance conferred at all stages of growth and ease of selection in the field, but unfortunately, it is readily overcome by new pathogen races (Chen and Line, 1995a, b; Carter et al., 2009). Historically, the acquisition of virulence for *Yr2* in the 1970s, *YrA* in the mid-1980s, *Yr9* in the 1990s, *Yr27* in 1998 in Mexico, *Yr24/Yr26* in 2008 in China and *Yr31* in 2008 in Mexico led to regional and continental epidemics, and crop losses (Liu et al., 2010, Wellings et al., 2009, Wellings, 2011). In contrast, adult-plant resistance or race non-specific resistance is generally quantitatively inherited, and involves reduced frequencies of infection, longer latent periods, lower urediniospore production, and smaller uredinial size compared to susceptible genotypes (Chen and Line, 1995a, Liang et al., 2006, Lu et al., 2009). Such resistance has a higher probability of being stable and durable (Niks and Rubiales, 2002). It now seems likely that genes conferring resistance to multiple diseases maintain effectiveness for long time periods (Li et al., 2014). As a consequence, there is increased interest in the development of wheat cultivars with race non-specific or adult-plant resistance (APR).

APR is more complicated for handling in wheat breeding compared with race-specific resistance because of its polygenic nature with minor effect. Nevertheless, molecular markers can be used to identify and locate genes for APR and to estimate the effect of each resistance gene, greatly facilitating selection for APR in breeding programs. Many APR genes in wheat are associated with molecular markers, such as *Yr16* on chromosome 2DL (Devos et al., 1993). *Yr18/Lr34* on 7DS (Lagudah et al., 2006), *Yr29/Lr46* on 1BL (Rosewarne et al., 2006), *Yr30* on 3BS (Singh et al., 2000a), *Yr36* on 6BS (Uauy et al., 2005; Fu et al., 2009), *Yr39* on 7BL (Lin and Chen, 2007), *Yr46/Lr67* on 4DL (Herrera-Foessel et al., 2011), *Yr48* on 5AL (Lowe et al., 2011), *Yr49* on 3DS (McIntosh et al., 2011), *Yr52* on 7BL (Ren et al., 2012a) and *Yr54* on 2DL (Basnet et al., 2013). Among them, the loci *Yr18/Lr34*, *Yr29/Lr46* and *Yr46/Lr67*, initially identified to confer resistance to stripe rust and leaf rust, also provide partial resistances to powdery mildew and stem rust, with additional designations as *Pm38/Sr57*, *Pm39/Sr58* and *Pm46/Sr55*, respectively (Lillemo et al., 2008; Li et al., 2014; Herrera-Foessel et al., 2014). These genes confer low to moderate levels of resistance to rusts when present alone, but in combinations with other slow-rusting genes high levels of resistance can be achieved (Singh et al., 2000a; Herrera-Foessel et al., 2011). However, the levels of resistance can be variable across environments and genetic backgrounds. For example, *Yr29/Lr46* conferred lower levels of resistance to leaf rust and powdery mildew than *Yr18/Lr34* (Lagudah, 2011), but displayed larger effects in Avocet/Pastor and Avocet/Francolin #1 genetic backgrounds (Rosewarne et al., 2012; Lan et al., 2014). *Yr18/Lr34* was cloned and shown to encode a putative ATP-binding cassette (ABC) transporter. The cloned sequence enabled development of diagnostic molecular markers that are suitable for marker assisted selection (Fu et al., 2009, Krattinger et al., 2009, Lagudah et al., 2009, Yuan et al., 2012). In addition to these named genes, over 140

QTL for reduced stripe rust severity have been identified during the last 10 years (Rosewarne et al., 2013). Although APR is more durable than major gene resistance, some sources can be overcome by pathogenicity changes in the pathogen (McDonald and Linde, 2002). This has led to increased emphasis in breeding programs being placed on pleiotropic APR QTL with unique gene sequences and reputed durability. The Chinese common wheat cultivar Lumai 21 was a leading cultivar with outstanding yield potential and broad adaptation, covering an area of 300 000ha in the Yellow and Huai Valleys Winter Wheat Region from 1994 to 2007 (Lan et al., 2010a). It exhibits a moderate level of APR to Chinese *Pst* races in the field, but is susceptible to some pathotypes at the seedling stage. It also has APR to leaf rust and powdery mildew. QTL mapping of APR to powdery mildew in the Lumai 21 × Jingshuang 16 F_3 population was reported by Lan et al. (2010a). However, the inheritance of APR to stripe rust in this population has not been reported. The objectives of the present study were to detect QTL for stripe rust APR to in the Lumai 21 × Jingshuang 16 population using molecular markers, and to assess the stability of the detected QTL across environments.

Materials and Methods

Plant materials

One hundred and ninety-nine F_2-derived bulk lines (hereafter described as F_3 lines) were produced from Lumai 21 × Jingshuang 16. Lumai 21, developed by the Yantai Academy of Agricultural Sciences, Shandong province, in 1991 is susceptible to Chinese *Pst* races CYR29 and CYR32 at the seedling stage, but is moderately resistant in the field. Jingshuang 16 (pedigree: Lovrin 10×Youmanghong 7), released in Beijing in 1985, is highly susceptible to *Pst* races CYR29 and CYR32 at the seedling stage, but shows a moderate level of APR to stripe rust (Wang et al., 2006). *Pst* races CYR29 and CYR32 was provided by Dr. Gangming Zhan, College of Plant Protection, Northwest Agricultural and Forestry University, Yangling, Shaanxi.

Field trials

The F_3 lines and parents were evaluated for stripe rust response during the 2009-2010 and 2010-2011 cropping seasons in Gansu and Sichuan provinces. Both locations are hotspots for stripe rust with ideal conditions for infection and spread. Field trials were conducted in randomized complete blocks with three replicates at each location. Plots consisted of single 1.5m rows with 25cm spacing. Approximately 50 seeds were sown in each row. Every tenth row was planted with the highly susceptible cultivar Mingxian 169 to aid spread of the pathogen within the trial. To ensure ample inoculum infection rows of Mingxian 169 were also planted perpendicularly and adjacent to the test rows. Inoculations were carried out using mixtures of predominantly *Pst* races CYR29 and CYR32 at the beginning of January in Sichuan. In Gansu, mixed *Pst* races CYR29 and CYR32 were inoculated around April 20. The parameter for analysis was maximum disease severity (MDS), scored as the percentage of leaf area covered by stripe rust when the disease severities on Mingxian 169 reached maximum levels around April 15 in Sichuan and June 10 in Gansu. The field data from Gansu 2010-2011 were excluded from the statistical analysis and QTL detection due to low stripe rust development.

Statistical analysis

Mean MDS was used in analysis of variance performed with PROC GLM in the statistical analysis system (SAS) software package (SAS institute, V8). The information in the ANOVA table was used to calculate broad sense heritability (h_b^2) of stripe rust resistance based on the formula $h_b^2 = \sigma_g^2 / (\sigma_g^2 + \sigma_{ge}^2/r + \sigma_\varepsilon^2/re)$ (Allard 1960), where $\sigma_g^2 = (MS_f - MS_{fe})/re$, $\sigma_{ge}^2 = (MS_{fe} - MS_e)/r$ and $\sigma_\varepsilon^2 = MS_e$; in this formula, σ_g^2 = genetic variance, σ_{ge}^2 = genotype × environment interaction variance, σ_ε^2 = error variance, MS_f = mean square of genotype, MS_{fe} = mean square of genotype × environment interaction, MS_e = mean square of error, r = number of replications and e = number of environments.

Genotyping

A total of 1, 375 simple sequence repeat (SSR) markers were screened on Lumai 21 and Jingshuang 16. Then pooled DNA representing contrasting bulks for stripe rust response were screened for polymorphism with 422 SSR markers polymorphic between the parents; these included BARC (Song et al., 2002), CFA and CFD (Sourdille et al., 2004), CFE (Zhang et al., 2005), CWEM (Peng et al., 2005), GDM (Pestsova et al., 2000), GWM (Röder et al., 1998) and WMC (Gupta et al., 2002) markers. SSR markers that showed similar patterns of polymorphism between the bulks and between parents were used to genotype all 199 F_3 lines from the Lumai 21×Jingshuang 16 cross.

Map construction and QTL analysis

Genetic linkage maps were constructed usingMap Manager QTX20 (Manly et al., 2001). Markers were assigned to linkage groups with a minimum logarithm of the odds (LOD) threshold of 3.0, and genetic distances between markers were estimated for recombination values based on the Kosambi mapping function (Kosambi, 1944). The ordering of markers and assignment of linkage groups to chromosomes were checked against a public wheat consensus map (Somers et al., 2004).

Stripe rust resistance QTL were detected by Inclusive composite interval mapping (ICIM) using the software QTL IciMapping V3.1 (Liet al., 2007, 2008; Wang, 2009), based on the MDS for each environment and those averaged from all environments combined. After performing a 1000 permutation test, a LOD threshold of 2.5 was set to declare QTL as significant. A walk speed of 1.0cM was chosen for all QTL detections. QTL effects were estimated as the phenotypic variance explained (PVE) and additive effects explained by the QTL. In this study, the genotype of Lumai 21 was defined as 2, and the genotype of Jingshuang 16 was defined as 0. Thus, the allele from Lumai 21 reduces stripe rust MDS when the additive effect is negative.

Results

Phenotypic evaluations

Stripe rust developed well across environments, except for Gansu in 2010-2011. The frequency distributions of stripe rust MDS for the 199 F_3 lines in three environments and averaged MDS for three environments revealed continuous distributions (Fig. 1), indicating polygenic inheritance. The averaged MDS of the F_3 lines across three environments was 36.5%, ranging from 8.1 to 80.6%. The MDS of the susceptible control Mingxian 169 ranged from 60 to 100%, from 90 to 100%, and from 50 to 90% in Sichuan 2009-2010, Sichuan 2010-2011 and Gansu 2009-2010 seasons, respectively. Lumai 21 was rated with a mean MDS of 45.0%, 56.7% and 24.0% in the Sichuan 2009-2010, Sichuan 2010-2011 and Gansu 2009-2010 seasons, respectively, whereas Jingshuang 16 had a mean MDS of 26-59% across all three environments.

The MDS data were significantly correlated across the three environments, ranging from 0.65 to 0.71 ($P < 0.01$) (Table 1), and the broad-sense heritability of MDS

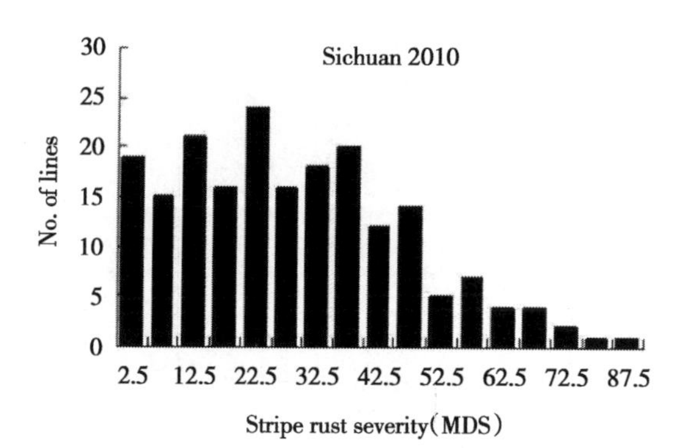

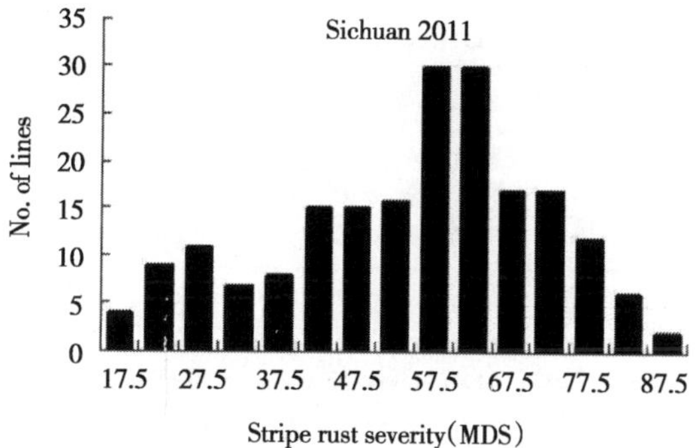

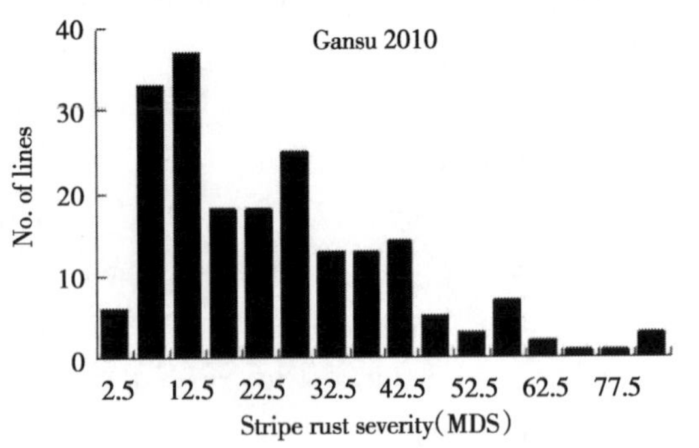

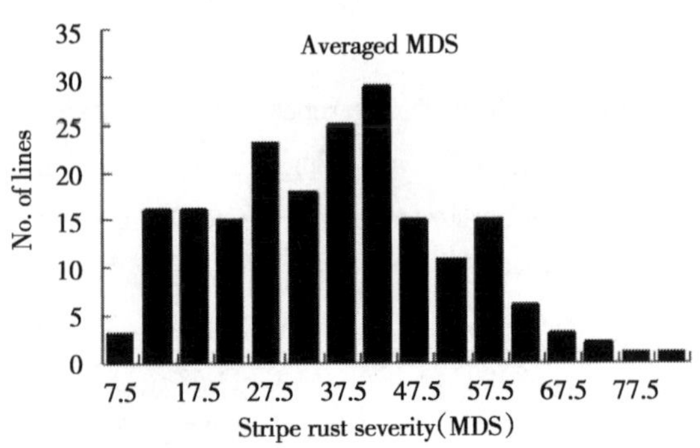

Fig. 1 Frequency distribution of stripe rust MDS for F_3 lines from the wheat cross Lumai 21×Jingshuang 16 in three environments and averaged MDS.

Table 1 Analysis of variance of MDS scores for F_3 lines generated from Lumai 21×Jingshuang 16

Source of variation	*df*	Mean square	*F* value	*P* value
Lines	198	2,158.3	13.7**	< 0.0001
Environments	2	154,726.9	984.2**	< 0.0001
Replicates (Environment)	6	7,689.7	48.9**	< 0.0001
Lines× environments	396	299.7	1.9**	< 0.0001
Error	1 188	157.2		

** Significant at $P=0.01$.

was 0.86. ANOVA revealed significant differences ($P=0.01$) in MDS among F_3 lines, environments, replicates within environments and line× environment interactions (Table 2).

Table 2 Quantitative trait loci (QTL) for adult plant resistance (APR) to stripe rust detected by inclusive composite interval mapping (ICIM) in the Lumai 21×Jingshuang 16 F_3 population across three environments

Location and year	QTL[a]	Marker interval	LOD[b]	PVE (%)[c]	Add[d]
Sichuan 2010	*QYr. caas-2DS. 2*	*Xbarc168-Xwmc18*	3.7	6.5	−5.2
	QYr. caas-5DS	*Xgwm190-Xgwm182*	5.0	18.0	14.0
Sichuan 2011	*QYr. caas-2DS. 2*	*Xbarc168-Xwmc18*	5.8	18.2	−5.0
	QYr. caas-5DS	*Xgwm190-Xgwm182*	3.3	5.1	6.6
Gansu 2010	*QYr. caas-2BL. 2*	*Xcfd73-Xgwm47*	3.0	6.2	−2.5
	QYr. caas-2DS. 2	*Xbarc168-Xwmc18*	2.7	2.3	−2.6
	QYr. caas-5DS	*Xgwm190-Xgwm182*	3.4	9.5	10.0
Averaged MDS in 3 environments	*QYr. caas-4DL. 2*	*Xbarc98-Xbarc1148*	3.8	8.0	−7.5
	QYr. caas-2BS. 2	*Xbarc156-Xbarc1147*	2.9	7.4	−3.0
	QYr. caas-2DS. 2	*Xbarc168-Xwmc18*	5.5	13.8	−5.4
	QYr. caas-5DS	*Xgwm190-Xgwm182*	3.9	9.2	8.6

a. QTL were detected with a minimum LOD score of 2.5 in at least one environment.

b. LOD, logarithm of odds score.

c. PVE, percentages of the phenotypic variance explained by individual QTL.

d. Add, additive effect of resistance allele.

Detection of QTL for stripe rust resistance

Five QTL for APR to stripe rust were identified on chromosomes 2B (two loci), 2DS, 4DL and 5DS based on ICIM using the MDS in each environment and the averaged value from all three environments (Fig. 2). They were designated *QYr. caas-2BS. 2*, *QYr. caas-2BL. 2*, *QYr. caas-2DS. 2*, *QYr. caas-4DL. 2*, and *QYr. caas-5DS*, respectively.

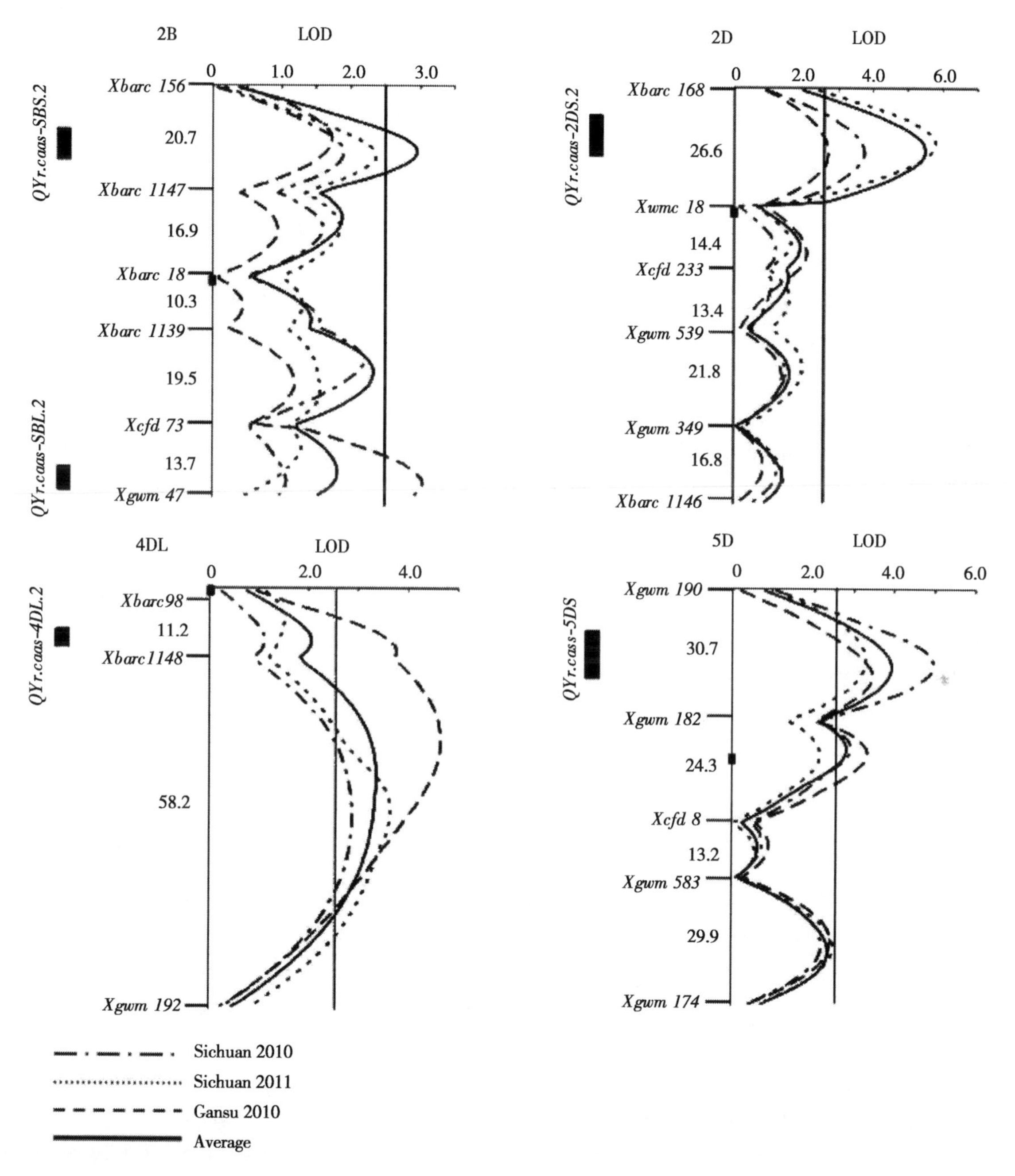

Fig. 2 LOD contours for QTL on chromosomes 2B (2 QTL), 2DS, 4DL and 5DS that reduce stripe rust severity in the F_3 population from Lumai 21×Jingshuang 16. The short arms are toward the top and approximate positions of centromeres are indicated by solid squares in the vertical axis. Genetic distances are shown in centiMorgans (cM) to the left of the vertical axes. The approximate positions of QTL are indicated by black rectangles to the left of markers and their heights represent the interval distances of flanking markers. LOD thresholds of 2.5 are indicated by dashed vertical lines.

A QTL for stripe rust response was located on chromosome 2DS, flanked by *Xbarc168* and *Xwmc18*, explaining 6.5%, 18.2%, 2.3% and 13.8% of the phenotypic variance in Sichuan 2009-2010, Sichuan 2010-2011, Gansu 2009-2010 seasons and the averaged MDS, respectively. The second consistently detected QTL with large effect, *QYr. caas-5DS*, was located on 5DS between *Xgwm190* and *Xgwm180*, and explained 5.1-18.0% of the phenotypic variance in the three environments as well as the overall mean. The remaining three QTL were detected in single environments. *QYr. caas-2BL. 2* and *QYr. caas-4DL. 2* were in marker intervals *Xcfd73-Xgwm47* and *Xbarc98-Xbarc1148*, and explained 6.2% and 7.6%, respectively, of the phenotypic variance in the Gansu 2009-2010 season. *QYr. caas-2BS. 2* was located between *Xbarc156* and *Xbarc1147* on chromosome 2BS, explaining of 7.4% the phenotypic variance in the averaged MDS. The allele on 5DS, derived from Jingshuang 16 also significantly reduced stripe rust severity. The total phenotypic variance explained by all five QTL ranged from 23.3 to 30.4% in a simultaneous fit across the three environments, suggesting an obvious effect of these QTL in reducing YR severity. Interactions among different QTL were not identified across the three environments.

Discussion

The Chinese cultivar Lumai 21 not only shows good and stable slow mildewing resistance, but also is moderately resistant to stripe rust at the adult plant stage. Four APR QTL for stripe rust were detected in Lumai 21. Based on the pedigree of Lumai 21 (Lumai 13/Yumai 2), the four APR QTL are likely derived from Yumai 2, a leading cultivar and a core breeding parent in Henan. The other parent Jingshuang 16 showed a moderately resistant reaction with MDS from 12.7% to 45% in field tests, and contributed one APR QTL (*QYr. caas-5DS*) to the cross.

Nineteen stripe rust resistance QTL have been mapped on chromosome 2B by molecular markers (Boukhatem et al., 2002; Ramburan et al., 2004; Mallard et al., 2005; Christiansen et al., 2006; Guo et al., 2008; Rosewarne et al., 2008; Carter et al., 2009; Dedryver et al., 2009; Lan et al., 2010b; Lowe et al., 2011; Chen et al., 2012; Vazquez et al., 2012; Ren et al., 2012b; Xu et al., 2013), and these QTL are concentrated in four chromosomal regions (Rosewarne et al., 2013). In the present study, the cultivar Lumai 21 contained two QTL, viz. *QYr. caas-2BS. 2* and *QYr. caas-2BL. 2*, more than 30cM apart, on chromosome 2B. *QYr. caas-2BS. 2* was flanked by *Xbarc156* and *Xbarc1147*, and was in the *QRYr2B. 2* region, whereas *QYr. caas-2BL. 2*, flanked by *Xcfd73* and *Xgwm47*, coincides with the *QRYr2B. 3* region (Somers et al., 2004; Rosewarne et al., 2013). However, these two QTL were detected only in a single environment, and they showed lower effects and stability than the other reported QTL. This indicated that the expression of the two QTL in Lumai 21 might be influenced by environmental factors.

Bariana et al. (2001) reported a QTL flanked by SSR markers *Xwmc111* and *Xwmc25* on chromosome 2DS in cultivar Katepwa. This region was also identified in wheat cultivars Oligoculm and Libellula (Suenaga et al., 2003; Lu et al., 2009). That QTL was approximately 24cM from *QYr. caas-2DS. 2* based on the consensus map (Somers et al., 2004), suggesting that it is different. Agenbag et al. (2012) identified a QTL (*QYr. ufs-2D*) flanked by SSR markers *Xgwm102* and *Wpt-664520* on 2DS in the French cultivar Cappelle-Desprez. *QYr. caas-2DS. 2* mapped to the same region, but contributed more to phenotypic variation (maximum of 18.2%). Mallard et al. (2005) identified a QTL (*QYr. inra-2DS*) in a similar chromosomal region in the French cultivar Camp Remy, explaining 24.0-69.0% of the phenotypic variance across years, indicating an obviously stronger effect across environments than both *QYr. ufs-2D* and *QYr. caas-2DS. 2*. However, the ancestral relationship between these French cultivars and Lumai 21 is not as clear as for Cappelle-Desprez and Camp Remy. There is a large region in the proximal part of chromosome 2DL that contains QTL from 4 parents, viz., Fukuho-komugi, Guardian, Alcedo and Naxos. This region is more than 20cM proximal to *QYr. caas-2DS. 2* (Suenaga et al., 2003; Somers et al., 2004; Melichar et al., 2008; Jagger et al., 2011; Ren et al., 2012c). These QTL seem to be different from *QYr. caas-2DS. 2*.

Suenaga et al. (2003) reported a QTL for stripe rust resistance on chromosome 4DL in the Israeli wheat Oligoculm. This QTL was closely linked to *Xwmc399* and contributed only a low level of resistance (variance range 2.5-8.0%). This QTL was approximately 24cM from *QYr. caas-4DL. 2* (He et al., 2011) indicating it is different from *QYr. caas-4DL. 2*. Hiebert et al. (2010) and Herrera-Foessel et al. (2011) located an APR gene (*Yr46/Lr67*) on 4DL in the Thatcher derivative RL6077, that was closely linked to markers *Xgwm165* and *Xgwm192* and fell within the marker interval carrying *QYr. caas-4DL. 2* based on the consensus map (Somers et al., 2004). Ren et al., (2012b) and Rosewarne et al. (2012) also identified a minor stripe rust resistance QTL at the same chromosomal region in Chinese wheat cultivar Bainong 64 and CIMMYT line Pastor, explaining 3.9-8.0% and 2.8-4.9% of the phenotypic variances, respectively. Of the two, *QYr. caas-4DL* contributed by Bainong 64 and *Yr46* might be the same since both confer partial resistance to stripe rust, leaf rust and powdery mildew (Liu et al., 2014). However, it is clear that Pastor did not have associated QTL for stripe rust and leaf rust resistances at this locus (Rosewarne et al., 2012) and therefore appear to be different. Although the locus in Lumai 21 conferred resistance to both stripe rust and leaf rust there was no significant effect in reducing the severity of powdery mildew (Lan et al., 2010a; Liu et al., 2014). We tested Lumai 21 and Jingshuang 16 using the SNP marker *csSNP856* that was co-segregating with *Lr67/Yr46* locus (Forrest et al., 2014), and didn't find any polymorphisms between two parents indicating that *QYr. caas-4DL. 2* is different from *Yr46*. The effects of the potential pleiotropic APR QTL might therefore be influenced by different genetic background.

QYr. caas-5DS in cultivar Jingshuang 16 is close to the centromere. Singh et al. (2000b) identified a QTL for APR to stripe rust in Mexican cultivar Opata 85 on chromosome 5DS, flanked by *Xfbb238* and *Xfba114* and located in the same region as *QYr. caas-5DS* based on the consensus map (Somers et al., 2004). However, the QTL explained 8.0% of the phenotypic variance in only one environment which showed a lower effect and stability than *QYr. caas-5DS*.

The same Lumai 21 × Jingshuang 16 population was previously used for QTL mapping of powdery mildew (PM) resistance and three QTL for APR to PM were identified on chromosomes 2B (2 QTL) and 2D (Lanet al., 2010a). In comparison to the present study, *QYr. caas-2BS. 2* and *QYr. caas-2BL. 2* co-located with QTL for PM resistance with the resistance alleles contributed by Lumai 21. This is consistent with a report by Liu et al. (2014), who demonstrated that *QPm. caas-2BS. 2* and *QPm. caas-2BL. 2* from Lumai 21 provided significant levels of resistance to stripe rust ($P<0.01$). However, *QPm. caas-2DL* did not show a significant effect in reducing stripe rust severity (Liu et al., 2014). Although *QYr. caas-2DS. 2* was co-located with *QPm. caas-2DL*, they are likely different genes.

In the present study, we detected two QTL conferring stable resistance to stripe rust across environments, including *QYr. caas-2DS. 2* and *QYr. caas-5DS*, and corresponding linked SSR markers. These two QTL and their linked markers could be used to achieve stable stripe rust resistance in wheat breeding. *QYr. caas-2BS. 2* and *QYr. caas-2BL. 2* may confer resistance to multiple diseases and therefore might be useful for improving stripe rust and powdery mildew resistances by marker-assisted selection.

Acknowledgments

We are grateful for the critical review of this manuscript by Prof. R. A. McIntosh, Plant Breeding Institute, University of Sydney, Australia. This study was supported by the National Key Basic Research Program of

China (2013CB127700), National Natural Science Foundation of China (31301307, 31261140370), and the China Agriculture Research System (CARS-3-1-3).

References

Agenbag, G. M., Z. A. Pretorius, L. A. Boyd, C. M. Bender, and R. Prins, 2012: Identification of adult plant resistance to stripe rust in the cultivar Cappelle-Desprez. *Theor. Appl. Genet.* 125, 109-120.

Allard, R. W., 1960: Princilpes of plant breeding. John Wiley and Sons, Inc., New York, London.

Bariana, H. S., M. J. Hayden, N. U. Ahmed, J. A. Bell, P. J. Sharp, and R. A. McIntosh, 2001: Mapping of durable adult plant and seedling resistances to stripe rust and stem rust diseases in wheat. *Aust. J. Agri. Res.* 52, 1247-1255.

Basnet, B. R., R. P. Singh, A. M. H. Ibrahim, S. A. Herrera-Foessel, J. Huerta-Espino, C. Lan, and J. C. Rudd, 2013: Characterization of *Yr54* and other genes associated with adult plant resistance to yellow rust and leaf rust in common wheat Quaiu 3. *Mol. Breeding* 33, 385-399.

Boukhatem, N., P. V. Baret, D. Mingeot, and J. M. Jacquemin, 2002: Quantitative trait loci for resistance against yellow rust in two wheat derived recombinant inbred line populations. *Theor. Appl. Genet.* 104, 111-118.

Boyd, L. A., 2005: Can Robigus defeat an old enemy? — Yellow rust of wheat. *J. Agric. Sci.* 143, 233-243.

Carter, A. H., X. M. Chen, K. Garland-Campbell, and K. K. Kidwell, 2009: Identifying QTL for high-temperature adult-plant resistance to stripe rust (*Puccinia striiformis* f. sp. *tritici*) in the spring wheat (*Triticum aestivum* L.) cultivar 'Louise'. *Theor. Appl. Genet.* 119, 1119-1128.

Chen, J. L., C. G. Chu, E. J. Souza, M. J. Guttieri, X. M. Chen, S. Xu, D. Hole, and R. Zemetra. 2012: Genome-wide identification of QTL conferring high-temperature adult-plant (HTAP) resistance to stripe rust (*Puccinia striiformis* f. sp. *tritici*) in wheat. *Mol. Breeding* 29, 791-800.

Chen, X. M., 2005: Epidemiology and control of stripe rust [*Puccinia striiformis* f. sy6p. *tritici*] on wheat. *Can. J. Plant Pathol.* 27, 314-337.

Chen, X. M., and R. F. Line, 1995a: Gene action in wheat cultivars for durable high-temperature adult-plant resistance and interactions with race-specific, seedling resistance to stripe rust caused by *Puccinia striiformis*. *Phytopathology* 85, 567-572.

Chen, X. M., and R. F. Line, 1995b: Gene number and heritability of wheat cultivars with durable, high-temperature, adult-plant resistance and race-specific resistance to *Puccinia striiformis*. *Phytopathology* 85, 573-578.

Christiansen, M. J., B. Feenstra, I. M. Skovgaard, and S. B. Andersen, 2006: Genetic analysis of resistance to yellow rust in hexaploid wheat using a mixture model for multiple crosses. *Theor. Appl. Genet.* 112, 581-591.

Dedryver, F., S. Paillard, S. Mallard, O. Robert, M. Trottet, S. Nègre, G. Verplancke, and J. Jahier, 2009: Characterization of genetic components involved in durable resistance to stripe rust in the bread wheat 'Renan'. *Phytopathology* 99, 968-973.

Fu, D. L., C. Uauy, A. Distelfeld, A. Blechl, L. Epstein, X. M. Chen, H. Sela, T. Fahima, and J. Dubcovsky, 2009: A kinase-start gene confers temperature-dependent resistance to wheat stripe rust. *Science* 323, 1357-1360.

Guo, Q., Z. J. Zhang, Y. B. Xu, G. H. Li, J. Feng, and Y. Zhou, 2008: Quantitative trait loci for high-temperature adult-plant and slow-rusting resistance to*Puccinia striiformis* f. sp. *tritici* in wheat cultivars. *Phytopathology* 98, 803-809.

Gupta, P. K., H. S. Balyan, K. J. Edwards, P. Isaac, V. Korzun, M. Röder, M. -F. Gautier, P. Joudrier, A. R. Schlatter, J. Dubcovsky, R. C. Dela Pena, M. Khairallah, G. Penner, M. J. Hayden, P. Sharp, B. Keller, R. C. C. Wang, J. Hardouin, P. Jack, P. Leroy, 2002: Genetic mapping of 66 new microsatellite (SSR) loci in bread wheat. *Theor. Appl. Genet.* 105, 413-422.

He, Z. H., C. X. Lan, X. M. Chen, Y. C. Zou, Q. S. Zhang, and X. C. Xia, 2011: Progress and perspective in research of adult-plant resistance to stripe rust and powdery mildew in wheat. *Sci. Agric. Sin.* 44, 2193-2215.

Herrera-Foessel, S. A., E. S. Lagudah, J. Huerta-Espino, M. J. Hayden, H. S. Bariana, D. Singh, and R. P. Singh, 2011: New slow-rusting leaf rust and stripe rust resistance genes *Lr67* and *Yr46* in wheat are pleiotropic or closely linked. *Theor. Appl. Genet.* 122, 239-249.

Herrera-Foessel, S. A., R. P. Singh, M. Lillemo, J. Huerta-Espino, S. Bhavani, S. Singh, C. X. Lan, V. Calvo-Salazar, and E. S. Lagudah. 2014. *Lr67/Yr46* confers adult plant resistance to stem rust and powdery mildew in wheat. *Theor. Appl. Genet.* 127, 781-789.

Hiebert, C. W., J. B. Thomas, B. D. McCallum, D. G.

Humphreys, R. M. Depauw, M. J. Hayden, R. Mago, W. Schnippenkoetter, and W. Spielmeyer, 2010: An introgression on wheat chromosome 4DL in RL6077 (Thatcher * 6/PI 250413) confers adult plant resistance to stripe rust and leaf rust (*Lr67*). *Theor. Appl. Genet.* 121, 1083-1091.

Jagger, L. J., C. Newell, S. T. Berry, R. MacCormack, and L. A. Boyd. 2011: The genetic characterisation of stripe rust resistance in the German wheat cultivar Alcedo. *Theor. Appl. Genet.* 122, 723-733.

Kosambi, D. D., 1944: The estimation of map distance from recombination values. *Annu. Eugen.* 12, 172-175.

Krattinger, S. G., E. S. Lagudah, W. Spielmeyer, R. P. Singh, J. Huerta-Espino, H. McFadden, E. Bossolini, L. L. Selter, and B. Keller, 2009: A putative ABC transporter confers durable resistance to multiple fungal pathogens in wheat. *Science* 323, 1360-1363.

Lagudah, E. S., 2011: Molecular genetics of race non-specific rust resistance in wheat. *Euphytica* 179, 81-91.

Lagudah E. S., S. G. Krattinger, S. Herrera-Foessel, R. P. Singh, J. Huerta-Espino, W. Spielmeyer, G. Brown-Guedira, L. L. Selter, and B. Keller, 2009: Gene-specific markers for the wheat gene *Lr34/Yr18/Pm38* which confers resistance to multiple fungal pathogens. *Theor. Appl. Genet.* 119, 889-898.

Lagudah, E. S., H. McFadden, R. P. Singh, J. Huerta-Espino, H. S. Bariana, and W. Spielmeyer, 2006: Molecular genetic characterization of the *Lr34/Yr18* slow rusting resistance gene region in wheat. *Theor. Appl. Genet.* 114, 21-30.

Lan, C. X., X. W. Ni, J. Yan, Y. Zhang, X. C. Xia, X. M. Chen, and Z. H. He, 2010a: Quantitative trait loci mapping of adult-plant resistance to powdery mildew in Chinese wheat cultivar Lumai 21. *Mol. Breeding* 25, 615-622.

Lan, C. X., S. S. Liang, X. C. Zhou, G. Zhou, Q. L. Lu, X. C. Xia, and Z. H. He, 2010b: Identification of genomic regions controlling adult-plant stripe rust resistance in Chinese landrace Pingyuan 50 through bulked segregant analysis. *Phytopathology* 100, 313-318.

Lan, C. X., R. P. Singh, J. Huerta-Espino, V. Calvo-Salazar, and S. A. Herrera-Foessel, 2014: Genetic analysis of resistance to leaf rust and stripe rust in wheat cultivar Francolin#1. *Plant Dis.* 98, 1227-1234.

Li, H. H., G. Y. Ye, and J. K. Wang, 2007: A modified algorithm for the improvement of composite interval mapping. *Genetics* 175, 361-374.

Li, H. H., J. M. Ribaut, Z. L. Li, and J. K. Wang, 2008: Inclusive composite interval mapping (ICIM) for digenic epistasis of quantitative traits in biparental populations. *Theor. Appl. Genet.* 116, 243-260.

Li, Z. F., C. X. Lan, Z. H. He, R. P. Singh, G. M. Rosewarne, X. M. Chen, and X. C. Xia, 2014: Overview and application of QTL for adult plant resistance to leaf rust and powdery mildew in wheat. *Crop Science* 54, 1907-1921.

Liang, S. S., K. Suenaga, Z. H. He, Z. L. Wang, H. Y. Liu, D. S. Wang, R. P. Singh, P. Sourdille, and X. C. Xia, 2006: Quantitative trait loci mapping for adult-plant resistance to powdery mildew in bread wheat. *Phytopathology* 96, 784-789.

Lillemo, M., B. Asalf, R. P. Singh, J. Huerta-Espino, X. M. Chen, Z. H. He, and Á. Bjørnstad, 2008: The adult plant rust resistance loci *Lr34/Yr18* and *Lr46/Yr29* are important determinants of partial resistance to powdery mildew in bread wheat line Saar. *Theor. Appl. Genet.* 116, 1155-1166.

Lin, F., and X. M. Chen, 2007: Genetics and molecular mapping of genes for race-specific all-stage resistance and non-race-specific high-temperature adult-plant resistance to stripe rust in spring wheat cultivar Alpowa. *Theor. Appl. Genet.* 114, 1277-1287.

Line, R. F., 2002: Stripe rust of wheat and barley in North America: a retrospective historical review. *Annu. Rev. Phytopathology* 40, 75-118.

Line, R. F., and X. M. Chen, 1995: Success in breeding for and managing durable resistance to wheat rusts. *Plant Dis.* 79, 1254-1255.

Liu, J. D., X. M. Chen, Z. H. He, L. Wu, B. Bai, Z. F. Li, and X. C. Xia, 2014: Resistace of slow mildewing genes to stripe rust and leaf rust in common wheat. *Acta Agronomica Sinica* 40, 1157-1564.

Liu, T. G., Y. L. Peng, and Z. Y. Zhang, 2010: First detection of virulence in *Puccinia striiformis* f. sp. *tritici* in China to resistance genes *Yr24* (=*Yr26*) present in wheat cultivar Chuanmai 42. *Plant Dis.* 94, 1163.

Lowe, I., L. C. Jankuloski, S. M. Chao, X. M. Chen, D. See, and J. Dubcovsky, 2011: Mapping and validation of QTL which confer partial resistance to broadly virulent post-2000 North American races of stripe rust in hexaploid wheat. *Theor. Appl. Genet.* 123, 143-157.

Lu, Y. M., C. X. Lan, S. S. Liang, X. C. Zhou, D. Liu, G. Zhou, Q. L. Lu, J. X. Jing, M. N. Wang, X. C. Xia, and Z. H. He, 2009: QTL mapping for adult-plant resistance to stripe rust in Italian common wheat cultivars Libellula and Strampelli. *Theor. Appl. Genet.* 119, 1349-1359.

Mallard, S., D. Gaudet, A. Aldeia, C. Abelard, A. L. Besnard, P. Sourdille, and F. Dedryver, 2005: Genetic analysis of durable resistance to yellow rust in bread wheat. *Theor. Appl. Genet.* 110, 1401-1409.

Manly, K. F., J. R. H. Cudmore, and J. M. Meer, 2001: Map Manager QTX, cross-platform software for genetic mapping. *Genome* 12, 930-932.

McDonald, B. A., and C. Linde, 2002: Pathogen population genetics, evolutionary potential, and durable resistance. *Annu. Rev. Phytopathology* 40, 349-379.

McIntosh, R. A., J. Dubcovsky, W. J. Rogers, C. Morris, R. Appels, and X. C. Xia, 2010: Catalogue of gene symbols for wheat: 2010 supplement, http://www.shigen.nig.ac.jp/wheat/komugi/genes/macgene/supplement2010.pdf.

McIntosh, R. A., J. Dubcovsky, W. J. Rogers, C. Morris, R. Appels, and X. C. Xia, 2011: Catalogue of gene symbols for wheat: 2011 supplement, http://www.shigen.nig.ac.jp/wheat/komugi/genes/macgene/supplement2011.pdf.

McIntosh, R. A., J. Dubcovsky, W. J. Rogers, C. Morris, R. Appels, X. C. Xia, 2013: Catalogue of gene symbols for wheat: 2013-2014 supplement, http://www.shigen.nig.ac.jp/wheat/komugi/genes/macgene/supplement2013.pdf.

McIntosh, R. A., K. M. Devos, J. Dubcovsky, W. J. Rogers, C. Morris, R. Appels, and D. J. Somers 2008: Catalogue of gene symbols for wheat: 2008 supplement, http://www.shigen.nig.ac.jp/wheat/komugi/genes/macgene/supplement 2008.pdf.

Melichar, J. P. E., S. Berry, C. Newell, R. MacCormack, L. and A. Boyd. 2008: QTL identification and microphenotype characterization of the developmentally regulated yellow rust resistance in the UK wheat cultivar Guardian. *Theor. Appl. Genet.* 117, 391-399.

Niks, R. E., and D. Rubiales, 2002: Potentially durable resistance mechanisms in plants to specialised fungal pathogens. *Euphytica* 124, 201-216.

Peng, J. H., Nora L. V. Lapitan, 2005: Characterization of EST-derived microsatellites in the wheat genome and development of eSSR markers. Funct. Integr. *Genomics* 5, 80-96.

Pestsova, E., M. W. Ganal, and M. S. Röder, 2000: Isolation and mapping of microsatellite markers specific for the D genome of bread wheat. *Genome* 43, 689-697.

Ramburan, V. P., Z. A. Pretorius, J. H. Louw, L. A. Boyd, P. H. Smith, W. H. P. Boshoff, and R. Prins, 2004: A genetic analysis of adult plant resistance to stripe rust in the wheat cultivar Kariega. *Theor. Appl. Genet.* 108, 1426-1433.

Ren, R. S., M. N. Wang, X. M. Chen, and Z. J. Zhang, 2012a: Characterization and molecular mapping of *Yr52* for high-temperature adult-plant resistance to stripe rust in spring wheat germplasm PI 183527. *Theor. Appl. Genet.* 125, 847-857.

Ren, Y., Z. F. Li, Z. H. He, L. Wu, B. Bai, C. X. Lan, C. F. Wang, G. Zhou, H. Z. Zhu, and X. C. Xia, 2012b: QTL mapping of adult-plant resistance to stripe rust and leaf rust in Chinese wheat cultivar Bainong 64. *Theor. Appl. Genet.* 125, 1253-1262.

Ren, Y., Z. H. He, J. Li, M. Lillemo, L. Wu, B. Bai, Q. C. Lu, H. Z. Zhu, G. Zhou, J. Y. Du, Q. L. Lu, and X. C. Xia, 2012c: QTL mapping of adult-plant resistance to stripe rust in a population derived from common wheat cultivars Naxos and Shanghai3/Catbird. *Theor. Appl. Genet.* 125, 1211-1221.

Röder, M. S., V. Korzun, K. Wendehake, J. Plaschke, M. H. Tixier, P. Leroy, and M. W. Ganal 1998: A microsatellite map of wheat. *Genetics* 149, 2007-2023.

Rosewarne, G. M., S. A. Herrera-Foessel, R. P. Singh, J. Huerta-Espino, C. X. Lan, and Z. H. He, 2013: QTL trait loci of stripe rust resistance in wheat. *Theor. Appl. Genet.* 126, 2427-2449.

Rosewarne, G. M., R. P. Singh, J. Huerta-Espino, H. M. William, S. Bouchet, S. Cloutier, H. McFadden, and E. S. Lagudah, 2006: Leaf tip necrosis, molecular markers and β1-proteasome subunits associated with the slow rusting resistance genes *Lr46/Yr29*. *Theor. Appl. Genet.* 112, 500-508.

Rosewarne, G. M., R. P. Singh, J. Huerta-Espino, and G. J. Rebetzke, 2008: Quantitative trait loci for slow-rusting resistance in wheat to leaf rust and stripe rust identified with multi-environment analysis. *Theor. Appl. Genet.* 116, 1027-1034.

Rosewarne, G. M., R. P. Singh, J. Huerta-Espino, S. A. Herrera-Foessel, K. L. Forrest, M. J. Hayden, and G. J. Rebetzke. 2012: Analysis of leaf and stripe rust severities reveals pathotype changes and multiple minor QTLs associated with resistance in an Avocet× Pastor wheat population. *Theor. Appl. Genet.* 124, 1283-1294.

Singh, R. P., J. Huerta-Espino, and S. Rajaram, 2000a:

Achieving near-immunity to leaf and stripe rusts in wheat by combining slow rusting resistance genes. *Acta Phytopathologica et Entomologica Hungarica* . 35, 133-139.

Singh, R. P. , J. C. Nelson, M. E. Sorrells, 2000b: Mapping *Yr28* and other genes for resistance to stripe rust in wheat. *Crop Science.* 40, 1148-1155.

Somers, D. J. , P. Isaac, and K. Edwards, 2004: A high-density microsatellite consensus map for bread wheat (*Triticum aestivum* L.). *Theor. Appl. Genet.* 109, 1105-1114.

Song, Q. J. , E. W. Fickus, and P. B. Cregan, 2002: Characterization of trinucleotide SSR motifs in wheat. *Theor. Appl. Genet.* 104, 286-293.

Sourdille, P. , S. Singh, T. Cadalen, G. L. Brown-Guedira, G. Gay, L. Qi, B. S. Gill, P. Dufour, A. Murigneux, and M. Bernard, 2004: Microsatellite-based deletion bin system for the establishment of genetic-physical map relationships in wheat (*Triticum aestivum* L.). *Funct. Integr. Genomics.* 4, 12-25.

Stubbs, R. W. , 1985: Stripe rust. In: Roelfs AP, Bushnell WR (eds) The cereal rusts II. Academic Press, Orlando, FL pp, 61-101.

Suenaga, K. , R. P. Singh, J. Huerta-Espino, and H. M. William, 2003: Microsatellite markers for genes *Lr34/Yr18* and other quantitative trait loci for leaf rust and stripe rust resistance in bread wheat. *Phytopathology* . 93, 881-890.

Uauy, C. , J. C. Brevis, X. M. Chen, I. Khan, L. Jackson, O. Chicaiza, A. Distelfeld, T. Fahima, and J. Dubcovsky, 2005: High-temperature adult-plant (HTAP) stripe rust resistance gene*Yr36* from *Triticum turgidum* ssp. *dicoccoides* is closely linked to the grain protein content locus *Gpc-B1*. *Theor. Appl. Genet.* 112, 97-105.

Vazquez, M. D. , C. J. Peterson, O. Riera-Lizarazu, X. Chen, A. Heesacker, K. Ammar, J. Crossa, and C. C. Mundt, 2012: Genetic analysis of adult plant, quantitiative resistance to stripe rust in wheat cultivar 'Stephens' in multi-environment trials. *Theor. Appl. Genet.* 124, 1-11.

Wan, A. M. , Z. H. Zhao, X. M. Chen, Z. H. He, S. L. Jin, Q. Z. Jia, G. Yao, J. X. Yang, B. T. Wang, G. B. Li, Y. Q. Bi, and Z. Y. Yuan, 2004: Wheat stripe rust epidemic and virulence of *Puccinia striiformis* f. sp. *tritici* in China in 2002. *Plant Dis.* 88, 896-904.

Wang, J. K. , 2009: Inclusive composite interval mapping of quantitative trait genes. *Acta Agronomica Sinica.* 35, 239-245.

Wang, Z. L. , S. D. Liu, H. Wang, Z. H. He, X. C. Xia, X. M. Chen, and X. Y. Duan, 2006: Genetic analysis of the resistance of the wheat variety Bainong 64 to slow powdery mildew. *Acta Bot. Boreal Occident. Sin.* 26, 0332-0336.

Wellings, C. R. , 2011: Global status of stripe rust: a review of historical and current threats. *Euphytica* . 179, 129-141.

Wellings, C. R. , R. P. Singh, A. Yahyauoi, K. Nazari, and R. A. McIntosh, 2009: The development and application of near-isogenic lines for monitoring cereal rust pathogens. In: McIntosh RA (ed) Proc. Borlaug Global Rust Initiative Technical Workshop. BGRI, Cd Obregon, Mexico, pp. 77-87.

Xu, L. S. , M. N. Wang, P. Cheng, Z. S. Kang, S. H. Hulbert, and X. M. Chen, 2013: Molecular mapping of *Yr53*, a new gene for stripe rust resistance in durum wheat accession PI 480148 and its transfer to common wheat. *Theor. Appl. Genet.* 126, 522-533.

Yuan, C. L. , J. Hui, H. G. Wang, K. Li, H. Tang, X. B. Li, and D. L. Fu, 2012: Distribution, frequency and variation of stripe rust resistance loci *Yr10*, *Lr34/Yr18* and *Yr36* in Chinese wheat cultivars. *J. Genetics and Genomics.* 39, 587-592.

Zhang, L. Y. , M. Bernard, P. Leroy, C. Feuillet, and P. Sourdille, 2005: High transferability of bread wheat EST-derived SSRs to other cereals. *Theor. Appl. Genet.* 111, 677-687.

Identification of genomic regions controlling adult-plant stripe rust resistance in Chinese landrace Pingyuan 50 through bulked segregant analysis

C. X. Lan[1], S. S. Liang[1], X. C. Zhou[2], G. Zhou[2], Q. L. Lu[2], X. C. Xia[1], and Z. H. He[1,3]

[1] *Institute of Crop Science, National Wheat Improvement Center/The National Key Facility for Crop Gene Resources and Genetic Improvement, Chinese Academy of Agricultural Sciences (CAAS), Zhongguancun South Street* 12, *Beijing* 100081, *China*; *second author*; [2] *Gansu Wheat Research Institute, Gansu Academy of Agricultural Sciences, Lanzhou* 730070, *Gansu Province, China*;, *and third author*; [3] *International Maize and Wheat Improvement Center (CIMMYT), CIMMYT China Office, c/o CAAS, Zhongguancun South Street* 12, *Beijing* 100081, *China*

Abstract: Stripe rust, caused by *Puccinia striiformis* f. sp. *tritici*, is one of the most widespread and destructive wheat diseases worldwide. Growing resistant cultivars with adult-plant resistance (APR) is an effective approach for the control of the disease. In this study, 540 simple sequence repeat (SSR) markers were screened to map quantitative trait loci (QTLs) for APR to stripe rust in a doubled haploid (DH) population of 137 lines derived from the cross Pingyuan 50×Mingxian 169. The DH lines were planted in randomized complete blocks with three replicates in Gansu and Sichuan provinces during the 2005-06, 2006-07, and 2007-08 cropping seasons, providing data for four environments. Artificial inoculations were carried out in Gansu and Sichuan with the prevalent Chinese race CYR32. Broad-sense heritability of resistance to stripe rust for maximum disease severity (MDS) was 0. 91, based on the mean value averaged across four environments. Inclusive composite interval mapping (ICIM) detected three QTLs for APR to stripe rust on chromosomes 2BS, 5AL, and 6BS, designated *QYr. caas-2BS*, *QYr. caas-5AL*, and *QYr. caas-6BS*, respectively, separately explaining from 4. 5 to 19. 9% of the phenotypic variation. *QYr. caas-5AL*, different from QTLs previously reported, was flanked by microsatellite markers *Xwmc410* and *Xbarc261*, and accounted for 5. 0 to 19. 9% of phenotypic variance. Molecular markers closely linked to the QTLs could be used in marker-assisted selection for APR in wheat breeding programs.

Introduction

Stripe rust, caused by *Puccinia striiformis* f. sp. *tritici*, is a devastating wheat disease worldwide, particularly in the areas where wheat is grown under cool and temperate conditions or at high altitudes[30]. China is the largest stripe rust epidemic region in the world, and stripe rust is very destructive to autumn-sown wheat in northwestern and southwestern parts of the country, where weather conditions are conducive for disease development every year[28]. Resistant cultivars are the most efficient, economically viable, and environmentally friendly means of controlling the dis-

Published in Phytopathology, 2010, 100 (4): 313-318

ease[15].

Due to frequent changes in the pathogen population, cultivars with major resistance genes lose their resistance to stripe rust within a few years of cultivation[26]. As a consequence, adult-plant resistance (APR) with reputed durability is increasingly being used in wheat breeding and production[13,16,22,29]. APR is generally characterized by susceptibility at the seedling stage, but by resistance at the adult stage when inoculated with the same pathotypes.

To date, 47 stripe rust resistance genes at 42 loci have been catalogued, and about 33 have been assigned temporary designations[17]. Ten of them, viz. *Yr11*, *Yr12*, *Yr13*, *Yr14*, *Yr16*, *Yr18*, *Yr29*, *Yr30*, *Yr36*, and *Yr39* are APR genes. The first four are poorly defined and there is evidence for virulence on plants with *Yr11*[8] and likely virulence for resistances attributed to *Yr12*, *Yr13*, and *Yr14* from pathogenicity data published in reports by the National Institute of Agricultural Botany in the U. K. On the other hand APR genes *Yr16*, *Yr18*, *Yr29*, *Yr30*, *Yr36*, *Yr39* and others characterized only as QTLs are considered to be sources or components of non-specific durable resistances, although most individually do not confer adequate levels of protection and need to be deployed in combinations. Genetic studies on lines that have acceptable levels of APR usually indicate involvement of three or more such genes[3,16,22]. Therefore, it is very important to map APR genes and to identify closely linked molecular markers that are useful for combining APR genes in wheat breeding.

The Chinese wheat landrace Pingyuan 50 was an important cultivar in the Yellow and Huai Valleys of China, where it occupied about 270 000 to 400 000ha annually in 1950s. It is known to have maintained a moderate level of resistance to all races of *P. s.* f. sp. *tritici* in the field for over 60 years[31,32]. The aim of this study was to identify and map QTLs for APR to stripe rust in a doubled haploid (DH) population generated from the cross Pingyuan 50 × Mingxian 169. Knowledge about the genetic basis of stripe rust resistance in Pingyuan 50 will enable more efficient selection of its resistance genes in wheat breeding programs.

Materials and methods

Plant materials

The 137 DH lines used in this study were developed from a cross between Chinese landrace cultivars Pingyuan 50 and Mingxian 169 by the wheat × maize crossing technique[10]. Pingyuan 50 originated in northern Henan in 1950s, and then spread to all parts of Henan and western parts of Shandong province. Its area expanded rapidly following the stripe rust epidemic in 1950. Mingxian 169 is a landrace from Shanxi province, and is highly susceptible to all races of *P. s.* f. sp. *tritici* at all growth stages.

Field trials

The DH lines and their parents were evaluated for APR to stripe rust at Qingshui, Gansu province, and Chengdu, Sichuan province during the 2005-2006, 2006-2007, and 2007-2008 cropping seasons. These areas are considered to be very favorable for stripe rust infection and spread. Field trials were conducted in randomized complete blocks with three replicates. Plots consisted of single rows of 1.5m length and 25cm between rows. Approximately 50 seeds were sown in each row.

Seeds were sown in autumn in Qingshui, Gansu province. Inoculations were performed with race CYR32 at the three-leaf stage in spring. Mingxian169 was planted after every 10 rows as a susceptible check, and Huixianhong was planted around the experimental area to ensure ample inoculum in spring. Stripe rust severities were assessed for the first time 4 weeks after inoculation, and then at weekly intervals on two further occasions using the modified Cobb scale[18].

In the 2007-2008 cropping season, the DH lines and their parents were sown in winter in Chengdu, Sichuan province. Mingxian 169 was planted after every 10 rows as a susceptible check, and Chuanyu 12 and Sy95-71 were planted around the tested lines to ensure ample inoculum. Inoculation was also performed with race CYR32, the prevalent race in the region, at the three-leaf stage. Stripe rust severities were rated, when the disease severities on Mingxian 169 had reached a maximum level around April 20, 2008.

Statistical analysis

Maximum disease severity (MDS) was used for analysis of variance (ANOVA) and QTL detection. SAS software (SAS Institute, Cary, NC) was used to compute ANOVA. Broad-sense heritability (h^2) estimates were based on the formula: $h^2 = \sigma_g^2 / (\sigma_g^2 + \sigma_{ge}^2/e + \sigma_\varepsilon^2/re)$, where σ_g^2, σ_{ge}^2, and σ_ε^2 are estimates of genotype, genotype × environment interaction, and error variances, respectively, and e and r are the numbers of environments and replications per environment, respectively.

SSR analysis

Of 540 pairs of wheat SSR primers, 273 pairs were wheat microsatellite consortium (WMC) primers[6], 160 pairs were Beltsville Agriculture Research Center (BARC) primers[24], 101 pairs were Gatersleben Wheat Microsatellite (GWM) primers[20], 4 pairs were Clermont Ferrand A-genome (CFA) primers[25], and 2 pairs were Clermont Ferrand D-genome (CFD) primers[7].

Bulked segregant analysis

Based onmaximum disease severities, resistant and susceptible bulks were constructed by mixing equal amountsof DNA from the five most resistant and five most susceptible DH lines, respectively. SSRs showing polymorphisms between the two parents and between the two bulks were used to genotype all 137 DH lines to determine their associations with stripe rust responses. Additional SSRs around the loci associated with APR were selected from available linkage maps (http: //wheat. pw. usda. gov/GG2/index. shtml, 23) and those polymorphic were genotyped on the entire DH population.

Map construction and QTL detection

Mapmaker 3. 0b[14] software was used to establish linkage groups. The mapping function of Kosambi[9] was used to convert recombination values to genetic distances. The software IciMapping 2. 2 was used to detect QTLs by the inclusive composite interval mapping (ICIM) method[11]. A logarithm of odds (LOD) of 2. 0 was set to declare QTL as significant. The percentages of phenotypic variance explained (PVE) by individual QTL and additive effects at the LOD peaks were obtained through stepwise regression[11]. Each QTL was represented by a 20-centimorgan (cM) interval with the local LOD maximum at the central point. Adjacent QTLs on the same chromosome were considered different when the distances between the curve peaks were greater than 20cM or when the support intervals were non-overlapping.

Results

Phenotypic analysis of MDS, correlations and heritabilities

The MDS data from four environments were highly significantly correlated ($P < 0.0001$), with the correlation coefficients ranging from 0. 37 to 0. 71. The MDS frequency distribution indicated quantitative inheritance of APR to stripe rust (Fig. 1). The mean MDS of Pingyuan 50 and Mingxian 169 in Gansu were 5. 7 and 66. 7%, 0. 3 and 11. 3%, and 11. 0 and 31. 6% in 2006, 2007, and 2008, respectively, whereas the means of the resistant and susceptible parents were 6% and 90%, respectively, in Sichuan 2008. The average of MDS of the DH lines over four environments was 23. 3%, ranging from 0 to 100%. The broad-sense heritability of MDS was 0. 91 based on across-environment genotype means. ANOVA revealed significant variation among the DH lines (Table 1).

QTL analysis for APR to stripe rust

Based on the mean MDS in each environment and the means over four environments, three QTLs for APR to stripe rust were detected on chromosomes 2BS, 5AL, and

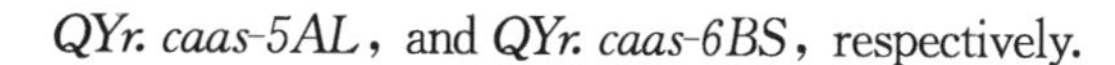

6BS (Table 2; Fig. 2), and designated *QYr. caas-2BS*, *QYr. caas-5AL*, and *QYr. caas-6BS*, respectively.

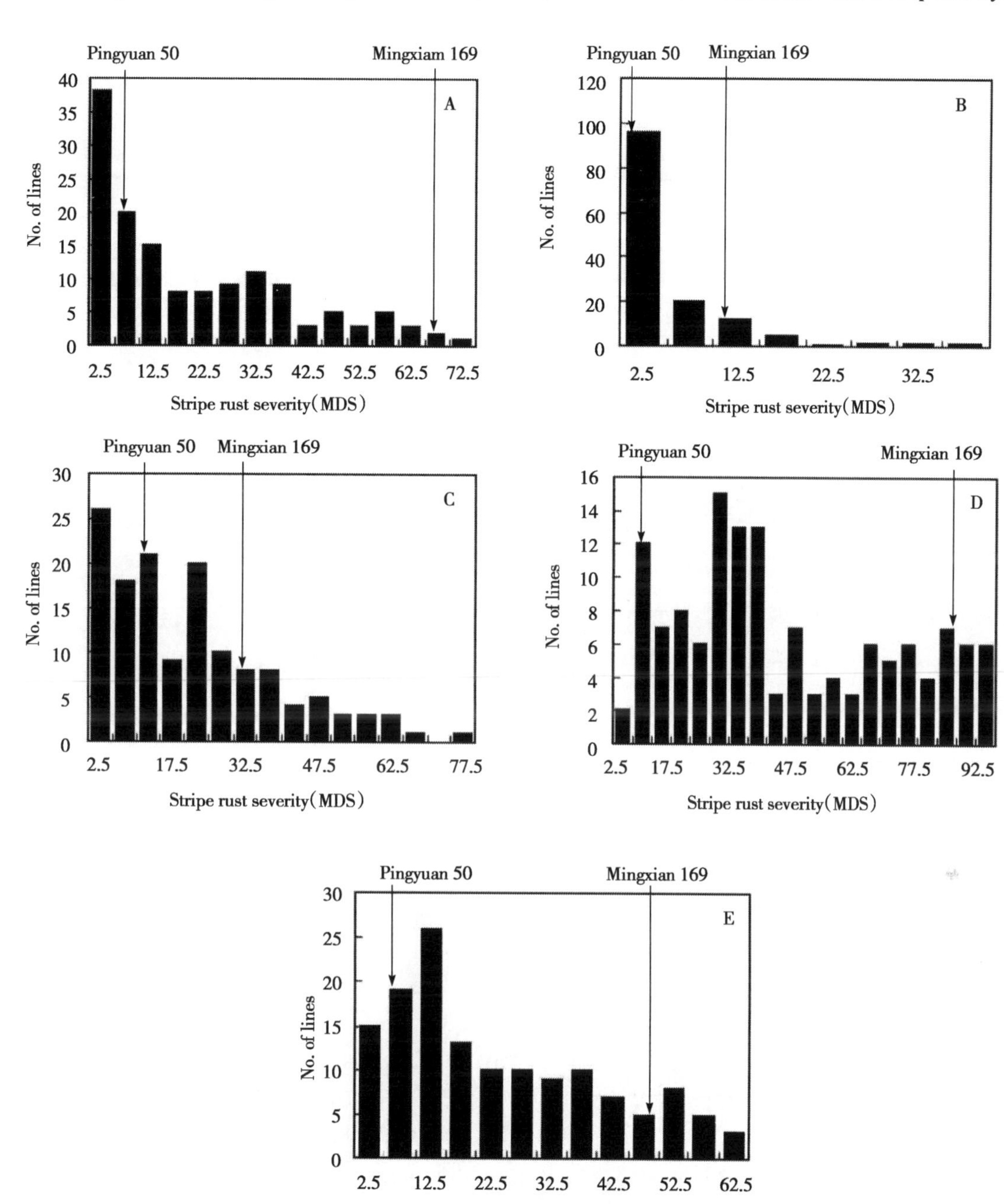

Fig. 1 A to E, Frequency distribution of stripe rust maximum disease severity (MDS) among doubled haploid (DH) lines derived from the cross Pingyuan 50×Mingxian 169. A, MDS, Gansu 2006; B, MDS, Gansu 2007; C, MDS, Gansu 2008; D, MDS, Sichuan 2008; and E, average MDS across all four environments. Mean values for the parents, Pingyuan 50 and Mingxian 169, are indicated by arrows.

Table 1 Analysis of variance of maximum disease severity (MDS) scores for the doubled haploid (DH) population derived from the cross Pingyuan 50×Mingxian 169

Source of variation	DF	Sum of squares	Mean square	*F* value
Replicate	2	1 882.06	941.03	11.70**
Environment	3	404 739.73	134 913.24	1 677.92**

（续）

Source of variation	DF	Sum of squares	Mean square	F value
Line	136	217 094.84	1 596.29	19.85**
Line×environment	408	231 426.17	571.42	7.11**
Error	1 069	86 194.29	80.41	

** $P<0.000\ 1$.

Table 2 Quantitative trait loci (QTLs) for maximum disease severity (MDS) to stripe rust by inclusive composition interval mapping (ICIM) in the doubled haploid (DH) population derived from the cross Pingyuan 50 × Mingxian 169

Location and year	QTL[a]	Marker interval	Position[b]	AE[c]	LOD[d]	PVE (%)[e]	Total PVE (%)
Gansu 2006	*QYr. caas-2BS*	*Xbarc13-Xbarc230*	14.2	−4.25	1.7	5.1	12.8
	QYr. caas-6BS	*Xgwm361-Xbarc136*	0.2	−5.22	2.5	7.7	
Gansu 2007	*QYr. caas-2BS*	*Xbarc13-Xbarc230*	3.0	−2.30	3.2	9.5	15.7
	QYr. caas-5AL	*Xwmc410-Xbarc261*	1.8	−1.86	2.0	6.2	
Gansu 2008	*QYr. caas-2BS*	*Xbarc13-Xbarc230*	2.0	−4.11	1.9	5.7	15.2
	QYr. caas-5AL	*Xwmc410-Xbarc261*	14.8	−3.86	1.7	5.0	
	QYr. caas-6BS	*Xgwm361-Xbarc136*	0.2	−3.66	1.6	4.5	
Sichuan 2008	*QYr. caas-5AL*	*Xwmc410-Xbarc261*	6.8	−12.48	5.6	19.9	19.9
Average MDS in four environments	*QYr. caas-5AL*	*Xwmc410-Xbarc261*	10.8	−5.93	4.8	16.5	24.1
	QYr. caas-6BS	*Xgwm361-Xbarc136*	0.2	−4.03	2.7	7.6	

a. Quantitative trait loci that overlap in the one-log support confidence intervals were assigned the same symbol.

b. Peak position in centimorgans from the first interval marker.

c. Additive effect of resistance allele.

d. Logarithm of odds (LOD) score.

e. PVE is the percentages of phenotypic variance explained by individual QTL.

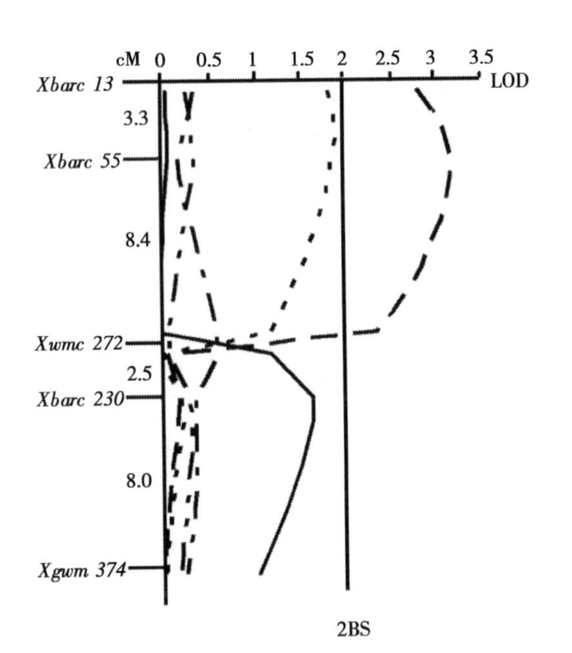

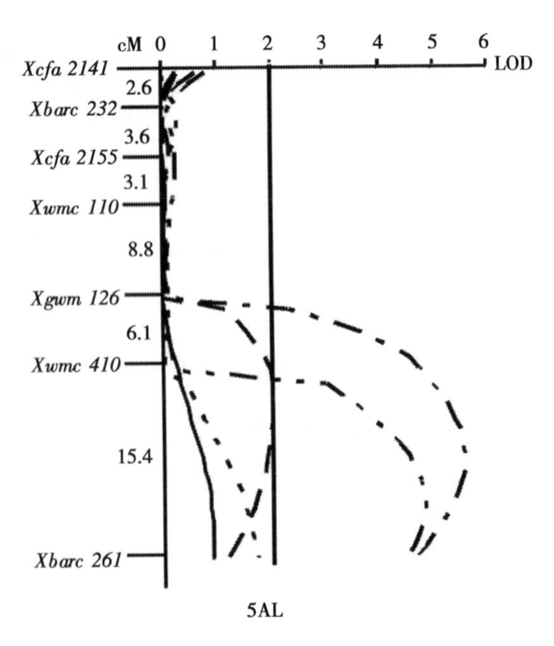

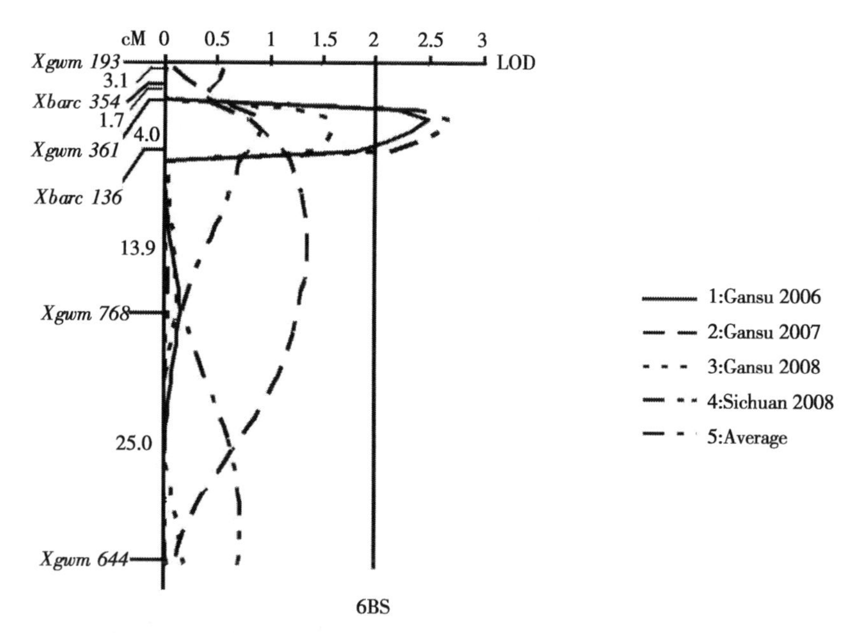

Fig. 2 Logarithm of odds (LOD) contours obtained by inclusive composite interval mapping (ICIM) for quantitative traits loci (QTLs) on chromosomes 2BS, 5AL, and 6BS that reduce stripe rust severity in Pingyuan 50 ×Mingxian 169 doubled haploid (DH) population. Gansu 2006, Gansu 2007, and Gansu 2008, maximum disease severity (MDS) in Gansu, 2006, 2007, and 2008, respectively. Sichuan 2008, maximum disease severity (MDS) in Sichuan 2008. Average, mean maximum disease severity (MDS) over four environments. The logarithm of odds (LOD) thresholds of 2.0 are indicated by a line.

The QTL *QYr. caas-2BS*, detected on chromosome 2BS in the SSR marker interval *Xbarc13* - *Xbarc230*, was identified in Gansu 2006, 2007, and 2008, accounting for 5.1%, 9.5%, and 5.7% of the phenotypic variance with additive effects of −4.25, −2.30, and −4.11, respectively. The most stable QTL, *QYr. caas-5AL*, was located between *Xwmc410* and *Xbarc261* on the long arm of chromosome 5A, explaining 6.2%, 5.0%, 19.9%, and 16.5% of the phenotypic variance in Gansu 2007, Gansu 2008, Sichuan 2008, and the mean MDS, respectively, with additive effects ranging from −12.48 to −1.86. *QYr. caas-6BS*, on the short arm of chromosome 6B, was flanked by *Xgwm361* and *Xbarc136*, and explained from 4.5 to 7.7% of the phenotypic variance in two environments as well as the mean data over all four environments. All QTLs were from Pingyuan 50.

Discussion

Chinese wheat landrace Pingyuan 50 was an excellent cultivar with durable resistance to stripe rust in addition to its relative high yield potential, strong stem, and broad adaptation[32]. Yuan et al. [31] reported that Pingyuan 50 was resistant to eight Chinese races of *P. s.* f. sp. *tritici* at the adult-plant stage, and demonstrated quantitative inheritance of APR to stripe rust, which is in agreement with the present study where three QTLs effective in two or more environments were detected.

Ramburan et al. (19) identified a QTL for stripe rust response flanked by markers *Xpsp3030* and *s16m40A* on chromosome 2BS in a DH population of Kariega× Avocet. Guo et al. [5] found a QTL at a similar position, located in the marker interval *Xgwm148*-*Xbarc167* in an $F_{2:3}$ population of Luke × Aquileja. Based on the wheat consensus map of Somers et al. [23], these two QTLs and *QYr. caas-2BS* may involve the same or closely linked loci. *QYr. caas-2BS* was identified in Gansu from 2006 to 2008. It showed a relatively stable genetic effect against stripe rust across different years. The SSR marker *Xbarc55* closely linked to this QTL could be used for improving wheat stripe rust resistance in breeding programs.

Boukhatem et al.[2] detected a QTL at the marker interval *Xfbb209-5A* - *Xabg391-5A* in the ITMI recombinant line population derived from Opata 85×Synthetic. Based on the wheat consensus map (http://wheat.pw.usda.gov/GG2/index.shtml, 23) this QTL is about 26.0cM distal to *QYr. caas-5AL* and therefore should be different. Chhuneja et al.[3] mapped a QTL on chromosome 5AL in the diploid A genome population *T. monococcum* (accession Paul 4087) × *T. boeoticum* (Pau5088), and this QTL was flanked by markers *Xbarc151* and *Xcfd12*, that are an estimated 68.7cM from *QYr. caas-5AL*[23]. One race-specific resistance gene has been mapped on chromosome 5AL, i.e. *Yr34* in the line WAWHT 2046[1]. This line is highly susceptible to race CYR32 at the seedling stage[12], and its position is about 34cM from *QYr. caas-5AL* based on the consensus map[23]. Therefore, *QYr. caas-5AL* appears to be a new gene conferring APR to stripe rust. It was detected with high LOD scores in three environments and in the mean MDS analysis. Selection for the SSR markers *Xwmc410* and *Xbarc261* closely linked to *QYr. caas-5AL* could be undertaken to improve APR to stripe rust in wheat.

Santra et al.[21] found a QTL for high temperature adult-plant resistance in cultivar Stephens. It was flanked by SSR markers *Xbarc101* and *Xbarc136* and located at a similar position to *QYr. caas-6BS*. Suenaga et al.[26] identified a QTL in cultivar Oligoculm, closely linked to microsatellite marker *Xgwm935* and with an estimated map distance of about 2cM from *QYr. caas-6BS* based on the consensus map[23]. Another QTL possibly at this locus was found by Christiansen et al.[4] in cultivars Solist and Kris. It was flanked by SSR markers *Xgwm518* and *Xwmc105b* and was estimated to be about 2.5cM away from *QYr. caas-6BS* (http://wheat.pw.usda.gov/GG2/index.shtml). Thus, these QTLs are at the same or closely linked loci. Uauy et al.[27] identified APR gene *Yr36* on chromosome 6BS in 'DIC' lines from *Triticum turgidum* ssp. *dicoccoides*. *Yr36* was completely linked to *Xbarc101*, and within 2cM interval defined by PCR-based markers *Xucw71* and *Xbarc136*. However, *QYr. caas-6BS* came from *Triticum aestivum*, and thus, *QYr. caas-6BS* and *Yr36* should be different stripe rust resistance genes. Although *QYr. caas-6BS* was identified only with relatively low LOD scores (1.6-2.7) and small effects, it was relatively stable across two environments and the averaged MDS. Reports of QTLs located at the same position in previous studies add credence to this conclusion.

Compared with themajor specific resistance genes, APR genes have relatively small effects on stripe rust response and it is necessary to incorporate three or more such genes in a single genotype to obtain acceptably high levels of resistance[22]. Molecular markers are obviously extremely useful for developing or maintaining such gene combinations in segregating breeding populations. Moreover, molecular markers are essential to detect such APR genes when major seedling resistance genes are present[27]. We previously identified five QTLs for stripe rust resistance in Italian wheat cultivars Libellula and Strampelli[16], and have found three additional QTLs in the present study. In order to validate the potential use of these genes we have made many crosses between these resistant parents and Chinese commercial cultivars with good agronomical traits. Molecular markers identified in this and earlier research will facilitate the selection of combined APR gene combinations among the segregating progeny.

Acknowledgments

The authors are very grateful to Prof. R. A. McIntosh, Plant Breeding Institute, University of Sydney, for his critical review of this manuscript. This study was supported by the National Science Foundation of China (30810214 and 30671294) and International Collaboration Project from the Ministry of Agriculture (2006-G2).

Reference

[1] Bariana, H. S., Parry, N., Barclay, I. R., Lough-

man, R., McLean, R. J., Shankar, M., Wilson, R. E., Willey, N. J., and Francki, M. 2006. Identification and characterization of stripe rust resistance gene *Yr34* in common wheat. Theor. Appl. Genet. 112: 1143-1148.

[2] Boukhatem, N., Baret, P. V., Mingeot, D., and Jacquemin, J. M. 2002. Quantitative trait loci for resistance against yellow rust in two wheat-derived recombinant inbred line populations. Theor. Appl. Genet. 104: 111-118.

[3] Chhuneja, P., Kaur, S., Garg, T., Ghai, M., Kaur, S., Prashar, M., Bains, N. S., Goel, R. K., Keller, B., Dhaliwal, H. S., and Singh, K. 2008. Mapping of adult plant stripe rust resistance genes in diploid A genome wheat species and their transfer to bread wheat. Theor. Appl. Genet. 116: 313-324.

[4] Christiansen, M. J., Feenstra, B., Skovgaard, I. M., and Andersen, S. B. 2006. Genetic analysis of resistance to yellow rust in hexaploid wheat using a mixture model for multiple crosses. Theor. Appl. Genet. 112: 581-591.

[5] Guo, Q., Zhang, Z. J., Xu, Y. B., Li, G. H., Feng, J., and Zhou, Y. 2008. Quantitative trait loci for high-temperature adult-plant and slow-rusting resistance to *Puccinia striiformis* f. sp. *tritici* in wheat cultivars. Phytopathology 98: 803-809.

[6] Gupta, P. K., Rustgi, S., Sharma, S., Singh, R., Kumar, N., and Balyan, H. S. 2003. Transferable EST-SSR markers for the study of polymorphism and genetic diversity in bread wheat. Mol. Genet. Genet. 270: 315-323.

[7] Guyomarc'h, H., Sourdille, P., Edwards, K. J., and Bernard, M. 2002. Studies of the transferability of microsatellites derived from *Triticum tauschii* to hexaploid wheat and to diploid related species using amplification, hybridization and sequence comparisons. Theor. Appl. Genet. 105: 736-744.

[8] Johnson, R., and Taylor, A. J. 1972. Isolates of *Puccinia striiformis* collected in England from wheat varieties Maris Beacon and Joss Cambier. Nature 238: 105-106.

[9] Kosambi, D. D. 1944. The estimation of map distance from recombination values. Annu. Eugen. 12: 172-175.

[10] Laurie, D. A., and Bennett, M. D. 1988. The production of haploid wheat plants from wheat × maize crosses. Theor. Appl. Genet. 76: 393-397.

[11] Li, H. H., Ye, G. Y., and Wang, J. K. 2007. A modified algorithm for the improvement of composite interval mapping. Genetics 175: 361-374.

[12] Li, Z. F., Xia, X. C., Zhou, C. X., Niu, Y. C., He, Z. H., Zhang, Y., Li, G. Q., Wan, A. M., Wang, D. S., Chen, X. M., Lu, Q. L., and Singh, R. P. 2006. Seedling and slow rusting resistance to stripe rust in Chinese common wheats. Plant Dis. 90: 1302-1312.

[13] Lin, F., and Chen, X. M. 2008. Quantitative trait loci for non-race-specific, high-temperature adult-plant resistance to stripe rust in wheat cultivar Express. Theor. Appl. Genet. 116: 797-806.

[14] Lincoln, S., Daly, M., and Lander, E. 1992. Constructing genetic maps with Mapmaker/EXP3.0. Whitehead Institute Tech. Rep, 3rd ed. Whitehead Institute, Cambridge.

[15] Line, R. F., and Chen, X. M. 1995. Success in breeding for and managing durable resistance to wheat rusts. Plant Dis. 79: 1254-1255.

[16] Lu, Y. M., Lan, C. X., Liang, S. S., Zhou, X. C., Liu, D., Zhou, G., Lu, L. Q., Jing, J. X., Wang, M. N., Xia, X. C., and He, Z. H. 2009. QTL mapping for adult-plant resistance to stripe rust in Italian common wheat cultivars Libellula and Strampelli. Theor. Appl. Genet. DOI: 10.1007/s00122-009-1139-6.

[17] McIntosh, R. A., Dubcovsky, J., Rogers, W. J., Morris, C. F., Appels, R., and Xia, C. X. 2009. Catalogue of gene symbols for wheat: 2009 (suppl) http://www.wheat.pw.usda.gov/GG2/pubs.shtml.

[18] Peterson, R. F., Campbell, A. B., and Hannah, A. E. 1948. A diagrammatic scale for estimating rust intensity of leaves and stems of cereals. Can. J. Res. Sect. C. 26: 496—500.

[19] Ramburan, V. P., Pretorius, Z. A., Louw, J. H., Boyd, L. A., Smith, P. H., Boshoff, W. H. P., and Prins, R. 2004. A genetic analysis of adult plant resistance to stripe rust in the wheat cultivar Kariega. Theor. Appl. Genet. 108: 1426-1433.

[20] Röder, M. S., Korzun, V., Wendehake, K., Plaschke, J., Tixier, M. H., Leroy, P., and Ganal, M. W. 1998. A microsatellite map of wheat. Genetics 149: 2007-2023.

[21] Santra, D. K., Chen, X. M., Santra, M., Campbell, K. G., and Kidwell, K. K. 2008. Identification and mapping QTL for high-temperature adult-plant re-

sistance to stripe rust in winter wheat (*Triticum aestivum* L.) cultivar Stephens. Theor. Appl. Genet. 117: 793-802.

[22] Singh, R. P., Huerta-Espino, J., and Rajaram, S. 2000. Achieving near-immunity to leaf and stripe rusts in wheat by combining slow rusting resistance genes. Acta. Phyto. Entom. Hung. 35: 133-139.

[23] Somers, D. J., Isaac, P., and Edwards, K. 2004. A high-density microsatellite consensus map for bread wheat (*Triticum aestivum* L.). Theor. Appl. Genet. 109: 1105-1114.

[24] Song, Q. J., Fickus, E. W., and Cregan, P. B. 2002. Characterization of trinucleotide SSR motifs in wheat. Theor. Appl. Genet. 104: 286-293.

[25] Sourdille, P., Singh, S., Cadalen, T., Brown-Guedira, G. L., Gay, G., Qi, L., Gill, B. S., Dufour, P., Murigneux, A., and Bernard, M. 2004. Microsatellite-based deletion bin system for the establishment of genetic-physical map relationships in wheat (*Triticum aestivum* L.). Funct. Int. Gen. 4: 12-25.

[26] Suenaga, K., Singh, R. P., Huerta-Espino, J., and William, H. M. 2003. Microsatellite markers for genes *Lr34/Yr18* and other quantitative trait loci for leaf rust and stripe rust resistance in bread wheat. Phytopathology 93: 881-890.

[27] Uauy, C., Brevis, J. C., Chen, X. M., Khan, I., Jackson, L., Chicaiza, O., Distelfeld, A., Fahima, T., and Dubcovsky, J. 2005. High-temperature adult-plant (HTAP) stripe rust resistance gene *Yr36* from *Triticum turgidum* ssp. *dicoccoides* is closely linked to the grain protein content locus *Gpc-B1*. Theor. Appl. Genet. 112: 97-105.

[28] Wan, A. M., Zhao, Z. H., Chen, X. M., He, Z. H., Jin, S. L., Jia, Q. Z., Yao, G., Yang, J. X., Wang, B. T., Li, G. B., Bi, Y. Q., and Yuan, Z. Y. 2004. Wheat stripe rust epidemic and virulence of *Puccinia striiformis* f. sp. *tritici* in China in 2002. Plant Dis. 88: 896-904.

[29] William, H. M., Singh, R. P., Huerta-Espino, J., Palacios, G., and Suenaga, K. 2006. Characterization of genetic loci conferring adult plant resistance to leaf rust and stripe rust in spring wheat. Genome 49: 977-990.

[30] Yahyaoui, A. H., Hovmøler, M. S., Ezzahiri, B., Jahoor, A., Maatougui, M. H., and Wolday, A. 2004. Survey of barley and wheat diseases in the central highlands of Eritrea. Phytopatholgy 43: 39-43.

[31] Yuan, W. H., Zhang, Z. J., and Zeng, S. M. 1994. Inheritance of durable resistance to stripe rust in wheat variety Pingyuan 50. Acta Phyto. Sin. 24 (1): 39-42. (In Chinese with English abstract)

[32] Zeng, S. M., Wang, F. Y., Wu, X. Y., Zhang, W. Q., Wang, J. Q., Song, W. Z., and Wang, S. Y. 1979. Analysis of horizontal resistance against stripe rust in wheat. Plant Protect 6 (1): 1-10. (In Chinese)

Identification of QTL for adult-plant resistance to powdery mildew in Chinese wheat landrace Pingyuan 50

M. A. Asad[1], B. Bai[2], C. X. Lan[1], J. Yan[3], X. C. Xia[1], Y. Zhang[1], and Z. H. He[1,4]

[1] *Institute of Crop Science, National Wheat Improvement Center/The National Key Facility for Crop Gene Resources and Genetic Improvement, Chinese Academy of Agricultural Sciences (CAAS), Beijing* 100081, *China*; [2] *Wheat Research Institute, Gansu Academy of Agricultural Sciences, Lanzhou* 730070, *China*; [3] *Cotton Research Institute, Chinese Academy of Agricultural Sciences (CAAS), Anyang* 455000, *China*; [4] *International Maize and Wheat Improvement Center (CIMMYT) China Office, c/o CAAS, Beijing* 100081, *China*

Abstract: Powdery mildew caused by *Blumeria graminis* f. sp. *tritici* is one of major wheat diseases worldwide. The Chinese wheat landrace Pingyuan 50 has expressed adult-plant resistance (APR) to powdery mildew in field over 60 years. To dissect the genetic basis of APR to powdery mildew in this cultivar, a mapping population of 137 double haploid (DH) lines derived from Pingyuan 50/Mingxian 169 was evaluated in replicated field trials for two years in Beijing (2009-2010 and 2010-2011) and one year in Anyang (2009-2010). A total of 540 polymorphic SSR markers were used to genotype the entire population for constructing linkage map and QTL analysis. Three QTL were mapped on chromosome 2BS (*QPm. caas-2BS. 2*), 3BS (*QPm. caas-3BS*) and 5AL (*QPm. caas-5AL*) with resistance alleles contributed by Pingyuan 50, explaining 5. 3, 10. 2 and 9. 1% of the phenotypic variance, respectively, and one QTL on chromosome 3BL (*QPm. caas-3BL*) derived from Mingxian 169 accounting for 18. 1% of the phenotypic variance. The ***QPm. caas-3BS***, ***QPm. caas-3BL*** and ***QPm. caas-5AL*** appear to be new powdery mildew APR loci. *QPm. caas-2BS. 2* and *QPm. caas-5AL* are possibly the pleiotropic or closely linked resistance loci to both stripe rust and powdery mildew. The Chinese landrace Pingyuan 50 could be a potential genetic resource to facilitate wheat breeding for improving APR to powdery mildew and stripe rust.

Key words: *Triticum aestivum* L., *Blumeria graminis* f. sp. *Tritici*, Disease resistance, Quantitative trait loci

Introduction

Powdery mildew, caused by *Blumeria graminis* f. sp. *tritici* (*Bgt*), is an important disease of wheat worldwide[1], resulting in significant reductions in both grain quality and yield in susceptible wheat cultivars[2,3], and substantial economic losses to wheat production annually on global scale[4]. Using powdery mildew resistance genes in elite cultivars is the most cost-effective and sustainable strategy to control this disease[5].

For the last three decades, most molecular disease resistance studies focused onmajor genes, known as qualitative or race specific resistance. These genes are genetically simple and easy to manipulate in breeding pro-

Published in The Crop Journal, 2014, 2: 308-314

grams, express complete resistance and follow the gene-for-gene interaction, and often associated with hypersensitive response that limit pathogen growth[6]. Race specific resistance is often transient due to evolution of new races of pathogen or increase in frequencies of previously rare variants[7], resulting in outbreaks of large epidemics, which 'burst' the once 'booming' cultivar like "E' an 1" in China[8]. Up to now, more than 70 powdery mildew resistance genes have been identified in wheat[9]. Most resistance genes to powdery mildew were ineffective in China due to their race specific nature and spread of new pathogen races.

One of the principal challenges in wheat breeding is to develop cultivars with durable disease resistance. Adult-plant resistance (APR) being race non-specific offers durable resistance based on the additive effects of several genes that delay infection, and reduces growth and reproduction of the pathogen at adult-plant stage[1]. Race non-specific or APR gene confers inadequate resistance for powdery mildew compared to race specific resistance genes. However, combining 4-5 APR genes into one genotype will achieve near-immunity or a high level of resistance in the field[10]. Advent of molecular markers made it easier to locate APR genes on chromosomes, and estimate the additive effect of each gene as compared with conventional techniques[11]. So far, more than 100 quantitative trait loci (QTL) for powdery mildew resistance have been identified and mapped on almost all wheat chromosomes in different genetic backgrounds (Z. F. Li, per. comm.), including the Swiss winter wheat cv. Forno[12], the French winter wheat lines RE714, Festin, Courtot, and RE9001[13-16], the North American winter wheats Massey and USG3209[10,17], the Japanese wheat cultivar Fukuho-komugi[18], the Israeli wheat cultivar Oligoculm[18], the CIMMYT wheat lines Opata 85, W7984, and Saar[19,20], the Australian wheat cultivar Avocet[20], and the Chinese wheat cultivars Bainong 64[21] and Lumai 21[11]. Unfortunately, only a few genotypes have good adaptability and associated agronomic traits in Chinese environments[22]. Wheat landraces are valuable genetic resources, and they carry multiple resistance genes for several diseases and are more adaptable to local environments[5]. It would, therefore, be more important to explore APR genes to powdery mildew in wheat landraces. Moreover, closely linked molecular markers to these genes will play an important role in management of APR genes in wheat breeding programs.

The Chinese wheat landrace Pingyuan 50 was one of leading cultivars in the Yellow and Huai Valley Autumn-sown Wheat Zone of China in the 1950s, and it has shown APR to stripe rust and powdery mildew in field for over 60 years. Previously, we mapped the QTL for APR to stripe rust in Pingyuan 50[22]. The main objectives of the present study were to detect the genetic location of powdery mildew resistance QTL in Pingyan 50 and determine whether they are pleiotropic or closely linked APR loci to both powdery mildew and stripe rust based on the detected chromosome regions.

Materials and methods

Plant materials

A doubled haploid (DH) population of 137 lines from Pingyuan 50/Mingxian 169 was used for QTL analysis. Pingyuan 50 showed APR to powdery mildew at adult-plant stage in the field trials. Mingxian 169, a landrace from Shanxi province is highly susceptible to all races of *P. striiformis* f. sp. *tritici* at all growth stages[22], whereas it is moderately resistant to powdery mildew at adult plant stage. The two parents were susceptible to *Bgt* isolates E20 at seedling stage. Jingshuang 16 was highly susceptible to powdery mildew, and used as a susceptible check cultivar.

Powdery mildew evaluation

The DH population was evaluated for powdery mildew responses over two years (2009-2010 and 2010-2011 wheat seasons) at two locations in the CAAS Experimental Station, Beijing, and CAAS Cotton Research Institute, Anyang, Henan province (herein referred to as Beijing 2010, Beijing 2011, and Anyang 2010). The hill plot fashion (50 seeds /hill) was used to plant the seeds in randomized complete blocks with three replicates. The highly susceptible cv. Jingshuang 16 was planted every tenth row as a check and around the experimental block as spreader. In Beijing, artificial inoculation with *Bgt* isolate E20 was performed before stem elongation. Disease severities were assessed as percentage of the infected penultimate leaves in five weeks after inoculation[23], and then scoring was repeated one week later when disease was at maximum level around May 20. In Anyang under natural infection, powdery mildew severity was rated once when cv. Jingshuang 16 expressed its maximum level of severity during the third week of May. Disease development was very low in Anyang in 2011 due to dry weather conditions. Therefore, disease scoring data in Anyang 2011 was not included in experimental analysis.

Statistical analysis

The frequency distribution of powdery mildew responses andcorrelation coefficients (r) based on maximum disease severity (MDS) in different environments were calculated in Microsoft Excel 2007. The area under the disease progress curve (AUDPC) was calculated according to Bjarko and Line[24]. Analysis of variance (ANOVA) was performed using the PROC GLM in the statistical analysis system (SAS Institute 1997). ANOVA information was then used to calculate broad-sense heritability (h^2) as: $h^2 = \sigma_g^2 / (\sigma_g^2 + \sigma_{ge}^2/e + \sigma_\varepsilon^2/re)$, where σ_g^2, σ_{ge}^2 and σ_ε^2 are estimates of genotypic, genotype × environment interaction and error variances, respectively, and e and r are the numbers of environments and replicates per environment, respectively.

SSR analysis

The 1528 pairs of simple sequence repeat (SSR) primers were used to scan the parents from published sources including WMC[25], BARC[26], GWM[27], CFA[28] and CFD[29] series (http: //wheat. pw. usda. gov). Bulked segregant analysis[30] was conducted, using equal amounts of ten resistant and ten susceptible lines based on MDS. Amplification of DNA, electrophoresis of PCR products on polyacrylamide gel and gel staining procedures were performed as described by Bryan et al.[31] and Bassam et al.[32]. The 540 polymorphic SSR markers were used to genotype the entire population for constructing linkage map and QTL analysis.

Map construction and QTL detection

Genetic linkage groups were constructed with the software Map Manager QTXb20[33], and map distances between markers were estimated by Kosambi mapping function[34]. Linkage groups were assigned to each chromosome according to published wheat consensus maps[35]. QTL analysis was performed with QTL Cartographer 2. 5 software by composite interval mapping method[36]. A logarithm of odds (LOD) were calculated from 2000 permutations for each trait to declare significant QTL at the $P = 0.01$ level. Estimates of phenotypic variance (R^2) explained by individual QTL and additive effects at LOD peaks were obtained by QTL Cartographer 2. 5. Two QTL on the same chromosome in different environments, having curve peak within the distance of 20cM, were considered as one common QTL, and different QTL when distances were over 20cM between the curve peaks.

Results

Phenotypic analysis

The MDS of the susceptible check Jingshuang 16 ranged from 80 to 100%, 60 to 90%, and 90 to 100%, whereas Pingyuan 50 and Mingxian 169 were 8. 5 and 7. 1%, 7. 7 and 6. 0%, and 12. 3 and 14. 5% in Anyang 2010, Beijing 2010, and Beijing 2011, respectively. Both parents showed mean MDS of powdery mildew less than 10% across three environments. Frequency distributions of MDS and AUDPC for DH lines showed continuous variation in all environments with clear transgressive segregation, indicating a quantitative nature of resistance to powdery mildew (Fig. 1). In addition, the MDS scores were significantly correlated across three environments ($r = 0.63$ to 0. 85). Analyses of variance of MDS and AUDPC showed significant variation among the DH lines (Table 1). The broad-sense heritabilities of MDS and AUDPC were 0. 80 and 0. 62, respectively, across three environments.

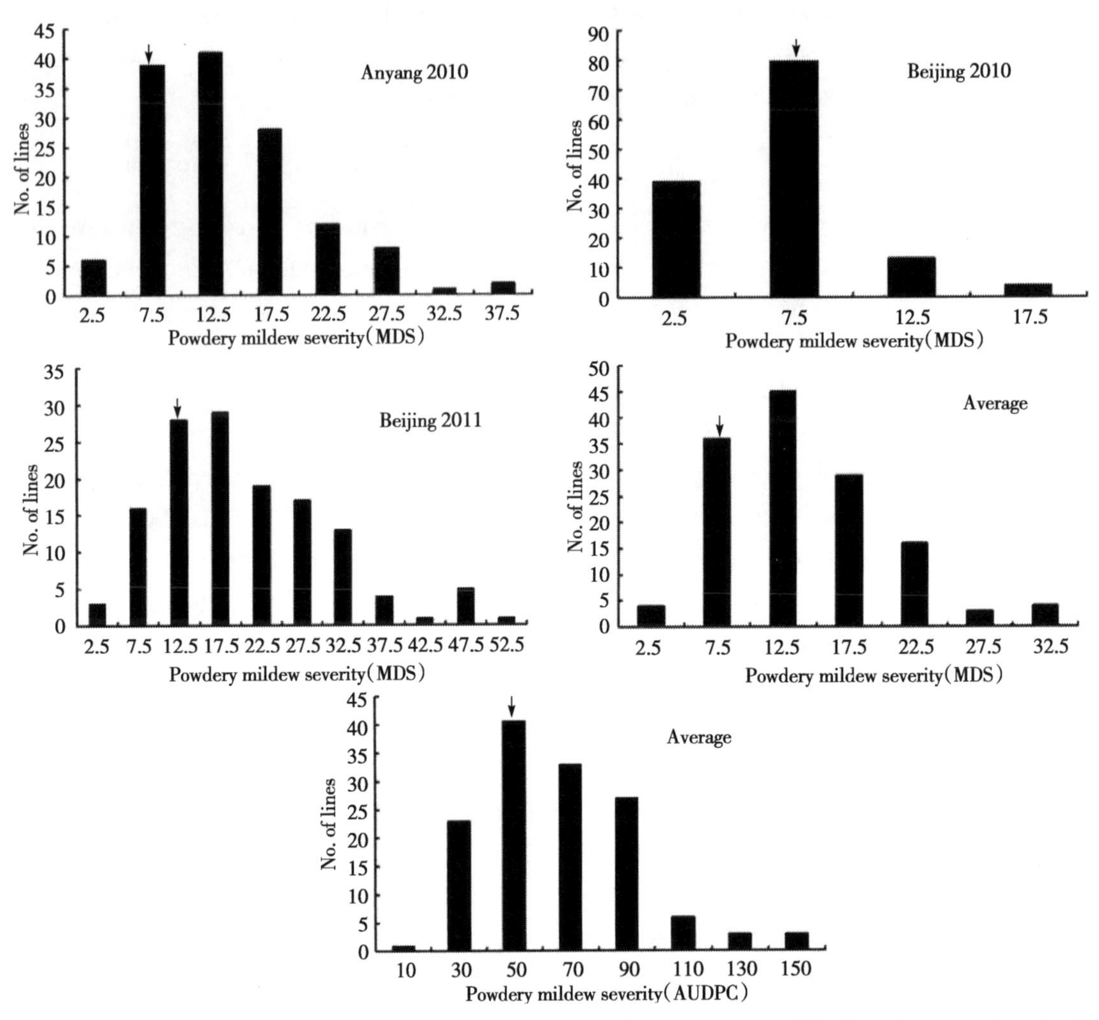

Fig. 1 Frequency distribution of powdery mildew maximum disease severity (MDS) and area under the disease progress curve (AUDPC) values in DH lines derived from Pingyuan 50/Mingxian 169. Average, mean MDS across three environments and AUDPC in Beijing over 2 years, respectively. Mean values for parents, Pingyuan 50 and Mingxian 169 are indicated by arrows.

Table 1 Analysis of variance of maximum disease severities (MDS) values for penultimate leaves and area under the disease progress curve (AUDPC) values for powdery mildew responses on doubled haploid (DH) lines from the Pingyuan 50/Mingxian 169 cross

Parameter	Source of variation	*df*	Sum of squares	Mean of squares	*F* values
MDS	Replicate	2	4 331.95	2 165.97	28.32**
	Environment	2	37 042.52	18 521.26	242.17**
	Line	136	48 714.61	358.19	4.68**
	Line×Environment	270	18 749.15	69.44	0.91
	Error	776	59 349.21	76.48	
AUDPC	Replicate	2	18 917.95	9 458.97	10.33**
	Environment	1	581 299.41	581 299.41	634.96**
	Line	136	653 914.62	4 808.19	5.25**
	Line×Environment	136	245 811.08	1 807.43	1.97**
	Error	545	498 938.04	915.48	

** Significant at $P < 0.0001$.

QTL analysis

Based on MDS, three QTL from Pingyuan 50 on chromosomes 2BS, 3BS, and 5AL, and one QTL on chromosome 3BL from Mingxian 169, respectively, were detected across environments (Table 2 and Fig. 2). They were designated as *QPm caas-2BS. 2*, *QPm caas-3BS*, *QPm caas-3BL* and *QPm caas-5AL*, respectively.

QTL on the short arm of chromosome 2B was detected in Beijing 2010, Beijing 2011, and averaged MDS across three environments and located in the marker interval of *Xbarcl3-Xgwm374*. It explained 4.0-9.1% of phenotypic variance across environments (Table 2).

Table 2 Quantitative trait loci (QTL) detected for adult-plant resistance (APR) to powdery mildew in the doubled haploid (DH) lines from the Pingyuan 50/Mingxian 169 cross

Environment	QTL[a]	Interval	LOD[b]	AE[c]	R^2 (%)[d]	Total R^2 (%)
Anyang 2010	*QPm caas-3BS*	*Xwmc366-Xgwm77*	3.2	−2.17	9.1	27.2
	QPm caas-3BL	*Xwmc527-Xwmc418*	4.4	2.83	18.1	
Beijing 2010	*QQPm caas-5AL*	*Xwmc410-Xbarc261*	2.5	−1.04	10.2	16.7
	QPm caas-2BS	*Xwmc272-Xgwm374*	2.0	−0.82	6.5	
Beijing 2011	*QPm caas-2BS. 2*	*Xbarc55-Xwmc272*	1.7	−2.64	5.3	9.3
Averaged MDS	*QPm caas-2BS*	*Xbarcl3-Xbarc55*	1.3	−2.24	4.0	
	QPm caas-2BS. 2	*Xbarc55-Xwmc272*	2.4	−1.57	8.6	17.7
	QPm caas-2BS	*Xbarcl3-Xbarc55*	3.0	−1.64	9.1	

a. QTL that extend across single one-log support confidence intervals were assigned the same symbol.

b. Logarithm of odds (LOD) score.

c. Additive effects.

d. R^2 is the proportion of phenotypic variance explained by the QTL.

The second QTL *QPm caas-3BS* was mapped on the short arm of chromosome 3B, flanked by SSR markers *Xwmc366* and *Xgwm77*, and accounted for 9.1% of the phenotypic variance with −2.17 of additive effect. The third QTL, *QPm caas-3BL*, was identified close to the centromere on the long arm of chromosome 3B linked to markers *Xwmc527* and *Xwmc418* with a LOD value of 4.4. This QTL explained 18.1% of phenotypic variation with 2.83 of additive effect, and only identified in Anyang 2010.

The fourth QTL, *QPm caas-5AL*, was detected on chromosome 5AL in the marker interval of *Xwmc410-Xbarc261*, and explained 10.2% of phenotypic variance with −1.04 of additive effect. The total phenotypic variance explained by the detected QTL for MDS ranged from 9.3 to 27.2% in single environments and was 17.7% for the mean across environments. Pingyuan 50 carries three QTL, while Mingxian 169 carries one (*QPm caas-3BL*).

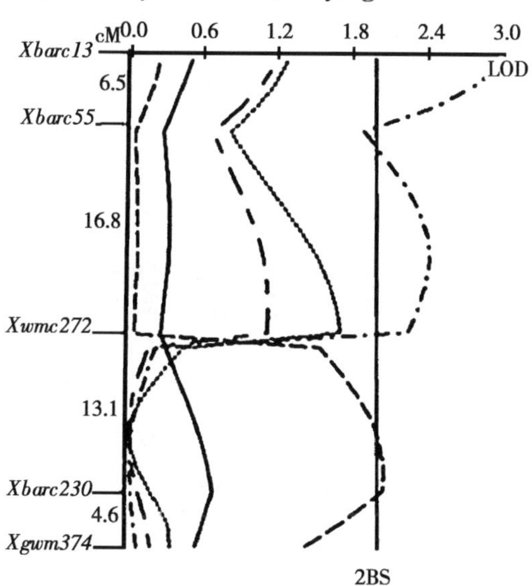

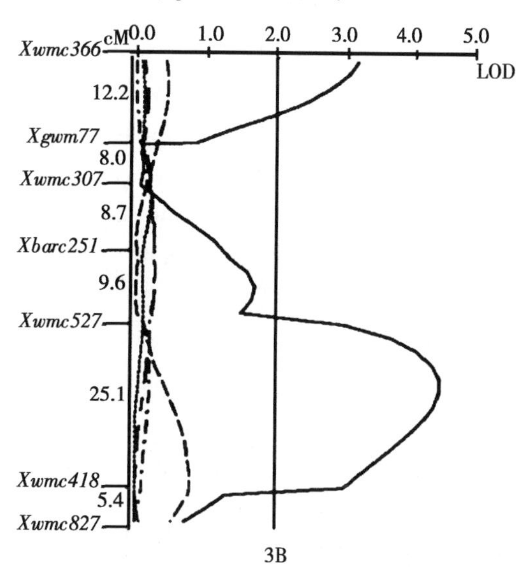

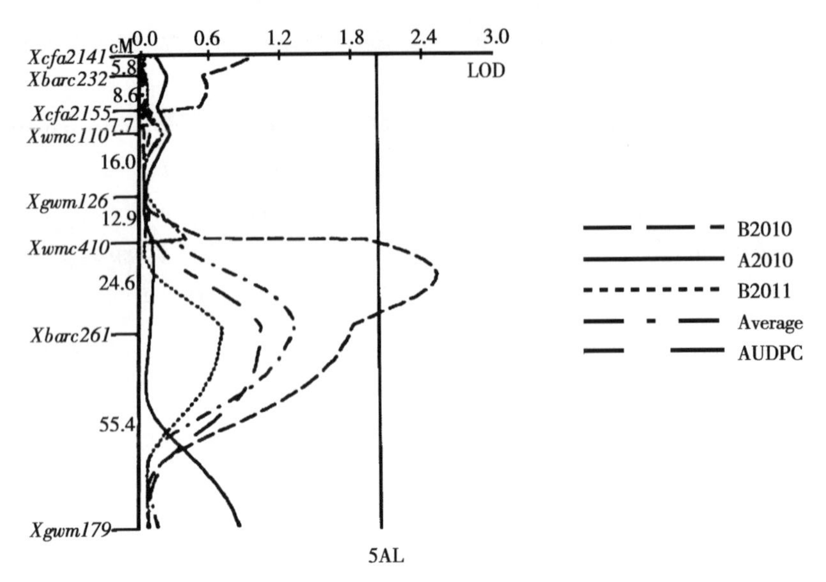

Fig. 2 Logarithm of odds (LOD) contours identified by composite interval mapping for QTL on chromosomes 2BS, 3B, and 5AL for powdery mildew APR in Pingyuan 50/Mingxian 169 double haploid population. A2010, maximum disease severities (MDS) in Anyang 2010; B2010 and B2011, maximum disease severities in Beijing 2010 and 2011, respectively; Average, mean MDS across three environments; AUDPC, averaged area under the disease progress curve in Beijing over two years. The LOD threshold for significance is 2.0.

Discussion

In the present study, the QTL detected in different environments on chromosome 2BS were within a genetic distance of less than 20cM. We considered them as one QTL according to hypothesis of declaration of significant QTL, and designated *QPm. caas-2BS. 2*. Previously, one QTL was also mapped on chromosome 2BS in Italian wheat cultivar Strampelli[37] and located around SSR marker *Xwmc25* with genetic distance about 32cM from *QPm. caas-2BS. 2* based on wheat consensus map[35]. In addition, *QPm. crag-2BS* [14] and *QPm. caas-2BS* [11], detected in Festin and Lumai 21, respectively, were mapped on short arm of chromosome 2B and located about 12cM on the both sides of *QPm. caas-2BS. 2*[35]. This QTL seems to be different from those in Strampelli, Festin and Lumai 21 based on their origins. Likely, two major powdery mildew resistance genes, *Pm26* and *Pm42* derived from wild emmer (*T. turgidum* var. *dicoccoides*) has also been mapped in the same vicinity less than 20cM from *QPm. caas-2BS. 2* [38,39]. This QTL should be different from the two seedling resistance genes, as it was mapped in Chinese landrace Pingyuan 50. Moreover, stripe rust resistance QTL, *QYr. caas-2BS*, was mapped in the same genomic region of *QPm. caas-2BS. 2* in the same population[22]. QTL for stripe rust resistance were identified at the same position in cv. Louise, Luke and Kariega[40-42]. This genomic region on chromosome 2BS seems to be pleiotropic effect on both powdery mildew and stripe rust. It can be used to breed durable resistance to powdery mildew and stripe rust by pyramiding with other resistance genes through marker assisted selection.

QPm. caas-3BS was identified in marker interval of *Xwmc366-Xgwm77* on chromosome 3BS contributed by Pingyuan 50, and explained 9.1% of phenotypic variation. Chen et al. [43] reported one QTL linked with *Xwms 533* on the short arm of chromosome 3B in wheat line 2174 with a genetic distance of about 56cM from *QPm. caas-3BS*[35]. Donini et al. [44] mapped *Pm13* on 3BS derived from *Ae. longissimum* and closely linked by RFLP markers. Therefore, it seems to be a new QTL for powdery mildew resistance on chromosome 3BS based on their chromosomal location and origins.

QPm. caas-3BL was mapped close to centromere on the long arm of chromosome 3B between SSR markers *Xwmc527-Xwmc418*, explained by 18.1% of the phenotypic variance and contributed by Mingxian 169. Race specific resistance genes *Pm41* in wild emmer (*T. turgidum* var. *dicoccoides*) has been mapped on chromosome 3BL with genetic distance of approximately 34cM from *QPm. caas-3BL*[45]. Although the genetic distance between *QPm. caas-3BS* and *QPm. caas-3BL* is less than 10cM[35], we considered them as two QTL due to their location on different arm of chromosome 3B. QTL for powdery mildew resistance on chromosome 3BL has not been yet reported, which appears to be a new locus for APR to powdery mildew.

QPm. caas-5AL detected on the long arm of chromosome 5A in the marker interval *Xwmc410* - *Xbarc261* explained 10.2% of phenotypic variance. Several QTL have been mapped on this chromosome in wheat lines Folke[1], Saar[20], *T. militinae*[46], and Forno[12] with genetic distances of 80, 80, 77, and 68cM away from *QPm. caas-5AL*, respectively, based on the wheat consensus map[35]. This appears to be new locus on chromosome 5AL for APR to powdery mildew. In addition, the stripe rust resistance QTL, *QYr. caas-5AL*[22], has also been mapped in the same genomic region on chromosome 5AL as *QPm. caas-5AL* in the Pingyuan 50/Mingxian 169 population, indicating that this region might be another pleiotropic APR loci to both powdery mildew and stripe rust. Moreover, *Yr48*, a partial stripe rust resistance gene derived from synthetic derivative PI 610750 has also been mapped at the same position of *QPm. caas-5AL* on chromosome 5AL[47]. This locus could be useful in improving APR to powdery mildew and stripe rust simultaneously.

The wheat landrace Pingyuan 50 is considered as a valuable germplasm of APR to both stripe rust and powdery mildew in local wheat breeding program, and three QTL for APR to stripe rust were mapped on Pingyuan 50[20]. In the present study, four QTL for APR to powdery mildew were mapped in the same population, and three of them derived from Pingyuan 50. Although these QTL were not stably detected across environments, *QPm. caas-2BS.2* and *QPm. caas-5AL* were mapped on the same chromosome regions as *QYr. caas-2BS* and *QYr. caas-5AL*, respectively, for APR to stripe rust, indicating pleiotropic APR loci to both powdery mildew and stripe rust were present in Pingyuan 50. QTL pyramiding is a very useful approach to maximize disease resistance by accumulating 4-5 minor genes in one genotype. We have successfully pyramided QTL for powdery mildew resistance derived from Bainong 64 and Lumai 21 through marker assisted selection[48]. The QTL detected in Pingyuan 50, particularly *QPm. caas-2BS.2* and *QPm. caas-5AL* in combination of previously identified 3 QTL including the locus *Pm38* in each of cv. Strampelli and Libellula for powdery mildew and stripe rust, should be useful for developing cultivars with potentially durable resistance to powdery mildew and stripe rust simultaneously.

Acknowledgments

This study was supported by the National Key Basic Research Program of China (2013CB127700), National Natural Science Foundation of China (31261140370 and 31260319), International Collaboration Projects from the Ministries of Science and Technology (2011DFG32990) and Agriculture (2011-G3), the National High Technology Research Program of China (2012AA101105), and the China Agriculture Research System (CARS-3-1-3). The author, M. A. Asad, gratefully acknowledges the full scholarship support for his Ph. D. studies from the China Scholarship Council (CSC), P. R. China.

References

[1] M. Lillemo, A. Bjørnstad, H. Skinnes, Molecular mapping of partial resistance to powdery mildew in winter wheat cultivar Folke. Euphytica 185 (2012) 47-59.

[2] K. L. Everts, S. Leath, P. L. Finney, Impact of powdery mildew and leaf rust on milling and baking quality of soft

red winter wheat. Plant Dis. 85 (2001) 423-429.

[3] R. L. Conner, A. D. Kuzyk, H. Su, Impact of powdery mildew on the yield of soft white spring wheat cultivars. Can. J. Plant Sci. 83 (2003) 725-728.

[4] A. Morgounov, H. A. Tufan, R. Sharma, B. Akin, A. Bagci, H. J. Braun, Y. Kaya, M. Keser, T. S. Payne, K. Sonder, R. McIntosh, Global incidence of wheat rusts and powdery mildew during 1969-2010 and durability of resistance of winter wheat variety Bezostaya 1. Eur. J. Plant Pathol. 132 (2012) 323-340.

[5] J. M. Wang, H. Y. Liu, H. M. Xu, M. Li, Z. S. Kang, Analysis of differential transcriptional profiling in wheat infected by *Blumeria graminis* f. sp *tritici* using gene chip. Mol. Biol. Rep. 39 (2012) 381-387.

[6] G. Gustafson, G. Shaner, Influence of plant age on the expression of slow-mildewing resistance in wheat. Phytopathology 72 (1982) 746-749.

[7] S. L. K. Hsam, F. J. Zeller, Breeding for powdery mildew resistance in common wheat (*Triticum aestivum* L.). *In*: R. R. Belanger, W. R. Bushnell, A. J. Dik, T. L. W. Carver (Eds), The powdery mildews, A comprehensive treatise, The American Phytopathological Society, St. Paul, 2002, pp. 219-238.

[8] D. Z. Yu, X. J. Yang, L. J. Yang, M. J. Jeger, J. K. M. Brown, Assessment of partial resistance to powdery mildew in Chinese wheat varieties. Plant Breed. 120 (2001) 279-284.

[9] R. A. McIntosh, W. J. Rogers, C. F. Morris, R. Appels, X. C. Xia, Catalogue of gene symbols for wheat: 2011 supplement. http://www.shigen.nig.ac.jp/wheat/komugi/genes /macgene/supplement 2011. pdf.

[10] S. X. Liu, C. A. Griffey, M. A. S. Maroof, Identification of molecular markers associated with adult plant resistance to powdery mildew in common wheat cultivar Massey. Crop Sci. 41 (2001) 1268-1275.

[11] C. X. Lan, X. W. Ni, J. Yan, Y. Zhang, X. C. Xia, X. M. Chen, Z. H. He, . Quantitative trait loci mapping for adult-plant resistance to powdery mildew in Chinese wheat cultivar Lumai 21. Mol. Breed. 25 (2010) 615-622.

[12] M. Keller, B. Keller, G. Schachermayr, M. Winzeler, J. E. Schmid, P. Stamp, M. M. Messmer, Quantitative trait loci for resistance against powdery mildew in a segregating wheat x spelt population. Theor. Appl. Genet. 98 (1999) 903-912.

[13] N. Chantret, D. Mingeot, P. Sourdille, M. Bernard, J. M. Jacquemin, G. Doussinault, A major QTL for powdery mildew resistance is stable over time and at two development stages in winter wheat. Theor. Appl. Genet. 103 (2001) 962-971.

[14] D. Mingeot, N. Chantret, P. V. Baret, A. Dekeyser, N. Boukhatem, P. Sourdille, G. Doussinault, J. M. Jacquemin, Mapping QTL involved in adult plant resistance to powdery mildew in the winter wheat line RE714 in two susceptible genetic backgrounds. Plant Breed. 121 (2002) 133-140.

[15] Y. Bougot, J. Lemoine, M. T. Pavoine, H. Guyomarch, V. Gautier, H. Muranty, D. Barloy, A major QTL effect controlling resistance to powdery mildew in winter wheat at the adult plant stage. Plant Breed. 125 (2006) 550-556.

[16] H. Muranty, M. T. Pavoine, B. Jaudeau, W. Radek, G. Doussinault, D. Barloy, Two stable QTL involved in adult plant resistance to powdery mildew in the winter wheat line RE714 are expressed at different times along the growing season. Mol. Breed. 23 (2009) 445-461.

[17] D. M. Tucker, C. A. Griffey, S. Liu, G. Brown-Guedira, D. S. Marshall, M. A. S. Maroof, Confirmation of three quantitative trait loci conferring adult plant resistance to powdery mildew in two winter wheat populations. Euphytica 155 (2007) 1-13.

[18] S. S. Liang, K. Suenaga, Z. H. He, Z. L. Wang, H. Y. Liu, D. S. Wang, R. P. Singh, P. Sourdille, X. C. Xia, Quantitative trait Loci mapping for adult-plant resistance to powdery mildew in bread wheat. Phytopathology 96 (2006) 784-789.

[19] A. Börner, E. Schumann, A. Furste, H. Coster, B. Leithold, M. S. Roder, W. E. Weber, Mapping of quantitative trait loci determining agronomic important characters in hexaploid wheat (*Triticum aestivum* L.). Theor. Appl. Genet. 105 (2002) 921-936.

[20] M. Lillemo, B. Asalf, R. P. Singh, J. Huerta-Espino, X. M. Chen, Z. H. He, A. Bjørnstad, The adult plant rust resistance loci *Lr34/Yr18* and *Lr46/Yr29* are important determinants of partial resistance to powdery mildew in bread wheat line Saar. Theor. Appl. Genet. 116 (2008) 1155-1166.

[21] C. X. Lan, S. S. Liang, Z. L. Wang, J. Yan, Y. Zhang, X. C. Xia, Z. H. He, Quantitative trait loci mapping for adult-plant resistance to powdery mildew in Chinese wheat cultivar Bainong 64. Phytopathology 99 (2009) 1121-1126.

[22] C. X. Lan, S. S. Liang, X. C. Zhou, G. Zhou, Q. L.

Lu, X. C. Xia, Z. H. He, Identification of genomic regions controlling adult-plant stripe rust resistance in Chinese landrace Pingyuan 50 through bulked segregant analysis. Phytopathology 100 (2010) 313-318.

[23] R. F. Peterson, A. B. Campbell, A. E. Hannah, A diagrammatic scale for estimating rust intensity of leaves and stems of cereals. Can. J. Re. 26 (1948) 496-500.

[24] M. E. Bjarko, R. F. Line, Heritability and number of genes controlling leaf rust resistance on four cultivars of wheat. Phytopathology 78 (1988) 457-461.

[25] P. K. Gupta, S. Rustgi, S. Sharma, R. Singh, N. Kumar, H. S. Balyan, Transferable EST-SSR markers for the study of polymorphism and genetic diversity in bread wheat. Mol. Gen. Genomics 270 (2003) 315-323.

[26] Q. J. Song, E. W. Fickus, P. B. Cregan, Characterization of trinucleotide SSR motifs in wheat. Theor. Appl. Genet. 104 (2000) 286-293.

[27] M. S. Röder, V. Korzun, K. Wendehake, J. Plaschke, M. H. Tixier, P. Leroy, M. W. Ganal, A microsatellite map of wheat. Genetics 149 (1998) 2007-2023.

[28] P. Sourdille, S. Singh, T. Cadalen, G. L. Brown-Guedira, G. Gay, L. Qi, Microsatellite based deletion bin system for the establishment of genetic physical map relationships in wheat (*Triticum aestivum* L.). Funct. Integr. Genomics 4 (2004) 12-25.

[29] H. Guyomarc'h, P. Sourdille, K. J. Edwards, M. Bernard, Studies of the transferability of microsatellites derived from *Triticum tauschii* to hexaploid wheat and to diploid related species using amplification, hybridization and sequence comparisons. Theor. Appl. Genet. 105 (2002) 736-744.

[30] R. W. Michelmore, I. Paran, R. V. Kesseli, Identification of markers linked to disease resistance genes by bulked segregant analysis: A rapid method to detect markers in specific genomic regions by using segregating populations. Proc. Natl. Acad. Sci. USA 88 (1991) 9828-9832.

[31] G. J. Bryan, A. J. Collins, P. Stephenson, A. Orry, J. B. Smith, M. D. Gale, Isolation and characterization of microsatellites from hexaploid bread wheat. Theor. Appl. Genet. 94 (1997) 557-563.

[32] B. J. Bassam, G. Caetano-Anolles, P. M. Gresshoff, Fast and sensitive silver staining of DNA in polyacrylamide gels. Anal. Biochem. 196 (1991) 80-83.

[33] K. F. Manly, R. H. Cudmore Jr, J. M. Meer, Map Manager QTX, cross-platform software for genetic mapping. Mamm. Genome 12 (2001) 930-932.

[34] D. D. Kosambi, The estimation of map distance from recombination values. Ann. Eugen. 12 (1943) 172-175.

[35] D. J. Somers, P. Isaac, K. Edwards, A high-density microsatellite consensus map for bread wheat (*Triticum aestivum* L.). Theor. Appl. Genet. 109 (2004) 1105-1114.

[36] S. Wang, C. J. Basten, Z. B. Zeng, Windows QTL Cartographer v2. 5, Statistical Genetics, North Carolina State University, 2005.

[37] M. A. Asad, B. Bai, C. X. Lan, J. Yan, X. C. Xia, Y. Zhang, Z. H. He, QTL Mapping for Adult Plant Resistance to Powdery Mildew in Italian Wheat cv. Strampelli. J. Integr. Agri. 5 (2013) 756-764.

[38] J. K. Rong, E. Millet, J. Manisterski, M. Feldman, A new powdery mildew resistance gene: Introgression from wild emmer into common wheat and RFLP-based mapping. Euphytica 115 (2000) 121-126.

[39] W. Hua, Z. Liu, J. Zhu, C. Xie, T. Yang, Y. Zhou, X. Duan, Q. Sun, Identification and genetic mapping of *Pm42*, a new recessive wheat powdery mildew resistance gene derived from wild emmer (*Triticum turgidum* var. *dicoccoides*). Theor. Appl. Genet. 119 (2009) 223-230.

[40] A. H. Carter, X. M. Chen, K. Garland-Campbell, K. K. Kidwell, Identifying QTL for high-temperature adult-plant resistance to stripe rust (*Puccinia striiformis* f. sp *tritici*) in the spring wheat (*Triticum aestivum* L.) cultivar 'Louise'. Theor. Appl. Genet. 119 (2009) 1119-1128.

[41] Q. Guo, Z. J. Zhang, Y. B. Xu, G. H. Li, J. Feng, Y. Zhou, Quantitative trait loci for high-temperature adult-plant and slow-rusting resistance to *Puccinia striiformis* f. sp *tritici* in wheat cultivars. Phytopathology 98 (2008) 803-809.

[42] V. P. Ramburan, Z. A. Pretorius, J. H. Louw, L. A. Boyd, P. H. Smith, W. H. P. Boshoff, R. Prins, A genetic analysis of adult plant resistance to stripe rust in the wheat cultivar Kariega. Theor. Appl. Genet. 108 (2004) 1426-1433.

[43] Y. H. Chen, R. M. Hunger, B. F. Carver, H. L. Zhang, L. L. Yan, Genetic characterization of powdery mildew resistance in U. S. hard winter wheat. Mol. Breed. 24 (2009) 141-152.

[44] P. Donini, R. M. D. Koebner, C. Ceoloni, Cytogenetic and molecular mapping of the wheat *Aegilops lon-*

gissima chromatin breakpoints in powdery mildew resistant introgression lines. Theor. Appl. Genet. 91 (1995) 738-743.

[45] G. Q. Li, T. L. Fang, H. T. Zhang, C. J. Xie, H. J. Li, T. M. Yang, E. Nevo, T. Fahima, Q. X. Sun, Z. Y. Liu, Molecular identification of a new powdery mildew resistance gene *Pm41* on chromosome 3BL derived from wild emmer (*Triticum turgidum* var. *dicoccoides*). Theor. Appl. Genet. 119 (2009) 531-539.

[46] I. Jakobson, H. Peusha, L. Timofejeva, K. Jarve, Adult plant and seedling resistance to powdery mildew in a *Triticum aestivum* × *Triticum militinae* hybrid line. Theor. Appl. Genet. 112 (2006) 760-769.

[47] I. Lowe, L. Jankuloski, S. Chao, X. Chen, D. See, J. Dubcovsky, Mapping and validation of QTL which confer partial resistance to broadly virulent post-2000 North American races of stripe rust in hexaploid wheat. Theor. Appl. Genet. 123 (2011) 143-157.

[48] B. Bai, Z. H. He, M. A. Asad, C. X. Lan, Y. Zhang, X. C. Xia, J. Yan, X. M. Chen, C. S. Wang, Pyramiding adult-plant powdery mildew resistance QTLs in bread wheat. Crop Pasture Sci. 63 (2012) 606-611.

Stripe rust resistance gene *Yr18* and its suppressor gene in Chinese wheat landraces

L. Wu[1], X. C. Xia[2], G. M. Rosewarne[3], H. Z. Zhu[1], S. Z. Li[1], Z. Y. Zhang, and Z. H[4]. He[2,5]

[1]*Crop Research Institute, Sichuan Academy of Agricultural Science (SAAS) /Key Laboratory of Biology and Genetic Breeding in Wheat (Southwest), Ministry of Agriculture, China, Chengdu, Sichuan* 610066, *China*; [2]*Institute of Crop Science, National Wheat Improvement Center/The National Key Facility for Crop Gene Resources and Genetic Improvement, Chinese Academy of Agricultural Sciences (CAAS)*, 12 *Zhongguancun South Street, Beijing* 100081, *China*; [3]*Department of Environment & Primary Industries*, 110 *Natimuk Rd, Horsham Victoria* 3400, *Australia*; [4]*Plant Protection Institute, Sichuan Academy of Agricultural Science (SAAS), Chengdu, Sichuan* 610066, *China*; [5]*International Maize and Wheat Improvement Center (CIMMYT) China Office, c/o CAAS*, 12 *Zhongguancun South Street, Beijing* 100081, *China*

Abstract: The slow rusting and mildewing gene *Yr18/Lr34/Pm38/Sr57* confers partial, durable resistance to multiple fungal pathogens and has its origins in China. A number of diagnostic markers were developed for this gene based on the gene sequence, but these markers do not always predict the presence of the resistant phenotype as some wheat varieties with the gene are susceptible to stripe rust in China. We hypothesized that these varieties have a suppressor of *Yr18*. The present study was undertaken to determine the presence of *Yr18*, the suppressor, and/or another resistance gene in 144 Chinese wheat landraces using molecular markers and stripe rust field data. Forty three landraces were predicted to have *Yr18* based on the presence of the markers, but had final disease severities higher than 70%, indicating that this gene may be under the influence of a suppressor. Four of these landraces, 'Sichuanyonggang 2', 'Baikemai', 'Youmai', and 'Zhangsihuang', were chosen for genetic studies. Crosses were made between the lines and 'Avocet S', with further crosses of Sichuanyonggang 2× 'Huixianhong' and Sichuanyonggang 2× 'Chinese Spring'. The F_1 plants of Sichuanyonggang 2/Chinese Spring was susceptible indicating the presence of a dominant suppressor gene. The results of genetic analyses of $F_{2:3}$ and BC_1F_2 families derived from these crosses indicated the presence of *Yr18*, a *Yr18* suppressor and another additive resistance gene. The *Yr18* region in Sichuanyonggang 2 was sequenced to ensure that it contained the functional allele. This is the first report of a suppressor of *Yr18/Lr34/Pm38/Sr57* gene with respect to stripe rust response.

Key words: *Puccinia striiformis*, yellow rust, *Triticum aestivum*

Introduction

Yr18/Lr34/Pm38/Sr57 (hereafter referred to as *Yr18*) is an important gene in wheat as it confers race non-specific, durable resistance to a number of diseases including stripe rust, caused by *Puccinia striiformis* Westend. f. sp. *tritici* Erikss. (*Yr18*) (McIntosh, 1992; Singh, 1992a), leaf rust (Dyck, 1987), powdery mildew (Spielmeyer et al., 2005, Lillemo et al.,

Published in Plant Breeding, 2015, 134 (60): 634-640

2008), stem rust (McIntosh, 2012), and barley yellow dwarf virus (Singh, 1993). It has been deployed for over 100 years, a testament to its durability (Kolmer et al., 2008). It is typically described as a slow rusting gene that is most effective at the adult plant stage; however seedlings can also exhibit resistance when exposed to low temperatures (Rubiales and Niks, 1995, Singh and Gupta, 1992). This gene also enhances the expression of some race-specific resistance genes and acts additively in combination with other slow rusting genes (German and Kolmer, 1992; Singh and Rajaram, 1992). Although the mechanism of quantitative resistance provided by *Yr18* is poorly understood, the response is non-hypersensitive and results in decreased infection frequency, increased latent period, higher rates of abortion of infection sites and smaller colonies (Rubiales and Niks, 1995; Singh and Huerta-Espino, 2003). Its expression also induces a leaf tip necrosis phenotype in flag leaves of adult wheat plants in some environments (Dyck, 1991; Singh, 1992b) and likely involves senescence-like processes (Krattinger et al., 2009).

Yr18 is located on wheat chromosome 7DS (Suanaga et al., 2003). Sequence analysis revealed an 11, 805 bp gene comprising of 24 exons and encoding an ABC transporter of the pleiotropic drug resistance class (Krattinger et al., 2009). Nine naturally occurring polymorphic sites in the gene sequence were reported, including an A/T in position 675 of intron 4 (Krattinger et al., 2009), a T/C in position 868 of intron 4, a T/C in position 1228 of intron 4, a T/C in position 48 of intron 6 and an 11 'A' /10 'A' in position 58～67 of intron 6 (Cao et al., 2010), an A deletion in exon 10 (Dakouri et al., 2010), a TTC deletion at positions 106～108 in exon 11 and a C/T in position 56 of exon 12 (Krattinger et al., 2009), and a G/T in position 136 of exon 22 (Lagudah et al., 2009). These mutations made up seven haplotypes with only one conferring resistance in, for example, Chinese Spring (A/T/T/T/11 'A' /N/N/C/G) (Krattinger et al., 2009; Lagudah, 2011), whereas the other six that led to susceptibility were described as T/T/T/T/11 'A' /N/TTC/T/G in 'Renan' (Krattinger et al., 2009), A/T/T/T/11 'A' /N/TTC/T/G in 'Galaxie' (Lagudah et al., 2009), A/T/T/T/11 'A' /N/N/C/T in 'Jagger' (Lagudah et al., 2009, Cao et al., 2010), A/C/C/C/10 'A' /N/N/C/G in '2174' (Cao et al., 2010), T/T/T/T/11 'A' /A/TTC/T/G in 'Invader' (Dakouri et al., 2010), and T/T/T/T/11 'A' /N/N/T/G in 'Odese Kaja13' (Dakouri et al., 2010).

Phenotype-based analysis of *Yr18* in wheat is challenging because the gene can interact with other resistance genes and its phenotype can be influenced by plant growth stages and environmental conditions. Molecular markers are of great value in this regard with early work identifying relatively useful markers such as *gwm1220* and *gwm295* (Spielmeyer et al., 2005), *swm10* (Bossolini et al., 2006), *csLV34* (Lagudah et al., 2006), *csLVMS1* (Spielmeyer et al., 2008), *csLVE17*, *csLVA1/SWSNP3* (Krattinger et al., 2009), and *cam1*, *cam2*, *cam8*, *cam11*, *cam16*, *cam23*, and *caISBP1* (Dakouri et al., 2010). However, these markers do not work in all germplasm as the previously listed SNP sequence mutations can affect gene function. Subsequently the *cssfr1-cssfr7* (Lagudah et al., 2009), and *caSNP4*, *caSNP12*, and *caIND11* (Dakouri et al., 2010) markers were developed with combinations being truly diagnostic on currently used germplasm.

It is assumed that the susceptible T/T/T/T/11 'A' /N/TTC/T/G haplotype in Renen is the original form of the gene as it is the only one thatoccurs in *Aegilops tauschii* (Krattinger et al., 2009; Dakouri et al., 2010, 2014). The resistance haplotype therefore likely formed in hexaploid wheat through a number of steps involving susceptible variants (Dakouri et al., 2010, 2014). From the geographical distribution of haplotypes, the resistant form has the highest frequencies in Asia, whereas the susceptible variants occur mostly in Europe. This supports the notion that the functional *Yr18* allele formed in Asia (Dakouri et al., 2010, 2014). The frequency of *Yr18* is extremely high in Chinese landraces at more than 80% (Yang et al.,

2008), a figure that is much higher than that in other countries and regions worldwide (Kolmer et al., 2008). Therefore, Chinese wheat landraces are important materials for studies on the origin and allelic variation of *Yr18*.

Wu et al. (2010) found that many Chinese wheat landraces that were predicted to possess the *Yr18* resistance allele by marker assays were highly susceptible to stripe rust in the field. This indicated that such landraces either contained susceptible allelic variations at the *Yr18* locus or a suppressor of gene function. This paper reports the distribution of *Yr18* in Chinese wheat landraces identified by markers *csLV34* and *cssfr1-cssfr7* along with field stripe rust scores. A suppressor in some Chinese wheat landraces containing the resistance gene was demonstrated by genetic analyses.

Materials and Methods

Plant materials

One hundred and forty four Chinese landraces were used to investigate the distribution of *Yr18*. Four landraces, namely Sichuanyonggang 2, Baikemai, Youmai, and Zhangsihuang, were used in crosses to test for suppressors of *Yr18*. Avocet S, Huixianhong and Chinese Spring were used as the other parents. Sichuanyonggang 2 was crossed with Avocet S, and crossed and backcrossed with Huixianhong and Chinese Spring. Baikemai, Youmai, and Zhangsihuang were crossed with Avocet S. The cross and backcross of Sichuanyonggang 2 and Huixianhong were tested in the field as $F_{2:3}$ and BC_1F_2 lines. The other crosses were tested in the field at the F_1 stage and in later generations (Table 1). *Yr18* was sequenced in Sichuanyonggang 2, 'Dabaimai', 'Dahongmang', and 'Sanheliuleng'. Segregating populations were not developed from crosses of the last three lines and Avocet S due to too few seed.

Table 1 Generations of crosses and backcrosses used in stripe rust tests

Cross	Generation	Abbreviation	Year	Number of lines
Sichuanyonggang 2/Avocet S	$F_{2:3}$	SAF	2013	399
		SAF	2014	399
Sichuanyonggang 2/Huixianhong	$F_{2:3}$	SHF	2013	363
Sichuanyonggang 2/2 * Huixianhong	BC_1F_2	SHB	2013	221
Sichuanyonggang 2/Chinese Spring	$F_{2:3}$	SCF	2014	431
Sichuanyonggang 2/2 * Chinese Spring	BC_1F_2	SCB	2014	229
Baikemai/Avocet S	$F_{2:3}$	BAF	2013	399
Youmai/Avocet S	$F_{2:3}$	YAF	2013	375
Zhangsihuang/Avocet S	$F_{2:3}$	ZAF	2013	395

Molecular marker analysis and *Yr18* gene sequencing

The SDS method was used to extract genomic DNA from four single seeds of each of the 144 landraces (Lagudah et al., 1991). The CTAB method (Sharp et al., 1988) was used to extract genomic DNA from young leaves of the parents and progenies of the populations. Tissue from approximately 10 F_1 plants was pooled prior to extraction of genomic DNA for each cross; single plants of the F_2 and BC_1 plants from crosses and backcrosses were used for DNA extraction. For molecular analysis, the use of *csLV34* and *cssfr1-cssfr7* primers, the PCR procedure, the CAPS procedure, gel electrophoresis and gel staining followed Lagudah et al. (2006, 2009). PCR primers used for sequencing the coding region of *Yr18* gene in Sichuanyonggang 2 were kindly provided by Dr. Evans Lagudah, CSIRO Plant Industry, Canberra, Australia. The software Primer Premier 5 was used to design PCR primers for sequencing the promoter region of *Yr18* in Sichuanyonggang 2. Sequencing of the coding and promoter regions were performed by Beijing Augct Biological Technology Co., Ltd (http://www.augct.com) and Shanghai Sangon Biological Engineering Technology & Service Co., Ltd (http://www.sangon.com). Sequence alignments of the coding and promoter regions were performed using the software DNAMAN (http://www.lynnon.com). The 3'-UTR region of *Yr18* in Sichuanyonggang 2 was also sequenced by BGI (http://www.genomics.cn).

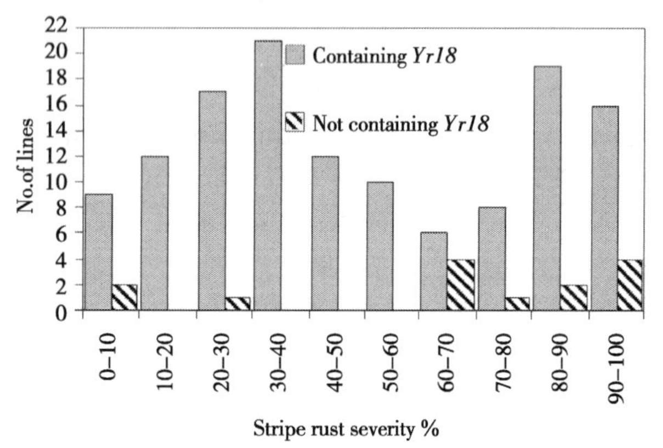

Fig. 1 Frequency distributions of 144 Chinese wheat landraces in field trials at Pi Xian in 2009 and 2010. Containing *Yr18* and not containing *Yr18* were according to the marker detection results by *csLV34* and *cssfr1*-*cssfr7*.

Field trials

The landraces were evaluated for stripe rust severity at the SAAS Pi Xian Field Station near Chengdu in the 2009 and 2010 cropping seasons. Progenies of the crosses were also evaluated for stripe rust reaction at the same location (Table 1). Field trials in each year were conducted as randomized complete blocks with three replicates. Plots consisted of single 1.5 m rows spaced 30 cm apart, with approximately 30 seeds sown in each row. The susceptible cultivar, 'Chuanyu 12', was planted after every 10 rows as a check, and planted in spreader rows surrounding the experimental area to ensure a uniform epidemic. The spreader rows at the three-leaf stage were inoculated with a mixture of *P. striiformis* f. sp. *tritici* (*Pst*) races in January of each year. The inocula comprised a mixture of *Pst* races CYR31, CYR32, CYR33 and Gui22-12, and avirulence/virulence patterns of these races were summarized in Table 2. Stripe rust severities were visually rated twice after the disease severities on Chuanyu 12 reached maximum levels (85-100%) between 10th and 20th April. The final disease severities (FDS) were used for analysis. For $F_{2:3}$ and BC_1F_2 populations, the mean, minimum, and maximum FDS of each line were recorded.

Genetic and statistical analyses

The average FDS of the landraces in 2009 and 2010 was used to calculate the mean FDS, minimum FDS, maximum FDS, SE, and CV% in two groups of germplasm, i.e. those containing *Yr18* and those not containing *Yr18*. The $F_{2:3}$ and BC_1F_2 families from Sichuanyonggang 2×Chinese Spring were classified into four phenotypic categories following Singh and Rajaram (1992).

Table 2 The statistical description of average final disease severity (FDS) of 144 landraces in 2009 and 2010 based on the marker test for *Yr18*/*Lr34*/*Pm38*/*Sr57*

	Sample size	Mean FDS*	Minimum FDS	Maximum FDS	SE	CV (%)	95% confidence level
Containing *Yr18*	130	51.7a	4.5	100	2.6	56.6	5.1
Not containing *Yr18*	14	67.7b	5.5	100	8.5	47.1	18.4

* Different letters following the mean indicate a significant difference at P=0.05.

These categories were homozygous parental type resistant (HPTR), homozygous parental type susceptible (HPTS), segregating for resistance with intermediate levels of response (SEGI) and segregating for intermediate types with some susceptible plants (SEGS). Because both parents were highly susceptible in other populations derived from the crosses between four landraces and Avocet S, and the resistance genes conferred only partial resistance, a cut-off value for the HPTR class could not be determined. Therefore, the four phenotypic categories were reduced to three. In these cases the classes were HPTS, SEGI and SEGS.

The parental score was used as a standard to classify the lines into different categories in each population. When the mean, minimum, and maximum FDS of a line was lower than those of the resistant parent, it was classified as HPTR; if the mean, minimum, and maximum FDS of a line was the same as or higher than those of susceptible parent, it was classified as HPTS; and when the mean FDS of a line was lower than that of susceptible parent, and the maximum score was higher than that of resistant parent and lower than that of susceptible parent, respectively, it was defined as SEGI; and if the minimum score of a line was lower than that of susceptible parent, but the maximum score was the same as or higher than that of susceptible parent, it was classified SEGS. The number of resistance and suppressor genes in the $F_{2:3}$ and BC_1F_2 population were estimated using Mendelian segregation analysis with expected frequencies determined empirically. Chi squared (χ^2) tests were used to determine the goodness of fit of the observed data to the theoretically expected segregation ratios. The Microsoft Excel analytical tool (CHITEST) was used for the statistical analysis.

Results

Marker detection

The *Yr18* alleles in 144 landraces were characterized by markers *csLV34* and *cssfr1-cssfr7* (Lagudah et al., 2009). All 144 landraces contained the sequence of the resistance allele in exon 22 as determined by *cssfr7*. The *cssfr6* marker detects polymorphisms in exon 12, and 14 landraces had the sequence of the susceptible allele in this region. An allele of susceptible genotypes in exon 11, determined by *cssfr1-cssfr5*, was present in a homogeneous state in 14 landraces, and was heterogeneous in a further 11. The *csLV34* marker was mostly accurate in predicting *cssfr1-7* genotypes, with the exception of one landrace that had the susceptible exon 11 allele being scored as resistant, one of the exon 11 resistant landraces being scored as susceptible, and four of the heterogeneous exon 11 landraces being scored as homozygous susceptible.

Sicchuanyonggang 2, Baikemai, Youmai, Zhangsihuang, Dabaimai, Dahongmang, Sanheliuleng, and Chinese Spring carried the *Yr18* resistance allele based on markers *csLV34* and *cssfr1-cssfr7*, whereas Avocet S and Huixianhong did not. There were no differences in marker results between *csLV34* and *cssfr1-cssfr7* in Sichuanyonggang 2, Baikemai, Youmai, and Zhansihuang. As *csLV34* was easier to detect, it was used in all subsequent marker work. Each family was classified according to the status of *csLV34*, either as homozygous with the resistance allele, homozygous with the susceptible allele, or heterozygous.

Distribution of *Yr18* in Chinese wheat landraces

According to the marker results, landraces with heterogeneous for *Yr18*, for example 'Jingchazhong' and 'Hulitou', were considered as containing *Yr18*. The 144 landraces were grouped into two categories based on molecular identification, either with or without the *Yr18* resistance allele of. Ninety percent (130) contained *Yr18* and 10% (14) did not (Fig. 1 and Table 3). Although the range and CV of the FDS of the two groups were large the mean FDS of landraces containing *Yr18* was significantly lower than that of landraces without *Yr18*. Eleven of the 14 landraces that did not contain *Yr18* scored between 60 and 100% severity, with only three scoring less than 30%.

Table 3 Final disease severities of parents and F_1 plants in 2009, 2010, 2013 and 2014 field trials

Parent/Population	2009	2010	2013 $F_{2:3}$ trial	2013 BC_1F_2 trial	2014 $F_{2:3}$ trial	2014 BC_1F_2 trial
Sichuanonggang 2	95	83.3	73.3		78.3	
Avocet S	100	98.3	100		91.7	
Sichuanyonggang 2/Avocet S F_1			80			

（续）

Parent/Population	2009	2010	2013 $F_{2:3}$ trial	2013 BC_1 F_2 trial	2014 $F_{2:3}$ trial	2014 BC_1 F_2 trial
Sichuanonggang 2	95	83.3	73.3	73.3		
Huixianhong	NA	NA	95	95		
Sichuanyonggang 2/Huixianhong F_1			NA			
Sichuanonggang 2	95	83.3			96.7	85
Chinese Spring	30	21.7			13.3	11.7
Sichuanyonggang 2/Chinese Spring F_1					77.5	
Baikemai	100	98.3	94.3			
Avocet S	100	98.3	98.3			
Baikemai/Avocet S F_1			80			
Youmai	100	66.7	81.7			
Avocet S	100	98.3	95			
Youmai/Avocet S F_1			80			
Zhangsihuang	100	66.7	86.7			
Avocet S	100	98.3	95			
Zhangsihuang/Avocet S F_1			70			

Re-sequencing of *Yr18*

The coding and promoter regions of the *Yr18* locus were sequenced in Sicchuanyonggang 2, Dabaimai, Dahongmang, Sanheliuleng, and Chinese Spring. The closest 9 kb in the 3' -UTR region was also sequenced in Sicchuanyonggang 2 and Chinese Spring. The resistant A/T/T/T/11 'A' /N/N/C/G haplotype of the *Yr18* locus coding region was present in all five genotypes and no sequence polymorphisms were identified. There were also no differences in the promoter regions among the four sequenced landraces and Chinese Spring, nor any differences in the adjacent 8.15 kb of the 3' -UTR region between Sichuanyonggang 2 and Chinese Spring.

FDS of parents and the progenies

In the 2009-2010 field tests, the mean FDS of Sichuanyonggang 2, Baikemai, Youmai, Zhangsihuang, Avocet S, and Chinese Spring were 89.2, 99.2, 83.4, 83.4, 99.2, and 25.9%, respectively. In the 2013/2014 field tests, the mean FDS of Sichuanyonggang 2 was 83.3%, whereas the FDS of Avocet S, Huixianhong, and Chinese Spring were 96, 95, and 12.5%, respectively. F_1 plants of Sichuanyonggang 2/Avocet S and Sichuanyonggang 2/Chinese Spring were susceptible with mean FDS of 80 and 77.5%, respectively. In the 2013 field tests, the mean FDS of Baikemai, Youmai, and Zhangsihuang were 94.3, 81.7, and 86.7%, respectively, and the FDS of Avocet S was 97.2%. F_1 plants of Baikemai/Avocet S, Youmai/Avocet S, and Zhangsihuang/Avocet S were susceptible with mean FDS of 80, 80, and 70%, respectively. All of the landraces and the susceptible lines Avocet S and Huixianhong had FDS between 66.7 and 100% (Table 4). Chinese Spring scores ranged between 11.7 and 30%. The mean FDS of lines homozygous for the resistance allele in each $F_{2:3}$ or BC_1F_2 line was always lower than lines within the same population that were homozygous for the susceptible allele. The heterozygous lines generally had an intermediate score (Table 5).

Table 4 Final disease severities of the $F_{2:3}$ or BC_1F_2 populations characterized according to *Yr18* status

Type	Item	SAF 2013	SAF 2014	SHF 2013	SHB 2013	SCF 2014	SCB 2014	BAF 2013	YAF 2013	ZAF 2013
Total	No. of lines	399	399	363	221			399	375	395
	Mean	77.7	70.4	72.8	84.3			56.0	56.6	62.0
	Minimum	8	15	5	15			0.3	10	6.7
	Maximum	100	100	100	100			100	95	100
	SD	21.4	21.0	21.7	13.9			27.6	18.7	24.5
	SE	1.1	1.0	1.1	0.9			1.4	1.0	1.2
Yr18 homo	No. of Lines	101	101	86		431	229	68	91	97
	Mean	63.9	54.9	54.1		59.5	37.8	42.0	44.4	47.8
	Minimum	16.7	15	15		7	5	3.3	15	6.7
	Maximum	98.3	100	95		100	85	95	78.3	100
	SD	20.3	19.8	20.8		24.2	19.6	25.1	16.0	21.1
	SE	2.0	2.0	2.2		1.2	1.3	3.0	1.7	2.1
Yr18 hete	No. of Lines	203	203	190	122			228	183	191
	Mean	77.3	70.7	74.8	78.8			53.6	55.4	60.2
	Minimum	8	18.3	21.7	15			0.3	10	12
	Maximum	100	98.3	100	100			100	93.3	100
	SD	20.4	18.3	18.1	14.7			25.9	16.8	23.2
	SE	1.4	1.3	1.3	1.3			1.7	1.2	1.7
No *Yr18*	No. of Lines	95	95	87	99			103	101	107
	Mean	93.1	86.1	86.8	91.0			70.5	69.9	78
	Minimum	23.3	15	5	45			2.5	27.5	17
	Maximum	100	100	100	100			100	95	100
	SD	12.8	14.5	16.7	9.2			26.8	15.4	20.5
	SE	1.3	1.5	1.8	0.9			2.6	1.5	2.0

Yr18 homo, homozygous for the resistance allele according to the *csLV34* marker,
Yr18 hete, heterozygous according to the *csLV34* marker,
No *Yr18*, homozygous a susceptible haplotype according to the *csLV34* marker.

Table 5 χ^2 test results of the genetic model[1] in $F_{2:3}$ and BC_1F_2 populations segregating for stripe rust reaction

Response category of populations	No. of lines in population								
	SAF[2] 2013	SAF[2] 2014	SHF[2] 2013	SHB[3] 2013	SCF[4] 2014	SCB[5] 2014	BAF[2] 2013	YAF[2] 2013	ZAF[2] 2013
HPTR					20	46			
HPTS	37	49	40	46	22		37	36	43
SEGI	118	112	120		147	67	114	124	107
SEGS	244	238	203	175	242	116	248	215	245
TOTAL	399	399	363	221	431	229	399	375	395
χ^2	1.36	0.95	2.07	2.18	4.19	3.91	5.31	2.37	4.11
P-value	0.54	0.60	0.35	0.15	0.28	0.14	0.07	0.31	0.12
Expected no. of lines									
HPTR					26.94	57.25			

（续）

Response category of populations	No. of lines in population								
	SAF[2] 2013	SAF[2] 2014	SHF[2] 2013	SHB[3] 2013	SCF[4] 2014	SCB[5] 2014	BAF[2] 2013	YAF[2] 2013	ZAF[2] 2013
HPTS	43.64	43.64	39.70	55.25	26.94		43.64	41.02	43.20
SEGI	118.45	118.45	107.77		134.69	57.25	118.45	111.33	117.27
SEGS	236.91	236.91	215.53	165.75	242.44	114.5	236.91	222.66	234.53
TOTAL	399	399	363	221	431	229	399	375	395

HPTR, homozygous parental type resistant,
HPTS, homozygous parental type susceptible,
SEGI, segregating for intermediate levels of resistance,
SEGS, segregating for intermediate types with some susceptible lines.
1. Tested genetic model is *Yr18*+*Yr18* suppressor + another partial resistance gene,
2. Ratio of tested genetic model HPTS : SEGI : SEGS is 7 : 19 : 38,
3. Ratio of tested genetic model of HPTS : SEGS is 1 : 3,
4. Ratio of tested genetic model of HPTR : HPTS : SEGI : SEGS is 1 : 1; 5 : 9,
5. Ratio of tested genetic model of HPTR : SEGI : SEGS is 1 : 1 : 2.

Genetic analyses of stripe rust resistance in Chinese wheat landraces

The main genetic model tested was for the landrace to contain *Yr18* (Y), a suppressor of *Yr18* (I) and another additive partial resistance gene (R) that was not affected by the suppressor. All genes were assumed to segregate independently. The expected ratio of HPTS : SEGI : SEGS in this genetic model is 7 : 19 : 38 in an $F_{2:3}$ population and this was tested in the populations of Sichuanyonggang 2/Avocet S (SAF2013 and SAF2014), Sichuanyonggang 2/Huixianhong (SHF2013), Baikemai/Avocet S (BAF2013), Youmai/Avocet S (YAF2013), and Zhangsihuang/Avocet S (ZAF2013). The expected ratios in a BC_1F_2 for the same model is 1 : 3 for HPTS : SEGS and this was tested in Sichuanyonggang 2/2 * Huixianhong (SHB2013). *Yr18* did not segregate in populations derived from Chinese Spring. Therefore the expected ratio of HPTR : HPTS : SEGI : SEGS was 11 : 5 : 9 in the $F_{2:3}$ of Sichuanyonggang 2/Chinese Spring (SCF2014), and 1 : 1 : 2 (HPTR : SEGI : SEGS) in the BC_1F_2 of Sichuanyonggang 2/2 * Chinese Spring (SCB2014). The *P* values for SAF2013, SAF2014, SHF2013, SHB2013, SCF2014, SCB2014, BAF2013, YAF2013, and ZAF2013 were 0.54, 0.6, 0.35, 0.15, 0.28, 0.14, 0.07, 0.31, and 0.12, respectively, indicating that the observed ratios of lines in the different classes were not significantly different from the expected ratios at *P*=0.05.

Discussion

Genetic analysis indicated *Yr18*, a *Yr18* suppressor and another partial resistance gene controlled stripe rust resistance in $F_{2:3}$ populations of Sichuanyonggang 2/Avocet S and Sichuanyonggang 2/Huixianhong, and in BC_1F_2 populations of Sichuanyonggang 2/2 * Huixianhong. The consistency of results across the three populations (SAF2013, SAF2014, and SHF2013) strongly supported the proposed genetic model of two partial resistance genes and a suppressor of one of the resistance genes segregating in these populations. The cross between Sichuanyonggang 2 and Chinese Spring was interesting as both parents contained *Yr18*. A key finding supporting a suppressor gene in Sichuanyonggang 2 is that the F_1 in this population was susceptible, presumably due to suppression of *Yr18*, and that the segregating generations also fitted the expected model. Given that the Chinese Spring populations fitted the expected ratio of a fixed *Yr18* plus two other genes, namely, a suppressor of *Yr18* and another partial resistance gene, we assumed that Sichuanyonggang 2 contained the suppressor. The other three crosses were made to Avocet S, a line that is generally considered not to contain effective genes for stripe rust resistance. The Sichuanyonggang 2/Avocet S population fitted the proposed model of *Yr18* + *Yr18* suppressor +

another partial resistance gene, indicating that Avocet S did not contain the suppressor. We concluded that the suppressor for the susceptible reactions in all other landrace populations were derived from the respective landrace parents, namely, Baikemai, Youmai, and Zhangsihuang. In addition, a partial resistance gene might be present in the four landraces since the area under the disease progress curve (AUDPC) of these landraces were a little lower than those of Avocet S and Huixianhong in 2009, 2010, 2013, 2014 and 2015 (Data not shown), and the FDS of landraces were lower than those of Avocet S and Huixianhong in 2010, 2013 and 2014, although the FDS of landraces were similar to those of Avocet S and Huixianhong in 2009 and 2015 (Table 4). In Chengdu, the FDS of Chinese Spring was much lower than that of Avocet S+*Yr18* during the past seven years, indicating the presence of additional resistance genes in Chinese Spring. In the crosses between Sichuanyonggang 2 and Chinese Spring, there might be one more minor gene from Chinese Spring; however, it is very difficult to distinguish all minor genes due to their minor and/or combined effects.

The resistance-conferring *Yr18* allele is widely present in Chinese wheat landraces. Kolmer et al. (2008) identified six of 20 (30%) Chinese landraces with *Yr18* using the *csLV34* marker. A more detailed study by Yang et al. (2008) indicated that 359 of 422 (85%) Chinese landraces contained this gene. The present results showed that 130 of 144 (90%) landraces contained *Yr18*. 'Strampelli', an Italian wheat, is considered the origin of most western cultivars containing *Yr18* (Kolmer et al., 2008). However, wheat genotypes obtained from China, including Chinese Spring, contain this gene and predate the development of Strampelli (Dyck, 1977). Also, the reference genetic stock, RL6058 ('Thatcher' + *Lr34*/*Yr18*/*Ltn1*) derived this gene from a Chinese wheat introduction (Dyck and Samborski, 1982). This was further supported by Dakouri et al. (2014) who also proposed that *Yr18* originated in Asia, highlighting the importance of working with Chinese wheat landraces to further elucidate the origin and allelic variation of *Yr18*. In the present study, there were 43 landraces for which widely validated markers predicted the presence of *Yr18*, but their stripe rust FDS was greater than 70%. Numerous QTL studies have shown that lines containing this gene rarely have such high stripe rust scores (Suenaga et al., 2003; Lillemo et al., 2008). Since the cloning of *Yr18* a number of susceptible haplotypes at this locus were identified. It would seem likely that this high number of susceptible landraces that contain this locus may indicate alternate susceptible alleles in the germplasm. However, sequencing results indicated that Sichuanyonggang 2 contained the *Yr*18 resistance haplotype. An alternate explanation was the presence of a suppressor gene, and our genetic studies of four landraces indicate that this was the case.

Partial resistance genes, when in combination, confer increased levels of resistance. A gene pyramiding strategy is considered to be effective in developing cultivars with high levels of durable resistance (Singh et al., 2000). Use of APR or partial resistance genes is becoming an important approach to control stripe rust in China, whereas major highly effective resistance genes were popularly used in the past. Consequently, in a survey of modern Chinese cultivars only 14 of 231 contained *Yr18* based on the *csLV34* marker (Yang et al., 2008). It will be useful to further characterize Chinese wheat landraces through mapping approaches. Future work will involve mapping not only the suppressor gene, but also the other additive gene (s) in the current populations. Three landraces that did not contain *Yr18* had low rust scores indicating the presence of other resistance genes. The identities of these genes and their interaction with the suppressor require further investigation. One of the major benefits of *Yr18* is its pleiotropic effect on other disease responses. The identification of similar multi-disease resistance genes in Chinese landraces would be extremely useful.

Suppressors of resistance are common in synthetic wheat. Suppressorswere reported for stripe rust (Kema et al., 1995; Ma et al., 1995; Cheng et al., 2013), leaf rust (Cheng et al., 2013; Nelson et al., 1997), stem rust (Kerber and Green, 1980; Assefa and Fehrmann, 2004), powdery mildew (Ren et al., 1997; McIntosh et al., 2011; Zeller and Hsam, 1996), and *Septoria tritici* (Arraiano et al., 2007), with loci identified on all three wheat genomes (Kema et al., 1995; Ma et al., 1995).

A well characterized, dominant suppressor affects the ability of *Pm8* to confer resistance to powdery mildew (Ren et al., 1997). This is a clear example of a suppressor having a specific interaction with a particular resistance gene as *Pm8* was suppressed by transcribed alleles at the orthologous *Pm3* locus (McIntosh et al., 2011; Hurni et al., 2014). Another suppressor is effective against the leaf rust resistance gene *Lr23* on chromosome 2BS, where a suppressor was mapped to chromosome 2DS between molecular markers *Xbsl28* and *Xcdo405* (Nelson et al., 1997). The suppressor of *Pm17* was localized to chromosome 7D (Zeller and Hsam, 1996).

Stripe rust resistance conferred by *Yr18* was suppressed in some Chinese wheat landraces. Four landrace populations confirmed the presence of a dominant suppressor of *Yr18*. Genetic analyses based on $F_{2:3}$ and BC_1F_2 families identified the presence of *Yr18*, a *Yr18* suppressor and another resistance gene in the Sichuanyonggang 2, Baikemai, Youmai, and Zhangsihuang populations. This is the first report of a suppressor stripe rust resistance conferred by *Yr18* in Chinese landrace wheats. Understanding the molecular basis of suppression will provide important information on the functioning of the durable rust resistance gene, *Yr18*. It also highlights the possibility that even well established functional markers based on the gene sequence may not always be effective in achieving predicted phenotypic effects.

Acknowledgments

The authors are grateful to Prof. R. A. McIntosh, Plant Breeding Institute, University of Sydney, for critical review of this manuscript. This study was supported by the National Natural Science Foundation of China (31271724), the National Key Basic Research Program of China (2013CB127700), Sichuan Province projects targeting research innovation and ability promotion (NSFC special supplemental grant), and the Sichuan Province Youth Science and Technology Innovation Team (2014TD0014).

References

Arraiano, L. S., J. Kirby, and J. K. M. Brown. 2007: Cytogenetic analysis of the susceptibility of the wheat line Hobbit sib (Dwarf A) to *Septoria tritici* blotch. *Theor. Appl. Genet.* 116, 113-122.

Assefa, S., and H. Fehrman. 2004: Evaluation of *Aegilops tauschii* Coss. for resistance to wheat stem rust and inheritance of resistance genes in hexaploid wheat. *Genet. Resour. Crop Evol.* 51, 663-669.

Bossolini, E., S. G. Krattinger, and B. Keller. 2006: Development of simple sequence repeat markers specific for the *Lr34* resistance region of wheat using sequence information from rice and *Aegilops tauschii*. *Theor. Appl. Genet.* 113, 1049-1062.

Cao, S. H., B. F. Carver, X. K. Zhu, T. L. Fang, Y. H. Chen, R. M. Hunger, and L. L. Yan. 2010: A single-nucleotide polymorphism that accounts for allelic variation in the *Lr34* gene and leaf rust reaction in hard winter wheat. *Theor. Appl. Genet.* 121, 385-392.

Cheng, W. Q., T. G. Liu, and L. Gao. 2013: Suppression of stripe rust and leaf rust resistances in interspecific crosses of wheat. *Euphytica* 192, 339-346.

Dakouri, A., B. D. McCallum, and S. Cloutier. 2014: Haplotype diversity and evolutionary history of the *Lr34* locus of wheat. *Mol. Breeding* 33, 639-655.

Dakouri, A., B. D. McCallum, A. Z. Walichnowski, and S. Cloutier. 2010: Fine-mapping of the leaf rust *Lr34* locus in *Triticum aestivum* (L.) and characterization of large germplasm collections support the ABC transporter as essential for gene function. *Theor. Appl. Genet.* 121, 373-384.

Dyck, P. L. 1977: Genetics of leaf rust reaction in three introductions of common wheat. *Can. J. Genet. Cytol.* 19, 711-716.

Dyck, P. L. 1987: The association of a gene for leaf rust

resistance with the chromosome 7D suppressor of stem rust resistance in common wheat. *Genome* 29, 467-469.

Dyck P. L. 1991: Genetics of adult plant leaf rust resistance in 'Chinese Spring' and 'Sturdy' wheats. *Crop Sci.* 31, 309-311.

Dyck, P. L., and D. J. Samborski. 1982: The inheritance of resistance to *Puccinia recondita* in a group of common wheat cultivars. *Can. J. Genet. Cytol.* 24, 273-283.

German, S. E., and J. A. Kolmer. 1992: Effect of gene *Lr34* in the enhancement of resistance to leaf rust of wheat. *Theor. Appl. Genet.* 84, 97-105.

Hurni, S., S. Brunner, D. Stirnweis, G. Herren, D. Peditto, R. A. McIntosh, and B. Keller. 2014: The powdery mildew resistance gene *Pm8* derived from rye is suppressed by its wheat ortholog *Pm 3*. *The Plant Journal* 79, 904-913.

Kema, G. H. J., W. Lange, and C. H. V. Silfhout. 1995: Differential suppression of stripe rust resistance in synthetic wheat hexaploids derived from *Triticum turgidum* subsp. *Diccocoides* and *Aegilops squarrosa*. *Phytopathology* 85, 425-429.

Kolmer, J. A., R. P. Singh, D. F. Garvin, L. Viccars, H. M. William, J. Huerta-Espino, F. C. Ogbonnaya, H. Raman, S. Orford, H. S. Bariana, and E. S. Lagudah. 2008: Analysis of the *Lr34/Yr18* rust resistance region in wheat germplasm. *Crop Sci.* 48, 1841-1852.

Krattinger, S. G., E. S. Lagudah, W. Spielmeyer, R. P. Singh, J. Huerta-Espino, H. McFadden, E. Bossolini, L. L. Selter, and B. Keller. 2009: A putative ABC transporter confers durable resistance to multiple fungal pathogens in wheat. *Science* 323, 1360-1363.

Lagudah, E. S. 2011: Molecular genetics of race non-specific rust resistance in wheat. *Euphytica* 179, 81-91.

Lagudah, E. S., R. Appels, and D. McNeil. 1991: The *Nor-D3* locus of *Triticum tauschii*: natural variation and genetic linkage to markers in chromosome 5. *Genome* 34, 387-395.

Lagudah, E. S., S. G. Krattinger, S. Herrera-Foessel, R. P. Singh, J. Huerta-Espino, W. Spielmeyer, G. Brown-Guedira, L. L. Selter, and B. Keller. 2009: Gene-specific markers for the wheat gene *Lr34/Yr18/Pm38* which confers resistance to multiple fungal pathogens. *Theor. Appl. Genet.* 119, 889-898.

Lagudah, E. S., H. McFadden, R. P. Singh, J. Huerta-Espino, H. S. Bariana, and W. Spielmeyer. 2006: Molecular genetic characterization of the *Lr34/Yr18* slow rusting resistance gene region in wheat. *Theor. Appl. Genet.* 114, 221-306.

Li, Z. F., Xia, X. C., Zhou, X. C., Niu, Y. C., He, Z. H., Zhang, Y., Li, G. Q., Wan, A. M., Wang, D. S., Chen, X. M., Lu, Q. L., and Singh, R. P. 2006: Seedling and slow rusting resistance to stripe rust in Chinese common wheats. *Plant Dis.* 90, 1302-1312.

Lillemo, M., B. Asalf, R. P. Singh, J. Huerta-Espino, X. M. Chen, Z. H. He, and Å. Bjørnstad. 2008: The adult plant rust resistance loci *Lr34/Yr18* and *Lr46/Yr29* are important determinants of partial resistance to powdery mildew in bread wheat line Saar. *Theor. Appl. Genet.* 116, 1155-1166.

Liu, T. G., Zhang, Z. Y., Liu, B., Gao, L., Peng, Y. L. and Chen, W. Q. 2015: Dectection of virulence to *Yr26* and pathogenicity to Chinese commercial winter wheat cultivars at seedling stage. *Acta Phytopathologica Sinica* 45 (1), 41-47. (in Chinese with English abstract)

Ma, H., R. P. Singh, and A. Mujeeb-Kazi. 1995: Suppression/expression of resistance to stripe rust in synthetic hexaploid wheat (*Triticum turgidum* × *T. tauschii*). *Euphytica* 83, 87-93.

McIntosh, R. A. 1992: Close genetic linkage of genes conferring adult-plant resistance to leaf rust and stripe rust in wheat. *Plant Pathol.* 41, 523-527.

McIntosh, R. A., J. Dubcovsky, W. J. Rogers, C. Morris, R. Appels, and X. C. Xia. Catalogue of gene symbols for wheat: 2012 supplement. http://www.shigen.nig.ac.jp/wheat/komugi/genes/symbolClassList.jsp 20140722

McIntosh, R. A., P. Zhang, C. Cowger, R. Parks, E. S. Lagudah, and, S. Hoxha. 2011: Rye-derived powdery mildew resistance gene *Pm8* in wheat is suppressed by the *Pm3* locus. *Theor. Appl. Genet.* 123, 359-367.

Nelson, J. C., R. P. Singh, J. E. Autrique, and M. E. Sorrels. 1997: Mapping genes conferring and suppressing leaf rust resistance in wheat. *Crop Sci.* 37, 1928-1935.

Ren, S. X., R. A. McIntosh, and Z. J. Lu. 1997: Genetic suppression of the cereal rye-derived gene *Pm8* in wheat. *Euphytica* 93, 353-360.

Rubiales, D., and R. E. Niks. 1995: Characterization of *Lr34*, a major gene conferring nonhypersensitive resistance to wheat leaf rust. *Plant Dis.* 79, 1208-1212.

Sharp, P. J., M. Kreis, P. R. Shewry, and M. D. Gale. 1988: Location of β-amylase sequence in wheat and its relatives. *Theor. Appl. Genet.* 75, 286-290.

Singh, R. P. 1992a: Genetic association of leaf rust resistance gene *Lr34* with adult plant resistance to stripe rust in bread wheat. *Phytopathology* 82, 835-838.

Singh, R. P. 1992b: Genetic association between gene *Lr34* for leaf rust resistance and leaf tip necrosis in wheats. *Crop Sci*. 32, 874-878.

Singh, R. P. 1993: Genetic association of gene *Bdv1* for tolerance to barley yellow dwarf virus with genes *Lr*34 and *Yr18* for adult plant resistance to rusts in bread wheat. *Plant Dis*. 77, 1103-1106.

Singh, R. P., and A. K. Gupta. 1992: Expression of wheat leaf rust resistance gene *Lr34* in seedlings and adult plants. *Plant Dis*. 76, 489-491.

Singh, R. P., J. Huerta-Espino, and S. Rajaram. 2000: Achieving near-immunity to leaf and stripe rusts in wheat by combining slow rusting resistance genes. *Acta Phytopathol. Entomol. Hungarica* 35, 133-139.

Singh, R. P., and J. Huerta-Espino. 2003: Effect of leaf rust resistance gene *Lr34* on components of slow rusting at seven growth stages in wheat. *Euphytica* 129, 371-376.

Singh, R. P., and S. Rajaram. 1992: Genetics of adult plant resistance of leaf rust in Frontana and three CIMMYT wheats. *Genome* 35, 24-31.

Spielmeyer, W., R. A. McIntosh, J. Kolmer, and E. S. Lagudah. 2005: Powdery mildew resistance and *Lr34*/*Yr18* genes for durable resistance to leaf and stripe rust cosegregate at a locus on the short arm of chromosome 7D of wheat. *Theor. Appl. Genet*. 111, 731-735.

Spielmeyer, W., R. P. Singh, H. McFadden, C. R. Wellings, J. Huerta-Espino, X. Kong, R. Appels, and E. S. Lagudah. 2008: Fine scale genetic and physical mapping using interstitial deletion mutants of *Lr34*/*Yr18*: a disease resistance locus effective against multiple pathogens in wheat. *Theor. Appl. Genet*. 116, 481-490.

Suenaga, K., R. P. Singh, J. Huerta-Espino, and H. M. William. 2003: Microsatellite markers for genes *Lr34*/*Yr18* and other quantitative trait loci for leaf and stripe rust resistance in bread wheat. *Phytopathology* 93, 881-890.

Wan, A. M., Wu, L. R., Jin, S. L., Yao, G. and Wan, B. T. 2003: Discovery and studies on CY32, a new race of *Puccinia striiformis* f. sp. *tritici* in China. *Acta Phytopathologica Sinica* 30, 347-352. (in Chinese with English abstract) Wu, L., H. Z. Zhu, S. Z. Li, Y. L. Zheng, X. C. Xia, and Z. H. He. 2010: Molecular characterization of *Lr34*/*Yr18*/*Pm38* in CIMMYT wheat lines using functional markers. *Scientia Agricultura Sinica* 43, 4553-4561. (in Chinese with English abstract)

Yang, W. X., F. P. Yang, D. Liang, Z. H. He, X. W. Shang, and X. C. Xia. 2008: Molecular characterization of slow-rusting genes *Lr*34/*Yr*18 in Chinese wheat cultivars. *Acta Agronomica Sinica* 34, 1109-1113. (in Chinese with English abstract)

Zeller, F. J., and S. L. K. Hsam. 1996: Chromosomal location of a gene suppressing powdery mildew resistance genes *Pm8* and *Pm17* in common wheat (*Triticum aestivum* L. em. Thell.). *Theor. Appl. Genet*. 93, 38-40.

A novel homeobox-like gene associated with reaction to stripe rust and powdery mildew in common wheat

D. Liu[1], X. C. Xia[1], Z. H. He[1,2], and S. C. Xu[3]

[1] *Institute of Crop Science, National Wheat Improvement Center/The National Key Facility for Crop Gene Resources and Genetic Improvement, Chinese Academy of Agricultural Sciences (CAAS), 12 Zhongguancun South Street, Beijing* 100081, *China*; [2] *International Maize and Wheat Improvement Center (CIMMYT) China Office, c/o CAAS, 12 Zhongguancun South Street, Beijing* 100081, *China*; [3] *Institute of Plant Protection, Chinese Academy of Agricultural Sciences (CAAS), 2 Yuanmingyuan West Road*, 100094 *Beijing, China*

Abstract: Stripe rust and powdery mildew, caused by *Puccinia striiformis* f. sp. *tritici* and *Blumeria graminis* f. sp. *tritici*, respectively, are severe diseases in wheat (*Triticum aestivum* L.) worldwide. In our study, differential amplification of a 201-bp cDNA fragment was obtained in a cDNA-AFLP analysis between near-isogenic lines *Yr10* NIL and Avocet S, inoculated with *P. striiformis* f. sp. *tritici* race CYR29. A full-length cDNA (1,357bp) of a homeobox-like gene, *TaHLRG* (GenBank accession EU385606), was obtained in common wheat based on the sequence of GenBank accession AW448633 with high similarity to the above fragment. The genomic DNA sequence (2,396bp) of *TaHLRG* contains three exons and two introns. *TaHLRG* appeared to be a novel homeobox-like gene, encoding a protein with a predicted 66-amino-acid homeobox domain. It was involved in race-specific responses to stripe rust in real-time quantitative PCR analyses with *Yr9* NIL, *Yr10* NIL and Avocet S. It was also associated with adult-plant resistance to stripe rust and powdery mildew based on the field trials of doubled haploid lines derived from the cross Bainong 64/Jingshuang 16 and two $F_{2:3}$ populations from the crosses Lumai 21/Jingshuang 16 and Strampelli/Huixianhong. A functional marker, *THR1*, was developed based on the sequence of *TaHLRG* and located on chromosome 6A using a set of Chinese Spring nulli-tetrasomic lines.

Intraduction

Stripe rust and powdery mildew, caused by *Puccinia striiformis* f. sp. *tritici* and *Blumeria graminis* f. sp. *tritici*, respectively, are serious wheat diseases worldwide, resulting in both yield losses and downgrading in quality[37]. Use of resistant cultivars is the most economical and environmentally sound method to control these diseases. Disease resistance in crop plants has been generally characterized as race-specific resistance and non-race specific or adult-plant resistance (APR)[14]. Race-specific resistance is often overcome by new pathogen races. In contrast, APR is often long-lived under field conditions and is therefore preferred by breeders aiming to achieve durable resistance, and has been emphasized to increase crop resistance in breeding[37]. For example, the wheat cultivar Knox and its derivative Massey showed stable and effective resistance against powdery mildew for 20 years[9]. Forty four stripe rust resistance and 49 powdery mildew resistance genes in wheat have been for-

Published in Phytopathology, 2008, 98 (12): 1291-1296

mally catalogued[17,26-28,30,33]. Some of them were transferred from wild relatives. A considerable number of them (*Yr11*, *Yr12*, *Yr13*, *Yr14*, *Yr16*, *Yr18*, *Yr29*, *Yr30*, *Yr31*, *Yr36*, *Yr39*) confer APR to stripe rust and two (*Pm38*, *Pm39*) provide APR to powdery mildew[18,19,21,28].

Previous studies indicated that APR genes involved in resistance to different diseases are clustered together on wheat chromosomes[35,39,41]. For example, *Yr18*, *Lr34*, and *Pm38* were located on the same position of chromosome 7DS conferring resistance to stripe rust, leaf rust and powdery mildew, respectively. They were also associated with leaf tip necrosis (*Ltn*)[28,37,41]. Similarly, *Yr29*, *Lr46*, and *Pm39*, which are associated with leaf tip necrosis, were mapped to the same location of chromosome 1BL and co-localize with genes encoding β1-proteasome subunits[35]. The association of APR genes is likely due to close linkage or the pleiotropic effects of a single locus for resistance to different diseases. However, few resistance genes have been cloned in wheat and none is of the APR type.

Many resistance genes contain conserved nucleotide-binding domains [31]. Homeobox genes encode DNA-binding domains with a helix-turn-helix structure, many of which are involved in gene regulation[8]. Typical homeodomains contain 60 amino acids and atypical domains have more or fewer amino acids[4,8]. In plants, homeobox proteins are divided into several families, such as Knox, WOX, Hd-Zip, PHD-finger, GLABRA-, and BEL1-like proteins, according to sequence conservation within the homeodomain and adjacent regions[5,11,12,20,23,34]. Most homeobox genes participate in plant development processes[4,5,7,8]. Some plant HD-Zip genes regulate a biochemical pathway for drought stress[1], and *H52* from tomato encoding a distinct HD-Zip protein was involved in cellular protection from programmed cell death[25]. *OCP3* in *Arabidopsis* mediates resistance to infection by necrotrophic pathogens[6]. Here we report a novel homeobox-like gene in common wheat, designated *TaHLRG*, which is associated with race-specific reaction to stripe rust and APR to both stripe rust and powdery mildew. The role of *TaHLRG* for resistance to wheat disease will be discussed.

Materials and methods

Plant materials

The near-isogenic lines *Yr9*/6 * Avocet S (*Yr9*NIL), *Yr10*/6 * Avocet S (*Yr10*NIL), and Avocet S, kindly provided by Dr. C. R. Wellings, Plant Breeding Institute Cobbitty, University of Sydney, Australia, were used for stripe rust tests in the greenhouse, cDNA-AFLP assay and cloning of the homeobox-like gene. Chinese Spring nulli-tetrasomic (NT) lines used for determining the location of the homeobox-like gene were kindly provided by Prof. R. A. McIntosh, University of Sydney. A doubled haploid (DH) population (128 lines) derived from the cross Bainong 64/Jingshuang 16 and two $F_{2:3}$ populations from the crosses Lumai 21/Jingshuang 16 (186 lines) and Strampelli/Huixianhong (226 lines) were used for the assessment of APR to stripe rust and powdery mildew in the field. Strampelli was introduced from Italy and showed good APR to stripe rust in China for the last 35 years, whereas Huixianhong is highly susceptible to all Chinese *P. striiformis* f. sp. *tritici* races. Bainong 64 and Lumai 21 were previous leading Chinese wheat cultivars with good APR to powdery mildew, whereas Jingshuang 16 is highly susceptible to current *B. graminis* f. sp. *tritici* races in China. The $F_{2:3}$ lines generated from F_2 plants were planted and harvested as bulks with over 50 plants of each line to produce $F_{2:3}$ populations that can be maintained as bulk populations, each deriving from a single F_2 plant.

Inoculation of *P. striiformis* f. sp. *tritici*.

The *Yr9*NIL, *Yr10*NIL, and Avocet S were inoculated with two *P. striiformis* f. sp. *tritici* races, CYR29 and/or culture 78028. *Yr9*NIL was susceptible to CYR29 and resistant to 78028, whereas *Yr10* was resistant, and Avocet S was susceptible to both isolates. Wheat seedlings were grown in a growth chamber at 17℃ with 50% relative humidity (RH), and inoculated with *P. striiformis* f. sp. *tritici* urediospores on the first fully expanded leaves as described by Stubbs[40] and Sun et al. [42]. After the inoculation, the seedlings were maintained in the dark at 9-11℃ with 100% RH. As a control, wheat seedlings were mock-inoculated with water and taken through the same process. Inoculated and corresponding mock-inoculated leaves for each treatment (*Yr9*-CYR29, *Yr9*-78028, *Yr10*-CYR29, and Avocet-CYR29) were collected at 12, 15, 18, 21, 24, 28, 32, and 36 h post-inoculation (hpi) for subsequent cDNA-AFLP and quantitative PCR (Q-PCR) analysis. Meanwhile, the seedlings of susceptible check cultivar, Mingxian169, in each treatment were transferred into a growth chamber with identical conditions, i. e., the day/night regime of 14h light (28,000lx) at 17℃ and 10h of darkness at 12℃, with 70% RH. All checks were fully susceptible (IT=4) two weeks after inoculation, indicating a successful inoculation in all treatments.

cDNA-AFLP assay

Total RNA was prepared from the first leaves of CYR29-inoculated and corresponding mock-inoculated *Yr10* NIL and Avocet S, collected at five time points (15, 18, 21, 24, and 36hpi) using TRIZOL reagent (Invitrogen, Carlsbad, CA, USA) according to the manufacturer's recommendations. DNase Ⅰ-treated (Qiagen, Hilden, Germany) total RNA was then used as a template for double-stranded cDNA amplification by a double-stranded cDNA amplification kit (*TaKaRa*, Dalian, China) following the manufacturer's recommendations. The double-stranded cDNA was digested with restriction enzymes *Pst* I and *Mse* I (New England Biolabs, Ipswich, CT, USA), and the 3' regions were captured on streptavidin Dynabeads (Invitrogen, Carlsbad, CA, USA). The pre-amplifications were performed with the *Pst* I and *Mse* I primers, with either T or C as the 3' nucleotide. From a 40-fold dilution of the pre-amplified samples, 5μl was used for the final selective amplifications using *Pst* I and *Mse* I primers with 3 selective nucleotides (CGA or TAC). Amplification products were separated on 5% polyacrylamide gels and visualized by silver staining[3]. The differential PCR fragments amplified from the cDNA-AFLP assay were cloned following the method described by Xu et al. [44].

Gene cloning by 3' RACE, 5' RACE and PCR amplification of genomic DNA

Oligonucleotides (Table 1) used for 3' and 5' RACE were designed based on the sequence of a wheat EST (GenBank accession AW448633). This EST was detected in a BLAST search, and was highly similar to a 201-bp differential cDNA-AFLP fragment. DNase I-treated (Qiagen, Hilden, Germany) total RNA from seedlings of *Yr10* NIL inoculated with CYR29 was used as a template for 3' and 5' rapid amplification of cDNA ends (RACE). For 3' RACE, an oligo-dT adapter primer (AP), 3'AP1, was used in RT-PCR for rapid amplification of the first-strand cDNA using a RT kit (Qiagen, Hilden, Germany) following the manufacturer' s recommendations. A gene-specific primer (GSP), 3'GSP1, and an adapter primer, 3'AP2, permitted capture of 3'-mRNA sequence for the 563 to 1357-bp region. PCR was performed using *LA Taq* DNA polymerase (*TaKaRa*, Dalian, China) with an annealing temperature of 60℃ according to the manufacturer's recommendation.

The 5' end of the cDNA was amplified using the 5'-RACE technique following the manufacturer's instructions (Invitrogen, Carlsbad, CA, USA). The primer 5'GSP1 was used to obtain the first-strand cDNA and a poly-C tail was added to the 3'end of the cDNA. In the second PCR, a cDNA fragment (1 to 286bp) was amplified with the primer set 5'GSP2/5'AP1 and sequenced. Then, 5'GSP3/5'AP2 were used in PCR amplification with the above 286-bp PCR product as template, and a 157-bp cDNA fragment (1 to 157bp) was amplified. The two fragments overlapped by 157bp with 100% sequence identity. The sequence of the 5'RACE fragment confirmed that the 286-bp cDNA fragment was the 5'-end sequence (1 to 286bp) of *TaHLRG*. The primer set OP4/OP5 was designed from the AW448633 sequence to obtain a 380-bp (8 to 387bp) cDNA fragment that overlapped with the 5'-end sequence, with 100% sequence identity in 279bp.

The primer set OP1/OP2 was used to amplify the region 266 to 665bp, overlapping with the OP4/OP5 amplification region and 3' RACE cDNA sequence in 122bp and 103bp, respectively, with 100% sequence identity. PCR was performed using *Pfu* DNA Polymerase (TIANGEN, Beijing, China) with an annealing temperature 55-65℃ following the manufacturer's recommendation.

Table 1 Oligonucleotides used for 3' RACE, 5' RACE, RT-PCR, and PCR amplification of genomic DNA

Primer	Primer sequence (5' -3')
3' AP1	CTGATCTAGAGGTACCGGATCCTTTTTTTTTTTTTTT
3' AP2	CTGATCTAGAGGTACCGGATCC
3' GSP1	TGGACCCGAGGAGGATGAATGG
5' AP1	GGCCACGCGTCGACTAGTACGGGIIGGGIIGGGIIG
5' AP2	GGCCACGCGTCGACTAGTAC
5' GSP1	TGGGGGCACCATCATCT
5' GSP2	GAGCGGTCCGGGGAGGCGTTGAA
5' GSP3	CTTGGCGCCTGCCCCGTCATACTTGTC
OP1	CAACGCCTCCCCGGACCGCTCC
OP2	GAGATCCCCAGGAAGCACAGC
OP3	GTAAGCTTCTCCTCCCACCCACCCCTCTCC
OP4	TCTCCTCCCACCCACCCCTCTCCA
OP5	CGGCCACTGGGGGCACCATCATCT
OP6	CAGCGGCACCCTCTTACATC
OP7	AGGCCCTCTATTTTGACGCTCGTGTTCT
OP8	AAGCGCACAGGGAAGCGAAACAACT
Tub2U	TGTGCCCCGTGCTGTTCTTATG
Tub2L	CCCTTGGCCCAGTTGTTACCC

Genomic DNA was extracted from *Yr10* NIL seedlings[36]. The genomic DNA sequence of the *TaHLRG* gene was assembled with PCR amplification of four primer sets (OP4/OP5, OP1/OP2, 3'GSP1/OP6 and OP7/OP8, Table 1), amplifying the regions covering 1 to 477, 259 to 755, 653 to 1985, and 1916 to 2396bp, respectively, with 100% identity in the overlapping sequences. PCR was performed using *Pfu* DNA Polymerase (TIANGEN, Beijing, China) with an annealing temperature of 55-65℃ according to the manufacturer's recommendation.

Real-time quantitative PCR

Total RNA samples were extracted using TRIZOL reagent (Invitrogen, Carlsbad, CA, USA) from the inoculated and corresponding mock-inoculated leaves of each treatment (*Yr9*-CYR29, *Yr9*-78028, *Yr10*-CYR29, and Avocet-CYR29) at eight time points (12, 15, 18, 21, 24, 28, 32, and 36hpi), following protocols recommended by the manufacturer. One-step RT-PCR (RT kit, Qiagen, Hilden, Germany) was performed using DNase I-treated (Qiagen, Hilden, Germany) total RNA as template.

In the Q-PCR assay, sequence-specific primer pairs, 3' GSP1/OP6 and Tub2U/Tub2L designed by DNAStar, were used to quantify accumulation of *TaHLRG* transcripts and to normalize the amount of cDNA samples, respectively. Tub2U/Tub2L was an endogenous control from the sequence of the *T. aestivum beta-tubulin* 2 gene (GenBank accession U76745), amplifying a 135-bp fragment (176 to 310bp). The cDNA region amplified with the primer set 3'GSP1/OP6 was 201bp (563 to 763bp), whereas the corresponding genomic region included an intron (771 to 1 902bp), facilitating a check of genomic DNA contamination in cDNA samples. All Q-PCR products from the two primer sets were sequenced to confirm the sequence identity with the template cDNA. Q-PCR experiments were performed in an ICycler iQ (Bio-Rad, USA) with high amplification efficiency using a SYBR GREEN kit (Qiagen, Hilden, Germany) according to the manufacturer's recommendations.

The $2^{-\Delta\Delta C_T}$ method[22] was used to normalize and calibrate the *TaHLRG* C_T-values relative to the *beta-tubulin* 2 endogenous control and the corresponding mock-inoculated controls. C_T-values included in the analyses were based on three replicates for each treatment at each time point. All statistical analyses were performed using Optical System Software, Version 3. 1.

Field trials for stripe rust and powdery mildew reactions

The 226 $F_{2:3}$ lines from cross Strampelli/Huixianhong were tested for stripe rust reaction in Qingshui following inoculation with *P. striiformis* f. sp. *tritici* race CYR32 in the 2004-2005, 2005-2006, and 2006-2007 cropping seasons. Qingshui, located in the Gansu province, is considered a stripe rust 'hot spot' in China, having ideal conditions for rust infection and spread. Field trials were conducted in a randomized complete block design with three replicates; each plot consisted of two 1. 5-m rows spaced 25cm apart. Approximately 150 seeds were sown in each plot. The highly susceptible line Tiaogan 601 was used as a susceptible check and was planted after every 10 plots. Spreader rows of Tiaogan 601 were planted perpendicular and adjacent to the rows of tested lines. Stripe rust severities were assessed around June 25 for the first time and then at weekly intervals for two further weeks using the modified Cobb scale[29,32].

The 128 DH lines from cross Bainong 64/Jingshuang 16 and 186 $F_{2:3}$ lines from cross Lumai 21/Jingshuang 16 were tested in the field for powdery mildew reaction in Beijing and Anyang located in Henan province, during the 2004-2005 and 2005-2006 cropping seasons. In Beijing, they were artificially inoculated with the prevalent *B. graminis* f. sp. *tritici* isolate E20, whereas in Anyang a mixture of pathotypes was used. Field trials were planted in randomized complete block designs with three replicates; each plot was comprised of two 1. 5-m rows spaced 25cm apart. Approximately 150 seeds were sown in each plot. Spreader rows of the susceptible cultivar Jingshuang 16 surrounded the tested cultivars and were also planted after every 10 plots. In Beijing, the infection severity on F-1 leaves (the leaf below the flag leaf) of 10 arbitrarily selected plants from each plot were scored, based on the actual percentage (0-100%) leaf area covered by powdery mildew[10]. Disease scoring was carried out for the first time three weeks after inoculation and then at weekly intervals until leaves were physiologically mature. Disease severities of 10 selected plants were averaged to obtain a mean severity for each line. In Anyang, powdery mildew severities on the F-1 leaf of each line were visually rated as percentage leaf area covered by powdery mildew, when the disease severities of the susceptible check cv. Jingshuang 16 reached a maximum level around May 18, in 2005 and 2006. The maximum stripe rust and powdery mildew severity were used to analyze their association with *TaHLRG* gene.

A functional marker, *THR1*, was developed from the sequence of *TaHLRG* using the software SSRHunter1. 3. One-way analysis of variance (ANOVA), using the software SAS 6. 12, was conducted to determine the association between the functional marker *THR1* and APR reaction to powdery mildew and stripe rust.

Results

Cloning a homeobox-like gene

In the cDNA-AFLP assay, 192 transcript derived fragments (TDF) were obtained and sequenced, ranging from 80 to 750bp in size. A 201-bp cDNA fragment was expressed at five time points (15, 18, 21, 24, and 36hpi) in *Yr10* NIL inoculated with *P. striiformis* f. sp. *tritici* race CYR29 and matched known genes involved in pathogen response. A wheat EST (GenBank accession AW448633) was detected through a BLAST search using the 201-bp cDNA sequence in a search against the wheat EST database in GenBank (http: // www. ncbi. nlm. nih. gov). Oligonucleotide primers were designed based on the AW448633 sequence, and a new *Triticum aestivum* homeobox-like resistance gene, designated *TaHLRG*, was cloned in *Yr10* NIL by 3' RACE, 5' RACE, and PCR amplification of genomic DNA, encoding a protein with a predicted HOMEOBOX _ 2 domain. In the genomic clone the sequence was 2,396bp (GenBank accession EU385606), with three exons from nucleotides 131 to 313, 411 to 770, and 1,903 to 2,373. The assembled full-length cDNA sequence (GenBank accession EU364815) was 1,357bp containing an open reading frame (ORF) of 1,014bp. It encoded a protein of 337 amino acid residues with a predicted 66-amino-acid HOMEOBOX _ 2 domain (PS500071, at positions 11 to 76) close to the N terminus that formed a conserved helix-turn-helix structure including a first helix (amino acids 11 to 43), a turn (amino acids 44 to 55) and a second helix (amino acids 56 to 76) as predicted by ScanProsite (http: // www. expasy. ch/tools/scanprosite/). *TaHLRG* was compared with 20 known genes with HOMEOBOX _ 2 domain sequences (Fig. 1) in maize, *Arabidopsis thaliana*, barley (*Hordeum vulgare* L.) and rice (*Oryza sativa* L.), which contained the conserved residues in alignment of Swiss-Prot true positive hits by PS500071 (http: //www. expasy. ch/cgi-bin/nicedoc. pl? PS50071) and were chosen from their representative sequence homology in each gene family. The predicted homeobox of *TaHLRG* contained six conserved residues (Leu-39, Phe-43, Trp-65, Phe-66, Asn-68, and Arg-70) and six frequently occurring residues (Arg-25, 69, 75; Gln-61, 67; Val-62). The other residues of *TaHLRG* were not homologous to any known homeodomain proteins, indicating that *TaHLRG* is a new homeobox-like gene.

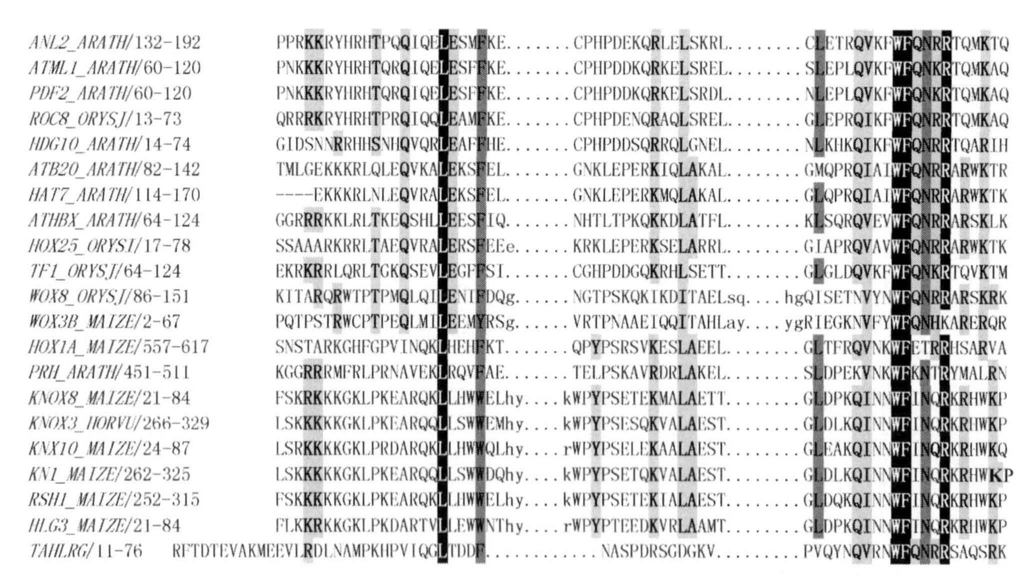

Fig. 1 HOMEOBOX _ 2 domain sequence comparison among *TaHLRG* and 20 known homeobox-like genes. The shaded amino acids are conserved residues. The numbers following the gene names indicate the position of helix-turn-helix domain in the deduced peptides.

Expression analysis of *TaHLRG*

Ina Q-PCR assay, the transcript level of *TaHLRG* was up-regulated in the compatible (*Yr9*-CYR29) interaction at 15hpi and no significant changes were found at seven other time points (12, 18, 21, 24, 28, 32, and 36hpi) (Fig. 2). In the incompatible interaction (*Yr10*-CYR29), the transcript level of *TaHLRG* was firstly down-regulated at 24hpi and then up-regulated at 28 hpi, respectively, and no significant changes were detected at six other time points (Fig. 2). These results indicated that *TaHLRG* was constitutively expressed in both inoculated and mock-inoculated samples, and might be involved in both compatible and incompatible race-specific responses.

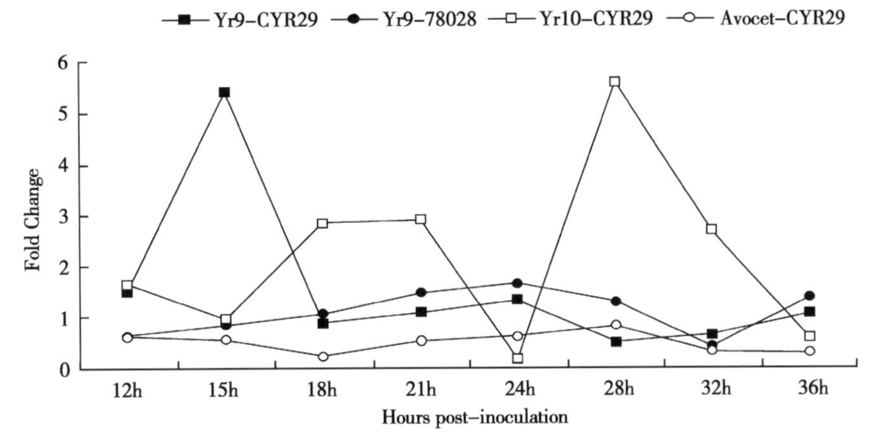

Fig. 2 *TaHLRG* transcript levels relative to corresponding mock-inoculated samples at eight sampling points. Interactionas were: *Yr9* NIL-CYR29 (compatible), *Yr9* NIL-78028 (incompatible), *Yr10* NIL-CYR29 (incompatible) and Avocet-CYR29 (compatible).

TaHLRG was associated with adult-plant resistance to stripe rust and powdery mildew

A functional marker, *THR1* (primer set OP3/5' GSP3), was developed from the sequence of *TaHLRG*, amplifying two PCR fragments of 145 and 157 bp in two parents of the populations, respectively. The sequence of the 157-bp PCR fragment is 100% identical to that of *TaHLRG*, whereas two simple sequence repeats were absent in the 145-bp PCR fragment.

THR1 was used to genotype three populations. One-way ANOVA analyses indicated that the gene explained 9.3-9.7% and 11.9-20.5% of the phenotypic variance for powdery mildew reaction in populations Bainong 64/Jingshuang 16 and Lumai 21/Jingshuang 16, respectively, and accounted for 11.8-22.5% of the phenotypic variance for stripe rust resistance in the population from Strampelli/Huixianhong across different environments (Table 2). The results indicated that *TaHLRG* was strongly associated with APR for both stripe rust and powdery mildew resistance in wheat.

Table 2 One-way ANOVA for the association between _TaHLRG_ and adult plant reaction to stripe rust and powdery mildew in three populations

Population	Disease	Location	Cropping season	R^{2} [a]	P [b]
Bainong 64/Jingshuang 16	Powdery mildew	Beijing	2004-05	9.3	0.000 5
		Anyang	2004-05	9.7	0.000 4
Lumai 21/Jingshuang 16	Powdery mildew	Beijing	2004-05	20.5	$<$0.000 1
		Beijing	2005-06	11.9	$<$0.000 1
		Anyang	2005-06	17.5	$<$0.000 1
Strampelli/Huixianhong	Stripe rust	Qingshui	2004-05	11.8	$<$0.000 1
		Qingshui	2005-06	22.5	$<$0.000 1
		Qingshui	2006-07	12.8	$<$0.000 1

a. Coefficient of determination: the maximum disease severities scored in the field trials were used to investigate the association between *TaHLRG* and adult-plant reaction to powdery mildew or stripe rust.

b. P=probability.

Chromosomal location of *TaHLRG*

In PCR amplification with the marker *THR1*, no PCR product was amplified from the Chinese Spring (CS) nulli-tetrasomic (NT) lines N6A/T6B and N6A/T6D (Fig. 3). However, a 145-bp fragment was amplified from CS and all other NT lines (data not shown), indicating that *TaHLRG* was located on wheat chromosome 6A.

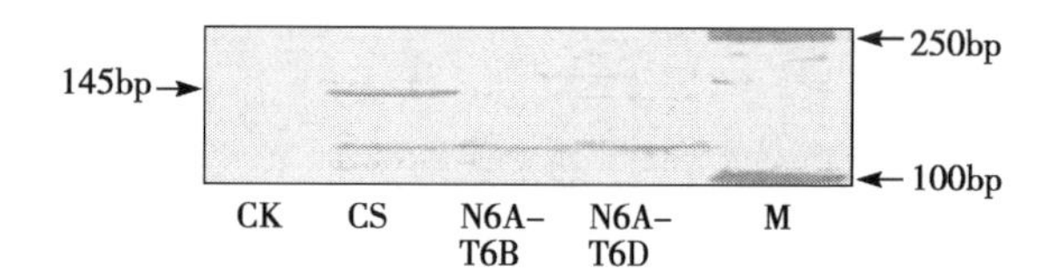

Fig. 3 Chromosomal location of *TaHLRG* by amplifying CS, N6A-T6B and N6A-T6D with the marker *THR1* and separating products on a 5% denaturing polyacrylamide gel. M, DNA ladder DL2000 (TaKaRa, Dalian, China); CK, check with no template DNA in the PCR reaction; CS, Chinese Spring; N6A-T6B, nullisomic 6A-tetrasomic 6B; N6A-T6D, nullisomic 6A-tetrasomic 6D.

Discussion

TaHLRG is a novel homeobox-like gene identified through a BLAST search against the nr database in GenBank. The specific motif, HOMEOBOX _ 2, contains 12 conserved residues, without a consensus pattern with other homeobox-like genes (http: //www.expasy.ch/cgi-bin/nicedoc.pl? PS 50071). Apart from the 12 conserved residues in the homeobox domain of *TaHLRG* (Fig. 1), other residues were not homologous to any known homeodomain proteins. Therefore, *TaHLRG* may encode a new homeobox protein.

A hypersensitive response (HR) is an efficient plant defense against biotrophic pathogens, resulting in localized cell death and formation of necrotic lesions. In incompatible interactions between resistant wheat cultivars and stripe rust races, the host cells start to become necrotic at 24hpi[16]. At the same time, the structural defense reactions such as formation of cell wall apposition, collar or papillae are more pronounced in the infected leaves of the resistant cultivars than in those of the susceptible ones[15]. After 24hpi, the infected cells become necrotic and collapse inhibiting the proliferation of the stripe rust fungus[16]. In the incompatible interaction (*Yr10*-CYR29), the expression of *TaHLRG* was detected in the above two phases. At the beginning of the structural defense reaction, 24 hpi, *TaHLRG* was down-regulated (Fig. 2). *TaHLRG* was strongly up-regulated at 28hpi, corresponding to a period of cell death, which may stop further development of pathogen. In contrast, no significant change of *TaHLRG* transcripts was found in the compatible interaction of Avocet-CYR29. This indicated that *TaHLRG* might play an important role in wheat line *Yr10* NIL against stripe rust race CYR29. In the compatible interaction of *Yr9*-CYR29, the transcript level of *TaHLRG* was strongly up-regulated as early as 15hpi, before host cells became necrotic. In contrast, the expression of *TaHLRG* showed no significant changes among different time points in the incompatible interaction of *Yr9*-78028. These reactions may reflect the specific incompatible and compatible interactions between resistance gene and stripe rust races. A number of DNA-binding proteins are important in the transmission of defense signals for the plant-pathogen interaction to either activate or suppress downstream gene expression, and in the regulation of cross-talk between different signaling pa thways[2,24]. *TaHLRG* has the characteristics of genes with a DNA-binding region. Possibly, *TaHLRG* is involved in cross-talk between different signaling pathways.

To date, three powdery mildew resistance genes[13] and five stripe rust resistance genes [26-28] are located on chromosome 6 in wheat. Among them, *Yr36*-6BS was characterized as a high-temperature APR gene that was susceptible to 15 different stripe rust races at the seedling stage and resistant to PST100 and PST101 at the adult plant stage under high temperature. *Pm11*, *Pm14*, *Pm31*, *Yr4a*, *Yr4b*, *Yr20*, and *Yr23* are major genes with specific resistance to different pathotypes. *Yr35*-6BS and *Yr38*-6A are dominant genes with resistance to a range of *P. striiformis* f. sp. *tritici* pathotypes. Based on its location, *TaHLRG* may be a new quantitative trait locus (QTL) associated with resistance to both powdery mildew and stripe rust. In addition, the EST AW448633 showed high homology (93%) to a wheat EST BG262981 on chromosomes 6A and 6B from a blast search of mapped wheat EST in the GrainGenes website (http://wheat.pw.usda.gov/GG2/blast.shtml), confirming the location of *TaHLRG*.

Previous studies indicated that APR genes against different pathogens are located on the same chromosomal regions and they may be involved in a similar biochemical pathway for APR in wheat[18,35,38,39,41]. The β1din subunit of the 20S proteasome, located in the genomic intervals carrying *Lr34/Yr18/Pm38/Ltn* and *Lr46/Yr29/Pm39/Ltn* [35], was shown to be involved in the establishment of systemic acquired resistance[43]. In our study, *TaHLRG* was characterized as a gene involved in race-specific responses. Highly significant associations between *TaHLRG* and stripe rust and powdery mildew resistance were found in three populations across two or three environments, indicating a stable effect associated with *TaHLRG* for resistance to stripe rust and powdery mildew.

❖ Acknowledgments

The authors are very grateful to the critical review of this manuscript by Prof. R. A. McIntosh, Plant Breeding Institute, University of Sydney. This study was supported by the National Science Foundation of China (30471083).

❖ References

[1] Agalou, A., Purwantomo, S., Övernäs, E., Johannesson, H., Zhu, X., Estiati, A, Kam, R., Engström, P., Slamet-Loedin, I., Zhu, Z., Wang, M., Xiong, L., Meijer, A., and Ouwerkerk, P. A genome-wide survey of HD-Zip genes in rice and analysis of drought-responsive family members. Plant Mol. Biol. 2008, 66: 87-103.

[2] Anderson, J. P., Badruzsaufari, E., Schenk, P. M., Manners, J. M., Desmond, O. J., Ehlert, C., Maclean, D. J., Ebert, P. R., and Kazan, K. Antagonistic interaction between abscisic acid and jasmonate-ethylene signaling pathways modulates defense gene expression and disease resistance in Arabidopsis. Plant Cell 2004, 16: 3460-3479.

[3] Bassam, B. J., Caetano-Anolles, G., and Gresshoff, P. M. Fast and sensitive silver staining of DNA in polyacrylamide gels. Anal. Biochem. 1991, 196: 80-83.

[4] Bürglin, T. R. A comprehensive classification of homeobox genes. Pages 25-71. in: Guidebook to the Homeobox Genes. D. Duboule, ed. Oxford University Press, Oxford U.K. 1994.

[5] Chan, R. L., Gago, G. M., Palena, C. M., and Gonzalez, D. H. Homeoboxes in plant development. Biochim. Biophys. Acta. 1998, 1442: 1-19.

[6] Coego, A., Ramirez, V., Gil, M., Flors, V., Mauch-Mani, B., and Vera, P. An Arabidopsis homeodomain transcription factor, *OVEREXPRESSOR OF CATIONIC PEROXIDASE* 3, mediates resistance to infection by necrotrophic pathogens. Plant Cell. 2005, 17: 2123-2137.

[7] Gehring, W. J., Affolter, M., and Bürglin, T. Homeodomain proteins. Annu. Rev. Biochem. 1994, 63: 487-526.

[8] Gehring, W. F., Qian, Y. Q., Billeter, M., Furukubo-Tokunaga, K., Schier, A. F., Resendez-Perez, D., Affolter, M., Otting, G., and Wüthrich, K. Homeodomain-DNA recognition. Cell. 1994, 78: 211-223.

[9] Griffey, C. A., and Das, M. K. Inheritance of adult-plant resistance to powdery mildew in Knox 62 and Massey winter wheats. Crop Sci. 1994, 34: 641-646.

[10] Gustafson, G. D., and Shaner, G. The influence of plant age on the expression of slow-mildewing resistance in wheat. Phytopathology. 1982, 72: 746-749.

[11] Haecker, A., Gross-Hardt, R., Geiges, B., Sarkar, A., Breuninger, H., Herrmann, M., and Laux, T. Expression dynamics of *WOX* genes mark cell fate decisions during early embryonic patterning in *Arabidopsis thaliana*. Development. 2004, 131: 657-668.

[12] Himmelbach, A., Hoffmann, T., Leube, M., Hohener, B., and Grill, E. Homeodomain protein ATHB6 is a target of the protein phosphatase ABI1 and regulates hormone responses in Arabidopsis. EMBO J. 2002, 21: 3029-3038.

[13] Huang, X. Q., Röder, M. S. Molecular mapping of powdery mildew resistance genes in wheat: A review.

Euphytica . 2004, 137: 203-223.

[14] Johnson, R. Durable resistance to yellow (stripe) rust in wheat and its implications in plant breeding. Pages 63-75. in: Breeding Strategies for Resistance to the Rusts of Wheat. N. W. Simmonds and S. Rajaram, eds. CIMMYT Mexico. 1988.

[15] Kang, Z. S. , Huang, L. L. , and Buchenauer, H. Ultrastructural changes and localization of lignin and callose in compatible and incompatible interactions between wheat and *Puccinia striiformis*. J. Plant Dis. Protect. 2002, 109: 25-37.

[16] Kang, Z. S. , Wang, Y. , Huang, L. L. , Wei, G. R. , and Zhao, J. Histology and ultrastructure of incompatible combination between *Puccinia striiformis* and wheat cultivars with low reaction type resistance. Scientia Agricultura Sinica. 2003, 2: 1102-1113.

[17] Kuraparthy, V. , Chhuneja, P. , Dhaliwal, H. S. , Kaur, S. , Bowden, R. L. , and Gill, B. S. Characterization and mapping of cryptic alien introgression from *Aegilops geniculata* with new leaf rust and stripe rust resistance genes *Lr57* and *Yr40* in wheat. Theor. Appl. Genet. 2007, 114: 1379-1389.

[18] Liang, S. S. , Suenaga, K. , He, Z. H. , Wang, Z. L. , Liu, H. Y. , Wang, D. S. , Singh, R. P. , Sourdile, P. , and Xia, X. C. Quantitative trait loci mapping for adult-plant resistance to powdery mildew in bread wheat. Phytopathology. 2006, 96: 784-789.

[19] Lin, F. , and Chen, X. M. Genetics and molecular mapping of genes for race specific and all-stage resistance and non-specific high-temperature adult-plant resistance to stripe rust in spring wheat cultivar Alpowa. Theoretical Applied Genetics . 2007, 114: 1277-1287.

[20] Lincoln, C. , Long, J. , Yamaguchi, J. , Serikawa, K. , and Hake, S. A *knotted 1*-like homeobox gene in Arabidopsis is expressed in the vegetative meristem and dramatically alters leaf morphology when overexpressed in transgenic plants. Plant Cell. 1994, 6: 1859-1876.

[21] Liu, S. X. , Griffey, C. A. , and Saghai-Maroof, M. A. Identification of molecular markers associated with adult plant resistance to powdery mildew in common wheat cultivar Massey. Crop Sci. 2001, 41: 1268-1275.

[22] Livak, K. J. , and Schmittgen, T. D. Analysis of relative gene expression data using realtime quantitative PCR and the $2^{-\Delta\Delta C}{}_{T}$ method. Methods. 2001, 25: 402-408.

[23] Long, J. A. , Moan, E. I. , Medford, J. I. , and Barton, M. K. A member of the KNOTTED class of homeodomain proteins encoded by the *STM* gene of Arabidopsis. Nature . 1996, 379: 66-69.

[24] Lorenzo, O. , Piqueras, R. , Sanchez-Serrano, J. J. , and Solano, R. ETHYLENE RESPONSE FACTOR 1 integrates signals from ethylene and jasmonate pathways in plant defense. Plant Cell. 2003, 15: 165-178.

[25] Mayda, E. , Tornero, P. , Conejero, V. , and Vera, P. A tomato homeobox gene (HD-Zip) is involved in limiting the spread of programmed cell death. Plant J. 1999, 20: 591-600.

[26] McIntosh, R. A. , Devos, K. M. , Dubcovsky, J. , Morris, C. F. , Appels, R. , and Anderson, O. A. Catalogue of gene symbols for wheat: 2005. (suppl) http: //www. wheat. pw. usda. gov/GG2/pubs. shtml, 2005.

[27] McIntosh, R. A. , Devos, K. M. , Dubcovsky, J. , and Rogers, W. J. Catalogue of gene symbols for wheat: 2004. (suppl) http: //www. wheat. pw. usda. gov/GG2/pubs. shtml, 2004.

[28] McIntosh, R. A. , Devos, K. M. , Dubcovsky, J. , Rogers, W. J. , Morris, C. F. , Appels, R. , Somers, D. J. , and Anderson, O. A. Catalogue of gene symbols: 2007 supplement. In: KOMUGI-Integrated Wheat Science Database. (http: //www. grs. nig. ac. jp/wheat/komugi) 2007.

[29] McIntosh, R. A. , Wellings, C. R. , and Park, R. F. Wheat rusts: an atlas of resistance genes. CSIRO Australia. 1995, pp 1-200.

[30] McIntosh, R. A. , Yamazaki, Y. , Devos, K. M. , Dubcovsky, J. , Rogers, W. J. , and Appels, R. Catalogue of gene symbols for wheat. *In*: Pogna NE, Romano M, Pogna EA, Galterio G (eds). Proc 10th Int Wheat Genet. Symp. Vol. 2003, 4: 1-34.

[31] Mondragon-Palomino, M. , Meyers, B. C. , Michelmore, R. W. , and Gaut, B. S. Patterns of positive selection in the complete NBS-LRR gene family of *Arabidopsis thaliana*. Genome Res. 2002, 12: 1305-1315.

[32] Peterson, R. F. , Campbell, A. B. , and Hannah, A. E. A diagrammatic scale for estimating rust intensity of leaves and stems of cereals. Can. J. Res. Sect. C. 1948, 26: 496-500.

[33] Ramburan, V. P. , Pretorius, Z. A. , Louw, J. H. , Boyd, L. A. , Smith, P. H. , Boshoff, W. H. P. , and Prins, R. A genetic analysis of adult plant resistance to stripe rust in the wheat cultivar Kariega. Theor. Appl. Genet. 2004, 108: 1426-1433.

[34] Reiser, L. , Modrusan, Z. , Margossian, L. , Sam-

ach, A., Ohad, N., Haughn, G. W., and Fischer, R. L. The *BEL1* gene encodes a homeodomain protein involved in pattern formation in the Arabidopsis ovule primordium. Cell. 1995, 83: 735-742.

[35] Rosewarne, G. M., Singh, R. P., Huerta-Espino, J., William, H. M., Bouchet, S., Cloutier, S., McFadden, H., and Lagudah, E. S. Leaf tip necrosis, molecular markers and β1-proteasome subunits associated with the slow rusting resistance genes *Lr46/Yr29*. Theor. Appl. Genet. 2006, 112: 500-508.

[36] Saghai-Maroof, M. A., Soliman, K. M., Jorgensen, R. A., and Allard, R. W. Ribosomal DNA spacer-length polymorphisms in barley: Mendelian inheritance, chromosomal location, and population dynamics. Proc. Natl. Acad. Sci. USA. 1984, 81: 8014-8018.

[37] Singh, R. P. Genetic association of leaf rust resistance gene *Lr34* with adult plant resistance to stripe rust in bread wheat. Phytopathology 82: 835-838.

[38] Singh, R. P., Nelson, J. C., and Sorrells, M. E. 2000. Mapping *Yr28* and other genes for resistance to stripe rust in wheat. Crop Sci. 1992, 40: 1148-1155.

[39] Spielmeyer, W., McIntosh, R. A., Kolmer, J., and Lagudah, E. S. Powdery mildew resistance and *Lr34/Yr18* genes for durable resistance to leaf and stripe rust cosegregate at a locus on the short arm of chromosome 7D of wheat. Theor. Appl. Genet. 2005, 111: 731-735.

[40] Stubbs, R. W. Pathogenicity analysis of yellow (stripe) rust of wheat and its significance in a global context. Pages 23-28. in: Breeding Strategies for Resistance to the Rusts of Wheat. N. W. Simmonds and S. Rajaram, eds. CIMMYT Mexico. 1988.

[41] Suenaga, K., Singh, R. P., Huerta-Espino, J., and William, H. M. Microsatellite markers for genes *Lr34/Yr18* and other quantitative trait loci for leaf rust and stripe rust resistance in bread wheat. Phytopathology. 2003, 93: 881-890.

[42] Sun, Q., Wei, Y., Ni, C., Xie, C., and Yang, T. Microsatellite marker for yellow rust resistance gene *Yr5* introgressed from spelt wheat. Plant Breed, 2002, 121: 539-541.

[43] Suty, L., Lequeu, J., Lancon, A., Etienne, P., Petitot, A. S., and Blein, J. P. Preferential induction of 20S proteasome subunits during elicitation of plant defense reactions: towards the characterization of "plant defense proteasomes". Int. J. Biochem. Cell Biol. 2003, 35: 637-650.

[44] Xu, M. L., Huaracha, E., and Korban, S. S. Development of sequence-characterized amplified regions (SCARs) from amplified fragment length polymorphism (AFLP) markers tightly linked to the *Vf* gene in apple. Genome. 2001, 44: 63-70.

中国小麦育成品种和农家种中慢锈基因 *Lr34/Yr18* 的分子检测

杨文雄[1,2]，杨芳萍[1,2]，梁丹[2,3]，何中虎[2,4]，尚勋武[5]，夏先春[2]

[1]甘肃省农业科学院作物研究所，兰州 730070；[2]中国农业科学院作物科学研究所/国家小麦改良中心，北京 100081；[3]安徽农业大学农学院，合肥 230036；[4]CIMMYT 中国办事处，北京 100081；[5]甘肃农业大学农学院，兰州 730070

摘要：*Lr34/Yr18* 是重要的慢叶锈和慢条锈基因，携带该连锁基因的小麦品种被广泛种植于世界许多国家。利用 STS 标记 *csLV34* 对慢叶锈和慢条锈基因 *Lr34/Yr18* 进行分子检测的结果表明，我国 231 份育成品种（系）中仅有 14 份材料携带 *Lr34/Yr18* 基因，占 6.1%。不同麦区分布频率不同，其中北部冬麦区为零，黄淮冬麦区、长江中下游冬麦区、西南冬麦区和西北春麦区分别为 3.0%、21.4%、16.7%和 33.3%。在 422 份农家种中，359 份含有 *Lr34/Yr18* 基因，占 85.1%。*Lr34/Yr18* 基因在不同麦区的分布频率也存在差异，北部冬麦区、黄淮冬麦区、长江中下游冬麦区、西南冬麦区、南部冬麦区和西北春麦区分别为 89.6%、77.4%、93.1%、93.8%、96.6%和 61.1%。*csLV34* 标记扩增产物为 150 和 229 bp 的片段，能有效鉴别品种是否携带 *Lr34/Yr18* 基因，是一个重复性好、准确率高的分子标记，可用于小麦 *Lr34/Yr18* 基因的鉴定与选择。

关键词：普通小麦；*Lr34/Yr18* 基因；STS 标记 *csLV34*；分子标记辅助选择

Molecular Characterization of Slow-Rusting Genes *Lr34/Yr18* in Chinese Wheat Cultivars

Yang Wenxiong[1,2], Yang Fangping[1,2], Liang Dan[2,3], He Zhonghu[2,4], Shang Xunwu[5], Xia Xianchun[2]

[1]*Crop Research Institute, Gansu Academy of Agricultural Sciences, Lanzhou* 730070 *Gansu*; [2]*Institute of Crop Sciences/ National Wheat Improvement Center, Chinese Academy of Agricultural Sciences*; [3]*College of Agronomy, Anhui Agricultural University, Hefei* 230036, *Anhui*; [4]*CIMMYT China Office, Beijing* 100081; [5]*College of Agronomy, Gansu Agricultural University, Lanzhou* 730070, *Gansu, China*

Abstract: *Lr34* and *Yr18* are important slow rusting resistance genes that have been widely used in many countries. The aim of this study was to identify *Lr34/Yr18* genes in 231 Chinese wheat cultivars and 422 landraces using an STS marker *csLV34*, which can amplify a 150-bp and a 229-bp fragment in the genotypes with and without *Lr34/Yr18*, respectively. The results indicated that only 14 genotypes carried the resistance genes *Lr34/Yr18* in the improved wheat cultivars, with a frequency of 6.1%. Different frequencies were found in wheat cultivars from different zones, which were 0, 3.0%, 21.4%, 16.7%, and 33.3% in North Plain Winter Wheat Region (NPWWR), Yellow-Huai Facultative Winter

原文发表在《作物学报》，2008，34（7）：1109-1113

Wheat Region (YHFWWR), Middle-Lower Yangtze Valley Winter Wheat Region (MLYVWWR), Southwest Winter Wheat Region (SWWR), and Northwest Spring Wheat Region (NSWR), respectively. Whereas, 85.1% of 422 landraces contained *Lr34/Yr18*, with the frequencies of 89.6%, 77.4%, 93.1%, 93.8%, 96.6%, and 61.1% in NPWWR, YHFWWR, MLYVWWR, SWWR, Southern China Winter Wheat Region (SCWWR) and NSWR, respectively. The marker *csLV34* exhibited a good repeatability, and can be therefore used for the identification of *Lr34/Yr18* in wheat breeding programs.

Key words: Common wheat (*Triticum aestivum* L.); *Lr34/Yr18*; STS marker *csLV34*; Marker-assisted selection

条锈病和叶锈病是世界小麦的重要病害。我国是世界最大的小麦条锈病流行区，尤其在西北和西南地区流行频繁；一般流行年份减产 10%～20%，较大流行年份减产超过 30%。20 世纪 50 年代以来，我国小麦条锈病曾发生 15 次中度以上流行，其中 1950 年、1964 年、1990 年和 2002 年四次全国性大流行造成严重产量损失[1,2]。小麦叶锈病在我国各地均有分布，过去在西南地区发生较重，近年来在华北地区也逐渐加重，成为小麦生产的重要问题。

育种与生产实践证明，种植和培育抗病品种是防治小麦锈病最经济有效的措施。然而，由于病原菌小种的频繁变异，大部分含主效基因的抗病品种的抗性丧失很快，因此培育持久抗性品种越来越重要。利用主效基因聚合、多系品种和抗病品种合理布局等在防治条锈病流行方面是行之有效的。但在现有小麦抗条锈主效基因中，只有 *Yr5*、*Yr10*、*Yr15* 和 *Yr26* 等少数基因对我国小麦条锈菌优势小种条中 32 号表现高抗[3,4]，仅仅利用这几个抗病基因很难实现条锈病的持久防治。

近年来慢锈基因的研究和利用越来越受重视，培育慢锈品种成为国际抗病育种的主流方向。慢锈性通常苗期表现感病，成株期表现中到高抗，故又称为成株抗性，具有持久抗性作用[5]。国际玉米小麦改良中心（CIMMYT）在小麦慢锈品种选育方面取得了很大成就[5]，育成一大批慢锈品种，并在世界各地推广，携带 *Lr34/Yr18* 的小麦品种在印度和巴基斯坦等发展中国家广泛种植。慢叶锈基因 *Lr34* 和慢条锈基因 *Yr18* 被定位在 7DS 的同一位置，二者要么紧密连锁，要么是“一因多效”[6,7]。*Lr34* 最初可能来源于巴西的农家种 Alfredo Chaves 或乌拉圭的农家种选系 Americano 44D[8]。Dyck 等[9]最早报道了 *Lr34*（最初定名为 *LrT2*）的抗性水平，并将其定位于 7DS，其载体品种有 Thatcher、Jupeteco 73R 和 Anza 等。中国农家种中国春也携带该基因[10]，TH3929 是 20 世纪 30 年代悉尼大学从南京引进的品种，兰天 8 为甘肃冬小麦研究所育成的品种，二者均携带 *Lr34*，说明中国小麦品种中 *Lr34/Yr18* 早已存在（与 McIntosh 私人交流，2007）。*Lr34/Yr18* 慢锈性表现稳定，迄今还没有发现其小种专化性。在高病害压力下 *Lr34/Yr18* 单独存在时抗性还不够强，但结合2～4 个其他慢锈基因则可获得高抗或近免疫抗性[5,11]。许多研究表明，*Lr34/Yr18* 还与其他病害的抗性有关。Singh 报道[12,13]大麦黄矮病毒病抗性基因（*Bdv1*）和控制叶干尖的 *Ltn* 基因与 *Lr34/Yr18* 紧密连锁，Spielmeyer 等[14]认为 *Lr34/Yr18* 与白粉病抗性基因紧密连锁，Liang 等[15]在该位点也发现一个慢白粉病的 QTL。可见，*Lr34/Yr18* 基因位点与多种病害的抗性相关，因而在小麦育种中应该受到足够重视。

Suenaga 等[16]利用 Fukuho-komugi × Oligoculm 的 DH 群体对 *Lr34/Yr18* 基因定位，发现 7DS 上的微卫星标记 *Xgwm295* 与 *Lr34/Yr18* 紧密连锁。Spielmeyer 等[14]用 RL6058 的重组自交系 F_6 群体定位了 *Lr34/Yr18*，也表明 *Xgwm295* 与 *Lr34/Yr18* 紧密连锁，遗传距离为 2.7cM。Lagudah 等[10]利用覆盖 *Lr34/Yr18* 基因区域的小麦 EST 序列，开发了与该基因紧密连锁的 STS 标记 *csLV34*，遗传距离为 0.4cM，并利用 3 个重组自交系群体和 24 个已知抗病基因的澳大利亚品种验证了该标记的有效性。

本研究旨在利用分子标记 *csLV34* 对我国小麦育成品种（系）和农家种的 *Lr34/Yr18* 基因进行检测，了解 *Lr34/Yr18* 基因的分布状况，筛选含有目标基因的优异小麦种质，为我国慢锈性小麦新品种的选育提供材料和方法。

1 材料和方法

1.1 供试材料

包括 231 份国内主要麦区育成品种（系）及 422

份农家种（品种名称略）。育成品种为 20 世纪 50 年代至目前的主栽品种，其中北部冬麦区 20 份，黄淮冬麦区 167 份，长江中下游冬麦区 14 份，西南冬麦区 24 份，西北春麦区 6 份；农家种中北部冬麦区 77 份，黄淮冬麦区 133 份，长江中下游冬麦区 131 份，西南冬麦区 16 份，南方冬麦区 29 份，西北春麦区 36 份。育成品种由本实验室保存，农家种由国家种质库提供。

1.2 基因组 DNA 提取

每个品种（系）选取能代表该品种（系）特征的 3 粒种子，按 Lagudah 等[17]方法分别提取籽粒基因组 DNA。

1.3 STS 标记检测

利用 Lagudah 等[10]开发的共显性 STS 标记 *csLV34* 检测 *Lr34/Yr18* 基因。上游引物序列为 5′-GTTGGTTAAGACTGGTGATGG-3′，下游引物序列为 5′-TGCTTGCTATTGCTGAATAGT-3′。

PCR 反应体系 20μl，含 20mmol L^{-1} Tris-HCl (pH8.4)，20mmol L^{-1} KCl，1.5mmol L^{-1} $MgCl_2$，dNTP 各 200μmol L^{-1}，每条引物 10pmol，*Taq* DNA 聚合酶 1U，模板 DNA 50ng。反应程序为 94℃变性 1min，57℃退火 1min，72℃延伸 2min，5 个循环；94℃变性 30s，57℃退火 30s，72℃延伸 1min，30 个循环；94℃变性 30s，57℃退火 30s，72℃延伸 5min。扩增产物以 2.0%的琼脂糖凝胶电泳分离检测，缓冲液体系为 1×TAE 溶液，150V 电压电泳 40min，溴化乙锭染色后，用 Gel Doc XR System 扫描成像并存入计算机。根据每个品种 3 粒种子 DNA 的检测结果判断该品种（系）的等位变异类型。

2 结果分析

2.1 *csLV34* 标记 PCR 多态性类型

利用 *csLV34* 标记扩增获得两种片段，凡含有 *Lr34/Yr18* 基因的材料均能扩增出 150bp 片段，不含 *Lr34/Yr18* 的材料均扩增出 229bp 片段，扩增条带清晰，重复性好（图 1）。

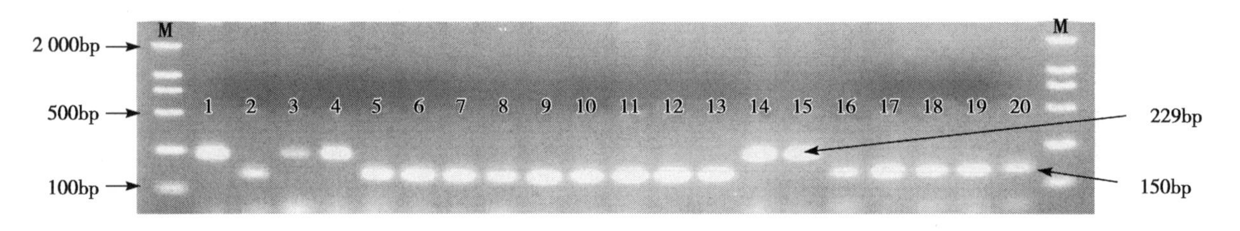

图 1 小麦品种慢锈性基因 *csLV34* 标记的多态性检测

Fig. 1 Polymorphic test of PCR fragments amplified with *csLV34* in Chinese wheat landraces

M：DL 2000；1：白壳红（ZM84）；2：白芒红（ZM33）；3：蚰子麦（ZM92）；4：红粒麦（ZM610）；5：沧州红（ZM654）；6：关东白麦（ZM651）；7：白葫芦头（ZM564）；8：光头白（ZM582）；9：马莲朵（ZM585）；10：光葫芦头（ZM533）；11：白秋麦（ZM539）；12：大白芒（ZM252）；13：小红芒（ZM253）；14：老芒麦（ZM3560）；15：红蜷芒（ZM2849）；16：红芒红（ZM3008）；17：红蚰子头（ZM2662）；18：蚰子麦（ZM2793）；19：紫秆白（ZM2720）；20：三月黄（ZM2685）。

M：DL 2000；1：Baikehong (ZM84)；2：Baimanghong (ZM33)；3：Youzimai (ZM92)；4：Honglimai (ZM610)；5：Cangzhouhong (ZM654)；6：Guandongbaimai (ZM651)；7：Baihulutou (ZM564)；8：Guangtoubai (ZM582)；9 Malianduo (ZM585)；10：Guanghulutou (ZM533)；11：Baiqiumai (ZM539)；12：Dabaimang (ZM252)；13：Xiaohongmang (ZM253)；14：Laomangmai (ZM3560)；15：Hongquanmang (ZM2849)；16：Hongmanghong (ZM3008)；17：Hongyouzitou (ZM2662)；18：Youzimai (ZM2793)；19：Ziganbai (ZM2720)；20：Sanyuehuang (ZM2685).

2.2 育成品种（系）及农家种 *Lr34/Yr18* 基因的分布状况

从表 1 可知，中国育成小麦品种（系）中仅 14 份材料含 *Lr34/Yr18* 基因，占 6.1%。其中 12 个为国内品种（系），分别是兰天 12、01-018、R97、济南 9 号、内乡 5 号、内乡 19、雅安早、宜麦 1 号、定西 29、郑州 6 号、万年 2 号和甘麦 7 号；另外两个是从意大利引进的南大 2419 和矮立多，曾在我国大面积推广。

从表 2 可知，我国不同麦区农家种中 *Lr34/Yr18* 基因的分布频率普遍很高，但在不同区域存在一定差异，黄淮冬麦区和西北春麦区分布频率较其他麦区低。

表 1 中国小麦育成品种（系）*Lr34/Yr18* 基因的分布频率

Table 1 Frequency of *Lr34/Yr18* gene in Chinese improved wheat cultivars and lines

麦区 Region	检测品种数 Number of cultivars tested	含 *Lr34/Yr18* 的品种数 Number of cultivars with *Lr34/Yr18*	频率 Frequency（%）
北部冬麦区 NPWWR	20	0	0
黄淮冬麦区 YHFWWR	167	5	3.0
长江中下游冬麦区 MLYVWWR	14	3	21.4
西南冬麦区 SWWR	24	4	16.7
西北春麦区 NSWR	6	2	33.3
合计 Total	231	14	6.1

NPWWR：North Plain Winter Wheat Region；YHFWWR：Yellow-Huai Facultative Winter Wheat Region；MLYVWWR：Middle-Lower Yangtze Valley Winter Wheat Region；SWWR：Southwest Winter Wheat Region；NSWR：Northwest Spring Wheat Region.

表 2 中国小麦农家种 *Lr 34/Yr18* 基因的分布频率

Table 2 Frequency of *Lr 34/Yr18* genes in Chinese wheat landraces

麦区 Region	检测品种数 Number of landraces tested	含 *Lr34/Yr18* 的品种数 Number of landraces with *Lr34/Yr18*	频率 Frequency（%）
北部冬麦区 NPWWR	77	69	89.6
黄淮冬麦区 YHFWWR	133	103	77.4
长江中下游冬麦区 MLYVWWR	131	122	93.1
西南冬麦区 SWWR	16	15	93.8
南部冬麦区 SCWWR	29	28	96.6
西北春麦区 NSWR	36	22	61.1
合计 Total	422	359	85.1

SCWWR：Southern China Winter Wheat Regine. Abbreviations of the other wheat regions are the same as in Table 1.

2.3 育成品种含 *Lr34/Yr18* 基因材料的亲本分析

231 份育成品种（系）中仅 14 份材料含 *Lr34/Yr18* 基因（表 3）。矮立多、南大 2419、内乡 5 号及内乡 19 的抗源相同，为 Wilhelmina、Rieti 或 Akagomughi；济南 9 号的抗病基因来自辛石 3 号而不是早洋麦（不携带 *Lr34/Yr18*），辛石 3 号是用农家种蚰子麦做亲本育成的品种，而蚰子麦含有 *Lr34/Yr18*；01-018 的抗源来自川 89-8（因 R131 和川 89-107 不含 *Lr34/Yr18*），R97 的抗源可能来自绵阳 92-8 或 91S-5-7；兰天 12 的抗源来自咸农 4 号，因为兰天 12 在田间的抗性表现与咸农 4 号相同（与杜久元私人交流，2007），咸农 4 号为慢锈性材料[18]。其他材料分别为矮立多、内乡 5 号和南大 2419 的衍生后代。可见，推广品种中 *Lr34/Yr18* 主要来源于外引材料和我国农家种。

表 3 含 *Lr34/Yr18* 基因的中国小麦育成品种系谱

Table 3 Pedigree of Chinese wheat cultivars with *Lr34/Yr18* genes

品种 Cultivar	系谱 Pedigree
矮立多 Ardito	Wilhelmina/Rieti//Akagomughi
济南 9 号 Jinan 9	辛石 3 号/早洋麦 Xinshi 3/ Early Piemium
内乡 5 号 Neixiang 5	南大 2419/（白火麦+碧玉麦+白芒麦）Nanda 2419/（Baihuomai+Quality+Baimangmai）
内乡 19 Neixiang 19	南大 2419/（白火麦+碧玉麦+白芒麦）Nanda 2419/（Baihuomai+ Quality +Baimangmai）
南大 2419 Nanda 2419	Rieti/Wilhelmina//Akagomughi
01-018	R131/川 89-8//川 89-107 R131/Chuan 89-8//Chuan 89-107

（续）

品种 Cultivar	系谱 Pedigree
R97	绵阳 92-8/91S-5-7 Mianyang 92-8/91S-5-7
兰天 12 Lantian 12	76-89-4/咸农 4 号 76-89-4/Xiannong 4
雅安早 Ya'anzao	矮立多衍生后代 Derivatives from Ardito
宜麦 1 号 Yimai 1	矮立多衍生后代 Derivatives from Ardito
定西 29 Dingxi 29	内乡 5 号衍生后代 Derivatives from Neixiang 5
郑州 6 号 Zhengzhou 6	内乡 5 号衍生后代 Derivatives from Neixiang 5
万年 2 号 Wannian 2	南大 2419 衍生后代 Derivatives from Nanda 2419
甘麦 7 号 Ganmai 7	南大 2419 衍生后代 Derivatives from Nanda 2419

3 讨论

csLV34 标记只有两种多态型，在含有*Lr34/Yr18*基因的小麦品种中可稳定扩增出150bp片段，在不含该基因的材料中可扩增出229bp片段。本研究验证了这一标记的有效性，它可用于目标基因*Lr34/Yr18*的鉴定，进行慢锈基因聚合和慢锈品种选育的分子标记辅助选择，以加快抗锈育种进程，提高育种效率。

除*Lr34/Yr18*外，Rosewarne等[19]报道*Lr46/Yr29*也为慢锈基因。2对连锁基因分别位于不同染色体，但却具有相似的反应机制，病害发展均表现为潜育期延长、孢子堆变小和产孢量减少[10,19]。在7DS上，*Lr34*、*Yr18*和慢白粉基因*Pm38*紧密连锁或位于同一位点；在1BL上，*Lr46*和*Yr29*也与慢白粉基因*Pm39*紧密连锁或位于同一位点；此外，这些基因均与叶干尖性状紧密相连，表明抗不同病害的慢病基因也许具有相同的作用机制。在抗性表型鉴定的基础上，利用分子标记使多个抗病基因的鉴定和聚合成为可能。利用*Lr34/Yr18*和*Lr46/Yr29*等基因的相关分子标记，聚合慢病基因和主效基因，培育含有多个抗病基因的小麦品种，可极大地提高小麦品种的抗性水平，促进持久抗性品种的选育进程。

为了充分利用*Lr34/Yr18*基因的慢锈性，可在育种中使用本研究发现的有关材料，如济南9号、矮立多、内乡5号、南大2419等。我国推广品种中*Lr34/Yr18*基因的来源除了农家种外，主要来自意大利，如矮立多和南大2419。此外，在甘肃种植多年的意大利品种Strampelli和Libellula也携带*Lr34/Yr18*基因（与McIntosh私人交流）。CIMMYT[20]开展慢锈性抗病育种将近30年，培育出一大批具有高产潜力和高水平抗性的品种，*Lr34/Yr18*在CIMMYT小麦品种中分布较广泛[21]。因此，在今后小麦育种中，除利用中国农家种外，还要大量引进并应用意大利和CIMMYT等国外优质抗源。

中国小麦育成品种（系）中*Lr34/Yr18*基因的频率极低，主要原因是主效抗病基因的抗性较强，田间容易选择，过去在品种选育中过分强调高抗或免疫的主效基因的选择，忽视了慢病基因的利用，致使慢病基因流失。因此，今后在兼顾高产、优质等育种目标的前提下，应加大慢锈性基因的选择力度，利用分子标记对慢锈性基因跟踪检测，确保多个慢锈性基因的有效聚合，以提高品种的抗性水平和持久性。

本研究结果也表明，422份小麦农家品种中85.1%携带*Lr34/Yr18*基因，加之该位点与锈病、白粉病和抗大麦黄矮病基因（*Bdv1*）等有关，致使中国农家种种植了很长时间，仍保持着对锈病和白粉病的抗性。此外，该基因与其他慢锈性基因和主效基因结合，会大大提高其抗锈病和白粉病的能力[5,20]。我国农家种抗逆性强，而且具有早熟性、多粒性、高度适应性等，是非常宝贵的种质资源。今后应加大对其研究力度，继续从中挖掘有价值的基因，为小麦新品种选育奠定基础。

参考文献

[1] Niu Y-C（牛永春），Wu L-R（吴立人）. Variation and strategies of resistance to stripe rust on wheat cultivars derived from Fan 6 and Mianyang-series. *Acta Phytopathol Sin*（植物病理学报），1997，27（1）：5-8.（in Chinese with English abstract）

[2] Wan A-M（万安民），Zhao Z-H（赵中华），Wu L-R

(吴立人). Reviews of occurrence of wheat stripe rust disease in 2002 in China. *Plant Prot* (植物保护), 2003, 29 (2): 5-8. (in Chinese with English abstract)

[3] Ma J X, Zhou R H, Dong Y S, Wang L F, Wang X M, Jia J Z. Molecular mapping and detection of the yellow rust resistance gene *Yr26* in wheat transferred from *Triticun turgidum* L. using microsatellite markers. *Euphytica*, 2001, 120: 219-226.

[4] Yang Z-M (杨作民), Xie C-J (解超杰), Sun Q-X (孙其信). Situation of the sources of stripe rust resistance of wheat in the post-CY32 era in China. *Acta Agron Sin* (作物学报), 2003, 29 (2): 161-168. (in Chinese with English abstract)

[5] Singh R P, Huerta-Espino J, Rajaram S. Achieving near-immunity to leaf and stripe rusts in wheat by combining slow rusting resistance genes. *Acta Phytopathol Hung*, 2000, 35: 133-139.

[6] McIntoch R A. Close genetic linkage of genes conferring adult-plant resistance to leaf rust and stripe rust in wheat. *Plant Pathol*, 1992, 41: 523-527.

[7] Singh R P. Genetic association of leaf rust resistance gene *Lr34* with adult plant resistance to stripe rust in bread wheat. *Phytopathology*, 1992, 82: 835-838.

[8] Roelfs A P. Resistance to leaf and stem rusts in wheat. In: Simmonds N W, Rajaram S eds. CIMMYT Breeding for Resistance to the Rusts of Wheat. Mexico, D. F.: CIMMYT, 1988. pp10-22.

[9] Dyck P L Genetics of leaf rust reaction in three introductions of common wheat. *Can J Genet Cytol*, 1977, 19: 711-716.

[10] Lagudah E S, McFadden H, Singh R P, Huerta-Espino J, Bariana H S, Spielmeyer W. Molecular genetic characterization of the*Lr34/Yr18* slow rusting resistance gene region in wheat. *Theor Appl Genet*, 2006, 114: 21-30.

[11] Singh R P, Gupta A K. Expression of wheat leaf rust resistance gene *Lr34* in seedlings and adult plants. *Plant Dis*, 1992, 76: 489-491.

[12] Singh R P. Association between gene *Lr34* for leaf rust resistance and leaf tip necrosis in wheat. *Crop Sci*, 1992, 32: 874-878.

[13] Singh R P. Genetic association of gene *Bdv*1 for tolerance to barley yellow dwarf virus with genes *Lr34* and *Yr18* for adult plant resistance to rusts in bread wheat. *Plant Dis*, 1993, 77: 1103-1106.

[14] Spielmeyer W, McIntosh R A, Kolmer J, Lagudah E S. Powdery mildew resistance and *Lr34/Yr18* genes for durable resistance to leaf and stripe rust cosegregate at a locus on the short arm of chromosome 7D of wheat. *Theor Appl Genet*, 2005, 111: 731-735.

[15] Liang S S, Suenaga K, He Z H, Wang Z L, Liu H Y, Wang D S Singh R P, Sourdille P, Xia X C. Quantitative trait loci mapping for adult-plant resistance to powdery mildew in bread wheat. *Phytopathology*, 2006, 96: 784-789.

[16] Suenaga K, Singh R P, Huerta-Espino J, William M. Microsatellite markers for genes *Lr34/Yr18* and other quantitative trait loci for leaf rust and stripe rust resistance in bread wheat. *Phytopathology*, 2003, 93: 881-890.

[17] Lagudah E S, Appels R, McNeil D. The *Nor-D3* locus of *Triticum tauschii*: natural variation and genetic linkage to markers in chromosome 5. *Genome*, 1991, 34: 387-395.

[18] Li Z F, Xia X C, Zhou X C, Niu Y C, He Z H, Zhang Y, Li G Q, Wan A M, Wang D S, Chen X M., Lu Q L, Singh R P. Seedling and slow rusting resistance to stripe rust in Chinese common wheats. *Plant Dis*, 2006, 90: 1302-1312.

[19] Rosewarne G M, Singh R P, Huerta-Espino J H, William M, Bouchet S, Cloutier S, McFadden H, Lagudah E S. Leaf tip necrosis, molecular markers and β1-proteasome subunits associated with the slow rusting resistance genes *Lr46/Yr29*. *Theor Appl Genet*, 2005, 112: 500-508.

[20] Singh R P, Huerta-Espino J, William H M. Genetics and breeding for durable resistance to leaf and stripe rusts in wheat. *Turk J Agric For*, 2005, 29: 121-127.

[21] Singh R P. Resistance to leaf rust in 26 Mexican wheat cultivars. *Crop Sci*, 1993, 33: 633-637.

利用STS标记检测CIMMYT小麦品种（系）中*Lr34/Yr18*、*Rht-B1b*和*Rht-D1b*基因的分布

梁丹[1,2]，杨芳萍[2,3]，何中虎[2,4]，姚大年[1]，夏先春[2]

[1]安徽农业大学农学院，合肥 230036；[2]中国农业科学院作物科学研究所/国家小麦改良中心/国家农作物基因资源与基因改良重大科学工程，北京 100081；[3]甘肃农业科学院作物研究所，兰州 730070；[4]CIMMTY 中国办事处，北京 100081

摘要：明确慢锈基因*Lr34/Yr18*和矮秆基因*Rht-B1b*（*Rht-1*）、*Rht-D1b*（*Rht-2*）在CIMMYT小麦中的分布，有助于慢病性品种选育和株高改良。利用三个STS标记检测慢锈基因*Lr34/Yr18*和矮秆基因*Rht-B1b*、*Rht-D1b*在263个CIMMYT小麦品种和高代品系中的分布。*csLV34*标记在含*Lr34/Yr18*的材料中扩增出一条150bp的特异带，在不含*Lr34/Yr18*的材料中扩增出229bp的特异带；利用两对互补引物NH-BF.2/WR1.2和NH-BF.2/MR1对*Rht-B1a*和*Rht-B1b*基因进行检测，在携带*Rht-B1a*和*Rht-B1b*的材料中分别扩增出一条400bp的特异带；利用DF/MR2标记检测*Rht-D1b*基因，在携带*Rht-D1b*的材料中扩增出一条280bp的片段。在263个品种中，57个品种携带*Lr34/Yr18*基因，占总数的21.7%；216个品种含*Rht-B1b*，占总数的82.1%；38个品种含*Rht-D1b*，占总数的14.4%。双矮秆基因型（*Rht-B1b*+*Rht-D1b*）12个，不含这两个矮秆基因的品种（*Rht-B1a*+*Rht-D1a*）21个。这三个STS标记可以方便、快速、准确地检测*Lr34/Yr18*、*Rht-B1b*和*Rht-D1b*基因。在CIMMYT材料中，*Rht-B1b*基因频率很高，*Rht-D1b*较低。

关键词：普通小麦；慢锈基因；矮秆基因；分子标记

Characterization of *Lr34/Yr18*, *Rht-B1b*, *Rht-D1b* genes in CIMMYT wheat cultivars and advanced lines using STS markers

Liang Dan[1,2], Yang Fangping[2], He Zhonghu[2,4], Yao Danian[1], Xia Xianchun[1,2]

[1]*College of Agronomy, Anhui Agricultural University, Hefei, Anhui* 230036, *China*; [2]*Institute of Crop Science/ The National Key Facility for Crop Gene Resources and Genetic Improvement/National Wheat Improvement Center, Chinese Academy of Agricultural Sciences* (*CAAS*), *Zhongguancun South Street* 12, *Beijing* 100081, *China*; [3]*Crop Research Institute, Gansu Academy of Agronomy Agricultural Science, Lanzhou, Gansu* 730070, *China*; [4]*International Maize and Wheat Improvement Center* (*CIMMYT*) *China Office, c/o CAAS, Zhongguancun South Street* 12, *Beijing* 100081, *China*

原文发表在《中国农业科学》，2009，42（1）：17－27

Abstract: Characterization of genes *Lr34/Yr18*, *Rht-B1b* (*Rht1*) and *Rht-D1b* (*Rht2*) in 263 CIMMYT wheat cultivars and advanced lines will benefit the improvement of rust resistance and plant height in Chinese wheat breeding program. A total of 263 CIMMYT wheat cultivars and advanced lines were tested by STS markers to understand the distribution of resistance genes *Lr34/Yr18* and dwarfing genes *Rht-B1b* and *Rht-D1b*. The marker *csLV34* could amplify a 150-bp fragment in the lines with *Lr34/Yr18*, and a 229-bp fragment in those lines without *Lr34/Yr18*. Two complementary markers, *NH-BF.2/WR1.2* and *NH-BF.2/MR1*, were used to investigate the alleles *Rht-B1a* and *Rht-B1b*, respectively. They could amplify a 400-bp fragment in the genotypes with *Rht-B1a* and *Rht-B1b*, respectively. The marker *DF/MR2* could generate a 280-bp fragment in the genotype with *Rht-D1b*. Of the 263 lines, 57 generated a 150-bp fragment with the marker *csLV34*, indicating the presence of *Lr34/Yr18* in these lines, with a frequency of 21.7%. Two hundred and sixteen lines (82.1%) were detected to have the allele *Rht-B1b*, and 38 lines (14.4%) contained the allele *Rht-D1b*. Twenty-one lines were detected to possess both the wild-type alleles *Rht-B1a* and *Rht-D1a*, while 12 had two dwarfing genes *Rht-B1b* and *Rht-D1b*. These STS markers could be useful to detect the gene *Lr34/Yr18*, *Rht-B1b* and *Rht-D1b*. The frequency of *Rht-B1b* was much higher than that of *Rht-D1b* in CIMMYT lines.

Key words: *Triticum aestivum* L.; Slow rusting resistance; Dwarfing gene; Molecular marker

小麦条锈病和叶锈病是我国重要病害。20 世纪 50 年代以来，小麦条锈病先后 15 次中等规模以上流行，造成大面积减产；小麦叶锈病过去在西南地区发生较重，近年来在华北地区也有逐渐加重的趋势[1]。利用抗病品种是控制上述两种病害的主要方法，但由于条锈菌和叶锈菌小种变异频繁，品种抗性丧失很快[2]，因此抗病育种始终处于被动状态。在国际上已正式定名的抗条锈主效基因中，只有 *Yr5*、*Yr10*、*Yr15* 和 *Yr26* 等少数基因抗目前的流行小种[3]，在已定名的 60 个小麦抗叶锈基因中，对中国小麦有效的抗叶锈基因也只有 *Lr9*、*Lr19*、*Lr24*、*Lr38* 等少数基因（李在峰，私人交流）。因而培育持久抗性品种成为小麦育种的重要目标，慢病性的利用是实现持久抗性的方法之一[4]。自 20 世纪 60 年代以来，矮秆和半矮秆小麦品种的育成和推广对产量的提高起到了关键作用[5,6]，进一步改良株高仍是今后提高产量的途径之一，如何实现矮秆基因最佳组合是育种中面临的重要问题。

小麦慢病性一般由几个到多个基因控制，小种专化性弱或无专化性，较少引起病原菌的变异，抗病性持久稳定。含慢病基因的品种虽然苗期感病，但成株期发病缓慢，病害严重度轻，减产幅度较小。基因 *Lr34* 和 *Yr18* 位于同一位点或紧密连锁，是目前应用最广泛的慢锈基因。Singh[7,8] 发现基因 *Lr34/Yr18* 与叶尖坏死基因 *Ltn1*、抗小麦黄矮病基因 *Bdv1* 均紧密连锁。近期研究发现，基因 *Lr34/Yr18* 还与抗白粉基因 *Pm38* 紧密连锁（Lillemo 等，私人交流）。Singh 等[4] 认为，把 *Lr34/Yr18* 与 2～4 个微效慢病基因聚合到一起，能选育出近免疫的抗条锈小麦品种。Imtiaz 等[9] 使用 140 个 Tiritea×Otane DH 系发现 *Lr34/Yr18* 基因位点在 7DS 染色体上，距 SSR 标记 *Xgwm44* 的遗传距离为 7cM。Spielmeyer 等[10] 利用 110 个重组自交系研究表明 *Lr34/Yr18* 基因与 7DS 上的微卫星标记 *Xgwm1220* 和 *Xgwm295* 紧密连锁，遗传距离分别为 0.9cM 和 2.7cM。Lagudah 等[11] 将一个 RFLP 标记转化成 STS 标记（*csLV34*），利用 768 株 Lalbahadur（不含 *Lr34/Yr18* 基因）× Lalbahadur（含 *Lr34/Yr18* 基因）的 F_2 代群体进行高密度作图，发现标记 *csLV34* 与 *Lr34/Yr18* 位点遗传距离为 0.4cM。利用该引物检测 24 个已知抗病基因的澳大利亚品种，所得带型与已知抗病基因信息完全吻合。矮秆基因 *Rht-B1b*（*Rht1*）和 *Rht-D1b*（*Rht2*）在小麦高产育种中起到重要作用。含矮秆基因 *Rht-B1b* 和 *Rht-D1b* 的品种的穗粒数增加，总体产量表现增加[12]。Ellis 等[13] 根据矮秆基因 *Rht-B1* 和 *Rht-D1* 突变体（*Rht-B1b* 和 *Rht-D1b*）中单个碱基对的差异分别设计两对互补引物来检测 *Rht-B1* 和 *Rht-D1* 位点的等位变异，经 19 个具有已知矮秆基因型的小麦验证，PCR 检测结果与品种的矮秆基因型完全吻合，利用该标记检测 157 个来自 Sunco（含 *Rht-B1b*）× Tasman（含 *Rht-D1b*）的 DH 系，*Rht-B1* 位点可以解释 23% 的株高表型变异，*Rht-D1* 位点可以解释 44% 的表型变异。杨松杰等[6] 利用该标记对中国主要麦区 239 个品种的 *Rht-B1b* 和 *Rht-D1b* 基因进行检测，*Rht-B1b* 的频率为 24.3%，*Rht-D1b* 的频率为 46.9%。CIMMYT 小麦在我国育种中发挥了重要作用，对提高产量、改善品质和抗病性做出了贡献，在新疆、甘肃、云南及四川广为利用[14]，近年来在黄淮麦区也育成了邯 6172 和郑麦 044 等主栽品种。Lagudah 等[11]开发的 STS 标记 *csLV34* 能有效地鉴定 *Lr34/Yr18* 基因，Ellis 等[13]开发的 STS 引物可以准确鉴定 *Rht-B1b* 和 *Rht-D1b* 基因，这为快速准确检测 CIMMYT 小麦品种中 *Lr34/Yr18*、*Rht-B1b* 和 *Rht-D1b* 的分布提供了可能。本研究对 263 个 CIMMYT 小麦品种（系）的 *Lr34/Yr18*、*Rht-B1* 和 *Rht-D1* 基因进行分子检测，明确这些位点不同等位基因的分布规律，为我国小麦育种提供有用的材料、信息以及分子标记辅助选择方法。

1 材料与方法

1.1 材料

1.1.1 供试材料 263 份，由 CIMMYT 小麦项目提供，2006-2007 年度种植于墨西哥。这些材料来自于 CIMMYT 水地育种项目，包括常用亲本、苗头高代品系以及近年来大面积推广的品种。利用 Janz（*Rht-B1b*）和 Kukri（*Rht-D1b*）作为分子标记检测 *Rht-B1b* 和 *Rht-D1b* 的对照品种，种子由澳大利亚 CSIRO 植物研究所提供。

1.2 DNA 提取

采用 SDS 法提取小麦基因组 DNA[15]，每份材料提取两粒种子的 DNA，利用紫外分光光度计检测 DNA 浓度，终浓度调整至 20ng/μl。

1.3 *Lr34/Yr18*、*Rht-B1b* 和 *Rht-D1b* 基因的分子标记检测

1.3.1 ***Lr34/Yr18* 基因的分子标记检测** *Lr34/Yr18* 基因特异性 STS 标记根据 Lagudah 等发表的序列[11]，由奥科生物技术有限公司合成。

引物序列为：上游引物 csLV34F：5’- GTTGGTTAAGACTGGTGATGG - 3’，下游引物 csLV34R：5’- TGCTTGCTATTGCTGAATAGT - 3’。

PCR 反应体系为：20μl 总体积中含 20mmol L^{-1} KCl，20mmol L^{-1} Tris-HCl（pH 8.4），1.5mmol L^{-1} $MgCl_2$，*Taq* DNA 聚合酶（天为时代）1U，dNTP（A、T、C、G）各 200μmol L^{-1}，每条引物 10pmol，模板 DNA 50ng。

PCR 反应条件为：首先 94℃ 变性 1min，然后 57℃退火 1min，72℃延伸 1min，4 个循环；再 94℃变性 30s，57℃退火 30s，72℃延伸 30s，29 个循环；最后 94℃ 变性 1min，57℃ 退火 1min，72℃ 延伸 5min。扩增产物以 2.0%的琼脂糖凝胶电泳检测，缓冲体系为 1×TAE 溶液，150V 电压电泳 30min，溴化乙锭染色后，用 GelDoc XR System 扫描成像并存入计算机。

1.3.2 ***Rht-B1b* 和 *Rht-D1b* 基因的分子标记检测** *Rht-B1b* 和 *Rht-D1b* 特异性 STS 引物根据 Ellis 等[13]发表并作部分修改的序列（Ellis，私人交流），由奥科生物技术有限公司合成。引物 NH-BF.2 与 WR1.2 用于检测 *Rht-B1a*（野生型）基因，NH-BF.2 与 MR1 用于检测 *Rht-B1b*（突变型）基因；引物 DF 与 MR2 用于检测 *Rht-D1b*（突变型）基因。

引物序列为：NH-BF.2：5’ - TCTCCTCCCTCCCCACCCCAAC - 3’；WR1.2：5’ - CCATGGCCATCTCGAGCTGC - 3’；MR1：5’ - CATCCCCATGGCCATCTCGAGCTA - 3’。

Rht-D1b 特异性引物序列为：DF：5’ - CGCGCAATTATTGGCCAGAGATAG - 3’；MR2：5’ - CCCCATGGCCATCTCGAGCTGCTA - 3’。

PCR 反应体系、反应条件以及电泳检测均参照杨松杰等发表论文[6]。

2 结果与分析

2.1 *Lr34/Yr18* 基因检测

Lr34/Yr18 基因标记 *csLV34* 是共显性标记（图 1），在含和不含 *Lr34/Yr18* 基因的材料中分别扩增出一条 150bp 和 229bp 的片段。对 263 份小麦品种（系）的检测表明（表 1），57 个品种（系）扩增出 150bp 片段，含 *Lr34/Yr18* 基因，占总数的 21.7%；206 个品种（系）扩增出 229bp 片段，不含 *Lr34/Yr18* 基因，占总数的 78.3%。

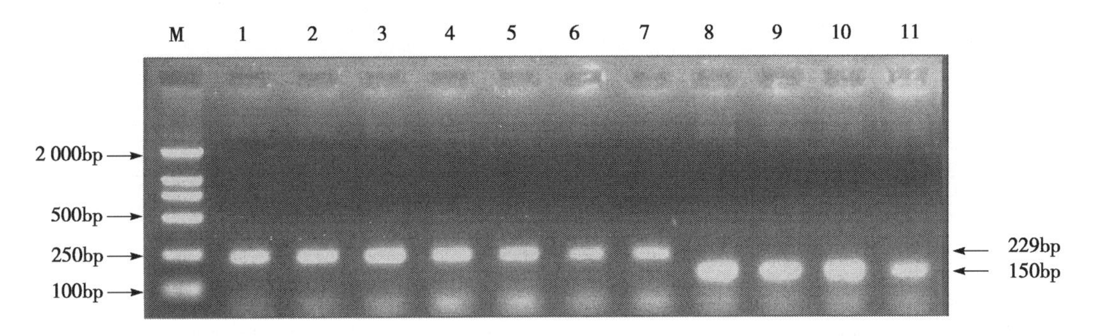

图 1 部分 CIMMYT 小麦品种（系）*Lr34/Yr18* 基因的检测结果

Fig. 1 PCR product amplified with the STS marker *csLV34* for *Lr34/Yr18* gene in some CIMMYT cultivars and lines

M. 2 000bp DNA ladder; 1. SERI/RAYON; 2. KRONSTAD F2004; 3. KAMBARA1; 4. WHEATEAR; 5. WEEBILL1; 6. ATTILA * 2/PBW65; 7. WAXWING; 8. SERI. 1B * 2/3/KAUZ * 2/BOW//KAUZ; 9. FRET2 * 2/BRAMBLING; 10. FRET2 * 2/4/SNI/TRAP # 1/3/KAUZ * 2/TRAP//KAUZ; 11. KAMB1 * 2/BRAMBLING

第 8、9、10、11 材料含 *Lr34/Yr18* 基因。Entries8，9，10 and 11 contain genes *Lr34/Yr18*.

表 1 263 份 CIMMYT 小麦品种系谱及其 *Lr34/Yr18*、*Rht-B1* 和 *Rht-D1* 基因位点的等位变异

Table 1 Allelic variation at the loci *Lr34/Yr18*, *Rht-B1* and *Rht-D1* in 263 CIMMYT wheat cultivars and advanced lines

序号 Code	系谱/名称 Pedigree	*Rht-B1*[1)]	*Rht-D1*[1)]	*csLV34*[2)]
1	SERI/RAYON	b	a	—
2	KRONSTAD F2004	b	a	—
3	KAMBARA1	b	a	—
4	WHEATEAR	b	a	—
5	WEEBILL1	b	a	—
6	WEEBILL1	b	a	—
7	SERI. 1B * 2/3/KAUZ * 2/BOW//KAUZ	b	a	+
8	ATTILA * 2/PBW65	b	a	—
9	WAXWING	b	a	—
10	PRL/2 * PASTOR	a	a	—
11	PBW65/2 * PASTOR	a	a	—
12	ALTAR 84/AE. SQUARROSA (221) //3 * BORL95/3/URES/JUN//KAUZ/4/WBLL1	b	a	—
13	ATTILA/3 * BCN * 2//BAV92	b	a	—
14	ATTILA/3 * BCN//BAV92/3/PASTOR	b	a	—
15	BABAX//IRENA/KAUZ/3/HUITES	a	a	—
16	BABAX/LR42//BABAX * 2/3/KURUKU	b	a	—
17	BABAX/LR42//BABAX * 2/3/PAVON 7S3，+LR47	b	a	—
18	BABAX/LR42//BABAX * 2/3/TUKURU	b	a	—
19	BABAX/LR42//BABAX * 2/3/VIVITSI	b	a	—
20	BABAX/LR42//BABAX * 2/3/VIVITSI	b	a	—
21	BABAX/LR42//BABAX * 2/3/VIVITSI	b	a	—
22	BABAX/LR42//BABAX * 2/4/SNI/TRAP # 1/3/KAUZ * 2/TRAP//KAUZ	b	a	—

（续）

序号 Code	系谱/名称 Pedigree	Rht-B1[1)]	Rht-D1[1)]	csLV34[2)]
23	BABAX/LR42//BABAX * 2/4/SNI/TRAP # 1/3/KAUZ * 2/TRAP//KAUZ	a	a	—
24	BL2064//SW89-5124 * 2/FASAN/3/TILHI	b	a	+
25	CAL/NH//H567.71/3/SERI/4/CAL/NH//H567.71/5/2 * KAUZ/6/PASTOR	b	a	—
26	CAR//KAL/BB/3/NAC/4/VEE/PJN//2 * TUI/5/MILAN	b	a	—
27	CHIBIA//PRLII/CM65531/3/FISCAL	b	a	—
28	CHIBIA//PRLII/CM65531/3/SKAUZ/BAV92	b	a	—
29	CHIBIA//PRLII/CM65531/3/SW89.5181/KAUZ	b	a	—
30	CNO79//PF70354/MUS/3/PASTOR/4/BABAX	b	a	—
31	ELVIRA/5/CNDO/R143//ENTE/MEXI75/3/AE. SQ/4/2 * OCI	b	a	—
32	FRET2 * 2/4/SNI/TRAP # 1/3/KAUZ * 2/TRAP//KAUZ	b	a	+
33	FRET2 * 2/4/SNI/TRAP # 1/3/KAUZ * 2/TRAP//KAUZ	b	a	+
34	FRET2 * 2/BRAMBLING	b	a	+
35	FRET2 * 2/BRAMBLING	b	a	+
36	FRET2 * 2/BRAMBLING	b	a	+
37	FRET2/KUKUNA//FRET2	b	a	+
38	FRET2/TUKURU//FRET2	b	a	—
39	FRET2/WBLL1//KAMB1	b	a	—
40	GAN/AE. SQUARROSA (408) //2 * OASIS/5 * BORL95	b	a	—
41	INQALAB 91 * 2/TUKURU	b	a	—
42	IRENA/2 * PASTOR	b	a	—
43	KAMB1 * 2/BRAMBLING	b	a	—
44	KAMB1 * 2/BRAMBLING	b	a	+
45	KAMB1 * 2/KIRITATI	b	a	—
46	KAUZ * 2/MNV//KAUZ/3/MILAN/4/BABAX	b	a	—
47	KAUZ//ALTAR 84/AOS/3/MILAN/KAUZ/4/HUITES	b	a	—
48	KAUZ//ALTAR 84/AOS/3/MILAN/KAUZ/4/HUITES	b	a	—
49	KAUZ//ALTAR 84/AOS/3/MILAN/KAUZ/4/HUITES	b	a	—
50	KAUZ//ALTAR 84/AOS/3/MILAN/KAUZ/4/HUITES	b	a	—
51	KAUZ/PASTOR//PBW343	b	a	—
52	KAUZ/PASTOR//PBW343	b	a	—
53	KIRITATI//ATTILA * 2/PASTOR	b	a	+
54	KIRITATI//PBW65/2 * SERI. 1B	a	a	—
55	KIRITATI//PBW65/2 * SERI. 1B	b	a	+
56	KIRITATI//PBW65/2 * SERI. 1B	b	b	—
57	KIRITATI//PBW65/2 * SERI. 1B	a	b	—
58	KIRITATI//PRL/2 * PASTOR	a	b	+
59	KIRITATI//PRL/2 * PASTOR	a	b	+
60	KIRITATI//SERI/RAYON	b	a	—
61	KIRITATI/3/HUW234+LR34//PRL/VEE # 10	b	a	+

（续）

序号 Code	系谱/名称 Pedigree	Rht-B1[1)]	Rht-D1[1)]	csLV34[2)]
62	KIRITATI/4/SERI. 1B * 2/3/KAUZ * 2/BOW//KAUZ	b	a	+
63	KIRITATI/4/SERI. 1B * 2/3/KAUZ * 2/BOW//KAUZ	b	a	+
64	KIRITATI/WBLL1	b	a	+
65	KIRITATI/WBLL1	b	a	—
66	MILAN/S87230//BABAX	b	a	—
67	MINO	b	a	—
68	NAC/TH. AC//3 * PVN/3/MIRLO/BUC/4/2 * PASTOR	b	a	—
69	OASIS/SKAUZ//4 * BCN * 2/3/PASTOR	b	a	—
70	OASIS/SKAUZ//4 * BCN/3/2 * PASTOR	b	a	—
71	OASIS/SKAUZ//4 * BCN/3/PASTOR/4/KAUZ * 2/YACO//KAUZ	b	a	+
72	PBW343/WBLL1//PANDION	b	b	—
73	PF74354//LD/ALD/4/2 * BR12 * 2/3/JUP//PAR214 * 6/FB6631/5/HP 1731	a	b	—
74	PFAU/SERI. 1B//AMAD/3/WAXWING	b	b	—
75	PFAU/WEAVER * 2//BRAMBLING	b	b	+
76	PFAU/WEAVER * 2//KIRITATI	b	b	—
77	PFAU/WEAVER * 2//KIRITATI	b	b	—
78	PFAU/WEAVER * 2//KIRITATI	b	b	+
79	PFAU/WEAVER * 2//TRANSFER#12，P88. 272. 2	b	a	—
80	PICUS/3/KAUZ * 2/BOW//KAUZ/4/TILHI	b	a	—
81	PRINIA/PASTOR	b	a	—
82	SITE/MO//PASTOR/3/TILHI	b	a	—
83	SKAUZ/BAV92//CHUM18/7 * BCN	b	a	+
84	TAM200/PASTOR//TOBA97	b	a	—
85	THELIN#2//ATTILA * 2/PASTOR/3/PRL/2 * PASTOR	b	a	—
86	THELIN//2 * ATTILA * 2/PASTOR	b	a	—
87	THELIN//2 * ATTILA * 2/PASTOR	b	a	—
88	THELIN/2 * WBLL1	b	a	—
89	THELIN/2 * WBLL1	b	a	—
90	THELIN/3/2 * BABAX/LR42//BABAX	b	a	—
91	THELIN/3/BABAX/LR42//BABAX/4/BABAX/LR42//BABAX	b	a	—
92	TOBA97/PASTOR	b	a	—
93	TOBA97/PASTOR	b	a	—
94	TUKURU//BAV92/RAYON	b	a	+
95	TUKURU//BAV92/RAYON	b	a	+
96	VEE/PJN//KAUZ/3/PASTOR/4/FISCAL	b	a	—
97	VORB/FISCAL	b	a	—
98	WAXWING * 2/4/SNI/TRAP#1/3/KAUZ * 2/TRAP//KAUZ	b	a	—
99	WAXWING * 2/KIRITATI	b	a	—
100	WAXWING * 2/KIRITATI	b	a	—

（续）

序号 Code	系谱/名称 Pedigree	*Rht-B1*[1)]	*Rht-D1*[1)]	*csLV34*[2)]
101	WAXWING * 2/KIRITATI	b	a	—
102	WAXWING * 2/KUKUNA	b	a	—
103	WAXWING * 2/KUKUNA	b	a	—
104	WAXWING * 2/TUKURU	b	a	—
105	WAXWING * 2/VIVITSI	b	a	—
106	WAXWING * 2/VIVITSI	b	a	—
107	WAXWING * 2/VIVITSI	b	a	—
108	WAXWING/4/SNI/TRAP#1/3/KAUZ * 2/TRAP//KAUZ	b	a	—
109	WBLL1 * 2/4/SNI/TRAP#1/3/KAUZ * 2/TRAP//KAUZ	b	a	—
110	WBLL1 * 2/4/YACO/PBW65/3/KAUZ * 2/TRAP//KAUZ	b	a	—
111	WBLL1 * 2/BRAMBLING	b	a	—
112	WBLL1 * 2/BRAMBLING	b	a	—
113	WBLL1 * 2/BRAMBLING	b	a	—
114	WBLL1 * 2/BRAMBLING	b	a	—
115	WBLL1 * 2/BRAMBLING	b	a	—
116	WBLL1 * 2/BRAMBLING	b	a	—
117	WBLL1 * 2/BRAMBLING	b	a	—
118	WBLL1 * 2/BRAMBLING	b	a	—
119	WBLL1 * 2/CHAPIO	b	a	—
120	WBLL1 * 2/KIRITATI	b	a	—
121	WBLL1 * 2/KIRITATI	b	a	—
122	WBLL1 * 2/KIRITATI	b	a	—
123	WBLL1 * 2/KIRITATI	b	a	—
124	WBLL1 * 2/KIRITATI	b	a	—
125	WBLL1 * 2/KKTS	b	a	—
126	WBLL1 * 2/KKTS	b	a	—
127	WBLL1 * 2/KUKUNA	b	a	—
128	WBLL1 * 2/KUKUNA	b	a	—
129	WBLL1 * 2/TUKURU	b	a	—
130	WBLL1/3/STAR//KAUZ/STAR/4/BAV92/RAYON	b	a	—
131	WBLL1/KUKUNA//KAMB1	b	a	—
132	WBLL4/KUKUNA//WBLL1	b	a	—
133	WEAVER/3/SAPI/TEAL//HUI/4/CROC _ 1/AE. SQUARROSA (213) //PGO/5/ SKAUZ * 2/SRMA	b	a	+
134	BL 1496/MILAN/3/CROC _ 1/AE. SQUARROSA (205) //KAUZ	b	a	—
135	BOW/NKT//CBRD/3/CBRD	b	a	—
136	KIRITATI//HUW234+LR34/PRINIA	b	b	+
137	RABE/LAJ3302	b	b	+
138	THELIN#2/TUKURU	a	a	+
139	V763. 2312/V879. C8. 11. 11. 11 (36) //STAR/3/STAR	b	a	—

（续）

序号 Code	系谱/名称 Pedigree	*Rht-B1* 1)	*Rht-D1* 1)	*csLV34* 2)
140	VK237/2 * PASTOR	b	a	—
141	WAXWING * 2/KIRITATI	b	a	—
142	WAXWING * 2/VIVITSI	b	a	+
143	PAVON F 76	a	b	—
144	FRET2 * 2/KUKUNA	b	a	+
145	HE1/3 * CNO79//2 * SERI/3/ATTILA/4/WH 542	b	a	+
146	JUCHI F2000	b	a	+
147	KAMB1 * 2/KHVAKI	b	a	+
148	KIRITATI	a	b	+
149	PFAU/WEAVER * 2//KIRITATI	a	b	—
150	PGO//CROC _ 1/AE. SQUARROSA (224) /3/2 * BORL95/4/CIRCUS	a	b	—
151	PGO//CROC _ 1/AE. SQUARROSA (224) /3/2 * BORL95/4/CIRCUS	b	b	—
152	PGO/SERI//BAV92	b	a	+
153	TAM200/TUI/6/PVN//CAR422/ANA/5/BOW/CROW//BUC/PVN/3/YR/4/TRAP#1	b	a	+
154	TAM200/TUI/6/PVN//CAR422/ANA/5/BOW/CROW//BUC/PVN/3/YR/4/TRAP#1	b	a	+
155	WL6736/5/2 * BR12 * 3/4/IAS55 * 4/CI14123/3/IAS55 * 4/EG, AUS//IAS55 * 4/ALD/6/OASIS/5 * BORL95/7/BORL95	a	b	—
156	TAM200/TUI	b	a	—
157	TAM200/TUI/3/BABAX/LR42//BABAX	b	a	—
158	CHIL/CHUM18/4/BUC/BJY/3/CNDR/ANA//CNDR/MUS	b	a	—
159	HAR3116	a	a	—
160	KAUZ//ALTAR 84/AOS/3/KAUZ/4/SW94. 15464	a	b	+
161	MILAN/SHA7/3/THB/CEP7780//SHA4/LIRA/4/SHA4/CHIL	a	b	—
162	NING MAI 9415. 16//SHA4/CHIL/3/NING MAI 50	b	a	—
163	NING MAI 9558//CHIL/CHUM18	a	a	—
164	YANG87-158 * 2//MILAN/SHA7	b	a	—
165	ZHENGZHOU 872//WIZZA _ 23/CONA-D/3/SUNSU	a	b	—
166	CHEN/AE. SQ//2 * WEAVER/3/OASIS/5 * BORL95	a	b	—
167	CHEN/AEGILOPS SQUARROSA (TAUS) //BCN/3/CMH81. 38/2 * KAUZ	b	a	—
168	CNDO/R143//ENTE/MEXI _ 2/3/AEGILOPS SQUARROSA (TAUS) /4/WEAVER/5/PASTOR	a	b	—
169	CROC _ 1/AE. SQUARROSA (205) //FCT/3/PASTOR	b	a	+
170	CHIBIA/DULUS	b	b	—
171	K6295. 4A	a	a	+
172	YANAC	b	a	+
173	KENYA NYANGUMI	b	b	+
174	KENYA SWARA	a	a	+
175	K4208	b	a	+

（续）

序号 Code	系谱/名称 Pedigree	*Rht-B1*[1)]	*Rht-D1*[1)]	*csLV34*[2)]
176	PVN//CAR422/ANA/5/BOW/CROW//BUC/PVN/3/YR/4/TRAP#1	a	b	—
177	SD3746	b	a	—
178	STOA	a	a	—
179	HEILO	b	a	—
180	BAU/TNMU	b	a	—
181	CATBIRD	b	a	+
182	GONDO	b	a	—
183	GUAM92//PSN/BOW	b	a	—
184	IVAN/6/SABUF/5/BCN/4/RABI//GS/CRA/3/AE. SQUARROSA (190)	b	a	—
185	KAUZ//TRAP#1/BOW	b	a	+
186	NG8675/CBRD	b	a	—
187	NING MAI 9558	a	b	+
188	SHA3/CBRD	b	a	—
189	SHA3/SERI//SHA4/LIRA	a	a	+
190	SHA5/WEAVER	a	a	—
191	SHA8/GEN	b	a	—
192	TINAMOU	b	a	—
193	WUH1/VEE#5//CBRD	b	a	—
194	WAXWING*2/KIRITATI	b	a	—
195	WAXWING*2/TUKURU	b	a	—
196	WAXWING*2/BRAMBLING	b	a	—
197	ACHTAR*3//KANZ/KS85-8-4	b	a	—
198	ACHTAR*3//KANZ/KS85-8-5	b	a	—
199	ALTAR 84/AEGILOPS SQUARROSA (TAUS) //OPATA	b	a	—
200	KANZ*4/KS85-8-4	a	a	—
201	PRL/SARA//TSI/VEE#5	b	a	—
202	BH1146*3/ALD//BUC/3/DUCULA/4/DUCULA	b	a	+
203	PF839197/BR35//BR23/3/PASTOR	b	a	—
204	TNMU/6/PEL74144/4/KVZ//ANE/MY64/3/PF70354/5/BR14/7/BR35	b	a	—
205	ALTAR 84/AE. SQ//OPATA/3/2*WH 542	b	a	—
206	CHUM18/BORL95//CBRD	b	a	+
207	CROC_1/AE. SQUARROSA (205) //KAUZ/3/SASIA	b	a	—
208	CROC_1/AE. SQUARROSA (205) //KAUZ/3/SASIA	b	a	—
209	KAUZ//ALTAR 84/AOS/3/MILAN/KAUZ	b	a	+
210	SW89. 3064//CMH82. 17/SERI	a	b	—
211	W462//VEE/KOEL/3/PEG//MRL/BUC	b	a	—
212	W485/HD29	a	b	—
213	CNDO/R143//ENTE/MEXI_2/3/AEGILOPS SQUARROSA (TAUS) /4/WEAVER/5/2*PASTOR	b	a	—
214	VOROBEY	b	a	—

（续）

序号 Code	系谱/名称 Pedigree	*Rht-B1* 1)	*Rht-D1* 1)	*csLV34* 2)
215	SOROCA	b	a	+
216	HEILO	b	a	−
217	HEILO	b	a	−
218	MILAN/AMSEL	b	a	−
219	OR791432/VEE#3. 2//MILAN	b	a	−
220	OR791432/VEE#3. 2//MILAN	b	a	−
221	PASTOR//MUNIA/ALTAR 84	b	a	−
222	PASTOR/TEERI	b	a	−
223	PASTOR/TEERI	b	a	−
224	ENEIDA F94	a	b	−
225	HUW468	b	a	+
226	INQALAB 91	b	a	−
227	INQALAB 91 * 2/KUKUNA	b	a	+
228	INQALAB 91 * 2/TUKURU	b	a	−
229	KANCHAN	a	b	−
230	PBW343	b	a	−
231	PBW343 * 2/KHVAKI	b	a	+
232	PBW343 * 2/KUKUNA	b	a	+
233	PBW343 * 2/TUKURU	b	a	+
234	58769	a	a	−
235	SW 8488 (W)	a	b	−
236	SW00-91382	a	b	−
237	SW02-90137	a	a	−
238	SW03-81497	a	a	−
239	SW22725	a	a	−
240	YUNMAI 47	b	a	−
241	SW1231	b	a	−
242	86715	a	b	−
243	R131	b	a	−
244	80. 8	a	a	−
245	SW2148	a	b	−
246	3570	a	b	−
247	CHUANMAI 42	b	a	−
248	CHUANMAI 47	a	b	−
249	CHUANMAI 43	b	a	−
250	SC SHINE	b	a	−
251	STALLION	b	a	−
252	SMART	b	a	−
253	KLEIN DON ENRIQUE	b	a	−

（续）

序号 Code	系谱/名称 Pedigree	*Rht-B1*[1)]	*Rht-D1*[1)]	*csLV34*[2)]
254	ONIX	b	a	—
255	CEP8880/3/BOW//BUC/BUL/4/EMB27	b	a	—
256	ITAPUA 40-OBLIGADO	b	a	—
257	ITAPUA 50 - AMISTAD	a	a	—
258	INIA CABURE	a	a	+
259	INIA CHURRINCHE	b	a	—
260	KLEIN DON ENRIQUE	b	a	—
261	OR 1	b	a	—
262	ITAPUA 40-OBLIGADO	b	a	—
263	PANDORA	b	a	—

1）“a”在基因位点*Rht-B1*、*Rht-D1*分别代表*Rht-B1a*、*Rht-D1a*；“b”在基因位点*Rht-B1*、*Rht-D1*分别代表*Rht-B1b*、*Rht-D1b*。

2）“+”表示品种（系）中含*Lr34/Yr18*基因；“—”表示品种（系）中不含*Lr34/Yr18*基因。

1) “a” for *Rht-B1* and *Rht-D1* loci represents the alleles *Rht-B1a* and *Rht-D1a*, respectively; “b” for *Rht-B1* and *Rht-D1* loci represents the alleles *Rht-B1b* and *Rht-D1b*, respectively.

2) “+” indicates the lines with the genes *Lr34/Yr18*, and “—” indicates those without *Lr34/Yr18*.

2.2 *Rht-B1b* 基因检测

利用已知矮秆基因品种Janz和Kukri作为对照，前者含*Rht-B1b*基因，后者含*Rht-D1b*基因[6]。利用Ellis提供的两对引物，检测*Rht-B1*基因位点（图2），在携带*Rht-B1a*的材料中，用NH-BF.2/WR1.2引物可扩增出一条400bp的片段；在携带*Rht-B1b*的材料中，用NH-BF.2/MR1引物可扩增出一条400bp的片段；两对引物PCR产物互补出现，可以相互验证。在检测的263个品种（系）中（表1），47个品系含*Rht-B1a*基因，占总数的17.9%；216个品系携带*Rht-B1b*基因，占总数的82.1%，两对引物检测结果完全互补。

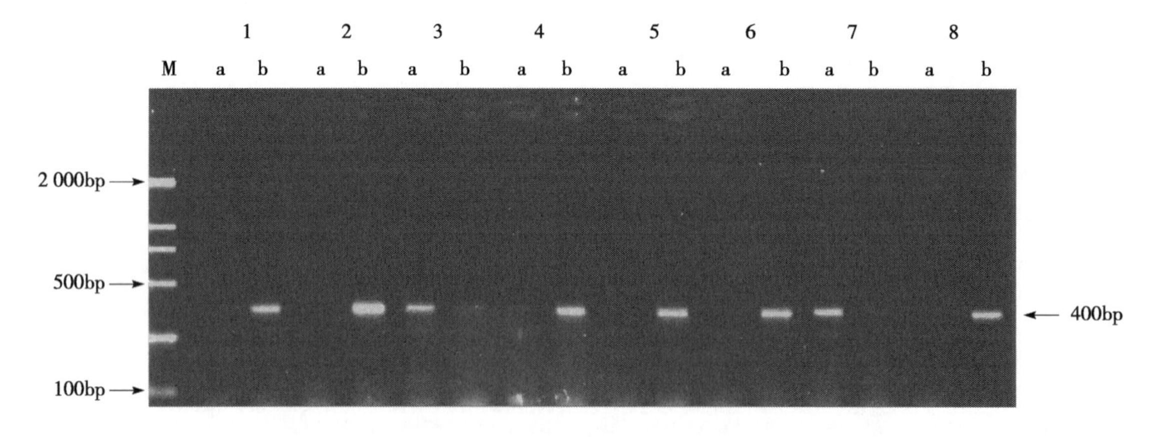

图2 部分CIMMYT小麦品种（系）*Rht-B1*基因位点的检测结果

Fig. 2 PCR product amplified with the STS marker for *Rht-B1* gene in some CIMMYT cultivars and lines

M. 2000bp DNA ladder. 1. BABAX/LR42//BABAX * 2/3/VIVITSI; 2. ATTILA/3 * BCN * 2//BAV92; 3. PRL/2 * PASTOR; 4. ATTILA * 2/PBW65; 5. SERI. 1B * 2/3/KAUZ * 2/BOW//KAUZ; 6. WEEBILL1; 7. Kukri; 8. Janz.

a. *Rht-B1a*基因特异性引物的PCR扩增产物，b. *Rht-B1b*基因特异性引物的PCR扩增产物。

a. PCR product amplified with the primer set NH-BF. 2/WR1. 2 for the allele *Rht-B1a*, b. PCR product amplified with the primer set NH-BF. 2/MR1 for the allele *Rht-B1b*.

2.3 *Rht-D1b* 基因检测

利用已知矮秆基因品种 Janz 和 Kukri 作为对照，用 Ellis 提供的一对引物 DF/MR2 检测 *Rht-D1* 基因位点，在携带 *Rht-D1b* 的材料中可扩增出一条 280bp 的片段（图 3）。在 263 份 CIMMYT 品种（系）中（表 1），通过 PCR 扩增，38 个品种（系）得到一条与 *Rht-D1b* 基因相对应的 280 bp 左右的片段，占总数的 14.4%；210 个品种（系）没有扩增产物，占总数的 85.6%。

分子检测表明，在 263 份 CIMMYT 材料中，双矮秆基因型（*Rht－B1b*＋*Rht－D1b*）共 12 个，占总数的 4.5%；高秆基因型（*Rht－B1a* ＋*Rht－D1a*）21 个，占总数的 7.9%。

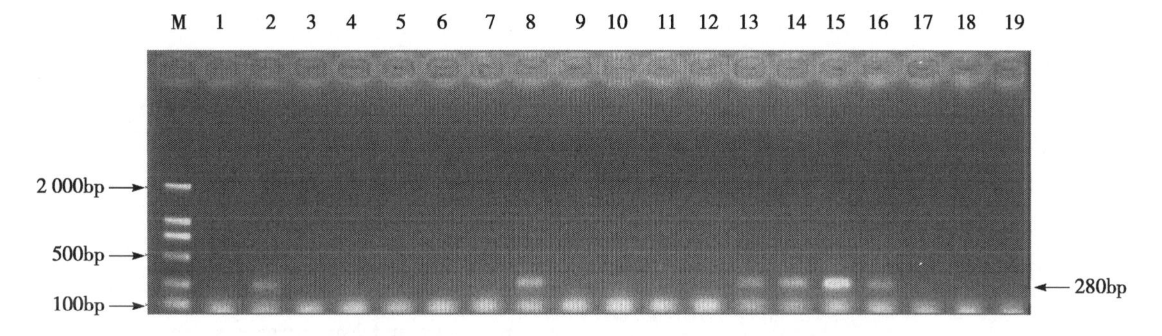

图 3 部分 CIMMYT 小麦品种（系）*Rht-D1b* 基因的检测结果

Fig. 3 PCR product amplified with the STS marker for *Rht-D1b* gene in some CIMMYT cultivars and lines

M. 2000bp DNAladder; 1. Janz; 2. Kukri; 3. WBLL1 * 2/KUKUNA; 4. WBLL1/KUKUNA//KAMB1; 5. WBLL4/KUKUNA//WBLL1; 6. HEILO; 7. WBLL1 * 2/KKTS; 8. RABE/LAJ3302; 9. PAVON F 76; 10. FRET2 * 2/KUKUNA; 11. HE1/3 * CNO79//2 * SERI/3/ATTILA/4/WH 542; 12. JUCHI F2000; 13. KIRITATI; 14. PFAU/WEAVER * 2//KIRITATI; 15. PGO//CROC _ 1/AE. SQUARROSA (224) /3/2 * BORL95/4/CIRCUS; 16. KENYA NYANGUMI; 17. BAU/TNMU; 18. CATBIRD; 19. GONDO

第 2、8、13、14、15、16 材料含 *Rht-D1b* 基因。Entries 2, 8, 13, 14, 15 and 16 contain the allele *Rht-D1b*.

3 讨论

分子标记辅助选择在育种中越来越受到重视，筛选并验证可靠的分子标记有助于快速、准确聚合抗病基因。在澳大利亚已将 *Lr34/Yr18* 基因及其紧密连锁的分子标记 *csLr34* 成功应用于小麦育种[11]。但迄今为止，利用分子标记检测 *Lr34/Yr18* 基因在国内还没有报道，CIMMYT 也未对其小麦品种进行系统检测。本试验利用 *csLV34* 标记检测了 263 份 CIMMYT 小麦品种（系）中 *Lr34/Yr18* 基因的等位变异，结果表明该标记扩增带型清晰易读，稳定性好，可用于大规模品种鉴定。

Singh 等[16]对 27 份印度和巴基斯坦小麦进行抗病鉴定，至少有 15 个品种含 *Lr34/Yr18* 基因，在 20 世纪 90 年代 CIMMYT 小麦品种中，60%左右含 *Lr34/Yr18* 基因[17]，印度和巴基斯坦推广的小麦品种绝大多数从 CIMMYT 引进，因而含 *Lr34/Yr18* 基因的比例较高。在本试验检测的 263 个 CIMMYT 品种（系）中，仅有 57 个品种（系）含 *Lr34/Yr18* 基因，占总数的 21.7%，频率与 20 世纪 90 年代差别较大，原因需进一步研究。在中国 216 个推广品种中，含 *Lr34/Yr18* 基因的品种仅占总数的 3.7%；但在 422 个农家品种中，85.1%的品种含 *Lr34/Yr18* 基因（本实验室资料，待发表）。与农家种相比，我国小麦推广品种含 *Lr34/Yr18* 基因很少，可能原因是在国内小麦育种中偏重主效抗病基因的选择，忽视了慢病性的利用。尽管育成主效抗病基因的品种在一定时间内高抗，但随着新小种的出现，其抗性迅速丧失，因而造成很大损失。所以在育种中要重视慢病性基因的选择，利用分子标记聚合慢病基因和主效基因，有计划地培育一批持久抗病品种。

在这批 CIMMYT 材料中，*Rht-B1b* 基因频率很高，而 *Rht-D1b* 基因比例较低，且含双矮秆基因型（*Rht-B1b*＋*Rht-D1b*）的品系较少。这些品种（系）矮秆基因型 *Rht-B1b* 和 *Rht-D1b* 的比例与先前报道的中国小麦的矮秆基因*Rht-B1b*和*Rht-D1b*比例差别较大[6]，在中国小麦中，*Rht-B1b*基因频率明显低于*Rht-D1b*基因的频率，其主要原因是育种目标和亲本来源不同。杨松杰等[6]研究发现，在中国品种中，*Rht-B1b*基因来自农林10（*Rht-B1b*＋*Rht-D1b*）和St2422/464，*Rht-D1b*基因来自农林10号、水源86、辉县红和蚰包麦，其中来自于农林10号和St2422/464的*Rht-B1b*矮秆基因型仅24个品种，说明在我国小麦育种中矮秆基因*Rht-B1b*利用较少；而CIMMYT的绝大多数品种的矮源都是农林10号或者其衍生系[18]，*Rht-B1b*基因频率较高的原因还有待进一步研究。这些CIMMYT品种（系）含双矮秆基因型（*Rht-B1b*＋*Rht-D1b*）的频率与中国小麦品种类似[6]。先前的研究表明，含双矮秆基因（*Rht-B1b*＋*Rht-D1b*）品种的产量稍高于高秆基因型（*Rht-B1a*＋*Rht-D1a*），但低于含单个矮秆基因的品种（*Rht-B1b*或*Rht-D1b*）[12]，这或许是导致含双矮秆基因（*Rht-B1b*＋*Rht-D1b*）的品系较少的原因。

4 结论

抗病性和丰产性是小麦育种的重要目标。本文报道263份CIMMYT品种（系）慢锈性基因*Lr34*/*Yr18*和矮秆基因*Rht-B1b*、*Rht-D1b*等位基因的分布，可以为我国和CIMMYT小麦育种提供有用的材料、信息以及分子标记辅助选择方法。

参考文献

[1] 李振岐，曾士迈．中国小麦锈病．北京：中国农业出版社，2002：2-3.
Li Z Q, Zeng S M. *Wheat Rust in China*. Beijing: China Agriculture Press, 2002: 2-3. (in Chinese)

[2] 王凤乐，吴立人，万安民．中国小麦条锈菌群体毒性变异研究．中国农业科学，1995，28：8-14.
Wang F L, Wu L R, Wan A M. Studies on virulence variation of wheat stripe rust population in China. *Scientia Agriculture Sinica*, 1995, 28: 8-14. (in Chinese)

[3] 闫红飞，杨文香，张维宏，刘大群．小麦抗叶锈基因*Lr38*差异表达的初步研究．中国农业科技导报，2007，9：118-120.
Yan H F, Yang W X, Zhang W H, Liu D Q. Preliminary study on differential expression of wheat leaf rust resistance gene *Lr38*. *Journal of Agricultural Science and Technology*, 2007, 9: 118-120.

[4] Singh R P, Huerta-Espino J, Rajaram S. Achieving near-immunity to leaf and stripe rusts in wheat by combining slow rusting resistance genes. *Acta Phytopathologica Hungarica*, 2000, 35 (1-4): 133-139.

[5] Hedden P. The genes of the Green Revolution. *Trends in Genetics*, 2003, 19: 5-9.

[6] 杨松杰，张晓科，何中虎，夏先春，周阳．用STS标记检测矮秆基因*Rht-B1b*和*Rht-D1b*在中国小麦中的分布．中国农业科学，2006，39：1680-1687.
Yang S J, Zhang X K, He Z H, Xia X C, Zhou Y. Distribution of dwarfing genes *Rht-B1b* and *Rht-D1b* in Chinese bread wheats detected by STS marker. *Scientia Agriculture Sinica*, 2006, 39: 1680-1687. (in Chinese)

[7] Singh R P. Association between gene *Lr34* for leaf rust resistance and leaf tip necrosis in wheat. 1992, *Crop Science*, 32: 874-878.

[8] Singh R P. Genetic association of gene *Bdvl* for barley yellow dwarf virus with genes *Lr34* and *Yr18* for adult plant resistance to rusts in bread wheat. *Plant Disease*, 1993, 77: 1103-1106.

[9] Imtiaz M, Ahmad M, Cromey M G, Griffin W B, Hampton J G. Detection of molecular markers linked to the durable adult plant stripe rust resistance gene *Lr34*/*Yr18* in bread wheat (*Triticum aestivum* L.). *Plant Breeding*, 2004, 123: 401-404.

[10] Spielmeyer W, McIntosh R A, Kolmer J, Lagudah E S. Powdery mildew resistance and *Lr34*/*Yr18* genes for durable resistance to leaf and stripe rust cosegregate at a locus on the short arm of chromosome 7D of wheat. *Theoretical and Applied Genetics*, 2005, 111: 731-735.

[11] Lagudah E S, McFadden H, Singh R P, Huerta E J, Bariana H S, Spielmeyer W. Molecular genetic characterization of the *Lr34*/*Yr18* slow rusting resistance gene region in wheat. *Theoretical and Applied Genetics*, 2006, 114: 21-30.

[12] Allan R E. Agronomic comparisons between *Rht1* and *Rht2* semidwarf genes in winter wheat. *Crop Science*, 1989, 29: 1103-1108.

[13] Ellis M H, Spielmeyer W, Rebetzke G J, Richards R A. "Perfect" markers for the *Rht-B1b* and *Rht-D1b* dwarfing genes in wheat. *Theoretical and Applied Genetics*, 2002, 105: 1038-1042.

[14] 何中虎，庞家智．CIMMYT麦类改良进展．北京：

中国农业科技出版社，1995：32-33.
He Z H，Pang J Z. *Progress of CIMMYT Wheat Improvement*. Beijing：China Agriculture and Science Press. 1995：32-33.（in Chinese）

[15] Devos K M，Gale M D. The use of random amplified polymorphic DNA markers in wheat. *Theoretical and Applied Genetics*，1992，84：567-572.

[16] Singh R P，Gupta A K. Genes for leaf rust resistance in Indian and Pakistani wheats tested with Mexican pathotypes of *Puccinia recondita* f. sp. *tritici*. *Euphytica*，1991，57：27-36.

[17] Bahl P N，Salimath P M，Mandal A K. *Genetics，Cytogenetics and Breeding of Crop Plants*. Oxford & IBH Publishing Co. PVT LTD，New Delhi and Calcutta，1997：75-144.

[18] Perkings J H. *This Translation of Ethics and Geopolitics and the Green Revolution*. Oxford University Press，1997：303-313.

CIMMYT 273 个小麦品种抗病基因 *Lr34/Yr18/Pm38* 的分子标记检测

伍玲[1,2,3]，夏先春[2]，朱华忠[3]，李式昭[3]，郑有良[1]，何中虎[2,4]

[1]四川农业大学小麦研究所，成都 611130；[2]中国农业科学院作物科学研究所/国家小麦改良中心，北京 100081；[3]四川省农业科学院作物研究所，成都 610066；[4]国际玉米与小麦改良中心北京办事处，北京 100081

摘要： 明确 CIMMYT 观察圃 273 个小麦品种（系）在慢病基因 *Lr34/Yr18/Pm38* 位点的等位变异类型及其对条锈病、叶锈病和白粉病的抗性，进一步验证抗病基因功能标记的有效性。利用 *Lr34/Yr18/Pm38* 紧密连锁的 STS 标记 *csLV34* 和基于该基因第 11 外显子（exon 11）等位变异开发的 5 对功能标记 *cssfr1*-*cssfr5* 检测新引进的 273 个 CIMMYT 小麦品种（系），同时在成都和北京分别对其进行田间条锈病和白粉病的抗性鉴定，并结合 CIMMYT 的田间条锈病和叶锈病抗性鉴定结果进行分析。STS 标记 *csLV34* 与 5 对功能标记 *cssfr1*-*cssfr5* 检测结果的一致性为 96.7%；在 273 份 CIMMYT 材料中有 43 份材料含有 *Lr34/Yr18/Pm38*，在不同地点对条锈病、叶锈病和白粉病具有不同程度的抗病性。功能标记 *cssfr1*-*cssfr5* 可准确鉴定 *Lr34/Yr18/Pm38* 位点 exon 11 中的等位变异，*cssfr3*、*cssfr4*、*cssfr5* 可用于这些含有 *Lr34/Yr18/Pm38* 材料的杂交后代的分子标记辅助选择。

关键词： 普通小麦；*Lr34/Yr18/Pm38*；功能标记；慢病性

Molecular Characterization of *Lr34/Yr18/Pm38* in 273 CIMMYT Wheat Cultivars and Lines for Yield Potential Trials Using Functional Markers

Wu Ling[1,2,3], Xia Xianchun[2], Zhu Huazhong[3], Li Shizhao[3], Zhang Youliang[1], He Zhonghu[2,4]

[1]*Triticeae Research Institute, Sichuan Agricultural University, Chengdu* 611130; [2]*Institute of Crop Sciences, Chinese Academy of Agricultural Sciences/National Wheat Improvement Center, Beijing* 100081; [3]*Crop Research Institute, Sichuan Academy of Agricultural Science, Chengdu* 610066; [4]*International Maize and Wheat Improvement Center (CIMMYT) China Office, Beijing* 100081

Abstract: The objectives of this study were to characterize the allelic variation at *Lr34/Yr18/Pm38* locus in 273 CIMMYT wheat cultivars (lines), evaluate their resistance against stripe rust, leaf rust and powdery mildew, and validate the functional markers for the locus. A total of 273 CIMMYT wheat genotypes were tested using a STS marker *csLV34* and 5 functional markers *cssfr1*-*cssfr5* which were developed based on a 3-bp deletion in the exon 11 of *Lr34/Yr18/Pm38* gene, and were tested for stripe rust resistance in the field in

原文发表在《中国农业科学》，2010，43（22）：4553－4561

Chengdu and CIMMYT respectively, leaf rust in CIMMYT, and powdery mildew in Beijing. The results tested by the STS marker *csLV34* showed a 96.7% identity to those of *cssfr1*-*cssfr5*. Of the 273 lines, 43 were characterized to have *Lr34/Yr18/Pm38* resistance gene, with partial resistance against stripe rust, leaf rust and powdery mildew in the field across different environments. The 5 functional markers (*cssfr1*-*cssfr5*) could be used to detect the allelic variants in *Lr34/Yr18/Pm38* locus, and 43 CIMMYT lines with *Lr34/Yr18/Pm38* may be used in wheat breeding targeting for stripe rust, leaf rust and powdery mildew resistance with marker-assisted selection of *cssfr3*, *cssfr4* and *cssfr5*.

Key words: *Triticum aestivum* L.; *Lr34/Yr18/Pm38*; functional marker; slow rusting and mildewing resistance

小麦条锈病、叶锈病和白粉病是制约小麦生产的重要因素。*Lr34/Yr18/Pm38* 位点是一个对小麦锈病、白粉病等多种病害具有部分抗性的重要慢病基因位点。明确 CIMMYT 小麦在 *Lr34/Yr18/Pm38* 位点的等位变异，及其对条锈病、叶锈病和白粉病的抗性，对 *Lr34/Yr18/Pm38* 位点分子标记在育种中的应用和 CIMMYT 小麦抗性基因的更好利用均具有重要意义。*Lr34/Yr18/Pm38* 位点的抗性很早就在小麦生产中发挥作用[1]。1977 年，Dyck[2]最早在加拿大品种 PI58548 叶锈病的鉴定中报道了该位点，之后更多的研究发现该位点同时对小麦叶锈病[2,3]、条锈病[4,5]和白粉病[6]均具有抗性，同时还与大麦黄矮病毒病[7]和小麦秆锈病[8]的抗性有关。*Lr34/Yr18/Pm38* 位点的抗性具有成株抗性、非小种专化性等特点，抗性持久稳定。Dyck[3]最先将 *Lr34* 定位在小麦 7D 染色体上，此后许多学者利用 QTL 分析技术将该位点定位在小麦 7DS 染色体上[9-14]，并将其对条锈病的抗性基因命名为 *Yr18*[15]，对白粉病的抗性基因命名为 *Pm38*[16]。由于有研究发现叶尖坏死基因 *Ltn1* 与 *Lr34/Yr18/Pm38* 位点紧密连锁[5,17]，所以在早期的研究中叶尖坏死作为 *Lr34/Yr18/Pm38* 的生理标记得以应用。但由于 *Lr34/Yr18/Pm38* 位点的抗性是在成株期表达，在田间容易被有效的主基因抗性掩盖；而生理标记叶尖坏死性状是受多基因作用的结果，在不同环境下有不同形式和不同程度的表达，并且容易受品种遗传背景的影响，因此小麦叶尖坏死不能完全代表 *Lr34/Yr18/Pm38* 的存在与否，而找到该位点可用的分子标记一直是研究者努力的目标。QTL 分析发现，SSR 标记 *Xgwm295* 和 *Xgwm1220* 与 *Lr34/Yr18/Pm38* 位点紧密连锁[12-14]，遗传距离分别为 0.9cM 和 2.7cM。为了找到更近的标记，两个实验室根据水稻基因组信息找到了与之同源的小麦 EST 和 D 组的 BAC 克隆，Bossolini 等[18]对其所得的 BAC 克隆序列进行分析，得到了 10 个 SSR 标记，其中标记 *SWM10* 与 *Lr34/Yr18/Pm38* 位点的连锁距离为 0.7cM，在 3 个独立来源的小麦“Frontana”、中国春、“Forno”中具有相同的等位性，在源于 CIMMYT 的春小麦中稳定存在，但在欧洲冬小麦中却存在不同的类型。Lagudah 等[19]以 EST 得到的克隆作为探针进行 RFLP 分析，找到了一个能清晰区分含有和不含 *Lr34/Yr18* 的 RFLP 标记，并成功地将该 RFLP 标记转化为共显性的 STS 标记 *csLV34*。他们通过对源于组合 Lalbahadur（不含 *Lr34*）× Lalbahadur（Parula7D）（含 *Lr34*）的 768 个 F_2 植株的精细作图分析，发现标记 *csLV34* 与 *Lr34/Yr18/Pm38* 位点间的遗传距离为 0.4cM，并用 60 个源于组合 Jupateco73R×Jupateco73S 的重组近交系和 24 个已知 *Lr34/Yr18/Pm38* 位点基因型的材料验证了标记 *csLV34* 的有效性。因为标记 *csLV34* 为共显性标记，其扩增产物稳定，两个片段大小差异为 79bp，容易用琼脂糖胶分离，因而得到广泛应用[1,20-22]。然而标记 *csLV34* 与位点间较低的重组和标记检测中对等位变异的不确定判断的存在限制了它的应用。2009 年 Krattinger 等[23]克隆了 *Lr34/Yr18/Pm38*，并预测 *Lr34/Yr18/Pm38* 编码的是一个多向耐药型（pleiotropic drug resistance，PDR）ABC 转运蛋白（ATP-binding cassette transporter）。这个基因控制了基于 *Lr34*、*Yr18* 和 *Pm38* 的抗性。该基因编码区大小为 11 805bp，由 24 个 exon 组成。他们比较了不同的小麦品种，发现仅存在两个等位基因，一个是感病的等位基因-*Lr34*，另一个是抗病的等位基因＋*Lr34*。这两个等位基因在序列上的差异存在于 intron 4 的 A/T SNP、exon11 中的 TTC 缺失和 exon 12 的 C/T SNP。Lagudah 等[24]根据＋*Lr34* 等位基因在 exon11 上的 TTC 缺失设计了 5 条引物，并与 *csLV34* 结合开发了 *cssfr1*-*cssfr5* 五对功能标记。这五对功能标记能准确区分在 *Lr34/Yr18/Pm38* 位点已知基因型的 34 份品种，可以准确鉴定小麦在 *Lr34/Yr18/Pm38* 位点的等位变异类型。

CIMMYT 小麦在中国小麦育种中发挥了重要作用，对提高产量、改善品质和抗病性做出了重要贡献[25]。CIMMYT 每年有 1 000 多份种质资源在中国各地的观察圃中种植[26]，这些种质资源的利用，丰富了中国小麦的遗传基础。在四川，CIMMYT 小麦是重要的条锈病抗源[27]。Singh 等[28]认为将包含 *Lr34/Yr18/Pm38* 在内的 2～4 个慢病基因聚合，可育出近免疫的抗条锈品种。Kolmer 等[1]用 *csLV34* 检测 127 份 CIMMYT 小麦，发现 62 份材料含有 *Lr34/Yr18/Pm38*，表明 CIMMYT 小麦是重要的慢病性资源。目前在中国国内还未见用 *Lr34/Yr18/Pm38* 位点的功能标记 *cssfr1-cssfr5* 对小麦品种检测的报道，还没有对新引进的 CIMMYT 41IBWSN 和 19HRWSN 系列小麦材料的条锈病和白粉病进行抗性鉴定和分子标记检测。本研究通过明确新引进的 CIMMYT 材料在 *Lr34/Yr18/Pm38* 位点的等位变异类型，鉴定这些小麦材料对条锈病、叶锈病、白粉病的抗性，比较分子标记检测结果与田间抗病鉴定结果，以进一步验证 *Lr34/Yr18/Pm38* 位点功能标记的有效性，为中国小麦抗病育种提供材料和技术。

1 材料与方法

1.1 试验材料

1.1.1 小麦品种 包括 2008 年引进的 CIMMYT 19HRWSN 和 41IBWSN 系列的小麦品种和品系。19HRWSN 系列是 CIMMYT 第 19 轮多雨环境小麦观察圃的高代品种（系），41IBWSN 系列是 CIMMYT第 41 轮国际观察圃的高代品种（系）。全部供试品种（系）273 份。

表 1 *Lr34/Yr18/Pm38* 位点紧密连锁的 STS 标记 *csLV34* 和功能标记 *cssfr1-cssfr5*

Table 1 STS marker *csLV34* closely linked to *Lr34/Yr18/Pm38* and functional markers *cssfr1-cssfr5* of *Lr34/Yr18/Pm38* locus

标记名称 Marker name	引物代号 Primer name	引物序列（5′-3′） Primer sequence	引物用量a) Amount of primers used（μl）	PCR 程序 PCR program	PCR 产物 PCR products（bp）	
					+*Lr34*	−*Lr34*
csLV34	csLV34F	GTTGGTTAAGACTGGTGATGG	0.5	5cycles of 94℃ 1min，57℃ 1min，72℃ 2min；30cycles of 94℃ 30s，57℃ 30s，72℃ 1min；94℃ 30s，57℃ 30s，72℃ 5min	150	229
	csLV34R	TGCTTGCTATTGCTGAATAGT	0.5			
cssfr1	Lr34DINT9F	TTGATGAAACCAGTTTTTTTTCTA	0.5	94℃ 5min；30cycles of 94℃ 1min，58℃ 1min，72℃ 1min；72℃ 10min	517	—
	Lr34PLUSR	GCCATTTAACATAATCATGATGGA	0.5			
cssfr2	Lr34DINT9F	TTGATGAAACCAGTTTTTTTTCTA	0.5	94℃ 5min，30cycles of 94℃ 1min，58℃ 1min，72℃ 1min；72℃ 10min	—	523
	Lr34MINUSR	TATGCCATTTAACATAATCATGAA	0.5			
cssfr3	Lr34DINT9F	TTGATGAAACCAGTTTTTTTTCTA	0.5	94℃ 5min，30cycles of 94℃ 1min，58℃ 1min，72℃ 1min；72℃ 10min	517 + 150	229
	Lr34PLUSR	GCCATTTAACATAATCATGATGGA	0.5			
	csLV34F	GTTGGTTAAGACTGGTGATGG	0.25			
	csLV34R	TGCTTGCTATTGCTGAATAGT	0.25			
cssfr4	Lr34DINT9F	TTGATGAAACCAGTTTTTTTTCTA	0.5	94℃ 5min，30cycles of 94℃ 1min，58℃ 1min，72℃ 1min；72℃ 10min	150	523 + 229
	Lr34MINUSR	TATGCCATTTAACATAATCATGAA	0.5			
	csLV34F	GTTGGTTAAGACTGGTGATGG	0.25			
	csLV34R	TGCTTGCTATTGCTGAATAGT	0.25			
cssfr5	Lr34DINT9F	TTGATGAAACCAGTTTTTTTTCTA	0.5	94℃ 5min，30cycles of 94℃ 1min，58℃ 1min，72℃ 1min；72℃ 10min	751	523
	Lr34MINUSR	TATGCCATTTAACATAATCATGAA	0.5			
	Lr34SPF	GGGAGCATTATTTTTTTCCATCATG	0.5			
	Lr34DINT13R2	ACTTTCCTGAAAATAATACAAGCA	0.5			

a）引物用量指浓度为 10μmol L^{-1}的引物在 20μl 反应体系中的用量；"—" 没有 PCR 产物。

a）The amount of primers used means the volume of primers used in 20μl total volume of PCR reaction，when the primer concentration is 10mol L^{-1}；"—" No PCR product.

1.1.2 *Lr34/Yr18/Pm38* 位点分子标记 包括 *Lr34/Yr18/Pm38* 位点紧密连锁的 STS 标记 *csLV34* 和基于 *Lr34/Yr18/Pm38* 位点 exon 11 的 TTC 缺失开发的功能标记 *cssfr1-cssfr5*，其引物序列和 PCR 反应程序列于表 1。引物由北京奥科生物技术有限责任公司（http：//www.augct.com）合成。

Lagudah 等[24]在 *Lr34/Yr18/Pm38* 的 intron 9 设计了特异的上游引物 L34DINT9F，根据 exon 11 的 TTC 缺失设计了两条下游引物 L34PLUSR 和 L34MINUSR，引物 L34DINT9F/L34PLUSR 组合形成了功能标记 *cssfr1*，引物 L34DINT9F/L34MINUSR 组合形成了功能标记 *cssfr2*；标记 *cssfr1* 和 *cssfr2* 为显性标记，二者互补；为了保证每个扩增反应的真实性，将它们分别加上标记 *csLV34* 的引物组成多重 PCR 体系形成了标记 *cssfr3* 和 *cssfr4*；而根据 exon 11 的 TTC 缺失设计的上游引物 L34SPF 与在 intron13 设计的下游引物 L34DINT13R2，并加上标记 *cssfr2* 的两条引物组成的多重 PCR 体系形成了共显性标记 *cssfr5*。

1.2 试验方法

1.2.1 **DNA 提取及分子标记检测** 每品种选取 4 粒有代表性的籽粒混合，采用 Lagudah 等[29]的方法提取籽粒基因组 DNA。根据表 1 的 PCR 程序在 PTC 200 扩增仪上进行 PCR 扩增。PCR 反应体系为 20μl，包括表 1 中的引物用量，2μl 10×PCR buffer，0.4μl 浓度各为 10mmol L^{-1}的 dNTP，标记 *csLV34* 检测用一般 *Taq* DNA 聚合酶 1U（天根公司，北京），其余标记均用 HotMaster *Taq* DNA 聚合酶 1U（天根公司，北京），模板 DNA 50ng。扩增产物以 1.5%的琼脂糖凝胶电泳分离检测，缓冲液体系为 1×TAE 溶液，200V 电压电泳 20min，溴化乙锭染色后，用 Gel Doc XR System 扫描成像并存入计算机。根据每个品种 6 个标记的检测结果判断该品种（系）在 *Lr34/Yr18/Pm38* 位点的等位变异类型。

1.2.2 **田间病害人工接种鉴定** 试验采用随机区组设计，3 次重复。条锈病田间人工接种鉴定试验于 2008—2009 年度在成都四川省农业科学院郫县试验站进行，白粉病人工接种鉴定试验同年在北京中国农业科学院东圃场试验地进行。

条锈病人工接种鉴定试验采用单行区，行长 150cm，行距 26.7cm，每 10 行设置 1 行感病对照品种 SY95-71，在鉴定材料行一端垂直种植 1 行混合的感病品种川育 12 和 SY95-71 做诱发行。2008 年 12 月下旬至 2009 年 1 月初，在诱发行中移栽已接种混合条锈菌小种并已发病的铭贤 169 幼苗。分别于 4 月初和 4 月中旬条锈病盛发期分两次记载条锈病严重度（旗叶和倒二叶条锈菌孢子堆面积占叶片面积的百分比），个别晚熟品种于 5 月初再记载一次。

白粉病人工接种鉴定试验每重复每品种 1 穴，200cm 行种 5 穴，穴距 45cm，行距 20cm，每 10 行设置 1 行感病对照品种京双 16，在鉴定材料行一端垂直种植 1 行感病对照品种京双 16 做诱发行。2009 年 3 月底，在诱发行中移栽已接种白粉菌小种 E20 并已发病的京双 16 幼苗。分别于 5 月 15 号和 5 月 22 号左右白粉病盛发期分两次记载供试材料倒二叶白粉病严重度（倒二叶白粉菌孢子堆面积占叶片面积的百分比）。

此外，从 CIMMYT 网站下载 19HRWSN 和 41IBWSN 材料在 CIMMYT 的条锈病和叶锈病鉴定结果。

2 结果

2.1 分子标记检测结果

csLV34 和 *cssfr1-cssfr5* 这 6 个标记扩增产物大小列于表 1，图 1 是这 6 个标记的扩增产物电泳图。标记 *csLV34* 有两个扩增片段，扩增出 150bp 片段的材料可能含有 *Lr34/Yr18/Pm38*，扩增出 229bp 片段的材料可能不含 *Lr34/Yr18/Pm38*[19]；有时 *csLV34* 的扩增产物中会有一个分子量大于 229bp 的片段，但该片段与 *Lr34/Yr18/Pm38* 无关[1]。标记 *cssfr1* 在 *Lr34/Yr18/Pm38* 位点 exon11 等位变异为 +*Lr34/Yr18/Pm38* 类型的材料中扩增出 517bp 片段；标记 *cssfr2* 在 exon11 等位变异为 −*Lr34/Yr18/Pm38* 类型的材料中扩增出 523bp 片段；标记 *cssfr3* 在 exon11 等位变异为 +*Lr34/Yr18/Pm38* 类型的材料中扩增出 517bp 和 150bp 两个片段，在 exon 11 等位变异为 −*Lr34/Yr18/Pm38* 类型的材料中扩增出 229bp 片段；标记 *cssfr4* 在 exon 11 等位变异为 +*Lr34/Yr18/Pm38* 类型的材料中扩增出 150bp 片段，在 exon 11 等位变异为 −*Lr34/Yr18/Pm38* 类型的材料中扩增出 523bp 和 229bp 两个片段；标记 *cssfr5* 在 exon 11 等位变异为 +*Lr34/Yr18/Pm38* 类型的材料可扩增出 751bp 片段，在 exon 11 等位变异为 −*Lr34/Yr18/Pm38* 类型的材料则扩增出 523bp 片段。

利用 *csLV34* 和 *cssfr1-cssfr5* 这 6 个标记检测了

供试的 273 份 CIMMYT 小麦在 *Lr34/Yr18/Pm38* 位点的等位变异。根据 6 个标记的检测结果，在 264 份材料中 6 个标记检测结果完全相同，占供试材料的 96.7%；其中 43 份材料为＋*Lr34/Yr18/Pm38* 类型，占供试材料的 15.8%；203 份材料为－*Lr34/Yr18/Pm38* 类型，占供试材料的 74.4%；18 份材料为杂合类型，占供试材料的 6.6%。43 份＋*Lr34/Yr18/Pm38* 类型材料的标记鉴定结果见表 2，其余 9 份材料 6 个标记的检测结果不一致，结果见表 3。分析它们的检测结果发现，因为标记 *csLV34* 的检测结果与功能标记引物检测结果不同，从而导致不能根据标记 *cssfr3*、*cssfr4* 判断被测材料的等位变异类型，所以最终不能确定这 9 份材料的等位变异类型。如果仅根据功能标记检测结果，这 9 份材料中有 1 份为＋*Lr34/Yr18/Pm38* 类型，4 份为－*Lr34/Yr18/Pm38* 类型，3 份为杂合类型。19HRWSN-88 在标记 *cssfr1* 和 *cssfr3* 检测中没有扩增出 517bp 片段，但在 *cssfr5* 的检测中扩增出了 751bp 片段，所以不能判断其等位变异类型。

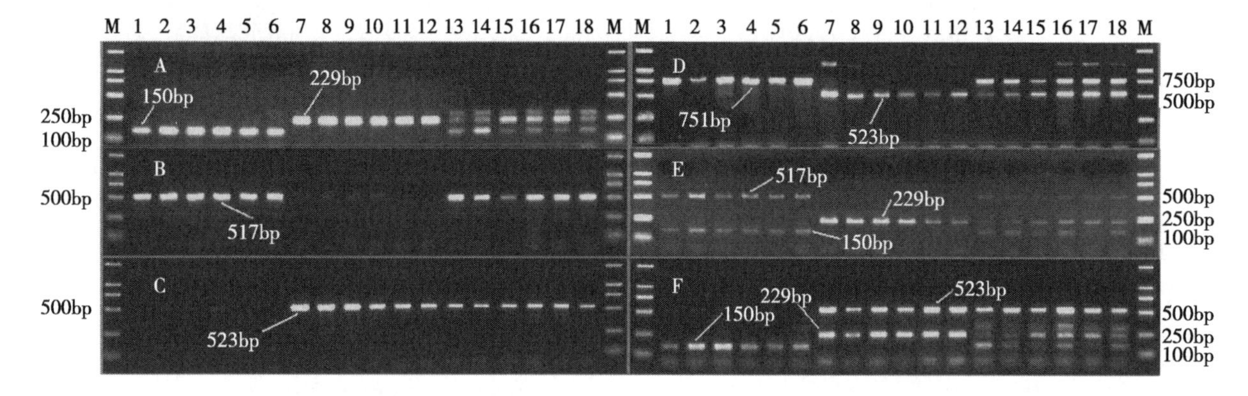

图 1 标记 *csLV34* 和 *cssfr1-cssfr5* PCR 扩增产物电泳图

Fig. 1 The electrophoretogram of PCR products amplified by markers *csLV34* and *cssfr1-cssfr5*

A：标记 *csLV34* 扩增产物电泳图；B：标记 *cssfr1* 扩增产物电泳图；C：标记 *cssfr2* 扩增产物电泳图；D：标记 *cssfr5* 扩增产物电泳图；E：标记 *cssfr3* 扩增产物电泳图；F：标记 *cssfr4* 扩增产物电泳图；M：DL2000；1：19HRWSN-73；2：19HRWSN-112；3：41IBWSN-46；4：41IBWSN-54；5：41IBWSN-78；6：41IBWSN-82；7：19HRWSN-75；8：19HRWSN-109；9：19HRWSN-110；10：19HRWSN-111；11：41IBWSN-5；12：41IBWSN-6；13：19HRWSN-121；14：41IBWSN-8；15：41IBWSN-14；16：41IBWSN-26；17：41IBWSN-109；18：41IBWSN-111。

A: Electrophoretogram of PCR products amplified by the marker *csLV34*; B: Electrophoretogram of PCR products amplified by the marker *cssfr1*; C: Electrophoretogram of PCR products amplified by the marker *cssfr2*; D: Electrophoretogram of PCR products amplified by the marker *cssfr5*; E: Electrophoretogram of PCR products amplified by the marker *cssfr3*; F: Electrophoretogram of PCR products amplified by the marker *cssfr4*; M: DL2000; 1: 19HRWSN-73; 2: 19HRWSN-112; 3: 41IBWSN-46; 4: 41IBWSN-54; 5: 41IBWSN-78; 6: 41IBWSN-82; 7: 19HRWSN-75; 8: 19HRWSN-109; 9: 19HRWSN-110; 10: 19HRWSN-111; 11: 41IBWSN-5; 12: 41IBWSN-6; 13: 19HRWSN-121; 14: 41IBWSN-8; 15: 41IBWSN-14; 16: 41IBWSN-26; 17: 41IBWSN-109; 18: 41IBWSN-111.

表 2 43 份＋*Lr34/Yr18/Pm38* 类型小麦材料的标记检测结果及田间病害严重度

Table 2 The results of marker test and disease severities of 43 CIMMYT wheat lines with ＋*Lr34/Yr18/Pm38* allele

品系编号 Line accession number	标记检测结果 The results of marker test						成都条锈病严重度 Stripe rust severity in Chengdu (%)	北京白粉病严重度 Powdery mildew severity in Beijing (%)	CIMMYT 叶锈病严重度 Leaf rust severity in CIMMYT (%)	CIMMYT 条锈病严重度 Stripe rust severity in CIMMYT (%)
	csLV34	*cssfr1*	*cssfr2*	*cssfr3*	*cssfr4*	*cssfr5*				
19HRWSN-2	＋	＋	＋	＋	＋	＋	15.0	7.7	20	5
19HRWSN-3	＋	＋	＋	＋	＋	＋	1.0	12.7	20	5
19HRWSN-6	＋	＋	＋	＋	＋	＋	26.7	11.0	20	20

（续）

品系编号 Line accession number	标记检测结果 The results of marker test						成都条锈病严重度 Stripe rust severity in Chengdu (%)	北京白粉病严重度 Powdery mildew severity in Beijing (%)	CIMMYT 叶锈病严重度 Leaf rust severity in CIMMYT (%)	CIMMYT 条锈病严重度 Stripe rust severity in CIMMYT (%)
	csLV34	*cssfr1*	*cssfr2*	*cssfr3*	*cssfr4*	*cssfr5*				
19HRWSN-8	+	+	+	+	+	+	3.7	11.0	10	1
19HRWSN-10	+	+	+	+	+	+	10.0	3.0	5	15
19HRWSN-14	+	+	+	+	+	+	16.7	8.7	10	1
19HRWSN-16	+	+	+	+	+	+	3.7	13.3	20	5
19HRWSN-17	+	+	+	+	+	+	10.0	11.0	5	1
19HRWSN-18	+	+	+	+	+	+	2.3	4.7	10	5
19HRWSN-21	+	+	+	+	+	+	16.7	3.0	5	1
19HRWSN-31	+	+	+	+	+	+	0.7	10.0	5	0
19HRWSN-32	+	+	+	+	+	+	1.0	11.7	5	10
19HRWSN-33	+	+	+	+	+	+	6.7	5.3	5	20
19HRWSN-37	+	+	+	+	+	+	30.0	12.5	10	5
19HRWSN-55	+	+	+	+	+	+	5.0	7.5	1	0
19HRWSN-56	+	+	+	+	+	+	2.3	10.0	30	1
19HRWSN-57	+	+	+	+	+	+	8.3	13.3	20	1
19HRWSN-58	+	+	+	+	+	+	10.0	11.7	5	5
19HRWSN-59	+	+	+	+	+	+	18.3	25.0	20	10
19HRWSN-61	+	+	+	+	+	+	8.3	9.3	20	0
19HRWSN-62	+	+	+	+	+	+	10.0	3.7	20	0
19HRWSN-71	+	+	+	+	+	+	6.7	15.0	10	20
19HRWSN-73	+	+	+	+	+	+	5.3	4.7	1	5
19HRWSN-112	+	+	+	+	+	+	2.0	5.3	10	1
19HRWSN-117	+	+	+	+	+	+	10.0	3.0	5	10
19HRWSN-118	+	+	+	+	+	+	80.0	6.3	10	30
19HRWSN-119	+	+	+	+	+	+	15.0	6.0	20	20
41IBWSN-4	+	+	+	+	+	+	16.7	11.0	0	30
41IBWSN-16	+	+	+	+	+	+	2.3	6.5	1	10
41IBWSN-17	+	+	+	+	+	+	1.0	5.3	1	10
41IBWSN-28	+	+	+	+	+	+	6.7	3.7	0	1
41IBWSN-46	+	+	+	+	+	+	11.7	2.0	1	10
41IBWSN-54	+	+	+	+	+	+	13.3	5.5	0	30
41IBWSN-77	+	+	+	+	+	+	26.7	11.5	30	1
41IBWSN-78	+	+	+	+	+	+	16.7	9.0	30	20
41IBWSN-82	+	+	+	+	+	+	76.7	8.7	0	30
41IBWSN-104	+	+	+	+	+	+	23.3	10.3	0	20
41IBWSN-106	+	+	+	+	+	+	11.7	3.0	0	30

（续）

品系编号 Line accession number	标记检测结果 The results of marker test						成都条锈病严重度 Stripe rust severity in Chengdu (%)	北京白粉病严重度 Powdery mildew severity in Beijing (%)	CIMMYT 叶锈病严重度 Leaf rust severity in CIMMYT (%)	CIMMYT 条锈病严重度 Stripe rust severity in CIMMYT (%)
	csLV34	*cssfr1*	*cssfr2*	*cssfr3*	*cssfr4*	*cssfr5*				
41IBWSN-107	+	+	+	+	+	+	26.7	4.7	1	30
41IBWSN-117	+	+	+	+	+	+	18.3	9.0	0	10
41IBWSN-138	+	+	+	+	+	+	70.0	6.3	0	30
41IBWSN-139	+	+	+	+	+	+	20.0	6.7	0	0
41IBWSN-153	+	+	+	+	+	+	23.3	4.3	0	0

"+"标记鉴定为+*Lr34*/*Yr18*/*Pm38* 类型。"+" Indicates+*Lr34*/*Yr18*/*Pm38* type.

表3 9份CIMMYT小麦的STS标记 *csLV34* 和功能标记 *cssfr1*-*cssfr5* 检测结果

Table 3 The results of 9 CIMMYT wheat lines tested with the STS marker *csLV34* and 5 functional markers *cssfr1*-cssfr5

品系编号 Line accession number	*csLV34*		*cssfr1*		*cssfr2*		*cssfr3*		*cssfr4*		*cssfr5*		*csLV34*	exon11 functional marker
	229 bp	150 bp	517 bp	523 bp	517 bp	229 bp	150 bp	523 bp	229 bp	150 bp	751 bp	523 bp		
19HRWSN-4	1	1	0	1	0	1	1	1	1	1	0	1	+/−	−
19HRWSN-13	1	0	1	1	1	1	0	1	1	0	1	1	−	+/−
19HRWSN-23	1	1	1	0	1	1	1	0	1	1	1	0	+/−	+
19HRWSN-54	1	1	0	1	0	1	1	1	1	1	0	1	+/−	−
19HRWSN-88	1	1	0	0	0	1	1	0	1	1	1	0	+/−	?
41IBWSN-25	0	1	0	1	0	0	1	1	0	1	0	1	+	−
41IBWSN-44	1	0	1	1	1	1	0	1	1	0	1	1	−	+/−
41IBWSN-81	0	1	1	1	1	0	1	1	0	1	1	1	+	+/−
41IBWSN-119	0	1	0	1	0	0	1	1	0	1	0	1	+	−

"+" +*Lr34*/*Yr18*/*Pm38*；"−" −*Lr34*/*Yr18*/*Pm38*。"+/−" 杂合类型 Heterozygote.

从上述结果可知5对功能标记仅在材料19HRWSN-88的检测中，标记*cssfr5*的751bp片段与标记*cssfr1*、*cssfr3*的517bp片段的结果不一致，其余结果完全相同，功能标记检测结果一致程度达到99.8%。标记*cssfr3*的结果是标记*csLV34*的结果与标记*cssfr1*结果的相加，而标记*cssfr4*的结果是标记*csLV34*的结果与标记*cssfr2*结果的相加。所以认为用*cssfr3*、*cssfr4*和*cssfr5*三个标记来检测小麦中*Lr34*/*Yr18*/*Pm38*位点exon11的等位基因即可。

2.2 田间病害鉴定结果

将供试的273份CIMMYT小麦品种（系）按标记检测分类，统计全部被测材料、+*Lr34*/*Yr18*/*Pm38*类型材料和−*Lr34*/*Yr18*/*Pm38*类型材料的病害严重度变幅和平均值，结果列于表4。

表4 三类小麦材料在4个环境下病害严重度变幅和平均值

Table 4 The range and average of disease severity for three types of wheat lines in 4 environments

小麦类型 Type of wheat lines	项目 Item	成都条锈病严重度 Stripe rust severity in Chengdu	北京白粉病严重度 Powdery mildew severity in Beijing	CIMMYT条锈病严重度 Stripe rust severity in CIMMYT	CIMMYT叶锈病严重度 Leaf rust severity in CIMMYT
所有材料 All 273 wheat lines	变幅 Range	0.3%～100%	1%～60%	0～60%	0～80%
	平均值 Average	22.3%	11.2%	12%	9.1%

（续）

小麦类型 Type of wheat lines	项目 Item	成都条锈病严重度 Stripe rust severity in Chengdu	北京白粉病严重度 Powdery mildew severity in Beijing	CIMMYT 条锈病严重度 Stripe rust severity in CIMMYT	CIMMYT 叶锈病严重度 Leaf rust severity in CIMMYT
+*Lr34*/*Yr18*/*Pm38* 类型 +*Lr34*/*Yr18*/*Pm38* type	变幅 Range	0.7%～30%（40份），70%～80%（3份）	2%～25%	0～30%	0～30%
	平均值 Average	16.1%	8.2%	10.7%	9%
−*Lr34*/*Yr18*/*Pm38* 类型 −*Lr34*/*Yr18*/*Pm38* type	变幅 Range	0.3%～100%	1%～60%	0～60%	0～80%
	平均值 Average	24%	12.1%	11.7%	9.1%

从表 4 可以发现，在不同地点对不同病害的抗性鉴定中，标记检测为+*Lr34*/*Yr18*/*Pm38* 类型材料的病害最高严重度均低于−*Lr34*/*Yr18*/*Pm38* 类型材料的最高严重度；在成都条锈病鉴定和北京白粉病鉴定中，标记检测为+*Lr34*/*Yr18*/*Pm38* 类型的材料的平均值也低于−*Lr34*/*Yr18*/*Pm38* 类型的材料的平均值。从表 2 可知除了材料 19HRWSN-118、41IBWSN-82 和 41IBWSN-138 等 3 份材料在成都条锈病严重度在 70%～80%之间外，+*Lr34*/*Yr18*/*Pm38* 类型材料在成都条锈病、北京白粉病和CIMMYT的条锈病与白粉病的病害严重度均在 30%以下。病害鉴定结果说明，标记检测为+*Lr34*/*Yr18*/*Pm38* 类型的材料在不同地点对条锈病、叶锈病和白粉病具有不同程度的抗性。

3　讨论

在开发出功能标记 *cssfr1*-*cssfr5* 之前，标记 *csLV34* 是检测 *Lr34*/*Yr18*/*Pm38* 最稳定有效的标记，得到了广泛应用[1,20-22]。而标记 *cssfr1*-*cssfr5* 是基于 *Lr34*/*Yr18*/*Pm38* 位点 exon 11 中 TTC 缺失开发的功能标记，它们可以准确鉴定小麦在 *Lr34*/*Yr18*/*Pm38* 位点的等位变异[24]。本试验利用 *cssfr1*-*cssfr5* 检测了 273 份 CIMMYT 小麦，结果表明，5 个标记的扩增带型均清晰易读、稳定性好，其中标记 *cssfr3*、*cssfr4* 和 *cssfr5* 可用于大规模品种鉴定和杂种后代选择。

尽管标记 *csLV34* 与 *Lr34*/*Yr18*/*Pm38* 位点紧密连锁，但仍有 0.4cM 的遗传距离[19]，所以标记 *csLV34* 与 *Lr34*/*Yr18*/*Pm38* 位点间有低频率的重组发生。Kolmer 等[1]利用 *csLV34* 检测了全球不同地方上千份栽培品种和地方品种，发现标记 *csLV34* 与 *Lr34*/*Yr18*/*Pm38* 位点间存在较低频率的重组；McCallum 等[20]在利用 *csLV34* 检测加拿大西部红春麦品种中也发现了标记 *csLV34* 与 *Lr34*/*Yr18*/*Pm38* 位点间的重组。Lagudah 等[24]在对功能标记 *cssfr1*-*cssfr5* 的验证中，证实在加拿大品种 AC Domain 和北美品种 Newton 中发生了标记 *csLV34* 与 *Lr34*/*Yr18*/*Pm38* 位点间的重组。在本试验中，有 9 份材料由 STS 标记 *csLV34* 检测的结果与功能标记 *cssfr1*-*cssfr5* 检测结果不一致，进一步说明标记 *csLV34* 与 *Lr34*/*Yr18*/*Pm38* 位点等位变异间的重组是存在的。

Krattinger 等[23]克隆了 *Lr34*/*Yr18*/*Pm38* 基因，认为在该位点只存在两个等位基因，分别是+*Lr34*/*Yr18*/*Pm38* 和−*Lr34*/*Yr18*/*Pm38*，这两个等位基因间有三个序列差异，及 intron 4 的 A/T SNP、exon 11 中 TTC 的缺失和 exon 12 的 C/T SNP，即这三个序列差异是共分离的。Lagudah 等[24]根据 exon 11 中 TTC 的缺失开发了 *cssfr1*-*cssfr5* 等 5 个功能标记，exon 12 的 C/T SNP 开发了 CAPS 标记 *cssfr6*。在他们的研究中发现了 5 份材料在 intron 4 的 A/T SNP 与 exon 11、exon 12 的序列差异不共分离，还有一个新的等位基因；而 exon 11 与 exon 12 的序列差异是共分离的，也就是说可以用根据 exon 11 的 TTC 缺失开发的 *cssfr1*-*cssfr5* 这 5 个功能标记检测小麦品种（系）是否为+*Lr34*/*Yr18*/*Pm38* 类型。由于到目前未发现 exon 11 的等位变异与 exon 12 的等位变异非共分离的材料，而 *cssfr6* 为 CAPS 标记，使用时较费时费力，其需要的内切酶也很昂贵，所以本试验中没用 CAPS 标记 *cssfr6* 来检测品种。

那么在该位点是否存在更多的等位变异或新的序列差异呢？Lagudah 等[24]用标记 *csLV34* 和功能标记 *cssfr1*、*cssfr2* 检测美国品种 Jagger 均表明它含有 *Lr34*，但它在田间却表现完全感病；序列分析发现

Jagger 存在一个 G/T 突变点，在 DNA 序列上形成了提前终止密码，从而导致合成的蛋白在 C 末端缺失了 185 个氨基酸而没有功能；针对 Jagger 突变，开发了 CAPS 标记 *cssfr7*。用标记 *csLV34*、*cssfr1-cssfr5*、*cssfr6*、*cssfr7* 检测 151 份中国小麦地方品种，并同时鉴定这些地方品种对条锈病抗性，发现有 82.1%的地方品种标记检测含有 *Lr34/Yr18/Pm38*，但其中 25.8%的品种在田间却表现完全感病，而这些感病的品种用 *cssfr7* 检测不是 Jagger 突变类型（国家小麦改良中心实验室资料）。这说明在中国地方品种中 *Lr34/Yr18/Pm38* 位点可能存在新的等位变异，也可能存在类似抑制基因等调控机制导致 *Lr34/Yr18/Pm38* 对条锈病等病害的抗性丧失。因此，在用上述标记检测中国小麦材料时，必须要结合表型鉴定，以确定哪些材料含有 *Lr34/Yr18/Pm38*。

由于在 *Lr34/Yr18/Pm38* 位点有多个功能标记，且这些标记均可用于标记辅助选择，所以在育种初期用这些标记检测亲本，明确亲本的等位变异类型并选出合适的标记应用于育种程序则可开展标记辅助选择工作。在选用本研究检测出的 43 份+*Lr34/Yr18/Pm38* 类型的 CIMMYT 小麦作为抗病亲本的育种工作中，功能标记 *cssfr3*、*cssfr4* 和 *cssfr5* 则非常适用于后代的辅助选择。

4 结论

利用标记 *csLV34* 和基于慢病基因 *Lr34/Yr18/Pm38* exon 11 中 TTC 缺失开发的 5 个功能标记 *cssfr1-cssfr5* 检测了 273 份 CIMMYT 小麦品种（系），并对这 273 份小麦材料进行田间条锈病、叶锈病和白粉病抗性鉴定；发现 43 份材料携带 *Lr34/Yr18/Pm38*，可用于小麦抗病育种。研究结果表明，功能标记 *cssfr1-cssfr5* 可准确鉴定 CIMMYT 材料慢病基因 *Lr34/Yr18/Pm38* 的等位变异类型；*cssfr3*、*cssfr4* 和 *cssfr5* 适合于这些 CIMMYT 小麦材料杂交后代的分子标记辅助选择。

参考文献

[1] Kolmer J A, Singh R P, Garvin D F, Viccars L, William H M, Huerta-Espino J, Ogbonnaya F C, Raman H, Orford S, Bariana H S, Lagudah E S. Analysis of the *Lr34/Yr18* rust resistance region in wheat germplasm. *Crop Science*, 2008, 48: 1841-1852.

[2] Dyck P L. Genetics of leaf rust reaction in three introductions of common wheat. *Canada Journal of Genetics and Cytology*, 1977, 19: 711-716.

[3] Dyck P L. The association of a gene for leaf rust resistance with the chromosome 7D suppressor of stem rust resistance in common wheat. *Genome*, 1987, 29: 467-469.

[4] McIntosh R A. Close genetic linkage of genes conferring adult-plant resistance to leaf rust and stripe rust in wheat. *Plant Pathology*, 1992, 41: 523-527.

[5] Singh R P. Genetic association of leaf rust resistance gene *Lr34* with adult plant resistance to stripe rust in bread wheat. *Phytopathology*, 1992, 82: 835-838.

[6] Spielmeyer W, McIntosh R A, Kolmer J, Lagudah E S. Powdery mildew resistance and *Lr34/Yr18* genes for durable resistance to leaf and stripe rust cosegregate at a locus on the short arm of chromosome 7D of wheat. *Theoretical and Applied Genetics*, 2005, 111: 731-735.

[7] Singh R P. Genetic association of gene *Bdv1* for tolerance to barley yellow dwarf virus with genes *Lr34* and *Yr18* for adult plant resistance to rusts in bread wheat. *Plant Disease*, 1993, 77: 1103-1106.

[8] Dyck P L. Inheritance of leaf rust and stem rust resistance in 'Roblin' wheat. *Genome*, 1993, 36: 289-293.

[9] Nelson J C, Singh R P, Autrique J E, Sorrells M E. Mapping genes conferring and suppressing leaf rust resistance in wheat. *Crop Science*, 1997, 37: 1928-1935.

[10] Singh R P, Nelson J C, Sorrells M E. Mapping *Yr28* and other genes for resistance to stripe rust in wheat. *Crop Science*, 2000, 40: 1148-1155.

[11] Bariana H S, Hayden M J, Ahmed N U, Bell J A, Sharp P J, McIntosh R A. Mapping of durable adult plant and seedling resistances to stripe rust and stem rust diseases in wheat. *Australian Journal of Agricultural Research*, 2001, 52: 1247-1255.

[12] Suenaga K, Singh R P, Huerta-Espino J, William H M. Microsatellite markers for genes *Lr34/Yr18* and other quantitative trait loci for leaf rust and stripe rust resistance in bread wheat. *Phytopathology*, 2003, 93: 881-890.

[13] Schnurbusch T, Paillard S, Schori A, Messmer M, Schachermayr G, Winzeler M, Keller B. Dissection of quantitative and durable leaf rust resistance in Swiss winter wheat reveals a major resistance QTL in the

Lr34 chromosomal region. *Theoretical and Applied Genetics*, 2004, 108: 477-484.

[14] Schnurbusch T, Bossolini E, Messmer M, Keller B. Tagging and validation of a major quantitative trait locus for leaf rust resistance and leaf tip necrosis in winter wheat cultivar Forno. *Phytopathology*, 2004, 94: 1036-1044.

[15] Lillemo M, Asalf B, Singh R P, Huerta-Espino J, Chen X M, He Z H, Bjørnstad Å. The adult plant rust resistance loci *Lr34/Yr18* and *Lr46/Yr29* are important determinants of partial resistance to powdery mildew in bread wheat line Saar. *Theoretical and Applied Genetics*, 2008, 116: 1155-1166.

[16] McIntosh R A, Dubcovsk, J, Rogers W J, Morris C, Appels R, XiaX C Catalogue of gene symbols for wheat: 2009 supplement. *Annual Wheat Newsletter*, 2010-06-08. http://www.shigen.nig.ac.jp/wheat/komugi/genes/macgene/supplement 2009.pdf.

[17] Dyck P L. Genetics of adult-plant leaf rust resistance in 'Chinese Spring' and 'Sturdy' wheats. *Crop Science*, 1991, 31: 309-311.

[18] Bossolini E, Krattinger S G, Keller B. Development of simple sequence repeat markers specific for the *Lr34* resistance region of wheat using sequence information from rice and *Aegilops tauschii*. *Theoretical and Applied Genetics*, 2006, 113: 1049-1062.

[19] Lagudah E S, McFadden H, Singh R P, Huerta-Espino J, Bariana H S, Spielmeyer W. Molecular genetic characterization of the *Lr34/Yr18* slow rusting resistance gene region in wheat. *Theoretical and Applied Genetics*, 2006, 114: 21-30.

[20] McCallum B D, Somers D J, Humphreys D G, Cloutier S. Molecular marker analysis of *Lr34* in Canada Western Red spring wheat cultivars//Appels Retal. *Proceedings of the 11th International Wheat Genetics Symposium*. Sydney: Sydney University Press, 2008,

[21] 杨文雄，杨芳萍，梁丹，何中虎，尚勋武，夏先春．中国小麦育成品种和农家种中慢锈基因 *Lr34/Yr18* 的分子检测．作物学报，2008，34 (7)：1109-1113.
Yang W X, Yang F P, Liang D, He Z H, Shang X W, Xia X C. Molecular characterization of slow-rusting genes *Lr34/Yr18* in Chinese wheat cultivars. *Acta Agronomica Sinica*, 2008, 34 (7): 1109-1113.

[22] 梁丹，杨芳萍，何中虎，姚大年，夏先春．利用 STS 标记检测 CIMMYT 小麦品种（系）中 *Lr34/Yr18*、*Rht-B1b* 和 *Rht-D1b* 基因的分布．中国农业科学，2009，42 (1)：17-27.
Liang D, Yang F P, He Z H, Yao D N, Xia X C. Characterization of *Lr34/Yr18*, *Rht-B1b*, *Rht-D1b* Genes in CIMMYT wheat cultivars and advanced lines using STS markers. *Scientia Agricultura Sinica*, 2009, 42 (1): 17-27. (in Chinese)

[23] Krattinger S G, Lagudah E S, Spielmeyer W, Singh R P, Huerta-Espino J, McFadden H, Bossolini E, Selter L L, Keller B. A putative ABC transporter confers durable resistance to multiple fungal pathogens in wheat. *Science*, 2009, 323: 1360-1363.

[24] Lagudah E S, Krattinger S G, Herrera-Foessel S, Singh R P, Huerta-Espino J, Spielmeyer W, Brown-Guedira G, Selter L L, Keller B. Gene-specific markers for the wheat gene *Lr34/Yr18/Pm38* which confers resistance to multiple fungal pathogens. *Theoretical and Applied Genetics*, 2009, 119: 889-898.

[25] 何中虎，庞家智．CIMMYT 麦类改良进展．北京：中国农业科技出版社，1995：115-124.
He Z H, Pang J Z. Progress of CIMMYT Wheat Improvement. Beijing: China Agricultural Science and Technology Press, 1995: 115-124. (in Chinese)

[26] Rajaram S, Borlaug N E, Ginkel M. CIMMYT International Wheat Breeding//Bread Wheat Improvement and Production. Curtis BC, Rajaram S, Macpherson HG (eds) Plant Production and Protection Series. No. 30 FAQ, Rome, 2002 . 103-117.

[27] 邹裕春，杨武云，朱华忠，杨恩年，蒲宗君，伍玲，张颙，汤永禄，黄钢，李跃建，何中虎，Singh R P, Rajaram S. CIMMYT 种质及育种技术在四川小麦品种改良中的利用．西南农业学报，2007，20：183-190.
Zou Y C, Yang W Y, Zhu H Z, Yang E N, Pu Z J, Wu L, Zhang Y, Tang Y L, Huang G, Li Y J, He Z H, Singh R P, Rajaram S. Utilization of CIMMYT germplasm and breeding technologies in wheat improvement in Sichuan, China. *Southwest China Journal of Agricultural Sciences*, 2007, 20: 183-190. (in Chinese)

[28] Singh R P, Huerta-Espino J, Rajaram S. Achieving near-immunity to leaf and stripe rusts in wheat by combining slow rusting resistance genes. *Acta Phytopathologica et Entomologica Hungarica*, 2000, 35 (1-4): 133-139.

[29] Lagudah E S, Appels R, McNeil D. The *Nor-D3* locus of *Triticum tauschii*: natural variation and genetic linkage to markers in chromosome 5. *Genome*, 1991, 34: 387-395.

兼抗型成株抗性
种质创新与育种应用

Pyramiding adult-plant powdery mildew resistance QTLs in bread wheat

B. Bai[1,2,3], M. A. Asad[2], C. X. Lan[2], Y. Zhang[2], X. C. Xia[2], Z. H. He[2,3,5], J. Yan[2], J. C. Wang[4], X. M. Chen[2], and C. S. Wang[1]

[1] *State Key Laboratory of Crop Stress Biology in Arid Areas/College of Agronomy, Northwest A & F University, Yangling* 712100, *Shaanxi, China*; [2] *Institute of Crop Science, National Wheat Improvement Center/The National Key Facility for Crop Gene Resources and Genetic Improvement, Chinese Academy of Agricultural Sciences (CAAS), Zhongguancun South Street* 12, *Beijing* 100081, *China*; [3] *International Maize and Wheat Improvement Center (CIMMYT), CIMMYT China Office, c/o CAAS, Zhongguancun South Street* 12, *Beijing* 100081, *China*; [4] *Wheat Research Institute, Gansu Academy of Agricultural Sciences, Nongkeyuanxincun* 1, *Lanzhou* 730070, *China*; [5] *Institute of Wheat, Yantai Academy of Agricultural Sciences*, 26 *Gangcheng Western Street, Yantai* 265500, *Shandong, China*

Abstract: Quantitative trait loci (QTL) pyramiding can be an effective approach for durable resistance to powdery mildew in wheat (*Triticum aestivum* L.). The Chinese wheat cultivars Bainong 64 and Lumai 21 with good agronomic traits conferred adult-plant resistance (APR) to powdery mildew possess four and three QTLs, respectively. To achieve optimal durable resistance, 21 F_6 lines combining two to five powdery mildew APR QTLs were developed from the cross Bainong 64/Lumai 21 using modified pedigree selection. The lines were planted in a randomized complete block design with two replicates in Beijing during the 2009-2010 and 2010-2011 cropping seasons, and were evaluated for powdery mildew response using the highly virulent *Blumeria graminis* f. sp. *tritici* isolate E20. Based on the phenotypic data of both maximum disease severity (MDS) and the area under the disease progress curve (AUDPC), analysis of variance indicated that there were highly significant effects of QTLs combinations on reducing powdery mildew MDS and AUDPC, respectively. Six pyramided QTL combinations possessing *QPm. caas-1A* and *QPm. caas-4DL* in common along with one or more of the others expressed better APR to powdery mildew than the more resistant parent Bainong 64. Thus pyramiding these two QTLs with one or more of *QPm. caas-2BS*, *QPm. caas-2BL*, and *QPm. caas-2DL* from Lumai 21 could be a desirable strategy to breed cultivars with high levels of durable resistance to powdery mildew. These results should be useful for pyramiding APR QTLs in wheat in order to develop cultivars with durable resistance to this important disease.

Key words: *Triticum aestivum*, *Blumeria graminis* f. sp. *tritici.*, adult-plant resistance gene, gene combinations, molecular markers

Introduction

Powdery mildew, caused by *Blumeria graminis* f. sp. *tritici* (*Bgt*), is an important air-borne disease of wheat (Bennett, 1984), causing yield losses of 1.5 to 45% (Conner et al., 2003), particularly in regions with moderate temperatures and moist conditions (Luo

Published in Crop & Pasture Science, 2012, 63 (7): 606 - 611 et al., 2009). In China, powdery mildew is a major disease in all winter wheat growing regions, especially in the Northern China Plain and Yellow and Huai River Valleys, but also in southwestern China and the Yangtze River area. From 2004-2009, the average area of wheat affected by powdery mildew was 6.85 million ha per year (He et al., 2011).

Host resistance is the most effective and sustainable approach to control crop disease and has no additional cost to growers (Singh et al., 2005). Generally, host resistance can be classified as qualitative or quantitative. Qualitative resistance, usually effective at all growth stages, is race specific and cultivars with such resistance may become susceptible some time after release. In contrast, quantitative resistance, often referred to as adult-plant resistance (APR), partial resistance, or slow disease development, tends to be race non-specific and is therefore potentially durable (Bennett, 1984; Niks and Rubiales, 2002). To date, more than 70 powdery mildew resistance genes (*Pm1*-*Pm45* and including multiple resistance alleles and genes not formally designated) have been described in wheat (Ma et al., 2011; McIntosh et al., 2011), but most of them confer race-specific resistance (Huang et al., 1997; Clarkson 2000).

Quantitative resistance is usually polygenic or oligogenic and controlled by minor genes with additive effects. For example, the pleiotropic resistance gene *Yr18*/*Lr34*/*Pm38* (Krattinger et al., 2009), confers durable protection against several diseases, including stem rust (Bhavani et al., 2011). The level of resistance conferred by this gene alone to each disease is not sufficient, but its combination with other minor genes, or other combinations of four to five slow-rusting genes, resulted in high levels of resistance (Singh et al., 2000, 2011). Italian wheats Libellula and Strampelli that remain resistant to stripe rust after 40 years of exposure to the pathogen in Gansu province have *Yr18* combined with other QTLs (Lu et al., 2009). With molecular markers closely linked to QTL, breeders can focus on pyramiding QTLs in a single genotype to increase both the level of resistance and durability (Castro et al., 2003). However, QTL pyramiding has limitations, especially in regard to the contributions of each QTL in different genetic backgrounds. There are few examples of successful MAS (Dekkers and Hospital, 2002), and few studies have examined the outcomes of combining QTLs from different sources. For example, Miedaner et al. (2006) combined three QTLs for *Fusarium* head blight (FHB) resistance into an elite European spring wheat, and found that each QTL reduced the deoxynivalenol (DON) concentration, but a chromosome 3A QTL, either alone or in combination with 3B and 5A QTLs, had no significant effect. The combination of both the 5A QTL and *Fhb1* achieved an adequate level of improvement to counteract the negative effect of dwarfing allele *Rht-D1* on response to FHB (Lu et al., 2011).

Bainong 64 and Lumai 21 each have three to four QTLs for APR to powdery mildew (Lan et al., 2009, 2010). We developed a set of advanced lines from the cross Bainong 64/Lumai 21 by conventional phenotypic selection and the aims of the present study were to determine if the most resistant selected lines carried QTLs from both parents, and if so, which ones.

Materials and methods

Plant materials

The parental cultivars Bainong 64 and Lumai 21 with APR to powdery mildew were leading cultivars in the Yellow and Huai Valleys Winter Wheat Zone during 1993-2000 and 1994-2010, respectively. Twenty-one F_6 lines (named BFB5-BFB25) were developed from the cross Bainong 64/Lumai 21 by a modified pedigree method in conjunction with field phenotyping of powdery mildew response and agronomic traits before identification of the QTLs in the two parents.

Phenotypic assessment of powdery mildew response

Bgt isolate E20 was used to evaluate seedling responses of the 21 lines in the greenhouse, and the parents and selected lines along with susceptible check cv. Jingshuang 16 were evaluated for APR in the field. Field trials were conducted in Beijing during wheat cropping seasons 2009-2010 and 2010-2011 using randomized complete blocks with two replicates. The lines under assessment were surrounded by Jingshuang 16 as spreader to ensure ample inoculum. Inoculations were carried out in Beijing using isolate E20 before stem elongation, and disease severities on penultimate leaves were scored as percentages of leaf area covered by powdery mildew at six weeks after inoculation, then at weekly intervals on further occasions, and the last time when the disease severity reached its maximum level.

Genotypic screening of pyramided lines

Genomic DNA was extracted using fresh leaves of each cultivar or line. Molecular markers closely linked to the QTLs were used to screen the F_5 pyramided lines firstly, then to screen the selected lines again at F_6. We used flanking markers *Xbarcl48* and *Xwmc550* for genotyping the 1A chromosome QTL region *QPm. caas-1A*, SSRs *Xgwml65*, *Xcfd23* and *Xwmc331* for genotyping *QPm. caas-4DL*, *Xbarc79* and *Xgwm518* for *QPm. caas-6BS*, *Xbarcl27* and *Xbarcl74* for *QPm. caas-7A*, *Xbarc 98* and *Xbarcl147* for *QPm. caas-2BS*, *Xbarcl139* and *Xgwm47* for *QPm. caas2BL*, and *Xbarcl8*, *Xgwm539* and *Xcfd233* for *QPm. caas-2DL*. All markers were described by Lan et al. (2009, 2010). The presence or absence of the corresponding QTLs in each line was inferred from the SSR genotypes.

For marker analyses PCR were performed in 15μl mixtures containing 50ng of template DNA, 1×PCR buffer, 4pmol of each primer, 200μM of each dNTP, and 1U of *Taq* DNA polymerase (Tiangen, Beijing). Amplification involved initial denaturing at 94℃ for 5 min followed by 35 cycles of 45s denaturation at 94℃, 45s annealling at 55-60℃ and 1 min of extension at 72℃, with a final step of 8 min at 72℃ for completion of primer extension. PCR products were separated on 6% denaturing polyacrylamide gels and visualized by silver staining.

Data analysis

Disease severity data were used to calculate the area under the disease progress curve (AUDPC) for each genotype using the following formula: $AUDPC = \sum_i [(x_i + x_{i+1})/2]t_i$, where x_i is disease severity on assessment date i, and t_i is the time in days between dates i and $i+1$ (Lin and Chen 2009). MDS and AUDPC data were used for the subsequent analysis of variance (ANOVA) by the Statistical Analysis System (SAS Institute, 2000).

Results

Powdery mildew resistance QTL in tested lines

The 21 lines were genotyped for powdery mildew response QTLs on chromosomes 1A, 4DL, 7A, 6BS, 2DL, 2BS, and 2BL by means of 16 co-dominant flanking SSR markers (Table 1). Seventeen lines combined QTLs donated by both parents and only four lines had three QTLs derived from Bainong 64. Ten lines had common QTL alleles on chromosome 1A, 4DL, and 7A derived from the parent Bainong 64; three genotypes had common QTL alleles on chromosome 2BS and 2BL, and two genotypes had common QTL alleles on chromosome 2BL, 2BS and 2DL, contributed by Lumai 21. Five QTL alleles presented in five lines, respectively, and 15 lines possessed three to four QTL alleles, and one line possessed two QTL alleles. However, the QTL on 6BS was absent in all lines.

Table 1 Composition of QTLs, and average MDS and AUDPC for powdery mildew response in parents and selected lines derived from Bainong 64/Lumai 21[1)]

Genotype	Selection history	QTL combination[2)]	No. of QTLs	Distribution of QTL alleles[3)]	Mean MDS[4)]	Mean AUDPC[5)]
Bainong 64	—	1A, 4DL, 6BS, 7A	4	B (1A, 4DL, 6BS, 7A)	7.0[cde]	45.0[defg]
Lumai 21	—	2BS, 2BL, 2DL	3	L (2BS, 2BL, 2DL)	12.0[a]	75.3[a]
BFB5	BS0410-2-2-1-1	1A, 4DL, 2BS	3	B (1A, 4DL), L (2BS)	3.5[f]	22.8[hig]
BFB6	BS0410-8-1-2-1	1A, 4DL, 2DL	3	B (1A, 4DL), L (2DL)	4.5[ef]	33.3[fghi]
BFB7	BS0410-8-1-2-2	1A, 4DL, 2DL	3	B (1A, 4DL), L (2DL)	4.0[ef]	31.5[fghi]
BFB8	BS0410-3-4-1-3	1A, 4DL, 2DL, 2BS, 2BL	5	B (1A, 4DL), L (2DL, 2BS, 2BL)	4.3[ef]	39[efghi]
BFB9	BS0410-3-4-5-4	1A, 4DL, 2BS, 2BL	4	B (1A, 4DL), L (2BS, 2BL)	9.5[abc]	63.5[abcd]
BFB10	BS0410-3-3-1-3	1A, 4DL, 2DL, 2BS, 2BL	5	B (1A, 4DL), L (2DL, 2BS, 2BL)	2.8[f]	20.8[hi]
BFB11	BS0410-3-3-2-1	1A, 4DL, 2DL, 2BS	4	B (1A, 4DL), L (2DL, 2BS)	2.8[f]	17.3[i]
BFB12	BS0410-10-2-1-2	4DL, 2BS	2	B (4DL), L (2BS)	11.3[ab]	72.8[ab]
BFB13	BS0410-17-1-3-1	1A, 4DL, 2DL	3	B (1A, 4DL), L (2DL)	4.0[ef]	23.5[hig]
BFB14	BS0410-17-1-3-3	1A, 4DL, 2DL	3	B (1A, 4DL), L (2DL)	3.8[ef]	22.8[hig]
BFB15	BS0410-18-9-1-2	1A, 4DL, 2DL, 2BS	4	B (1A, 4DL), L (2DL, 2BS)	8.5[bcd]	59.3[abcde]
BFB16	BS0410-42-3-3-1	1A, 4DL, 2BL, 7A	4	B (1A, 4DL, 7A), L (2BL)	5.8[def]	51.0[abcdef]
BFB17	BS0410-42-3-3-2	1A, 4DL, 2BL, 7A	4	B (1A, 4DL, 7A), L (2BL)	5.0[ef]	35.5[fghi]
BFB18	BS0410-42-3-3-4	1A, 4DL, 2DL, 2BL, 7A	5	B (1A, 4DL, 7A), L (2DL, 2BL)	5.8[def]	43.8[defgh]
BFB19	BS0410-42-3-3-5	1A, 4DL, 2DL, 2BL, 7A	5	B (1A, 4DL, 7A), L (2DL, 2BL)	5.8[def]	32.5[fghi]
BFB20	BS0410-42-3-3-7	1A, 4DL, 2BL, 7A	4	B (1A, 4DL, 7A), L (2BL)	4.0[ef]	32.8[fghi]
BFB21	BS0410-42-3-4-2	1A, 4DL, 2DL, 2BL, 7A	5	B (1A, 4DL, 7A), L (2DL, 2BL)	5.3[def]	47.8[cdef]
BFB22	BS0410-48-1-7-2	1A, 4DL, 7A	3	B (1A, 4DL, 7A)	9.8[abc]	68.8[abc]
BFB23	BS0410-48-1-7-3	1A, 4DL, 7A	3	B (1A, 4DL, 7A)	11.3[ab]	76.5[a]
BFB24	BS0410-48-1-7-4	1A, 4DL, 7A	3	B (1A, 4DL, 7A)	10.0[abc]	59.0[abcde]
BFB25	BS0410-48-1-7-8	1A, 4DL, 7A	3	B (1A, 4DL, 7A)	11.0[ab]	68.8[abc]

1) MDS and AUDPC of the susceptible check Jingshuang 16 averaged in two environments was 67.5% and 589.5, respectively.

2) 1A, 4DL, 7A, 6BS, 2DL, 2BS, and 2BL represent QTLs previously mapped in Bainong 64/Jingshuang 16 and Lumai 21/Jingshuang 16 (Lan et al., 2009, 2010).

3) QTL alleles distribution, B and L indicated the QTL alleles derived from Bainong 64 and Lumai 21, respectively.

4), 5) Average data for Beijing 2009-2010 and 2010-2011.

Validation of pyramided lines

The parents and 21 lines were all susceptible to isolate E20 at the seedling stage. The mean MDS of the two resistant parents were 7.0% and 12.0%, and the mean AUDPC was 45.0 and 75.3, respectively (Table 1). Twelve pyramided lines had lower MDS and AUDPC than both parents. BFB10 expressed the lowest mean MDS (2.8%) and the lower AUDPC (20.8) in Beijing over two years. Disease severity was significantly ($P < 0.01$) influenced by genotype, environment, and genotype by environment interaction effects (Table 2).

Table 2 Analysis of variance of MDS and AUDPC for powdery mildew responses on parents, susceptible check and pyramided lines selected from Bainong 64/Lumai 21

Source	*DF*	MDS	AUDPC
		Sum of squares	Sum of squares
Environments[1)]	1	41.3*	50 646.1**
Replications	1	5.51	776.3
Genotypes	23	15 066.5**	1 167 712.2**
Environment×genotypes	23	371.4	276 219.2**
Error	47	459.0	34 225.2

1) Environments include Beijing 2009-2010 and 2010-2011.

Effects of QTL combinations on powdery mildew response

The pyramided lines and parents were categorized into 11 genotypes (Table 3). QTL combination (*QPm. caas-2DL*, *QPm. caas-2BS*, and *QPm. caas- 2BL*) present in Lumai 21 had a lower effect on disease response than other combinations, but there was no significant difference between the combinations (*QPm. caas-1A* + *QPm. caas-4DL* + *QPm. caas-7A*) and (*QPm. caas-4DL* + *QPm. caas-2BS*). Six combinations had lower MDS than the combination of QTLs present in Bainong 64, but only three combinations had significantly lower MDS value than that of Bainong 64 ($P < 0.05$), and one combination had significantly lower AUDPC value than that of Bainong 64 ($P < 0.05$). No significant differences were found among the six combinations, indicating that the contributions of these six genotypes to APR were significantly higher than the other four QTL combinations. In addition, both MDS and AUDPC evaluations showed that the lines possessing higher numbers of QTLs were more resistant (Fig. 1).

Table 3 Effects of different QTL combinations on powdery mildew response

Genotype	No. of lines[1)]	QTL combinations	Mean MDS[2)]	Mean AUDPC[3)]
Lumai21	1	2BS+2BL+2DL	12.0^{a}	75.3^{a}
Bainong 64	1	1A+4DL+6BS+7A	7.0^{b}	45.0^{b}
Line	1	4DL+2BS	11.3^{a}	72.8^{a}
	4	1A+4DL+7A	10.5^{a}	68.3^{a}
	1	1A+4DL+2BS+2BL	9.5^{a}	63.5^{a}
	3	1A+4DL+2DL+2BL+7A	5.6^{bc}	41.1^{bc}
	3	1A+4DL+2BL+7A	5.5^{bc}	39.8^{bc}
	2	1A+4DL+2DL+2BS	5.2^{bc}	38.3^{bc}
	4	1A+4DL+2DL	4.1^{c}	29.9^{bc}
	1	1A+4DL+2BS	3.5^{c}	27.8^{bc}
	2	1A+4DL+2DL+2BS+2BL	3.5^{c}	22.8^{c}

1) Numbers of lines with same combination of QTLs.

2), 3) Values followed by the same letter in the same column are not significantly different at $P = 0.05$.

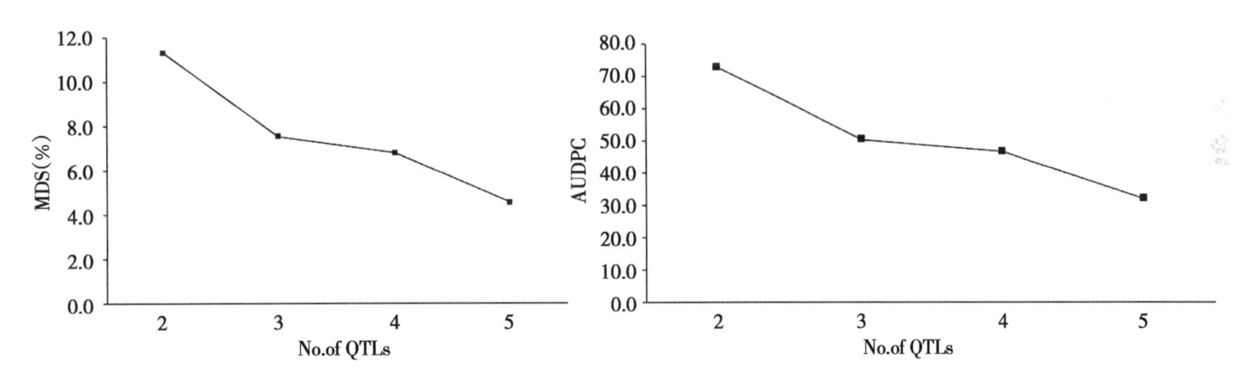

Fig. 1 Effects of QTL number on reducing MDS and AUDPC. The number in the abscissa indicates the number of QTLs combined in each subset of lines; for example, 2 means lines possessing two QTLs. The averaged values of MDS and AUDPC in each class are indicated on the vertical axes.

Discussion

More than 80 powdery mildew APR QTLs were mapped on all 21 chromosomes. Most QTL mapping studies have focused on identifying and estimating of effects on phenotypic variation within mapping populations (He et al., 2011). The effects of QTLs on disease response in breeding populations remain largely unclear, and studies are required to determine which QTL combinations produce superior additive gene effects when they are transferred to different genetic backgrounds. The present study provides useful information on the effect of QTL on powdery mildew response in breeding populations. QTLs, *QPm. caas-1A*, *QPm. caas-7A*, *QPm. caas-4DL*, *QPm. caas-2BS*, *QPm. caas-2BL*, and *QPm. caas-2DL* derived from two parents, two to five of them were combined in different lines, respectively, and the phenotypic data showed that QTL combinations had significant effects toward resistance across all environments. In earlier studies, the dominance and epistatic effects on FHB response were significant in some crosses of Chinese resistant and susceptible cultivars, and these effects increased resistance in some crosses, but decreased resistance in others (Bai et al., 2000). Digenic epistatic interaction between non-allelic QTLs for APR to stripe rust was also reported by Lu et al. (2009). Genotype × environment interaction affected expression of resistance to FHB in wheat (Miedaner et al., 2001).

The selection history showed that lines BFB22, BFB23, BFB24 and BFB25 and BFB16, BFB17, BFB18, BFB19 and BFB20 came from single F_4 plants (BFB21 came from the same F_3 plant as the latter genotypes); BFB6 and BFB7 were from the same F_4 plant; BFB8, BFB9 and BFB10, and BFB11 were progenies of the same F_3 or F_4 plants; and BFB13 and BFB14 were progenies of one F_4 plant. Thus many of the lines had similar genetic backgrounds and were more likely to respond in similar ways. We therefore grouped QTL combinations according to QTL composition and analyzed the effects of various QTLs on powdery mildew response. The six combinations involving genes from both parents had lower mean MDS and AUDPC than other combinations, and could be the optimal QTL combinations from this cross. Furthermore, these QTL combinations shared *QPm. caas-1A and QPm. caas-4DL* derived from Bainong 64, and other QTLs derived from Lumai 21 such as *QPm. caas-2BL*, *QPm. caas-2BS*, or *QPm. caas-2DL*, indicating that pyramiding these QTLs in a line would be an ideal strategy to obtain high to near-immune resistance. This result confirms that pyramiding QTLs can be a suitable strategy for breeding highly and durably resistant wheat cultivars.

We also found that lines with higher numbers of QTL alleles had lower MDS and AUDPC values (Fig. 1), which indicated that additive effects of the QTLs for powdery mildew resistance played an important role in lines. BFB10, followed by BFB8, possessing five QTLs, was highly resistant and had the lowest MDS and lower AUDPC values among all lines. Lines containing four QTLs also had low MDS and AUDPC values, the lines BFB22, BFB23, BFB24, and BFB25 with three QTLs derived from Bainong 64, and Lumai 21, had higher MDS values than other lines except BFB12, and four of them also had higher AUDPC than other lines except BFB12 accordingly. This is in agreement with the previous reports that four to five minor genes for resistance to leaf rust and stripe rust were sufficient to confer near immunity (Singh et al., 2011). However, the effects of pyramided QTLs on powdery mildew response in some lines did not always coincide with the number of QTLs. For example, BFB11 possessing four QTLs was more resistant than lines BFB18, BFB19, and BFB21 with five QTLs, but there is no significant difference between them, and BFB5, BFB6, BFB7, BFB13, and BFB14 with three QTLs were significantly resistant than BFB9 and BFB15, each with four QTLs ($P<0.05$, Table 1). This difference could be caused by the different effect of each QTL, or by the interaction of QTLs (data not shown). Thus specific combinations of lower numbers of QTL may be as effective as random combinations of larger numbers.

In the QTL pyramiding for powdery mildew resistance, six QTLs, *QPm. caas-1A*, *QPm. caas-7A*, *QPm. caas-4DL*, *QPm. caas-2BS*, *QPm. caas-2BL*, and *QPm. caas-2DL* derived from Bainong 64 and Lumai 21, were pyramided in 21 wheat lines, respectively. However, only the QTL *QPm. caas-6BS* derived from Bainong 64 was absent in all these lines, which could be caused by our phenotype selection of plants. Forty eight plants were selected in the F_2 breeding population derived from cross of Bainong 64/Lumai 21, but the 21 F_6 lines with high resistance to powdery mildew and good agronomic traits were progenies of eight F_2 plants, so the QTL on 6BS could be absent in these lines because of its lower resistance to the disease than other QTL alleles or its negative interaction with other QTL alleles or its linkage with other agronomic traits.

Based on the current results, it is likely that high levels of APR can be achieved by traditional field selection, but a large population size and selection in the early generations, such as F_2 or BC_1F_1 and F_3, will be needed and major genes must be avoided (Singh et al., 2005). It will be desirable to apply molecular markers if they are available to identify any known and potentially useful genes in the parents, but as the QTL genetic and environmental interactions cannot be predicted, it is difficult to decide the approximate number of QTLs needed for a higher level of APR. One option is to select 3-5 QTLs with relatively large effects in early generations by molecular markers, but multilocational testing, or at least two generations of testing in a high disease pressure nursery, will still be needed. Experienced breeders with a good knowledge of minor genes can achieve APR resistance by phenotypic selection such as that currently being done in the CIMMYT wheat breeding program for rust resistance; selection by means of molecular markers will still require powdery mildew nurseries and disease phenotyping for validation.

Acknowledgments

We are grateful to the critical review of this manuscript by Prof. R. A. McIntosh, Plant Breeding Institute, University of Sydney, Australia. This study was supported by the National Program on Key Basic Research Project of China (2009CB118301) and the National Key Technology R&D Program of China (2011BAD35B03).

References

Bai GH, Shaner G, and Ohm H (2000) Inheritance of resistance to *Fusarium graminearum* in wheat. Theoretical and Applied Genetics 100, 1-8. doi: 10.1007/ PL00002902

Bennett F (1984) Resistance to powdery mildew in wheat: a review of its use in agriculture and breeding programmes. Plant Pathology 33, 297-300. doi: 10.1111/j. 1365-3059. 1984. tb01324.x

Bhavani S, Singh R P, Argillier O, Huerta-Espino J, Singh S, Njau P, Brun S, Lacam S, and Desmouceaux N (2011) Mapping durable adult plant stem rust resistance to the race Ug99 group in six CIMMYT wheats. BGRI 2011 Technical Workshop, St. Paul, Minnesota, pp: 44-53.

Castro AJ, Chen XM, Hayes PM, Johnston M (2003) Pyramiding quantitative trait locus (QTL) alleles determining resistance to barley stripe rust: effects on resistance at the seedling stage. Crop Science 43, 651-659. doi: 10.2135/cropsci2003.2234

Conner RL, Kuzyk AD, Su H (2003) Impact of powdery mildew on the yield of soft white spring wheat cultivars. Canadian Journal of Plant Science 83, 725-728. doi: 10.4141/P03-043

Clarkson JDS (2000) Virulence survey report for wheat powdery mildew in Europe, 1996-1998. Cereal Rusts and Powdery Mildews Bulletin http://www.crpmb.org/2000/1204clarkson

Dekkers JCM and Hospital F (2002) The use of molecular genetics in the improvement of agricultural populations. Nature reviews, genetics 3, 22-32. doi: 10.1038/nrg701

He ZH, Lan CX, Chen XM, Zou YC, Zhuang QS and Xia XC (2011) Progress and perspective in research of adult-plant resistance to stripe rust and powdery mildew in wheat. Scientia Agricultura Sinica 44, 2193-2215. (in Chinese). doi: 10.3864/j.issn.0578-1752.2011.11.001

Huang XQ, Hsam SLK and Zeller FJ (1997) Identification of powdery mildew resistance genes in common wheat (*Triticum aestivum* L. em Thell.). IX: cultivars, land races and breeding lines grown in China. Plant Breeding 116, 233-238. doi: 10.1111/j. 1439-0523. 1997. tb00988.x

Krattinger SG, Lagudah ES, Spielmeyer W, Singh RP, Huerta-Espino J, McFadden H, Bossolini E, Selter LL and Keller B (2009) A putative ABC transporter confers durable resistance to multiple fungal pathogens in wheat. Science 323, 1360-1363. doi: 10. 1126/science. 1166453

Lan CX, Liang SS, Wang ZL, Yan J, Zhang Y, Xia XC, He ZH (2009) Quantitative trait loci mapping for adult-plant resistance to powdery mildew in Chinese wheat cultivar Bainong 64. Phytopathology 99, 1121-1126. doi: 10.1094/PHYTO-99-10-1121

Lan CX, Ni XW, Yan J, Zhang Y, Xia XC, Chen XM, He ZH (2010) Quantitative trait loci mapping of adult-plant resistance to powdery mildew in Chinese wheat cultivar Lumai 21. Molecular Breeding 25, 615-622. doi: 10.1007/s11032-009-9358-8

Lin F and Chen XM (2009) Quantitative trait loci for non-race-specifc, high-temperature adult-plant resistance to stripe rust in wheat cultivar Express. Theoretical and Applied Genetics 118, 631-642. doi: 10. 1007/s00122-008-0894-0

Lu QX, Szabo-Hever A, Bjørnstad Åsmund, Lillemo M, Semagn K, Mesterhazy A, Ji F, Shi JR, Skinnes H (2011) Two major resistance quantitative trait loci are required to counteract the increased susceptibility to *Fusarium* head blight of the *Rht-D1b* dwarfing gene in wheat. Crop Science 51, 2430-2438. doi: 10. 2135/cropsci2010. 12. 0671

Lu YM, Lan CX, Liang SS, Zhou XC, Liu D, Zhou G, Lu QL, Jing JX, Wang MN, Xia XC, He ZH (2009) QTL mapping for adult-plant resistance to stripe rust in Italian common wheat cultivars Libellula and Strampelli. Theoretical and Applied Genetics 119, 1349-1359. doi: 10. 1007/s00122-009-1139-6

Luo PG, Luo YH, Chang ZJ, Zhang HY, Zhang M, Ren ZL (2009) Characterization and chromosomal location of *Pm40* in common wheat: a new gene for resistance to powdery mildew derived from *Elytrigia intermedium*. Theoretical and Applied Genetics 118, 1059-1064. doi: 10. 1007/s00122-009-0962-0

MaHQ, Kong ZX, Fu BS, Li N, Zhang LX, Jia HY, Ma ZQ (2011) Identification and mapping of a new powdery mildew resistance gene on chromosome 6D of common wheat. Theoretical and Applied Genetics 123, 1099-1106. doi: 10. 1007/s00122-011-1651-3

McIntosh RA, Dubcovsky J, Rogers WJ, Morris CF, Appels R, Xia XC (2011) Catalogue of gene symbols for wheat: 2011 supplement. http: //www. shigen. nig. ac. jp/wheat/komugi/genes/macgene/supplement2011. pdf.

Miedaner T, Schilling AG, Geiger HH (2001) Molecular genetic diversity and variation of aggressiveness in population of *Fusarium graminearum* and *F. culmurum* sampled from wheat fields in different countries. Journal of Phytopathology. 149, 641-648. doi: 10. 1046/j. 1439-0434. 2001. 00687. x

Miedaner T, Wilde F, Steiner B, Buerstmayr H, Korzun V, Ebmeyer E (2006) Stacking quantitative trait loci (QTL) for *Fusarium* head blight resistance from non-adapted sources in an European elite spring wheat background and assessing their effects on deoxynivalenol (DON) content and disease severity. Theoretical and Applied Genetics 112, 562-569. doi: 10. 1007/s00122-005-0163-4

Niks REand Rubiales D (2002) Potentially durable resistance mechanisms in plants to specialized fungal pathogens. Euphytica 124, 201-216. DOI: 10. 1023/A: 1015634617334

Singh RP, Huerta-Espino J and William HM (2005) Genetics and breeding for durable resistance to leaf and stripe rusts in wheat. Turkish journal of agriculture & forestry 29, 121-127. doi: 10. 1023/A: 1015634617334

Singh RP, Huerta-Espino J and Rajaram S (2000) Achieving near-immunity to leaf and stripe rusts in wheat by combining slow rusting resistance genes. Acta Phytopathol. Entomol. Hungarica 35, 133-139.

Singh, RP, Huerta-Espino J, Bhavani S, Herrera-Foessel SA, Singh D, Singh PK, Velu G, Mason RE, Jin Y, Njau P, Crossa J (2011) Race non-specific resistance to rust diseases in CIMMYT spring wheats. Euphytica 179, 175-186. doi: 10. 1007/s10681-010-0322-9

Breeding adult plant resistance to stripe rust in spring bread wheat germplasm adapted to Sichuan Province of China

E. N. Yang[1], Y. C. Zou[1], W. Y. Yang[1],
Y. L. Tang[1], Z. H. He[2], and R. P. Singh[3]

[1]*Crop Research Institute, Sichuan Academy of Agriculture Sciences, Sichuan Chengdu*, 610066;
[2]*CAAS-CIMMYT, Beijing, China*; [3]*CIMMYT, Apdo. Postal* 6-641, 06600, *Mexico DF, Mexico.*

Abstract: Sichuan is an important wheat producing province of China where severe stripe rust epidemics occur annually. Developing high-yielding wheat varieties with good and stable stripe rust resistance is foremost breeding objective of all breeding programs. Because minor genes based adult-plant resistance (APR) is considered durable, a shuttle breeding program between Sichuan Academy of Agricultural Sciences (SAAS) and CIMMYT was initiated in 2000 to transfer APR identified in CIMMYT wheats to wheat germplasm adapted in Sichuan. During 2007-2009, a total of 669 advanced generation lines obtained from this shuttle breeding effort were provided to Plant Protection Research Institute, SAAS for official multi-environment stripe rust tests, and 231 elite lines were characterized for yield performance by the agronomists in Crop Research Institute, SAAS. Between 11-39% lines were highly resistant depending on the year of testing and 17 (7.3%) lines had 5% or higher yields than the checks mean. The adapted resistant lines are being used by various breeding programs to enhance resistance diversity and three lines are being tested in National or Provincial Yield Trials for possible releases.

Key words: *Puccinia striiformis*, *Triticum aestivum*, yellow rust, durable resistance

Introduction

Stripe, or yellow, rust caused by fungus *Puccinia striiformis tritici* is considered to be the most important disease of wheat (*Triticum aestivum*) in China. Sichuan is an important wheat producing province where wheat crop is annually affected from severe stripe rust epidemics. Several resistant varieties have been released over years however resistance of many varieties was short-lived due to evolution of new races (Wan et al., 2007). Race CYR32 identified in late 1990s caused severe epidemics since then in Sichuan and several other provinces and the use of fungicides has increased in the last decade. Several studies conducted in recent years have shown that varieties that have shown resistance durability carry adult plant resistance (APR) based on combinations of multiple minor genes. Slow rusting genes *Yr18*, *Yr29* and *Yr46* are often involved in durable adult plant resistance complex (Singh. 2007). Objective of study was to transfer the high level of resistance bred into CIMMYT wheats to wheat germplasm adapted in Sichuan province using a Mexico-Sichuan shuttle-breeding scheme aimed towards the release of wheat varieties with adult plant resistance.

Materials and methods

To implement a shuttle breeding scheme between

Mexico and China, 16 high yielding but stripe rust susceptible (or moderately susceptible) Sichuan wheat varieties and high yielding lines were crossed in Mexico during 2000 in various combinations with 14 CIMMYT wheats that carry high levels of APR. The resulting 143 F_1 s were then backcrossed with Sichuan parents and approximately 400 seeds per combination obtained. The BC_1 and BC_1F_2 populations of about 1 200 plants were grown and selected in Mexico. About 500 BC_1F_3 bulked seeds from selected plants in each population were grown in Chengdu, Sichuan province from 2003 onwards for continuing single plant selections under stripe rust pressure created with local predominant races. During 2007-2009, a total of 669 advanced lines were evaluated by the Plant Protection Research Institute (PPRI), Sichuan Academy of Agricultural Sciences (SAAS) under artificially inoculated, multi-environment stripe rust official trials in Sichuan. Stripe rust severity was used to group lines in the following four categories: R or resistant (0-15%), MR or moderately resistant (20-35%), MS or moderately susceptible (40-55%) and S or susceptible (60-100%).

A set of 'Avocet' near-isogenic lines (NILs) and tester varieties were also planted to determine the effectiveness of various resistance genes to the yellow rust races used in the field trials. Yield performance of 231 lines was determined in large plot, two-replicate trials by the agronomists in the Crop Research Institute (CRI), SAAS.

Results and discussion

Response of Avocet NILs and other testers to yellow rust

Race analyses conducted in Sichuan and China have shown that races CYR29, CYR30, CYR31 and CYR32 have dominated from mid 1990s onwards and CYR32 was has been the most important race in the last decade (Wan et al., 2007). In 2008 a new race CYR33 with additional virulence was identified. Performance of Avocet NILs and other testers during, 2007, 2008 and 2009 is summarized in Table 1. Resistance genes that were not effective in all three years were *Yr1*, *Yr2*, *Yr6*, *Yr7*, *Yr8*, *Yr9*, *Yr17*, *Yr31*, *YrA* and *YrSp* (Table 1). Resistance genes such as *Yr18*, *Yr24*, *Yr27*, *Yr28*, *Yr29* and *YrCV* either were ineffective in some years or did not confer adequate resistance and grouped in the MS category. *Yr5*, *Yr10* and *Yr15* were highly effective in all years.

Table 1 Performance of Avocet NILs and other tester varieties in Sichuan for three years

Line	Yellow rust response		
	2007	2008	2009
Avocet	S	S	S
Avocet+*Yr1*	S	S	S
Kalyansona (*Yr2*, *Yr29*)	MS	MS	S
Avocet+*Yr5*	R	R	R
Avocet+*Yr6*	S	S	S
Avocet+*Yr7*	S	S	S
Avocet+*Yr8*	S	MS	MS
Avocet+*Yr9*	S	S	S
Avocet+*Yr10*	R	R	R
Avocet+*Yr15*	R	R	R
Avocet+*Yr17*, *Yr18*	MS	MS	MS
Avocet+*Yr18*	MS	MR	MS
Avocet+*Yr24*	MS	R	MS
Avocet+*Yr26*	MR	R	MR
Avocet+*Yr27*	R	MS	S
Avocet+*Yr28*	S	MR	S
Avocet+*Yr29*	MS	MS	MS
Avocet+*Yr31*	S	S	S
Avocet+*YrA*	S	S	S
Avocet+*YrCV*	S	MR	S
Avocet+*YrSp*	S	MS	S
Opata 85 (*Yr27*, *Yr18*, *Yr30*, +)	R	R	MR
Pavon 76 (*Yr6*, *Yr7*, *Yr29*, *Yr30*, +)	MS	MS	MS
PBW 343 (*Yr9*, *Yr27*, *Yr29*, +)	R	MR	MS
Seri 82 (*Yr2*, *Yr9*, *Yr29*, *Yr30*, +)	R	R	MR
Super Kauz (*Yr9*, *Yr27*, *Yr18*, *Yr30*, +)	R	R	MR
Tatara (*Yr3*+*Yr29*+)	MR	R	MR

Stripe rust resistance of advanced breeding lines

Between 11 to 39% advanced breeding lines were highly resistant (R), and 26-46% carried moderate resistance (MR) to stripe rust (Table 2). Eighty-nine lines were tested for two years, 2007 and 2008, to determine the stability of their performance. Lines in R category in 2007 maintained their resistance in 2008. A majority of lines with MR response were also classified in the same category. In 2008 frequency of lines with MR response was much higher and 2 additional lines were grouped in the R category.

Table 2 Stripe rust resistance of 89 advanced lines in 2007 and 2008 Disease Identification Nursery in Sichuan Province

Year tested	Response to stripe rust and frequency of advanced lines							
	Resistant		Moderately resistant		Moderately susceptible		Susceptible	
	No.	%	No.	%	No.	%	No.	%
2007	34	38.2	27	30.3	22	24.7	6	6.7
2008	36	40.5	45	50.6	6	6.7	2	2.3

Grain yield performance of advanced breeding lines

Seventeen (7.3%) of the 231 lines tested during 2007 and 2009 had 5% or higher yields than the checks. Five and seven of these 17 lines belonged to R and MR categories, respectively to stripe rust. New variety proposed for release in Sichuan province must have at least 5% higher yields than the checks mean for 3 years in Provincial Yield Trial. In addition they must also be classified as R or MR for stripe rust resistance by the PPRI pathologists.

Twelve highly resistant lines with good agronomic traits and adaption in Sichuan environments were provided to various breeding programs in Sichuan. These lines are being used as resistant parents in our and other breeding programs to enhance genetic diversity and durability of resistance to stripe rust. Additional advanced shuttle breeding lines were provided to wheat breeding programs in Beijing and Gansu provinces for utilization in their breeding programs and other studies. One elite line, 08RC2525 (CHUANMAI 32 * 2/CHAPIO), is in National Yield Trial and two lines, 07RC3929 (SW119 * 2/TUKURU) and 07RC3941 (SW119 * 2/TUKURU), in Sichuan Provincial Yield Trial of 2010. Ten best performing entries in resistant category R from various crosses are listed in Table 3.

Table 3 Ten best yielding entries from different crosses that were classified in the resistant category (R)

Sichuan designation	Cross	Response to stripe rust	Yield (% of checks)
08RC2525	CHUANMAI 32 * 2/CHAPIO	MR	+15.55
07RC3929	SW119 * 2/TUKURU	R	+10.46
06RC4117	SW119 * 2/TUKURU	R	+7.42
08RC2329	CHUANMAI 32 * 2/CHAPIO	R	+6.51
08RC2338	CHUANMAI 32 * 2/CHAPIO	R	+4.83
08RC1177	CHUANMAI 32 * 2/CHAPIO	R	+4.56
07RC3941	SW119 * 2/TUKURU	R	+4.43
08RC1202	CHUANMAI 32 * 2/JARU	R	−4.29
08RC1796-1	SW00-60165 * 2/TUKURU	R	−5.11
08RC2724	CHUANMAI 32 * 2/JARU	R	−5.36

Selecting complex adult plant resistance has given us good experience and confidence that such resistance can be utilized successfully in developing new wheat varieties that may show resistance durability in coming years. This will help Sichuan farmers to rely more on genetic resistance than depend on chemical control. We aim to conduct studies such as: a) resistance tracing-determine the response to a new race to determine durability; b) methodology to transfer minor genes based resistance to varieties with major genes- several new wheat varieties and elite breeding lines in Sichuan carry a major gene and therefore molecular markers are required for this work; c) determining the role of known slow rusting resistance genes such as *Yr18* and *Yr29* in high level of resistance of some lines- *Yr18* and *Yr29* alone are not very effective in Sichuan and our initial molecular marker studies indicate that only some highly resistant APR lines carried *Yr18*, hence some other uncharacterized minor genes are present; and d) utilization of APR present in land-races- Sichuan province is rich in land-races and yellow rust resistance in some has survived for a long time, however genetic basis of resistance is not known.

Acknowledgments

This shuttle breeding program was supported by CIMMYT Wheat Program, National 863 program, the Important International Cooperative Program of the Ministry of Science and Technology of China and the International Cooperative Program of the Science and Technology Department of Sichuan Province. The authors are grateful to Dr. N. Borlaug, Dr. S. Rajaram and Prof. R. A. McIntosh for their suggestions and encouragements.

References

SINGH R. P. (2007) Increasing genetic yield and mitigating effects of key biotic and abiotic constraints to wheat production in India through international wheat resources and partnerships. J. Wheat Research, 1: 13-18.

WAN A. M., CHEN X. M., HE Z. H. (2007) Wheat stripe rust in China. Australian Journal of Agricultural Research, 58: 605-619.

小麦慢白粉病 QTL 对条锈病和叶锈病的兼抗性

刘金栋[1]，陈新民[1]，何中虎[1,2]，伍玲[3]，白斌[4]，
李在峰[5]，夏先春[1]

[1]中国农业科学院作物科学研究所/国家小麦改良中心，北京 100081；
[2]CIMMYT 中国办事处，北京 100081；[3]四川省农业科学院作物研究所，成都 610066；
[4]甘肃省农业科学院小麦研究所，兰州 730070；[5]河北农业大学植物保护学院植物病理系，保定 071001

摘要：聚合兼抗白粉病、条锈病和叶锈病的慢病性基因，是培育持久多抗小麦品种的重要措施。百农 64 和鲁麦 21 均为慢白粉病品种，分别含有 4 个和 3 个慢白粉病抗性 QTL。将百农 64 与鲁麦 21 杂交，获得 21 个聚合 2～5 个慢白粉病抗性 QTL 的 F_6 株系，于 2012—2013 年度分别在四川郫县和甘肃天水进行条锈病田间抗性鉴定，在河北保定和河南周口进行叶锈病田间抗性鉴定。分析 21 个株系条锈和叶锈病的最大严重度和病程曲线下面积，检测单个 QTL 和 QTL 聚合体对条锈病和叶锈病的抗性效应。结果表明，*QPm. caas-4DL*、*QPm. caas-6BS* 和 *QPm. caas-2BL* 对条锈病均有显著的抗性，分别解释表型变异的 16.9%、14.1%和 17.3%；*QPm. caas-4DL* 对叶锈病也有显著抗性，可解释表型变异的 35.3%；*QPm. caas-1A/QPm. caas-4DL/QPm. caas-2DL/QPm. caas-2BS/QPm. caas-2BL* 和 *QPm. caas-1A/QPm. caas-4DL/QPm. caas-2BS/QPm. caas-2BL* 聚合体对条锈病和叶锈病的抗性显著高于两亲本，它们均含有来自百农 64 的 *QPm. caas-4DL* 以及来自鲁麦 21 的 *QPm. caas-2BL* 和 *QPm. caas-2BS*，表明这些 QTL 具有明显的兼抗性效应。在小麦抗病育种中，聚合慢病性 QTL 越多，慢病性越强，聚合 4～5 个慢病性 QTL 时，株系可达到高抗甚至接近免疫的水平，是选育持久抗性小麦品种的重要手段。

关键词：普通小麦；慢病性；持久抗性；基因聚合；QTL

Resistance of Slow Mildewing Genes to Stripe Rust and Leaf Rust in Common Wheat

Liu Jindong[1], Chen Xinmin[1], He Zhonghu[1,2], Wu Ling[3],
Bai Bin[4], Li Zaifeng[5], Xia Xianchun[1]

[1]*Institute of Crop Science/ National Wheat Improvement Center*, *Chinese Academy of Agricultural Sciences (CAAS)*, *Beijing* 100081, *China*; [2]*CIMMYT-China Office*, *c/o CAAS*, *Beijing* 100081, *China*; [3]*Crop Research Institute*, *Sichuan Academy of Agricultural Sciences*, *Chengdu* 610066, *China*; [4]*Wheat Research Institute*, *Gansu Academy of Agricultural Sciences*, *Lanzhou* 730070, *China*; [5]*Department of Plant Pathology*, *College of Plant Protection*, *Agricultural University of Hebei*, *Baoding* 071001, *China*

原文发表在《作物学报》，2014，40（9）：1557－1564

Abstract: Pyramiding quantitative trait loci (QTLs) is an effective method to improve resistance to powdery mildew, stripe rust, and leaf rust in common wheat. We have developed 21 lines (F_6) carrying 2-5 slow mildewing QTLs by crossing slow powdery mildew cultivars Bainong 64 and Lumai 21 possessing four and three slow mildewing QTLs, respectively. These F_6 lines were evaluated in the field in Pianxian, Sichuan and Tianshui, Gansu for stripe rust resistance and in Baoding, Hebei and Zhoukou, Henan for leaf rust resistance during the 2012-2013 cropping season. According to the maximum disease severities (MDS) and the area under the disease progress curve (AUDPC), QTLs *Qpm. caas-4DL*, *QPm. caas-6BS*, and *QPm. caas-2BL* were highly resistant to stripe rust ($P<0.01$), which explained 16.9%, 14.1%, and 17.3% of phenotypic variance, respectively. Locus *Qpm. caas-4DL* also showed high resistance to leaf rust ($P<0.01$) with phenotypic contribution of 35.3%. Lines that pyramided five (*QPm. caas-1A*/*QPm. caas-4DL*/*QPm. caas-2DL*/*QPm. caas-2BS*/*QPm. caas-2BL*) and four QTLs (*Qpm. caas-1A*/*QPm. caas-4DL*/*QPm. caas-2BS*/*QPm. caas-2BL*) exhibited higher resistance to both stripe and leaf rust compared to their parents. This result indicates that the combination of *Qpm. caas-4DL* (from Bainong 64) and *QPm. caas-2BS* and *QPm. caas-2BL* (Lumai 21) has a marked effect on improving adult resistance to powdery mildew, strip rust and leaf rust, and the more QTLs are pyramided, the stronger slow disease resistance can be achieved. In breeding practice, the combination of 4-5 slow mildewing or rusting QTLs can result in durable resistance to multiple diseases.

Key words: *Triticum aestivum* L.; Slow mildewing and slow rusting resistance; Durable resistance; Gene pyramiding; QTL

小麦条锈病、叶锈病和白粉病是全球性重要病害，分别由小麦条锈菌（*Puccinia striiformis* f. sp. *tritici*）、小麦白粉菌（*Blumeria graminis* f. sp. *tritici*）和小麦叶锈菌（*P. recondita* f. sp. *tritici*）引起，具有发生频率高、流行范围广和暴发性强的特点，可导致3.0%～49.0%的产量损失[1-3]。长期以来，利用寄主抗性防治这些病害取得了显著成效[4]。

寄主抗性分为垂直抗性和水平抗性。垂直抗性又称为生理小种专化性抗性、苗期抗性、全生育期抗性或主效基因抗性，是由一个或少数几个主效基因控制，对特定病原菌生理小种表现出高抗或免疫，具有病原菌生理小种专化性，常因病原菌生理小种变异而丧失抗性；水平抗性亦称慢病性、成株抗性或非小种专化抗性，由多个微效基因控制，对病原菌无小种专化性或专化性弱，苗期表现为感病，成株期表现为中抗或高抗，且抗性持久[5,6]。慢病性由多个微效基因的加性效应控制，基因聚合是获得慢病性、选育兼抗多种病害品种的重要方式。Singh 等[7]发现，聚合4～5 个微效慢病基因的小麦材料对叶锈和条锈病呈高抗至免疫。意大利小麦品种 Strampelli 和 Libellula 及我国农家种平原 50 至今仍然高抗条锈病。抗性遗传分析表明，Strampelli 和 Libellula 含有 *Yr18*/*Lr34*/*Pm38*/*Sr57* 和 2～4 个其他慢病基因[8]；平原 50 含有 3 个慢条锈基因和 3 个慢白粉基因[9]。*Yr18*/*Lr34*/*Pm38*/*Sr57* 对小麦条锈病、叶锈病、白粉病和秆锈病均具有抗性[10,11]。

随着 QTL 定位和分子标记研究工作的不断深入，育种家可以利用分子标记聚合抗病 QTL，培育持久抗性品种[7,12,13]。从 20 世纪 70 年代开始，国际玉米小麦改良中心（CIMMYT）就选育出一批抗性持久且兼抗多种病害的慢病性品种，例如 Amadina、Chapio、Cook、Kukuna、Parula、Pavon 76、Sonoita 81、Tonichi 81 和 Tukuru，均含有 *Yr18*/*Lr34*/*Pm38*/*Sr57* 或 *Yr29*/*Lr46*/*Pm39*/*Sr58* 位点及 2～3 个微效基因[14,15]。因此，聚合慢病性基因是获得兼抗多种病害的持久抗性小麦品种的重要手段。

虽然小麦抗病性 QTL 定位研究很多，但是由于不同遗传背景下的 QTL 聚合到同一遗传背景存在诸多困难[16]，不同来源的 QTL 可能存在互作，QTL 聚合育种实践的报道很少。Miedaner 等[17]将 3 个来源不同的赤霉病抗性 QTL 聚合到同一种质，发现单个 QTL 均可降低赤霉病发病率和脱氧雪腐镰刀菌烯醇（DON）含量，但 3A 上的一个 QTL 单独存在或与 3B 和 5A 上 QTL 聚合时对赤霉病没有显著遗传效应。Lu 等[18]将 5A 上的 QTL 与 *Fhb1* 聚合在一起，发现可以降低矮秆基因 *Rht-D1b* 对赤霉病引起的效应。可见，QTL聚合体遗传效应分析具有重要的应用价值。

百农64和鲁麦21均具有优良的农艺性状，分别在20世纪90年代和21世纪初为我国黄淮麦区主推品种。它们苗期对白粉菌[19]、条锈菌[20]和叶锈菌流行小种均感病，田间成株期表现抗病，具有典型的慢病性特征。在百农64中检测到4个慢白粉病抗性QTL（*QPm. caas-1A*、*QPm. caas-4DL*、*QPm. caas-6BS* 和 *QPm. caas-7A*）[21,22]，在鲁麦21中发现3个慢白粉病抗性QTL（*QPm. caas-2BS*、*QPm. caas-2BL* 和 *QPm. caas-2DL*）。Bai 等[23]采用杂交和改良系谱法，对百农64和鲁麦21所含慢白粉病抗性QTL进行聚合，创制了21个 F_6 聚合株系，并对这些材料的白粉病抗性进行了分析。在此基础上，本研究将分析这21个聚合不同慢白粉病基因的株系对条锈病和叶锈病的遗传效应，明确这些慢白粉病抗性QTL及QTL聚合体对条锈病和叶锈病的抗性，发掘高抗条锈、叶锈和白粉病的慢病性材料和QTL组合，为选育兼抗多种病害的持久抗性小麦品种提供材料和方法。

1 材料与方法

1.1 聚合株系基因型及其分子标记

百农64与鲁麦21杂交采用改良系谱法，根据田间白粉病抗性和综合农艺性状选择，在早代淘汰白粉病严重度高的株系，选择低严重度株系继续种植，经过连续多代鉴定和选择，获得21个 F_6 抗病株系（编号：BFB5～BFB25）[23]；利用16个分子标记[21,22]进行检测，明确这些株系携带2～5个慢白粉病QTL。这些慢白粉病QTL中，*QPm. caas-1A* 的连锁标记为 *Xbarc148* 和 *Xwmc550*，*QPm. caas-4DL* 的连锁标记为 *Xgwm165*、*Xcfd23* 和 *Xwmc331*，*QPm. caas-6BS* 的连锁标记为 *Xbarc79* 和 *Xgwm518*，*QPm. caas-7A* 的连锁标记为 *Xbarc127* 和 *Xbarc174*，*QPm. caas-2BS* 的连锁为 *Xbarc98* 和 *Xbarc1147*，*QPm. caas2BL* 的连锁标记为 *Xbarc1139* 和 *Xgwm47*，*QPm. caas-2DL* 的连锁标记为 *Xbarc18*、*Xgwm539* 和 *Xcfd233*。

1.2 田间抗病性鉴定

将21个 F_6 聚合基因株系及其亲本于2012—2013年度种植于四川郫县（30°05′N，102°54′E）和甘肃天水（34°05′N′，104°35′E）进行条锈病抗性鉴定；在河北保定（113°40′N，38°10′E）和河南周口（114°38′N，33°37′E）进行叶锈病抗性鉴定。采用完全随机区组设计，3次重复，单行区，行长1.5m，行距0.25m，每行种植50粒。在郫县和天水点，每10行种植1行高感条锈病品种对照辉县红，小区周围种植高感品种作为诱发行；在保定和周口，每10行种植1行高感叶锈病对照郑州5389。以保证充分接种，实际观察所有对照均发病完全。

1.3 抗病性评价

接种后6周左右，当对照品种充分发病时开始调查发病严重度，记录旗叶和倒二叶上条锈菌或叶锈菌孢子堆面积占总叶片面积的百分数。每隔7d调查一次，共调查2～3次，直到叶片上孢子堆不再增加为止。用最大严重度（maximum disease severity，MDS）和病程曲线下面积（area under the disease progress curve，AUDPC）作为抗病性评价指标，$AUDPC = \sum[(x_i + x_i + 1)/2](t_i + 1 - t_i)$，式中 x_i 表示第i次调查的严重度，t_i 表示第i次调查距接种后的天数[24]。用SAS 9.2软件进行统计分析和显著性比较。

2 结果与分析

2.1 21个基因聚合株系条锈病和叶锈病抗性鉴定

MDS和AUDPC在株系间和环境间的差异均达极显著水平（$P<0.01$）。百农64和鲁麦21在郫县和天水两个环境下条锈病平均MDS分别为36.8%和50.0%，平均AUDPC分别为210.6和290.5；12个聚合基因株系的条锈病平均MDS均低于百农64和鲁麦21。其中，株系BFB14在两个环境下的条锈病平均MDS值（16.2%）和平均AUDPC值（92.1）最低；在保定和周口的叶锈病平均MDS分别为15.8%和35.0%，平均AUDPC分别为91.6和212.3。其他11个株系的叶锈病平均MDS低于百农64和鲁麦21。株系BFB9在两个环境下叶锈病的平均MDS（3.6%）和平均AUDPC（18.9）最低（表1）。

表 1 21 个聚合基因株系及其亲本 3 种病害的最大严重度（MDS）和病程曲线下面积（AUDPC）

Table 1 Composition of slow mildewing QTL, and averaged MDS and AUDPC for powdery mildew, stripe rust and leaf rust response in 21 F_6 lines from the Bainong 64/Lumai 21 cross[1] and their parents

品系 Line	慢白粉病 QTL Slow mildewing resistance QTLs	QTL 数 No. of QTL	白粉病 Powdery mildew		条锈病 Stripe rust		叶锈病 Leaf rust	
			MDS MDS	AUDPC AUDPC	MDS MDS	AUDPC AUDPC	MDS MDS	AUDPC AUDPC
百农 64 Bainong 64	1A/4DL/6BS/7A	4	7.0 cde	45.0 defg	36.8 cdefgh	210.6 efgh	15.8 ef	91.6 ef
鲁麦 21 Lumai 21	2BS/2BL/2DL	3	12.0 a	75.3 a	50.0 abcd	290.5 bcde	35.0 a	212.9 a
BFB5	1A/4DL/2BS	3	3.5 f	22.8 hig	33.7 fghi	195.5 fghi	9.4 fgh	52.0 fgh
BFB6	1A/4DL/2DL	3	4.5 ef	33.3 fghi	35.3 fghi	198.6 fghi	9.0 fgh	50.1 fgh
BFB7	1A/4DL/2DL	3	4.0 ef	31.5 fghi	48.0 abcd	288.3 cde	8.7 fgh	48.0 fgh
BFB8	1A/4DL/2DL/2BS/2BL	5	4.3 ef	39 efghi	27.5 ghijk	160.5 ghijk	4.6 h	25.4 gh
BFB9	1A/4DL/2BS/2BL	4	9.5 abc	63.5 abcd	24.7 hijk	143.3 hijk	3.6 h	18.9 h
BFB10	1A/4DL/2DL/2BS/2BL	5	2.8 f	20.8 hi	21.0 hijk	123.7 jk	8.9 gh	48.5 fgh
BFB11	1A/4DL/2DL/2BS	4	2.8 f	17.3 i	18.2 jk	101.6 jk	5.7 gh	31.6 gh
BFB12	4DL/2BS	2	11.3 ab	72.8 ab	36.5 efgh	186.1 fghij	35.7 a	200.7 a
BFB13	1A/4DL/2DL	3	4.0 ef	23.5 hig	31.5 ghij	172.5 ghijk	15.4 ef	87.2 ef
BFB14	1A/4DL/2DL	3	3.8 ef	22.8 hig	16.2 k	91.7 k	10.0 fgh	54.8 fgh
BFB15	1A/4DL/2DL/2BS	4	8.5 bcd	59.3 abcde	54.3 ab	324.2 abcd	26.7 cd	145.8 cd
BFB16	1A/4DL/2BL/7A	4	5.8 def	51.0 abcdef	26.8 ghijk	168.0 ghijk	15.5 ef	83.3 ef
BFB17	1A/4DL/2BL/7A	4	5.0 ef	35.5 fghi	29.7 ghijk	177.7 fghij	14.3 efg	72.9 efg
BFB18	1A/4DL/2DL/2BL/7A	5	5.8 def	43.8 defgh	27.3 ghijk	162.7 ghijk	32.9 bc	182.1 bc
BFB19	1A/4DL/2DL/2BL/7A	5	5.8 def	32.5 fghi	39.5 cdefg	241.6 defg	21.3 de	114.3 de
BFB20	1A/4DL/2BL/7A	4	4.0 ef	32.8 fghi	39.2 cdefgh	227.1 efgh	21.0 de	113.2 de
BFB21	1A/4DL/2DL/2BL/7A	5	5.3 def	47.8 cdef	45.3 cdef	257.9 cdef	27.9 cd	155.8 cd
BFB22	1A/4DL/7A	3	9.8 abc	68.8 abc	53.5 ab	327.1 abc	31.9 bc	175.9 bc
BFB23	1A/4DL/7A	3	11.3 ab	76.5 a	61.7 a	374.5 a	29.2 bcd	158.8 cd
BFB24	1A/4DL/7A	3	10.0 abc	59.0 abcde	57.7 ab	349.7 ab	32.8 bc	179.0 bc
BFB25	1A/4DL/7A	3	7.0 cde	45.0 defg	50.8 abc	316.2 abcd	27.9 cd	157.0 cd

慢白粉病 QTL 由 Lan 等[21,22]鉴定，*QPm. caas-4DL* 表中缩写为 4DL，余同，其中 *QPm. caas-4DL* 和 *QPm. caas-6BS* 兼抗白粉病、条锈病和叶锈病，*QPm. caas-2BS* 和 *QPm. caas-2BL* 兼抗白粉病和条锈病[20,25]。白粉病鉴定结果为 2009—2010 和 2010—2011 年度北京和河南安阳两点的平均值[23]。条锈病数据为 2012—2013 年度四川郫县和甘肃天水两点的平均值。叶锈病的数据为 2012—2013 度年河北保定和河南周口两点的平均值。平均值后不同字母表示品系间差异显著（$P<0.05$）。

Slow powdery mildew QTL werepreviously mapped[21,22], *QPm. caas-4DL* is short for 4DL, so as other QTLs, *QPm. caas-4DL* and *QPm. caas-6BS* have significant resistance to powdery mildew, stripe rust and leaf rust; *QPm. caas-2BS* and *QPm. caas-2BL* have significant resistance to powdery mildew and stripe rust[20, 25]; Averaged data of powdery mildew for 2009-10 and 2010-11 cropping seasons in Beijing and Anyang of Henan province[23]; Averaged data of stripe rust for 2012-13 cropping season in Pixian of Sichuan province and Tianshui of Gansu province; Averaged data for 2012-13 cropping season in Baoding of Hebei province and Zhoukou of Henan province. The values followed by the different letters behind the average data are significantly different at the $P = 0.05$ level.

2.2 慢白粉病 QTL 对条锈和叶锈病的抗性效应与互作分析

QPm. caas-4DL、*QPm. caas-2BS* 和 *QPm. caas-2BL* 对条锈病具有显著的抗性遗传效应（$P<0.01$），分别解释条锈病平均表型变异的 16.9%、14.1%和 17.3%，*QPm. caas-6BS* 可解释 1.8%的表型变异。*QPm. caas-2BS* 和 *QPm. caas-2BL* 之间存在互作，但效应较低，环境与 QTL 间也存在互作。*QPm. caas-4DL* 对叶锈病也具有显著抗性，可解释表型变异的

35.3%，而 *QPm. caas-6BS* 可解释 8.3%的表型变异 （表 2）。

表 2 4 个环境下慢白粉病抗性 QTL 对条锈病和叶锈病的效应及其互作

Table 2 The effects of slow mildewing QTL on stripe rust and leaf rust, and their interactions based on MDS data from four environments

变异来源 Source	*df*	条锈病 Stripe rust		叶锈病 Leaf rust		变异来源 Source	*df*	条锈病 Stripe rust		叶锈病 Leaf rust	
		SS	%	*SS*	%			*SS*	%	*SS*	%
4DL	1	2 951.7**	16.9	1 514.0**	35.3	Env×*6BS*	1	172.9	1.0	22.5	0.5
6BS	1	306.5	1.8	357	8.3	Env×*2BS*	1	330.3	1.9	—	—
2BS	1	2 469.2**	14.1	—	—	Env×*2BL*	1	879.1	5.0	—	—
2BL	1	3 032.1**	17.3	—	—	Env	1	174.9	1.0	104.1	0.1
2BS × 2BL	1	15.3	0.1	—	—	E	102	5 652.3		1 8591	
Env× 4DL	1	82.3	0.5	—	—						

QPm. caas-4DL 缩写为 4DL，余同；Env：环境；E：误差；**表示 0.01 的显著水平；%表示部分平方和占总平方和的百分比，即可以解释的表型变异，"—" 表示数据不存在。

QPm. caas-4DL is short for 4DL, so as other QTLs; Env: envriment; E: error; ** Significant at 0.01 propability; Partial sum of squares as % of total sum of squares, which can be interpreted as an indication of phenotypic variance explained; "—" indicates that the data does not exist.

2.3 慢白粉病 QTL 聚合体对条锈和叶锈病的抗性效应

百农 64 和鲁麦 21 以及 21 个慢病性 QTL 聚合株系中共有 11 种 QTL 组合（表 3）。*QPm. caas-1A/QPm. caas-4DL/QPm. caas-2DL/QPm. caas-2BS/QPm. caas-2BL*、*QPm. caas-1A/QPm. caas-4DL/QPm. caas-2BS/QPm. caas-2BL* 聚合体的条锈病 MDS 及 AUDPC 值最低，远低于两亲本及其他 7 种 QTL 聚合体，且差异显著（*P*<0.05），对条锈和叶锈病有较好的抗性。*QPm. caas-1A/QPm. caas-4DL/QPm. caas-7A* 组合的条锈病 MDS 和 AUDPC 最高，与鲁麦 21 差异不显著，但显著高于百农 64 及其他 8 种 QTL 聚合体（*P*<0.05），21 个聚合株系中其余 6 种 QTL 聚合体抗性与百农 64 差异不显著。

表 3 不同慢白粉病抗性 QTL 聚合体对白粉病、条锈病和叶锈病抗性效应

Table 3 Effects of different slow mildewing QTL combinations on powdery mildew, stripe rust and leaf rust response

QTL 聚合体 QTL combination for powdery mildew	株系数 No. of lines	白粉病 Powdery mildew		条锈病 Stripe rust		叶锈病 Leaf rust	
		MDS	AUDPC	MDS	AUDPC	MDS	AUDPC
1A/4DL/6BS/7A	百农 64	7.0 b	45.0 b	36.8 b	210.8 b	15.8 c	91.8 c
2BS/2BL/2DL	鲁麦 21	12.0 a	75.3 a	50.0 a	290.8 a	35.0 a	213.3 a
1A/4DL/7A	4	10.5 a	68.3 a	56.0 a	341.8 a	30.5 ab	167.8 b
1A/4DL/2DL/2BL/7A	3	5.6 bc	41.1 bc	37.3 b	220.8 b	27.2 b	151.0 b
1A/4DL/2DL/2BS	2	5.2 bc	38.3 bc	36.5 b	212.8 b	16.5 c	88.8 c
4DL/2BS	1	11.3 a	72.8 a	36.5 b	186.3 bc	35.6 a	201.0 a
1A/4DL/2BL/7A	3	5.5 bc	39.8 bc	31.8 bc	190.8 bc	16.8 c	89.8 c
1A/4DL/2BS	1	3.5 c	27.8 bc	33.7 bc	195.8 bc	9.3 de	52.2 de
1A/4DL/2DL	4	4.1 c	29.9 bc	32.8 bc	187.8 bc	10.8 cd	60.0 cd
1A/4DL/2BS/2BL	1	9.5 a	63.5 a	24.7 c	143.7 c	3.5 e	18.8 e
1A/4DL/2DL/2BS/2BL	2	3.5 c	22.8 c	24.5 c	142.2 c	6.8 de	37.0 de

QPm. caas-4DL 表中缩写为 4DL，余同；白粉病（MDS）和（AUDPC）为 2009—2010 和 2010—2011 年度北京和安阳两点数据平均值[23]；条锈病（MDS）和（AUDPC）为 2012—2013 年度郫县和天水两点数据平均值；叶锈病（MDS）和（AUDPC）为 2012—2013 年度保定和周口两点数据平均值；同一列数值后不同字母表示 QTL 聚合体之间差异显著（*P*<0.05）。

QPm. caas-4DL is short for 4DL, so as other QTLs; [1]Number of lines with common QTL combinations; [2]Averaged data of powdery mildew for 2009-2010 and 2010-2011 cropping seasons in Beijing and Anyang[23]; Averaged data of stripe rust for 2012-2013 cropping seasons in Pianxian and Tianshui; Averaged data of leaf rust for 2012-2013 cropping seasons in Baoding and Zhoukou; Values followed by different letters in the same column are significantly different (*P*=0.05).

QPm caas-1A/*QPm caas-4DL*/*QPm caas-2DL*/*QPm caas-2BS*/*QPm caas-2BL*、*QPm caas- 1A*/*QPm caas- 4DL*/*QPm caas-2BS*/*QPm caas-2BL* 和 *QPm caas-1A*/*QPm Caas-4DL*/*QPm caas-2BS* 组合的叶锈病 MDS 及 AUDPC 值较低，显著低于双亲及其他 6 种 QTL 聚合体，且差异显著（$P<0.05$），对叶锈病有较好抗性。*QPm caas-1A*/*QPm caas-4DL*/*QPm caas-7A* 和 *QPm caas-4DL*/*QPm caas-2BS* 组合的叶锈病 MDS 和 AUDPC 较高，与鲁麦 21 差异不显著，但显著高于 21 个聚合株系中其他 7 种 QTL 聚合体（$P<0.05$），其余 4 种 QTL 聚合体抗性与百农 64 无显著差异。

除株系 BFB12 外，随着含有白粉病抗性 QTL 数量的增多，聚合株系的条锈病和叶锈病 MDS 和 AUDPC呈现逐渐降低趋势（图 1）。

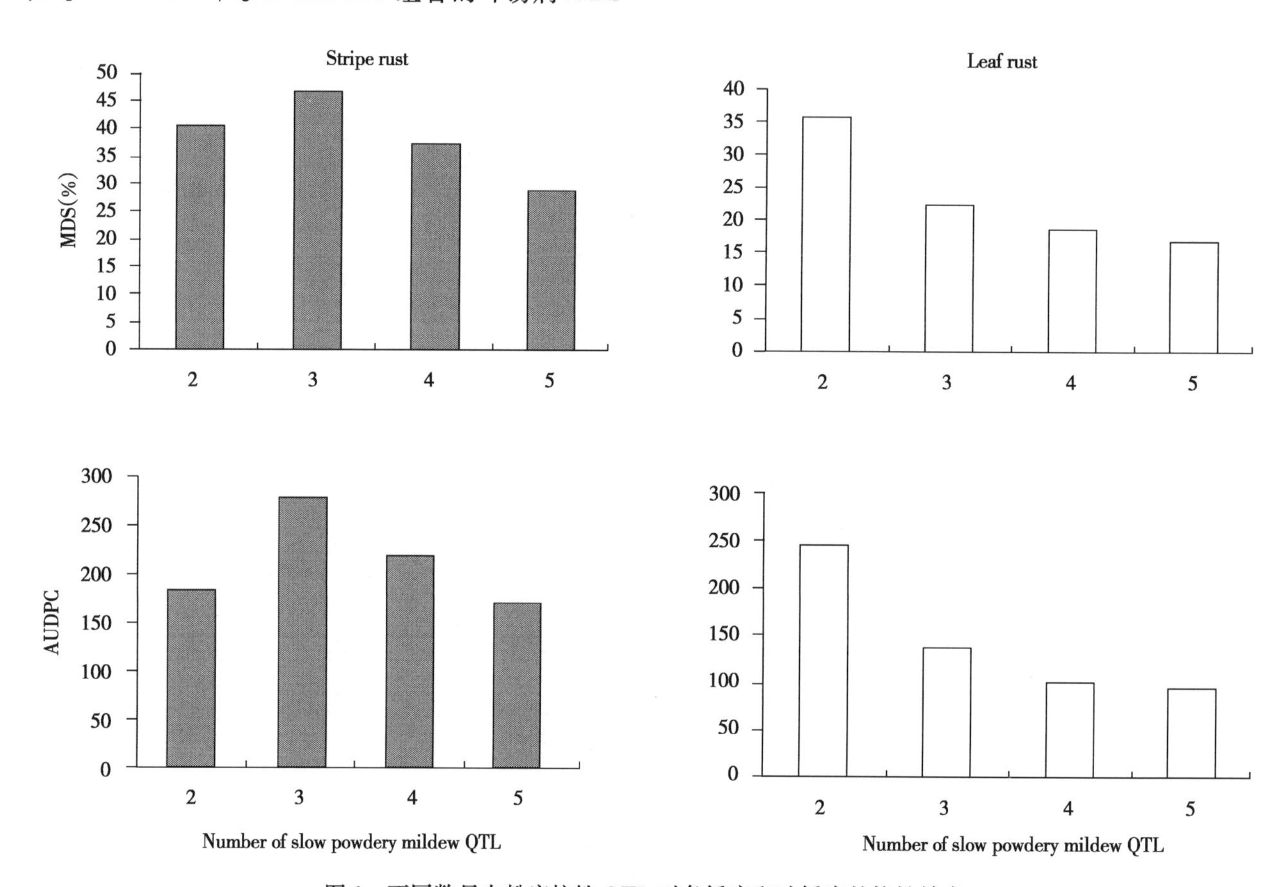

图 1 不同数量白粉病抗性 QTL 对条锈病和叶锈病的抗性效应

Fig. 1 Effects of slow powdery mildew QTL number for reducing stripe rust and leaf rust MDS and AUDPC

横坐标表示株系所含有的慢白粉病 QTL 数目，如 2 代表株系含有两个 QTL；纵坐标表示含有相同 QTL 个数株系的平均 MDS 和 AUDPC 值。

The number in the abscissa indicates the number of slow powdery mildew QTL combined in each subset of lines; for example, 2 means lines possessing two QTL. The averaged values of MDS and AUDPC in each class are indicated on the vertical axis.

3 讨论

利用分子标记辅助选择聚合慢病性基因是培育兼抗白粉病、条锈病和叶锈病品种的重要途径。迄今为止，已证实 *Yr18*/*Lr34*/*Pm38*/*Sr57*、*Yr29*/*Lr46*/*Pm39*/*Sr58* 和 *Yr46*/*Lr67*/*Pm46*/*Sr55* 等多个小麦慢病性基因兼抗条锈病、叶锈病、白粉病和秆锈病[26-29]。同时，还在 1BL、2BS、2BL、3BS、6BS 和 7DS 上发现了多个兼抗条锈病、叶锈病和白粉病的基因簇[30]。这些基因的发现为培育兼抗多种病害的小麦品种提供了可能。CIMMYT 约 60%的小麦品种聚合多个慢病性基因，并已建立通过聚合 *Yr18*/*Lr34*/*Pm38*/*Sr57*、*Yr29*/*Lr46*/*Pm39*/*Sr58* 和 *Yr46*/*Lr67*/*Pm46*/*Sr55* 等几个慢病性基因，选育兼抗几种病害的持久抗性品种的主要育种策略，运用该策略成功选育出一批兼抗型小麦品种[4]。携带 *Yr18/Lr34/Pm38/Sr57* 基因的材料在CIMMYT小麦种质资源中广泛存在，已保持70多年的抗性，发展中国家含 *Yr18/Lr34/Pm38/Sr57* 的小麦品种种植面积约有2600万 hm^2[31]，在病害流行年份发挥着重要作用。美国、澳大利亚和欧洲的研究重点近年也从垂直抗性逐步转向慢病性[32-34]。

百农64含有的白粉病抗性QTL *QPm. caas-4DL* 与 *Yr46/Lr67/Pm46/Sr55* 很可能为同一基因，虽然 *Yr46/Lr67/Pm46/Sr55* 抗性效应不及 *Yr18/Lr34/Pm38/Sr57*，但百农64的农艺性状良好，该位点将是小麦持久抗病育种的另一重要基因资源。Ren等[20]在 *QPm. caas-6BS* 同一位置分别发现条锈病和叶锈病抗性QTL，该位点是一个新发现的兼抗多种病害的位点。Ren等[20]在鲁麦21中定位了两个条锈病抗性QTL *QYr. caas-2BS* 和 *QYr. caas-2BL*，分别与白粉病抗性QTL *QPm. caas-2BS* 和 *QPm. caas-2BL* 位置一致，这两个位点也是小麦持久兼抗育种的重要基因。利用慢病性基因容易做到兼抗与持久抗性的结合，研究表明兼抗几种病害的慢病基因都具有持久抗性（Ravi Singh，个人交流）。已经定名的几个慢病性基因（如 *Yr18/Lr34/Pm38/Sr57*、*Yr29/Lr46/Pm39/Sr58* 和 *Yr46/Lr67/Pm46/Sr55* 等）都具有这种特点。我国小麦白粉病、条锈病和叶锈病发生严重，选育兼抗多种病害小麦品种对于我国小麦生产具有重要意义。来自百农64的 *QPm. caas-4DL*、*QPm. caas-6BS* 和来自鲁麦21的 *QPm. caas-2BL* 和 *QPm. caas-2BS* 等对条锈病、叶锈病和白粉病皆表现慢病性，利用这些基因选育兼抗多种病害的小麦品种可取得事半功倍的效果。虽然过去国内也有育成兼抗且具有持久抗性的小麦品种，但有目的地进行这种育种的尚不多。

在21个聚合株系中，含有抗病QTL数量较多的株系的MDS和AUDPC值均较低，表明QTL加性效应起了重要作用。Singh等[7]发现聚合4～5个微效基因的小麦材料对条锈和叶锈病呈高抗至免疫。但是，QTL聚合体的遗传效应并不总是与所含抗病QTL的数量成正比。例如，株系BFB12和BFB14分别含有2个和3个抗性QTL，但条锈病MDS和AUDPC值均较低。其原因可能是不同的QTL聚合体效应不同，部分QTL间互作会产生加性效应，或者某些株系含有未知的抗性QTL，导致条锈病和叶锈病MDS和AUDPC值较低，抗性增强。

Ren等[24]在鲁麦21中还发现另外两个条锈病抗性QTL *QYr. caas-4DL. 2* 和 *QPm. caas-2DS. 2*，推测部分株系可能含有这两个QTL，导致BFB12和BFB14等含有较少兼抗型QTL的株系抗性较强。相同QTL聚合体的株系抗性并不完全一致，如BFB11和BFB15均含有 *QPm. caas-1A/QPm. caas-4DL/QPm. caas-2DL/QPm. caas-2BS* 基因组合，但BFB11抗性远大于BFB15，估计BFB11含有其他条锈病抗性QTL或存在QTL间互作。但综合21个株系数据来看，株系含有QTL越多，对多种病害的抗性越强。

结合Bai等[20]对21个聚合株系的白粉病鉴定结果进行分析，株系BFB5、BFB7、BFB8、BFB10、BFB11和BFB14对白粉病、条锈病和叶锈病均具有较高的抗性，可作为育种亲本材料用于小麦抗锈病和白粉病育种。*QPm. caas-1A/QPm. caas-4DL/QPm. caas-2DL/QPm. caas-2BS/QPm. caas-2BL*、*QPm. caas-1A/QPm. caas-4DL/QPm. caas-2DL* 和 *QPm. caas-1AQPm. caas-4DL/QPm. caas-2BS* 组合对条锈病、叶锈病和白粉病均具有较好的抗性，是选育兼抗多种病害小麦慢病性品种的良好材料。*QPm. caas-1A/QPm. caas-4DL/QPm. caas-2BS/QPm. caas-2BL* 对条锈病和叶锈病抗性较好，可用于条锈病和叶锈病多发区的抗病育种。

4 结论

6个聚合百农64和鲁麦21慢白粉病抗性QTL的 F_6 株系对白粉病、条锈病和叶锈病均具有较好的慢病性，且农艺性状优良，是选育抗病高产品种的优良材料。证实了利用分子标记进行慢病性QTL聚合的可行性及其有效性，进一步说明聚合4～5个慢病性QTL足以在田间表现高水平抗性，为实现抗病育种思路转变提供了材料和方法。

参考文献

[1] Wellings C R, Mcintosh R A, Hussain M. A new source of resistance to *Puccinia striiformis* f. sp. *tritici* in spring wheats (*Triticum aestivum*). *Plant Breed*, 1988, 100: 288-296.

[2] Komer J A. Genetics of resistance to wheat leaf rust. *Annu Rev Phytopathol*, 1996, 34: 435-455.

[3] Conner R L, Kuzyk A D, Su H. Impact of powdery mildew on the yield of soft white spring wheat cultivars. *Can J Plant Sci*, 2003, 83: 725-728.

[4] Singh R P, Huerta-Espino J, William H M. Genetics and breeding for durable resistance to leaf and stripe rusts in wheat. *Turk J Agric For*, 2005, 29: 121-127.

[5] Roberts J, Caldwell R M. General resistance (slow mildewing) to *Erysiphe graminis* f. sp. *tritici* in Knox wheat. *Mol Gen Genet*, 1970, 60: 1310.

[6] Gustafson G D, Shaner G. Influence of plant age on the expression of slow-mildewing resistance in wheat (*Triticum aestivum*). *Phytopathology*, 1982, 72: 746-749.

[7] Singh R P, Huerta-Espino J, Bhavani S, Herrera-Foessel S A, Singh D, Singh P K, Velu G, Mason R E, Jin Y, Njau P, Crossa J. Race non-specific resistance to rust diseases in CIMMYT spring wheats. *Euphytica*, 2011, 179: 175-186.

[8] Lu Y M, Lan C X, Liang S S, Zhou X C, Liu D, Zhou G, Lu Q L, Jing J X, Wang M N, Xia X C, He Z H. QTL mapping for adult-plant resistance to stripe rust in Italian common wheat cultivars Libellula and Strampelli. *Theor Appl Genet*, 2009, 119: 1349-1359.

[9] Lan C X, Liang S S, Zhou X C, Zhou G, Lu Q L, Xia X C, He Z H. Identification of genomic regions controlling adult-plant stripe rust resistance in Chinese landrace Pingyuan 50 through bulked segregant analysis. *Phytopathology*, 2010, 100: 313-318.

[10] Krattinger S G, Lagudah E S, Spielmeyer W, Singh R P, Huerta-Espino J, Mcfadden H, Bossolini E, Selter L L, Keller B. A putative ABC transporter confers durable resistance to multiple fungal pathogens in wheat. *Science*, 2009, 323: 1360-1363.

[11] Bhavani S, Singh R P, Argillier O, Huerta-Espino J, Singh S, Njau P, Brun S, Lacam S, and Desmouceaux N. Mapping durable adult plant stripe rust resistance to the race Ug99 group in six CIMMYT wheats. BGRI 2011 Technical Workshop, St. Paul, Minnesota, 2011. pp: 44-53.

[12] Singh R P, Huerta-Espino J, Rajaram S. Achieving near-immunity to leaf and stripe rusts in wheat by combining slow rusting resistance genes. *Acta Phytopathol Entomol Hungarica*, 2000, 35: 133-139.

[13] Castro A J, Chen X M, Hayes P M, Johnston M. Pyramiding quantitative trait locus (QTL) alleles determining resistance to barley stripe rust: effects on resistance at the seedling stage. *Crop Sci*, 2003, 43: 651-659.

[14] Marasas C N, Smale M, Singh R P. The impact of agricultural maintenance research: The case of leaf rust resistance breeding in CIMMYT-related spring bread wheat//CD-ROM Proceeding Internal Congress on Impacts of Agricultural Research and Development. San Jose, Costa Rica, 2002.

[15] Singh R P, William H M, Huerta-Espino J, Rosewarne G. Wheat rust in Asia: Meeting the challenges with old and new technologies//New Directions for a Diverse Planet. Proceedings of the 4th International Crop Science Congress. Brisbane, Australia, 2004, 26.

[16] Dekkers J C M, Hospital F. The use of molecular genetics in the improvement of agricultural populations. *Nat Rev Genet*, 2003, 3: 22-32.

[17] Miedaner T, Wilde F, Steiner B, Buerstmayr H, Korzun V, Ebmeyer E. Stacking quantitative trait loci (QTL) for Fusarium head blight resistance from non-adapted sources in an European elite spring wheat background and assessing their effects on deoxynivalenol (DON) content and disease severity. *Theor Appl Genet*, 2006, 112: 562-569.

[18] Lu Q X, Szabo-Hever A, Bjørnstad Åsmund, Lillemo M, Semagn K, Mesterhazy A, JiF, Shi J R, Skinnes H. Two major resistance quantitative trait loci are required to counteract the increased susceptibility to *Fusarium* head blight of the *Rht-D1b* dwarfing gene in wheat. *Crop Sci*, 2011, 51: 2430-2438.

[19] Wang Z L, Li L H, He Z H, Duan X Y, Zhou Y L, Chen X M, Lillemo M, Singh R P, Wang H, Xia X C. Seeding and adult-plant resistance to powdery mildew in Chinese bread wheat cultivars and lines. *Plant Dis*, 2005, 89: 457-463.

[20] 任妍. 普通小麦抗条锈病基因分子定位. 中国农业科学院博士论文, 2012. pp. 56-63.
Ren Y. Molecular Mapping of Stripe Rust Resistance Genes in Common Wheat. PhD Dissertation of Chinese Academy of Agricultural Sciences, Beijing, China, 2012. pp 56-63. (in Chinese with English abstract)

[21] Lan C X, Liang S S, Wang Z L, Yan J, Zhang Y, Xia X C, He Z H. Quantitative trait loci mapping for adult-plant resistance to powdery mildew in Chinese wheat cultivar Bainong 64. *Phytopathology*, 2009, 99: 1121-1126.

[22] Lan C X, Ni X W, Yan J, Zhang Y, Xia X C, Chen X M, He Z H. Quantitative trait loci mapping of adult-plant resistance to powdery mildew in Chinese wheat cultivar Lumai 21. *Mol Breeding*, 2010, 25: 615-622.

[23] Bai B, He Z H, Asad M A, Lan C X, Zhang Y, Xia X C, Yan J, Chen X M, Wang C S. Pyramiding adult-plant powdery mildew resistance QTLs in bread wheat. *Crop Pasture Sci*, 2011, 63: 606-611.

[24] Lin F and Chen X M. Quantitative trait loci for non-race-specific, high-temperature adult-plant resistance to stripe rust in wheat cultivar Express. *Theor Appl Genet*, 2009, 118: 631-642.

[25] Ren Y, Li Z F, He Z H, Wu L, Bai B, Lan C X, Wang C F, Zhou G, Zhu H Z, Xia X C. QTL mapping of adult-plant resistance to stripe rust and leaf rust in Chinese wheat cultivar Bainong 64. *Theor Appl Genet*, 2012, 125: 1253-1262.

[26] Lillemo M, Asalf B, Singh R P, Huerta-Espino J, Chen X M, He Z H, Bjørnstad Å. The adult plant rust resistance loci *Lr34/Yr18* and *Lr46/Yr29* are important determinants of partial resistance to powdery mildew in bread wheat line Saar. *Theor Appl Genet*, 2008, 116: 1155-1166.

[27] Singh R P. Genetic association of leaf rust resistance gene *Lr34* with adult plant resistance to stripe rust in bread wheat. *Phytopathology*, 1992, 82: 835-838.

[28] Dyck P L, Kerber E R, Aung T. An interchromosomal reciprocal translocation in wheat involving leaf rust resistance gene *Lr34*. *Genome*, 1994, 37: 556-559.

[29] Herrera-Foessel S A, Lagudah E S, Huerta-Espino J, Hayden M, Bariana H S, Singh R P. New slow-rusting leaf rust and stripe rust resistance gene *Lr67* and *Yr46* in wheat are pleiotropic or closely linked. *Theor Appl Genet*, 2011, 122: 239-249.

[30] 何中虎，兰彩霞，陈新民，邹裕春，庄巧生，夏先春．小麦条锈病和白粉病成株抗性研究进展和展望．中国农业科学，2011，44：2193-2215.
He Z H, Lan C X, Chen X M, Zou Y C, Zhuang Q S, Xia X C. Progress and perspective in research of adult-plant resistance to stripe rust and powdery mildew in wheat. *Sci Agric Sin*, 2011, 44: 2193-2215. (in Chinese with English abstract)

[31] Marasas C N, Smale M, Singh R P. The economic impact of productivity maintenance research: breeding for leaf rust resistance in modern wheat. *Agric Eco*, 2003, 29: 253-263.

[32] Chen X M, Line R F. Gene action in wheat cultivars for durable high-temperature adult-plant resistance and interactions with race-specific, seedling resistance to stripe rust caused by *Puccinia striiformis*. *Phytopathology*, 1995, 85: 567-572.

[33] Bariana H S, Kailasapillai S, Brown G N, Sharp P J. Marker assisted identification of *Sr2* in the National Cereal Rust Control Program in Australia, In: Slinkard A E ed. Proc 9th Intl Wheat Genet Symp. Vol. 5. University of Saskatchewan, Saskatoon, SK, Canada: Univ. Extension Press, 1998. pp. 83-91.

[34] Keller M, Keller B, Schachermayr G, Winzeler M, Schmid J E, Stamp P, Messmer M M. Quantitative trait loci for resistance against powdery mildew in a segregating wheat × spelt population. *Theor Appl Genet*, 1999, 98: 903-912.

兼抗型成株抗性小麦品系的培育、鉴定与分子检测

刘金栋[1]，杨恩年[2]，肖永贵[1]，陈新民[1]，伍玲[2]，
白斌[3]，李在峰[4]，G. M. Rosewarne[2,5]，夏先春[1]，何中虎[1,5]

[1]中国农业科学院作物科学研究所/国家小麦改良中心，北京 100081；
[2]四川省农业科学院作物研究所，成都 610066；[3]甘肃省农业科学院
小麦研究所，兰州 730070；[4]河北农业大学植物保护学院，保定 071001；
[5]CIMMYT 中国办事处，北京 100081

摘要：小麦条锈病、叶锈病和白粉病是我国小麦的重要真菌病害，培育兼抗型成株抗性品种是控制病害最为经济有效和持久安全的方法。本研究选用由成株抗性育种方法培育的 21 份冬小麦高代品系和 96 份春小麦高代品系，在多个环境下进行这 3 种病害的成株期抗性鉴定，并利用紧密连锁的分子标记检测了兼抗型基因 *Lr34*/*Yr18*/*Pm38*、*Lr46*/*Yr29*/*Pm39* 和 *Sr2*/*Yr30* 的分布。田间鉴定表明，21 份冬小麦品系中有 17 份兼抗 3 种病害，占 80.9%；96 份春小麦品系中有 85 份兼抗 3 种病害，占 88.5%。分子标记检测发现，21 份冬小麦品系均含 *QPm caas-4DL*，其中 7 份还含 *QPm caas-2BS*，9 份还含 *QPm caas-2BL*；96 份春小麦品系中，18 份含 *Lr34*/*Yr18*/*Pm38*，37 份含 *Lr46*/*Yr29*/*Pm39*，29 份含 *Sr2*/*Yr30*。以上结果表明，分子标记与常规育种相结合，可有效培育兼抗型成株抗性品种，为我国小麦抗病育种提供了新思路。

关键词：普通小麦；兼抗型；成株抗性；条锈病；叶锈病；白粉病

Development, Field and Molecular Characterization of Advanced Lines with Pleiotropic Adult-Plant Resistance in Common Wheat

Liu Jindong[1], Yang Ennian[2], Xiao Yonggui[1], Chen Xinmin[1],
Wu Ling[2], Bai Bin[3], Li Zaifeng[4], Xa Xianchun[1], Garry M. Rosewarne[2,5], He Zhonghu[1,5]

[1]*Institute of Crop Science / National Wheat Improvement Center, Chinese Academy of Agricultural Sciences (CAAS), Beijing* 100081, *China*; [2]*Crop Research Institute, Sichuan Academy of Agricultural Sciences, Chengdu* 610066, *China*; [3]*Wheat Research Institute, Gansu Academy of Agricultural Sciences, Lanzhou* 730070, *China*; [4]*College of Plant Protection, Agricultural University of Hebei, Baoding* 071001, *China*; [5]*CIMMYT-China Office, c/o CAAS, Beijing* 100081, *China*

原文发表在《作物学报》，2015，41（10）：1472－1480

Abstract: Stripe rust, leaf rust, and powdery mildew are devastating fungal diseases of common wheat (*Triticum aestivum* L.) in China, and cultivars with pleiotropic adult-plant resistance are believed to be the most important solution to control these diseases effectively and environmental friendly. A total of 21 winter wheat advanced lines and 96 spring wheat advanced lines collected from adult-plant resistance breeding programs were used to estimate the level of resistance against the stripe rust, leaf rust and powdery mildew across several environments. Simultaneously, the distribution of pleiotropic resistance genes *Lr34/Yr18/Pm38*, *Lr46/Yr29/Pm39*, and *Sr2/Yr30* were also detected using molecular marker closely linked to the target genes. The field test showed that 17 winter wheat lines (80.9%) and 85 spring wheat lines (88.5%) performed acceptable resistance against the three diseases. All the 21 winter wheat lines tested contain *QPm. caas-4DL*, of which seven contain *QPm. caas-2BS* and nine contain *QPm. caas-2BL*. Among the 96 spring wheat lines, 18 carry *Lr34/Yr18/Pm38*, 37 carry *Lr46/Yr29/Pm39*, and 29 lines possess *Sr2/Yr30*. These results indicate that molecular-marker-assistant selection in combination with conventional breeding is effective and applicable in developing pleiotropic adult-plant resistance cultivars, which provides a new thought for wheat resistance breeding.

Key words: *Triticum aestivum* L.; Pleiotropic resistance; Adult-plant resistance; Stripe rust; Leaf rust; Powdery mildew

小麦条锈病、叶锈病和白粉病分别由 *Puccinia striiformis* f. sp. *tritici*、*P. triticina* f. sp. *Tritici* 和 *Blumeria graminis* f. sp. *tritici* 引起，具有流行范围广、发生频率高、暴发性强的特点[1,2]，严重威胁小麦生产。我国是世界上最大的小麦条锈病流行区域，主要分布在西北和西南地区，包括陕西、甘肃、河南南部、湖北、四川和云南等地，年发生面积约420万 hm^2[3]。20世纪80年代以来，随着矮秆品种的推广、水肥条件的改善和1B/1R易位系抗性的丧失，白粉病已逐渐成为我国冬麦区的主要病害，且常年发生，年发生面积约680万 hm^2[3]。近年来，由于全球气候变暖等原因，叶锈病在我国有加重的趋势，流行区域已从西南、西北和长江流域的部分地区扩展到华北和东北地区[3,4]。因此，培育兼抗3种病害的品种已成为我国小麦育种的重要目标。

寄主抗性可分为苗期抗性和成株期抗性。苗期抗性又称主效基因抗性或全生育期抗性，由一个或少数主效基因控制，对病害表现出高抗或免疫，对病原菌生理小种选择压力较大，易引起生理小种变异，过度应用苗期抗性容易造成品种抗性频繁丧失。成株期抗性也称部分抗性或慢病性，苗期表现感病，成株期表现抗病，由多个微效基因控制，对病原菌小种专化性较弱，可减缓病原菌生理小种的变异，抗性持久且稳定[5]。我国及世界多数国家都以苗期主效基因抗性利用为主，培育高抗甚至免疫的品种，导致品种抗性频繁丧失，抗病育种始终处于被动状态。早期研究认为，成株抗性基因作用机制复杂，这在一定程度上阻碍了成株抗性基因的应用。2005年，国际玉米小麦改良中心（CIMMYT）将4～5个成株抗性基因聚合在同一小麦品种中，从而培育出高抗且具持久抗性的成株抗性材料[6]。近年来，培育成株抗性品种已成为国际抗病育种的主要方向。

迄今，国际已正式命名的小麦条锈病抗性基因有70个，其中 *Yr18*、*Yr29*、*Yr30*、*Yr36*、*Yr46*、*Yr54*、*Yr59* 和 *Yr62* 为成株抗性基因[7,8]；正式命名的小麦叶锈病抗性基因有72个，其中 *Lr12*、*Lr13*、*Lr22a*、*Lr22b*、*Lr34*、*Lr35*、*Lr37*、*Lr46*、*Lr48*、*Lr49*、*Lr67* 和 *Lr68* 为成株抗性基因[8,9]；正式命名的小麦白粉病抗性基因有47个，其中 *Pm38*、*Pm39* 和 *Pm46* 为成株抗性基因[8,9]。Lillemo 等[10]将位于7DS上的成株抗性基因 *Lr34*、*Yr18* 和 *Pm38* 定义为同一个兼抗型基因 *Lr34/Yr18/Pm38*。现已开发了 *Lr34/Yr18/Pm38* 的共显性 STS 标记 *csLv34*[11]，广泛应用于分子标记辅助育种[12]，并最终克隆了该基因[13]。Lillemo 等[10]还将1BL染色体上的 *Lr46*、*Yr29* 和 *Pm39* 定义为同一个兼抗型基因 *Lr46/Yr29/Pm39*；随后开发了其共显性 CAPS 标记 *csLV46G2*，并逐步应用于基因检测及分子标记辅助育种（Lagudah，私人交流）。*Lr34/Yr18/Pm38* 和 *Lr46/Yr29/Pm39* 除对上述三种病害表现成株抗性外，还对秆锈病具有一定的成株抗性，且常伴随叶尖坏死现象，可作为田间选择的形态标记[10,14]。Singh[15]发现 *Lr34/*

Yr18/Pm38 与成株期耐大麦黄矮病基因 *Bdv1* 有高度相关性，7DS 染色体区域还存在一个控制小麦斑枯病的 *Sb1* 基因[16]，它们可能与 *Lr34/Yr18/Pm38* 是同一基因。兼抗型基因 *Sr2/Yr30* 位于 3BS 染色体上[17]，其 CAPS 标记 *csSr2* [18]已得到广泛应用。Herrera-Foessel 等[19]在小麦品系 RL6077 中检测到条锈病、叶锈病和白粉病的兼抗基因 *Lr67/Yr46/Pm46* [20]；Forrest 等[21]借助 KASP 技术开发了其 SNP 标记 *csSNP856*，但由于稳定性差、受群体遗传背景影响较大等原因，不适用于检测 *Lr67/Yr46/Pm46* 基因。

类似 *Lr34/Yr18/Pm38*、*Lr46/Yr29/Pm39*、*Sr2/Yr30* 和 *Lr67/Yr46/Pm46* 的兼抗型基因的抗性持久稳定，因其具有一因多效性而倍受育种家重视，发掘和利用兼抗型成株抗性基因，并将其用于新品种培育已成为国际发展趋势。自 20 世纪 60 年代以来，CIMMYT 就开始了成株抗性育种工作，并成功选育出一批兼抗型品种，如 Chapio、Cook、Pavon 76、Sonoita 81 和 Tukuru，在发展中国家大面积推广[22]。为了从根本上解决我国西南麦区抗病品种抗性频繁丧失的突出问题，2000 年四川省农业科学院与 CIMMYT合作开展了条锈病成株抗性育种。中国农业科学院作物科学研究所则以白粉病为主攻方向，采用成株抗性策略，在 QTL 定位的基础上，培育成株抗性品种，为抗白粉病育种提供优良材料。本研究选用通过兼抗型成株抗性育种方法培育的冬小麦和春小麦高代品系进行田间成株期条锈病、叶锈病和白粉病抗性鉴定，并利用分子标记检测兼抗型基因 *Lr34/Yr18/Pm38*、*Lr46/Yr29/Pm39* 和 *Sr2/Yr30* 的分布，为我国小麦抗病育种提供可行的兼抗型成株抗性育种方法和优良抗病材料。

1 材料与方法

1.1 试验材料

试验材料包括 27 份冬小麦和 97 份春小麦。冬小麦包括 21 份来自百农 64/鲁麦 21 的高代品系及其亲本，兼抗型冬小麦品种 Strampelli 和平原 50，以及用分子标记辅助育种育成的高代品系 SD11P421 和 SD11P423。在对百农 64 和鲁麦 21 抗性 QTL 定位的基础上[23,24]，将二者杂交，进行白粉病成株抗性基因聚合，在 F_2 和 F_3 代淘汰成株期高感病株系，选择中感至抗病株系继续种植，F_4 和 F_5 代选择抗病株系，最终育成 21 份含有 1～3 个兼抗型成株抗性 QTL 的 F_6 抗病品系[25]，分别命名为 BFB5-BFB25。SD11P421 和 SD11P423 是平原 50 和烟农 19 杂交并通过回交，培育的兼抗型成株抗性品系。

春小麦包括 96 份高代品系和品种 Wheatear。2006 年四川农业科学院用四川品种与 CIMMYT 已鉴定的兼抗型成株抗性亲本杂交并回交，采用混合选择法（bulked selection），将抗性基因导入农艺性状优良、广适性的四川小麦品种，于 2012 年育成 96 份稳定的春小麦高代品系。

1.2 田间抗病性鉴定

27 份冬小麦于 2012—2013 和 2013—2014 年度种植于四川郫县和甘肃天水进行抗条锈病鉴定，以辉县红做感病对照；2012—2013 和 2013—2014 年度种植于河北保定和河南周口用于抗叶锈病鉴定，以郑州 5389 做感病对照；2013—2014 年度种植于北京和贵州贵阳进行抗白粉病鉴定，以京双 16 做感病对照。96 份春小麦于 2013—2014 年度种植于四川郫县和甘肃天水进行抗条锈病鉴定，在河北保定及河南周口进行抗叶锈病鉴定，在北京及贵州贵阳进行抗白粉病鉴定，采用与冬小麦相同的对照品种。各试验环境采用相同的田间设计，均为单行区，3 次重复，行长 1.5m,行距 0.2m，每行 50 粒。田间管理按当地常规。

采用国内外已报道的方法分别进行条锈菌[23,26]、叶锈菌[27]和白粉菌[25]田间接种。其中，条锈病菌采用混合小种，郫县点包括 CYR29、CYR31 和 CYR32，天水点包括 CYR29、CYR31、CYR32 和 CYR33；叶锈病菌两试点均为强毒性生理小种 THTT；白粉病菌两试点均为强毒性菌株 E20。

当对照品种发病完全后开始调查严重度，分别记录旗叶和倒二叶上病菌孢子堆面积占总叶面积的百分数，以最大病害严重度（maximum disease severity, MDS）作为抗病性评价指标[28]。采用 SAS 9.2 计算基本统计量，进行方差分析。

1.3 兼抗基因的分子标记检测

由于 *Lr67/Yr46/Pm46* 无合适的功能标记，仅对兼抗基因 *Lr34/Yr18/Pm38*、*Lr46/Yr29/Pm39* 和 *Sr2/Yr30* 进行检测。在小麦苗期，采用 CTAB 法按系混提基因组 DNA，终浓度 30ng μl^{-1}左右。

根据 Lagudah 等[11]发表的序列合成 *Lr34/Yr18/*

Pm38 共显性STS标记*csLv34*的引物（*csLV34*F：5′-GTTGGTTAAGACTGGTGATGG - 3′；*csLV34*R：5′- TGCTTGCTATTGCTGAATAGT - 3′），预期扩增产物为150bp（含目的基因）和229bp（不含目的基因）。PCR反应体系共20μl，包括2μl 10×buffer，1U *Taq* DNA聚合酶（2.5U μl^{-1}，北京天根公司），0.4μl dNTPs（各10mmol L^{-1}），0.5μl引物（4μmol L^{-1}），50ng模板DNA。PCR反应程序为：94℃变性1min，57℃退火1min，72℃延伸2min，5个循环；94℃变性30s，57℃退火30s，72℃延伸1min，30个循环；最后94℃变性1min，57℃退火30s，72℃延伸5min。

根据Mago等[18]报道的序列信息合成*Sr2/Yr30*共显性CAPS标记*csSr2*（*csSr2*F：5′- CAAGGGTT-GCTAGGATTGGAAAAC - 3′；*csSr2*R：5′- AGATA-ACTCTTATGATCTTACATTTTTCTG -3′）[18]，含有*Sr2/Yr30*的纯合株系扩增产物经*Bsp*HI酶切呈现3条带（172+112+53bp），而不含该基因的纯合株系扩增后酶切产物为2条带（225+112bp）。PCR反应体系20μl，包括2μl 10×buffer，0.5U *Taq* DNA聚合酶（2.5U μl^{-1}），0.2μl dNTPs（各10mmol L^{-1}），0.5μl引物（4μmol L^{-1}），100ng模板DNA。PCR反应程序为：95℃预变性2min；95℃变性30s，60℃退火40s，72℃延伸50s，30个循环；最后72℃延伸5min。

以CAPS标记*csLv46G22*检测兼抗基因*Lr46/Yr29/Pm39*（Lagudah，私人交流）。

所有引物均由北京奥科公司合成。扩增产物在2.0%琼脂糖凝胶中电泳检测，缓冲液体系为1×TAE溶液，150V电压电泳30min，溴化乙锭染色。

2 结果与分析

2.1 成株期3种病害的抗性鉴定

方差分析表明，冬小麦和春小麦基因型间、年度间和重复间MDS差异显著，基因型与环境互作显著，但明显小于基因型间的方差（表1）。27份冬小麦材料的条锈病、叶锈病和白粉病MDS平均值分别为35.5、20.2和10.6，变异范围分别为3～58、3～36和4～40（部分数据见表2）；97份春小麦3种病害的MDS平均值分别为10.7、17.6和13.4，变异范围分别为2～33、0～70和1～82（部分数据见表3）；3种病害的对照品种辉县红、郑州5389和京双16，其MDS值在冬小麦组分别为70～100、90～100和75～100，在春小麦组分别为75～100、90～100和80～100，说明田间发病充分，鉴定结果可靠。

表1 高代品系及品种条锈病、叶锈病和白粉病MDS方差分析

Table 1 Analysis of variance of stripe rust, leaf rust, and powdery mildew in MDS in wheat advanced lines and cultivars

变异来源 Source of variance	条锈病 Stripe rust			叶锈病 Leaf rust			白粉病 Powdery mildew		
	df	*MS*	*F*	*df*	*MS*	*F*	*df*	*MS*	*F*
冬小麦 Winter wheat									
重复 Replicate	2	2016.7	20.7**	2	1 506.8	23.1*	2	895.6	15.3*
环境 Environment (E)	3	29 305.	242.0**	3	19 856.3	192.8**	1	13 526.8	156.1*
基因型 Genotype (G)	26	1 856.9	19.8**	26	1 342.5	18.5**	26	1 246.3	17.6**
基因型×环境 G×E	78	282.6	2.1**	78	242.5	1.9**	26	125.3	1.0**
误差 Error	214	125.6		214	76.6		106	102.3	
春小麦 Spring wheat									
重复 Replicate	2	1 989.3	26.6*	2	1 756.3	28.5*	2	1 328.5	25.3*
环境 Environment (E)	1	15 796.3	123.1**	1	13 258.3	118.2**	1	14 263.2	91.5*
基因型 Genotype (G)	96	1 564.3	15.3**	96	1 428.5	20.6**	96	1 250.1	10.6**
基因型×环境 G×E	96	369.5	2.6**	96	258.6	1.9*	96	258.3	1.3*
误差 Error	386	189.3		386	152.3		386	166.6	

*和**分别表示在0.05和0.01概率水平显著。

* and** indicate significance at 0.05 and 0.01 probability levels, respectively.

在冬小麦21份BFB育种品系中，BFB10、BFB14和BFB19等17份兼抗3种病害，占81.0%，BFB9、BFB12、BFB20和BFB24兼抗叶锈病和白粉病，但不抗条锈病，占19.0%；此外，SD11P421和SD11P423对叶锈病有较好的抗性，白粉病抗性稍差，中感条锈病。在冬小麦育成品种中，Strampelli兼抗3种病害；百农64和鲁麦21对叶锈病和白粉病抗性较好，但对条锈病抗性略差；平原50对条锈病和叶锈病有较好抗性，但白粉病抗性较差（表2）。

表2 冬小麦主要品系条锈病（SR）、叶锈病（LR）和白粉病（PM）的MDS值及兼抗型成株抗性基因

Table 2 Maximum disease severity (MDS) values of major winter wheat lines in response to stripe rust (SR), leaf rust (LR), and powdery mildew (PM) and the presence of pleiotropic resistance genes

品系 Line	来源 Origin	系谱 Pedigree	QTL	MDS			*Lr34/Yr18/Pm38*	*Lr46/Yr29/Pm39*	*Sr2/Yr30*
				SR	LR	PM			
Strampelli	意大利 Italy	Libero//SanPastore-14/Jacometti-49		3	11	23	+	–	–
平原50 Pingyuan 50	中国河南 Henan, China	不详 Unknown		5	20	37	–	+	–
百农64 Bainong 64	中国河南 Henan, China	百农8717/3/偃大2-629-52/石82-5594//百农84-4046-1 Bainong8717/3/Yanda 2-629-52/Shi 82-5594//Bainong 84-4046-1	4DL	32	18	12	–	–	–
鲁麦21 Lumai 21	中国山东 Shandong, China	鲁麦17/豫麦54 Lumai 17/Yumai 54	2BS/2BL	48	8	5	–	–	–
BFB9	中国北京 Beijing, China	鲁麦21/百农64 Lumai 21/Bainong 64	4DL/2BS/2BL	47	3	8	–	–	–
BFB10	中国北京 Beijing, China	鲁麦21/百农64 Lumai 21/Bainong 64	4DL/2BS/2BL	23	12	5	–	–	–
BFB12	中国北京 Beijing, China	鲁麦21/百农64 Lumai 21/Bainong 64	4DL/2BS	46	33	5	–	–	–
BFB13	中国北京 Beijing, China	鲁麦21/百农64 Lumai 21/Bainong 64	4DL/2BS	32	17	4	–	–	–
BFB14	中国北京 Beijing, China	鲁麦21/百农64 Lumai 21/Bainong 64	4DL/2BS	25	10	10	–	–	–
BFB15	中国北京 Beijing, China	鲁麦21/百农64 Lumai 21/Bainong 64	4DL	34	29	4	–	–	–
BFB16	中国北京 Beijing, China	鲁麦21/百农64 Lumai 21/Bainong 64	4DL	20	20	4	–	–	–
BFB19	中国北京 Beijing, China	鲁麦21/百农64 Lumai 21/Bainong 64	4DL/2BL	16	21	4	–	–	–
BFB20	中国北京 Beijing, China	鲁麦21/百农64 Lumai 21/Bainong 64	4DL/2BL	50	26	7	–	–	–
BFB24	中国北京 Beijing, China	鲁麦21/百农64 Lumai 21/Bainong 64	4DL	43	36	6	–	–	+
BFB25	中国北京 Beijing, China	鲁麦21/百农64 Lumai 21/Bainong 64	4DL/2BL	40	30	4	–	–	–
SD11P421	中国山东 Shandong, China	平原50/烟农19*3 Pingyuan 50/Yannong 19*3	4DL	58	26	40	–	–	–
SD11P423	中国山东 Shandong, China	平原50/烟农19*3 Pingyuan 50/Yannong 19*3	4DL	52	18	38	–	–	–

QTL指兼抗型QTL，由Lan等[23,24]鉴定，其中*QPm.caas-4DL*缩写为4DL，余此类推。MDS数据分别为4个（SR）、4个（LR）和2个（PM）环境的平均值，抗病性等级分为高抗（MDS<20）、中抗（20≤MDS<50）、中感（50≤MDS<70）和高感（MDS≥70）。+和–分别表示兼抗基因存在和不存在。

QTL stands for pleiotropic resistance QTL, which were mapped by Lan et al.[23, 24]. QTL 4DL is an abbreviation for *QPm.caas-4DL*, and the rest may be inferred by analogy. The MDS values were the averages over four, four, and two environments for SR, LR, and PM diseases, respectively. Resistance were graded with high resistance (MDS<20), moderate resistance (20≤MDS<50); moderate susceptibility (50≤MDS<70), and high susceptibility (MDS≥70). Symbols "+" and "–" indicate the presence and absence of the pleiotropic resistance gene, respectively.

在96份春小麦品系中，13EW331、13EW359和13EW419等85份兼抗条锈病、叶锈病和白粉病，占88.5%；13AYT45、13EW269和13EW271等6份兼抗条锈病和叶锈病，但不抗白粉病，占6.3%；13EW306、13EW348和13EW435等5份兼抗条锈病和白粉病，但不抗叶锈病，占5.2%（表3）。Wheatear也兼抗3种病害。

表3 主要春小麦品系条锈病（SR）、叶锈病（LR）和白粉病（PM）的最大病害严重度（MDS）及兼抗型成株抗性基因

Table 3 Maximum disease severity（MDS）values of major spring wheat lines in response to stripe rust（SR），leaf rust（LR），and powdery mildew（PM）and the presence of pleiotropic resistance genes

品系 Line	系谱 Pedigree	MDS			*Lr34/Yr18/Pm38*	*Lr46/Yr29/Pm39*	*Sr2/Yr30*
		SR	LR	PM			
13AYT3	1522 * 2/Sunco	8	9	18	+	h	—
13AYT37	Bv98s-3849/G. C. W. /Seri/Mian 20/Chuanmai 56	3	16	11	—	—	—
13AYT4	1522 * 2/Sunco	7	5	17	+	+	—
13AYT45	Ravi1546/Ravi1043	8	37	58	—	+	—
13EW264	Sw05Rc470/3570 * 2	21	27	82	—	—	—
13EW269	Singh/1231 * 2	12	13	47	—	+	—
13EW271	Singh/1231 * 2	13	16	56	—	h	—
13EW274	Bv98s-3849/G. C. W. /Seri/Mian 20/Chuanmai 56	11	18	4	—	—	—
13EW293	1522 * 2/Sunco	7	5	11	+	+	—
13EW294	1522 * 2/Sunco	10	5	10	—	+	—
13EW306	Ms Song * 2/Nl682//2 * 1522	6	48	8	—	—	—
13EW329	07GH173/1522	9	0	2	+	+	—
13EW331	07GH304/Chuanmai 56	11	17	11	h	h	—
13EW332	07GH304/Chuanmai 56	8	13	31	h	+	—
13EW341	Chuanmai42 * 2/3/Pfau/Weaver * 2//Transfer #12，P88. 272. 2	13	19	2	+	—	—
13EW346	3570 * 2/3/Kiritati//Prl/Pastor * 2	13	5	6	+	+	—
13EW348	3570 * 2/3/Kiritati//Prl/Pastor * 2	5	45	14	—	+	—
13EW350	Chuanmai42 * 2/3/Pbw343/Wbll1//Pandion	12	32	6	—	—	+
13EW353	Chuanmai47 * 2/4/Thelin/3/Babax * 2/*Lr42*//Babax	8	33	11	—	—	+
13EW356	Chuanmai47 * 2//Toba97/Pastor	10	7	11	—	—	+
13EW359	3570 * 2/3/Kiritati//Attila * 2/Pastor	6	3	1	+	+	+
13EW361	Yunmai47 * 2/4/Babax/Lr42//Babax * 2/3/Tukuru	23	15	20	—	—	—
13EW362	Sw1231 * 2/3/Kiritati//Prl/Pastor * 2	18	8	2	+	—	+
13EW365	3570 * 2/3/Gan/*Ae. squarrosa*（408）//Oasis * 2/Borl95 * 2	14	20	8	+	+	—
13EW367	3570 * 2/3/Gan/*Ae. squarrosa*（408）// Oasis * 2/Borl95 * 2	9	8	7	+	+	—
13EW374	Chuanmai42 * 2/3/Pbw343/Wbll1//Pandion	9	26	11	—	—	+
13EW376	Rc5193-12	9	8	7	—	—	—
13EW383	Rc4981-1	8	6	2	—	—	—
13EW403	Chumai47 * 2//Toba97/Pastor	33	23	25	—	+	+
13EW419	Chuanmai47 * 2//Waxwing * 2/Vivitsi	3	16	13	—	h	—
13EW435	Chuanmai47 * 2//Waxwing * 2/Vivitsi	2	70	10	—	+	+
Wheatear		13	23	1	—	—	+

三种病害的MDS数据均为两个环境的平均值，抗病性等级分为高抗（MDS<20）、中抗（20≤MDS<50）、中感（50≤MDS<70）和高感（MDS≥70）。+和一分别表示兼抗基因存在和不存在；h表示杂合基因型。

The MDS values werethe averages over two environments for the three diseases. Resistance were graded with high resistance（MDS<20），moderate resistance（20≤MDS<50）；moderate susceptibility（50≤MDS<70），and high susceptibility（MDS≥70）. Symbols "+" and "–" indicate the presence and absence of the pleiotropic resistance gene，respectively，whereas the letter "h" indicates heterozygous genotype.

2.2 兼抗型基因分子标记检测

用分子标记检测 27 份冬小麦材料，在 Strampelli 中检测到 *Lr34/Yr18/Pm38*，在平原 50 中检测到 *Lr46/Yr29/Pm39*（表 2），而在其他冬小麦中未检测到兼抗型基因，这与前人的研究结果[23,24,30,31]基本一致。Bai 等[25]利用分子标记检测 21 份冬小麦 BFB 品系，发现所有 BFB 品系均含 *QPm. caas-4DL*，BFB9、BFB12 和 BFB14 等 7 份还含 *QPm. caas-2BS*，BFB9、BFB19 和 BFB20 等 9 份还含 *QPm. caas-2BL*。

在本研究的 96 份春小麦品系中，13EW329、13EW359 和 13EW362 等 18 份含 *Lr34/Yr18/Pm38*，13EW331 和 13EW332 含 *Lr34/Yr18/Pm38* 杂合基因型，13AYT45、13EW269 和 13EW332 等 37 份含 *Lr46/Yr29/Pm39*，13EW271、13EW331 和 13EW419 等 6 份含 *Lr46/Yr29/Pm39* 杂合基因型，13EW252、13EW300 和 13EW343 等 29 份含 *Sr2/Yr30*，13EW356、13EW335 和 13EW435 等 25 份含两个兼抗型基因，13EW359 含三个兼抗型基因（表 3）。其余 33 份未检测到目标兼抗基因。

3 讨论

3.1 抗性基因来源分析

在兼抗型成株抗性育种中，选择含有成株抗性基因的亲本进行杂交至关重要。若能明确其抗性基因的分布与效应，则可显著提高育种效率。百农 64 和鲁麦 21 是我国黄淮麦区主推品种，农艺性状优良，来自百农 64 的 *QPm. caas-4DL* 和鲁麦 21 的 *QPm. caas-2BS*、*QPm. caas-2BL* 兼抗白粉病、条锈病和叶锈病[23,24,29]。由百农 64 衍生出来的冬小麦品系 BFB5 至 BFB25 在田间兼抗条锈病、叶锈病和白粉病，均含有 1～3 个兼抗型 QTL[25]，是农艺性状优良的兼抗型成株抗性育种材料，其中 BFB10、BFB13、BFB14、BFB16 和 BFB19 等株系的抗性和农艺性状尤为突出，应加大其在育种中的应用。平原 50 携带 *QPm. caas-2BS* 和 *QPm. caas-5AL*，兼抗条锈病和白粉病[30]，SD11P421 和 SD11P423 由平原 50 和烟农 19 回交转育而成，对三种病害均有一定抗性，但由于抗性基因间可能存在互作，SD11P421 和 SD11P423 对条锈病和白粉病抗性相对较差。Strampelli 从意大利引入我国后，50 多年来依然保持较好的抗性水平，主要因为其携带有 *Lr34/Yr18/Pm38* 等基因[31]，本试验中也表现高抗三种病害，但其农艺性状较 BFB 品系差，育种上使用相对困难。

春小麦品系的兼抗型基因 *Lr34/Yr18/Pm38* 主要来自亲本 Sunco[32]、Kiritati 和 Borlaug 95，*Lr46/Yr29/Pm39* 来自亲本 Sunco 和 Pastor[33]，*Sr2/Yr30* 来自亲本 Prl（Singh，私人交流）。13AYT3、13AYT 4、13EW293 和 13EW294 来自 1522 * 2/Sunco，其亲本之一 Sunco 为澳大利亚优质品种，携带兼抗型基因 *Lr34/Yr18/Pm38* [32]和 *Lr46/Yr29/Pm39*（Singh，私人交流）；13EW359 和 13EW362 来自不同组合，但均含有来自 Kiritati 的 *Lr34/Yr18/Pm38*；13EW365 和 13EW367 含有来自 Borlaug 95 的 *Lr34/Yr18/Pm38*。13EW346 和 13EW359 等品系来自不同组合，均含有来自 Pastor 的 *Lr46/Yr29/Pm39* [33]。13EW346 和 13EW362 等均含有来自 Prl 的 *Sr2/Yr30*。依据系谱推测，部分品系可能还含有其他抗性基因，例如 13AYT37 和 13EW274 可能含有来自 Seri 的兼抗型成株抗性基因，13EW350 和 13EW374 等可能含有来自 Pandion 的 *Yr17* 和 PBW343 的 *Yr9* 和 *Yr27*，13EW346 和 13EW359 等还可能含有来自 Pastor 的 *Yr31*，13EW341、13EW353 和 13EW361 可能含有来自 Quaiu 3 的 *Yr54*，13EW410 等可能含有来自 Fancolin 的 *YrF* [34]。此外，13EW419 等品系中检测到 *Lr46/Yr29/Pm39* 和 *Sr2/Yr30*，由于缺乏资料，目前还难以确定这些品系中兼抗型基因的来源。

成株抗性基因的加性效应起重要作用[35]，但我们发现部分品系的抗性效应并不一定与成株抗性基因的数量成正比，比如冬小麦品系 BFB15 和 BFB16 仅含有 *QPm. caas-4DL*，但其抗性高于含有多个兼抗 QTL 的 BFB9 和 BFB25。春小麦 13EW383 和 13EW376 品系中未检测到兼抗型基因，但其抗病性显著高于含有两个兼抗型基因的 13EW330 和 13EW435。究其原因，可能是目前已发现的兼抗型基因和基因簇中，只有 *Lr34/Yr18/Pm38*、*Lr46/Yr29/Pm39* 和 *Sr2/Yr30* 具有可以应用的分子标记，而 13EW383 和 13EW376 等品系中可能存在未检测或尚未发现的抗性基因，同时部分基因间可能存在互作效应，影响品系抗性。因此，在未来的抗病育种中，发掘新的兼抗基因并开发可用的分子标记十分重要。

3.2 兼抗型成株抗性育种方法

本研究表明，无论是在北京以白粉病为目标还是在四川以条锈病为目标的成株抗性育种皆取得了成功，由于对带有成株抗性亲本的杂交后代选择指标把握得当，即使群体处理方法不同，但都育成了兼抗三种病害的成株抗性品系，并且它们农艺性状优良，接近生产应用水平，建议作为育种亲本利用。国际上只有CIMMYT育成并推广了兼抗型成株抗性品种，本文为国内首次报道有目的的成株抗性育种工作。国内外经验说明兼抗型成株抗性育种方法已经成熟，可在更多单位推广。我国农家种中成株抗性基因分布频率较高，但从20世纪50年代开始，由于追求高抗、免疫，造成成株抗性基因丢失，导致现有品种以主效抗性基因为主，抗性频繁丧失。鉴于目前兼抗型成株抗性育种方法已经成熟，育种家应改变传统育种策略，从高抗免疫转向兼抗型成株抗性育种。*Lr34*/*Yr18*/*Pm38*、*Lr46*/*Yr29*/*Pm39*、*Sr2*/*Yr30* 和 *Lr67*/*Yr46*/*Pm46* 已被证实兼抗多种病害，并且前三个已有可应用于育种的分子标记，同时还在1B、2B、2D、3A、3B、4A、4B、5A、5B、6A、6B和7B染色体上发现了多个兼抗型基因簇[3,9]，为培育兼抗多种病害的成株抗性品种提供了抗性基因。

由于兼抗型成株抗性基因的效应相对较小，需采用与主效基因育种不同的方法，用当地高产感病品种与本文报道的兼抗型品系杂交（也可用农艺性状亲本回交或三交），主要是扩大分离世代群体规模，改变选择标准，在F_2、BC_1F_1或F_3代选择中感或中抗类型，淘汰含主效基因的高抗单株，可结合分子标记辅助选择技术提高选择效率，本课题组已报道过具体实施方案[3,7,36]。在高代品系稳定后，需在多个环境下进行田间表型鉴定，或者至少在高感病压力病圃中连续鉴定两年，并利用分子标记检测，以确认其遗传基础[3,7]。

我国兼抗型成株抗性育种研究工作起步较晚，与主效抗性基因相比，国内外发掘的成株抗性基因有限，在很大程度上制约了小麦抗病育种发展。近年来SNP技术发展迅速，小麦90K和630K芯片相继问世，其衍生的KASP技术逐步应用到实践中，使成株抗性基因的鉴定更为快速和准确。应充分利用SNP芯片技术的高通量和高密度优势，发掘与兼抗型基因连锁更为紧密的SNP标记，以提升筛选效率，在加大对 *Lr34*/*Yr18*/*Pm38*、*Lr46*/*Yr29*/*Pm39*、*Sr2*/*Yr30* 和 *Lr67*/*Yr46*/*Pm46* 等兼抗型基因利用的同时，加强对来自于CIMMYT等国外和我国已有优良抗性品种的抗性鉴定和基因发掘，并将其用于育种工作中。

4 结论

在多个环境下对由兼抗型成株抗性育种方法培育的高代品系进行田间成株期条锈病、叶锈病和白粉病抗性鉴定，并检测其含有的兼抗基因。在选育出的21份冬小麦品系和96份春小麦品系中，分别有17份和85份兼抗三种病害，且具有优良农艺性状，有望成为抗病育种中的重要材料。本研究为我国小麦兼抗型成株抗性育种提供了可行的育种方法和优良的育种材料。

参考文献

[1] Wellings C R, McIntosh R A, Hussain M. A new source of resistance to *Puccinia striiformis* f. sp. *tritici* in spring wheats (*Triticum aestivum*). *Plant Breed*, 1988, 100: 288-296.

[2] Conner R L, Kuzyk A D, Su H. Impact of powdery mildew on the yield of soft white spring wheat cultivars. *Can J Plant Sci*, 2003, 83: 725-728.

[3] 何中虎，兰彩霞，陈新民，邹裕春，庄巧生，夏先春．小麦条锈病和白粉病成株抗性研究进展和展望．中国农业科学，2011，44：2193-2215.
He Z H, Lan C X, Chen X M, Zou Y C, Zhuang Q S, Xia X C. Progress and perspective in research of adult-plant resistance to stripe rust and powdery mildew in wheat. *Sci Agric Sin*, 2011, 44: 2193-2215. (in Chinese with English abstract)

[4] Zhao X L, Zheng T C, Xia X C, He Z H, Liu D Q, Yang W X, Yin G H, Li Z F. Molecular mapping of leaf rust resistance gene *LrZH84* in Chinese wheat line Zhou 8425B. *Theor Appl Genet*, 2008, 117: 1069-1075.

[5] Gustafson G D, Shaner G. Influence of plant age on the expression of slow-mildewing resistance in wheat (*Triticum aestivum*). *Phytopathology*, 1982, 72: 746-749.

[6] Singh R P, Huerta-Espino J, Bhavani S, Herrera-Foessel S A, Singh D, Singh P K, Velu G, Mason R E, Jin Y, Njau P, Crossa J. Race non-specific resistance to rust diseases in CIMMYT spring wheats. *Euphytica*, 2011, 179: 175-186.

[7] Rosewarne G M, Herrera-Foessel S A, Singh R P, Huerta-Espino J, Lan C X, He Z H. Quantitative trait loci of stripe rust resistance in wheat. *Theor Appl*

Genet, 2013, 126: 2427-2449.

[8] McIntosh R A, Dubcovsky J, Rogers W J, Morris C F, Appels R, Xia X C. Catalogue of gene symbols for wheat: 2013-2014 supplement. 2014. http://www.shigen.nig.ac.jp/wheat/komugi/genes/macgene/supplement2013-2014.pdf

[9] Li Z F, Lan C X, He Z H, Singh R P, Rosewarne G M, Chen X M, Xia X C. Overview and application of QTL for adult plant resistance to leaf rust and powdery mildew in wheat. *Crop Sci*, 2014, 54: 1907-1925.

[10] Lillemo M, Asalf B, Singh R P, Huerta-Espino J, Chen X M, He Z H, Bjørnstad Å. The adult plant rust resistance loci *Lr34/Yr18* and *Lr46/Yr29* are important determinants of partial resistance to powdery mildew in bread wheat line Saar. *Theor Appl Genet*, 2008, 116: 1155-1166.

[11] Lagudah E S, McFadden H, Singh R P, Huerta-Espino J, Bariana H S, Spielmeyer W. Molecular genetic characterization of the *Lr34/Yr18* slow rusting resistance gene region in wheat. *Theor Appl Genet*, 2006, 114: 21-30.

[12] 梁丹，杨芳萍，何中虎，姚大年，夏先春．利用 STS 标记检测 CIMMYT 小麦品种（系）中 *Lr34/Yr18*，*Rht-B1b* 和 *Rht-D1b* 基因的分布．中国农业科学，2009，42：17-27.
Liang D, Yang F P, He Z H, Yao D N, Xia X C. Characterization of *Lr34/Yr18*, *Rht-B1b*, *Rht-D1b* genes in CIMMYT wheat cultivars and advanced lines using STS markers. *Sci Agric Sin*, 2009, 42: 17-27. (in Chinese with English abstract)

[13] Krattinger S G, Lagudah E S, Spielmeyer W, Singh R P, Huerta-Espino J, McFadden H, Bossolini E, Selter L L, Keller B. A putative ABC transporter confers durable resistance to multiple fungal pathogens in wheat. *Science*, 2009, 323: 1360-1363.

[14] Rosowarne G M, Singh R P, Huerta-Espino J, William H M, Bouchet S, Cloutier S, McFadden H, Lagudah E S. Leaf tip necrosis, molecular markers and β-proteasome subunits associated with the slow rusting resistance gene *Lr46/Yr29*. *Theor Appl Genet*, 2006, 112: 500-508.

[15] Singh R P. Genetic association of leaf rust resistance gene *Lr34* with adult-plant resistance to stripe rust in bread wheat. *Phytopathology*, 1992, 82: 835-838.

[16] Lillemo M, Joshi A K, Prasad R, Chand R, Singh R P. QTL for spot blotch resistance in bread wheat line Saar co-locate to the biotrophic disease resistance loci *Lr34* and *Lr46*. *Theor Appl Genet*, 2013, 126: 711-719.

[17] Skovmand B, Wilcoxson R D, Shearer B L, Stucker R E. Inheritance of slow rusting to stem rust in wheat. *Euphytica*, 1978, 27: 95-107.

[18] Mago R, Brown-Guedira G, Dreisigacker S, Breen J, Jin Y, Singh R, Appels R, Lagudah E S, Ellis J, Spielmeyer W. An accurate DNA marker assay for stem rust resistance gene *Sr2* in wheat. *Theor Appl Genet*, 2011, 122: 735-744.

[19] Herrera-Foessel S A, Lagudah E S, Huerta-Espino J, Hayden M, Bariana H S, Singh R P. New slow-rusting leaf rust and stripe rust resistance genes *Lr67* and *Yr46* in wheat are pleiotropic or closely linked. *Theor Appl Genet*, 2011, 122: 239-249.

[20] McIntosh R A, Dubcovsky J, Rogers W J, Morris C F, Appels R, Xia X C. Catalogue of gene symbols for wheat: 2012 supplement. 2012. http://www.shigen.nig.ac.jp/wheat/komugi/genes/macgene/supplement 2012.pdf

[21] Forrest K, Pujol V, Bulli P, Pumphrey M, Wellings C, Herrera-Foessel S, Huerta-Espino J, Singh R, Lagudah E, Hayden M, Spielmeyer W. Development of a SNP marker assay for the *Lr67* gene of wheat using a genotyping by sequencing approach. *Mol Breed*, 2014, 34: 2109-2118.

[22] Singh R P, William H M, Huerta-Espino J, Rosewarne G. Wheat rust in Asia: Meeting the challenges with old and new technologies. In: New Directions for a Diverse Planet. Proceedings of the 4th International Crop Science Congress, Brisbane, Australia. 2004, 26.

[23] Lan C X, Liang S S, Wang Z L, Yan J, Zhang Y, Xia X C, He Z H. Quantitative trait loci mapping for adult-plant resistance against powdery mildew in Chinese wheat cultivar Bainong 64. *Phytopathology*, 2009, 99: 1121-1126.

[24] Lan C X, Ni X W, Yan J, Zhang Y, Xia C X, Chen X M, He Z H. Quantitative trait loci mapping of adult-plant resistance to powdery mildew in Chinese wheat cultivar Lumai 21. *Mol Breed*, 2010, 25: 615-622.

[25] Bai B, He Z H, Asad M A, Lan C X, Zhang Y, Xia X C, Yan J, Chen X M, Wang C S. Pyramiding adult-plant powdery mildew resistance QTLs in bread wheat. *Crop Pasture Sci*, 2011, 63: 606-611.

[26] 伍玲，夏先春，朱华忠，李式昭，郑有良，何中虎．CIMMYT 273 个小麦品种抗病基因 *Lr34/Yr18/Pm38*

的分子标记检测．中国农业科学，2010，43：4553-4561.

Wu L, Xia X C, Zhu H Z, Li S Z, Zheng Y L, He Z H. Molecular characterization of *Lr34/Yr18/Pm38* in 273 CIMMYT wheat cultivars and lines using functional markers. *Sci Agric Sin*, 2010, 43: 4553-4561. (in Chinese with English abstract)

[27] Li Z F, Xia X C, He Z H, Li X, Zhang L J, Wang H Y, Meng Q F, Yang W X, Li G Q, Liu D Q. Seedling and slow rusting resistance to leaf rust in Chinese wheat cultivars. *Plant Dis*, 2010, 94: 45-53.

[28] Peterson R F, Campbell A B, Hannah A E. A diagrammatic scale for estimating rust intensity of leaves and stems of cereals. *Can J Res*, 1948, 26: 496-500.

[29] Ren Y, Li Z F, He Z H, Wu L, Bai B, Lan C X, Wang C F, Zhou G, Zhu H Z, Xia X C. QTL mapping of adult-plant resistances to stripe rust and leaf rust in Chinese wheat cultivar Bainong 64. *Theor Appl Genet*, 2012, 125: 1253-1262.

[30] Lan C X, Liang S S, Zhou X C, Zhou G, Lu Q L, Xia X C, He Z H. Identification of genomic regions controlling adult plant stripe rust resistance to in Chinese wheat landrace Pingyuan 50 through bulked segregant analysis. *Phytopathology*, 2010, 100: 313-318.

[31] Lu Y M, Lan C X, Liang S S, Zhou X C, Liu D, Zhou G, Lu L Q, Jing J X, Wang M N, Xia X C, He Z H. QTL mapping for adult-plant resistance to stripe rust in Italian common wheat cultivars Libellula and Strampelli. *Theor Appl Genet*, 2009, 119: 1349-1359.

[32] Kaur J, Bariana H S. Inheritance of adult plant stripe rust resistance in wheat cultivars Kukri and Sunco. *J Plant Pathol*, 2010, 92: 391-394.

[33] Rosewarne G M, Singh R P, Huerta-Espino J, Herrera-Foessel S A, Forrest K L, Hayden M J, Rebetzke G J. Analysis of leaf and stripe rust severities reveals pathotype changes and multiple minor QTLs associated with resistance in an Avocet × Pastor wheat population. *Theor Appl Genet*, 2012, 124: 1283-1294.

[34] Lan C X, Rosewarne G M, Singh R P, Herrera-Foessel S A, Huerta-Espino J, Basnet B R, Yang E N. QTL characterization of resistance to leaf rust and stripe rust in the spring wheat line Francolin# 1. *Mol Breed*, 2014, 34: 789-803.

[35] 刘金栋，陈新民，何中虎，伍玲，白斌，李在峰，夏先春．小麦慢白粉病QTL对条锈病和叶锈病的兼抗性．作物学报，2014，40：1557-1564.

Liu J D, Chen X M, He Z H, Wu L, Bai B, Li Z F, Xia X C. Resistance of slow mildewing genes to stripe rust and leaf rust in common wheat. *Acta Agron Sin*, 2014, 40: 1557-1564. (in Chinese with English abstract)

[36] 张勇，申小勇，张文祥，陈新民，阎俊，张艳，王德森，王忠伟，刘悦芳，田宇兵，夏先春，何中虎．高分子量谷蛋白5+10亚基和1B/1R易位分子标记辅助选择在小麦品质育种中的应用．作物学报，2012，38：1743-1751.

Zhang Y, Shen X Y, Zhang W X, Chen X M, Yan J, Zhang Y, Wang D S, Wang Z W, Liu Y F, Tian Y B, Xia X C, He Z H. Marker-Assisted Selection of HMW-Glutenin 1Dx5 + 1Dy10 Gene and 1B/1R Translocation for Improving Industry Quality in Common Wheat. *Acta Agron Sin*, 2012, 38: 1743-1751. (in Chinese with English abstract)

杂交与诱变相结合改良 CIMMYT 种质效果分析

吴振录[1]，樊哲儒[1]，李剑峰[1]，张跃强[1]，王岩军[1]，何中虎[2,3]

[1]新疆农业科学院核技术生物技术研究所，乌鲁木齐 830000；
[2]中国农业科学院作物科学研究所，北京 100081；[3]国际玉米小麦改良中心中国办事处，北京 100081

摘要： 本文总结了 40 多年杂交与诱变相结合改良 CIMMYT 种质的育种进展。从 1974 年至今，每年将各杂交组合当代种子均分为两份，一份用^{60}Coγ-射线 80Gy 辐照，另一份种子不辐照，为单纯杂交材料，两部分材料同步筛选，育成了 11 个新品种，其中 4 个成为新疆春小麦种植面积最大的主栽品种。研究表明，中低剂量γ-射线辐照春小麦杂种当代干种子，能够扩大杂种后代多数农艺性状的变异范围，株高与籽粒饱满度的相关系数有所降低，为选育矮秆兼抗旱的品种提供了可能。本项目育成的新春 2 号、新春 6 号等抗旱高产品种，连续 30 年在新疆春小麦中种植面积最大，近年又育成了优质品种新春 26 号等。CIMMYT 种质是重要的高产和优质源，杂交与诱变结合改良 CIMMYT 种质，可显著提高育种效率。

关键词： 春小麦；杂交；诱变；CIMMYT 种质

Development of Spring Wheat Varieties by CIMMYT Germplasm through Combination of Hybridization and Radiation

Wu Zhenlu[1], Fan Zheru[1], Li Jianfeng[1], Zhang Yueqiang[1], Wang Yanjun[1], He Zhonghu[2,3]

[1]*Institute of Nuclear & Biological Technologies, Xinjiang Academy of Agricultural Sciences, Urumqi* 830000; [2]*Crop Research Institute, Chinese Academy of Agricultural Science, Beijing* 100081; [3]*CIMMYT China Office, Beijing* 100081

Abstract: The progress of spring wheat breeding through the combination of hybridization and irradiation and utilization of CIMMYT in the last 40 years was reviewed. From 1974 up to now, F_1 seeds were equally divided into two parts, one part was irradiated by γ-ray, and the other part was planted without any treatment. And these two parts were equally selected by pedigree method. Eleven varieties were developed, and four of them became leading varieties with largest acreage in Xinjiang. Our data showed that the middle or low dose γ-ray irradiation on F_1 dry seeds could enlarge the variation of most agronomic traits in hybrid progenies and reduce the correlation coefficient between plant height and kernel plumpness, which may increase the opportunity to breed varieties with combination of drought resistance

---

原文发表在《麦类作物学报》2010，30（5）976-980；本文补充了 2010—2015 年的新资料，并将题目做了相应修改

and lodging resistance. Varieties such as Xinchun 2 and Xinchun 6 with high yield potential and drought resistance were developed, which were successively leading varieties with the largest area from 1986 to 2015 in Xinjiang. New varieties, such as Xinchun 26 with high quality, were released in recent years. Our experiences showed that hybridization combined with irradiation treatment could significantly improve breeding efficiency. In addition, utilization of CIMMYT germplasm has played a key role for the variety development.

Key words: Spring wheat; Hybridization; Radiation; CIMMYT germplasm

小麦是新疆最重要的粮食作物，近年每年种植面积约 90 万公顷，春小麦占新疆小麦面积 40%左右。新疆地处中亚干旱气候区，由于降水量少，一般麦田必须灌溉。新疆各麦区都依据可利用水量确定小麦种植面积，麦田灌溉基本有保证，因此小麦单产高于全国平均水平。新疆小麦育种始于 20 世纪 50～60 年代，50 年代主要是整理农家品种和系统选育，60 年代开始小麦杂交育种，随之开展了辐射诱变育种的探索[1-3]。从 20 世纪 70 年代起，新疆春小麦育种实施了杂交与诱变结合的方法，以改良 CIMMYT种质为主要目标，明显提高了育种效果。本文对这一工作进行了系统总结，以期为今后育种工作提供参考。

1　总体思路

20 世纪 60 年代，新疆春小麦杂交育种和诱变育种是在不同课题组分别进行的。在辐射诱变育种中，对 γ-射线辐射诱变采用传统的高剂量（半致死剂量约 250Gy）处理小麦风干种子，原品种被辐照后农艺性状变劣的机率很大，而出现优良变异的机率很小。当时国内外已有人把诱变处理同杂交育种结合使用，认为可增加变异性以扩大育种的选择基础[4]。20 世纪 70 年代，新疆农业科学院核技术生物技术研究所辐射育种课题组开始研究使用中低剂量 γ-射线处理杂种材料。具体研究路线是在大量原始材料中精选亲本配制杂交组合，再把杂交当代种子分成两等份，一份用^{60}Coγ-射线 80Gy 剂量辐射处理（F_1M_1），另一份不做处理用做单纯杂交材料（F_1），两部分种子同时播种，然后逐世代按系谱法选育。当时已经清楚 γ-射线 80Gy 的处理剂量只会造成轻微的辐射损伤，几乎不抑制出苗，基本不会死苗。我们只希望辐射处理能在杂交变异的基础上再增加一些微小变异，或者产生一些单纯杂交未获得的基因重组，或者能打断一些基因连锁等。如果万一辐射的效果不好，同时进行的单纯杂交育种也会有所收获。我们对中低剂量 γ-射线辐照春小麦杂种当代干种子的效果进行了研究，证明辐射处理能够扩大杂种后代大多数农艺性状的变异范围。例如 4 个杂交组合的 F_2代和 F_2M_2代群体的试验表明，株高的变异系数 F_2为 14.1%，F_2M_2为 15.0%；单株穗数的变异系数 F_2为 43.2%，F_2M_2为 46.5%；主穗粒数的变异系数 F_2为 25.0%，F_2M_2为 27.7%；单株粒数的变异系数 F_2为 50.8%，而 F_2M_2为 54.8%；单株粒重的变异系数 F_2为 52.6%，F_2M_2为 55.2%；籽粒饱满度的变异系数 F_2为 26.2%，而 F_2M_2为 29.0%；单株百粒重的变异系数 F_2为 15.7%，而 F_2M_2为 16.6%。另外，这 4 个杂交组合总计 F_2M_2的株高与籽粒饱满度的相关系数比 F_2有所降低，由 F_2的 0.62 下降到 F_2M_2的 0.55；同时，这 4 个杂交组合Ⅰ类优株（株高<100cm，籽粒饱满度中上）的出现频率 F_2为 3.52%，而 F_2M_2增加到 8.58%；Ⅱ类优株（株高<90cm，籽粒饱满度中上）的出现频率 F_2为 0%，而 F_2M_2增加到 1.49%。由于籽粒饱满度是新疆春小麦品种抗旱性强弱的最重要指标，株高与籽粒饱满度相关系数降低，为选育矮秆兼抗旱的品种提供了较大可能[5,6]。在辐射扩大杂种后代分离变异范围方面，我们与王琳清[7]和孙光祖[8]的结果基本一致。

国际玉米小麦改良中心（CIMMYT）培育的春小麦品种以高产、矮秆、抗病、广适闻名于世，CIMMYT种质引到新疆后表现矮秆、高产、抗病、早熟、多花、优质和对日照长度反映迟钝等特点，适应性较好，引起我们高度重视。从 1974 年至今，我们一直采用杂交与诱变结合世代同步育种的做法，持续不断改良 CIMMYT 种质，选育出的新春 2 号、新春 3 号、新春 6 号和新春 17 号先后成为新疆春小麦生产中种植面积最大的主栽品种，并且作为主栽品种的种植时间接连长达 30 年。实践证明这个技术路线确实提高了春小麦育种效果。

2 抗旱高产育种

2.1 新春2号和新春3号

1974年1月，我们在海南岛南繁中选配了塞洛斯/奇春4号杂交组合，其中塞洛斯是CIMMYT绿色革命的主要品种之一。当年4月初将这批100多个组合杂种当代种子各均分为两份，一份种子用$^{60}Co\gamma$-射线80Gy辐照，另一份种子不辐射处理。两部分种子同时春播，此后按系谱法世代同步选育。1984年审定了新春2号，1986年审定了新春3号（表1）。两个新品种都来源于塞洛斯/奇春4号的辐射处理（F_1M_1）后代，而未经过辐射的此组合单纯杂交材料（F_1）后代中未选出新品种。

表1 新春2号和新春3号在1981—1983年新疆春小麦区域试验中的产量表现

Table 1 Yield data in Xinjiang spring wheat regional trials from 1981 to 1983

品种 Variety	产量 Yield (t/hm^2)	增产幅度 Yield increase (%)	产量位次 Yield Rank
新春2号 Xinchun 2	5.064	12.1	1
新春3号 Xinchun 3	4.965	9.9	2
巴春1号（CK）Bachun 1	4.517	0	7

新春2号是新疆春小麦干热麦区大面积单产超$7.5t/hm^2$的突破性品种，在哈密盆地、准噶尔盆地和塔里木盆地相继大面积突破$7.5t/hm^2$，种植面积迅速扩大，1986年达到6.5万hm^2，成为新疆春小麦种植面积最大的品种。新春2号中秆、中熟，同时具有抗病性中等、不易落粒和中筋品质等特点，其突出优点是抗旱且高产，每推广到一个新地方就在当地创造高产记录，改变了过去抗旱品种不高产、高产品种不抗旱的局面。

通过对基部叶片功能期和籽粒饱满度的持续筛选，我们从新春2号中选出了改良型新春2号。其抗旱性比原新春2号明显提高，籽粒饱满度和容重都提高一个等级，进而成为新疆干热风最严重的吐鲁番盆地托克逊县的主栽品种[9]。经过对新春2号的全面更新，新春2号成为1986—1995年十年新疆春小麦累计种植面积最大的品种（七年种植面积第一，三年第二）。

新春3号在此年中曾有三年在新疆种植面积居第一，其丰产性亦好，但抗旱、耐高温性不及新春2号。

2.2 新春6号和新春7号

1985年6月，我们用来自中国农业科学院作物科学研究所的半矮秆、高蛋白品系中7906（引自CIMMYT，组合为CNO _ GalloxBb4A/K4496）为母本，以改良型新春2号为父本配制杂交组合。仍然将杂种当代种子均分两份，一份种子用$^{60}Co\gamma$-射线80Gy辐照，另一份种子不辐射处理，用做单纯杂交材料。经过多世代筛选，1993年审定了新春6号，1997年审定了新春7号（表2）。这两个新品种都来源于中7906/改良型新春2号杂交当代未进行$^{60}Co\gamma$-射线辐射处理（F_1）的后代材料，而经过辐射的材料（F_1M_1）后代未选育出新品种。

表2 新春6号和新春7号在1990—1992年新疆春小麦区域试验中的产量表现

Table 2 Yield data in Xinjiang spring wheat regional trials from 1990 to 1992

品种 Variety	产量 Yield (t/hm^2)	增产幅度 Yield increase (%)	产量位次 Yield Rank
新春6号 Xinchun 6	6.366	4.4	2
新春7号 Xinchun 7	6.450	5.7	1
新春2号（CK）Xinchun 2	6.101	0	3

新春6号是新疆春小麦大面积单产突破$9t/hm^2$的超高产品种。最初在新疆焉耆盆地，然后在新疆大多数春麦区相继实现了大面积单产$9t/hm^2$，推广速度很快，1996年成为新疆春小麦种植面积最大的品种。新春6号品种为半矮秆、早熟、生长势强、籽粒灌浆快、抗旱而且千粒重高，其叶片短、宽，单位面积成穗数多，超高产栽培条件下穗数、穗粒数和千粒重三个产量构成因素可以同步超亲。1998年通过国家审定并开始在内蒙古自治区和甘肃省推广。1998和1999年该品种的种植面积占新疆春小麦的40%，是新疆1996—2006年、2008—2015年种植面积最大的春小麦品种。新春6号生育期105 d，产量构成因素为600穗/m^2、穗粒数40粒、千粒重50g，单产可达$10.6t/hm^2$，同时适应范围较广，是一个理想的超高产品种[10]。2004年和2005年的抗旱性试验表明，新春2号、新春6号和新春7号等都具有抗旱性强、对土壤水分改善敏感和水分高效利用特性[11,12]。

2.3 新春17

2005年审定的新春17来源于新春6号/NS64杂交组合的单纯杂交后代（F_1），NS64为埃及品种Giza

163，也是来自CIMMYT的品种。新春17也具丰产性突出的特点（表3），2007年曾成为新疆春小麦种植面积最大的品种。

表3　新春17在2002—2003年新疆春小麦区域试验中的产量表现

Table 3　Yield data in Xinjiang spring wheat regional trials from 2002 to 2003

品种 Variety	产量 Yield (t/hm^2)	增产幅度 Yield increase (%)	产量位次 Yield Rank
新春17 Xinchun 17	6.371	2.3	1
新春6号（CK）Xinchun 6	6.226	0	3

3　优质育种

3.1　新春6号、新春7号和新春17的品质优于新春2号

新春6号和新春7号的母本中7906是优质品系，籽粒蛋白质含量可达17.3%。新春6号和新春7号的籽粒蛋白质含量及面筋强度均超过当时新疆春小麦主栽品种新春2号。新春7号的品质比新春6号更好，同时新春7号具有5+10优质高分子量麦谷蛋白亚基（HMW-GS）。新春17号的籽粒品质也优于新春6号，并且也具有5+10优质亚基。新春6号、新春7号、新春17号的品质均优于新春2号。

3.2　新春26和新春30

1996年6月配制了“新春9号/新春6号”等杂交组合，新春9号为埃及品种Giza 164（即来自CIMMYT的Veery 5）。仍然将杂种当代种子各均分两份，一份种子用$^{60}Co\gamma$-射线80Gy辐照，另一份种子不辐射处理，用做单纯杂交材料。经过多世代筛选，2007年审定了新春26，2009年审定了新春30（表4）。两个新品种都来源于新春9号/新春6号的辐射处理（F_1M_1）后代，而未经过辐射的此组合单纯杂交材料（F_1）后代未选出新品种。

表4　2004—2007年新疆春小麦区域试验中品种的产量表现

Table 4　Yield data in Xinjiang spring wheat regional trials from 2004 to 2007

试验年份 Year	品种 Variety	产量 Yield (t/hm^2)	增产幅度 Yield increase (%)	产量位次 Yield Rank
2004—2005	新春26 Xinchun 26	6.571	1.4	6
	新春6号（CK）Xinchun 6	6.481	0	7
2006—2007	新春30 Xinchun 30	6.555	2.4	1
	新春6号（CK）Xinchun 6	6.404	0	3

新春26属于半矮秆、中早熟品种，新春30属于半矮秆、早熟品种，它们的共同突出特点是籽粒品质优于优质亲本新春9号。这两个新品种的HMW-GS组合都为2*、7+9、5+10，都具有优质亚基。

经农业部农产品质量监督检验测试中心（乌鲁木齐）检测，新春26粗蛋白（干基）17.3%，湿面筋34.1%，面团形成时间10.7min，面团稳定时间30.5min，面团最大拉伸阻力705E.U，面团延伸度192mm，拉伸曲线面积170.4cm^2，面包体积745ml，面包评分87.5。经农业部谷物及制品质量监督检验测试中心（哈尔滨）对新春26的面条和面包试验，面条总评分89.3，面包体积850ml，面包评分90，烘烤面包品质与对照金像粉持平，面条品质超过澳大利亚小麦。表明新春26既能烘烤出很好的面包，又能制作出优质面条，是一个多用途优质小麦[13]。2008年，我们对新疆40年来春小麦种植面积居第一位曾达三年以上品种的主要产量性状与新春26同时进行了比较（表5）。新春26不仅熟期、株高，而且其他主要产量性状都与新春6号接近。新春6号和新春26，比喀什白皮、新春2号和新春3号的株高明显降低，熟期提早，千粒重和容重明显提高。从新疆农家品种系统选育出的喀什白皮品质很好，新春2号和新春3号品质不及喀什白皮，而新春26品质比喀什白皮更好，说明春小麦育种不仅大幅度提高了新疆小麦的产量，而且显著改善了加工品质。

表5　新疆四十年来最重要春小麦品种的主要产量性状，2008年

Table 5　Characters on yield of leading varieties in Xinjiang in 2008 crop season

品　种 Variety	抽穗期（月/日）Heading (month/day)	成熟期（月/日）Maturity (month/day)	株高 Height (cm)	穗数 Spike number (m^2)	穗粒数 Grain number (spike)	千粒重 1 000-kernel weight (g)	收获指数 Harvest index	产量 Yield (t/hm^2)
喀什白皮 Kashibaipi	5/31	7/28	102.3	456.1	31.7	34.8	0.24	4.775

（续）

品　种 Variety	抽穗期（月/日）Heading (month/day)	成熟期（月/日）Maturity (month/day)	株高 Height (cm)	穗数 Spike number (m^2)	穗粒数 Grain number (spike)	千粒重 1 000-kernel weight (g)	收获指数 Harvest index	产量 Yield (t/hm^2)
新春2号 Xinchun 2	5/30	7/26	91.0	477.5	36.0	45.6	0.40	8.004
新春3号 Xinchun 3	5/31	7/28	90.3	458.6	38.8	38.1	0.33	7.240
新春6号 Xinchun 6	5/25	7/19	83.3	486.8	34.3	51.2	0.43	8.195
新春26 Xinchun 26	5/28	7/21	85.0	546.2	35.4	45.3	0.43	9.064

在2009年大面积高产栽培试验中，新春26在4.6hm^2获得9t/hm^2的产量，表明其具有大面积高产潜力，是新疆小麦优质育种的突破。新疆最大的面粉加工企业——新疆天山面粉集团公司已经把新春26确定为加价收购品种，是新疆现在唯一获得面粉加工企业加价收购的小麦品种。虽然每千克加价只有0.1元，但由于新春26产量不低，农户还是积极性的。

4　近五年育成的新品种

4.1　新春33

2010年审定的新春33（表6）来源于新春9号/新春6号杂交组合的辐射处理（F_1M_1）后代，是新春26和新春30的姐妹品种，因此新春9号/新春6号组合的辐射处理（F_1M_1）后代实际选育出三个春小麦品种。而这个杂交组合未经过辐射的单纯杂交材料（F_1）后代未选出新品种。新春33和新春30同属于中强筋类型。

表6　新春33在2007—2008年新疆春小麦区域试验中的产量表现

Table 6　Yield data in Xinjiang spring wheat regional trials from 2007 to 2008

品种 Variety	产量 Yield (t/hm^2)	增产幅度 Yield increase (%)	产量位次 Yield Rank
新春33 Xinchun 33	6.213	0.6	1
新春6号（CK）Xinchun 6	6.176	0	3

4.2　新春37

2012年审定的新春37（表7）是2002年以高产优质春小麦品系49-5（新春26姐妹系）为母本，与加拿大优质强筋春小麦品种野猫杂交，当代杂交种子经$^{60}Co\gamma$-射线80Gy辐照（F_1M_1），再经过多代筛选和南繁北育选育而成的优质、强筋、高产、抗病春小麦新品种。农业部农产品质量监督检验测试中心（乌鲁木齐）检测表明，新春37蛋白质含量16.0%，湿面筋含量34.5%，面团稳定时间25.8min，最大拉伸阻力612E.U，延伸度206mm，面包评分85.5分，属优质强筋类型。这个杂交组合未经过辐射的单纯杂交材料（F_1）后代未选出新品种。

表7　新春37在2009—2010年新疆春小麦区域试验中的产量表现

Table 7　Yield data in Xinjiang spring wheat regional trials from 2009 to 2010

品种 Variety	产量 Yield (t/hm^2)	增产幅度 Yield increase (%)	产量位次 Yield Rank
新春37 Xinchun 37	7.313	0.6	1
新春6号（CK）Xinchun 6	7.270	0	2

4.3　新春40

2013年审定的新春40（表8）属于中强筋类型。1999年以新春6号为母本，以引进美国春小麦种质UC1041为父本杂交，当代杂交种子经$^{60}Co\gamma$-射线80Gy辐照（F_1M_1），通过多代单株、单穗选择和南繁加代育成的春小麦新品种。而这个杂交组合未经过辐射的单纯杂交材料（F_1）后代未选出新品种。

表 8 新春 40 在 2011—2012 年新疆春小麦区域试验中的产量表现

Table 8 Yield data in Xinjiang spring wheat regional trials from 2011 to 2012

品种 Variety	产量 Yield (t/hm²)	增产幅度 Yield increase (%)	产量位次 Yield Rank
新春 40 Xinchun 40	5.773	1.7	1
新春 6 号（CK）Xinchun 6	5.676	0	4

4.4 新春 44

2015 年审定的新春 44（表 9）是我们 2004 年以自育新品系 17-11 为母本，CIMMYT 小麦种质 YN-76 为父本杂交，当代杂交种子经 ^{60}Coγ-射线 80Gy 辐照（F_1M_1），通过多代单株、单穗选择，南繁北育，于 2015 年育成的新品种。而这个杂交组合未经过辐射的单纯杂交材料（F_1）后代未选出新品种。经农业部农产品质量监督检验测试中心（乌鲁木齐）检测，新春 44 品种籽粒蛋白质（干基）含量为 16.0%，湿面筋含量（湿基）为 34.6%，面团形成时间为 9.5min，面团稳定时间为 17.4min，最大抗延阻力为 503E.U，面团延伸度为 191mm，拉伸曲线面积为 119.6cm^2，面包总评分 95.5 分，为优质强筋小麦。

表 9 新春 44 在 2013—2014 年新疆春小麦区域试验中的产量表现

Table 9 Yield data in Xinjiang spring wheat regional trials from 2013 to 2014

品种 Variety	产量 Yield (t/hm²)	增产幅度 Yield increase (%)	产量位次 Yield Rank
新春 40 Xinchun 40	7.053	3.1	2
新春 6 号（CK）Xinchun 6	6.839	0	3

2010—2015 年期间，我们选育出新春 37、新春 40 和新春 44 三个新品种的三个杂交组合，都是从当代杂交种子经 ^{60}Coγ-射线 80Gy 辐照（F_1M_1）的后代中选育出新品种的，而这三个杂交组合未经过辐射的单纯杂交材料（F_1）后代都未选出新品种。

5 讨论

从 20 世纪 70 年代初期至现在 40 多年间，我们把品种间杂交和辐射诱变相结合，选育的春小麦新品种在生产上发挥了重要作用，先后成为主栽品种的有新春 2 号、新春 3 号、新春 6 号、新春 7 号和新春 17。有连续约十年时间，这些品种每年的种植面积覆盖了本地春小麦面积的 50%～70%，现在新春 26 和新春 30 也已开始在生产上发挥作用。分析这些新品种的来源，新春 2 号、新春 3 号、新春 6 号、新春 7 号、新春 26 和新春 30 六个品种分别出自三个杂交组合，新春 2 号和新春 3 号选自塞洛斯/奇春 4 号的辐射处理（F_1M_1）后代，新春 6 号和新春 7 号选自中 7906/改良型新春 2 号的单纯杂交（F_1）的后代，新春 26 和新春 30 选自新春 9 号/新春 6 号的辐射处理（F_1M_1）后代。这六个品种基本上属于三个批次，即三个时间段，都大体适合了不同时间段新疆小麦生产的实际需求，即 20 世纪 80 年代小麦生产需要高产品种，90 年代以来需要超高产品种，21 世纪初需要优质品种。值得注意的是每批次两个品种都是来自同一个杂交组合的同一个处理，或者都来自经过辐射处理的后代，或者都来自单纯杂交的后代；而且往往都能出两个新品种，似乎具有可重复性。但新春 17 就其主要特点应归属我们育成新品种的第二个批次，它来源于新春 6 号/NS64 的单纯杂交（F_1）后代材料，此杂交组合的辐射处理（F_1M_1）后代未育成新品种。如果当年不对杂交种子进行辐射处理，我们大概就不能获得第一个和第三个时间段的成果。杂交与辐射诱变相结合的技术成为新疆春小麦育种工作获得丰硕成果的关键。总结过去 40 多年杂交与辐射诱变结合的育种工作，可以形成以下三点认识。

（1）十分重视 CIMMYT 小麦种质的利用与改良。虽然我们原先是辐射育种课题组，但要辐照杂交材料就必须像从事杂交育种一样重视杂交亲本的研究。在 20 世纪 70 年代初，CIMMYT 种质被引进新疆，引进种质的矮秆、高产、抗病、早熟、多花、优质和对日照长度反映迟钝等特点，引起我们高度重视。此前新疆的春小麦育种不论是杂交育种还是辐射育种，都以中等产量水平为目标。我们引进 CIMMYT种质后，很快把春小麦育种的目标调整为高产育种[14]。CIMMYT 种质与新疆种质的生态类型差异较大、亲缘关系较远，而且优良性状互补性好，因而成为改良新疆春小麦的重点外来亲本。40 多年来配制了数千个杂交组合，在辐照杂种与单纯杂交育种中效果最好的外来亲本是来源于 CIMMYT 的塞洛斯、新春 9 号和中 7906。

（2）辐射杂交育种与单纯杂交育种应当并重。新春 2 号、新春 3 号、新春 26 和新春 30 分别来源于两个杂交组合的辐射处理（F_1M_1）后代，新春 6 号、新春 7 号和新春 17 分别来源于另两个杂交组合的单纯杂交（F_1）后代。育种实践表明，中低剂量 γ-射线辐照春小麦杂种当代干种子对杂种后代的分离会产生深远影响，虽然辐照小麦杂种材料能扩大后代的变异范围，产生新的微小变异，育成单纯杂交不能得到的新品种，然而在育种成效上既表现出正效应也出现负效应。可能有些通过杂交重组能够出现的变异类型经过辐射又失去了。若将辐照杂交育种和单纯杂交育种同步进行，可以显著增加成功的机会[15]。

（3）外来品种做母本育种效果好的现象值得注意。纵观 40 多年的育种工作，我们在三个时间段里分别各育出两个重要品种的三个重要杂交组合（塞洛斯/奇春 4 号、中 7906/改良型新春 2 号、新春 9 号/新春 6 号），都是以外来品种为母本而以新疆品种为父本的。这并不是我们事前有预见、有计划的结果，因为在配制杂交组合时我们对外来品种或本地品种做父母本是随机的，这一现象值得进一步研究。

参考文献

[1] 吴振录．新疆春小麦主栽品种的生态类型初步分析［J］．新疆农业科学，1982（1）：3-6.

[2] 吴振录．新疆农作物辐射育种成就［J］．核农学通报 1994，15（5）：201-204.

[3] 吴振录．就新春 2 号的育成谈春小麦育种的几个问题［J］．新疆农业科学，1985（2）：8-10.

[4] 中国科学院遗传研究所．突变育种手册［M］．北京：科学出版社，1972，5.

[5] 吴振录．γ-射线辐照春小麦杂种扩大变异范围的研究初报［J］．原子能农业应用，1983（3）：25-29.

[6] 吴振录．中低剂量 γ-射线辐照春小麦杂种提高育种效果的研究［J］．原子能农业应用，1985，增刊：128-133.

[7] 王琳清，张维强，范庆霞，等．辐射与杂交结合提高冬小麦辐射育种效果的探讨［J］．作物学报，1980，6（4）：237-244.

[8] 孙光祖，陈义纯．应用辐射与杂交相结合的方法选育春小麦品种的体会［J］．原子能农业应用，1981，（4）：15-21.

[9] 吴振录，胡伯洪．春小麦良种新春 2 号的更新、高产示范与推广［J］．新疆农业科学，1995（2）：64-67.

[10] 吴振录，樊哲儒，陈玉魁，等．超高产优质中筋春小麦品种-新春 6 号［J］．麦类作物学报，2005，25（2）147.

[11] 吴振录，黄光宏，樊哲儒，等．小麦水分高效利用种质的筛选方法探讨［J］．麦类作物学报，2005，25（5）143-146.

[12] 吴振录，卢运海，范玲，等．小麦抗旱性、水敏感性和水分高效利用特性的研究//第九届全国作物生理学研讨会论文集［M］．北京：中国农业出版社，2006，127-134.

[13] 樊哲儒，吴振录，李剑峰，等．多用途优质强筋春小麦新品种-新春 26［J］．麦类作物学报，2008，28（2）356.

[14] Wu Zhenlu. Breeding spring wheat for high yield potential and utilization of CIMMYT germplasm in Xinjiang，CIMMYT. WPSR 1997，No46：36-39 .

[15] 吴振录，陈玉魁，樊哲儒，等．γ-射线辐照小麦杂种的亲本组配研究［J］．核农学报，1995，9（4）：193-199.

CIMMYT 种质与育种技术在四川小麦品种改良中的应用

邹裕春[1]，杨武云[1]，朱华忠[1]，杨恩年[1]，伍玲[1]，黄钢[1]，李跃建[1]，
何中虎[2,3]，R. P. Singh[3]，S. Rajaram[3]，G. M. Rosewarne[3]

[1]四川省农业科学院作物研究所，成都 610066；[2]中国农业科学院作物科学研究所，北京 100081；
[3]International Maize and Wheat Improvement Center (CIMMYT)，
Apdo. Postal 6-641，06600 Mexico，D. F.，Mexico

四川省农业科学院与国际玉米小麦改良中心(CIMMYT)开展小麦科技合作已有30多年的历史，引进CIMMYT材料5000多份，四川育种单位利用含CIMMYT种质材料育成新品种51个。大致可分为四个阶段：第一阶段1984—1988年，主要进行人员互访、信息交流和种质资源交换；第二阶段1988—1997年，四川省农业科学院参加了中国农业科学院组织的中国与CIMMYT小麦穿梭育种合作项目，成为该项目国内首轮4个穿梭点之一；第三阶段1997—2010年，1997年CIMMYT中国办事处成立后，双方合作进一步加强，并取得重要进展；第四阶段从2011年至今，CIMMYT与四川省农业科学院签署了“关于小麦育种、病害及农艺学研究的国际合作研究协议”，派高级科学家Garry M Rosewarne博士长驻四川，进行持久抗病育种研究，并建立“中国-CIMMYT南方联合实验站”。通过长期合作，极大地提高了四川小麦育种和生产水平，并培养了一批优秀研究和管理人才，推动国际合作迈上新台阶，扩大了四川小麦育种的国际影响，同时也使CIMMYT的小麦育种受益匪浅。

1 四川—CIMMYT 小麦育种合作概况

1.1 摸索认识阶段

从1966年第一份CIMMYT品种Penjamo 62引进四川，到1988年底中国和CIMMYT签订小麦穿梭育种合作协议，是四川育种家对CIMMYT及CIMMYT品种、育种资源从完全不了解到逐步摸索认识的时期。20世纪70年代前后，CIMMYT育成的矮秆、耐肥、抗倒和光周期不敏感的高产小麦品种在印度、巴基斯坦带来的绿色革命，使四川育种家看到了新希望。至20世纪80年代初的20多年，先后有20余份CIMMYT小麦品种引入四川，但没有一份能直接推广，它们的生态适应性和产量潜力都远不及四川在70年代末、80年代初育成的繁六和绵阳11。

1984年，邹裕春研究员与美国俄勒冈州立大学小麦育种家一起访问CIMMYT，这是四川第一位小麦育种家与CIMMYT的第一次直接接触。同年，CIMMYT小麦项目副主任Klatt博士访问中国，专程到四川进行考察，四川小麦生产及育种给他留下了深刻印象。1984年底，邹裕春回国后即开展了与CIMMYT较为系统的合作，每年选择3～5个CIMMYT国际圃在四川种植，一方面了解CIMMYT材料在四川的表现，同时也希望从中筛选出适合四川使用的杂交亲本材料。后来育成了四川审定的第一个具有CIMMYT血缘的小麦新品种川麦25，川麦30的CIMMYT亲本也是在这一阶段筛选积累下来的。1985-1987年，CIMMYT每年都派专家来四川考察。1985年，四川省农业科学院作物所小麦专家赴CIMMYT开展合作研究，加深了双方的合作和了解。

1.2 参加中国—CIMMYT 小麦穿梭育种

1988年，中国—CIMMYT小麦穿梭育种合作协议正式签订。加强了小麦种质资源交换，推动科学家定期互访、交流及参加田间选种，四川青年专家赴

原文发表在《西南农业学报》，2007，20(2)：183-190；本文由杨恩年研究员进行补充修改

CIMMYT 培训人数显著增加。为了让更多中国专家了解 CIMMYT 小麦育种理念、方法、田间设计、田间记载以及系谱的记录方法等，在CIMMYT资助下，我们翻译出版了《CIMMYT 的小麦育种》一书，该书被指定为在 CIMMYT 培训的中国学员必读参考书。期间，100 余份四川品种及选系交换到 CIMMYT，包括获繁六、绵阳 11 和后来在 CIMMYT 育成抗印度粒腥黑穗病的抗源川麦 18。为了便于中国三个春麦穿梭点的小麦专家赴 CIMMYT 执行穿梭育种任务，CIMMYT 还专门设立了南京圃、哈尔滨圃和成都圃。四川省农业科学院作物所专家分别于 1989、1991、1992 和 1995 年赴 CIMMYT 进行短期穿梭选种，每年都有数百份 CIMMYT 材料带回四川，包括育成强筋新品种川麦 36、川麦 39 的亲本 Milan 选系。多年的利用结果表明，这几份 Milan 选系在四川是极为难得的优质与高抗条锈病相结合的优良亲本，估计在今后一个时期内，还将继续在育种中发挥作用。杨武云博士在培训期间从 CIMMYT 引进人工合成种，回国后育成了影响很大的川麦 42。

1.3 CIMMYT 种质应用取得重要进展

1997 年 CIMMYT 北京办事处成立，使CIMMYT能更及时、准确地了解各个穿梭点的动态，从而促进合作向更广泛、更深入的方向发展。四川省农业科学院作物研究所近年来与 CIMMYT 开展的小麦条锈病微效多基因持久性抗性育种、垄作栽培、南亚稻麦耕作制度等项目都是在这一阶段发展起来的[1,2]。以这些项目的成功经验及培养的人才为基础，我们承担的中—澳四川小麦改良项目、与巴基斯坦的中-巴节水项目都得以顺利实施。在这期间，四川利用 CIMMYT 种质育成小麦新品种 28 个。

1.4 进行持久抗病育种

2011—2015 年，高级科学家 Garry M Rosewarne 博士长驻四川省农业科学院，进行持久抗病育种研究，并建立“中国—CIMMYT 南方联合实验站”。在这期间，四川利用 CIMMYT 种质育成小麦新品种 22 个。

2 合作主要成果

2.1 种质资源交换与品种改良

通过引进 CIMMYT 的各种国际圃和参加穿梭育种的专家在 CIMMYT 田间选种，约 5 000 多份 CIMMYT 小麦品种/选系和各种育种材料被引入四川并在以成都为主的生态条件下观察筛选。用 CIMMYT 小麦种质在四川已育成小麦新品种 51 个，并在条锈新抗源、强筋优质和高产育种上都有新的突破。其中，川麦 30 曾在四川创造连续三年验收亩*产超 500kg 的小面积单产纪录[1,2]，累计推广 177 万 hm^2，获四川省科技进步二等奖；川麦 32 是 90 年代中期条锈病新生理小种爆发时全省唯一大面积推广的高抗条锈病的新品种，还通过了国家审定；川麦 36、川麦 39 都是 CIMMYT 育成的 Milan 选系的衍生后代，是四川仅审定的两个强筋新品种，实现了品质育种的突破；川麦 42 是世界上首次利用 CIMMYT 硬粒小麦-节节麦人工合成种 Syn-CD768（Altar84/*Aegilops tauschii188*）育成的推广品种，创造了省区试最高单产纪录（6.13t/hm^2），是长江上游麦区第一个超过 6 000kg/hm^2 的国家审定小麦品种[1-3]。

2.2 科学家交流与青年专家培训

四川省农业科学院先后有 14 位小麦专家赴 CIMMYT考察、参加短期合作研究、穿梭选种等，有 5 位青年专家在 CIMMYT 参加 1～6 个月的专题培训，回国后分别担任院长、副院长、所长、副所长等职，他们又都成为四川省的学术带头人。四川省农业科学院翻译出版了介绍 CIMMYT 的专著 2 册，与 CIMMYT合作发表论文 50 余篇，扩大了 CIMMYT 及四川小麦育种的影响。

2.3 与 CIMMYT 合作获得的荣誉与奖励

1989 年，CIMMYT 授予邹裕春“合作研究成功荣誉证书”。

1989 年，由 CIMMYT 国际大麦产量圃筛选出的品系品种威 24 获四川省政府科技进步三等奖。

1996 年，由于与 CIMMYT 等国际合作的突出贡献，邹裕春被科技部、农业部、国家计委和国家引智办联合授予“全国农业引智先进个人”称号。

1999 年，CIMMYT 授予黄钢、邹裕春“中国 CIMMYT 小麦穿梭合作突出贡献荣誉证书”。

2000 年，CIMMYT 小麦育种项目首席科学家 Rajaram 博士获四川省政府金顶奖。

---

* 亩为非法定计量单位。15 亩＝1 公顷——编者注

2000 年，四川省政府授予四川省农业科学院作物所“四川省农业引智基地”。

2001 年，用 CIMMYT 种质育成的小麦品种川麦 30 获四川省政府科技进步二等奖。

2007 年，CIMMYT 小麦育种项目首席科学家 Ravi Singh 博士获四川省政府金顶奖。

2007 年，利用 CIMMYT 种质 Milan 育种的强筋小麦品种川麦 36、川麦 39 获四川省科技进步二等奖。

2009 年，利用 CIMMYT 人工合成种育成的川麦 42 获四川省科技进步一等奖。

2010 年，利用 CIMMYT 人工合成种育成的川麦 42 获国家科技进步二等奖。

2013 年，CIMMYT 派驻四川省农业科学院的小麦育种高级科学家 Garry M Rosewarne 博士获四川省政府金顶奖。

2014 年，四川省农业科学院获 CIMMYT 颁发的“杰出合作奖”，4 位科学家获“杰出校友奖”。

2.4 合作育种新进展

利用 CIMMYT 人工合成种育成的突破性小麦品种川麦 42 连续 7 年被列为国家重点推广品种，并于 2010 年在江油市大堰乡泉水村创造了亩产 710kg 的高产新纪录。国审品种川麦 104 在两年国家区域试验中平均亩产 408.7kg，比对照川麦 42 增产 8.4%。在继续利用 CIMMYT 人工合成种的同时，我们还利用高产硬小麦及四川地方四倍体小麦与节节麦合成了新的人工合成六倍体小麦。

从 2000 年开始，四川省农业科学院与 CIMMYT 合作开展了以条锈病成株抗性为主、兼抗叶锈和白粉病的育种工作，在三个方面取得重要进展：(1) 利用三个条锈成株抗性 RILs 群体发现了 12 个成株抗条锈 QTLs 及 10 个成株抗叶锈 QTLs，并利用分子标记将 CIMMYT 兼抗材料 Chapio 中的三个兼抗性基因 *Yr31*、*Yr30*、*Yr18* 转入四川高产品种，育成一批高抗条锈病、兼抗叶锈病、白粉病且农艺性状优良的材料[4-6]。(2) 利用成株抗性育种方法育成 96 份高代抗病品系，经多环境田间鉴定及 Lr34/Yr18/Pm38、Lr46/Yr29/Pm39 和 Sr2/Yr30 分子标记检测表明，85 份兼抗三种病害，18 份含 Lr34/Yr18/Pm38，37 份含 Lr46/Yr29/Pm39，29 份含 Sr2/Yr30；部分品系可能还含有未知成株抗性基因[7,8]。(3) 育成三个兼抗型成株抗性高产品系，分别参加国家和四川省区试，其中川麦 82 兼抗条锈、叶锈、白粉病，在四川省 2014 年度区试中亩产 420.8kg，比对照增产 13.5%，居第四组首位。通过分子标记与常规育种相结合，可有效培育兼抗型成株抗性品种，为我国小麦抗病育种提供了新思路[8]。

3 经验总结与讨论

3.1 加强合作的针对性

了解 CIMMYT 各种圃的特点和主要用途是我们选择、引种的重要依据。主要通过轮流引进各种国际圃以及在 CIMMYT 对各个圃的实地观察和根据资料查阅点名引进，然后将引进材料在成都观察并试配组合。已育成品种的几份 CIMMYT 材料都是这样筛选出来的，特别是育成了我省首个强筋品种和多个高抗品种的 Milan 选系，就是在查阅 CIMMYT 品质实验室资料的基础上点名引进的。培训是认识 CIMMYT 和选种的最好机会，人工合成种是杨武云博士在 CIMMYT 培训时引回的。

根据在 CIMMYT 和成都同时对 CIMMYT 和成都材料的多年观察，双方材料特点十分明显。多数四川品种在 CIMMYT 表现矮秆、抗倒、多花、大粒、大穗、早熟和灌浆快、落黄好、抗印度粒黑穗病，受到 CIMMYT 育种家的高度重视，并已被大量用于育种；但四川小麦品种秆粗、叶大、分蘖少、群体差、加工品质差，感锈病（抗条锈也仅为主基因抗性）、白粉病、纹枯病等，并易落粒。多数 CIMMYT 材料在四川表现秆细（但韧性好）、叶小（多为狭长）、分蘖好、群体好，抗多种病害（但赤霉病重），不断穗轴，籽粒整齐、皮薄、光亮、商品品质及加工品质好；但穗稀尖长、小花结实少、籽粒小。在四川冬春易干旱条件下，CIMMYT 种质苗期长势弱，不利于根系生长及生物产量积累，而春季光照弱、雾露多，且拔节至抽穗灌浆期多阵雨伴大风的天气特点又造成多数 CIMMYT 种质在四川植株偏高、秆细易倒伏及易早衰等缺点。

针对四川小麦生产和育种现状，以及我们积累的种质资源与育种经验，在 CIMMYT 众多小麦育种技术、策略及经验中，我们有针对性地选择了冬/春杂交、穿梭育种、混合选择和微效多基因持久抗性等育种新技术及新思路，用于改进我们的育种技术，并结合分子标记辅助选择，已形成了新的育种体系，取得了很好效果。

3.2 坚持引进创新

在认识CIMMYT小麦育种资源特点的基础上，针对我们育种当前和长远需要的类型，选择各类材料在成都再筛选、分类。选择亲本试配组合时，国内亲本应以丰产性和地方适应性为主，CIMMYT亲本则必须能给我们引进新性状、新基因，使育成的新品种具有与我省现有品种不同的新特点。用人工合成种育成世界首例商用小麦品种川麦42的亲本选择就是一例，很好掌握了人工合成种的表型与遗传特点，再与矮秆、早熟、大穗的川麦30杂交育成；川麦42在两年省区试中产量均列小组第一，平均超过对照12%。

针对CIMMYT的育种技术及经验，更应根据自身基础和现有条件，瞄准国际发展趋势进行消化创新。我们已将CIMMYT的冬/春杂交、穿梭育种、改良混合选择和微效多基因持久抗性等育种新技术及新思路，结合国际上已经成熟的DH群体技术、重组自交系群体技术，共同融于我们自己的育种技术，形成了新的育种体系，大大加快了育种进度，提高了育种效率。比如，为了加强地方适应性，已将“冬/春杂交”改成“冬/春//春”和CIMMYT/四川//四川模式，两次春性或四川材料，可以是一次回交，也可以用另一材料做三交。穿梭可以根据情况选择部分穿梭，但主要选择应在四川进行。近年新开展的微效多基因持久抗性育种合作，更是CIMMYT/四川//四川回交模式与部分穿梭、成都选择为主相结合的成功典例。

CIMMYT材料的遗传背景和选择环境与四川材料有很大差异，因而不少四川育种家深感CIMMYT材料不好利用。我们在早期的教训也是深刻的，都希望能从引进材料中直接筛选出品种。1995年，邹裕春根据多年与CIMMYT育种合作的经验，特别是对大量CIMMYT材料在CIMMYT和在成都两地不同环境种植的变化观察，就明确提出引进CIMMYT材料在全国不同生态条件下利用的两种不同模式，在新疆、青海、宁夏等地可直接推广利用或一次杂交就可能选出品种；而在长江流域直接利用的可能性极小，应渐进利用，逐步杂交，不断选择，才能选育出有突出新性状的新品种。我们用CIMMYT材料一次杂交成功的品种极少，而且双亲性状必须互补性很强。其余基本上都是逐步杂交或用已具CIMMYT血缘的品种或中间材料再杂交获得的。这也是四川虽较早就与CIMMYT合作，但用CIMMYT材料育成的审定品种则主要在2001年以后的原因。根据我们的经验，在利用CIMMYT材料选育品种时，还要重视对高代选系间的筛选和从高代选系中再分离出来的单株选拔与鉴定，川麦30、川麦32、川麦36都是在高代选系中再分离出来的单株选拔、鉴定育成的。

3.3 依靠合作团队

对CIMMYT材料及四川育种、生产都有较好认识的专家，与经CIMMYT培训的青年专家组成的合作团队，是双方合作成功的重要保证。自1989年四川省农业科学院派出第一位青年专家李跃建赴CIMMYT培训，先后有5位青年专家到CIMMYT学习。四川省农业科学院与CIMMYT的成功合作，就是依靠了这批曾在CIMMYT培训或短期合作考察的专家，他们对CIMMYT材料及育种方法都有较好了解，通过分工合作，吸取CIMMYT的成功经验及有用资源，实现了提高四川育种水平的共同目标。

4 前景展望与建议

建议在原有工作基础上重点加强以下几个领域的合作：(1) 提高合作层次和质量。双方签署长期合作协议，建立清晰的战略合作框架，进一步在小麦育种和农艺技术研究上明确具体合作内容和形式，形成长期合作共赢机制。(2) 深化和丰富合作方式，特别要继续坚持种质资交换和利用。进一步加强小麦条锈病持久性抗性、品质改良、白皮抗穗发芽、高产广适品种选育及近缘种质的交换和利用，加强生物技术育种合作。(3) 继续加强对青年科技人员的培训。加大人员互派，促进学术和信息交流，特别要通过多种方式培养青年科技人员，促进他们成为作物研究、创新与合作的生力军。

参考文献

[1] 黄钢，邹裕春，汤永禄，杨恩年．四川-CIMMYT小麦穿梭育种中的农艺学合作研究．西南农业学报，2007，20 (2)：191-198.

[2] 李跃建，张颙，陈沧桑．四川省农业科学院-CIMMYT国际科技合作回顾与展望．西南农业学报，2007，20 (2)：173-177.

[3] 张颙，杨武云，胡晓蓉，等．源于硬粒小麦——节节麦人工合成种的高产抗病小麦新品种川麦42主要农艺性状分析．西南农业学报，2004，17 (2)：141-145.

[4] Wu Ling, Xia Xian-chun, Zheng You-liang, Zhang Zheng-yu, Zhu Hua-zhong, Liu Yong-jian, Yang En-nian, Li Shi-zhao and He Zhong-hu. QTL Mapping for adult-plant resistance to stripe rust in a common wheat RIL population derived from Chuanmai 32/Chuanyu 12. Journal of Integrative Agriculture, 2001, 11 (11): 1775-1782.

[5] Yang E N, Rosewarne G M, Herrera-Foessel S A, Huerta-Espino J, Tang Z X, Sun C F, Ren Z L, Singh R P. QTL analysis of the spring wheat "Chapio" identifies stable stripe rust resistance despite inter-continental genotype x environment interactions. Theor Appl Genet, 2013, 126: 1721-1732.

[6] Rosewarne G M, Li Z F, Singh R P, Yang E N, Herrera-Foessel S A, Huerta-Espino J. Different QTLs are associated with leaf rust resistance in wheat between China and Mexico. Mol Breeding, 2015, 35: 127.

[7] Yang E N, Zou Y C, YANG W Y, Tang Y G, He Z H, Singh R P. Breeding adult plant resistance to stripe rust in spring bread wheat germplasm adapted to Sichuan Province of China. Czech Journal of Genetics and Plant Breeding, 2011 (Special Issue), 47: S165-S168.

[8] 刘金栋，杨恩年，肖永贵，陈新民，伍玲，白斌，李在峰，夏先春，Garry M. ROSEWARNE，何中虎．普通小麦兼抗型成株抗性品系的培育、鉴定与分子检测．作物学报，2015，41 (10)：1472-1480.

CIMMYT种质对四川、云南、甘肃和新疆春性小麦产量遗传进展的贡献

张勇[1]，李式昭[2]，吴振录[3]，杨文雄[4]，
于亚雄[5]，夏先春[1]，何中虎[1,6]

[1]中国农业科学院作物科学研究所/国家小麦改良中心/国家农作物基因资源与基因改良重大科学工程，北京 100081；[2]四川省农业科学院作物研究所，成都 610066；[3]新疆农业科学院核技术生物技术研究所，乌鲁木齐 830000；[4]甘肃省农业科学院作物研究所，兰州 730070；[5]云南省农业科学院粮食作物研究所，昆明 650205；[6]CIMMYT 中国办事处，北京 100081

摘要： 研究历史品种产量潜力变化规律有助于提高小麦育种水平。来自四川、云南、甘肃和新疆的代表性品种连续两年分别种植在四川成都、云南丽江、甘肃武威和新疆昌吉，在肥水供应充足、控制病虫害和倒伏的条件下分析了产量和相关农艺性状的变化趋势。结果表明，四川、云南、甘肃和新疆品种的产量随育成年份显著增加，年遗传进展分别为 0.73%、0.34%、0.58%和 1.43%。四川品种的产量遗传进展与产量构成因子变化关系不密切；云南主要来自于减少穗数和增加穗粒数；甘肃主要来自于增加穗粒数；新疆主要来自于增加主穗粒重和收获指数，并与成熟期提早及株高降低有一定关系。各地区品种中 *Rht-B1b* 和 *Rht-D1b* 矮秆基因均来自于 CIMMYT 种质，其产量潜力的提高主要得益于 CIMMYT 种质的引进和有效利用，其中四川和云南主要利用 CIMMYT 种质对条锈病的抗性；甘肃和新疆主要利用其矮秆、高产、穗粒数多及广泛适应性特性。

关键词： 普通小麦；产量潜力；CIMMYT 种质

Contribution of CIMMYT wheat germplasm to the genetic improvement of grain yield in Sichuan, Yunnan, Gansu and Xinjiang provinces

Zhang Yong[1], Li Shi-zhao[2], Wu Zhen-lu[3], Yang Wen-xiong[4],
Yu Ya-xiong[5], Xia Xian-chun[1], He Zhong-hu[1,6]

[1]*Crop Science Institute / National Wheat Improvement Center, Chinese Academy of Agriculture Sciences* (*CAAS*), 100081, *Beijing*; [2]*Crop Research Institute, Sichuan AAS*, 610066, *Chengdu*; [3]*China Institute of Nuclear & Biological Technology, Xinjiang AAS*, 830000, *Urumqi*; [4]*Crop Research Institute, Gansu AAS*, 730070, *Lanzhou*; [5]*Crop Research Institute, Yunnan Academy of Agricultural Sciences*, 650205, *Kunming*; [6]*CIMMYT-China, C/O, CAAS*, 100081, *Beijing*

Abstract: Information on advances in wheat (*Triticum aestivum* L.) productivity is essential for genetic improvement on yield potential. Four yield potential trials with totally 59 leading cultivars from Sichuan, Yunnan, Gansu, and Xinjiang, were conducted using a randomized complete block design with three replications under controlled environments in two successive seasons from 2007 to 2009, planted in Chengdu in Sichuan province, Lijiang in Yunnan province, Wuwei in Gansu province, and Changji in Xinjiang, respectively. Molecular markers were used to detect the presence of dwarfing genes and 1B/1R translocation. The results indicated that the annual genetic gain in yield in Sichuan, Yunnan, Gansu, and Xinjiang was 0.73%, 0.34%, 0.58%, and 1.43%, respectively. There were no obvious trend of yield component improvement for yield increase in Sichuan province; while reduced spikes per square meter and increased kernels per spike were the main approach for yield increase in Yunnan province; increased kernels per spike were the main approach for yield increase in Gansu province; and increased kernel weight of main spike and harvest index were the main approach for yield increase in Xinjiang province, together with the contribution from reduced plant height and earlier maturity. It also indicated that the dwarfing genes *Rht-B1b* and *Rht-D1b* were all from CIMMYT lines, and the significant progresses of genetic gain in yield in the four provinces were mainly due to the direct and indirect use of CIMMYT germplasm. Stripe rust resistance was the main factor for using CIMMYT wheat in Sichuan and Yunnan; while CIMMYT wheats contributes to high yield potential with high kernel number per spike, short plant height, and wide adaptability in Xinjiang and Gansu.

Key words: *T. aestivum*; yield potential; CIMMYT germplasm

不断提高单位面积产量是我国小麦育种最重要的目标[1-3]。单产的显著提高和新品种的大面积推广为小麦的持续发展提供了重要保障，而品种产量潜力的提高对增加单产有重要作用[3]。对历史主栽品种和新育成品系进行系统分析有助于了解品种更换过程中产量潜力构成因素的变化情况，从而为下一步的品种改良提供参考依据[1-3]。

国外已对小麦品种产量遗传潜力进行了大量研究，不同国家的产量潜力遗传进展介于 0.2%和 1.4%之间，单位面积穗数和粒数增加、收获指数提高和株高降低是单产增加的主要原因，矮秆基因和 1B/1R 易位系的利用在产量遗传改良中发挥了重要作用[4-10]。我国也开展了部分相关研究，其中 Zhou 等的研究较为系统，认为 1B/1R 易位系和矮秆基因的利用对我国北方冬麦区品种改良贡献较大，株高降低、穗粒重和收获指数的显著提高是产量提高的关键[1,2]。但国内有关该方面的研究主要集中于北方冬麦区以及长江流域的四川和江苏，对云南、甘肃和新疆等地区尚没有涉及[1,2,11-16]。

小麦是四川、云南、甘肃和新疆的重要作物，这些地区的品种改良在很大程度上得益于对国际玉米小麦改良中心（CIMMYT）种质的有效改造和利用[17-20]。CIMMYT 以春小麦育种举世闻名，所育成品种丰产性好，植株较矮、株型紧凑、较抗倒伏、品质较好、抗病性强[21]，可在云南和新疆等地直接推广种植，并在甘肃等地用作杂交亲本或直接推广应用[18]。自 20 世纪 70 年代以来，CIMMYT 种质在四川、云南、甘肃和新疆经直接引进或改造利用育成了 127 个品种[3,19,20]，其中云南、甘肃和新疆直接推广利用的品种达 21 个（本实验室内部资料）。本文在肥水供应充足、严格防病虫和防倒伏条件下对四川、云南、甘肃、新疆地区 CIMMYT 种质育成品种和历史主栽品种及部分近期育成品系的产量潜力及主要农艺性状的遗传进展进行研究，并对矮秆基因和 1B/1R 易位系的利用状况进行分析，旨在探讨 CIMMYT 种质对上述地区小麦产量潜力的贡献和品种产量潜力进一步提高的途径，为今后育种提供理论依据。

1 材料与方法

1.1 试验材料

来自四川、云南、甘肃、新疆共 59 份 CIMMYT 种质育成品种和各时期主栽品种及近期新育成品系于 2007—2009 年度分别种植在四川成都、云南丽江、甘肃武威和新疆昌吉，这些地点的产量水平高，能较好地反映当地品种的产量潜力。试验材料名称及其育成年份等信息详见表 1。供试材料基本反映了各地区主栽品种的演变和发展概况。四川省试验是在 Zhou 等[2]对 2000 年前育成品种研究的基础上进行，因此只选用了 2000 年后育成的品种。

表 1 参试品种名称、育成年份和矮秆基因携带等情况

Table 1 Cultivars tested, year of release, dwarfing genes, and status of 1B/1R translocation

品种 Cultivar	育成年份 Year of release	系谱 Pedigree	矮秆基因 Dwarfing gene	1B/1R
四川				
川麦 107 Chuanmai 107	2000	2469/80-28-7	—	N
川农 16 Chuannong 16	2002	川育 12/87-422	*Rht-B1b*	P
川麦 39 Chuanmai 39	2003	Milan 's'/90-7	*Rht-B1b*	N
川麦 42 Chuanmai 42	2003	Syn-CD768/SW89-3243//川 6415	*Rht-B1b*	N
川麦 43 Chuanmai 43	2004	Syn-CD768/SW89-3243//川 6415	*Rht-B1b*	N
川麦 44 Chuanmai 44	2004	96 夏 440/贵农 21	*Rht-D1b*, *Rht8c*	N
资麦 1 号	2006	绵阳 29/川麦 25	*Rht-B1b*, *Rht-D1b*	N
川育 20 Chuanyu 20	2006	SW3243//35050/21530	*Rht8c*	P
西科麦 4 号	2007	墨 460/9601-3	*Rht-B1b*	N
川 07005 Chuan 07005	新品系	川农 16//贵农 21/3295	*Rht8c*	N
云南				
滇西洋麦 Dianxiyangmai	1945	地方品种	*Rht8c*	N
阿勃 Abbondanza	1957	Abondaza	—	N
云南 778 Yunnan 778	1965	南大 2419 选系	—	N
南原 1 号 Nanyuan 1	1972	南大 2419/Minn 2-50-25	—	N
风麦 13 Fengmai 13	1975	云南 778/ Orofen	—	N
查平戈 Chapingo	1975	CIMMYT 种质 Chapingo F74	*Rht-D1b*	N
墨波	1975	CIMMYT 种质 Potam S70	*Rht-D1b*	N
墨沙	1975	CIMMYT 种质 Saric F70	*Rht-D1b*	N
0230	1980	PAKF4/6313/3/TOB/CTFN//BB/4/BMAN/ON//CAL/5/MAYA74 "S"	*Rht-D1b*	N
0103	1985	CIMMYT 种质 Veery "S"	*Rht-D1b*	N
0483	1985	EMU "S" /MRS//KAL/BB	*Rht-D1b*	N
精选 9 号	1988	CIMMYT 种质 Veery "S"	*Rht-D1b*	N
国际 13 Guoji 13	1990	CIMMYT 种质 Veery "S"	*Rht-B1b*	P
风麦 24 Fengmai 24	1992	云麦 36/Mexico 965	—	N
E001	1994	CIMMYT 种质 Attila	*Rht-B1b*	N
云麦 39 Yunmai 39	1994	云麦 29/Flicker	*Rht-B1b*, *Rht8c*	P
R101	1997	AGA/4* HORKS	*Rht-B1b*	N
德麦 4 号 Demai 4	1997	毕麦 5 号/II8156 (墨) //中引 1022	*Rht-B1b*	P
绵阳 20 Mianyang 20	1997	AGA/4* HORKS	*Rht8c*	N
云麦 42 Yunmai 42	1999	抗锈 782/云麦 29//YR70-PAM	*Rht8c*	N
靖麦 7 号 Jingmai 7	2000	436/ 092-1 (墨)	—	P
云选 11-12 Yunxuan 11-12	2003	Ning8391//SHA4/LIRA	*Rht8c*	N
云麦 47 Yunmai 47	2004	79213-194/92B-4074	*Rht-B1b*	P

（续）

品种 Cultivar	育成年份 Year of release	系谱 Pedigree	矮秆基因 Dwarfing gene	1B/1R
云麦 57 Yunmai 57	2008	PFAU/Milan	*Rht-B1b*，*Rht8c*	N
甘肃				
甘麦 8 号 Ganmai 8	1964	51 麦/ Abondaza	—	N
临农 14 Linnong 14	1975	Funo/新疆大颗子	—	N
张春 9 号 Zhangchun 9	1975	民选 116/ Abondaza	—	N
陇春 8 号 Longchun 8	1976	甘麦 8 号选系	—	N
宁春 4 号 Lingmai 4	1980	Sonora64/宏图	*Rht-D1b*	N
晋 2148 Jin 2148	1981	晋江赤仔/华东 5 号//Orofen/3/瑞梯	—	N
陇春 10 号 Longchun 10	1981	70-84-2-1/墨西哥 27 号	*Rht8c*	N
武春 121 Wuchun 121	1985	甘麦 8 号/Nuri F70	*Rht8c*	P
花培 764 Huapei 764	1988	Cajeme F71/东乡大头兰麦//阿 4	*Rht-D1b*	N
甘春 16 Ganchun 16	1989	单 357/甘春 11 号	—	N
张春 20 Zhangchun 20	1989	7606/抗黄矮病二体异附加系 L1	*Rht8c*	N
高原 602 Gaoyuan 602	1991	高原 182/3987-88（3）	—	N
陇春 8139 Longchun 8139	1993	陇春 7 号/68-73-20-3	—	N
临麦 30 Linmai 30	1996	74503-1-7-1-2/07802	—	N
陇春 15 Longchun 15	1996	750025/山前麦	—	N
甘春 20 Ganchun 20	1997	中作 8131-1/甘 630（88-862/630）	—	N
陇春 20 Longchun 20	2001	832-748/0103（Veery "S"）	—	P
武春 3 号 Wuchun 3	2001	石 1269 系选	*Rht-B1b*	P
陇春 23 Longchun 23	2004	CMBW90M4860-0TOPY-16M-1Y-010M-010Y-1M-0	*Rht-B1b*	P
新疆				
喀什白皮 Kashibaipi	1975	地方品种	—	N
新春 2 号 Xinchun 2	1984	Siete Cerros/奇春 4 号	*Rht-B1b*	N
新春 3 号 Xinchun 3	1986	Siete Cerros/奇春 4 号	*Rht-B1b*	N
新春 6 号 Xinchun 6	1993	中 7906/新春 2 号	*Rht-B1b*	N
新春 23 号 Xinchun 23	2006	CIMMYT 种质 Kambara	*Rht-B1b*，*Rht8c*	N
新春 26 号 Xinchun 26	2007	CM33027/新春 6 号	*Rht-B1b*	N

— 表示不含已知矮秆基因，P 和 N 分别表示有和没有。

“—”, indicates no known dwarfing gene; “P”, present; N, not present.

1.2 田间设计

采用随机区组设计，3 次重复，小区面积 7～10m^2。播期和播量同当地品种比较试验，播种时撒毒谷防治地下害虫。各试点试验地肥力水平均较高，并于播种前施足底肥，拔节期浇水时结合追肥，能充分满足试验要求。小区人工除草，孕穗、灌浆期喷氧化乐果和粉锈宁各一次，防止蚜虫和病害发生。孕穗至成熟期间用竹竿搭架编织防倒网防止倒伏，网眼直径 20cm，防倒网随植株的生长逐渐抬高，使网眼始终处于植株中上部。抽穗至成熟期间防止鸟类危害，减少产量损失。

1.3 性状调查

包括抽穗期（播种至抽穗天数）、成熟期（播种至成熟天数）、株高（cm）、穗数（穗/m^2）、穗粒数（粒/穗）、千粒重（g）和产量（t/hm^2）。收获前每小区中间选取两个长 50cm 的样段，贴地面收割，随机选取 30 个主穗分析主穗粒重，同时查穗数后放入纱袋，晒干脱粒，茎秆剪碎后置 60℃烘箱 24h 后称重，计算收获指数，并根据穗数和千粒重计算穗粒数。收获时去除两边行，晒干后称重，并折算公顷产量。其中云南和甘肃未进行主穗粒重和收获指数调查，甘肃未记载抽穗期和收获期。

1.4 1B/1R 易位系和矮秆基因分析

Rht-B1b、*Rht-D1b*、*Rht8c* 和 1B/1R 易位系检测方法按 Zhou 等[1,2]。

1.5 统计分析

用 Statistical Analysis System[22]统计分析软件进行方差分析和品种间各性状多重比较，计算基本统计量。按下列公式计算产量等性状遗传进展：$\ln(y_i)=a+bx_i+u$，其中 y_i 和 x_i 分别代表品种 i 的产量等性状值和育成年份，$\ln(y_i)$ 为 y_i 的自然对数，a 为方程截距，b 为遗传进展，u 为残差。性状间遗传相关分析按 Becker[23]方法计算。

2 结果与分析

方差分析表明（表略），各试点产量和穗数、穗粒数、主穗粒重、收获指数、株高，四川、云南和新疆点千粒重以及四川、新疆点抽穗期和成熟期的品种效应均达 0.05 或 0.01 显著水平，年度相关效应和年度内重复效应不显著。由于品种性状演变趋势在各点两年试验中表现基本一致，因此以各点试验平均值进行品种各性状分析。将上述 4 点品种间各性状均值多重比较和遗传进展分析结果分列于表 2 至表 5 和图 1。

表 2 四川品种农艺性状遗传进展

Table 2 Genetic gain of cultivars on agronomic traits in Sichuan Province

品种	产量 GY	穗数 SN	穗粒数 GPS	千粒重 TKW	主穗粒重 MSKW	收获指数 HI	株高 PH	抽穗期 DH	成熟期 DM
川麦 107	7.68ab	385de	36.5b	47.3c	1.73ab	44.7b	97c	127f	183d
川农 16	7.65ab	483ab	28.4e	47.5c	1.35d	46.2ab	85e	122h	180f
川麦 39	6.90b	372e	29.2d	47.7c	1.39cd	36.2d	98c	130a	188a
川麦 42	8.13a	433bcd	31.6c	54.4a	1.72ab	47.3a	95c	124g	183d
川麦 43	8.34a	486a	30.8d	51.2ab	1.56bc	47.7a	90d	127ef	185c
川麦 44	7.79a	393cde	40.6a	42.7cd	1.74ab	48.0a	86e	127de	183d
资麦 1 号	8.18a	470ab	34.5c	43.5d	1.50cd	47.5a	89d	128cd	183d
川育 20	7.76a	395cde	36.8b	48.5bc	1.78a	44.0b	99b	128c	185c
西科麦 4 号	8.13a	441abc	32.2c	48.9bc	1.56bc	41.0c	101a	129b	187b
川 07005	8.04a	474ab	28.1e	51.8ab	1.45cd	46.0ab	95c	122h	182e
均值	7.86±0.9	433±60.2	32.9±4.9	48.3±4.8	1.58±0.2	44.9±4.2	93.8±7.1	126±3.1	184±6.7
遗传进展	0.73*	17.63	−0.58	0.04	−0.59	0.10	0.57	−0.05	0.05
R^2	0.16	0.14	0.04	0.02	0.03	0.01	0.06	0.01	0.01

DH, DM, PH, SN, GPS, TKW, MSKW, GY and HI each represents days to heading, days to maturity, plant height, spike number, grains per spike, thousand kernel weight, main spike kernel weight, grain yield, and harvest index, respectively. The same as below.

同列中不相同字母表示 0.05 水平差异显著。

Different letters in the same column indicate significant difference at $P=0.05$.

*，**和*** 分别表示 0.05，0.01 和 0.001 显著水平。

*, **, and *** each indicates significance at 0.05, 0.01, and 0.001 probability level, respectively.

2.1 四川品种产量及主要农艺性状演变趋势

四川供试品种平均产量为 7.86t/hm²，2000 年后不同年份育成品种的产量随推广时期显著增加，从 2003 年川麦 39 的 6.90hm² 到 2004 年川麦 43 的 8.34hm²，平均年遗传进展 0.73%（表 2，图 1a），这与 Zhou 等[2]对 2000 年前育成品种的研究结果基本一致。穗数、穗粒数、千粒重、主穗粒重、收获指数、株高、抽穗期和成熟期遗传进展均不显著。并可看出矮秆基因 *Rht-B1b* 得到了广泛利用，其频率高达 60%，川农 16 和川育 20 还携带 1B/1R 易位系（表 2）。川麦 107 于 2000 年分别通过四川和长江上游审定，表现广适、中抗条锈病，是2000—2006年的主栽品种，年最大推广面积近500万亩*，至今仍在百万亩以上，并在云南和贵州有较大面积，于2003年开始作为四川省和国家区域试验对照。川农16和川麦39的产量均低于川麦107，但差异不显著，表现株高较矮、中抗至高抗条锈病；其中川农16是CIMMYT种质Alondra的衍生后代，川麦39是CIMMYT种质Milan的衍生后代，为四川审定的第一个强筋品种。之后育成品种的产量较川麦107均有一定程度的提高，尽管未达显著水平；其中川麦42、川麦43、资麦1号、西科麦4号和川07005的产量均超过8.00t/hm^2，显著高于川麦39；川麦42和川麦43为姊妹系，是CIMMYT小麦人工合成种的后代。川麦42是四川省迄今为止在区域试验中产量最高的品种，平均比对照川麦107增产16%，对条锈病免疫，但感白粉和赤霉病，近三年推广面积均在百万亩以上，已成为当地的主栽品种，于2010年成为长江上游区域试验的对照品种。川麦43的产量潜力最高，但与川麦42差异不显著，表现株高较矮、对条锈病免疫。西科麦4号、资麦1号和川07005对条锈病高抗至免疫，其中西科麦4号和川07005为CIMMYT种质衍生后代，资麦1号的父本川麦25是四川审定的第一个含CIMMYT血缘的品种。从以上分析可看出，四川省2000年后育成品种的产量潜力有了明显提高，但其产量的遗传进展与产量构成因子变化关系并不密切，这与育种者采用不同的产量育种技术路线有关。但从以上分析仍可看出，其产量遗传进展与CIMMYT种质的贡献密切相关，主要得益于人工合成六倍体小麦的应用和抗条锈病性的显著提高[24]。除川麦107外，其余品种均是CIMMYT材料的后代。

2.2 云南品种产量及主要农艺性状演变趋势

云南供试品种平均产量为6.10t/hm^2，不同年份育成品种的产量随推广时期显著增加，从20世纪50年代阿勃的4.82t/hm^{2}到1994年S001的6.94t/hm^2，平均年遗传进展0.34%（表3，图1b）；穗粒数显著增加，平均年遗传进展0.45%；抽穗期显著推迟，平均年遗传进展0.06%；千粒重、株高和成熟期遗传进展不显著。还可看出，随着20世纪70年代中期CIMMYT种质的引进，*Rht-B1b*和*Rht-D1b*等矮秆基因得到了广泛利用，国际13、靖麦7号和云麦47等的1B/1R易位系均来自于CIMMYT种质（表1）。该地区品种产量演变大致可分为四个阶段。第一阶段的滇西洋麦是新中国成立前云南中西部种植面积最大的地方品种，分布较广。第二阶段是意大利种质的广泛应用，1957年引入的阿勃表现广泛适应性，在20世纪60年代中期开始大面积推广，最高时播种面积近90万亩；云南778是意大利品种南大2419的选系，但更耐肥、抗倒伏、耐锈病力较强，年最大推广面积超过100万亩；南原1号是南大2419的衍生后代，产量达6.20t/hm^2，但株高较高，表现高产、广适、高抗三种锈病，适于在中等肥力水平下种植，是20世纪70年代的主栽品种；凤麦13结合了当时两个主栽品种的优点，是20世纪70年代末和80年代初的主栽品种，年播种面积曾在100万亩以上，之后由于感条锈、且株高较高而淘汰。第三个阶段是CIMMYT种质的引进与直接利用，包括查平戈、墨波、墨沙、0230、0103、0483、精选9号、国际13等品种，产量有所提高，介于5.53～6.85t/hm^2，矮秆基因*Rht-D1b*随之引入，株高降低到90cm左右，适于水地种植；其中精选9号的产量最高，高抗条锈病，于1988年通过审定，1991年秋播面积达42万亩。第四个阶段是CIMMYT种质的改造，包括1990年后用CIMMYT材料育成的凤麦24、云麦39、德麦4号、靖麦7号、云麦42和云麦47，产量介于5.77～6.67t/hm^2；其中云麦42和云麦47产量超过6.50t/hm^2，分别于1999年和2003年通过审定；云麦42株高较高，为优质强筋品种，分蘖力强、抗病性好、灌浆速度快，适于旱地种植；云麦47株高较矮，大穗大粒、耐肥抗倒，高抗白粉病、中抗条锈病，适于在高肥水条件下种植。而同期经引进审定的CIMMYT种质E001、R101、云选11-12和云麦57等产量介于6.22～6.94t/hm^2，其中E001产量最高，表现株高较矮、适应性广、高抗条锈病；云麦57其次，表现高抗条锈病、中抗白粉病，为优质强筋品种，于2008年通过审定。由于墨西哥的气候条件与云南相似性很高，所以CIMMYT种质在此可以直接利用。从以上分析可看出，云南品种产量潜力的显著提高与穗粒数增加和穗数减少有关，主要得益于对国外种质特别是20世纪70年代后CIMMYT种质的有效利用以及由此带来的品种抗条锈病性的提高。

---

* 亩为非法定计量单位。15亩=1公顷——编者注

表 3　云南品种农艺性状遗传进展

Table 3　Genetic gain of cultivars on agronomic traits in Yunnan Province

品种	产量	穗数	穗粒数	千粒重	株高	抽穗期	成熟期
滇西洋麦	5.32de	633abc	28.6cd	35.4cdefg	95cdefg	125a	177a
阿勃	4.82e	474bcde	46.2ab	34.4efghij	108bc	127a	181a
云南 778	5.55cde	464bcde	32.8bcd	35.2defgh	123a	127a	174a
南原 1 号	6.20abcd	519abcd	42.8ab	34.9efgh	106bcd	129a	180a
凤麦 13	6.11bcd	498bcde	44.4ab	31.8ghijk	102bcdef	127a	177a
查平戈	6.42abcd	619abcd	28.4d	39.6ab	95cdefg	122a	171a
墨波	6.35abcd	436cde	38.5abcd	39.4ab	89fg	124a	175a
墨沙	5.53cde	706a	28.6cd	37.4bcde	73h	125a	171a
0230	6.47abcd	620abcd	38.1abcd	36.1bcdef	89fg	127a	176a
0103	5.68bcde	512abcd	32.9bcd	33.7fghijk	93defg	126a	173a
0483	5.51cde	540abcd	40.6abcd	31.6hijk	89fg	128a	176a
精选 9 号	6.85ab	546abcd	34.3abcd	36.2bcdef	95cdefg	130a	178a
国际 13	5.70bcde	667ab	40.5abcd	30.4k	88fg	128a	174a
凤麦 24	5.77bcde	497bcde	46.8ab	38.8abcd	104bcde	129a	178a
E001	6.94a	550abcd	38.9abcd	31.1ijk	90fg	127a	174a
云麦 39	6.35abcde	443cde	42.3abcd	33.6fghijk	95cdefg	127a	180a
R101	6.22abcd	570abcd	38.8abcd	32.2ghijk	92efg	128a	175a
德麦 4 号	6.22abcd	434cde	41.5abcd	34.5efghi	94cdefg	127a	175a
绵阳 20	6.31abcd	460cde	37.8abcd	38.1bcde	87g	130a	180a
云麦 42	6.75abc	520abcd	42.2abcd	42.0a	99cdefg	130a	178a
靖麦 7 号	5.82bcde	417de	42.6abc	37.6bcde	114ab	133a	181a
云选 11-12	6.29abcd	458cde	40.5abcd	32.2ghijk	90fg	125a	174a
云麦 47	6.67abc	297e	47.3a	39.0abc	84g	129a	174a
云麦 57	6.50abcd	455cde	44.2ab	30.7jk	92defg	130a	177a
均值	6.10±1.0	514±108.6	39.1±6.9	35.2±3.4	94.9±11.7	127±4.6	176±4.2
遗传进展	0.34**	−6.03*	0.45*	0.00	−0.24	0.06*	0.00
R^2	0.37	0.16	0.19	0.01	0.10	0.22	0.01

2.3　甘肃品种产量及主要农艺性状演变趋势

甘肃供试品种平均产量为 7.35t/hm²。不同年份育成品种的产量随推广时期显著增加，从 1964 年甘麦 8 号的 6.22t/hm² 到 2001 年武春 3 号的 8.40t/hm²，平均年遗传进展 0.58%（表 4，图 1c）；穗粒数显著增加，平均年遗传进展 0.33%；穗数、千粒重和株高遗传进展不显著；还可看出，随着 20 世纪 80 年代 CIMMYT 种质的引进，*Rht-B1b* 和 *Rht-D1b* 等矮秆基因随之引入，*Rht-D1b* 最早出现在宁春 4 号，来自其亲本 Sonora 64。该地区品种产量演变大致可分为三个阶段。第一阶段育成的甘麦 8 号较好结合了双亲的优良性状，高抗条锈和叶锈病、适应性广，最大年推广面积曾超过 1 000 万亩，在甘肃、宁夏、陕西汉中及新疆阿尔泰、哈密等地大面积种植。临农 14 和张春 9 号产量有所提高，但在产量潜力和综合性状方面并无突破；1976 年从甘麦 8 号中系选得到的陇春 8 号产量显著提高，最大年推广面积曾在 160 万亩以上。第二个阶段始于宁夏于 1981 年育成的品种宁春 4 号，其产量高达 8.12t/hm²，穗粒数显著高于之前推广的甘麦 8 号、陇春 8 号等品种，并基本保留了母本即 CIMMYT 引进种质 Sonora 64 矮秆、高产、广适、品质优良的特性，适于在宁夏、内蒙古、甘肃河西灌区、新疆等省区水地种植，目前年播种面积仍在300万亩以上，已累计推广近1亿亩，是我国春小麦种植面积最大、年限最长的品种。晋2148（来自福建）和陇春10号的产量潜力显著低于宁春4号。武春121是利用CIMMYT资源Nuri F70改造甘麦8号育成的，表现高产、矮秆、抗倒，但产量潜力并没有显著提高，1991年在河西走廊推广面积140万亩。此后育成和推广品种的产量介于7.35～7.80t/hm²，显著低于宁春4号，只有小面积推广种植。第三个阶段包括近期育成的武春3号和陇春23。这两个品种都携带1B/1R易位系和*Rht-B1b*矮秆基因，产量和穗粒数比之前育成品种均有所提高，株高显著降低，其中武春3号表现高产、优质、高抗条锈病、适应性广，于2001年审定，适宜在甘肃、宁夏、内蒙古、青海、新疆等省区水地种植；陇春23来自于CIMMYT种质CM4860，表现抗倒伏、高抗条锈病、中抗白粉病，于2004年审定，已在甘肃大面积推广，并已成为甘肃东片水地区域试验的对照品种。综合以上分析可知，甘肃品种产量潜力的显著提高与穗粒数增加有关，主要得益于对国外种质的引进和利用，1980年前以意大利品种为主，1980年后则主要来自于CIMMYT种质的贡献。

表4 甘肃品种农艺性状遗传进展

Table 4 Genetic gain of cultivars on agronomic traits in Gansu Province

品种	产量	穗数	穗粒数	千粒重	株高
甘麦8号	6.22g	510bcd	31.0efg	39.0a	100bcde
临农14	6.27g	512bcd	30.5fg	40.0a	102abcd
张春9号	6.54fg	504bcd	30.5fg	42.5a	92defg
陇春8号	6.90e	507bcd	30.5fg	44.5a	95def
宁春4号	8.12ab	559a	35.0abcd	42.5a	85g
晋2148	6.40g	516b	30.0g	41.0a	95def
陇春10号	6.77ef	499bcd	31.5defg	43.5a	96cdef
武春121	8.10ab	513bcd	34.5abcde	43.0a	99cdef
花培764	7.35d	515bc	33.5abcdefg	42.5a	96cdef
甘春16	7.42d	504bcd	31.5defg	39.5a	98bcde
张春20	7.80bc	491bcd	35.5abc	43.0a	89efg
高原602	7.55cd	507bcd	34.0abcdef	40.5a	107ab
陇春8139	7.64cd	489bcd	33.5abcdefg	42.5a	112a
临麦30	7.40d	486d	32.0cdefg	41.5a	106abc
陇春15	7.79bc	486cd	33.5abcdefg	44.0a	96cdef
甘春20	7.60cd	502bcd	32.5bcdefg	41.5a	96cdef
陇春20	7.69cd	507bcd	36.5a	38.5a	101bcd
武春3号	8.40a	503bcd	36.0ab	45.5a	87fg
陇春23	8.05ab	555a	36.0ab	43.5a	83g
均值	7.35±0.7	508±20.7	32.9±2.4	42.0±2.4	96.5±7.8
遗传进展	0.64***	−0.45	0.33**	0.01	−0.04
R^2	0.58	0.01	0.48	0.05	0.01

2.4 新疆品种产量及主要农艺性状演变趋势

新疆供试品种平均产量为7.24t/hm²，不同年份育成品种的产量随推广时期显著增加，从1975年喀什白皮的4.87t/hm²到2007年新春26的9.32t/hm²，平均年遗传进展1.43%（表5）；主穗粒重显著增加，平均年遗传进展0.98%；穗数、穗粒数、千粒重、收获指数、株高、抽穗期和成熟期遗传进展不显著。喀什白皮是20世纪70到80年代中期新疆春小麦推广面积最大的主栽品种（农家种），表现主穗粒重、穗粒数和收获指数较低、株高较高（不含任何已知的矮秆基因）、晚熟。随着20世纪80年代中期新春2号、新春3号的选育成功，新疆春小麦大面积单产迅速突破7.0t/hm²。新春2号和新春3号是姊妹系，为CIMMYT材料Siete Cerros的后代，其穗粒数、主穗粒重和收获指数显著提高，株高大幅降低（均含*Rht-B1b*矮秆基因）、抽穗期和成熟期显著提早，是1986到1995年新疆春小麦种植面积最大的主栽品种，其中新春2号在干热的准噶尔、哈密和塔里木盆地大面积种植单产多次超7.50t/hm²，1992年最大推广面积151万亩；新春3号主要分布在新疆较冷凉春麦区，1991年最大推广面积140万亩。新春6号选自CIMMYT材料中7906（CNO-Gallox Bb4A/K4496）和改良新春2号的后代，其主穗粒重和收获指数进一步提高，高产潜力大、株高较矮、秆硬、抗倒伏能力强、早熟，是20世纪90年代后期到2005年新疆种植面积最大的春小麦品种，大面积单产曾突破9.00t/hm²。新春23号为CIMMYT引进种质Kambara，在生产试验中比对照新春6号增产12.3%，于2006年通过审定，适合在中等或偏低肥力条件下种植。新春26选自CIMMYT种质CM33027和新春6号的后代，产量潜力显著提高，于2007年通过自治区审定，是一个高产优质面包面条兼用型品种，表现矮秆、抗倒伏能力较强，适于在北疆春麦区较高肥力条件下种植。综合以上分析可知，新疆品种产量潜力的显著提高与主穗粒重和收获指数增加、株高降低有关，主要得益于对CIMMYT种质的引进和有效利用。

表5 新疆主栽品种农艺性状遗传进展

Table 5 Genetic gain of cultivars on agronomic traits in Xinjiang Province

品种	产量	穗数	穗粒数	千粒重	主穗粒重	收获指数	株高	抽穗期	成熟期
喀什白皮	4.87e	506b	29.5c	37.5c	1.10d	30.3d	108a	53a	110a
新春2号	7.18c	488b	35.9a	41.9b	1.51b	42.3b	92c	51c	107c
新春3号	6.75d	517ab	37.3a	33.8d	1.27c	37.5c	94b	52b	108b
新春6号	7.44c	514ab	32.9b	49.2a	1.63a	45.7a	81d	46f	101e
新春23	7.87b	509ab	37.6a	42.0b	1.57ab	43.0b	92c	50d	105d
新春26	9.32a	544a	35.8a	46.3a	1.66a	42.7b	88c	49e	107c
均值	7.24±1.5	513±46.7	34.8±3.6	41.8±6.6	1.46±0.3	40.3±6.4	92.3±9.3	50±2.8	106±3.4
遗传进展	1.43*	2.52	0.35	0.05	0.98*	0.76	−0.48	−0.22	−0.10
R^2	0.82	0.37	0.36	0.30	0.66	0.50	0.40	0.28	0.22

3 讨论

品种改良的主要目标是进一步提高产量潜力，增强其抗主要病虫害的能力并改善品质[3]。在Zhou等[1,2]对我国冬麦区小麦产量潜力研究基础上，本研究表明四川、云南、甘肃和新疆小麦产量潜力均有显著提高，遗传进展分别为0.73%、0.34%、0.58%和1.43%，其中新疆地区最高，云南地区最低，这与各地区的生态环境和品种类型密切相关。四川盆地地势较低，小麦主产区分布在海拔300～700m的地区，虽以雨养为主，但生长季温度高、湿度大、日照少、降水较多，水肥条件较好；云南小麦主产区分布在海拔1 000～2 400m的地区，有田麦（水地）和地麦（旱地）之分，肥力总体较低，基本没有灌溉条件；甘肃境内地势复杂，水地主要依赖于灌溉，产量高，但旱地面积较大，产量较低；新疆春小麦约90%为水浇地[3,19,20]；因此四川和新疆品种类型以水地为主，如本研究中所选用的川麦42等和新春2号、新春6号等；而云南和甘肃品种类型较多，水地和旱地兼顾，如云麦42和陇春8139为典型的旱地品种，云麦47以及CIMMYT引进种质云麦57和武春3号、陇春23为典型的水地品种，这可能是导致四川和新疆品种产量遗传进展较高，而云南和甘肃品种产量遗传进展相对较低的主要原因。

Zhou等[1,2]研究表明，我国北方冬麦区和长江流域小麦主产区的产量年遗传进展介于0.31%～1.23%，其中北京和山东较高，在1.0%以上；河北、河南、江苏和四川等较低；并指出，上述麦区的产量遗传进展与洛夫林10等1B/1R品种以及意大利品种的引进和利用有关，主要来自于1B/1R易位系和矮秆基因的利用以及与之相关的穗粒重和收获指数的提高和株高的降低。与之相比，本文中新疆春麦区的产量遗传进展较高，四川居中，甘肃和云南较低。地点间产量与各性状的变化趋势关系不密切。新疆品种的产量与主穗粒重（r=0.91，P<0.05）和收获指数（r=0.82，P<0.05）显著正相关，与株高（r=−0.76，P<0.05）显著负相关，主穗粒重与千粒重（r=0.85，P<0.05）和收获指数（r=0.96，P<0.01）、抽穗期与株高（r=0.90，P<0.01）显著正相关，且主穗粒重随推广时期显著增加，其产量遗传进展主要来自于增加主穗粒重和收获指数，并与株高降低和早熟有一定关系，进一步分析表明其产量潜力主要来自于改造CIMMYT种质，伴以直接推广，如新春2号和新春6号等都是CIMMYT材料的衍生后代，新春23号为CIMMYT种质Kambara。四川品种的产量遗传进展与产量构成因子变化关系不密切，其产量潜力主要来自于改造CIMMYT种质，并与品种的抗条锈病性提高有关，川麦42和川麦43等都是CIMMYT种质的后代，表现高抗至免疫条锈病。云南品种的产量与穗数（r=−0.53，P<0.01）、穗数与穗粒数（r=−0.60，P<0.01）显著负相关，且穗数随推广时期显著降低，穗粒数显著增加，其产量遗传进展主要来自于减少穗数和增加穗粒数；进一步分析表明其产量潜力主要来自于直接利用CIMMYT种质，伴以改造利用其条锈病抗性，如精选9号等直接引自CIMMYT，云麦42等是CIMMYT材料的后代，表现中抗至高抗条锈病。甘肃品种的产量与穗粒数（r=0.97，P<0.001）显著正相关，且穗粒数随推广时期显著增加，其产量遗传进展主要来自于增加穗粒数；进一步分析表明其产量潜力主要来自于改造CIMMYT材料，伴以直接利用，如宁春4号和武春3号均为CIMMYT材料的后代，陇春23直接来自CIMMYT种质CM4860。综合以上分析可知，四川和云南主要利用CIMMYT种质对条锈病的抗性，甘肃和新疆则主要利用其矮秆、高产、穗粒数多、适应性广泛的特性。更为重要的是，本文分析表明，上述麦区的*Rht-B1b*和*Rht-D1b*矮秆基因都来自CIMMYT，因而为这些地区降低株高、提高品种抗倒伏能力发挥了重要作用。CIMMYT种质不仅过去在我国春麦区小麦品种改良上起了主导作用[17-20,25-28]，今后还将发挥关键作用。

与Zhou等[2]的研究结果有所区别，本研究发现四川新育成品种中含有大量的矮秆基因，其中以*Rht-B1b*频率较高，占60%；云南则以*Rht-B1b*和*Rht-D1b*为主；甘肃品种中的矮秆基因利用频率则较低，19个品种仅7个含已知矮秆基因；而新疆约83%的品种含*Rht-B1b*矮秆基因，这与CIMMYT种质在上述地区的利用是直接相关的。川农16、川麦44和资麦1号分别含*Rht-B1b*等矮秆基因，株高低于90cm；川麦42和川麦43等分别含*Rht-B1b*等矮秆基因，株高介于90～101cm；川麦107株高97cm，不含任何已知的矮秆基因。云南的南原1号和靖麦7号等株高较高（>100cm），不含任何已知的矮秆基因；墨沙和云麦47含*Rht-B1b*等矮秆基因，株高低于75cm；E001和云麦39等含*Rht-B1b*等矮秆基因，株高介于88～95cm。甘肃的甘麦8号和陇春20等株高在100cm以上，不含任何已知的矮秆基因；宁春4号、武春3号和陇春23含*Rht-B1b*等矮秆基因，株高低于90cm。新疆的喀什白皮株高较高，不含任何已知的矮秆基因；新春2号等较低，分别含*Rht-B1b*等矮秆基因。说明利用已知的矮秆基因在四川进行矮化育种有一定作用，但效果不显著，其当前品种很可能含有一些未知的矮秆基因；而在云南、甘肃和新疆效果明显。此外，川麦16、云麦39和云麦47、武春3号和陇春23等都携带1B/1R易位系，说明1B/1R易位系在上述地区也得到了一定应用，可能对产量改良也起到了一定作用。

值得一提的是，由于四川是我国条锈病生理小种变化十分频繁的热点地区，其主栽品种川麦42已于2008年丧失条锈病抗性（杨武云，个人通讯），因此，我们主张利用国际上采用的微效多基因持久抗性方法来培育抗病品种。自2000年开始，CIMMYT与四川开展了该方面的合作研究，将四川育成品种或稳定高代选系送往CIMMYT与其具有微效多基因持久抗性且品质较好的选系进行杂交，在四川和CIMMYT穿梭选择，目前已获得了部分表现慢条锈性且产量较高的高代品系[28]，08RC2525和07RC391等在区域试验中表现突出。CIMMYT与云南也开展了类似合作，云麦60已于2010年通过云南省品种审定。因此，通过穿梭育种项目的实施，在病害重发区培育具有条锈病持久抗性品种是可能的。

需要说明的是，本研究中云南和甘肃两点所选材料包括了水地和旱地两种，对产量遗传进展影响较大，把历史上育成的代表性水地和旱地品种种植在同一种环境中分析其产量遗传进展具有一定的局限性。

由于工作量巨大，本研究未从生理性状方面对产量遗传进展进行解析，而当前小麦产量瓶颈的突破将取决于生理性状的进一步改良，灌浆期冠层温差等指标的应用将促进产量遗传潜力的进一步提高[29,30]。

4 结论

四川、云南、甘肃和新疆品种的产量随育成年份显著增加，年遗传进展0.34%～1.43%，其中新疆地区较高，云南地区较低。四川品种的产量遗传进展与产量构成因子变化关系不密切；云南主要来自于减少穗数和增加穗粒数；甘肃主要来自于增加穗粒数；新疆主要来自于增加主穗粒重和收获指数，并与株高降低和早熟有一定关系。各地区育成品种中*Rht-B1b*和*Rht-D1b*矮秆基因均来自于CIMMYT种质，其产量潜力的提高主要得益于CIMMYT种质的引进和有效利用，其中四川和云南主要利用CIMMYT种质对条锈病的抗性；甘肃和新疆主要利用其矮秆、高产、穗粒数多和适应性广泛的特性。

参考文献

[1] Zhou Y, He Z H, Sui X X, Xia X C, Zhang X K, Zhang G S. Genetic improvement of grain yield and associated traits in the Northern China Winter Wheat Region from 1960 to 2000. *Crop Sci*, 2007a, 47: 245-253.

[2] Zhou Y, Zhu H Z, Cai S B, He Z H, Zhang X K, Xia X C, Zhang G S. Genetic improvement of grain yield and associated traits in the southern China winter wheat region: 1949 to 2000. *Euphytica*, 2007b, 157: 465-473.

[3] Zhuang Q-S (庄巧生). Chinese Wheat Improvement and Pedigree Analysis (中国小麦品种改良及系谱分析). Beijing: China Agriculture Press, 2003. (in Chinese)

[4] Brancourt Hulmel M, Doussinault G, Lecomte C, Berard P, Buanec B L, Trottet M. Genetic improvement of agronomic traits of winter wheat cultivars released in France from 1946 to 1992. *Crop Sci*, 2003, 43: 37-45.

[5] Donmez E, Sears R G, Shroyer J P, Paulsen G M. Genetic gain in yield attributes of winter wheat in the Great Plains. *Crop Sci*, 2001, 41: 1412-1419.

[6] McCaig T N, DePauw R M. Breeding hard red spring wheat in western Canada: historical trends in yield and related variables. *Can J Plant Sci*, 1995, 75: 387-393.

[7] Perry M, DÒntuono M. Yield improvement and associated characteristics of some Australian spring wheat cultivars introduced between 1860 and 1982. *Austr J Agric Res*, 1989, 40: 457-472.

[8] Calderni D F, Dreccer M F, Slafer G A. Genetic improvement in wheat yield and associated traits: a reexamination of previous results and the latest trends. *Plant Breed*, 1995, 114: 108-112.

[9] Sayre K D, Rajaram S, Fischer R F. Yield potential progress in short bread wheat in northwest Mexico. *Crop Sci*, 1997, 37: 36-42.

[10] Ortiz Monasterio R, Sayre K D, Rajaram S, McMahon M. Genetic progress in wheat yield and nitrogen use efficiency under four nitrogen rates. *Crop Sci*, 1997, 37: 898-904.

[11] Wu Z-S (吴兆苏), Wei X-Z (魏燮中). Genetic improvement and trends in yield and its associated traits of wheat cultivars in middle and low reaches of Yangtse River. *Sci Agric Sin* (中国农业科学), 1984, (3): 14-20. (in Chinese with English abstract)

[12] Yu S-R (俞世蓉), Wu Z-S (吴兆苏), Yang Z-P (杨竹平). Genetic improvement in yield and its associated traits of wheat cultivars in northern Jiangsu Province since 1970s. *Sci Agric Sin* (中国农业科学), 1988, 21 (4): 15-21. (in Chinese with English abstract)

[13] Tian X-M (田笑明). Genetic improvement and trends in yield and its associated traits of wheat cultivars in Xinjiang. *Acta Agron Sin* (作物学报), 1991, (4): 297-320. (in Chinese with English abstract)

[14] Chen H-B (陈化榜), Li Q-Q (李晴祺). Genetic improvement in yield and its associated traits of wheat cultivars in Shandong Province since 1950s. *J Shandong Agric Unvi* (山东农业大学学报), 1991, 22 (1): 95-98. (in Chinese)

[15] Lei Z-S (雷振声), Lin Z-J (林作楫). Genetic improvement in yield and its associated traits of wheat cultivars in Henan Province. *Sci Agric Sin* (中国农业科学), 1995, 28 (suppl): 28-33. (in Chinese with English abstract)

[16] Xu W-G (许为钢), Hu L (胡琳), Wu Z-S (吴兆苏), Gai J-Y (盖钧镒). Studies on genetic improvement of yield and yield components of wheat cultivars in mid-Shaanxi area. *Acta Agron Sin* (作物学报), 2000, 26 (3): 352-358. (in Chinese with English ab-

stract)

[17] He Z-H (何中虎), Zhang A-M (张爱民). Advance of Wheat Breeding in China (中国小麦育种研究进展). Beijing: China Science and Technology Press, 2002. (in Chinese)

[18] Zhang Y (张勇), Wu Z-L (吴振录), Zhang A-M (张爱民), Maarten van Ginkel, He Z-H (何中虎). Adaptation of CIMMYT wheat germplasm in Chines spring wheat regions. *Sci Agric Sin* (中国农业科学), 2006, 39 (4): 655-663.

[19] He Z H, Rajaram S, Xin Z Y, Huang G Z (eds). A History of Wheat Breeding in China. Mexico, D. F.: CIMMYT, 2001.

[20] He Z H, Rajaram S. China/CIMMYT collaboration on wheat breeding and germplasm exchange: results of 10 years of shuttle breeding (1984-1994). *Wheat Special Report No. 46*. Mexico, D. F.: CIMMYT, 1997.

[21] Pingali P L (Eds). CIMMYT 1989-1999 world wheat facts and trends. Global wheat research in a changing world: challenges and achievements. Mexico, D. F.: CIMMYT. 1999.

[22] SAS Institute. SAS User's Guide: Statistics. SAS Institute, Cary, NC. 2000.

[23] Becker, W A. Manuel for quantitative genetics. 4th ed. Academic Enterprises Pullman, WA. 1984.

[24] Yang Wuyun, Liu Dengcai, Li Jun, Zhang Lianquan, Wei Huiting, Hu Xiaorong, Zheng Youliang, He Zhonghu, Zou Yuchun. Synthetic hexaploid wheat and its utilization for wheat genetic improvement in China. *J Genet Genomics*, 2009, 36: 539-546.

[25] Wan X-L (宛秀兰). A preliminary study on the adaptability of Mexican wheat varieties in China. *Acta Agro Sin* (作物学报), 1981, 17 (4): 249-257. (in Chinese with English abstract)

[26] Yao J-B (姚金保), Zhou C-F (周朝飞), Qian C-M (钱存鸣), Yao G-C (姚国才), Yang X-M (杨学明). The progress of the shuttling breeding program between Jiangsu and CIMMYT. *Tritical Crops* (麦类作物学报), 1998, 18 (5): 14-16. (in Chinese with English abstract)

[27] Yuan H-M (袁汉民), Wu S-J (吴淑筠), Zhang F-G (张富国), Qian X-X (钱晓曦). Studies on genetic resources of Mexican wheat in Ningxia. *Ningxia Agronomy and Forest* (宁夏农林科技), 1998, (4): 8-12. (in Chinese with English abstract)

[28] Zou Y-C (邹裕春), Yang W-Y (杨武云), Zhu H-Z (朱华忠), Yang E-N (杨恩年), Pu Z-J (蒲宗君), Wu L (伍玲), Zhang Y (张颙), Tang Y-L (汤永禄), Huang G (黄钢), Li Y-J (李跃建), He Z-H (何中虎), Ravi Singh, Rajaram S. Utilization of CIMMYT germplasm and breeding technologies in wheat improvement in Sichuan, China. *Southwest China J Agric Sci* (西南农业学报) 2007, 20 (2): 183-190. (in Chinese with English abstract)

[29] Fischer R A, Edmeades G O. Breeding and cereal yield progress. *Crop Sci*, 2010, 50: 85-98.

[30] Reynolds M, Foulkes M J, Slafer G A, Berry P, Parry M A J, Snape J W, Angus W J. Raising yield potential in wheat. *J. Exp. Bot*, 2009, 60: 1899-1918.

附　　录

附录 1　利用 CIMMYT 种质育成审定品种目录

（带 * 者为主栽品种）

品　名	组　合	审定年份与地区
北部与黄淮冬麦区		
中优 9507	中作 8131-1 选系	2001，北京
兰天 1 号	洛夫林 13/墨西哥 30	1988，甘肃
兰天 3 号*	洛夫林 13/墨西哥 30//天农 1 号	1994，甘肃
兰天 25*	92-72/ Mo（s）311	2009，甘肃
邯 6172*	邯 4032/中引 1 号	2001，河北、国家
鲁麦 17	山前麦/3/安徽 9 号/红壳殴柔白//拜尼莫 62	1990，山东
济南 17*	临汾 5064/鲁麦 13	1999，山东
济麦 19*	临汾 5064/鲁麦 13	2003，国家
临汾 5064	临汾 5694/SaricF74//临汾 5054	1988，山西
豫麦 62*	周 8425A/SW73295	1999，河南
郑麦 004*	豫麦 13/90M434//冀麦 38	2004，国家
豫农 949	郑太育 9215/90M434//90（232）	2005，河南
平安 7 号	洛麦 4 号//990111//WS89-5422	2008，河南
徐州 2962	叶考拉/徐州 32331	1985，江苏
徐洲 22	叶考拉/徐州 32331	1991，江苏
皖麦 2	St2422/464/那纳瑞 60	1983，安徽
皖麦 4	St2422/464/那纳瑞 60	1985，安徽
皖麦 33*	安农 8326/中作 8131-1	1997，安徽
长江中下游冬麦区		
宁麦 7	扬麦 3/SERI	1993，江苏
宁麦 10	上海 7//PRL"S"/VEERY"S"	1999，江苏
生抗 2	Alondra"S"/繁 60096	2000，江苏
鄂麦 12*	滇 750025-12/鄂麦 6 号	1992，湖北
鄂麦 18*	SKUA（USA）/865146（Mexico）/鄂麦 11	2002，湖北
湘麦 12	墨西哥 120/大理 63//万雅 2 号	1991，湖南
西南冬麦区		
绵农 1 号*	绵阳 11/Alondra "S"	1991，四川
绵农 2 号	（75-21-1×75-19）F4 ×（绵阳 11/ALD′S′）F3	1992，四川
绵农 3 号	（75-21-4×75-19）F4 ×（绵阳 11/ALD′S′）F3	1993，四川
绵农 4 号*	绵农 3 号姊妹系	1993，四川
川麦 25	1414/川育 5//Genaro 80	1994，四川
川麦 30*	1426/4/IR68-77/YAA//ALDS/3/YAZ/ST2022/983	1998，四川
川麦 32*	1900 "s" /宁 8439//1900	2001，四川、国家
川麦 35	SW1862/2469	2002，四川
川麦 36*	Milan ′s′/SW5193	2002，四川

（续）

品　名	组　合	审定年份与地区
西南冬麦区		
川麦 38	Syn-CD769/SW89-3243//川 6415	2003，四川
川麦 42*	Syn-CD769/SW89-3243//川 6415	2003，四川、国家
川育 18	川育 5/Milan 's'//94F2-4	2003，四川
川育 19	川育 5/Milan 's'//绵阳 26	2003，四川
绵阳 35	05363-8-1/绵优 2 号	2003，四川
西科麦 1 号	绵阳 88-304×墨西哥 M-212	2003，四川
川麦 39*	Milan 's'/90-7	2003，四川、国家
西科麦 2 号	川育 11/Milan 's'	2004，四川
绵麦 37*	SW2148/90-100	2004，四川
川麦 43*	Syn-CD769/SW89-3243//川 6415	2004，四川、国家
川麦 45	GH430/SW1862	2005，四川
川麦 47	Syn786/MY26//MY26	2005，四川
宜麦 8 号	宜 98-53/SW8188	2005，四川
川麦 48	SW8188/SW8688	2006，四川
资麦 1 号	绵阳 29/川麦 25	2006，四川
川育 20*	SW3243//35050/21530	2007，国家
西科麦 4 号	Milan/9601-3	2008，国家
川麦 51	1275-1/99-1522	2008，国家
川麦 52	川麦 36/ SW1862	2008，四川
川麦 53	477/绵农 4 号//Y314	2009，四川
川麦 55*	SW3243/SW8688	2009，四川
川麦 56	川麦 30/川麦 42	2009，四川
博麦 1 号	绵阳 29/川麦 30	2009，四川
绵麦 367*	1275-1/99-1522	2010，国家
绵麦 228	1275-1/内 2938//99-1522	2011，四川
川麦 58	川麦 42/03 间 3//川麦 42	2011，四川
绵麦 51	1275-1/99-1522	2012，国家
川麦 104*	川麦 42/川农 16	2012，四川
川麦 61	郑 9023/间 3/2/间 3/3/1522	2012，四川
西科麦 7 号	0105-2/SW8688	2012，四川
绵麦 1618	1275-1//内 2938/99-1522	2013，四川
蜀麦 969	SHW-L1/SW8188//川育 18/3/川麦 42	2013，四川
川麦 64	川麦 42/川农 16	2013，四川
川麦 91	内麦 8/郑 9023//00062/3/川麦 42	2014，四川
川麦 90	间 38/99116//川麦 42	2014，四川
川麦 80	郑 005/2 * 1522	2014，四川
川麦 66	99-1572/98-266//01-3570	2014，四川
川麦 67	99-1572//SW8688//01-3570	2014，四川
资麦 2 号	R25/川麦 30	2014，四川

（续）

品　名	组　合	审定年份与地区
西南冬麦区		
中科麦 138	川麦 42/川育 16	2014，四川
川麦 68	99-1572/98-266//01-3570	2015，四川
川麦 92	内麦 8 号/间 3//川麦 42	2015，四川
川麦 81	SW8019/99-1572//99-1572	2015，四川
绵麦 285	1275-1/99-1522	2015，四川
川麦 69	川麦 104×B2183	2015，四川
国豪麦 3 号	1227-185/99-1522//99-1572	2015，国家
滇 0483*	EMU "S"/MRS//KAL/BB，CM38199	1985，云南
滇 0103*	Veery "S"，CM38199	1985，云南
春 980*	Alondra-Pichibuila，CM32566	1985，云南
750025-12*	矮绒穗 2-18-1/墨沙//7312	1986，云南
精选 9 号*	CIMMYT 引进种质	1988，云南
S001*	CIMMYT 引进种质	1991，云南
云麦 36	墨巴 65/早熟阿金//墨沙/3/75002-12	1991，云南
84-420	卡捷姆/684-214 40kr	1991，云南
凤麦 24*	云麦 36/Mexico 965	1992，云南
凤麦 26	39491/云植 803	1992，云南
云麦 38	87P（0）-1-1-7//洛夫 231/802-736/3/75-14/4/822-852	1993，云南
云麦 39*	云麦 29/Flicker	1994，云南
凤麦 27*	7902/39491	1994，云南
E001*	ATTILA	1994，云南
靖麦 4 号*	中 7906/80B230/4/毕麦 5 号/3/墨查/库斯曼//内乡 5 号	1994，云南
德麦 3 号	陇春 2 号/查平戈	1997，云南
德麦 4 号*	毕麦 5 号/II8156（墨）//中引 1022	1997，云南
R101*	AGA/4 * HORKS	1997，云南
凤麦 29	0483/4/凤麦 10 号/墨西 120//110-705/3/综抗矮 2 号/5/82S-1667	1998，云南
楚麦 4 号	繁 6/墨沙	1998，云南
云麦 42*	抗锈 782/云麦 29//YR70-PAM	1999，云南
德麦 5 号	绵阳 11/Yun 80-1（CIMMYT）	1999，云南
靖麦 7 号	436/墨西哥高代材料 092-1	2000，云南
凤麦 30	39491/云植 803//丰优/抗锈 784	2000，云南
凤麦 31 号	S242/882-805	2000，云南
靖麦 8	477/3845（CIMMYT line）	2000，云南
凤麦 32	0483/4/凤麦 10 号/墨西哥 120＋32/3/综抗矮 2 号/5/882-182	2002，云南
楚麦 6 号*	842-849/精选 7 号	2002，云南
云选 11-12*	Ning8391//SHA4/LIRA	2003，云南
靖麦 10 号	TB78-212/引 48//092-1	2003，云南
德麦 7 号	云植 437/892-17	2003，云南
云麦 46	VEERY "S"/82B-477//83D4-1	2004，云南

（续）

品　名	组　合	审定年份与地区
西南冬麦区		
靖麦 11	（高加索 78-3845/墨 980）/多父本混合花粉	2004，云南
云麦 47*	79213-194/92B-4074	2004，云南
凤麦 33	882V-2232/Rikaze//兴麦 775	2004，云南
云麦 57	CIMMYT 引进	2008，云南
云麦 60	CIMMYT 引进，含慢病性基因	2010，云南
云麦 62	NG8391/2/SHA4/LIRA/3/902-41	2012，云南
云麦 63	NG8391/2/SHA4/LIRA/3/902-41	2011，云南
云麦 64	云麦 39/云麦 42	2012，云南
云麦 67	豫麦 35/云麦 29	2013，云南
云麦 69	云麦 42/陕 623	2014，云南
兴麦 5	高加索/66-36-3-6（CIMMYT）	1978，贵州
综矮抗 2	大山洞 1/Orofen//索诺拉 64/3/贵农 1/4/Kavkaz	1978，贵州
801	72-120/叶考拉	1978，贵州
毕麦 26	雅安矮 2 号/卡捷姆//白兔 2 号	1985，贵州
华南冬麦区		
福红壳 13	波他姆 S70/苏麦 3	1976，福建
福繁 904	福矮麦 2 号/75-5283//红芒 22	1988，福建
龙溪 6 号	龙溪 35/Alondra "S"//福繁 16	1994，福建
安麦 74-5	红芒麦/伊尼亚 F66	1983，广东
粤麦 2 号	M87/白芒	1989，广东
粤麦 6 号	M87/B-5（M87 来自 CIMMYT）	1989，广东
东北春麦区		
垦北 1 号	北新 4 号/墨巴 66	1982，黑龙江
克丰 3 号*	克 $71F_4$-370-7/那达多列斯 63	1982，黑龙江
龙麦 11	沈 68-71/他诺瑞 F71	1983，黑龙江
黑春 3 号	尖麦 302/叶考拉	1984，黑龙江
克丰 4 号	克 71F4-370-7/墨巴 66	1985，黑龙江
垦北 2 号	北 75-616//北新 4/墨巴 66/3/克 72 远 308 矮	1986，黑龙江
龙麦 13	墨巴 66/松 71-175//克 74-202	1986，黑龙江
垦红 6 号*	克丰 1 号/中引 432	1987，黑龙江
龙麦 12	墨巴 66/松 71-175//克 74-202	1987，黑龙江
合春 13	佳 72-819/L9//M87	1988，黑龙江
龙麦 15	克 76-686// 科春 14/他那瑞	1989，黑龙江
克旱 13	克旱 8 号//克 71F4-370-7/那达多列斯 63	1992，黑龙江
垦红 9 号*	东农 120//克丰一号/中引 321	1992，黑龙江
龙麦 19*	那达多列斯 63/克 70F3-49//龙 74-5778	1994，黑龙江
克旱 14	克 80-10/克 $81F_4$-80-0-1	1995，黑龙江
克丰 6 号*	克 85F3-868//克 73-402/克 $71F_4$-370-7/那大多列斯 63	1995，黑龙江
垦红 12	Fulchutor/克 76-250（墨巴 66 衍生系）	1995，黑龙江

（续）

品　名	组　合	审定年份与地区
东北春麦区		
垦红14*	钢82-122（墨巴66衍生系）/东农120	1997，黑龙江
垦红16	CM5434（CIMMT引入）	1997，黑龙江
北麦1号	垦红14号（墨巴66衍生系）/九三3U108	2005，国家
北麦7号	垦红6号（墨麦中引432衍生系）/克丰6号	2008，国家
龙麦33	克丰1号/中71432	2008，黑龙江
北麦8号	垦大1号/克71F4.370-7/墨巴66	2009，黑龙江
龙麦37	人工合成六倍体小麦/龙辐91B569//龙26/3/01D1572-2	2014，黑龙江，
龙春1号	CROC-1/A. SQ//2 * OPATA/9273	2014，黑龙江
铁春1号*	科春14/他诺瑞	1982，辽宁
铁春2号	科春14/他诺瑞//辽29	1990，辽宁
辽春10号*	克71F4370-10/墨巴66//UP321/3/辽春6号/京红1号	1990，辽宁、国家
铁岭3号	科春14/他诺瑞	1999，辽宁
辽春18	克71F4370-10/墨巴66//UP321/3/辽春6号/京红1号	2005，辽宁、国家
丰强3号*	他诺瑞/新曙光1号//吉7136-1-3	1983，吉林
丰强4号	他诺瑞/新曙光1号//吉7136-1-3	1983，吉林
长春1号	辽春6号/他诺瑞	1987，吉林
长春2号	波他姆/科春14//60169	1990，吉林
白麦1号	新曙光3号/他诺瑞	1991，吉林
北部春麦区		
内麦11	波他姆S70/文革1号	1984，内蒙古
内麦12	原农61/京红5//墨巴65	1984，内蒙古
内麦13	他诺瑞F71/辽春7	1984，内蒙古
内麦14	京红9号选系	1986，内蒙古
哲春3号	科春14/纽瑞	1986，内蒙古
内麦16	克丰1号/那达多列斯63	1986，内蒙古
内麦17	查平戈/内麦3	1987，内蒙古
内麦19*	81NS10-1/宁春4号*2	1991，内蒙古
赤麦1号	辽春5号/3/辽鉴28/伊利亚//美乐兰	1987，内蒙古
哲春4号	他诺瑞/辽春5号	1989，内蒙古
乌麦5号	克群/沙瑞克//内麦11	1989，内蒙古
蒙优1号*	红芒/叶考拉	1991，内蒙古
内麦18	中引198（CIMMYT）/晋麦2148	1991，内蒙古
蒙麦28*	宁春4/中7606/4/叶考拉/斗地1/3/宏图/塞洛斯//宏图	1997，内蒙古
蒙麦29	墨巴65/新曙光3//克津	1997，内蒙古
呼麦4号	克69-69/那达多列斯63	1997，内蒙古
巴优1号	冀84-5418/宁春4号	2004，国家
农麦2号	宁1608/蒙鉴3号	2006，国家
冀春1号	墨巴66/科春5号	1978，河北
冀张春2号	冀春1号/约瑞C69	1987，河北

（续）

品　名	组　合	审定年份与地区
北部春麦区		
晋春 3 号	咸农 39/墨巴 66	1974，山西
晋春 4 号	墨巴 66/克春 14	1976，山西
晋春 5 号	京红 5 号/塞洛斯	1981，山西
晋春 6 号	云麦 26/墨巴 66+京春 69-736	1982，山西
晋春 7 号	无芒 1 号/沙瑞克 F70	1982，山西
晋麦 37	沙瑞克 F70/临汾 3029/3/74100/2/蚰包 036/小偃 759	1991，山西
晋春 45	沙瑞克 F70/临汾 3029/3/74100//蚰包 036/小偃 759	1993，山西
晋春 12	波他姆 S70/京红 1 号	1995，山西
麦 878-80	叶考拉/晋春 9 号	2004，山西
京红 7 号	京红 1 号/那林诺 59	1982，北京
京红 8 号	京红 4 号/墨巴 66	1982，北京
京 711	叶考拉/科春 14	1983，北京
京红 9 号	京红 4/墨巴 66	1988，北京
中作 8131-1	京 771/中 7606//引 1053	1989，北京
京红 10 号	京 772/Alondra "S"-Pima 77	1991，北京
津春 1 号	墨巴/科春 5 号	1993，天津
秦麦 5 号	榆春 1 号/依尼亚 F66	1985，陕西
榆春 4 号	榆 7741-1//榆春 3 号/叶考拉	1995，陕西
西北春麦区		
民勤 7586	6836-3-5/卡捷姆	1978，甘肃
临麦 26	陇春 7 号/68-1076//叶考拉	1980，甘肃
陇春 10	70-84-2-1/墨西哥 27 号	1981，甘肃
民勤 732	墨巴 66/甘麦 24	1982，甘肃
武春 1 号*	甘麦 23/（哈什白皮+墨巴 66）	1985，甘肃
武春 121*	甘麦 8 号/墨纽 F70	1985，甘肃
甘春 15	（拜尼莫 62/甘麦 42）F_2//新曙光 1 号	1986，甘肃
陇花 2 号*	卡提姆//东乡大头兰麦/阿 4	1988，甘肃
甘 630*	715/70（CIMMYT line）	1989，甘肃
民勤 78152	甘麦 8 号/7586	1991，甘肃
甘春 18	1059/甘麦 23//墨巴 66	1992，甘肃
陇春 20	832-748/0103，0103 引自 CIMMYT	2000，甘肃
陇春 22	MY94-9（CIMMYT 材料 CHIL/BUC）	2004，甘肃
陇春 23*	CMBW90M4860-0TOPY-16M-1Y-010M-010Y-1M-0	2004，甘肃
陇春 28	9807-2-14/CM7033//CM7015	2011，甘肃，
宁春 304	墨巴 65 /宏图	1979，宁夏
宁春 4 号*	索诺拉 64//阿勃/碧玉	1981，宁夏
宁春 7 号	墨巴 65/宏图，辐射	1983，宁夏
宁春 11	叶考拉/京红 5 号+科春 14 号+斗地 1 号［F_4］//索诺拉 64［F_2］/3/雁 804［F_1］/4/宁春 4 号	1988，宁夏

（续）

品　名	组　合	审定年份与地区
西北春麦区		
宁春 14	永良 5 号/MO77	1990，宁夏
宁春 15	永良 4 号/中 7906	1990，宁夏
宁春 16	SG（81rs10）/宁春 4 号//宁春 4 号	1992，宁夏
宁春 18	叶考拉/榆 293//卡捷姆/榆 293	1994，宁夏
宁春 23	永良 4 号/中 7906//陕农 7859	1995，宁夏
宁春 24	79N121/墨卡	1995，宁夏
宁春 30	CIMMYT 材料系选	2000，宁夏
宁春 31	宁春 43/JUNCO"S"	2000，宁夏
宁春 39	永 833（T2739/农院 G89）/宁春 4 号	2006，国家
宁春 41	宁春 4 号/永旱 2 号	2005，宁夏
宁冬 10*	NZT/302//ALD/4/NAD//TMP/CI12426/3/EMU/5/北农 2 号	2007，宁夏
宁春 50*	宁春 4 号 2/ChamI（法国硬粒小麦）	2011，宁夏
宁春 51	永 3002/宁春 4 号	2011，宁夏
宁春 53	宁春 39 号/M7021	2014，宁夏
新疆冬春麦区		
墨巴 65	CIMMYT 种质	新疆，无审定制度
塞洛斯	CIMMYT 种质	新疆，无审定制度
卡捷姆	CIMMYT 种质	新疆，无审定制度
约瑞	CIMMYT 种质	新疆，无审定制度
哈垦 1 号	墨巴 66/喀什白皮	1980，新疆
伊春 5 号	伊春 1/奥帕尔//塞洛斯/3/新曙光 1 号	1982，新疆
哈垦 2 号	墨巴 65/喀什白皮	1983，新疆
巴春 3 号	红星/墨巴 65	1984，新疆
巴春 7 号	74-16/波他姆 S70	1984，新疆
昌春 2 号	奇春 1 号/墨巴 65	1984，新疆
昌春 3 号	喀什白皮/那达多列士//奇春 4 号	1984，新疆
阿春 2 号	阿春 1 号/沙瑞克	1984，新疆
阿春 3 号	阿春 1 号/沙瑞克	1984，新疆
解放 5 号	墨巴 66/喀什白皮	1984，新疆
塔春 1 号	昌春 1 号//解放 4 号/沙瑞克	1984，新疆
新春 2 号*	（塞洛斯/奇春 4 号）＋辐射处理	1984，新疆、国家
新春 3 号*	（塞洛斯/奇春 4 号）＋辐射处理	1986，新疆
石春 1 号	青春 5 号/塞洛斯	1986，新疆
哈春 1 号	哈密大头郎/卡捷姆	1987，新疆
哈春 2 号	阿勃/塞洛斯	1990，新疆
哈春 3 号	哈密红大头郎/塞洛斯	1990，新疆
哈春 4 号	新春 1 号/新大头黄/塞洛斯	1992，新疆
新春 6 号*	中 7906/改良型新春 2 号	1993，新疆、国家
新春 7 号	中 7906/改良型新春 2 号	1997，新疆

（续）

品　名	组　合	审定年份与地区
新疆冬春麦区		
新春 9 号	Giza164（Kvz/Buho "s"//Kal/Bb，CM33027）	1999，新疆
新春 11	新春 2 号/86-7	2002，新疆
新春 14*	CIMMYT 种质 Attila	2003，新疆
新春 17*	新春 6 号/Giza163	2005，新疆
新春 23	CIMMYT 种质，Kambaral	2006，新疆
新春 26*	（CM33027/新春 6 号）＋辐射处理	2007，新疆
新春 30*	（CM33027/新春 6 号）＋辐射处理	2009，新疆
新春 33	（CM33027/新春 6 号）＋辐射处理	2010，新疆
新春 37*	49-5/野猫，49-5 为 CM33027/新春 6 号	2012，新疆
新春 39	NS64/新春 8 号，NS64 即 Giza163	2012，新疆
新春 40	新春 6 号/UC1041（来自美国）	2013，新疆
新春 42	2001-29/新春 26	2014，新疆
新春 43	90-33/新春 6 号	2015，新疆
新春 44	17-11/YN-76（引自 CIMMYT）	2015，新疆
青藏春冬麦区		
高原 56	7020/墨巴 65	1978，青海
青春 533	367B/Alondra "S"-76	1988，青海
高原 158	[（Abbondanza/6508）/（墨巴 66//索诺拉 62/红大 2）]/[（矮秆/Kavkaz）/高原 472] F_2	1994，青海
高原 175	[（矮单/Kavkaz）/高原 472]///[（Abbondanza/6508）/（墨巴 66//索诺拉 62/红大 2）]/多年生 1]//（矮单大 2 高-2/高原 338）	1994，青海
高原 365	兰粒///{（Abbondanza/6508）F_4/[墨巴 65/（索诺拉 62/红大 2）F_4]//（矮大 2-2/M47）} F_5	1994，青海
高原 V028	HER/SAP "S"//VEE	1998，青海
墨引 1 号	CIMMYT 高代系（系谱不详）	2004，青海
墨引 2 号	CIMMYT 高代系（系谱不详）	2004，青海
日喀则 18	叶考拉 F70/72013	1983，西藏
白春 1	72013/沙瑞克 71	1989，西藏
藏春 667	日喀则 12//天蓝冰草/纽瑞 F70	1996，西藏

附录2　发表论文和专（译）著目录

1. He Z. H., and Rajaram, S.. 1994. Differential response of bread wheat characters to high temperature, Euphytica, 72: 197-203.
2. 邹裕春，刘仲齐，李有春.1994. CIMMYT的小麦育种.成都：四川科学技术出版社.
3. 何中虎，庞家智.1995. CIMMYT麦类改良进展.北京：中国农业科技出版社.
4. Singh R. P., Chen W. Q., and He Z. H.. 1999. Leaf rust resistance of spring, facultative and winter wheat cultivars from China, Plant Disease, 83: 644-651.
5. 何中虎，黄钢，肖世和，译.1999. 提高小麦产量潜力—突破增产屏障.北京：中国科学技术出版社.
6. He Zhonghu, S. Rajaram, Xin Z. Y., and Huang J. Z.. 2001. A History of wheat breeding in China, Mexico, D. F.: CIMMYT.
7. 胡英考，辛志勇.2001. 小麦合成种M53抗白粉病基因的RAPD和SSR标记.作物学报，27(4)：415-419.
8. Anmin Wan, Zhonghua Zhao, Xianming Chen, Zhonghu He, Shelin Jin, Qiuzhen Jia, Ge Yao, Jiaxiu Yang, Baotong Wang, Gaobao Li, Yunqing Bi, and Zhongying Yuan. 2004. Wheat stripe rust epdemic and virulence of Puccinia striiformis f. sp. tritici in China in 2002, Plant Disease, 88 (8): 896-904.
9. Chen X. M., Luo Y. H., Xia X. C., Xia L. Q., Chen X., Ren Z. L., He Z. H., and Jia J. Z.. 2005. Chromosomal location of powdery mildew resistance gene Pm 16 in wheat using SSR molecular marker analysis, Plant Breeding, 124: 225-228.
10. Wang Z. L., Li L H., He Z. H., Duan X. Y., Zhou Y. L., Chen X. M., Lillemo M. Singh R. P., Wang H., and Xia X. C.. 2005. Seedling and adult plant resistance to powdery mildew in Chinese bread wheat cultivars and lines, Plant Disease, 89 (5): 457-463.
11. 罗瑛皓，陈新民，夏兰芹，陈孝，何中虎，任正隆.2005. 小麦抗白粉病基因聚合体DH材料的分子标记鉴定.作物学报，31(5)：565-570.
12. Genying Li, Zhonghu He, Roberto Javier Pena, Xianchun Xia, Morten Lillemo, and Qixin Sun. 2006. Identification of novel secaloindoline-a and secaloindoline-b alleles in CIMMYT hexaploid triticale lines, Journal of Cereal Science, 43: 378-386.
13. Morten Lillemo, Chen Feng, Xianchun Xia, Manilal William, Roberto J. Pena, Richard Trethowan, and Zhonghu He. 2006. Puroindoline grain hardness alleles in CIMMYT bread wheat germplasm, Journal of Cereal Science, 44: 86-92.
14. R. M. Trethowan, Alexi Morgunov, Zhonghu He, R. De. Pauw, J. Cross, M. Warburton, Arman Baytasov, Chunli Zhang, M. Mergoum, and G. Alvarado. 2006. The global adaptation of bread wheat at high latitudes, Euphytica, 152: 303-316.
15. S. S. Liang, K. Suenaga, Z. H. He, Z. L. Wang, H. Y. Liu, D. S. Wang, R. P. Singh, P. Sourdile, and X. C. Xia. 2006. Quantitative trait loci mapping for adult-plant resistance to powdery mildew in bread wheat, Phytopathology, 96 (7): 784-789.
16. Yong Zhang, Zhonghu He, Aimin Zhang, Maarten van Ginkel, Roberto J. Peña, and Guoyou Ye. 2006. Pattern analysis on protein properties of Chinese and CIMMYT spring wheat cultivars sown in China and CIMMYT, Australian Journal of Agricultural Sciences, 57: 811-822.
17. Yong Zhang, Zhonghu He, Aimin Zhang, Maarten van Ginkel, and Guoyou Ye. 2006. Pattern analysis on grain yield performance of Chinese and CIMMYT spring wheat cultivars sown in China and CIMMYT, Eu-

phytica，147：409-420.

18. Z. F. Li，X. C. Xia，X. C. Zhou，Y. C. Niu，Z. H. He，Y. Zhang，G. Q. Li，A. M. Wan，D. S. Wang，X. M. Chen，Q. L. Lu，and R. P. Singh. 2006. Seedling and slow rusting resistance to stripe rust in Chinese common wheats，Plant Disease，90（10）：1302-1312.
19. Li GQ，Li ZF，Yang WY，Zhang Y，He ZH，Xu SC，Singh RP，Qu YY，and Xia XC. 2006. Molecular tagging of stripe rust resistance gene YrCH42 in Chinese wheat cultivar Chuanmai 42 and its allelism with Yr 24 and Yr 26，TAG，112：1434-1440.
20. 陈锋，夏先春，王德森，Morten Lillemo，何中虎 . 2006. CIMMYT 人工合成小麦与普通小麦杂交后代籽粒硬度 puroindoline 基因等位变异检测 . 中国农业科学，39（3）：440-447.
21. 刘慧远，Kazuhiro Suenaga，何中虎，马均，Michel Bernard Pierre Sourdille，夏先春 . 2006. 普通小麦白粉病成株抗性的 QTL 分析 . 作物学报，32（2）：197-202.
22. 李根英，Susanne Dreisigacker，Marilyn Warburton，夏先春，何中虎，孙其信 . 2006. 小麦指纹图谱数据库的建立及 SSR 分子标记试剂盒的研发 . 作物学报，32（12）：1771-1778.
23. 穆培源，何中虎，徐兆华，王德森，张艳，夏先春 . 2006. IMMYT 普通春小麦品种 Waxy 蛋白及淀粉特性研究 . 作物学报，32（7）：1071-1075.
24. 肖永贵，阎俊，何中虎，张勇，张晓科，刘丽，李天富，曲延英，夏先春 . 2006. 1BL/1RS 易位对小麦产量性状和白粉病抗性的影响及其 QTL 分析 . 作物学报，32（11）：1636-1641.
25. 王竹林，王德森，何中虎，王辉，陈新民，段霞瑜，周益林，夏先春 . 2006. 小麦品种百农 64 的慢白粉病 QTL 分析 . 中国农业科学，39（10）：1956-1961.
26. 张晓科，夏先春，何中虎，周阳 . 2006. 用 STS 标记检测春化基因 Vrn-A1 在中国小麦中的分布 . 作物学报，32（7）：1038-1043.
27. 张勇，吴振录，张爱民，Maarten van Ginkel，何中虎 . 2006. CIMMYT 小麦在中国春麦区的适应性分析 . 中国农业科学，39（4）：655-663.
28. A. M. Wan，X. M Chen，and Z. H. He. 2007. Wheat stripe rust in China，Australian Journal of Agricultural Research，58：605-619.
29. 李根英，夏先春，何中虎，孙其信 . 2007. CIMMYT新型人工合成小麦 *Pina* 和 *Pinb* 基因等位变异 . 作物学报，33（2）：242-249.
30. 张勇，何中虎，吴振录，张爱民，Maarten van Ginkel. 2007. CIMMYT 和中国硬质春麦在 4 种 IMMYT 不同处理环境中产量和蛋白品质性状分析 . 作物学报，33（7）：1182-1186.
31. D. Liu，X. C. Xia，Z. H. He，and S. C. Xu. 2008. A novel homeobox-like gene associated with reaction to stripe rust and powdery mildew in common wheat，Phytopathology，98（12）：1291-1296.
32. Morten Lillemo，Belachew Asalf，Ravi Singh，Julio Huerta-Espino，Xinmin Chen，Zhonghu He，and Åsmund Bjørnstad. 2008. The adult plant rust resistance loci *Lr34/Yr18* and *Lr46/Yr29* are important determinants of partial resistance to powdery mildew in bread wheat line Saar，TAG，116：1155-1166.
33. X. K. Zhang，X. C. Xia，Y. G. Xiao，J. Dubcovsky，and Z. H. He. 2008. Allelic variation at the vernalization genes *Vrn-A1*，*Vrn-B1*，*Vrn-D1* and *Vrn-B3* in Chinese common wheat cultivars and their association with growth habit，Crop Science，48：458-470.
34. 穆培源，刘丽，陈峰，夏先春，张艳，王德森，何中虎 . 2008. CIMMYT 人工合成小麦改良品系的 HMW-GS 和 LMW-GS 组成及其对面筋品质的影响 . 麦类作物学报，28：607-612.
35. 穆培源，王亮，陈峰，何中虎，韩新年，徐红军，夏先春 . 2008. CIMMYT 普通冬小麦品种的籽粒硬度及 Puroindoline 基因变异 . 麦类作物学报，28：41-46.
36. 穆培源，刘丽，陈峰，夏先春，张艳，王德森，何中虎 . 2008. CIMMYT 人工合成小麦改良品系的 HMW-GS 和 LMW-GS 组成及其对面筋品质的影响 . 麦类作物学报，28：607-612.

37. 倪小文，阎俊，陈新民，夏先春，何中虎，张勇，王德森，Morten Lillemo. 2008. 鲁麦 21 慢白粉病抗性基因数目和遗传力分析. 作物学报，34：1317-1322.

38. 杨文雄，杨芳萍，梁丹，何中虎，尚勋武，夏先春. 2008. 中国小麦育成品种和农家种中慢锈基因 *Lr34*/*Yr18* 的分子检测. 作物学报，34：1109-1113.

39. Caixia Lan, Shanshan Liang, Zhulin Wang, Jun Yan, Yong Zhang, Xianchun Xia, and Zhonghu He. 2009. Quantitative trait loci mapping for adult-plant resistance to powdery mildew in Chinese wheat cultivar Bainong 64, Phytopathology, 99: 1121-1126.

40. Dongyun Ma, Zhanga Yan, Xianchun Xia, Craig F. Morris, and Zhonghu He. 2009. Milling and Chinese raw white noodle qualities of common wheat near-isogenic lines differing in puroindoline b alleles, Journal of Cereal Sciences, 50: 126-130.

41. R. Ortiz, H. J. Bruan, J. Crossa, J. H. Crouch, G. Davenport, J. Dixon, S. Dreisigacker, E. Duveiller, Z. H. He, J. Huerta, A. K. Joshi, M. Kishii, P. Kosina, Y. Manes, M. Mezzalama, A. Morgounov, J. Murakami, J. Nocol, G. O. Ferrara, J. I. Ortiz-Monasterio, T. S. Payne, R. J. Pena, M. P. Reynolds, K. D. Sayre, R. C. Sharam, R. Singh, J. K. Wang, M. Warburton, H. X. Wu, and M . Iwanaga. 2009. Wheat genetics resources enhancement by the International Maize and Wheat Improvement Center (CIMMYT), Genet Resour Crop Evol, 55: 1095-1140.

42. X. Y. He, Z. H. He, W. Ma, R. Appels, and X. C. Xia. 2009. Allelic variants of PSY1 genes in Chinese and CIMMYT wheat cultivars and development of functional markers, Molecular Breeding, 23: 553-563.

43. Xinyao He, Jianwu Wang, Zhonghu He, Karim Ammar, Roberto Javier Peña, and Xianchun Xia. 2009. Allelic variants at the Psy-A1 and Psy-B1 loci in durum wheat and their associations with grain yellowness, Crop Science, 49: 2058-2064.

44. F. P. Yang, X. K. Zhang, X. C. Xia, D. A. Laurie, W. X . Yang, and Z. H. He. 2009. Distribution of the photoperiod insensitive Ppd-D1a allele in Chinese wheat cultivars, Euphytica, 165: 445-452.

45. Y. M. Lu, C. X. Lan, S. S. Liang, X. C. Zhou, D. Liu, Xia Xianchun, and He Zhonghu. 2009. QTL mapping for adult-plant resistance to stripe rust in Italian common wheat cultivars Libellula and Strampelli, TAG, 119: 1349-1359.

46. Yang Wuyun, Liu Dengcai, Li Jun, Zhang Lianquan, Wei Huiting, Hu Xiaorong, Zheng Youliang, He Zhoughu, Zou Yuchun. 2009. Synthetic hexaploid wheat and its utilization for wheat genetic improvement in China, J. Genet. Genomics, 36: 539-546.

47. 梁丹，杨芳萍，何中虎，姚大年，夏先春. 2009. 利用 STS 标记检测 CIMMYT 小麦品种（系）中 *Lr34*/*Yr18*、*Rht-B1b* 和 *Rht-D1b* 基因的分布. 中国农业科学，42（1）：17-27.

48. 任妍，梁丹，张平平，何中虎，陈静，傅体华，夏先春. 2009. 中国和 CIMMYT 小麦品种 Bx7 亚基超量表达基因（$Bx7^{OE}$）的分子检测. 作物学报，35（3）：403-411.

49. 唐怀君，殷贵鸿，夏先春，冯建军，曲延英，何中虎. 2009. 1BL・1RS 特异性分子标记的筛选及其对不同来源小麦品种 1RS 易位染色体的鉴定. 作物学报，35（11）：2107-2115.

50. Liu L, Ikeda T M, Branlard G, Peña R J, Rogers W J, Lerner S E, María DI, Kolman A, XIA X C, Linhai Wang, Wujun Ma, Rudi Appels, Hisashi Yoshida, Aili Wang, Yueming Yan, and He Z H. 2010. Comparison of low molecular weight glutenin subunits identified by SDS-PAGE, 2-DE, MALDI-TOF-MS and PCR in common wheat, BMC Plant Biology, 10: 124.

51. Dan Liang, Zhonghu He, Jianwei Tang, Roberto Javier Peña, Ravi Singh, Xinyao He, Xiaoyong Shen, DanianYao, and Xianchun Xia. 2010. Characterization of CIMMYT bread wheats for high- and low-molecular-weight glutenin subunits and other quality-related genes with SDS-PAGE, RP-HPLC and molecular markers, Euphytica, 172: 235-250.

52. He Zhonghu, and Alain Bonjean. 2010. Cereals in China, Mexico, D. F., CIMMYT.
53. Lan CX, Ni XW, Yan J, Zhang Y, Xia CX, Chen XM, and He ZH. 2010. Quantitative trait loci mapping of adult-plant resistance to powdery mildew in Chinese wheat cultivar Lumai 21, Molecular Breeding, 25: 615-622.
54. Lan CX, Liang SS, Zhou XC, Zhou G, Lu QL, Xia XC, and He ZH. 2010. Identification of genetic regions controlling stripe rust adult-plant resistance in Chinese landrace Pingyuan 50, Phytopathology, 100: 313-318.
55. Z. F. Li, X. C. Xia, Z. H. He, L. J. Zhang, X. Li, H. Y. Wang, Q. F. Meng, W. X Yang, G. Q Li, and D. Q Liu. 2010. Seedling and slow rusting resistance to leaf rust in Chinese wheat cultivars, Plant Disease, 94: 45-53.
56. Sui Xinxia, Xia Xianchun, and He Zhonghu. 2010. Molecular mapping of a non-host resistance gene Yrpst in barley (*hordeum vulgare* L.) for resistance to wheat stripe rust, Hereditas, 147: 176-128.
57. 武玲，夏先春，朱华忠，李式昭，郑有良，何中虎 . 2010. CIMMYT 小麦品种抗病基因 *Lr34/Yr18/Pm38* 的分子标记检测 . 中国农业科学，43 (22): 4553-4561.
58. 朱华忠，王忠伟，伍玲，Ravi P Singh，J Huerta-Espino，何中虎，胡嘉，陈放，夏先春 . 2010. 小麦品种川麦 107 对条锈病成株抗性的 QTL 定位 . 中国农业科学，43 (4): 706-712.
59. Hui Jin, Jun Yan, Roberto J. Peña, Xianchun Xia, Alexey Morgounov, Liming Han, Yong Zhang, and Zhonghu He. 2011. Molecular detection of high- and low-molecular-weight glutenin subunit genes in common wheat cultivars from 20 countries using allele-specific markers, Crop and Pasture Science, 62: 746-754.
60. Tao Li, Zengyan Zhang, Yingkao Hu, Xiayu Duan, and Zhiyong Xin. 2011. Identification and molecular mapping of a resistance gene to powdery mildew from the synthetic wheat line M53, J. Appl. Genetics, 52: 137-143.
61. Yang E N, Zou Y C, Yang W Y, Tang Y L, He Z H, and Singh R. 2011 (special issue). Breeding adult plant resistance to stripe rust in spring bread wheat germplasm adapted to Sichuan Province of China, Czech J. Genet. Plant Breed., 47: S165-S168.
62. 韩烨，何中虎，夏先春，李星，李在峰，刘大群 . 2011. CIMMYT 小麦材料的苗期和成株抗叶锈病鉴定 . 作物学报，37 (07): 1125-1133.
63. 何中虎，兰彩霞，陈新民，邹裕春，庄巧生，夏先春 . 2011. 小麦条锈病和白粉病成株抗性研究进展与展望 . 中国农业科学，44 (11): 2193-2215.
64. 杨芳萍，韩利明，阎俊，夏先春，张勇，曲延英，王忠伟，何中虎 . 2011. 春化和光周期基因等位变异在 23 个国家小麦品种中的分布 . 作物学报，37 (11): 1917-1925.
65. 张勇，李式昭，吴振录，杨文雄，于亚雄，夏先春，何中虎 . 2011. CIMMYT 种质对四川、云南、甘肃和新疆春性小麦产量遗传增益的贡献 . 作物学报，37 (10): 1752-1762.
66. B Bai, Z H He, M A Asad, C X Lan, Y Zhang, X C Xia, J Yam, X M Chen, and C S Wang. 2012. Pyramiding adult plant powdery mildew resistance QTL in bread wheat, Crop and Pasture Science, 63: 606-611.
67. Bai Bin, Ren Yan, Xia Xianchun, Du Jiuyuan, Zhou Gang, Wu Ling, Zhu Huazhong, He Zhonghu, and Wang Chengshe. 2012. Mapping of quantitative trait loci for adult plant resistance to stripe rust in German wheat cultivar Ibis, Journal of Integrative Agriculture, 11 (4): 528-536.
68. Ling Wu, Youliang Zheng, Xianchun Xia, Yunliang Peng, Huazhong Zhu, Yongjian Liu, Yu Wu, Shizhao Li, and Zhonghu He. 2012. QTL mapping of adult-plant resistance to stripe rust in Chinese wheat cultivar Chuanyu 16, Journal of Agricultural Science, 4 (3): 57-70.
69. Muhammad Azeem Asad, Xianchun Xia, Chengshe Wang, and Zhonghu He. 2012. Molecular mapping of stripe rust resistance gene *YrSN104* in Chinese wheat line Shaannong 104, Hereditas, 149: 146-152.
70. M A Asad, B Bai, C X Lan, J Yan, X C Xia Y Zhang, and Z H He. 2012. Molecular mapping of quantitative

trait loci for adult plant resistance to powdery mildew in Italian wheat cultivar Libellula, Crop and Pasture Science, 63: 539-546.

71. Wu Ling, Xia Xianchun, Zheng Youliang, Zhang Zhengyu, Zhu Huazhong, Liu Yongjian, Yang Ennian, Li Shizhao, and He Zhonghu. 2012. QTL mapping for adult-plant resistance to stripe rust in a common wheat RIL population derived from Chuanmai 32/Chuanyu 12, Journal of Integrative Agriculture, 11 (11): 1775-1782.
72. Yan Ren, Zhonghu He, Jia Li, Morten Lillemo, Ling Wu, Bin Bai, Qiongxian Lu, Huazhong Zhu, Gang Zhou, Jiuyuan Du, Qinglin Lu, and Xianchun Xia. 2012. QTL mapping of adult-plant resistance to stripe rust in a population derived from common wheat cultivars Naxos and Shanghai 3/Catbird, TAG, 125: 1211-1221.
73. Yan Ren, Zaifeng Li, Zhonghu He, Ling Wu, Bin Bai, Caixia Lan, Cuifen Wang, Gang Zhou, Huazhong Zhu, and Xianchun Xia. 2012. QTL mapping of adult-plant resistances to stripe rust and leaf rust in Chinese wheat cultivar Bainong 64, TAG, 125: 1253-1262.
74. Zhang Xiaofei, Jin Hui, Zhang Yan, Liu Dongcheng, Li Genying, Xia Xianchun, He Zhonghu, and Zhang Aimin. 2012. Composition and functional analysis of low-molecular-weight glutenin alleles with Aroona near-isogenic lines of bread wheat, BMC Plant Biology, 12: 243.
75. 杨芳萍，夏先春，张勇，张晓科，刘建军，唐建卫，杨学明，张俊儒，刘茜，李式昭，何中虎. 2012. 春化、光周期和矮秆基因在不同国家小麦品种中的分布及其效应. 作物学报，38 (6): 1-12.
76. Asad Muhammad Azeem, Bin Bai, Caixia Lan, Jun Yan, Xianchun Xia, Yong Zhang, and Zhonghu He. 2013. QTL Mapping for adult plant resistance to powdery mildew in Italian wheat cv. Strampelli, Journal of Integrate Agriculture, 12: 756-764.
77. Rosewarne G M, Herrera-Foessel S A, Singh R P, Huerta-Espino J, Lan C X, and He Z H. 2013. Quantitative trait loci of stripe rust resistance in wheat, TAG, 126: 2427-2449.
78. Yang E N, Rosewarne G M, Herrera-Foessel S A, Huerta-Espino J, Tang Z X, Sun C F, Ren Z L, and Singh R P. 2013. QTL analysis of the spring wheat “Chapio” identifies stable stripe rust resistance despite inter-continental genotype x environment interactions, TAG, 126: 1721-1732.
79. Zhonghu He and Daowen Wang (eds). 2013. Proceedinds: 11th International Gluten Workshop, Beijing, China, August 12-15, 2012. Mexico, D. F., CIMMYT.
80. Awais Rasheed, Xianchun Xia, Francis Ogbonnaya, Tariq Mahmood, Zongwen Zhang, Abdul Mujeeb-Kazi, and Zhonghu He. 2014. Genome-wide association for grain morphology in synthetic hexaploid wheats using digital imaging analysis, BMC Plant Biology, 14: 128.
81. Bai B., Du J. Y., Lu Q. L., He C. Y., Zhang L. J., Zhou G., Xia X. C., He Z. H., and Wang C. S. 2014. Effective resistance to wheat stripe rust in a region with high disease pressure, Plant Dis., 98: 891-897.
82. Caixia Lan, Garry M. Rosewarne, Ravi P. Singh, Sybil A. Herrera-Foessel, Julio Huerta-Espino, Bhoja R. Basnet, Yelun Zhang, Ennian Yang. 2014. QTL characterization of resistance to leaf rust and stripe rust in the spring wheat line Francolin#1. Molecular Breeding, 34: 789-803.
83. Muhammad Azeem Asad, Bin Bai, Caixia Lan, Jun Yan, Xianchun Xia, Yong Zhang, Zhonghu He. 2014. Identification of QTL for adult-plant resistance to powdery mildew in Chinese wheat landrace Pingyuan 50, The Crop Journal, 2, 308-314.
84. Yue Zhou, Yan Ren, Morten Lillemo, Zhanjun Yao, Peipei Zhang, Xianchun Xia, ZhonghuHe, Zaifeng Li, and Daquan Liu. 2014. QTL mapping of adult-plant resistance to leaf rust in a RIL population derived from a cross of wheat cultivars Shanghai 3/Catbird and Naxos, TAG, 127: 1873-1883.
85. Yong Ren, Shengrong Li, Xianchun Xia, Qiang Zhou, Yuanjiang He, Yuming Wei, Youliang Zheng, and

Zhonghu He. 2015. Molecular mapping of a recessive stripe rust resistance gene *yrMY37* in Chinese wheat cultivar Mianmai 37，Molecular Breeding，35：1-9.

86. Zaifeng Li，Caixia Lan，Zhonghu He，Ravi P. Singh，Garry M. Rosewarne，Xinmin Chen，and Xianchun Xia. 2014. Overview and application of QTL for adult plant resistance to leaf rust and powdery mildew in wheat，Crop Science，54：1-19.
87. Zhonghu He，Xianchun Xia，Shaobing Peng，and Thomas Lumpkin. 2014. Meeting demands for increased cereal production in China，Journal of Cereal Sciences，59：235-244.
88. 刘金栋，陈新民，何中虎，伍玲，白斌，李在峰，夏先春 . 2014. 小麦慢白粉病 QTL 对条锈病和叶锈病的兼抗性 . 作物学报，40 (9)：1157-1564.
89. 刘金栋，杨恩年，肖永贵，陈新民，伍玲，白斌，李在峰，Garry M. Rosewarne，夏先春，何中虎 . 2015. 兼抗型成株抗性小麦品系的培育、鉴定与分子检测 . 作物学报，41 (10)：1472-1480.
90. Ren Yan，Liu Li，He Zhonghu，Wu Ling，Bai Bin，and Xia Xianchun. 2015. QTL mapping of adult-plant resistance to stripe rust in a 'Lumai 21 × Jingshuang 16' wheat population，Plant Breeding，134：501-507.
91. Ling Wu，Xianchun Xia，Garry M. Rosewarne，Huazhong Zhu，Shizhao Li，Zhengyu Zhang，and Zhonghu. He. 2015. Stripe rust resistance gene *Yr18* and its suppressor gene in Chinese wheat landraces，Plant Breeding，134：634-640.
92. Ai-yong Qi，Pei-pei Zhang，Xian-chun Xia，Zhong-hu He，Julio Huerta-Espino，Zai-feng Li，Da-qun Liu. 2015. Molecular mapping and markers for leaf rust resistance gene *Lr24* in CIMMYT wheat line 19HRWSN-122，Euphytica，206：57-66.